ALTERNATING CURRENT MACHINES

(Electrical Machines—Part II)

IN

S.I. UNITS

By the Same Author :

- A Text Book of Electrical Machines
- Direct Current Machines
- A Textbook of Power Plant Engineering
- Electrical Engineering Materials
- Elements of Electrical Engineering and Electronics
- Thermal Engineering
- A Textbook of Engineering Thermodynamics
- Applied Mechanics

ALTERNATING CURRENT MACHINES

(Electrical Machines—Part II)

IN

S.I. UNITS

for

B.Sc. (Engg.) ; U.P.S.C. ; GATE ; A.M.I.E. (India) Sec. B ;
Diploma and other Competitive Examinations

By

R.K. RAJPUT

M.E. (*Heat Power Engg.) Hons.–Gold Medallist ; Grad. (Mech. Engg. & Elect. Engg.); M.I.E. (India); M.S.E.S.I.; M.I.S.T.E.; C.E. (India)*

Principal (*Formerly*):
Punjab College of Information Technology
PATIALA (PUNJAB)

LAXMI PUBLICATIONS (P) LTD

(An ISO 9001:2015 Company)

BENGALURU • CHENNAI • GUWAHATI • HYDERABAD • JALANDHAR
KOCHI • KOLKATA • LUCKNOW • MUMBAI • RANCHI • NEW DELHI
BOSTON (USA) • NAIROBI (KENYA)

ALTERNATING CURRENT MACHINES

Printed and bound in India

Typeset at : Goswami Associates, Delhi.

First Edition : 1992, *Second Edition* : 1998, *Third Edition* : 2002, *Reprint* : 2003, 2008, 2009, 2011, 2013, Edition : 2018

ISBN : 978-81-7008-222-4

Branches

© **Bengaluru**	080-26 75 69 30
© **Chennai**	044-24 34 47 26
© **Guwahati**	0361-254 36 69, 251 38 81
© **Hyderabad**	040-27 55 53 83, 27 55 53 93
© **Jalandhar**	0181-222 12 72
© **Kochi**	0484-237 70 04, 405 13 03
© **Kolkata**	033-22 27 43 84
© **Lucknow**	0522-220 99 16
© **Mumbai**	022-24 93 12 61
© **Ranchi**	0651-220 44 64

PUBLISHED IN INDIA BY

LAXMI PUBLICATIONS (P) LTD.

(An ISO 9001:2015 Company)

113, GOLDEN HOUSE, DARYAGANJ, NEW DELHI-110002, INDIA

Telephone : 91-11-4353 2500, 4353 2501

Fax : 91-11-2325 2572, 4353 2528

www.laxmipublications.com info@laxmipublications.com

C—00297/018/02

Printed at : Sanjeev Offset Press, Delhi.

PREFACE TO THE THIRD EDITION

I am pleased to present the third edition of this book. The warm reception, which the previous editions and reprint of this book have enjoyed all over India and abroad, has been a matter of great satisfaction to me.

The entire book has been thoroughly revised and enlarged ; large number of solved examples (questions having been selected from various universities and competitive examinations) and ample additional text have been added. Besides this a "**Questions Bank**" *with Answers* has also been added to make the book a more comprehensive and useful unit for the students preparing for various examinations.

Any suggestions for the improvement of this book will be thankfully acknowledged and incorporated in the new edition.

R.K. RAJPUT

PREFACE TO THE FIRST EDITION

A large number of books are available on the subject 'Electrical Machines' in the market but they lack in proper subject treatment, solved examples are not large in number and are too difficult to understand without guidance. The purpose of writing this book is to bridge these gaps and to present such a book to the students which should be easy to understand and should assimilate comprehensive treatment of the subject matter in simple, lucid and direct language.

The book comprises two parts. Part I contains D.C. machines whereas Part II assimilates polyphase circuits and A.C. machines. There are 650 neat line diagrams, 520 solved examples and 550 unsolved problems, in the book. Besides theoretical questions, a large number of objective type questions have been added at the end of each chapter. In Part I a chapter namely 'Miscellany' has been added which includes those topics which could not be dealt/accommodated in other chapters. A.M.I.E. questions have been added at the end of each chapter for the students appearing in section B of Institution of Engineers (India) examination.

The book will prove to be of great help to the students preparing for engineering degree, A.M.I.E. section B, diploma and other competitive examinations.

The author's grateful thanks are due to Smt. Ramesh Rajput (wife) for extending full co-operation during the preparation of the manuscript.

The author is also thankful to Mr. Rakesh Syal for drawing neat diagrams for this book.

In the end the author wishes to express his gratitude to M/s Laxmi Publications for taking extra pains in printing the book at a rapid pace and very systematically.

Although every care has been taken to make the book free of errors both in text as well as solved examples, yet the author shall feel obliged, if any error present is brought to his notice. Constructive criticism of the book will be warmly received.

R.K. RAJPUT

Compiled by :
Smt. RAMESH RAJPUT

CONTENTS

PART II

(A.C. MACHINES)

Chapter *Page*

0

Electromechanical Energy Conversion

1. Introduction. 2. Principle of energy conversion. 3. Faraday's laws of electromagnetic induction. 4. Singly and multiply-excited magnetic field systems. 5. Torque production in rotating machines. 6. General analysis of electromechanical system. Worked examples. Highlights—Objective Type Questions—Theoretical Questions—Unsolved Examples.

1. INTRODUCTION

- The main advantage of electric energy over other forms of energy is that it can be *transmitted over long distances with ease and high efficiency*. Its main use is in the form of a transmitting link for transporting other forms of energy, *e.g.*, mechanical, sound, light etc. from one physical location to another.
- **An electromechanical energy conversion device** is one *which converts electrical energy into mechanical energy and mechanical energy into electrical energy*. Electromechanical energy conversion takes place *via* the medium of a magnetic or electric field the *magnetic field being most suited for practical conversion devices*. The conversion process is basically a *reversible one* though practical devices may be designed and constructed to particularly suit one mode of conversion or other.
  — From the *view-point of magnitudes of energy involved rotating or linear electrical machines are the most important energy converters*. These are primarily used *for bulk energy conversion and utilisation*.

Examples : *Motors and generator* (*continuous energy-conversion equipment*).

— *A "second category"* includes devices for measurement and control, frequently referred to as *transducers*. These generally operate under linear-output conditions and *with relatively small signals*.

Examples : *Microphones, pictures, sensors,* and *loudspeakers*.

— A *"third category"* of devices encompasses *force-producing devices*.

Examples : *Solenoids, relays and electromagnets*.

2. PRINCIPLE OF ENERGY CONVERSION

For converting energy from one form to another the **"principle of conservation of energy"** (*which states that energy can neither be created nor destroyed, it can merely be converted from one form to another*) can be invoked. In an energy conversion device, out of the total input energy, a larger part of the energy is converted into useful output energy, some energy is stored and rest is converted to heat (called energy loss). Thus *energy balance*, which should include the above terms, can be written for a motor and a generator as follows :

Motor (*action*) :

$$\begin{bmatrix}\text{Total electrical}\\ \text{energy input}\end{bmatrix} = \begin{bmatrix}\text{Mechanical energy}\\ \text{output}\end{bmatrix} + \begin{bmatrix}\text{Total energy stored}\\ \text{(in magnetic field)}\end{bmatrix} + \begin{bmatrix}\text{Energy converted}\\ \text{into heat}\end{bmatrix} \quad ...(1)$$

Eqn. (1) is written so that the electrical and mechanical energy terms have *positive* values for motor action. The equation applies well to generator action ; these terms then simply have *negative* values.

Generator (*action*) :

$$\begin{bmatrix}\text{Total mechanical}\\ \text{energy input}\end{bmatrix} = \begin{bmatrix}\text{Electrical energy}\\ \text{output}\end{bmatrix} + \begin{bmatrix}\text{Total energy stored}\\ \text{(in magnetic field)}\end{bmatrix} + \begin{bmatrix}\text{Energy converted}\\ \text{into heat}\end{bmatrix} \quad ...(2)$$

In the above cases, the *sign of the heat generation term(s) is such that heat generation within the system results in a flow of thermal energy out of the system.* In the systems, to be considered here, the conversion of energy into heat occurs by such mechanisms as ohmic heating due to current flow in the windings of the electric terminals and mechanical friction due to the motion of the system components forming the mechanical terminals.

Thus, energy balance equation may be written as :

$$\int dW_{\text{input}} = \int dW_{\text{output}} + \int dW_{\text{magnetic field}} + \int dW_{\text{heat}} \quad ...(3)$$

Fig. 1 shows the flow of energy in electromechanical energy conversion *via* a coupling field.

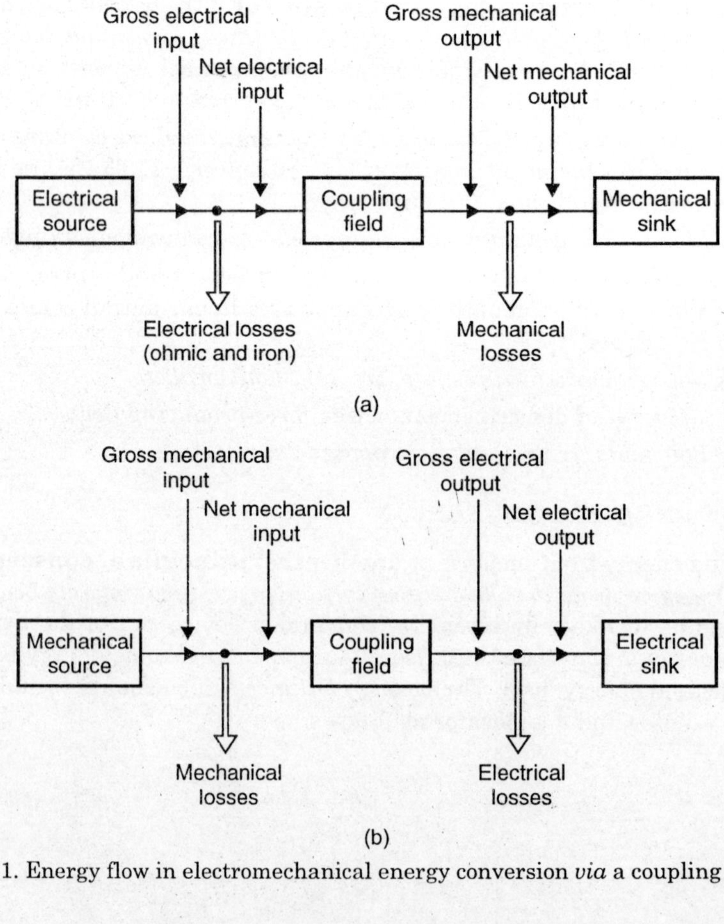

Fig. 1. Energy flow in electromechanical energy conversion *via* a coupling field.

- Owing to the inertia associated with mechanical components, electromechanical energy conversion devices are *slow-moving*. Therefore, the coupling field necessary for energy conversion *must vary slowly* and as such this field is *quasi-static in nature.* Electromagnetic radiation from the coupling field is almost negligible.

3. FARADAY'S LAWS OF ELECTROMAGNETIC INDUCTION

"The phenomenon whereby an e.m.f. and hence current is induced in any conductor which is cut across or is cut by a magnetic flux is known as ***electromagnetic induction****".*

Faraday's First Law. It states as follows :

Whenever the magnetic flux linked with a circuit changes, an e.m.f. is always induced in it".

Or

"Whenever a conductor cuts magnetic flux, an e.m.f. induced in that conductor".

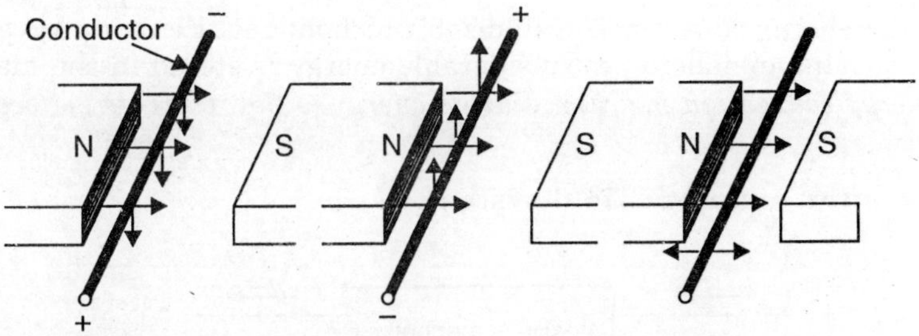

(*a*) Voltage induced across a wire moving downward.

(*b*) Voltage induced across a wire moving upward.

(*c*) No voltage is induced in a wire moving parallel to the field.

Fig. 2. When a conductor is moved across a magnetic field a voltage is induced in the conductor.

Faraday's Second Law. It states as follows :

"The magnitude of induced e.m.f. is equal to the rate of change of flux-linkages".

Mathematically, $$e = -N\frac{d\phi}{dt} \text{ volts} \qquad ...(4)$$

where, e = induced e.m.f.,

$\frac{d\phi}{dt}$ = rate of change of flux, and

N = number of turns of the coil.

[Usually, a *minus sign* is given to the right-hand side expression to signify the fact that the *induced e.m.f. sets up current in such a direction that magnetic effect produced by it opposes the very cause producing it.*]

Induced e.m.f. :

Induced e.m.f. may be of the following two types :

1. Dynamically induced e.m.f.
2. Statically induced e.m.f.

Dynamically Induced e.m.f. **:**

Refer Fig. 2. The e.m.f. induced (e) in the conductor is given by :

$$e = Blv \text{ volt} \qquad ...(5)$$

where, B = flux density of the magnetic field in *tesla*,
l = length of the conductor in *metres*, and
v = velocity of the conductor in *m/s*.

If the conductor moves at an angle θ with the direction of flux then the induced e.m.f.

$$e = Blv \sin \theta \text{ volts} \quad ...(6)$$

The direction of the induced e.m.f. is given by *Fleming's Right hand rule.*

Statically Induced e.m.f. **:**

The e.m.f. induced by *variation of flux is termed as "statically induced e.m.f.".*

Statically induced e.m.f. can be further subdivided as follows :

(*i*) Self-induced e.m.f.

(*ii*) Mutually induced e.m.f.

4. SINGLY AND MULTIPLY-EXCITED MAGNETIC FIELD SYSTEMS

Let us consider energy-conversion systems in which magnetic circuits have air-gaps between the stationary and moving members ; a considerable energy is stored in the magnetic field. *This field acts as the energy-conversion medium and its energy is the reservoir between the electric and mechanical systems.*

(*a*) Singly-excited magnetic field system

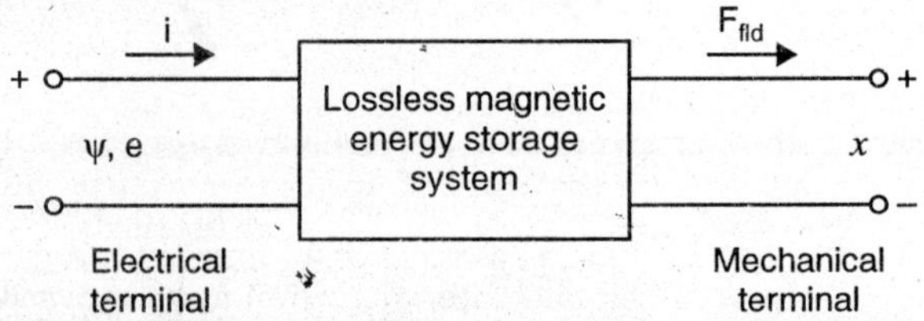

(a) Schematic magneto-field electromechanical-energy conversion device.

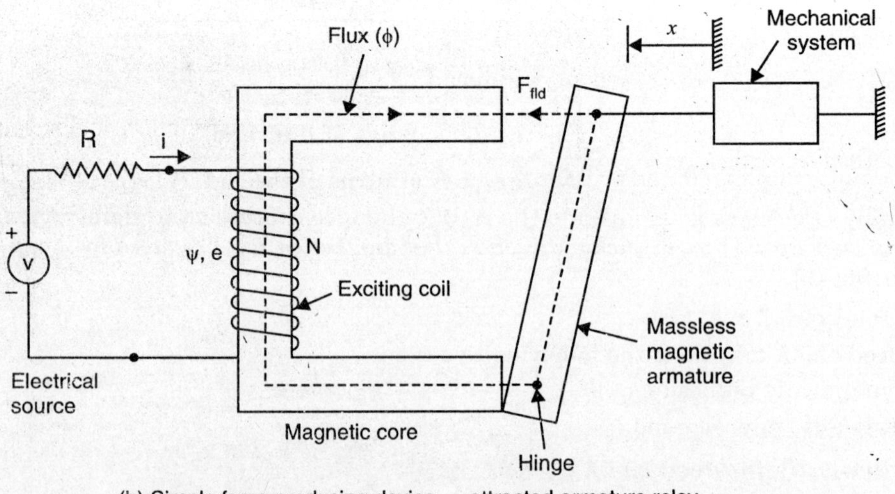

(b) Simple force-producing device — attracted armature relay.

Fig. 3. Singly-excited magnetic field system.

Fig. 3 shows a singly-excited magnetic field system. A mechanical force F_{fld} acts on the armature of the relay [Fig. 3 (*b*)] which is connected to a mechanical system comprising an active and/or passive element.

Assumptions made :

(*i*) No loss of energy in the magnetic core.

(*ii*) The whole of flux created by the core is confined to it.

(*iii*) The resistance of the exciting coil is represented outside in a lumped fashion.

Let us assume that the armature moves a differential distance dx in the direction of F_{fld}. Then, as per the law of conservation of energy, the energy balance equation may be written as :

Electrical energy input = mechanical energy output + increase in field energy.

or, $$i d\psi = F_{fld} \cdot dx + dW_{fld}$$ ...[7 (*a*)] (fld stands for field)

[where ψ = flux linkages = $N\phi$, N being number of terms of the exciting coil.]

or, $$F_{fld} \cdot dx = i.d\psi - dW_{fld}$$ ...[7 (*b*)]

Neglecting iron reluctance, for a linear case the inductance of the coil is a function $L(x)$ of linear distance of the armature.

$$\therefore \quad dW_{fld} = \frac{1}{2} dL(x)\, i^2$$

and, $$d\psi = i \times dL(x)$$

Thus, $$F_{field} \cdot dx = i^2 dL(x) - \frac{1}{2} i^2 dL(x) = \frac{1}{2} i^2 dL(x)$$ ...(8)

Eqn. (8) indicates that out of the total electrical energy input, half of energy input appears as mechanical output and half of it gets stored in the magnetic field. This statement is also applicable in the case when the armature undergoes finite movement, subject to the condiiion that, it (armature) moves very slowly such that current in the exciting coil remains unaltered.

Eqn. (8) can be written as :

$$F_{fld} = \frac{1}{2} i^2 \frac{dL(x)}{dx}$$ ...(9)

$$= \frac{dW_{fld}(i, x)}{dx}$$ ...(10)

Now, $$L(x) = \frac{N^2}{S(x)}$$ ...(11)

(where, S is the reluctance)

Substituting in Eqn. (9), we get

$$F_{fld} = \frac{1}{2} i^2 \frac{d}{dx}\left[\frac{N^2}{S(x)}\right]$$

$$= -\frac{1}{2} \frac{(Ni)^2}{S^2(x)} \frac{dS(x)}{dx} = -\frac{1}{2} \phi^2 \frac{d}{dx} S(x)$$ ...(12)

$$\left(\because \phi = \frac{Ni}{S}\right)$$

Eqn. (10) indicates that the force acts in a direction so as *to increase coil inductance* and according to eqn. (12) it acts in a direction to *reduce the reluctance of the magnetic circuit.*

Similarly, in a rotary case,

Torque, $$T_{fld} = \frac{dW_{fld}(i, \theta)}{d\theta}.$$

The cases mentioned above are usually employed for motion through a *limited distance or rotation through prescribed angles* and the expressions derived for singly excited system. Here the electromechanical conversion takes place through only one A.C. or D.C. source and as such they are known as "*Singly-excited magnetic field systems*"; *examples* being *electromagnetic relay, reluctance motor*.

(*b*) Multiply-excited magnetic field system

Some devices, which have the special requirement of producing an electrical signal proportional to forces or velocities or producing force proportional to electrical signal (current or voltage), require **two excitations**. *One excitation establishes a magnetic field of specified strength* while the other excitation produces the desired signal (electrical or mechanical). Such a system is called as *doubly excited or multiply-excited magnetic field system* ; **examples** being *synchronous motor, and alternator on which stator and rotor have A.C. and D.C. excitation respectively.*

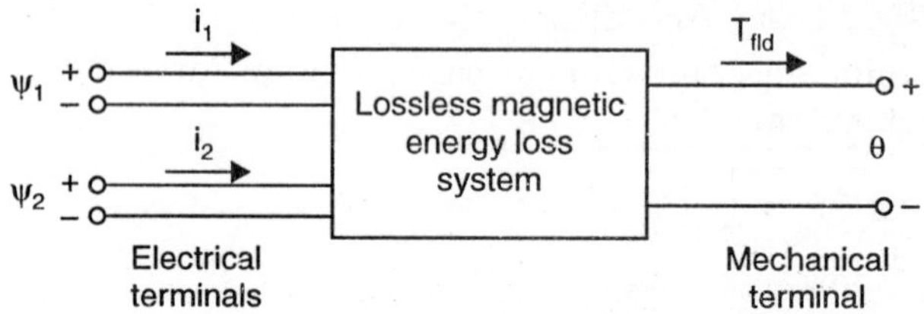

(a)

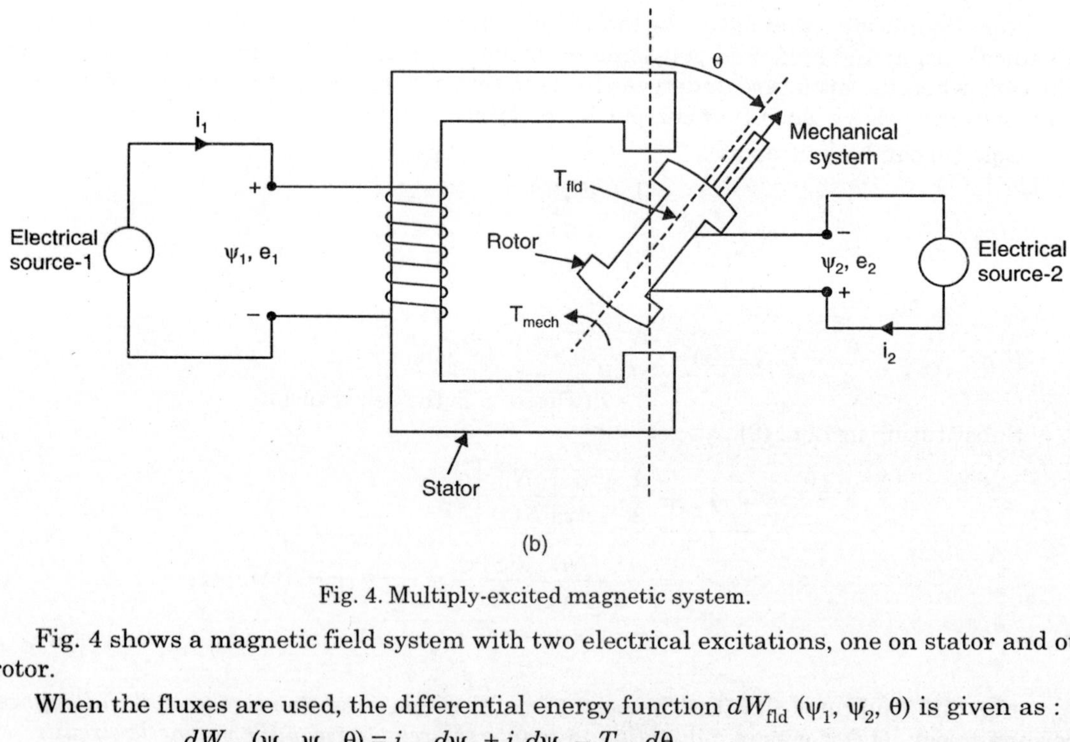

(b)

Fig. 4. Multiply-excited magnetic system.

Fig. 4 shows a magnetic field system with two electrical excitations, one on stator and other on rotor.

When the fluxes are used, the differential energy function $dW_{fld}\ (\psi_1, \psi_2, \theta)$ is given as :

$$dW_{fld}\ (\psi_1, \psi_2, \theta) = i_1 \,.\, d\psi_1 + i_2 d\psi_2 - T_{fld}\ d\theta \qquad ...(13)$$

Also, $$i_1 = \left.\frac{\partial W_{fld}\ (\psi_1, \psi_2, \theta)}{\partial \psi_1}\right|_{\psi_2,\ \theta} \qquad ...(14)$$

$$i_2 = \left.\frac{\partial W_{fld}(\psi_1, \psi_2, \theta)}{\partial \psi_2}\right|_{\psi_1, \theta} \quad ...(15)$$

and,
$$T_{fld} = -\left.\frac{\partial W_{fld}(\psi_1, \psi_2, \theta)}{\partial \theta}\right|_{\psi_1, \psi_2} \quad ...(16)$$

5. TORQUE PRODUCTION IN ROTATING MACHINES

Refer Fig. 5. When currents flow in stator and rotor windings of a machine, these windings produce their own magnetic fields along their respective axes which are sinusoidally distributed along the air gap. These fields tend to align themselves, which results in the production of a torque. This torque is produced only if the two fields have the same number of poles and are stationary with respect to each other. The flux components set up by the stator and rotor currents cross the air-gap twice and complete their circuits *via* the stator and rotor iron ; these component fields cause the appearance of north (*N*) and south (*S*) poles on stator and rotor surfaces ; the field axes being along N—S and out of *N*-pole. As the two relatively rotating fields cross each other, they will produce alternating torque so that the average torque is zero. Therefore, all the varieties of electrical machines *viz.* Induction, synchronous, D.C. are devised to produce interacting fields with zero relative velocity.

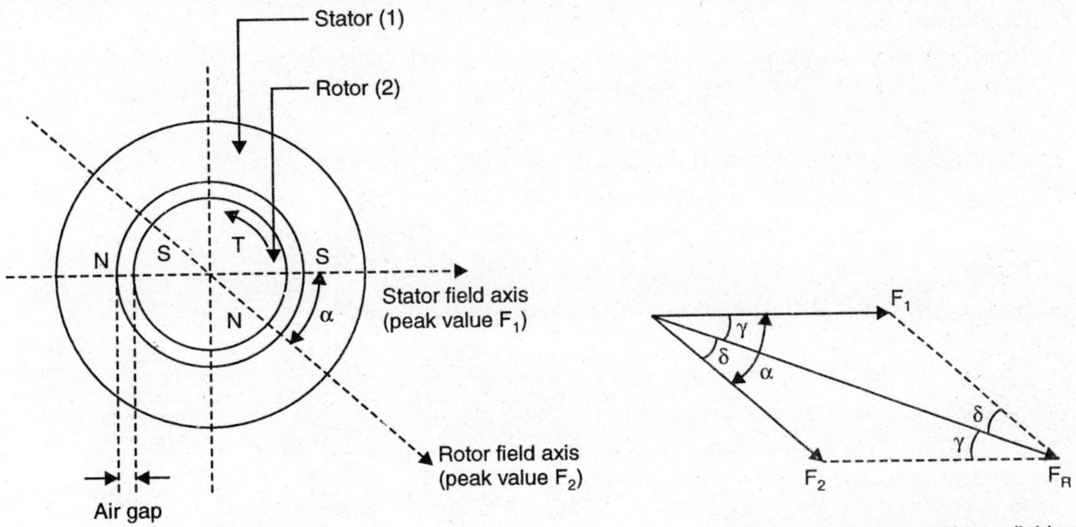

(a) Stator and rotor fields (2-pole machine)

(b) Vector sum of stator and rotor fields

Fig. 5. Torque production in rotating machines.

Assumptions made :

- Rotor is cylindrical (non-salient pole) so that air-gap is uniform throughout.
- Stator and rotor mmfs are sinusoidal space waves.
- The air gap is narrow.
- Most of the resultant flux is common to stator and rotor windings.
- Reluctance of iron path of flux is negligible.

Let F_1, F_2 = peak values of the spatial sinusoidal mmfs of the stator and rotor respectively (See Fig. 5).

α = angle between F_1 and F_2.

The peak value of the resultant mmf F_R is given as : Refer Fig. 5 (*b*)

$$F_R = \sqrt{F_1^2 + F_2^2 + 2F_1F_2 \cos \alpha}$$

The torque developed (due to interaction of fields) can be expressed as :

$$T = K\, F_1 F_2 \sin \alpha \quad ...(17)$$

or,

$$T = K'\, F_R\, F_2 \sin \delta \quad ...(18)$$

where δ is the angle between the axes of F_R and F_2.

The resultant field establishes air gap flux of

$$\phi_R = \frac{F_R}{S} \text{ per pole}$$

where, S = reluctance per pole,

and, $F_2 \propto I_2$ (rotor current)

Hence, torque can be also expressed, in more useful form, as

$$T \propto \phi_R\, I_2 \sin \delta. \quad ...(19)$$

6. GENERAL ANALYSIS OF ELECTROMECHANICAL SYSTEM

Refer to Fig. 3 again.

We know that,

Mechanical energy output = Electrical energy input–increase in field energy

$$F_{fld} \,.\, dx = id\psi - dW_{fld} \quad ...(20)$$

Also,

$$W_{fld} = i\psi - W_{fld}\,(i, x)$$

(where $\psi = N\phi$ and W_{fld} = co-energy)

Then

$$dW_{fld} = d(i\psi) - dW'_{fld} - (i, x)$$

$$= id\psi + \psi di - \left[\frac{\partial W'_{fld}}{\partial i}\, di + \frac{dW'_{fld}}{\partial x} \,.\, dx\right] \quad ...(21)$$

Substituting for dW_{fld} from eqn. (21) in eqn. (20), we get

$$F_{fld} \cdot dx = id\psi - \left[i.\,d\psi + \psi.\,di - \left\{\frac{dW'_{fld}}{\partial i}\, di + \frac{dW'_{fld}}{\partial x} \,.\, dx\right]\right.$$

or,

$$F_{fld} \,.\, dx = i.d\psi - i.d\psi - \psi.di + \frac{dW'_{fld}}{\partial i} di + \frac{dW'_{fld}}{\partial x} \,.\, dx$$

or,

$$F_{fld} \,.\, dx = \left(\frac{\partial W'_{fld}}{\partial i} - \psi\right) di + \frac{\partial W'_{fld}}{\partial x} \,.\, dx \quad ...(22)$$

Since the incremental changes *di* and *dx* are independent and *di* is not present on L.H.S. of eqn. (22), therefore, its coefficient of R.H.S. must be zero.

$$\therefore \quad \frac{\partial W'_{fld}}{\partial i} - \psi = 0$$

or,

$$\psi = \frac{\partial W'_{fld}}{\partial i}\, (i, x)$$

Thus, from eqn. (22), we have

$$F_{fld} = \frac{\partial W'_{fld}\,(i, x)}{\partial x} \quad ...(23)$$

This expression (for mechanical force developed) applies when *i* is an independent variable *i.e.*, it is a *current excited system*.

WORKED EXAMPLES

Example 1. *Fig. 6 shows a relay made from infinitely-permeable magnetic material with a movable plunger, also of infinite-permeable material. The height of the plunger is much greater than the air-gap length ($h >> l_g$).*

Calculate the magnetic stored energy W_{fld} as a function of plunger position ($0 < x < d$) for $N = 1200$ turns, $l_g = 1.8$ mm, $d = 0.12$ m, $l = 0.08$ m and $i = 12$ A.

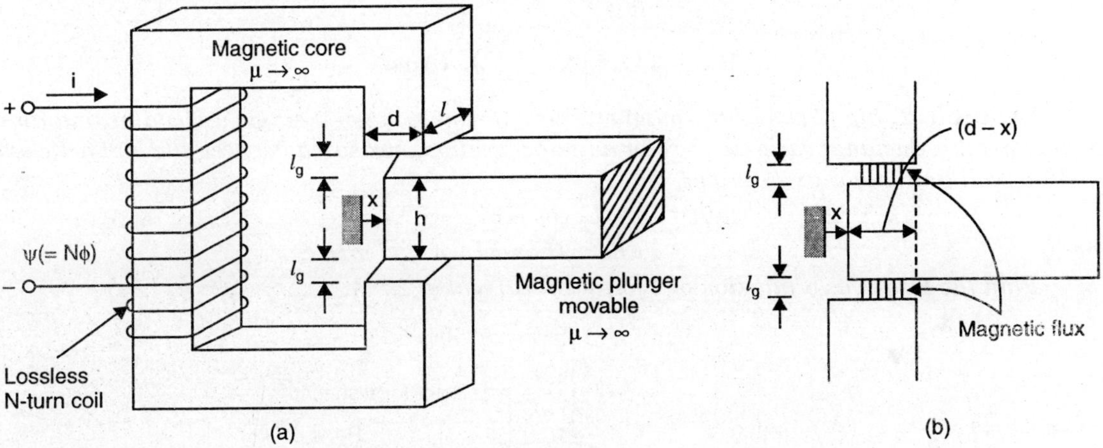

Fig. 6. (*a*) Relay with movable plunger.
(*b*) Detail showing air-gap configuration with the plunger partially removed.

Solution. *Given* : $N = 1200$ turns ; $l_g = 1.8$ mm ; $d = 0.12$ m ; $l = 0.08$ m ; $i = 12$ A.

Magnetic stored energy W_{fld} :

We know that, magnetic stored energy,

$$W_{fld} = \frac{1}{2} L(x) i^2 \quad ...(i)$$

where L is the inductance.

Now,
$$L(x) = \frac{\mu_0 N^2 A_g}{2 l_g}$$

where, μ_0 = absolute permeability = $4\pi \times 10^{-7}$ *H/m* ;
N = number of turns ;
A_g = gap cross-sectional area, and
l_g = length of air gap each, above and below the plunger.

Here,
$$A_g = l(d - x) = ld\left(1 - \frac{x}{d}\right)$$

Thus,
$$L(x) = \frac{\mu_0 N^2 ld\left(1 - \frac{x}{d}\right)}{2 l_g}$$

Substituting in eqn. (*i*), we have

$$\therefore \qquad W_{fld} = \frac{1}{2}\left[\frac{\mu_0 N^2 l d\left(1-\frac{x}{d}\right)}{2l_g}\right] i^2$$

$$= \frac{1}{2}\left[\frac{(4\pi\times10^{-7})\,(1200)^2\,(0.08)\,(0.12)\left(1-\frac{x}{d}\right)}{2(1.8\times10^{-3})}\right]\times 12^2$$

or, $$W_{fld} = \mathbf{347.4\left(1-\frac{x}{d}\right)\ J.\ (Ans.)}$$

Example 2. *Fig. 7 shows the magnetic circuit consisting of a single-coil stator and an oval rotor. The coil inductance varies with rotor angular position, measured between the magnetic axis of the stator coil and major axis of rotor, as*

$$L(\theta) = L_0 + L_2 \cos(2\theta)$$

where, $$L_0 = 11.66\ mH \text{ and } L_2 = 2.97\ mH.$$

Find the torque as a function of θ for a coil current of 2.2 A.

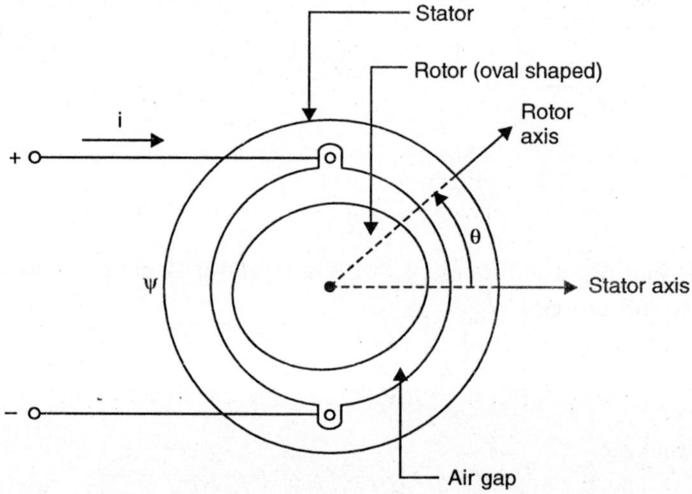

Fig. 7

Solution. *Given* : $L(\theta) = L_0 + L_2 \cos(2\theta)$

$L_0 = 11.66$ mH ; $L_2 = 2.97$ mH ;

$i = 2.2$ A

We know that, $$T_{fld}(\theta) = \frac{1}{2} i^2 \frac{dL(\theta)}{d\theta}$$

$$= \frac{i^2}{2}\frac{d}{d\theta}[L_0 + L_2\cos(2\theta)] = \frac{i^2}{2}[-2L_2\sin(2\theta)]$$

$$= -\frac{(2.2)^2}{2}[-2\times(2.97\times10^{-3})\sin(2\theta)]\ \text{Nm}$$

or, $$T_{fld}(\theta) = \mathbf{-0.0144\sin(2\theta)\ Nm.\quad (Ans.)}$$

It may be noted that, in this case the *torque acts in such a direction as to pull the rotor axis in alignment with the coil axis and hence to maximise the coil inductance.*

Example 3. *The following data relate to the electromagnetic relay shown in Fig. 3.*

Number of terms, N = 1200

Cross-sectional area of the core, A_{core} *= 2.7 cm × 2.7 cm.*

Permeability of the core = infinite

Neglecting fringing effects, find the following :

(i) The inductance of coil for an air gap length (l_g) *of 0.45 cm. The energy stored in the field and force acting on the armature when the coil carries a current of 2.25 A.*

(ii) The mechanical energy output when the air gap is decreased from 0.45 cm to 0.225 cm, assuming that the current through the coil remains constant at 2.25 A.

(iii) The expression for force on armature as a function of l_g*, coil current being constant at 2.25 A. Also calculate the work done by the magnetic field when* l_g *changes from 0.45 cm to 0.225 cm from the expression* $\int F_{fld}\, dl_g$ *and thereby verify the result of part (ii).*

Solution. Refer Fig. 3.

(i) **L ; W$_{fld}$; F$_{fld}$:**

$$\mathbf{L} = \frac{\mu_0 N^2 A_{core}}{l_g} = \frac{(4\pi \times 10^{-7}) \times (1200)^2 \times (2.7 \times 2.7 \times 10^{-4})}{(0.45 \times 10^{-2})} = \mathbf{0.293\ H.} \quad \textbf{(Ans.)}$$

$$\mathbf{W_{fld}} = \frac{1}{2} Li^2 = \frac{1}{2} \times 0.293 \times (2.25)^2 = \mathbf{0.742\ J.} \quad \textbf{(Ans.)}$$

$$\mathbf{F_{fld}} = \frac{W_{fld}}{l_g} = \frac{0.742}{(0.45 \times 10^{-2})} (-) = \mathbf{164.89\ Nm\ (-).} \quad \textbf{(Ans.)}$$

Negative sign indicates that force acts in a direction to reduce air gap length (l_g).

(ii) **W$_{mech}$:**

$$dW_{mech} = \frac{1}{2}\, dW_{elect} = \frac{1}{2}\, i_0 (\psi_2 - \psi_1) = \frac{1}{2} i_0^2 (L_2 - L_1)$$

As, $$L = \frac{\mu_0 N^2 A_{core}}{l_g} = \frac{(4\pi \times 10^{-7}) \times (1200)^2 \times (2.7 \times 2.7 \times 10^{-4})}{l_g} = \frac{0.001319}{l_g}$$

$$L_1 (l_g = 0.45 \times 10^{-2}\ \text{m}) = \frac{0.001319}{0.45 \times 10^{-2}} = 0.293\ \text{H}$$

$$L_2 (l_g = 0.225 \times 10^{-2}\ \text{m}) = \frac{0.001319}{0.225 \times 10^{-2}} = 0.586\ \text{H}$$

Substituting the values, we have

$$\mathbf{W_{mech}} = \frac{1}{2} \times (2.25)^2 (0.586 - 0.293) = \mathbf{0.742\ J.} \quad \textbf{(Ans.)}$$

(iii) **Expression for force (F$_{fld}$) as a function of l$_g$ and W$_{mech}$:**

$$F_{fld} = \frac{W_{fld}}{l_g} = -\frac{\frac{1}{2} Li^2}{l_g} = -\frac{1}{2}\left[\frac{\mu_0 N^2 A_{core}}{l_g \times l_g}\right] i^2$$

$$= -\frac{1}{2}\left[\frac{4\pi \times 10^{-7} \times 1200^2 \times (2.7 \times 2.7 \times 10^{-4})}{l_g^2}\right] \times (2.25)^2$$

or $$F_{fld} = -\frac{0.00334}{l_g^2} \quad \text{... Required expression. (Ans.)}$$

$$dW_{mech} = \int_{l_{g_1}}^{l_{g_2}} F_{fld}\, dl_g = \int_{0.45\times10^{-2}}^{0.225\times10^{-2}} -\frac{0.00334}{l_g^2} = \left[\frac{0.00334}{l_g}\right]_{0.45\times10^{-2}}^{0.225\times10^{2}}$$

$$= 0.00334\left[\frac{1}{0.225\times10^{-2}} - \frac{1}{0.45\times10^{-2}}\right] = \mathbf{0.742\ J}$$

$= \mathbf{0.742\ J}$ (Result of part (*ii*) verified). **(Ans.)**

Example 4. *The self and mutual inductances of the two exciting coils of a multiply-excited translatory system are :*

$$L_{11} = L_{22} = \frac{3.6}{1+2x},\ L_{12} = L_{21} = \frac{1.8}{1+2x}$$

Calculate the time average force and coil currents at x = 0.4 m when :

(i) Both the coils are connected in series across a voltage source of 100 cos 314 t ;

(ii) Both the coils are connected in parallel across a voltage source of 100 cos 314 t.

Solution. *Given* : $L_{11} = L_{22} = \frac{3.6}{1+2x}$; $L_{12} = L_{21} = \frac{1.8}{1+2x}$; $x = 0.4$ m

Time average force $F_{fld(av.)}$; i_1, i_2 :

(*i*) ***When the coils are connected in series* :**

$$\text{At } x = 0.4 \text{ m} ;\ L_{11} = L_{22} = \frac{3.6}{1+2\times0.4} = 2,$$

and $$L_{12} = L_{21} = \frac{1.8}{1+2\times0.4} = 1$$

With armature held at $x = 0.4$ m, the circuit equations for the two coils are :

$$v_1 = L_{11}\frac{di_1}{dt} + L_{12}\frac{di_2}{dt} = 2\frac{di_1}{dt} + \frac{di_2}{dt} \quad ...(1)$$

$$v_2 = L_{22}\frac{di_2}{dt} + L_{21}\frac{di_1}{dt} = 2\frac{di_2}{dt} + \frac{di_1}{dt} \quad ...(2)$$

As both the coils are connected in series (*i.e.*, $i_1 = i_2$)

$\therefore$ $$v_1 + v_2 = 100 \cos 314\, t$$

or $$3\frac{di_1}{dt} + 3\frac{di_2}{dt} = 100 \cos 314\, t \quad \text{[From (i) and (ii)]}$$

or $$\frac{di_1}{dt} = \frac{di_2}{dt} = \frac{100}{6}\cos 314\, t = 16.67 \cos 314\, t \quad (\because\ i_1 = i_2)$$

By integrating, $$\mathbf{i_1 = i_2} = \frac{16.67}{314}\sin 314\, t = \mathbf{0.053 \sin 314\, t.} \quad \textbf{(Ans.)}$$

Now, $$W_{fld}(i_1, i_2, x) = \frac{1}{2}L_{11}i_1^2 + \frac{1}{2}L_{22}i_2^2 + L_{12}i_1 i_2$$

$$= \frac{1}{2}\left(\frac{3.6}{1+2x}\right)i_1^2 + \frac{1}{2}\left(\frac{3.6}{1+2x}\right)i_2^2 + \left(\frac{1.8}{1+2x}\right)i_1 i_2$$

$$= \frac{1.8}{1+2x}(i_1^2 + i_2^2 + i_1 i_2)$$

$$F_{fld} = -\frac{\partial W_{fld}}{\partial x} = \frac{3.6}{(1+2x)^2}(i_1^2 + i_2^2 + i_1 i_2)$$

At $$x = 0.4 \text{ m} ; F_{fld}(x = 0.4 \text{ m}) = \frac{3.6}{(1+2\times 0.4)^2}(i_1^2 + i_2^2 + i_1 i_2)$$

$$= 1.111 (i_1^2 + i_2^2 + i_1 i_2)$$

$\therefore$ $$F_{fld}(t) = 3 \times 1.111 (0.053 \sin 314\, t)^2 \qquad (\because i_1 = i_2)$$

$$= 3 \times 1.111 \times (0.053)^2 \sin^2 314\, t$$

$$= 0.00936 \sin^2 314\, t$$

$$F_{fld}\text{(avg.)} = \frac{3 \times 1.111}{2} \times (0.053)^2 = \mathbf{0.00468\ N.} \quad \textbf{(Ans.)}$$

(*ii*) ***When the coils are connected in parallel :***

In this case, $v_1 = v_2 = 100 \cos 314\, t$

From eqns. (1) and (2), we have

$$v_1 = v_2 = \left[\frac{2di_1}{dt} + \frac{di_2}{dt} = 2\frac{di_2}{dt} + \frac{di_1}{dt}\right] = 100 \cos 314\, t$$

or $$\frac{di_1}{dt} = \frac{di_2}{dt} = \frac{100}{3} \cos 314\, t = 33.33 \cos 314\, t$$

By integrating, we have

$$i_1 = i_2 = \frac{33.33}{314} \sin 314\, t = \mathbf{0.106 \sin 314\, t.} \quad \textbf{(Ans.)}$$

$\therefore$ $$F_{fld}(t) = 3 \times 1.111 (0.106)^2 \sin^2 314\, t$$

or $$F_{fld}\text{(avg.)} = \frac{3 \times 1.111}{2}(0.106)^2 = \mathbf{0.0187\ N.} \quad \textbf{(Ans.)}$$

Example 5. *Fig. 8 shows a system, in which the inductances are as given below :*

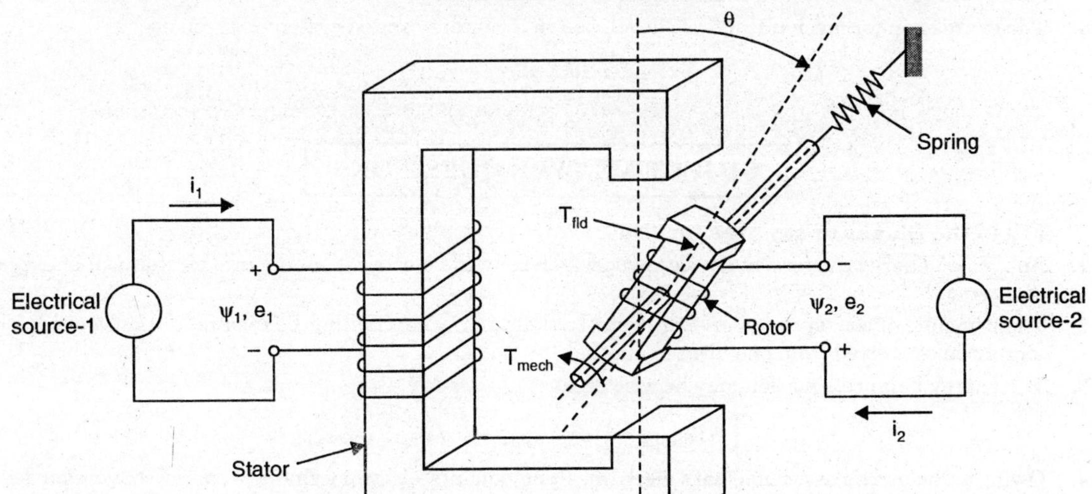

Fig. 8

$L_{11} = (2 + \cos 2\theta) \times 10^{-3}\, H$;
$L_{12} = 0.2 \cos \theta\, H$;
$L_{22} = 15 + 5 \cos 2\theta\, H$
Determine the torque $T_{fld}(\theta)$ *for current* $i_1 = 0.96\,A$ *and* $i_2 = 0.012\,A$.

Solution. The torque (T_{fld}) can be determined as follows :

$$T_{\text{fld}} = \frac{i_1^2}{2}\frac{dL_{11}(\theta)}{d\theta} + \frac{i_2^2}{2}\frac{dL_{22}(\theta)}{d\theta} + i_1 i_2 \frac{dL_{12}(\theta)}{d\theta}$$

$$= \frac{i_1^2}{2}\frac{d}{d\theta}(2 + \cos 2\theta) \times 10^{-3} + \frac{i_2^2}{2}\frac{d}{d\theta}(15 + 5\cos 2\theta) + i_1 i_2 \frac{d}{d\theta}(0.2 \cos\theta)$$

$$= \frac{i_1^2}{2}(-2 \times 10^{-3}) \sin 2\theta + \frac{i_2^2}{2}(-10 \sin 2\theta) + i_1 i_2 (-0.2 \sin\theta)$$

Substituting the values of i_1 (= 0.96 A) and i_2 (= 0.012 A), we have

$$T_{\text{fld}} = \frac{(0.96)^2}{2}(-2 \times 10^{-3}) \sin 2\theta + \frac{(0.012)^2}{2}(-10 \sin 2\theta) + (0.96 \times 0.012)(-0.2 \sin\theta)$$

$$= -0.922 \times 10^{-3} \sin 2\theta - 0.72 \times 10^{-3} \sin 2\theta - 2.3 \times 10^{-3} \sin\theta$$

or $$\mathbf{T_{fld} = -1.642 \times 10^{-3} \sin 2\theta - 2.3 \times 10^{-3} \sin\theta.} \quad \textbf{(Ans.)}$$

HIGHLIGHTS

1. An *electromechanical energy conversion device* is one which converts electrical energy into mechanical energy and mechanical energy into electrical energy.
2. The phenomenon where by an e.m.f. and hence current is induced in any conductor which is cut across or cut by a magnetic flux is known as *electromagnetic induction.*
3. The examples of *singly-excited* magnetic field system are *electromagnetic relays, reluctance motor.*
4. The examples of *multiply-excited magnetic field systems are synchronous motor, alternator* on which stator and rotor have A.C. and D.C. excitation respectively.
5. The expression for mechanical force developed in a current excited system is given as :

$$F_{\text{fld}} = \frac{\partial W'_{\text{fld}}(i, x)}{\partial x}.$$

OBJECTIVE TYPE QUESTIONS

Fill in the Blanks or say "Yes" or "No" :

1. An .... energy conversion device is one which converts electrical energy into mechanical energy and *vice versa.*
2. The principle of conservation of energy states that energy can neither be created nor destroyed, it can merely be converted from one form to another.
3. The energy balance equation may be written as :

$$\int dW_{\text{input}} = \int dW_{\text{output}} + \int dW_{\text{magnetic field}} + \text{..........} .$$

4. Owing to the inertia associated with mechanical components, electromechanical energy conversion devices (EMEC) are ....
5. Electromagnetic radiation from the coupling field in EMEC device is ....
6. Faraday's .... law states that the magnitude of induced e.m.f. is equal to the rate of change of flux-linkages.
7. The e.m.f. induced by variations of flux is termed as dynamically induced e.m.f.

8. Electromagnetic relay is an example of singly-excited magnetic field system.
9. An altenator is an example of .... excited magnetic field system.
10. Reluctance motor is an example of excited magnetic field system.

ANSWERS

1. electromechanical	2. Yes	3. $\int dW_{heat}$	4. slow moving	5. negligible
6. second	7. No	8. Yes	9. multiply	10. singly.

THEORETICAL QUESTIONS

1. What is an "Electromechanical energy conversion device" ?
2. Discuss briefly "Principle of energy conversion".
3. Describe briefly Faraday's laws of electromagnetic induction.
4. Explain briefly an electromechanical energy conversion device with the help of a block diagram.
5. For a singly excited magnetic field system, derive the relation for the magnetic stored energy.
6. Show that the torque developed in a doubly excited system is equal to the rate of increase of field energy with respect to displacement at constant.
7. Derive an expression for the magnetic force developed in a multiply-excited translational magnetic system.
8. Discuss briefly about torque production in rotating machines.
9. Discuss briefly general analysis of electromechanical system, and derive an expression for the mechanical force developed in a current excited system.

UNSOLVED EXAMPLES

1. The relay shown in Fig. 6 is made from infinitely-permeable magnetic material with a movable plunger, also of infinitely permeable material. The height of the plunger is much greater than the airgap length ($h >> l_g$).
Calculate the magnetic stored energy W_{fld} as a function of plunger position ($0 < x < d$) for $N = 1000$ turns;
$l_g = 2.0$ mm ; $d = 0.15$ m, $l = 0.1$ m, and $i = 10$ A. $\left[\textbf{Ans. } 236\left(1-\frac{x}{d}\right) \text{J}\right]$
2. Fig. 7 shows the magnetic circuit consisting of a single-coil stator and an oval rotor. The coil inductance varies with rotor angular position, measured between the magnetic axis of the stator coil and major axis of rotor, as
$$L(\theta) = L_0 + L_2 \cos (2\theta)$$
where $L_0 = 10.6$ mH and $L_2 = 2.7$ mH.
Find the torque as a function of θ for a coil current of 2 A. [**Ans.** $T_{fld}(\theta) = -0.0108 \sin (2\theta)$ Nm]
3. The following data relate to the electromagnetic relay shown in Fig. 3.
Number of turns = 1000
Cross-sectional area of the core = 3 cm × 3 cm
Permeability of the core = infinite
Neglecting fringing effects, find the following :
(*i*) The inductance of coil for air gap length (l_g) of 0.5 cm, the energy stored in the field, and force acting on the armature when the coil carries a current of 2.5 A.
(*ii*) The mechanical energy output when the air gap is decreased from 0.5 cm to 0.25 cm, assuming that the current through the coil remains constant at 2.5 A.
[**Ans.** (*i*) 226.1 mH ; 0.706 J ; – 141.37 Nm (*ii*) 0.706 J]

4. The self and mutual inductances of the two exciting coils of a multiply-excited translatory systems are :

$$L_{11} = L_{22} = \frac{4}{1+2x},\ L_{12} = L_{21} = \frac{2}{1+2x}$$

Calculate the time average force and coil currents at $x = 0.5$ m when :

(*i*) Both the coils are connected in series across a voltage source of 100 cos 314 t.

(*ii*) Both the coils are connected in parallel across a voltage source of 100 cos 314 t.

[**Ans.** (*i*) 0.053 sin 314 t ; 0.00 421 (*ii*) 0.106 sin 314 t ; 0.0169 N]

5. Fig. 8 shows a system, in which the inductances are as given below :

$L_{11} = (3 + \cos 2\theta) \times 10^{-3}$; $L_{12} = 0.3 \cos \theta$; $L_{22} = 30 + 10 \cos 2\theta$.

Determine the torque $T_{fld}(\theta)$ for current $i_1 = 0.8$ A and current $i_2 = 0.01$ A.

[**Ans.** $T_{fld} = 0.00164 \sin 2\theta - 0.0024 \sin \theta$]

ALTERNATING CURRENT MACHINES

(Electrical Machines—Part II)

1

Polyphase Circuits

1.1. INTRODUCTION

- Generation, transmission and heavy-power utilisation of A.C. electric energy almost invariably involve a type of system or circuit called a *polyphase system* or *polyphase circuit*. In such a system, each voltage source consists of a group of voltages having relative magnitudes and phase angles. Thus, a *m-phase system will employ voltage sources which, conventionally, consist of m voltages substantially equal in magnitude and successively displaced by a phase angle of 360°/m.*
- A *3-phase system* will employ voltage sources whicn, conventionally, consist of *three voltages substantially equal in magnitude* and *displaced by phase angles of 120°*. Because it *possesses definite economic and operating advantages, the 3-phase system is by far the most common,* and consequently emphasis is placed on 3-phase *circuits*.

1.2. ADVANTAGES OF POLYPHASE SYSTEMS

The advantages of polyphase systems *over single-phase systems* are :

1. A polyphase transmission line requires less conductor material than a single-phase line for transmitting the same amount power at the same voltage.

2. For a given frame size a polyphase machine gives a higher output than a single-phase machine. For example, output of a 3-phase motor is 1.5 times the output of single-phase motor of same size.

3. Polyphase motors have a uniform torque where most of the single-phase motors have a pulsating torque.

4. Polyphase induction motors are self-starting and are more efficient. On the other hand single-phase induction motors are not self-starting and are less efficient.

5. Per unit of output, the polyphase machine is very much cheaper.

6. Power factor of a single-phase motor is lower than that of polyphase motor of the same rating.

7. Rotating field can be set up by passing polyphase current through stationary coils.

8. Parallel operation of polyphase alternators is simple as compared to that of single-phase alternators because of pulsating reaction in single-phase alternator.

It has been found that the above advantages are *best realised in the case of three-phase systems*. Consequently, the electric power is generated and transmitted in the form of three-phase system.

1.3. GENERATION OF THREE-PHASE VOLTAGES

- Let us consider an elementary 3-phase 2-pole generator as shown in Fig. 1.1. On the armature are three coils, *ll'*, *mm'*, and *nn'* whose axes are displaced 120° in space from each other.

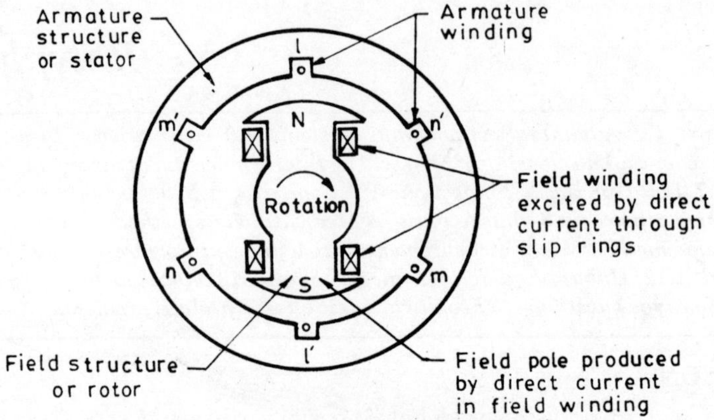

Fig. 1.1. Elementary 3-phase 2-pole generator.

- When the field is excited and rotated, voltages will be generated in the three phases in accordance with Faraday's law. If the field structure is so designed that the flux is distributed sinusoidally over the poles, the flux linking any phase will vary sinusoidally with time and sinusoidal voltages will be induced in three-phases. These three waves will be displaced 120 electrical degrees (Fig. 1.2) in time as a result of the phases being

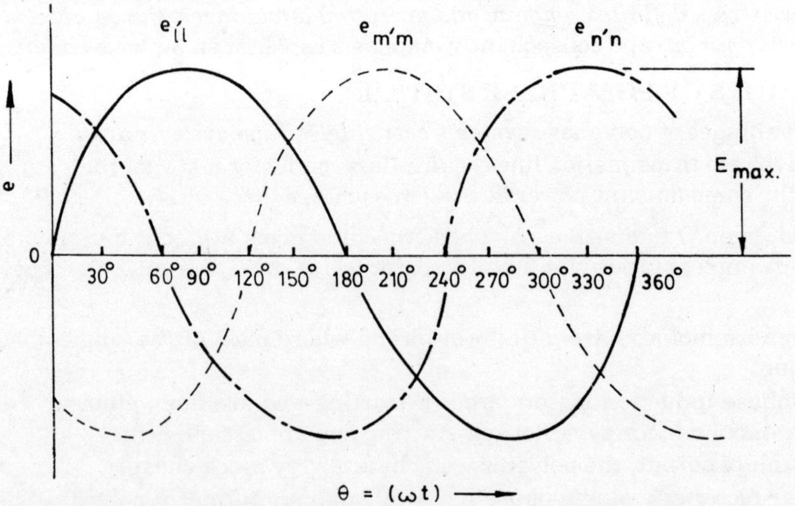

Fig. 1.2. Voltage waves generated in windings of Fig. 1.1.

displaced 120° in space. The corresponding phasor diagram is shown in Fig. 1.3. The equations of the instantaneous values of the three voltages (given by Fig. 1.2) are :

$$e_{l'l} = E_{max.} \sin \omega t$$
$$e_{m'm} = E_{max.} \sin (\omega t - 120°)$$
$$e_{n'n} = E_{max.} \sin (\omega t - 240°)$$

The sum of the above three e.m.fs. is always zero as shown below :

Resultant instantaneous e.m.f.

$$= e_{l'l} + e_{m'm} + e_{n'n}$$
$$= E_{max.} \sin \omega t + E_{max.} \sin (\omega t - 120°) + E_{max.} \sin (\omega t - 240°)$$
$$= E_{max.} [\sin \omega t + (\sin \omega t \cos 120° - \cos \omega t \sin 120° + \sin \omega t \cos 240° - \cos \omega t \sin 240°)]$$
$$= E_{max.} [\sin \omega t + (-\sin \omega t \cos 60° - \cos \omega t \sin 60° - \sin \omega t \cos 60° + \cos \omega t \sin 60°)]$$
$$= E_{max.} (\sin \omega t - 2 \sin \omega t \cos 60°)$$
$$= E_{max.} (\sin \omega t - \sin \omega t) = 0.$$

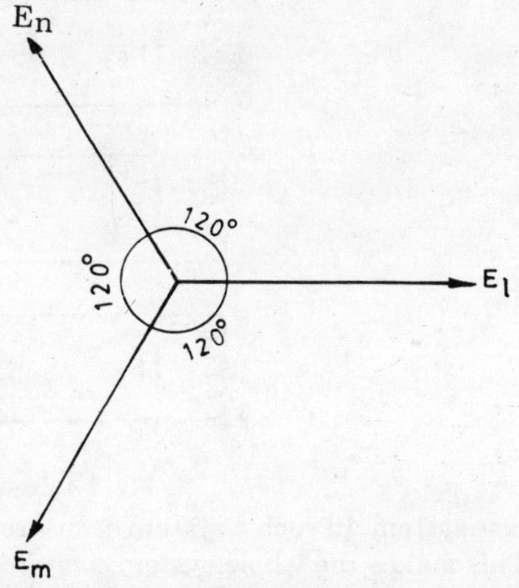

Fig. 1.3. Phasor diagram of generated voltages.

1.4. PHASE SEQUENCE AND NUMBERING OF PHASES

- *By phase sequence is meant the order in which the three phases attain their peak or maximum.*

In the generation of three-phase e.m.fs. in Fig. 1.2 clockwise rotation of the field system in Fig. 1.1 was assumed. This assumption made the e.m.f. of phase '**m**' lag behind that of '**l**' by 120° and in a similar way, made that of '**n**' lag behind that of '**m**' by 120° (or that of **l** by 240°). Hence, the order in which the e.m.fs. of phases **l, m** and **n** attain their maximum value is **lmn**. It is called the *phase order* or phase sequence **l** → **m** → **n**. If now the rotation of field structure of Fig. 1.1 is reversed *i.e.* made counter-clockwise, then the order in which three phases would attain their corresponding maximum voltages would also be reversed. The phase sequence would become **l** → **n** → **m**. This means that e.m.f. of phase '**n**' would now lag behind that of phase '**l**' by 120° instead of 240° as in the previous case.

The phase sequence of the voltages applied to a load, in general, is determined by the order in which the 3-phase lines are connected. The *phase sequence can be reversed* by *interchanging any pair of lines.* (In the case of an induction motor, reversal of sequence results in the reversed direction of motor rotation).

- The three-phases may be *numbered* **l, m, n** or 1, 2, 3 or they may be given three *colours* (as is customary).

The colours used commercially are *red, yellow* (or sometimes white) and *blue.* In this case sequence is RYB.

Evidently in any three-phase system, there are two possible sequences, in which three coils or phase voltages may pass through their maximum value *i.e.* red → yellow → blue (RYB) or red → blue → yellow (RBY).

By convention :

RYB ...... taken as *positive.*

RBY ...... taken as *negative.*

1.5. INTER-CONNECTION OF THREE PHASES

Each coil of three phases has two terminals [one 'start' (*S*) and another 'finish' (*F*)] and if individual phase is connected to a separate load circuit, as shown in Fig. 1.4, we get a non-interlinked

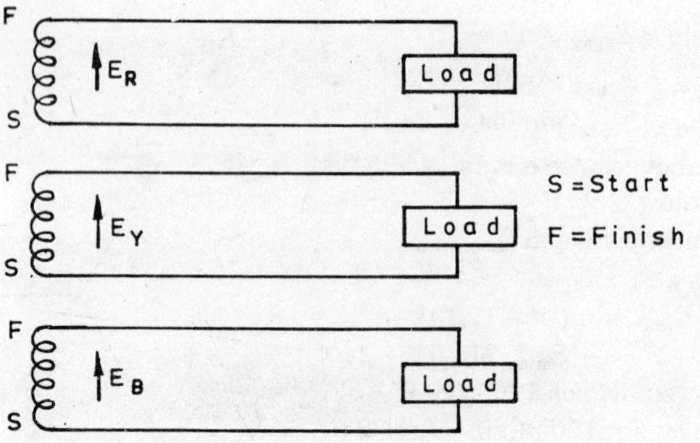

Fig. 1.4. Non-interlinked 3-phase system.

3-phase system. In such a system each circuit will require two conductors, therefore, 6 conductors in all. This makes the whole system *complicated* and *expensive*. Hence the *three phases are generally interconnected which results in substantial saving of copper*.

The general method of inter-connections are :

1. Star or Wye (Y) connection.
2. Mesh or delta (Δ) connection.

1.6. STAR OR WYE (Y) CONNECTION

- In this method of inter-connection the similar ends either the 'start' or 'finish' are joined together at point N. This common point N [Fig. 1.5 (a)] is called *star point* or *neutral point*. Ordinarily only three wires are carried to the external circuit giving 3-phase, 3-wire start connected system but sometimes a *fourth-wire*, known as *neutral wire* is carried to the neutral point of the external load circuit giving 3-phase, 4 wire star connected system.
- The *voltage between any line and the neutral point* (*i.e.* voltage across the phase winding) is called the **'phase voltage' (E_{ph})** ; while the *voltage available between any pair of terminals* (or outers) is called the **'line voltage' (E_L)**.

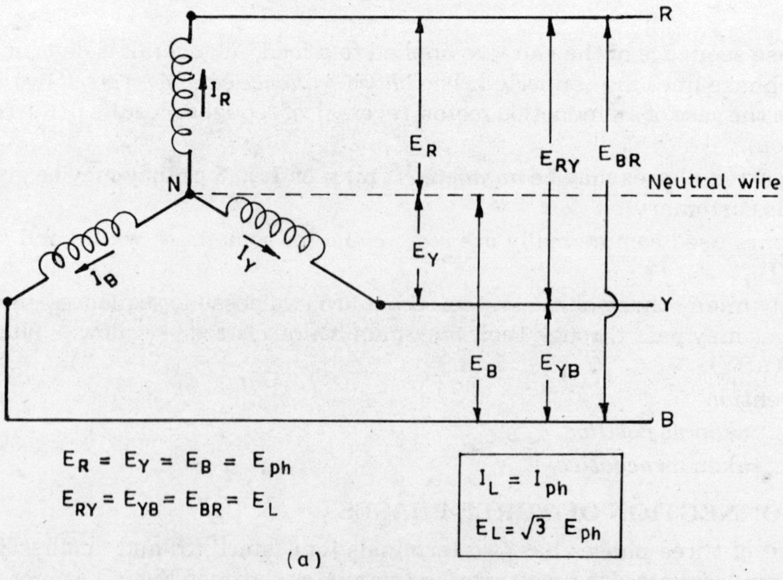

Fig. 1.5 (a)

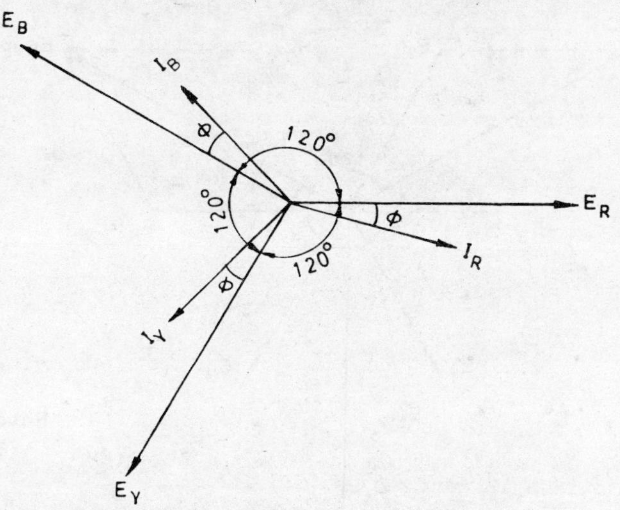

Fig. 1.5. (*b*) Star-connected three-phase net work.

- In star connection, as is evident in Fig. 1.5 (*a*) there are two-phase windings between each pair of terminals, but since their *similar* ends have been joined together, they are in *opposition.* Obviously, the instantaneous value of potential difference between any two terminals is the *arithmetic difference* of the two-phase e.m.fs. concerned. However, the r.m.s. value of this potential difference is given by the *vector difference* of the two-phase e.m.fs.

- Fig. 1.5 (*b*) shows the vector diagram for phase voltages and currents in a star connection where a *balanced system has been assumed.* [A balanced system is one in which (*i*) the voltages in all phases are equal in magnitude and differ in phase from one another by equal angles, in this case, the angle = 360/3 = 120°, (*ii*) the currents in the three phases are equal in magnitude and also differ in phase from one another by equal angles. A three-phase balanced load is that in which the loads connected across the three-phases are *identical*]. Thus, we have

$$E_R = E_Y = E_B = E_{ph} \text{ (phase e.m.f.)}$$

Line voltage, $E_{RY} (= E_L)$ = Vector difference of E_R and E_Y

$$= E_R - E_Y$$

Line voltage, $E_{YB} = E_Y - E_B$

Line voltage, $E_{BR} = E_B - E_R$.

(*a*) **Relation Between Line Voltages and Phase Voltages.** Refer Fig. 1.6.

The potential difference between outers R any Y is

$$E_{RY} = E_R - E_Y \qquad \text{[vector difference]}$$

or

$$E_{RY} = E_R + (-E_Y) \qquad \text{[vector sum]}$$

Hence, E_{RY} is found by compounding E_R and E_Y reversed and its value is given by the diagonal of the parallelogram (Fig. 1.6). Obviously the angle between E_R and E_Y reversed is 60° and the value of

$$E_{RY} (\text{or } E_L) = \sqrt{E_R^2 + E_Y^2 + 2E_R E_Y \cos 60°} = \sqrt{E_{ph}^2 + E_{ph}^2 + 2E_{ph} \times E_{ph} \times \frac{1}{2}} = \sqrt{3} E_{ph}$$

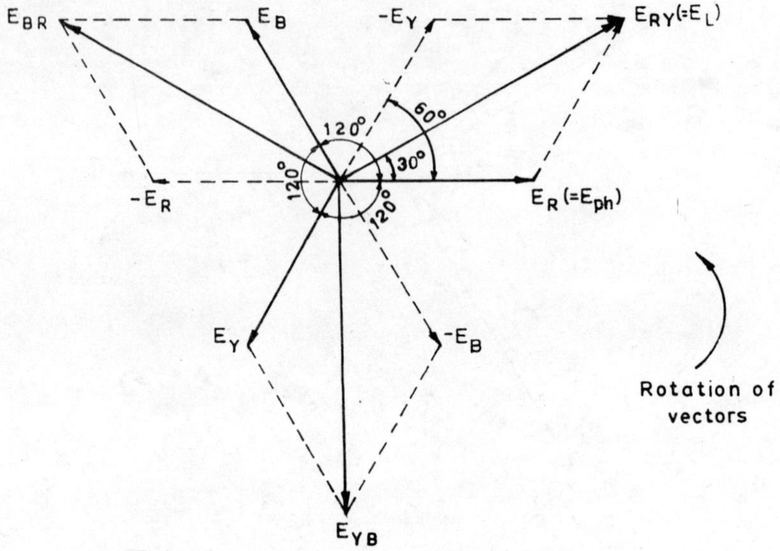

Fig. 1.6. Vector diagram for star connected net-work.

Similarly $\quad E_{YB}(= E_L) = E_Y - E_B = \sqrt{3}E_{ph}$

and $\quad E_{BR}(= E_L) = E_B - E_R = \sqrt{3}E_{ph}$

i.e. $\quad E_{RY} = E_{YB} = E_{BR} = E_L = \sqrt{3}E_{ph}$

Hence, $\quad E_L = \sqrt{3}E_{ph}$ ...(1.2)

(*i.e. Line voltage* = $\sqrt{3}$ *phase voltage*).

(*b*) **Relation between Line Currents and Phase Currents.** Since in star-connected system each line conductor is connected to separate phase, so the current flowing through the line and phase are same.

Current in outer (or line) $\quad R = I_R$

Current in outer $\quad Y = I_Y$

Current in outer $\quad B = I_B$

Since $\quad I_R = I_Y = I_B =$ say, I_{ph}—the phase current

∴ Line current, $\quad I_L = I_{ph}$ ...(1.3)

(*c*) **Power.** If the phase current has a phase difference of ϕ with phase voltage,

Power per phase $\quad = E_{ph}I_{ph} \cos \phi$

Total power (true), $\quad P = 3 \times$ power per phase

$$P = 3 \times E_{ph}I_{ph} \cos \phi \quad ...[1.4\ (a)]$$

Now $\quad E_{ph} = \dfrac{E_L}{\sqrt{3}}$ and $I_{ph} = I_L$

Hence in terms of line values, the above expression becomes

$$P = 3 \times \frac{E_L}{\sqrt{3}} I_L \cos \phi$$

or

$$P = \sqrt{3}E_L I_L \cos \phi \quad ...[1.4\ (b)]$$

(Apparent power $= \sqrt{3}E_L I_L$).

In a *balanced star-connected net work* the following points are worthnoting :

(*i*) *Line voltages are* $\sqrt{3}$ *times the phase voltages.*

(*ii*) *Line currents are equal to phase currents.*

(*iii*) Line voltages are 120° apart.

(*iv*) Line voltages are 30° ahead of the respective phase voltages.

(*v*) The angle between line currents and the corresponding line voltages is (30° ± ϕ) + for lagging currents – ve for leading currents.

(*vi*) True power = $\sqrt{3}\, E_L I_L \cos \phi$, where ϕ is the angle between respective phase current and phase voltage, not between the line current and line voltage.

(*vii*) Apparent power = $\sqrt{3} E_L I_L$.

(*viii*) In balance system, the potential of the neutral or star point is zero.

∴ Potential at neutral (or star) point $= E_{NR} + E_{NY} + E_{NB} = 0$.

STAR OR WYE CONNECTION

Example 1.1. *Three equal impedances each having a resistance of 25 Ω and reactance of 40 Ω are connected in star to a 400 V, 3-phase, 50 Hz system. Calculate :*

(*i*) *The line current* (*ii*) *Power factor, and*

(*iii*) *Power consumed.*

Solution. Resistance per phase, $R_{ph} = 25\ \Omega$

Reactance per phase, $X_{ph} = 40\ \Omega$

Line voltage, $E_L = 400$ V

Line current, I_L **:**

Power factor, $\cos\phi$ **:**

Power consumed, P **:**

Refer Fig. 1.7.

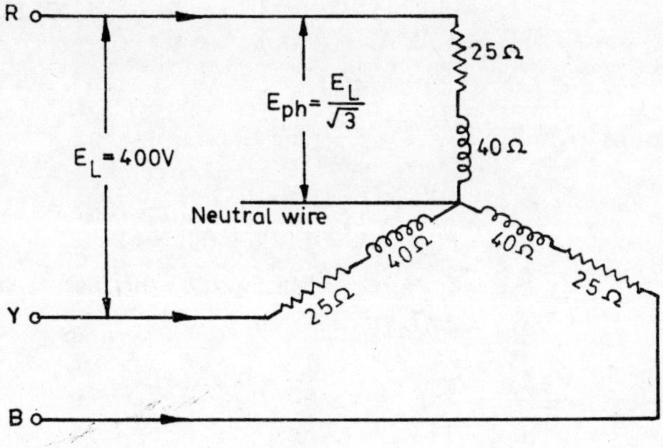

Fig. 1.7

Impedance per phase, $Z_{ph} = \sqrt{R_{ph}^2 + X_{ph}^2}$

∴ $Z_{ph} = \sqrt{25^2 + 40^2} = 47.17\ \Omega$

Phase voltage, $E_{ph} = \frac{E_L}{\sqrt{3}} = \frac{400}{\sqrt{3}} = 231$ V

Phase current, $I_{ph} = \frac{E_{ph}}{Z_{ph}} = \frac{231}{47.17} = 4.9$ A (app.)

(*i*) **Line current,** I_L = phase current, I_{ph}

$\therefore$ $\mathbf{I_L = 4.9\ A.}$ **(Ans.)**

(*ii*) **Power factor,** $\cos\phi = \frac{R_{ph}}{Z_{ph}} = \frac{95}{47.17} = \mathbf{0.53\ (lag).}$ **(Ans.)**

(*iii*) **Power consumed,** $P = \sqrt{3}E_L I_L \cos\phi = \sqrt{3} \times 400 \times 4.9 \times 0.53$

$= \mathbf{1800\ W\ (app.).}$ **(Ans.)**

[or $P = 3I_{ph}^2 R_{ph} = 3 \times 4.9^2 \times 25 = 1800$ W].

Example 1.2. *Three identical coils are connected in star to a 400 V (line voltage), 3-phase A.C. supply and each coil takes 300 W. If the power factor is 0.8 (lagging). Calculate :*

(*i*) *The line current,* (*ii*) *Impedance, and*

(*iii*) *Resistance and inductance of each coil.*

Solution. Line voltage, $E_L = 400$ V

Power taken by each coil, $P_{ph} = 300$ W

Power factor, $\cos\phi = 0.8$ (lagging)

$\mathbf{I_L ; Z ; R_{ph} ; L_{ph}}$ **:**

Phase voltage, $E_{ph} = \frac{E_L}{\sqrt{3}} = \frac{400}{\sqrt{3}}$ V

Also $P_{ph} = E_{ph} I_{ph} \cos\phi$

$300 = \frac{400}{\sqrt{3}} \times I_{ph} \times 0.8$

$\therefore$ $I_{ph} = \frac{300 \times \sqrt{3}}{400 \times 0.8} = 1.62$ A.

(*i*) **Line current,** I_L = phase current, I_{ph}

$\therefore$ $\mathbf{I_L = 1.62\ A.}$ **(Ans.)**

(*ii*) **Coil impedance,** $Z_{ph} = \frac{E_{ph}}{I_{ph}} = \frac{\frac{400}{\sqrt{3}}}{1.62} = 142.5\ \Omega$

$\therefore$ $\mathbf{Z_{ph} = 142.5\ \Omega.}$ **(Ans.)**

(*iii*) $R_{ph} = Z_{ph} \cos\phi = 142.5 \times 0.8 = 114\ \Omega$

Coil reactance, $X_{ph} = Z_{ph} \sin\phi = 142.5 \times 0.6 = \mathbf{85.5\ \Omega}$ **(Ans.)**

But $X_{ph} = 2\pi f L_{ph}$

$\therefore$ $L_{ph} = \frac{X_{ph}}{2\pi f} = \frac{85.5}{2\pi \times 50} = 0.272$ H.

Hence, $\mathbf{R_{ph} = 114\ \Omega.}$ **(Ans.)**

and $\mathbf{L_{ph} = 0.272\ H.}$ **(Ans.)**

Example 1.3. *In a 3-phase, 3-wire system with star-connected load the impedance of each phase is (3 + j4) Ω. If the line voltage is 230 V, calculate :*

(*i*) *The line current, and* (*ii*) *The power absorbed by each phase.*

Solution. Line voltage, $E_L = 230$ V

Resistance per phase, $R_{ph} = 3\ \Omega$

Reactance per phase, $X_{ph} = 4\ \Omega$

$\mathbf{I_L ;\ P_{ph}}$ **:**

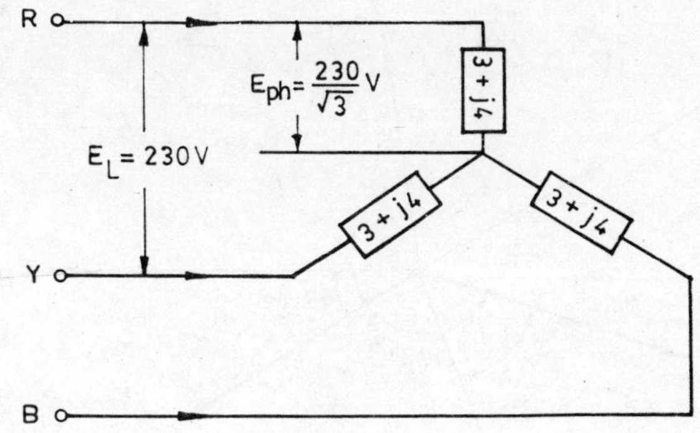

Fig. 1.8

Phase voltage, $E_{ph} = \frac{E_L}{\sqrt{3}} = \frac{230}{\sqrt{3}}$ V

Impedance per phase, $Z_{ph} = \sqrt{R_{ph}^2 + X_{ph}^2} = \sqrt{3^2 + 4^2} = 5\ \Omega$

Power factor, $\cos\phi = \frac{R}{Z} = \frac{3}{5} = 0.6$

Phase current, $I_{ph} = \frac{E_{ph}}{Z_{ph}} = \frac{230}{\sqrt{3} \times 5} = 26.56$ A

Line current, $\mathbf{I_L = I_{ph} = 26.56}$ **A. (Ans.)**

Power absorbed by each phase,

$$\mathbf{P_{ph}} = E_{ph} I_{ph} \cos\phi = \frac{230}{\sqrt{3}} \times 26.56 \times 0.6 = \mathbf{2116\ W\ (Ans.)}$$

$$[P_{ph} = I_{ph}^2 R_{ph} = 26.56^2 \times 3 = 2116\ \text{W}]$$

Solution by Symbolic Notation. In Fig. 1.9 E_R, E_Y and E_B are the phase voltages whereas I_R, I_Y and I_B are phase currents.

Taking E_R *as the reference vector,* we get

$$E_R = \frac{230}{\sqrt{3}} \angle 0° = 133 \angle 0° = 133 + j0 \text{ volt}$$

$$E_Y = 133 \angle -120° = 133\,(-0.5 - j0.866) = (-66.5 - j115) \text{ volts}$$

$$E_B = 133 \angle 120° = 133\,(-0.5 + j0.866) = (-66.5 + j115) \text{ volts}$$

$$Z = 3 + j4 = 5 \angle 53° \, 8'$$

$$I_R = \frac{E_R}{Z} = \frac{133 \angle 0°}{5 \angle 53° \, 8'} = 26.6 \angle -53° \, 8'$$

This current lags behind the reference voltage (E_R) by 53° 8′ (Fig. 1.9).

$$I_Y = \frac{E_Y}{Z} = \frac{133 \angle -120°}{5 \angle 53° \, 8'} = 26.6 \angle -173° \, 8'$$

It lags the reference vector *i.e.,* E_R by 173° 8′ which amounts to lagging behind its phase voltage E_Y by 53° 8′,

$$I_B = \frac{E_B}{Z} = \frac{133 \angle 120°}{5 \angle 53° \, 8'} = 26.6 \angle 66° \, 52'$$

This current leads E_R by 66° 52′ which is the same as *lagging* behind its phase voltage by 53° 8′.

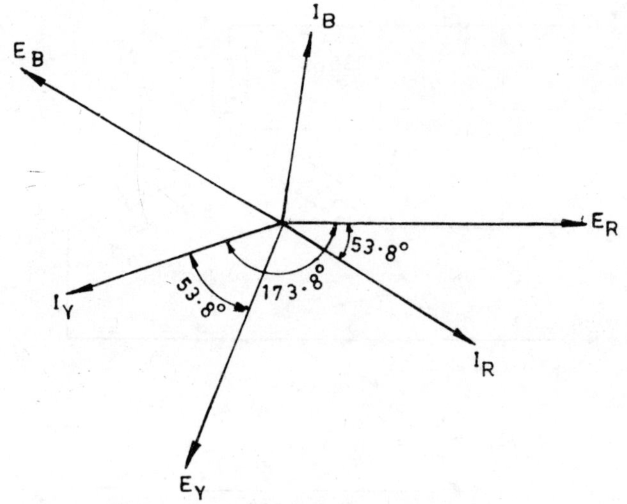

Fig. 1.9

Let us consider R-phase for calculation of power

$$E_R = (133 + j0) \; ; \; I_R = 26.6\,(0.6 - j0.8) = (15.96 - j21.28)$$

Using method of conjugates, we get

$$P_{VA} = (133 - j0)\,(15.96 - j21.28) = 2116 - j2830$$

∴ Real power absorbed/phase = 2116 W ...(As before)

Example 1.4. *A star-connected, 6000 V, 3-phase alternator is supplying 4000 kW at a power factor of 0.8. Calculate the active and reactive components of the current in each phase.*

Solution. Line voltage, $E_L = 6000$ V

Power supplied, $P = 4000$ kW

Power factor, $\cos\phi = 0.8$

Active and reactive components of current :

We know that, $P = \sqrt{3}\, E_L I_L \cos\phi$

$$4000 \times 1000 = \sqrt{3} \times 6000 \times I_L \times 0.8$$

i.e.,
$$I_L = \frac{4000 \times 1000}{\sqrt{3} \times 6000 \times 0.8} = 481 \text{ A}$$

∴ $I_{ph} = I_L = 481$ A

Active component $= I_{ph} \cos\phi = 481 \times 0.8 =$ **384.8 A. (Ans.)**

Reactive component $= I_{ph} \sin\phi = 481 \times 0.6 =$ **288.6 A. (Ans.)**

Example 1.5. *In a 3-phase, 4-wire system, two phases have currents of 20 A and 12 A in lagging power factors of 0.8 and 0.6 respectively, while the third phase is open-circuited. Calculate the current in the neutral and draw the vector diagram.*

Solution. Refer Figs. 1.10 and 1.11.

$$\cos\phi_1 = 0.8$$

∴ $\phi_1 = \cos^{-1} 0.8 = 36°\ 52'$ (lag)

$$\cos\phi_2 = 0.6$$

∴ $\phi_2 = \cos^{-1} 0.6 = 53°\ 8'$ (lag)

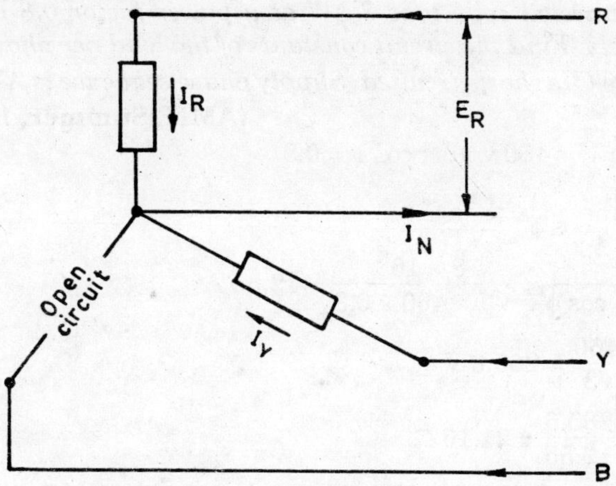

Fig. 1.10

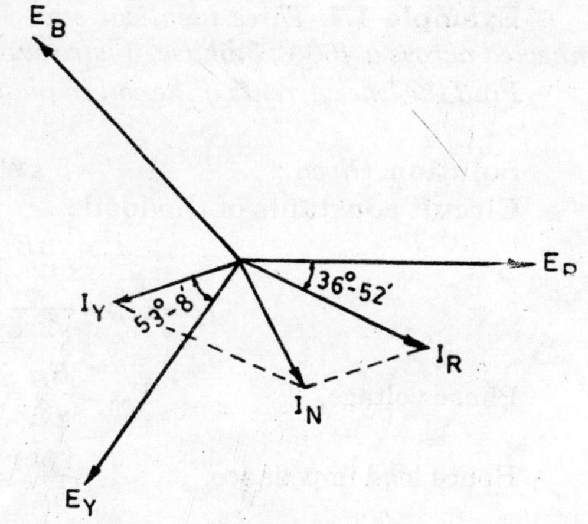

Fig. 1.11

Let E_R be the reference vector.

Then $\qquad I_R = 20 \angle -36° 52' = 20\,(0.8 - j0.6) = (16 - j12)$

and $\qquad I_Y = 12 \angle -173° 8' = 12(-1 - j0.12) = (-12 - j1.44)$

The current through the neutral

$$I_N = I_R + I_Y$$

$$= (16 - j12) + (-12 - j1.44) = 4 - j13.44 = 14 \angle -73° 24'$$

Hence, **current in the neutral = 14 ∠– 73° 24′. (Ans.)**

Example 1.6. *A balancea 3-phase star connected load of 100 kW takes a leading current of 80 A, when connected across a 3-phase, 1100 V, 50 Hz supply. Find the circuit constants of the load per phase.* **(AMIE Winter, 1997)**

Solution. *Given* : $P = 100$ kW ; $I_{ph}\ (= I_L) = 80$ A ; $E_L = 1100$ V ; $f = 50$ Hz

Circuit constants of the load per phase, R, C :

As the 3-φ load is balanced and star connected, line or phase current,

$$I_L = (I_{ph}) = \frac{P}{\sqrt{3}\, E_L \cos\phi} = \frac{100 \times 10^3}{\sqrt{3} \times 1100 \times \cos\phi}$$

or $$\cos\phi = \frac{100 \times 10^3}{\sqrt{3} \times 1100 \times 80} = 0.656$$

Load impedance, $$Z = \frac{E_{ph}}{I_{ph}} = \frac{(1100/\sqrt{3})}{80} = 7.94\ \Omega$$

∴ $$\mathbf{R} = Z\cos\phi = 7.94 \times 0.656 = \mathbf{5.2\ \Omega}.\ \textbf{(Ans.)}$$

Now $X_C = \dfrac{1}{2\pi f C}$ as the current given is *leading* current.

∴ $$C = \frac{1}{2\pi f X_C}$$

But $$X_C = Z\sin\phi = 7.94 \times 0.755 = 5.99\ \Omega$$

∴ $$\mathbf{C} = \frac{1}{2\pi \times 50 \times 5.99}\ \text{F}$$

$$= \frac{10^6}{2\pi \times 50 \times 5.99}\ \mu\text{F} = \mathbf{531.4\ \mu F}.\ \textbf{(Ans.)}$$

Example 1.7. *Three identical star-connected coils take 8 kW at a power factor 0.8 when connected across a 460 V, 3-phase, 3-wire supply. Find the circuit constants of the load per phase.*

Find the line currents if the coil in phase C is short circuited. Supply phase sequence is A-B-C.

(AMIE Summer, 1997)

Solution. *Given* : $P = 8$ kW ; $V = 460$ volts ; $\cos\phi = 0.8$

Circuit constants of the load :

$$P = \sqrt{3} E_L I_L \cos\phi$$

$$I_L(= I_{ph}) = \frac{P}{\sqrt{3}\, E_L \cos\phi} = \frac{8 \times 10^3}{\sqrt{3} \times 460 \times 0.8} = 12.55 \text{ A}$$

Phase voltage, $$E_{ph} = \frac{E_L}{\sqrt{3}} = \frac{460}{\sqrt{3}} = 265.6 \text{ V}$$

Hence load impedance, $$Z = \frac{E_{ph}}{I_{ph}} = \frac{265.6}{12.55} = 21.16\ \Omega$$

Load resistance, **R** $= Z\cos\phi = 21.16 \times 0.8 =$ **16.93** Ω. **(Ans.)**

Load reactance, **X** $= Z\sin\phi = 21.16 \times 0.6 =$ **12.7** Ω. **(Ans.)**

Line currents if the coil in phase C is short circuited :

The connection arrangement is shown in Fig. 1.12 when coil C is shorted.

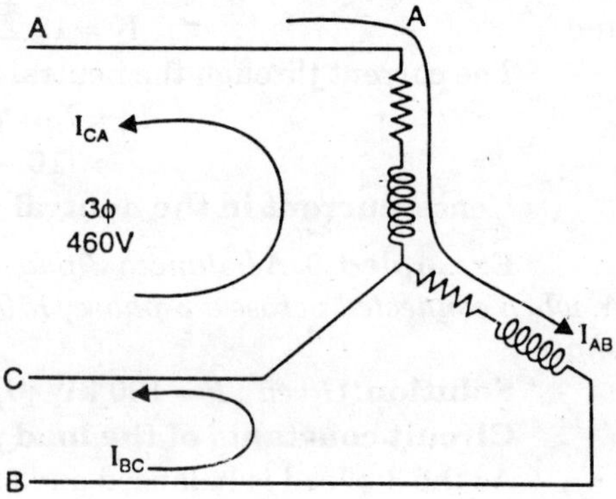

Fig. 1.12

Given that phase sequence is A-B-C.

Hence, $E_{AB} = 460 \angle 0°$

$E_{BC} = 460 \angle -120°$

$E_{CA} = 460 \angle +120°$

Also, $Z = 21.6 \angle 36.8°$

$$I_{AB} = \frac{E_{AB}}{2Z} = \frac{460 \angle 0°}{2 \times 21.16 \angle 36.8°} = 10.87 \angle -36.8°$$

$$I_{BC} = \frac{E_{BC}}{Z} = \frac{460 \angle -120°}{21.16 \angle 36.8°} = 21.74 \angle -156.8°$$

$$I_{CA} = \frac{E_{CA}}{Z} = \frac{460 \angle 120°}{21.16 \angle 36.8°} = 21.74 \angle 83.2°$$

Line current in line A $= I_{AB} - I_{CA}$

$= 10.87 \angle -36.8° - 21.74 \angle 83.2°$

$= 10.87(0.8 - j0.6) - 21.74(0.118 + j0.993)$

$= 8.70 - j6.52 - 2.56 - j21.59 =$ **(6.14 – j28.11)A. (Ans.)**

Line current in line B $= I_{BC} - I_{AB}$

$= 21.74 \angle -156.8° - 10.87 \angle -36.8°$

$= 21.74(-0.92 - j0.394) - (8.70 - j6.52)$

$= -20.0 - j8.56 - 8.70 + j6.52 =$ **(– 28.70 – j2.04)A. (Ans.)**

Line current in line $C = I_{CA} - I_{BC}$

$= (-2.56 - j21.59) - (-20 - j8.56) = -2.56 - j21.59 + 20 + j8.56$

$=$ **(17.44 – j13.03)A. (Ans.)**

Example 1.8. *In a star-connected load each phase consists of a resistance of 50 Ω in parallel with a capacitor of capacitance 16 μF. When it is connected to a 400 V, 3-phase, 5 Hz supply, calculate :*

(i) The line current, *(ii) The power factor,*

(iii) The power absorbed, and *(iv) The total kVA.*

Solution. Resistance per phase, $R = 50\ \Omega$

Capacitance, $C = 16\ \mu F = 16 \times 10^{-6}$ F

The circuit is shown in Fig. 1.13.

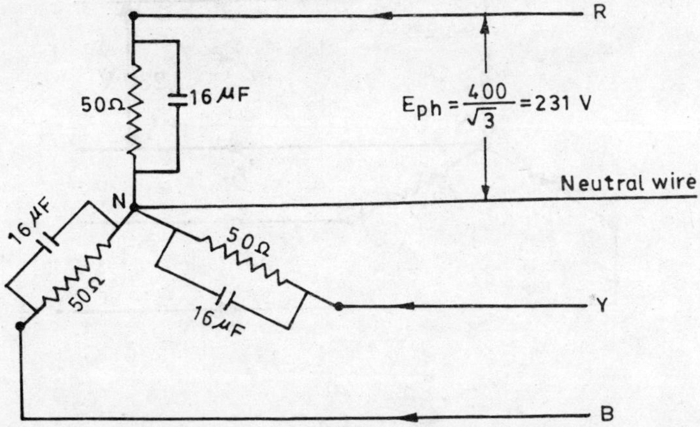

Fig. 1.13

(i) $$E_{ph} = \left(\frac{400}{\sqrt{3}}\right) \angle 0° = 231 \angle 0° = (231 + j0)$$

Admittance of each phase,

$$Y_{ph}\left(= \frac{1}{Z_{ph}}\right) = \frac{1}{R} + j\omega C$$

$$= \frac{1}{50} + j(2\pi \times 50 \times 16 \times 10^{-6}) = 0.02 + j0.005$$

$\therefore$ $$I_{ph} = E_{ph} \cdot Y_{ph} = 231\ (0.02 + j0.005) = 4.62 + j1.155 = 4.76 \angle 14°$$

For a star-connection, $I_{ph} = I_L$

$\therefore$ $$I_L = 4.76 \text{ A}$$

Hence, **line current = 4.76 A. (Ans.)**

(ii) **Power factor** $= \cos 14° =$ **0.97 (leading). (Ans.)**

(iii) Now $E_{ph} = (231 + j0)$; $I_{ph} = (4.62 + j1.155)$

$\therefore$ $$P_{VA} = (231 + j0)\ (4.62 - j1.155)$$

$$= 231 \times 4.62 - j1.155 \times 231 = 1067.22 - j266.8$$

$$= 1100 \angle -14° \quad \text{(per phase)}$$

$\therefore$ **Total power absorbed** $= 3 \times 1067.22 = 3201.66$ W $=$ **3.201 kW. (Ans.)**

(iv) **Total volt-amperes** $= 3 \times 1100 = 3300$ VA $=$ **3.3 kVA. (Ans.)**

Example 1.9. *A 3-phase, star-connected system with 230 V between each phase and neutral has resistances of 8, 10 and 20 Ω respectively in three phases, calculate :*

(i) The current flowing in each phase, *(ii) The neutral current, and*

(iii) The total power absorbed.

Solution. Refer Figs. 1.14 (*a*) and (*b*).

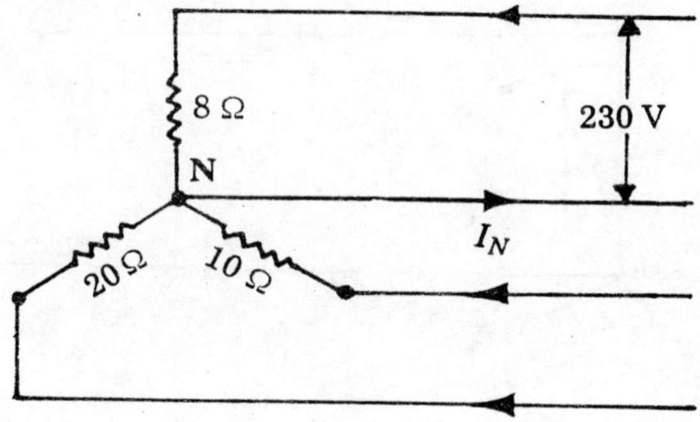

(*a*)

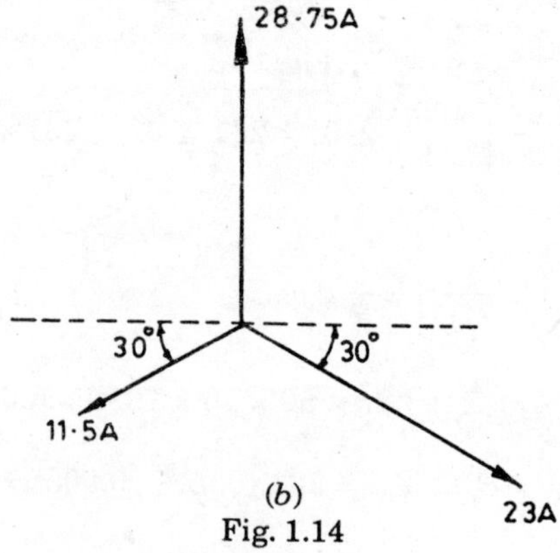

(*b*)
Fig. 1.14

Phase voltage, $E_{ph} = 230$ V

(*i*) Current in 8 Ω resistor $= \frac{230}{8} =$ **28.75 A. (Ans.)**

Current in 10 Ω resistor $= \frac{230}{10} =$ **23 A. (Ans.)**

Current in 12 Ω resistor $= \frac{230}{20} =$ **11.5 A. (Ans.)**

(*ii*) The above currents are mutually displaced by 120°. The neutral current I_N is the vector sum of these three currents.

I_N can be found by splitting up these three-phase currents into their *X*-components and *Y*-components and then by combining them together.

Σ*X*-components $= 23 \cos 30° - 11.5 \cos 30° = 11.5 \cos 30° = 9.96$ A

Σ*Y*-components $= 28.75 - 23 \sin 30° - 11.5 \sin 30° = 28.75 - 34.5 \sin 30° = 11.5$ A

∴ **Neutral current,** $I_N = \sqrt{(9.96)^2 + (11.5)^2} =$ **15.21 A. (Ans.)**

(*iii*) **Total power absorbed,**

$P = 230\ (28.75 + 23 + 11.5) =$ **14547.5 W. (Ans.)**

Example 1.10. *In a 3-phase, 4-wire system, there is a balanced 3-phase motor load taking 10 kW at a power factor of 0.8 lagging while lamps connected between phase conductors and the neutral are taking 2.5 kW, 2 kW and 5 kW respectively. Voltage between line conductors is 430 V. Calculate :*

(*i*) *The current in the neutral wire.* (*ii*) *The current in each conductor.*

Solution. Motor load = 10 kW or 10000 W

Power factor of motor load, $\cos \phi = 0.8$

Lamp loads : 2.5 kW, 2 kW and 5 kW

Voltage between line conductors, $E_L = 430$ V.

Refer Figs. 1.15 and 1.16.

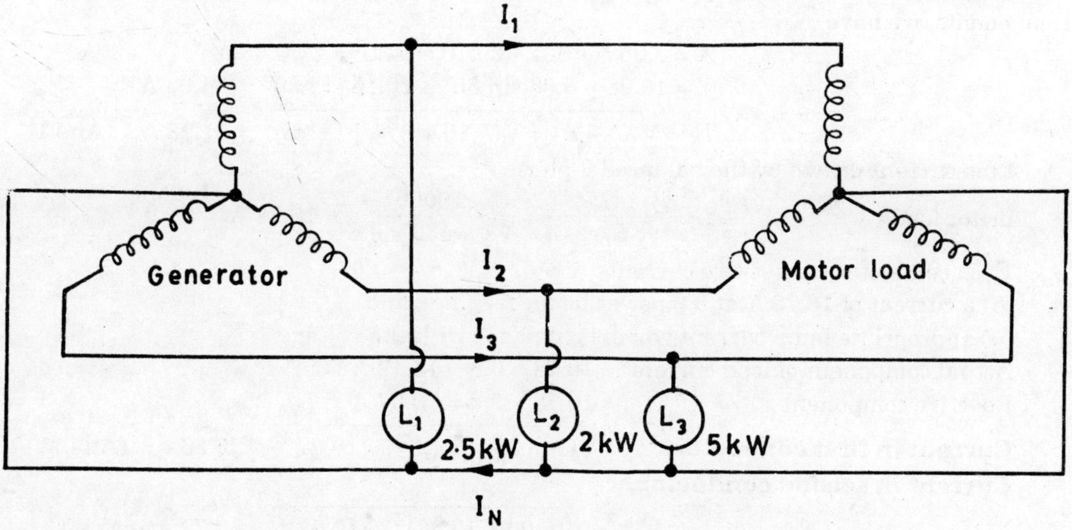

Fig. 1.15

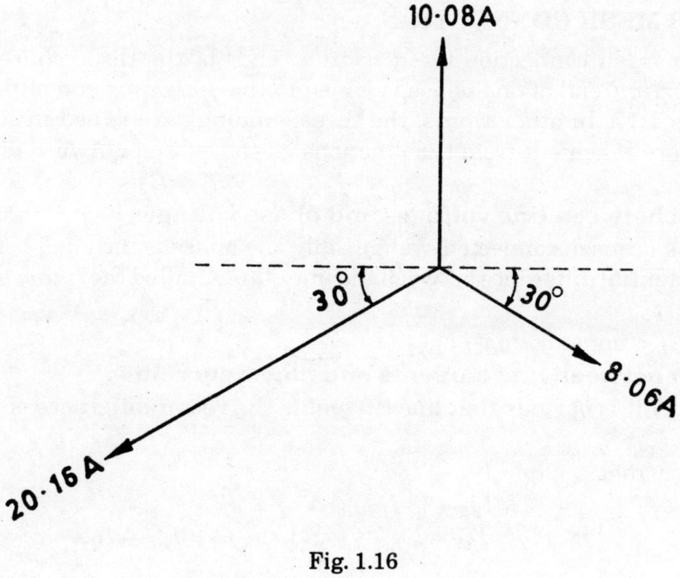

Fig. 1.16

Let us first find the current in the neutral wire due to lamp loads L_1 (2.5 kW), L_2 (2 kW) and L_3 (5 kW) respectively.

Phase voltage, $E_{ph} = \frac{E_L}{\sqrt{3}} = \frac{430}{\sqrt{3}} = 248$ V

Current through lamp L_1 $= \frac{2.5 \times 1000}{248} = 10.08$ A

Current through lamp L_2 $= \frac{2 \times 1000}{248} = 8.06$ A

Current through lamp L_3 $= \frac{5 \times 1000}{248} = 20.16$ A.

The neutral current is the vector sum of three lamp currents. Resolving them into their X and Y components, we have

$$X = 8.06 \cos 30° - 20.16 \cos 30° = -10.48 \text{ A}$$

$$Y = 10.08 - 8.06 \sin 30° - 20.16 \sin 30° = -4.03 \text{ A}$$

$$\therefore \quad I_N = \sqrt{X^2 + Y^2} = \sqrt{(-10.48)^2 + (-4.03)^2} = \textbf{11.23 A. (Ans.)}$$

Line current drawn by the balanced 3-phase

motor load, $I_L = \frac{P}{\sqrt{3} E_L \cos \phi} = \frac{10000}{\sqrt{3} \times 430 \times 0.8} = 16.78$ A

Each conductor carries two currents :

(*i*) a current of 16.78 A at 0.8 power factor lagging ; and

(*ii*) appropriate lamp current which is in phase with the voltage.

Actual component of load current $= 16.78 \times 0.8 = 13.42$ A

Reactive component $= 16.78 \times 0.6 = 10.07$ A

Current in first conductor, $I_1 = \sqrt{(10.08 + 13.42)^2 + (10.07)^2} = \textbf{25.56 A. (Ans.)}$

Current in second conductor,

$$I_2 = \sqrt{(8.06 + 13.42)^2 + (10.07)^2} = \textbf{23.72 A (Ans.)}$$

Current in third conductor, $I_3 = \sqrt{(20.16 + 13.42)^2 + (10.07)^2} = \textbf{35.06 A. (Ans.)}$

1.7. DELTA (Δ) OR MESH CONNECTION

In a delta or mesh connection the *dissimilar* ends of the three-phase windings are joined together *i.e.* the '*starting*' end of one phase is joined to the '*finishing*' end of the other phase and so on as shown in Fig. 1.17. In other words, the three windings are joined in series to form a closed mesh. Three leads are taken out from the junctions as shown and outward directions are taken as positive.

(*a*) **Relation between line voltages and phase voltages :**

Since in delta or mesh connected system, only one phase is included between any pair of line outers, therefore potential difference between the line outers, called the *line voltage*, is equal to phase voltage.

i.e., *Line voltage*, E_L = *phase voltage*, E_{ph}.

(*b*) **Relation between line currents and phase currents :**

From Fig. 1.17 it is obvious that line current is the vector difference of phase currents of two phases concerned.

Thus, line current, $I_R = I_{YR} - I_{RB}$ (Vector difference)

$= I_{YR} + (-I_{RB})$ (Vector sum)

Similarly, $I_Y = I_{BY} - I_{YR}$ and $I_B = I_{RB} - I_{BY}$

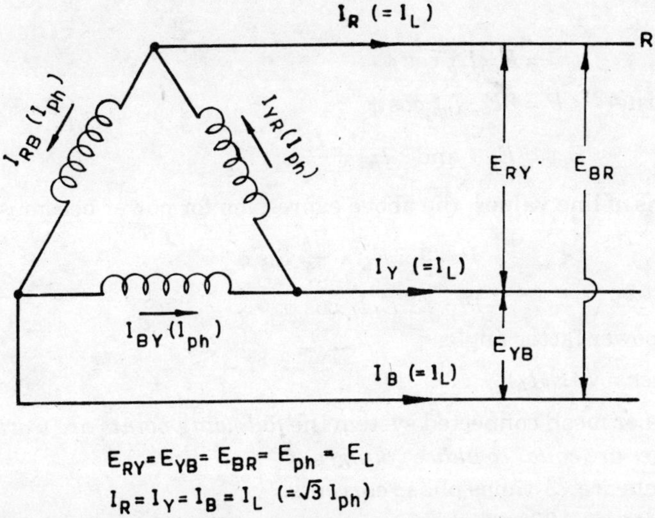

Fig. 1.17. Delta or mesh connected diagram.

Refer Fig. 1.18. Since phase angle between phase current I_{YR} and $-I_{RB}$ is 60°,

$$\therefore \quad I_R = \sqrt{I_{YR}^2 + I_{RB}^2 + 2I_{YR}I_{RB}\cos 60°}$$

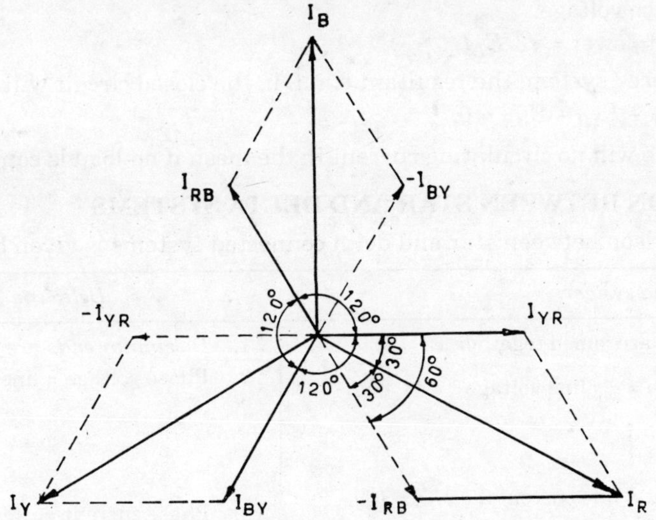

Fig. 1.18. Vector diagram for delta connected net work.

Assuming the delta connected system or network be balanced, the phase current in each winding is equal and let each be equal to I_{ph} (*i.e.*, $I_{YR} = I_{BY} = I_{RB} = I_{ph}$)

$$\therefore \quad I_R (= I_L) = \sqrt{I_{ph}^2 + I_{ph}^2 + 2I_{ph}I_{ph}\cos 60°} = \sqrt{2I_{ph}^2 + 2I_{ph}^2 \times \frac{1}{2}} = \sqrt{3}I_{ph}$$

Similarly, $\quad I_Y = I_B = \sqrt{3}I_{ph}$

Hence, $\quad I_L = \sqrt{3}\ I_{ph}$ ...(1.5)

(*i.e.*, *line current* = $\sqrt{3}$ *phase current*)

(*c*) **Power :**

Power/phase $= E_{ph} I_{ph} \cos \phi$

Total power (true) $P = 3E_{ph} I_{ph} \cos \phi$

But $E_{ph} = E_L$ and $I_{ph} = \dfrac{I_L}{\sqrt{3}}$

Hence, in terms of line values, the above expression for power becomes

$$P = 3 \times E_L \times \frac{E_L}{\sqrt{3}} \cos \phi$$

or

$$P = \sqrt{3}\, E_L I_L \cos \phi \quad ...(1.6)$$

where ϕ = the phase power factor angle.

(*Apparent power* $= \sqrt{3} E_L I_L$)

In case of delta or mesh connected system the *following points are worthnoting* :

(*i*) *Line voltages are equal to phase voltages.*

(*ii*) Line currents are $\sqrt{3}$ times phase currents.

(*iii*) Line currents are 120° apart.

(*iv*) Line currents are 30° behind their respective phase currents.

(*v*) The angle between line currents and corresponding line voltages is $(30° \pm \phi)$ as in the star system.

(*vi*) True power $= \sqrt{3}\, E_L I_L \cos \phi$, where ϕ is the phase angle between respective phase current and phase voltage.

(*vii*) Apparent power $= \sqrt{3}\, E_L I_L$.

(*viii*) In balanced system, the resultant e.m.f. in the closed circuit will be zero.

i.e., $E_{RY} + E_{YB} + E_{BR} = 0$.

Hence, there will no circulating current in the mesh if no-load is connected to the lines.

1.8. COMPARISON BETWEEN STAR AND DELTA SYSTEMS

The comparison between star and delta connected systems is given below :

Star connected system	*Delta connected system*
1. *Similar ends* are joined together.	1. *Dissimilar ends* are joined.
2. Phase voltage $= \dfrac{1}{\sqrt{3}}$ line voltage $\left(i.e., E_{ph} = \dfrac{E_L}{\sqrt{3}}\right)$.	2. Phase voltage = line voltage (*i.e.*, $E_{ph} = E_L$).
3. Phase current = line current (*i.e.*, $I_{ph} = I_L$).	3. Phase current $= \dfrac{1}{\sqrt{3}} \times$ line current $\left(i.e., I_{ph} = \dfrac{I_L}{\sqrt{3}}\right)$.
4. Possible to carry neutral to the load.	4. Neutral wire not available.
5. Provides 3-phase 4-wire arrangement.	5. Provides 3-phase 3-wire arrangement.
6. Can be used for lighting as well as power load.	6. Can be used for power loads only.
7. Neutral wire of a star connected alternator can be connected to earth, so relays and protective devices can be provided in the star connected alternators for safety.	7. Not possible. Delta connected system is mostly used in transformer for running of small low voltage 3-phase motors and *best suited for rotary converters*.

DELTA OR MESH CONNECTION

Example 1.11. *Three identical coils connected in delta across 400 V, 50 Hz, 3-phase supply take a line current of 15 A at a power factor 0.8 lagging. Calculate :*

(*i*) *The phase current, and*

(*ii*) *The impedance, resistance and inductance of each winding.*

Solution. Line voltage, $E_L = 400$ V

Line current, $I_L = 15$ A

Power factor, $\cos\phi = 0.8$ lagging

$\mathbf{I_{ph}}$; $\mathbf{Z_{ph}}$; $\mathbf{R_{ph}}$; **L :**

Phase voltage, $E_{ph} = E_L = 400$ V

(*i*) **Phase current,** $I_{ph} = \dfrac{I_L}{\sqrt{3}} = \dfrac{15}{\sqrt{3}} = \mathbf{8.66\ A.}$ **(Ans.)**

(*ii*) **Impedance of each phase,** $\mathbf{Z_{ph}} = \dfrac{E_{ph}}{I_{ph}} = \dfrac{400}{8.66} = \mathbf{46.19\ \Omega.}$ **(Ans.)**

Resistance of each phase, $\mathbf{R_{ph}} = Z_{ph}\cos\phi = 46.19 \times 0.8 = \mathbf{36.95\ \Omega.}$ **(Ans.)**

Reactance of each phase, $X_{ph} = Z_{ph}\sin\phi = 46.19\sqrt{(1-\cos^2\phi)}$

$$= 46.19\sqrt{1-(0.8)^2} = 27.71\Omega$$

$\therefore$ **Inductance,** $\mathbf{L} = \dfrac{X_{ph}}{2\pi f} = \dfrac{27.71}{2\pi \times 50} = \mathbf{0.088\ H.}$ **(Ans.)**

Example 1.12. *A 220 V, 3-phase voltage is applied to a balanced delta-connected 3-phase load of phase impedance (6 + j8).*

(*i*) *Find the phasor current in each line.*

(*ii*) *What is the power consumed per phase ?*

(*iii*) *What is the phasor sum of the three line currents ? What does it have this value ?*

Solution. Refer Fig. 1.19.

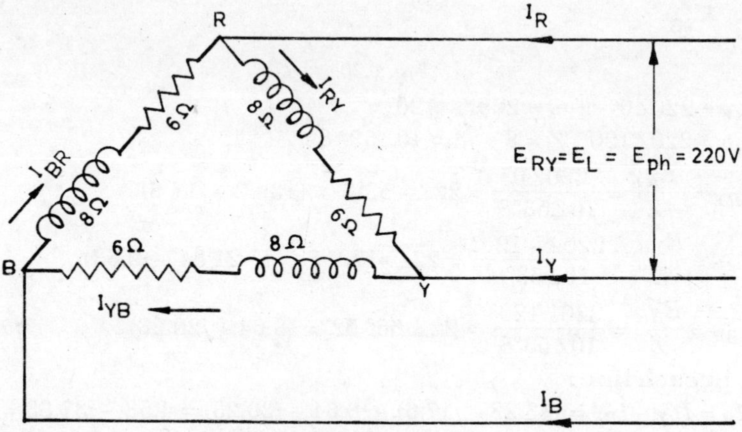

Fig. 1.19

Resistance per phase, $R_{ph} = 6\ \Omega$

Reactance per phase, $X_{ph} = 8\ \Omega$

$E_L = E_{ph} = 220$ V

Impedance per phase, $Z_{ph} = \sqrt{R_{ph}^2 + X_{ph}^2} = \sqrt{6^2 + 8^2} = 10\ \Omega$

(*i*) Phase current, $I_{ph} = \dfrac{E_{ph}}{Z_{ph}} = \dfrac{220}{10} = 22$ A

∴ **Line current,** $I_L = \sqrt{3} \times 22 =$ **38.1 A. (Ans.)**

(*ii*) **Power consumed per phase,**

$$P_{ph} = I_{ph}^2 \times R_{ph} = 22^2 \times 6 = \mathbf{2904\ W.}\ \textbf{(Ans.)}$$

(*iii*) Phasor sum would be zero because the three currents are equal in magnitudes and have a mutual phase difference of 120°.

Solution by Symbolic Notation. Let E_{RY} is taken as a reference vector (Fig. 1.20).

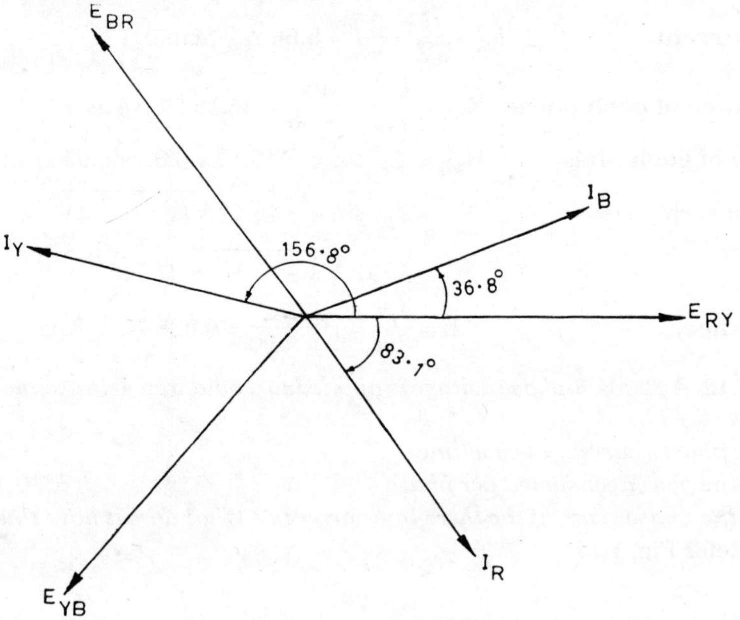

Fig. 1.20

$E_{RY} = 220\angle 0°$, $E_{YB} = 220\angle -120°$

$E_{BR} = 220\angle 120°$, $Z = 6 + j8 = 10\angle 53°\ 8'$,

$$I_{RY} = \frac{E_{RY}}{Z} = \frac{220\angle 0°}{10\angle 53°8'} = 22\angle -53°8' = (13.22 - j17.6)\ \text{A}$$

$$I_{YB} = \frac{E_{YB}}{Z} = \frac{220\angle -120°}{10\angle 53°8'} = 22\angle -173°\ 8' = (-21.84 - j2.63)$$

$$I_{BR} = \frac{E_{BR}}{Z} = \frac{220\angle 120°}{10\angle 53°8'} = 22\angle\ 66°\ 52' = (8.64 + j20.23)$$

(*i*) **Current in each line :**

$I_R = I_{RY} - I_{BR} = (13.22 - j17.6) - (8.64 + j20.23) = 4.58 - j37.83 =$ **38.1∠83.1°. (Ans.)**

$I_Y = I_{YB} - I_{RY} = (-21.84 - j2.63) - (13.22 - j17.6)$

$= -21.84 - j2.63 - 13.22 + j17.60 = -35.06 + j14.97 =$ **38.12∠156.8°. (Ans.)**

$I_B = I_{BR} - I_{YB} = (8.64 + j20.23) - (-21.84 + j2.63) = 8.64 + j20.23 + 21.84 + j2.63$

$= 30.48 + j22.86 =$ **38.1∠36.8°. (Ans.)**

(*ii*) **Power consumed per phase :**

Using conjugate of voltage, we get for R-phase

$$P_{VA} = E_{RY} \cdot I_{RY} = (220 - j0)(13.22 - j17.6) = (2908.4 - j3872) \text{ volt ampere}$$

True power per phase = **2.908 kW. (Ans.)**

(*iii*) **Phase sum of the three line currents**

$$= I_R + I_Y + I_B$$
$$= (4.58 - j37.83) + (-35.06 + j14.96) + (30.48 + j22.86) = 0$$

Hence, **the phasor sum of three line currents drawn by a 'balanced load' is zero. (Ans.)**

Example 1.13. *A delta-connected balanced 3-phase load is supplied from a 3-phase, 400 V supply. The line current is 30 A and the power taken by the load is 12 kW. Find :*

(*i*) *Impedance in each branch ; and*

(*ii*) *The line current, power factor and power consumed if the same load is connected is star.*

Solution. Delta-connection :

$$E_{ph} = E_L = 400 \text{ V}$$
$$I_L = 30 \text{ A}$$
$$\therefore \quad I_{ph} = \frac{I_L}{\sqrt{3}} = \frac{30}{\sqrt{3}} = 17.32 \text{ A.}$$

(*i*) **Impedance per phase**

$$\mathbf{Z_{ph}} = \frac{E_{ph}}{I_{ph}} = \frac{400}{17.32} = \mathbf{23.09\ \Omega.\ (Ans.)}$$

Now
$$P = \sqrt{3} E_L I_L \cos\phi$$
$$12000 = \sqrt{3} \times 400 \times 30 \times \cos\phi.$$

or
$$\cos\phi \text{ (power factor)} = \frac{12000}{\sqrt{3} \times 400 \times 30} = 0.577$$

(*ii*) **Star-connection**

$$E_{ph} = \frac{E_L}{\sqrt{3}} = \frac{400}{\sqrt{3}} = 231 \text{ V}$$
$$\mathbf{I_L} = I_{ph} = \frac{E_{ph}}{Z_{ph}} = \frac{231}{23.09} = \mathbf{10\ A\ \ (Ans.)}$$

Power factor, $\cos\phi = 0.577$ (since impedance is same)

Power consumed $= \sqrt{3}\, E_L I_L \cos\phi = \sqrt{3} \times 400 \times 10 \times 0.577 =$ **3997.6 W (Ans.)**

Example 1.14. *Three 50 Ω non-inductive resistances are connected in (i) star, (ii) delta across a 400 V, 50 Hz., 3-phase mains. Calculate the power taken from the supply system in each case. In the event of one of the three resistances getting opened, what would be the value of the total power taken from the mains in each of the two cases.*

Solution. Star connection :

Phase voltage, $E_{ph} = \dfrac{E_L}{\sqrt{3}} = \dfrac{400}{\sqrt{3}} = 231$ V

Phase current, $I_{ph} = \dfrac{E_{ph}}{R_{ph}} = \dfrac{231}{50} = 4.62$ A

Power consumed, $P = 3 I_{ph}^2 R_{ph} = 3 \times 4.62^2 \times 50 =$ **3200 W. (Ans.)**

[or $P = \sqrt{3} E_L I_L \cos\phi = \sqrt{3} \times 400 \times 4.62 \times 1 = 3200$ W]

Delta connection :

Phase voltage, $E_{ph} = E_L = 400$ V

Phase current, $I_{ph} = \dfrac{E_{ph}}{R_{ph}} = \dfrac{400}{50} = 8$ A

Power consumed, $P = 3I_{ph}^2 R_{ph} = 3 \times 8^2 \times 50 =$ **9600 W. (Ans.)**

When one of the resistances is disconnected :

(*i*) **Star connection.** Refer Fig. 1.21.

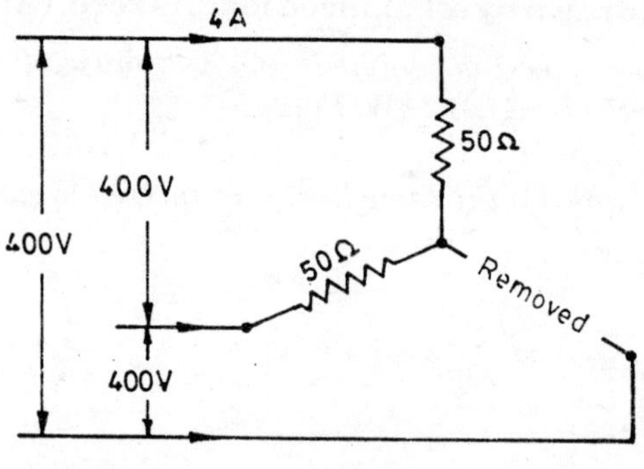

Fig. 1.21

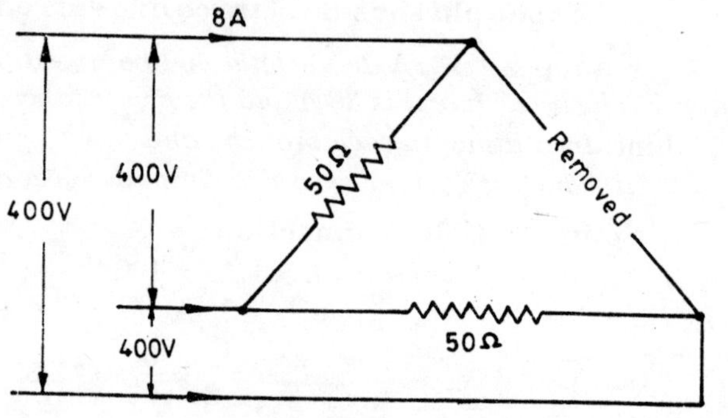

Fig. 1.22

When one of the resistances is disconnected, the circuit is no longer 3-phase but converted into single-phase circuit, having two resistances each of 50 ohm connected in series across supply of 400 V.

Hence line current, $I_L = \dfrac{E_L}{2R_{ph}} = \dfrac{400}{2 \times 50} = 4$ A

Power consumed, $P = 4^2 (50 + 50) =$ **1600 W. (Ans.)**

[or $P = VI \cos \phi = 400 \times 4 \times 1 = 1600$ W].

(*ii*) **Delta connection.** Refer Fig. 1.22.

Potential difference across each resistance, $E_L = 400$ V

Current in each resistance $= \dfrac{400}{50} = 8$ A

Power consumed in both resistances $= 2 \times 8^2 \times 50 =$ **6400 W. (Ans.)**

[or $P = 2 \times E_{ph} I_{ph} \cos \phi = 2 \times 400 \times 8 \times 1 = 6400$ W].

Example 1.15. *The secondary of a 3-phase star-connected transformer, which has a phase voltage of 230 V feeds a 3-phase delta connected load ; each phase of which has a resistance of 30 Ω and an inductive reactance of 40 Ω. Draw the circuit diagram of the system and calculate :*

(*i*) *The voltage across each phase of load,*

(*ii*) *The current in each phase of load,*

(*iii*) *The current in the transformer secondary windings, and*

(*iv*) *The total power taken from the supply and its power factor.*

Solution. Refer Fig. 1.23.

Resistance per phase, $R_{ph} = 20$ Ω

Reactance per phase, $X_{ph} = 40$ Ω

Phase voltage across transformer secondary,

$E_{ph} = 230$ V

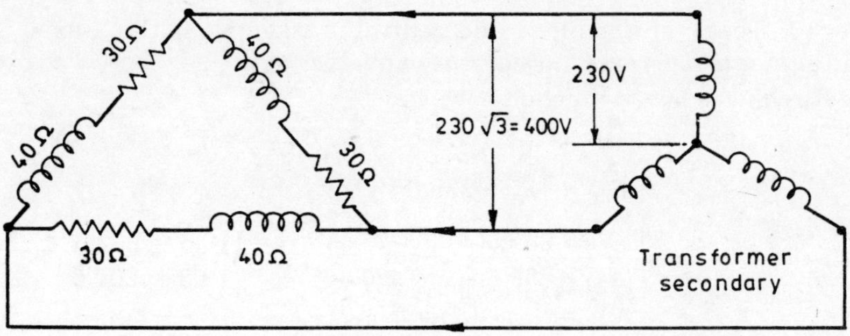

Fig. 1.23

Line voltage across delta connected load = line voltage across transformer secondary

$$= \sqrt{3} \times 230 = 400 \text{ V (app.)}$$

(*i*) **Voltage across each phase of the load,**

$$E_{ph} = E_L = \mathbf{400\ V.}\quad \textbf{(Ans.)}$$

(*ii*) **Current in each phase of the load,**

$$I_{ph} = \frac{E_{ph}}{Z_{ph}} = \frac{400}{\sqrt{R_{ph}^2 + X_{ph}^2}} = \frac{400}{\sqrt{30^2 + 40^2}} = \frac{400}{50} = \mathbf{8\ A.}\quad \textbf{(Ans.)}$$

(*iii*) **Current in the transformer secondary**

$$= \text{line current of load} = \sqrt{3} \times I_{ph} = \sqrt{3} \times 8 = \mathbf{13.86\ A.}\ \textbf{(Ans.)}$$

(*iv*) **Power factor,** $\cos\phi = \frac{R_{ph}}{Z_{ph}} = \frac{30}{50} = \mathbf{0.6.}$ **(Ans.)**

Total power consumed, $P = \sqrt{3} E_L I_L \cos\phi = \sqrt{3} \times 400 \times 13.86 \times 0.6 = \mathbf{5761.5\ W.}$ **(Ans.)**

Example 1.16. *Three non-inductive resistances, each of 50 ohms are connected in star to a 3-phase 415 volts supply. Three inductive coils of 100 ohms each connected in delta fashion are also connected to the same supply. Calculate :*

(*i*) *Line current* (*ii*) *Power factor of the system.*

(AMIE Winter, 1999)

Solution. The connection diagram and the vector diagram of the star and delta connected impedances are shown in Fig. 1.24 (*a*) and (*b*).

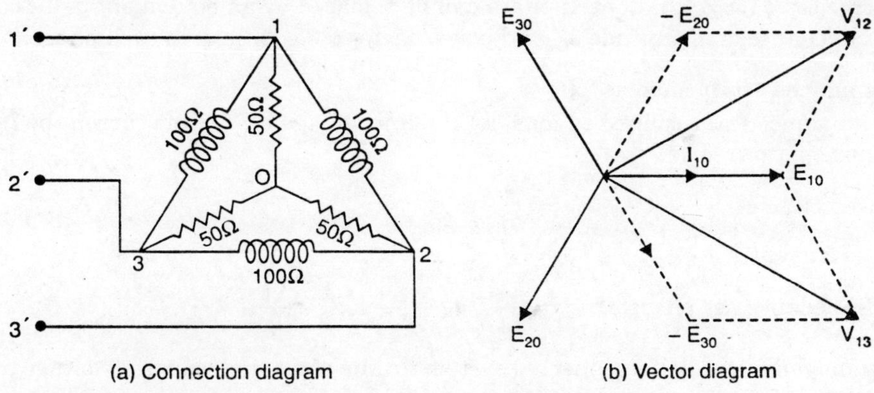

Fig. 1.24

The voltage E_{10} between the line 1 and neutral is taken along the X-axis. Since the load is **balanced**, it will *suffice to determine the current in one line only*.

Applying Kirchhoff's law to junction 1, we have

$$I_{11'} = I_{10} + I_{12} + I_{13}$$

Now $$E_{10} = \frac{415}{\sqrt{3}} = 239.6 \text{ V}$$

$$E_{20} = 239.6(-0.5 - j0.866) = -119.8 - j207.5$$

$$E_{30} = 239.6(-0.5 + j0.866) = -119.8 + j207.5$$

Now $$V_{12} = E_{10} + E_{02} = E_{10} + (-E_{20}) = E_{10} - E_{20}$$

$$= 239.6 - (-119.8 - j207.5) = 359.4 + j207.5$$

$$V_{13} = E_{10} + E_{03} = E_{10} - E_{30}$$

$$= 239.6 - (-119.8 + j207.5) = 359.4 - j207.5$$

(*i*) **Line current :**

Now, $$I_{10} = \frac{E_{10}}{50} = \frac{239.6}{50} = 4.792 + j0$$

$$I_{12} = \frac{V_{12}}{j100} = \frac{359.4 + j207.5}{j100} = 2.075 - j3.594$$

$$I_{13} = \frac{V_{13}}{j100} = \frac{359.4 - j207.5}{j100} = -2.075 - j3.594$$

$\therefore$ $$I_{11}' = I_{10} + I_{12} + I_{13}$$

$$= (4.792 + j0) + (2.075 - j3.594) + (-2.075 - j3.594)$$

$$= 4.792 - j7.188 = \mathbf{8.639 \angle -56.3°}.$$

Hence line current $= \mathbf{8.639\ A.}$ **(Ans.)**

Power factor $= \cos 56.3° = \mathbf{0.5548.}$ **(Ans.)**

☞**Example 1.17. (Balanced parallel loads)** *Three star-connected impedances $Z_1 = (20 + j37.7)\ \Omega$ per phase are in parallel with three delta-connected impedances $Z_2 = (30 - j159.3)\ \Omega$ per phase. The line voltage is 398 V. Find the line current, power and reactive volt-amperes taken by the combination.* **(AMIE Winter, 1998)**

Solution. Any balanced star-connected system can be completely replaced by an equivalent delta-connected system or vice-versa because of their relationship between phase and line voltages and currents. For example, a balanced star-connected load having an impedance of magnitude Z with a power factor $\cos \phi$ (or $Z \angle\phi$) in each phase can be replaced by an equivalent delta-connected load having an impedance of magnitude $3Z$ and power factor ϕ (*i.e.* $3Z \angle\phi$) in each phase.

This may be established as follows :

For a balanced star-connected load, let the line voltage be V_L, line current be I_L and impedance per phase be Z.

Phase voltage, $$V_{ph} = \frac{V_L}{\sqrt{3}}$$

Phase current, $$I_{ph} = I_L$$

$\therefore$ Impedance per phase, $$Z_{star} = \frac{V_{ph}}{I_{ph}} = \frac{V_L/\sqrt{3}}{I_L} = \frac{1}{\sqrt{3}} \cdot \frac{V_L}{I_L} \quad ...(i)$$

Now in equivalent delta-connected system for the same line values of voltage and current as in case of star-connected system,

Phase voltage, $$V_{ph} = V_L$$

Phase current, $I_{ph} = \frac{I_L}{\sqrt{3}}$

$\therefore$ Impedance per phase, $Z_{delta} = \frac{V_{ph}}{I_{ph}} = \frac{V_L}{I_L/\sqrt{3}} = \sqrt{3}\,\frac{V_L}{I_L}$ ...(ii)

Comparing (i) and (ii), we have

$$Z_{star} = \frac{1}{3} \times Z_{delta} \quad \text{or} \quad Z_{delta} = 3Z_{star}$$

The equivalent delta load of the star-connected load is given as 3Z, *i.e.* $(60 + j113.1)\ \Omega$

Line voltage, $V_L = 398$ V

Phase voltage, $V_{ph} = 398$ V in case of delta-connections.

Taking phase voltage as reference phasor

$$V_{ph} = 398 \angle 0° \text{ V}$$

Phase current for impedance Z_1, $I_{ph1} = \frac{V_{ph}}{Z_1} = \frac{398 \angle 0°}{(60 + j113.1)}$

$$= \frac{398 \angle 0°}{128 \angle 62.05°} = 3.109 \angle -62.05° \text{ A} \quad \text{or} \quad \mathbf{(1.46 - j2.746)\ A}$$

Phase current to impedance Z_2, $I_{ph2} = \frac{398 \angle 0°}{(30 - j159.3)}$

$$= \frac{398 \angle 0°}{162.1 \angle -79.33°}$$

$$= 2.455 \angle 79.33° \text{ A} \quad \text{or} \quad (0.4545 + j2.412) \text{ A}$$

Total phase current, $I_{ph} = I_{ph1} + I_{ph2}$

$$= (1.46 - j2.746) + (0.4545 + j2.412)$$

$$= (1.914 - j0.334) \text{ A} \quad \text{or} \quad 1.943 \angle -9.9°$$

Total line current, $\mathbf{I_L} = \sqrt{3} I_{ph} = \sqrt{3} \times 1.943 = $ **3.36 A. (Ans.)**

Power, $\mathbf{P} = \sqrt{3} V_L I_L \cos \phi$

$= \sqrt{3} \times 398 \times 3.36 \cos(-9.9°) =$ **2281.7 W. (Ans.)**

Reactive power $= \sqrt{3}\, V_L I_L \sin \phi$

$= \sqrt{3} \times 398 \times 3.36 \sin(-9.9°)$

$= -398.2$ *VAR* or **398.2 VAR** (inductive). **(Ans.)**

1.9. MEASUREMENT OF POWER IN 3-PHASE CIRCUIT

- The power in 3-phase load can be measured by using the following methods :
  1. Three wattmeters method.
  2. Two wattmeters method.
  3. One wattmeter method.
- A wattmeter consists of two coils : Refer Fig. 1.25.
  1. *Current coil*—possesses a *low resistance.*
  2. *Pressure or potential coil*—possesses a *high resistance.*

The '*current coil*' is connected in *series with the line carrying the current* and the '*pressure coil*' *is connected across the two points whose potential*

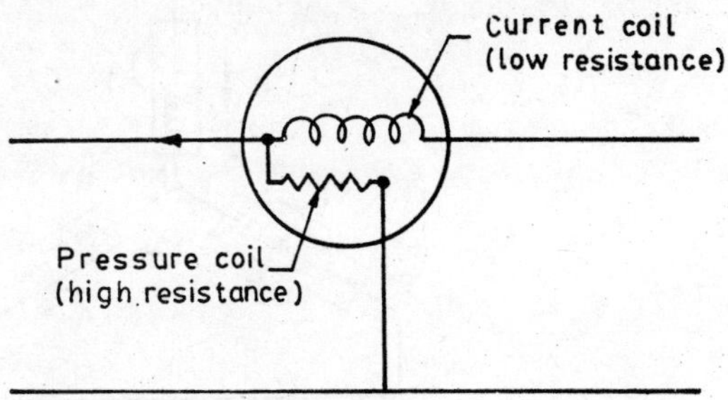

Fig. 1.25. Connections of a wattmeter.

difference is to be measured. A wattmeter shows a reading which is proportional to the *product* of the *current* through its current coil, the *potential difference* across its pressure coil and *cosine of the angle* between this voltage and current.

1.9.1. Three Wattmeters Method. Figs. 1.26 and 1.27 show the connection diagram for star-connected and delta-connected loads respectively. As indicated in the figures three wattmeters are connected in each of the three phases of the load whether star or delta connected. The current

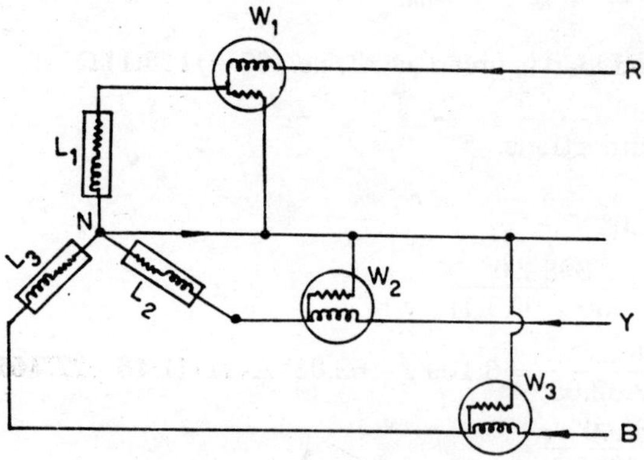

Fig. 1.26. Star-connected load.

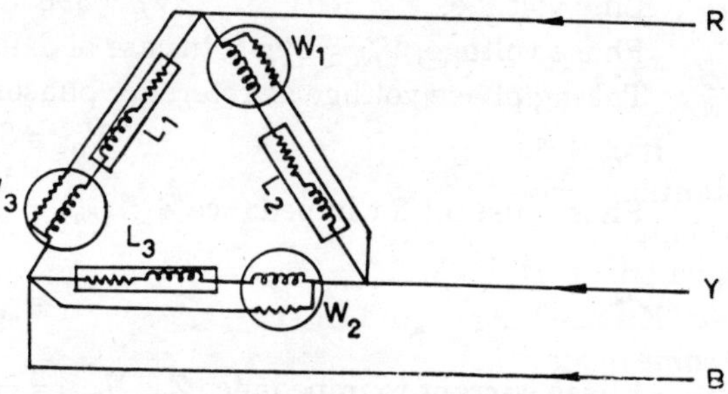

Fig. 1.27. Delta-connected load.

coil of each wattmeter carries the current of one phase only and the pressure coil measures the phase-voltage of the phase. Hence, *each wattmeter measures the power in a single phase. The total power in the load is given by the algebraic sum of the readings of the three wattmeters.*

While using this method following *difficulty* is met with :

- In case of star-connected load it is not always possible to get at neutral point which is required for connections (Fig. 1.26).
- In case of delta-connected load, under ordinary conditions it is not generally feasible to break into the phases of the load.

To measure power it is not necessary to use three wattmeters, two wattmeters can be used for the purpose as explained in the Article 1.9.2.

1.9.2. Two Wattmeters Method. Balanced or unbalanced load, Figs. 1.28 and 1.29 show connection diagrams for star-connected and delta-connected loads respectively. In this method the current coils of the two wattmeters are inserted in *any two* lines and the pressure (or potential) coil of each joined to the *third line.*

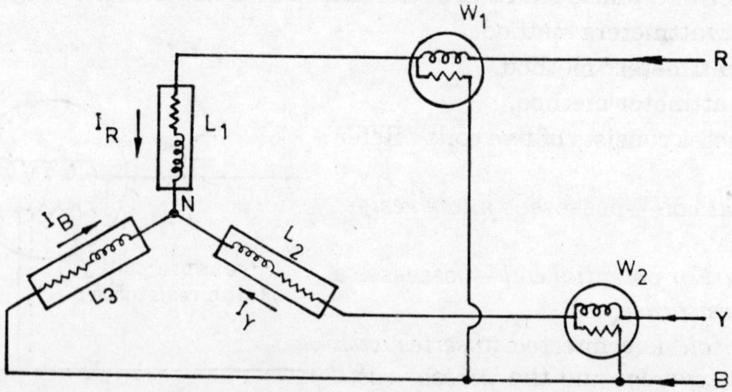

Fig. 1.28. Star-connected load.

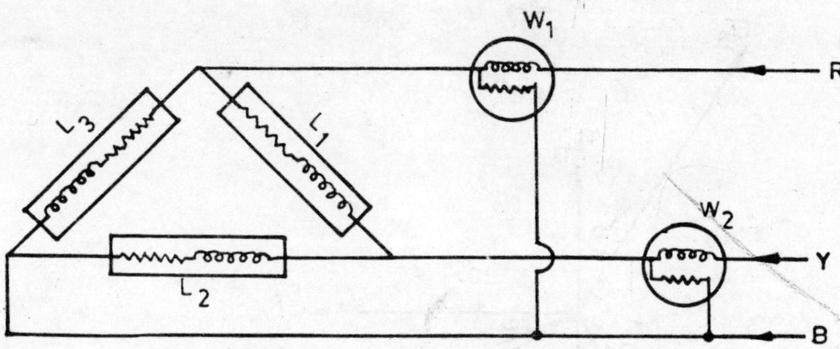

Fig. 1.29. Delta-connected load.

It can be proved that the sum of the instantaneous powers indicated by W_1 and W_2 gives the instantaneous power absorbed by the three loads L_1, L_2 and L_3. Let us consider a star-connected load (although it can be equally applied to a delta-connected load which can always be replaced by an equivalent star-connected load).

Keeping in mind that it is important to take *the direction of the voltage through the circuit as the same as that taken for the current when establishing the readings of the two wattmeters.*

Instantaneous current through $W_1 = i_R$

Instantaneous potential difference across $W_1 = e_{RB} = e_R - e_B$

Instantaneous power read by $W_1 = i_R(e_R - e_B)$

Instantaneous current through $W_2 = i_Y$

Instantaneous potential difference across $W_2 = e_{YB} = e_Y - e_B$

Instantaneous power read by $W_2 = i_Y(e_Y - e_B)$

$$\therefore \quad W_1 + W_2 = i_R(e_R - e_B) + i_Y(e_Y - e_B) = i_R e_R + i_Y e_Y - e_B(i_R + i_Y)$$

Now according to Kirchhoff's point law

$$i_R + i_Y + i_B = 0$$

$$\therefore \quad i_R + i_Y = -i_B$$

$$\therefore \quad W_1 + W_2 = i_R e_R + i_Y e_Y + i_B e_B = p_1 + p_2 + p_3$$

where p_1 = power absorbed by load L_1, p_2 = power absorbed by load L_2, and

p_3 = power absorbed by load L_3.

$$\therefore \quad W_1 + W_2 = \text{total power absorbed.}$$

This proof is true whether the load is *balanced* or *unbalanced*.

In case the load is star-connected, then it should have no neutral connection (*i.e.*, 3-phase, 3-wire connected) and if it has a neutral connection (*i.e.*, 3-phase, 4-wire connected) that it should be exactly balanced so that in each case there is no neutral current i_N otherwise Kirchhoff's point law will give $i_R + i_Y + i_B + i_N = 0$.

In the above derivation we have considered the *instantaneous* readings. In fact the moving system wattmeter, due to its inertia, cannot quickly follow the variations taking place in cycle, hence it indicates the *average power.*

$\therefore$ Total power $= W_1 + W_2$

$$= \frac{1}{T}\int_0^T i_R e_{RB}\, dt + \frac{1}{T}\int_0^T i_Y e_{YB}\, dt. \qquad \text{...(1.7)}$$

Two Wattmeters Method—Balanced load. The total power consumed by a *balanced load* can be found by using two wattmeters (Figs. 1.28 and 1.29). When load is assumed inductive in Fig. 1.26, the vector diagram for such a balanced star-connected load is shown in Fig. 1.30.

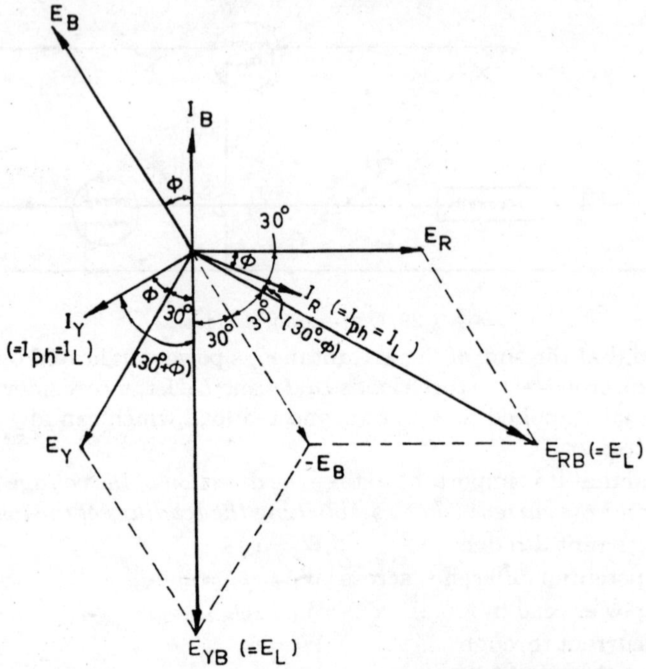

Fig. 1.30. Vector diagram—two wattmeters method.

Let us consider the problem in terms of r.m.s. values (instead of instantaneous values).

Let E_R, E_Y, E_B = r.m.s. values of the three phase voltages;

and I_R, I_Y, I_B = r.m.s. values of the currents.

Since these voltages and currents are assumed sinusoidal, they can be represented by vectors, the currents lagging behind their respective phase voltages by φ.

Refer Fig. 1.30.

Current through wattmeter $W_1 = I_R$

Potential difference across pressure coil of wattmeter

$$W_1 = E_{RB} = E_R - E_B \qquad \text{(Vectorially)}$$

The value of E_{RB} is found by compounding E_R and E_B reversed as shown in Fig. 1.30. It may be observed that phase difference between E_{RB} and I_R = (30° – φ).

$\therefore$ Reading of wattmeter $W_1 = E_{RB} I_R \cos (30° - \phi)$ ...(i)

Similarly, current through wattmeter $W_2 = I_Y$

Potential difference across pressure coil of wattmeter

$$W_2 = E_{YB} = E_Y - E_B \qquad \text{(Vectorially)}$$

The value of E_{YB} is found by compounding E_Y and E_B reversed as shown in Fig. 1.30. The phase difference between E_{YB} and I_Y = (30° + φ).

$\therefore$ Reading of wattmeter $W_2 = E_{YB} I_Y \cos (30° + \phi)$ ...(ii)

Since the load is balanced, $E_{RB} = E_{YB} = E_L$ (Line voltage)

and $I_R = I_Y = I_L$ (Line current)

$\therefore \qquad W_1 = E_L I_L \cos (30° - \phi)$

and $\qquad W_2 = E_L I_L \cos (30° + \phi)$

$\therefore$ Total power,

$$
\begin{aligned}
P &= W_1 + W_2 \\
&= E_L I_L \cos (30° - \phi) + E_L I_L \cos (30° + \phi) \\
&= E_L I_L [\cos (30° - \phi) + \cos (30° + \phi)] \\
&= E_L I_L [\cos 30° \cos \phi + \sin 30° \sin \phi + \cos 30° \cos \phi - \sin 30° \sin \phi] \\
&= E_L I_L (2 \cos 30° \cos \phi) = E_L I_L \times 2 \times \frac{\sqrt{3}}{2} \cos \phi = \sqrt{3} E_L I_L \cos \phi
\end{aligned}
$$

i.e., $\qquad P = \sqrt{3} E_L I_L \cos \phi$

Hence the *sum of the readings of the two wattmeters give the total power consumption in the 3-phase load.*

It is worth noting that in the above case the phase sequence of *RYB* has been assumed, the readings of the two wattmeters will change if the phase sequence is reversed.

Variations in wattmeter readings. As shown above that for a *lagging power factor*

$W_1 = E_L I_L \cos (30° - \phi)$

and $\qquad W_2 = E_L I_L \cos (30° + \phi)$

From above it is evident that individual readings of the wattmeters *not only depend on the load but also upon its power factor*. Let us take up the following cases :

(*i*) When $\phi = 0$

i.e., power factor is *unity* (*i.e.,* load is *resistive*)

Then $\qquad W_1 = W_2 = E_L I_L \cos 30°$

The reading of each wattmeter will be equal and opposite (*i.e.,* up-scale reading).

(*ii*) When $\phi = 60°$

i.e., power factor = 0.5 (lagging)

Then $\qquad W_2 = E_L I_L \cos (30° + 60°) = 0$

Hence, the power is measured by W_1 above.

(*iii*) When $\quad 90° > \phi > 60°$

i.e., $\quad 0.5 > \text{p.f.} > 0$

Then W_1 is still *positive* but reading of W_2 is *reversed*. For a leading p.f., conditions are just the opposite of this. In that case, W_1 will read *negative* because the phase angle between the current and voltage is *more than* 90°. For getting the total power, the reading of W_2 is to be *subtracted* from that of W_1. Under this condition, W_2 will read '*down scale*' *i.e., backwards*. Hence, *to obtain a reading on* W_2, *it is necessary to reverse either its pressure coil or current coil, usually the former.*

All readings taken after reversal of pressure coil are to be taken as negative.

(*iv*) When $\phi = 90°$

(*i.e.,* p.f. = 0 *i.e.,* pure inductive or capacitive load)

Then $\qquad W_1 = E_L I_L \cos (30° - 90°) = E_L I_L \sin 30°$

and $\qquad W_2 = E_L I_L \cos (30° + 90°) = - E_L I_L \sin 30°$

These two readings are equal in magnitude but opposite in sign

$\therefore \qquad W_1 + W_2 = 0.$

So far we have considered *lagging angles* (taken as *positive*). Now let us discuss how the readings of wattmeters change when the power factor is *leading* one.

— For $\phi = +60°$ (lag) : $W_2 = 0$

— For $\phi = -60°$ (lead) : $W_1 = 0$

Thus we find that for angles of lead the readings of the two wattmeters are interchanged. Hence, when the power is *leading* :

$$W_1 = E_L I_L \cos(30° + \phi)$$
$$W_2 = E_L I_L \cos(30° - \phi).$$

Power Factor—When the load is 'balanced'. When load is balanced with a *lagging power factor* and the voltage and currents are sinusoidal :

$$W_1 + W_2 = E_L I_L \cos(30° - \phi) + E_L I_L \cos(30° + \phi) = \sqrt{3} E_L I_L \cos\phi \quad ...(i)$$

Similarly,
$$W_1 - W_2 = E_L I_L \cos(30° - \phi) - E_L I_L \cos(30° + \phi)$$
$$= E_L I_L (2 \sin\phi \sin 30°) = E_L I_L \sin\phi \quad ...(ii)$$

Dividing (*ii*) by (*i*), we get

$$\tan\phi = \frac{\sqrt{3}(W_1 - W_2)}{(W_1 + W_2)} \quad ...(1.8)$$

For a *leading power*, this expression becomes

$$\tan\phi = -\frac{\sqrt{3}(W_1 - W_2)}{(W_1 + W_2)} \quad ...(1.9)$$

After finding tan ϕ, hence ϕ, the value of power factor cos ϕ can be found (from trigonometrical tables).

One important point which must be kept in mind is that if W_2 reading has been taken after reversing the pressure coil *i.e.*, if W_2 is *negative*, then the eqn. (1.8) becomes

$$\tan\phi = \sqrt{3}\,\frac{W_1 - (-W_2)}{W_1 + (-W_2)}$$

or
$$\tan\phi = \sqrt{3}\,\frac{W_1 + W_2}{W_1 - W_2} \quad ...(1.10)$$

The power factor may also be expressed in terms of ratio of the readings of the two wattmeters.

Let
$$\frac{\text{Smaller reading}}{\text{Larger reading}} = \frac{W_2}{W_1} = \alpha$$

Then from eqn. (1.8) above, we have

$$\tan\phi = \frac{\sqrt{3}\left[1 - \left(\frac{W_2}{W_1}\right)\right]}{\left[1 + \left(\frac{W_2}{W_1}\right)\right]} = \frac{\sqrt{3}(1 - \alpha)}{(1 + \alpha)}$$

We know that,
$$\sec^2\phi = 1 + \tan^2\phi$$

or
$$\frac{1}{\cos^2\phi} = 1 + \tan^2\phi \quad \text{or} \quad \cos^2\phi = \frac{1}{1 + \tan^2\phi}$$

$$\cos\phi = \frac{1}{\sqrt{1 + \tan^2\phi}} = \frac{1}{\sqrt{1 + \left[\frac{\sqrt{3}(1 - \alpha)}{1 + \alpha}\right]^2}}$$

or
$$= \frac{1}{\sqrt{1 + 3\left(\frac{1-\alpha}{1+\alpha}\right)^2}} = \frac{1+\alpha}{\sqrt{(1+\alpha)^2 + 3\,(1-\alpha)^2}}$$

$$= \frac{1+\alpha}{\sqrt{1+\alpha^2+2\alpha+3\,(1+\alpha^2-2\alpha)}} = \frac{1+\alpha}{\sqrt{4+4\alpha^2-4\alpha}}$$

$$= \frac{1+\alpha}{2\sqrt{1-\alpha+\alpha^2}}$$

i.e.,
$$\cos\phi = \frac{1+\alpha}{2\sqrt{1-\alpha+\alpha^2}} \qquad \text{...(1.11)}$$

If a curve is plotted between α and cos φ, then the curve obtained will be as shown in Fig. 1.31, this curve is called **watt-ratio curve.**

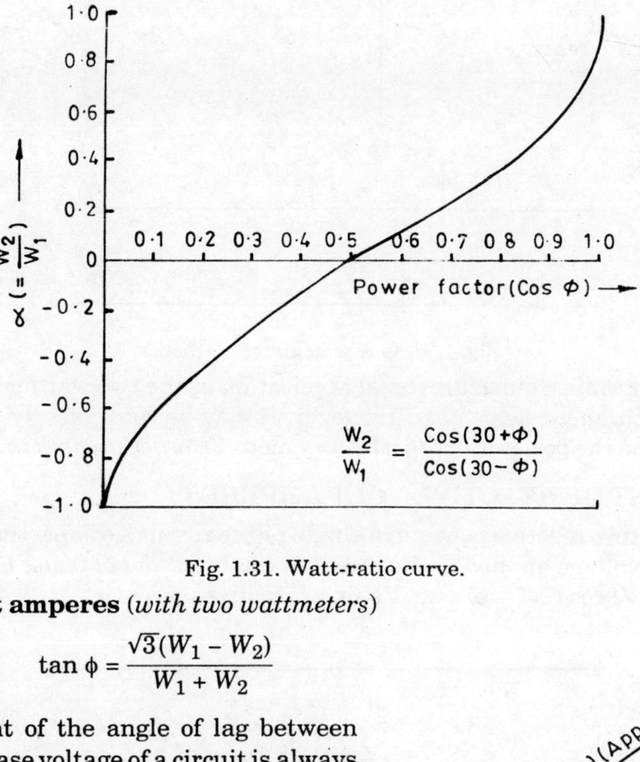

Fig. 1.31. Watt-ratio curve.

Reactive volt amperes (*with two wattmeters*)

We know that $\tan\phi = \frac{\sqrt{3}(W_1 - W_2)}{W_1 + W_2}$

As the tangent of the angle of lag between phase current and phase voltage of a circuit is always equal to the ratio of reactive power to the true power (Fig. 1.32). Hence, in case of a balanced load, the reactive power is given by √3 *times the difference of the readings of the two wattmeters* used to measure the power of a 3-phase circuit by two wattmeter method.

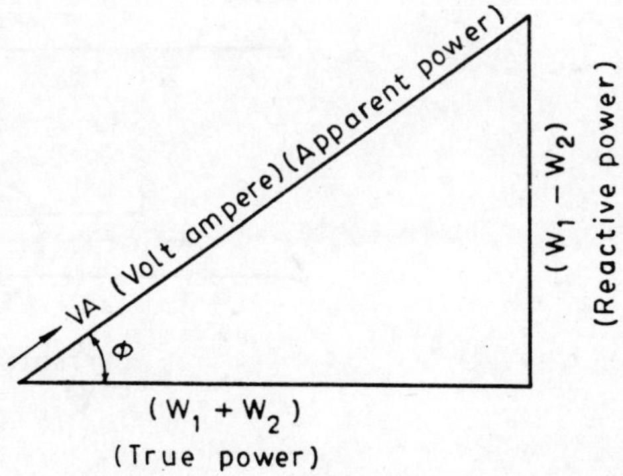

Fig. 1.32

Mathematical proof is as follows :

$$\sqrt{3}\,(W_1 - W_2) = \sqrt{3}\,[(E_L I_L \cos(30° - \phi) - E_L I_L \cos(30° + \phi)]$$
$$= \sqrt{3}\,E_L I_L\,[(\cos 30° \cos\phi + \sin 30° \sin\phi) - (\cos 30° \cos\phi - \sin 30° \sin\phi)]$$
$$= \sqrt{3}\,E_L I_L\,[\cos 30° \cos\phi + \sin 30° \sin\phi - \cos 30° \cos\phi + \sin 30° \sin\phi]$$
$$= \sqrt{3}\,E_L I_L\,(2 \sin 30° \sin\phi) = \sqrt{3}\,E_L I_L \sin\phi$$

i.e., $$\sqrt{3}\,(W_1 - W_2) = \sqrt{3}\,E_L I_L \sin\phi.$$

1.9.3. One Wattmeter Method. In this method the current coil is connected in any one line and the pressure coil is connected alternately between this and the other two lines as shown in Fig. 1.33. The two readings so obtained, for a balanced load, correspond to those obtained by normal two wattmeter method.

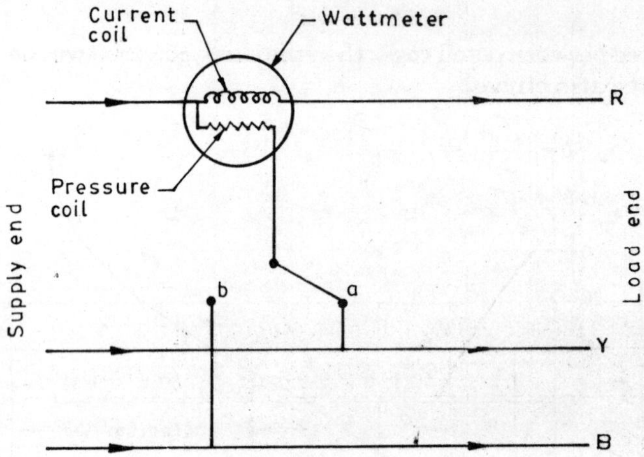

Fig. 1.33. One wattmeter method.

This method is not of as much universal application as the two wattmeter method because it is restricted to *fairly balanced loads only*. However, it may be conveniently applied, for instance, when it is desired to find the power input to a factory motor *in order to check the load upon the motor*.

1.10. MEASUREMENT OF REACTIVE VOLT AMPERES

In order to measure reactive power in a single phase circuit a *compensated wattmeter* is used. In this wattmeter the voltage applied to the pressure coil is *90° out of phase with the actual voltage* an˙ ence it will read $VI \cos(90° - \phi)$ *i.e.,* $VI \sin\phi$, reactive power.

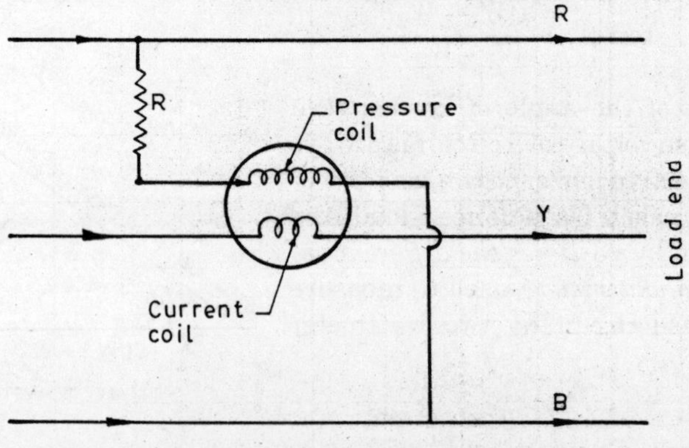

Fig. 1.34. Measurement of reactive volt-amperes.

In balanced three-phase circuit the reactive power can be determined by using one wattmeter. The necessary connections are shown in Fig. 1.34. The current coil is inserted in one line and the pressure coil is connected across the other two lines.

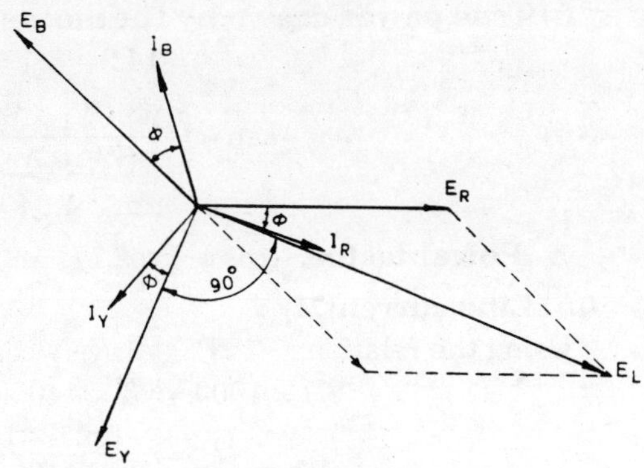

Fig. 1.35

The current following through the current coil of the wattmeter

$$= I_Y = I_{ph} \text{ (say)}$$

The potential difference across the potential coil of wattmeter,

$$E = E_R - E_B$$
$$= \sqrt{3}\, E_{ph} \text{ leading vector } E_Y \text{ by } 90°$$
$$= \sqrt{3}\, E_{ph} \text{ leading the vector of current } I_Y \text{ by } (90° + \phi)$$

∴ Reading of the wattmeter (Fig. 1.35)

$$= \sqrt{3}\, E_{ph} I_{ph} \cos (90° + \phi)$$
$$= -\sqrt{3}\, E_{ph} I_{ph} \sin \phi = -W_r$$

Total reactive power of the circuit,

$$= 3\, E_{ph} I_{ph} \sin \phi = -\sqrt{3}\, W_r$$

Reactive power (as earlier started) can also be determined from two wattmeter readings connected for measurement of power.

$$W_1 - W_2 = E_L I_L \cos (30° - \phi) - E_L I_L \cos (30° + \phi)$$

or $$W_1 - W_2 = E_L I_L \times 2 \sin 30° \sin \phi$$

or $$E_L I_L \sin \phi = W_1 - W_2$$

Reactive power of load circuit.

$$W_r = \sqrt{3}\, E_L I_L \sin \phi = \sqrt{3}\, (W_1 - W_2).$$

POWER MEASUREMENT IN 3-PHASE CIRCUITS

Example 1.18. *The power input to a 3-phase induction motor is read by two wattmeters. The readings are 920 W and 300 W. Calculate the power factor of the motor.*

Solution. Reading of wattmeter, $W_1 = 920$ W

Reading of wattmeter, $W_2 = 300$ W

Power factor of the motor, cos φ :

Using the relation, $$\tan \phi = \frac{\sqrt{3}\,(W_1 - W_2)}{(W_1 + W_2)} = \frac{\sqrt{3}\,(920 - 300)}{(920 + 300)} = \frac{\sqrt{3} \times 620}{1220} = 0.88$$

∴ $$\phi = \tan^{-1} 0.88 = 41.35°$$

Power factor of the motor,

$$\cos \phi = \cos 41.35° = \mathbf{0.75 \text{ (lag)}}. \quad \textbf{(Ans.)}$$

Example 1.19. *While performing a load test on a 3-phase wound-rotor induction motor by two wattmeters method, the readings obtained on two wattmeters were + 14.2 kW and – 6.1 kW and the line voltage was 440 V. Calculate :*

(i) True power drawn by the motor *(ii) Power factor, and*

(iii) Line current.

Solution. Reading of wattmeter, $W_1 = 14.2$ kW

Reading of wattmeter, $W_2 = -6.1$ kW

Line voltage, $E_L = 440$ V

(*i*) **True power drawn by the motor**

$$= 14.2 - 6.1 = \mathbf{8.1\ kW.}\ \mathbf{(Ans.)}$$

(*ii*) $$\tan \phi = \frac{\sqrt{3}\,(W_1 - W_2)}{(W_1 + W_2)} = \frac{\sqrt{3}\,[14.2 - (-6.1)]}{[14.2 + (-6.1)]} = \frac{\sqrt{3} \times 20.3}{8.1} = 4.34$$

$\therefore$ $$\phi = \tan^{-1} 4.34 = 77°$$

$\therefore$ **Power factor,** $\cos \phi = \cos 77° = \mathbf{0.2249\ (lag).}$ **(Ans.)**

(*iii*) **Line current, I_L :**

Using the relation, $$P = \sqrt{3}\, E_L I_L \cos \phi$$

$$8.1 \times 1000 = \sqrt{3} \times 440 \times I_L \times 0.2249$$

$\therefore$ $$I_L = \frac{8.1 \times 1000}{\sqrt{3} \times 440 \times 0.2249} = 47.26\ \text{A}$$

Hence, **line current = 47.26 A. (Ans.)**

Example 1.20. *A 3-phase, 440 V motor load has a power factor of 0.6. Two wattmeters connected to measure the power show the input to be 25 kW. Find the reading on each instrument.*

Solution. Input power = 25 kW

Line voltage, $E_L = 440$ V

Power factor of the motor load, $\cos \phi = 0.6$

W_1 ; W_2 :

Using the relation : $$\tan \phi = \frac{\sqrt{3}\,(W_1 - W_2)}{W_1 + W_2} \quad ...(i)$$

Now, $\cos \phi = 0.6, \quad \therefore \quad \phi = 53.13°$

$\therefore$ $\tan \phi = \tan 53.13° = 1.333$

Also $W_1 + W_2 = 25$ kW (*given*) ...(*ii*)

Substituting these values in (*i*), we get

$$1.333 = \frac{\sqrt{3}\,(W_1 - W_2)}{25}$$

$\therefore$ $$W_1 - W_2 = \frac{1.333 \times 25}{\sqrt{3}} = 19.24\ \text{kW} \quad ...(iii)$$

From (*ii*) and (*iii*), we get **W_1 = 22.12 kW**

and **W_2 = 2.88 kW. (Ans.)**

☞**Example 1.21.** *In a 3-phase circuit two wattmeters used to measure power indicate 1200 W and 600 W respectively. Find the power factor of the circuit :*

(*i*) *When both wattmeter readings are positive.*

(*ii*) *When the latter is obtained by reversing the current coil connections.*

(AMIE Summer, 1999)

Solution. (*i*) *When both wattmeter readings are positive :*

Reading of wattmeter, $W_1 = 1200$ W

Reading of wattmeter, $W_2 = 600$ W

We know that, $$\tan \phi = \frac{\sqrt{3}\,(W_1 - W_2)}{W_1 + W_2} = \frac{\sqrt{3}\,(1200 - 600)}{(1200 + 600)} = 0.577$$

$$\phi = \tan^{-1} 0.577 = 30°$$

Power factor, $\cos \phi = \cos 30° = \mathbf{0.866\ (lag).}$ **(Ans.)**

(ii) When the reading of wattmeter W_2 is obtained by reversing the coil connection :

Reading of wattmeter, $W_1 = 1200$ W

Reading of wattmeter, $W_2 = -600$ W

We know that, $$\tan \phi = \frac{\sqrt{3}\,(W_1 - W_2)}{(W_1 + W_2)} = \frac{\sqrt{3}\,[1200 - (-600)]}{[1200 + (-600)]} = \frac{\sqrt{3} \times 1800}{600} = 5.196$$

or $$\phi = \tan^{-1} 5.196 = 79.1°$$

Hence, **power factor** $= \cos \phi = \cos 79.1° = $ **0.1889. (Ans.)**

Example 1.22. *In order to measure the power input and the power factor of an over-excited synchronous motor two wattmeters are used. If the meters indicate (– 3.5 kW) and (+ 8.0 kW) respectively. Calculate :*

(i) Power factor of the motor. *(ii) Power input to the motor.*

Solution. *(i) Since an over-excited synchronous motor runs with a leading power factor*, we should use the relation,

$$\tan \phi = -\frac{\sqrt{3}\,(W_1 - W_2)}{(W_1 + W_2)}$$

Moreover it is W_1 that gives negative reading and not W_2. Hence $W_1 = -3.5$ kW.

$$\therefore \quad \tan \phi = -\frac{\sqrt{3}\,(-3.5 - 8)}{(-3.5 + 8)} = -\frac{\sqrt{3} \times 11.5}{5.5} = 3.62$$

$$\therefore \quad \phi = \tan^{-1} 3.62 = 74.6° \text{ (lead)}$$

∴ **Power factor,** $\cos \phi = \cos 74.6° = $ **0.2655 (lead). (Ans.)**

(ii) **Power input** $= W_1 + W_2 = -3.5 + 8 = $ **4.5 kW. (Ans.)**

Example 1.23. *Two wattmeters are used to measure power input to a 1.5 kV, 50 Hz, 3-phase motor running on full-load at an efficiency of 85 per cent. Their readings are 250 kW and 80 kW respectively. Calculate :*

(i) Input, *(ii) Power factor,*

(iii) Line current, and *(iv) Output.*

Solution. *Since the motor is running at full-load, its power factor must be greater than 0.5. Hence W_2 reading is positive*

$$\therefore \quad W_1 = +250 \text{ kW} \quad \text{and} \quad W_2 = +80 \text{ kW}$$

(i) **Input** $= W_1 + W_2 = 250 + 80 = $ **330 kW. (Ans.)**

(ii) $$\tan \phi = \frac{\sqrt{3}\,(W_1 - W_2)}{W_1 + W_2} = \frac{\sqrt{3}\,(250 - 80)}{(250 + 80)} = 0.892$$

$$\therefore \quad \phi = \tan^{-1} 0.892 = 41.74°$$

and **power factor** $= \cos \phi = \cos 41.74° = $ **0.746 (lag). (Ans.)**

(iii) **Power,** $$P = \sqrt{3}\, E_L I_L \cos \phi$$

$$\therefore \quad I_L = \frac{P}{\sqrt{3}\, E_L \cos \phi} = \frac{330 \times 1000}{\sqrt{3} \times 1.5 \times 1000 \times 0.746} = 170.27$$

Hence, **line current = 170.27 A. (Ans.)**

(iv) **Output** = input × efficiency = 330 × 0.85 = **280.5 kW. (Ans.)**

☞**Example 1.24.** *Two wattmeters are used to measure power input to a synchronous motor. Each of them indicates 60 kW. If the power factor be changed to 0.866 leading, determine the readings of the two wattmeters, the total input power remaining the same. Draw the vector diagram for the second condition of the load.*

Solution. Reading of wattmeter, $W_1 = 60$ kW

Reading of wattmeter, $W_2 = 60$ kW.

First case. In the first case, *both wattmeters read equal and positive. Hence, motor must be running at unity power factor.*

Second case. *Power factor is 0.866 leading.*

In this case : $W_1 = E_L I_L \cos (30° + \phi)$

$W_2 = E_L I_L \cos (30° - \phi)$

$\therefore \quad W_1 + W_2 = \sqrt{3}\, E_L I_L \cos \phi$

$W_1 - W_2 = - E_L I_L \sin \phi$

$\therefore \quad \tan \phi = - \dfrac{\sqrt{3}\,(W_1 - W_2)}{(W_1 + W_2)}$

Since $\cos \phi = 0.866$ (*given*)

$\therefore \quad \phi = \cos^{-1} 0.866 = 30°$

and $\tan \phi = \tan 30° = \dfrac{1}{\sqrt{3}}$

$\therefore \quad \dfrac{1}{\sqrt{3}} = - \dfrac{\sqrt{3}\,(W_1 - W_2)}{120}$

[$\because$ $W_1 + W_2 = 120$ kW (*given*)]

$\therefore \quad W_1 - W_2 = - \dfrac{120}{3} = - 40$ ...(*i*)

But $W_1 + W_2 = 120$ ...(*ii*)

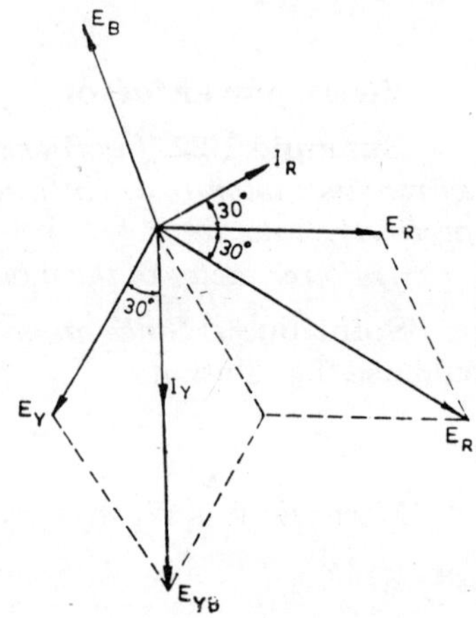

Fig. 1.36

From (*i*) and (*ii*), we get $\mathbf{W_1 = 40}$ **kW**, $\mathbf{W_2 = 80}$ **kW. (Ans.)**

For connection diagram, please refer to Fig. 1.29. The vector or phasor diagram is shown in Fig. 1.36.

Example 1.25. *Two wattmeters connected to read the total power in a 3-phase system supplying a balanced load read 10.5 kW and – 2.5 kW respectively. Calculate the total power and power factor.*

Also, explain the significance of (i) equal wattmeter readings and (ii) a zero reading on one wattmeter. **(AMIE Winter, 1998)**

Solution. *Given* : $W_1 = 10.5$ kW ; $W_2 = - 2.5$ kW.

For two wattmeter method,

total power = $W_1 + W_2 = 10.5 + (-2.5) =$ **8.0 kW. (Ans.)**

We know that, $\tan \phi = \dfrac{\sqrt{3}\,(W_1 - W_2)}{W_1 + W_2} = \dfrac{\sqrt{3}\,[10.5 - (-2.5)]}{[10.5 + (-2.5)]} = 2.8145$

or $\phi = 70.44°$

$\therefore$ **Power factor,** $\cos \phi = \cos 70.44° =$ **0.335. (Ans.)**

(*i*) For readings of the two wattmeters to be equal,

$W_1 = W_2$

or $\sqrt{3}\, E_L I_L \cos (30° - \phi) = \sqrt{3}\, E_L I_L \cos (30° + \phi)$

or $\cos (30° - \phi) = \cos (30° + \phi)$

$\therefore \quad \phi = 0°$ or $\cos \phi = 1$

i.e., for *unity power factor, the readings of two wattmeters are equal.* **(Ans.)**

(*ii*) As readings of wattmeters are

$W_1 = \sqrt{3} E_L I_L \cos (30° - \phi)$, and

$W_2 = \sqrt{3}\, E_L I_L \cos (30° + \phi)$,

for *reading of one of the wattmeters to be zero ϕ must be 60°, which makes $W_2 = \sqrt{3}\, E_L I_L \cos 90°$* = **0.**

(Ans.)

Example 1.26. *A star connected 3-phase load consists of three similar impedances. When load is connected to a 3-phase 400 V, 50 Hz supply it takes 30A line current at a power factor of 0.8 lagging.*

(*i*) *Calculate the total power taken by the load.*

(*ii*) *Calculate the resistance and reactance of each phase of the load.*

(*iii*) *If power is measured by two wattmeters method, then calculate the reading of each wattmeter.* **(AMIE Winter, 1996)**

Solution. *Given* : $E_L = 400$ V ; $f = 50$ Hz ; $I_L = (I_{ph}) = 30$ A, $\cos\phi = 0.8$ lagging.

(*i*) **Total power taken by the load, P :**

$$P = \sqrt{3} E_L I_L \cos\phi$$

$$= \sqrt{3} \times 400 \times 30 \times 0.8 = 16627 \text{ W} \quad \text{or} \quad \mathbf{16.627\ kW.}\ \ \textbf{(Ans.)}$$

(*ii*) **Resistance and reactance of each phase of the load, R, X :**

$$E_{ph} = \frac{E_L}{\sqrt{3}} = \frac{400}{\sqrt{3}} = 230.95 \text{ V}$$

∴ Load impedance, $Z = \frac{E_{ph}}{I_{ph}} = \frac{230.95}{30} = 7.7\ \Omega$

∴ Load resistance, $\mathbf{R} = Z \cos\phi = 7.7 \times 0.8 = \mathbf{6.16\ \Omega}$. **(Ans.)**

and, load reactance, $\mathbf{X} = Z \sin\phi = 7.7 \times 0.6 = \mathbf{4.62\ \Omega}$. **(Ans.)**

(*iii*) **Reading of each wattmeter, $\mathbf{W_1, W_2}$:**

Reading of wattmeter 1, $W_1 = E_L I_L \cos(30° + \phi)$

Reading of wattmeter 2, $W_2 = E_L I_L \cos(30° - \phi)$

where $\phi = \cos^{-1} 0.8 = 36.9°$

∴ $\mathbf{W_1} = 400 \times 30 \cos(30° + 36.9°) = 4708$ W or **4.708 kW. (Ans.)**

and $\mathbf{W_2} = 400 \times 30 \cos(30° - 36.9°) = 11913$ W or **11.913 kW. (Ans.)**

Example 1.27. *A 3-phase, 3-wire, 415 V system supplies a balanced load of 20 A at a power factor 0.8 lag. The current coil of wattmeter (1) is in phase R and of wattmeter 2 in phase B. Calculate :*

(*i*) *The reading on 1 when its voltage coil is across R and Y ;*

(*ii*) *The reading on 2 when its voltage coil is across B and Y ;*

(*iii*) *The reading on 1 when its voltage coil is across Y and B.* **(AMIE Winter, 1997)**

Solution. *Given* : $E_L = 415$ V, $I_L = 20$ A, $\cos\phi = 0.8$ (lag) or $\phi = \cos^{-1} 0.8 = 36.87°$ (lag)

(*i*) *The reading on 1 when its voltage coil is across R and Y and current coil is in R phase*

$$= E_L I_L \cos(30° - \phi) = 415 \times 20 \times \cos(30° - 36.87°)$$

$$= 8240 \text{ W} \quad \text{or} \quad \mathbf{8.24\ kW.}\ \ \textbf{(Ans.)}$$

(*ii*) *Reading on wattmeter 2 when its voltage coil is across B and Y and current coil is in phase B*

$$= E_L I_L \cos(30° + \phi) = 415 \times 20 \cos(30° + 36.87°)$$

$$= 3260 \text{ W} \quad \text{or} \quad \mathbf{3.26\ kW.}\ \ \textbf{(Ans.)}$$

(*iii*) *When the current coil is in one phase (phasor R) and the voltage coil is across other two phases (Y and R),* the wattmeter reads $\frac{\text{total } kVAR}{\sqrt{3}}$ and

Total $kVAR = \sqrt{3}(W_1 - W_2)$

∴ Wattmeter reading $= \frac{\text{Total } kVAR}{\sqrt{3}} = \frac{\sqrt{3}(W_1 - W_2)}{\sqrt{3}}$

$= W_1 - W_2 = 8240 - 3260 =$ **4980 VAR. (Ans.)**

Example 1.28. *The ratio of the readings of two wattmeters connected to measure power in a balanced 3 – φ, 3 wire load is 5 : 3. The load is known to be inductive with a lagging power factor. Calculate the power factor of the load.* **(Punjab University)**

Solution. Let the ratio of wattmeter readings be r *i.e.*, $\frac{W_2}{W_1} = r$

Now,
$$\tan \phi = \sqrt{3}\left(\frac{W_1 - W_2}{W_1 + W_2}\right) = \sqrt{3}\left[\frac{1 - \frac{W_2}{W_1}}{1 + \frac{W_2}{W_1}}\right] = \sqrt{3}\left(\frac{1 - r}{1 + r}\right)$$

and power factor,
$$\cos \phi = \frac{1}{\sec \phi} = \frac{1}{\sqrt{1 + \tan^2 \phi}} = \frac{1}{\sqrt{1 + 3\left(\frac{1 - r}{1 + r}\right)^2}}$$

Substituting $r = \frac{W_2}{W_1} = \frac{3}{5} = 0.6$ in the above expression, we get

$$\cos \phi = \frac{1}{\sqrt{1 + 3\left(\frac{1 - 0.6}{1 + 0.6}\right)^2}} = \mathbf{0.918\ (lag).\ \ (Ans.)}$$

Example 1.29. *A symmetrical star-connected three-phase load is made of three identical coils having an internal resistance of 12 Ω and inductance of 142 mH. The power taken by this load when connected to a balanced 3-phase, three wire supply of 400 V, 50 Hz is measured by using the two wattmeters method. The wattmeters used are similar and are of type of 400 V, 5 A, UPF, class 1 accuracy.*

(i) Determine expected readings of W_1 and W_2.

(ii) Compute the total power. **(GATE. 1995)**

Solution. Line voltage, $E_L = 400$ V

Phase voltage, $E_{ph} = \frac{400}{\sqrt{3}} = 231$ V

Coil resistance, $R_{ph} = 12\ \Omega$

Coil inductive reactance, $X_{ph} = 2\pi f L_{ph} = 2\pi \times 50 \times 142 \times 10^{-3} = 44.61\ \Omega$

Coil impedance, $Z_{ph} = \sqrt{R_{ph}^2 + X_{ph}^2} = \sqrt{12^2 + 44.61^2} = 46.2\ \Omega$

Coil phase angle, $\phi = \cos^{-1}\frac{R_{ph}}{Z_{ph}} = \cos^{-1}\left(\frac{12}{46.2}\right) = 74.94°$ (lagging)

Phase current, $I_{ph} = \frac{E_{ph}}{Z_{ph}} = \frac{231}{46.2} = 5$ A

(i) **Readings of wattmeters W_1, W_2 :**

Readings of wattmeter, $W_1 = E_L I_L \cos(30° - \phi)$

$= 400 \times 5.0 \cos(30° - 74.94°) =$ **1415.7 W. (Ans.)**

($\because\ I_L = I_{ph} = 5.0$ A)

Reading of wattmeter, $W_2 = E_L I_L \cos(30° + \phi)$

$= 400 \times 5.0 \times \cos(30° + 74.24°) =$ **– 515.6 W. (Ans.)**

(ii) **Total power :**

Total power $= W_1 + W_2$

$= 1415.7 + (-515.6) =$ **900.1 W. (Ans.)**

Example 1.30. *Two wattmeters are being used to measure power of a balanced load of 30 A at power factor 0.8 lagging being supplied by a 3-phase, 3-wire, 440 V supply. Calculate :*

(*i*) *Power consumed.* (*ii*) *Reading of wattmeter No. 1*

(*iii*) *Reading of wattmeter No. 2.*

Solution. Line voltage, $E_L = 440$ V

Line current, $I_L = 30$ A

Power factor, $\cos \phi = 0.8$

Phase angle, $\phi = \cos^{-1} 0.8 = 36.87°$

(*i*) **Power consumed,** $P = \sqrt{3}\, E_L I_L \cos \phi = \sqrt{3} \times 440 \times 30 \times 0.8$

$= 18290$ W or **18.29 kW. (Ans).**

(*ii*) **Reading of wattmeter No. 1**

$= E_L I_L \cos (30° - \phi) = 440 \times 30 \cos (30° - 36.87°) =$ **13105 W. (Ans.)**

(*iii*) **Reading of wattmeter No. 2**

$= E_L I_L \cos (30° + \phi) = 440 \times 30 \times \cos (30° + 36.87°)$

= **5185 W (app.). (Ans.)**

☞**Example 1.31.** *A balanced star-connected load, each having a resistance of 10 Ω and inductive reactance of 20 Ω is connected to a 400 V, 50 Hz supply. The phase sequence is RYB. Two wattmeters connected to read total power have their current coils connected in the red and blue lines respectively. Calculate the reading of each wattmeter.*

Draw the circuit and vector diagram.

Solution. Resistance per phase, $R_{ph} = 10\ \Omega$

Reactance per phase, $X_{ph} = 20\ \Omega$

Line voltage, $E_L = 400$ V

Impedance per phase, $Z_{ph} = \sqrt{R_{ph}^2 + X_{ph}^2} = \sqrt{10^2 + 20^2} = 22.36\ \Omega$

Phase voltage, $E_{ph} = \dfrac{E_L}{\sqrt{3}} = \dfrac{400}{\sqrt{3}} = 231$ V

$$I_{ph}\ (= I_L) = \frac{E_{ph}}{Z_{ph}} = \frac{231}{22.36} = 10.33 \text{ A}$$

Power factor, $\cos \phi = \dfrac{R_{ph}}{Z_{ph}} = \dfrac{10}{22.36} = 0.447$

∴ $\phi = 63.43°$

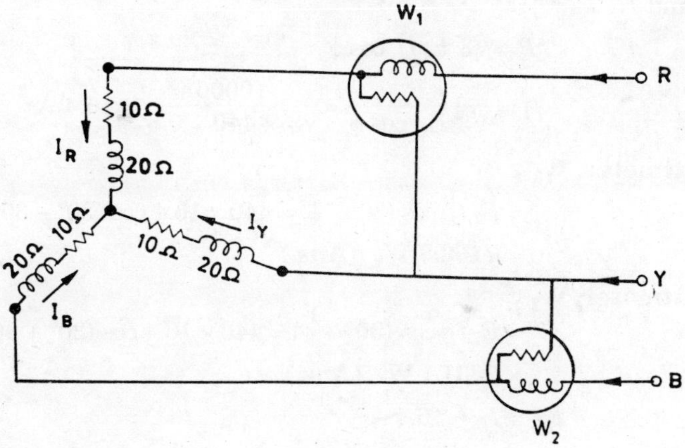

Fig. 1.37

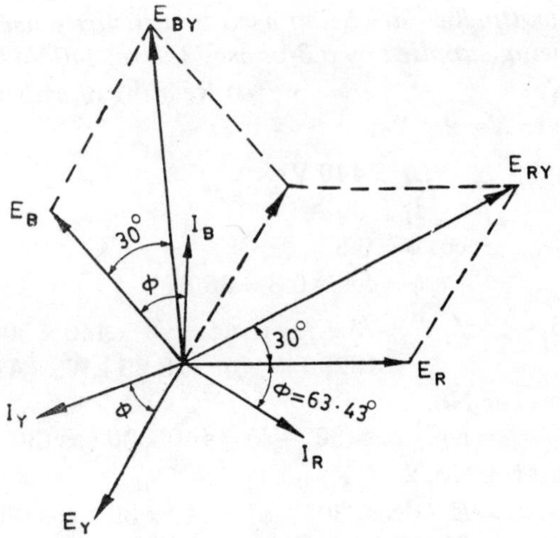

Fig. 1.38

The circuit and vector diagrams are shown in Figs. 1.37 and 1.38 respectively.

Reading of wattmeter, $W_1 = E_{RY} \times I_R \times \cos(30° + \phi) = E_L \times I_L \times \cos(30° + \phi)$

$= 400 \times 10.33 \cos(30° + 63.43°) =$ **– 247 W (app.). (Ans.)**

Reading of wattmeter, $W_2 = E_{BY} \times I_B \times \cos(30° - \phi) = E_L \times I_L \times \cos(30° - 63.43°)$

$= 400 \times 10.33 \times \cos(30° - 63.43°) =$ **3448 W (app.). (Ans.)**

☞**Example 1.32.** *Three equal star-connected inductors take 10 kW at power factor 0.8 when connected to a 440 V, 3 φ, 3-wire supply. Give the connection diagram of two single-phase wattmeters to measure the power in the circuit. Allocate the reading of each wattmeter. Find the line currents if one of the inductors is short-circuited. Draw the vector diagrams of the currents and voltages under the condition.*

Solution. Power, $P = 10 \text{ kW} = 10000 \text{ W}$

Line voltage, $E_L = 440 \text{ V}$

Power factor, $\cos\phi = 0.8$ (lag)

∴ $\phi = 36.9°$ (lag)

The connection diagram of two single-phase wattmeters to measure the power in the circuit is shown in Figs. 1.39 and 1.40.

$$P = \sqrt{3}\, E_L I_L \cos\phi$$

∴ Line current,
$$I_L = \frac{P}{\sqrt{3}\, E_L \cos\phi} = \frac{10000}{\sqrt{3} \times 440 \times 0.8} = 16.4 \text{ A}.$$

Reading of wattmeter, W_1

$= E_L I_L \cos(30° - \phi) = 440 \times 16.4 \cos(30° - 36.9°)$

= 7163.7 W. (Ans.)

Reading of wattmeter, W_2

$= E_L I_L \cos(30° + \phi) = 440 \times 16.4 \cos(30° + 36.9°)$

= 2831.1 W. (Ans.)

Phase current, $I_{ph} = I_L = 16.4 \text{ A}$

Phase impedance, $Z_{ph} = \dfrac{E_{ph}}{I_{ph}} = \dfrac{\frac{440}{\sqrt{3}}}{16.4} = 15.49\ \Omega$

Phase resistance, $R_{ph} = Z_{ph} \cos \phi = 15.49 \times 0.8 = 12.39\ \Omega$

Phase reactance, $X_{ph} = Z_{ph} \sin \phi = 15.49 \times 0.6 = 9.29\ \Omega$

The circuit diagram when one of the inductors is short-circuited is shown in Fig. 1.39.

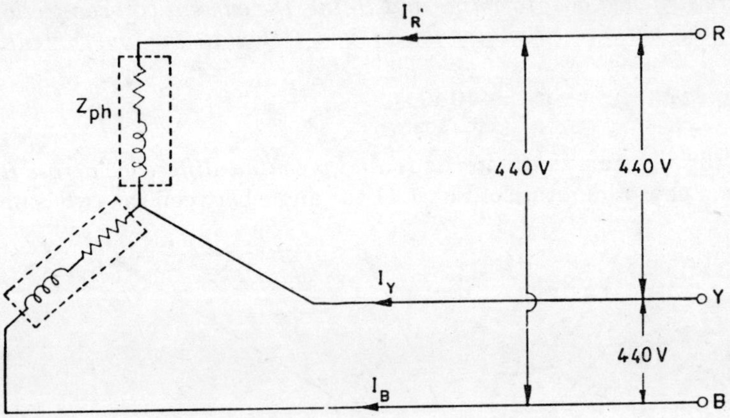

Fig. 1.39

Current in line R, $I_R = \dfrac{E_{RY}}{Z_{ph}} = \dfrac{440}{15.49} = 28.4$ A

Current in line B, $I_B = \dfrac{E_{BY}}{Z_{ph}} = \dfrac{440}{15.49} = 28.4$ A

The currents in line R and line B are equal in magnitude but differ in phase by 60°, as shown in Fig. 1.40.

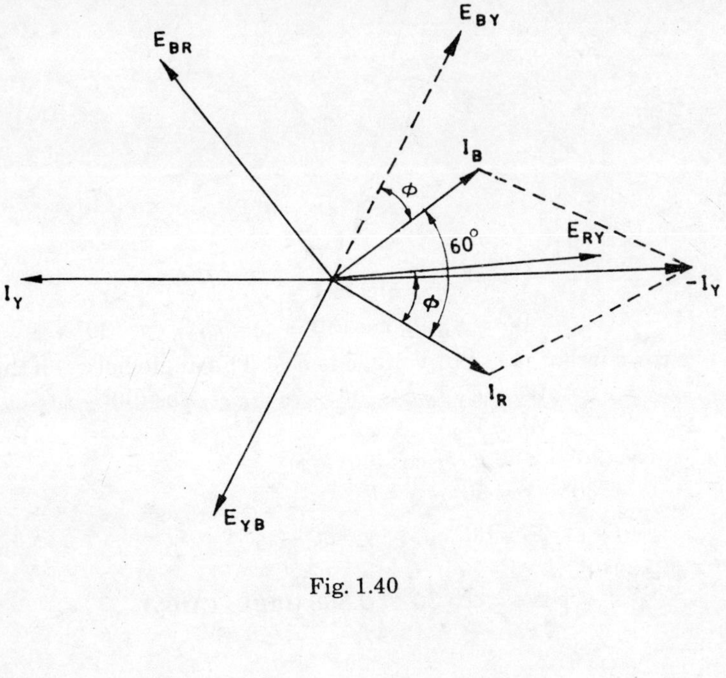

Fig. 1.40

Current in line Y, $I_Y = -(I_R + I_B)$

Magnitude of current flowing in line Y,

$$I_Y = 2I_R \cos 30° = 2 \times 28.4 \times \cos 30° = \mathbf{49.19\ A.}\ \textbf{(Ans.)}$$

Example 1.33. *A 3-phase, 3-wire balanced load with a lagging power factor is supplied at 440 V (between lines). A single phase wattmeter (scaled in kW) when connected with its current coil in the R-line and voltage coil between R and Y lines gives a reading of 8 kW. When the same terminals of the voltage coil are switched over to Y-line and B-line, the current coil connections remaining the same, the reading of the wattmeter remains unchanged. Calculate the line current and power factor of the load.*

Solution. Line voltage, $E_L = 440$ V

Phase sequence $R \to Y \to B$

The current through the wattmeter is I_R *and potential difference across its pressure coil is* E_{RY}. As seen from the phasor diagram of Fig. 1.41 the angle between the two is $(30° + \phi)$.

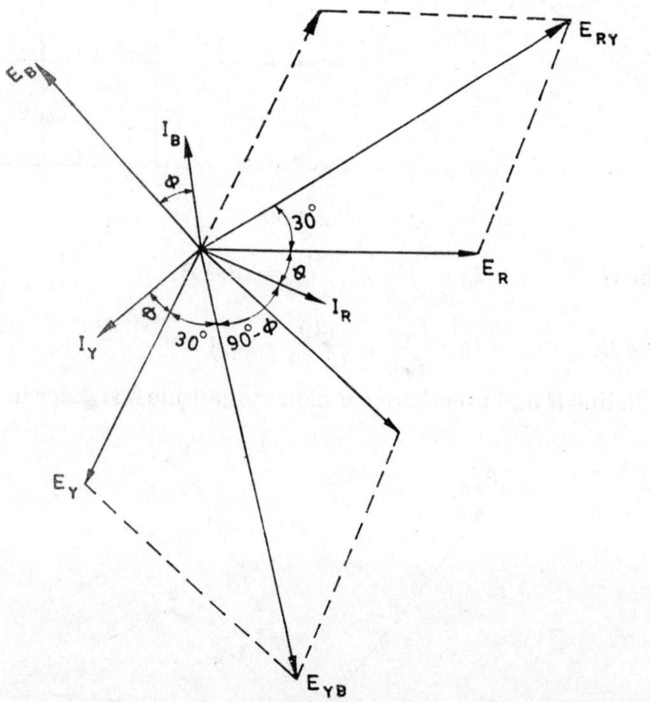

Fig. 1.41

$$\therefore \quad W_1 = E_{RY} I_R \cos (30° + \phi) = E_L I_L \cos (30° + \phi) \quad ...(i)$$

In the second case, current is I_R but voltage is E_{YB}. The angle between the two is $(90° - \phi)$.

$$\therefore \quad W_2 = E_Y I_R \cos (90° - \phi) = E_L I_L \cos (90° - \phi) \quad ...(ii)$$

Since $W_1 = W_2$ *(given)*

$$E_L I_L \cos (30° + \phi) = E_L I_L \cos (90° - \phi)$$

$$\therefore \quad 30° + \phi = 90° - \phi$$

or $2\phi = 60° \quad \therefore \quad \phi = 30°$

$\therefore$ **Load power factor,**

$$\cos \phi = \cos 30° = \mathbf{0.866\ (lag).}\ \textbf{(Ans.)}$$

Now, $W_1 = W_2 = 8 \text{ kW} = 8000 \text{ W}$

From (*ii*), we get $8000 = E_L I_L \cos (90° - \phi) = 440 \times I_L \cos (90° - 30°)$

$$\therefore \quad I_L = \frac{8000}{440 \times \cos 60°} = 36.36 \text{ A}$$

Hence, **line current = 36.36 A. (Ans.)**

Example 1.34. *A wattmeter indicates 6 kW when its current coil is connected in red phase and its potential coil is connected between the neutral and red phase of a symmetrical 3-phase system supplying a balanced load of 40 A at 440 V. What will be reading of the instrument if the current coil connections remain unchanged, the potential coil be connected between blue and yellow phases ?*

Solution. When the wattmeter current coil is connected in red phase and its pressure coil is connected between red phase and neutral, the wattmeter reads per phase $E_{ph} I_{ph} \cos \phi$

or $E_{ph} I_{ph} \cos \phi = 6 \text{ kW} = 6000 \text{ W}$

But $E_{ph} = \frac{440}{\sqrt{3}}$ V and $I_{ph} = 40$ A (*given*)

$$\therefore \text{ Power factor,} \quad \cos \phi = \frac{6000}{E_{ph} I_{ph}} = \frac{6000}{\frac{440}{\sqrt{3}} \times 40} = 0.59$$

and $\sin \phi = \sqrt{1 - (0.59)^2} = 0.807$

When the pressure coil is connected between blue and yellow phases and current coil is connected in red phase, the instrument will read $\frac{\text{Total kVAR}}{\sqrt{3}}$ or $\sqrt{3}$ kVAR per phase.

$$\therefore \textbf{ Reading of instrument} = \sqrt{3} E_{ph} I_{ph} \sin \phi = \sqrt{3} \times \frac{440}{\sqrt{3}} \times 40 \times 0.807$$

$= 14203$ or **14.203 kVAR. (Ans.)**

Example 1.35. *A wattmeter reads 5.54 kW when its current coil is connected in R phase and its voltage coil is connected between the neutral and R phase of a symmetrical 3-phase system supplying a balanced load of 30 A at 400 V. What will be the reading on the instrument if the connections to the current coil remain unchanged and the voltage coil is connected between B and Y phases ? Take phase sequence RYB. Draw corresponding phasor diagram.* **(Panjab University)**

Solution. Phase current, $I_{ph} = I_L = 30$ A

Phase voltage, $E_{ph} = \frac{E_L}{\sqrt{3}} = \frac{400}{\sqrt{3}} = 231$ V

Power supplied per phase, P = reading of wattmeter = 5.54 kW

Load power factor, $\cos \phi = \frac{P}{E_{ph} \cdot I_{ph}}$

$= \frac{5.54 \times 10^3}{231 \times 30} = 0.8$

or $\phi = \cos^{-1} 0.8 = 36.87°$

In the *second case* when pressure coil is connected across phases B and Y, the phase angle between current in phase R and voltage across voltage coil is $(90° - \phi)$ as obvious from the phasor diagram, shown in Fig. 1.42.

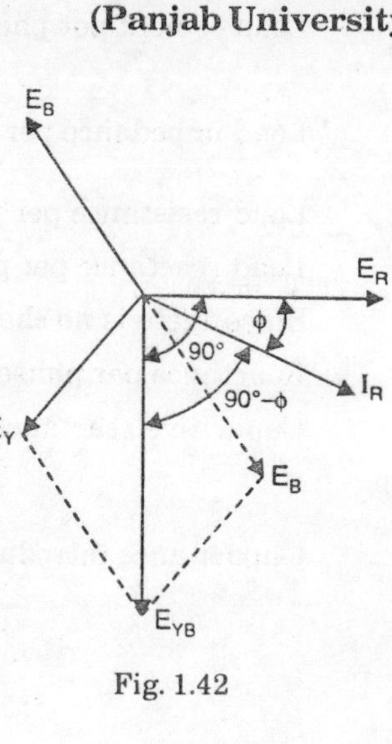

Fig. 1.42

∴ Wattmeter reading $= E_{YB}\, I_R \cos (90° - \phi)$

$= 400 \times 30 \times \cos (90° - 36.87°)$

$= 7200$ W or **7.2 kW. (Ans.)**

Example 1.36. *Two wattmeters A and B give readings as 5000 W and 1000 W respectively during the power measurement of 3 – φ, 3 wire balanced load system. (a) Calculate the power and power factor (i) when both meters read direct and (ii) when one of them reads in reverse (b) if the voltage of the circuit is 400 V, what is the value of the capacitance which must be introduced in each phase to cause the whole of the power to appear on A ? The frequency of supply is 50 Hz.*

(Nagpur University, 1998)

Solution. *Given* : $W_1 = 5000$ W ; $W_2 = 1000$ W ; $V = 400$ volts ; $f = 50$ Hz.

(*a*) **Power (P) and power factor (cos φ) :**

(*i*) *When both meters read direct :*

Total power, $P = W_1 + W_2 = 5000 + 1000 = 6000$ W or **6 kW. (Ans.)**

We know that, $\tan \phi = \dfrac{\sqrt{3}\,(W_1 - W_2)}{W_1 + W_2} = \dfrac{\sqrt{3}\,(5000 - 1000)}{(5000 + 1000)} = 1.1547$

or $\phi = \tan^{-1} (1.1547) = 49.1°$

∴ Power factor, $\cos \phi = \cos 49.1° =$ **0.655 (lagging). (Ans.)**

(*ii*) *When one of the meters reads in reverse :*

Here, $W_2 = -\,1000$ W

Total power, $P = W_1 + W_2 = 5000 + (-\,1000) = 4000$ W or **4 kW. (Ans.)**

Again, $\tan \phi = \dfrac{\sqrt{3}\,(W_1 - W_2)}{W_1 + W_2} = \dfrac{\sqrt{3}\,[5000 - (-\,1000)]}{(5000 - 1000)} = 2.598$

or $\phi = \tan^{-1} 2.598 = 68.95°$

∴ $\cos \phi = \cos 68.95° =$ **0.359 (lagging). (Ans.)**

(*b*) **Capacitance to be introduced in each phase, C :**

Reading of wattmeter, *B will be zero and whole of the power would appear on wattmeter A when power factor* $\cos \phi' = 0.5$ *or* $\phi' = 60°$. It means power factor is to be improved from 0.359 to 0.5.

Load cement per phase, $I_{ph}\, (= I_L) = \dfrac{P}{\sqrt{3}\, E_L \cos \phi} = \dfrac{4000}{\sqrt{3} \times 400 \times 0.359} = 16.08$ A

Load impedance per phase, $Z_{ph} = \dfrac{E_{ph}}{I_{ph}} = \dfrac{400/\sqrt{3}}{16.08} = 14.36\ \Omega$

Load resistance per phase, $R_{ph} = Z_{ph} \cos \phi = 14.36 \times 0.359 = 5.156\ \Omega$

Load reactance per phase, $X_{ph} = Z_{ph} \sin \phi = 14.36 \sin \cos^{-1} (0.359) = 13.4\ \Omega$

Since there is no change in resistance, therefore,

Reactance per phase in the circuit $X'_{ph} = R_{ph} \cdot \tan \phi' = 5.156 \tan 60° = 8.93\ \Omega$

Capacitive reactance introduced in each phase,

$$X_c = X_{ph} - X'_{ph} = 13.4 - 8.93 = 4.47\ \Omega$$

Capacitance introduced in each phase, $\mathbf{C} = \dfrac{1}{2\pi f X_L}$

$$= \frac{1}{2\pi \times 50 \times 4.47} \times 10^6\ \mu\text{F} = \mathbf{712\ \mu F}.\ \textbf{(Ans.)}$$

Example 1.37. *A 3-phase, 434 V, 50 Hz supply is connected to a 3-phase, star-connected induction motor and synchronous motor. Impedance of each phase of induction motor is 1.25 + j2.17 Ω. The three-phase synchronous motor is over-excited and it draws a current of 120 A at 0.87 leading power factor. Two wattmeters are connected in usual manner to measure power drawn by the two motors. Calculate :*

(i) Reading on each wattmeter, *(ii) Combined power factor.*

Solution. Line voltage, $E_L = 434$ V

Impedance of each phase of induction motor,

$$Z_1 = 1.25 + j2.17\ \Omega$$

Current drawn by the synchronous motor = 120 A at p.f. 0.87 leading

Reading on each wattmeter :

Combined power factor :

Let us assume that the synchronous motor is star-connected. *Since it is over-excited it has a leading power factor.* The connections and phasor diagrams are shown in Figs. 1.43 and 1.44.

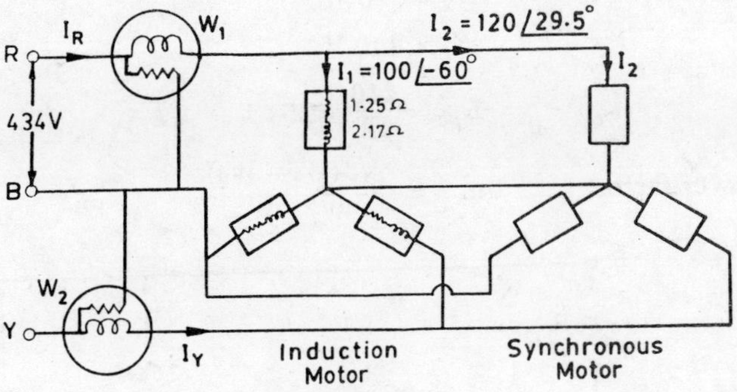

Fig. 1.43

Phase voltage in each case $= \dfrac{434}{\sqrt{3}} = 250$ V

$$Z_1 = 1.25 + j2.17 = 2.5\ \angle 60°$$

$$I_1 = \frac{250}{2.5} = 100 \text{ A}$$

lagging the reference vector E_R by 60°

$$\therefore \quad I_1 = 100\ \angle -60° = 50 - j86.6$$

$$I_2 = 120\ \angle 29.5° = 104.4 + j59.1$$

$$I_R = I_1 + I_2 = (50 - j86.6) + (104.4 + j59.1)$$

$$= (154.4 - j27.5) = 156.8\angle -10.1°$$

(*i*) As shown in Fig. 1.44, I_R lags E_R by 10.1°

Similarly, I_Y lags E_Y by 10.1°.

As seen from Fig. 1.43, current through W_1 is I_R and voltage across it is $E_{RB} = E_R - E_B$. As seen, $E_{RB} = 434$ V lagging by 30°. Phase difference between E_{RB} and I_R is $30 - 10.1 = 19.9°$.

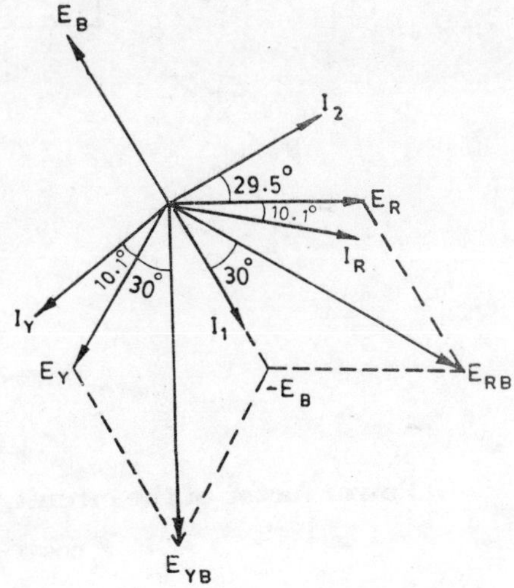

Fig. 1.44

Reading of W_1 = 434 × 156.8 × cos 19.9°

= **63988 W. (Ans.)**

Current I_Y is also (like I_R) the vector sum of the line currents draws by the two motors. It is equal to 156.8 A and lags behind its respective phase voltage E_Y by 10.1°. Current through W_2 is I_Y and voltage across it is $E_{YB} = E_Y - E_B$. As seen E_{YB} = 43 V.

Phase difference between E_{YB} and I_Y = 30° + 10.1° = 40.1° (lag)

∴ **Reading of wattmeter, W_2** = 434 × 156.8 × cos 40.1° = **52054 W. (Ans.)**

(*ii*) **Combined power factor** = cos 10.1° = **0.984 (lag). (Ans.)**

Example 1.38. *A 3-phase balanced supply with a line voltage of 440 V at a frequency of 50 Hz supplies a star-connected balance load. Each phase of the load consists of a resistance and a capacitor joined in series and the readings on two wattmeters connected to measure the total power supplied are 810 W and 2100 W, both positive. Calculate :*

(*i*) *Power factor of the circuit,* (*ii*) *Line current, and*

(*iii*) *Capacitance of each capacitor.*

Solution. Reading of wattmeter, W_1 = 2100 W

Reading of wattmeter, W_2 = 810 W

Line voltage, E_L = 440 V

Phase voltage, $$E_{ph} = \frac{440}{\sqrt{3}} = 254 \text{ V}$$

For a leading power factor, $$\tan \phi = -\frac{\sqrt{3}(W_1 - W_2)}{W_1 + W_2}$$

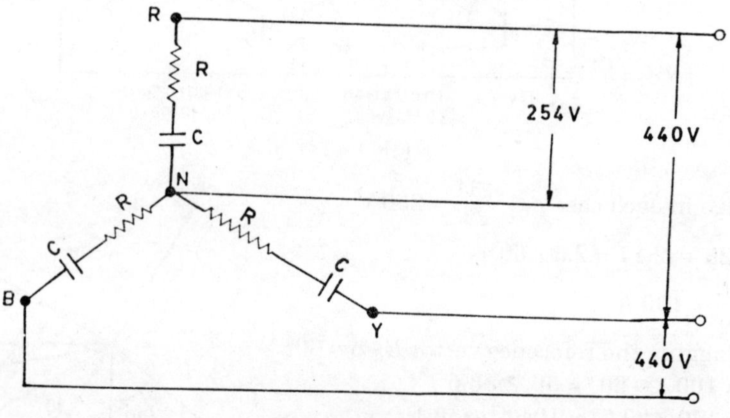

Fig. 1.45

$$\therefore \quad \tan \phi = -\sqrt{3} \times \frac{(2100 - 810)}{(2100 + 810)} = -0.767$$

$$\therefore \quad \phi = -37.48°.$$

(*i*) **Power factor of the circuit,**

$$\cos \phi = \cos(-37.48°) = \mathbf{0.793\ (leading).\ (Ans.)}$$

(*ii*) Power, $$P = \sqrt{3}\, E_L I_L \cos \phi$$

$$(W_1 + W_2) = \sqrt{3} E_L I_L \cos \phi \qquad [\because P = W_1 + W_2]$$

$$(2100 + 810) = \sqrt{3} \times 440 \times I_B \times 0.793$$

$\therefore \quad I_L = \dfrac{2910}{\sqrt{3} \times 440 \times 0.793} = 4.81$ A

Hence, **line current = 4.81 A. (Ans.)**

(*iii*) Impedance per phase, $Z_{ph} = \dfrac{E_{ph}}{I_{ph}} = \dfrac{254}{4.81}$ $[\because I_{ph} = I_L = 4.81 \text{ A}]$

$= 52.8\ \Omega$

$Z_{ph} = 52.8 \angle -37.48° = 41.9 - j32.12$

$\therefore$ Capacitive reactance, $X_C = 32.12\ \Omega$ or $\dfrac{1}{2\pi fC} = 32.12$

$\therefore \quad C = \dfrac{1}{2\pi f \times 32.12} \text{ F} = \dfrac{10^6}{2\pi \times 50 \times 32.12}\ \mu\text{F} = 99.1\ \mu\text{F}$

Hence, **capacitance of each capacitor = 99.1 μF. (Ans.)**

Example 1.39. *A 3-phase, 3-wire, CBA system supplies a Δ-connected load in which Z_{AB} = 25 ∠90° ; Z_{BC} = 15 ∠30° ; Z_{CA} = 20 ∠0° ohms. Line voltage is 230 V. Find the wattmeter readings when the two wattmeters method is used with meters in lines A and B.* **(A.M.I.E. 1998)**

Solution. *Given* : $Z_{AB} = 25 \angle 90°\ \Omega$; $Z_{BC} = 15 \angle 30°\ \Omega$; $Z_{CA} = 20 \angle 0°\ \Omega$; $E_L = 230$ V.

Wattmeter readings W_1, W_2 :

Let the voltage E_{AC} be the reference phasor, then

$E_{AC} = 230 \angle 0°$; $E_{CB} = 230 \angle -120°$ V and $E_{BA} = 230 \angle 120°$

Phase current,

$$I_{CA} = \frac{E_{CA}}{Z_{CA}} = \frac{-E_{AC}}{Z_{CA}} = \frac{-230 \angle 0°}{20 \angle 0°}$$

$= -11.5 \angle 0°$ A or $(-11.5 + j0)$ A

Phase current,

$$I_{AB} = \frac{E_{AB}}{Z_{AB}} = \frac{-E_{BA}}{Z_{AB}} = \frac{-230 \angle 120°}{25 \angle 90°}$$

$= -9.2 \angle 30°$ A or $(-7.967 - j4.6)$A

Phase current,

$$I_{BC} = \frac{E_{BC}}{Z_{BC}} = \frac{-E_{CB}}{Z_{BC}} = \frac{-230 \angle -120°}{15 \angle 30°}$$

$= -15.333 \angle -150°$

$= (13.279 + j7.666)$ A

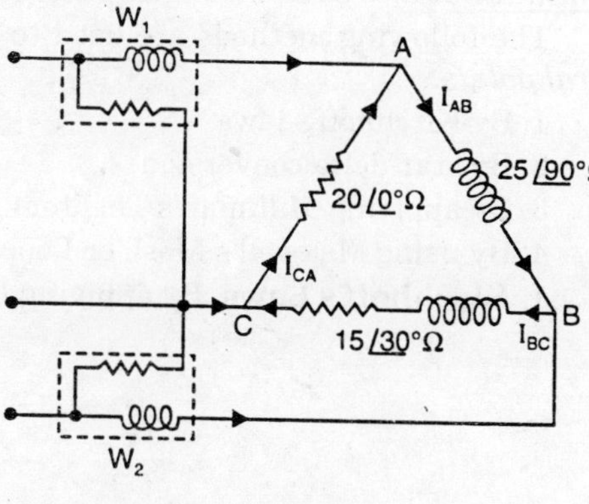

Fig. 1.46

Now line current, $I_A = I_{AB} - I_{CA} = (-7.967 - j4.6) - (-11.5 + j0)$

$= (3.533 - j4.6) = 5.8 \angle -52.47°$

Line current, $I_B = I_{BC} - I_{AB} = (13.279 + j7.666) - (-7.967 - j4.6)$

$= (21.246 + j12.266) = 24.533 \angle 30°$

Reading of wattmeter W_1 = Real part of $^*E_{AC} \times I_A$

= Real part of $230 \angle 0° \times 5.8 \angle -52.47°$

= **812.64 W. (Ans.)** (*means conjugate of)

Reading of wattmeter W_2 = Real part of $^*E_{BC} \times I_B$

$= -230 \angle -120° \times 24.533 \angle 30°$ = **2443.3 W. (Ans.)**

1.11. UNBALANCED STAR-CONNECTED LOAD

The following two cases will be considered :

1. Three-phase four-wire system.
2. Three-phase three-wire system.

1.11.1. Three Phase Four Wire System

- In this type of system a total of *four conductors* are used, three for three phases and one for neutral.
- This system is *very flexible* and is *used for distribution of electric power to domestic, commercial, industrial and other consumers.*
- The line voltage is $\sqrt{3}$ times the phase voltage. The domestic and other single-phase loads needing higher voltages are connected between lines. *The currents in three lines will not be equal and the phasor sum of line currents would return through neutral. The current in the neutral must be equal and opposite to the sum of line currents.*

1.11.2. Three-Phase Three-Wire System. In case a star-connected load is unbalanced and if there is *no neutral wire,* then its star point is isolated from the star-point of the generator. The potential of the load star-point is different from that of the generator star-point. The potential of the former is subject to variations according to the unbalance of the load and under certain conditions of loading, the potentials of the two star-points may differ considerably. Such an isolated load star-point or neutral point is called a *'floating'* neutral point *because its potential is always changing and not fixed.* All star-connected unbalanced load supplied from polyphase systems without a neutral wire, have floating neutral points. *Any unbalancing of the load causes variations not only of the potential of star-point but also of voltages across the different branches of the load.* Hence in that case, *phase voltage is not* $\frac{1}{\sqrt{3}}$ *of the line voltage.* In general, the phase voltages are not only unequal in magnitude but also subtend angles other than 120° with one another.

The following methods are used to tackle *unbalanced star-connected loads having isolated neutral points.*

1. By Kirchhoff's laws.
2. By star-delta conversion.
3. By applying Milliman's Theorem.
4. By using Maxwell's Mesh or Loop Current Method.

1. **Kirchhoff's Laws.** By applying Kirchhoff's laws, we get (Refer Fig. 1.47)

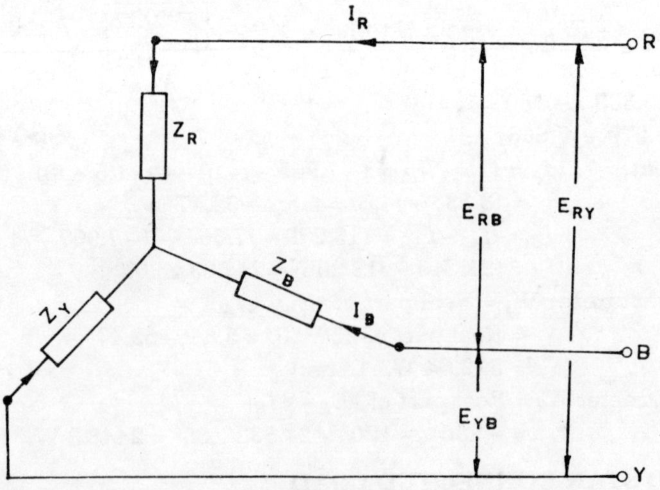

Fig. 1.47

$$E_{RY} = I_R Z_R - I_Y Z_Y \quad ...(i)$$
$$E_{YB} = I_Y Z_Y - I_B Z_B \quad ...(ii)$$

$$E_{BR} = I_B Z_B - I_R Z_R \qquad ...(iii)$$

and $$I_R + I_Y + I_B = 0 \qquad ....(iv)$$

Adding *(i)* and *(iii)*, we get

$$E_{RY} + E_{BR} = I_R Z_R - I_Y Z_Y + I_B Z_B - I_R Z_R$$
$$= I_B Z_B - I_Y Z_Y \qquad ...(v)$$

or $$I_B = \frac{E_{RY} + E_{BR} + I_Y Z_Y}{Z_B}$$

Substituting $I_Y + I_B = -I_R$ from *(iv)* in *(iii)*, we get

$$E_{BR} = I_B Z_B + (I_Y + I_B) Z_R = I_B(Z_B + Z_R) + I_Y Z_R \qquad ...(vi)$$

Substituting the value of I_B from *(iv)* in *(vi)*, we get

$$E_{BR} = \left(\frac{E_{RY} + E_{BR} + I_Y Z_Y}{Z_B}\right)(Z_B + Z_R) + I_Y Z_R$$

or $$E_{BR}Z_B = E_{RY}Z_B + E_{BR}Z_B + I_Y Z_Y Z_B + E_{RY}Z_R + E_{BR}Z_R + I_Y Z_Y Z_R + I_Y Z_R Z_B$$

or $$I_Y(Z_R Z_Y + Z_Y Z_B + Z_B Z_R) = -Z_R(E_{RY} + E_{BR}) - E_{RY}Z_B$$

or $$I_Y(Z_R Z_Y + Z_Y Z_B + Z_B Z_R) = Z_R E_{YB} - Z_B E_{RY} \qquad (\because\ E_{RY} + E_{YB} + E_{BR} = 0)$$

or $$I_Y = \frac{E_{YB}Z_R}{Z_R Z_Y + Z_Y Z_B + Z_B Z_R} - \frac{E_{RY}Z_B}{Z_R Z_Y + Z_Y Z_B + Z_B Z_R}$$

i.e., $$I_Y = \frac{E_{YB}}{Z_Y + Z_B + \dfrac{Z_Y Z_B}{Z_R}} - \frac{E_{RY}}{Z_R + Z_Y + \dfrac{Z_R Z_Y}{Z_B}}$$

Similarly $$I_B = \frac{E_{BR}}{Z_B + Z_R + \dfrac{Z_B Z_R}{Z_Y}} - \frac{E_{YB}}{Z_Y + Z_B + \dfrac{Z_Y Z_B}{Z_R}}$$

and $$I_R = \frac{E_{RY}}{Z_R + Z_Y + \dfrac{Z_R Z_Y}{Z_B}} - \frac{E_{BR}}{Z_B + Z_R + \dfrac{Z_B Z_R}{Z_Y}}.$$

2. Star-delta Conversion. In case of unbalanced star-connected load it is somewhat more convenient to convert it into equivalent delta-connected load and solve it.

If Z_R, Z_Y and Z_B be the impedances of the given star-connected load and Z_{RY}, Z_{YB}, Z_{BR} be the impedances of the equivalent delta-connected load (Fig. 1.48), then

$$Z_{RY} = Z_R + Z_Y + \frac{Z_R Z_Y}{Z_B} = \frac{Z_R Z_B + Z_{YB} + Z_{RY}}{Z_B}$$

$$Z_{YB} = Z_Y + Z_B + \frac{Z_Y Z_B}{Z_R} = \frac{Z_Y Z_R + Z_B Z_R + Z_Y Z_B}{Z_R}$$

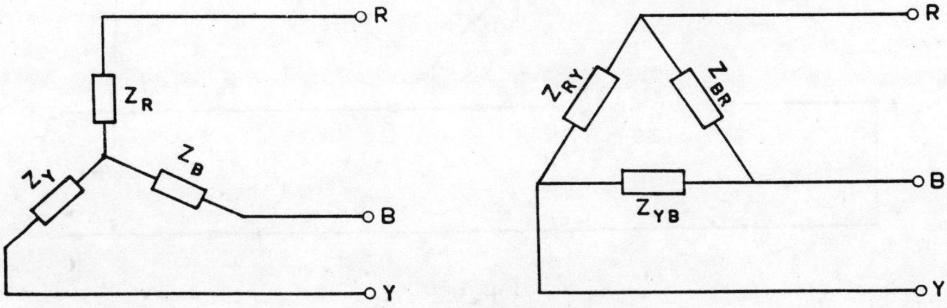

Fig. 1.48

$$Z_{BR} = Z_B + Z_R + \frac{Z_B Z_R}{Z_Y} = \frac{Z_B Z_Y + Z_R Z_Y + Z_B Z_R}{Z_Y}.$$

Now the phase currents and line currents can be determined. (Refer worked examples).

3. **Milliman's Theorem.** Refer Fig. 1.49. A number of linear bilateral admittances Y_1, Y_2, Y_3, ..., Y_n are connected to a common point (or mode) O'. The voltages of the free ends of these admittances with respect to another common point O are V_{1O}, V_{2O}, V_{3O}, ..., V_{nO}. Then according to this theorem, the voltage of O' with respect to A is given by

$$V_{O'O} = \frac{V_{1O}\,Y_1 + V_{2O}Y_2 + V_{3O}\,Y_3 + \ldots\ldots\, V_{nO}Y_n}{Y_1 + Y_2 + Y_2 + \ldots\ldots + Y_n}$$

$$V_{O'O} = \frac{\sum_{k=1}^{n} V_{kO}\,Y_k}{\sum_{k=1}^{n} Y_k}.$$

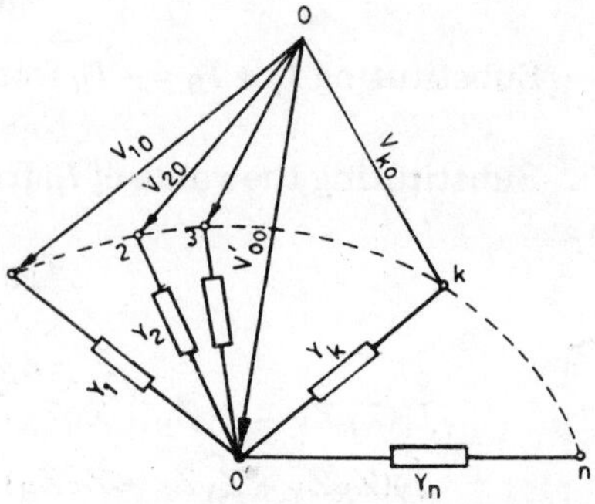

Fig. 1.49

1.12. UNBALANCED DELTA CONNECTED LOAD

When the load impedance in the three phases are not equal in magnitude or phase or both, the load is said to be *unbalanced.* If three unequal loads are connected to form a delta and connected across a 3-phase supply the currents in the three loads will not be equal in magnitude and/or phase. The three-phase currents and the line currents will also be unbalanced.

Example 1.40. *A 440 V, 50 Hz, 3-phase supply has delta-connected load having 50 Ω between R and Y, 159 mH between Y and B and 15.9 μF between B and R. Find :*

(i) The line current for the sequence RYB.

(ii) The value of star-connected balanced resistors for the same power.

Solution. Phase voltage (E_{ph}) = line voltage (E_L) = 440 V

The potential difference across three-phases are given by

$$E_{RY} = 440(1 + j0) = 440\angle 0° \text{ V}$$

$$E_{YB} = 440(-0.5 - j0.866) \text{ or } 440 \angle -120° \text{ V}$$

$$E_{BR} = 440(-0.5 + j0.866) \text{ or } 440 \angle 120° \text{ V}$$

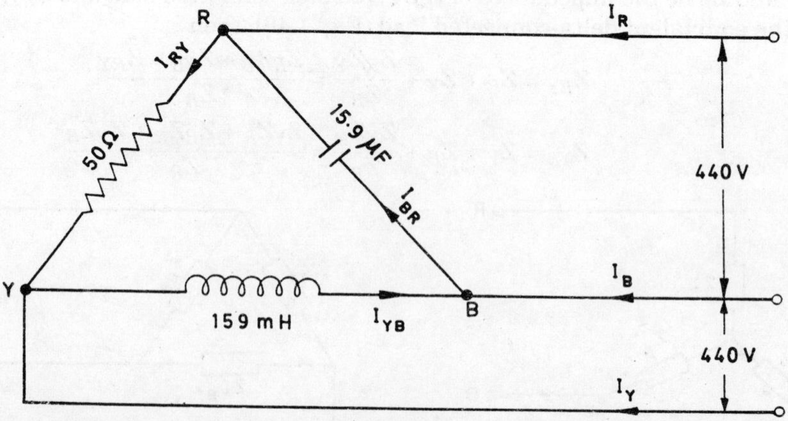

Fig. 1.50

Impedance, $Z_{RY} = (50 + j0) = 50 \angle 0° \ \Omega$

Impedance, $Z_{YB} = (0 + j2\pi fL) = (0 + j2\pi \times 50 \times 159 \times 10^{-3})$

$= (0 + j50)$ or $50 \angle 90° \ \Omega$

Impedance, $Z_{BR} = \left(0 - \dfrac{j}{2\pi fC}\right)$

$= \left(0 - \dfrac{j}{2\pi \times 50 \times 15.9 \times 10^{-6}}\right) = (0 - j200) = 200\angle -90°$

(*i*) **Phase currents :**

Phase current, $I_{RY} = \dfrac{E_{RY}}{Z_{RY}} = \dfrac{440\angle 0°}{50\angle 0°} = 8.8(1 + j0)$ A

Phase current, $I_{YB} = \dfrac{E_{YB}}{Z_{YB}} = \dfrac{440 \angle -120°}{50 \angle 90°} = 8.8 \angle -210°$

Phase current $I_{BR} = \dfrac{E_{BR}}{Z_{BR}} = \dfrac{440 \angle 120°}{200\angle -90°} = 2.2 \angle 210° = 2.2(-0.866 - j0.5)$ A.

Line currents :

Line current, $I_R = I_{RY} - I_{BR} = 8.8(1 + j0) - 2.2(-0.866 - j0.5)$

$= 8.8 + j0 + 1.905 + j1.1 = 10.7 + j1.1 = \mathbf{10.75 \angle 5.8°}$ **A. (Ans.)**

Line current, $I_Y = I_{YB} - I_{RY} = 8.8(-0.866 + j0.5) - 8.8(1 + j0)$

$= -7.62 + j4.4 - 8.8 - j0$

$= -16.42 + j4.4 = \mathbf{16.99 \angle 164.9°}$ **A. (Ans.)**

Line current, $I_B = I_{BR} - I_{YB} = 2.2(-0.866 - j0.5) - 8.8(-0.866 + j0.5)$

$= -1.905 - j1.1 + 7.62 - j4.4 = 5.715 - j5.5$

$= \mathbf{7.93 \angle -43.9°}$ **A. (Ans.)**

(*ii*) **Power supplied :**

$$P = I_{RY}^2 \times R = 8.8^2 \times 50 = 3872 \text{ W}$$

Let the resistance per phase be R_{ph}, which when connected in star across 440 V, 3-phase supply take power of 3872 W.

Line current, $I_L = I_{ph} = \dfrac{E_L/\sqrt{3}}{R_{ph}} = \dfrac{440/\sqrt{3}}{R_{ph}}$

Power supplied, $P = \sqrt{3}\, E_L I_L$

$\therefore \quad 3872 = \sqrt{3} \times 440 \times \dfrac{440/\sqrt{3}}{R_{ph}}$

$\therefore \quad \mathbf{R_{ph}} = \dfrac{\sqrt{3} \times 440 \times 440/\sqrt{3}}{3872} = \dfrac{440 \times 440}{3872} = \mathbf{50 \ \Omega}$. **(Ans.)**

☞**Example 1.41.** *Three loads (31 + j59), (30 – j40) and (80 + j60) Ω are connected in delta across 400 V, 3-phase supply. Find :*

(*i*) *Phase currents,* (*ii*) *Line currents, and*

(*iii*) *Power consumed.*

Solution. Phase voltage, E_{ph} = line voltage, E_L = 400 V

The potential differences across three-phases are given by :

$$E_{RY} = 400 \angle 0°$$
$$E_{YB} = 400 \angle -120°$$
$$E_{BR} = 400 \angle 120°$$

The phase impedances are :

$$Z_{RY} = 31 + j59 = 66.6 \angle 62.3°$$
$$Z_{YB} = 30 - j40 = 50 \angle -53.1°$$
$$Z_{BR} = 80 + j60 = 100 \angle 36.9°$$

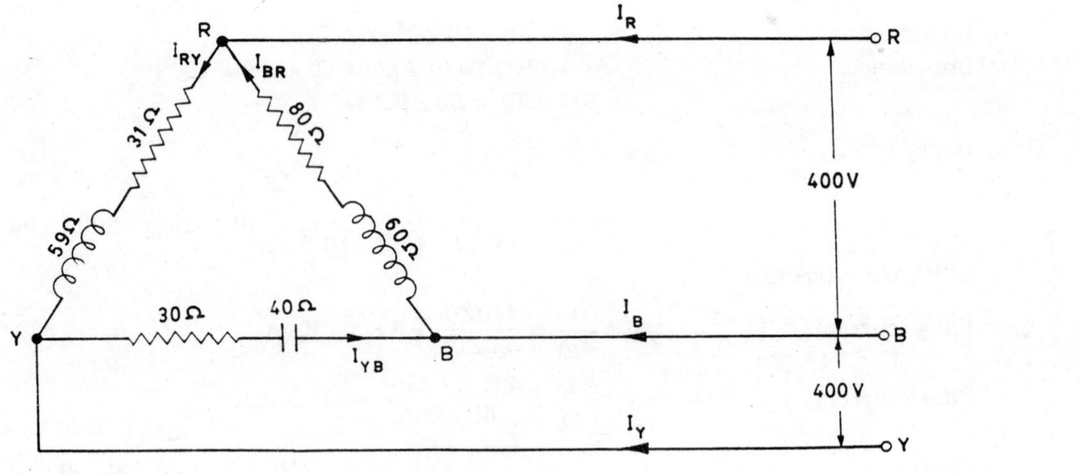

Fig. 1.51

Phase currents :

$$\mathbf{I_{RY}} = \frac{E_{RY}}{Z_{RY}} = \frac{400\ \angle 0°}{66.6\angle 62.3°} = \mathbf{6\ \angle -62.3°\ A.\ \ (Ans.)}$$

$$\mathbf{I_{YB}} = \frac{E_{YB}}{Z_{YB}} = \frac{400\ \angle -120°}{50\angle -53.1°} = \mathbf{8\ \angle -66.9°\ A.\ \ (Ans.)}$$

$$\mathbf{I_{BR}} = \frac{E_{BR}}{Z_{BR}} = \frac{400\ \angle 120°}{100\angle 36.9°} = \mathbf{4\ \angle 83.1\ A.\ \ (Ans.)}$$

Line currents :

$$\mathbf{I_R} = I_{RY} - I_{BR} = 6\ \angle -62.3° - 4\ \angle 83.1°$$
$$= 6(0.465 - j0.885) - 4(0.12 + j0.993)$$
$$= 2.79 - j5.31 - 0.48 - j3.97 = 2.31 - j\ 9.28 = \mathbf{9.56\ \angle -76°\ A.\ \ (Ans.)}$$

$$\mathbf{I_Y} = I_{YB} - I_{RY} = 8\ \angle -66.9° - 6\ \angle -62.3°$$
$$= 8(0.392 - j0.92) - 6(0.465 - j0.885)$$
$$= (3.136 - j7.36) - (2.79 - j5.31)$$
$$= 0.346 - j2.05 = \mathbf{2.07\ \angle -80.4°\ A.\ \ (Ans.)}$$

$$\mathbf{I_B} = I_{BR} - I_{YB} = 4\ \angle 83.1° - 8\ \angle -66.9°$$
$$= 4(0.12 + j0.993) - 8(0.392 - j0.92)$$
$$= 0.48 + j3.97 - 3.136 + j7.36$$
$$= -2.656 + j11.33 = \mathbf{11.637\ \angle 103.2°\ A.\ \ (Ans.)}$$

Total power is the sum of power in different phases

$$P = E_{RY} \times I_{RY} + E_{YB} \times I_{YB} + E_{BR} \times I_{BR}$$
$$= 400\ \angle 0° \times 6\ \angle -62.3° + 400\ \angle -120° \times 8\ \angle -66.9° + 400\ \angle 120° \times 4\angle 83.1°$$
$$= 400 \times 6 \cos(-62.3°) + 400 \times 8 \cos(53.1°) + 400 \times 4 \cos(-36.9°)$$
$$= 1115.6 + 1921.3 + 1279.5 = 4316.4\ \text{W}$$

[OR
$$P = I_{RY}^2\, R_{RY} + I^2_{YB} \times R_{YB} + I^2_{BR} \times R_{BR}$$
$$= 6^2 \times 31 + 8^2 \times 30 + 4^2 \times 80 = 1116 + 1920 + 1280 = 4316\ \text{W}$$]

Hence, **total power = 4316.4 W. (Ans.)**

Example 1.42. *For the unbalanced delta-connected load of Fig. 1.52, find the phase currents, line currents and total power consumed by the load, when phase sequence is :*

(*i*) *RYB* (*ii*) *RBY.*

(AMIE Advanced Electrical M/Cs Winter, 1998)

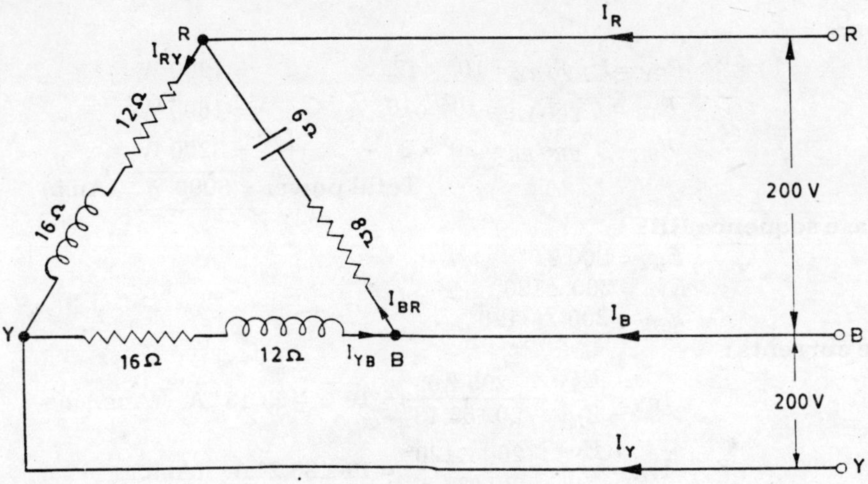

Fig. 1.52

Solution. (*i*) **Phase sequence RYB :**

The potential differences across three phases are :

$$E_{RY} = 200 \angle 0°$$
$$E_{YB} = 200 \angle -120°$$
$$E_{BR} = 200 \angle 120°$$

Impedance, $Z_{RY} = 12 + j16 = 20 \angle 53.13°$

Impedance, $Z_{YB} = 16 + j12 = 20 \angle 36.8°$

Impedance, $Z_{BR} = 8 - j6 = 10 \angle -36.8°$

Phase currents :

Phase current, $$\mathbf{I_{RY}} = \frac{E_{RY}}{Z_{RY}} = \frac{200 \angle 0°}{20 \angle 53.13°} = \mathbf{10 \angle -53.13° \text{ A.}} \quad \textbf{(Ans.)}$$

Similarly, $$\mathbf{I_{YB}} = \frac{E_{YB}}{Z_{YB}} = \frac{200 \angle -120°}{20 \angle 36.8°} = \mathbf{10 \angle -156.8° \text{ A.}} \quad \textbf{(Ans.)}$$

$$\mathbf{I_{BR}} = \frac{E_{RB}}{Z_{BR}} = \frac{200 \angle 120°}{10 \angle -36.8°} = \mathbf{20 \angle 156.8° \text{ A.}} \quad \textbf{(Ans.)}$$

Line currents :

Line current, $$\mathbf{I_R} = I_{RY} - I_{BR} = 10 \angle -53.13° - 12 \angle 156.8°$$
$$= 10(0.60 - j0.8) - 20(-0.72 + j0.394) = 6.0 - j8 + 18.4 - j7.88$$
$$= 24.4 - j15.88 = \mathbf{29.11 \angle -33° \text{ A.}} \quad \textbf{(Ans.)}$$

Similarly, $$\mathbf{I_Y} = I_{YB} - I_{RY} = 10 \angle -156.8° - 10 \angle -53.13°$$
$$= 10(-0.919 - j0.394) - 10(0.6 - j0.8) = -9.19 - j3.94 - 6 + j8$$
$$= -15.19 + j4.06 = \mathbf{15.72 \angle 165° \text{ A.}} \quad \textbf{(Ans.)}$$

$$\mathbf{I_B} = I_{BR} - I_{YB} = 20 \angle 156.8° - 10 \angle -156.8°$$
$$= 20(0.919 + j0.394) - 10(-0.919 - j0.394)$$
$$= -18.38 + j7.88 + 9.19 + j3.94 = -9.19 + j11.82 = 14.97 \angle 127.9°$$

[Check $\Sigma I = (24.4 - j15.88) + (-15.19 + j4.06) + (-9.19 + j11.82) = 0$]

Power :

$$P_{RY} = I^2_{RY}R_{RY} = 10^2 \times 12 = 1200 \text{ W}$$
$$P_{YB} = I^2_{YB}R_{YB} = 10^2 \times 16 = 1600 \text{ W}$$
$$P_{BR} = I^2_{BR}R_{BR} = 20^2 \times 8 = 3200 \text{ W}$$

Total power = 6000 W. (Ans.)

(*ii*) **Phase sequence RBY :**

Here,
$$E_{RY} = 200 \angle 0°$$
$$E_{YB} = 200 \angle 120°$$
$$E_{BR} = 200 \angle -120°$$

Phase currents :

$$\mathbf{I_{RY}} = \frac{E_{RY}}{Z_{RY}} = \frac{200 \angle 0°}{20\angle 53.13°} = \mathbf{10 \angle -53.13° \ A.} \ \textbf{(Ans.)}$$

$$\mathbf{I_{YB}} = \frac{E_{YB}}{Z_{YB}} = \frac{200 \angle 120°}{20\angle 36.8°} = \mathbf{10 \angle 83.2° \ A.} \ \textbf{(Ans.)}$$

$$\mathbf{I_{BR}} = \frac{E_{BR}}{Z_{BR}} = \frac{200 \angle -120°}{10\angle -36.8°} = \mathbf{20 \angle -83.2° \ A.} \ \textbf{(Ans.)}$$

Line currents :

$$\mathbf{I_R} = I_{RY} - I_{BR} = 10 \angle -53.13° - 20 \angle -83.2°$$
$$= 10(0.6 - j0.8) - 20(0.118 - j0.993) = 6 - j8 - 2.36 + j19.86$$
$$= 3.64 + j11.86 = \mathbf{12.4 \angle 72.9° \ A.} \ \textbf{(Ans.)}$$

$$\mathbf{I_Y} = I_{YB} - I_{RY} = 10 \angle 83.2° - 10 \angle -53.13°$$
$$= 10(0.118 + j\,0.993) - 10(0.6 - j0.8)$$
$$= 1.18 + j9.93 - 6 + j8 = -4.82 + j17.93 = \mathbf{18.56 \angle 105° \ A.} \ \textbf{(Ans.)}$$

$$\mathbf{I_B} = I_{BR} - I_{YB} = 20 \angle -83.2° - 10 \angle 83.2°$$
$$= 20(0.118 - j0.993) - 10(0.118 + j0.993)$$
$$= 2.36 - j19.86 - 1.18 - j9.93 = 1.18 - j29.79 = \mathbf{29.81 \angle -87.7°.} \ \textbf{(Ans.)}$$

Power :

$$P_{RY} = 10^2 \times 12 = 1200 \text{ W}$$
$$P_{YB} = 10^2 \times 16 = 1600 \text{ W}$$
$$P_{BR} = 20^2 \times 8 = 3200 \text{ W}$$

Total power = **6000 W. (Ans.)**

Example 1.43. *A 3-phase, 400 V, 50 Hz system has the following load connected in delta.*

Between R and Y lines : a non-reactive resistor of 50 Ω.

Between Y and B lines : a coil having a resistance of 10 Ω and reactance of 30 Ω.

Between B and R lines : a loss free capacitor of 15 μ F.

Calculate : (i) The phase currents (ii) The line currents.

Assume phase sequence to be RYB.

Solution. $E_{ph} = E_L = 400$ V

The potential difference across the three phases are :

$$E_{RY} = 400 \angle 0° \text{ V}$$
$$E_{YB} = 400 \angle -120° \text{ V}$$
$$E_{BR} = 400 \angle 120° \text{ V}$$

Impedances

$$Z_{RY} = 50 + j0 = 50 \angle 0° \ \Omega$$
$$Z_{YB} = 10 + j30 = 31.62 \angle 71.56° \ \Omega$$

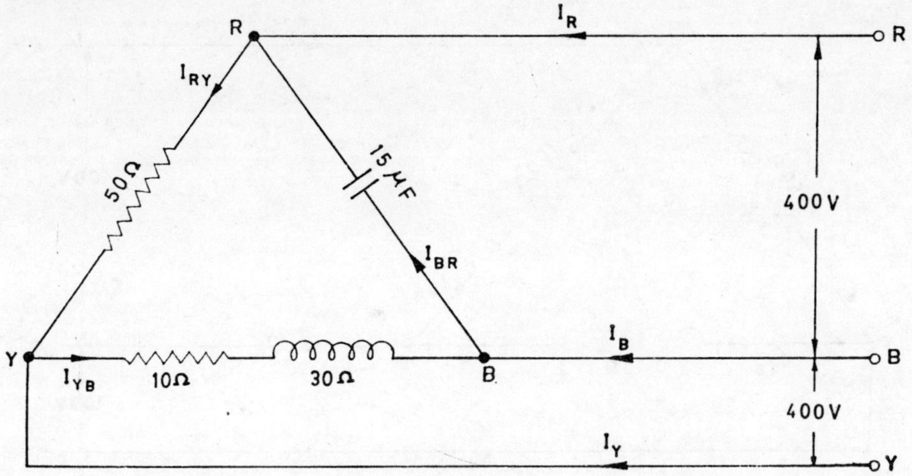

Fig. 1.53

$$Z_{BR} = \left(0 - \frac{j}{2\pi fC}\right) = \left(0 - \frac{j}{2\pi \times 50 \times 15 \times 10^{-6}}\right)$$
$$= (0 - j212) = 212 \angle -90° \ \Omega.$$

Phase currents :

$$\mathbf{I_{RY}} = \frac{E_{RY}}{Z_{RY}} = \frac{400 \angle 0°}{50\angle 0°} = \mathbf{8 \angle 0° \ A. \ \ (Ans.)}$$

$$\mathbf{I_{YB}} = \frac{E_{YB}}{Z_{YB}} = \frac{400 \angle -120°}{31.62\angle 71.56°} = \mathbf{12.65 \angle -191.56° \ A. \ \ (Ans.)}$$

$$\mathbf{I_{BR}} = \frac{E_{BR}}{Z_{BR}} = \frac{400 \angle 120°}{212\angle -90°} = \mathbf{1.886 \angle 210° \ A. \ \ (Ans.)}$$

Line currents :

$$\mathbf{I_R} = I_{RY} - I_{BR} = 8 \angle 0° - 1.886 \angle 210°$$
$$= 8 - 1.886\,(-0.866 - j0.5) = 8 + 1.633 + j0.943$$
$$= 9.633 + j0.943 = \mathbf{9.68 \angle 5.6° \ A. \ \ (Ans.)}$$

$$\mathbf{I_Y} = I_{YB} - I_{RY} = 12.65 \angle -191.56° - 8 \angle 0°$$
$$= 12.65(-0.979 + j0.2) - 8 = -12.38 + j2.53 - 8$$
$$= -20.38 + j2.53 = \mathbf{20.53 \angle 7.07° \ A. \ \ (Ans.)}$$

$$\mathbf{I_B} = I_{BR} - I_{YB} = 1.886 \angle 210° - 12.65 \angle -191.56°$$
$$= 1.886\,(-0.866 - j0.5) - 12.65(-0.979 + j0.2)$$
$$= -1.633 - j\,0.943 + 12.38 - j2.53$$
$$= 10.747 - j3.473 = \mathbf{11.29 \angle 17.4°. \ \ (Ans.)}$$

☞**Example 1.44.** *A balanced 3-phase system supplies an unbalanced 3-phase delta-connected load made up of resistors 50 Ω and 100 Ω and a reactor having an inductance of 0.15 H with negligible resistance. Line voltage = 100 V at 50 Hz. Calculate :*

(*i*) *The total power in the system.*

(*ii*) *The total volt amperes reactive.*

Solution. The delta-connected load and its phasor diagram are shown in Figs. 1.54 and 1.55 respectively.

Reactance, $X_{YB} = 2\pi fL = 2\pi \times 50 \times 0.15 = 47.12 \ \Omega$

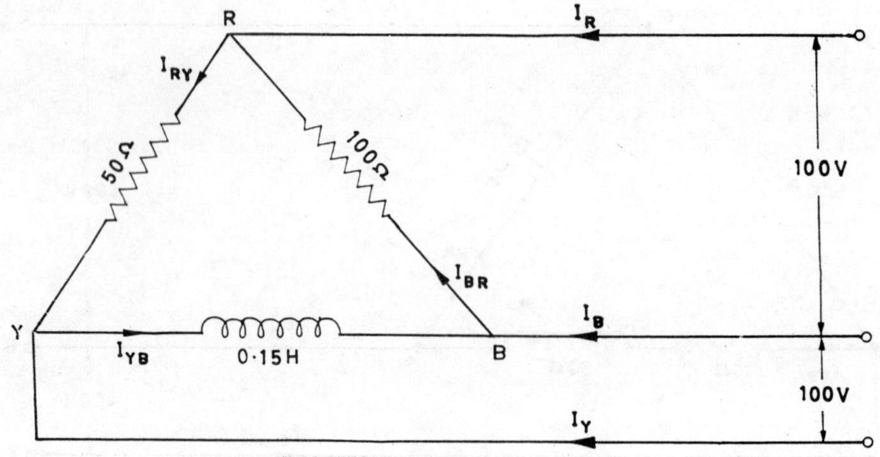

Fig. 1.54

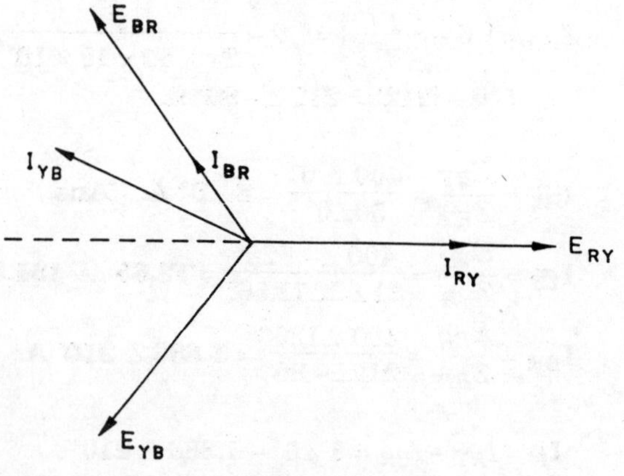

Fig. 1.55

The potential differences across three phases are

$$E_{RY} = 100 \angle 0°$$
$$E_{YB} = 100 \angle -120°$$
$$E_{BR} = 100 \angle 120°$$

Impedances :

$$Z_{RY} = 50 + j0 = 50 \angle 0°$$
$$Z_{YB} = 0 + j47.12 = 47.12 \angle 90°$$
$$Z_{BR} = 100 + j0 = 100 \angle 0°.$$

Phase currents :

$$I_{RY} = \frac{E_{RY}}{Z_{RY}} = \frac{100 \angle 0°}{50 \angle 0°} = 2 \angle 0° = 2 + j0$$

$$I_{YB} = \frac{E_{YB}}{Z_{YB}} = \frac{100 \angle -120°}{47.12 \angle 90°} = 2.12 \angle -210°$$

$$= 2.12\,(-0.866 + j0.5) = -1.836 + j1.06$$

$$I_{BR} = \frac{E_{BR}}{Z_{BR}} = \frac{100 \angle 120°}{100 \angle 0°} = 1 \angle 120° = -0.5 + j0.866$$

Watts in branch RY $= \dfrac{E^2_{RY}}{R_{RY}} = \dfrac{100^2}{50} = 200$ W

VARs = 0

Watts in branch YB = 0

VARs = 100 × 2.12 = 212 (lag)

Watts in branch BR $= \dfrac{E^2_{BR}}{R_{BR}} = \dfrac{100^2}{100} = 100$ W

VARs = 0

(*i*) **Total power** = 200 + 100 = **300 W. (Ans.)**

(*ii*) **Total VARs** (*volt-amperes reactive*) = **212 (lag). (Ans.)**

Example 1.45. *A 400 V, 3-phase 4 wire system has the loads 6 – j8, 6 + j0 and 4 + j3 Ω connected in star. Find :*

(*i*) *The line currents,* (*ii*) *Neutral current, and*

(*iii*) *Total power.*

Solution. The connections are shown in Fig. 1.56.

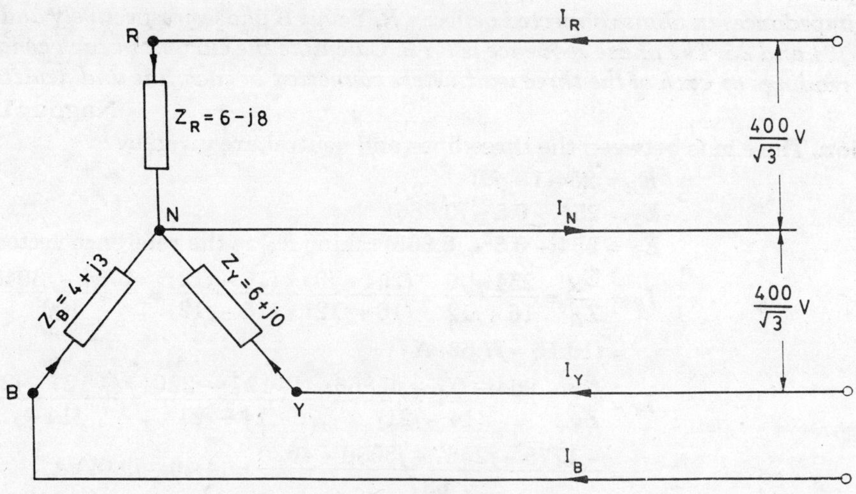

Fig. 1.56

The potential differences across the three phases are :

$$E_{RN} = \frac{400}{\sqrt{3}} \angle 0° = 230.9 \angle 0° \text{ V}$$

$$E_{YN} = 230.9 \angle -120° \text{ V}$$

$$E_{BN} = 230.9 \angle 120° \text{ V}$$

Impedances :

$$Z_R = 6 - j8 = 10 \angle -53.13°$$

$$Z_Y = 6 + j0 = 6 \angle 0°$$

$$Z_B = 4 + j3 = 5 \angle 36.9°$$

(*i*) **Line currents** :

$$\mathbf{I_R} = I_{RN} = \frac{E_R}{Z_R} = \frac{230.9 \angle 0°}{10 \angle -53.13°} = \mathbf{23.09 \angle 53.13° \text{ A.}} \quad \textbf{(Ans).}$$

$$\mathbf{I_Y} = I_{YN} = \frac{E_Y}{Z_Y} = \frac{230.9 \angle -120°}{6 \angle 0°} = \mathbf{38.48 \angle -120° \text{ A.}} \quad \textbf{(Ans.)}$$

$$\mathbf{I_B} = I_{BN} = \frac{E_B}{Z_B} = \frac{230.9 \angle 120°}{5 \angle 36.9°} = \mathbf{46.18 \angle 83.1° \ A. \ \ (Ans.)}$$

(*ii*) **The current in the neutral wire :**

$$\mathbf{I_N} = -(I_R + I_Y + I_B)$$
$$= -[23.09 \angle 53.13° + 38.48 \angle -120° + 46.18 \angle 83.1°]$$
$$= -[23.09(0.6 + j0.8) + 36.48(-0.5 - j0.866) + 46.18(0.12 + j0.99)]$$
$$= -[13.85 + j18.47 - 18.24 - j31.59 + 5.54 + j45.72]$$
$$= -(1.15 + j32.6) = -1.15 - j32.6 = \mathbf{32.62 \angle -92° \ A. \ \ (Ans.)}$$

Power :

Power taken by phase *R*,

$$P_B = I_R^2 \times R_R = 23.09^2 \times 6 = 3199 \text{ W (app.)}.$$

Similarly,
$$P_Y = I_Y^2 \times R_Y = 38.48^2 \times 6 = 8884 \text{ W (app.)}.$$

and
$$P_B = I_B^2 \times R_B = 46.18^2 \times 4 = 8530 \text{ W (app.)}.$$

Total power taken $= P_R + P_Y + P_B = 3199 + 8884 + 8530 =$ **20613 W. (Ans.)**

Example 1.46. *A 440/254 volt, 3 phase, 4-core supplies an unbalanced load represented by the following impedances in ohms connected between R, Y and B phases respectively and the neutral 16 + j12 ; 14 – j21 and 25. The phase sequence is RYB. Calculate the current in each conductor of the cable and the readings on each of the three waitmeters connected in each line and neutral.*

(Nagpur University)

Solution. The e.m.fs between the three lines and neutral are given by :

$$E_R = 254(1 + j0)$$
$$E_Y = 254(-0.5 - j0.866)$$
$$E_B = 254(-0.5 + j0.866) \text{ taking } E_R \text{ as the reference vector}$$

$$I_R = \frac{E_R}{Z_R} = \frac{254 + j0}{16 + j12} = \frac{(254 + j0) \times (16 - j12)}{(16 + j12) \times (16 - j12)} = \frac{4064 - j3048}{400}$$
$$= (10.16 - j7.62) \text{ A}$$

$$I_Y = \frac{E_Y}{Z_Y} = \frac{254(-0.5 - j0.866)}{(14 - j21)} = \frac{-127 - j220}{14 - j21} = \frac{(-127 - j220) \times (14 + j21)}{(14 - j21)(14 + j21)}$$
$$= \frac{-1778 - j2667 - j3080 + 4620}{637} = (4.46 - j9.02) \text{ A}$$

$$I_B = \frac{E_B}{Z_B} = \frac{254(-0.5 + j0.866)}{25} = \frac{-127 + j220}{25} = (-5.08 + j8.8) \text{ A}$$

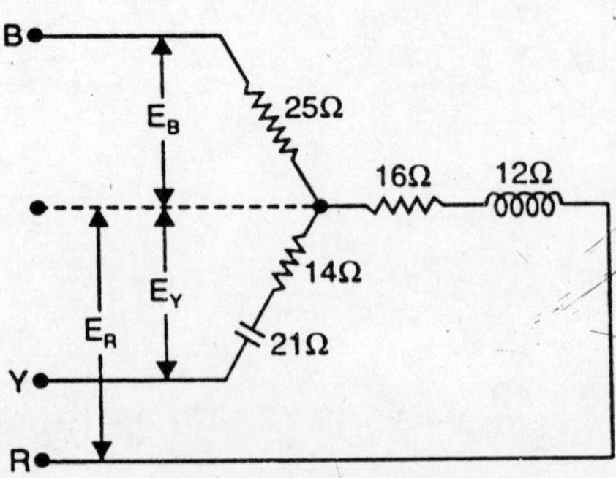

(a) Circuit Diagram

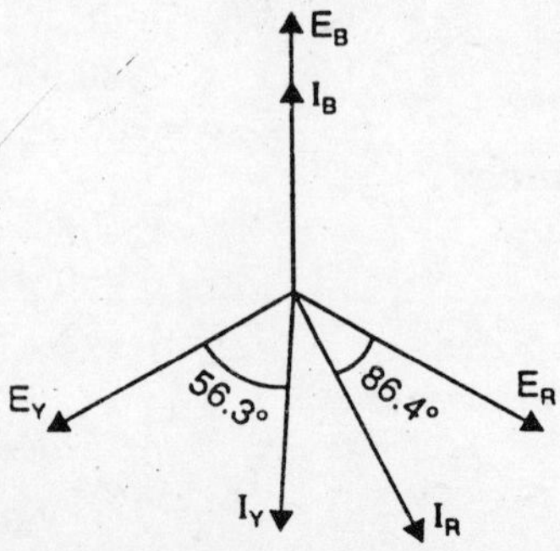

(b) Phasor Diagram

Fig. 1.57

Now current in neutral wire,

$$\mathbf{I_N} = -(I_R + I_Y + I_B)$$
$$= -[(10.16 - j7.62) + (4.46 - j9.02) + (-5.08 + j8.8)]$$
$$= -9.54 + j7.84$$

Evaluating these currents, we get

$$\mathbf{I_R} = \sqrt{10.16^2 + (-7.62)^2} = \mathbf{12.7\ A.}\ \ \textbf{(Ans.)}$$
$$\mathbf{I_Y} = \sqrt{4.46^2 + (-9.02)^2} = \mathbf{10.06\ A.}\ \ \textbf{(Ans.)}$$
$$\mathbf{I_B} = \sqrt{(-5.08)^2 + (8.8)^2} = \mathbf{10.16\ A.}\ \ \textbf{(Ans.)}$$
$$\mathbf{I_N} = \sqrt{(-9.54)^2 + (7.84)^2} = \mathbf{12.35\ A.}\ \ \textbf{(Ans.)}$$

Power in R-phase $= I_R^2 \times R_R = (12.7)^2 \times 16 = \mathbf{2580.6\ W.}$ **(Ans.)**

Power in Y-phase $= I_Y^2 \times R_Y = (10.06)^2 \times 14 = \mathbf{1416.8\ W.}$ **(Ans.)**

Power in B-phase $= I_B^2 \times R_B = (10.16)^2 \times 25 = \mathbf{2580.6\ W.}$ **(Ans.)**

☞**Example 1.47.** *A 3-phase, 4-wire system having a 440 V line to line has the following loads connected between the respective lines and neutral :*

$$Z_R = 5\ \angle 0°\ \Omega,\ Z_Y = 5\ \angle 37°\ \Omega,\ Z_B = 5\ \angle -53°\ \Omega.$$

Calculate the current in neutral wire and power taken by each load when phase sequence is :

(i) RYB *(ii) RBY.*

Solution. (*i*) **Phase sequence RYB** (Figs. 1.58 and 1.59).

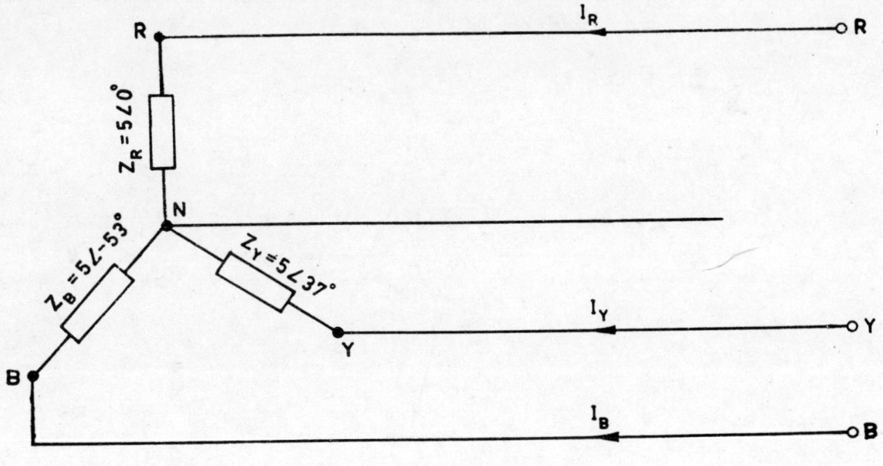

Fig. 1.58

The potential differences across the three phases are :

$$E_{RN} = \frac{440}{\sqrt{3}} \angle 0° = 254\ \angle 0°,\quad E_{YN} = 254\ \angle -120°$$
$$E_{BN} = 254\ \angle 120°$$

Impedances :

$$Z_R = 5\ \angle 0°,\ Z_Y = 5\ \angle 37°$$
$$Z_B = 5\ \angle -53°$$

Line currents :

$$I_R = I_{RN} = \frac{E_{RN}}{Z_R} = \frac{254\ \angle 0°}{5\ \angle 0°} = 50.8\ \angle 0°$$

$$I_Y = I_{YN} = \frac{E_{YN}}{Z_Y} = \frac{254 \angle -120°}{5 \angle 37°} = 50.8 \angle -157°$$

$$I_B = I_{BN} = \frac{E_{BN}}{Z_B} = \frac{254 \angle 120°}{5 \angle -53°} = 50.8 \angle 173°$$

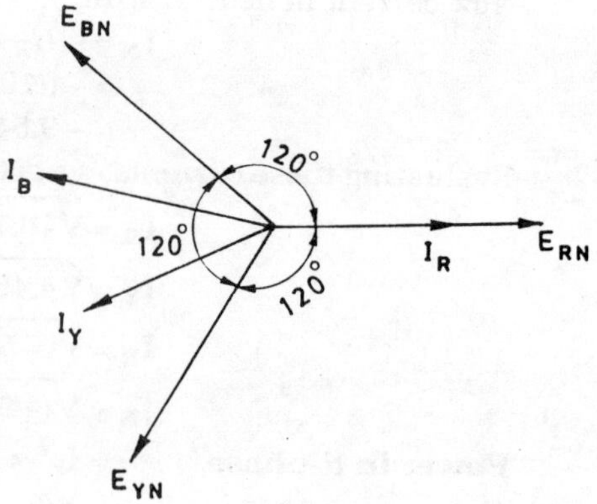

Fig. 1.59

Current in the neutral :

$$\mathbf{I_N} = -(I_R + I_Y + I_B)$$
$$= -(50.8 \angle 0° + 50.8 \angle -157° + 50.8 \angle 173°)$$
$$= -[50.8 + 50.8(-0.92 - j0.39) + 50.8(-0.99 + j0.122)]$$
$$= -[50.8 - 46.74 - j19.81 - 50.3 + j6.2]$$
$$= -(-46.24 - j13.61) = 46.24 + j13.61$$
$$= \mathbf{48.2 \angle 16.4°.} \textbf{ (Ans.)}$$

Now $R_R = 5\ \Omega$, $R_Y = 5 \cos 37° = 4\ \Omega$,
$R_B = 5 \cos(-53°) = 3\ \Omega$

Power taken by phase R,

$$P_R = 50.8^2 \times 5 = \mathbf{12903\ W.} \textbf{ (Ans.)}$$

Similarly, $P_Y = 50.8^2 \times 4 = \mathbf{10322\ W.}$ **(Ans.)**

and $P_B = 50.8^2 \times 3 = \mathbf{7742\ W.}$ **(Ans.)**

Phase Sequence RBY (Fig. 1.60) :

$$E_{RN} = 254 \angle 0°, E_{YN} = 254 \angle 120°$$
$$E_{BN} = 254 \angle -120°$$

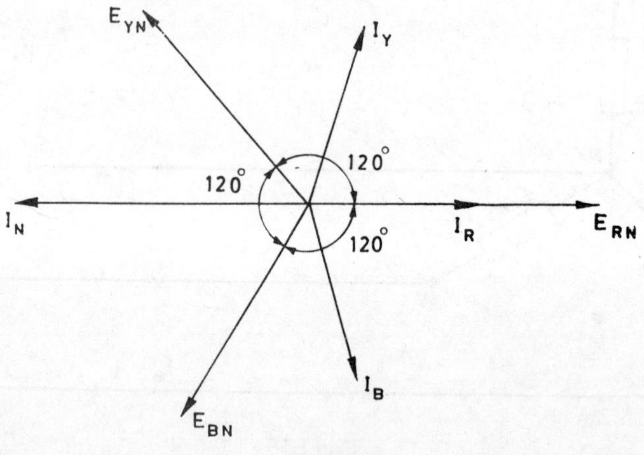

Fig. 1.60

Line currents :

$$I_R = I_{RN} = \frac{E_{RN}}{Z_R} = \frac{254 \angle 0°}{5 \angle 0°} = 50.8 \angle 0°$$

$$I_Y = I_{YN} = \frac{E_{YN}}{Z_Y} = \frac{254 \angle 120°}{5 \angle 37°} = 50.8 \angle 83°$$

$$I_B = I_{BN} = \frac{E_{BN}}{Z_B} = \frac{254 \angle -120°}{5 \angle -53°} = 50.8 \angle -67°$$

Current in the neutral :

$$\mathbf{I_N} = -(I_R + I_Y + I_B)$$
$$= -(50.8 \angle 0° + 50.8 \angle 83° + 50.8 \angle -67°)$$
$$= -[50.8 + 50.8(0.122 + j0.992) + 50.8(0.39 - j0.92)]$$
$$= -(50.8 + 6.197 + j50.4 + 19.81 - j46.74)$$
$$= -(76.81 + j3.66) = -76.81 - j3.66$$
$$= \mathbf{76.89 \angle -177.3° \text{ A.}} \quad \textbf{(Ans.)}$$

Power would remain the same because magnitude of branch currents is unaltered.

Example 1.48. *In a 3-phase, 4-wire system, the line voltage is 400 V and non-inductive loads of 5, 4 and 2.5 kW are connected between the three lines conductors and the neutral, as shown in Fig. 1.61. Calculate :*

(*i*) *The current in lines ; and*

(*ii*) *The current in the neutral conductor.*

Solution. Refer Fig. 1.61.

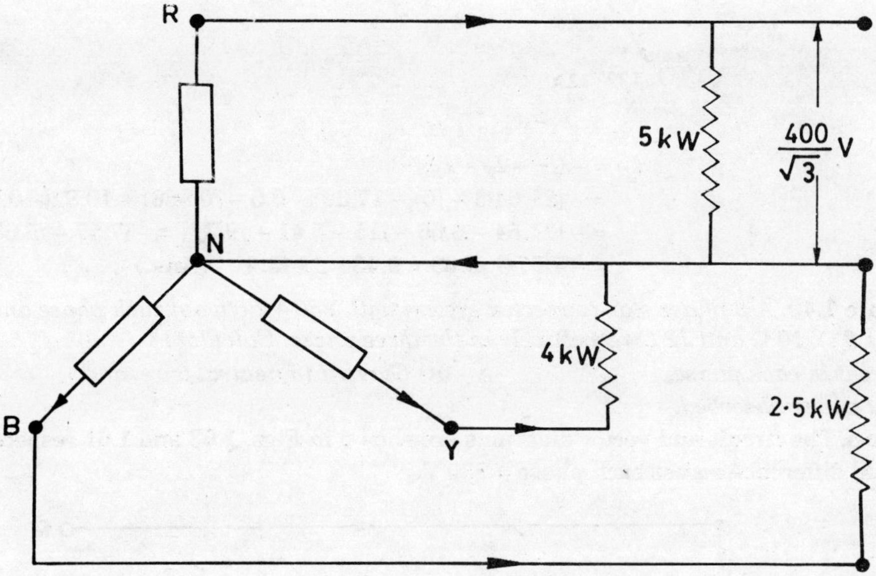

Fig. 1.61

Phase voltage across each load,

$$E_{ph} = \frac{E_L}{\sqrt{3}} = \frac{400}{\sqrt{3}} = 231 \text{ V}$$

(*i*) Current in phase R,

$$I_R = \text{(line current)} = \frac{\text{kW} \times 1000}{E_{ph} \cos \phi} = \frac{5 \times 1000}{231 \times 1} = \mathbf{21.64 \text{ A.}} \quad \textbf{(Ans.)}$$

Current in phase Y, $\quad I_Y = \dfrac{4 \times 1000}{231 \times 1} = \mathbf{17.32 \text{ A.}}$ **(Ans.)**

Current in phase B, $\quad I_B = \dfrac{2.5 \times 1000}{231 \times 1} = \mathbf{10.82 \text{ A.}}$ **(Ans.)**

These currents are mutually 120° out of phase as the circuit is purely resistive. Taking I_R as the reference vector, line currents may be given as

$$\mathbf{I_R} = 21.64 \angle 0° = 21.64\ (1 + j0)$$
$$\mathbf{I_Y} = 17.32 \angle -120° = 17.32\ (-0.5 - j0.866)$$
$$\mathbf{I_B} = 10.82 \angle 120° = 10.82\ (-0.5 + j0.866)$$

(*ii*) **Current in the neutral wire** (Refer Fig. 1.62) :

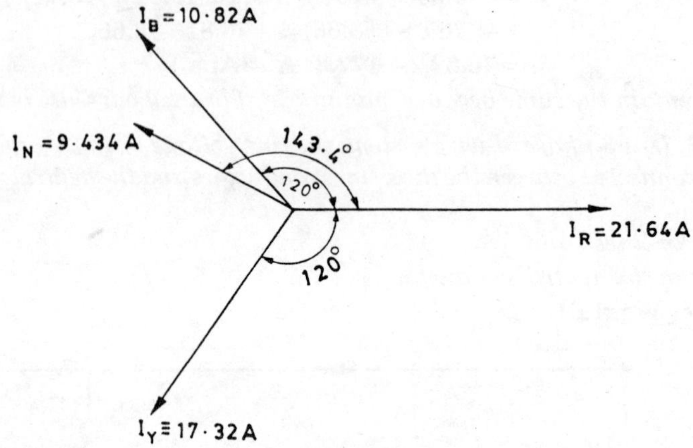

Fig. 1.62

$$I_N = -(I_R + I_Y + I_B)$$
$$= -[21.64(1 - j0) + 17.32\ (-0.5 - j0.866) + 10.82(-0.5 + j0.866)]$$
$$= -[21.64 - 8.66 - j15 - 5.41 + j9.37] = -(7.57 - j5.63)$$
$$= -7.57 + j5.63 = \mathbf{9.434 \angle 143.4°}.\ \textbf{(Ans.)}$$

Example 1.49. *A 3-phase star-connected system with 230 V between each phase and neutral has resistances 8 Ω, 10 Ω and 12 Ω respectively in the three phase. Calculate :*

(*i*) *Current in each phase,* (*ii*) *Current in neutral wire, and*

(*iii*) *Total power absorbed.*

Solution. The circuit and vector diagrams are shown in Figs. 1.63 and 1.64 respectively.

Potential difference across each phase = 230 V.

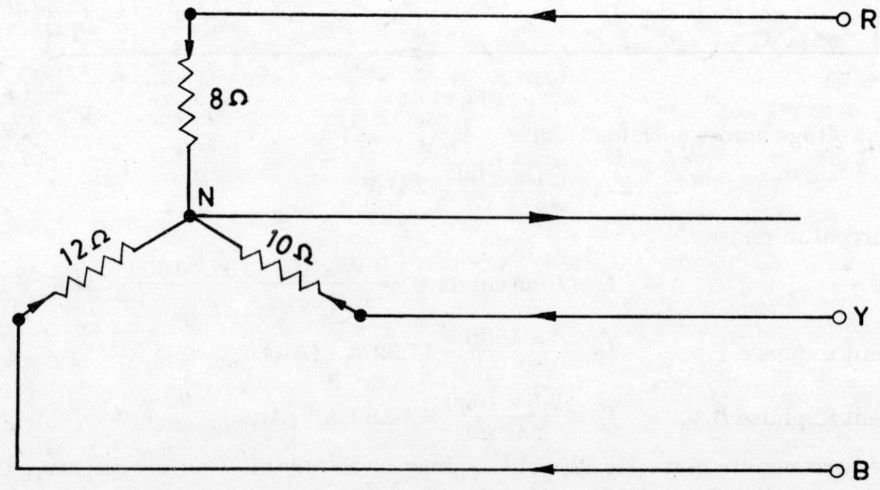

Fig. 1.63

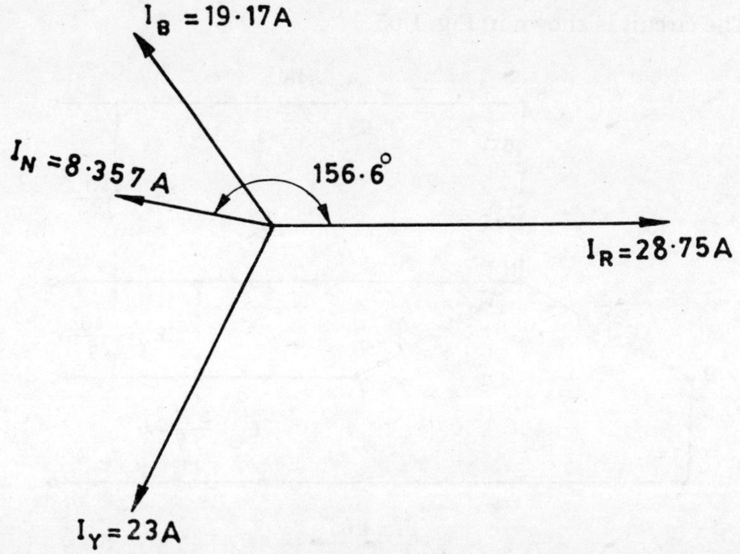

Fig. 1.64

(*i*) **Phase currents :**

Current in 8 Ω resistance $= \frac{230}{8} = $ **28.75 A. (Ans.)**

Current in 10 Ω resistance $= \frac{230}{10} = $ **23 A. (Ans.)**

Current in 12 Ω resistance $= \frac{230}{12} = $ **19.17 A. (Ans.)**

Since the circuit is purely resistive, so currents will be in phase with their respective voltages and will be mutually 120° apart. Taking I_R as the reference vector (Fig. 1.64) the line currents are given as :

$$I_R = 28.75 \angle 0° = 28.75(1 + j0)$$
$$I_Y = 23 \angle - 120° = 23 (- 0.5 - j0.866)$$
$$I_B = 19.17 \angle 120° = 19.17 (- 0.5 + j0.866).$$

(*ii*) **Current in neutral wire,**

$$I_N = -(I_R + I_Y + I_B) = -[28.75(1 + j0) + 23(- 0.5 - j0.866) + 19.17(- 0.5 + j0.866)$$
$$= -[28.75 - 11.5 - j19.92 - 9.58 + j16.6]$$
$$= -[7.67 - j3.32] = -7.67 + j3.32 = \mathbf{8.357 \angle 156.6° \text{ A.}} \text{ (Ans.)}$$

(*iii*) **Total power absorbed**

$$= I_R^2 \times 8 + I_Y^2 \times 10 + I_B^2 \times 12 = 28.75^2 \times 8 + 23^2 \times 10 + 19.17^2 \times 12$$
$$= 6612.5 + 5290 + 4409.8 = \mathbf{16312.3 \text{ W.}} \text{ (Ans.)}$$

☞**Example 1.50.** *A 440 V (line-to-line), 3-phase, 4-core supplies an unbalanced load represented by the following impedances in ohms connected between R, Y and B phases respectively and the neutral 8 + j6, 7 – j10.5 and 12.5. Calculate :*

(i) Current in each conductor of the cable, (ii) Current in the neutral wire, and

(iii) Readings of the three wattmeters connected in each line and neutral.

The phase sequence is RYB.

Solution. The circuit is shown in Fig. 1.65.

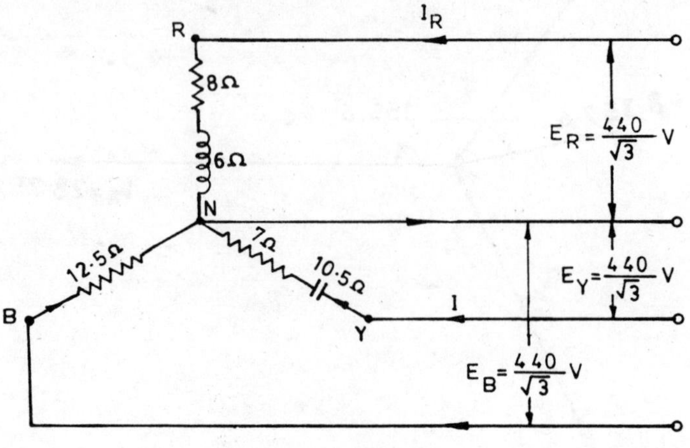

Fig. 1.65

The potential differences across the three phases are :

$$E_{RN} = \frac{440}{\sqrt{3}} = 254 \angle 0°$$
$$E_{YN} = 254 \angle -120°$$
$$E_{BN} = 254 \angle 120°.$$

Impedances :

$$Z_R = 8 + j6 = 10 \angle 36.9°$$
$$Z_Y = 7 - j10.5 = 12.62 \angle -56.3°$$
$$Z_B = 12.5 = 12.5 \angle 0°.$$

(*i*) **Phase currents :**

$$\mathbf{I_R} = I_{RN} = \frac{E_{RN}}{E_R} = \frac{254 \angle 0°}{10 \angle 36.9°}$$
$$= \mathbf{25.4 \angle -36.9° \text{ A.}} \quad \textbf{(Ans.)}$$

$$\mathbf{I_Y} = I_{YN} = \frac{E_{YN}}{Z_Y} = \frac{254 \angle -120°}{12.62 \angle -56.3°}$$
$$= \mathbf{20.12 \angle -63.7° \text{ A.}} \quad \textbf{(Ans.)}$$

$$\mathbf{I_B} = I_{BN} = \frac{E_{BN}}{Z_B} = \frac{254 \angle 120°}{12.5 \angle 0°}$$
$$= \mathbf{20.32 \angle 120° \text{ A.}} \quad \textbf{(Ans.)}$$

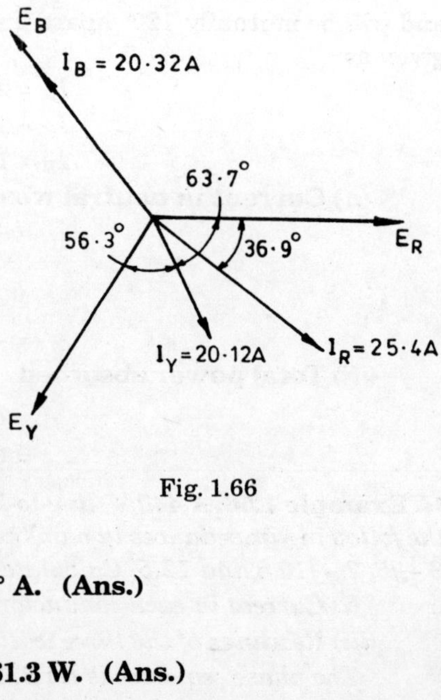

Fig. 1.66

(*ii*) **Current in the neutral wire :**

$$I_N = -(I_R + I_Y + I_B)$$
$$= -[25.4 \angle -36.9° + 20.12 \angle -63.7° + 20.32 \angle 120°]$$
$$= -[25.4(+0.8 - j0.6) + 20.12(0.44 - j0.896) + 20.32(-0.5 + j0.866)]$$
$$= -(20.32 - j15.24 + 8.85 - j18.03 - 10.16 + j17.6)$$
$$= -(19.01 - j15.67) = -19.01 + j15.67 = \mathbf{24.64 \angle 39.5° \text{ A.}} \quad \textbf{(Ans.)}$$

(*iii*) **Readings of wattmeters :**

Power in phase, $R = I_R^2 \times R_R = 25.4^2 \times 8 = \mathbf{5161.3 \text{ W.}}$ **(Ans.)**

Power in phase, $Y = I^2_Y \times R_Y = 20.12^2 \times 7 =$ **2833.7 W. (Ans.)**

Power in phase, $B = I^2_B \times R_B = 20.32^2 \times 12.5 =$ **5161.3 W. (Ans.)**

KIRCHHOFF'S LAWS (Article 1.11)

Example 1.51. *Calculate the voltage drop across each of the following three star-connected impedances when they are connected to a 400 V, 3-phase system :*

$$Z_R = j40\ \Omega,\ Z_Y = j2\ \Omega,\ Z_B = -j2\ \Omega.$$

Also find the potential of the star-point above ground.

The phase sequence is RYB.

Solution. Refer Fig. 1.67.

Considering E_{RY} as a reference vector. The vector expressions for the three line voltages are :

$E_{RY} = 400(1 + j0)$ V $= (400 + j0)$ V

$E_{YB} = 400(-0.5 - j0.866) = (-200 - j346.4)$ V

$E_{BR} = 400(-0.5 + j0.866) = (-200 + j346.4)$ V

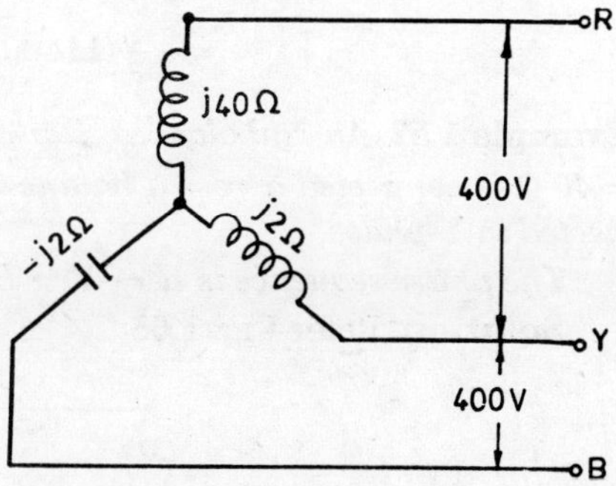

Fig. 1.67

Impedances :

$Z_R = (0 + j40)$, $\qquad Z_Y = (0 + j2)$

$Z_B = (0 - j2)$

Now for finding I_R, I_Y and I_B, using the relations :

$$I_R = \frac{E_{RY}}{Z_R + Z_Y + \frac{Z_R Z_Y}{Z_B}} - \frac{E_{BR}}{Z_B + Z_R + \frac{Z_B Z_R}{Z_Y}}$$

$$= \frac{400 + j0}{(0 + j40) + (0 + j2) + \frac{(0 + j40)(0 + j2)}{(0 - j2)}} - \frac{(-200 + j346.4)}{(0 - j2) + (0 + j40) + \frac{(0 - j2)(0 + j40)}{(0 + j2)}}$$

$$= \frac{(400 + j0)}{(0 + j42) - (j40)} - \frac{(-200 + j346.4)}{(0 + j38) - j40} = -\frac{400}{j2} - \frac{(-200 + j346.4)}{-j2}$$

$$= -j200 + j100 + 346.4 = (173.2 - j100) \text{ A or } 200 \angle -30° \text{ A}$$

$$I_Y = \frac{E_{YB}}{Z_Y + Z_B + \frac{Z_Y Z_B}{Z_R}} - \frac{E_{RY}}{Z_R + Z_Y + \frac{Z_R Z_Y}{Z_B}}$$

$$= \frac{(-200 - j346.4)}{(0 + j2) + (0 - j2) + \frac{(0 + j2)(0 - j2)}{(0 + j40)}} - \frac{(400 + j0)}{(0 + j40) + (0 + j2) + \frac{(0 + j40)(0 + j2)}{(0 - j2)}}$$

$$= \frac{(-200 - j346.4)}{-j0.1} - \frac{(400 + j0)}{(0 + j42) - j40}$$

$$= \frac{-200 - j346.4}{-j0.1} - \frac{(400 + j0)}{j2} = -j2000 + 3464 + j200 = 3464 - j1800$$

$$= 3904 \angle -27.45° \text{ A}$$

$$I_B = \frac{E_{BR}}{Z_B + Z_R + \frac{Z_B Z_R}{Z_Y}} - \frac{E_{YB}}{Z_Y + Z_B + \frac{Z_Y Z_B}{Z_R}}$$

$$= \frac{(-200 + j346.4)}{(0 - j2) + (0 + j40) + \frac{(0 - j2)(0 + j40)}{(0 + j2)}} - \frac{(-200 - j346.4)}{(0 + j2) + (0 - j2) + \frac{(0 + j2)(0 - j2)}{(0 + j40)}}$$

$$= \frac{(-200 + j346.4)}{j38 - j40} - \frac{(-200 - j346.4)}{-j0.1} = \frac{(-200 + j346.4)}{-j2} - \frac{(-200 - j346.4)}{-j0.1}$$

$$= -j100 - 173.2 + j2000 - 3464 = -3637.2 + j1900 = 4104 \angle 152.4° \text{ A.}$$

Voltage drops across impedances :

Voltage drop, $E_R = I_R Z_R = (173.2 - j100) \times j40 =$ **(4000 + j6928) V. (Ans.)**

Voltage drop, $E_Y = I_Y Z_Y = (3464 - j1800)(j2) =$ **(3600 + j6928) V. (Ans.)**

Voltage drop, $E_B = I_B Z_B = (-3637.2 + j1900)(-j2) =$ **(3800 + j7274.4) V. (Ans.)**

Potential of star point above ground :

$$= -\frac{1}{3}(E_R + E_Y + E_B)$$

$$= -\frac{1}{3}[(4000 + j6928) + (3600 + j6928) + (3800 + j7274.4)]$$

$$= -\frac{1}{3}(11400 + j17656) = -(3800 + j7043.5) = \mathbf{-8003 \text{ V.}} \text{ (Ans.)}$$

☞**Example 1.52.** *An unbalanced star-connected load shown in Fig. 1.68, $Z_R = j40\ \Omega$, $Z_Y = -j40\ \Omega$, $Z_B = 40\ \Omega$ is connected across a balanced 400 V, 3-phase supply. Find the reading of the wattmeter connected in Y-phase.*

The phase sequence is R → Y → B.

Solution. Refer Fig. 1.68.

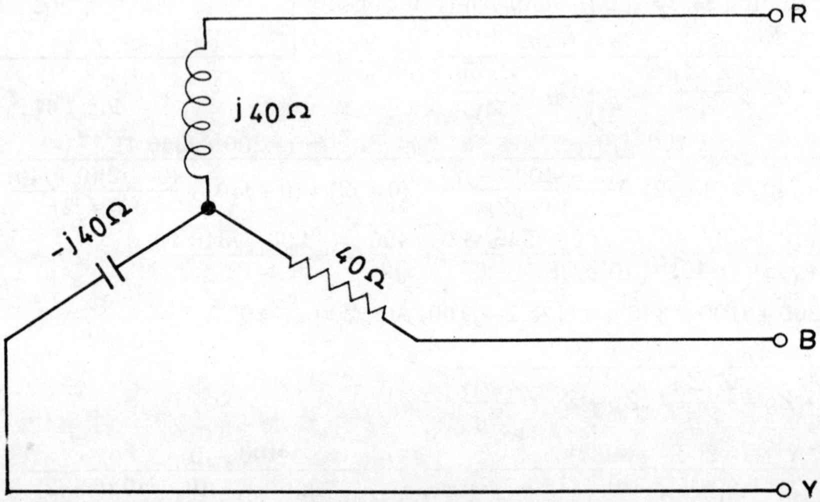

Fig. 1.68

Taking E_{RY} as reference vector

$$E_{RY} = 400 \angle 0° = 400(1 + j0)$$
$$E_{YB} = 400 \angle -120° = 400(-0.5 - j0.866)$$
$$E_{BR} = 400 \angle 120° = 400(-0.5 + j0.866)$$

Impedances are

$$Z_R = (0 + j40), \quad Z_Y = (0 - j40), \quad Z_B = (40 + j0)$$

Now,
$$I_R = \frac{E_{RY}}{Z_R + Z_Y + \dfrac{Z_R Z_Y}{Z_B}} - \frac{E_{BR}}{Z_B + Z_R + \dfrac{Z_B Z_R}{Z_Y}}$$

$$= \frac{(400 + j0)}{j40 - j40 + \frac{j40 \times (-j40)}{40}} - \frac{400(-0.5 + j0.866)}{40 + j40 + \frac{40 \times j40}{-j40}}$$

$$= \frac{400}{40} - \frac{-200 + j346.4}{40 + j40 - 40}$$

$$= 10 - \frac{-200 + j346.4}{j40} = 10 - \frac{+j200 - 346.4}{40}$$

$$= 10 - j5 + 8.66 = (1.34 - j5) \text{ A}$$

$$I_Y = \frac{E_{YB}}{Z_Y + Z_B + \frac{Z_Y Z_B}{Z_R}} - \frac{E_{RY}}{Z_R + Z_Y + \frac{Z_R Z_Y}{Z_B}}$$

$$= \frac{400(-0.5 - j0.866)}{-j40 + 40 + \frac{(-j40) \times 40}{j40}} - \frac{400 + j0}{j40 - j40 + \frac{j40(-j40)}{40}}$$

$$= \frac{-200 - j346.4}{-j40 + 40 - 40} - \frac{400 + j0}{40}$$

$$= -j5 + 8.66 - 10 = (-1.34 - j5) \text{ A}$$

$$I_B = \frac{E_{BR}}{Z_B + Z_R + \frac{Z_B Z_R}{Z_Y}} - \frac{E_{YB}}{Z_Y + Z_B + \frac{Z_Y Z_B}{Z_R}}$$

$$= \frac{400(0.5 + j0.866)}{40 + j40 + \frac{40 \times j40}{-j40}} - \frac{400(-0.5 - j0.866)}{-j40 + 40 + \frac{(-j40) \times 40}{j40}}$$

$$= \frac{200 + j346.4}{40 + j40 - 40} - \frac{-200 - j346.4}{-j40 + 40 - 40} = \frac{200 + j346.4}{j40} - \frac{-200 - j346.4}{-j40}$$

$$= 8.66 + j5 + j5 - 8.66 = j10 \text{ A}$$

Voltage drop, $E_R = I_R Z_R = (1.34 - j5) \times j40 = (200 + j53.6)$ V

Voltage drop, $E_Y = I_Y Z_Y = (-1.34 - j5) \times (-j40) = (-200 + j53.6)$ V

Voltage drop, $E_B = I_B Z_B = j10 \times 40 = (0 + j400)$ V

Voltage across potential coil of wattmeter

= voltage between supply neutral and load neutral

$$= \frac{1}{3}(E_R + E_Y + E_B) = \frac{1}{3}(200 + j53.6 - 200 + j53.6 + j400) = j169 \text{ V}$$

Current through wattmeter current coil × conjugate of voltage across wattmeter potential coil

$$= I_Y \times (-j169) = (-1.34 - j5) \times (-j169) = (-845 + j226.5)$$

∴ **Wattmeter reading = – 845 W. (Ans.)**

STAR-DELTA CONVERSION

☞**Example 1.53.** *An unbalanced star-connected load has branch impedances of $Z_R = 10 \angle 30°$ Ω, $Z_Y = 20 \angle 45°$ Ω and $Z_B = 20 \angle 60°$ Ω and is connected across a balanced 3-phase, 3-wire supply of 440 V. Find :*

(i) Line currents, and *(ii) Voltage across each impedance.*

Use star-delta conversion method.

Solution. Taking E_{RY} as reference vector

Then $E_{RY} = 440 \angle 0°$, $E_{YB} = 440 \angle -120°$

$E_{BR} = 440 \angle 120°$

Refer Fig. 1.69.

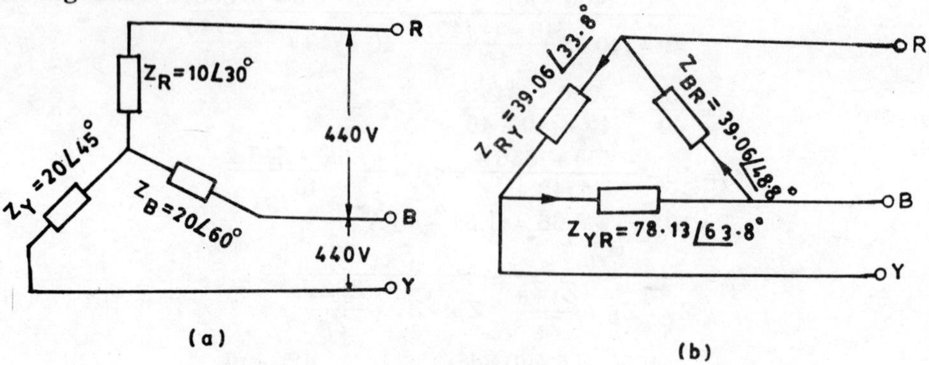

Fig. 1.69

Also $Z_R = 10 \angle 30°$ *(given)*

$Z_Y = 20 \angle 45°$ *(given)*

$Z_B = 20 \angle 60°$ *(given)*

Impedances of equivalent delta-connected load (Refer Fig. 1.69) :

Now
$$Z_{RY} = Z_R + Z_Y + \frac{Z_R R_Y}{Z_B} = \frac{Z_R Z_B + Z_Y Z_B + Z_R Z_Y}{Z_B}$$

$$= \frac{(10 \angle 30°)(20 \angle 60°) + (20 \angle 45°)(20 \angle 60°) + (10 \angle 30°)(20 \angle 45°)}{20 \angle 60°}$$

$$= \frac{200 \angle 90° + 400 \angle 105° + 200 \angle 75°}{20 \angle 60°}$$

$$= \frac{200(0 + j1) + 400(-0.26 + j0.966) + 200(0.26 + j0.966)}{20 \angle 60°}$$

$$= \frac{j200 - 104 + j386.4 + 52 + j193.2}{20 \angle 60°}$$

$$= \frac{-52 + j779.6}{20 \angle 60°} = \frac{781.3 \angle 93.8°}{20 \angle 60°} = 39.06 \angle 33.8° \ \Omega$$

$$Z_{YB} = Z_Y + Z_B + \frac{Z_Y Z_B}{Z_R} = \frac{Z_Y Z_R + Z_B Z_R + Z_Y Z_B}{Z_R}$$

$$= \frac{(20 \angle 45°)(10 \angle 30°) + (20 \angle 60°)(10 \angle 30°) + (20 \angle 45°)(20 \angle 60°)}{10 \angle 30°}$$

$$= \frac{200 \angle 75° + 200 \angle 90° + 400 \angle 105°}{10 \angle 30°}$$

$$= \frac{200(0.62 + j0.966) + j200 + 400(-0.26 + j0.966)}{10 \angle 30°}$$

$$= \frac{52 + j193.2 + j200 - 104 + j386.4}{10 \angle 30°}$$

$$= \frac{-52 + j779.6}{10 \angle 30°} = \frac{781.3 \angle 93.8°}{10 \angle 30°} = 78.13 \angle 63.8° \ \Omega$$

$$Z_{BR} = Z_B + Z_R + \frac{Z_B Z_R}{Z_Y} = \frac{Z_B Z_Y + Z_R Z_Y + Z_B Z_R}{Z_Y}$$

$$= \frac{(20 \angle 60°)(20 \angle 45°) + (10 \angle 30°)(20 \angle 45°) + (20 \angle 60°)(10 \angle 30°)}{20 \angle 45°}$$

$$= \frac{400 \angle 105° + 200 \angle 75° + 200\angle 90°}{20 \angle 45°}$$

$$= \frac{-52 + j779.6}{20 \angle 45°} = \frac{781.3 \angle 93.8}{20 \angle 45°} = 39.06 \angle 48.8° \ \Omega$$

$$\mathbf{I_{RY}} = \frac{E_{RY}}{Z_{RY}} = \frac{440 \angle 0°}{39.06 \angle 33.8°} = 11.26 \angle -33.8° \text{ A}$$

$$= 11.26(0.83 - j0.56) = \mathbf{(9.34 - j6.3) \ A.} \quad \textbf{(Ans.)}$$

$$\mathbf{I_{YB}} = \frac{E_{YB}}{Z_{YB}} = \frac{440 \angle -120°}{78.13 \angle 63.8°} = 5.63 \angle -183.8°$$

$$= 5.63(-0.998 + j0.066) = \mathbf{(-5.62 + j0.37) \ A.} \quad \textbf{(Ans.)}$$

$$\mathbf{I_{BR}} = \frac{E_{BR}}{Z_{BR}} = \frac{440 \angle 120°}{39.06 \angle 48.8°} = 11.26 \angle 71.2°$$

$$= 11.26(0.322 + j0.947) = \mathbf{(3.625 + j10.66) \ A.} \quad \textbf{(Ans.)}$$

(*i*) **Line currents :**

$$\mathbf{I_R} = I_{RY} - I_{BR} = (9.34 - j6.3) - (3.625 + j10.66)$$

$$= 5.715 - j16.96 = \mathbf{17.89 \angle -71.4° \ A.} \quad \textbf{(Ans.)}$$

$$\mathbf{I_Y} = I_{YB} - I_{RY} = (-5.62 + j0.37) - (9.34 - j6.3)$$

$$= -14.96 + j6.67 = \mathbf{16.4 \angle 156° \ A.} \quad \textbf{(Ans.)}$$

$$\mathbf{I_B} = I_{BR} - I_{YB} = (3.625 + j10.66) - (-5.62 + j0.37)$$

$$= 9.245 + j10.29 = \mathbf{13.833 \angle 48° \ A.} \quad \textbf{(Ans.)}$$

These are the currents in the phases of the star-connected unbalanced load. Let us find voltage drop across each star-connected branch impedance.

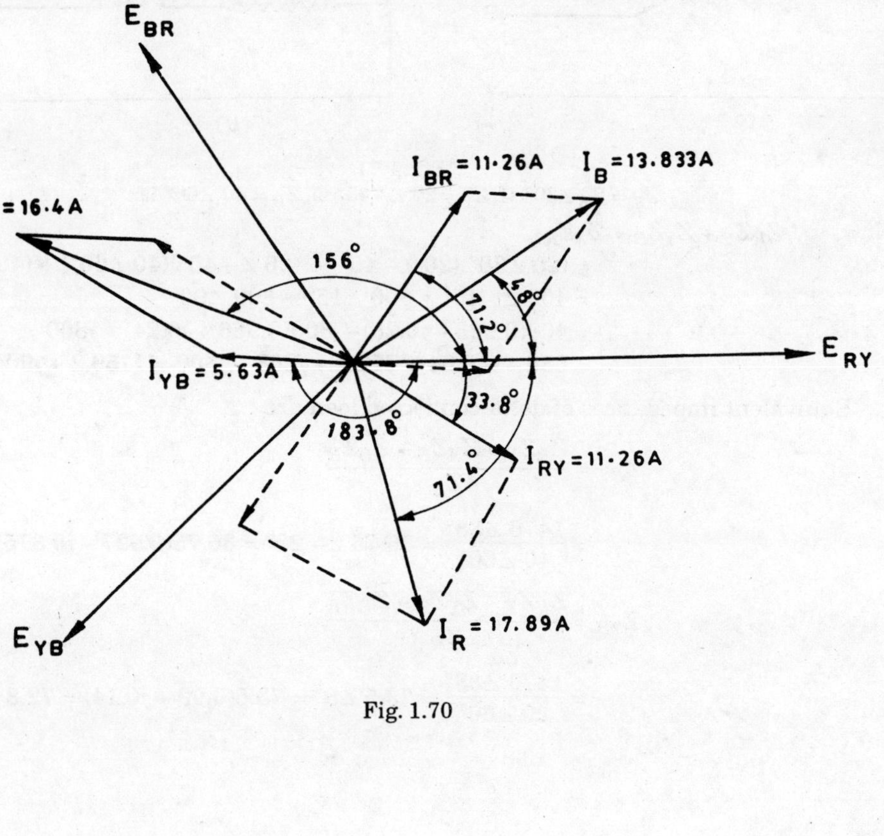

Fig. 1.70

(*ii*) **Voltage across impedances :**

Voltage across $\mathbf{Z_R} = I_R \times Z_R$

$= 17.89 \angle -71.4° \times 10 \angle 30° = \mathbf{178.9 \angle -41.4° \text{ V.}}$ **(Ans.)**

Voltage across $\mathbf{Z_Y} = I_Y \times Z_Y = 16.4 \angle 156° \times 20 \angle 45°$

$= \mathbf{328 \angle 210° \text{ V.}}$ **(Ans.)**

Voltage across $\mathbf{Z_B} = I_B \times Z_B = 13.833 \angle 48° \times 20 \angle 60°$

$= \mathbf{276.67 \angle 108° \text{ V.}}$ **(Ans.)**

The phasor diagram is shown in Fig. 1.70.

☞**Example 1.54.** *The branch impedances of an unbalanced star-connected load are :* $Z_R = 20 \angle 30°\ \Omega$, $Z_Y = 20 \angle -45°\ \Omega$, $Z_B = 40 \angle 60°\ \Omega$. *It is connected across a balanced 3-phase, 3-wire supply of 200 V. Find* :

(*i*) *Line currents, and* (*ii*) *Voltage across each impedance.*

Use star-delta conversion method.

Solution. The unbalanced star-connected load and its equivalent delta-connected load are shown in Fig. 1.71.

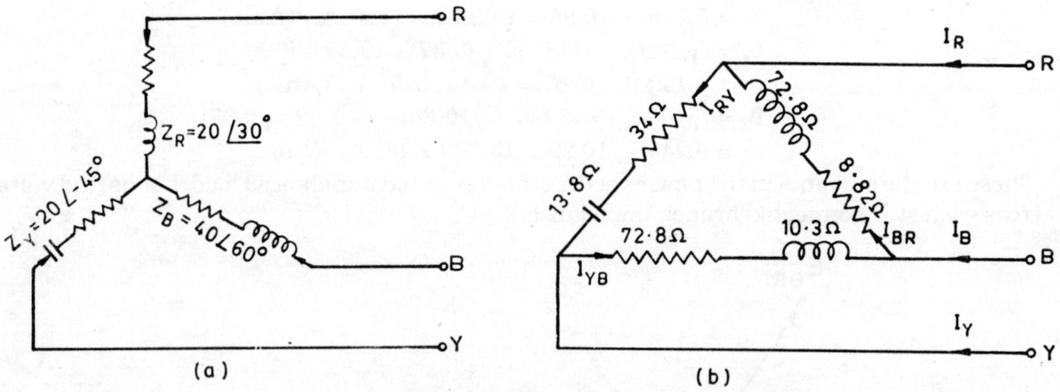

Fig. 1.71

$Z_R = 20 \angle 30°\ \Omega,\ Z_Y = 20 \angle -45°\ \Omega,\ Z_B = 40 \angle 60°\ \Omega$

Now, $Z_R Z_Y + Z_Y Z_B + Z_B Z_R$

$= (20 \angle 30°)(20 \angle -45°) + (20 \angle -45°)(40 \angle 60°) + (40 \angle 60°)(20 \angle 30°)$

$= 400 \angle -15° + 800 \angle 15° + 800 \angle 90°$

$= 400(0.966 - j0.26) + 800(0.966 + j0.24) + j800$

$= 386.4 - j104 + 772.8 + j208 + j800 = 1159.2 + j904 = 1470 \angle 38°$

∴ Equivalent impedances of delta-connected load are :

$$Z_{RY} = \frac{Z_R Z_Y + Z_Y Z_B + Z_B Z_R}{Z_B}$$

$$= \frac{1470 \angle 38°}{40 \angle 60°} = 36.75 \angle -22° = 36.75(0.927 - j0.375) = 34 - j13.8$$

$$Z_{YB} = \frac{Z_R Z_Y + Z_Y Z_B + Z_B Z_R}{Z_R}$$

$$= \frac{1470 \angle 38°}{20 \angle 30°} = 73.5 \angle 8° = 73.5(0.99 + j0.14) = 72.8 + j10.3$$

$$Z_{BR} = \frac{Z_R Z_Y + Z_Y Z_B + Z_B Z_R}{Z_Y} = \frac{1470 \angle 38^\circ}{20 \angle -45^\circ}$$

$$= 73.5 \angle 83^\circ = 73.5(0.12 + j0.99) = 8.82 + j72.8$$

Taking E_{RY} as the difference vector and assuming sequence as $R \to Y \to B$,

$$E_{RY} = 200 \angle 0^\circ, E_{YB} = 200 \angle -120^\circ, E_{BR} = 200 \angle 120^\circ$$

$$I_{RY} = \frac{E_{RY}}{Z_{RY}} = \frac{200 \angle 0^\circ}{36.75 \angle -22^\circ} = 5.44 \angle 22^\circ$$

$$= 5.44(0.927 + j0.375) = 5.04 + j2.04$$

$$I_{YB} = \frac{E_{YB}}{Z_{YB}} = \frac{200 \angle -120^\circ}{73.5 \angle 8^\circ} = 2.72 \angle -128^\circ$$

$$= 2.72\,(-0.616 - j0.79) = -1.67 - j2.15$$

$$I_{BR} = \frac{E_{BR}}{Z_{BR}} = \frac{200 \angle 120^\circ}{73.5 \angle 83^\circ} = 2.72 \angle 37^\circ = 2.72(0.8 + j0.6) = 2.17 + j1.63$$

(*i*) **Line Currents :**

$$\mathbf{I_R} = I_{RY} - I_{BR}$$

$$= (5.04 + j2.04) - (2.17 + j1.63) = 2.87 + j0.41 = \mathbf{2.9 \angle 8.1^\circ\ A.}\ \ \textbf{(Ans.)}$$

$$\mathbf{I_Y} = I_{YB} - I_{RY} = (-1.67 - j2.15) - (5.04 + j2.04)$$

$$= -6.71 - j4.19 = \mathbf{7.91 \angle -148^\circ\ A.}\ \ \textbf{(Ans.)}$$

$$\mathbf{I_B} = I_{BR} - I_{YB} = (2.17 + j1.63) - (-1.67 - j2.15)$$

$$= 3.84 + j3.78 = \mathbf{5.39 \angle 44.5^\circ\ A.}\ \ \textbf{(Ans.)}$$

These are the currents in the phases of star-connected unbalanced load. Let us find the drop across each star-connected branch impedance.

(*ii*) **Voltage drop across impedances** :

Voltage drop across $\mathbf{Z_R} = I_R Z_R = 2.9 \angle 8.1^\circ \times 20 \angle 30^\circ = \mathbf{58 \angle 38.1^\circ\ V.}$ **(Ans.)**

Voltage drop across $\mathbf{Z_Y} = I_Y Z_Y = 7.91 \angle -148^\circ \times 20 \angle -45^\circ = \mathbf{158.2 \angle -193^\circ\ V.}$ **(Ans.)**

Voltage across $\mathbf{Z_B} = I_B Z_B = 5.39 \angle 44.5^\circ \times 40 \angle 60^\circ = \mathbf{215.6 \angle 104.5^\circ\ V.}$ **(Ans.)**

Example 1.55. *A star-connected load consisting of non-inductive r esistors of 30, 12 and 20 ohms connected to the R, Y and B lines respectively is fed by a 300 V (line), 3-phase supply. The phase sequence is RYB. Calculate the voltage across each resistor.*

Solution. Refer Fig. 1.72.

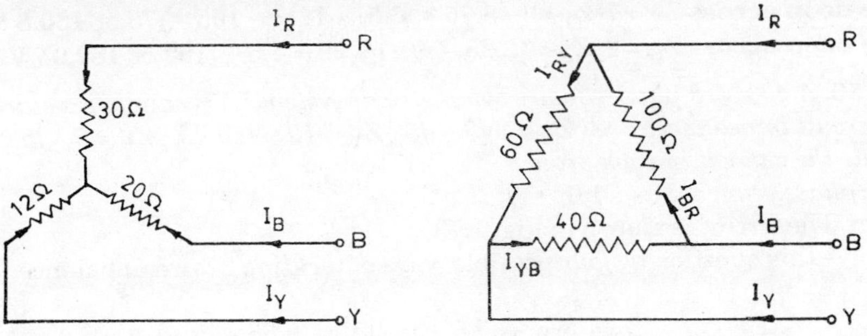

Fig. 1.72

Using star-delta conversion method,

Impedances are :

$$Z_R = 30\ \Omega, \quad Z_Y = 12\ \Omega, \quad Z_B = 20\ \Omega$$

Now $Z_R Z_Y + Z_Y Z_B + Z_{BR} = 30 \times 12 + 12 \times 20 + 20 \times 30 = 360 + 240 + 600 = 1200\ \Omega$

Equivalent impedances of delta-connected load are :

$$Z_{RY} = \frac{Z_R Z_Y + Z_Y Z_B + Z_{BR}}{Z_B} = \frac{1200}{20} = 60\ \Omega$$

$$Z_{YB} = \frac{Z_R Z_Y + Z_Y Z_B + Z_B Z_R}{Z_R} = \frac{1200}{30} = 40\ \Omega$$

$$Z_{BR} = \frac{Z_R Z_Y + Z_Y Z_B + Z_B Z_R}{Z_Y} = \frac{1200}{12} = 100\ \Omega$$

Phase currents are :

$$I_{RY} = \frac{E_{RY}}{Z_{RY}} = \frac{300}{60} = 5\text{ A}$$

$$I_{YB} = \frac{E_{YB}}{Z_{YB}} = \frac{300}{40} = 7.5\text{ A}$$

$$I_{BR} = \frac{E_{BR}}{Z_{BR}} = \frac{300}{100} = 3\text{ A}$$

Each current is in phase with its own voltage because the load is purely resistive.

To find line currents for delta-connection using vector algebra and choosing E_{RY} as the reference axis, we get

$$I_{RY} = 5 + j0$$

$$I_{YB} = 7.5(-0.5 - j0.866) = -3.75 - j6.5$$

$$I_{BR} = 3(-0.5 + j0.866) = -1.5 + j2.6$$

Line currents for delta-connection are :

$$I_R = I_{RY} - I_{BR}$$

$$= (5 + j0) - (-1.5 + j2.6) = (6.5 - j2.6) \text{ or } 7\text{ A in magnitude}$$

$$I_Y = I_{YB} - I_{RY}$$

$$= (-3.75 - j6.5) - (5 + j0) = -8.75 - j6.5 \text{ or } 10.9\text{ A in magnitude}$$

$$I_B = I_{BR} - I_{YB} = (-1.5 + j2.6) - (-3.75 - j6.5)$$

$$= 2.25 - j9.1 \text{ or } 9.37\text{ A in magnitude}$$

These line currents for delta-connection are the phase currents for star-connection.

Voltage drop across each limb of star-connected load is :

Voltage drop across $\mathbf{Z_R}$ $= I_R Z_R = (6.5 - j2.6) \times 30 = 195 - j78$ or **210 V. (Ans.)**

Voltage drop across $\mathbf{Z_Y}$ $= I_Y Z_Y = (-8.75 - j6.5) \times 12 = -105 - j78$ or **130.8 V. (Ans.)**

Voltage drop across $\mathbf{Z_B}$ $= I_B Z_B = (2.25 - j9.1) \times 20 = 4.5 - j182$ or **182.05 V. (Ans.)**

☞**Example 1.56.** *A 400 V, 3-phase, 3-wire symmetrical system RYB supplies a star-connected load whose branch circuit impedances are :* $Z_R = 5\sqrt{3} + j5$; $Z_Y = 10 + j10\sqrt{3}$; $Z_B = 0 - j5$.

Determine the current in each branch.

Take the phase sequence as RYB.

Solution. The circuit is shown in Fig. 1.73.

Let us solve this question by *four* possible ways of tackling 3-wire unbalanced star-connected load problems.

Now

$$Z_R = 5\sqrt{3} + j5 = 10\ \angle 30° = (8.66 + j5)$$

$$Z_Y = 10 + j10\sqrt{3} = 20\ \angle 60° = (10 + j17.32)$$

$$Z_B = 0 - j5 = 5\ \angle -90° = -j5$$

Also, let

$$E_{RY} = 400\ \angle 0° = (400 + j0)$$

$$E_{YB} = 400\ \angle -120° = (-200 - j346)$$

$$E_{BR} = 400\ \angle 120° = (-200 + j346)$$

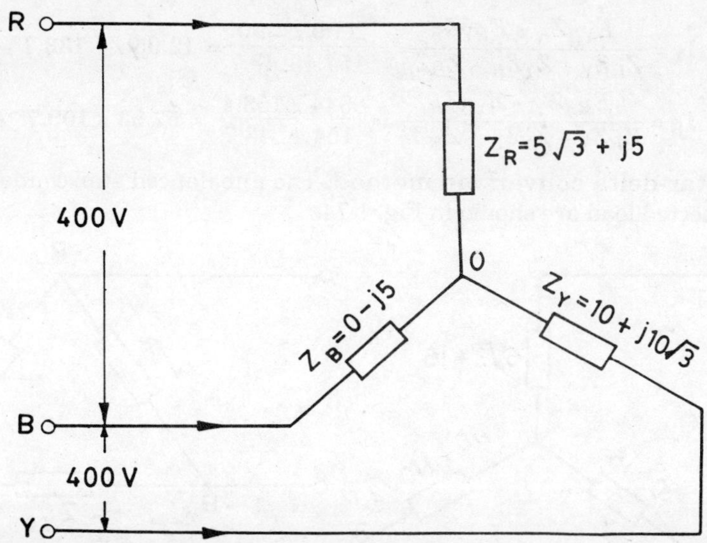

Fig. 1.73

1. By using Kirchhoff's law :

We know that

$$I_R = \frac{E_{RY}}{Z_R + Z_Y + \dfrac{Z_R Z_Y}{Z_B}} - \frac{E_{BR}}{Z_B + Z_R + \dfrac{Z_B Z_R}{Z_Y}}$$

$$= \frac{E_{RY} Z_B}{Z_R Z_B + Z_Y Z_B + Z_R Z_Y} - \frac{E_{BR} Z_Y}{Z_{BY} + Z_{RY} + Z_{BR}} \quad ...(i)$$

$$I_Y = \frac{E_{YB} Z_R}{Z_R Z_Y + Z_Y Z_B + Z_B Z_R} - \frac{E_{RY} Z_B}{Z_{RY} + Z_{YB} + Z_{BR}} \quad ...(ii)$$

and

$$I_B = \frac{E_{BR} Z_Y}{Z_R Z_Y + Z_Y Z_B + Z_B Z_R} - \frac{E_{YB} \times Z_R}{Z_{RY} + Z_{YB} + Z_{BR}} \quad ...(iii)$$

Now $Z_R Z_Y + Z_Y Z_B + Z_B Z_R$

$$= (10 \angle 30°)(20° \angle 60°) + (20 \angle 60°)(5 \angle -90°) + (5 \angle -90°)(10 \angle 30°)$$

$$= 200 \angle 90° + 100 \angle -30° + 50 \angle -60°$$

$$= j200 + (86.6 - j50) + (25 - j43.3)$$

$$= 111.6 + j106.7 = 154.4 \angle 43.7°$$

$$E_{RY} Z_B - E_{BR} Z_Y = 400 \times 5 \angle -90° - 400 \angle 120° \times 20 \angle 60°$$

$$= 2000 \angle -90° - 8000 \angle 180° = -j2000 - (-8000)$$

$$= 8000 - j2000 = 8246 \angle -14°$$

$$E_{YB} Z_R - E_{RY} Z_B = 400 \angle -120° \times 10 \angle 30° - 400 \angle 0° \times 5 \angle -90°$$

$$= 4000 \angle -90° - 2000 \angle -90°$$

$$= -j4000 + j2000 = -j2000 = 2000 \angle -90°$$

$$E_{BR} Z_Y - E_{YB} Z_R = 400 \angle 120° \times 20 \angle 60° - 400 \angle -120° \times 10 \angle 30°$$

$$= 8000 \angle 180° - 4000 \angle -90° = -8000 + j4000 = 8944 \angle 153.4°$$

Currents in each branch [from (*i*), (*ii*) and (*iii*)]

$$\mathbf{I_R} = \frac{E_{RY} Z_B - E_{BR} Z_Y}{Z_R Z_Y + Z_Y Z_B + Z_B Z_R} = \frac{8246 \angle -14°}{154.4 \angle 43.7°} = \mathbf{53.4 \angle -57.7° \ A.} \quad \textbf{(Ans.)}$$

$$\mathbf{I_Y} = \frac{E_{YB}Z_R - E_{RY}Z_B}{Z_RZ_Y + Z_YZ_B + Z_BZ_R} = \frac{2000 \angle -90°}{154.4 \angle 43.7} = \mathbf{12.95 \angle -133.7° \ A. \ (Ans.)}$$

$$\mathbf{I_B} = \frac{E_{BR}Z_Y - Z_{YB}Z_R}{Z_RZ_Y + Z_YZ_B + Z_BZ_R} = \frac{8944 \angle 153.4°}{154.4 \angle 43.7} = \mathbf{57.93 \angle 109.7° \ A. \ (Ans.)}$$

2. By using Star-delta conversion method. The unbalanced star-connected load and its equivalent delta-connected load are shown in Fig. 1.74.

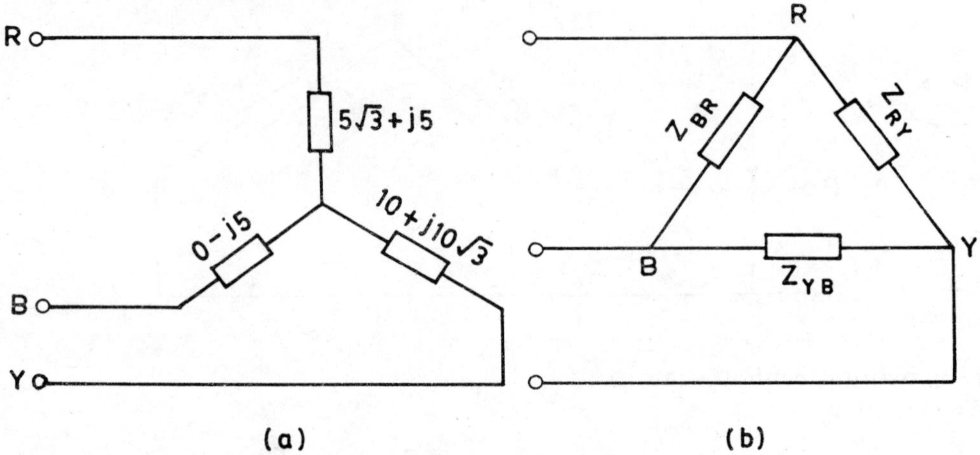

Fig. 1.74

Now, $Z_RZ_Y + Z_YZ_B + Z_BZ_R = 154.4 \angle 43.7°$ [Already calculated]

Equivalent delta-connected impedances are :

$$Z_{RY} = \frac{Z_RZ_Y + Z_YZ_B + Z_BZ_R}{Z_B} = \frac{154.4 \angle 43.7°}{5 \angle -90°} = 30.88 \angle 133.7°$$

$$Z_{YB} = \frac{Z_RZ_Y + Z_YZ_B + Z_BZ_R}{Z_R} = \frac{154.4 \angle 43.7°}{10 \angle 30°} = 15.44 \angle 13.7°$$

$$Z_{BR} = \frac{Z_RZ_Y + Z_YZ_B + Z_BZ_R}{Z_Y} = \frac{154.4 \angle 43.7°}{20 \angle 60°} = 7.72 \angle -16.3°$$

Phase currents are :

$$I_{RY} = \frac{E_{RY}}{Z_{RY}} = \frac{400 \angle 0°}{30.88 \angle 133.7} = 12.95 \angle -133.7° = (-8.95 - j9.36)$$

$$I_{YB} = \frac{E_{YB}}{Z_{YB}} = \frac{400 \angle -120°}{15.44 \angle 13.7} = 25.9 \angle -133.7° = (-17.89 - j18.72)$$

$$I_{BR} = \frac{E_{BR}}{Z_{BR}} = \frac{400 \angle 120°}{7.72 \angle -16.3°} = 51.8 \angle 136.3° = (-37.45 + j35.78)$$

Line currents for delta-connection are :

$$\mathbf{I_R} = I_{RY} - I_{BR}$$
$$= (-8.95 - j9.36) - (-37.45 + j35.78)$$
$$= 28.5 - j45.14 = \mathbf{53.38 \angle 57.7° \ A. \ (Ans.)}$$
$$\mathbf{I_Y} = I_{YB} - I_{RY} = (-17.89 - j18.72) - (-8.95 - j9.36)$$
$$= -8.94 - j9.36 = 12.94 \angle -133.7°$$

$$\mathbf{I_B} = I_{BR} - I_{YB}$$
$$= (-37.45 + j35.78) - (-17.89 - j18.72) = -19.56 + j54.5 = 57.9 \angle 109.7°$$
$$\Sigma I = I_R + I_Y + I_B = 0 \text{ ...... as a check.}$$

3. By using Maxwell's Loop Current Method :

Fig. 1.75 shows the loop or mesh currents.

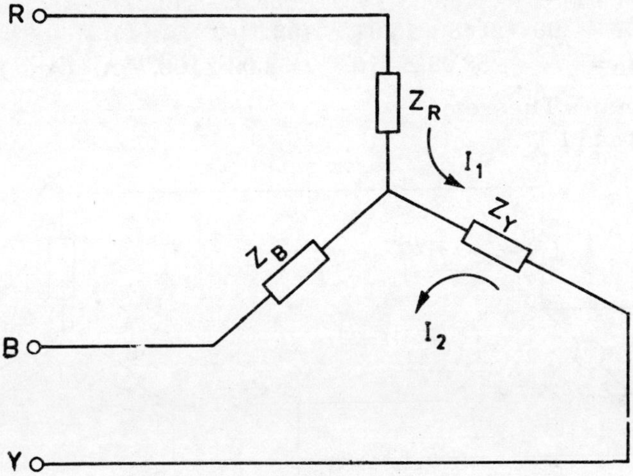

Fig. 1.75

It may be noted that $I_R = I_1$

$$I_Y = I_2 - I_1$$

and $I_B = -I_2.$

Considering drops across R and Y arms, we get

$$I_1 Z_R + Z_Y(I_1 - I_2) = E_{RY}$$

or $$I_1(Z_R + Z_Y) - I_2 Z_Y = E_{RY} \quad ...(i)$$

Similarly, considering the arms Y and B, we have

$$(I_2 - I_1) Z_Y + I_2 Z_B = E_{YB}$$
$$-I_1 Z_Y + I_2(Z_Y + Z_B) = E_{YB} \quad ...(ii)$$

Solving for I_1 and I_2, we get

$$I_1 = \frac{E_{RY}(Z_Y + Z_B) + E_{YB} Z_Y}{(Z_R + Z_Y)(Z_Y + Z_B) - Z_Y^2}$$

$$I_2 = \frac{E_{YB}(Z_R + Z_Y) + E_{RY} Z_Y}{(Z_R + Z_Y)(Z_Y + Z_B) - Z_Y^2}$$

$$I_1 = \frac{400(10 + j12.32) + 400 \angle -120 \times 20 \angle 60°}{(18.66 + j22.32)(10 + j12.32) - 400 \angle 120°}$$

$$= \frac{4000 + j4928 + 4000 - j6928}{(186.6 + j229.9 + j223.2 - 275) - (-200 + j346.4)}$$

$$= \frac{8000 - j2000}{111.6 + j106.7} = \frac{8246.2 \angle -14°}{154 \angle 43.7°}$$

$$= 53.5 \angle -57.7° = (28.59 - j45.22)$$

$$I_2 = \frac{(-200 - j346)(18.66 + j22.32) + 400(10 + j17.32)}{111.6 + j106.7}$$

$$= \frac{-373.2 - j4464 - j6456.4 + 7723 + 4000 + j6928}{111.6 + j106.7}$$

$$= \frac{8000 - j4000}{111.6 + j106.7} = \frac{8944 \angle -26.6°}{154 \angle 43.7°} = 58.08 \angle -70.3° = 19.58 - j54.7$$

$\therefore$ $\quad$ $\mathbf{I_R} = I_1 = \mathbf{53.5 \angle -57.7° \ A.}$ **(Ans.)**

$\mathbf{I_Y} = I_2 - I_1 = (19.58 - j54.7) - (28.59 - j45.22)$

$= -90 - j9.48 = \mathbf{13.07 \angle -133.5° \ A.}$ **(Ans.)**

$\mathbf{I_B} = -I_2 = -58.08 \angle -70.3° = \mathbf{58.08 \angle 109.7° \ A.}$ **(Ans.)**

4. Using Milliman's Theorem :

Refer Figs. 1.76 and 1.77.

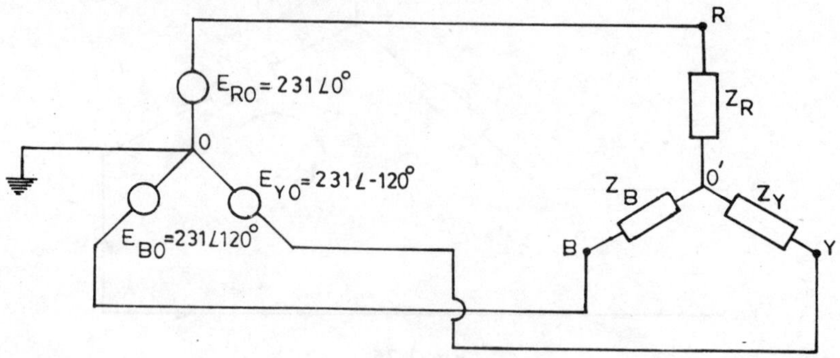

Fig. 1.76

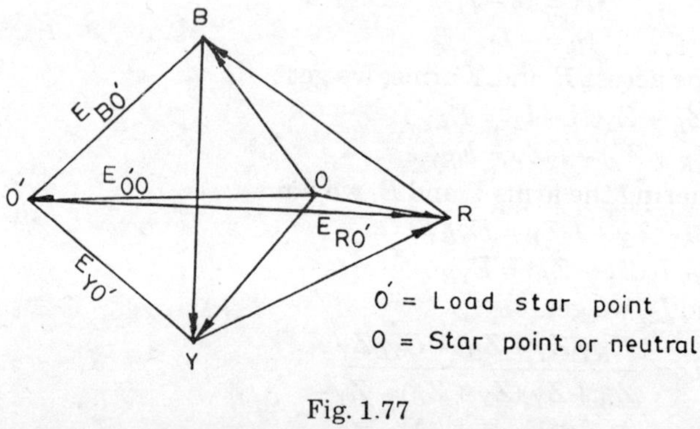

Fig. 1.77

According to Milliman's theorem, the voltage of the *load* star point O' with respect to the star-point or neutral O of the generator or supply (normally zero potential) is given by

$$E_{O'O} = \frac{E_{RO}Y_R + E_{YO}Y_Y + E_{BO}Y_B}{Y_R + Y_Y + Y_B}$$

where E_{RO}, E_{YO} and E_{BO} are the phase voltages of the generator or 3-phase supply.

Voltage across each phase of the load is

$$E_{RO'} = E_{RO} - E_{O'O}$$

$$E_{YO'} = E_{YO} - E_{O'O}$$

$$E_{BO'} = E_{BO} - E_{O'O}$$

Now, $\quad I_{RO'} = (E_{RO} - E_{O'O})\,Y_R, \quad I_{YO'} = (E_{YO} - E_{O'O})Y_Y$

$$E_{BO'} = (E_{BO} - E_{O'O})Y_B$$

Here, $$Y_R = \frac{1}{10\angle 30°} = 0.1\angle -30° = (0.0866 - j0.05)$$

$$Y_Y = \frac{1}{20\angle 60°} = 0.05\angle -60° = (0.025 - j0.0433)$$

$$Y_B = \frac{1}{5\angle -90°} = 0.2\angle 90° = (0 + j0.2)$$

$\therefore$ $$Y_R + Y_Y + Y_B = (0.0866 - j0.05) + (0.025 - j0.0433) + (0 + j0.2)$$
$$= 0.1116 + j0.1067 = 0.1544\angle 43.7°$$

Let $$E_{RO} = \frac{400}{\sqrt{3}}\angle 0° = (231 + j0)$$

$$E_{YO} = 231\angle -120° = (-115.5 - j200)$$
$$E_{BO} = 231\angle 120° = (-115.5 + j200)$$
$$E_{O'O} = \frac{231 \times 0.1\angle -30° + 231\angle -120° \times 0.05\angle -60° + 231\angle 120° \times 0.2\angle 90°}{0.1544\angle 43.7°}$$
$$= \frac{23.1\angle -30° + 11.55\angle -180° + 46.2\angle 210°}{0.1544\angle 43.7°}$$
$$= \frac{20 - j11.55 - 11.55 - 40 - j23.1}{0.1544\angle 43.7°}$$
$$= \frac{-31.55 - j34.65}{0.1544\angle 43.7°} = \frac{46.86\angle -132.3°}{0.1544\angle 43.7°}$$
$$= 303.5\angle -176° = (-302.7 - j21.2)$$
$$E_{RO'} = E_{RO} - E_{O'O} = 231 - (-302.7 - j21.2)$$
$$= 533.7 + j21.2 = 534.1\angle 2.27°$$
$$E_{YO'} = E_{YO} - E_{O'O} = -115.5 - j200 + 302.7 + j21.2$$
$$= 187.2 - j178.8 = 258.9\angle -43.7°$$
$$E_{BO'} = E_{BO} - E_{O'O} = -115.5 + j200 + 302.7 + j21.2$$
$$= 187.2 + j221.2 = 289.8\angle 49.7°$$

$\therefore$ $$I_{RO'} = 534.1\angle 2.27 \times 0.1\angle -30° = 53.41\angle -27.7°$$
$$I_{YO'} = 258.9\angle -43.7° \times 0.05\angle -60° = 12.94\angle -103.7°$$
$$I_{BO'} = 289.8\angle 49.7° \times 0.2\angle 90° = 57.96\angle 139.7°.$$

1.13. POWER FACTOR IMPROVEMENT

The power factors of different loads are as follows :

Type of load	Power factor (range)
Heating and lighting	0.95 to unity
Motor loads	0.5 to 0.9
Single-phase motors	as low as 0.4
Electric welding units	0.2 to 0.3

The power factor is given by :

$$\cos\phi = \frac{\text{kW}}{\text{kVA}} \quad \text{or} \quad \text{kVA} = \frac{\text{kW}}{\cos\phi}$$

In case of *single-phase supply*,

$$\text{kVA} = \frac{VI}{1000} \quad \text{or} \quad I = \frac{1000\ \text{kVA}}{V}$$

$\therefore$ $$I \propto \text{kVA}$$

In case of *three phase supply*,

$$\text{kVA} = \frac{\sqrt{3}\, E_L I_L}{1000} \quad \text{or} \quad I_L = \frac{1000\ \text{kVA}}{\sqrt{3}\, E_L}$$

$$\therefore \quad I_L \propto \text{kVA}$$

In each case, the *kVA is directly proportional to current.*

The higher current due to poor power factor affects the system and lead to the following undesirable results :

1. Rating of alternators and transformers are proportional to their output current hence inversely proportional to power factor, therefore large *generators and transformers are required to deliver same load but at a low power factor.*

2. When a load having a low lagging power is switched on, there is a large voltage drop in the supply voltage because of the increased voltage drop in the supply lines and transformers. This drop in voltage adversely affects the starting torque of motors and necessitates *expensive voltage stabilizing equipment* for keeping the consumer's voltage fluctuations within the statutory limits.

3. The cross-sectional area of the bus-bar, and the contact surface of switch gear is required to be *enlarged* for the same power to be delivered but at a low power factor.

4. To transmit same power at low power factor, more current will have to be carried by the transmission line or the distributor or cable. If the current density in the line is to be kept constant *the size of the conductor will have to be increased.* Thus *more copper is required to deliver the same load but at a low power factor.*

5. Copper losses are proportional to the square of the current hence inversely proportional to the square of the power factor *i.e.* more copper losses at low power factor and hence *poor efficiency.*

Thus we find that the capital cost of transformers, alternators, distributors, transmission lines etc. *increase with the low power factor.*

Causes of Low Power Factor :

(*i*) All A.C. motors (except overexcited synchronous motors and certain types of commutators motors) and transformers operate at lagging power factor.

(*ii*) Due to typical characteristic of the arc, arc lamps operate at low power factor.

(*iii*) When there is increase in supply voltage, which usually occurs during low load periods (such as lunch hours, night hours etc.) the magnetizing current of inductive reactances increases and power factor of the electrical plant as a whole decreases.

(*iv*) Arc and induction furnaces etc. operate at a very low lagging power factor.

(*v*) Due to improper maintenance and repairs of motors the power factor at which motors operate fall.

Methods of Improving Power Factor. The power factor may improve by using the following methods :

1. Use of high power factor motors.
2. Use of induction motors with phase advancers.
3. Use of static capacitors.
4. Use of capacitance boosters.
5. Use of synchronous condensers.

POWER FACTOR IMPROVEMENT

Example 1.57. *A 40-MVA, 11 kV 3-phase alternator supplies full-load at a lagging power factor of 0.6. Find the percentage increase in earning capacity if the power factor is increased to 0.9.*

Solution. Rating of the alternator = 40 MVA

Supply voltage = 11 kV

Initial power factor = 0.6 (lagging)

Final power factor after increase = 0.9 (lagging)

Percentage increase in earning capacity :

The earning capacity is proportional to the power (in MW or kW) supplied by the alternator.

MW supplied at 0.6 lagging = 40 × 0.6 = 24

MW supplied at 0.9 lagging = 40 × 0.9 = 36

Increase in MW = 36 – 24 = 12

The increase in earning capacity is proportional to 12.

∴ **Percentage increase in earning capacity** $= \frac{12}{24} \times 100 =$ **50%. (Ans.)**

Example 1.58. *A 3-phase, 40 kW, 440 V, 50 Hz induction motor operates on full-load with an efficiency of 90 per cent and at a power factor of 0.8 lagging. Calculate the total kVA rating of capacitors required to raise the full-load power factor to 0.9 lagging. What will be capacitance per phase if the capacitances are :*

(i) Delta connected *(ii) Star-connected.*

Solution. Motor output = 40 kW

Supply voltage = 440 V

Efficiency of motor = 90%

Motor power input, $P = \frac{\text{output}}{\text{efficiency}} = \frac{40}{0.9} = 44.44$ kW

Power factor 0.8 (lagging)

$$\cos \phi_1 = 0.8$$

$$\therefore \quad \phi_1 = \cos^{-1} 0.8 = 36.9°$$

$$\tan \phi_1 = 0.75$$

$$\text{Motor kVAR}_1 = P \tan \phi_1 = 44.44 \times 0.75 = 33.33$$

Power factor 0.9 (lagging)

Motor power input = 44.44 kW [As before]

[Power is same as before since the capacitors are *loss free i.e.*, they *do not absorb any power*]

$$\cos \phi_2 = 0.9$$

$$\therefore \quad \phi_2 = 25.8°$$

$$\tan \phi_2 = 0.484$$

$$\therefore \quad \text{Motor kVAR}_2 = P \tan \phi_2 = 44.44 \times 0.484 = 21.5.$$

The difference in values of kVAR is *due to the capacitors which supply leading kVAR to partially neutralize the lagging kVAR of the machine.*

∴ **Leading kVAR supplied by capacitors is**

$= \text{kVAR}_1 - \text{kVAR}_2 = 33.33 - 21.5$

= 11.83 ...... MQ or NT.

Since capacitors are loss-free, their kVAR is same as kVA

∴ Total kVA rating of capacitors = **11.83 (Ans.)**

$$\text{kVAR/capacitor} = \frac{11.83}{3} = 3.943$$

∴ VAR/capacitor = 3943.

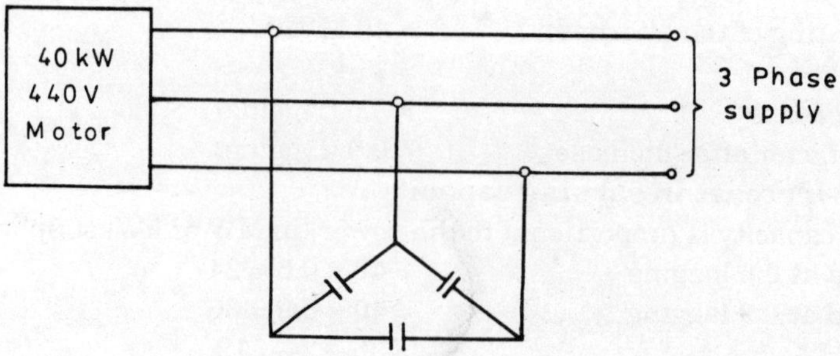

Fig. 1.78

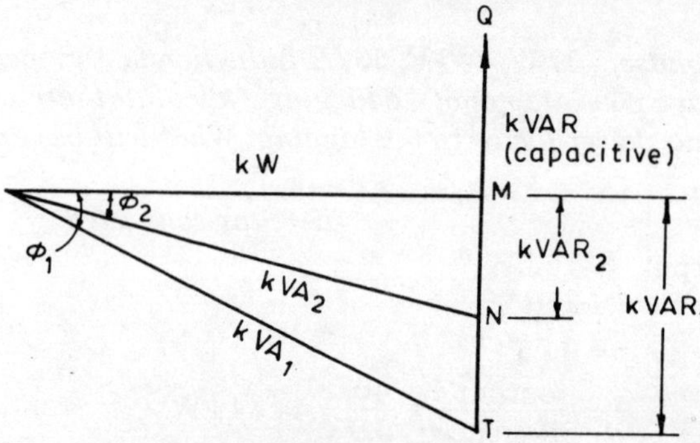

Fig. 1.79

(*i*) **Delta connection.** In delta-connection, voltage across each capacitor is 440 V.

Current drawn by each capacitor, $I_C = \dfrac{3943}{440} = 8.96$ A

Now, $$I_C = \frac{V}{X_C} = \frac{V}{\dfrac{1}{2\pi fC}} = 2\pi fCV$$

$$\therefore \quad C = \frac{I_C}{2\pi fV} = \frac{6.96}{2\pi \times 50 \times 440} = 50.35 \times 10^{-6} \text{ F} = 50.35\ \mu\text{F}$$

Hence, **capacitance of each capacitor = 50.35 μF. (Ans.)**

(*ii*) **Star-connection.** In star-connection, voltage across each capacitor

$$= \frac{440}{\sqrt{3}} = 254 \text{ V}$$

Current drawn by each capacitor,

$$I_C = \frac{3943}{254} = 15.52 \text{ A}$$

Now, $$I_C = \frac{V}{X_C} = \frac{V}{\dfrac{1}{2\pi fC}} 2\pi fCV$$

$$\therefore \quad C = \frac{I_C}{2\pi fV} = \frac{15.52}{2\pi \times 50 \times 440} = 112.3 \times 10^{-6} \text{ F} = 112.3\ \mu\text{F}$$

Hence, **capacitance of each capacitor = 112.3 μF. (Ans.)**

Example 1.59. *Three impedance coils each having a resistance of 10 Ω and a reactance of 7.5 Ω are connected in star to a 400 V, 3-phase 50 Hz supply. Calculate :*

(*i*) *Power factor* (*ii*) *Line current, and*

(*iii*) *Power supplied.*

If three capacitors, each of the same capacitance, are connected in delta to the same supply so as to from parallel circuit with the above impedance coils, calculate the capacitance of each capacitor to obtain resultant power factor of 0.9 lagging.

Solution. Supply voltage, $E_L = 400$ V [Line-to-line]

Phase resistance, $R_{ph} = 10\ \Omega$

Phase reactance, $X_{ph} = 7.5\ \Omega$

Phase voltage, $E_{ph} = \dfrac{400}{\sqrt{3}} = 231$ V

Phase impedance, $Z_{ph} = \sqrt{R_{ph}^2 + X_{ph}^2} = \sqrt{10^2 + (7.5)^2} = 12.5\ \Omega$

(*i*) **Power factor,** $\cos \phi_1 = \dfrac{R_{ph}}{Z_{ph}} = \dfrac{10}{12.5} = \mathbf{0.8.}$ **(Ans.)**

Phase current, $I_{ph} = \dfrac{E_{ph}}{Z_{ph}} = \dfrac{231}{12.5} = \mathbf{18.48\ A.}$ **(Ans.)**

(*ii*) Line current, (I_L) = phase current, (I_{ph})

∴ $I_L = \mathbf{18.48\ A.}$ **(Ans.)**

(*iii*) **Power supplied,** $P = \sqrt{3} E_L I_L \cos \phi = \sqrt{3} \times 400 \times 18.48 \times 0.8 = \mathbf{10243\ W.}$ **(Ans.)**

Motor $\text{VAR}_1 = \sqrt{3} \times E_L I_L \sin \phi_1 = \sqrt{3} \times 400 \times 18.48 \times 0.6 = \mathbf{7682.}$ **(Ans.)**

When capacitors are connected :

Power factor, $\cos \phi_2 = 0.9$

∴ $\phi_2 = 25.8°$

and $\tan \phi_2 = 0.4834$

Since capacitors themselves do not absorb any power, power remains the same *i.e.*, 10243 W even when capacitors are connected. The only thing that changes is the VAR.

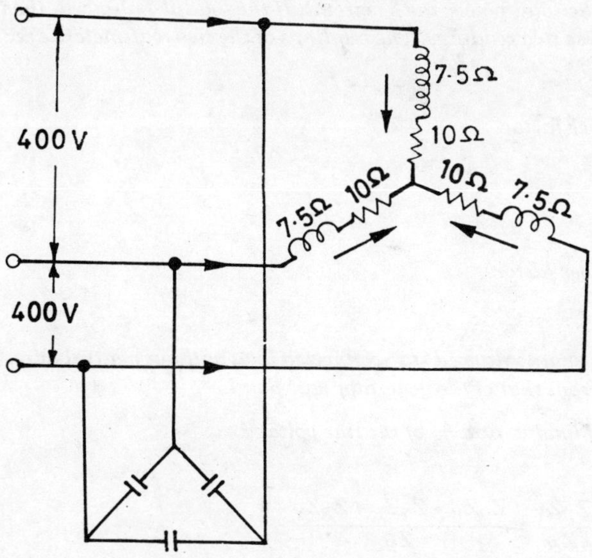

Fig. 1.80

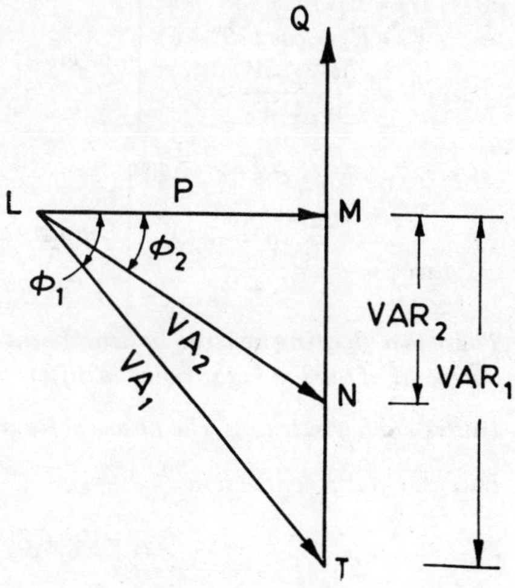

Fig. 1.81

Now $\quad VAR_2 = P \tan \phi_2 = 10243 \times 0.4834 = 4951$

Leading VAR supplied by three capacitors is

$$= VAR_1 - VAR_2 = 7682 - 4951 = 2731 = MQ \text{ or } NT \text{ (Fig. 1.81)}$$

$$VAR/capacitor = \frac{2731}{3} = 910.3$$

For delta connection voltage across each capacitor is 400 V

$$\therefore \quad I_C = \frac{910.3}{400} = 2.275 \text{ A}$$

Also

$$I_C = \frac{V}{\frac{1}{2\pi fC}} = V \times 2\pi fC$$

$$\therefore \quad C = \frac{I_C}{V \times 2\pi f} = \frac{2.275}{400 \times 2 \times 3.142 \times 50} = 18.10 \times 10^{-6} \text{ F or } 18.10 \ \mu\text{F}$$

Hence, **capacitance of each capacitor = 18.10 μF. (Ans.)**

HIGHLIGHTS

1. *In a star-connected system*

$$E_{ph} = \frac{E_L}{\sqrt{3}}$$

$$I_{ph} = I_L$$

$$P = \sqrt{3} E_L I_L \cos \phi$$

2. *In a delta-connected system*

$$E_{ph} = E_L$$

$$I_{ph} = \frac{I_L}{\sqrt{3}}$$

$$P = \sqrt{3} E_L I_L \cos \phi.$$

3. *Two-wattmeters method is generally used to measure 3-phase power. In this method, the current coils of the two wattmeters are connected in any two lines and their potential coils to the remaining third line. The sum of the two wattmeters readings gives the total power in the circuit. If the load is balanced, then its power factor can also be calculated from these two readings. The readings of the two wattmeters are :*

(*i*) *Lagging power factor*

$$W_1 = E_L I_L \cos (30° - \phi)$$

$$W_2 = E_L I_L \cos (30° + \phi)$$

$$\tan \phi = \frac{\sqrt{3}(W_1 - W_2)}{W_1 + W_2}$$

(*ii*) *Leading power factor*

$$W_1 = E_L I_L \cos (30° + \phi)$$

$$W_2 = E_L I_L \cos (30° - \phi)$$

$$\tan \phi = -\frac{\sqrt{3}(W_1 - W_2)}{W_1 + W_2}$$

4. *There is a 'floating neutral' or 'star' point in an unbalanced star-connected load without neutral wire. The potential of the load star point is different from that of the generator star point.*

Under such conditions, the phase voltage of load is **not** $\frac{1}{\sqrt{3}}$ *of the line voltage.*

5. *In a star-delta connection*

$$Z_{RY} = Z_R + Z_Y + \frac{Z_R Z_Y}{Z_B} = \frac{Z_R Z_B + Z_Y Z_B + Z_R Z_Y}{Z_B}$$

$$Z_{YB} = Z_Y + Z_B + \frac{Z_Y Z_B}{Z_R} = \frac{Z_Y Z_R + Z_B Z_R + Z_Y Z_B}{Z_R}$$

$$Z_{BR} = Z_B + Z_R + \frac{Z_B Z_R}{Z_Y} = \frac{Z_B Z_Y + Z_R Z_Y + Z_B Z_R}{Z_Y}.$$

6. *In a single-phase as well as in a three-phase system, the kVA is directly proportional to the current I. The disadvantage of a lower power factor is that the current required for a given power is very high, which fact leads to many undesirable results. The power factor may be improved by the following :*
   - *Static capacitors*
   - *Phase advancers*
   - *Synchronous capacitors*
   - *Capacitor boosters*
   - *High power factor motors.*

OBJECTIVE TYPE QUESTIONS

A. Choose the Correct Answer :

1. In two wattmeters method of power measurement, one of the wattmeters will show negative reading when the load power factor angle is strictly
   (*a*) less than 30° (*b*) less than 60°
   (*c*) greater than 30° (*d*) greater than 60°. **[U.P.S.C., 1996]**
2. Three identical impedances are connected in delta to a 3-phase supply of 400 V. The line-current is 34.65 A and the total power taken is 14.4 kW. The resistance of the load in each phase (in ohm) is
   (*a*) 20 (*b*) 16
   (*c*) 12 (*d*) 10. **[U.P.S.C., 1993]**
3. The power in a 3-phase system is given by $\sqrt{3} V_L I_L \cos \phi$ where ϕ is the phase angle between
   (*a*) line-voltage and line current (*b*) phase voltage and phase current
   (*c*) line voltage and phase current (*d*) phase voltage and line current.
4. Which of the following statements associated with 3-phase delta connected circuits is *true* ?
   (*a*) Line voltage is equal to phase voltage (*b*) Line current is equal to phase current
   (*c*) Line voltage is $\sqrt{3}$ time of phase voltage (*d*) Line currents are 60° apart.
5. In a 3-phase supply, floating neutral is undesirable because it may result in ...... across the load.
   (*a*) unequal line voltages (*b*) high voltage
   (*c*) low voltage.
6. Phase reversal in a 4-wire unbalanced load supplied from a balanced 3-phase supply causes change in
   (*a*) the power consumed (*b*) magnitude of phase currents
   (*c*) only the magnitude of the neutral current
   (*d*) magnitude as well as phase angle of the neutral current.
7. A 3-phase star connected symmetrical load consumes *P* watts of power from a balanced supply. If the same load is connected in delta to the same supply, the power consumption will be
   (*a*) *P* (*b*) $\sqrt{3}P$
   (*c*) 3*P* (*d*) not determined from the given data.
   [U.P.S.C., 1995]
8. Three unequal impedances are connected in star to a 3-phase system. The sum of three line currents will be
   (*a*) equal to the each line current (*b*) zero
   (*c*) none of these.

9. Three equal impedances are first connected in delta across a 3 – ϕ balanced supply. If the same impedances are connected in star across the same supply

(*a*) phase currents will be one-third (*b*) line currents will be one-third

(*c*) power consumed will be one-third (*d*) none of the above.

10. Which of the following is a four wire system ?

(*a*) Delta (*b*) Star

(*c*) Both delta and star (*d*) Neither delta nor star.

11. In a 3-phase balanced star-connected load, neutral current is equal to

(*a*) zero (*b*) I_P

(*c*) I_L (*d*) unpredictable.

12. Which of the following equations is valid for a 3-phase 4-wire balanced star-connected load ?

(*a*) $I_R + I_Y + I_B = I_N = 0$ (*b*) $I_R + I_Y - I_B = I_N$

(*c*) $I_R - I_Y + I_B = I_N$ (*d*) $\frac{V_B + V_R + V_Y}{Z} = I_N$.

13. Three unequal impedances are connected in delta to a 3-phase, 3-wire system.

(*a*) The voltages across the three phases will be different

(*b*) Both of the phase currents and line currents will be unbalanced

(*c*) Phase currents will be unbalanced but the line currents will be balanced

(*d*) None of the above.

14. The relationship between the line and phase voltages of a delta-connected circuit is given by

(*a*) $V_L = V_P$ (*b*) $V_L = \sqrt{3} V_P$

(*c*) $V_L = \frac{V_P}{\sqrt{2}}$ (*d*) $V_L = \frac{2}{\pi} V_P$. **[B.T.E. M.P. 1996]**

15. In case of a delta connected load, if one resistor is removed, the power will become

(*a*) zero (*b*) one-third

(*c*) two-third (*d*) none of the above.

16. Readings of 1154 and 577 watts are obtained when two wattmeters method was used on a balanced load. The delta connected load impedance for a system of 100 V will be

(*a*) 15 ∠± 30° (*b*) 15 ∠+ 30°

(*c*) 15 ∠– 30° (*d*) 15 ∠+ 90°. **[U.P.S.C., 1993]**

17. Which of the following statements is true about two-wattmeters method for power measurement in 3-phase circuit ?

(*a*) Power can be measured using two wattmeters method only for star-connected 3-phase circuits

(*b*) When two wattmeters show identical readings, the power factor is 0.5

(*c*) When power factor is unity, one of the wattmeter reads zero

(*d*) When the readings of the two wattmeters are equal but of opposite sign, the power factor is zero. **[U.P.S.C., 1992]**

18. The minimum number of wattmeters to measure power in a 3-phase unbalanced star-connected load is

(*a*) one (*b*) two

(*c*) three (*d*) four.

19. While measuring power in a three phase load by two-wattmeters method, the readings of two wattmeters will be equal and opposite when

(*a*) *pf* is unity (*b*) load is balanced

(*c*) phase angle is between 60° and 90° (*d*) the load is purely inductive. **[U.P.S.C., 1995]**

20. In the measurement of 3-phase power by two-wattmeters method, if the two wattmeter readings are equal the power factor of the circuit is

(*a*) 0.8 lagging (*b*) 0.8 leading

(*c*) zero (*d*) unity. **[A.M.I.E., 1998]**

21. Which of the following are the necessary conditions for an entire 3-ϕ system to be balanced ?
1. The line voltages are equal in magnitude
2. The phase differences between successive line voltages are equal
3. The impedances in each of the phases are identical.

Select the correct answer using the codes given below :

Codes :

(*a*) 1, 2 and 3 (*b*) 1 and 3
(*c*) 1 and 2 (*d*) 2 and 3. **[U.P.S.C., 1996]**

22. The power delivered to a 3-phase load can be measured by the use of 2 wattmeters only when the
(*a*) load is balanced (*b*) load is unbalanced
(*c*) 3-phase load is connected to the source through 3-wires
(*d*) 3-phase load is connected to the source through 4-wires. **[U.P.S.C., 1995]**

23. The ratio of the readings of two wattmeters connected to measure power in a balanced 3-phase load is 5 : 3 and the load is inductive. The power factor of the load is
(*a*) 0.917 lead (*b*) 0.917 lag
(*c*) 0.6 lead (*d*) 0.6 lag. **[U.P.S.C., 1994]**

24. In two phase system phase voltages differ by
(*a*) 60° (*b*) 90°
(*c*) 120° (*d*) 180°.

25. W_1 and W_2 are the readings of two wattmeters used to measure power of a 3-phase balanced load. The reactive power drawn by the load is
(*a*) $W_1 + W_2$ (*b*) $W_1 - W_2$
(*c*) $\sqrt{3}(W_1 + W_2)$ (*d*) $\sqrt{3}(W_1 - W_2)$. **[A.M.I.E., 1997]**

26. In a two wattmeters method of measuring power in a 3-phase system one of the wattmeters reads negative implying
(*a*) wattmeter connection is faulty (*b*) load is unbalanced
(*c*) power flow is in the reverse direction (*d*) power factor is less than 0.5. **[A.M.I.E., 1998]**

27. v_{RN}, v_{YN} and v_{BN} are the instantaneous line to neutral voltages and i_R, i_Y and i_B are instantaneous line currents in a balanced three phase circuit, the computation $v_{RN}(i_Y - i_B) - (v_{YN} - v_{BN})i_R$ will yield a quantity proportional to the
(*a*) active power (*b*) power factor
(*c*) reactive power (*d*) complex power. **[GATE, 1993]**

28. If the R-phase of a 3-phase star connected alternator becomes reversely connected by mistake, it will affect
(*a*) E_{RY} and E_{YB} (*b*) E_{YB} and E_{BR}
(*c*) E_{RY} and E_{BR} (*d*) E_{RY}, E_{YB} and E_{BR}.

29. When phase sequence at the 3-phase load is reversed
(*a*) phase currents are changed in magnitude
(*b*) phase currents change in phase angle but not in magnitude
(*c*) total power consumed is changed
(*d*) phase powers are changed.

30. Lamp load is being supplied from a 3-phase, 4-wire, 230/400 V ac supply system. If a 3-phase motor is now switched on across the same supply, then neutral current will
(*a*) increase (*b*) decrease
(*c*) remain unchanged (*d*) unpredictable.

31. Lamps of 40 W, 100 W and 200 W ratings are connected in different phases of a 3-phase, 4-wire supply. If the neutral wire breaks, then the lamp likely to fuse first will be
(*a*) 200 W (*b*) 100 W
(*c*) 40 W (*d*) all lamps.

32. The instantaneous values of currents in phases R and Y of a 3-phase system are 25 A each. For a phase sequence of BRY, the instantaneous value of current in phase B is

(*a*) 25 A (*b*) 50 A

(*c*) 12.5 A (*d*) 43.3 A.

33. The power measurement in balanced 3-ϕ circuit can be done by

(*a*) one wattmeter method only (*b*) two wattmeters method only

(*c*) three wattmeters method only (*d*) any one of the above. **[B.T.E. M.P., 1998]**

ANSWERS

1. (*d*)	**2.** (*c*)	**3.** (*b*)	**4.** (*a*)	**5.** (*a*)	**6.** (*d*)	**7.** (*c*)
8. (*b*)	**9.** (*c*)	**10.** (*b*)	**11.** (*a*)	**12.** (*a*)	**13.** (*b*)	**14.** (*a*)
15. (*c*)	**16.** (*b*)	**17.** (*d*)	**18.** (*b*)	**19.** (*d*)	**20.** (*d*)	**21.** (*a*)
22. (*c*)	**23.** (*b*)	**24.** (*b*)	**25.** (*d*)	**26.** (*d*)	**27.** (*c*)	**28.** (*c*)
29. (*b*)	**30.** (*c*)	**31.** (*c*)	**32.** (*b*)	**33.** (*d*).		

B. Say 'Yes' or 'No' :

1. Polyphase system means the system which consists of numerous windings or circuits.

2. Electrical displacement = $\dfrac{\text{180 electrical degrees}}{\text{number of phases}}$.

3. A load circuit is said to be balanced when the loads (impedances) connected in various phases are same in magnitude as well as in phase.

4. Rotating magnetic field can be set up by passing polyphase current through stationary coils.

5. Polyphase system is less capable and reliable than single phase system.

6. Parallel operation of polyphase alternators is simple as compared to that of single phase alternators because of pulsating reaction in single-phase alternators.

7. Double script notation is usually preferred in A.C. electrical circuits.

8. Phase sequence is the order or sequence in which the currents or voltages in different phases attain their maximum values one after the other.

9. If the phase sequence is reversed by interchanging any two terminals of the 3-phase supply, the motor would rotate in the same direction.

10. Star-connected system is obtained by joining together similar ends either the start or the finish, the other ends are joined to the line wires.

11. The common point at which similar (start or finish) ends, are connected is called the neutral or star point.

12. In a balanced star-connected system line voltage = phase voltage.

13. In a balanced delta-connected system line current is equal to phase current.

14. Delta-connected system is best suited for rotary converters.

15. Three wattmeters method is useful for measuring power in 3-phase, 3-wire load circuits.

16. Power in a 3-phase circuit = $\sqrt{3}\, E_L I_L \cos\phi$.

17. In the case of balanced three-phase circuit the reactive power can be determined by use of one wattmeter.

18. A load having a low lagging power factor gives a poor voltage regulation.

19. Static capacitors are installed to improve the power factor of a group of A.C. motors.

20. Static capacitors are practically loss free.

ANSWERS

1. Yes	**2.** No	**3.** Yes	**4.** Yes	**5.** No	**6.** Yes	**7.** Yes
8. Yes	**9.** No	**10.** Yes	**11.** Yes	**12.** No	**13.** No	**14.** Yes
15. No	**16.** Yes	**17.** Yes	**18.** Yes	**19.** Yes	**20.** Yes.	

THEORETICAL QUESTIONS

1. State the advantages of A.C. polyphase supply system over single-phase system.
2. Why is the number of phases in a polyphase system always three rather than any other number ?
3. Explain clearly what is meant by 'phase sequence' of 3-phase voltages.
4. What are the two systems in which three-phases can be connected ? What are the advantages and disadvantages of each system ?
5. What are the advantages of inter-connecting the 3-phases ?
6. Derive an expression for power in a 3-phase star-connected system in terms of (*i*) phase values and (*ii*) line values of voltages and currents.
7. Derive the numerical relationship between line and phase currents for a balanced 3-phase delta-connected load.
8. Deduce an expression for power in a 3-phase balanced load circuit. Show that it is the same irrespective of the load being connected in star or delta.
9. Compare the star and delta connections in a 3-phase system.
10. Enumerate various methods for 3-phase power measurement, and describe in detail two wattmeters method for 3-phase power measurement.
11. Prove that the power in a balanced 3-phase circuit can be deduced from the readings of two wattmeters. Draw the relevant connection diagram and vector diagram.
12. Discuss two-wattmeters method for power measurement in 3-phase system and obtain a relation for the power factor.
13. Derive an expression for star-connected impedances equivalent to three delta-connected impedances.
14. What does the following signify in case of measurement of balanced 3-phase power by two-wattmeters method :
    (*i*) When the reading of one wattmeter is zero.
    (*ii*) When the readings of two wattmeters are equal.
    (*iii*) When the readings of two wattmeters are equal and opposite.
    In each case give reasons to your answer.
15. Explain what will be the consequences of disconnecting the neutral in a 3-phase, 4-wire system.
16. What are the disadvantages of having a low lagging power factor in a system.
17. Enumerate the various equipments which can be employed to improve the power factor.

UNSOLVED EXAMPLES

Star/Delta Connections

1. Three equal impedances each having a resistance of 20 Ω and reactance of 15 Ω are connected in star to a 400 V, 3-phase, 50 Hz system. Calculate :
   (*i*) The line current, (*ii*) The power factor, and
   (*iii*) The power consumed. [**Ans.** 9.24 A ; 0.8 (lag) ; 5120 W]
2. Three equal impedances are star-connected to a 3-phase, 50 Hz supply. If the resistance and reactance of each branch are 25 Ω and 38 Ω respectively, calculate :
   (*i*) The line current, and (*ii*) The power consumed. [**Ans.** 5.28 A ; 2086 W]
3. Three resistances of 20 Ω each are connected in star across the 400 V, 3-phase A.C. supply. Calculate :
   (*i*) The line and phase currents, (*ii*) Phase voltages, and
   (*iii*) Total power taken. [**Ans.** (*i*) 11.55 A ; (*ii*) 231 V ; (*iii*) 8.0 kW]

4. A star-connected, 3-phase load consists of three similar impedances. When the load is connected to a 3-phase, 500 V, 50 Hz supply, the line current is 28.85 A and the power factor is 0.8 lagging, calculate :

(*i*) The total power taken by the load, and (*ii*) The resistance of each phase of the load.

[**Ans.** 20 kW ; 8 Ω]

5. Three non-inductive resistances each of 50 Ω are connected in star across 400 V, 3-phase A.C. supply. Calculate the current through each. Calculate the current if they were connected in delta across the same supply.

[**Ans.** 4.62 A ; 8 A]

6. Three identical coils are connected in star to a 200 V (line voltage), 3-phase, A.C. supply and each coil takes 200 W. The power factor is 0.8 (lagging), calculate :

(*i*) The line current, (*ii*) Impedance, and

(*iii*) Resistance and inductance of each coil. [**Ans.** 2.165 A ; 53.334 Ω, 42.662 Ω, 0.102 H]

7. In a three-phase, 3-wire system with star-connected load the impedance of each phase is $(6 + j8)$ Ω. If the line voltage is 230 V, calculate :

(*i*) The line current, and (*ii*) The power absorbed by each phase.

[**Ans.** 13.3 A ; 1067 W]

8. A balanced star-connected load of $(8 + j6)$ Ω per phase is connected to 3-phase, 230 V supply. Find :

(*i*) Line current, (*ii*) Power factor,

(*iii*) Power, (*iv*) Reactive volt-amperes, and

(*v*) Total volt-amperes. [**Ans.** 13.28 A ; 0.8 (lag) ; 4.232 W ; 3174 ; 5290]

9. A star-connected, 5000 V, 3-phase alternator is supplying 3,000 kW at power factor of 0.8. Calculate the active and reactive components of the current in each phase. [**Ans.** 346.2 A ; 260 A]

10. In a 3-phase, 4-wire system, two phases have currents of 10 A and 6 A in lagging power factors of 0.8 and 0.6 respectively, while the third phase is open-circuited. Calculate the current in the neutral and sketch the vector diagram.

[**Ans.** $7 \angle -73° 24'$]

11. In a star-connected load each phase consists of a resistance of 100 Ω in parallel with a capacitor of capacitance 31.8 μF. When it is connected to a 416 V, 3-phase, 50 Hz supply, calculate :

(*i*) The line current, (*ii*) The power factor,

(*iii*) The power absorbed, and (*iv*) The total kVA.

[**Ans.** 3.39 A ; 0.707 (leading) ; 1.728 kW ; 2.443 kVA]

12. Three identical coils connected in delta across 400 V, 50 Hz, 3- phase supply takes a line current of 17.32 A at power factor 0.5 lagging. Determine :

(*i*) The current in each phase, and

(*ii*) Resistance, inductance and impedance of each phase winding. [**Ans.** 10 A ; 20 Ω ; 0.11 H, 40 Ω]

13. A 220 V, 3-phase voltage is applied to a balanced delta-connected 3-phase load of phase impedance $(15 + j20)$ Ω. Find :

(*i*) The phasor current in each line. (*ii*) What is the power consumed per phase ?

(*iii*) What is the phasor sum of the three line currents ? Why does it have this value ?

[**Ans.** 15.24 A ; 1161.6 W ; zero]

14. Calculate (*i*) line current and (*ii*) the total power absorbed when three coils each having a resistance of 10 Ω and reactance of 7 Ω are connected (*a*) in star and (*b*) in delta across a 400 V, 3-phase supply.

[**Ans.** 18.93 A, 10750 W ; 56.7 A, 32250 W]

15. A delta-connected balanced 3-phase load is supplied from a 3-phase 400 V supply. The line current is 20 A and power taken by the load is 10000 W. Find :

(*i*) Impedance in each branch ; and

(*ii*) The line current, power factor and power consumed if the same load is connected in star.

[**Ans.** 34.6 Ω, 20/3 A, 0.7217, 3330 W]

16. A balanced 3-phase load consists of resistances of 4 Ω each. Determine the total power when the resistances connected to 400 V supply are :
(*i*) Star connected (*ii*) Delta connected. [**Ans.** 40 kW ; 120 kW]

17. Three non-inductive resistances, each of 100 Ω are connected in star across 400 V supply. Calculate the current through each. What would be the current through each if they are connected in delta across the same supply. [**Ans.** 2.31 A ; 4 A]

18. Three 100 Ω non-inductive resistances are connected in (*i*) star (*ii*) delta across a 400 V, 50 Hz, 3-phase mains. Calculate the power taken from the supply system in each case. In the event of one of the three resistances opened, what would be the value of the total power taken from the mains in each of the two cases. [**Ans.** 1600 W ; 4800 W]

19. A 3-phase delta connected load ; each phase of which has an inductive reactance of 40 Ω and a resistance of 25 Ω, is fed from the secondary of a 3-phase star-connected transformer, which has phase voltage of 240 V. Draw the circuit diagram of the system and calculate :
(*i*) The current in each phase of the load, (*ii*) The voltage across each phase of the load,
(*iii*) The current in the transformer secondary windings, and
(*iv*) The total power taken from the supply and its power factor. [**Ans.** 8.8 A ; 415.7 V ; 15.3 A ; 5820 W]

20. Three similar resistors are connected in star across 400 V, 3-phase supply. The line current is 5 A. Calculate the value of each resistor. To what value line voltage be changed to obtain the same current with the resistors connected in delta ? [**Ans.** 46.2 Ω ; 133.3 V]

21. A 3-phase, star-connected system with 230 V between each phase and neutral has a resistance of 4, 5 and 6 Ω respectively in three phases. Calculate :
(*i*) The current flowing in each phase, (*ii*) The neutral current, and
(*iii*) The total power absorbed. [**Ans.** 57.5 A, 46 A, 38.3 A ; 16.71 A ; 32610 W]

22. In a 3-phase, 4 wire system there is a balanced 3-phase motor load taking 20 kW at a power factor of 0.8 lagging while lamps connected between phase conductors and the neutral are taking 5, 4 and 10 kW respectively. Voltage between line conductors is 430 V. Calculate the current in each conductor and in the neutral wire of the feeder supplying these loads. [**Ans.** 51.2 A ; 47.5 A ; 80 A ; 22.4 A]

Power Measurement in 3-phase Circuits

23. The power input to a 3-phase induction motor is read by two wattmeters. The readings are 860 W and 240 W. What is the power factor of the motor ? [**Ans.** 0.7155 (lag)]

24. While performing a load test on a 3-phase wound rotor induction motor by two wattmeters method, the readings obtained on two wattmeters were + 12.5 kW and – 4.8 kW and the line voltage was 440 V. Calculate :
(*i*) Power drawn by the motor, (*ii*) Power factor, and
(*iii*) Line current. [**Ans.** 7.7 kW ; 0.2334 ; 43.3 A]

25. The input power to a 3-phase delta-connected motor was measured by two wattmeters method. The readings were 20.8 kW and – 6.8 kW and the line voltage was 400 V. Calculate :
(*i*) Input power, (*ii*) Power factor, and
(*iii*) Line current. [**Ans.** 14 kW ; 0.281 ; 72 A]

26. A 3-phase, 500 V motor load has a power factor of 0.4. Two wattmeters connected to measure the power show the input to be 30 kW. Find the reading on each instrument. [**Ans.** 35 kW ; – 5 kW]

27. The power in a 3-phase circuit is measured by two wattmeters. If the total power is 100 kW and power factor is 0.66 leading, what will be the reading of each wattmeter ? For what power factor will one of the wattmeters read zero ? [**Ans.** 17.15 kW ; 82.85 kW ; p.f. = 0.5]

28. Two wattmeters used to measure the power input in a 3-phase circuit indicate 1000 W and 500 W respectively. Find the power factor of the circuit
(*i*) when both wattmeter readings are positive ; and
(*ii*) when the latter is obtained by reversing the current coil connections.
[**Ans.** 0.866 (lag) ; 0.1889 (lag)]

29. Two wattmeters connected to read the total power in a 3-phase system supplying a balanced load read 10.5 kW and – 2.5 kW respectively. Calculate the total power and the power factor. Draw suitable phasor diagram, explain the significance of

(*i*) equal wattmeter readings ; and (*ii*) a zero reading on one wattmeter.

[**Ans.** 8 kW ; 0.3348 (lag) (*i*) unity p.f., (*ii*) 0.5 p.f.]

30. The power input to a 1000 V, 3-phase induction motor running on full-load is measured by two wattmeters which indicate 300 kW and 100 kW respectively. Determine :

(*i*) Input, (*ii*) Line current,
(*iii*) Power factor, and (*iv*) Output if the efficiency of the motor is 92%.

[**Ans.** 400 kW ; 303.5 A ; 0.756 (lag) ; 368 kW]

31. Two wattmeters are used for measuring the power input and the power factor of an over-excited synchronous motor. If the readings of the meters are – 2.0 kW and + 7.0 kW respectively, calculate the input and power factor of the motor. [**Ans.** 5 kW ; 0.3057 (leading)]

32. The power input to a 2-kV, 50 Hz, 3-phase motor running on full-load at an efficiency of 90 per cent is measured by two wattmeters which indicate 300 kW and 100 kW respectively. Calculate :

(*i*) Input, (*ii*) Power factor,
(*iii*) Line current, and (*iv*) Output. [**Ans.** 400 kW ; 0.756 (lag) ; 153 A ; 360 kW]

33. In a balanced three-phase system power is measured by two-wattmeters method and the ratio of the two wattmeter readings is 2 : 1. Find the power factor of the system. [**Ans.** 0.866 (log)]

34. The power input of a synchronous motor is measured by two wattmeters both of which indicate 50 kW. If the power factor of the motor be changed to 0.866 leading, determine the readings of the two wattmeters, the total input power remaining the same. Draw the vector diagram for the second condition of load. [**Ans.** 33.33 kW ; 66.67 kW]

35. Each phase of a 3-phase delta-connected load consists of an impedance $Z = 20 \angle 60° \ \Omega$. The line voltage is 440 V at 50 Hz. Calculate the power consumed by each phase impedance and the total power. What will be the readings of the two wattmeters connected ? [**Ans.** 4840 W ; 14520 W ; 14520 W. zero]

36. A 3-phase 3-wire, 415 V supplies a balanced load of 20 A at power factor 0.8 lagging. Two wattmeters are used to measure power. Calculate :

(*i*) Power, (*ii*) Reading of wattmeter No 1, and
(*iii*) Reading of wattmeter No. 2. [**Ans.** 11.5 kW ; 8240 W ; 3260 W]

37. A balanced load is supplied from a 3-phase, 400 V, 3-wire system whose power is measured by two wattmeters. If the total power supplied is 26 kW at 0.75 power factor lagging, find the readings of each of the two wattmeters. [**Ans.** 19.62 kW ; 6.38 kW]

38. A 3-phase 4-wire, star-connected system supplies only non-inductive load. The current in line R is 8 A, the current in line Y is 10 A and the current in line B is 6 A. The voltage from each line to neutral is 120 V. Find :

(*i*) Wattage shown by each of the three wattmeters, and
(*ii*) Power taken by the lighting load. [**Ans.** 960 W, 1200 W, 720 W ; 2880 W]

39. Three identical coils, each having a resistance of 20 Ω and reactance of 20 Ω are connected in (*i*) star ; and (*ii*) delta across 440 V, 3-phase supply. Calculate for each method of connection the line current and reading on each of the two wattmeters connected to measure power.

[**Ans.** 8.98 A ; 3817.5 W, 1022.5 W ; 26.95 A ; 12452.5 W ; 3067.5 W]

40. A balanced star-connected load, each having a resistance of 10 Ω and inductive reactance of 30 Ω is connected to a 400 V, 50 Hz supply. The phase sequence is RYB. Two wattmeters connected to read total power have their current coils connected in the red and blue lines respectively. Calculate the reading of each wattmeter. Draw the circuit and vector diagrams. [**Ans.** W_1 = – 585 W ; W_2 = 2185 W]

41. Three equal star-connected inductors take 8 kW at power factor 0.8 when connected to a 460 V, 3ϕ, 3-wire supply. Give the connection diagram of 2 single-phase wattmeters to measure the power in the circuit. Allocate the reading of each wattmeter. Find the line currents if one of the inductor is short-circuited. Draw the vector diagrams of the currents and voltages under the condition.

[**Ans.** 5732 W, 2268 W ; 21.74 A, 21.74 A]

42. A 3-phase, 3-wire balanced load with a lagging power factor is supplied at 400 V (between lines). A single-phase wattmeter (scaled in kW) when connected with its current coil in the *R*-line and the voltage coil between *R* and *Y* lines gives a reading of 6 kW. When the same terminals of the same voltage coil are switched over to *Y*-line and *B*-line, the current-coil connections remaining the same, the reading of the wattmeter remains unchanged. Calculate the line current and power factor of the load. Phase sequence is $R \rightarrow Y \rightarrow B$. [**Ans.** 30 A ; 0.866 (lag)]

43. A wattmeter reads 5.54 kW when its current coil is connected in red phase and its potential coil is connected between the neutral and red phase of a symmetrical 3-phase system supplying a balanced load of 30 A at 400 V. What will be the reading of the instrument if the current coil connections remains unchanged but the potential coil be connected between blue and yellow phase ? What does this figure represent ? [**Ans.** 7 kVAR]

44. A 3-phase, 434 V, 50 Hz supply is connected to a 3-phase star-connected induction motor and synchronous motor. Impedance of each phase of induction motor is $1.25 + j2.17\ \Omega$. The 3-phase synchronous motor is over-excited and it draws a current of 120 A at 0.87 leading power factor. Two wattmeters are connected in usual manner to measure power drawn by the two motors. Calculate :

(*i*) Reading on each wattmeter, and (*ii*) Combined power factor.

[**Ans.** 63970 W, 52050 W ; 0.9845 (lag)]

45. A star-connected balanced load is supplied from a 3-phase balanced supply with a line voltage of 416 V at a frequency of 50 Hz. Each phase of load consists of a resistance and a capacitor joined in series and the readings on two wattmeters connected to measure the total power supplied are 782 W and 1980 W, both positive. Calculate :

(*i*) The power factor of the circuit, (*ii*) The line current, and

(*iii*) The capacitance of each capacitor. [**Ans.** 0.8 (leading) ; 4.8 A ; 106 μF]

Unbalanced Load Circuits

46. A 400 V, 50 Hz, 3-phase supply has delta-connected load having 100 Ω between *R* and *Y*, 318 mH between *Y* and *B* and 31.8 μF between *B* and *R*. Find the line currents for the phase sequence *RYB*. Calculate the value of star-connected balanced resistor for the same power.

[**Ans.** $7.73 \angle 15°$ A ; $7.73 \angle 165°$ A ; $4 \angle 270°$ A ; 100 Ω]

47. Three loads $(31 + j59)$, $(30 - j40)$ and $(80 + j60)$ Ω are connected in delta across a 400 V, 3-phase supply. Find the phase currents, line currents and power.

[**Ans.** $3 \angle -62.3°$ A ; $4 \angle -66.9°$ A ; $\angle 83.1°$ A ; $4.78 \angle -76°$ A ; $1.038 \angle -80.35°$ A ; $5.819 \angle -256.8°$; 1078.9 W]

48. A 3-phase, 400 V, 50 Hz system has the following load connected in delta : between *R* and *Y* lines a non-reactive resistor of 100 Ω, between *Y* and *B* lines a coil having a reactance of 66 Ω and a resistance of 20 Ω, between *B* and *R* lines a loss free capacitor of 30 μF. Calculate :

(*i*) The phase currents, and (*ii*) The line currents.

Assume the phase sequence to be *RYB*.

[**Ans.** $4 \angle 0°$ A ; $6.324 \angle -191.56°$ A ; $3.774 \angle 210°$ A ; $7.5 \angle 14.6°$ A ; $10.275 \angle 173°$ A ; $4.3 \angle -47.13°$ A]

49. A balanced 3-phase system supplies an unbalanced 3-phase *delta-connected* load made up of two resistors 100 Ω and 200 Ω and a reactor having an inductance of 0.3 H with negligible resistance. $E_L = 100$ V at 50 Hz. Calculate :

(*i*) The total power in the system, and (*ii*) The total volt-amperes reactive.

[**Ans.** 150 W ; 106 (lag)]

50. A 400 volt 3-phase, 4-wire system has the loads $12 - j16$, $12 + j0$ and $8 + j6\ \Omega$ connected in star. Find the line currents, neutral current and total power.

[**Ans.** $11.55\ \angle 53.1°$ A ; $19.25\ \angle -120°$ A ; $23.09\ \angle 83.1°$ A ; $I_N = 15.47\ \angle -90.3°$ A ; 10310.11 W]

51. A 3-phase 3-wire system having a 254 V line-to-neutral has the following loads connected between the respective lines and neutral : $Z_R = 10\ \angle 0°\ \Omega$, $Z_Y = 10\ \angle 37°\ \Omega$, $Z_B = 10\ \angle -53°\ \Omega$. Calculate the current in the neutral wire and power taken by each load when phase sequence is

(*i*) *RYB* (*ii*) *RBY*.

[**Ans.** (*i*) $24.46\ \angle 16°\ 15'$; 6452 W ; 5162 W ; 3871 (*ii*) $38.5\ \angle -177.3°$; Power same as in (*i*)]

52. A 3-phase star connected system with 230 V between each phase and neutral has resistances of 4 Ω, 5 Ω and 6 Ω respectively in the three phases. Calculate :

(*i*) Current flowing in each phase, (*ii*) Neutral current, and

(*iii*) Total power absorbed. [**Ans.** 57.5 A, 46 A, 38.33 A ; 16.71 A ; 32620 W]

53. A 440/254 V, 3-phase, 4-core supplies an unbalanced load represented by the following impedances in ohms between *R*, *Y* and *B* phases respectively and the neutral $16 + j12$, $14 - j21$ and 25 Ω. The phase sequence is *RYB*. Calculate :

(*i*) The current in each conductor of the cable, and

(*ii*) Readings of three wattmeters connected in each line and neutral.

[**Ans.** 12.7 A, 10.06 A, 10.16 A ; 2580 W, 1413 W ; 2580 W]

54. Calculate the voltage across each of the following star-connected impedances when they are connected to a 400 V, 3-phase system, $Z_R = j20\ \Omega$, $Z_Y = j1\ \Omega$, $Z_B = -j1\ \Omega$.

Also find the potential of the star point above ground.

The phase sequence is *RYB*. [**Ans.** $(4000 + j6928)$ V ; $(3600 + j6928)$ V ; $(3800 + j7274.4)$ V ; – 8000 V]

55. An unbalanced star-connected load comprising of the impedances $Z_R = (10 + j0)$, $Z_Y = (3 + j4)$, $Z_B = (0 - j10)$ is connected across a 3-phase 200 V circuit with balanced voltages. Find :

(*i*) Three line currents, and (*ii*) Voltage across each impedance.

The phase sequence is *RYB*. [**Ans.** $(-20.2 + j4.54)$; $(7.24 + j5.48)$; $(12.78 - j10.02)$; $(-200.2 + j45.4)$; $(-0.2 + j45.4)$; $(-100.2 - j127.8)$]

56. An unbalanced star-connected load is supplied from a symmetrical 3-phase, 440 V, 3-wire system. The branch impedances of the load are $Z_1 = 5\ \angle 30°\ \Omega$, $Z_2 = 10\ \angle 45°\ \Omega$ and $Z_3 = 10\ \angle 60°\ \Omega$. Assuming the positive phase sequence of *RYB* determine the line currents and voltage across each impedance. Draw the phasor diagram. [**Ans.** $35.75\ \angle -71.3°$ A ; $32.77\ \angle 156.1°$ A ; $27.67\ \angle 48°$ A ; $178.7\ \angle -41.3°$ V, $327.7\ \angle 201.1°$ V, $276.7\ \angle 108°$ V]

57. An unbalanced star-connected load has branch impedances of $Z_1 = 10\ \angle 30°\Omega$, $Z_2 = 30\ \angle -45°\ \Omega$, Z_3, $= 20\ \angle 60°\ \Omega$ and is connected across a balanced 3-phase, 3-wire supply of 200 V. Find the line currents and the voltage across each impedance using star-delta conversion method.

[**Ans.** $5.39\ \angle 8°\ 32'$ A, $15.79\ \angle -148°\ 6'$ A, $10.73\ \angle 44°\ 16'$ A, $53.9\ \angle 38°\ 32'$ V, $157.9\ \angle -193.6°$, $214.6\ \angle 104°\ 16'$]

58. A 300 V (line) 3-phase supply feeds a star-connected load consisting of non-inductive resistors of 15, 6 and 10 Ω connected to the *R*, *Y* and *B* lines respectively. The phase sequence is *RYB*. Calculate the voltage across each resistor. [**Ans.** $(195 - j78)$ V, $(-105 - j78)$ V, $(4.5 + j182)$ V]

59. A 400 V, 3-phase, 3-wire symmetrical system *RYB* supplies a star-connected load whose branch circuit impedances are : $Z_R = 10\sqrt{3} + j10$; $Z_Y = 20 + j20\sqrt{3}$; $Z_B = 0 - j10$.

Determine the current in each branch.

Take the phase sequences as $R \rightarrow Y \rightarrow B$. [**Ans.** $25.55\ \angle -57°\ 45'$; $6.48\ \angle -133.7°$; $29\ \angle 109°\ 45'$]

Power Factor Improvement

60. A 50-MVA, 11 kV 3-phase alternator supplies full-load at a lagging power factor of 0.7. What would be percentage increase in earning capacity if the power factor is increased to 0.95. [**Ans.** 35.7 per cent]

61. A 3-phase, 37.3 kW, 440 V, 50 Hz induction motor operates on full-load with an efficiency of 89% and a power factor of 0.85 lagging. Calculate the total kVA rating of capacitors required to raise the full-load power factor to 0.95 lagging. What will be the capacitance per phase if the capacitors are :

(*i*) Delta-connected, and (*ii*) Star-connected.

[**Ans.** 12.19 kVA ; 66.8 μF ; 200.4 μF]

62. Three impedance coils, each having a resistance of 20 Ω and a reactance of 15 Ω are connected in star to a 400 V, 3-phase, 50 Hz supply. Calculate :

(*i*) The line current, (*ii*) Power supplied, and (*iii*) The power factor.

If three capacitors, each of the same capacitance are connected in delta to the same supply as to form parallel circuit with the above impedance coils, calculate the capacitance of each capacitor to obtain a resultant power factor of 0.95 lagging. [**Ans.** 9.24 A, 5120 W, 0.8 (lag) ; 14.32 μF]

63. A balanced 3-phase load, each element of which consist of a 5 Ω resistance in series with an inductance of reactance 3 Ω, is connected in star across mains whose line potential difference is 400 V and frequency 50 Hz. What is the power consumed by the system ?

Three equal capacitors are connected across the mains in the mesh. What should their capacitance be to cause the total line current to be in phase with the line voltage ? [**Ans.** 23.55 kW ; 93.8 μF/phase]

2
Transformers

2.1. GENERAL ASPECTS

Although the transformer is not classified as an electric machine, the principles of its operation are fundamental for the induction motor and synchronous machines. Since A.C. electric machines are normally built for low frequencies only the low-frequency power transformer will be considered in this text.

Function. *The function of a transformer, as the name implies, is to transform alternating current energy from one voltage into another voltage.* The transformer has *no rotating parts*, hence it is often called a *static transformer.*

When energy is transformed into a higher voltage the transformer is called a *step-up transformer* but when the case is otherwise it is called a *step-down transformer*. Most power transformers operate at constant voltage, *i.e.* if the power varies the current varies while the voltage remains fairly constant.

Applications. A transformer performs many important functions in prominent areas of electrical engineering.

- In *electrical power engineering* the transformer makes it possible to convert electric power from a generated voltage of about 11 kV (as determined by generator design limitations) to higher values of 132 kV, 220 kV, 400 kV, 500 kV and 765 kV thus permitting transmission of huge amounts of power along long distances to appropriate distribution points at tremendous savings in the cost of transmission lines as well as in power losses.
- At *distribution points* transformers are used to reduce these high voltages to a safe level of 400/230 volts for use in homes, offices etc.
- In *electric communication circuits* transformers are used for a variety of purposes *e.g.*, as an impedance transformation device to allow maximum transfer of power from the input circuit to the output device.

- In *radio and television circuits* input transformers, interstage transformers and output transformers are widely used.
- Transformers are also used in *telephone circuits, instrumentation circuits* and *control circuits.*

2.2. BASIC DEFINITIONS

- A *transformer* is a *static electromagnetic device designed for the transformation of the (primary) alternating current system into another (secondary) one of the same frequency with other characteristics, in particulars, other voltage and current.*
- As a rule a transformer consists of a core assembled of sheet transformer steel and two or several windings coupled *electromagnetically*, and in the case of *autotransformer, also electrically.*
- A transformer with two windings is called *double-wound transformer* ; a transformer with three or more windings is termed a *triple wound* or *multi-winding* one.
- According to the kind of current, transformers are distinguished as single-phase, three-phase and poly-phase ones. A *poly-phase transformer winding is a group of all phase windings of the same voltage*, connected to each other in a definite way.
- **Primary and secondary windings.** The transformer winding to which the energy of the alternating current is delivered is called the *primary winding* ; the other winding from which energy is received is called the *secondary winding*.
- In accordance with the names of the windings, all quantities pertaining to the primary winding as, for example, power, current, resistance etc., are also primary, and those pertaining to the secondary winding secondary.
- **h.v. and l.v. windings.** The winding connected to the circuit with the higher voltage is called the *high-voltage winding* (h.v.), the winding connected to the circuit with the lower voltage is called the *low-voltage winding*. (l.v.). If the secondary voltage is *less* than the primary one, the transformer is called a *step-down transformer* and if *more-a step-up transformer.*
- *A tapped transformer* is one whose windings are fitted with special taps for changing its voltage or current ratio.
- **Oil and dry transformers.** To avoid the deterimental effect of the air on the winding insulation and improve the cooling conditions of the transformer its core together with the windings assembled on it is immersed in a tank filled with transformer oil. Such transformers are called **oil transformers**. Transformers not immersed in oil are called **dry transformers.**

2.3. WORKING PRINCIPLE OF A TRANSFORMER

A transformer operates on the principle of *mutual inductance,* between two (and sometimes more) inductively coupled coils. It consists of two windings in close proximity as shown in Fig. 2.1. *The two windings are coupled by magnetic induction.* (There is no conductive connection between the windings). One of the windings called *primary* is energised by a sinusoidal voltage. The second winding, called *secondary* feeds the load. The alternating current in the primary winding sets up an alternating flux (ϕ) in the core. The secondary winding is linked by most of this flux and e.m.fs are induced in the two windings. The e.m.f. induced in the secondary winding drives a current through the load connected to the winding. Energy is transferred from the primary circuit to the secondary circuit through the medium of the magnetic field.

In brief, a transformer is a *device* that :

(*i*) *transfers electric power from one circuit to another* ;

(*ii*) *it does so without change of frequency ; and*

(*iii*) *it accomplishes this by electromagnetic induction (or mutual inductance).*

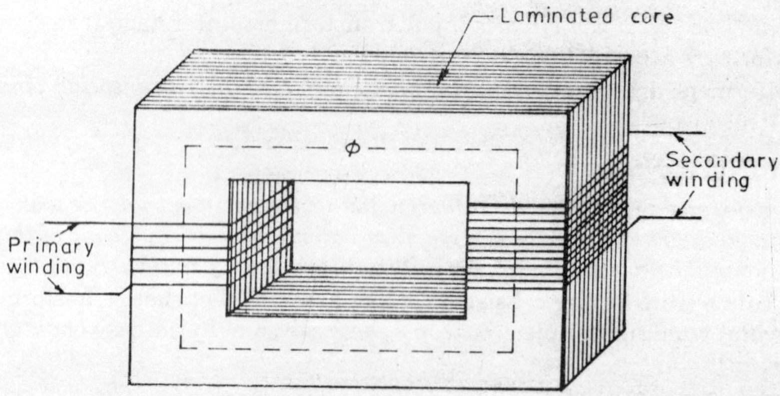

Fig. 2.1. Two winding transformer.

2.4. TRANSFORMER RATINGS

The rated quantities of a transformer, its power, voltage, frequency, etc. are given in Manufacturer's name plate, which should always be arranged so as to be accessible. But the term 'rated' can also be applied to quantities not indicated on the name plate, but relating to the rated duty, as for example, the rated efficiency, rated temperature conditions of the cooling medium, etc. :

- *The rated duty* of a transformer is determined by the quantities given in the name plate.
- The *rated power* of the transformer is the power at the secondary terminals, indicated in the name plate and expressed in kVA.
- *The rated primary voltage* is the voltage indicated in the transformer name plate ; if the primary is provided with taps, the rated tapped voltage is specially noted.
- *The rated secondary voltage* is the voltage across the transformer *secondary terminals at no-load* and with the rated voltage across the primary terminals ; if the secondary winding has taps, then their rated voltage is specially indicated.
- *The rated currents of the transformer*, primary and secondary, are the currents indicated in the name plate of the transformer and calculated by using the corresponding rated values of power and voltage.

2.5. KINDS OF TRANSFORMERS

The following kinds of transformers are the most important ones :

1. **Power transformers.** *For the transmission and distribution of electric power.*

2. **Auto-transformers.** *For converting voltages within relatively small limits to connect power systems of different voltages, to start A.C. motors etc.*

3. **Transformer for feed installations with static convertors.** (Mercury arc rectifiers, ignitrons, semi-conductor valves, etc.) *When converting A.C. into D.C.* (rectifying) *and convering D.C. into A.C. (inverting).*

4. **Testing transformers.** *For conducting tests at high and ultra-high voltages.*

5. **Power transformers for special applications.** Furnace, welding etc.

6. **Radio-transformers.** *It is used in radio engineering etc.*

Note. *Distribution transformers* should be designed to have maximum efficiency at a load much *lower than full-load* (*about 50 per cent*).

Power transformers should be designed to have maximum efficiency *at or near full-load.*

2.6. TRANSFORMER CONSTRUCTION

All transformers have the following essential elements :

1. Two or more **electrical windings** insulated from each other and from the core (except in auto-transformers).

2. A **core,** which in case of a single-phase distribution transformers usually comprises *cold-rolled silicon-steel strip* instead of an assembly of punched silicon-steel laminations such as are used in the larger power-transformer cores. The *flux path in the assembled core is parallel to the directions of steel's grain or 'orientation'.* This results in a *reduction in core losses* for a given flux density and frequency, or *it permits the use of higher core densities* and *reduced size of transformers for given core losses.*

Other necessary parts are :

- A *suitable container* for the assembled core and windings.
- A *suitable medium* for insulating the core and its windings from each other and from the container.
- Suitable *bushings* for insulating and bringing the terminals of the windings out of the case.

The two basic types of transformer construction are :

1. *The core type.*
2. *The shell type.*

The above two types differ in their relative arrangements of copper conductors and the iron cores. In the '*core type*', the *copper virtually surrounds the iron core,* while *in the 'shell type', the iron surrounds the copper winding.*

2.6.1. Core Type Transformer. The completed magnetic circuit of the core-type transformer is in the shape of the hollow rectangle, exactly as shown in Fig. 2.2 in which I_0 is the no-load current and ϕ is the flux produced by it. N_1 and N_2 are the number of turns on the primary and secondary sides respectively.

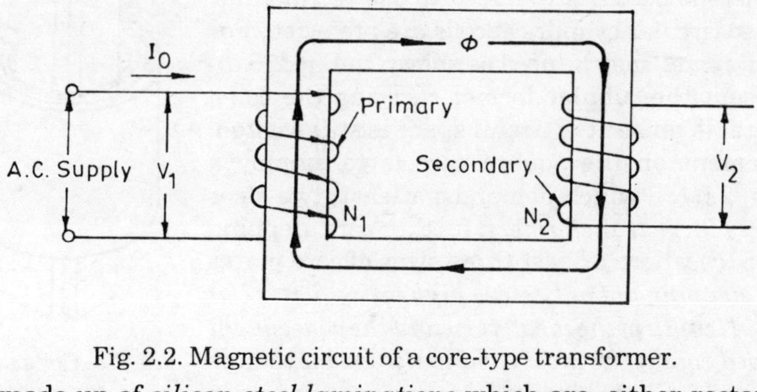

Fig. 2.2. Magnetic circuit of a core-type transformer.

The core is made up of *silicon-steel laminations* which are, either rectangular or L-shaped. With the coils wound on two legs the appearance is that of Fig. 2.3. If the two coils shown were the respective high and low-side coils as in Fig. 2.3, the *leakage reactance would be much too great.* In order to provide maximum *linkage* between windings, the group on *each leg is made up of both high-tension and low-tension coils.* This may be seen in Fig. 2.4, where a cross-sectional cut is taken across the legs of the core. By placing the high-voltage winding around the low-voltage winding, only one layer of high-voltage insulation is required, that between the two coils. *If the high-voltage coils were adjacent to the core, an additional high-voltage insulation layer would be necessary between the coils and the iron core.*

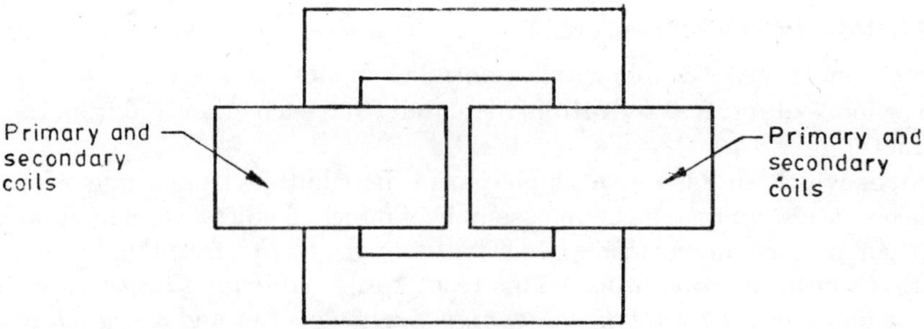

Fig. 2.3. Core-type transformer.

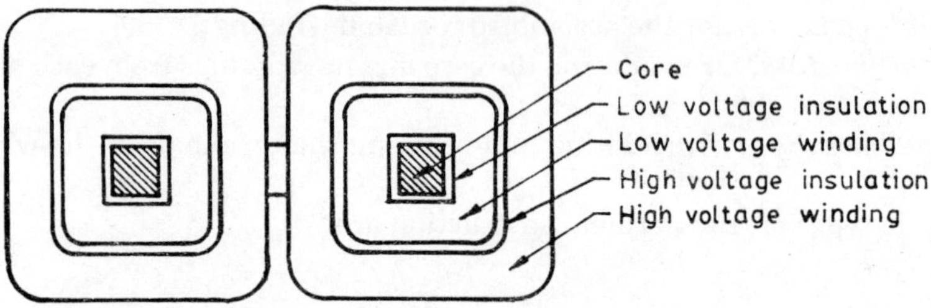

Fig. 2.4. Cross-section of a core-type transformer.

Fig. 2.5 shows the coils and laminations of a core-type transformer with a cruciform core and circular coils.

- Fig. 2.6 shows *the different types of cores used in core type transformers.*

Rectangular cores [Fig. 2.6 (*a*)] with rectangular cylindrical coils can be used for small size core-type transformers. For large size transformers it becomes wasteful to use rectangular cylindrical coils and so circular cylindrical coils are preferred. For such purposes, '*square cores*' may be used as shown in Fig. 2.6 (*b*) where circles represent the tubular former carrying the coils. Evidently a considerable amount of useful space is still wasted. A common improvement on the square core is to employ a '*cruciform core*' [Fig. 2.6 (*c*)] which demands, atleast, two sizes of core strips. For *very large transformers*, further core stepping is done as in Fig. 2.6 (*d*) where atleast three sizes of core plates are necessary. *Core stepping not only gives high space factor but also results in reduced length of the mean turn and the consequent* I^2R *loss. Three stepped core is the most commonly* used although more steps may be used for very large transformers as shown in Fig. [2.6 (*e*)].

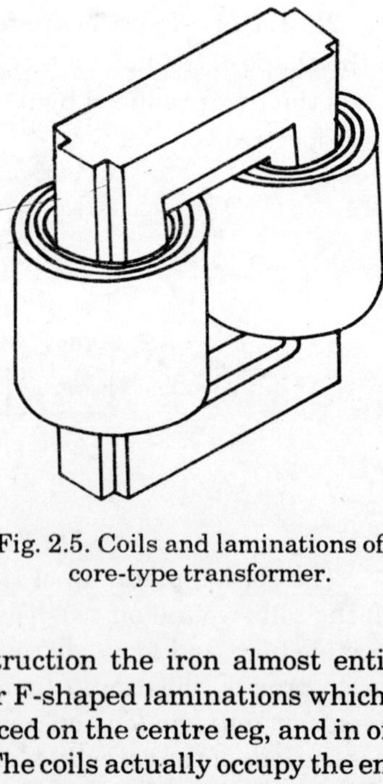

Fig. 2.5. Coils and laminations of a core-type transformer.

2.6.2. Shell Type Transformer. In the shell-type construction the iron almost entirely surrounds the copper (Fig. 2.7). The core is made up of E-shaped or F-shaped laminations which are stacked to give a rectangular figure eight. All the windings are placed on the centre leg, and in order to reduce leakage, each high-side coil is adjacent to a low-side coil. The coils actually occupy the entire space of both windows, are flat or pencake in shape, and are usually constructed of strip copper. Again, *to reduce the amount the high-voltage insulation required, the low-voltage coils are placed adjacent to the iron core.*

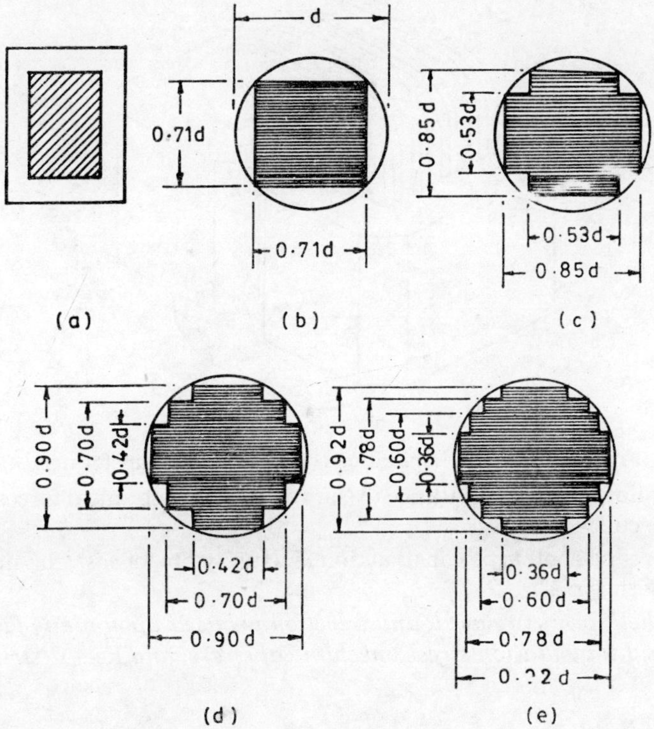

Fig. 2.6. Various types of cores.

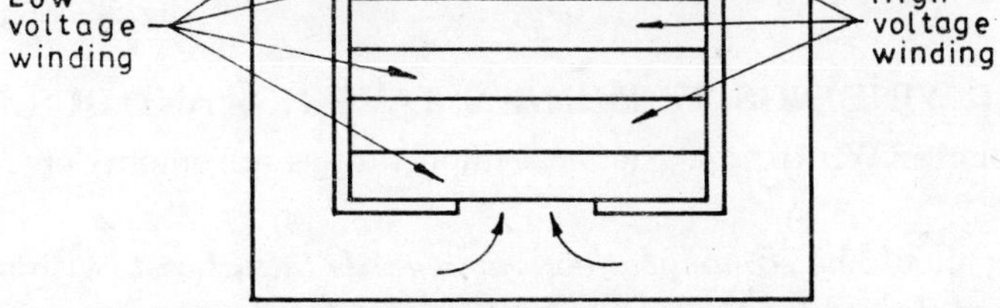

Fig. 2.7. Shell-type transformer.

Fig. 2.8 shows the coils and laminations of a typical shell-type transformer.

Choice of Core- or Shell Type Construction. In general, the core-type has a longer mean length of core and a shorter mean length of coil turn. The core type also has a smaller cross-section of iron and so will need a greater number of turns of wire, since, in general, not as high a flux may be reached in the core. However, *core type is better adopted for some high-voltage service since there is more room for insulation*. The *shell type has better provision for mechanically supporting and*

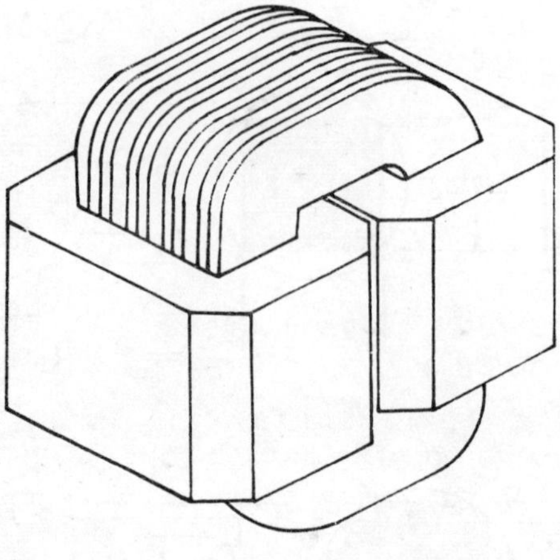

Fig. 2.8. Coils and laminations of a shell-type transformer.

bracing the coils. This allows better resistance to the very high mechanical forces that develop during a high-current short circuit.

The choice of core- or shell type construction is usually one of *cost*, for similar characteristics can be obtained with both types.

Both core and shell forms are used, and selection is based upon many factors such as voltage rating, kVA rating, weight, insulation stress, mechanical stress, and heat distribution.

2.6.3. Spiral Core Transformer. The typical spiral core is shown in Fig. 2.9. The core is assembled either of a *continuous* strip of transformer steel wound in the form of a circular or elliptical cylinder or of a group of short strips assembled to produce the same elliptical-shaped core. By using this construction the core flux always follows along the grain of the iron. Cold-rolled steel of high silicon content enables the designer to use higher operating flux densities with lower loss per kg. *The higher flux density reduces the weight per kVA.*

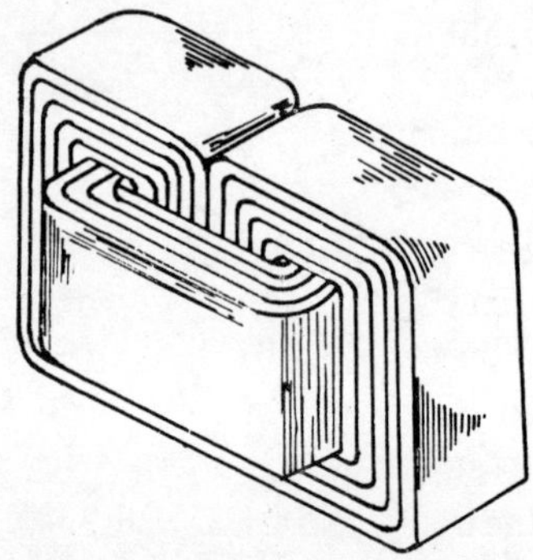

Fig. 2.9. Spiral-core transformer.

2.7. TRANSFORMER WINDINGS, TERMINALS, TAPPINGS AND BUSHINGS

2.7.1. Transformer Windings. The most important requirements of transformer windings are :

1. The winding should be *economical both as regards initial cost*, with a view to the market availability of copper, and the efficiency of the transformer in service.

2. The heating conditions of the windings should *meet standard requirements*, since departure from these requirements towards allowing higher temperature will drastically shorten the service life of the transformer.

3. The winding should be *mechanically stable* in respect to the forces appearing when sudden short circuit of the transformer occur.

4. The winding should have the *necessary electrical strength in respect to over voltages.*

The different *types of windings* are *classified* and briefly discussed below :

1. Concentric windings :

(*i*) Cross-over (*ii*) Helical

(*iii*) Disc.

2. Sandwich windings

1. Concentric windings. Refer Fig. 2.10. These windings are *used for core type transformers.* Each limb is wound with a group of coils consisting of both primary and secondary turns which may be concentric cylinders. The l.v. winding is placed next to the core and h.v. winding on the outside. But the two windings can be sub-divided, *and interlaced with high tension and low tension section alternately to reduce leakage reactance.* These windings can be further divided as follows :

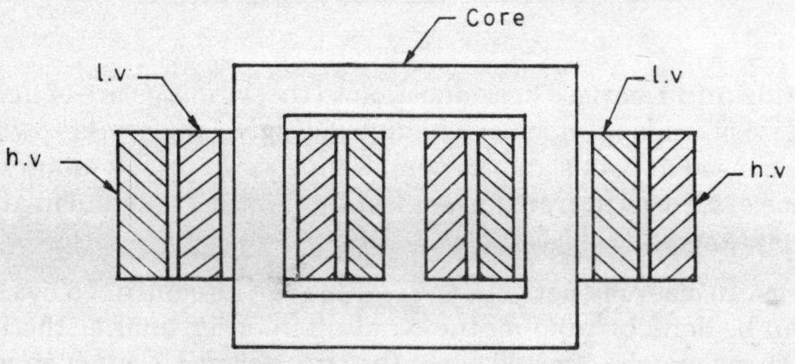

Fig. 2.10. Concentric coils.

(*i*) **Cross-over windings.** Cross-over windings are used for currents up to 20 A and so they are suitable for h.v. winding of small transformers. The conductors are either cotton covered round wires or strips insulated with paper. Cross-over coils are wound over formers and each coil consists of a number of layers with a number of turns per layer. The complete winding consists of a number of coils connected in series. Two ends of each coil are brought out, one from inside and one from outside. The inside end of a coil is connected to the outside end of the adjacent coil.

(*ii*) **Helical winding.** A helical winding consists of rectangular strips wound in the form of a helix. The strips are wound in parallel radially and each turn occupies the total radial depth of winding.

Helical coils are well suited for *l.v. windings of large transformers*. They can also be used for h.v. windings by putting extra insulation between layers in addition to insulation of conductors.

(*iii*) **Continuous disc winding.** This type of winding consists of a number of flat strips wound spirally from inside (radially) outwards. The conductor is used in such lengths as are sufficient for complete winding or section of winding between tappings. The conductor can either be a single strip or a number of strips in parallel, wound on the flat. This gives a robust construction for each disc. The discs are wound on insulating cylinders spaced from it by strips along the length of cylinder. The discs are separated from each other with press board sectors attached to the vertical strips. The vertical and horizontal spacers provide ducts for free circulation of oil which is in contact with every turn.

2. Sandwich coils. Sandwich coils (Fig. 2.11) are employed in transformers of *shell type.* Both high and low voltage windings are split into a number of sections. Each high voltage section lies between the low voltage sections.

The *advantage of sandwich coils is that their leakage can be easily controlled and so any desired value of leakage reactance can be had by the division of windings.*

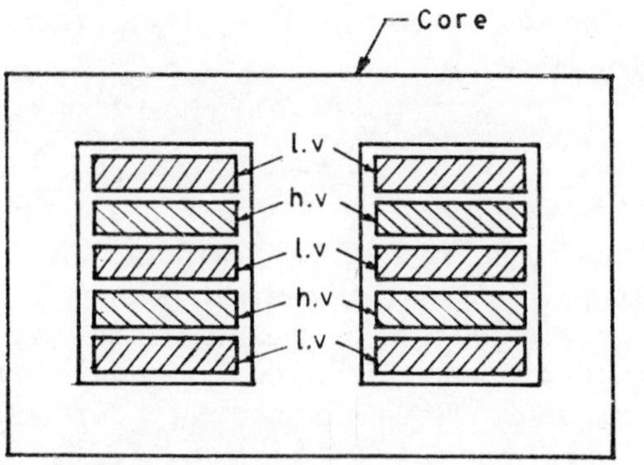

Fig. 2.11. Sandwich coils.

2.7.2. Terminals and Leads. The connection to the windings are of insulated copper rods or bars. The shape and size of leads is important in high voltage transformers owing to dielectric stress and corona which are caused at bends and corners. Connections from windings are directly taken to the busbars in the case of air-cooled transformers while they are taken to insulated bushings in the case of oil-cooled transformers.

2.7.3. Tappings. In a supply network the voltage can be controlled by changing the transformation ratio. This can be done by tapping the winding in order to alter the number of turns. The change in number of turns may be effected when the transformer is out of circuit (known as off load tap changing) or when on load (known as on load tap changing). *The tappings are provided on the high voltage winding because a fine voltage variation is obtained owing to large number of turns. It is difficult to obtain voltage variation within close percentage limits in low voltage winding as there are few turns and voltage per turn is a large percentage of the total voltage.*

In transformers, the tappings can be provided at :

(*i*) phase ends ; and

(*ii*) neutral point or in the middle of the windings.

- The advantage of providing tappings at *phase ends* is that the number of bushing insulators is reduced, this is important where the cover space is limited. Some transformers have reinforced insulation at the phase ends. It is essential that in such cases either the tapping should not be provided at end turns or the reinforcement should be carried beyond the lower tap.
- When the tappings are made at the *neutral point* the insulation between various parts is small. *This arrangement is economical especially in the case of high voltage transformers.*

2.7.4. Bushings. The bushings *are employed for insulating and bringing out terminals of the winding from the container to the external circuit.* For low-voltage transformers this is accomplished by employing *bushings of porcelain around the conductor at the point of entry.* For high voltages it is necessary to employ bushings of larger sizes. In modern transformers the problem is met by using *large porcelain or composition bushings for voltages as high as 33 kV, above that oil filled* or *condenser type bushings are used.*

2.8. TRANSFORMER COOLING

2.8.1. Cooling Methods. The transformers get heated due to iron and copper losses occurring in them. It is necessary to dissipate this heat so that the temperature of the windings is kept below the value at which the insulation begins to deteriorate. The *cooling of transformers is more difficult than that of rotating machines because the rotating machines create a turbulent air flow which assists in removing the heat generated due to losses.* Luckily the losses in transformers are comparatively small. Nevertheless the elaborate cooling arrangements have been devised to deal with the whole range of sizes.

As far as cooling methods are concerned, the transformers are of following two types :

1. Dry type.
2. Oil immersed type.

Dry Type Transformers. Small transformers upto *25 kVA size* are of the *dry type* and have the following cooling arrangements :

(*i*) **Air natural.** In this method the natural circulation of surrounding air is utilized to carry away the heat generated by losses. A sheet metal enclosure protects the winding from mechanical injury.

(*ii*) **Air blast.** Here the transformer is cooled by a continuous blast of cool air forced through the core and windings (Fig. 2.12). The blast is produced by a fan. The air supply must be filtered to prevent accumulation of dust in ventilating ducts.

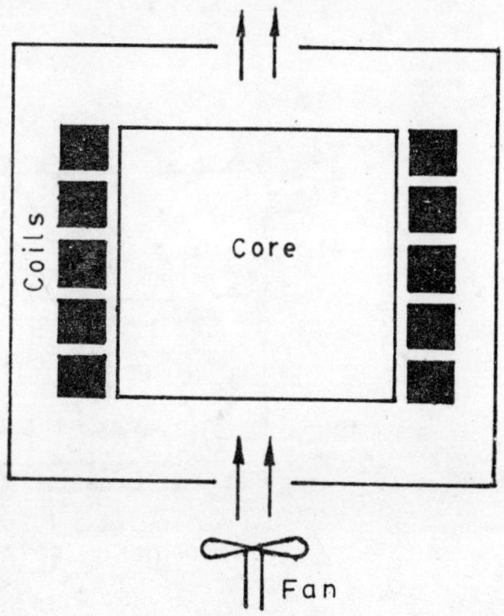

Fig. 2.12

Oil Immersed Transformers. In general most transformers are of oil immersed types. The *oil provides better insulation than air and it is a better conductor of heat than air*. Mineral oil is used for this purpose.

Oil immersed transformers are classified as follows :

(*i*) **Oil immersed self-cooled transformers.** The transformer is immersed in oil and heat generated in cores and windings is passed to oil by conduction. Oil in contact with heated parts rises and its place is taken by cool oil from the bottom. The natural oil transfers its heat to the tank walls from where heat is taken away by the ambient air. The oil gets cooler and falls to the bottom from where it is dissiptated into the surroundings. The tank surface is the best dissipator of heat but a plain tank will have to be excessively large, if used without any auxiliary means for high rating transformers. As both space and oil are costly, these auxiliary means should not increase the cubic capacity of the tank. The heat dissipating capacity can be increased by providing (*i*) corrugations, (*ii*) fins, (*iii*) tubes (Fig. 2.13) and (*iv*) radiator tanks.

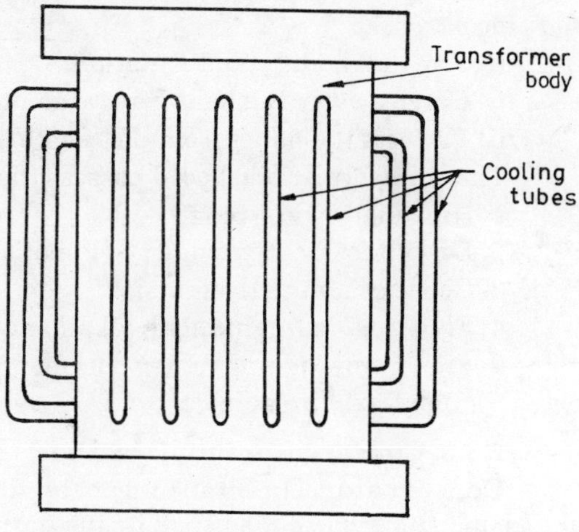

Fig. 2.13. Transformer with cooling tubes.

The *advantages of 'oil natural' cooling is that it does not clog the ducts and the windings are free from effects of moisture.*

(*ii*) **Oil immersed forced air-cooled transformers.** In this type of cooling, air is directed over the outer surfaces of the tank of the transformer immersed in oil.

(*iii*) **Oil immersed water-cooled transformers.** Heat is extracted from the oil by means of a stream of water pumped through a metallic coil immersed in the oil just below the top of the tank. The heated water is in turn cooled in a spray pond or a cooling tower.

(*iv*) **Oil immersed forced oil cooled transformers.** In such transformers heat is extracted from the oil by pumping the oil itself upward through the winding and then back by way of external radiators which may themselves be cooled by fans. The *extra cost of oil pumping equipment must of* *course be economically justified but it has incidentally the advantage of reducing the temperature difference between the top and bottom of enclosing tank.*

Fig. 2.14 shows the cooling of transformers having capacities from 10000 kVA and higher. In such cases air blast cooling of radiator is used.

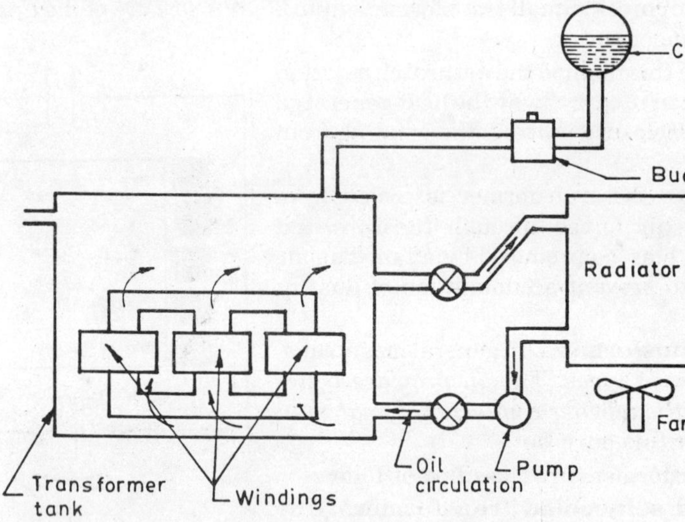

Fig. 2.14. Air blast cooling of radiator.

2.8.2. Transformer Oil. It is a *mineral oil obtained by refining crude petroleum*. It serves the following *purposes* :

(*i*) Provides additional insulation.

(*ii*) Carries away the heat generated in the core and coils.

(*iii*) Protects the paper from dirt and moisture.

The transformer oil should possess the following *properties* :

1. High dielectric strength.
2. Low viscosity to provide good heat transfer.
3. Good resistance to emulsion.
4. Free from inorganic acid, alkali and corrosive sulphur.
5. Free from sludging under normal operating conditions.
6. High flash/fire point.

2.8.3. Conservator and Breather

Conservator. The oil should not be allowed to come in contact with atmospheric air as it may take up moisture which may spoil its insulating properties. Also air may cause acidity and sludging of oil. To prevent this, many transformers are provided with conservators. The *function of a conservator (Fig. 2.14) is to take up contraction and expansion of oil without allowing it to come in contact with outside air*. The conservator consists of an air tight metal-drum fixed above the level of the top of the tank and connected with it by a pipe. The main tank is completely filled with oil when cold. The *conservator is partially filled with oil*. So the oil surface in contact with air is greatly reduced. The sludge thus formed remains in the conservator itself and does not go to the main tank.

Breather. When the temperature changes, the oil expands or contracts and there is a displacement of air. When the transformer cools, the oil level goes down, and air is drawn in. This is known as *breathing*. The air, coming in, is passed through an apparatus called *breather* for the purpose of extracting moisture. The *breather consists of a small vessel which contains a drying agent like silica get crystal impregnated with cobalt crystal.*

Note. *Sludging* means the slow formation of solid hydrocarbons due to heating and oxidation. The sludge deposit itself on the windings and cooling ducts producing overheating. This makes transformer still hotter producing more sludge. This process may continue till the transformer becomes unusable due to overheating. So the contact of oil with air should be avoided as the air contains oxygen.

2.9. SINGLE PHASE TRANSFORMER

2.9.1. Elementary Theory of an Ideal Transformer. The basic theory of a transformer is not difficult to understand. To simplify matters as much as possible, let us first consider *an ideal transformer*, that is, *one in which the resistance of the windings is negligible and the core has no losses.*

Let the secondary be open (Fig. 2.15), and let a sine wave of potential difference v_1 (Fig. 2.16) be impressed upon the primary. The impressed potential difference causes an alternating current to

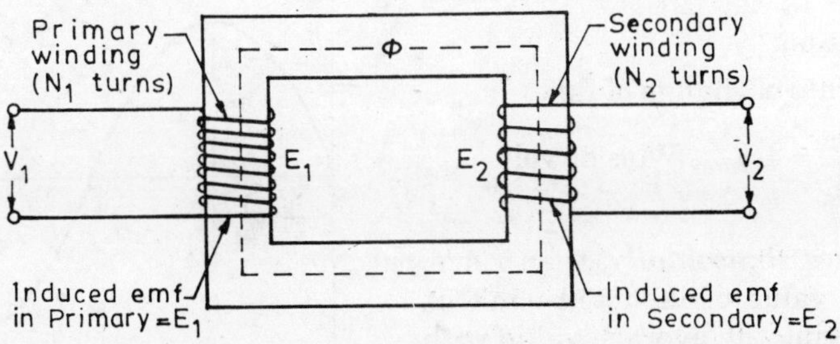

Fig. 2.15. Elementary diagram of an ideal transformer with an open secondary winding.

flow in the primary winding. Since the primary resistance is negligible and there are no losses in the core, the effective resistance is zero and the circuit is purely reactive. Hence the current wave i_m lags the impressed voltage wave v_1 by 90 time degrees, as shown in Fig. 2.17. The reactance of circuit is very high and the magnetizing current is very small. This current in the N_1 turns of the primary magnetizes the core and produces a flux ϕ that is at all times proportional to the current (if the permeability of the circuit is assumed to be constant), and therefore in time phase with the current. The flux, by its rate of change, induces in the primary winding E_1 which at every instant of time is equal in value and opposite in direction to V_1. It is called *counter e.m.f.* of the primary. The value which the primary current attains must be such that the flux which it produces in the core is of sufficient value to induce in the primary the required counter e.m.f.

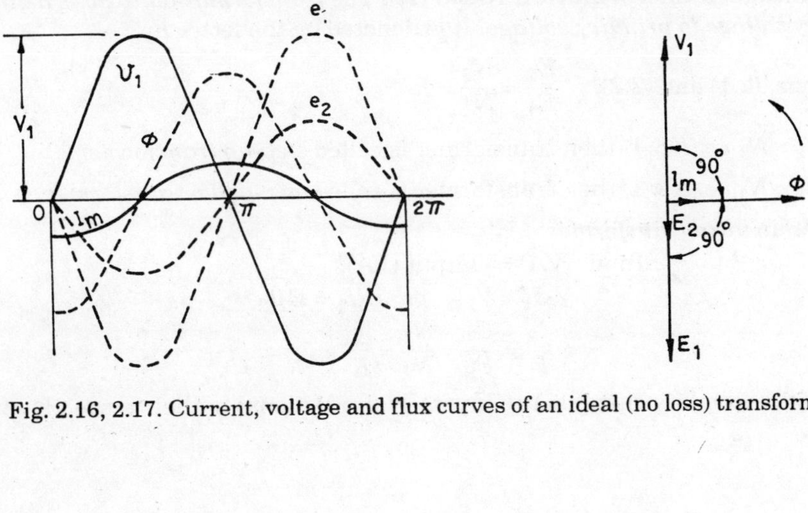

Fig. 2.16, 2.17. Current, voltage and flux curves of an ideal (no loss) transformer.

Since the flux also threads (or links) the secondary winding a voltage e_2 is induced in the secondary. This voltage is likewise proportional to the rate of change of flux and so is in time phase with e_1, but it may have any value depending upon the number of turns N_2 in the secondary.

2.9.2. E.M.F. Equation of a Transformer

Let N_1 = number of turns in primary,

N_2 = number of turns in secondary,

ϕ_m = maximum flux in the core, Wb.

$= B_m \times A$, [where B_m is the maximum flux density in the core and A is the core area], and

f = frequency of a.c. input, Hz.

Refer Fig. 2.18. Since the flux increases from its zero value to maximum value ϕ_m in one quarter of the cycle *i.e.* in $\frac{T}{4}$ or $\frac{1}{4f}$ second (T being time-period of the cycle),

∴ Average rate of change of flux

$$= \frac{\phi_m}{\frac{1}{4f}} = 4f\phi_{max} \text{ Wb/s or volt}$$

If flux ϕ varies *sinusoidally,* then r.m.s. (root mean square) value of inducted e.m.f. is obtained by multiplying the average value with form factor.

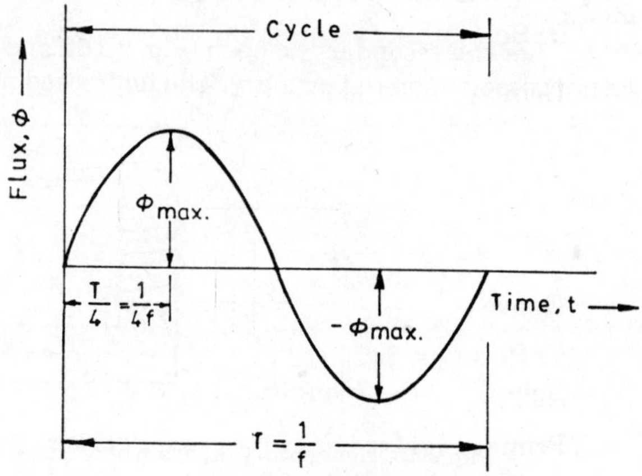

Fig. 2.18

But, form factor $= \frac{\text{r.m.s. value}}{\text{average value}} = 1.11$

∴ r.m.s. value of e.m.f./turn

$$= 1.11 \times 4f\phi_{max} = 4.44f\phi_{max} \text{ volt}$$

Now, r.m.s. value of induced e.m.f. in the whole of primary winding,

$$E_1 = 4.44f\phi_{max} N_1 \quad ...(2.1)$$

Similarly r.m.s. value of induced e.m.f. in secondary is,

$$E_2 = 4.44f\phi_{max} N_2 \quad ...(2.2)$$

In an ideal transformer on no-load $V_1 = E_1$ and $V_2 = E_2$ (Fig. 2.15).

2.9.3. Voltage Transformation Ratio (K). The *transformation ratio is defined as the ratio of the secondary voltage to primary voltage.* It is denoted by the letter K.

From eqns. (2.1) and (2.2), $$\frac{E_2}{E_1} = \frac{N_2}{N_1} = K \quad ...(2.3)$$

- If $N_2 > N_1$ *i.e.* $K > 1$, then transformer is called *step-up transformer.*
- If $N_2 < N_1$ *i.e.* $K < 1$, then transformer is called *step-down transformer.*

Again for an *ideal transformer*

$$\text{Input } (VA) = \text{Output } (VA)$$

$$V_1 I_1 = V_2 I_2 \text{ or } E_1 I_1 = E_2 I_2$$

or $$\frac{I_2}{I_1} = \frac{E_1}{E_2} = \frac{N_1}{N_2} = \frac{1}{K} \quad ...(2.4)$$

i.e., Primary and secondary currents are inversely proportional to their respective turns.

Example 2.1. *A 40 kVA, single phase transformer has 400 turns on the primary and 100 turns on the secondary. The primary is connected to 2000 V, 50 Hz supply. Determine :*

(i) The secondary voltage on open circuit.

(ii) The current flowing through the two windings on full-load.

(iii) The maximum value of flux.

Solution. Rating $= 40$ kVA

Primary turns, $N_1 = 400$

Secondary turns, $N_2 = 100$

Primary induced voltage, $E_1 = V_1 = 2000$ V

(i) Secondary voltage on open, circuit V_2 :

Using the relation, $$\frac{E_2}{E_1} = \frac{N_2}{N_1}$$

$$E_2 = E_1 \times \frac{N_2}{N_1}$$

$$\therefore \quad E_2 = V_2 = 2000 \times \frac{100}{400} = 500 \text{ V}$$

Hence, $\mathbf{V_2 = 500}$ **V. (Ans.)**

(ii) Primary current, I_1 :

Secondary current I_2 :

Primary full-load current, $$I_1 = \frac{\text{kVA} \times 100}{V_1} = \frac{40 \times 1000}{2000} = \mathbf{20 \text{ A.}} \text{ (Ans.)}$$

Secondary full-current, $$I_2 = \frac{\text{kVA} \times 1000}{V_2} = \frac{40 \times 1000}{500} = \mathbf{80 \text{ A}} \text{ (Ans.)}$$

(iii) Maximum value of flux, ϕ_{max} :

Using e.m.f. equation, $$E_1 = 4.44 f \phi_{max} N_1$$

$$\therefore \quad 2000 = 4.44 \times 50 \times \phi_{max} \times 400$$

$$\therefore \quad \phi_{max} = \frac{2000}{4.44 \times 50 \times 400} = 0.0225 \text{ Wb.}$$

Hence, $\phi_{max} = \mathbf{0.0225}$ **Wb. (Ans.)**

Example 2.2. *The no-load ratio required in a single-phase 50 Hz transformer is 6600/600 V. If the maximum value of flux in the core is to be about 0.08 Wb, find the number of turns in each winding.*

Solution. Primary, $E_1 = V_1 = 6600$ V

Secondary, $E_2 = V_2 = 600$ V

Maximum value of flux $\phi_{max} = 0.08$ Wb.

Primary turns, N_1 :

Secondary turns, N_2 :

Using the relation, $$E_1 = 4.44 f \phi_{max} N_1$$

$$6600 = 4.44 \times 50 \times 0.08 \times N_1$$

$$\therefore \quad N_1 = \frac{6600}{4.44 \times 50 \times 0.08} \simeq 372$$

Hence, $\mathbf{N_1 = 372.}$ **(Ans.)**

Also $\frac{E_2}{E_1} = \frac{N_2}{N_1}$

$\therefore$ $N_2 = \frac{E_2 N_1}{E_1} = \frac{600 \times 372}{6600} \simeq 34$

Hence, $\mathbf{N_2 = 34.}$ **(Ans.)**

Example 2.3. *A single-phase transformer is connected to a 230 V, 50 Hz supply. The net cross-sectional area of the core is 60 cm². The number of turns in the primary is 500 and in the secondary 100. Determine :*

(*i*) *Transformation ratio.* (*ii*) *E.m.f. induced in secondary winding.*

(*iii*) *Maximum value of flux density in the core.*

Solution. Primary turns, $N_1 = 500$

Secondary turns, $N_2 = 100$

Primary, $E_1 = V_1 = 230$ V

Core area, $a = 60 \text{ cm}^2 = 60 \times 10^{-4} \text{ m}^2$

(*i*) **Transformation ratio, K :**

$$K = \frac{N_2}{N_1} = \frac{100}{500} = 0.2$$

Hence, **K = 0.2. (Ans.)**

(*ii*) **Maximum value of flux density, $\mathbf{B_{max}}$:**

Using the e.m.f. equation, $E_1 = 4.44 f \phi_{max} N_1$

$\therefore$ $230 = 4.44 \times 50 \times \phi_{max} \times 500$

or $$\phi_{max} = \frac{230}{4.44 \times 50 \times 500} = 0.00207 \text{ Wb}$$

Now, $$B_{max} = \frac{\phi_{max}}{A} = \frac{0.00207}{60 \times 10^{-4}} = 0.345 \text{ T}$$

[where T stands for tesla (Wb/m²)]

Hence, $\mathbf{B_{max} = 0.345}$ **T. (Ans.)**

(*iii*) **E.m.f. induced in the secondary winding, $\mathbf{E_2}$:**

Using the relation, $\frac{E_2}{E_1} = \frac{N_2}{N_1}$

$$\frac{E_2}{230} = \frac{100}{500}$$

$\because$ $\mathbf{E_2 = 46}$ **V. (Ans.)**

Example 2.4. *3300/300 V single-phase 300 kVA transformer has 1100 primary turns. Find :*

(*i*) *Transformation ratio.* (*ii*) *Secondary turns.*

(*iii*) *Voltage/turn.*

(*iv*) *Secondary current when it supplies a load of 200 kW at 0.8 power factor lagging.*

Solution. Primary, $E_1 = 3300$ V

$N_1 = 1100$

Secondary, $E_2 = 300$ V

Rating of the transformer = 300 kVA

Output = 200 kW

(*i*) **Transformation ratio, K = ?**

$$K = \frac{N_2}{N_1} = \frac{E_2}{E_1} = \frac{300}{3300} = \frac{1}{11}$$

Hence, $\mathbf{K = \frac{1}{11}}$. **(Ans.)**

(*ii*) **Secondary turns, N_2 = ?**

Using the relation, $$\frac{E_2}{E_1} = \frac{N_2}{N_1}$$

∴ $$N_2 = \frac{E_2 N_1}{E_1} = \frac{300 \times 1100}{3300} = 100$$

Hence, $\mathbf{N_2 = 100}$. **(Ans.)**

(*iii*) **Voltage/turn = ?**

Voltage/turn $$\frac{E_1}{N_1} = \frac{3300}{1100} = \mathbf{3\ V}.\ \textbf{(Ans.)}$$

(*iv*) **Secondary current,** $$I_2 = \frac{\text{Output}}{V_2 \cos \phi} = \frac{200 \times 1000}{300 \times 0.8} = 833.33 \text{ A}$$

Hence, $\mathbf{I_2 = 833.33\ A}$. **(Ans.)**

Example 2.5. *The voltage per turn of a single-phase transformer is 1.1 V. When the primary winding is connected to a 220 V, 50 Hz A.C. supply, the secondary voltage is found to be 550 V. Find :*

(*i*) *Primary and secondary turns.*

(*ii*) *Core area if the maximum flux density is 1.1 T.*

Solution. Voltage per turn = 1.1 V

Primary, $E_1 = 220$ V

Secondary, $E_2 = 550$ V

Max. flux density, $B_{max} = 1.1$ T

(*i*) **Primary turns,** $$\mathbf{N_1} = \frac{E_1}{1.1} = \frac{220}{1.1} = \mathbf{200}.\ \textbf{(Ans.)}$$

Secondary turns, $$\mathbf{N_2} = \frac{E_2}{1.1} = \frac{550}{1.1} = \mathbf{500}.\ \textbf{(Ans.)}$$

(*ii*) **Core area A :**

Using the relation, $$E_1 = 4.44 f \phi_{max} N_1$$

$$220 = 4.44 \times 50 \times \phi_{max} \times 200$$

∴ $$\phi_{max} = \frac{220}{4.44 \times 50 \times 200} = 0.004955 \text{ Wb}$$

Core area, $$A = \frac{\phi_{max}}{B_{max}} = \frac{0.004955}{1.1} = 0.004504 \text{ m}^2 = \mathbf{45.04\ cm^2}.\ \textbf{(Ans.)}$$

Example 2.6. *The core of 1000 kVA, 11000/550 V, 50 Hz, single-phase transformer has a cross-section of 20 cm × 20 cm. If the maximum core density is not to exceed 1.3 tesla, calculate :*

(*i*) *The number of h.v. and l.v. turns per phase ;*

(*ii*) *The e.m.f. per turns.*

Assume a stacking factor of 0.9. **(AMIE Winter, 1997)**

Solution. *Given :* $A = 20 \text{ cm} \times 20 \text{ cm} = 400 \times 10^{-4} \text{ m}^2$; $B_{max} = 1.3 \text{ Wb/m}^2$, $f = 50$ Hz ; $E_1 = 11000$ V ; $E_2 = 550$ V ; Stacking factor = 0.9

(*i*) **Number of turns N_1 (h.v.) ; N_2 (l.v.) :**

We know that, Flux = flux density × area of cross-section × stacking factor

$$\therefore \quad \phi_{max} = 1.3 \times 400 \times 10^{-4} \times 0.9 = 0.0468 \text{ Wb}$$

$$E_1 = 4.44\, f\phi_{max}\, N_1$$

or $$11000 = 4.44 \times 50 \times 0.0468 \times N_1$$

$$\therefore \quad N_1 = \frac{11000}{4.44 \times 50 \times 0.0468} = 1058.75 \simeq \mathbf{1059.}\ \textbf{(Ans.)}$$

and $$N_2 = \frac{E_2}{E_1} \times N_1 = \frac{550}{11000} \times 1059 = 52.95 \simeq \mathbf{53.}\ \textbf{(Ans.)}$$

(*ii*) **The e.m.f. per turn :**

E.M.F. per turn $$= \frac{E_1}{N_1} = \frac{11000}{1059} = \mathbf{10.387\ V.}\ \textbf{(Ans.)}$$

2.9.4. Transformer with Losses but no Magnetic Leakage. We shall consider the following two cases :

1. When transformer is on no-load.
2. When transformer is on load.

2.9.4.1. Transformer on No-load. A transformer is said to be on *no-load* if its secondary side is open and primary is connected to a sinusoidal alternating voltage V_1. The alternating applied voltage will cause *flow of alternating current in the primary winding which will create alternating flux*. This primary input current (I_0) under no-load conditions supply :

(*i*) *Iron losses in the core* (*i.e.* hysteresis loss and eddy current loss).

(*ii*) A very *small amount of copper losses in primary* (there being no copper loss in the secondary as it is open).

Thus I_0 is not at 90° behind V_1, but lags it by an angle $\phi_0 < 90°$.

No-load power input, $P_0 = V_1 I_0 \cos \phi_0$ ...(2.5)

where $\cos \phi_0$ = primary power factor under no-load conditions.

As is evident from Fig. 2.19, primary current I_0 has the following two components :

(*i*) *Active* or *working or iron loss component* I_w. This component is in *phase with* V_1 and mainly supplies the iron loss plus small quantity of primary copper loss.

$$I_w = I_0 \cos \phi_0 \qquad ...(i)$$

(*ii*) *Magnetising component* I_m. This component is *in quadrature with* V_1 and its function is to *sustain the alternating flux in the core.* It is *wattless.*

$$I_m = I_0 \sin \phi_0 \qquad ...(ii)$$

Also $$I_0 = \sqrt{I_w^2 + I_m^2} \qquad ...(iii)$$

The following points are *worthnoting* :

- The no-load primary current I_w is *very small as compared to the full-load primary current.*
- As I_0 is very small, the no-load primary copper loss is negligibly small which means that *no-load primary input is practically equal to the iron loss in the transformer.*

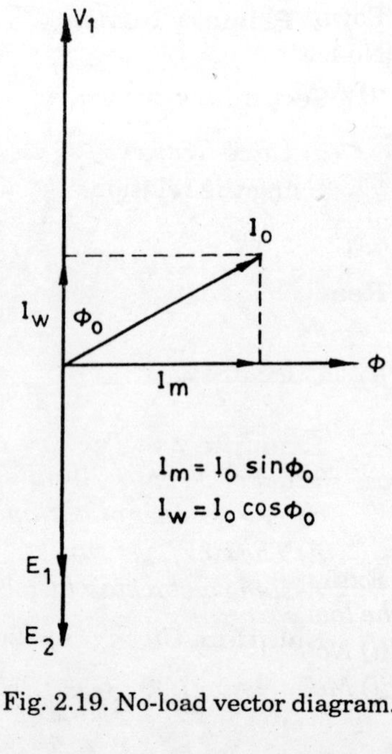

Fig. 2.19. No-load vector diagram.

- Since, it is primarily the core loss which is responsible for shift in the current vector, angle ϕ_0 is known as **hysteresis angle of advance.**

☞**Example 2.7.** *A 3300/300 V single-phase transformer gives 0.6 A and 60 W as ammeter and wattmeter readings when supply is given to the low voltage winding and high voltage winding is kept open, find :*

(i) Power factor of no-load current, *(ii) Magnetising component, and*

(iii) Iron loss component.

Solution. No-load losses, $P_0 = 60$ W

No-load current, $I_0 = 0.6$ A

(*i*) $P_0 = VI_0 \cos \phi_0$

$\therefore$ $60 = 300 \times 0.6 \times \cos \phi_0$

or $$\cos \phi_0 = \frac{60}{300 \times 0.6} = 0.33 \text{ (lagging)}$$

Hence, **power factor of no-load current = 0.33. (Ans.)**

(*ii*) Magnetising component, $I_m = I_0 \sin \phi_0 = 0.6\sqrt{1 - \cos^2 \phi_0} = 0.6\sqrt{1 - (0.33)^2} = 0.566$ A

$\therefore$ $I_m =$ **0.566 A. (Ans.)**

(*iii*) Iron loss component, $I_w = I_0 \cos \phi_0 = 0.6 \times 0.33 =$ **0.198 A. (Ans.)**

[or $$I_w = \sqrt{I_0^2 - I_m^2} = \sqrt{(0.6)^2 - (0.566)^2} = 0.198 \text{ A}].$$

☞**Example 2.8.** *Find (i) active and reactive components of no-load current ; and (ii) no-load current of a 440/220 V single-phase transformer if the power input on no- load to the high voltage winding is 80 W and power factor of no- load current is 0.3 lagging.*

Solution. Primary, $E_1 = 440$ V

Secondary, $E_2 = 220$ V

Power factor, $\cos \phi_0 = 0.3$ (lagging)

No-load losses, $P_0 = 80$ W

(*i*) **Active component** (or *wattful component*),

$$I_w = (I_0 \cos \phi_0) = \frac{P_0}{V_1} = \frac{80}{440} = \mathbf{0.182\ A.\ (Ans.)}$$

$$\cos \phi_0 = 0.3 \ ; \ \phi_0 = \cos^{-1} (0.3) = 72.54°$$

$\therefore$ $\tan \phi_0 = 3.18$

Reactive component (or *magnetising components*)

$$I_m = I_w \tan \phi_0 = 0.182 \times 3.18 = \mathbf{0.578\ A.\ (Ans.)}$$

(*ii*) $I_w = I_0 \cos \phi_0$

$\therefore$ $$I_0 = \frac{I_w}{\cos \phi_0} = \frac{0.182}{0.3} = \mathbf{0.606.\ (Ans.)}$$

or $$[I_0 = \sqrt{I_w^2 + I_m^2} = \sqrt{(0.182)^2 + (0.578)^2} = 0.606 \text{ A}].$$

Example 2.9. *A 3300/220 V, 30 kVA, single-phase transformer takes a no-load current of 1.5 A when the low voltage winding is kept open. The iron loss component is equal to 0.4 A find :*

(i) No-load input power.

(ii) Magnetising component and power factor of no-load current.

Solution. Rating of transformer = 30 kVA

Primary, $E_1 = 3300$ V

Secondary, $E_2 = 220$ V

No-load current, $I_0 = 1.5$ A

Iron loss component, $I_m = 0.4$ A

(*i*) **No-load input power,**

$$P_0 = V_1 I_0 \cos \phi_0 = V_1 I_w \qquad (\because \ I_w = I_0 \cos \phi_0)$$
$$= 3300 \times 0.4 = \mathbf{1320\ W.\ (Ans.)}$$

(*ii*) **Magnetising component,**

$$I_m = \sqrt{I_0^2 - I_w^2} = \sqrt{1.5^2 - 0.4^2} = \mathbf{1.44\ A.\ \ (Ans.)}$$

No-load power factor,

$$\cos \phi_0 = \frac{I_w}{I_0} = \frac{0.4}{1.5} = \mathbf{0.267.\ \ (Ans.)}$$

2.9.4.2. Transformer on Load. *The transformer is said to be loaded when the secondary circuit of a transformer is completed through an impedance or load.* The magnitude and phase of secondary current I_2 with respect to secondary terminal voltage will depend upon the characteristic of load, *i.e.* current I_2 will be in phase, lag behind and lead the terminal voltage V_2 respectively when the load is purely resistive, inductive and capacitive.

The secondary current I_2 sets up its own ampere-turns (= $N_2 I_2$) and creates its own flux ϕ_2 opposing the main flux ϕ_0 created by no-load current I_0. The opposing secondary flux ϕ_2 weakens the primary flux ϕ_0 momentarily hence primary counter or back e.m.f. E_1 tends to be reduced. V_1 gains the upper hand over E_1 momentarily and hence causes more current to flow in primary. Let this additional primary current be I_2'. It is known as *load component of primary current*. The additional primary m.m.f. $N_1 I_2'$ sets up its own flux ϕ_2' which is in opposition to ϕ_2 (but is in the same direction as ϕ_0) and is equal to it in magnitude. Hence they *cancel* each other. Thus we find that the magnetic effects of secondary current I_2 are immediately neutralised by the additional primary current I_2' which is brought into existence exactly at the same instant as I_2.

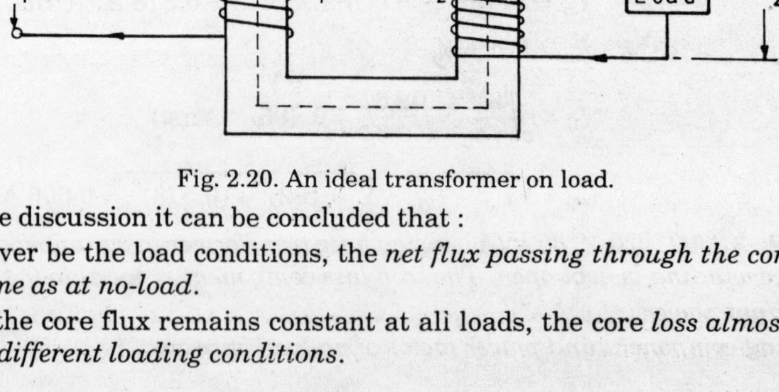

Fig. 2.20. An ideal transformer on load.

From above discussion it can be concluded that :

- Whatever be the load conditions, the *net flux passing through the core is approximately the same as at no-load*.
- Since the core flux remains constant at all loads, the core *loss almost remains constant under different loading conditions*.

Since $\phi_2 = \phi_2'$

$\therefore$ $$N_2 I_2 = N_1 I_2'$$

i.e., $$I_2' = \frac{N_2}{N_1} \times I_2 = K I_2 \qquad \left(\because \frac{N_2}{N_1} = K\right)$$

The total primary current is the vector sum of I_0 and I_2' ; the current I_2' is in *antiphase* with I_2 and K times in magnitude.

The components of primary current can be shown as below :

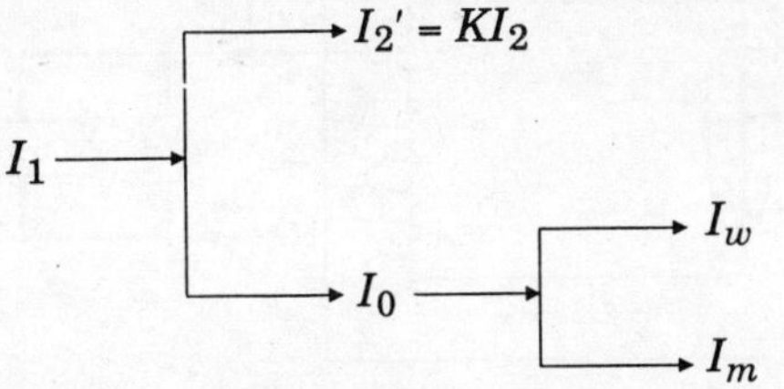

The vector diagrams for transformer on non-inductive, inductive and capacitive loads are shown in Fig. 2.21 (*a, b, c*).

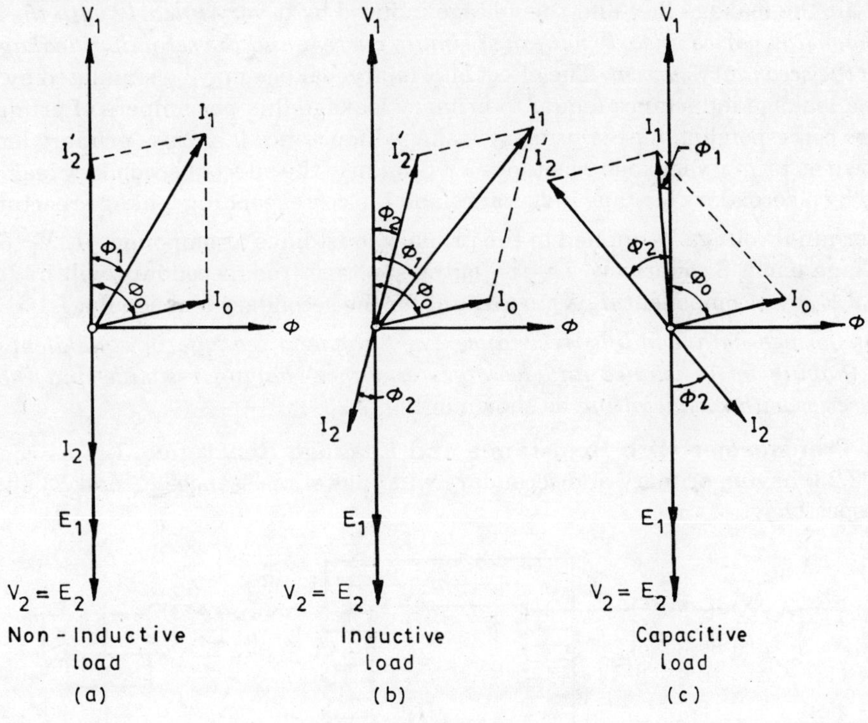

Fig. 2.21

2.9.5. Resistance and Magnetic Leakage. In an ideal transformer it is assumed that windings have got no resistance and there is no leakage flux. But in actual practice it is not possible to have an ideal transformer.

Resistance. In an actual transformer the primary as well secondary winding possess some resistance due to which some voltage drop takes place in them.

Magnetic leakage. Magnetic flux cannot be confined into a desired path. The greater portion of the flux (*i.e.*, the mutual flux) remains confined to the core and links both the windings but a small portion, called the *leakage flux,* completes its path through the air surrounding the coils. As shown in Fig. 2.22 each of the winding is associated with a leakage flux. Since the path of the leakage flux

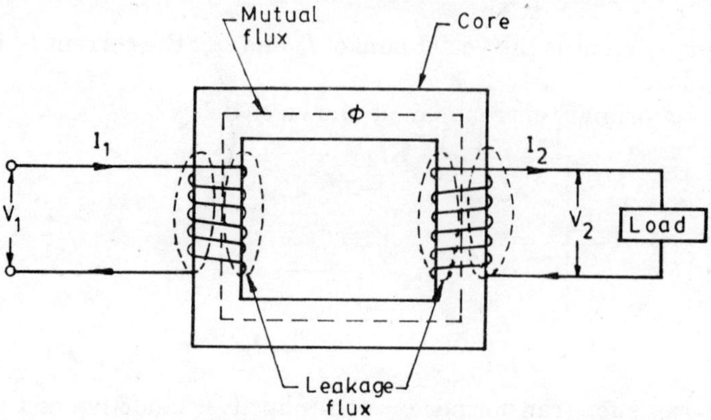

Fig. 2.22. Magnetic flux in a transformer.

is largely in air the leakage flux and the voltage induced by it vary *linearly with the current. The primary leakage flux varies linearly with the primary current and the secondary leakage flux varies linearly with the secondary current.* The effect of primary leakage may be simulated by assigning to the primary a leakage inductance (equal to primary leakage flux per ampere of primary current). The reactance corresponding to this primary leakage inductance (*i.e.* $2\pi f \times$ primary leakage inductance) is known as primary leakage reactance X_1. Similarly the effect of secondary leakage flux may be simulated by a secondary leakage inductance and the corresponding leakage reactance X_2.

The terminal voltage V_1 applied to the primary must have a component $jI_1 X_1$ to balance the primary leakage e.m.f. Similarly V_2 *i.e.* the output voltage from secondary will be less than the induced e.m.f. E_2 by a component $jI_2 X_2$ to account for the secondary leakage flux.

A transformer with winding resistance and magnetic leakage is equivalent to an ideal transformer (having no resistance and leakage resistance) having resistive and inductive coils connected in series with each winding as shown in Fig. 2.23.

2.9.6. Transformer with Resistance and Leakage Reactance. Let us consider transformer (Fig. 2.23) having primary and secondary windings of resistances R_1 and R_2 and reactances X_1 and X_2 respectively.

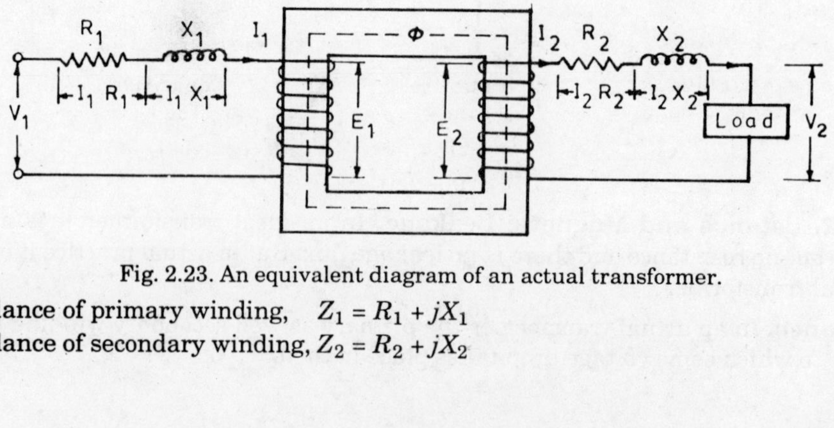

Fig. 2.23. An equivalent diagram of an actual transformer.

Impedance of primary winding, $Z_1 = R_1 + jX_1$

Impedance of secondary winding, $Z_2 = R_2 + jX_2$

The *vector diagrams* for different kinds of loads for such a transformer are shown in Fig. 2.24 (*a, b, c*). *In these diagrams, vectors for resistive drops are drawn parallel to current vectors* whereas *reactive drops are drawn perpendicular to the current vectors*. The angle ϕ_1 between V_1 and I_1 gives the power factor angle of the transformer.

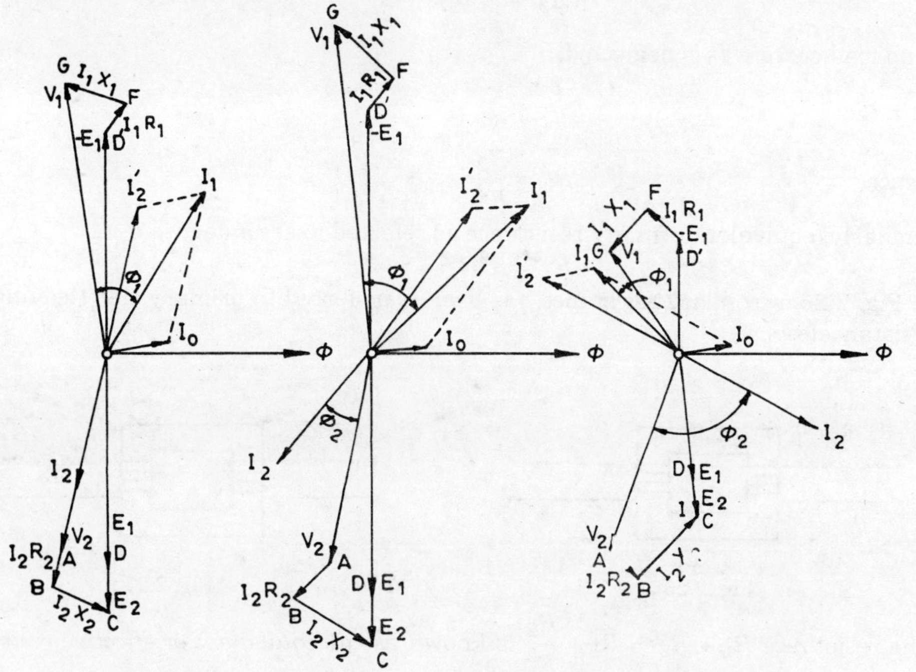

Fig. 2.24. (*a*) Unity p.f., (*b*) lagging p.f. and (*c*) leading p.f.

2.9.7. Equivalent Resistance and Reactance

Equivalent Resistance. In Fig. 2.25 is shown a transformer whose primary and secondary windings have resistances of R_1 and R_2 respectively (shown external to the windings). The resistances of the two windings can be transferred to any one of the two windings. If both the resistances are concentrated in one winding the calculations become simple since then we can work in one winding only.

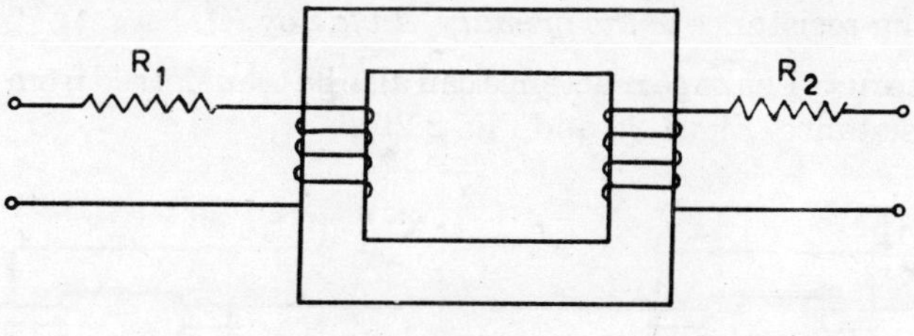

Fig. 2.25

We shall now prove that a resistance of R_2 in secondary is equivalent to $\dfrac{R_2}{K^2}$ in primary and its value will be denoted by R_2', — *the equivalent secondary resistance referred to primary.*

In the secondary, copper loss is $I_2^2R_2$. This loss is supplied by primary which takes a current of I_1. Hence, if R_2' is the equivalent resistance referred to primary which would have caused the same loss as R_2 in secondary, then

$$I_1^2R_2' = I_2^2R_2 \quad \text{or} \quad R_2' = \left(\frac{I_2}{I_1}\right)^2 R_2$$

If no-load current I_0 is neglected,

then, $$\frac{I_2}{I_1} = \frac{1}{K}$$

Hence, $$R_2' = \frac{R_2}{K^2}$$

Similarly, equivalent primary resistance as referred to secondary is

$$R_1' = K^2R_1$$

In Fig. 2.26, secondary resistance has been transferred to primary side (leaving secondary circuit resistanceless).

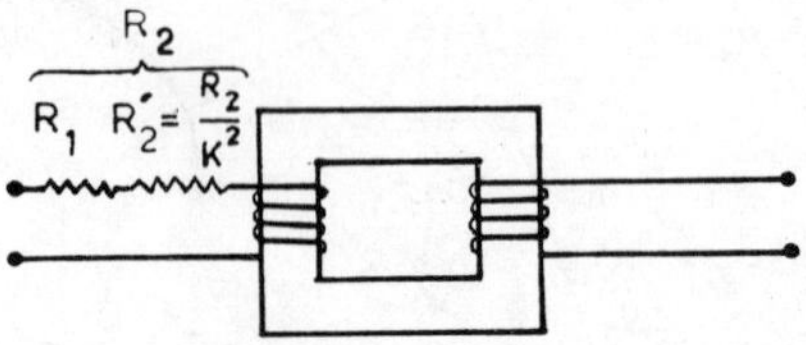

Fig. 2.26

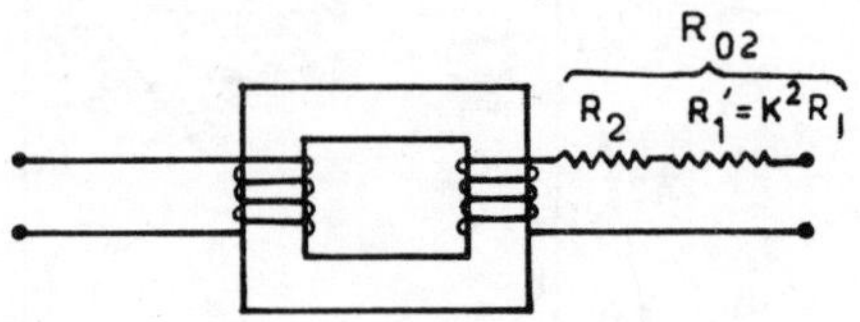

Fig. 2.27

The resistance $R_1 + R_2' = R_1 + \frac{R_2}{K^2}$ is known as the *equivalent or effective resistance of the transformer as referred to the primary* and may be designated as R_{01}.

Thus $$R_{01} = R_1 + R_2' = R_1 + \frac{R_2}{K^2}. \quad \text{...(2.6)}$$

Similarly, the equivalent resistance of the transformer as referred to the secondary (Fig. 2.27) is

$$R_{02} = R_2 + R_1' = R_2 + K^2R_1 \quad \text{...(2.7)}$$

The following points are worth remembering :

(*i*) When shifting resistance *to the secondary, multiply* it by K^2.

(*ii*) When shifting resistance *to the primary, divide* by K^2.

Leakage reactance. Leakage reactance can also be transferred from one winding to the other in the same way as resistance (Fig. 2.28 and Fig. 2.29).

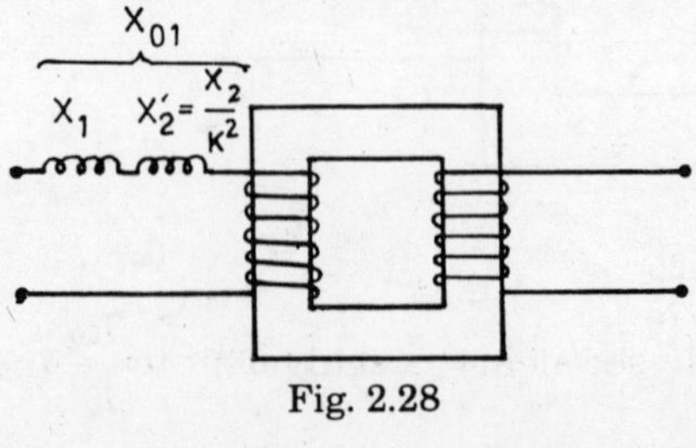

Fig. 2.28

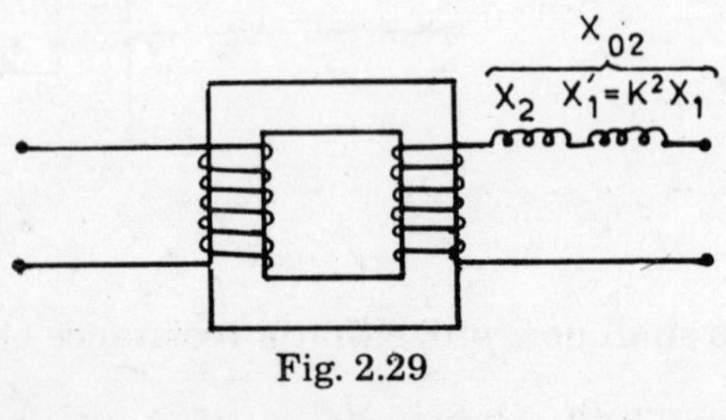

Fig. 2.29

$$X_2' = \frac{X_2}{K^2} \quad \text{and} \quad X_1' = K^2 X_1$$

and $$X_{01} = X_1 + X_2' = X_1 + \frac{X_2}{K^2} \qquad \text{...(2.8)}$$

$$X_{02} = X_2 + X_1' = X_2 + K^2 X_1 \qquad \text{...(2.9)}$$

Total impedance. It is obvious that total impedance of the transformer as referred to primary is given by

$$Z_{01} = \sqrt{R_{01}^{\ 2} + X_{01}^{\ 2}} \qquad \text{(Fig. 2.30) ...(2.10)}$$

and $$Z_{02} = \sqrt{R_0^{\ 2} + X_{02}^{\ 2}} \qquad \text{(Fig. 2.31) ...(2.11)}$$

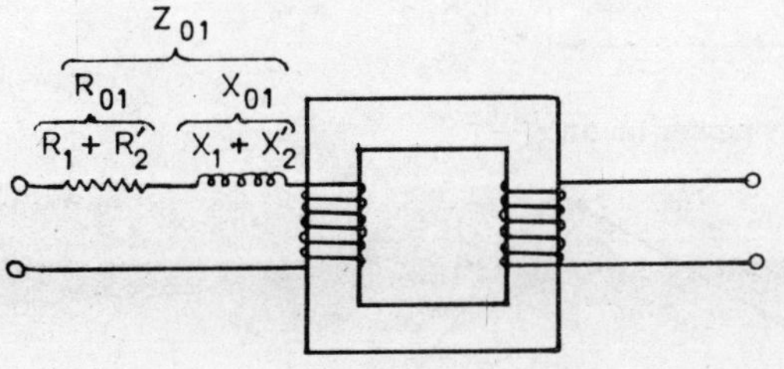

Fig. 2.30

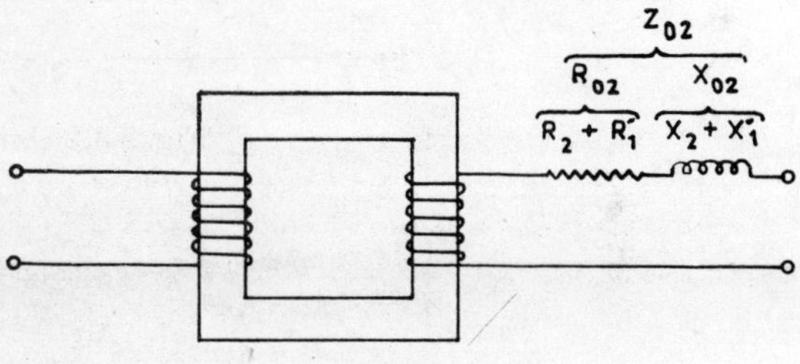

Fig. 2.31

2.9.8. Total Voltage Drop in a Transformer

(*i*) **Approximate voltage drop**. When there is no-load on the transformer, then,

$$V_1 = E_1 \text{ (approximately)}.$$

and $$E_2 = KE_1 = KV_1$$

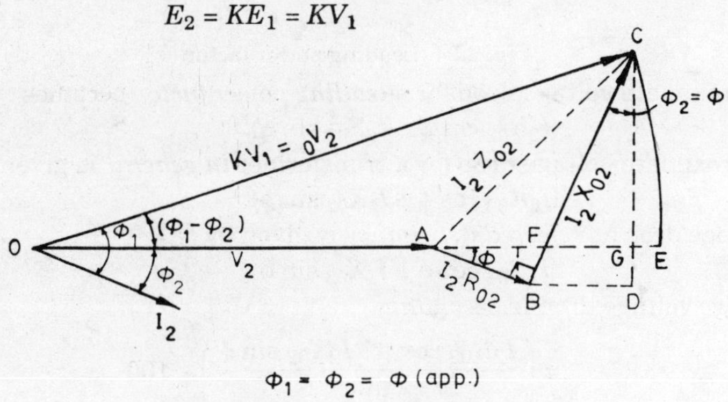

Fig. 2.32. Lagging power factor.

Also $E_2 = {}_0V_2$, where ${}_0V_2$ is secondary terminal voltage on no-load

or $E_2 = {}_0V_2 = KV_1$

V_2 = secondary voltage on load.

Refer Fig. 2.32. The procedure of finding the approximate voltage drop of the transformer as *referred to secondary* is given below :

- Taking O as centre, radius OC draw an arc cutting OA produced at E.
  The total voltage drop $I_1 Z_{02} = AC = AE$ which is approximately equal to AG.
- From B draw BF perpendicular on OA produced. Draw CG perpendicular to OE and draw BD parallel to OE.

Approximate voltage drop $= AG = AF + FG = AF + BD$ [$\because$ $FG = BD$]

$= I_2R_{02} \cos \phi + I_0X_{02} \sin \phi$

This is the value of approximate voltage drop for a *lagging power factor*.

Figs. 2.33 and 2.34 refer to *unity* and *leading power factor* respectively.

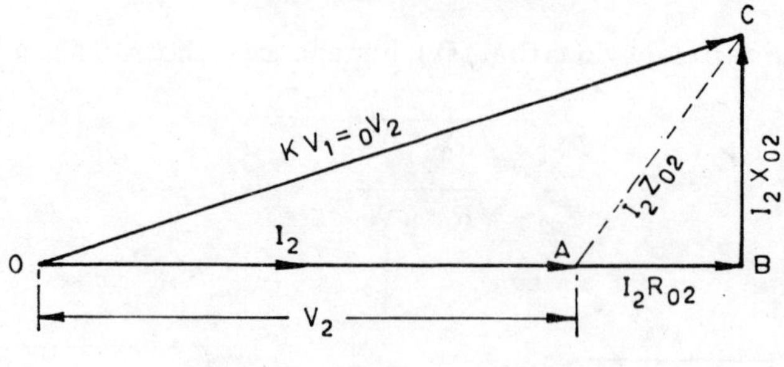

Fig. 2.33. Unity power factor.

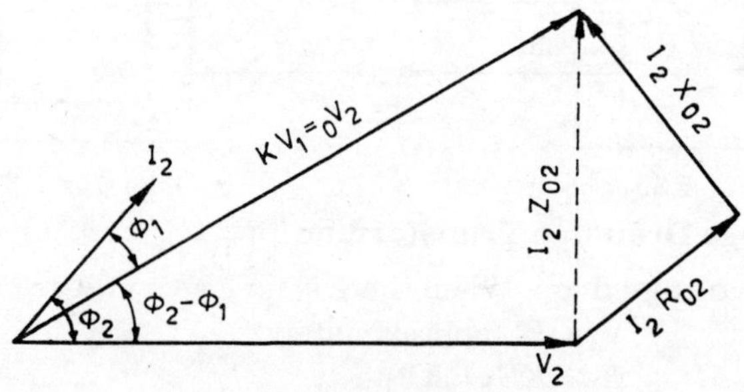

Fig. 2.34. Leading power factor.

- The approximate voltage drop for a *leading power factor* becomes :

$$(I_2R_{02} \cos \phi - I_2X_{02} \sin \phi)$$

- The approximate voltage drop for a transformer *in general* is given by :

$$(I_2R_{02} \cos \phi \pm I_1X_{02} \sin \phi) \qquad \text{...(2.12)}$$

- The voltage drop as *referred to primary* is given by :

$$(I_1R_{01} \cos \phi \pm I_1X_{02} \sin \phi)$$

- *Percentage* voltage drop in *secondary*

$$= \left(\frac{I_2R_{02} \cos \phi \pm I_2X_{02} \sin \phi}{{}_0V_2} \right) \times 100 \qquad \text{...(2.13)}$$

(*ii*) **Exact voltage drop.** When we refer to Fig. 2.32 we find that exact drop is AE and not AG. If quantity GE is added to AG the exact value of voltage drop will be obtained.

From right angled triangle OCG, we get

$$CG^2 = OC^2 - OG^2 = (OC + OG)\,(OC - OG)$$

$$= (OC + OG)\,(OE - OG) = GE \times 2OC$$

$$\therefore \qquad GE = \frac{CG^2}{2OC}$$

Now $\qquad CG = I_2X_{02} \cos \phi - I_2R_{02} \sin \phi$

$$\therefore \qquad GE = \frac{(I_2X_{02} \cos \phi - I_2R_{02} \sin \phi)^2}{2\,{}_0V_2}$$

∴ The exact voltage drop for a *lagging* power factor is

$$= AG + GE$$

$$= (I_2R_{02}\cos\phi + I_2X_{02}\sin\phi) + \frac{(I_2X_{02}\cos\phi - I_2R_{02}\sin\phi)^2}{2{}_0V_2} \quad ...(2.14)$$

For a *leading* power, the expression becomes

$$= (I_2R_{02}\cos\phi - I_2X_{02}\sin\phi) + \frac{(I_2X_{02}\cos\phi + I_2R_{02}\sin\phi)^2}{2{}_0V_2} \quad ...(2.15)$$

In *general*, the exact voltage drop is

$$= (I_2R_{02}\cos\phi \pm I_2X_{02}\sin\phi) + \frac{(I_2X_{02}\cos\phi \mp I_2R_{02}\sin\phi)^2}{2{}_0V_2}$$

Percentage drop is

$$= \frac{(I_2R_{02}\cos\phi \pm I_2X_{02}\sin\phi)\times 100}{{}_0V_2} + \frac{(I_2X_{02}\cos\phi \mp I_2R_{02}\sin\phi)^2\times 100}{2{}_0V_2^2} \quad ...(2.16)$$

[Upper signs ... for *lagging* power factor
Lower signs ... for *leading* power factor]

2.9.9. Equivalent Circuit. In transformers, the problems concerning voltages and currents can be solved by the use of phasor diagrams. However, it is more convenient to represent the transformer by an equivalent circuit. If an equivalent circuit is available the computations can be done by the direct application of circuit theory. An *equivalent circuit is merely a circuit interpretation of the equations which describe the behaviour of the device.*

The transformer windings, in the equivalent circuit, are shown as ideal. The resistance and leakage reactance of the primary and secondary are shown separately in the primary and secondary circuits. The effect of magnetising current is represented by X_0 connected in parallel across the winding. The effect of core loss is represented by a non-inductive resistance R_0 as shown in Fig. 2.35.

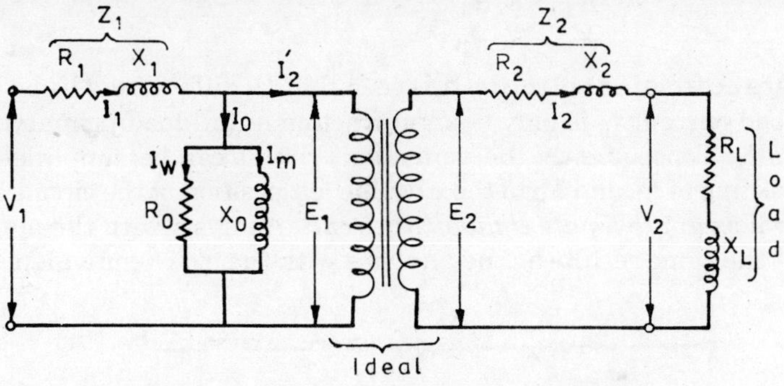

Fig. 2.35. Equivalent circuit of a transformer.

The equivalent circuit can be simplified by transferring the secondary resistances and reactances to the primary side in such a way that the ratio of E_2 to E_1 is not affected in magnitude or phase. Let R_2' be the resistance which must be placed in the primary circuit to produce the same drop as produced by R_2 in the secondary. Then R_2' causes a voltage drop in primary equal to $I_2'R_2'$.

The ratio of $I_2'R_2'$ and I_2R_2 must be the same as the turn ratio $\frac{N_1}{N_2}$.

Thus $$\frac{I_2' R_2'}{I_2 R_2} = \frac{N_1}{N_2} = \frac{1}{K} \quad \text{or} \quad R_2' = R_2 \times \frac{I_2}{I_2'} \times \frac{1}{K}$$

But $$\frac{I_2}{I_2'} = \frac{N_1}{N_2} = \frac{1}{K}$$

$\therefore$ $$R_2' = \frac{R_2}{K^2}$$

Similarly $$X_2' = \frac{X_2}{K^2}.$$

The load resistance and reactance can be transferred to the primary side in the same way. When all the secondary impedances have been transferred to the primary side, the winding need not be shown in the equivalent circuit. The exact equivalent circuit of the transformer with all impedances transferred to the primary side is shown in Fig. 2.36.

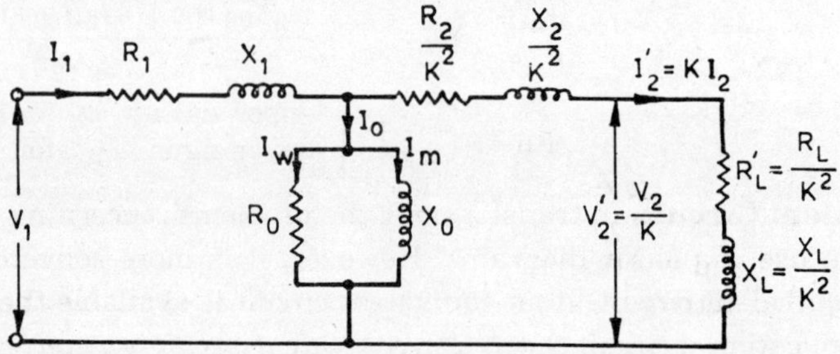

Fig. 2.36. Equivalent circuit with secondary impedances transferred to primary.

Just as the secondary impedances have been transferred to the primary it is possible to transfer the primary winding resistance and reactance to the secondary side. The primary resistance R_1 and leakage reactance X_1 when transferred to secondary are denoted by R_1' and X_1' and are given by

$$R_1' = K^2 R_1$$
$$X_1' = K^2 X_1.$$

Approximate equivalent circuit. It is seen that E_1 differs from V_1 by a very small amount. Moreover, the no-load current I_0 is only a small fraction of full-load primary current so that I_2' is practically equal to I_1. Consequently the equivalent circuit can be simplified by transferring the parallel branch consisting of R_0 and X_0 to the extreme left position of the circuit as shown in Fig. 2.37. This circuit is known as *approximate equivalent circuit*. Analysis with the approximate equivalent circuit gives almost the same results as the analysis with the exact equivalent circuit. *However, the*

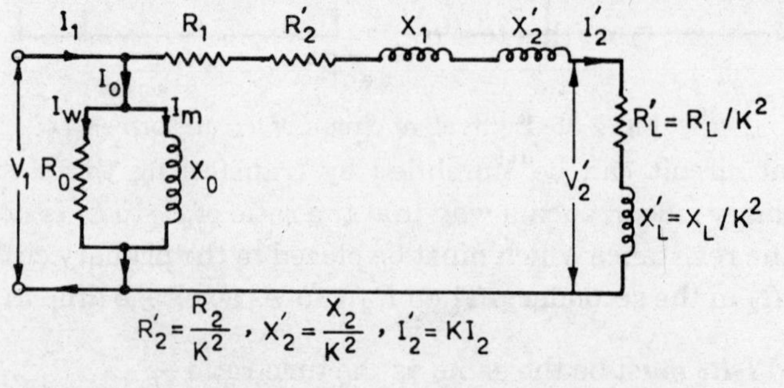

Fig. 2.37. Approximate equivalent circuit of transformer.

analysis with the approximate equivalent circuit is simple because the resistances R_1 and R_2' and leakage reactances X_1 and X_2' can be combined.

2.9.10. Transformer Tests. The performance of a transformer can be calculated on the basis of its equivalent circuit which contains the following four main parameters :

(*i*) Equivalent resistance R_{01} as referred to primary (or secondary R_{02}).

(*ii*) Equivalent leakage resistance X_{01} as referred to primary (or secondary X_{02}).

(*iii*) Core loss conductance G_0 (or resistance R_0).

(*iv*) Magnetising susceptance B_0 (or reactance X_0).

These parameters or constants can be determined by the following *two tests* :

1. Open-circuit or no-load test.

2. Short-circuit or impedance test.

The above two tests are *convenient to perform and very economical because they furnish the required information without actually loading the transformer.*

2.9.10.1. Open-Circuit or No-Load Test (O.C. Test). An open-circuit or no-load test is conducted to find :

(*i*) No-load loss or core loss.

(*ii*) No-load current I_0 which is helpful in finding R_0 and X_0.

The connections for this test are made as shown in Fig. 2.38. One winding of the transformer (*usually high voltage winding*) is *left open* and the other is connected to its supply of *normal voltage and frequency*. Ammeter A and wattmeter W are connected to measure *no-load current* (I_0) *and no-load input power* (P_0) respectively.

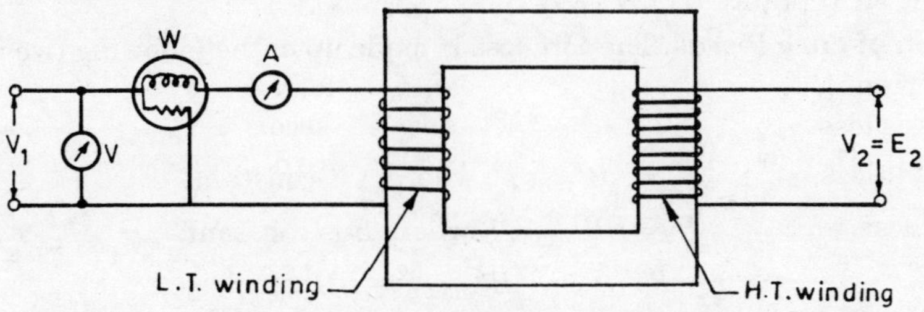

Fig. 2.38. Circuit diagram for open-circuit test.

As the primary no-load current I_0 (as measured by ammeter) is small (usually 3 to 10% of rated load current) copper loss is negligibly small in primary (L.T. winding) and nil in secondary winding (it being open). Hence the *wattmeter reading represents practically the core-loss under no-load conditions* (and *this loss is same for all loads*).

From the data available from this test R_0, X_0, $\cos \phi_0$ (no-load power factor), I_w and I_m can be calculated as follows :

Now, Iron loss $= P_i =$ input power on no-load

$= P_0$ watts (say)

No-load current $= I_0$

Applied primary voltage $= V_1$

Also $P_0 = V_1 I_0 \cos \phi_0$ (where $\cos \phi_0$ = no-load power factor)

$$\therefore \quad \cos \phi_0 = \frac{P_0}{V_1 I_0} \qquad ...(2.17)$$

or
$$\phi_0 = \cos^{-1}\frac{P_0}{V_1 I_0}$$

No-load current *wattful component*,

$$I_w = I_0 \cos\phi_0 = \frac{P_0}{V_1} \quad ...(2.18)$$

No-load current *magnetising component*.

$$I_m = \sqrt{I_0^2 - I_w^2} \quad ...(2.19)$$

The no-load resistance, $R_0 = \frac{V_1}{I_w} = \frac{V_1^2}{P_0}$ ...(2.20)

The no-load reactance, $X_0 = \frac{V_1}{I_m} = \frac{V_1}{\sqrt{I_0^2 - I_w^2}}$ ...(2.21)

The no-load vector diagram is shown in Fig. 2.19.

Since the current is practically all-exciting current when a transformer is on no-load (*i.e.* $I_0 = I_m$) and the voltage drop in primary leakage impedance is small, hence the exciting admittance Y_0 of the transformer is given by

$$I_0 = V_1 Y_0 \quad \text{or} \quad Y_0 = \frac{V_1}{I_0} \quad ...(2.22)$$

The exciting conductance $G_0 = \frac{P_0}{V_1^2}$ ...(2.23)

The exciting susceptance $B_0 = \sqrt{Y_0^2 - G_0^2}$ ...(2.24)

Separation of core losses. The core loss is made up of the following two parts :

(*i*) Eddy current loss.

(*ii*) Hysteresis loss.

Eddy current loss, $P_e = AB^2_{max} f^2$, where A is constant.

Hysteresis loss, $P_h = BB^{1.6}_{max} f$, where B is constant.

Total loss $= P_e + P_h = AB^2_{max} f^2 + BB^{1.6}_{max} f$

The values of constants A and B can be found out by conducting two experiments using two different frequencies but the same maximum flux density ; thereafter eddy current and hysteresis loss can be found separately.

2.9.10.2. Short-Circuit or Impedance Test (S.C. Test). This test is conducted to determine the following :

(*i*) Full-load copper loss.

(*ii*) Equivalent resistance and reactance *referred to metering side.*

In this test (Fig. 2.39) the terminals of the secondary winding (*usually low voltage winding*) are short-circuited by a thick conductor or through an ammeter which may serve the additional purpose of indicating rated load current. A low voltage, usually 5 to 10% of normal primary voltage, at correct frequency is applied to the primary and *is continuously increased till full-load currents flow in the primary as well as secondary windings* (as indicated by the respective ammeters).

Since applied voltage is very low so flux linking with the core is very small and therefore, iron losses are so small that these can be neglected, the reading of the wattmeter gives total copper losses at full-load.

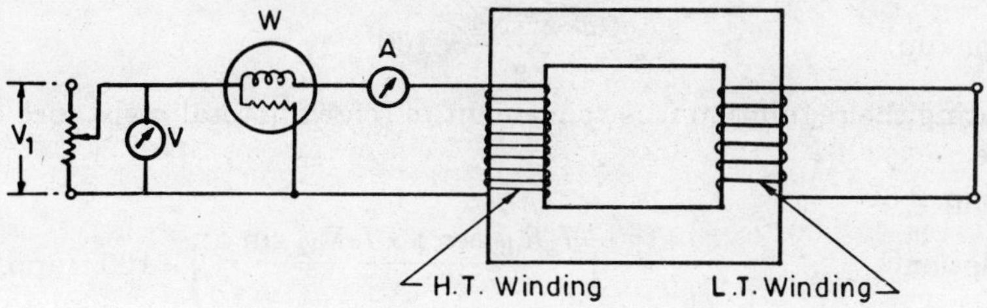

Fig. 2.39. Short-circuit test.

The equivalent circuit of the transformer under short-circuit condition is shown in Fig. 2.40.

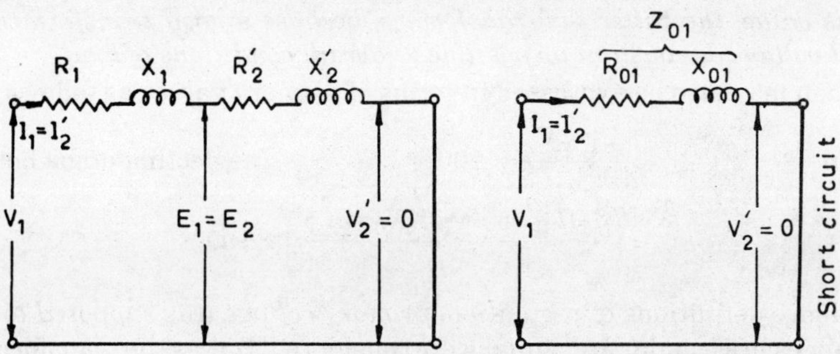

Fig. 2.40. Equivalent circuit of transformer under short-circuit condition.

Let V_{SC} = voltage required to circulate rated load currents

I_1 = reading of the ammeter on the primary side

Z_{01} = total impedance as referred to primary side

R_{01} = total resistance as referred to primary side

X_{01} = total reactance as referred to primary side.

Then, equivalent impedance as referred to primary side,

$$Z_{01} = \frac{V_{SC}}{I_1} \qquad ...(2.25)$$

Also $$P = {I_1}^2 R_{01}$$

$\therefore$ $$R_{01} = \frac{P}{{I_1}^2} \qquad ...(2.26)$$

and $$X_{01} = \sqrt{{Z_{01}}^2 - {R_{01}}^2} \qquad ...(2.27)$$

2.9.11. Regulation of a Transformer

- Due to the resistances of the windings and leakage reactances voltage drop takes place in a transformer. Accordingly the output voltage under load conditions is different from the output voltage under no-load conditions. **Voltage regulation** is defined as :

"The change in secondary voltage when rated load at a specified power is removed".

It is specified as a percentage of the rated secondary voltage.

Thus, if $_0V_2$ = secondary terminal voltage at *no-load*

$= E_2 = KE_1 = KV_1$ because at no-load the impedance drop is negligible

V_2 = secondary terminal voltage on *full-load*

Then, % regulation (down) $$= \frac{_0V_2 - V_2}{_0V_2} \times 100 \qquad ...(2.28)$$

% regulation (up) $= \dfrac{{}_0V_2 - V_2}{V_2} \times 100$ ...(2.29)

For calculating the regulation it is convenient to refer the total resistance and reactance to the secondary side.

- Refer Fig. 2.32.

%age regulation $= \left(\dfrac{I_2R_{02} \cos \phi \pm I_2X_{02} \sin \phi}{{}_0V_2} \right) \times 100$ (app.) ...(230)

[+ sign for *lagging* power factor
– sign for *leading* power factor]

The less this value, the better the transformer, because a good transformer should keep its secondary terminal voltage as constant as possible under all conditions of load.

The regulation may also be expressed in terms of primary values as follows :

% regulation $= \dfrac{V_1 - V_2'}{V_1} \times 100$ (neglecting angle between V_1 and V_2')

$= \left(\dfrac{I_1R_{01} \cos \phi \pm I_1X_0 \sin \phi}{V_1} \right) \times 100$ ...(2.31)

- In the above definitions of regulation, *primary voltage was supposed to be kept constant* and changes in secondary terminal were considered. But as the transformer is loaded, the secondary terminal voltage falls (for lagging power factor) ; hence to keep the circuit voltage constant, the primary voltage must be increased. The rise *in primary voltage required to maintain rated output voltage from no-load to full-load at a given power factor expressed as percentage of rated primary voltage gives the regulation of the transformer.*

If the primary voltage has to be raised from its rated value V_1 to V_1', then

% regulation $= \dfrac{V_1' - V}{V_1} \times 100$ ...[2.31 (*a*)]

2.9.12. Percentage Resistance and Reactance

- Invariably the equivalent resistance and reactance of a transformer are expressed in percent.

Percentage resistance is the *resistance drop in volts at rated current and frequency as a percentage of the rated voltage i.e.,*

$$\text{Percent } R = \frac{IR}{V} \times 100 \quad ...(i)$$

Percentage reactance is the *reactance drop in volts at rated current and frequency expressed as a percentage of the rated voltage i.e.,*

$$\text{Percent } X = \frac{IX}{V} \times 100 \quad ...(ii)$$

and

$$\text{Percent } Z = \frac{IZ}{V} \times 100 \quad ...(iii)$$

The important **advantage** of expressing resistance and reactance of a transformer in percentage is that the *percentage resistance and percentage reactance have same values whether determined referred to primary or secondary whereas when expressed in ohms they have different values when referred to primary and secondary.*

Further, *percent reactances and impedances are convenient for computing the current that will flow in a transformer when the secondary is accidentally shorted.* When the secondary is shorted, I_1

$= \frac{E_1}{Z_0}$ or $E_1 = I_1Z_0$ *i.e.*, when secondary is shorted, I_1Z_0 is 100 percent of E_1. Therefore, the short-circuit current is given by the equation,

$$\text{Short circuit current}\,;\, I_{SC} = \frac{100}{\text{Percent impedance}} \times \text{full load rated current} \quad ...(iv)$$

Unless the transformer is quite small, equivalent resistance is negligible and percent reactance may be substituted for percent impedance in the above equation (*iv*).

- **Another method of designating the resistances and reactances** is by the **per unit values.** The per unit values are equal to the percentage values divided by 100.

$$\therefore \qquad \text{Per unit } R = \frac{IR}{V} \quad ...(v)$$

and

$$\text{Per unit } X = \frac{IX}{V} \quad ...(vi)$$

2.9.13. Kapp Regulation Diagram. It is seen that the secondary terminal *voltage falls* as the load on the transformer is increased when power factor is lagging and it increases when the power factor is *leading*. Thus the secondary terminal voltage not only depends on the load but on power factor also.

Dr. Kapp proposed the graphical method of determining the voltage drop in a transformer and the graphical construction is called *Kapp regulation diagram*.

In order to draw Kapp regulation diagram it is necessary to know the equivalent resistance and reactance as referred to secondary i.e. R_{02} *and* X_{02}.

Following is the **procedure** of drawing the Kapp regulation diagram :

(*i*) Draw a vector *OL* to represent the *secondary terminal voltage on load,* V_2.

(*ii*) Draw *OX* at an angle ϕ_2 to *OL*, where cos ϕ_2 is the power factor of load. Then *OX* represents the phase of secondary current.

(*iii*) Draw *LM* parallel to *OX* and equal to resistance drop referred to secondary, *i.e.* I_2R_{02}, and perpendicular *MN* equal to reactance drop referred to the secondary *i.e.* I_2X_{02}. Then *NLM* is the impedance triangle referred to the secondary, and *NL* is the total voltage drop.

Hence *NL* is the secondary *no-load voltage* ${}_0V_2$ and the arithmetic difference (*ON* – *OL*) is the secondary drop from no-load to full-load.

(*iv*) Transfer triangle *NLM* to *OO'P*, as shown ; then $O'L = ON = {}_0V_2$. Hence for a given secondary current, the locus of *N* is a circle with centre *O* and radius ${}_0V_2$, while the locus of *L* is a circle of radius ${}_0V_2$ with center *O'*.

(*v*) To find the voltage drop on full-load at *any power factor* cos ϕ, draw a radius *OQS* inclined at angle ϕ to *OX*. If the impedance triangle is drawn in position *UQT*, then *OU* = *OS*, the secondary no-load voltage, ${}_0V_2$ and *OQ* = the secondary voltage on load, V_2. Hence length *QS* represents the voltage drop.

It will be seen that ; for all phase angle ϕ inclined between $\angle XOV$ (which is leading) and $\angle WOX$ (which is lagging), there is a voltage drop on load. For phase angles equal to either of these two limits the voltage on load is equal to that on no-load. For phase angles between $\angle WOX$ and $\angle XOV$ (reckoning in an anti-clockwise direction) there is *actually increase of voltage on no-load.*

The Kapp diagram *is very helpful in determining the variation of regulation with power factor.* However, its disadvantage is that since the lengths of the sides of the impedance triangle are very small as compared to radii of the circles, the *diagram has to be drawn on a very large scale for getting sufficiently accurate results.*

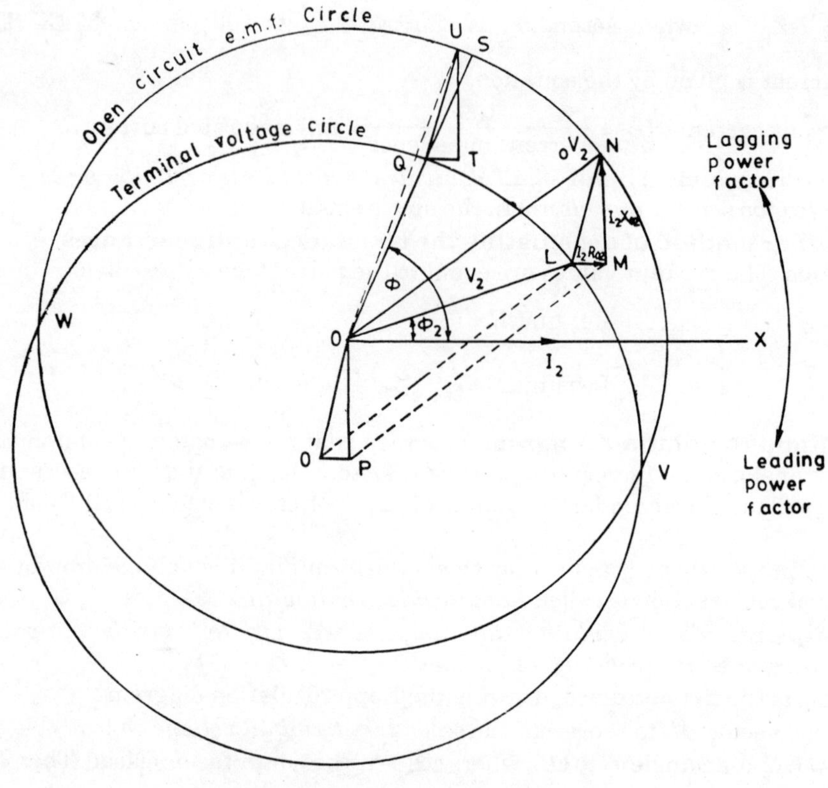

Fig. 2.41. Kapp regulation diagram.

2.9.14. Sumpner's or Back to Back Test

- Although open circuit and short-circuit tests can be used to find the efficiency and regulation of the transformer accurately yet for determination of *temperature rise* it becomes necessary to put the transformer to full-load for several hours. Transformers of smaller capacity can be loaded artificially by means of lamp loads or water loads but it may be a very difficult to arrange suitable loads for large rating transformers. Moreover there is ample wastage of energy.
- In this test *two similar transformers are required.* These are loaded fully (in similar way as two D.C. machines in a regenerative test) and the power required from the supply is that necessary for supplying the iron and copper losses of both transformers.

Fig. 2.42 shows the *connection diagram for Sumpner's test.*

The primary windings, usually low voltage windings, of the two transformers are connected in parallel across the same A.C. supply. With switch S open, the wattmeter W_1 reads the core loss for the two transformers. The secondaries are so connected that their potentials are in opposition to each other. This would be so if $V_{LM} = V_{NP}$ and L is joined to N whilst M is joined to P. In that case, there would be no secondary current flowing around the loop formed by the two secondaries.

A is an auxiliary low-voltage transformer (regulating transformer) which can be adjusted to give a variable voltage and hence current in the secondary loop circuit. By proper adjustment of A, full-load secondary current can be made to flow as shown. It is seen that I_2 flows from P to N and then from L to M. Flow of I_1 is confined to $RQJYTUVQ$ and it does not pass through W_1. Hence

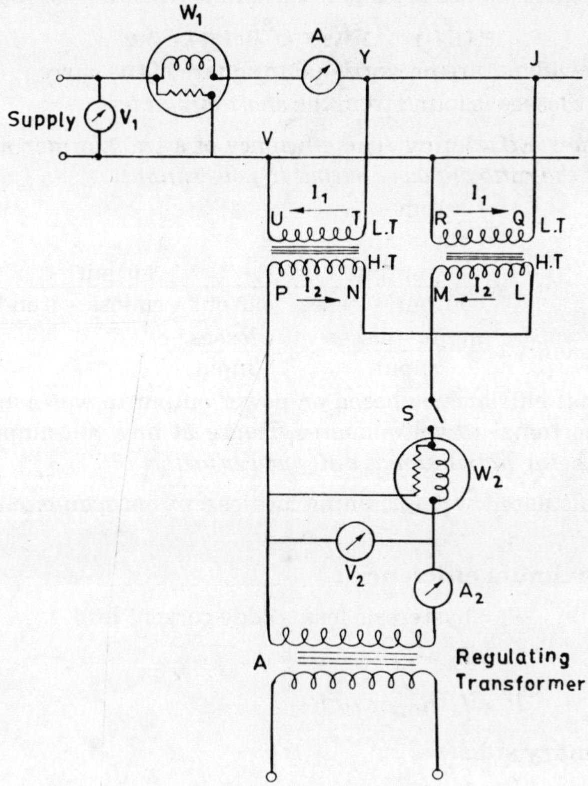

Fig. 2.42. Sumpner's test.

wattmeter W_2 measures full-load copper loss (or at any other load current I_2) and wattmeter W_1 continues to read the core loss. Evidently, the *power taken is twice the losses of a single transformer.*

2.9.15. Transformer Losses. The losses in a transformer are classified as follows :

1. Iron losses (or core losses).
2. Copper losses.

1. **Iron or core losses.** It includes *hysteresis loss* and *eddy current loss.*

(*i*) **Hysteresis loss.** Since the flux in a transformer core is alternating, power is required for the continuous reversals of the elementary magnets of which the iron is composed. This loss is known as *hysteresis loss.*

Hysteresis loss $= K_h f B^{1.6}_{max}$ ...(2.32)

where f is the frequency in Hz, B_{max} is the maximum flux density in core and K_h is a constant.

(*ii*) **Eddy current loss.** This is due to the flow of eddy currents in the core. Thin laminations, insulated from each other, reduce the eddy current loss to small proportion.

Eddy current loss $= K_e f^2 B^2_{max}$ ...(2.33)

where K_e is a constant.

Iron or core loss is found from open circuit test. The input of the transformer when on no-load measures the core loss.

2. **Copper losses.** These losses are due to the ohmic resistance of the transformer windings.

Total copper loss $= I_1^2R_1 + I_2^2R_2 = I_1^2R_{01} = I_2^2R_{02}$

These losses, as is evident, are proportional to square of the current (or kVA)^2.

The value of copper losses is found from the *short-circuit test*.

2.9.16. Transformer Efficiency. The efficiency of a transformer at a particular load and power factor is defined as *the ratio of power output to power input*.

$$\therefore \text{ Efficiency} = \frac{\text{output}}{\text{input}}$$

$$= \frac{\text{output}}{\text{output + losses}} = \frac{\text{output}}{\text{output + cu loss + iron loss}} \quad ...(2.34)$$

or
$$\text{Efficiency} = \frac{\text{input} - \text{losses}}{\text{input}} = 1 - \frac{\text{losses}}{\text{input}} \quad ...(2.35)$$

It may be noted that efficiency is based on power output in watts and not in volt-amperes, although losses are proportional to volt-amperes. Hence at any volt-ampere load, the efficiency depends on power factor, being *maximum at unity power factor*.

Efficiency can be calculated by determining core losses from open-circuit test and copper losses from short-circuit test.

Condition for maximum efficiency :

Iron losses, P_i = hysteresis loss + eddy current loss

$= P_h + P_e$

Copper losses, $P_c = I_1^2R_{01}$ or $I_2^2R_{02}$

Considering primary side :

Input to primary $= V_1I_1 \cos\phi_1$

$$\text{Efficiency, } \eta = \frac{V_1I_1\cos\phi_1 - \text{losses}}{V_1I_1\cos\phi_1} = \frac{V_1I_1\cos\phi_1 - I_1^2R_{01} - P_i}{V_1I_1\cos\phi_1}$$

$$= 1 - \frac{I_1R_{01}}{V_1\cos\phi_1} - \frac{P_i}{V_1I_1\cos\phi_1}$$

Differentiating both sides w.r.t. I_1, we get

$$\frac{d\eta}{dI_1} = 0 - \frac{R_{01}}{V_1\cos\phi_1} + \frac{P_i}{V_1I_1^2\cos\phi_1}$$

For η to be maximum,

$\dfrac{d\eta}{dI_1} = 0$. Hence the above equation reduces to

$$\frac{R_{01}}{V_1\cos\phi_1} = \frac{P_i}{V_1I_1^2\cos\phi_1}$$

or
$$P_i = I_1^2R_{01} \text{ or } I_2^2R_{02} \quad ...(2.36)$$

or **Copper losses = Iron losses**

The output current corresponding to maximum efficiency is

$$I_2 = \sqrt{\frac{P_i}{R_{02}}} \quad ...(2.37)$$

By proper design it is possible to make the maximum efficiency occur at any desired load.

Variation of efficiency with power factor. We know that transformers efficiency,

$$\eta = \frac{\text{output}}{\text{input}} = \frac{\text{input} - \text{losses}}{\text{input}}$$

$$= 1 - \frac{\text{losses}}{\text{input}} = 1 - \frac{\text{losses}}{(V_2 I_2 \cos\phi + \text{losses})}$$

Let, $\frac{\text{losses}}{V_2 I_2} = \beta$

$\therefore$ $$\eta = 1 - \left(\frac{\text{losses}/V_2 I_2}{\cos\phi + \text{losses}/V_2 I_2}\right) \quad \text{or} \quad = 1 - \frac{\beta}{(\cos\phi + \beta)}$$

or $$\eta = 1 - \frac{(\beta/\cos\phi)}{1 + (\beta/\cos\phi)} \qquad ...(2.38)$$

The variations of efficiency with power factor at different loadings on a typical transformer are shown in Fig. 2.43.

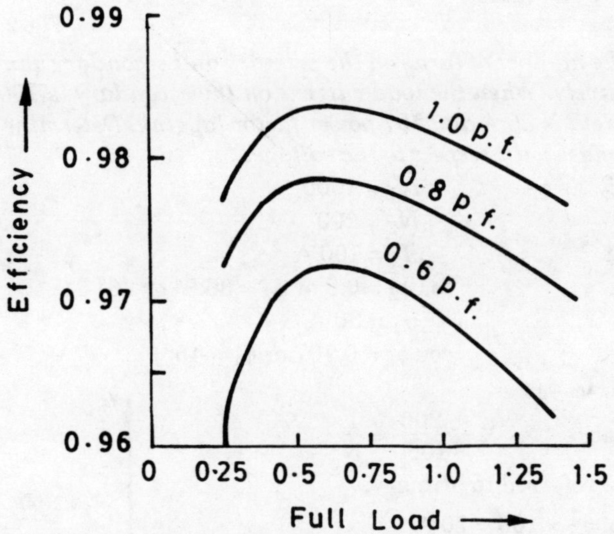

Fig. 2.43. Variations of efficiency with power factor at different loadings.

TRANSFORMER ON LOAD

☞**Example 2.10.** *A 230 V/115 V single-phase transformer takes a no-load current of 2 A at a power factor of 0.2 lagging with low voltage winding kept open. If the low voltage winding is now loaded to take a current of 15 A at 0.8 power factor lagging find the current taken by high voltage winding.*

Solution. Primary, $E_1 = V_1 = 230$ V

Secondary, $E_2 = V_2 = 115$ V

No load current $I_0 = 2$ A

No-load power factor, $\cos\phi_0 = 0.2$ or $\phi_0 = 78.46°$ or $78° \, 29'$

Load power factor, $\cos\phi_2 = 0.8$ or $\phi_2 = 36.9°$ or $36° \, 54'$

Current taken by h.v. winding, I_1 :

Now, transformation ratio, $K = \frac{V_2}{V_1} = \frac{E_2}{E_1} = \frac{115}{230} = \frac{1}{2}$

∴ Secondary current referred to primary,

$$I_2' = KI_2 = \frac{1}{2} \times 15 = 7.5 \text{ A}$$

Phase angle, between I_0 and I_2'

$= 78.46° - 36.9°$

$= 41.56°$ or $41° \, 34'$

Using parallelogram law of forces (Fig. 2.44), we get

$$I_1 = \sqrt{2^2 + 7.5^2 + 2 \times 2 \times 7.5 \times \cos 41.56°}$$

$= 9.09$ A

Hence, **current taken by h.v. winding,**

$$\mathbf{I_1 = 9.09 \text{ A.}} \quad \textbf{(Ans.)}$$

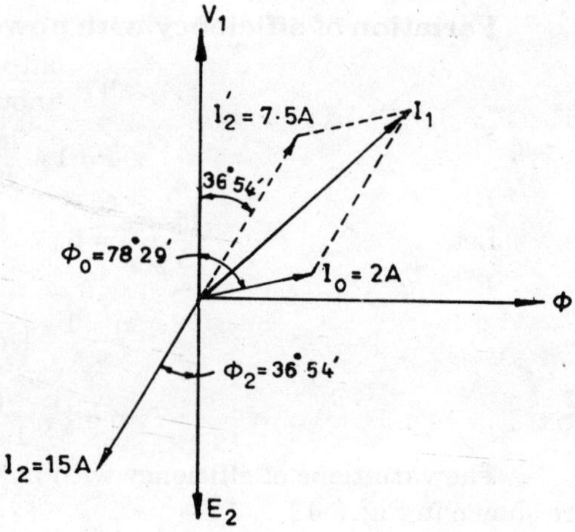

Fig. 2.44

Example 2.11. *The number of turns on the primary and secondary windings of a transformer are 1000 and 200 respectively. When the load current on the secondary is 100 A at 0.8 power factor lagging, the primary current is 30 A at 0.707 power factor lagging. Determine the no-load current of the transformer and its phase with respect to the voltage.*

Solution. Primary turns, $N_1 = 1000$

Secondary turns, $N_2 = 200$

Secondary current. $I_2 = 100$ A

Power factor, $\cos \phi_2 = 0.8$ or $\phi_2 = 36.9°$ or $36° \, 54'$

Primary current, $I_1 = 30$ A

Power factor, $\cos \phi_1 = 0.707$ or $\phi_1 = 45°$

No load current, I_0, ϕ_0 :

Transformation ratio, $K = \frac{200}{1000} = \frac{1}{5}$

Secondary current referred to primary,

$$I_2' = KI_2 = \frac{1}{5} \times 100 = 20 \text{ A}$$

Refer Fig. 2.45. I_1 is the vector sum of I_0 and I_2'. Let I_0 lag behind V_1 by an angle ϕ_0.

Resolving currents into their X-and Y-components, we get

$$I_0 \cos \phi_0 + 20 \cos 36.9° = 30 \cos 45°$$

$$\therefore \; I_0 \cos \phi_0 = 30 \cos 45° - 20 \cos 36.9°$$

$$= 21.21 - 16 = 5.21 \text{ A} \qquad ...(i)$$

$$I_0 \sin \phi_0 + 20 \sin 36.9° = 30 \sin 45°$$

$$\therefore \; I_0 \sin \phi_0 = 30 \sin 45° - 20 \sin 36.9°$$

$$= 21.21 - 12 = 9.21 \text{ A} \qquad ...(ii)$$

From (*i*) and (*ii*), we get

$$\tan \phi_0 = \frac{9.21}{5.21} = 1.767$$

$$\therefore \qquad \phi_0 = 60.5°$$

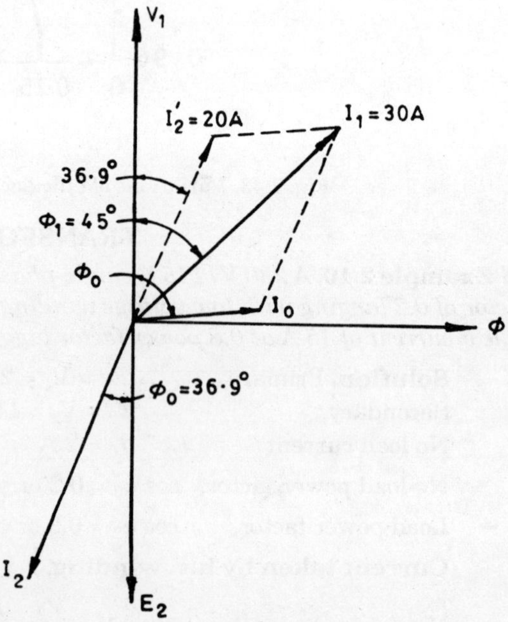

Fig. 2.45

Putting the value of ϕ_0 in *(i)*, we get

$$I_0 \cos 60.5° = 5.21$$

$$\therefore \quad I_0 = \frac{5.21}{\cos 60.5°} = 10.58 \text{ A}$$

Hence, **no-load current = 10.58 A**

and $\quad \phi_0 = \mathbf{60.5°}$. **(Ans.)**

Example 2.12. *A 30 kVA, 2000/200 V, single-phase, 50 Hz transformer has a primary resistance of 3.5 Ω and reactance of 4.5 Ω. The secondary resistance and reactance are 0.015 Ω and 0.02 Ω respectively. Find :*

(i) Equivalent resistance, reactance and impedance referred to primary.

(ii) Equivalent resistance, reactance and impedance referred to secondary.

(iii) Total copper loss of the transformer.

Solution. Primary resistance, $R_1 = 3.5\ \Omega$

Primary reactance, $X_1 = 4.5\ \Omega$

Secondary resistance, $R_2 = 0.015\ \Omega$

Secondary reactance, $X_2 = 0.02\ \Omega$

Transformation ratio, $K = \frac{200}{2000} = \frac{1}{10}$

(i) Equivalent resistance *referred to primary*,

$$R_{01} = R_1 + R_2' = R_1 + \frac{R_2}{K^2} = 3.5 + \frac{0.015}{\left(\frac{1}{10}\right)^2} = \mathbf{5.0\ \Omega}. \quad \textbf{(Ans.)}$$

Equivalent reactance *referred to primary*,

$$X_{01} = X_1 + X_2' = X_1 + \frac{X_2}{K^2} = 4.5 + \frac{0.02}{\left(\frac{1}{10}\right)^2} = \mathbf{6.5\ \Omega}. \quad \textbf{(Ans.)}$$

Equivalent impedance *referred to primary*,

$$Z_{01} = \sqrt{(R_{01})^2 + (X_{01})^2} = \sqrt{(5)^2 + (6.5)^2} = \mathbf{8.2\ \Omega}. \quad \textbf{(Ans.)}$$

(ii) Equivalent resistance *referred to secondary*,

$$R_{02} = R_2 + R_1' = R_2 + K^2 R_1$$

$$= 0.015 + \left(\frac{1}{10}\right)^2 \times 3.5 = \mathbf{0.05\ \Omega}. \quad \textbf{(Ans.)}$$

Equivalent reactance *referred to secondary*,

$$X_{02} = X_2 + X_1' = X_2 + K^2 X_1 = 0.02 + \left(\frac{1}{10}\right)^2 \times 4.5 = \mathbf{0.065\ \Omega}. \quad \textbf{(Ans.)}$$

Equivalent impedance *referred to secondary*,

$$Z_{02} = \sqrt{(R_{02})^2 + (X_{02})^2} = \sqrt{(0.05)^2 + (0.065)^2} = \mathbf{0.082\ \Omega}. \quad \textbf{(Ans.)}$$

(iii) **Total copper loss :**

Secondary current $= \frac{30 \times 1000}{200} = 150$ A

Total copper loss $= I_2^2 R_{02} = 150^2 \times 0.05 = \mathbf{1125\ W}$. **(Ans.)**

☞**Example 2.13.** *A single-phase transformer has the following data :*

Turn ratio 20 : 1 ; R_1 = 20 Ω, X_1 = 80 Ω ; R_2 = 0.04 Ω ; X_2 = 0.2 Ω. No-load current = 1.2 A leading the flux by 30°.

The secondary delivers 180 A at a terminal voltage of 400 V and at a power factor of 0.8 lagging. Determine by the aid of a vector diagram :

(i) The primary applied voltage. *(ii) The primary power factor.*

(iii) The efficiency.

Solution. Refer Fig. 2.46 :

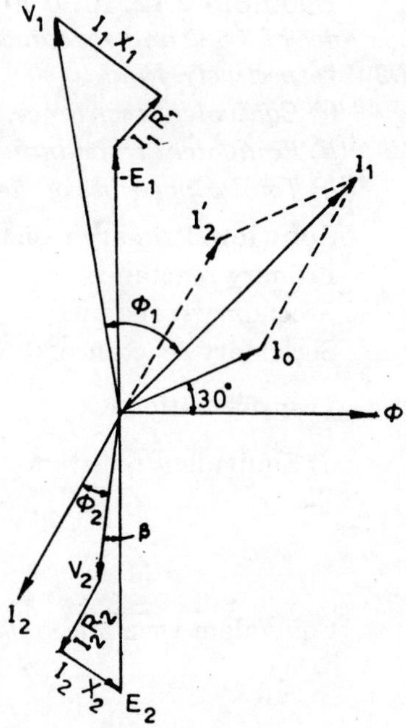

Fig. 2.46

(*i*) **Primary applied voltage, V_1 :**

Taking V_2 as the reference vector

$$\therefore \quad V_2 = 400 \angle 0° = 400 + j0$$

$$I_2 = 180(0.8 - j0.6) = 144 - j108$$

$$Z_2 = (0.04 + j0.2)$$

$$E_2 = V_2 + I_2Z_2$$

$$= (400 + j0) + (144 - j108) \times (0.04 + j0.2)$$

$$= 400 + (5.76 + j28.8 - j4.32 + 21.6)$$

$$= (427.36 + j24.48) = 428.1 \angle 3.28°.$$

Obviously,

$$\beta = 3.28°$$

$$E_1 = \frac{E_2}{K} = 20E_2 = 20(427.36 + j24.48) = 8547 + j490$$

$$\therefore \quad -E_1 = -8547 - j490 = 8561 \angle 183.28°$$

Secondary current referred to primary,

$$I_2' = -KI_2 = \frac{(-144 + j\,108)}{20} = -7.2 + j5.4$$

As seen from Fig. 2.46, I_0 leads V_2 by an angle

3.28° + 90° + 30° = 123.28°

$$\therefore \quad I_0 = 1.2 \angle 123.28° = 1.2(\cos 123.28° + j \sin 123.28°)$$

$$= 1.2(-0.548 + j0.836) = -0.657 + j1.003.$$

Primary current,

$$I_1 = -I_2' + I_0 = (-7.2 + j5.4) + (-0.657 + j1.003) = -7.857 + j6.403 = 10.14 \angle 140.8°$$

$$V_1 = -E_1 + I_1Z_1 = (-8547 - j490) + (-7.857 + j6.403)(20 + j80)$$

$$= -8547 - j490 + (-157.14 - j628.56 + j128.06 - 512.24)$$

$$= -9216.38 - j990.5 = \mathbf{9269 \angle 186.13°}. \quad \textbf{(Ans.)}$$

(*ii*) **Primary power factor, cos ϕ_1 :**

Phase angle between V_1 and I_1,

$$\phi_1 = 186.13° - 140.8° = 45.33°$$

∴ Primary power factor = cos 45.33° = **0.703 (lag). (Ans.)**

(*iii*) **Efficiency :**

No-load primary input power

$$= V_1I_1 \cos \phi_0 = 9269 \times 1.2 \times \cos 60° = 5561.4 \text{ W}$$

$$R_{02} = R_2 + K^2R_1 = 0.04 + \left(\frac{1}{20}\right)^2 \times 20 = 0.09\ \Omega$$

Total copper losses as referred to secondary

$$= I_2^2 R_{02} = (180)^2 \times 0.09 = 2916 \text{ W}$$

Output $= V_2 I_2 \cos\phi_2 = 400 \times 180 \times 0.8 = 57600$ W

Total losses $= 5561.4 + 2916 = 8477.4$ W

Input $=$ Output + Losses $= 57600 + 8477.4 = 66077.4$ W

$\therefore$ Efficiency, $\eta = \dfrac{\text{output}}{\text{input}} = \dfrac{57600}{66077.4} = 0.872$ or **87.2%. (Ans.)**

Example 2.14. *The high voltage and low voltage windings of a 2200/220 V single-phase 50 Hz. transformer has resistances of 4.8 Ω and 0.04 Ω and reactances 2 Ω and 0.018 Ω respectively. The low voltage winding is connected to a load having an impedance of (6 + j4) Ω. Determine :*

(i) Current in l.v. winding, *(ii) Current in h.v. winding,*

(iii) Load voltage, and *(iv) Power consumed by the load.*

Solution. Primary resistance, $R_1 = 4.8\ \Omega$

Primary reactance, $X_1 = 2\ \Omega$

Secondary resistance, $R_2 = 0.04\ \Omega$.

Secondary reactance, $X_2 = 0.018\ \Omega$

Impedance of load, $Z_L = (6 + j4)$

Transformation ratio, $K = \dfrac{E_2}{E_1} = \dfrac{220}{2200} = \dfrac{1}{10}$

Equivalent resistance referred to l.v. side,

$$R_{02} = R_2 + K^2 R_1 = 0.04 + \left(\frac{1}{10}\right)^2 \times 4.8 = 0.088\ \Omega$$

Equivalent reactance referred to l.v. side,

$$X_{02} = X_2 + K^2 X_1 = 0.018 + \left(\frac{1}{10}\right)^2 \times 2 = 0.038\ \Omega$$

Equivalent impedance referred to l.v. side,

$$Z_{02} = R_{02} + jX_{02} = (0.088 + j0.038)\ \Omega$$

(i) **Current in l.v. winding,**

$$I_2 = \frac{220}{Z_{02} + Z_L} = \frac{220}{(0.088 + j0.038) + (6 + j4)}$$

$$= \frac{220}{6.088 + j4.038} = \frac{220}{\sqrt{(6.088)^2 + (4.038)^2}} = \textbf{30.11 A. (Ans.)}$$

(ii) **Current in h.v. winding,**

$$I_1 \simeq I_2' = KI_2 = \frac{1}{10} \times 30.11 = \textbf{3.011 A. (Ans.)}$$

(iii) **Load voltage,** $= I_2 Z_L = 30.11(6 + j4) = 30.11 \times \sqrt{(6)^2 + (4)^2} =$ **217.12 V. (Ans.)**

(iv) **Power consumed by the load,**

$$= I^2 R_L = 30.11^2 \times 6 = \textbf{5439.6 W. (Ans.)}$$

☞**Example 2.15.** *A single-phase transformer has $Z_1 = 1.4 + j5.2\ \Omega$ and $Z_2 = 0.0117 + j0.0465\ \Omega$. The input voltage is 6600 V and the turn ratio is 10.6 : 1. The secondary feeds a load which draws 300 A at 0.8 power factor lagging. Find the secondary terminal voltage and the kW output. Neglect no-load current I_0.*

Solution. Impedance $Z_1 = 1.4 + j5.2\ \Omega$

Impedance $Z_2 = 0.0117 + j0.0465\ \Omega$

Input voltage $V_1 = 6600$ V

Turn ratio, $K = 10.6 : 1$

Secondary load current, $I_2 = 300$ A

Power factor, $\cos \phi_2 = 0.8$

Secondary terminal voltage, V_2 :

kW output :

Let the secondary terminal voltage $= V_2 \angle 0°$

$$I_2 = 300 \angle -36.87° \text{ A}$$

$$E_2 = V_2 \angle 0° + I_2 Z_2$$

$$= V_2 \angle 0° + (300 \angle -36.87°)(0.0117 + j0.0465) = V_2 \angle 0° + 14.4 \angle 39°$$

$$E_1 = 10.6 E_2 = 10.6 V_2 \angle 0° + 152.64 \angle 39°$$

$$I_1 = \frac{300}{10.6} \angle -36.87° = 28.3 \angle -36.87°$$

$$V_1 = E_1 + I_1 Z_1$$

$$= 10.6 V_2 \angle 0° + 152.64 \angle 39° + (28.3 \angle -36.87°)(1.4 + j5.2)$$

$$= 10.6 V_2 + 152.64 \angle 39° + 152.4 \angle 38°$$

$$= 10.6 V_2 + 118.62 + j96.06 + 120 + j93.83$$

$$= 10.6 V_2 + 238.62 + j189.89$$

Since $|V_1| = 6600$ V

$\therefore$ $(10.6 V_2 + 238.62)^2 + (189.89)^2 = 6600^2$

or $V_2 \simeq 600$ **V**

Hence, **secondary terminal voltage = 600 V. (Ans.)**

Output :

Output $= V_2 I_2 \cos \phi_2 = 600 \times 300 \times 0.8 = 144000$ W or **144 kW. (Ans.)**

Example 2.16. *The full load copper loss on h.v. side of 100 kVA, 11000/317 V 1-phase transformer is 0.62 kW and on the l.v. side is 0.48 kW.*

(i) Calculate R_1, R_2 and R_2' in ohms :

(ii) The total reactance is 4 percent, find X_1, X_2 and X_2' in ohms if the reactance is divided in the same proportion as resistance. **(AMIE Winter, 1997)**

Solution. *Given* : Rated kVA = 100 ; $V_1 = 11000$ V ; $V_2 = 317$ V ; loss on *h.v.* side = 0.62 kW ; loss of *l.v.* side = 0.48 kW ; total reactance = 4%.

(i) **R_1, R_2 and R_2' :**

h.v. side full load current, $I_1 = \dfrac{\text{rated kVA} \times 1000}{V_1} = \dfrac{100 \times 1000}{11000} = 9.09$ A

l.v. side full load current, $I_2 = \dfrac{\text{rated kVA} \times 1000}{V_2} = \dfrac{100 \times 1000}{317} = 315.46$ A

Primary winding resistance, $\mathbf{R_1} = \dfrac{\text{full-load copper loss in primary winding}}{I_1^2}$

$$= \frac{0.62 \times 1000}{(9.09)^2} = \mathbf{7.5\ \Omega}. \ \textbf{(Ans.)}$$

Secondary winding resistance, $\mathbf{R_2} = \dfrac{\text{full-load copper loss in secondary winding}}{I_2^2}$

$$= \frac{0.48 \times 1000}{(315.46)^2} = \mathbf{0.00482\ \Omega.}\ \ \textbf{(Ans.)}$$

Transformation ratio, $K = \frac{V_2}{V_1} = \frac{317}{11000} = 0.02882$

Secondary resistance referred to primary,

$$R_2' = \frac{R_2}{K^2} = \frac{0.00482}{(0.02882)^2} = \mathbf{5.8\ \Omega.}\ \ \textbf{(Ans.)}$$

(*ii*) $\mathbf{X_1, X_2}$ **and** $\mathbf{X_2'}$ **:**

Percentage reactance referred to *h.v.* side $= \frac{I_1 X_{01}}{V_1} \times 100 = 4$

or transformer total reactance referred to primary,

$$X_{01} = \frac{4 \times V_1}{100 I_1} = \frac{4 \times 11000}{100 \times 9.09} = 48.4\ \Omega.$$

Now, $\frac{R_1}{R_2'} = \frac{X_1}{X_2'}$ ...(Given)

or $\frac{R_1 + R_2'}{R_2'} = \frac{X_1 + X_2'}{X_2'}$

or $\frac{7.5 + 5.8}{5.8} = \frac{48.4}{X_2'}$ $(\because\ X_1 + X_2' = X_{01})$

or $\mathbf{X_2'} = \frac{48.4 \times 5.8}{(7.5 + 5.8)} = \mathbf{21.1\ \Omega.}$ **(Ans.)**

Primary winding reactance, $\mathbf{X_1} = X_{01} - X_2' = 48.4 - 21.1 = \mathbf{27.3\ \Omega.}$ **(Ans.)**

Secondary winding reactance, $\mathbf{X_2} = K^2 X_2' = \left(\frac{377}{11000}\right)^2 \times 21.1 = \mathbf{0.0175\ \Omega.}$ **(Ans.)**

HYSTERESIS AND EDDY CURRENT LOSSES

Example 2.17. *A 230 V, 2.5 kVA single-phase transformer has an iron loss of 100 W at 40-Hz and 70 W at 30-Hz. Find the hysteresis and eddy current losses at 50-Hz.*

Solution. Iron loss at 40-Hz = 110 W

Iron loss at 30-Hz = 75 W

Hysteresis and eddy current loss at 50-Hz :

We know that hysteresis loss,

$$P_h \propto f = Af$$

Eddy current loss, $P_e \propto f^2 = Bf^2$

Iron loss, $P_i = P_h + P_e = Af + Bf^2$

At 40-Hz :

$$100 = 40A + (40)^2 B$$

i.e., $40A + 1600B = 100$

or $A + 40B = 2.5$ ...(*i*)

At 30-Hz :

$$70 = 30A + (30)^2 B$$

i.e., $30A + 900B = 70$

or $3A + 90B = 7$ ...(*ii*)

Solving (*i*) and (*ii*), we get $A = \frac{11}{6}$, $B = \frac{1}{60}$.

Hysteresis loss at 50-Hz

$$= Af = \frac{11}{6} \times 50 = \mathbf{91.67\ W.}\ \ \textbf{(Ans.)}$$

Eddy current loss at 50-Hz

$$= Bf^2 = \frac{1}{60} \times 50^2 = \mathbf{41.67\ W.}\ \ \textbf{(Ans.)}$$

Example 2.18. *When a transformer is supplied at 400 V, 50-Hz the hysteresis loss is found to be 310 W and eddy current loss is found to be 260 W. Determine the hysteresis loss and eddy current loss when the transformer is supplied at 800 V, 100-Hz.* **(AMIE Summer, 1996)**

Solution. Hysteresis loss at 400 V, 50 Hz = 310 W

Eddy current loss at 400 V, 50-Hz = 260 W

We know that,

Induced e.m.f., $E = 4.44 f \phi_{max} N = 4.44 (B_{max} A) f N$

i.e. $E \propto B_{max} f$

or $B_{max} \propto \frac{E}{f} = a \frac{E}{f}$ [a = constant]

In the **Ist case,** $B_{max_1} = a \cdot \frac{E_1}{f_1} = \frac{400}{50} a = 8a.$

In the **IInd case,** $B_{max_2} = a \frac{E_2}{f_2} = \frac{800}{100} a = 8a$

Hence, $B_{max_1} = B_{max_2}$

[$\because$ Core area A, and number of turns N remain the same]

Hence, $P_h = Af$ and $P_e = Bf^2$

$P_{h1} = 310$ W at $f_1 = 50$-Hz

$\therefore$ $A = \frac{310}{50} = 6.2$

Hysteresis loss at 100-Hz

$$= Af_2 = 6.2 \times 100 = \mathbf{620\ W.}\ \ \textbf{(Ans.)}$$

$P_{e1} = 260$ W at $f_1 = 50$-Hz

$\therefore$ $B = \frac{260}{50^2} = 0.104$

Eddy current loss at 100-Hz

$$= Bf_2^2 = 0.104 \times 100^2 = \mathbf{1040\ W.}\ \ \textbf{(Ans.)}$$

Example 2.19. *A 4400 V, 50-Hz transformer has a hysteresis loss of 1200 W, eddy current loss of 1800 W and full-load copper loss of 4000 W. If the transformer is supplied at 6600 V, 75-Hz. what will be the losses ?*

Assume that the full-load current remains the same.

Solution. Hysteresis loss (at 4400 V, 50-Hz), $P_{h1} = 1200$ W

Eddy current loss, $P_{e1} = 1800$ W

Full-load copper loss = 4000 W

Losses at 6600 V, 75 Hz :

We known that, $B_{max} \propto \frac{E}{f} = a \frac{E}{f}$

At 4400 V, 50-Hz

$$B_{max_1} = \frac{4400}{50} a = 88a$$

At 6600 V, 75-Hz

$$B_{max_2} = \frac{6600}{75} a = 88a$$

$$\therefore \quad B_{max_1} = B_{max_2}$$

Flux density remains the same is both cases.

$$\therefore \quad \frac{P_{h2}}{P_{h1}} = \frac{f_2}{f_1}$$

$$\frac{P_{h2}}{1200} = \frac{75}{50}$$

$$\therefore \quad \mathbf{P}_{h2} = \frac{1200 \times 75}{50} = 1800 \text{ W}$$

Hence, **hysteresis loss at 6600 V, 75-Hz = 1800 W. (Ans.)**

Also,
$$\frac{P_{e2}}{P_{e1}} = \frac{(f_2)^2}{(f_1)^2} = \frac{75^2}{50^2} = 2.25$$

$$\therefore \quad P_{e2} = P_{e1} \times 2.25 = 1800 \times 2.25 = 4050 \text{ W}$$

Hence, **eddy current loss at 6600 V, 75-Hz = 4050 W. (Ans.)**

Since the full-load current remains the same, full-load copper loss = 4000 W (as before).

☞**Example 2.20.** *A transformer with normal voltage impressed has a flux density of 1.2 T and a core loss comprising 1200 W eddy current loss and 3500 W hysteresis loss. What do these losses become under the following conditions :*

(i) Increasing the applied voltage by 5% at rated frequency ;

(ii) Reducing the frequency by 5% with normal voltage impressed ; and

(iii) Increasing both impressed voltage and frequency by 5%.

Solution. Flux density (at normal voltage) = 1.2 T

Eddy current loss = 1200 W

Hysteresis loss = 3500 W

We known that hysteresis loss,

$$P_h \propto B_{max}^{1.6} f = AB_{max}^{1.6} f$$

$$P_e \propto B_{max}^2 f^2 = BB_{max}^2 f^2$$

From the relation, $E = 4.44(B_{max} \times \text{area})Nf$, we get

$$B_{max} \propto \frac{E}{f}$$

Putting this value of B_{max} in the above equations, we have

$$P_h = A\left(\frac{E}{f}\right)^{1.6} f = AE^{1.6} f^{-0.6}$$

and
$$P_e = B\left(\frac{E}{f}\right)^2 f^2 = BE^2$$

From the given data $\quad P_h = 3500 = AE^{1.6} f^{-0.6}$ ...(i)

and $\quad P_e = 1200 = BE^2$ ...(ii)

where E and f are the normal values of primary voltage and frequency.

(i) Voltage becomes $E + 5\%E = 1.05E$

New hysteresis loss, $P_h = A(1.05E)^{1.6} f^{-0.6}$ ...(iii)

Dividing (iii) by (i), we get

$$\frac{P_h}{3500} = \frac{A(1.05E)^{1.6} f^{-0.6}}{AE^{1.6} f^{-0.6}} = (1.05)^{1.6} = 1.081$$

$\therefore$ $P_h = 3500 \times 1.081 = \mathbf{3783.5\ W.}$ **(Ans.)**

New eddy current loss, $P_e = B(1.05E)^2$ ...(iv)

Dividing (iv) by (ii), we get

$$\frac{P_e}{1200} = \frac{B(1.05E)^2}{BE^2} = 1.1$$

$\therefore$ $\mathbf{P_e = 1200 \times 1.1 = 1320\ W.}$ **(Ans.)**

(ii) As seen from (i), eddy current loss would not be affected by change of frequency. Thus $P_e = 1320$ W

New hysteresis loss, $P_h = AE^{1.6}(0.95f)^{-0.6}$ ...(v)

From (i) and (v), we have

$$\frac{P_h}{3500} = \frac{AE^{1.6}(0.95f)^{-0.6}}{AE^{1.6} f^{-0.6}} = (0.95)^{-0.6}$$

$\therefore$ $P_h = 3500 \times \frac{1}{(0.95)^{0.6}} = \mathbf{3609.4\ W.}$ **(Ans.)**

(iii) In this case both E and f are increased by 5%.

New hysteresis loss, $P_h = A(1.05E)^{1.6}(1.05f)^{-0.6}$ ...(vi)

From (i) and (ii), we get

$$\frac{P_h}{3500} = \frac{A(1.05E)^{1.6}\,(1.05f)^{-0.6}}{AE^{1.6}\,(f)^{-0.6}} = (1.05)^{0.6}(1.05)^{-0.6} = \frac{(1.05)^{1.6}}{(1.05)^{0.6}} = 1.05$$

$\therefore$ $P_h = 3500 \times 1.05 = \mathbf{3675\ W.}$ **(Ans.)**

As P_e is unaffected by changes in f its value is the same as found in (i) above (*i.e.* 1320 W).

Example 2.21. *The core loss for a given single phase transformer is found to be 2000 W at 50 Hz. Keeping the flux density constant, the frequency at the supply is raised to 75 Hz resulting in core loss of 3200 W. Compute separately hysteresis and eddy current losses at both the frequencies.* **(GATE, 1994)**

Solution. *Given* : Core loss at 50 Hz = 2000 W ; core loss at 75 Hz = 3200 W.

Hysteresis and eddy current losses at both the frequencies ; P_e, P_h ;

If flux density is maintained constant, then eddy current loss, $P_e = k_e f^2$,

and hysteresis loss, $P_h = k_h f$

Thus, $2000 = k_e \times 50^2 + k_h \times 50$

or $40 = 50k_e + k_h$ ...(i)

and $3200 = k_e \times 75^2 + k_h \times 75$

or $\frac{3200}{75} = 75k_e + k_h$ ...(ii)

Subtracting (i) from (ii), we get

$$\left(\frac{3200}{75} - 40\right) = 25k_e$$

$\therefore \qquad k_e = \dfrac{42.67 - 40}{25} = 0.1068$

and $\qquad k_h = 40 - 50 \times 0.1068 = 34.66$

Hence, *At 50 Hz :*

$$P_e = k_e f^2 = 0.1068 \times 50^2 = \mathbf{267\ W.}\ \ \textbf{(Ans.)}$$

$$P_h = k_h f = 34.66 \times 50 = \mathbf{1733\ W.}\ \ \textbf{(Ans.)}$$

At 75 Hz :

$$P_e = 0.1068 \times 75^2 = \mathbf{600\ W.}\ \ \textbf{(Ans.)}$$

$$P_h = 34.66 \times 75 = \mathbf{2599.5\ W.}\ \ \textbf{(Ans.)}$$

TRANSFORMER TESTS (O.C. AND S.C.) AND EQUIVALENT CIRCUIT

☞**Example 2.22.** *A 50-Hz, single-phase transformer has a turn ratio of 5. The resistances are 0.8 Ω, 0.02 Ω and reactances are 4 Ω and 0.12 Ω for high-voltage and low-voltage windings respectively. Find :*

(i) The voltage to be applied to the h.v. side to obtain full-load current of 180 A in the l.v. winding on short-circuit.

(ii) The power factor on short-circuit.

Draw the equivalent circuit and vector diagram.

Solution. Turn ratio, $\quad \dfrac{N_1}{N_2} = 5,\ K = \dfrac{1}{5}$

Primary resistance, $\quad R_1 = 0.8\ \Omega$

Primary reactance, $\quad X_1 = 4\ \Omega$

Secondary resistance, $\quad R_2 = 0.02\ \Omega$

Secondary reactance, $\quad X_2 = 0.12\ \Omega$

Total resistance referred to primary,

$$R_{01} = R_1 + R_2' = R_1 + \frac{R_2}{K^2} = 0.8 + 0.02 \times (5)^2 = 0.8 + 0.5 = 1.3\ \Omega$$

Total reactance referred to secondary,

$$X_{01} = X_1 + X_2' = X_1 + \frac{X_2}{K^2} = 4 + 0.12(5)^2 = 4 + 3 = 7\ \Omega$$

Total impedance referred to primary,

$$Z_{01} = \sqrt{R_{01}^2 + X_{01}^2} = \sqrt{(1.3)^2 + (7)^2} = 7.12\ \text{ohm}$$

Also $\quad I_1 = KI_2 = \dfrac{1}{5} \times 180 = 36\ \text{A}$

(i) **Voltage to be applied** on h.v. side to obtain full-load current I_1,

$$V_{SC} = I_1 Z_{01} = 36 \times 7.12 = \mathbf{256.3\ V.}\ \ \textbf{(Ans.)}$$

(ii) The **power factor** on short-circuit,

$$\cos \phi_0 = \frac{R_{01}}{Z_{01}} = \frac{1.3}{7.12} = \mathbf{0.1825.}\ \ \textbf{(Ans.)}$$

Equivalent circuit and vector diagram are shown in Figs. 2.47, 2.48 and 2.49.

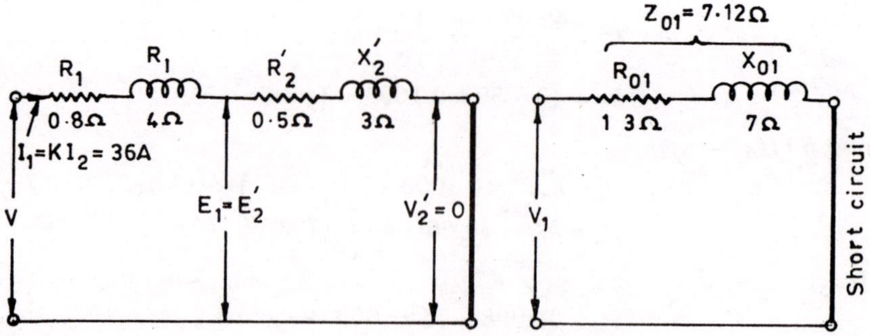

Fig. 2.47

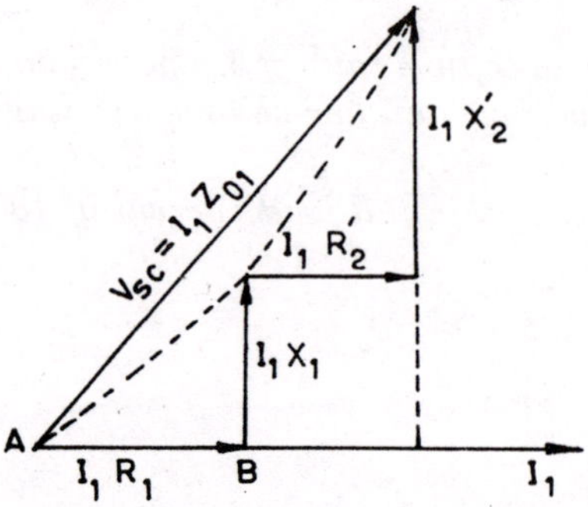

Fig. 2.48

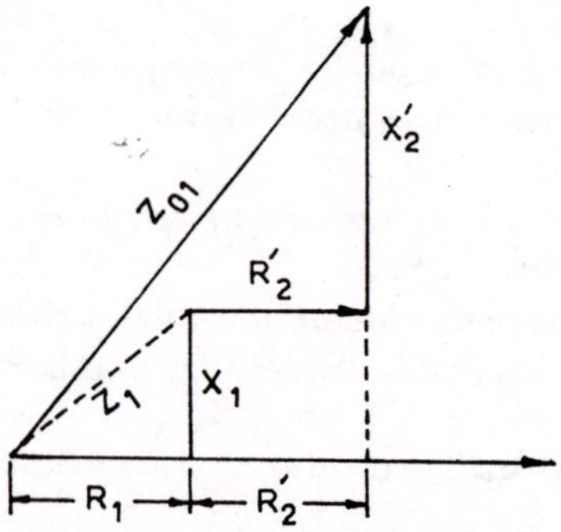

Fig. 2.49

Example 2.23. *A 12 kVA 4000/400 V transformer has primary and secondary winding resistances of 13 Ω and 0.15 Ω and leakage reactances of 20 Ω and 0.25 Ω respectively. The magnetising reactance is 6000 Ω and the resistance representing core loss is 12000 Ω. Determine :*

(*i*) *Equivalent resistance and reactance as referred to primary.*

(*ii*) *Input current with secondary terminals open circuited.*

(*iii*) *Input current when the secondary load current is 30 A at 0.8 power factor lagging.*

Solution. Transformation ratio, $K = \frac{400}{4000} = \frac{1}{10}$

Primary resistance, $R_1 = 13\ \Omega$

Primary reactance, $X_1 = 20\ \Omega$

Secondary resistance, $R_2 = 0.15\ \Omega$

Secondary reactance, $X_2 = 0.25\ \Omega$

Magnetising reactance, $X_0 = 6000\ \Omega$

Resistance representing the core loss, $R_0 = 12000\ \Omega$

(*i*) **Equivalent resistance referred to primary,**

$$R_{01} = R_1 + R_2' = R_1 + \frac{R_2}{K^2}$$

$$= 13 + \frac{0.15}{\left(\frac{1}{10}\right)^2} = 13 + 15 = \mathbf{28\ \Omega}. \quad \textbf{(Ans.)}$$

Equivalent reactance referred to primary,

$$X_{01} = X_1 + X_2' = X_1 + \frac{X_2}{K^2}$$

$$= 20 + \frac{0.25}{\left(\frac{1}{10}\right)^2} = 20 + 25 = \mathbf{45\ \Omega.}\ \ \textbf{(Ans.)}$$

(*ii*) **Input current with secondary terminals open-circuited (or input current at no-load) :**

Magnetising component, $I_m = \frac{4000}{6000} = 0.667$ A

Wattful component, $I_w = \frac{4000}{12000} = 0.333$ A

Hence input current at no-load,

$$I_0 = 0.333 - j0.667 = \mathbf{0.745 \angle - 63.5°\ A.}\ \ \textbf{(Ans.)}$$

(*iii*) **Input current $\mathbf{I_1}$:**

Secondary output current, $I_2 = 30$ A (*given*)

Power factor, $\cos \phi_2 = 0.8$ (lagging)

i.e., $I_2 = 30 \angle - 36.9°$

Secondary current referred to primary,

$$I_2' = KI_2 = \frac{30}{10} \angle - 36.9° = 3 \angle - 36.9° = 2.4 - j1.8$$

$\therefore$ Input current, $I_1 = I_2' + I_0 = (2.4 - j1.8) + (0.333 - j0.667)$

$$= 2.733 - j2.467 = \mathbf{3.68 \angle - 42.07°\ A.}\ \ \textbf{(Ans.)}$$

Example 2.24. *A 200/2000 V transformer is fed from a 200 V supply. The total winding resistance and leakage reactance as referred to low voltage side are 0.15 Ω and 0.6 Ω respectively. The resistance representing core loss is 450 Ω and magnetising reactance is 250 Ω. A load of impedance (600 + j400 Ω) is connected across the secondary terminals. Calculate :*

(i) Input current, *(ii) Secondary terminal voltage, and*

(iii) Primary power factor.

Solution. Transformation ratio, $K = \frac{2000}{200} = 10$

Total winding resistance referred to primary side (l.v.)

$$R_{01} = 0.15\ \Omega$$

Total winding reactance referred to primary side (l.v.),

$$X_{01} = 0.6\ \Omega$$

Resistance representing the core loss, $R_0 = 450\ \Omega$

Magnetising reactance, $X_0 = 250\ \Omega$

Load impedance connected across secondary terminals,

$$Z_L = (600 + j400)$$

i.e. $R_L = 600\ \Omega,\ X_L = 400\ \Omega$

Now, load resistance referred to primary,

$$R_L' = \frac{R_L}{K^2} = \frac{600}{(10)^2} = 6\ \Omega,$$

and load reactance referred to primary,

$$X_L' = \frac{X_L}{K^2} = \frac{400}{(10)^2} = 4\ \Omega$$

The magnetising component of no-load current I_0,

$$I_m = \frac{200\ \angle 0°}{j250} = -j0.8\ \text{A}$$

The wattful component of no-load current I_0,

$$I_w = \frac{200\ \angle 0°}{450} = 0.444\ \text{A}$$

(*i*) **Input current, $\mathbf{I_1}$:**

Secondary current referred to primary,

$$I_2' = \frac{V_1}{(R_1 + R_2') + R_L' + j(X_1 + X_2' + X_L')}$$

$$= \frac{200\ \angle 0°}{R_{01} + R_L' + j(X_{01} + X_L')} = \frac{200\ \angle 0°}{0.15 + 6 + j(0.6 + 4)}$$

$$= \frac{200\ \angle 0°}{6.15 + j4.6} = \frac{200(6.15 - j4.6)}{(6.15 + j4.6)(6.15 - j4.6)}$$

$$= \frac{200(6.15 - j4.6)}{58.98} = 20.85 - j15.6\ \text{A}$$

$$\therefore \quad I_1 = I_2' + I_m + I_w$$

$$= (20.85 - j15.6) + (j0.8) + 0.444 = 21.294 - j16.44$$

$$= \mathbf{26.9\ \angle - 37.67°\ A.}\ \textbf{(Ans.)}$$

(*ii*) **Secondary terminal voltage, $\mathbf{V_2}$:**

Secondary terminal voltage referred to primary,

$$V_2' = I_2'(R_L' + jX_L') = (20.85 - j15.6)(6 + j4)$$

$$= (26.04\ \angle - 36.8°)(7.21\ \angle 33.7°) = 187.75\ \angle - 3.1°\ \text{V}$$

Secondary terminal voltage,

$$V_2 = KV_2' = 10 \times 187.75 = \mathbf{1877.5\ V.}\ \textbf{(Ans.)}$$

(*iii*) **Primary power factor, $\cos \phi_1$:**

$$\cos \phi_1 = \cos (37.67°) = \mathbf{0.791\ (lagging).}\ \textbf{(Ans.)}$$

☞**Example 2.25.** *Obtain the approximate equivalent circuit of a given 200/2000 V single-phase 30 kVA transformer having the following test results :*

O.C. test : 200 V, 6.2 A, 360 W on l.v. side

S.C. test : 75 V, 18 A, 600 W on h.v. side.

Solution. O.C. test (l.v. side)

Primary voltage, $V_1 = 200$ V

No-load current, $I_0 = 6.2$ A

No-load loss, $P_0 = 360$ W

Now, $P_0 = V_1 I_0 \cos \phi_0$

$$360 = 200 \times 6.2 \times \cos \phi_0$$

$$\therefore \quad \cos \phi_0 = \frac{360}{200 \times 6.2} = 0.29$$

$$\sin \phi_0 = 0.957$$

Wattful component of no-load current,

$$I_w = I_0 \cos \phi_0 = 6.2 \times 0.29 = 1.8\ \text{A}.$$

Resistance representing the core loss,

$$R_0 = \frac{V_1}{I_w} = \frac{200}{1.8} = \mathbf{111.11\ \Omega.\ (Ans.)}$$

Magnetising component of no-load current,

$$I_m = I_0 \sin \phi_0 = 6.2 \times 0.95 = 5.93\ \text{A}$$

Magnetising reactance, $X_0 = \frac{V_1}{I_m} = \frac{200}{5.93} = \mathbf{33.7\ \Omega.\ (Ans.)}$

S.C. test (h.v. side) :

Short-circuit voltage, $V_{SC} = 75$ V

Short-circuit current, $I_{SC} = 18$ A

Losses, $P_{SC} = 600$ W

Impedance of transformer referred to h.v. side,

$$Z_{02} = \frac{V_{SC}}{I_{SC}} = \frac{75}{18} = 4.167\ \Omega$$

$$P_{SC} = {I_{SC}}^2 \times R_{02}$$

$$600 = 18^2 \times R_{02}$$

$$\therefore \quad R_{02} = \frac{600}{18^2} = 1.85\ \Omega$$

Transformation ratio, $K = \frac{2000}{200} = 10$

Referred to 200 V side :

$$Z_{01} = \frac{Z_{02}}{K^2} = \frac{4.167}{(10)^2} = 0.04167\ \Omega$$

$$R_{01} = \frac{R_{02}}{K^2} = \frac{1.85}{(10)^2} = 0.0185\ \Omega$$

$$\therefore \quad X_{01} = \sqrt{{Z_{01}}^2 - {R_{01}}^2} = \sqrt{(0.04167)^2 - (0.0185)^2} = \mathbf{0.0373\ \Omega.\ (Ans.)}$$

Approximate equivalent circuit is shown in Fig. 2.50.

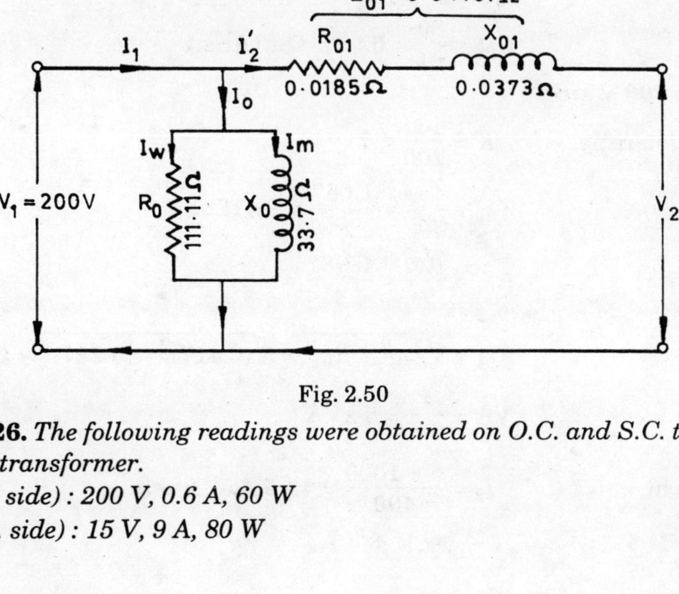

Fig. 2.50

Example 2.26. *The following readings were obtained on O.C. and S.C. tests on a 200/400 V, 50 Hz, single-phase transformer.*

O.C. test (l.v. side) : 200 V, 0.6 A, 60 W

S.C. test (h.v. side) : 15 V, 9 A, 80 W

Calculate the secondary voltage when delivering 4 kW at 0.8 power factor lagging, the primary voltage being 200 V. **(AMIE Winter, 1996)**

Solution. O.C. test—l.v. side (*Instruments in the primary side* and secondary open)

Primary voltage, $V_1 = 200$ V

No-load current, $I_0 = 0.6$ A

No-load loss, $P_0 = 60$ W

Now,
$$P_0 = V_1 I_0 \cos \phi_0$$
$$60 = 200 \times 0.6 \times \cos \phi_0$$

$$\therefore \quad \cos \phi_0 = \frac{60}{200 \times 0.6} = 0.5$$
$$\sin \phi_0 = 0.866$$

Wattful component of no-load current
$$I_w = I_0 \cos \phi_0 = 0.6 \times 0.5 = 0.3 \text{ A}$$

Magnetising component of no-load current,
$$I_m = I_0 \sin \phi_0 = 0.6 \times 0.866 = 0.52 \text{ A}$$

$\therefore$ Resistance representing the core loss,
$$R_0 = \frac{V_1}{I_w} = \frac{200}{0.3} = \mathbf{666.67\ \Omega.\ (Ans.)}$$

Magnetising reactance, $X_0 = \dfrac{V_1}{I_m} = \dfrac{200}{0.52} = \mathbf{384.6\ \Omega.\ (Ans.)}$

S.C. test—h.v. side (*Instruments in the secondary side* and primary short-circuited)

Short circuit voltage, $V_{SC} = 15$ V

Short circuit current, $I_{SC} = 9$ A

Losses, $P_{SC} = 80$ W

Impedance of transformer referred to secondary,
$$Z_{02} = \frac{V_{SC}}{I_{SC}} = \frac{15}{9} = 1.667\ \Omega$$

Also,
$$P_{SC} = I_{SC}^2 \times R_{02}$$
$$80 = 9^2 \times R_{02}$$

$$\therefore \quad R_{02} = \frac{80}{81} = \mathbf{0.987\ \Omega.\ (Ans.)}$$

Referred to 200 V side :

Transformation ratio, $K = \dfrac{400}{200} = 2$

$$Z_{01} = \frac{Z_{02}}{K^2} = \frac{1.667}{2^2} = 0.417\ \Omega$$

$$R_{01} = \frac{R_{02}}{2^2} = \frac{0.987}{4} = 0.247\ \Omega$$

$$\therefore \quad X_{01} = \sqrt{Z_{01}^2 - R_{01}^2} = \sqrt{(0.417)^2 - (0.247)^2} = 0.336\ \Omega$$

Output, $\text{kVA} = \dfrac{4}{0.8} = 5$ kVA

Output current, $I_2 = \dfrac{5 \times 1000}{400} = \mathbf{12.5\ A.\ (Ans.)}$

This value of I_2 is *approximate* because V_2 (which is to be calculated as yet) has bo n taken equal to 400 V (which in fact is equal to E_2 or ${}_0V_2$)

Now, $Z_{02} = 1.667\ \Omega$

$R_{02} = 0.987\ \Omega$

$\therefore \quad X_{02} = \sqrt{Z_{02}^2 - R_{02}^2} = \sqrt{1.667^2 - 0.987^2} = 1.343\ \Omega$

Total drop as referred to secondary

$= I_2 (R_{02} \cos \phi_2 + X_{02} \sin \phi_2)$

$= 12.5(0.987 \times 0.8 + 1.343 \times 0.6) = 19.94$ V

$\therefore \quad V_2 = 400 - 19.94 \simeq 380$ V

Hence, **secondary voltage = 380 V. (Ans.)**

Approximate equivalent circuit is shown in Fig. 2.51.

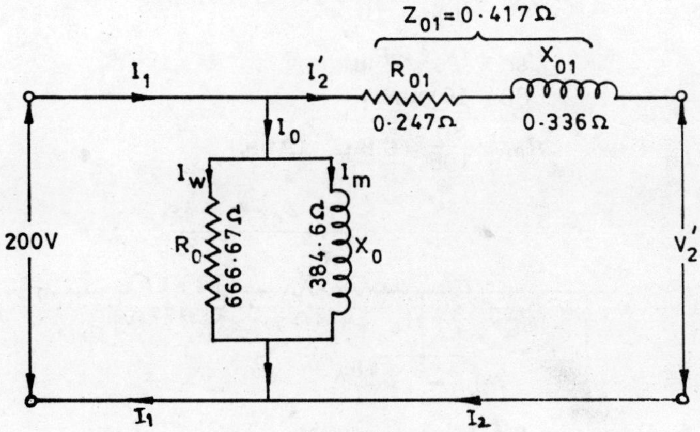

Fig. 2.51

Example 2.27. *A 4 kVA, 400/200 V, 50-Hz single-phase transformer has the following test data :*

O.C. test (l.v. side) 200 V, 1 A, 64 W

S.C. test (h.v. side) 15 V, 10 A, 80 W

Determine :

(i) Equivalent circuit referred to l.v. side, and

(ii) Secondary load voltage on full-load at 0.8 power factor lagging.

Solution. (*i*) **O.C. test—l.v. side :**

Voltage, $V_0 = 200$ V

No-load current, $I_0 = 1$ A

No-load loss, $P_0 = 64$ W

Now, $P_0 = V_0 I_0 \cos \phi_0$

$64 = 200 \times 1 \times \cos \phi_0$

$\therefore \quad \cos \phi_0 = 0.32$

$\sin \phi_0 = 0.9474$

Wattful component of no-load current,

$$I_w = I_0 \cos \phi_0 = 1 \times 0.32 = 0.32 \text{ A}$$

Magnetising component of no-load current,

$$I_m = I_0 \sin \phi_0 = 1 \times 0.9474 = 0.9474 \text{ A}$$

∴ Resistance representing the core loss,

$$R_0 = \frac{V_0}{I_w} = \frac{200}{0.32} = \mathbf{625\ \Omega.\ (Ans.)}$$

Magnetising reactance, $X_0 = \frac{V_0}{I_m} = \frac{200}{0.9474} = \mathbf{211.1\ \Omega.\ (Ans.)}$

S.C. test—h.v. side :

Short-circuit voltage, $V_{SC} = 15$ V

Short-circuit current, $I_{SC} = 10$ A

Losses, $P_{SC} = 80$ W

Impedance of the circuit referred to h.v. side,

$$Z_{01} = \frac{V_{SC}}{I_{SC}} = \frac{15}{10} = \mathbf{1.5\ \Omega.\ (Ans.)}$$

Also $P_{SC} = I_{SC}^2 \times R_{01}$

∴ $80 = 10^2 \times R_{01}$

∴ $R_{01} = \frac{80}{100} = \mathbf{0.8\ \Omega.\ (Ans.)}$

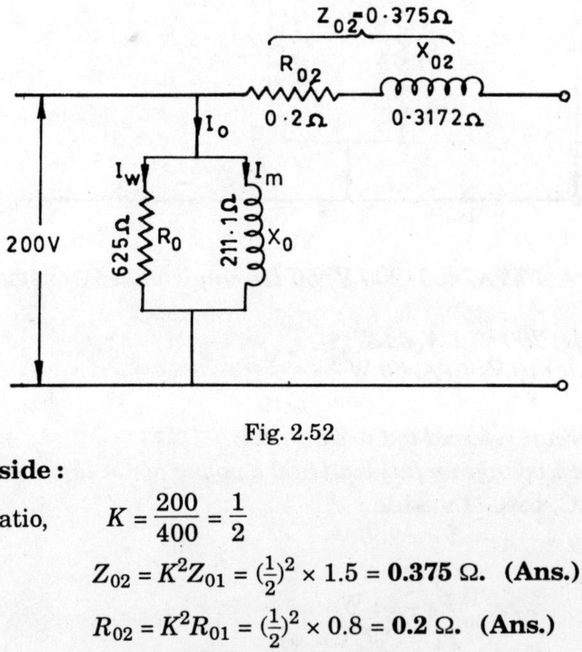

Fig. 2.52

Referred to l.v. side :

Transformation ratio, $K = \frac{200}{400} = \frac{1}{2}$

$$Z_{02} = K^2 Z_{01} = (\tfrac{1}{2})^2 \times 1.5 = \mathbf{0.375\ \Omega.\ (Ans.)}$$

$$R_{02} = K^2 R_{01} = (\tfrac{1}{2})^2 \times 0.8 = \mathbf{0.2\ \Omega.\ (Ans.)}$$

∴ $X_{02} = \sqrt{Z_{02}^2 - R_{02}^2} = \sqrt{(0.375)^2 - (0.2)^2} = \mathbf{0.3172\ \Omega.\ (Ans.)}$

The approximate equivalent circuit is shown in Fig. 2.52.

(*ii*) **Secondary load voltage, V_2 :**

Secondary full-load current, $I_2 = \frac{4 \times 1000}{200} = 20$ A

$\cos \phi_2 = 0.8$

$\sin \phi_2 = 0.6$

$$\therefore \quad I_2 = 20(0.8 - j0.6) = (16 - j12)$$

$$Z_{02} = R_{02} + jX_{02} = 0.2 + j0.3172$$

Secondary load voltage, $V_2 = 200 - I_2 Z_{02} = 200 - (16 - j12)(0.2 + j0.3712)$

$$= 200 - (3.2 + j5.075 - j2.4 + 3.806)$$

$$= 200 - (7 + j2.675) = 193 - j2.675 \simeq \textbf{193 V. (Ans.)}$$

Example 2.28. *A single-phase step-down transformer has a turn ratio of 3. The resistance and reactance of the primary winding are 1.2 Ω and 6 Ω and those of the secondary winding are 0.05 Ω and 0.03 Ω respectively. If the h.v. winding is supplied at 230 V, 50-Hz with l.v. winding short circuited, find :*

(*i*) *Current in the l.v. winding,* (*ii*) *Copper loss in the transformer, and*
(*iii*) *Power factor.*

Solution. Turn ratio $= 3$, *i.e.*, $K = \frac{1}{3}$

Resistance of primary, $R_1 = 1.2\ \Omega$
Reactance of primary, $X_1 = 6\ \Omega$
Resistance of secondary, $R_2 = 0.05\ \Omega$
Reactance of secondary, $X_2 = 0.3\ \Omega$
Referred to *h.v. side*

$$R_{01} = R_1 + R_2' = R_1 + \frac{R_2}{K^2} = 1.2 + 0.05 \times (3)^2 = 1.65\ \Omega$$

$$X_{01} = X_1 + X_2' = X_1 + \frac{X_2}{K^2} = 6 + 0.3 \times (3)^2 = 8.7\ \Omega$$

$$Z_{01} = \sqrt{R_{01}^2 + X_{01}^2} = \sqrt{(1.65)^2 + (8.7)^2} = 8.855\ \Omega$$

Current in h.v. winding when l.v. winding is short-circuited,

$$I_{SC} = \frac{V_{SC}}{Z_{01}} = \frac{230}{8.855} = 25.97\text{ A}$$

Neglecting I_0, $I_1 = I_2' = 25.97$ A

(*i*) **Current in the l.v. winding,**

$$I_2 = \frac{I_2'}{K} = \frac{25.97}{\left(\frac{1}{3}\right)} = \textbf{77.91 A. (Ans.)}$$

(*ii*) **Total copper loss** $= I_1^2 R_{01} = (25.97)^2 \times 1.65 = $ **1112.8 W. (Ans.)**

(*iii*) **Primary power factor :**

Power input on short-circuit,

$$P_{SC} = V_{SC} I_{SC} \cdot \cos \phi_{SC}$$

$$1112.8 = 230 \times 25.97 \times \cos \phi_{SC}$$

$$\therefore \quad \cos \phi_{SC} = \frac{1112.8}{230 \times 25.97} = \textbf{0.1863. (Ans.)}$$

☞**Example 2.29.** *A single-phase, 3 kVA, 230/115 V, 50-Hz transformer has the following constants :*

Resistance : Primary 0.3 Ω, secondary 0.09 Ω.
Reactance : Primary 0.4 Ω, secondary 0.1 Ω.
Resistance of equivalent exciting circuit referred to primary, $R_0 = 600\ \Omega$.
Reactance of equivalent exciting circuit-referred to primary, $X_0 = 200\ \Omega$.

What would be the readings of the instruments when the transformer is connected for (i) O.C. test, (ii) S.C. test. In both tests supply is given to h.v. side.

Solution. Primary resistance, $R_1 = 0.3\ \Omega$

Primary reactance, $X_1 = 0.4\ \Omega$

Secondary resistance, $R_2 = 0.09\ \Omega$

Secondary reactance, $X_2 = 0.1\ \Omega$

$R_0 = 600\ \Omega$

$X_0 = 200\ \Omega$

(*i*) **O.C. test :**

Wattful component of no-load current,

$$I_w = \frac{V_1}{R_0} = \frac{230}{600} = 0.383 \text{ A}$$

Magnetising component of no-load current,

$$I_w = \frac{V_1}{X_0} = \frac{230}{200} = 1.15 \text{ A}$$

$\therefore$ No-load current, $I_0 = \sqrt{I_w^2 + I_m^2} = \sqrt{(0.383)^2 + (1.15)^2} = 1.212 \text{ A}$

Input on no-load to h.v. winding

$$= V_1 I_w = 230 \times 0.383 = 88.09 \text{ W}$$

Hence, the readings of the instruments are :

Voltmeter reading = 230 V. (Ans.)

Ammeter reading = 1.212 A. (Ans.)

Wattmeter reading = 88.09 W. (Ans.)

(*ii*) **S.C. test :**

Transformation ratio, $K = \frac{115}{230} = \frac{1}{2}$

Referred to *h.v. side*, $R_{01} = R_1 + R_2' = R_1 + \frac{R_2}{K^2} = 0.3 + \frac{0.09}{(\frac{1}{2})^2} = 0.66\ \Omega$

$$X_{01} = X_1 + X_2' = X_1 + \frac{X_2}{K^2} = 0.4 + \frac{0.1}{(\frac{1}{2})^2} = 0.8\Omega$$

$\therefore$ $Z_{01} = \sqrt{R_{01}^2 + X_{01}^2} = \sqrt{(0.66)^2 + (0.8)^2} = 1.037\ \Omega$

Full-load current in the h.v. winding

$$I_1 = \frac{3 \times 1000}{230} = 13.04 \text{ A}$$

Voltage to be applied to h.v. winding on short-circuiting l.v. winding to pass full-load current of 13.04 A through h.v. winding.

$$V_{SC} = I_1 Z_{01} = 13.04 \times 1.037 = 13.52 \text{ V}$$

Power supplied to h.v. winding on short-circuit

$$= I_1^2 \,.\, R_{01} = (13.04)^2 \times 0.66 = 112.2 \text{ W}$$

Hence, the readings of the instruments are :

Voltmeter reading = 13.52 V. (Ans.)

Ammeter reading = 13.04 A. (Ans.)

Wattmeter reading = 112.2 W. (Ans.)

Example 2.30. *The primary of a transformer is rated at 10 A. On open circuit the readings are $V_1 = 1000$ V ; $V_2 = 500$ V, $I = 0.42$ A and $P_{OC} = 100$ W. On short circuit the readings are $I_1 = 10$ A, $V_1 = 126$ V and $P_{SC} = 400$ W. Draw an equivalent circuit for the transformer and determine the parameters. Predict the output voltage across a load impedance $Z_L = (19 + j12)\ \Omega$.*

(AMIE Summer, 1994)

Solution. O.C. test :

$$I_0 = 0.42 \text{ A}, V_1 = 1000 \text{ V}, P_0 = 100 \text{ W}$$

No-load power factor, $\cos \phi_0 = \dfrac{P_0}{V_1 I_0} = \dfrac{100}{1000 \times 0.42} = 0.238$

$$I_w = I_0 \cos \phi_0 = 0.42 \times 0.238 = 0.1 \text{ A}$$

$$I_m = I_0 \sin \phi_0 = 0.42 \sin (\cos^{-1} \phi_0) = 0.42 \times 0.971 = 0.408 \text{ A}$$

$\therefore$ $$\mathbf{R_0} = \frac{V_1}{I_w} = \frac{1000}{0.1} = 10000\ \Omega \text{ or } \mathbf{10\ k}\ \Omega. \ \textbf{(Ans.)}$$

and $$X_0 = \frac{V_1}{I_m} = \frac{1000}{0.408} = 2450\ \Omega \text{ or } \mathbf{2.45\ k}\ \Omega. \ \textbf{(Ans.)}$$

S.C. test :

$$I_{SC} = 10 \text{ A}, V_{SC} = 126 \text{ V}, P_{SC} = 400 \text{ W}$$

$$Z_{01} = \frac{V_{SC}}{I_{SC}} = \frac{126}{10} = 12.6\ \Omega$$

$$\mathbf{R_{01}} = \frac{P_{SC}}{I_{SC}^2} = \frac{400}{10^2} = \mathbf{4\ \Omega.} \ \textbf{(Ans.)}$$

$$\mathbf{X_{01}} = \sqrt{Z_{01}^2 - R_{01}^2} = \sqrt{12.6^2 - 4^2} = \mathbf{11.95\ \Omega.} \ \textbf{(Ans.)}$$

The equivalent circuit is shown in Fig. 2.53.

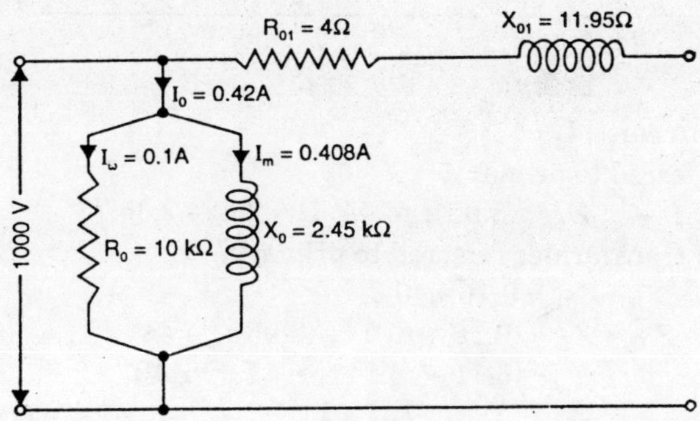

Fig. 2.53

It is given that on open circuit voltage on secondary side is 500 V, hence rated voltage of secondary, $V_2 = 500$ V

Transformation ratio, $K = \dfrac{V_2}{V_1} = \dfrac{500}{1000} = 0.5$

Equivalent resistance referred to secondary,

$$R_{02} = K^2 R_{01} = 0.5^2 \times 4 = 1\ \Omega$$

Equivalent reactance referred to secondary,

$$X_{02} = K^2 X_{01} = 0.5^2 \times 11.95 = 2.987\ \Omega$$

$\therefore \quad Z_{02} = (1 + j2.987)\ \Omega$

Load impedance, $Z_L = (19 + j12)\ \Omega$

Load voltage $= \dfrac{V_2 \times Z_L}{(Z_{02} + Z_L)}$

$$= \frac{500 \times (19 + j12)}{(1 + j2.987) + (19 + j12)} = \frac{500 \times 22.47 \angle 32.3°}{(20 + j14.987)}$$

$$= \frac{500 \times 22.47 \angle 32.3°}{25 \angle 36.85°} = \mathbf{449.4 \angle -4.55° \ V.} \quad \textbf{(Ans.)}$$

Example 2.31. *The diagram (Fig. 2.54) shows the equivalent circuit for a single phase transformer. Figures give resistances and reactances in ohms in terms of primary side. The ratio of secondary to primary turns is 10 and the load is inductive. Find :*

(*i*) *The primary current* ; (*ii*) *The secondary terminal voltage.*

(S.C.E. & T.E.W.B. 1994)

Solution. Transformation ratio, $K = \dfrac{N_2}{N_1} = 10$

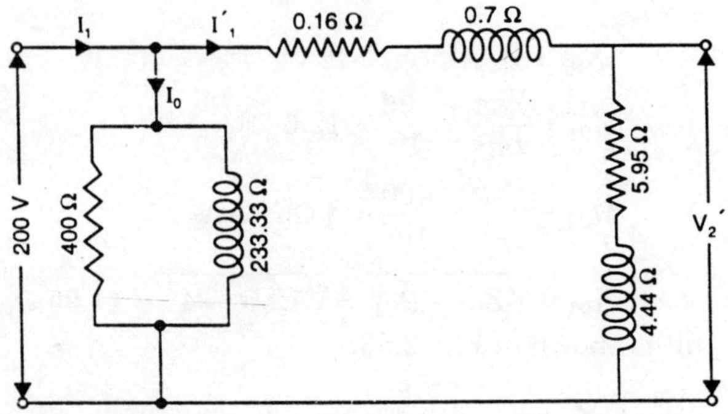

Fig. 2.54

(*i*) **The primary current, I_1 :**

Load impedance referred to primary,

$$Z_L = (5.95 + j4.44)\ \Omega = 7.424 \angle 36.73°$$

Total impedance of transformer referred to primary,

$$Z_{01} = 0.16 + j0.7$$

$$Z_{01} + Z_L = (0.16 + j0.7) + (5.95 + j4.44)$$

$$= (6.11 + j5.14) \text{ or } 7.98 \angle 40°$$

$$Y_0 = \frac{1}{R_0} - j\frac{1}{X_0} = \frac{1}{400} - \frac{j}{233.3} = (0.0025 - j0.00429)$$

No-load current, $I_0 = V \times Y_0 = (200 + j0)(0.0025 - j0.00429) = (0.5 - j0.858)$ A

Load current referred to primary,

$$I_1' = \frac{Y}{Z_{01} + Z_L} = \frac{200 + j0}{7.98 \angle 40°} = 25.063 \angle -40° = (19.2 - j16.11) \text{ A}$$

Primary current, $\mathbf{I_1} = I_0 + I_1' = (0.5 - j0.858) + (19.2 - j16.11)$

$= (19.7 - j16.968)$ or $\mathbf{26 \angle -40.74°}$. **(Ans.)**

(*ii*) **The secondary terminal voltage :**

Secondary voltage referred to primary,

$$V_2' = I_1' Z_L = 25.063 \angle -40° \times 7.424 \angle 36.73° = 186.07 \angle -3.27°$$

Secondary terminal voltage, $\mathbf{V_2} = KV_2' = 10 \times 186.07 \angle -3.27°$
$= \mathbf{1860.7 \angle -3.27°}$. **(Ans.)**

REGULATION AND EFFICIENCY OF A TRANSFORMER

☞**Example 2.32.** *The following readings were obtained from O.C. and S.C. tests on 8 kVA 400/120 V, 50-Hz transformer.*

O.C. test (l.v. side) : 120 V ; 4 A ; 75 W.

S.C. test (h.v. side) : 9.5 V ; 20 A ; 110 W.

Calculate :

(i) The equivalent circuit (approximate) constants,

(ii) Voltage regulation and efficiency for 0.8 lagging power factor load, and

(iii) The efficiency at half full-load and 0.8 power factor load.

Solution. (*i*) Transformation ratio, $K = \frac{120}{400} = \frac{3}{10}$

O.C. test :

It is seen from the O.C. test, that *primary open*, the secondary draws a no-load current of 4 A.

Since $K = \frac{3}{10}$, the corresponding no-load current $I_0 = 4 \times \frac{3}{10} = 1.2$ A.

Also $P_0 = V_1 I_0 \cos \phi_0$

$\therefore$ $75 = 400 \times 1.2 \times \cos \phi_0$

i.e., $\cos \phi_0 = \frac{75}{400 \times 1.2} = 0.156$

and $\sin \phi_0 = 0.987$

Now, wattful component of no-load current, I_0,

$$I_w = I_0 \cos \phi_0 = 1.2 \times 0.156 = 0.187 \text{ A}$$

and magnetising component of no-load current I_0,

$$I_m = I_0 \sin \phi_0 = 1.2 \times 0.987 = 1.184 \text{ A}$$

$\therefore$ Resistance representing the core loss,

$$R_0 = \frac{V_1}{I_w} = \frac{400}{0.187} = \mathbf{2139\ \Omega}. \quad \textbf{(Ans.)}$$

Magnetising reactance, $X_0 = \frac{V_1}{I_m} = \frac{400}{1.184} = \mathbf{337.8\ \Omega}$. **(Ans.)**

S.C. test :

During S.C. test the *instruments have been placed in primary.*

Here, $V_{SC} = 9.5$ V

$I_{SC} = 20$ A

$P_{SC} = 110$ W

Now, $Z_{01} = \frac{V_{SC}}{I_{SC}} = \frac{9.5}{20} = \mathbf{0.475\ \Omega}$. **(Ans.)**

$$R_{01} = \frac{P_{SC}}{I_{SC}^2} = \frac{110}{(20)^2} = \mathbf{0.25\ \Omega}. \quad \textbf{(Ans.)}$$

$\therefore$ $X_{01} = \sqrt{Z_{01}^2 - R_{01}^2} = \sqrt{(0.475)^2 - (0.25)^2} = \mathbf{0.404\ \Omega}$. **(Ans.)**

The equivalent circuit is shown in Fig. 2.55.

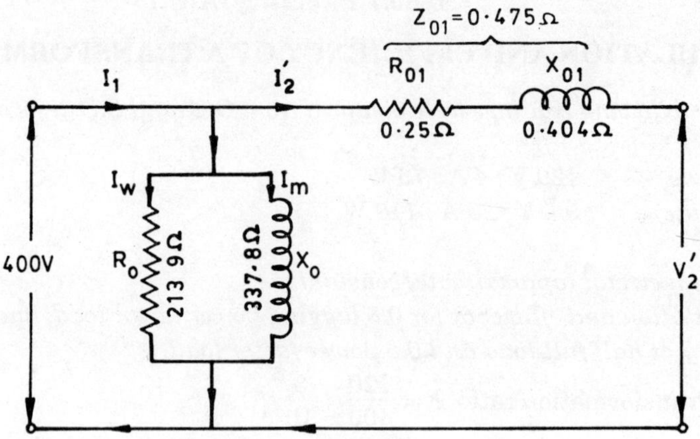

Fig. 2.55

(*ii*) Total *approximate voltage drop* as referred to primary is

$= I_1 (R_{01} \cos \phi + X_{01} \sin \phi)$

Now, full-load, $I_1 = \dfrac{8 \times 1000}{400} = 20$ A

∴ Voltage drop $= 20 (0.25 \times 0.8 + 0.404 \times 0.6) = 8.85$ V

% Voltage regulation $= \dfrac{8.85}{400} \times 100 =$ **2.21%. (Ans.)**

Iron loss, $P_i = 75$ W

Copper loss, $P_c = 110$ W

Full-load losses $= P_i + P_c = 75 + 110 = 185$ W

∴ Full-load output $= 8 \times 1000 \times 0.8 = 6400$ W

∴ **Full-load efficiency,** $\eta = \dfrac{\text{output}}{\text{output + losses}} = \dfrac{6400}{6400 + 185} = 0.972$ or **97.2%. (Ans.)**

(*iii*) **Half full-load :**

Iron loss, $P_i = 75$ W

Copper loss, $P_c = (\frac{1}{2})^2 \times 110 = 27.5$ W

Total losses $= P_i + P_c = 75 + 27.5 = 102.5$ W

Output at half full-load $= \frac{1}{2} \times 8 \times 1000 \times 0.8 = 3200$ W

∴ **Efficiency** $= \dfrac{\text{output}}{\text{output + losses}} = \dfrac{3200}{3200 + 102.5} = 0.969$ or **96.9%. (Ans.)**

Example 2.33. *Consider a 20 kVA, 2200/220 V, 50 Hz transformer. The OC/SC test results are as follows :*

O.C. Test : 220 V, 4.2 A, 148 W (l.v. side)

S.C. Test : 86 V, 10.5 A, 360 W (h.v. side)

(i) Determine the regulation at 0.8 p.f. lagging at full load.

(ii) What is the p.f. on short-circuit. **(Nagpur University, 1998)**

Solution. *Given* :

S.C. Test :

$$V_{SC} = 86 \text{ V}, I_{SC} = 10.5 \text{ A}, P_C = 360 \text{ W}$$

(*i*) **Regulation :** Equivalent impedance referred to primary, $Z_{01} = \dfrac{V_{SC}}{I_{SC}} = \dfrac{86}{10.5} = 8.19\ \Omega$

($\because$ Short-circuit test has been conducted on the h.v. (primary) side)

Equivalent resistance referred to primary, $R_{01} = \dfrac{P_{SC}}{I_{SC}^2} = \dfrac{360}{(10.5)^2} = 3.27\ \Omega$

Equivalent reactance referred to primary, $X_{01} = \sqrt{Z_{01}^2 - R_{01}^2} = \sqrt{8.19^2 - 3.27^2} = 7.51\ \Omega$

Full load primary current, $I_1 = \dfrac{20 \times 1000}{2200} = 9.09$ A

Percentage regulation $= \dfrac{I_1 (R_{01} \cos \phi + X_{01} \sin \phi)}{V_1} \times 100$

$= \dfrac{9.09(3.27 \times 0.8 + 7.51 \times 0.6)}{2200} \times 100 =$ **2.94%. (Ans.)**

(*ii*) **P.f. on short-circuit, cos ϕ_{SC} :**

$$\cos \phi_{SC} = \frac{R_{01}}{Z_{01}} = \frac{3.27}{8.19} = \mathbf{0.399\ (lag).\ \ (Ans.)}$$

Example 2.34. *High voltage side short circuit test data for 20 kVA, 2300/230 V transformer are :*

Power = 250 watts ; Current = 8.7 A ; Voltage = 50 V.

Calculate equivalent impedance, resistance, reactance referred to h.v. side. Find the transformer regulation at 0.7 lagging power factor. **(AMIE Summer, 1997)**

Solution. *Given* : **h.v. side :** $P_{SC} = 250$ W ; $I_{SC} = 8.7$ A ; $V_{SC} = 50$ V ; $\cos \phi = 0.7$.

From S.C. test with measurements performed on h.v. side the various parameters referred to *h.v.* side are :

$$R_{01} = \frac{P_{SC}}{I_{SC}^2} = \frac{250}{8.7^2} = 3.303\ \Omega$$

$$Z_{01} = \frac{V_{SC}}{I_{SC}} = \frac{50}{8.7} = 5.747\ \Omega$$

$$\therefore \quad X_{01} = \sqrt{Z_{01}^2 - R_{01}^2} = \sqrt{5.747^2 - 3.303^2} = 4.703\ \Omega$$

We know that, % Regulation $= \dfrac{I_1 R_{01} \cos \phi + I_1 X_{01} \sin \phi}{V_1} \times 100$

where $I_1 = \dfrac{20 \times 10^3}{2300} = 8.696$ A (rated current) ; and

$\cos \phi = 0.7$ (given) and hence $\sin \phi = \sin (\cos^{-1} 0.7) = 0.714$.

$\therefore$ **% Regulation** $= \dfrac{8.696 \times 3.303 \times 0.7 + 8.696 \times 4.703 \times 0.714}{2300} \times 100 =$ **2.14%. (Ans.)**

Example 2.35. *A 10 kVA, 2300/230 V, 50 Hz, distribution transformer has h.v. winding resistance 3.96 Ω and leakage resistance of 15.8 Ω. The l.v. winding has corresponding value of 0.0396 Ω and 0.158 Ω respectively. The transformer has a core loss of 58 W under normal operating conditions. Find :*

(*i*) *Load terminal voltage when transformer delivers rated current at 0.8 p.f. lagging to a load on l.v. side, with h.v. side voltage held at rated value and compute efficiency at this load.*

(*ii*) *The h.v. side voltage necessary to maintain rated voltage at load terminals when the transformer is loaded as above. Is efficiency same as in (i) ?* **(AMIE Summer, 1999)**

Solution. *Given* : $R_1 = 3.96\ \Omega$; $X_1 = 15.8\ \Omega$; $R_2 = 0.0396\ \Omega$; $X_2 = 0.158\ \Omega$; $P_i = 58$ W ; *p.f.* = 0.8 lagging

(*i*) **Load terminal voltage :**

Rated secondary current, $I_2 = \dfrac{10 \times 10^3}{230} = 43.48$ A

$$R_{02} = R_2 + K^2R_1 = 0.0396 + \left(\frac{230}{2300}\right)^2 \times 3.96 = 0.0792\ \Omega$$

$$X_{02} = X_2 + K^2X_1 = 0.158 + \left(\frac{230}{2300}\right)^2 \times 15.8 = 0.316\ \Omega$$

∴ Voltage drop on load side = $I_2 (R_{02} \cos \phi + X_{02} \sin \phi)$

= 43.48 (0.0792 × 0.8 + 0.316 × 0.6) = 11 V

∴ **Load terminal voltage** = 230 – 11 = **219 V. (Ans.)**

Copper loss $(P_C) = I_2^2R_{02} = 43.48^2 \times 0.0792 = 149.73$ W

$$\therefore \quad \eta = \frac{\text{Output}}{\text{Output} + \text{Core loss } (P_i) + \text{Copper loss } (P_C)}$$

$$= \frac{10 \times 10^3 \times 0.8}{(10 \times 10^3 \times 0.8) + 58 + 149.73} = 0.9747 \text{ or } \mathbf{97.47\%.} \text{ (Ans.)}$$

(*ii*) To maintain rated voltage at load terminals of 230 V, the O.C. voltage required at secondary terminals would be 230 + 11 = 241 V. Hence primary induced voltage required

$$= 241 \times \frac{2300}{230} = \mathbf{2410\ V.}\ \textbf{(Ans.)}$$

Example 2.36. *The following test results were obtained for a 1000 / 100 V, 100 kVA single-phase transformer :*

O.C. test :

Primary volts = 1000, secondary volts = 100, watts in primary = 950.

S.C. test :

Primary volts for full-load current = 20, watts in primary = 1000.

Determine the regulation and efficiency of the transformer at full-load and at 0.8 power factor lagging.

Solution. Transformation ratio, $K = \dfrac{100}{1000} = \dfrac{1}{10}$

O.C. test :

Primary volts, $V_1 = 1000$ V

Iron loss, $P_i = 950$ W = 0.95 kW

S.C. test :

Full-load copper loss, $P_c = 1000$ W = 1 kW

In S.C. test 20 V in primary are equivalent to $20 \times \dfrac{1}{10} = 2$ V in secondary.

Now, full-load secondary current,

$$I_2 = \frac{100 \times 1000}{100} = 1000 \text{ A}$$

Also $$Z_{02} = \frac{2}{1000} = 0.002\ \Omega$$

and $$R_{02} = \frac{1000}{(1000)^2} = 0.001\ \Omega$$

$\therefore$ $$X_{02} = \sqrt{Z_{02}^2 - R_{02}^2} = \sqrt{0.002^2 - 0.001^2} = 0.00173\ \Omega$$

Approximate voltage drop at 0.8 power factor lagging

$$= I_2(R_{02} \cos \phi + X_{02} \sin \phi)$$
$$= 1000(0.001 \times 0.8 + 0.00173 \times 0.6) = 1.838 \text{ V}$$

$\therefore$ **% regulation down** $= \frac{1.838}{100} \times 100 =$ **1.838%. (Ans.)**

Full-load output $= 100 \times 0.8 = 80$ kW

Total losses $= P_i + P_c = 0.95 + 1 = 1.95$ kW

$\therefore$ **Efficiency,** $\eta = \frac{\text{output}}{\text{output + losses}} = \frac{80}{80 + 1.95} = 0.976$ or **97.6%. (Ans.)**

Example 2.37. *The primary and secondary winding resistances of a 30 kVA, 6600/250 V single-phase transformer are 8 Ω and 0.015 Ω respectively. The equivalent leakage reactance as referred to the primary winding is 30 Ω. Find the full-load regulation for load power factors of :*

(i) unity *(ii) 0.8 lagging* *(iii) 0.8 leading.*

Solution. Transformer ratio, $K = \frac{250}{6600} = \frac{1}{26.4}$

Primary resistance, $R_1 = 8\ \Omega$

Secondary resistance, $R_2 = 0.015\ \Omega$

Equivalent leakage reactance as referred to primary,

$$X_{01} = 30\ \Omega$$

Now, $$R_{02} = R_2 + K^2R_1 = 0.015 + \left(\frac{1}{26.4}\right)^2 \times 8 = 0.02648\ \Omega$$

$$X_{02} = K^2X_{01} = \left(\frac{1}{26.4}\right)^2 \times 30 = 0.04304\ \Omega$$

Secondary full-load current, $I_2 = \frac{30 \times 1000}{250} = 120$ A.

(i) **Unity power factor,** $\cos \phi_2 = 1$:

Voltage drop $= I_2R_{02} \cos \phi_2 + I_2X_{02} \sin \phi_2 = 120 \times 0.02648 \times 1 + 0 = 3.177$ V

% regulation $= \frac{3.177}{250} \times 100 =$ **1.27%. (Ans.)**

(ii) **0.8 power factor (lagging) :**

Voltage drop $= I_2R_{02} \cos \phi_2 + I_2X_{02} \sin \phi_2$

$= 120 \times 0.02648 \times 0.8 + 120 \times 0.04304 \times 0.6 = 5.64$ V

% regulation $= \frac{564}{250} \times 100 =$ **2.256%. (Ans.)**

(iii) **0.8 power factor (leading) :**

Voltage drop $= I_2R_{02} \cos \phi_2 - I_2X_{02} \sin \phi_2$

$= 120 \times 0.02648 \times 0.8 - 120 \times 0.04304 \times 0.6 = -0.557$ V

% regulation $= -\frac{0.557}{250} \times 100 =$ **– 0.223%. (Ans.)**

Example 2.38. *Short-circuit test is conducted on a 5 kVA, 400 V/100 V single phase transformer with 100 V winding shorted. The input voltage at full load current is 40 V. The wattmeter, on the input reads 250 W. Find the power factor for which regulation at full load is zero.*

(GATE, 1994)

Solution. *Given* : Rating of transformer : 5 kVA, 400 V/100 V

Input power, on short-circuit, P_{SC} = 250 W

The input voltage at full load current, V_{SC} = 40 V

Power factor for which regulation at full load is zero :

Input current I_{SC} = full load current, $I_1 = \frac{5 \times 1000}{400} = 12.5$ A

Equivalent resistance as referred to primary, $R_{01} = \frac{P_{SC}}{I_{SC}^2} = \frac{250}{(12.5)^2} = 1.6\ \Omega$

Equivalent impedance referred to primary, $Z_{01} = \frac{V_{SC}}{I_{SC}} = \frac{40}{12.5} = 3.2\ \Omega$

Equivalent reactance referred to primary, $X_{01} = \sqrt{Z_{01}^2 - R_{01}^2} = \sqrt{3.2^2 - 1.6^2} = 2.77\ \Omega$

Regulation will be zero when phase angle,

$$\phi = \tan^{-1}\frac{R_{02}}{X_{02}} = \tan^{-1}\frac{R_{01}}{X_{01}} = \tan^{-1}\left(\frac{1.6}{2.77}\right) = \mathbf{30°}$$

or **power factor** $= \cos\phi = \cos(30°) =$ **0.866. (Ans.)**

(Refer Example 2.41 also)

Example 2.39. *A 50 MVA, 76.2 V/33 kV, 1-phase, 50 Hz, two-winding transformer with tap changer has percentage impedance of 0.5 + j7.0. What tapping must be used to maintain rated voltage at the secondary on*

(i) full load at 0.8 lagging power factor, and

(ii) 40 MVA load at 0.6 lagging power factor.

Assume that the tap changer is provided on the h.v. side. **(AMIE Summer, 1995)**

Solution. *Given* : Rated MVA = 50 ; f = 50 Hz ; E_1 = 76.2 V ; E_2 = 33 kV ; percentage impedance = 0.5 + j7.0.

(*i*) Percentage regulation $= \frac{L_2R_{02}\cos\phi + I_2X_{02}\sin\phi}{E_2} \times 100$

$$= \frac{I_2R_{02}}{E_2} \times 100 \cos\phi + \frac{I_2X_0}{E_2} \times 100 \sin\phi$$

$$= \%R\cos\phi + \%X\sin\phi$$

$$= 0.5 \times 0.8 + 7 \times 0.6 = 4.6\%$$

i.e., voltage to be raised for maintaining rated voltage = 4.6%

So tap setting required on h.v. side = – 4.6% or **4.6% down. (Ans.)**

(*ii*) At load of 40 MVA *i.e.* 80% of full load or 0.8 times full load,

voltage regulation $= 0.8\ (\%R\cos\phi + \%X\sin\phi)$

$$= 0.8(0.5 \times 0.6 + 7 \times 0.8) = 4.72\%$$

Tap setting required on h.v. side = **4.72% down. (Ans.)**

Note. *If the tap changer is on the primary side, tap setting will be down and if it is to be provided on secondary side, tap setting will be up for raising the secondary voltage on load.*

Example 2.40. *The percentage resistance and reactance of a transformer are 2% and 4% respectively. Find the approximate regulation on full-load at :*

(i) Unity power factor, *(ii) 0.8 power factor lagging, and*

(iii) 0.8 power factor leading.

Solution. Percentage resistance = 2%

Percentage reactance = 4%

Approximate % regulation = % resistance × cos ϕ ± % reactance × sin ϕ

(i) **Unity power factor :**

Approximate % regulation = % resistance cos ϕ ($\because$ sin ϕ = 0)

= 2 × 1 = **2%. (Ans.)**

(ii) **0.8 power factor lagging :**

Approximate % regulation = % resistance × cos ϕ + % reactance × sin ϕ

= 2 × 0.8 + 4 × 0.6 = 1.6 + 2.4 = **4%. (Ans.)**

(iii) **0.8 power factor leading :**

Approximate % regulation = % resistance × cos ϕ – % reactance × sin ϕ

= 2 × 0.8 – 4 × 0.6 = 1.6 – 2.4 = **– 0.8%. (Ans.)**

Example 2.41. *A single-phase 80 kVA, 2000/200 V, 50-Hz transformer has impedance drop of 8% and resistance drop of 4% :*

(i) Find the regulation at full-load 0.8 power factor lagging.

(ii) At what power factor is the regulation zero.

Solution. Impedance drop = 8%

Resistance drop = 4%

i.e.,
$$\frac{I_2Z_{02}}{V_2} = \frac{8}{100} \quad \text{or} \quad I_2Z_{02} = \frac{200 \times 8}{100} = 16 \text{ V}$$

and
$$\frac{I_2R_{02}}{V_2} = \frac{4}{100} \quad \text{or} \quad I_2R_{02} = \frac{200 \times 4}{100} = 8 \text{ V}$$

$\therefore$
$$I_2X_{02} = \sqrt{(I_2Z_{02})^2 - (I_2R_{02})^2} = \sqrt{16^2 - 8^2} = 13.86 \text{ V.}$$

(i) % regulation
$$= \frac{I_2R_{02}\cos\phi_2 + I_2X_{02}\sin\phi_2}{V_2}$$
$$= \frac{8 \times 0.8 + 13.86 \times 0.6}{200} = \frac{6.4 \times 8.316}{200} = \mathbf{7.36\%.}\ \textbf{(Ans.)}$$

(ii) Regulation can be zero only when the power factor is leading. Then

$$\frac{I_2R_{02}\cos\phi_2 - I_2X_{02}\sin\phi_2}{V_2} = 0$$

or
$$I_2R_{02}\cos\phi_2 - I_2X_{02}\sin\phi_2 = 0$$

or
$$I_2R_{02}\cos\phi_2 = I_2X_{02}\sin\phi_2$$

or
$$\tan\phi_2 = \frac{I_2R_{02}}{I_2X_{02}} = \frac{8}{13.86} = 0.577 \quad \text{or} \quad \phi_2 = 30°$$

$\therefore$ **Power factor** = cos ϕ_2 = **0.866 leading. (Ans.)**

Example 2.42. *The high voltage of a single-phase 200 kVA 4400/220 V transformer takes a current of 35 A and power of 1250 W at 80 V when the low voltage winding is short-circuited. Determine :*

(i) The voltage to be applied to the high voltage winding on full-load at 0.8 power factor lagging if the full-load secondary terminal voltage is to be kept at 220 V.

(ii) % regulation.

Solution. (*i*) Transformation ratio, $K = \frac{220}{4400} = \frac{1}{20}$

Voltage applied to h.v. winding in S.C. test

$$V_{SC} = 80 \text{ V}$$

and

$$I_{SC} = 35 \text{ A}$$

$$P_{SC} = 1250 \text{ W}$$

Referred to *h.v. side,*

$$Z_{01} = \frac{V_{SC}}{I_{SC}} = \frac{80}{35} = 2.28\ \Omega$$

$$R_{01} = \frac{P_{SC}}{I_{SC}^2} = \frac{1250}{35^2} = 1.02\ \Omega$$

$$X_{01} = \sqrt{Z_{01}^2 - R_{01}^2} = \sqrt{2.28^2 - 1.02^2} = 2.04\ \Omega$$

Full-load current in h.v. winding,

$$I_1 = \frac{200 \times 1000}{4400} = 45.45 \text{ A}$$

Voltage drop in resistance $= I_1R_{01} = 45.45 \times 1.02 = 46.36$ V

Voltage drop in reactance $= I_1X_{01} = 45.45 \times 2.04 = 92.72$ V

Full-load secondary voltage referred to primary,

$$V_2' = \frac{V_2}{K} = \frac{220}{\frac{1}{20}} = 4400 \text{ V}$$

To get full-load secondary voltage equal to 220 V, supply voltage V_1 must be equal to the vector sum of 4400 V, I_1R_{01} and IX_{01}.

From Fig. 2.56,

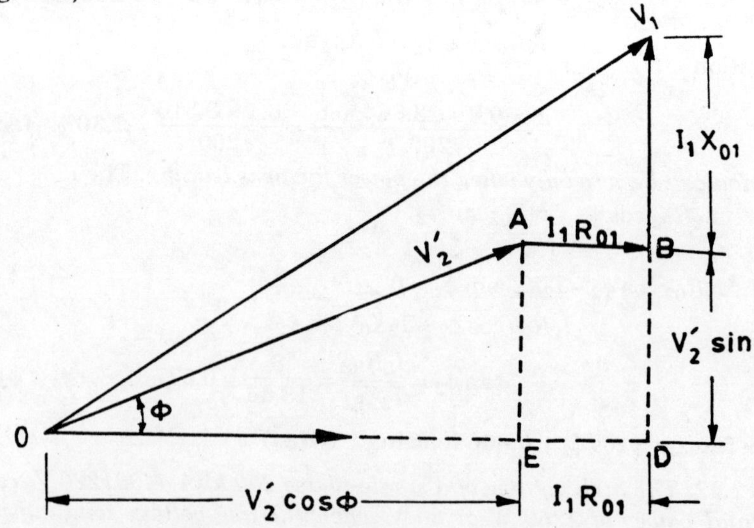

Fig. 2.56

$$OC^2 = OD^2 + CD^2 = (OE + ED)^2 + (BD + BC)^2$$

$$V_1^{\,2} = (V_2' \cos \phi + I_1 R_{01})^2 + (V_2' \sin \phi + I_1 X_{01})^2$$

$$= (4400 \times 0.8 + 46.36)^2 + (4400 \times 0.6 + 92.72)^2$$

or $$V_1 = 4493 \text{ V}$$

Voltage to be applied to h.v. winding to get a full-load secondary voltage of 220 V = 4493 V. (Ans.)

(*ii*) % **regulation** $= \dfrac{V_1 - V_2'}{V_1} \times 100 = \dfrac{4493 - 4400}{4493} \times 100 =$ **2.07%. (Ans.)**

Example 2.43. *The high voltage and low voltage windings of a 50 kVA, 4400/220 V, 50-Hz transformer have resistances of 2.2 Ω and 0.005 Ω respectively. The full-load current is obtained when 160 V at 50-Hz is applied to h.v. winding with l.v. winding short-circuited. Find :*

(*i*) *The equivalent resistance and reactance of the transformer referred to h.v. side, and*

(*ii*) *Reactance of each winding.*

Assume that the ratio of resistance to reactance is the same for each winding and the full-load efficiency of transformer is 0.98.

Solution. (*i*) Transformation ratio, $K = \dfrac{220}{4400} = \dfrac{1}{20}$

Primary resistance, $R_1 = 22\ \Omega$

Secondary resistance, $R_2 = 0.005\ \Omega$

Full-load output $= 50 \text{ kVA} = 50 \times 10^3 \text{ VA}$

Full-load input $= \dfrac{\text{output}}{\text{efficiency}} = \dfrac{50 \times 10^3}{0.98} \text{ VA}$

Full-load input current, $I_1 = \dfrac{\text{input in VA}}{\text{supply voltage}} = \dfrac{50 \times 10^3}{0.98 \times 4400} = 11.6 \text{ A}$

Impedance referred to h.v. side

$$Z_{01} = \frac{V_{SC}}{I_{SC}} = \frac{160}{11.6} = 13.8\ \Omega$$

Total resistance referred to h.v. side,

$$R_{01} = R_1 + R_2' = R_1 + \frac{R_2}{K^2} = 2.2 + \frac{0.005}{\left(\frac{1}{20}\right)^2} = 2.2 + 2 = \mathbf{4.2\ \Omega.\ (Ans.)}$$

$$X_{01} = \sqrt{Z_{01}^{\,2} - R_{01}^{\,2}} = \sqrt{(13.8)^2 - (4.2)^2} = \mathbf{13.15\ \Omega.\ (Ans.)}$$

(*ii*) Since the ratio of resistance to reactance remains same (given),

$$\therefore \quad \frac{R_1}{X_1} = \frac{R_2}{X_2} = \frac{R_2'}{X_2'}$$

i.e., $$\frac{R_1}{R_2'} = \frac{X_1}{X_2'}$$

or $$\frac{R_1 + R_2'}{R_2'} = \frac{X_1 + X_2'}{X_2'} \quad \text{or} \quad \frac{R_{01}}{R_2'} = \frac{X_{01}}{X_2'}$$

or $$X_2' = \frac{R_2' X_{01}}{R_{01}} = \frac{2 \times 13.15}{4.2} = 6.26\ \Omega$$

Reactance of primary winding,

$$X_1 = X_{02} - X_2' = 13.15 - 6.26 = \mathbf{6.85\ \Omega}. \quad \textbf{(Ans.)}$$

Reactance of secondary winding,

$$X_2 = K^2 X_2' = \left(\frac{1}{20}\right)^2 \times 6.26 = \mathbf{0.01565\ \Omega}. \quad \textbf{(Ans.)}$$

☞**Example 2.44.** *The following test results were obtained in a 250/500 V transformer :*

O.C. test (l.v. side) : 250 V, 1 A, 80 W.

S.C. test (l.v. winding short-circuited) : 20 V, 12 A, 100 W.

Determine :

(i) The circuit constants.

(ii) The applied voltage and efficiency when the output is 10 A at 500 V and 0.8 power factor lagging.

Solution. Transformation ratio, $K = \dfrac{500}{250} = 2$

(*i*) **Circuit constants :**

O.C. test *(l.v. side)*

Voltage, $V_1 = 250$ V

No-load current, $I_0 = 1$ A

No-load loss, $P_0 = 80$ W

Also, $P_0 = V_1 I_0 \cos\phi_0$

$$80 = 250 \times 1 \times \cos\phi_0$$

$$\cos\phi_0 = \frac{80}{250} = 0.32$$

Wattful component of no-load current I_0,

$$I_w = I_0 \cos\phi_0 = 1 \times 0.32 = 0.32 \text{ A}$$

Magnetising component of no-load current, I_0,

$$I_m = \sqrt{I_0^2 - I_w^2} = \sqrt{1^2 - 0.32^2} = 0.95 \text{ A}$$

Now, resistance representing the core loss,

$$R_0 = \frac{V_1}{I_w} = \frac{250}{0.32} = \mathbf{781.25\ \Omega}. \quad \textbf{(Ans.)}$$

Magnetising reactance, $X_0 = \dfrac{V_1}{I_m} = \dfrac{250}{0.95} = \mathbf{263.16\ \Omega}$. **(Ans.)**

S.C. test *(l.v. winding short-circuited)*

Short-circuit voltage, $V_{SC} = 20$ V

Short-circuit current, $I_{SC} = 12$ A

Losses, $P_{SC} = 100$ W

As the primary is short-circuited all *values refer to secondary winding.*

$$\therefore \quad R_{02} = \frac{P_{SC}}{I_{SC}^2} = \frac{100}{(12)^2} = 0.694\ \Omega$$

$$Z_{02} = \frac{V_{SC}}{I_{SC}} = \frac{20}{12} = 1.677\ \Omega$$

and

$$X_{02} = \sqrt{Z_{02}^2 - R_{02}^2} = \sqrt{(1.667)^2 - (0.694)^2} = 1.516\ \Omega$$

As R_0 and X_0 refer to primary, let us transfer these values to primary as follows :

$$R_{01} = \frac{R_{02}}{K^2} = \frac{0.694}{(2)^2} = \mathbf{0.174\ \Omega}$$

$$X_{01} = \frac{X_{02}}{K^2} = \frac{1.516}{(2)^2} = \mathbf{0.38\ \Omega}$$

$$Z_{01} = \frac{Z_{02}}{K^2} = \frac{1.669}{(2)^2} = \mathbf{0.417\ \Omega}$$

The equivalent circuit is shown in Fig. 2.57.

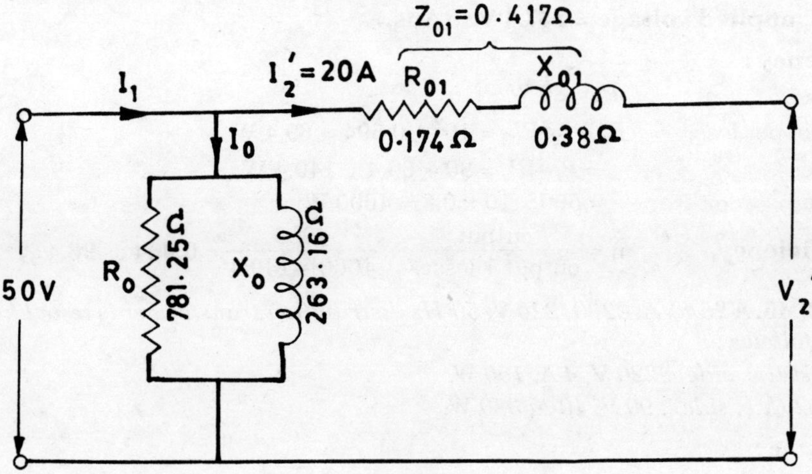

Fig. 2.57

(*ii*) **Applied voltage, V_1' :**

The applied voltage V_1' is the vector sum of V_1 and I_1Z_{01} (Fig. 2.58).

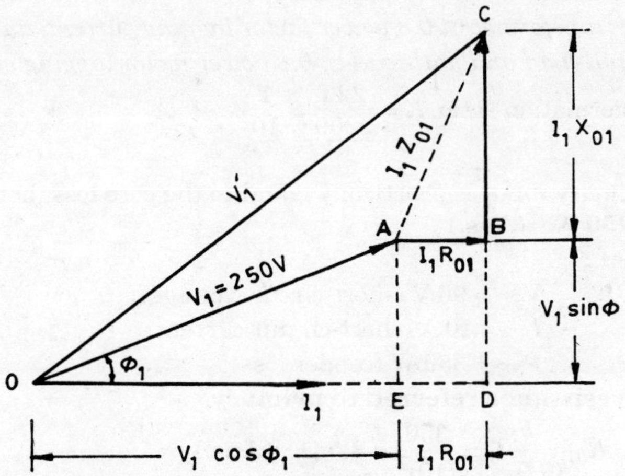

Fig. 2.58

Output current, $I_2 = 10$ A

$\therefore$ $I_1 = KI_2 = 2 \times 10 = 20$ A

Now, $I_1R_{01} = 20 \times 0.174 = 3.48$ V

$I_1X_{01} = 20 \times 0.38 = 7.6$ V

Refer Fig. 2.58. Neglecting the angle between V_1 and V_1', we have

$$OC^2 = OD^2 + DC^2$$

$$OC = V_1' = \sqrt{(OE + ED)^2 + (DB + BC)^2}$$

$$= \sqrt{(V_1 \cos \phi_1 + I_1R_{01})^2 + (V_1 \sin \phi_1 + I_1X_{01})^2}$$

$$= \sqrt{(250 \times 0.8 + 3.48)^2 + (250 \times 0.6 + 7.6)^2} = 257.4 \text{ V}$$

Hence, **applied voltage = 257.4 V. (Ans.)**

Efficiency:

Iron loss, $P_i = 80$ W

Total copper loss, $P_c = I_2^2R_{02} = 10^2 \times 0.694 = 69.4$ W

Total loss, $= P_i + P_c = 80 + 69.4 = 149.4$ W

Output $= 500 \times 10 \times 0.8 = 4000$ W

∴ **Efficiency,** $\eta = \dfrac{\text{output}}{\text{output + losses}} = \dfrac{4000}{4000 + 149.4} =$ **0.964** or **96.4%. (Ans.)**

☞**Example 2.45.** *A 25-kVA, 2200/220 V, 50-Hz distribution transformer is tested for efficiency and regulation as follows :*

O.C. test (l.v. side) : 220 V, 4 A, 150 W.

S.C. test (h.v. side) : 90 V, 10 A, 350 W.

Determine :

(*i*) *Core loss,*

(*ii*) *Equivalent resistance referred to primary,*

(*iii*) *Equivalent resistance referred to secondary,*

(*iv*) *Equivalent reactance referred to primary,*

(*v*) *Equivalent reactance referred to secondary,*

(*vi*) *Regulation of transformer at 0.8 power factor lagging current, and*

(*vii*) *Efficiency at full-load and half-load at 0.8 power factor lagging current.*

Solution. Transformation ratio, $K = \dfrac{220}{2200} = \dfrac{1}{10}$

(*i*) **Core loss :**

Since no-load primary input is practically equal to the core loss, hence, **core loss as found from no-load test, is 150 W. (Ans.)**

(*ii*) **From S.C. test :**

$V_{SC} = 90$ V (short-circuit voltage)

$I_{SC} = 10$ A (short-circuit current)

$P_{SC} = 350$ W (copper loss)

∴ **Equivalent resistance referred to primary,**

$$R_{01} = \frac{P_{SC}}{I_{SC}^2} = \frac{350}{(10)^2} = \mathbf{3.5\ \Omega.\ (Ans.)}$$

(*iii*) **Equivalent resistance referred to secondary,**

$$R_{02} = K^2R_{01} = \left(\frac{1}{10}\right)^2 \times 3.5 = \mathbf{0.035\ \Omega.\ (Ans.)}$$

(*iv*) Also $$Z_{01} = \frac{V_{SC}}{I_{SC}} = \frac{90}{10} = 9\ \Omega$$

∴ **Equivalent reactance referred to primary,**

$$X_{01} = \sqrt{Z_{01}^2 - R_{01}^2} = \sqrt{9^2 - 3.5^2} = \mathbf{8.29\ \Omega.}\ \textbf{(Ans.)}$$

(*v*) **Equivalent reactance referred to secondary,**

$$X_{02} = K^2 X_{01} = \left(\frac{1}{10}\right)^2 \times 8.29 = \mathbf{0.0829\ \Omega.}\ \textbf{(Ans.)}$$

(*vi*) **% regulation :**

Let us find the rise in voltage necessary to maintain the output terminal voltage constant from no-load to full-load.

Rated primary current $= \frac{25 \times 1000}{2200} = 11.36$ A

Now, using the relation,

$$V_1' = \sqrt{(V_1 \cos\phi + I_1 R_0)^2 + (V_1 \sin\phi + I_1 X_{01})^2}$$

$$= \sqrt{(2200 \times 0.8 + 11.36 \times 3.5)^2 + (2200 \times 0.6 + 11.36 \times 8.29)^2}$$

$$= 2289 \text{ V (app.)}$$

∴ **% regulation** $= \frac{V_1' - V_1}{V_1} \times 100$ (Art. 2.9.11)

$$= \frac{2289 - 2200}{2200} \times 100 = \mathbf{4.045\%.}\ \textbf{(Ans.)}$$

We can get the same result by working in the secondary :

Rated secondary current,

$$I_2 = \frac{I_1}{K} = \frac{11.36}{1/10} = 113.6 \text{ A}$$

$$_0V_2 = \sqrt{(V_2 \cos\phi + I_2 R_{02})^2 + (V_2 \sin\phi + I_2 X_{02})^2}$$

$$= \sqrt{(220 \times 0.8 + 113.6 \times 0.035)^2 + (220 \times 0.6 + 113.6 \times 0.0829)^2}$$

$$= 228.9 \text{ V}$$

∴ % regulation $= \frac{_0V_2 - V_2}{V_2} \times 100 = \frac{228.9 - 220}{220} \times 100 = \mathbf{4.045\%.}$ **(Ans.)**

(*vii*) **Efficiency :**

Core loss, $P_i = 150$ W

Copper loss at full-load,

$$P_c = I_1^2 R_{01} = 11.36^2 \times 3.5 = 451.7 \text{ W}$$

Copper loss at half-load,

$$P_c = \left(\frac{11.36}{2}\right)^2 \times 3.5 = 112.9 \text{ W}$$

∴ Efficiency at *full-load*

$$= \frac{\text{output}}{\text{output} + P_i + P_c} = \frac{25 \times 1000 \times 0.8}{(25 \times 1000 \times 0.8) + 150 + 451.7}$$

$$= \mathbf{0.9707} \text{ or } \mathbf{97.07\%.}\ \textbf{(Ans.)}$$

Efficiency at half *full-load*

$$= \frac{(25 \times 1000/2) \times 0.8}{(25 \times 1000/2) \times 0.8 + 150 + 112.9} = \mathbf{0.9744 \text{ or } 97.44\%.} \quad \textbf{(Ans.)}$$

☞**Example 2.46.** *Two similar 100 kVA, single-phase transformers gave the following test readings when tested by Sumpner's test.*

Supply power = 2.4 kW.

Power supplied to secondary circuit in passing full-load current through it = 3.2 kW.

Find the efficiency and regulation of each transformer at unity power factor.

Solution. Core loss (or iron loss) of the transformers

= 2.4 kW

Iron loss of each transformer,

$$P_i = \frac{2.4}{2} = 1.2 \text{ kW}$$

Full-load copper loss of both transformers

= 3.2 kW

Full-load copper loss of each transformer

$$P_c = \frac{3.2}{2} = 1.6 \text{ kW}$$

Total full-load losses of each transformer

$= P_i + P_c = 1.2 + 1.6 = 2.8$ kW

Output at full-load at unity power factor

$= 100 \times 1 = 100$ kW

$$\therefore \textbf{ Full-load efficiency, } \eta = \frac{\text{output}}{\text{output + losses}} = \frac{100}{100 + 2.8} = \mathbf{0.9727 \text{ or } 97.27\%.} \quad \textbf{(Ans.)}$$

Example 2.47. *A 10 kVA, 2500/250 V, single phase transformer gave the following test results :*

Open circuit test : 250 V, 0.8 A, 50 W

Short circuit test : 60 V, 3 A, 45 W

(*i*) *Calculate the efficiency of half full load at 0.8 p.f.*

(*ii*) *Calculate the load kVA at which maximum efficiency occurs and also the maximum efficiency at 0.8 p.f.*

(*iii*) *Compute the voltage regulation at 0.8 p.f. leading.* **(AMIE Winter, 1998)**

Solution. Full rated current $= \dfrac{10 \times 1000}{2500} = 4$ A

Hence reading of wattmeter corresponding to full load current of 4 A

$$= 45 \times \left(\frac{4}{3}\right)^2 = 80 \text{ W}$$

∴ Full load copper losses = 80 W, and

Full load iron losses = 50 W

(*i*) **Efficiency at half load at 0.8 p.f. :**

$$\eta = \frac{\text{half load output}}{\text{half load output + losses}}$$

$$= \frac{(10 \times 1000 \times 0.5) \times 0.8}{[(10 \times 1000 \times 0.5) \times 0.8] + 50 + 80 \times (\frac{1}{2})^2}$$

$$= \frac{4000}{4000 + 50 + 20} = 0.983 \text{ or } \mathbf{98.3\%.} \ \textbf{(Ans.)}$$

(*ii*) **Load kVA for maximum efficiency, and its value :**

For maximum efficiency :

Copper loss = Iron loss = 50 W

$\therefore$ Current at which maximum efficiency occurs $= \frac{50 \times 4}{80} = 2.5$ A

$\therefore$ **Load kVA** $= 10 \times \frac{2.5}{4} = \mathbf{6.25\ kVA.}$ **(Ans.)**

$$\eta_{max} \text{ (at 0.8 p.f.)} = \frac{(6.25 \times 1000 \times 0.8)}{(6.25 \times 1000 \times 0.8) + 50 + 50} = 0.98 \text{ or } \mathbf{98\%.} \ \textbf{(Ans.)}$$

(*iii*) **Voltage regulation at 0.8 p.f. leading :**

From short circuit test :

$$V_{SC} = 60 \text{ V} ; I_{SC} = 3 \text{ A} ; P_{SC} = 45 \text{ W}$$

Now,

$$R_{01} = \frac{P_{SC}}{I_{SC}^2} = \frac{45}{3^2} = 5\ \Omega$$

$$Z_{01} = \frac{V_{SC}}{I_{SC}} = \frac{60}{3} = 20\ \Omega$$

$$\therefore \quad X_{01} = \sqrt{Z_{01}^2 - R_{01}^2} = \sqrt{20^2 - 5^2} = 19.36\ \Omega$$

$\therefore$ Voltage drop at leading p.f. of 0.8

$$= I_1 R_{01} \cos\phi - I_1 X_{01} \sin\phi$$

$$= 4 \times 5 \times 0.8 - 4 \times 19.36 \times 0.6 = -30.46 \text{ V}$$

Hence % **voltage regulation** $= \frac{-30.46}{2500} \times 100 = \mathbf{-1.218\%.}$ **(Ans.)**

Example 2.48. *A 50 kVA single-phase transformer has a full-load primary current of 250 A and total resistance referred to primary is 0.006 ohm. If the iron loss amounts to 200 W, find the efficiency on full-load and half-load at (i) unity power factor and (ii) 0.8 power factor.*

Solution. Full-load primary current,

$$I_1 = 250 \text{ A}$$

Total resistance referred to primary,

$$R_{01} = 0.006\ \Omega$$

Iron loss, $P_i = 200$ W

Full-load copper loss, $P_c = I_1^2 R_{01} = (250)^2 \times 0.006 = 375$ W

Total full-load loss $= P_i + P_c = 200 + 375 = 575 \text{ W} = 0.575$ kW

Half full-load copper loss $= (\frac{1}{2})^2 P_c = (\frac{1}{2})^2 \times 375 = 93.75$ W

Total loss on half-load $= 200 + 93.75 = 293.75 \text{ W} = 0.294$ kW (say)

(*i*) **At unity power factor :**

Full-load output, $= 50 \times 1 = 50$ kW

Efficiency at full-load $= \frac{\text{output}}{\text{output + losses}} = \frac{50}{50 + 0.575} = \mathbf{0.9886}$ or **98.86%.** **(Ans.)**

Half-full-load output $= 50 \times \frac{1}{2} \times 1 = 25$ kW

Efficiency at half full-load at unity power factor

$$= \frac{25}{25 + 0.294} = \mathbf{0.9984} \text{ or } \mathbf{99.84\%.} \ \textbf{(Ans.)}$$

(*ii*) **At 0.8 power factor :**

Full-load output $= 50 \times 0.8 = 40$ kW

Efficiency on full-load at 0.8 power factor

$$= \frac{40}{40 + 0.575} = \mathbf{0.9858} \text{ or } \mathbf{98.58\%}. \quad \textbf{(Ans.)}$$

Half full-load output $= 50 \times \frac{1}{2} \times 0.8 = 20$ kW

Efficiency on half full-load at 0.8 power factor

$$= \frac{20}{20 + 0.294} = \mathbf{0.9855} \text{ or } \mathbf{98.55\%}. \quad \textbf{(Ans.)}$$

☞**Example 2.49.** *A 40 kVA, single-phase transformer has an iron loss of 300 W and full-load copper loss of 600 W,*

(*i*) *Find the load at which maximum efficiency occurs and the value of maximum efficiency at unity power factor.*

(*ii*) *If the maximum efficiency occurs at 80% of full-load, find the new core loss and full-load copper loss assuming that total full-load loss is a constant.*

Solution. Rating of transformer = 40 kVA

Iron loss, $P_i = 300$ W

Full-load copper loss, $P_c = 600$ W

(*i*) Let the maximum efficiency occurs at *x* times full-load, then

$$x^2 P_c = P_i$$

or

$$x = \sqrt{\frac{P_i}{P_c}} = \sqrt{\frac{300}{600}} = 0.707 = 70.7\%$$

Hence, **efficiency occurs at 70.7% of full-load. (Ans.)**

Maximum efficiency :

Output at unity power factor

$$= 0.707 \times 40 \times 1 = 28.28 \text{ kW}$$

Total losses $= P_i + P_c = 2P_i$ $\left[\text{Because when efficiency is maximum, } P_i = P_c\right]$

$= 2 \times 300 = 600$ W

$= 0.6$ kW

∴ **Maximum efficiency**

$$= \frac{\text{output}}{\text{output + losses}} = \frac{28.28}{28.28 + 0.6} = \mathbf{0.9792} \text{ or } \mathbf{97.92\%}. \quad \textbf{(Ans.)}$$

(*ii*) **New core loss, P_i' :**

New copper loss, P_c' :

Maximum efficiency occurs at 80% of full-load

Now, $P_i' + P_c' = P_i + P_c$ (*given*)

$= 300 + 600 = 900$ W

Also $0.8 = \sqrt{\dfrac{P_i'}{P_c'}}$ or $\dfrac{P_i'}{P_c'} = [0.8]^2 = 0.64$

or $\dfrac{P_i'}{P_i' + P_c'} = \dfrac{0.64}{1 + 0.64} = 0.3902$ or $\dfrac{P_i'}{900} = 0.3902$

∴ $P_i' = 351.2$ W

and $P_c' = 900 - P_i' = 900 - 351.2 = 548.8$ W

Hence, **new iron loss = 351.2 W. (Ans.)**

new copper loss = 548.8 W. (Ans.)

Example 2.50. *The primary and secondary resistances of a 1100/220 V transformer are 0.3 Ω and 0.02 Ω respectively. If iron loss amounts to 260 W determine the secondary current at which maximum efficiency occurs and find the maximum efficiency at 0.8 power factor.*

Solution. Transformation ratio, $K = \frac{220}{1100} = \frac{1}{5}$

Primary resistance, $R_1 = 0.3\ \Omega$

Secondary resistance, $R_2 = 0.02\ \Omega$

Iron loss, $P_i = 260$ W

Total resistance referred to secondary,

$$R_{02} = R_2 + K^2R_1 = 0.02 + \left(\frac{1}{5}\right)^2 \times 0.3 = \mathbf{0.032\ \Omega} \quad \textbf{(Ans).}$$

Let I_2 be the secondary current at maximum efficiency.

At maximum efficiency,

Copper loss = Iron loss

$$I_2^2R_{02} = 260$$

$$\therefore \quad I_2^2 = \frac{260}{R_{02}} = \frac{260}{0.032} = 8125$$

or $\quad I_2 = 90.14$ A

Hence, **secondary current at maximum efficiency**

= 90.14 A. (Ans.)

Maximum efficiency at 0.8 power factor :

Output at maximum efficiency at 0.8 power factor

$= V_2I_2 \cos \phi_2 = 220 \times 90.14 \times 0.8 = 15865$ W

Losses $= P_i + P_c = 2P_i \quad (\because\ P_i = P_c)$

$= 2 \times 260 = 520$ W

$$\therefore \quad \textbf{Maximum efficiency} = \frac{\text{output}}{\text{output + losses}} = \frac{15865}{15865 + 520} = \mathbf{0.9683} \text{ or } \mathbf{96.83\%}. \quad \textbf{(Ans.)}$$

Example 2.51. *A single-phase 200 kVA transformer has an efficiency of 95% on full-load at 0.8 power factor and on half full-load at 0.8 power factor. Find :*

(i) Iron loss *(ii) Full-load copper loss.*

Solution. Rating of transformer = 200 kVA

Efficiency = 95%

Full-load output at 0.8 power factor

$= 200 \times 0.8 = 160$ kW

Full-load input at 0.8 power factor

$$= \frac{\text{output}}{\text{efficiency}} = \frac{160}{0.95} = 168.4 \text{ kW}$$

Total losses on full-load = input – output = 168.4 – 160 = 8.4 kW

If P_i is iron loss and P_c full-load copper loss,

Then $\quad P_i + P_c = 8.4$ kW ...(*i*)

Output on half full-load at 0.8 power factor

$$= \frac{200}{2} \times 0.8 = 80 \text{ kW}$$

Input $= \dfrac{\text{output}}{\text{efficiency}} = \dfrac{80}{0.95} = 84.2$ kW

Total losses on half full-load $= 84.2 - 80 = 4.2$ kW

Copper losses at half full-load $= (\tfrac{1}{2})^2 P_c = \tfrac{1}{4} P_c$

Then total losses, $P_i + \tfrac{1}{4} P_c = 4.2$ ...(*ii*)

Subtracting (*ii*) from (*i*), we get

$$\tfrac{3}{4} P_c = 8.4 - 4.2 = 4.2 \quad \text{or} \quad P_c = \frac{4.2 \times 4}{3} = 5.6 \text{ kW}$$

$\therefore$ $P_i = 8.4 - 5.6 = 2.8$ kW

Hence, **iron loss** **= 2.8 kW. (Ans.)**

and **full-load copper loss** **= 5.6 kW. (Ans.)**

Example 2.52. *The efficiency of a 250 kVA, single-phase transformer is 96% when delivering full-load at 0.8 power factor lagging and 97.2% when delivering half full-load at unity power factor. Determine the efficiency at 75% of full-load at 0.8 power factor lagging.*

Solution. Rating of transformer = 250 kVA

Full-load efficiency at 0.8 p.f. lagging = 96%

Half full-load efficiency at unity p.f. = 97.2%

Efficiency at 0.75 full-load at 0.8 p.f. lagging :

At full-load :

Output at 0.8 p.f. $= 250 \times 0.8 = 200$ kW

Input $= \dfrac{\text{output}}{\text{efficiency}} = \dfrac{200}{0.96} = 208.33$ kW

Total losses = input – output

$= 208.33 - 200 = 8.33$ kW

i.e., $P_i + P_c = 8.33$ kW ...(*i*)

At half full-load :

Output at unity p.f. $= \dfrac{250}{2} \times 1 = 125$ kW

Input $= \dfrac{\text{output}}{\text{efficiency}} = \dfrac{125}{0.972} = 128.6$ kW

Total losses $= 128.6 - 125 = 3.6$ kW

Copper loss at $\tfrac{1}{2}$ load $= (\tfrac{1}{2})^2 P_c = \tfrac{1}{4} P_c$

$\therefore$ Total loss, $P_i + \tfrac{1}{4} P_c = 3.6$ kW ...(*ii*)

Subtracting (*ii*) from (*i*), we get

$$\tfrac{3}{4} P_c = 8.33 - 3.6$$

$\therefore$ $P_c = 6.31$ kW

and $P_i = 8.33 - 6.31 = 2.02$ kW

Efficiency at 75% of full-load at 0.8 p.f. lagging :

Copper loss at 75% of load

$= (0.75)^2 \times P_c = (0.75)^2 \times 6.31 = 0.5625 \times 6.31 = 3.55$ kW

Total loss at 75% of full-load

$= P_i + (0.75)^2 \times P_c = 2.02 + 3.55 = 5.57$ kW

Output at 75% of load at 0.8 power factor

$$= 250 \times 0.75 \times 0.8 = 150 \text{ kW}$$

Efficiency $$= \frac{\text{output}}{\text{output + losses}} = \frac{150}{150 + 5.57} = \mathbf{96.42\%.}\ \textbf{(Ans.)}$$

Example 2.53. *A 10 kVA, 400/200 V single phase 50-Hz transformer has a maximum efficiency of 92% at 80% of full-load at unity p.f. Determine the efficiency at full-load at 0.8 p.f. lagging.*

Solution. Rating of transformer = 10 kVA

Maximum efficiency = 92%

Efficiency at full-load at 0.8 p.f. lagging :

Output at 80% of full-load at unity p.f.

$$= 10 \times 0.8 \times 1 = 8 \text{ kW}$$

Input at 80% of full-load $$= \frac{\text{output}}{\text{efficiency}} = \frac{8}{0.92} = 8.69 \text{ kW}$$

Total losses = input – output = 8.69 – 8 = 0.69 kW

At maximum efficiency

$$P_i = (0.8)^2 P_c = 0.64 P_c \qquad ...(i)$$

Total loss at maximum efficiency

$$= 2P_i = 0.69 \qquad (\because P_i = P_c)$$

$$\therefore \quad P_i = \frac{0.69}{2} = 0.345 \text{ kW}$$

Also $$P_c = \frac{P_i}{0.64} \qquad \text{[from } (i)]$$

$$= \frac{0.345}{0.64} = 0.539 \text{ kW}$$

Total full-load loss $= P_i + P_c = 0.345 + 0.539 = 0.884$ kW

Full-load output at 0.8 p.f. = 10 × 0.8 = 8 kW

∴ Efficiency at full-load at 0.8 p.f. lagging

$$= \frac{\text{output}}{\text{output + losses}} = \frac{8}{8 + 0.884} = \mathbf{0.90} \text{ or } \mathbf{90\%.}\ \textbf{(Ans.)}$$

Example 2.54. *A 40 kVA, 2000/250 V single-phase transformer has resistances of 0.8 Ω, 0.012 Ω and reactances of 4 Ω and 0.04 Ω for the high voltage and low voltage windings respectively. It has a maximum efficiency of 96% at 80% of full-load at unity p.f. The magnetising current for the h.v. side at 2000 V is 1.2 A. Find the readings of suitable instruments for open circuit test and short-circuit test, supply being given to h.v. side in both cases.*

Solution. Rating of transformer = 40 kVA

Transformation ratio, $$K = \frac{250}{2000} = \frac{1}{8}$$

Primary resistance, $R_1 = 0.8\ \Omega$

Primary reactance, $X_1 = 4\ \Omega$

Secondary resistance, $R_2 = 0.012\ \Omega$

Secondary reactance, $X_2 = 0.04\ \Omega$

Maximum efficiency = 96%

Magnetising current, $I_m = 1.2$ A

Readings of instruments :

Load at maximum efficiency = 0.8 × 40 × 1 = 32 kW

Maximum efficiency $= \dfrac{\text{output}}{\text{output + losses}}$

$$0.96 = \frac{32}{32 + \text{losses}}$$

$\therefore$ $$\text{Losses} = \frac{32}{0.96} - 32 = 1.333 \text{ kW}$$

But losses at maximum efficiency

$$= 2P_i = 1.333$$

$\therefore$ Iron loss, $$P_i = \frac{1.333}{2} = 0.666 \text{ kW} = 666 \text{ W}$$

O.C. test :

Supply voltage, $V_1 = 2000$ V

Wattmeter reading = iron loss = 666 W

Wattful component of no-load current I_0,

$$I_w = \frac{666}{2000} = 0.333 \text{ A}$$

Magnetising component of no-load current I_0,

$$I_m = 1.2 \text{ A} \qquad \textit{(given)}$$

$\therefore$ No-load current $$I_0 = \sqrt{I_w^2 + I_m^2} = \sqrt{(0.333)^2 + (1.2)^2} = 1.245 \text{ A}$$

$\therefore$ Reading of voltmeter = **2000 V. (Ans.)**

Reading of ammeter = **1.245 A. (Ans.)**

Reading of wattmeter = **666 W. (Ans.)**

S.C. test :

The l.v. side is short circuited and supply is given to h.v. side to pass full-load current.

Referred to h.v. side :

$$R_{01} = R_1 + R_2' = R_1 + \frac{R_2}{K^2}$$

$$= 0.8 + \frac{0.012}{(1/8)^2} = 0.8 + 0.012 \times 64 = 1.57 \ \Omega$$

$$X_{01} = X_1 + X_2' = X_1 + \frac{X_2}{K^2} = 4 + \frac{0.04}{(1/8)^2} = 4 + 0.04 \times 64 = 6.56 \ \Omega$$

$$Z_{01} = \sqrt{R_{01}^2 + X_{01}^2} = \sqrt{(1.57)^2 + (6.56)^2} = 6.74 \ \Omega$$

Full-load current in h.v. winding $= \dfrac{40 \times 1000}{2000} = 20$ A

Voltage to be applied to h.v. winding to pass full-load current in S.C. test,

$$V_{SC} = I_1 Z_{01} = 20 \times 6.74 = 134.8 \text{ V}$$

Full-load copper loss $= I_1^2 R_{01} = 20^2 \times 1.57 = 628$ W

$\therefore$ Reading of voltmeter = **134.8 V. (Ans.)**

Reading of ammeter = **20 A. (Ans.)**

Reading of wattmeter = **628 W. (Ans.)**

☞**Example 2.55.** *A 120 kVA 3300/300 V single-phase transformer has an efficiency of 97% both on full-load at unity p.f. and on half full-load at unity power factor. The power factor of the no-load current is 0.3 lagging and the regulation on full-load at 0.8 p.f. lagging is 2.5 per cent. Draw the equivalent circuit referred to the l.v. side.*

Solution. Ratting of transformer = 120 kVA

Transformation ratio $= \frac{300}{3300} = \frac{1}{11}$

Efficiency = 97%.

Power factor of no-load current, $\cos \phi = 0.3$ lagging

Regulation = 2.5%

Let P_i = iron loss

and P_c = full-load copper loss

Full-load output a unity p.f. $= 120 \times 1 = 120$ kW

Full-load losses $= P_i + P_c$

Efficiency (full-load) $= \frac{\text{output}}{\text{output + losses}}$

$$0.97 = \frac{120}{120 + P_i + P_c}$$

$$\therefore \quad P_i + P_c = \frac{120}{0.97} - 120 = 3.71 \text{ kW} = 3710 \text{ W} \quad ...(i)$$

At half full-load, losses $= P_i + (\tfrac{1}{2})^2 P_c = P_i + \tfrac{1}{4} P_c$

Output on half full-load at unity p.f.

$$= \frac{120}{2} \times 1 = 60 \text{ kW}$$

$\therefore$ Efficiency (half full-load) $= \frac{\text{output}}{\text{output + losses}}$

$$0.97 = \frac{60}{60 + \left(P_i + \frac{P_c}{4}\right)}$$

$$\therefore \quad P_i + \frac{P_c}{4} = \frac{60}{0.97} - 60 = 1.856 \text{ kW} = 1856 \text{ W} \quad ...(ii)$$

Subtracting (*ii*) from (*i*), we get

$$\tfrac{3}{4} P_c = 3710 - 1856 = 1854 \text{ W}$$

$$\therefore \quad P_c = 2472 \text{ W}$$

and $$P_i = 3710 - 2472 = 1238 \text{ W}$$

Referring to l.v. side,

$$V_2 I_0 \cos \phi_0 = P_i$$

$$\therefore \quad I_0 = \frac{P_i}{V_2 \cos \phi_0} = \frac{1238}{300 \times 0.3} = 13.75 \text{ A}$$

$$I_w = I_0 \cos \phi_0 = 13.75 \times 0.3 = 4.125 \text{ A}$$

$$\sin \phi_0 = \sqrt{1 - 0.3^2} = 0.954$$

$$\therefore \quad I_m = I_0 \sin \phi_0 = 13.75 \times 0.954 = 13.12 \text{ A}$$

$$R_0 = \frac{V_2}{I_w} = \frac{300}{4.125} = 72.7\ \Omega$$

$$X_0 = \frac{V_2}{I_m} = \frac{300}{13.12} = 22.86\ \Omega$$

Current in l.v. winding on full-load,

$$I_2 = \frac{120 \times 1000}{300} = 400\ \text{A}$$

Full-load copper loss $= I_2^2 R_{02} = 2472$

$$\therefore \qquad R_{02} = \frac{2472}{(400)^2} = 0.01545\ \Omega$$

Let X_{02} be the equivalent reactance of the transformer referred to l.v. side.

$$\%\ \text{regulation} = \frac{I_2(R_{02}\cos\phi_2 + X_{02}\sin\phi_2)}{V_{02}} \times 100$$

$$2.5 = \frac{400 \times 100(0.01545 \times 0.8 + X_{02} \times 0.6)}{300}$$

$$(0.01545 \times 0.8 + 0.6\,X_{02}) = \frac{2.5 \times 300}{400 \times 100} = 0.01875$$

$$\therefore \qquad X_{02} = \frac{0.01875 - 0.01545 \times 0.8}{0.6} = 0.01065\ \Omega$$

The equivalent circuit referred to l.v. side is shown in Fig. 2.59.

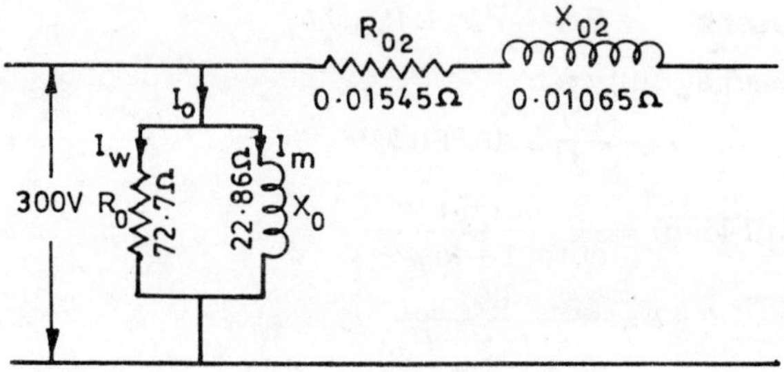

Fig. 2.59

Example 2.56. *The maximum efficiency of a single-phase 250 kVA, 2000/250 V transformer occurs at 80% of full-load and is equal to 97.5% at 0.8 p.f. Determine the efficiency and regulation on full-load at 0.8 p.f. lagging if the impedance of the transformer is 9 per cent.*

Solution. Rating of transformer = 250 kVA

Transformation ratio, $K = \frac{250}{2000} = \frac{1}{8}$

Maximum efficiency, $= 97.5\%$

Impedance of transformer $= 9\%$

Efficiency and regulation on full-load :

Let P_i and P_c be the iron loss and full-load copper loss respectively.

Maximum efficiency occurs at 80% of full-load

$$\therefore \qquad 0.8 = \sqrt{\frac{P_i}{P_c}} \quad \text{or} \quad 0.64 = \frac{P_i}{P_c}$$

$$\therefore \qquad P_i = 0.64 P_c \qquad ...(i)$$

At maximum efficiency

Output at 0.8 p.f. $= 0.8 \times 250 \times 0.8$ kW = 160 kW

Maximum efficiency = 97.5%

Input $= \dfrac{160}{0.975} = 164.1$ kW

Total losses = input – losses = 164.1 – 160 = 4.1 kW

But at maximum efficiency,

Total loss, $= 2P_i = 4.1$

$\therefore \quad P_i = \dfrac{4.1}{2} = 2.05$ kW

From (*i*), we have $P_c = \dfrac{P_i}{0.64} = \dfrac{2.05}{0.64} = 3.2$ kW

Efficiency :

Output at full-load at 0.8 p.f. $= 250 \times 0.8 = 200$ kW

Total full-load losses $= P_i + P_c = 2.05 + 3.2 = 5.25$ kW

$\therefore$ **Efficiency at full-load at 0.8 p.f.**

$$= \frac{\text{output}}{\text{output + losses}} = \frac{200}{200 + 5.25} = \mathbf{0.9744} \text{ or } \mathbf{97.44\%}. \quad \textbf{(Ans.)}$$

Regulation :

Full-load copper loss $= I_2{}^2 R_{02} = 3.2$

$V_2 I_2$ = kVA of the transformer

= 250 kVA, where V_2 is in kV and I_2, full-load current

% resistance (%*R*) $= \dfrac{I_2 R_{02}}{V_2} \times 100 = \dfrac{I_2{}^2 R_{02}}{V_2 I_2} \times 100 = \dfrac{3.2}{250} \times 100 = 1.28$ per cent

% impedance (%*Z*) = 9 per cent

% reactance (%*X*) $= \sqrt{(\%Z)^2 - (\%R)^2} = \sqrt{(9)^2 - (1.28)^2} = 8.908$ per cent

% regulation at 0.8 p.f. lagging $= \%R \cos\phi + \%X \sin\phi$

$= 1.28 \times 0.8 + 8.908 \times 0.6 =$ **6.368 per cent. (Ans.)**

☞**Example 2.57.** *A 5 kVA, 230/115 V transformer takes 1.1 A and 70 W when 230 V is applied to h.v. winding and l.v. winding is kept open. It takes 20 A and 140 W when the l.v. winding is short-circuited and 17 V is applied to h.v. winding. Find :*

(i) The no-load current as a fraction of full-load input current at 0.8 p.f. lagging,

(ii) Percentage regulation on full-load at 0.8 p.f. lagging,

(iii) The load at which maximum efficiency occurs, and

(iv) The maximum efficiency at unity p.f.

Solution. Rating of transformer = 5 kVA

Transformation ratio, $K = \dfrac{115}{230} = \dfrac{1}{2}$

Core loss, $P_0 = 70$ W

No-load current, $I_0 = 1.1$ A

Full-load current in h.v. winding,

$$I_2' = \frac{5 \times 1000}{230} = 21.74 \text{ A}$$

Also, $P_0 = V_1 I_0 \cos\phi_0$

$70 = 230 \times 1.1 \times \cos\phi_0$

$\therefore \quad \cos \phi_0 = \dfrac{70}{230 \times 1.1} = 0.276$

$\sin \phi_0 = 0.961$

$I_w = I_0 \cos \phi_0 = 1.1 \times 0.276 = 0.3036$ A

$I_m = I_0 \sin \phi_0 = 1.1 \times 0.961 = 1.0571$ A

$I_0 = I_w - jI_m = (0.3036 - j1.0571)$A

Full-load output current referred to primary,

$I_2' = 21.74$ A (Calculated above)

$\cos \phi_2 = 0.8$

$\sin \phi_2 = 0.6$

$I_2' = 21.74(0.8 - j0.6) = (17.39 - j13.04)$ A

$I_1 = I_2' + I_0 = (17.39 - j13.04) + (0.3036 - j1.0571)$

$= 17.7 - j14.1 = 22.63$ A

$I_0 = 1.1$ A

$\dfrac{I_0}{I_1} = \dfrac{1.1}{22.63} = 0.0486$

$I_0 = 4.86\% \, I_1$

(*i*) **No-load current = 4.86% of full-load input current. (Ans.)**

(*ii*) **Referred to h.v. side,**

$V_{SC} = 17$ V

$I_{SC} = 20$ A

$P_{SC} = 140$ W

$$Z_{01} = \frac{V_{SC}}{I_{SC}} = \frac{17}{20} = 0.85 \ \Omega$$

$$R_{01} = \frac{P_{SC}}{I_{SC}^2} = \frac{140}{(20)^2} = 0.350 \ \Omega$$

$$X_{01} = \sqrt{Z_{01}^2 - R_{01}^2} = \sqrt{(0.85)^2 - (0.35)^2} = 0.775 \ \Omega$$

% regulation at 0.8 p.f. lagging

$$= \frac{I_2' (R_{01} \cos \phi + X_{01} \sin \phi)}{V_1} \times 100$$

$$= \frac{21.74(0.35 \times 0.8 + 0.775 \times 0.6)}{230} \times 100 = \mathbf{7.04\%.} \ \textbf{(Ans.)}$$

(*iii*) Core loss, $P_0 = P_i = 70$ W

Full-load copper loss, $P_c = 140$ W

Let maximum efficiency occurs at x times full-load,

$$x = \sqrt{\frac{P_i}{P_c}} = \sqrt{\frac{70}{140}} = 0.707 = 70.7\%$$

Maximum efficiency occurs at 70.7% of full-load. (Ans.)

(*iv*) **At maximum efficiency,**

Output at unity p.f. $= 0.707 \times 5000 \times 1 = 3535$ W

Copper loss, P_c = core or iron loss, P_i

$\therefore$ Total loss $= 70 + 70 = 140$ W

Maximum efficiency (at unity power factor)

$$= \frac{\text{output}}{\text{output + losses}} = \frac{3535}{3535 + 140} = \mathbf{96.19\%.} \ \textbf{(Ans.)}$$

Example 2.58. *The maximum efficiency of a 500 kVA, 3300/500 V, 50 Hz single phase transformer is 97% and occurs at $\frac{3}{4}$ full load, unity power factor. If the impedance is 10%, calculate the regulation at full load ; power factor 0.8 lagging.* **(AMIE Summer, 2000)**

Solution. *Given* : Rating of the transformer = 500 kVA ; $\eta_{max} = 97\% \left(\text{at } \frac{3}{4} \text{ full load} \right)$;

$\cos \phi = 1$; Z_{01} (impedance) = 10%.

Regulation at full load, p.f. 0.8 lagging :

At maximum efficiency ; Iron losses = Copper losses

Let, P_c = full load copper losses

Then, copper losses at $\frac{3}{4}$ full load $= \left(\frac{3}{4}\right)^2 \times P_c = \frac{9}{6} P_c$

Now, $$\eta = \frac{\text{output}}{\text{output + losses}}$$

∴ For maximum η of 97%, we have

$$0.97 = \frac{(500 \times 3/4)}{500 \times \frac{3}{4} + 2 \times \frac{9}{16} P_c} = \frac{375}{375 + \frac{9}{8} P_c}$$

or $$0.97 \left(375 + \frac{9}{8} P_c \right) = 375$$

or $$P_c = \frac{8}{9} \left(\frac{375}{0.97} - 375 \right) = 10.31 \text{ kW}$$

If resistance referred to primary is R_{01},

Then, $$I_1^2 R_{01} = 10.31 \times 10^3 \text{ W}$$

or $$R_{01} = \frac{10.31 \times 10^3}{(151.5)^2} = 0.449\ \Omega \qquad \left(\because I_1 = \frac{500 \times 10^3}{3300} = 151.5 \text{ A} \right)$$

% Impedance $$= \frac{I_1 Z_{01}}{V_1} \times 100 = 10$$

∴ $$Z_{01} = \frac{10}{100} \times \frac{V_1}{I_1} = \frac{10}{100} \times \frac{3300}{151.5} = 2.178\ \Omega$$

∴ $$X_{01} = \sqrt{Z_{01}^2 - R_{01}^2} = \sqrt{2.178^2 - 0.449^2} = 2.13\ \Omega$$

% Regulation $$= \frac{I_1 R_{01} \cos \phi + I_1 X_{01} \sin \phi}{V_1} \times 100$$

$$= \frac{151.5 \times 0.449 \times 0.8 + 151.5 \times 2.13 \times 0.6}{3300} \times 100$$

$$= \mathbf{7.52\%}. \quad \textbf{(Ans.)}$$

Example 2.59. *A 200 kVA transformer has an efficiency of 98% at full-load. If the maximum efficiency occurs at three quarters of full-load, calculate the efficiency at half load. Assume negligible magnetising current and p.f. of 0.8 at all loads.* **(Punjab Univ. 1998)**

Solution. *Given* : Rating of the transformer = 200 kVA

Efficiency, η = 98% at full-load

Power factor, $\cos \phi$ = 0.8 (at all loads)

Efficiency at half-load :

Full-load input $= \dfrac{200 \times 0.8}{0.98} = 163.26$ kW

Total losses on full-load $= P_i + P_c =$ Input – Output

$= 163.26 - 160 = 3.26$ kW ...(*i*)

Since maximum efficiency occurs at three quarters of full load, so

$$\left(\frac{3}{4}\right)^2 P_c = P_i$$

or $$\frac{9}{16} P_c = P_i \qquad ...(ii)$$

Solving (*i*) and (*ii*), we get

$$P_c = 2.09 \text{ kW} \quad \text{and} \quad P_i = 1.17 \text{ kW}$$

At half-load and 0.8 p.f. :

Power output $= \dfrac{1}{2} \times 200 \times 0.8 = 80$ kW

Total losses $= P_i + x^2 P_c = 1.17 + \left(\dfrac{1}{2}\right)^2 \times 2.09 = 1.69$ kW

Efficiency, $\eta = \dfrac{\text{output}}{\text{output + total losses}}$

$= \dfrac{80}{80 + 1.69} = 0.9793$ or **97.93%. (Ans.)**

Example 2.60. *A 600 kVA, 1-phase transformer has an efficiency of 92% both at full-load and half-load at unity power factor. Determine its efficiency at 60% of full load at 0.8 p.f. lag.*

(AMIE Winter, 1997)

Solution. Let P_i and P_c be the iron loss and full-load copper loss respectively.

At full-load and unity power factor :

Output, $P = 600 \times 1.0 = 600$ kW

Transformer efficiency, $\eta = 92\%$ or 0.92

Power input $= \dfrac{P}{\eta} = \dfrac{600}{0.92} = 652.17$ kW

Total losses $(P_i + P_c)$ = Input – output $= 625.17 - 600 = 52.17$ kW ...(*i*)

At half full-load and unity power factor :

Power output $= \dfrac{1}{2} \times 600 \times 1.0 = 300$ kW

Power input $= \dfrac{300}{0.92} = 326.08$ kW

Total losses $= P_i + \left(\dfrac{1}{2}\right)^2 P_c = 326.08 - 300 = 26.08$ kW ...(*ii*)

Solving (*i*) and (*ii*), we have

$$P_c = 34.78 \text{ kW and } P_i = 17.38 \text{ kW}$$

At 60% full-load and 0.8 p.f. lag :

Power output $= 600 \times 0.6 \times 0.8 = 288$ kW

Total losses $= P_i + (0.6)^2 P_c = 17.39 + (0.6)^2 \times 34.78 = 29.91$ kW

$\therefore$ **Efficiency,** $\eta = \dfrac{288}{288 + 29.91} = 0.9059$ or **90.59%. (Ans.)**

☞**Example 2.61.** *The maximum efficiency of a 3-phase 11000/400 V, 500 kVA transformer is 98.8% and occurs at 80% full load, unity power factor. Its percentage impedance is 4.5%. Load power factor is now varied while the load current and the supply voltage are held constant at their rated values. Determine the load power factor at which the secondary terminal voltage is minimum and find the value of the latter.* **(GATE, 1993)**

Solution. Rating of transformer : 11000/400 V, 500 kVA

Maximum efficiency = 98.8% at 80% of load

Percentage impedance = 4.5%

Load at maximum efficiency $= \text{Full load kVA} \times \sqrt{\dfrac{\text{Iron losses}}{\text{Full load copper losses}}}$

$\therefore$ 80% full load $= \text{Full load} \times \sqrt{\dfrac{\text{Iron losses}}{\text{Full load copper losses}}}$

$$\therefore \quad 0.64 = \frac{\text{Iron losses}}{\text{Full load copper losses}}$$

$\therefore$ Iron losses = 0.64 × Full load copper losses

Input at unity power factor = 500 kW

Output = 500 × 0.988 = 494 kW

Total losses = Copper losses at maximum losses + Iron losses

$= (500 - 494) \times 10^3 = 6000$ W

At maximum efficiency, copper losses = iron losses

$\therefore$ Copper losses at maximum efficiency $= \dfrac{6000}{2} = 3000$ W

Since maximum efficiency is at 80% of full load

$\therefore$ Full load copper losses $= \dfrac{3000}{(0.8)^2} = 4687.5$ W

$$\therefore \quad 3I_1^2R_{01} = 4687.5$$

where, $I_1 = \dfrac{500 \times 10^3}{\sqrt{3} \times 11000} =$ **26.243 A. (Ans.)**

$$\therefore \quad R_{01} = \frac{4687.5}{3 \times 26.243^2} = 2.27\ \Omega$$

$$\therefore \quad R_{02} = K^2R_{01} = \left(\frac{400}{11000}\right)^2 \times \frac{4687.5}{3 \times (26.243)^2} \qquad \left(\because K = \frac{400}{11000}\right)$$

$= 0.003\ \Omega$

Percentage impedance = 4.5%

Actual impedance $= \left(\dfrac{11000}{\sqrt{3} \times 26.43}\right) \times \dfrac{4.5}{100} = 10.89\ \Omega.$

$$\therefore \quad X_{01} = \sqrt{Z_{01}^2 - R_{01}^2} = \sqrt{10.89^2 - 2.27^2} = 10.65\ \Omega$$

$$\therefore \quad X_{02} = K^2X_{01} = \left(\frac{400}{11000}\right)^2 \times 10.65 = 0.014\ \Omega$$

Drop due to this resistance and reactance $= I_2(R_{02} \cos \phi + X_{02} \sin \phi)$

For terminal voltage to be minimum, the drop will be maximum,

$$\frac{d(drop)}{d\phi} = I_2(-R_{02} \sin \phi + X_{02} \cos \phi) = 0$$

$\therefore \qquad R_{02} \sin \phi = X_{02} \cos \phi$

or $\qquad \dfrac{\sin \phi}{\cos \phi} = \dfrac{X_{02}}{R_{02}} = \dfrac{0.014}{0.003} = 4.667$ or $\tan \phi = 4.667 \quad \therefore \ \phi = 78°$

Thus load power factor for minimum voltage = cos 78° $\simeq$ **0.21 (lag). (Ans.)**

Example 2.62. *In a 50 kVA, 11 kV/400 V transformer, the iron and copper losses are 500 W and 600 W respectively under rated conditions.*

(i) Calculate the efficiency on unity power factor at full load.

(ii) Find the load for maximum efficiency and the iron and copper losses corresponding to this load. **(GATE, 1998)**

Solution. *Given* : Rating of transformer = 50 kVA ; P_i = 500 W ; P_c = 600 W ; cos ϕ = 1.

(*i*) **Efficiency on unity power factor at full load :**

$$\eta = \frac{\text{full load kVA} \times \text{p.f.}}{\text{full load kVA} \times \text{p.f.} + P_i + P_c}$$

$$= \frac{50 \times 10^3 \times 1}{50 \times 10^3 \times 1 + 500 + 600} = 0.9785 \text{ or } \mathbf{97.85\%.\ (Ans.)}$$

(*ii*) **Load for max. η and iron and copper losses :**

Load for η_{max} $\quad = 50 \text{ kVA} \sqrt{\dfrac{P_i}{P_c}} = 50 \times \sqrt{\dfrac{500}{600}} = \mathbf{45.64\ kVA.\ (Ans.)}$

Iron loss $\quad$ = copper loss = **500 W. (Ans.)**

Example 2.63. *A 25 kVA, 2000/200 V, 50 Hz transformer offers its maximum efficiency at 80% of full load. Its per unit resistance and impedance are 0.012 and 0.05 respectively. The no load current is 4% of its full current and 0.2 power factor. Determine its regulation and efficiency at $\frac{1}{2}$ of the full load and at 0.8 power factor lagging.* **(AMIE Winter, 1996)**

Solution. Output at unity power factor = 25 kW

Maximum efficiency = 80%

$\therefore$ Input power $= \dfrac{25}{0.8} = 31.25$ kW

$\therefore$ Full load losses = 31.25 – 25 = 6.25 kW

At maximum efficiency, copper loss = iron loss

$\therefore$ Iron loss $= \dfrac{6.25}{2} = 3.125$ kW

Percentage resistance, % R = 0.012 × 100 = 1.2% *(given)*

Percentage impedance, % X = 0.05 × 100 = 5% *(given)*

$\therefore$ % Reactance $= \sqrt{(\% Z)^2 - (\% R)^2}$

$= \sqrt{5^2 - 1.2^2} = 4.854\%$

Full load primary current $= \dfrac{25 \times 10^3}{2000} = 12.5$ A

$\therefore$ No load current, $\quad I_0 = \dfrac{4}{100} \times 12.5 = 0.5$ A

Half full load current $= I_0 + I_{F/2}$

$= 0.5(0.2 + j0.98) + \dfrac{12.5}{2}(0.8 + j0.6)$

$$= (0.1 + j0.49) + (5 + j3.75)$$
$$= 5.1 + j4.24 = 6.63 \angle 39.7°$$

We know that, % **Regulation** $= \dfrac{I_1R_{01}\cos\phi + I_1X_{01}\sin\phi}{V_1} \times 100$

$$= V_R \cos\phi + V_X \sin\phi$$

$\therefore$ % Regulation at half load and 0.8 p.f. $= 1.2 \times 0.8 + 4.854 \times 0.6$

$= 0.96 + 2.91 =$ **3.87%. (Ans.)**

Efficiency (at half load and 0.8 p.f.) $= \dfrac{\text{output}}{\text{output + iron loss + copper loss}}$

$$= \frac{\frac{1}{2} \times 25 \times 0.8}{\frac{1}{2} \times 25 \times 0.8 + 3.125 + 3.125 \times \left(\frac{6.63}{12.5}\right)^2}$$

$= \dfrac{10}{10 + 3.125 + 0.879} = 0.714$ or **71.4%. (Ans.)**

Example 2.64. *A single phase transformer has percentage regulation of 4 and 4.4 for lagging power factors of 0.8 and 0.6 respectively. The full load copper loss is equal to iron loss. Calculate :*

(*i*) *The lagging power factor at which full load regulation is maximum.*

(*ii*) *The full load efficiency at unity power factor.* **(UPSC, 1996)**

Solution. Let percentage resistance and reactance drop be v_R and v_X respectively.

Percentage regulation $= v_R \cos\phi + v_X \sin\phi$ where $\cos\phi$ is load power factor

For 0.8 lagging p.f., % age regulation $= v_R \times 0.8 + v_X \times 0.6 = 4$ ...(*i*)

For 0.6 lagging p.f., % age regulation $= v_R \times 0.6 + v_X \times 0.8 = 4.4$ ...(*ii*)

Solving equations (*i*) and (*ii*), we get

$$v_R = 2.0 \text{ and } v_X = 4$$

Percentage copper loss = Percentage resistance drop = 2% of the output

Percentage iron loss = Percentage full-load copper loss = 4% of the output

(*i*) **The lagging p.f. at which full load regulation is maximum :**

Full-load voltage regulation will be maximum for load p.f. of

$\cos\phi = \cos\left(\tan^{-1}\dfrac{X_{02}}{R_{02}}\right) = \cos\left(\tan^{-1}\dfrac{4}{2}\right) =$ **0.4472 lag. (Ans.)**

(*ii*) **The full-load efficiency at unity power factor :**

Full-load efficiency, $\eta = \dfrac{\text{output}}{\text{output + losses}}$

$= \dfrac{100}{100 + 2 + 2} = 0.9615$ or **96.15%. (Ans.)**

Example 2.65. *The full load voltage drops in a transformer are 2% and 4% due to resistance and leakage reactance respectively. The full load copper loss is equal to iron loss. Calculate :*

(*i*) *The efficiency on half load at unity power factor* ;

(*ii*) *The lagging power factor on full load output at which the voltage drop is a maximum* ;

(*iii*) *This maximum percentage voltage drop.* **(AMIE Winter, 1999)**

Solution. % resistance drop $= \dfrac{I_2R_{02} \times 100}{V_2} = 2$

% reactance drop $= \dfrac{I_2X_{02} \times 100}{V_2} = 4$

$$\therefore \quad \frac{I_2^2 R_{02}}{V_2 I_2} = 0.02$$

$\therefore$ Full load ohmic loss $= 0.02 \times$ rated kVA = iron loss

(i) The efficiency at half load at unity power factor

$$= \frac{\left[\text{Rated VA} \times \left(\frac{1}{2}\right)\right] \text{(output)}}{\underset{\text{(output)}}{\left[\text{Rated VA} \times \left(\frac{1}{2}\right)\right]} + \underset{\text{(copper loss)}}{\left[\frac{1}{4} \times 0.02 \text{ rated VA}\right]} + \underset{\text{(iron loss)}}{[0.02 \text{ rated VA}]}}$$

$$= \frac{\frac{1}{2}}{\frac{1}{2} + \frac{1}{4} \times 0.02 + 0.02} \times 100 = \frac{0.5}{0.5 + 0.005 + 0.02} \times 100$$

$= \mathbf{95.24\%}$. **(Ans.)**

(ii) The lagging power factor :

Maximum voltage drop means maximum voltage regulation, therefore, full load power factor at which voltage drop is maximum,

$$\cos \phi = \frac{R_{02}}{Z_{02}} = \frac{\text{percentage resistance}}{\text{percentage impedance}}$$

$$= \frac{2}{\sqrt{2^2 + 4^2}} = \mathbf{0.447 \text{ lagging.}} \quad \textbf{(Ans.)}$$

(iii) Maximum percentage voltage drop :

Percentage voltage drop $= V_R \cos \phi + V_X \sin \phi$

$= 2 \times 0.447 + 4 \times 0.894 = \mathbf{4.47\%}$. **(Ans.)**

Example 2.66. *(a) A 30-kVA, single phase transformer has an iron loss of 457 W and copper loss of 125 W, when delivering half the full load. At what percent of the full load will the transformer have maximum efficiency ?*

(b) A 1-kVA, 220 V/110 V, 400 Hz transformer is designed to be used at a frequency of 60 Hz. What will be the kVA rating of the transformer at the reduced frequency ? **(AMIE Winter, 1998)**

Solution. (a) *Given* : Rating of the transformer = 30 kVA ; $P_i = 457$ W ;

$P_c = 125$ W at half full load.

Percentage of full load for maximum efficiency :

Let, x = full load

Then, half load $= \frac{x}{2}$

Further let, y = the load at which maximum efficiency occurs.

Then, copper loss at maximum efficiency = iron loss = 457 W

and $y^2 \propto 457$ ...(i)

Also, $\left(\frac{x}{2}\right)^2 \propto 125$ ...(ii)

From (i) and (ii), we have

$$\frac{y^2}{(x/2)^2} = \frac{457}{125} = 3.656 \quad \text{or} \quad \frac{y}{x/2} = 1.912$$

or $$y = \frac{x}{2} \times 1.912 = 0.956x$$

Hence, % of full load at which maximum efficiency occurs is **95.6%. (Ans.)**

(*b*) *Given* : Rating of the transformer = 1 kVA ; V_1 = 220 V ; f_1 = 400 Hz, f_2 = 60 Hz

kVA rating at the reduced frequency :

We know that for a transformer,

$$E(= V) = 4.44 f\phi_{max} Nf = 4.44\, B_{max} A_i Nf$$

Assuming that flux density in the core remains constant, we have

$$V \propto f$$

Hence $$\frac{V_1}{V_2} = \frac{f_1}{f_2}$$

$$\therefore \quad \frac{220}{V_2} = \frac{400}{60} \quad \text{or} \quad V_2 = \frac{220 \times 60}{400} = 33 \text{ V}$$

Hence kVA rating being proportional to voltage will change to

$$1 \times \frac{33}{220} = \mathbf{0.15\ kVA.}\ \mathbf{(Ans.)}$$

☞**Example 2.67.** (*a*) *What is meant by similar polarity ends of the two windings of a single phase transformer ? How are these ends identified ?*

(*b*) *A 15 kVA, 2300/230 V, 50 Hz single phase transformer gave the following test data :*

Open Circuit Test $V_0 = 2300$ V, $I_0 = 0.21$ A

$P_0 = 50$ W

Short Circuit Test $V_{SC} = 47$ V, $I_{SC} = 6.0$ A,

$P_{SC} = 160$ W

(*i*) *Find the equivalent circuit referred to high voltage side.*

(*ii*) *Calculate the full load voltage regulation at 0.8 p.f. lagging when the load voltage is held at 220 volts.*

(*iii*) *What is the efficiency at half the rated load at unity p.f. ?*

(*iv*) *Find the maximum efficiency and corresponding output power.*

(*c*) *State reasons for using tertiary windings in a transformer.* **(UPSC, 1997)**

Solution. (*a*) Refer Fig. 2.60.

For polarity, **dots** are provided on the diagram. *These dots help to determine the polarity of the voltages and currents in the core.* The physical significance of the dot convention is that a current flowing into the dotted end of a winding produces a positive m.m.f., while a current flowing into the undotted end of a winding produces a negative m.m.f. Therefore, two currents, both flowing into the dotted end of their respective windings, produce m.m.fs that add together. If one current flows into a dotted end of a winding and flows out of a dotted end, then m.m.fs will subtract from each other.

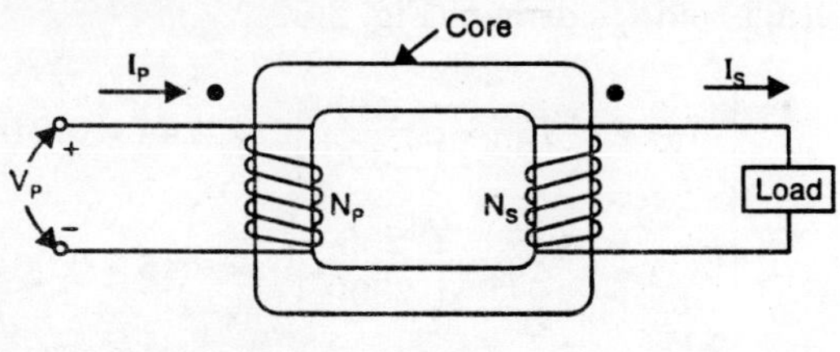

Fig. 2.60

(*b*) From the given data, open circuit test and short circuit test are both carried out on high voltage side and hence results will directly lead to equivalent circuit referred to high voltage side.

(*i*) **Equivalent circuit (h.v. side)**

Open circuit test :

$$V_0 = V_1 = 2300 \text{ V}, I_0 = 0.21 \text{ A}, P_0 = 50 \text{ W}$$

$$\mathbf{R_0} = \frac{V_1^2}{P_0} = \frac{2300^2}{50} = \mathbf{105800\ \Omega.\ \ (Ans.)}$$

$$I_w = \frac{V_1}{R_0} = \frac{2300}{105800} = 0.02174\text{ A}$$

$$I_m = \sqrt{I_0^2 - I_w^2} = \sqrt{0.21^2 - 0.02174^2} = 0.20887\text{ A}$$

$$\therefore \quad \mathbf{X_0} = \frac{V_1}{I_m} = \frac{2300}{0.20887} = \mathbf{11011.6\ \Omega.\ \ (Ans.)}$$

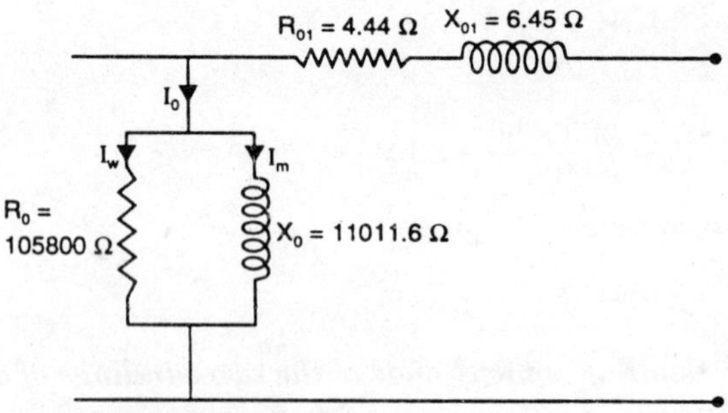

Fig. 2.61. Equivalent circuit referred to h.v.

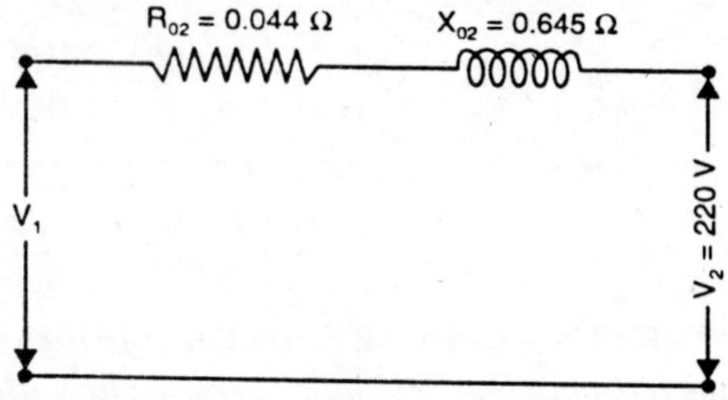

Fig. 2.62

Short circuit test :

$$V_{SC} = 47\text{ V},\ I_{SC} = 6.0\text{ A},\ P_{SC} = 160\text{ W}$$

$$\mathbf{R_{01}} = \frac{P_{SC}}{I_{SC}^2} = \frac{160}{(6.0)^2} = \mathbf{4.44\ \Omega.\ \ (Ans.)}$$

$$Z_{01} = \frac{V_{SC}}{I_{SC}} = \frac{47}{6.0} = 7.833\ \Omega$$

$$\therefore \quad \mathbf{X_{01}} = \sqrt{Z_{01}^2 - R_{01}^2} = \sqrt{7.833^2 - 4.44^2} = \mathbf{6.45\ \Omega.\ \ (Ans.)}$$

(*ii*) **Full load voltage regulation at 0.8 p.f. lagging when the load voltage is held at 220 V :**

For finding the regulation, the *shunt branch is omitted* and circuit referred to secondary (low voltage) side is drawn (Fig. 2.62)

$$\left[\begin{array}{l} R_{02} = K^2 R_{01} = \left(\dfrac{230}{2300}\right)^2 \times 4.44 = 0.044\ \Omega \\[2ex] X_{02} = K^2 X_{02} = \left(\dfrac{230}{2300}\right)^2 \times 6.45 = 0.0645\ \Omega \end{array}\right]$$

Load current $= \dfrac{15 \times 1000}{220} = 68.18\text{ A}$

Voltage regulation $= \dfrac{I_2\,(R_{02}\cos\phi + X_{02}\sin\phi)}{220}$

$$= \frac{68.18(0.044 \times 0.8 + 0.065 \times 0.6)}{220} = 0.023 \text{ or } \mathbf{2.3\%.\ \ (Ans.)}$$

(*iii*) **Efficiency at half the rated load at unity p.f. :**

Output power $= \frac{1}{2} \times 15 \times 1 = 7.5\text{ kW or }7500\text{ W},\ P_i = 50\text{ W}$

Full load copper losses, $P_c = 160 \times \left[\dfrac{(15000/2300)}{6} \right]^2 = 189$

Copper losses at half load $= \left(\dfrac{1}{2}\right)^2 \times 189 = 47.25$ W.

$$\therefore \quad \eta = \frac{\text{output power}}{\text{output power + iron losses + copper losses}}$$

$$= \frac{7500}{7500 + 50 + 47.25} = 0.9872 \text{ or } \mathbf{98.72\%}. \text{ (Ans.)}$$

(*iv*) **Maximum efficiency and corresponding output power :**

Maximum efficiency occurs at $\sqrt{\dfrac{50}{189}} = 0.5143$ times the rated load.

Then, losses = 50 + 50 = 100 W or 0.1 kW

Considering unity p.f. ; load efficiency,

$$\eta = \frac{0.5143 \times 15 \times 1}{0.5143 \times 15 \times 1 + 0.1} = 0.9872 \text{ or } 98.72\%.$$

Hence, the **maximum efficiency is 98.72%** and the corresponding load is nearly **51.43% of the rated load** (= 7.714 kW). **(Ans.)**

(c) **Tertiary winding** *is a delta connected winding added to the transformer, so that the third harmonic components of voltage in the delta causing a circulating current flow within the winding, are suppressed. This suppresses the third-harmonic components of voltage in the same manner as grounding the transformer neutrals.*

Example 2.68. *A 1-phase transformer is rated at 100 kVA, 2300/230 V, 50 Hz. The maximum flux density in the core is 1.0 T and the net cross-sectional area is 0.048 m². Determine :*

(i) The number of primary and secondary turns needed.

(ii) Magnetising current if the mean length of the magnetic circuit is 2.6 metres and the relative permeability is 1,200. Neglect the current drawn for the core loss.

(iii) The power input on open-circuit with rated voltage was 400 watts and power input on short circuit with full load current flowing was 1,200 watts. Determine the efficiency of the transformer at 75% of full load with 0.8 p.f. lag.

(iv) What will be the effect on its efficiency if it is connected to a supply of same voltage but double the frequency (i.e. 100 Hz). **(Nagpur University)**

Solution. *Given* : Rating of transformer = 100 kVA, 2300 V/230 V ; f = 50 Hz ;
B_{max} = 1.0 T (Wb/m²) ; A = 0.048 m² ; l = 2.6 m ; μ_r = 1200.

(*i*) **Number of turns, N_1, N_2 :**

Transformer ratio, $K = \dfrac{E_2}{E_1} = \dfrac{230}{2300} = \dfrac{1}{10}$

Maximum value of flux, $\phi_{max} = B_{max} \times A = 1.0 \times 0.048 = 0.048$ Wb

We know that, $E_1 = 4.44 f \phi_{max} N_1$

$$\therefore \quad N_1 = \frac{E_1}{4.44 f\phi_{max}} = \frac{2300}{4.44 \times 50 \times 0.048} \simeq \mathbf{218}. \text{ (Ans.)}$$

$$N_2 = KN_1 = \frac{1}{10} \times 218 \simeq \mathbf{22}. \text{ (Ans.)}$$

(*ii*) **Magnetising current, I_m :**

Magnetising current is given by,

$$I_m = \frac{B}{\mu_0\mu_r} \times \frac{l}{N_1}$$

$$= \frac{1.0}{4\pi \times 10^{-7} \times 1200} \times \frac{2.6}{218} = \mathbf{7.91\ A.\ (Ans.)}$$

$$\left[\because\ \mu = \mu_0\mu_r = \frac{B}{H} = \frac{B}{NI/l} \text{ or } \mu_0\mu_r = \frac{Bl}{NI} \text{ or } I = \frac{B}{\mu_0\mu_r} \times \frac{l}{N} \right]$$

(*iii*) **Efficiency of transformer at 75% of full load with 0.8 p.f. lag :**

Given : $P_0 = P_i = 400$ W = 0.4 kW ; $P_{SC} = 1200$ W = 1.2 kW, with full load current flowing.

Output at 75% of full load and 0.8 p.f. = 0.75 × 100 × 0.8 = 60 W

Input at 75% of full load and 0.8 p.f.

$$= \text{output} + P_i + x^2 P_c$$

$$= 60 + 0.4 + \left(\frac{3}{4}\right)^2 \times 1.2 = 61.075 \text{ kW} \quad (\because\ P_c = P_{SC}, \text{ in this case})$$

∴ Efficiency,

$$\eta = \frac{\text{output}}{\text{input}} = \frac{60}{61.075} = 0.9824 \text{ or } \mathbf{98.24\%.\ (Ans.)}$$

(*iv*) When supply frequency is *doubled, keeping supply voltage same, the iron losses will increase and efficiency will decrease.* **(Ans.)**

2.9.17. All-day Efficiency. *All-day efficiency* is the *ratio of energy (kWh) delivered in a 24 hour period divided by the energy (kWh) input in the same length of time.*

$$\therefore \qquad \eta_{\text{all-day}} = \frac{\text{output in kWh}}{\text{input in kWh}} \text{ (for 24 hours)} \qquad ...(2.39)$$

Transformers used on residence-lighting circuits (and distribution circuits generally) are either idle or only lightly loaded during much of 24-hour period. However, they must at all times be connected to the line and ready to serve, so that the core losses are being supplied continually. *It is therefore very important that such transformers be designed for minimum core loss.* The copper losses are relatively less important, since they depend on the load. Because they are lightly loaded much of the time, *distribution transformers are designed for relatively large full-load copper loss and have their maximum power efficiencies at light loads.* This design results in improved all-day efficiency for these transformers. *Power transformers,* on the other hand are loaded more or less continuously and *are designed for full-load copper losses equal to about twice the no-load losses.*

To calculate all-day efficiency, it is necessary to know how the load on the transformer varies from hour to hour. The quotient obtained by dividing the energy output by the energy output plus energy losses over a 24-hour period yields the efficiency expressed as a decimal fraction.

The use of a *load factor* facilitates practical calculations.

Example 2.69. *A 15 kVA, 2000/200 V transformer has an iron loss of 250 W and full-load copper loss 350 W. During the day it is loaded as follows :*

No. of hours	*Load*	*Power factor*
9	$\frac{1}{4}$ *load*	0.6
7	*full-load*	0.8
6	$\frac{3}{4}$ *load*	1.0
2	*no-load*	—

Calculate the all-day efficiency.

Solution. Rating of transformer = 15 kVA

Iron loss, P_i = 250 W = 0.25 kW

Full-load copper loss, P_c = 350 W = 0.35 kW

Iron loss / day = 0.25 × 24 = 6 kWh

Copper loss at $\frac{1}{4}$ *load* $= \left(\frac{1}{4}\right)^2 \times P_c = \frac{1}{16} \times 0.35 = 0.0218$ kW

Copper loss for 9 hours at $\frac{1}{4}$ load

= 9 × 0.0218 = 0.196 kWh

Copper loss at *full-load* = P_c = 0.35 kW

Copper loss for 7 hours on full-load

= 7 × 0.35 = 2.45 kWh

Copper loss at $\frac{3}{4}$ *load* $= \left(\frac{3}{4}\right)^2 \times P_c = \frac{9}{16} \times 0.35 = 0.197$ kW

Copper loss for 6 hours at $\frac{3}{4}$ load = 0.197 × 6 = 1.18 kWh

Copper loss/day = 0.196 + 2.45 + 1.18 = 3.826 kWh

Total loss/day = iron loss/day + copper loss/day = 6 + 3.826 = 9.826 kWh

Total output/day $= \frac{1}{4} \times 15 \times 0.6 \times 9 + 15 \times 0.8 \times 7 + \frac{3}{4} \times 15 \times 1.0 \times 6$

= 20.25 + 84 + 67.5 = 171.75 kWh

All-day efficiency $= \frac{\text{output}}{\text{output + losses}} = \frac{171.75}{171.75 + 9.826}$ = **0.9459** or **94.59%. (Ans.)**

☞Example 2.70. *A 40 kVA transformer has got a maximum efficiency of 97% at 80% of load at unity p.f. During the day it is loaded as follows :*

No. of hours	*Load*	*Power factor*
9	*6 kW*	*0.6 lag*
8	*25 kW*	*0.8 lag*
7	*30 kW*	*0.9 lag*

Find the all-day efficiency.

Solution. Rating of transformer = 40 kVA

Maximum efficiency = 97%

At maximum efficiency

Iron loss = copper loss

Output at maximum efficiency at unity p.f.

= 0.8 × 40 × 1 = 32 kW

Efficiency $= \frac{\text{output}}{\text{output + losses}}$

$$0.97 = \frac{32}{32 + \text{losses}}$$

∴ Losses $= \frac{32}{0.97} - 32 = 0.989$ kW

Iron loss $= \frac{0.989}{2} = 0.4945$ kW

Copper loss at 80% of load = 0.4945 kW

Full-load copper loss $= \frac{0.4945}{(0.8)^2} = 0.7726$ kW

Iron loss for 24 hours (i.e., per day)

$$= 0.4945 \times 24 = 11.87 \text{ kWh}$$

Let P_{c1} be the copper loss at 6 kW

$$P_c \propto (\text{kVA of transformer})^2$$

$$P_{c1} \propto (\text{kVA of load})^2$$

$$\propto \left(\frac{\text{kW of load}}{\text{power factor}}\right)^2$$

$$\therefore \quad \frac{P_{c1}}{P_c} = \frac{\left(\frac{\text{kW of load}}{\text{p.f.}}\right)^2}{(\text{kVA of tansformer})^2}$$

$$\therefore \quad P_{c1} = \frac{\left(\frac{6}{0.6}\right)^2}{(40)^2} \times P_c = \frac{\left(\frac{6}{0.6}\right)^2}{(40)^2} \times 0.7726 = 0.04828 \text{ kW}$$

$$= 0.04828 \times 9 = 0.435 \text{ kWh}$$

Copper loss at 25 kW $= \dfrac{\left(\frac{25}{0.8}\right)^2}{(40)^2} \times 0.7726 = 0.4715$ kW

Copper loss for 8 hours at 25 kW

$$= 0.4715 \times 8 = 3.772 \text{ kWh}$$

Copper loss at 30 kW $= \dfrac{\left(\frac{30}{0.9}\right)^2}{(40)^2} \times 0.7726 = 0.5365$ kW

Copper loss for 7 hours at 30 kW

$$= 0.5365 \times 7 = 3.755 \text{ kWh}$$

Total losses/day = Copper loss/day + Iron loss/day

$$= (0.435 + 3.772 + 3.755) + 11.87 = 19.832 \text{ kWh}$$

Total output/day $= 6 \times 9 + 25 \times 8 + 30 \times 7 = 54 + 200 + 210 = 464$ kWh

$\therefore$ **All-day efficiency** $= \dfrac{\text{output}}{\text{output + losses}} \times 100$

$$= \frac{464}{464 + 19.832} \times 100 = \mathbf{95.9\%}. \textbf{ (Ans.)}$$

Example 2.71. *The all-day efficiency of 250 kVA transformer is 96.5% when it is loaded as follows :*

No. of hours	*Load*	*Power factor*
10	*150 kW*	*0.8*
9	*180 kW*	*0.9*
5	*No-load*	—

If maximum efficiency of this transformer occurs at 80% of full load, find iron loss and full-load copper loss.

Solution. Rating of transformer = 250 kVA

All-day efficiency = 96.5%

Output/day $= 150 \times 10 + 180 \times 9 = 1500 + 1620 = 3120$ kWh

All-day efficiency $= \dfrac{\text{output in kWh}}{\text{(output + losses) in kWh}}$

$$0.965 = \frac{3120}{3120 + \text{losses}}$$

$\therefore$ Total losses in kWh $= 113.16$ kWh

Let P_i be the iron loss in kW and P_c be the full-load copper loss in kW

Copper loss at 150 kW $= \dfrac{\left(\dfrac{150}{0.8}\right)^3}{(250)^2} \times P_c = 0.5625\ P_c$

Copper loss for 10 hours at 150 kW

$= 0.5625\ P_c \times 10 =$ **5.625 P_c kWh. (Ans.)**

Copper loss at 180 kW $= \dfrac{\left(\dfrac{180}{0.9}\right)^2}{(250)^2} \times P_c = 0.64\ P_c$

Copper loss at 180 kW for 9 hours

$= 0.64\ P_c \times 9 =$ **5.76 P_c kWh. (Ans.)**

Iron loss for 24 hours $= P_i \times 24$ kWh

Total losses/day $= (5.625\ P_c + 5.76\ P_c) + 24\ P_i = 11.385\ P_c + 24\ P_i$

But total losses/day $= 113.16$ kWh

$\therefore$ $11.385\ P_c + 24\ P_i = 113.16$ ...(*i*)

Copper loss at 80% of full-load

$= (0.8)^2\ P_c = 0.64\ P_c$

At maximum efficiency $P_i = 0.64\ P_c$ ...(*ii*)

Putting this in (*i*), we get

$$11.385\ P_c + 24 \times 0.64\ P_c = 113.16$$
$$26.745\ P_c = 113.16$$
$$P_c = \mathbf{4.23\ kW}$$

$\therefore$ **Full-load copper loss** $=$ **4.23 kW. (Ans.)**

Iron loss, $P_i = 0.64\ P_c = 0.64 \times 4.23 =$ **2.707 kW. (Ans.)**

Example 2.72. *The daily variation of load on a 100 kVA transformer is as follows :*

8.00 AM to 1.00 PM — 65 kW, 35 kVAR

1.00 PM to 6.00 PM — 80 kW, 50 kVAR

6.00 PM to 1.00 AM — 30 kW, 30 kVAR

1.00 AM to 8.00 AM — No-load

This transformer has no-load core loss of 270 watts and a full-load ohmic loss of 1200 watts. Determine the all-day efficiency of the transformer. **(Punjab University)**

Solution. All day output $= 65 \times 5 + 80 \times 5 + 30 \times 7 = 935$ kWh

Iron loss for 24 hours $= \dfrac{270 \times 24}{1000} = 6.48$ kWh

Load in kVA from 8.00 AM to 1.00 PM $= \sqrt{\text{kW}^2 + \text{kVAR}^2} = \sqrt{65^2 + 35^2} = 73.82$ kVA

Load in kVA from 1.00 PM to 6.00 PM $= \sqrt{80^2 + 50^2} = 94.34$ kVA

Load in kVA from 6.00 PM to 1.00 AM $= \sqrt{30^2 + 30^2} = 42.43$ kVA

Copper losses for 24 hours $= \left(\frac{73.82}{100}\right)^2 \times 1200 \times 5 + \left(\frac{94.34}{100}\right)^2 \times 1200 \times 5 + \left(\frac{42.43}{100}\right)^2 \times 1200 \times 7$

$= 3269.63 + 5340.02 + 1512.26 = 10122$ Wh or 10.122 kWh

$\therefore$ All-day efficiency, $\eta = \frac{\text{output}}{\text{output + iron losses + copper losses}}$

$= \frac{935}{935 + 6.48 + 10.122} = 0.9826$ or **98.26%. (Ans.)**

2.9.18. Transformer Noise. The **"hum"** caused by energized power transformer, under *no-load conditions*, originates in the *core where the laminations tend to vibrate by magnetic forces.* The noise is transmitted through the oil to the tank side and thence to the surroundings.

The following are the *main factors which produce noise in transformers* :

1. *Magnetostriction* (occurrence of dimensional changes both parallel to, and perpendicular to the direction of magnetisation).
2. The mechanical vibrations caused by the laminations, depending upon the tightness of clamping, size, gauge, associated structural parts, etc.
3. The mechanical vibration of trank walls.
4. The damping.

The noise emission may be reduced by the following methods/means :

1. Prevention of vibration of core-plate by the use of a lower flux density and giving attention to constructional feature (such as clamping bolts, proportions and dimensions of the *'steps'* in plate width, tightness of clamping and uniformity of plates).
2. Using cushions, padding, or oil barriers to sound insulate the transformer from tank.
3. Designing suitably the tank and stiffeners to check tank wall vibration.
4. Sound insulating the tank from the ground or surrounding air.

However, the *noise problem cannot b2 solved completely.*

2.9.19. Auto-transformer. *A transformer in which part of the winding is common to both the primary and secondary circuits is known as an* **auto-transformer.** *The primary is electrically connected to the secondary, as well as magnetically coupled to it.*

Refer Fig. 2.64. *LM* is primary winding having N_1 turns and *MS* is secondary winding having N_2 turns. If no-load current and iron losses are neglected.

$$\frac{V_2}{V_1} = \frac{N_2}{N_1} = K.$$

The current in the section *MS* is vector difference of I_2 and I_1 ; but since the two currents are practically in phase opposition, the resultant is $(I_2 - I_1)$ where $I_2 > I_1$.

Saving of copper (in comparison to conventional two winding transformer) :

The volume and hence weight of copper is proportional to the length and area of cross-section of the conductors. But the *length of conductor is proportional to the number of turns and cross-section depends on current.* Hence the weight of copper is proportional to the *product of number of turns and current to be carried.*

Weight of copper in conventional two winding transformer

$$\propto (N_1 I_1 + N_2 I_2).$$

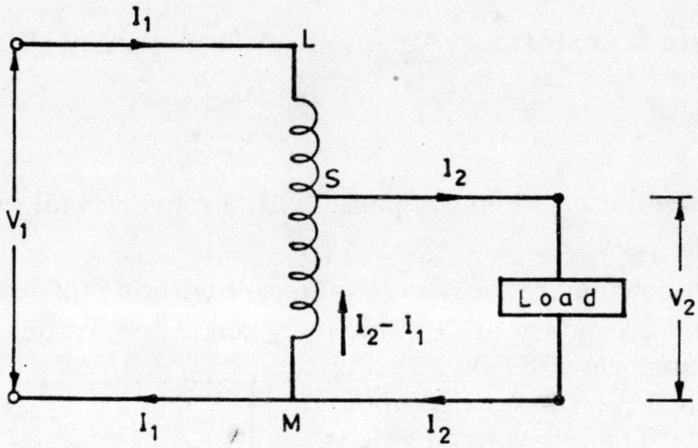

Fig. 2.63. Auto-transformer.

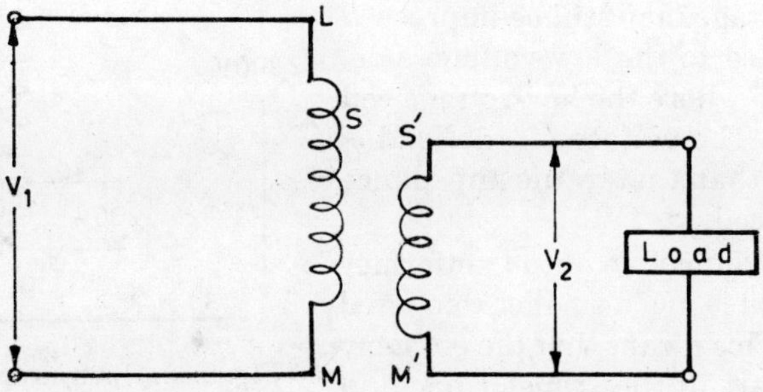

Fig. 2.64. Conventional two winding transformer.

Weight of copper in auto-transformer

= weight of copper in section LS + weight of copper is section MS.

But weight of copper in section $LS \propto (N_1 - N_2)I_1$

and weight of copper in section $MS \propto N_2\,(I_2 - I_1)$

$\therefore$ Weight of copper in auto-transformer

$$\propto (N_1 - N_2)\,I_1 + N_2\,(I_2 - I_1)$$

$$\therefore \quad \frac{\text{Weight of copper in auto-transformer } (W_a)}{\text{Weight of copper in ordinary transformer } (W_0)}$$

$$= \frac{(N_1 - N_2)\,I_1 + N_2\,(I_2 - I_1)}{N_1 I_1 + N_2 I_2} = \frac{(N_1 - 2N_2)\,I_1 + N_2 I_2}{N_1 I_1 + N_2 I_2}$$

$$= \frac{\dfrac{N_1}{N_2} - 2 + \dfrac{I_2}{I_1}}{\dfrac{N_1}{N_2} + \dfrac{I_2}{I_1}} = \frac{\dfrac{1}{K} - 2 + \dfrac{1}{K}}{\dfrac{1}{K} + \dfrac{1}{K}}$$

$$= 1 - K \qquad \text{...(2.40)}$$

$$\left[\because \frac{N_2}{N_1} = K, \frac{I_2}{I_1} = \frac{1}{K}\right]$$

$\therefore$ Saving in copper $= W_0 - W_a = W_0 - (1 - K)\,W_0 = KW_0$

$\therefore$ Saving in copper $= K \times$ weight of copper in ordinary tranformer

It can be proved that power transformed = input $(1 - K)$

The rest of the power is conducted *directly from the source to the load.*

Advantages of auto-transformer. An auto-transformer entails the following *advantages* :

(*i*) Higher efficiency

(*ii*) Small size

(*iii*) Lower cost

(*iv*) Better voltage regulation when compared with a conventional two-winding transformer of the same rating.

Disadvantages. Following are the *disadvantages / limitations* of auto-transformers :

(*i*) The primary and secondary are conductively connected, rather than isolated as in the conventional (ordinary) transformer. Because of this, *both sides are subject to any stresses set up by disturbances on either side.* The low-voltage side is subject to high-voltage stress and should be insulated for the higher voltage. In the case of a step down transformer, the high voltage may still be impressed upon equipment connected to the low-voltage side. This is shown in Fig. 2.65 where the low-voltage coil has accidentally opened. The voltage on the load is nearly 2300 V, being less than that by the impedance volt drop in the primary coil.

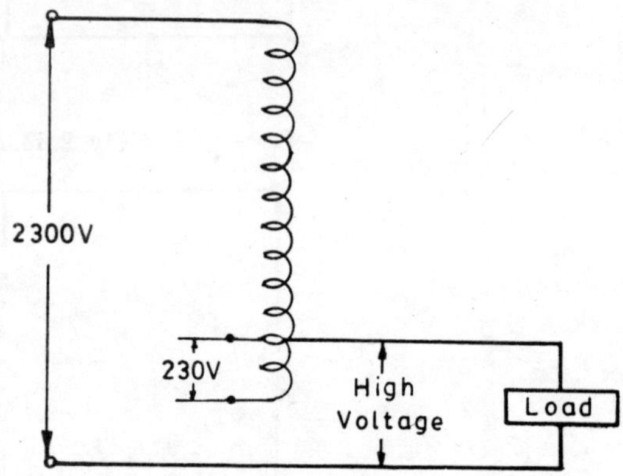

Fig. 2.65. Open circuit on auto-transformer.

(*ii*) As the voltage ratio of an auto-tranformer increases, the common coil is much smaller compared with the entire winding. This means that the economy gained is only a small part of the transformer and therefore this advantage is minimized.

Thus, because of the above disadvantages, lack of isolation and decreased economy, *auto-transformers are rarely used in ratios greater than 4 : 1, except for low-power devices on low- voltage systems.*

Uses. The auto-transformers find the following *applications* :

1. To tie together transmission or distribution circuits of slightly different voltages (*e.g.* 11000 V system with a 13200 V system).

2. To obtain partial line voltages for starting induction and synchronous motors with squirrel-cage windings.

3. To give a small boost to a distribution cable to correct for the voltage drop.

4. As furnace transformers for getting a convenient supply to suit the furnace winding from a 230 V supply.

5. As regulating transformers.

6. To obtain a neutral in a 3-wire A.C distribution system in the same way as a balancer set is used in a 3-wire D.C. distribution system.

7. A continuously variable auto-transformer finds useful applications in electrical testing laboratory.

2.9.20. Polarity of Transformers. In a transformer each of the primary terminals becomes alternately positive and negative with respect to the other and the same is true about the secondary terminals. If the transformer is to be used alone, the polarity is not important but if the transformer is to be used in parallel with another transformer the *instantaneous polarity is important* because the terminals having identical instantaneous polarity have to be connected together.

Refer Figs. 2.66 and 2.67. The terminals of the high voltage (h.v.) winding (*i.e.* H_1 and H_2) and the terminals of the low voltage (l.v.) winding (*i.e.* X_1 and X_2) are so marked that when the instantaneous voltage is directed from H_1 to H_2 in the high voltage winding, it is directed from X_1 to X_2 in the low voltage winding. In other words if the high voltage terminal H_1 is positive with respect to H_2 at any instant when the low voltage terminal X_1 will be positive with respect to X_2 at the same instant. It follows that when the terminals are arranged as in Fig. 2.66 the polarity is subtractive whereas the additive polarity is represented as in Fig. 2.67.

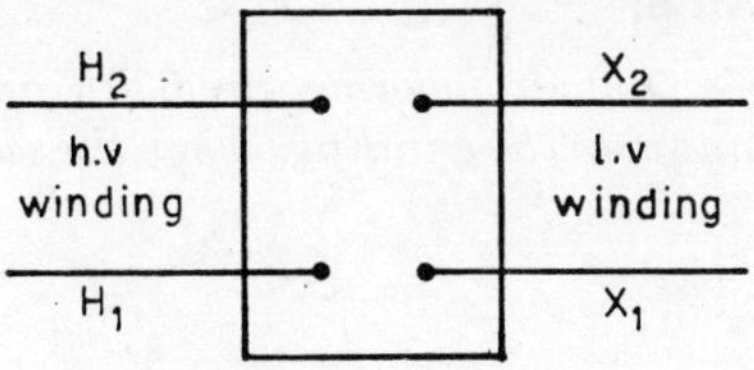

Fig. 2.66. Subtractive polarity.

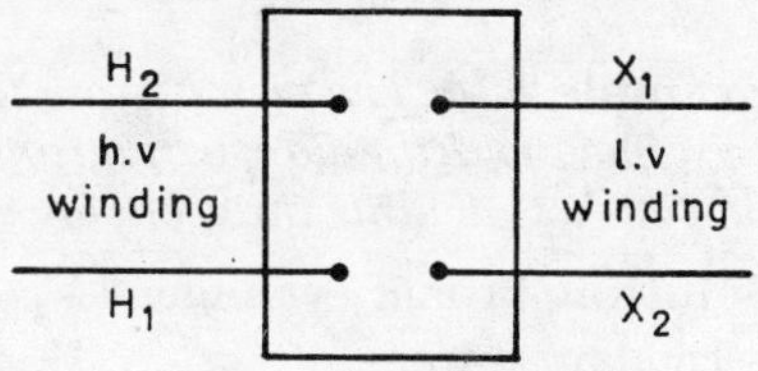

Fig. 2.67. Additive polarity.

Polarity Test. A polarity test is carried out to *find out the terminals having the same instantaneous polarity* assuming that the terminals are not marked. The connections are shown in Fig. 2.68.

- One h.t. and one l.t. terminals are joined together. A voltmeter is placed between the remaining two terminals.
- A convenient moderate voltage is impressed on the h.t. winding.

If the voltage V′, is 'greater' than the applied voltage V, then the transformer has 'additive' polarity.

If V′ is 'less' than V, the transformer has 'subtractive' polarity.

The terminals are then marked accordingly.

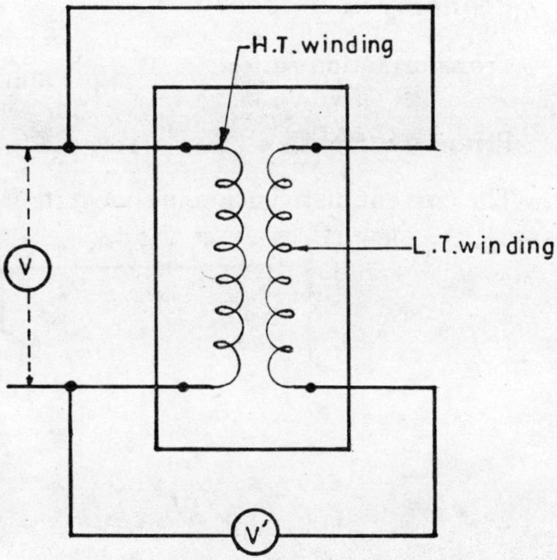

Fig. 2.68. Polarity test.

Example 2.73. *A load of 6 kW is supplied by an anto-transformer at 120 V and at unity power factor. If the primary voltage is 240 V, determine :*

(*i*) *Transformation ratio,* (*ii*) *Secondary current,*

(*iii*) *Primary current,*

(*iv*) *Number of turns across secondary if the total number of turns is 280.*

(*v*) *Power transformed, and*

(*vi*) *Power conducted directly from supply mains to load.*

Solution. Load suppied = 6 kW

Output voltage, $V_2 = 120$ V

Primary voltage, $V_1 = 240$ V

Total number of turns, $N_1 = 280$

(*i*) **Transformation ratio,** $K = \dfrac{V_2}{V_1} = \dfrac{120}{240} = \dfrac{1}{2}$ **(Ans.)**

(*ii*) **Secondary current,** $I_2 = \dfrac{6 \times 1000}{V_2 \cos \phi} = \dfrac{6000}{120 \times 1} = \mathbf{50\ A.}$ **(Ans.)**

(*iii*) **Primary current,** $I_1 = KI_2 = \frac{1}{2} \times 50 =$ **25 A. (Ans.)**

(*iv*) **Truns across secondary,** $N_2 = kN_1 = \frac{1}{2} \times 280 =$ **140. (Ans.)**

(*v*) **Power transformed** $= \text{Load} \times (1 - K) = 6 \times (1 - \frac{1}{2}) =$ **3 kW. (Ans.)**

(*vi*) **Power conducted directly from supply mains**

$= 6 - 3 =$ **3 kW. (Ans.)**

Example 2.74. *The primary and secondary voltages of an auto-transformer are 600 V and 500 V respectively. Show with the aid of a diagram the current distribution in the windings when the secondary current is 210 A. Calculate the economy in copper.*

Solution. Primary voltage, $V_1 = 600$ V

Secondary voltage, $V_2 = 500$ V

Secondary current, $I_2 = 210$ A

Economy in copper :

Transformation ratio, $\frac{V_2}{V_1} = \frac{500}{600} = \frac{5}{6}$

Primary current, $I_1 = KI_2 = \frac{5}{6} \times 210 = 175$ A

The current distribution is shown in Fig. 2.69.

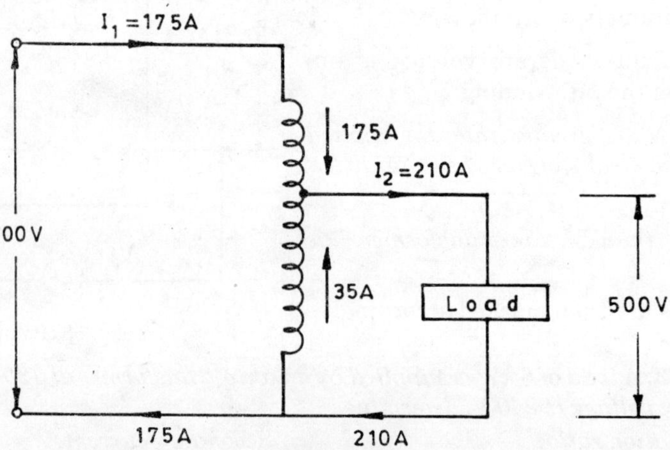

Fig. 2.69

We know that,

$$\frac{\text{Weight of copper on auto-transformer}}{\text{Weight of copper on ordinary transformer}} = 1 - K$$

Economy in copper

$$= \frac{\text{wt. of Cu on ordinary transformer} - \text{wt. of Cu on auto-transformer}}{\text{wt. of Cu on ordinary transformer}}$$

$$= 1 - \frac{\text{wt. of Cu on auto-transformer}}{\text{wt. of Cu on ordinary transformer}}$$

$$= 1 - (1 - K) = K = \frac{5}{6} \text{ or } \mathbf{83.33\%.} \textbf{ (Ans.)}$$

☞Example 2.75. *A 200/300 V auto-transformer draws power from a 200 V source and supplies a 5 kW load with a power factor of 0.8 lagging. A second load of 1.5 kW is supplied at unity power factor from 100 V winding. Neglecting losses, calculate the current drawn by the transformer from the 200 V line and its power factor.*

Solution. First load : 5 kW at 0.8 p.f. lagging.

Second load : 1.5 kW at unity p.f.

The connection diagram is shown in Fig. 2.70.

Transformation ratio for primary and 5 kW load $= \frac{300}{200} = 1.5$

Current drawn by the *first load*,

$$I_1 = \frac{5 \times 1000}{300 \times 0.8} = 20.8 \text{ A}$$

∴ Current drawn by primary to supply this load

$= 20.8 \times 1.5 = 31.2$ A ...... 0.8 p.f. lagging

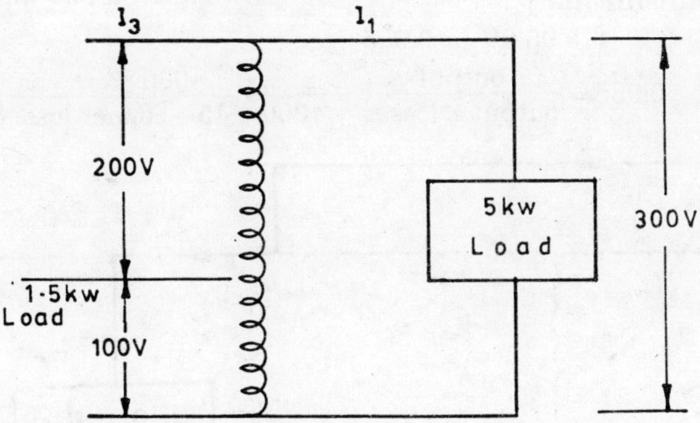

Fig. 2.70

Current drawn by *second load*

$$= \frac{1.5 \times 1000}{100 \times 1} = 15 \text{ A}$$

Transformation ratio for primary and 1.5 kW load

$$= \frac{100}{200} = \frac{1}{2}$$

∴ Current drawn by primary to supply this load

$= \frac{1}{2} \times 15 = 7.5$ A .... at unity power factor

Hence total primary current drawn from the 200 V supply is the vector sum of

(*i*) 31.2 A at 0.8 p.f. lagging, and

(*ii*) 7.5 A at unity p.f.

Resolving these currents into their *X*- and *Y*-components, we get

X-component $= 31.2 \times 0.8 + 7.5 = 32.5$ A

Y-component $= 31.2 \times 0.6 = 18.7$ A

∴ **Total primary current**

$I_3 = 32.5 + j18.7 =$ **37.5 A. (Ans.)**

Power factor $= \frac{32.5}{37.5} =$ **0.866. (Ans.)**

Example 2.76. *A 4-kVA single-phase 50-Hz transformer has a full-load efficiency of 95.5% and iron loss of 45 W. The transformer is now connected as an auto-transformer to 220 V supply. If it delivers 4 kW load at unity power factor to a 110 V circuit, calculate the efficiency of the operation and the current drawn by the high-voltage side.*

Solution. Fig. 2.71 shows the connections for a 2-winding transformer. In Fig. 2.72 the same unit has been connected as an auto-transformer to a 220 V supply. Since the two-windings are connected in series, hence voltage across each is 110 V.

In both connections the iron loss would remain the same. Since the auto-transformer windings will each carry but *half the current* as the conventional two-winding transformer, the copper losses will be $\frac{1}{4}$*th of previous value.*

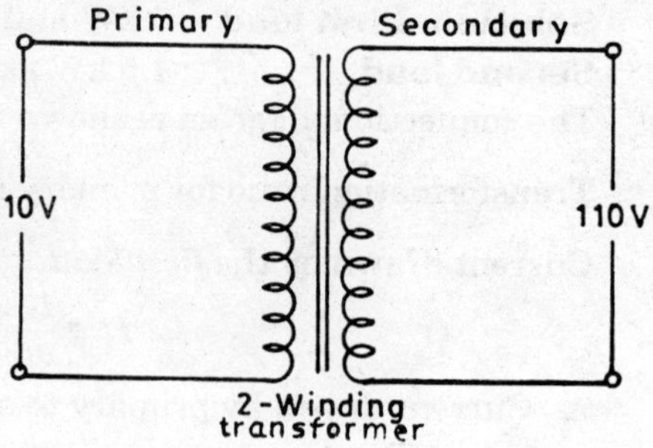

Fig. 2.71. Two-winding transformer.

Two-winding transformer :

Efficiency $= 95.5\%$ or 0.955

$$\therefore \quad 0.955 = \frac{\text{output}}{\text{output + losses}} = \frac{4000}{4000 + 45 + \text{copper loss}}$$

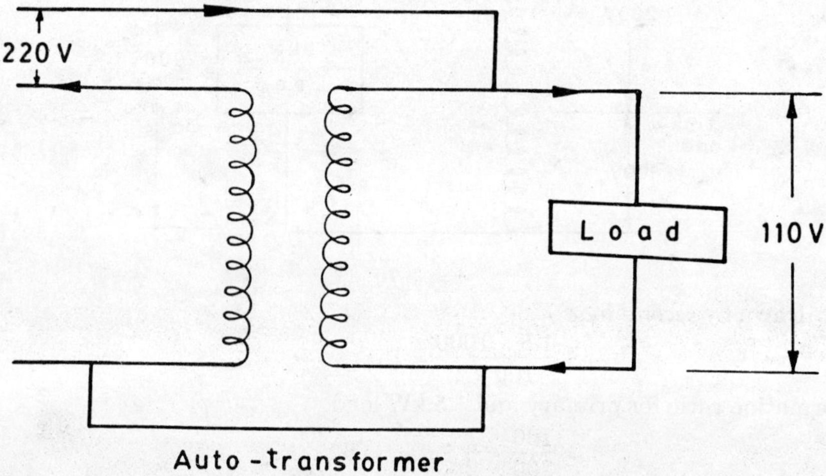

Fig. 2.72. Auto-transformer.

$$\therefore \quad \text{Copper loss} = \frac{4000}{0.955} - (4000 + 45) = 143.48 \text{ W}$$

Auto-transformer :

$$\text{Copper loss} = \frac{143.48}{4} = 35.87 \text{ W}$$

Iron loss $= 45$ W

$$\therefore \quad \textbf{Efficiency} = \frac{4000}{4000 + 35.87 + 45} = 0.9802 \text{ or } \mathbf{98.02\%}. \quad \textbf{(Ans.)}$$

Example 2.77. *A 200 kVA, 2300/460 V, 50 Hz, 2-winding transformer is to be used as an auto-transformer to step-up the voltage of 2300 V to 2760 V. If the transformer has an efficiency of 96% at 0.8 p.f. lagging, impedance of 4% and regulation of 3%, determine :*

(*i*) *Voltage and current ratings of each side* ;

(*ii*) *kVA rating* ;

(*iii*) *Efficiency at unity power factor* ;

(*iv*) *Percentage impedance* ;

(*v*) *Regulation* ;

(*vi*) *Short-circuit current of each side* ;

(*vii*) *kVA transformed and kVA conducted at full load, while it is used as an auto-transformer.*

(Nagpur University)

Solution. *Given* : Rating of 2-winding transformer = 200 kVA, 2300/460 V ;
η = 96% at 0.8 p.f. lagging ; Impedance = 4% ; regulation = 3%.

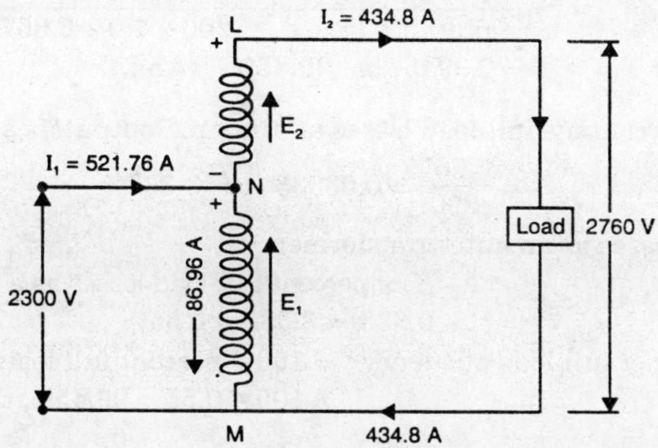

Fig. 2.73

(*i*) **Voltage and current ratings of each side :**

Rated voltage of the auto-transformer on l.v. side is 2300 V and on h.v. side is

= 2300 + 460 = **2760 V. (Ans.)**

The rated currents of h.v. and l.v. windings are respectively,

$$I_{\text{h.v.}} = \frac{200 \times 1000}{2300} = 86.96 \text{ A}$$

$$I_{\text{l.v.}} = \frac{200 \times 1000}{460} = 434.8 \text{ A}$$

Rated current of *h.v.* side of the auto-transformer

= rated current of 460 V side of the two winding transformer

= **434.8 A. (Ans.)**

Rated current of 434.8 A in the l.v. winding needs, by transformer action, a rated current of 86.96 A in the 2300 V winding acting as the primary.

Applying Kirchhoff's current law at point *N*, we get line current ;

$$I_1 = 434.8 + 86.96 = \mathbf{521.76\ A.}\ \ \textbf{(Ans.)}$$

(*ii*) **kVA rating :**

$$\text{kVA rating of the auto-transformer} = \frac{2760 \times 434.8}{1000} = \mathbf{1200\ kVA.}\ \ \textbf{(Ans.)}$$

$$\text{Transformation ratio, } K = \frac{\text{l.v. side voltage}}{\text{h.v. side voltage}} = \frac{2300}{2760} = 0.833$$

(*iii*) **Efficiency at unity power factor :**

Power output of two-winding transformer at 0.8 p.f. lagging

$$= 200 \times 0.8 = 160 \text{ kW}$$

Full-load losses $= \text{input} - \text{output} = \dfrac{\text{output}}{\text{efficiency}} - \text{output}$

$$= \frac{160}{0.96} - 160 = 6.667 \text{ kW}$$

Since the auto-transformer operates at rated voltage and rated currents, the losses remain constant *i.e.*, 6.667 kW.

Efficiency of auto-transformer for an output of 1200×1.0 kW is

$$\eta = \frac{\text{output}}{\text{output} + \text{losses}} = \frac{1200 \times 1.0}{1200 \times 1.0 + 6.667}$$

$$= 0.9945 \text{ or } \mathbf{99.45\%. \text{ (Ans.)}}$$

Alternatively. Percentage full-load losses at unity p.f. output as a two-winding transformer $= \dfrac{6.667}{200} = 0.03333$ or 3.333%

Percent full-load losses as an auto-transformer

$= (1 - K) \times$ percent full-load losses as a two-winding transformer

$= (1 - 0.833) \times 3.333 = 0.55\%$

and auto-transformer full-load efficiency $= 100 -$ percent full load losses

$= 100 - 0.55 = 99.45\%$, the same as before.

(*iv*) **Percentage impedance :**

Percentage impedance of two-winding transformer = 4%

The ohmic drop at full load is the same in both losses. If the impedance drop is referred to h.v. side of the auto-transformer (Fig. 2.73), the per unit or percent drop in the auto-transformer is with respect to V_1 while that in the two-winding transformer it is with respect to

$$V_1 - V_2 = V_1(1 - K)$$

If the impedance drop is referred to h.v. side of the auto-transformer (Fig. 2.73), then the winding current is

$$I_2 - I_1 = I_2(1 - K)$$

so the impedance drop with respect to I_2 is to be reduced $(1 - K)$ times in the case of an auto-transformer.

Therefore, per unit or percent impedance drop in an auto-transformer

$= (1 - K) \times$ per unit or percent drop as a two-winding transformer

$= (1 - 0.833) \times 4 = \mathbf{0.668\%.}$ **(Ans.)**

(*v*) **Regulation :**

As regulation is proportional to per unit or percent impedance drop, so regulation in an auto-transformer

$= (1 - K) \times$ regulation as a two-winding transformer

$= (1 - 0.833) \times 3 = \mathbf{0.5\%.}$ **(Ans.)**

(*vi*) **Short-circuit current of each side :**

Short-circuit current as an auto-transformer

$$= \frac{1}{\text{per unit impedance of the auto-transformer}}$$

$$= \frac{1}{0.00668} = 149.7 \text{ per unit}$$

$\therefore$ Short-circuit current on h.v. side $= \dfrac{149.7 \times 434.8}{1000} =$ **65.09 kA. (Ans.)**

Short-circuit current on l.v. side $= \dfrac{149.7 \times 521.76}{1000} =$ **78.11 kA. (Ans.)**

(*vii*) kVA transformed and kVA conducted at full-load :

Here winding *MN* acts as the primary and winding *LN* as the secondary.

$\therefore$ **kVA transformed** $= \dfrac{86.96 \times 2300}{1000}$ or $\dfrac{434.8 \times 460}{1000} =$ **200 kVA. (Ans.)**

kVA conducted = 1200 – 200 = **1000 kVA. (Ans.)**

2.9.21. Parallel Operation of Single-Phase Transformers. The necessity of parallel operation of transformers is felt when the amount of power to be transformed is greater than that which can be handled by one transformer.

The *important conditions* which must be fulfilled if two or more transformers are to operate successfully in parallel to deliver a common load are :

1. The *voltage ratings* of both primaries and secondaries must be identical. This obviously implies that the transformation ratios are the same. *Small differences* are permissible if the resultant *circulating currents* can be tolerated.

2. The transformer *must be* properly connected with regard to *polarity*.

3. The equivalent impedances should be *inversely proportional* to the respective kVA ratings.

4. The ratio of the equivalent resistance to equivalent reactance of all transformers should be same.

These conditions are most easily met by paralleling transformers of identical ratings of the same make and model. Careful study must be performed with different kVA ratings of even the same make or the effects in steps 1, 3 and 4 may appear in undesirable amounts. Step 2 has to be *strictly* observed even if steps 1, 3 and 4 are *slightly* modified.

Figs. 2.74 and 2.75 show how two transformers should be connected in parallel when they have the *same* polarity (Fig. 2.74) and when they have *opposite* polarity (Fig. 2.75).

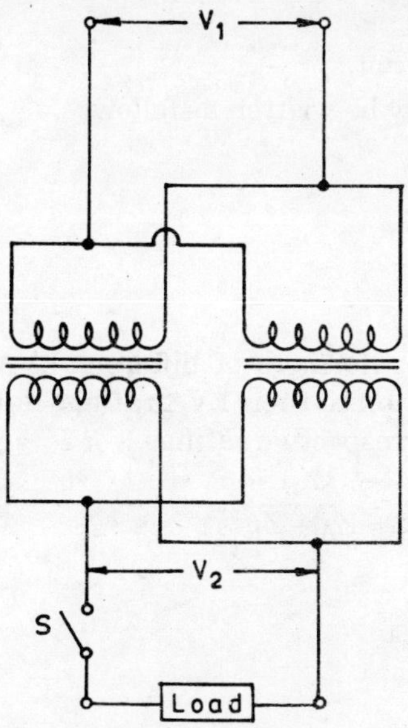

Fig. 2.74. Transformers with *same* polarity.

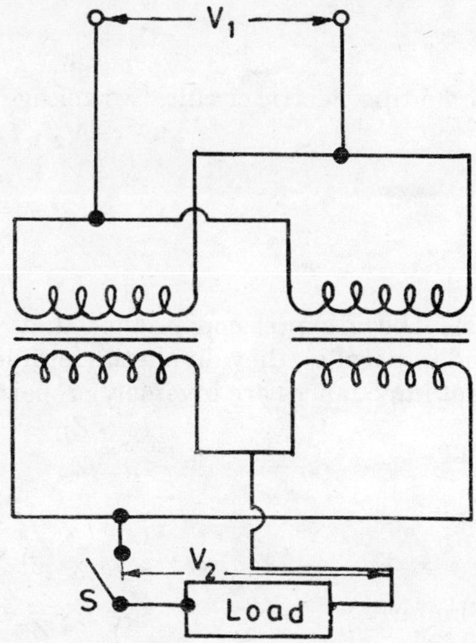

Fig. 2.75. Transformer with *opposite* polarity.

Load sharing by two transformers. We shall be considering following two cases :

(*i*) Equal voltage ratios

(*ii*) Unequal voltage ratios.

Case I. Equal Voltage Ratios. Let us assume that no-load voltage of both secondaries is the same *i.e.* $E_A = E_B = E_2$, and that the two voltages are coincident (*i.e.* there is no phase difference between E_A and E_B which would be true if the magnetising currents of the two transformers are not much different from each other) under these conditions, both primaries and secondaries of the two transformers can be connected in parallel and there will *circulate no current between them on 'no-load'*.

Fig. 2.76 represents two impedances in parallel. Let R_A, X_A and Z_A be the total equivalent resistance, reactance and impedance respectively of transformer *A* and R_B, X_B and Z_B be the total equivalent resistance, reactance and impedance of transformer *B* (referred to secondary).

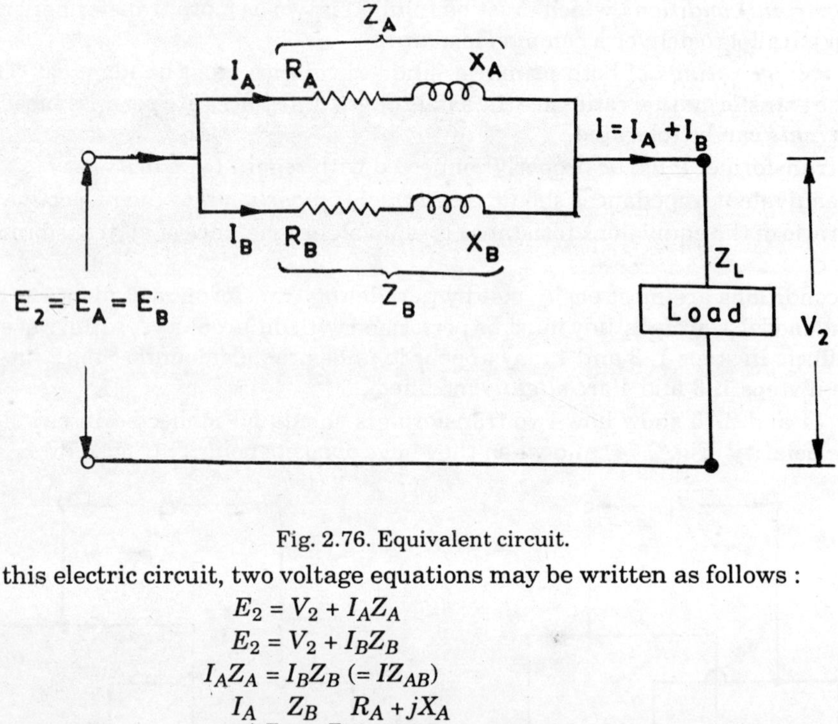

Fig. 2.76. Equivalent circuit.

From this electric circuit, two voltage equations may be written as follows :

$$E_2 = V_2 + I_A Z_A$$

and
$$E_2 = V_2 + I_B Z_B$$

or
$$I_A Z_A = I_B Z_B \; (= IZ_{AB})$$

or
$$\frac{I_A}{I_B} = \frac{Z_B}{Z_A} = \frac{R_A + jX_A}{R_B + jX_B} \quad \text{...(2.46)}$$

The above expression implies that when two transformers of different kVA ratings are connected in parallel, they divide the total load in proportion to their kVA ratings only when their equivalent impedances are inversely proportional to their respective ratings.

Again
$$\frac{I_A}{I_B} = \frac{Z_B}{Z_A} \quad \text{or} \quad \frac{I_A}{I_A + I_B} = \frac{Z_B}{Z_A + Z_B}$$

or
$$I_A = I \times \frac{Z_B}{Z_A + Z_B} \quad (\because \; I = I_A + I_B)$$

Similarly
$$I_B = I \times \frac{Z_A}{Z_A + Z_B}$$

Load shared by transformer A,

$$Q_A = V_2 I_A \times 10^{-3} \text{ kVA} = V_2 I \frac{Z_B}{Z_A + Z_B} \times 10^{-3} \text{ kVA} \quad ...(2.47)$$

Load shared by transformer B,

$$Q_B = V_2 I_B \times 10^{-3} \text{ kVA}$$

$$Q_B = V_2 I \frac{Z_A}{Z_A + Z_B} \times 10^{-3} \text{ kVA} \quad ...(2.48)$$

Total, $\quad Q = Q_A + Q_B = V_2 I \times 10^{-3}$ kVA

Hence, $$Q_A = \frac{Z_B}{Z_A + Z_B} Q \quad ...(2.49)$$

and $$Q_B = \frac{Z_A}{Z_A + Z_B} Q \quad ...(2.50)$$

Note. If the equivalent impedance Z_A and Z_B are given in per unit system or as percentage and ratings of the two transformers are different, then impedances of both should be transferred to common base kVA.

Case II. Unequal Voltage ratios. In case turn ratios are not exactly equal, then with the primaries connected to the same supply, the *secondary induced e.m.f. will be unequal and a circulating current will flow, even at no-load.* If the transformers are sharing a load, this *circulating current will be superimposed on the currents drawn by the load.*

Let V_1 = primary supply voltage,
K_A = voltage ratio of transformer A,
K_B = voltage ratio of transformer B,
Z_A = equivalent impedance of transformer A $(= R_A + jX_A)$ referred to secondary,
Z_B = equivalent impedance of transformer B $(= R_B + jX_B)$ referred to secondary,
I_A = output current of transformer A, and
I_B = output current of transformer B.

Now, induced e.m.f. in the secondary of transformer A

$$E_A = \frac{V_1}{K_A} = V_2 + I_A Z_A \quad ...(2.51)$$

Similarly, $$E_B = \frac{V_1}{K_B} = V_2 + I_B Z_B \quad ...(2.52)$$

Also, $V_2 = IZ_L$, where, Z_L is the impedance of the load

$$= (I_A + I_B) Z_L \quad ...(2.53)$$

$\therefore$ $$E_A = I_A Z_A + (I_A + I_B) Z_L \quad ...(2.54)$$

and $$E_B = I_B Z_B + (I_A + I_B) Z_L \quad ...(2.55)$$

$$E_A - E_B = I_A Z_A - I_B Z_B$$

or $$I_A = \frac{(E_A - E_B) + I_B Z_B}{Z_A} \quad ...(2.56)$$

Substituting the value of I_A in eqn. (2.55), we get

$$E_B = I_B Z_B + \frac{(E_A - E_B) + I_B Z_B}{Z_A} \times Z_L + I_B Z_L$$

or $$I_B \left(Z_B + Z_L + \frac{Z_B}{Z_A} \times Z_L \right) = \frac{E_B Z_A - (E_A - E_B) Z_L}{Z_A}$$

or
$$I_B = \frac{E_B Z_A - (E_A - E_B) Z_L}{Z_A Z_B + Z_L (Z_A + Z_B)} \quad ...(2.57)$$

PARALLEL OPERATION OF SINGLE-PHASE TRANSFORMERS

Transformers with 'Equal Voltage Ratios'

Example 2.78. *Two single-phase transformers with equal turns have impedances of (0.6 + j4) and (0.8 + j10) Ω with respect to the secondary. If they operate in parallel, determine how they will share total load of 120 kW at 0.8 power factor lagging.*

Solution. Impedance of transformer A, $Z_A = 0.6 + j4$

Impedance of transformer B, $Z_B = 0.8 + j10$

Total load to be shared = 120 kW at 0.8 power factor lagging

Load shared by transformer A, P_A :

Load shared by transformer B, P_B :

Now $\cos \phi = 0.8$

$\therefore$ $\phi = -36.9°$

Total load in kVA $= \frac{120}{0.8} = 150$

Hence, load to be shared, $P = 150 \angle -36.9°$ kVA

$$Z_A = (0.6 + j4)\ \Omega = 4.04 \angle 81.5°\ \Omega$$
$$Z_B = (0.8 + j10)\ \Omega = 10.03 \angle 85.4°\ \Omega$$
$$Z_A + Z_B = (0.6 + j4) + (0.8 + j10) = (1.4 + j14) = 14.07 \angle 84.3°\ \Omega$$

Load shared by transformer A,

$$P_A = P \times \frac{Z_B}{Z_A + Z_B} \text{ kVA}$$
$$= 150 \angle -36.9° \times \frac{10.03 \angle 85.4°}{14.07 \angle 84.3°}$$
$$= 106.93 \angle -39.7° \text{ kVA} = \mathbf{86.72\ kW.} \quad \textbf{(Ans.)}$$

Load shared by transformer B,

$$P_B = P \times \frac{Z_A}{(Z_A + Z_B)} \text{ kVA}$$
$$= 150 \angle -36.9° \times \frac{4.04 \angle 81.5°}{14.0 \angle 84.3}$$
$$= 43.07 \angle -39.7° \text{ kVA} = \mathbf{33.14\ kW.} \quad \textbf{(Ans.)}$$

[As a check $P = P_A + P_B = 86.72 + 33.14 = 119.86$ kW. The small error is due to approximations made in the calculations.]

☞**Example 2.79.** *Two single-phase transformers are connected in parallel on both primary and secondary. One transformer has a full-load resistance drop of 1% and reactance drop of 6%. For the other transformer the corresponding values are 1.5% and 5% respectively. How will the following loads be shared by the two transformers :*

(i) 160 kW at 0.8 power factor lagging

(ii) 155 kW at 0.9 power factor leading.

Solution. Impedance of transformer A, $Z_A = (1 + j6)$

Impedance of transformer B, $Z_B = (1.5 + j5)$

$$Z_A = (1 + j6) = 6.08 \angle 80.54°\ \Omega$$

$$Z_B = (1.5 + j5) = 5.22 \angle 73.3° \ \Omega$$

$$Z_A + Z_B = (1 + j6) + (1.5 + j5) = (2.5 + j11) = 11.28 \angle 72.2° \ \Omega$$

(*i*) **160 kW at 0.8 p.f. lagging :**

Total load in kW $= 160$ kW

$$\cos \phi = 0.8 \text{ lagging}$$

$$\phi = -36.9°$$

Total load in kVA $= \dfrac{160}{0.8} = 200$ kVA

Hence total load to be shared, P

$$= 200 \angle -36.9° \text{ kVA}$$

Load shared by transformer A,

$$P_A = P \times \frac{Z_B}{Z_A + Z_B} = 200 \angle -36.9° \times \frac{5.22 \angle 73.3°}{11.28 \angle 77.2°}$$

$= 92.55 \angle -40.8°$ kVA $=$ **70.06 kW. (Ans.)**

Load shared by transformer B,

$$P_B = P \times \frac{Z_A}{Z_A + Z_B} = 200 \angle -36.9° \times \frac{6.08 \angle 80.54°}{11.28 \angle 77.2°}$$

$= 107.8 \angle -33.56°$ kVA $=$ **89.83 kW. (Ans.)**

(*ii*) **155 kW at 0.9 power factor leading :**

Total load in kW $= 155$ kW

$$\cos \phi = 0.9 \text{ leading}$$

or $\phi = 25.84°$

Total load in kVA $= \dfrac{155}{0.9} = 172.2$ kVA

Hence total load to be shared, $P = 172.2 \angle 25.84°$

Load shared by transformer A,

$$P_A = P \times \frac{Z_B}{Z_A + Z_B} = 172.2 \angle 25.84° \times \frac{5.22 \angle 73.3°}{11.28 \angle 77.2°}$$

$= 79.69 \angle 21.94°$ kVA $=$ **73.9 kW. (Ans.)**

Load shared by transformer B,

$$P_B = 172.2 \angle 25.84° \times \frac{Z_A}{Z_A + Z_B} = 172.2 \angle 25.84° \times \frac{6.08 \angle 80.54°}{11.28 \angle 77.2°}$$

$= 92.82 \angle 29.2° =$ **81.1 kW. (Ans.)**

Example 2.80. *Two single phase transformers are connected in parallel on both primary and secondary. They have resistances of 0.2 Ω and 0.1 Ω and reactances of 0.35 Ω and 0.6 Ω respectively referred to secondary side. The no-load secondary voltage of each transformer is 550 V. Determine :*

(i) The output current of each transformer ; and

(ii) The ratio of kW outputs if the secondary load is 200 A at 0.8 p.f. lagging.

Solution. (*i*) Impedance of transformer A

$$Z_A = (0.2 + j0.35) \ \Omega = 0.403 \angle 60.25° \ \Omega$$

Impedance of transformer B

$$Z_B = (0.1 + j0.6) \ \Omega = 0.608 \angle 80.53° \ \Omega$$

Let I_A and I_B be the currents delivered by transformers A and B respectively.

Then $\dfrac{I_A}{I_B} = \dfrac{Z_B}{Z_A} = \dfrac{0.608 \angle 80.53°}{0.403 \angle 60.25°} = 1.509 \angle 20.28 = (1.416 + j0.523)$

$\therefore$ $I_A = I_B\,(1.415 + j0.523)$ ...(i)

Total current, $I = 200\,(0.8 - j0.6)\,A = (160 - j120)\,A$

Also, $I = I_A + I_B$

$(160 - j120) = I_B\,(1.415 + j0.523) + I_B$

or $(160 - j120) = I_B\,(2.415 + j0.523)$

or $200 \angle -36.9° = I_B\,(2.47 \angle 12°)$

$\therefore$ $I_B = \dfrac{200 \angle -36.9°}{2.47 \angle 12°} = 80.97 \angle -48.9°$

$\therefore$ **Current delivered by transformer B = 80.97 A. (Ans.)**

Now, $I_A = I_B\,(1.415 + j0.523)$ [eq. (i)]

$= 80.97 \angle -48.9° \times 1.509 \angle +20.28° = 122.1 \angle -28.62°\ A$

Current delivered by transformer A = 122.1 A. (Ans.)

(ii) $\dfrac{\text{Output of transformer } A}{\text{Output of transformer } B} = \dfrac{\text{active component of } I_A}{\text{active component of } I_B}$

$= \dfrac{122.1 \cos(-28.62°)}{80.97 \cos(-48.9°)} = \dfrac{107.18}{53.22} =$ **2.014. (Ans.)**

☞**Example 2.81.** *A 1000 kVA transformer having a percentage impedance of (1 + j4) and a 500 kVA transformer having a percentage impedance of (2 + j6) are connected in parallel. The no-load secondary voltage of each transformer is 550 V. Find the load shared by each transformer and its power factor if the total load is 750 kVA at 0.8 power factor lagging.*

Solution. Impedance of transformer A (1000 kVA)

$= (1 + j4)\ \Omega$

Impedance of transformer B (500 kVA)

$= (2 + j6)\ \Omega$

No-load secondary voltage of each transformer = 550 V

Total load to be shared = 750 kVA at 4.8 p.f. lagging.

It may be noted that impedances given above *refer to different ratings*, these should be adjusted to the same basic kVA say 1000 kVA.

$\therefore$ Impedance of 500 kVA referred to basic kVA of 1000

$$Z_B = (2 + j6) \times \frac{1000}{500} = (4 + j12)\ \Omega = 12.65 \angle 71.6°$$

$$Z_A = (1 + j4) = 4.12 \angle 76°$$

$$Z_A + Z_B = (1 + j4) + (4 + j12) = 5 + j16 = 16.76 \angle 72.6°$$

Total load to be shared, $P = 750 \angle -36.9°$

Load shared by transformer A,

$$P_A = P \times \frac{Z_B}{Z_A + Z_B} = (750 \angle -36.9°) \times \frac{12.65 \angle 71.6°}{16.76 \angle 72.6°}$$

$= 566.08\angle -37.9°$ kVA = **446.7 kW at 0.789 lagging. (Ans.)**

Load shared by transformer B,

$$P_B = P \times \frac{Z_A}{Z_A + Z_B} = (750 \angle -36.9°) \times \frac{4.12 \angle 76°}{16.76 \angle 72.6°}$$

$= 184.36 \angle -33.5°$ = **153.7 kW at 0.834 lagging. (Ans.)**

Example 2.82. *Two single-phase transformers A and B have the following parameters : Transformer A : 10 kVA, 400 V/200 V, percentage resistance and percentage reactance 3% and 4% respectively ; Transformer B : 5 kVA, 400 V/200 V, percentage resistance and percentage reactance are 4% and 3% respectively. These two transformers are connected in parallel and they share a common load of 12 kW at a power factor of 0.8 lagging. Determine the active and reactive power delivered by transformer A.* **(GATE, 1999)**

Solution.

Transformer A :	**Transformer B :**
Rating = 10 kVA	Rating = 5 kVA
400 V/200 V	400 V/200 V
$\% R = 3 ; \% X = 4$	$\% R = 4, \% X = 3$
$\% Z_A = 3 + j4$	$\% Z_B = 4 + j3$

Load = 12 kW, 0.8 p.f. (lag)
$= 15 \angle -36.87°$ kVA

$\% Z_A$ and $\% Z_B$ on common base kVA of 10

$$\% Z_A = 3 + j4, \; \% Z_B = \frac{10}{5}(4 + j3) = 8 + j6$$

Now power delivered by transformer *A*

$$= (15 \angle -36.87°) \times \frac{Z_B}{Z_A + Z_B}$$

$$= (15 \angle -36.87°) \times \frac{8 + j6}{(3 + j4) + (8 + j6)}$$

$$= (15 \angle -36.87°) \times \frac{8 + j6}{11 + j10}$$

$$= (15 \angle -36.87°) \times \frac{(8 + j6)}{(11 + j10)} \times \frac{(11 - j10)}{(11 - j10)}$$

$$= (15 \angle -36.87°) \times \left[\frac{(88 - j80 + j66 + 60)}{121 + 100}\right]$$

$$= (15 \angle -36.87°) \times \frac{148 - j14}{221}$$

$$= (15 \angle -36.87°) \times (0.6697 - j0.0633)$$

$$= (15 \angle -36.87°) \times [0.6727 \angle -5.4°]$$

$$= 10 \angle -42.27° = 7.4 - j6.73$$

∴ **Active power** delivered by transformer A = **7.4 kW. (Ans.)**

Reactive power delivered by transformer A = **6.73 kVAR. (Ans.)**

☞**Example 2.83.** *Two single-phase 1100/220 V step down transformers with equal voltage ratios have the following short-circuit data :*

Transformer A : Primary voltage = 60 V, primary current = 30 A, power input to primary = 540 W

Transformer B : Primary voltage = 45 V, primary current = 30 A, power input to primary = 720 W

The two transformers connected in parallel, feed a total load of 20 kW at 0.8 power factor lagging. Find the secondary load current, power factor and power output of each transformer.

Solution. Total impedance of transformer *A*, referred to primary,

$$Z_A = \frac{V_{SC}}{I_{SC}} = \frac{60}{30} = 2\ \Omega$$

Total resistance referred to primary,

$$R_A = \frac{P_{SC}}{I_{SC}^2} = \frac{540}{30^2} = 0.6\ \Omega$$

Total reactance, $X_A = \sqrt{Z_A^2 - R_A^2} = \sqrt{2^2 - (0.6)^2} = 1.908\ \Omega$

$$Z_A = (0.6 + j1.908)\ \Omega = 2\ \angle 72.54°\ \Omega$$

Total impedance of transformer B referred to primary,

$$Z_B = \frac{V_{SC}}{I_{SC}} = \frac{45}{30} = 1.5\ \Omega$$

Total resistance referred to primary,

$$R_B = \frac{P_{SC}}{I^2_{SC}} = \frac{720}{30^2} = 0.8\ \Omega$$

Total reactance, $X_B = \sqrt{1.5^2 - (0.8)^2} = 1.269\ \Omega$

$\therefore$ $Z_B = (0.8 + j1.269)$ ohm $= 1.5\ \angle 57.8°\ \Omega$

Let I_A and I_B be the secondary currents delivered by transformers A and B.

Then $\dfrac{I_A}{I_B} = \dfrac{Z_B}{Z_A} = \dfrac{1.5\ \angle 57.8°}{2\ \angle 72.54°} = 0.75\ \angle -14.74° = (0.725 - j0.191)$

$\therefore$ $I_A = I_B\,(0.725 - j0.191)$

Secondary load current, $I = \dfrac{20 \times 1000}{220 \times 0.8} = 113.6$ A

where $V_2 = 220$ volts, p.f. of total current, $I = 0.8$ lagging and total power delivered = 20 kW

$$I = 113.6\,(0.8 - j0.6) = 90.88 - j68.16$$

Also $I = I_A + I_B$

$$90.88 - j68.16 = I_B(0.725 - j0.191) + I_B = I_B\,(1.725 - j0.191)$$

$\therefore$ $I_B = \dfrac{90.88 - j68.16}{1.725 - j0.191} = \dfrac{113.6\ \angle -36.9°}{1.736 \angle -6.32°}$

$= 65.44\ \angle -30.58°$ = **65.44 A, 0.861 lagging. (Ans.)**

$\therefore$ $I_A = (65.44\ \angle -30.58°)\,(0.725 - j0.191)$

$= (65.44\ \angle -30.58°)\,(0.75\ \angle -14.76°) = 49.1\text{ A}\ \angle -45.3°$

= **49.1 A, 0.703 lagging. (Ans.)**

Power output of transformer A

= Voltage × Active component of I_A

= $220 \times 49.1 \times \cos(-45.3°) \times 10^{-3}$ kW = **7.598 kW. (Ans.)**

Power output of transformer B

= Voltage × Active component of I_B

= $220 \times 65.44 \cos(-30.58°) \times 10^{-3}$ = **12.394 kW. (Ans.)**

Transformers with 'Unequal Voltage Ratios'

☞**Example 2.84.** *Two single-phase transformers are connected in parallel to supply a load having a resistance of 10 Ω and reactance of 5 Ω. The effective resistance and reactance of the transformer A referred to secondary are 0.1 Ω and 1.2 Ω and the corresponding values for transformer B are 0.16 Ω and 1.5 Ω. The no-load e.m.f. of transformer A is 2200 V and that of transformer B is 2100 V. Determine :*

(*i*) *Circulating current on no-load, and*

(*ii*) *Current output of each transformer and its power factor.*

Solution. No-load e.m.f. of transformer *A*, $E_A = 2200$ V

No-load e.m.f. of transformer *B*, $E_B = 2100$ V

$$R_A = 0.1\ \Omega,\ X_A = 1.2\ \Omega$$

$$R_B = 0.16\ \Omega,\ X_B = 1.5\ \Omega$$

$\therefore$ $$Z_A = 0.1 + j1.2 = 1.204\ \angle 85.24°$$

and $$Z_B = 0.16 + j1.5 = 1.508\ \angle 83.9°$$

Now $$Z_A + Z_B = (0.1 + j1.2) + (0.16 + j1.5) = (0.26 + j2.7) = 2.712\ \angle 84.5°$$

Load impedance, $$Z_L = 10 + j5 = 11.18\ \angle 26.6°$$

(*i*) **Circulating current,**

$$I_c = \frac{E_A - E_B}{Z_A + Z_B} = \frac{2200 - 2100}{2.712\ \angle 84.5°} = \mathbf{36.87\ A.}\ \ \textbf{(Ans.)}$$

(*ii*) **Current output of transformer A,**

$$I_A = \frac{E_A Z_B + (E_A - E_B) Z_B}{Z_A Z_B + Z_L (Z_A + Z_B)}$$

$$= \frac{2200 \times 1.508\ \angle 83.9° + (2200 - 2100) \times 11.18\ \angle 26.6°}{(1.204\ \angle 85.24°)(1.508\ \angle 83.9°) + (11.18\ \angle 26.6°)(2.712 \angle 84.5°)}$$

$$= \frac{3317.6\ \angle 83.9° + 1118\ \angle 26.6°}{1.815 \angle 169.14° + 30.32 \angle 111.1°}$$

$$= \frac{352.54 + j3298.8 + 999.66 + j500.6}{-1.78 + j0.342 + (-10.92 + j28.29)}$$

$$= \frac{1352.2 + j3799.4}{-12.7 + j28.63} = \frac{4032\ \angle 70.4°}{31.32\ \angle 113.9°} = 128.7\ \angle -43.5°$$

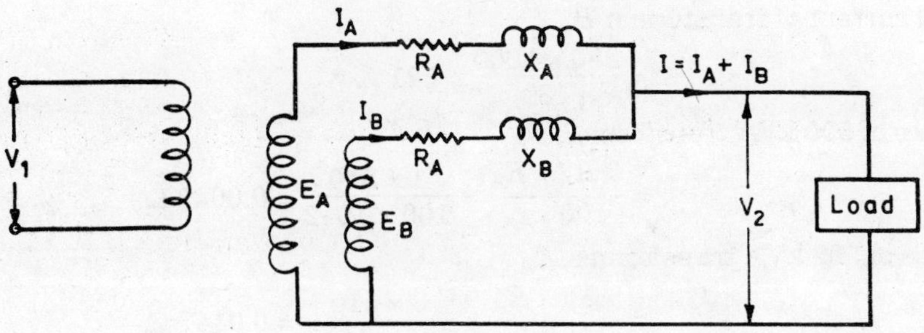

Fig. 2.77

$\therefore$ **Current delivered by transformer A = 128.7 A. (Ans.)**

Power factor $= \cos(-43.5°) = \mathbf{0.725}$ **lagging. (Ans.)**

Current output of transformer *B*,

$$I_B = \frac{E_B Z_A - (E_A - E_B) Z_L}{Z_A Z_B + Z_L (Z_A + Z_B)}$$

$$= \frac{2100 \times 1.204\ \angle 85.24° - (2200 - 2100) \times 11.18\ \angle 26.6°}{(1.204\ \angle 85.24°)(1.508\ \angle 83.9°) + (11.18\ \angle 26.6°)(2.712\ \angle 84.5°)}$$

$$= \frac{2528.4\ \angle 85.24° - 1118\ \angle 26.6°}{1.815\ \angle 169.14° + 30.32\ \angle 111.1°}$$

$$= \frac{(209.8 + j2519.68) - (999.66 + j500.6)}{-1.78 + j0.342 + (-10.92 + j28.29)}$$

$$= \frac{-789.86 + j2019.08}{-12.7 + j28.63} = \frac{2168.01 \angle 111.36^\circ}{31.32 \angle 113.9^\circ} = 69.22 \angle -2.54^\circ$$

∴ **Current delivered by transformer B = 69.22 A. (Ans.)**

Power factor = cos (– 2.54°) = **0.999 lagging. (Ans.)**

Example 2.85. *Two transformers A and B of ratings 500 kVA and 250 kVA are supplying a total load of 700 kVA at unity power factor. Their open circuit voltages are 510 V and 500 V respectively. Transformer A has 1% resistance and 3% reactance and the transformer B has 1.5% resistance and 6% reactance. Find :*

(i) Cross-current in the secondaries on no-load, and

(ii) Load shared by each transformer.

Solution. Rating of transformer A = 500 kVA

Rating of transformer B = 250 kVA

Total load = 700 kVA at unity power factor

Open circuit voltage of transformer A,

$$E_A = 510 \text{ V}$$

Open circuit voltage of transformer B,

$$E_B = 500 \text{ V}$$

It is more convenient to work with ohmic resistances and for that purpose, we will convert percentage value into numerical values by assuming 480 V as the terminal voltage (this value is arbitrary but this assumption will not introduce appreciable error).

Full-load current of transformer A

$$= \frac{500 \times 1000}{480} = 1042 \text{ A}$$

Full-load current of transformer B

$$= \frac{250 \times 1000}{480} = 521 \text{ A}$$

Resistance of 500 kVA transformer, R_A

$$= \frac{\%R_A \times E_A}{100 \times I_A} = \frac{1 \times 510}{100 \times 1042} = 0.0049 \ \Omega$$

Reactance of 500 kVA transformer, X_A

$$= \frac{\%X_A \times E_A}{100 \times I_A} = \frac{3 \times 510}{100 \times 1042} = 0.0147 \ \Omega$$

Resistance of 250 kVA transformer, R_B

$$= \frac{\%R_B \times E_B}{100 \times I_B} = \frac{1.5 \times 500}{100 \times 521} = 0.0144 \ \Omega$$

Reactance of 250 kVA transformer, X_B

$$= \frac{\%X_B \times E_B}{100 \times I_B} = \frac{6 \times 500}{100 \times 521} = 0.0576 \ \Omega$$

∴ Impedance of transformer A,

$$Z_A = 0.0049 + j0.0147 = 0.0155 \angle 71.56^\circ \ \Omega$$

Impedance of transformer B,

$$Z_B = 0.0144 + j0.0576 = 0.0594 \angle 75.96^\circ \ \Omega$$

$$Z_A + Z_B = (0.0049 + j0.0147) + (0.0144 + j0.0576)$$

$$= (0.0193 + j0.0723)\ \Omega = 0.075\ \angle 75^\circ\ \Omega$$

Load, $P = 700\ \angle 0^\circ$ kVA

Load impedance, $$Z_B = \frac{V_2^2}{P \times 1000} = \frac{(480)^2}{700\angle 0^\circ \times 1000} = 0.329\ \angle 0^\circ = 0.329\ \Omega + j0$$

(*i*) **Cross-current in the secondaries on no-load,**

$$I_c = \frac{E_A - E_B}{Z_A + Z_B} = \frac{510 - 500}{0.075\ \angle 75^\circ} = \frac{10}{0.075\ \angle 75^\circ} = \mathbf{133.3\ \angle -75^\circ\ A.}\ \textbf{(Ans.)}$$

(*ii*) **Load shared by each transformer :**

$$I_A = \frac{E_A Z_B + (E_A - E_B) Z_B}{Z_A Z_B + (Z_A + Z_B) Z_B}$$

$$= \frac{510 \times (0.0594\ \angle 75.96^\circ) + (510 - 500) \times (0.329\ \angle 0^\circ)}{(0.0155\ \angle 71.56^\circ) \times (0.0594\ \angle 75.96^\circ) + (0.075\ \angle 75^\circ) \times (0.329\ \angle 0^\circ)}$$

$$= \frac{10.634 + j29.376}{(0.00092\ \angle 147.52^\circ) + (0.024675\ \angle 75^\circ)}$$

$$= \frac{10.634 + j29.376}{0.000776 + j0.00049 + 0.006386 + j0.023834}$$

$$= \frac{10.634 + j29.376}{0.007162 + j0.02432} = \frac{31.24\ \angle 70^\circ}{0.025\ \angle 73.6^\circ} = 1250\ \angle -3.6^\circ\ \text{A}$$

$$I_B = \frac{E_B Z_A + (E_A - E_B) Z}{Z_A Z_B + (Z_A + Z_B) Z}$$

$$= \frac{500 \times (0.0155\ \angle 71.56^\circ) + (510 - 500) \times (0.329 \angle 0^\circ)}{(0.0155 \angle 71.56^\circ) \times (0.0594\ \angle 75.96^\circ) + (0.075\ \angle 75^\circ) \times (0.329\ \angle 0^\circ)}$$

$$= \frac{7.75\ \angle 71.56^\circ - 3.29}{0.025\ \angle 73.6^\circ} = \frac{(2.45 + j7.35) - 3.29}{0.025\ \angle 73.6^\circ}$$

$$= \frac{-0.84 + j7.35}{0.025\ \angle 73.6^\circ} = \frac{7.4\ \angle 96.5^\circ}{0.025\ \angle 73.6^\circ} = 296\ \angle 22.9^\circ\ \text{A}$$

∴ **Load shared by transformer A,**

$$P_A = 480 \times (1250\ \angle -3.6^\circ)\ \text{kVA} = \mathbf{600\ \angle -3.6^\circ\ kVA.}\ \textbf{(Ans.)}$$

Load shared by transformer B,

$$P_B = 480 \times 296\ \angle 22.9^\circ = \mathbf{142.08\ \angle 22.9^\circ\ kVA.}\ \textbf{(Ans.)}$$

☞**Example 2.86.** *A 100 kVA, 250 V single-phase transformer, A having a percentage impedance of (0.6 + j3) is connected in parallel with a 50 kVA, 250 V single-phase transformer, B having a percentage impedance of (0.8 + j4). The no-load induced e.m.fs. in the secondaries of the transformers are 260 V and 255 V respectively. If the total load on the set is 120 kW at 0.8 power lagging, determine :*

(*i*) *The secondary current of each transformer and its power factor.*

(*ii*) *The voltage across the load terminals.*

(*iii*) *Load in kW taken by each transformer.*

Solution. Rating of transformer, $A = 100$ kVA

Rating of transformer, $B = 50$ kVA

Percentage impedance of transformer A, $Z_A = (0.6 + j3)$

Percentage impedance of transformer B, $Z_B = (0.8 + j4)$

Total load on the set, $P = 120$ kW at 0.8 p.f. lagging

Full-load current of transformer A,

$$I_A = \frac{100 \times 1000}{250} = 400 \text{ A}$$

Full-load current of transformer B,

$$I_B = \frac{50 \times 1000}{250} = 200 \text{ A}$$

Resistance of 100 kVA transformer,

$$R_A = \frac{\%R_A \times E_A}{100 \times I_A} = \frac{0.6 \times 250}{100 \times 400} = 0.0038 \ \Omega$$

Reactance of 100 kVA transformer,

$$X_A = \frac{\%X_A \times E_A}{100 \times I_A} = \frac{3 \times 250}{100 \times 400} = 0.0188 \ \Omega$$

Resistance of 50 kVA transformer,

$$R_B = \frac{\%R_B \times E_B}{100 \times I_B} = \frac{0.8 \times 250}{100 \times 200} = 0.01 \ \Omega$$

Reactance of 50 kVA transformer,

$$X_B = \frac{\%X_B \times E_B}{100 \times I_B} = \frac{4 \times 250}{100 \times 200} = 0.05 \ \Omega$$

Hence, $$Z_A = R_A + jX_A = 0.0038 + j0.0188 = 0.0192 \angle 78.6°$$

$$Z_B = R_B + jX_B = (0.01 + j0.05) = 0.051 \angle 78.7°$$

$$Z_A + Z_B = (0.0038 + j0.0188) + (0.01 + j0.05)$$

$$= (0.0138 + j0.0688) = 0.07 \angle 78.6°$$

Let 250 V be the terminal voltage and Z_L be the load impedance

kVA of load $$= \frac{\text{kW}}{\text{p.f.}} = \frac{120}{0.8} = 150 \text{ kVA}$$

Total load current, $$I = \frac{250}{Z_B}$$

kVA supplied $$= 250 \times \frac{250}{Z_B} \times 10^{-3} = \frac{250^2}{Z_B} \times 10^{-3}$$

But kVA supplied = 150 at 0.8 p.f. lagging

$$\cos \phi_2 = 0.8$$

$$\phi_2 = -36.87°$$

$\therefore$ kVA supplied $= 150 \angle -36.87°$

$\therefore$ $$\frac{250^2}{Z_B} \times 10^{-3} = 150 \angle -36.87°$$

$\therefore$ $$Z_B = \frac{250^2 \times 10^{-3}}{150 \angle -36.87°} = 0.417 \angle 36.87°$$

(*i*) Let I_A and I_B be the currents delivered by transformers A and B respectively

$$I_A = \frac{E_A Z_B + (E_A - E_B) Z_B}{Z_A Z_B + (Z_A + Z_B) Z_L}$$

$$= \frac{260 \times (0.051 \angle 78.7°) + (260 - 255) 0.417 \angle 36.87°}{(0.0192 \angle 78.6°) \times (0.051 \angle 78.7°) + (0.07 \angle 78.6°) \times (0.417 \angle 36.87°)}$$

$$= \frac{13.26 \angle 78.7° + 2.085 \angle 36.87°}{0.00098 \angle 157.3° + 0.029 \angle 115.5°}$$

$$= \frac{2.6 + j13 + 1.67 + j1.25}{-0.0009 + j0.00038 - 0.0125 + j0.026}$$

$$= \frac{4.27 + j14.25}{-0.0134 + j0.0264} = \frac{14.87 \angle 73.3°}{0.0296 \angle 116.9°} = 502 \angle -43.6° \text{ A}$$

Hence, $\mathbf{I_A}$ **= 502 A, 0.724 power factor lagging. (Ans.)**

$$I_B = \frac{E_B Z_A - (E_A - E_B) Z_B}{Z_A Z_B + (Z_A + Z_B) Z_B}$$

$$= \frac{255 \times (0.0192 \angle 78.6°) - (260 - 255) \times (0.417 \angle 36.87°)}{(0.0192 \angle 78.6°) \times (0.051 \angle 78.7°) + (0.07 \angle 78.6°) \times (0.417 \angle 36.87°)}$$

$$= \frac{4.896 \angle 78.6° - 2.085 \angle 36.87°}{0.00098 \angle 157.3° + 0.029 \angle 115.5°}$$

$$= \frac{0.967 + j4.8 - (1.67 + j1.25)}{-0.0009 + j0.00038 - 0.0125 + j0.026}$$

$$= \frac{-0.703 + j3.55}{-0.0134 + j0.0264} = \frac{3.62 \angle 101.2°}{0.0296 \angle 116.9°} = 122.3 \angle -15.7°$$

Hence, $\mathbf{I_B}$ **= 122.3 A, 0.963 power factor lagging. (Ans.)**

(*ii*) **Voltage across the load terminals,**

$$= E_B - I_B Z_L = 255 - [(122.3 \angle -15.7°) \times (0.051 \angle 78.7°)]$$

$$= 255 - 6.24 \angle 63° = 255 - (2.83 + j5.56)$$

$$= 252.17 - j5.56 = \mathbf{252.23 \text{ V}}. \textbf{ (Ans.)}$$

(*iii*) **Load delivered by transformer A,**

$$P_A = V_2 I_A \cos \phi_A \times 10^{-3} \text{ kW}$$

$$= 252.23 \times 502 \times 0.724 \times 10^{-3} \text{ kW} = \mathbf{91.7 \text{ kW}}. \textbf{ (Ans.)}$$

Load delivered by transformer B,

$$P_B = V_2 I_B \cos \phi_B \times 10^{-3} \text{ kW}$$

$$= 252.23 \times 122.3 \times 0.963 \times 10^{-3} = \mathbf{29.7 \text{ kW}}. \textbf{ (Ans.)}$$

2.10. THREE PHASE TRANSFORMER

2.10.1. Three-phase Transformer Connections. Virtually all power distribution is by poly-phase system of voltages. Three-phase transformations may be made with the use of properly connected single-phase transformers. These connections are in extensive commercial use. The most frequently used connections are the following :

(*i*) Primary *Y*—secondary *Y*. (*ii*) Primary Δ—secondary Δ.

(*iii*) Primary Δ—secondary *Y*, *or vice versa*. (*iv*) Primary and secondary open Δ.

(*v*) Primary *T* secondary *T* (Scott connection)

Thus the most common connections are *Y-Y*, Δ-Δ, *Y*-Δ, Δ-*Y*, open delta or *V-V* and Scott connection or *T-T* connection.

2.10.1.1. The Y-Y connection. Fig. 2.78 shows a bank of three transformers connected in *Y* on both the primary and secondary sides. If the ratio of transformation of each transformer is *K*, the same ratio will exist between the *line* voltages on the primary and secondary sides. This *connection will give satisfactory service only if the three-phase load is balanced* ; when the load is unbalanced, the electrical neutral will shift from its exact centre to a point that will make the line-to-neutral voltages unequal.

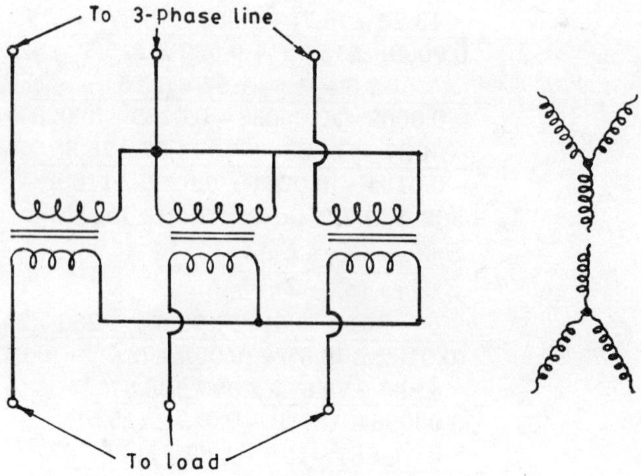

Fig. 2.78. The *Y-Y* connection of transformers.

Advantages :

1. This type of connection requires fewer turns per winding since the voltage across each is $\frac{1}{\sqrt{3}}$ times the line voltage ; hence it is *cheaper*.

2. The phase or winding current being equal to the line current, the cross-section of the winding wire is large, therefore the winding is stronger to bear stresses imposed upon it during heavy load or short circuit.

3. There is less dielectric stress on the insulating materials owing to lesser voltage *i.e.* $\frac{1}{\sqrt{3}}$ of line voltage.

The above advantages are enumerated with the understanding that, other things being equal, its rival is the delta-delta connection.

Disadvantages :

1. In case the load on the secondary side is unbalanced, as in the case of distributing network, the potential of the star-point will assume any value if the star-point is not earthed. This may impose full-line voltage on secondary windings. The shifting of the neutral point must be prevented by connecting the primary star-point to the star-point of the alternator winding.

2. In spite of grounding the star-point, if there is a third harmonic in the form of the alternator voltage, the third harmonic will appear in the voltages of the secondary side. This will cause triple frequency currents in the three-phase circuits. These currents when they flow in the neutral wire are additive and do not cancel out. Hence they will *cause interference to telephone lines located along the same route.*

3. The magnetising current in a transformer has third harmonic components. These currents will find a return path via the connection between the primary star-point of transformer and the neutral point of the alternator. However, if this connection is missing, these components will distort the flux wave which will produce a voltage having a third harmonic in each of the transformer, both on the primary and secondary sides. And, as before, if the star-point on the secondary is earthed, or grounded, triple harmonic currents will appear in the secondary circuit, and they will flow through the neutral wire causing interference to telephone lines in the vicinity.

4. If the star-points of both the primary and the secondary sides are not earthed, the regulation of the phases will be very poor if the load happen to be unbalanced as the case of distribution network.

2.10.1.2. The Δ-Δ Connection. Fig. 2.79 shows a bank of transformers connected in Δ on both the primary and secondary sides. *This arrangement is generally used in systems in which the voltages are not very high and especially when continuity of service must be maintained even though one of the transformers should fail.*

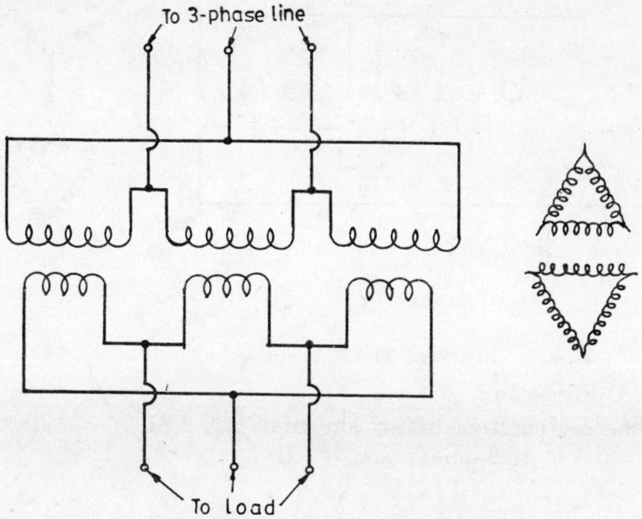

Fig. 2.79. The Δ-Δ connection of transformers.

Advantages :

1. The system voltages are more stable in relation to an unbalanced load.

2. If one transformer fails it may be switched out of the line and operation continued at a reduced power level. This is known as open-delta or *V-V* operation.

3. There is no distortion of flux, because the third harmonic component of magnetising current can flow in the delta connected primary windings without flowing in the line wires.

4. No difficulty is experienced due to unbalancing of loads on secondary side.

Disadvantages :

1. In comparison to *Y-Y* connections it requires more insulation.

2. The absence of star-point may be disadvantageous. If one line gets earthed due to fault, maximum voltage between windings and core will be full line voltage.

2.10.1.3. The Y-Δ Connection. The *Y*-Δ connection is shown in Fig. 2.80.

It is principally used where the voltage is to be stepped down, as for example, at the end of a transmission line. It is also employed in *moderately low-voltage distribution circuits for stepping down* from transmission voltages of 4000 or 8000 V to 230 (and 115 V).

- The *Y* connection takes advantage of the fact that one that leg of a *Y*, or the line-to-neutral voltage, is less than the line-to-line voltage by a $\sqrt{3}$ factor. This is especially important when the primary voltage is a few hundred thousand volts.
- The *Y*-Δ does have a phase shift between the primary and secondary voltages. This 30° phase shift means that a *Y*-Δ transformer bank *cannot* be paralleled with either a *Y-Y* or a Δ-Δ. The phasor voltage differences between the two systems would be around sin 30° = 0.5 times the secondary voltages. This would cause an excessive circulating current between transformer banks.

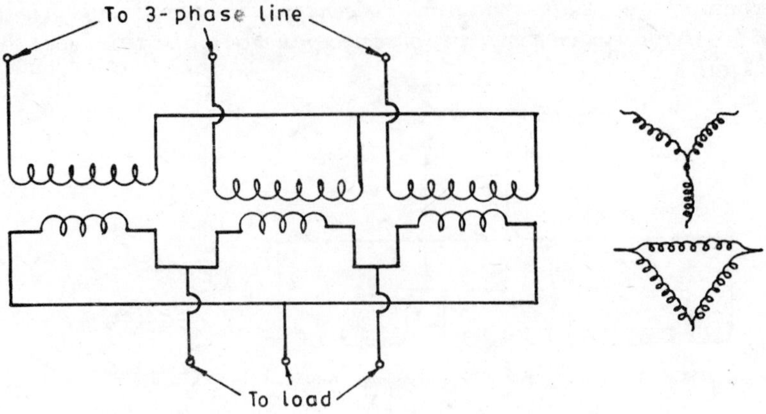

Fig. 2.80. The Y-Δ connection.

2.10.1.4. The Δ-Y Connection

- The three-phase Δ-Y connections are shown in Fig. 2.81.

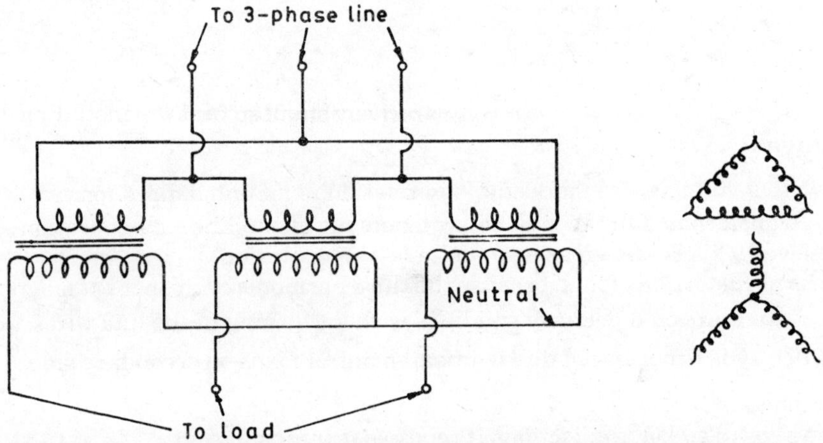

Fig. 2.81. The Δ-Y connection of transformers.

This type of connection is employed where it is necessary to *step up the voltage*, as for example, at the beginning of a high-tension transmission system.

- The ratio of secondary to primary voltage is $\sqrt{3}$ times the transformation ratio of each transformer.
- The neutral of the secondary is grounded for providing 3-phase 4-wire service. This connection is popular since it can be used to serve both the 3-phase power equipment and single-phase lighting circuit.
- This connection is not open to the objection of a floating neutral and voltage distortion because the existence of a Δ-connection allows a path for the third-harmonic currents. It would be observed that the primary and secondary line voltages and line currents are out of phase with each other by 30°. Because of this 30° shift, it is impossible to parallel such a bank with a Δ-Δ or Y-Y bank of transformers even though the voltage ratio are correctly adjusted.

2.10.1.5. The V-V (open-Δ) Connections. Fig. 2.82 shows the *V-V* connection.

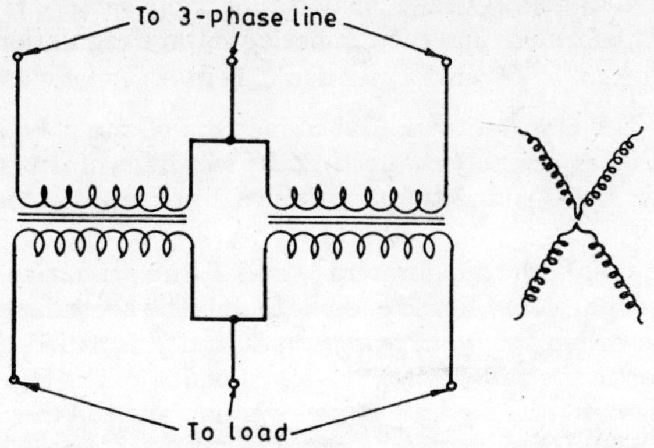

Fig. 2.82. The *V-V* (open Δ) connection of transformers.

If one of the transformers of a Δ-Δ bank is removed and a three-phase source is connected to the primaries (as shown in Fig. 2.82), three equal three-phase voltages will be measured at the secondary terminals at no-load. This method of transforming three-phase power, using two transformers, is called the open-delta, or *V-V* connection.

This type of connection is used in the following cases :

(*i*) When the three-phase load is comparatively small so that the installing does not warrant a three transformer bank.

(*ii*) When one of the transformers in a Δ-Δ bank fails, so that the service may be continued until the faulty transformer is repaired or a good one is substituted.

(*iii*) When it is anticipated that the future load will increase to warrant the chosing of the open Δ at some time later.

V-V connection has a number of *features* that are *advantageous* :

- Upon failure of the primary or secondary of one transformer of a complete Δ-Δ transformer circuit, the system reverts to a *V-V* circuit, so this is an *automatic standby*. The power-handling capacity of a *V-V* circuit is $\frac{1}{\sqrt{3}}$ times the capacity of a full Δ-Δ of the same transformers. This feature works both ways, so a circuit is sometimes installed as *V-V* with the understanding that its power handling may be multiplied by $\sqrt{3}$ by adding one more transformer.
- Open delta or *V-V* circuits do introduce some voltage unbalance due to the non-symmetry of the voltage regulation effects under load. However, the small degree of unbalance is not normally noticed by a motor load or other types of commercial load.

Disadvantages of V-V connection :

1. The secondary terminal voltages tend to become unbalanced to a great extent when the load is increased, this happens even when the load is perfectly balanced.

2. The average factor at which the *V*-bank operates is less than that of the load. This power factor is actually *86.6 per cent of the balanced load factor*. Another important point to note is that, except, for a balanced unity power factor load, the two transformers in the *V-V* bank operate at different power factors.

Uses of V-V connection :

(*i*) The *V-V* circuit is frequently used for two *auto-transformers.* Here advantage is taken of power handling of auto-transformers and their superior voltage regulation and efficiency.

(*ii*) Another major use of *V-V* transformer banks is in *A.C. motor starting.*

2.10.1.6. Scott or T-T Connection. The connection of one polyphase system into another polyphase system is possible by suitably connecting the windings of transformers. One of the early types that was used is Scott or *T-T* connection, by which a 2-phase system is available from a 3-phase system or *vice-versa.*

In Fig. 2.83 two single-phase transformers *M* and *T*, the primaries of which are connected to a 3-phase supply. The secondary of *M* forms one phase and the secondary of *T* the other phase of a true 2-phase system. *M* is called the *main transformer* and *T* is called the *teaser*. One end of the teaser primary is connected to the mid-point of the main primary. The two ends of the main primary are connected to two lines wires of a 3-phase, 3-wire system, and the third line wire is connected to a tapping *X* on the teaser primary.

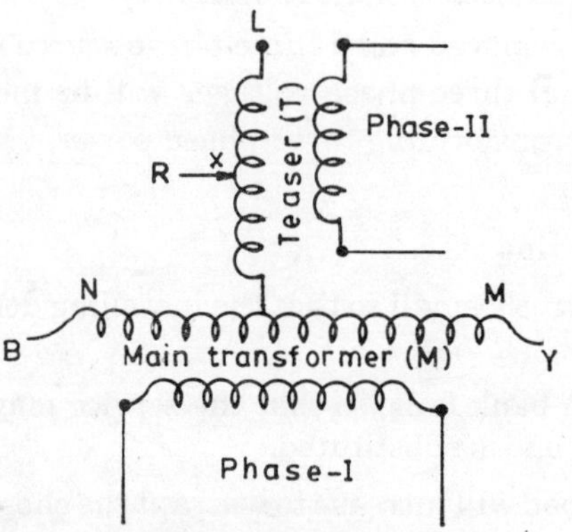

Fig. 2.83. Scott or *T-T* connection.

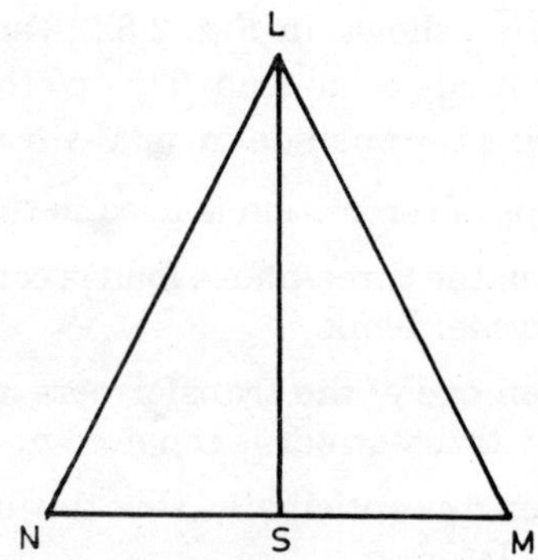

Fig. 2.84

If the supply voltages are assumed symmetrical, the triangle of voltages is *equilateral* as shown in Fig. 2.84. The vertical line $LS = \frac{\sqrt{3}}{2} NM$, so that it gives the relationship between the number of turns of two primaries, as

$$\frac{\text{Number of turns for the main transformer } M}{\text{Number of turns for the teaser } T} = \frac{100}{86.6}$$

Hence, if two identical single-phase transformers are to be used for Scott-connection, the primary one must have a tapping point brought out from its *mid-point of the primary,* and the secondary transformer must have a tapping *X* brought out from *86.6% of its primary turns.*

As the two-phase side is asymmetrical, there cannot be perfect balanced conditions on the three-phase side. However, neglecting the impedance of the windings, it can be shown that if load on the 2-phase side is balanced the 3-phase side is also balanced (Figs. 2.85, 2.86 and 2.87).

- Fig. 2.85 shows the connection diagram. The 2-phase side has equal load impedances.
- In Fig. 2.86, the vector diagram is drawn for *unity power factor load.* The secondary currents are not shown. *On* is the *teaser* primary current, and since it flows at the mid point of the *main transformer,* it divides itself equally in two halves. The resultant current in the lines *B* and *Y* is the vector sum of the balancing amperes of the main primary and

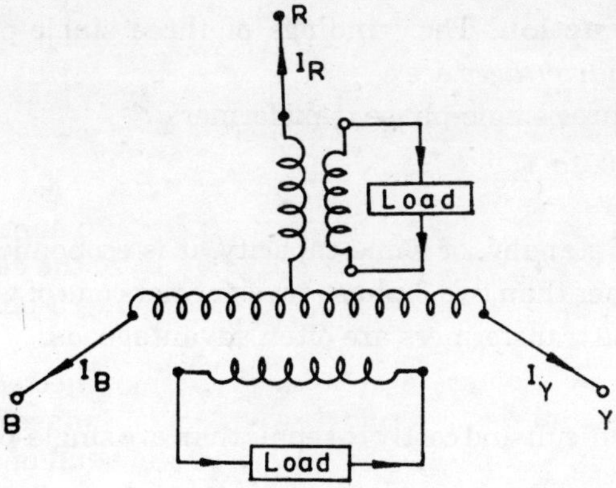

Fig. 2.85

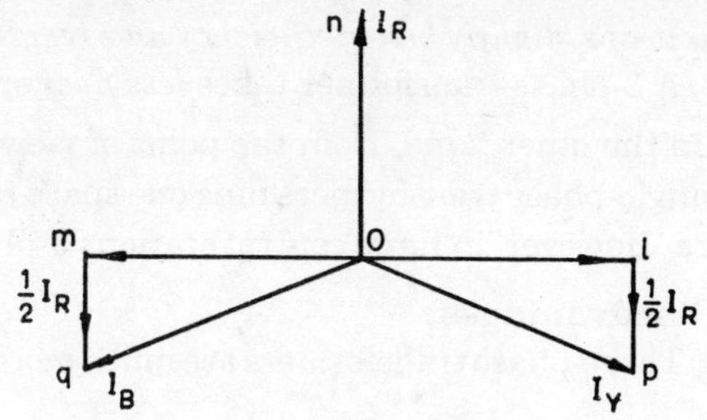

Fig. 2.86

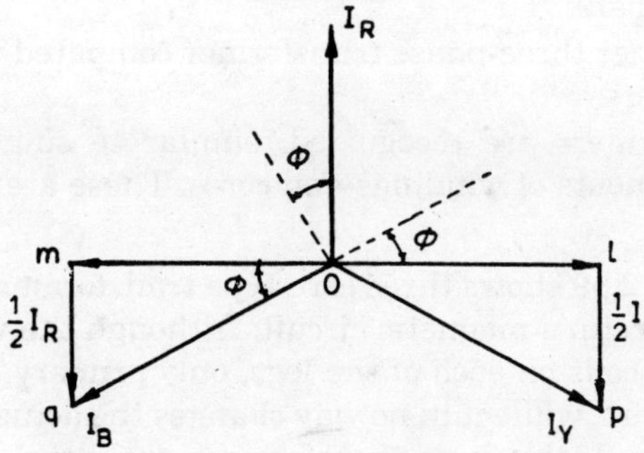

Fig. 2.87

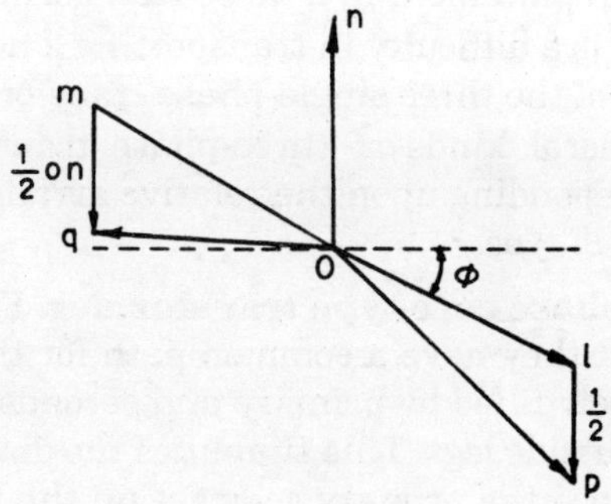

Fig. 2.88

$\frac{1}{2}$ the current of the teaser primary. Hence *op* and *oq* are the line currents. Below is the proof that *ol* = *om* = *on*, and that each one of these currents is at 120° from the other two.

Proof. Let us assume that the ratio of turns of the *main* transformer be 100 : 100 (primary to secondary) and that of teaser 86.6 : 100. If the secondary load-currents are 100 A at unity power factor, the primary balancing amperes of the main transformer will be 100 and lie along *ml*. The *teaser* balancing amperes will be $\frac{100}{86.6} \times 100 = 115.4$ and will lie along *on*. The *teaser* primary current splits into half at 0 and flows in the two halves of the main primary. Hence the line current in *Y* and *B* is given by $\left[(100)^2 + \left(\frac{115.4}{2}\right)^2\right]$ since these two components are at right angles to each other. Now

$$\left[(100)^2 + \left(\frac{115.4}{2}\right)^2\right]^{1/2} = 115.4 \text{ A}$$ and the angle between *ol* and *op* is ϕ such that $\phi = \tan^{-1}\frac{57.7}{100} = 30°$ (Fig. 2.86). Hence *op* lags *on* by 90° + 30° = 120° Q.E.D.

- In Fig. 2.87 the vector diagram is drawn for a condition when the *load power factor is* cos ϕ. The same proof can be applied to this case as was done for the case of unity power factor. If the loads are unbalanced and have different power factors, such as cos ϕ_1, and cos ϕ_2, primary side becomes unbalanced. The vector diagram of Fig. 2.88 shows the extent of unbalance when the *teaser* load is of unity power factor and the *main* transformer load has a power factor of cos ϕ (lag).

2.10.2. Three-phase Transformer Construction. The windings of three single-phase transformers can be wound on a common core. The *advantages* are :

1. One 3-phase transformer is *cheaper* than three single-phase transformers.
2. It has *slightly better efficiency and regulation.*
3. A 3-phase transformer takes *less floor space.*

On the other hand, from the point of view of standby, or same capacity, it is economical to have 3 single-phase transformers *plus* one spare rather than two 3-phase transformers one of which is a spare. However, in large central stations 3-phase transformers are often advantageous.

Disadvantages :

1. Three-phase transformers are much more difficult and costly to repair than are single-phase units.

2. When failure does occur and it becomes necessary to substitute a replacement unit to maintain service, the cost of spare is much greater than it would be were a single-phase transformer to be used as a replacement in a three-transformer bank.

3. There is a difficulty in transporting a heavier three-phase transformer compared with the moving of each of the three single-phase transformers.

Two general kinds of three-phase transformers are recognized, similar to single-phase transformers,depending upon the relative arrangements of windings and cores. These are the *core type* and the *shell type*.

Three-phase core type transformer. Fig. 2.89 shows three core-type transformers placed together so that they have a common path for the return magnetic circuit. Although the windows should be entirely filled by primary and secondary coils on each of the legs, only primary coils are shown on the outside legs. This simplifies the diagram, while it in no way changes the actual theory that follows, since the primary coils set up the flux. If the three transformers are identical in all respects, a balanced three-phase system of voltages will produce three fluxes in the cores which have the same maximum value, but differ in time phase by 120°. In the common leg of the three cores of Fig. 2.90, the three fluxes add, and the net flux is therefore always zero. The common leg may then be eliminated, with a subsequent saving in core material and size of transformer. A single polyphase transformer would be of impractical construction if it were the same as Fig. 2.89 with the centre leg omitted. Instead, the core-type polyphase transformer is manufactured so that it looks like that shown in Fig. 2.90. Actually, what we have done is, that axes of three coils have been moved into one plane. This causes the magnetic reluctance of coil M to differ somewhat from that of L and N. This

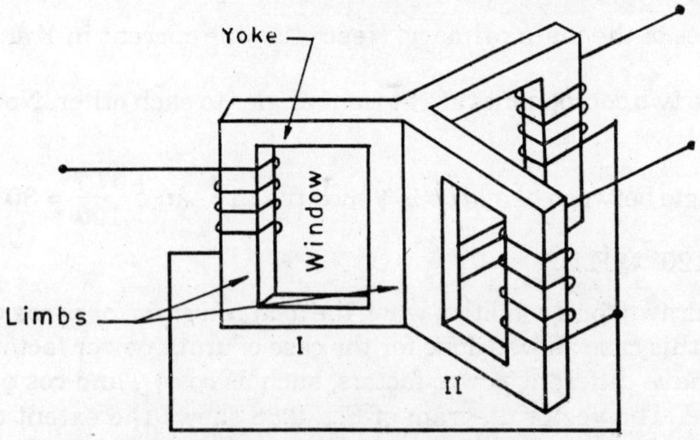

Fig. 2.89. Core-type transformers for polyphase transformation.

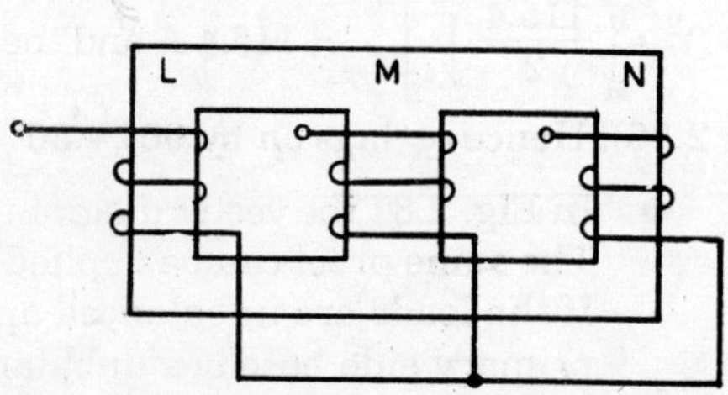

Fig. 2.90. Core-type three-phase transformer.

produces a slight unbalance in the three magnetizing currents, but the effect is not serious, especially under load.

Three-phase shell type transformer :

- In Fig. 2.91 are shown three shell-type transformers stacked one above the other, with only the primary coils shown.

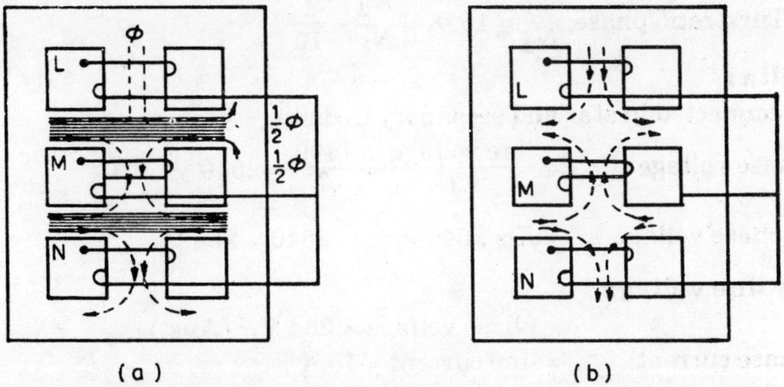

Fig. 2.91. Shell-type three-phase transformer (*a*) coil *M* wound in the same direction, (*b*) coil *M* wound in opposite direction.

- In Fig. 2.91 (*a*) the three coils are wound in the same direction. The flux in the common core area between adjacent phases such as the shaded portions is thus equal to the *difference* of two of the phase fluxes. Since the fluxes are 120° apart in time, this mutual flux is equal to $\sqrt{3} \times (\frac{1}{2}\phi)$, or 0.866 of the flux in the centre leg. If the same flux density is maintained throughout, less iron is required in the common legs. Now, however if the centre coil, phase *M* is reversed, the flux in the common core is equal to the *sum* of the fluxes of two adjacent phases. This is shown in Fig. 2.91 (*b*). As in any three-phase system, the sum of two fluxes is equal to either flux alone, which in this case is 0.5 ϕ. This represents a further saving in iron, and for this reason shell-type three-phase transformers are *usually wound with centre coil opposing the two outside ones*.

2.10.3. Parallel Operation of 3-phase Transformers. The conditions for paralleling 3-phase transformers are same as that required for parallel operation of single-phase transformers with the following *additions* :

(i) The voltage ratio must refer to the terminal voltage of primary and secondary.

(ii) The phase displacement between primary and secondary voltages must be the same for all transformers which are to be paralleled.

(iii) The phase sequence must be the same.

Following points are worthnoting while dealing with 3-phase transformers :

- *The calculations are made for one phase only*. The value of equivalent impedance used is the equivalent impedance per phase *referred to secondary*.
- When the impedances of primary and secondary windings are given separately then primary impedance *must be transferred to secondary by multiplying it with (transformation ratio)^2.*
- In case of Y/Δ or Δ/Y transformers the voltage ratios as given in the questions, *refer to terminal voltages and are quite difference from turn ratio.*

THREE-PHASE TRANSFORMERS

Star/Delta Connections

Example 2.87. *A 3-phase step down transformer takes 15 A when connected to 4400 V mains. The turn ratio per phase is 10. Neglecting losses find the secondary line voltage, line current and output if the transformer windings are connected (i) star/delta, and (ii) delta/star.*

Solution. Turn ratio/phase, $\frac{N_1}{N_2} = 10$, $K = \frac{N_2}{N_1} = \frac{1}{10}$

(*i*) **Star/Delta :**

Primary is connected in star and secondary in delta.

Primary phase voltage $= \frac{\text{line voltage}}{\sqrt{3}} = \frac{4400}{\sqrt{3}} = 2540$ V

Secondary phase voltage $= K \times 2540 = \frac{1}{10} \times 2540 = 254$ V

Secondary line voltage

= phase voltage = **254 V. (Ans.)**

Primary phase current = line current = 15 A

Secondary phase current $= 15 \times \frac{1}{K} = 15 \times 10 = 150$ A

Secondary line current

$= \sqrt{3} \times$ phase current $= \sqrt{3} \times 150 =$ **259.8 A. (Ans.)**

Output in kVA

$= \sqrt{3} \times 254 \times 2528 \times 10^{-3} =$ **114.29 kVA. (Ans.)**

(*ii*) **Delta/Star :**

Primary is connected in delta and secondary in star

Primary phase voltage = line voltage = 4400 V

Secondary phase voltage $= K \times 4400 = \frac{1}{10} \times 4400 = 440$ V

Secondary line voltage $= \sqrt{3} \times$ phase voltage $= \sqrt{3} \times 440 =$ **762 V. (Ans.)**

Primary phase current $= \frac{\text{line current}}{\sqrt{3}} = \frac{15}{\sqrt{3}} = 8.66$ A

Secondary phase current $= \frac{8.66}{K} = 8.66 \times 10 = 86.6$ A

∴ **Secondary line current** = phase current

= **86.6 A. (Ans.)**

Output in kVA $= \sqrt{3} \times 762 \times 86.6 \times 10^{-3} =$ **114.29 kVA. (Ans.)**

Example 2.88. *A 3-phase transformer has a delta-connected primary and is supplied at 11000 V. The terminal voltage (line voltage) of the star-connected secondary at 0.8 power factor lagging is 400 V. The effective resistance and reactance drops are 1.5% and 6% respectively. Determine the approximate transformation ratio.*

Solution. Supply voltage = 11000 V

Line voltage of star-connected secondary = 400 V

Effective resistance = 1.5%

Effective reactance = 6%

Approximate transformation ratio :

Primary phase voltage = line voltage = 11000 V

Secondary terminal voltage/phase

$$V_2 = \frac{440}{\sqrt{3}} = 254 \text{ V}$$

Let V_{02} be the no-load secondary voltage/phase

Then % regulation $= \frac{V_{02} - V_2}{V_{02}} \times 100$

Also approximate % regulation

$$= \% \text{ resistance} \times \cos \phi \pm \% \text{ reactance} \times \sin \phi$$

$$\cos \phi = 0.8 \text{ lagging}, \sin \phi = 0.6$$

∴ Approximate % regulation

$$= 1.5 \times 0.8 + 6 \times 0.6 = 1.2 + 3.6 = 4.8$$

Then approximately,

% regulation $= \frac{V_{02} - V_2}{V_{02}} \times 100 = 4.8$

$$V_{02} - V_2 = 0.048\, V_{02}$$

$$0.952\, V_{02} = V_2$$

$$V_{02} = \frac{V_2}{0.952} = \frac{254}{0.952} = 266.8 \text{ V}$$

Approximate transformation ratio

$$= \frac{V_{02}/\text{phase}}{V_1/\text{phase}} = \frac{266.8}{11000} = \mathbf{0.02425.}\ \textbf{(Ans.)}$$

Example 2.89. *A 3-phase 1000 kVA. 6600/1100 V transformer is delta-connected on the primary and star-connected on the secondary. The primary resistance/phase is 1.8 Ω and secondary resistance/phase is 0.025 Ω. Determine the efficiency on full-load at :*

(i) Unity power factor ; and

(ii) 0.8 power factor lagging if the iron loss is 15 kW.

Solution. Rating of three-phase transformer = 1000 kVA

Primary resistance/phase, $R_1 = 1.8\ \Omega$

Secondary resistance/phase, $R_2 = 0.025\ \Omega$

Iron loss, $P_i = 15$ kW

Primary phase voltage = 6600 V

Secondary phase voltage $= \frac{1100}{\sqrt{3}} = 635$ V

Transformation ratio, $K = \frac{635}{6600} = 0.096$

Equivalent resistance referred to secondary

$$R_{02} = R_2 + K^2 R_1 = 0.025 + (0.096)^2 \times 1.8 = 0.0416\ \Omega/\text{phase}$$

Secondary phase current = line current $= \frac{1000 \times 10^3}{\sqrt{3} \times 1100} = 524.8$ A

Total full-load copper loss,

$$P_c = 3 I_{ph}^2 R_{ph} = 3 \times (524.8)^2 \times 0.0416 = 34371 \text{ W} = 34.37 \text{ kW}$$

Total loss $= P_i + P_c = 15 + 34.37 = 49.37$ kW.

(*i*) **At unity p.f. :**

Output on full-load $= \text{kVA} \times \cos\phi = 1000 \times 1 = 1000 \text{ kW}$

Percentage efficiency $= \dfrac{\text{output}}{\text{output + losses}} \times 100 = \dfrac{1000}{1000 + 49.37} \times 100 =$ **95.29%. (Ans.)**

(*ii*) **At 0.8 p.f. :**

Output on full-load $= 1000 \times 0.8 = 800 \text{ kW}$

Percentage efficiency $= \dfrac{\text{output}}{\text{output + losses}} \times 100 = \dfrac{800}{800 + 49.37} \times 100 =$ **94.2%. (Ans.)**

Example 2.90. *The resistance/phase and leakage reactance/phase of the primary winding of a 3-phase, 1200 kVA, 11000/2200 V, 50 Hz, step down transformer are 0.8 Ω and 4.8 Ω and the corresponding values for the secondary winding are 0.01 Ω and 0.06 Ω respectively. If the primary winding is connected in delta and secondary in star, calculate :*

(*i*) *The voltage to be applied at normal frequency to the primary to circulate full-load current through it when the secondary is short-circuited.*

(*ii*) *The power input under this condition.*

Solution. Resistance/phase of primary winding, $R = 0.8 \ \Omega$

Leakage reactance/phase of primary winding, $X_1 = 4.8 \ \Omega$

Resistance/phase of secondary winding, $R_2 = 0.01 \ \Omega$

Leakage reactance/phase of secondary winding, $X_2 = 0.06 \ \Omega$

(*i*) Primary phase voltage = line voltage = 11000 V = 11 kV

Secondary phase voltage $= \dfrac{2.2}{\sqrt{3}}$ kV

Transformation ratio, $K = \dfrac{2.2}{\sqrt{3} \times 11} = \dfrac{1}{5\sqrt{3}} = 0.115$

Equivalent resistance/phase referred to primary,

$$R_{01} = R_1 + R_2' = R_1 + \frac{R_2}{K^2} = 0.8 + \frac{0.01}{(0.115)^2} = 1.55 \ \Omega$$

Equivalent reactance referred to primary,

$$X_{01} = X_1 + X_2' = X_1 + \frac{X_2}{K^2} = 4.8 + \frac{0.06}{(0.115)^2} = 9.3 \ \Omega$$

Equivalent impedance/phase referred to primary,

$$Z_{01} = \sqrt{R_{01}^2 + X_{01}^2} = \sqrt{(1.55)^2 + (9.3)^2} = 9.43 \ \Omega$$

Primary line current on full-load

$$= \frac{1200 \times 10^3}{\sqrt{3} \times 11000} = 62.98 \text{ A}$$

Primary phase current on full-load

$$= \frac{62.98}{\sqrt{3}} = 36.36 \text{ A}$$

Voltage to be applied to the primary to circulate full-load current taken when secondary is short-circuited

$$= I_{SC} Z_{01} = 36.36 \times 9.43 \text{ V/phase} = 342.87 \text{ V/phase}$$

Voltage (line value) to be applied to primary

= phase voltage (since delta-connected)

= **342.87 V/phase. (Ans.)**

(*ii*) **Power input** $= 3I_{sc}^2 R_{01} = 3 \times 36.36^2 \times 1.55$

$= \mathbf{6147\ W}$ or $\mathbf{6.147\ kW}$. **(Ans.)**

Open Delta Connection or V-V Connection

Example 2.91. *Three single-phase transformers are mesh-connected and are supplying a line current of 200 A. Determine the current in each transformer. If one unit is faulty and removed, what will be the capacity of the new set, if the current in each transformer remains the same ?*

Solution. Since the transformers are mesh-connected, *current in each transformer*

$$= \frac{\text{line current}}{\sqrt{3}} = \frac{200}{\sqrt{3}} = \mathbf{115.47\ A.}\ \textbf{(Ans.)}$$

In delta/mesh connection let V_2 be the terminal voltage and I be the line current.

Then Δ capacity $= \sqrt{3}\, V_2 I$.

In V connection (when one transformer is faulty) the line current will be the same as phase current.

Hence line current in V connection

$$= \frac{I}{\sqrt{3}}$$

$$V \text{ capacity} = \sqrt{3}\, V_2 \frac{I}{\sqrt{3}} = V_2 I$$

$$\frac{V \text{ capacity}}{\Delta \text{ capacity}} = \frac{V_2 I}{\sqrt{3}\, V_2 I} = \frac{1}{\sqrt{3}} = 0.578 = 57.8\%$$

V capacity = **57.8%** Δ capacity. **(Ans.)**

Example 2.92. *Three single-phase transformers are connected in mesh. If one transformer is found faulty and removed what will be the reduction in rating of each of the other transformers ? What must be rating of each transformer in a V-V connection to supply a 3-phase balanced load of 200 kVA ? If a third similar transformer is included what will be the rated capacity of the set ?*

Solution. Let the rating of each transformer be Q_1.

Total rating of the three transformers $= 3Q_1$

Total rating of the two transformer in V-V connection $= 2Q_1$

But we know that $\dfrac{V \text{ capacity}}{\Delta \text{ capacity}} = 0.578$

Δ capacity $= 3Q_1$

$\therefore$ V capacity $= 0.578 \times 3Q_1 = 1.734Q_1$

Reduction in rating for the V-V set

$= 2Q_1 - 1.734Q_1 = 0.266Q_1$

Reduction in rating of each transformer

$$= \frac{0.266Q_1}{2} = 0.133Q_1 = 0.133 \text{ times its rating}$$

= **13.3% of its rating. (Ans.)**

Total load = 200 kVA

Load to be shared by one transformer $= \dfrac{200}{2} = 100$ kVA

Let Q_2 be the rating of each transformer. Then

$$(Q_2 - 0.133Q_2) = 100 \text{ kVA}$$

$$0.867Q_2 = 100 \text{ kVA}$$

$$\therefore \quad Q_2 = \frac{100}{0.867} \text{ kVA} = 115.34 \text{ kVA}$$

Rating of each transformer in V-V connection

= **115.34 kVA. (Ans.)**

If a third transformer of 115.34 kVA is included,

Rating of the new set (Δ-Δ set)

= 3 × 115.34 = **346 kVA. (Ans.)**

Example 2.93. *Three single-phase transformers, connected in Δ-Δ, supply a balanced 3-phase load of 1500 kW at 4400 V at 0.8 power factor lagging. The transformers are supplied from 3-phase mains at 11000 V. Find the currents in the windings of each transformer. If one transformer is found faulty and the other two are connected in V-V determine the currents in the windings of each transformer and its power factor.*

Solution. Δ-Δ connection

$$\sqrt{3} V_2 I_2 \cos \phi = 1500 \times 10^3 \text{ W}$$

where, V_2 = secondary line voltage = 4400 V,

I_2 = secondary line current,

$\cos \phi$ = power factor of current = 0.8,

and, load delivered = 1500×10^3 W.

$$\therefore \quad \sqrt{3} \times 4400 \times I_2 \times 0.8 = 1500 \times 10^3$$

$$I_2 = \frac{1500 \times 10^3}{\sqrt{3} \times 4400 \times 0.8} = 246 \text{ A}$$

Secondary phase current $= \dfrac{\text{line current}}{\sqrt{3}} = \dfrac{246}{\sqrt{3}} = 142$ A

∴ **Current in the secondary winding = 142 A. (Ans.)**

Primary phase voltage = 11000 V

Secondary phase voltage = 4400 V

Transformation ratio, $K = \dfrac{4400}{11000} = 0.4$

Current in the primary winding

= $K \times 142 = 0.4 \times 142 =$ **56.8 A. (Ans.)**

V-V connection. In *V-V* connection, secondary line current is the same as secondary phase current

$$\sqrt{3} V_2 I_2 \cos \phi = 1500 \times 10^3$$

$$I_2 = \frac{1500 \times 10^3}{\sqrt{3} \times 4400 \times 0.8} = 246 \text{ A}$$

Secondary phase current = line current = 246 A

Current in the secondary winding = 246 A. (Ans.)

Current in the primary winding

= $K \times 246 = 0.4 \times 246 =$ **98.4 A. (Ans.)**

Power factor of transformer in V-V connection

= 0.866 times the power factor in Δ-Δ connection

= 0.866 × 0.8 = **0.6928 lagging. (Ans.)**

Scott or T-T Connection

Example 2.94. *Two single phase electric furnaces A and B are supplied at 220 V from a 3-phase 1100 V supply by means of a Scott-connected transformer combination. If the total output is 600 kW at 0.6 power factor lagging determine the currents in the winding and transformation ratio of each transformer.*

Solution. ***Main transformer* :**

Primary volts = 1100 V

Secondary volts = 220 V

Transformation ratio,

$$K = \frac{N_2}{N_1} = \frac{V_2}{V_1} = \frac{220}{1100} = \frac{1}{5}. \quad \textbf{(Ans.)}$$

Output of main transformer $= \frac{600}{2} = 300$ kW

Current in secondary winding,

$$I_2 = \frac{\text{output in watts}}{V_2 \cos \phi} = \frac{300 \times 1000}{220 \times 0.8} = \textbf{1704.5 A.} \quad \textbf{(Ans.)}$$

***Teaser Transformer* :**

Secondary voltage = 220 V

Primary voltage $= \frac{\sqrt{3}}{2} \times 1100 = 952.6$ V

Transformation ratio,

$$K = \frac{N_2}{N_1} = \frac{V_2}{V_1} = \frac{220}{952.6} = 0.231 = \frac{\mathbf{1}}{\mathbf{4.33}}. \quad \textbf{(Ans.)}$$

Output of teaser transformer $= \frac{600}{2} = 300$ kW

Current in secondary winding

$$= \frac{300 \times 10^3}{220 \times 0.8} = \textbf{1704.5 A.} \quad \textbf{(Ans.)}$$

Current in the primary winding

$$= K \times 1704.5 = 0.231 \times 1704.5 = \textbf{393.7 A.} \quad \textbf{(Ans.)}$$

☞**Example 2.95.** *A Scott-connected transformer set supplies two electric furnaces A and B at 110 V each from a 3-phase 4400 V system. If the furnace connected to teaser transformer secondary takes 270 kW and other one connected to main transformer secondary takes 450 kW at unity p.f. find the line current taken from the 3-phase mains.*

Solution. Refer Figs. 2.92 and 2.93.

Unity p.f. :

Load on teaser transformer = 270 kW

Teaser transformer secondary current $= I_{2A} = \frac{270 \times 1000}{1 \times 110} = 2454.5$ A

Transformation ratio of teaser transformer

$$= \frac{N_2}{\left(\frac{\sqrt{3}}{2} N_1\right)} = 1.155 \frac{N_2}{N_1} = 1.155 \text{ K}$$

where $K = \frac{N_2}{N_1} = \frac{V_2}{V_1} = \frac{110}{4400} = \frac{1}{40}$

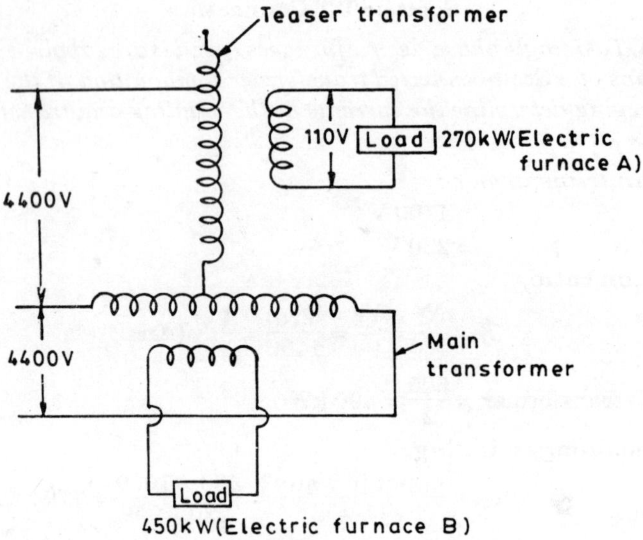

Fig. 2.92

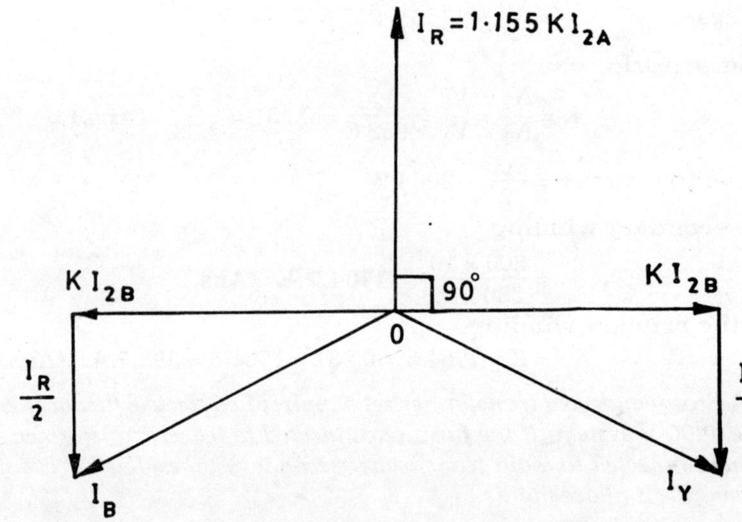

Fig. 2.93

Teaser primary current, $I_R = 1.155KI_{2A}$

$$= 1.155 \times \frac{1}{40} \times 2454.5 = 70.87 \text{ A}$$

Primary current of main transformer,

$$I_B \text{ or } I_Y = \text{Vector sum of } KI_{2B} \text{ and } \frac{1.155KI_{2A}}{2}$$

where $I_{2B} = \dfrac{450 \times 10^3}{110 \times 1}$

$$KI_{2B} = \frac{1}{40} \times \frac{450 \times 10^3}{110 \times 1} = 102.27 \text{ A}$$

$$\frac{I_R}{2} = \frac{1.155KI_{2A}}{2} = \frac{70.87}{2} = 35.43 \text{ A}$$

Main primary current, I_B or $I_Y = \sqrt{(KI_{2B})^2 + \left(\frac{I_R}{2}\right)^2} = \sqrt{(102.27)^2 + (35.43)^2} = 108.23$ A

Hence, **current in line, R = 70.87 A. (Ans.)**
Current in line, Y = 108.23 A. (Ans.)
Current in line, B = 108.23 A. (Ans.)

Example 2.96. *A Scott-connected transformer set supplies two single-phase furnaces A and B each at 120 V from a 3-phase 6600 V system. If the furnace connected to teaser transformer takes 300 kW at unity power factor and the other one connected to main transformer takes 495 kW at 0.8 power factor lagging determine the line currents taken from the 3-phase mains. Neglect losses.*

Solution. Load on teaser transformer = 300 kW
Secondary current of teaser transformer,

$$I_{2A} = \frac{300 \times 1000}{V_2 \times \cos\phi} = \frac{300 \times 1000}{120 \times 1} = 2500 \text{ A}$$

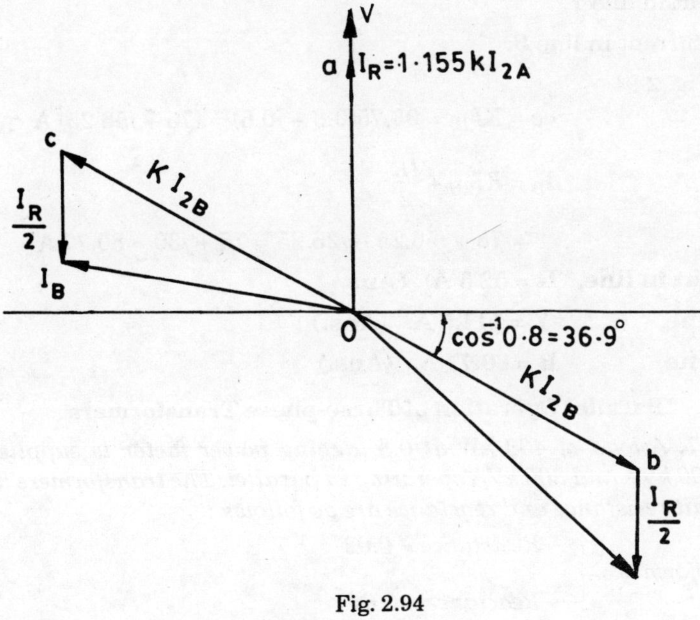

Fig. 2.94

Transformation ratio of teaser transformer

$$= \frac{N_2}{\left(\frac{\sqrt{3}}{2} N_1\right)} = 1.155 \text{ K}$$

where $K = \frac{N_2}{N_1} = \frac{V_2}{V_1} = \frac{120}{6600} = \frac{1}{55}$

Teaser primary current, $I_R = 1.155KI_{2A} = 1.155 \times \frac{1}{55} \times 2500 = 52.5$ A

$$\frac{I_R}{2} = \frac{52.5}{2} = 26..25 \text{ A}$$

Load on main transformer secondary = 495 kW

$$V_2 = 120 \text{ V}, \cos \phi = 0.8$$

Current in main transformer secondary

$$= \frac{495 \times 1000}{V_2 \cos \phi} = \frac{495 \times 1000}{120 \times 0.8} = 5156.25 \text{ A}$$

$$KI_{2B} = \frac{1}{55} \times 5156.25 = 93.75 \text{ A}$$

Referring to Fig. 2.94

$$ob = KI_{2B} = 93.75(0.8 - j0.6) = (75 - j56.25) \text{ A}$$

$$\frac{I_R}{2} = -j26.25 \text{ A}$$

$$I_Y = KI_{2B} + \frac{I_R}{2} = (75 - j56.25 - j26.25) = (75 - j82.5) = 111.5 \text{ A}$$

where I_Y is the current in line Y.

Let I_B be the current in line B.

Referring to Fig. 2.94

$$oc = KI_{2B} = 93.75(0.8 + j0.6) = (75 + j56.25) \text{ A}$$

$$I_B = KI_{2B} + \frac{I_R}{2}$$

$$= 75 + j56.25 - j26.25 = 75 + j30 = 80.77 \text{ A}$$

Hence, **current in line, R = 52.5 A. (Ans.)**

Current in line, Y = 111.5 A. (Ans.)

Current in line, B = 80.77 A. (Ans.)

Parallel Operation of Three-phase Transformers

Example 2.97. *A load of 400 kW at 0.8 lagging power factor is supplied by two 3-phase transformers rated 200 kVA and 300 kVA operating in parallel. The transformers have the same turn ratio and their per unit resistance and reactance are as follows :*

200 kVA transformer — *Resistance = 0.02*, *Reactance = 0.04*

300 kVA transformer — *Resistance = 0.01*, *Reactance = 0.05*

Solution. Load P in kVA $= \frac{\text{load in kW}}{\text{power factor}} = \frac{400}{0.8} = 500$ kVA

$\therefore$ $P = 500 \angle -36.9°$ kVA. [$\because$ $\cos \phi = 0.8$ lagging]

On the basis of 300 kVA :

Percentage impedance of 200 kVA transformer,

$$Z_A = \frac{300}{200}(2 + j4) = (3 + j6) = 6.7 \angle 63.4°$$

Percentage impedance of 300 kVA transformer,

$$Z_B = \frac{300}{300}(1 + j5) = 1 + j5 = 5.1 \angle 78.7°$$

$$Z_A + Z_B = (3 + j6) + (1 + j5) = (4 + j11) = 11.7 \angle 70°$$

Load shared by 200 kVA transformer,

$$P_A = P \times \frac{Z_B}{Z_A + Z_B} = \frac{500 \angle -36.9° \times 5.1 \angle 78.7°}{11.7 \angle 70°} = 217.9 \angle -28.2°$$

= **217.9 kVA at p.f. 0.881 (lag). (Ans.)**

Load shared by 300 kVA transformer,

$$P_B = P \times \frac{Z_A}{Z_A + Z_B}$$

$$= \frac{500 \angle -36.9° \times 6.7 \angle 63.4°}{11.7 \angle 70°} = 286.3 \angle -43.5°$$

= **286.3 kVA at p.f. 0.725 (lag). (Ans.)**

☞**Example 2.98.** *The following data pertain to two transformers A and B connected in parallel to a load of* $(2 + j1.5)\ \Omega$.

Transformer A :

Impedance in secondary, $Z_A = (0.15 + j0.5)\ \Omega$

Induced e.m.f., $E_A = 207 \angle 0°$ *volts*

Transformer B :

Impedance in secondary, $Z_B = (0.1 + j0.6)\ \Omega$

Induced e.m.f., $E_B = 205 \angle 0°$ *volts*

Find the power output and power factor of each transformer.

Solution. Equivalent impedance of transformer *A*,

$$Z_A = (0.15 + j0.5)\ \Omega$$

Equivalent impedance of transformer *B*,

$$Z_B = (0.1 + j0.6)\ \Omega$$

Load impedance, $Z_B = (2 + j1.5)\ \Omega$

Induced e.m.f. of transformer *A*,

$$E_A = 207 \angle 0° = (207 + j0)\text{ V}$$

Induced e.m.f. transformer *B*,

$$E_B = 205 \angle 0° = (205 + j0)\text{ V}$$

Let I_A = current supplied by transformer A,

I_B = current supplied by transformer B, and

V_2 = terminal voltage.

Now total load current, $I = I_A + I_B$

Voltage across load, $V_2 = IZ_L = (2 + j1.5)(I_A + I_B)$

Since induced e.m.f. in transformer *A* is given as,

$$E_A = V_2 + I_A Z_A$$

$$\therefore \qquad (207 + j0) = (2 + j1.5)(I_A + I_B) + (0.15 + j0.5)I_A$$

or $$(2.15 + j2.0)I_A + (2 + j1.5)I_B = 207 + j0 \qquad ...(i)$$

Similarly $$E_B = V_2 + I_B Z_B$$

or $$(205 + j0) = (2 + j1.5)(I_A + I_B) + (0.15 + j0.5)I_A$$

or $$(2 + j1.5)\, I_A + (2.1 + j2.1)I_B = 205 + j0 \qquad ...(ii)$$

Solving expressions (*i*) and (*ii*), we get

$$I_A = \frac{(2.1 + j2.1)(207 + j0) - (2 + j1.5)(205 + j0)}{(2.15 + j2.0)(2.1 + j2.1) - (2 + j1.5)(2 + j1.5)}$$

$$= \frac{24.7 + j127.2}{-1.435 + j2.715} = \frac{129.7 \angle 79°}{3.07 \angle 117.9°}$$

$$= 42.26 \angle -38.9° = (32.89 - j26.55) \text{ A}$$

$$I_B = \frac{(207 + j0)(2 + j1.5) - (205 + j0)(2.15 + j2.0)}{(2 + j1.5)(2 + j1.5) - (2.1 + j2.1)(2.15 + j2.0)}$$

$$= \frac{26.75 + j99.5}{-1.435 + j2.715} = \frac{103 \angle 75°}{3.07 \angle 117.9°}$$

$$= 33.56 \angle -42.9° = (24.58 - j22.84) \text{ A}$$

Terminal voltage, $$V_2 = (I_A + I_B)Z_L$$

$$= (32.89 - j26.55 + 24.58 - j22.84)(2 + j1.5)$$

$$= (57.47 - j49.39)(2 + j1.5) = 189 - j12.58 = 189.4 \angle -3.9° \text{ V}$$

Transformer A :

Power factor angle, $\phi_A = -3.9° - (-38.9°) = 35°$

Power factor, $\cos \phi_A = \cos 35° =$ **0.8192 (lag). (Ans.)**

Power output, $P_A = \dfrac{V_2 I_A \cos \phi_A}{1000} \text{ kW} = \dfrac{189.4 \times 42.26 \times 0.8192}{1000} =$ **6.56 kW. (Ans.)**

Transformer B :

Power factor angle, $\phi_B = -3.9° - (-42.9°) = 39°$

Power factor, $\cos \phi_B = \cos 39° =$ **0.7771 (lag). (Ans.)**

Power output, $P_B = \dfrac{V_2 I_B \cos \phi_B}{1000} \text{ kW}$

$= \dfrac{189.4 \times 33.56 \times 0.7771}{1000} \text{ kW} =$ **4.94 kW. (Ans.)**

2.11. INSTRUMENT TRANSFORMERS

It is not practicable to connect instruments and meters directly to the lines in high voltage circuits. Instead instrument transformers are used. The following are the *two basic advantages* inherent in this method :

(*i*) *Standard rated instruments may be used.*

(*ii*) Operating personnel coming in contact with the instruments are *not subjected to high voltage and current of the lines*, and so there is less danger to them. Even with a low-voltage system, instrument transformers are used for measuring large currents, so that heavy leads to the instrument panel and to the ammeter and other current terminals are avoided.

The principle of the instrument transformer is fundamentally the same as that of the power transformer. The instrument transformers are classified as follows :

1. Potential transformers.
2. Current transformers.

2.11.1. Potential Transformers (P.T.)

- A potential transformer is a *step down transformer* used along with a low range voltmeter for measuring a high voltage. The primary is connected across the high voltage supply and the secondary to the voltmeter or potential coil of the wattmeter. Since the voltmeter (or potential coil) impedance is very high, the secondary current is very small and the potential transformer behaves as an ordinary two winding transformer operating on no-load. Fig. 2.95 shows a potential transformer used to measure the voltage of a circuit. It may be noted that the *secondary is grounded*. This is done so that *if the insulation breaks down, the high voltage does not endanger personnel who may be reading the meters.*

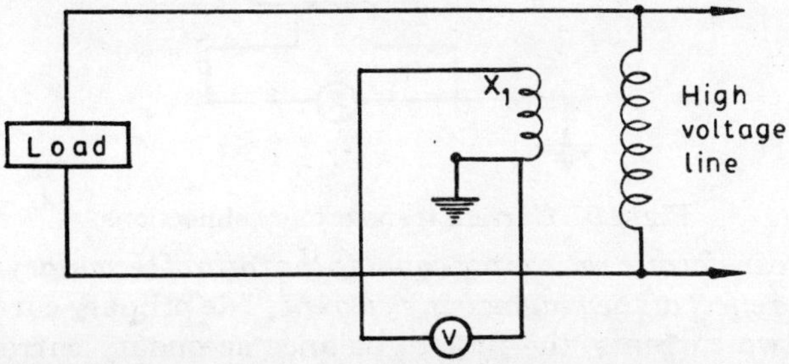

Fig. 2.95. Potential transformer connections.

- These transformers are *made with high quantity iron core operating at very low flux densities so that the magnetising current may be very small.* Careful design ensures minimum variation of voltage ratio with load and minimum phase shift between input and output voltages. Potential transformers secondary are commonly designed for an output of 110 V.

2.11.2. Current Transformers (C.T.)

- Just as a shunt extends the range of a D.C. ammeter, so does the current transformer perform the same function in A.C. circuits. Thus a *high magnitude alternating current can be easily measured by a combination of a current transformer and a low range ammeter.*
- The primary of a current transformer (C.T.) consists of a few turns of thick cross-section connected in *series* with the high current line. Very often the primary is just one turn formed by taking the line conductor through the secondary winding (Fig. 2.96). The secondary winding consists of a large number of turns of fine wire designed for either 5 A or 1 A rating. Thus a current transformer is *step-up* transformer. The current transformer

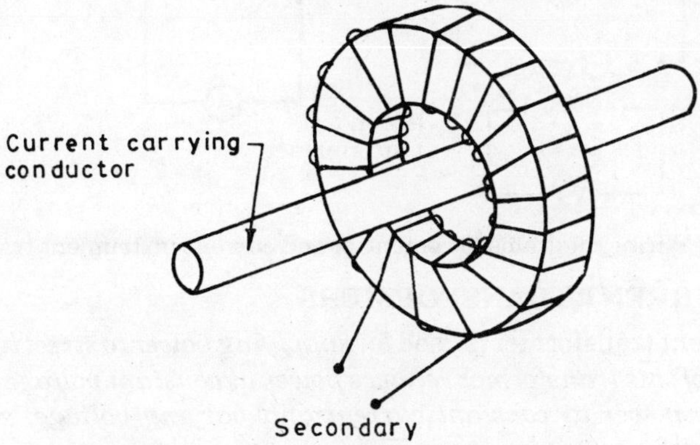

Fig. 2.96. Line conductor acting as primary.

has the secondary effectively short-circuited through the low impedance of the ammeter. Fig. 2.97 shows the current transformer connections.

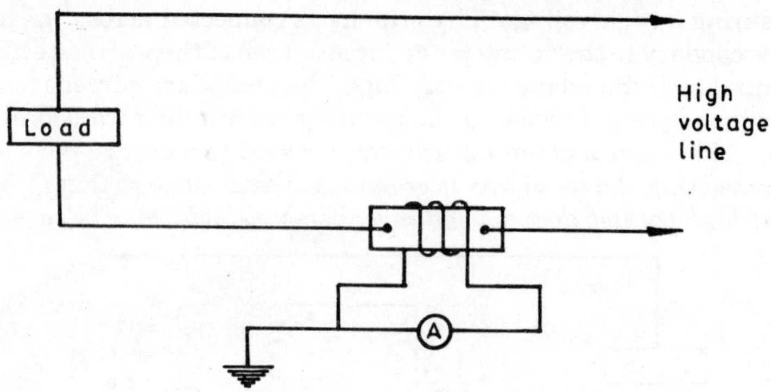

Fig. 2.97. Current transformer connections.

- The *current transformer ratio is not equal to the ratio of secondary to primary turns, mainly because of the effect of the magnetising current.* The primary current can be thought of as the sum of two currents, the first to balance secondary current so that primary and secondary m.m.fs. may balance and the second being the no-load current I_0. The component I_0 besides being responsible for a slight error in the current ratio, is also responsible for a phase angle error. The transformer *must be carefully designed to minimise the ratio and phase angle error.*
- It may be noted that current transformer *must never be operated on open-circuit* for the following two reasons :

(*i*) There will be no secondary m.m.f. and since the primary current (and m.m.f.) is fixed, the core flux will increase enormously. This will cause large eddy current and hysteresis losses and the *resulting high temperature may damage the insulation or even the core.*

(*ii*) A very high voltage will be induced in the multi-turn secondary and this high voltage may be dangerous both to life and to the insulation.

Fig. 2.98 shows the wiring diagrams for potential and current instrument transformers.

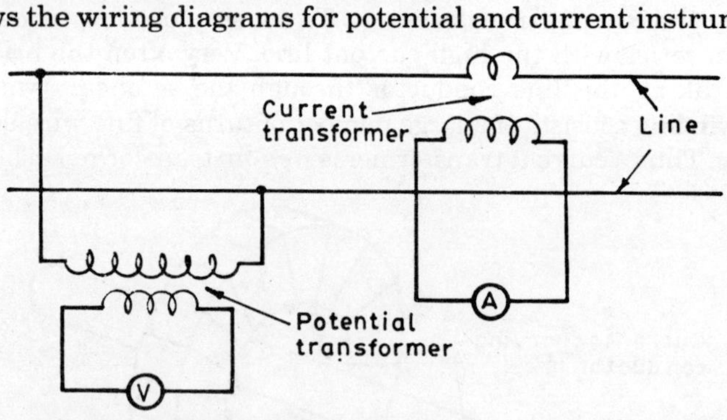

Fig. 2.98. Wiring diagrams for potential and current instrument transformers.

2.12. CONSTANT-CURRENT TRANSFORMERS

A constant-current transformer is used for *supplying power to street lights which are connected in series.* The *primary of this transformer receives power at constant voltage and varying current and the secondary supplies power at constant current and varying voltage,* which depends upon the number of lamps in series.

Principle of Operation. The principle of operation of a constant-current transformer is illustrated in Fig. 2.99.

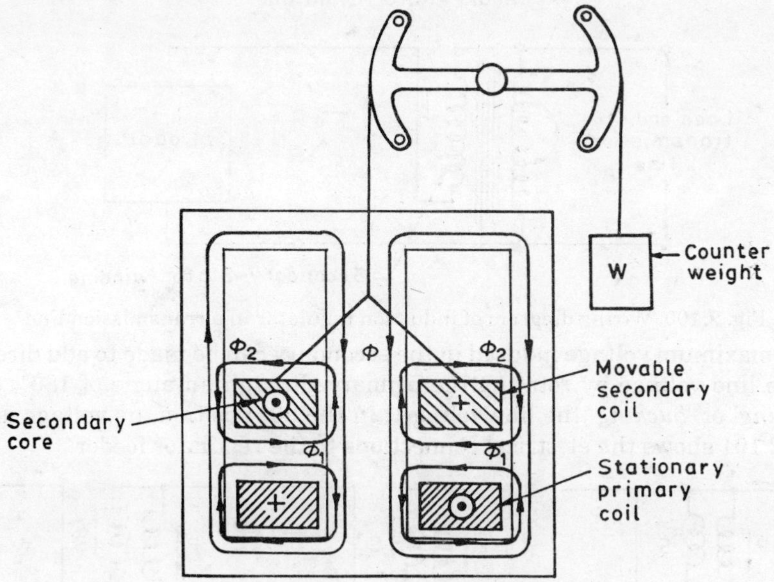

Fig. 2.99. Constant-current transformer.

- One coil is stationary while the other one is movable, and its weight is partly balanced by a counter-weight *W*. Either coil may be the movable coil.
- The openings through the coils are much longer than the core. This permits a large leakage flux to be set up through the coils when the currents are maintained in them.
- It is clear from Fig. 2.99 that a mechanical force of repulsion exists between the coils. This force tends to drive the movable coil upward on the core, and owing to the counterweight, a relatively small force produces motion. *At maximum load the movable coil is close to the stationary one and the voltage induced in the former is a maximum.* When the secondary load is *reduced* by cutting out some of the lamps in the series circuit, the current in the remaining units increases momentarily. This causes the force of repulsion to increase. The *floating coil then moves upward and thereby increases the equivalent leakage reactance of the transformer, reduces the secondary voltage,* and *brings the current back to normal. The coil comes to rest when the terminal voltage is properly reduced to suit the lesser number of lamps and the current attains its correct constant value.*
- The *rating* is usually given in kW, and the secondary current. The standard values of current are 6.6 ,7.5 and 20 A. The load power factor is almost 100%. The *regulation* is expressed as *the percentage of the current (secondary) from its rated value. The regulation should not exceed 1%.*

2.13. INDUCTION REGULATORS

- The induction regulator is really a special type of potential transformer with primary winding mounted on a cylindrical core that may be turned on its axis. The *secondary winding is stationary*. Both windings are mounted in slots, much as are the windings of an induction motor. A small air gap separates the stator from the rotor.
- Regulators of this type are *used to regulate the voltage on single-phase feeder circuits. Voltage drop in the feeder is compensated for by the voltage induced in the secondary of the regulator.*

A wiring diagram of the electrical connections is given in Fig. 2.100.

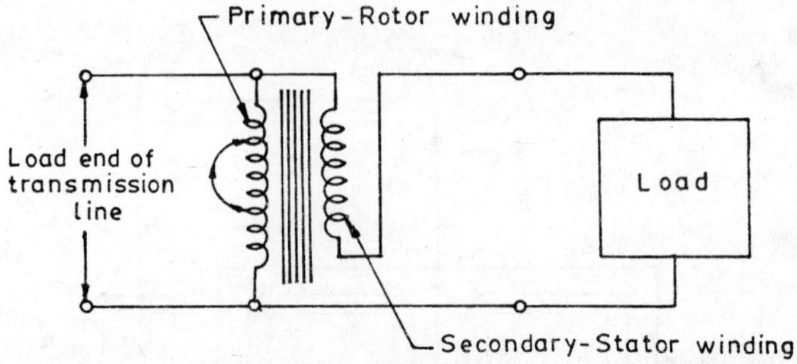

Fig. 2.100. Wiring diagram of induction regulator in a transmission line.

Since the maximum voltage induced in the secondary can be made to add directly or subtract directly from the line voltage by rotating the primary through an angle of 180°, the regulator is capable of '*boosting*' or '*bucking*' the line voltage ; in the 90° position, no voltage is induced in the secondary. Fig. 2.101 shows the electrical connections to the regulator feeder.

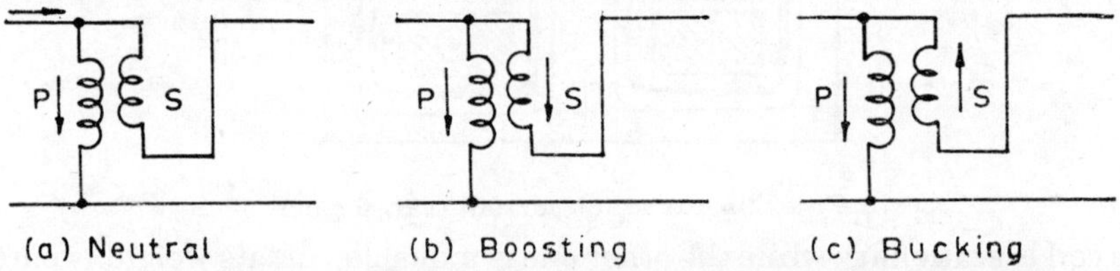

Fig. 2.101. Circuit connections for feeder regulator.

- The induction regulator is essentially a specially constructed transformer in which the *amount of flux set up by the primary is constant but that portion of the flux linking the secondary is variable.* It thus has a variable co-efficient of coupling, somewhat similar to that in the constant-current transformer. In the induction regulator, however, the spacing between the axis of the coils is varied (Fig. 2.102). The *primary is wound on a movable core,* while the *secondary is fixed on the stator.* In Fig. 2.102, the axes of both the coils coincide, and since maximum flux linkages is obtained, the secondary induced voltage is likewise a *maximum.* The primary may be rotated through 180°, thereby giving another maximum secondary voltage position 180° electrical degrees out of phase with the first one.

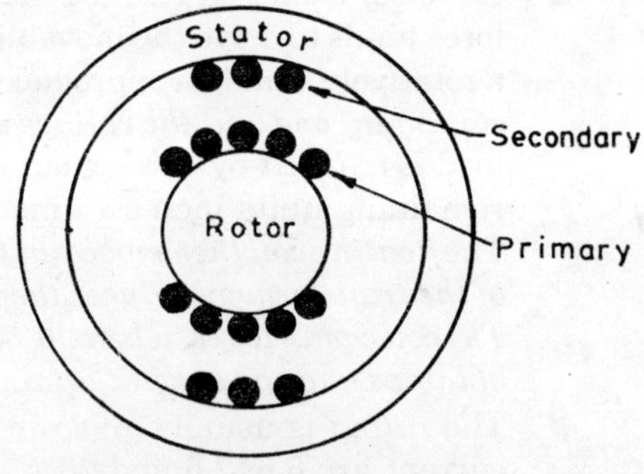

Fig. 2.102. The induction regulator maximum-voltage position.

When the axes of the coils are 90° apart, *little or no voltage is induced in the secondary,* since the flux linkage is at a maximum. However, the secondary coil may still carry considerable current due to the load circuit. With the little flux linking the coils, the secondary alone now sets up its own flux, and thus acts as a high-reactance coil in series with the load. In order to overcome this reactive volt drop, a short-circuited winding is placed on the rotor. The axis of the third winding is 90° from that of the primary winding Fig. (2.103). The secondary coil flux thus induces a voltage in the short circuited winding, and the resultant current produces a flux *opposing* that caused by load current. This reduces the net flux, and hence the series-reactive volt drop.

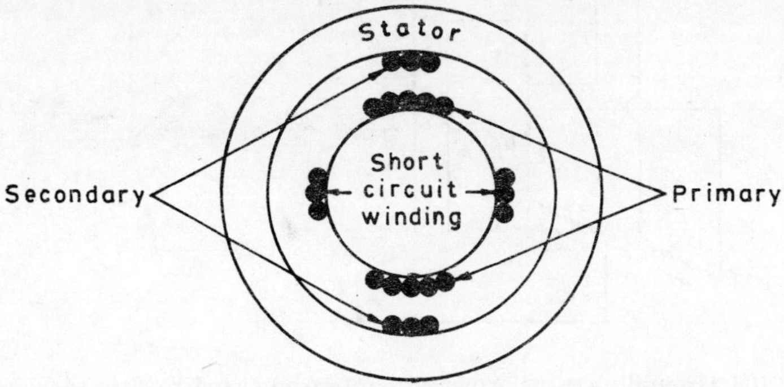

Fig. 2.103. Induction regulator with short circuited winding.

Advantages. The induction regulator :

(*i*) provides stepless voltage variation without any arcing or short-circuiting of turns as in the case of transformers ;

(*ii*) is reliable in operation because of the absence of any parts subject to wear ;

(*iii*) has simple and rugged construction ;

(*iv*) can withstand overload well ; and

(*v*) is not affected by load and power-factor variations.

Applications :

1. Automatic induction regulators, which can hold voltage within ± 1%, are ideally suited for applications *requiring constant voltage*, as in the field of *electronics and control circuits*.

2. Lighting circuits can be supplied through induction regulators, where lighting and power circuits are supplied from the same bus and are thus subject to wide fluctuations of voltage.

3. Induction regulators can also be used for speed control of variable speed devices.

Hand-operated induction regulators are used in laboratories for providing smooth variable voltage over a range of 0 to 200% of the normal voltage for calibration and testing work, and in theatres for light control.

2.14. GROUNDING TRANSFORMERS

A grounding transformer is a transformer intended primarily for the purpose of providing a neutral point for grounding purposes.

- It may be a two-winding unit with a delta-connected secondary winding and *Y*-connected primary winding which provides the neutral for grounding purposes, or it may be a single-winding 3-phase auto-transformer with windings in inter-connected *Y* or zig-zag. With the latter, the windings consist of six equal parts, each designed for one-third the line-to-line voltage, two of these parts are placed on each leg and connected as in Fig. 2.104.

In *case of a ground fault on any line, the ground current flows equally in three legs of the auto-transformer, and the inter-connection offers the minimum impedance to the flow of the single-phase fault current.*

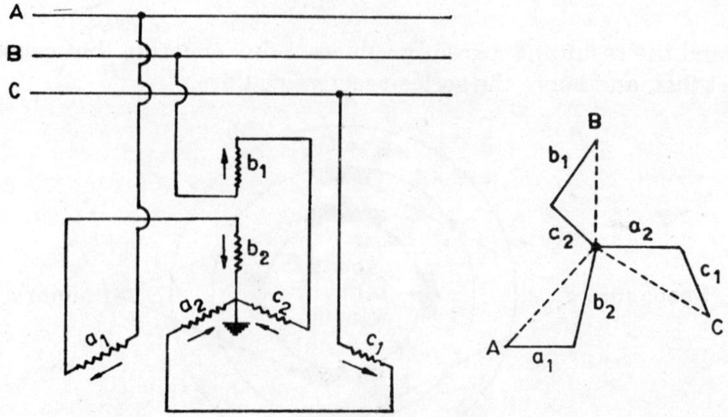

Fig. 2.104. Grounding transformer with inter-connected Y or zig-zag winding.

2.15. FURNACE TRANSFORMERS

- *These transformers supply power to electric furnaces of induction, resistance, open-arc, and submerged-arc types.*
- The secondary voltages are low, occasionally less than 100 V, but generally several hundred volts. Size range from a few kVA to over 50 MVA, with secondary currents over 60000 A. High currents are obtained by parallel connection of many winding sections. Current is collected by internal bus bars and brought through the transformer cover by the bus bars or by high-current bushings.
- The *power input to the furnace is controlled by adjusting the output voltage of the furnace transformer.* Optimum performance of the furnace may require adjustment of the secondary voltage over a range of 3 : 1 or more. This may be accomplished by a regulating transformer between the high-voltage power source and a fixed-rated furnace transformer. More frequently, regulation is obtained by *taps in high voltage winding.* In addition to taps in high voltage winding, *delta Y-switch* in the high-voltage winding is often used to extend the range of voltage by an additional ratio of 1.73.
- *Motor-operated off-load tap changers are usual,* but occasionally on-load tap-changing equipment is justified by the saving in melt time and reduced breaker maintenance. The load-tap-changing duty is more severe than on the usual power transformer, not only with respect to frequency of operation, but also because of the extreme range, which results in large kVA increments per tap.
- *Circuit reactance furnishes current stability for A.C. arc furnaces.* In the larger size the inherent impedance of the transformer and its associated secondary conductors is sufficient for adequate stability. This is not generally true for smaller arc furnaces. Consequently it is customary in furnace transformers rated 7500 kVA and below to *include a reactor in the tank with the transformer.* This reactor is connected in the high-voltage circuit and is furnished with taps to permit adjusting the total reactance to that required to maintain arc stability under the existing service conditions.

2.16. HIGH FREQUENCY TRANSFORMERS

The lamination thicknesses, beyond the normal power frequency range at frequencies higher than 50-60 Hz, have to be reduced to control the iron losses to an acceptable level. If the operating frequencies extend beyond a few *kilocycles per second, it becomes necessary to use iron-dust cores or the ferrites, special oxides which exhibit reasonably good magnetic properties with very small eddy-current losses.*

In such transformers, *variable frequency is the normal operating condition and as the frequency increases, the significance of the elements in the equivalent circuit changes.*

- At *very low frequencies,* for example, the magnetising reactance is small enough to be comparable with or even less than load impedance so that this tends to be *'shorted out'* and the secondary terminal voltage is greatly reduced.
- In the *intermediate frequency* range, X_m is high enough for its effects to be neglected and the leakage impedance absorbs only a moderate fraction of the available voltage.
- With *further increase of frequency,* however the leakage reactance drop tends to cause the output voltage to *fall appreciable again.*

Capacitance between turns and between the windings and earth can no longer be ignored since these are shunted across the circuit in a distributed fashion causing behaviour similar to a leading power-factor load.

The design problem here is to ensure the transfer, from primary to secondary, without excessive distortion, of an input signal having many frequency components ; for example a pulse of voltage or an input having the complex frequency spectrum of a musical note. In the high-power field too, with the advent of thyristor inverters, supply voltages may be far from sinusoidal and may appear as castellated or other compositive waveforms.

2.17. TRANSFORMER SPECIFICATIONS

The transformer specifications give the rating and performance expectations of the transformer. These are broadly as mentioned below :

1. kVA rating*;
2. Rated voltage ;
3. Number of phases (1-phase or 3-phase) ;
4. Rated frequency ;
5. Connections (Δ or Y for 3-phase transformer) ;
6. Tappings if any ;
7. Type of core (core or shell) ;
8. Type (power or distribution) ;
9. Ambient temperature (generally average 40°C) ;
10. Type of cooling.
    - (*i*) Cooling medium—air, oil or water
    - (*ii*) Circulation type—natural or forced
    - (*iii*) Simple or mixed cooling.
11. Temperature rise above ambient in °C depending upon the class of winding insulation.
12. Voltage regulation
    - (*i*) Percent or per unit at full-load at 75°C unity p.f. or 0-8 p.f. lag
    - (*ii*) Impedance percent or per unit
    - (*iii*) Reactance—percent or per unit.
13. No-load current in amperes or percent of rated current at rated voltage and rated frequency.
14. Efficiency in percent or per unit at full load, $\frac{1}{2}$ load, $\frac{3}{4}$ load at unity p.f. or 0.8 p.f.

kVA rating :

- The *kVA rating* is the *kVA output which the transformer can deliver at rated voltage and frequency under usual operating conditions without exceeding the standard limits of temperature rise.*

  The *kVA figures always refer to the* **output kVA** appearing at the secondary load terminals ; the **input kVA**, of course, is *slightly higher because of internal losses* (core and winding losses).
- *Rated secondary voltage* is the voltage that appears across the secondary terminals when rated current flows.
- The *rated primary voltage* is equal to the rated secondary voltage multiplied by the turn-ratio between primary and secondary *i.e.* V_1 (rated) = V_2 (rated) × $\frac{N_1}{N_2}$.

IS Specifications :

1. Outdoor type distribution transformers (IS : 1180-1964) :

- *Standard ratings* : 16, 25, 40, 50, 63, 80 and 100 kVA.
- *No-load voltage ratios* : 3300/433 V, 6600/433 V and 11000/433 V.
- *Tappings* : Shall be provided on h.v. side in 5 steps. The ranges shall be ± 2.5% and ± 5% off-load tap changers to be used.
- *Connections* : Δ/Y (*Dy* 11) with neutral brought out to a separate insulated terminal.
- *Cooling* : By low viscosity transformer oil.
- *Conservator tank* : To be provided on transformer of rating 50 kVA or above.
- *Limits of temperature rise* :

  The following temperature rises shall be permitted over the ambient temperature of 45°C :

  (*i*) Winding (temperature to be measured by resistance method) ...... 55°C

  (*ii*) Oil (temperature rise to be measured by thermometer in the top oil) ...... 45°C

  The above temperature rises are for ON, OB, OW type cooling.
- *Impedance* : 4.5% at 75°C, subject to the tolerance limit of ± 10%.

2. Power transformers (IS : 2026-1962) :

- *Standard ratings (for 3-phase transformers)* : 25, 40, 63, 100, 125, 160, 200, 250, 315, 400, 500, 630, 800, 1000, 1250, 1600, 2000, 2500, 3125, 4000, 6300, 8000, 10000, 12500, 16000, 20000, 25000, 31500, 40000, 50000, 63000 and 80000 kVA.

  Standard ratings (for 1-single phase transformers) : 1, 2, 5, 10, 16 and 25 kVA.

  Above 25 kVA, the standard rating for single phase transformers shall be one-third of the value given for 3-phase transformers.
- *Tappings* : The standard tapping ranges are $2\frac{1}{2}$ % ± 5%. Tap changing is carried out by means of an externally operated off-circuit switch capable of being locked in positions.
- *Limits of temperature rise* :

  (*i*) Windings (measured by resistance) ...... 55°C (*ON, OB, OW* cooling)
...... 60°C (*OFN, OFB* cooling)
...... 65°C (*OFW* cooling)

  (*ii*) Oil (measured by thermometer in top oil) ...... all types ...... 45°C

  (*iii*) Cores ...... core shall be designed so that the temperature rise on any part of the external surface does not exceed that of the winding.
- *Impedance* : 4.5% at 75°C, for transformers of rating upto and including 100 kVA, 11 kV and 4.75% for transformers of rating above 100 kVA including 1000 kVA, 11 kV.

HIGHLIGHTS

1. *The function of a transformer is to transform alternating current energy from one voltage into another voltage. It operates on the principle of mutual inductance (between two or more inductively coupled coils).*
2. **Distribution transformers** *should be designed to have maximum efficiency at a load much lower than full-load (about 50%).*
   Power transformers *should be designed to have maximum efficiency at or near full-load.*
3. *The transformation ratio (K) is defined as the ratio of the secondary voltage to primary voltage.*
4. **Approximate** *voltage drop* $= I_2R_{02} \cos \phi \pm I_2 X_{02} \sin \phi$

   Exact *voltage drop* $= (I_2R_{02} \cos \phi \pm I_2X_{02}) + \dfrac{(I_2X_{02} \cos \phi \mp I_2R_{02} \sin \phi)^2}{2 \; {}_0V_2}$
5. *Transfer of resistance or reactance from*
   Primary to secondary ...... $\times K^2$
   Secondary to primary ...... $\times \dfrac{1}{K^2}$
6. *The change in secondary voltage when rated load at a specified power is removed.*

$$\% \text{ regulation} = \frac{I_2R_{02} \cos \phi \pm I_2X_{02} \sin \phi}{{}_0V_2} \times 100$$

$$\eta_{all\text{-}day} = \frac{\text{output in kWh}}{\text{input in kWh}} \text{ (for 24 hours).}$$

7. *A transformer in which part of the winding is common to both the primary and secondary circuit is known as an* **auto-transformer.**
8. *The induction regulator is a special type of potential transformer with primary winding mounted on a cylindrical core that may be turned on its axis. The secondary winding is stationary.*

OBJECTIVE TYPE QUESTIONS

(A) Choose the Correct Answer :

1. Which of the following does not change in a transformer ?
   (*a*) Current (*b*) Voltage
   (*c*) Frequency (*d*) All of the above.
2. In a transformer the energy is conveyed from primary to secondary
   (*a*) through cooling coil (*b*) through air
   (*c*) by the flux (*d*) none of the above.
3. A transformer core is laminated to
   (*a*) reduce hysteresis loss (*b*) reduce eddy current losses
   (*c*) reduce copper losses (*d*) reduce all above losses.
4. Which loss is not common between a transformer and rotating machines ?
   (*a*) Eddy current loss (*b*) Copper loss
   (*c*) Windage loss (*d*) Hysteresis loss.
5. The degree of mechanical vibrations produced by the laminations of a transformer depends on
   (*a*) tightness of clamping (*b*) gauge of laminations
   (*c*) size of laminations (*d*) all of the above.
6. The no-load current drawn by transformer is usually what percent of the full-load current ?
   (*a*) 0.2 to 0.5 per cent (*b*) 2 to 5 per cent
   (*c*) 12 to 15 per cent (*d*) 20 to 30 per cent.

7. In case there are burrs on the edges of the laminations of the transformer, it is likely to result in
(*a*) vibrations (*b*) noise
(*c*) higher eddy current loss (*d*) higher hysteresis loss.

8. The path of a magnetic flux in a transformer should have
(*a*) high resistance (*b*) high reluctance
(*c*) low resistance (*d*) low reluctance.

9. No-load test on a transformer is carried out to determine
(*a*) copper loss (*b*) magnetising current
(*c*) magnetising current and loss (*d*) efficiency of the transformer.

10. The dielectric strength of transformer oil is expected to be
(*a*) 1 kV (*b*) 33 kV
(*c*) 100 kV (*d*) 330 kV.

11. Sumpner's test is conducted on transformers to determine
(*a*) temperature (*b*) stray losses
(*c*) all-day efficiency (*d*) none of the above.

12. The permissible flux density in case of cold rolled grain oriented steel is around
(*a*) 1.7 Wb/m^2 (*b*) 2.7 Wb/m^2
(*c*) 3.7 Wb/m^2 (*d*) 4.7 Wb/m^2.

13. During the short-circuit test on a small transformer the frequency is increased from 50 Hz to 200 Hz. The copper losses will increase by a factor of
(*a*) 16 (*b*) 4
(*c*) 1 (*d*) $\frac{1}{4}$.

14. The efficiency of a transformer will be maximum when
(*a*) copper losses = hysteresis losses (*b*) hysteresis losses = eddy current losses
(*c*) eddy current losses = copper losses (*d*) copper losses = iron losses.

15. No-load current in a transformer
(*a*) lags behind the voltage by about 75° (*b*) leads the voltage by about 75°
(*c*) lags behind the voltage by about 15° (*d*) leads the voltage by about 15°.

16. The purpose of providing an iron core in a transformer is to
(*a*) provide support to windings (*b*) reduce hysteresis loss
(*c*) decrease the reluctance of the magnetic path
(*d*) reduce eddy current losses.

17. Which of the following is not a part of transformer installation ?
(*a*) Conservator (*b*) Breather
(*c*) Buchholz relay (*d*) Exciter.

18. While conducting short-circuit test on a transformer the following side is short-circuited
(*a*) high voltage side (*b*) low voltage side
(*c*) primary side (*d*) secondary side.

19. In a transformer the toppings are generally provided on
(*a*) primary side (*b*) secondary side
(*c*) low voltage side (*d*) high voltage side.

20. The use of higher flux density in the transformer design
(*a*) reduces weight per kVA (*b*) reduces iron losses
(*c*) reduces copper losses (*d*) increases part load efficiency.

21. The chemical used in breather for transformer should have the quality of
(*a*) ionizing air (*b*) absorbing moisture
(*c*) cleansing the transformer oil (*d*) cooling the transformer oil.

22. The chemical used in breather is
(*a*) asbestos fibre (*b*) silica sand
(*c*) sodium chloride (*d*) silica gel.

23. If a pump motor is run on 2/3rd of its supply voltage, it will
(*a*) continue to deliver same power (*b*) burn
(*c*) stall (*d*) continue to run at lower speed.

24. An ideal transformer has infinite values of primary and secondary inductances. The statement is
(*a*) true (*b*) false.

25. The transformer ratings are usually expressed in terms of
(*a*) volts (*b*) amperes
(*c*) kW (*d*) kVA.

26. The noise resulting from vibrations of laminations set by magnetic force, is termed as
(*a*) magnetostriction (*b*) boo
(*c*) hum (*d*) zoom.

27. Hysteresis loss in a transformer varies as (B_{max} = maximum flux density)
(*a*) B_{max} (*b*) $B_{max}^{1.6}$
(*c*) B_{max}^{2} (*d*) $B_{max}^{2.4}$

28. Material used for construction of transformer core is usually
(*a*) wood (*b*) copper
(*c*) aluminium (*d*) silicon steel.

29. The thickness of laminations used in a transformer is usually
(*a*) 0.4 mm to 0.5 mm (*b*) 4 mm to 5 mm
(*c*) 14 mm to 15 mm (*d*) 25 mm to 40 mm.

30. The function of conservator in a transformer is
(*a*) to protect against internal fault (*b*) to reduce copper as well as core losses
(*c*) to cool the transformer oil
(*d*) to take care of the expansion and contraction of transformer oil due to variation of temperature of surroundings.

31. The highest voltage for transmitting electrical power in India is
(*a*) 33 kV (*b*) 66 kV
(*c*) 132 kV (*d*) 400 kV.

32. In a transformer the resistance between its primary and secondary is
(*a*) zero (*b*) 1 Ω
(*c*) 1000 Ω (*d*) infinite.

33. A transformer oil must be free from
(*a*) sludge (*b*) odour
(*c*) gases (*d*) moisture.

34. A Buchholz relay can be installed on
(*a*) auto-transformers (*b*) air-cooled transformers
(*c*) welding transformers (*d*) oil cooled transformers.

35. Gas is usually not liberated due to dissociation of transformer oil unless the oil temperature exceeds
(*a*) 50°C (*b*) 80°C
(*c*) 100°C (*d*) 150°C.

36. The main reason for generation of harmonics in a transformer could be
(*a*) fluctuating load (*b*) poor insulation
(*c*) mechanical vibrations (*d*) saturation of core.

37. Distribution transformers are generally designed for maximum efficiency around
(*a*) 90% load (*b*) zero load
(*c*) 25% load (*d*) 50% load.

38. Which of the following properties is not necessarily desirable in the material for transformer core ?
(*a*) Mechanical strength
(*b*) Low hysteresis loss
(*c*) High thermal conductivity
(*d*) High permeability.

39. Helical coils can be used on
(*a*) low voltage side of high kVA transformers
(*b*) high frequency transformers
(*c*) high voltage side of small capacity transformers
(*d*) high voltage side of high kVA rating transformers.

40. High frequency transformers sometimes make use of ferrite cores because it has
(*a*) high specific gravity
(*b*) high resistance
(*c*) high hysteresis
(*d*) low permeability.

41. Cross over windings are used on
(*a*) low voltage side of high kVA rating transformers
(*b*) current transformers
(*c*) high voltage side of high kVA rating transformers
(*d*) high voltage side of low kVA rating transformers.

42. While conducting short-circuit test on a transformer the following side is short-circuited.
(*a*) h.v. side
(*b*) l.v. side
(*c*) primary side
(*d*) secondary side.

43. A substance is placed in strong magnetic field. Which of the following will have maximum influence on the capacity of the substance to get magnetised ?
(*a*) Permeability
(*b*) Susceptibility
(*c*) Permittivity
(*d*) Resistivity.

44. Harmonics in transformer result in
(*a*) increased core losses
(*b*) increased I^2R losses
(*c*) magnetic interference with communication circuits
(*d*) all of the above.

45. Tertiary winding is provided in transformers having
(*a*) mesh/star winding
(*b*) mesh/mesh winding
(*c*) star/star winding
(*d*) any of the above.

46. The core used in high frequency transformer is usually
(*a*) copper core
(*b*) cost iron core
(*c*) air core
(*d*) mild steel core.

47. The full-load copper loss of a transformer is 1600 W. At half-load the copper loss will be
(*a*) 6400 W
(*b*) 1600 W
(*c*) 800 W
(*d*) 400 W.

48. The value of flux involved in the e.m.f. equation of a transformer is
(*a*) average value
(*b*) r.m.s. value
(*c*) maximum value
(*d*) instantaneous value.

49. Silicon steel used in laminations mainly reduces
(*a*) hysteresis loss
(*b*) eddy current losses
(*c*) copper losses
(*d*) all of the above.

50. Which winding of the transformer has less cross-sectional area ?
(*a*) Primary winding
(*b*) Secondary winding
(*c*) Low voltage winding
(*d*) High voltage winding.

51. Power transformers are generally designed to have maximum efficiency around
(*a*) no-load
(*b*) half-load
(*c*) near full-load
(*d*) 10% overload.

52. Which of the following is the main advantage of an auto-transformer over a two winding transformer ?
(*a*) Hysteresis losses are reduced
(*b*) Saving in winding material
(*c*) Copper losses are negligible
(*d*) Eddy losses are totally eliminated.

53. During short-circuit test iron losses are negligible because

(*a*) the current on secondary side is negligible (*b*) the voltage on secondary side does not vary
(*c*) the voltage applied on primary side is low (*d*) full-load current is not supplied to the transformer.

54. Two transformers are connected in parallel. These transformers do not have equal percentage impedance. This is likely to result in

(*a*) short-circuiting of the secondaries
(*b*) power factor of one of the transformers is leading while that of the other lagging
(*c*) transformers having higher copper losses will have negligible core losses
(*d*) loading of the transformers not in proportion to their kVA ratings.

55. The changes in volume of transformer cooling oil due to variation of atmospheric temperature during day and night is taken care of by which part of transformer ?

(*a*) Conservator (*b*) Breather
(*c*) Bushings (*d*) Buchholz relay.

56. The transformer laminations are insulated from each other by

(*a*) mica strip (*b*) thin coat of varnish
(*c*) paper (*d*) any of the above.

57. Which type of winding is used in 3-phase shell-type transformer ?

(*a*) Circular type (*b*) Sandwich type
(*c*) Cylindrical type (*d*) Rectangular type.

58. During open circuit test of a transformer

(*a*) primary is supplied rated voltage (*b*) primary is supplied full-load current
(*c*) primary is supplied current at reduced voltage
(*d*) primary is supplied rated kVA.

59. Open circuit test on transformers is conducted to determine

(*a*) hysteresis losses (*b*) copper losses
(*c*) core losses (*d*) eddy current losses.

60. Short circuit test on transformers is conducted to determine

(*a*) hysteresis losses (*b*) copper losses
(*c*) core losses (*d*) eddy current losses.

61. For the parallel operation of single-phase transformers it is necessary that they should have

(*a*) same efficiency (*b*) same polarity
(*c*) same kVA rating (*d*) same number of turns on the secondary side.

62. The transformer oil should have ...... voltatility and ...... viscosity

(*a*) low ...... low (*b*) high ...... high
(*c*) low ...... high (*d*) high ...... low.

63. The function of breather in a transformer is

(*a*) to provide oxygen inside the tank (*b*) to cool the coils during reduced load
(*c*) to filter the transformer cooling oil
(*d*) to arrest flow of moisture when outside air enters the transformer.

64. The secondary winding of which of the following transformers is always kept closed ?

(*a*) Step-up transformer (*b*) Step-down transformer
(*c*) Potential transformer (*d*) Current transformer.

65. The size of a transformer core will depend on

(*a*) frequency (*b*) area of the core
(*c*) flux density of the core material (*d*) (*a*) and (*b*) both.

66. Natural air cooling is generally restricted for transformers up to

(*a*) 1.5 MVA (*b*) 5 MVA
(*c*) 15 MVA (*d*) 50 MVA.

67. A shell-type transformer has
(*a*) high eddy current losses (*b*) reduced magnetic leakage
(*c*) negligible hysteresis losses (*d*) none of the above.

68. A transformer can have regulation closer to zero
(*a*) on full-load (*b*) on overload
(*c*) on leading power factor (*d*) on zero power factor.

69. A transformer transforms
(*a*) voltage (*b*) current
(*c*) current and voltage (*d*) power.

70. Which of the following is not the standard voltage for power supply in India
(*a*) 11 kV (*b*) 33 kV
(*c*) 66 kV (*d*) 122 kV.

ANSWERS

1. (*c*)	**2.** (*c*)	**3.** (*b*)	**4.** (*a*)	**5.** (*d*)	**6.** (*b*)	**7.** (*c*)
8. (*d*)	**9.** (*c*)	**10.** (*b*)	**11.** (*a*)	**12.** (*a*)	**13.** (*a*)	**14.** (*d*)
15. (*a*)	**16.** (*c*)	**17.** (*d*)	**18.** (*b*)	**19.** (*c*)	**20.** (*a*)	**21.** (*b*)
22. (*d*)	**23.** (*c*)	**24.** (*b*)	**25.** (*d*)	**26.** (*c*)	**27** (*b*)	**28.** (*d*)
29. (*a*)	**30.** (*d*)	**31.** (*d*)	**32.** (*d*)	**33.** (*d*)	**34.** (*d*)	**35.** (*d*)
36. (*d*)	**37.** (*d*)	**38.** (*c*)	**39.** (*a*)	**40.** (*b*)	**41.** (*d*)	**42.** (*b*)
43. (*b*)	**44.** (*d*)	**45.** (*c*)	**46.** (*c*)	**47.** (*d*)	**48.** (*c*)	**49.** (*a*)
50. (*d*)	**51.** (*c*)	**52.** (*b*)	**53.** (*c*)	**54.** (*d*)	**55.** (*a*)	**56.** (*b*)
57. (*b*)	**58.** (*a*)	**59.** (*c*)	**60.** (*b*)	**61.** (*b*)	**62.** (*a*)	**63.** (*d*)
64. (*d*)	**65.** (*d*)	**66.** (*a*)	**67.** (*b*)	**68.** (*c*)	**69.** (*d*)	**70.** (*d*)

(B) Say 'Yes' or 'No'

1. When a transformer raises the voltage it is called the step-up transformer.

2. A transformer must not be connected to D.C. source.

3. The ratio of primary voltage to secondary voltage is known as transformation ratio.

4. An ideal transformer is one in which the resistance of the windings is negligible and the core has no losses.

5. Primary and secondary currents are directly proportional to their respective turns.

6. The function of the magnetising component of no-load current is to sustain the alternating flux in the core.

7. The no-load primary input is practically equal to the iron loss in the transformer.

8. A transformer is said to be loaded when the secondary circuit of a transformer is completed through an impedance load.

9. Magnetic flux can be confined into a designed path.

10. When shifting resistance to the secondary, divide it by K^2

11. An open-circuit test is conducted to find no-load or core loss.

12. Short-circuit test is conducted to find full-load copper loss.

13. The change in voltage when rated load at a specified power is removed is termed as voltage regulation.

14. Iron or core losses include copper loss and eddy current loss.

15. Iron or core losses are found from short-circuit test.

16. The efficiency of a transformer at a particular load and power factor is defined as the ratio of power output to power input.

17. Copper losses = iron losses is the condition for maximum efficiency of a transformer.

18. $\eta_{\text{all-day}} = \dfrac{\text{output in kWh}}{\text{input in kWh}}$ (for 24 hours)

19. A transformer in which part of the winding is common to both the primary and secondary circuits is known as auto-transformer.

20. The Δ-Δ connection is generally used in systems in which the voltages are not very high and especially when continuity of service must be maintained even though one of the transformers should fail.
21. The *Y*-Δ connection is principally used where the voltage is to be stepped up.
22. The Δ-*Y* connection is employed where it is necessary to step-up the voltage.
23. The *V-V* circuit is frequently used for two auto-transformers.
24. It is practicable to connect instruments and meters directly to the lines in high voltage circuits.
25. A potential transformer is a step down transformer used along with a low range voltmeter for measuring a high voltage.
26. The current transformer ratio is not equal to the ratio of secondary to primary turns, mainly because of the effect of the magnetising current.
27. A constant-current transformer is used to supply power to street lights which are connected in series.
28. In case of an induction regulator primary winding is stationary.

ANSWERS

1. Yes	**2.** Yes	**3.** No	**4.** Yes	**5.** No	**6.** Yes	**7.** Yes
8. Yes	**9.** No	**10.** No	**11.** Yes	**12.** Yes	**13.** Yes	**14.** No
15. No	**16.** Yes	**17.** Yes	**18.** Yes	**19.** Yes	**20.** Yes	**21.** No
22. Yes	**23.** Yes	**24.** No	**25.** Yes	**26.** Yes	**27.** Yes	**28.** No.

THEORETICAL QUESTIONS

1. What is a transformer ? How does it transfer electric energy from one circuit to another ?
2. Explain the principle of operation of a transformer.
3. What is meant by *transformer action* ? Under what conditions will it take place ?
4. If an alternating current is impressed on one coil, what will be the frequency of the induced voltage in another coil with which it is coupled ?
5. Enumerate the various kinds of transformers.
6. Explain the need for stepping up and stepping down voltages in a power system. How does a transformer accomplish ?
7. Why are transformer windings divided into several coils ?
8. What properties should a good transformer oil possess ?
9. What purposes are served by placing transformers in oil-filled tanks ?
10. Why are the tanks of some large transformers corrugated ?
11. Write a short note on 'transformer cooling'.
12. Enumerate and explain briefly different types of windings.
13. Derive an expression for induced e.m.f. in a transformer in terms of frequency, the maximum value of flux and the number of turns on the windings.
14. Derive an expression for the e.m.f. of an ideal transformer winding.
15. Explain the function of the active and reactive components of the no-load current of a static transformer.
16. Why are some transformers constructed with primary and secondary windings divided into two halves ?
17. Distinguish between *power* and *distribution* transformers.
18. Draw and explain the no-load phasor diagram for a single-phase transformer.
19. Draw the vector diagram of a power transformer under full-load condition.
20. What is meant by *equivalent resistance* of a transformer ? How may it be calculated in primary terms and secondary terms ?
21. What is meant by the *equivalent reactance* of a transformer ? How may it be calculated in primary terms and secondary terms ?
22. How can the equivalent impedance of a transformer be determined ?

23. Outline carefully the procedure for performing the short-circuit test.
24. What useful information is obtained from the short-circuit test ?
25. Outline carefully the procedure for performing the open-circuit test.
26. What useful information is obtained from open-circuit test ?
27. What are the two components of the core loss ?
28. How is the hysteresis loss affected by a change in flux density ?
29. Develop the equivalent circuit of a single phase transformer.
30. Draw the equivalent circuit of a transformer and show how the constants of the primary and secondary windings may be combined to give a simplified equivalent circuit with the values of constants given in terms of the secondary winding.
31. Describe the method of calculating the regulation and efficiency of a single-phase transformer by open circuit and short circuit tests.
32. Distinguish between the 'efficiency' and the 'regulation' of a transformer. Show how power factor affects both of them.
33. Explain with circuit diagrams, the open-circuit and short-circuit tests to be carried out in the laboratory for the determination of the parameters of a single-phase transformer. Derive the condition for maximum efficiency.
34. What is Sumpner's test ? Draw a circuit diagram to conduct this test and explain its principle.
35. Define all-day efficiency.
36. What is an auto-transformer ? What advantages are possessed by auto-transformers ?
37. Derive an expression for saving of copper when an auto-transformer is used.
38. State the applications of auto-transformers. Why is this transformer not used as a distribution transformer ? Prove that for the same capacity and voltage ratio, an auto-transformer requires less copper than an ordinary transformer.
39. What are the sources of heat in a transformer ? Describe briefly the various methods used for cooling of transformers.
40. To what does the '*polarity*' of a transformer refer ?
41. List the conditions that must be fulfilled before two transformers can be operated successfully in parallel ?
42. Under what condition will there be no circulating current when two transformers are operated in parallel at no-load ?
43. Assuming that two transformers having equal ratios of transformation are connected in parallel, indicate how the total load divides between them ?
44. Assuming that two transformers having unequal ratios of transformation are connected in parallel, indicate how the total load divides between them.
45. List the four possible ways of connecting a bank of three transformers for three-phase service. State the applications of each of the connections listed above.
46. State with the help of connection and vector diagrams how a 2-phase supply can be obtained from a 3-phase supply.
47. Explain how *T-T* connection is made for three-phase to three-phase operation.
48. What is the advantage of *V-V* connection ?
49. What is the total load capacity of *V-V* bank as compared with a Δ-Δ bank ?
50. List the advantages and disadvantages of three-phase transformers.
51. What are the two general types of three-phase transformers ? Does either have any advantage over the other ?
52. What are the conditions for paralleling two 3-phase transformers ? Differentiate between the essential and desirable conditions.
53. Describe the general constructional features of the constant current transformer.
54. What are the instrument transformers ? Give the connection diagrams for current and potential transformers.
55. Describe the general constructional features of the induction regulator.
56. Explain the principle of operation of the induction regulator.

UNSOLVED EXAMPLES

E.m.f. Equation-Turn Ratio

1. The no-load ratio required in a single-phase 50-Hz transformer is 6600/300 V. If the maximum value of flux in the core is to be about 0.09 weber, find the number of turns in each winding. [**Ans.** N_1 = 330 ; N_2 = 15]

2. A 20 kVA, single-phase transformer has 200 turns on the primary and 40 turns on the secondary. The primary is connected to 1000 V, 50-Hz supply. Determine :
   (*i*) the secondary voltage on open circuit ;
   (*ii*) the currents flowing through the two windings on full-load ; and
   (*iii*) the maximum value of flux. [**Ans.** (*i*) 200 V ; (*ii*) 20 A, 100 A ; (*iii*) 0.0225 Wb]

3. A single-phase transformer is connected to a 230 V, 50-Hz supply. The net cross-sectional area of the core is 50 cm^2. The number of turns in the primary is 460 and in the secondary 80. Determine :
   (*i*) Transformation ratio ; (*ii*) Peak value of flux density in the core ; and
   (*iii*) E.m.f. in the secondary winding. [**Ans.** 0.1739 ; 0.4504 T ; 40 V]

4. A 6600/440 V single-phase 600 kVA transformer has 1200 primary turns. Find :
   (*i*) Transformation ratio ; (*ii*) Secondary turns ;
   (*iii*) Voltage/turn ; and
   (*iv*) Secondary current when it supplies a load of 400 kW at 0.8 power factor lagging.
   $\left[\textbf{Ans. } \frac{1}{15} \text{ ; 80 ; 5.5 V ; 1136 A}\right]$

5. Find the primary and secondary turns of a 3300/300 V, single-phase, 50-Hz, 30 kVA transformer if the flux in the core is to be about 0.06 Wb. Also determine the primary and secondary currents if the losses are to be neglected. [**Ans.** 264 ; 24 ; 100 A ; 9.09 A]

6. The voltage/turn of a single-phase transformer is 1.1 V, when the primary winding is connected to a 220 V, 50-Hz A.C. supply, the secondary voltage is found to be 660 V. Find :
   (*i*) Primary and secondary turns ; and
   (*ii*) Core area if the maximum flux density is 1.2 T. [**Ans.** 200 ; 600 ; 41.29 cm^2]

Transformer on No-load

7. A 2000/200 V single-phase transformer gives 0.5 A and 40 W as ammeter and wattmeter readings when supply is given to the low voltage winding and high voltage winding is kept open. Find :
   (*i*) The magnetising component, (*ii*) The iron loss component, and
   (*iii*) The power factor of no-load current. [**Ans.** 0.4582 A ; 0.2 A ; 0.4 lagging]

8. Find (*i*) active component and reactive components of no-load current and ; (*ii*) no-load current of a 230 V/115 V single-phase transformer if the power input on no-load to the high voltage winding is 70 W and power factor of no-load current is 0.25 lagging. [**Ans.** 0.3044 A ; 1.179 A ; 1.2176 A]

9. A single-phase transformer has 500 turns on the primary and 40 turns on the secondary winding. The mean length of the magnetic path in the core is 150 cm and the joints are equivalent to an air-gap of 0.1 mm. When a potential difference of 3000 V is applied to the primary, maximum flux density is 1.2 T. Calculate :
   (*i*) The cross-sectional area of the core, (*ii*) No-load secondary voltage,
   (*iii*) The no-load current drawn by the primary, and
   (*iv*) Power factor on no-load.
   Given that AT/cm for a flux density of 1.2 T in the iron to be 5, the corresponding iron loss to be 2 W/kg at 50 Hz and density of iron as 7.8 g/cm^3. [**Ans.** 225 cm^2 ; 240 V ; 1.208 A ; 0.1457]

Transformer on Load

10. A 230 V/115 V single phase transformer takes a no-load current of 1.7 A at a power factor of 0.18 lagging with low voltage winding kept open. If the low voltage winding is now loaded to take a current of 13 A at 0.8 power factor lagging find the current taken by high voltage winding [**Ans.** 7.834 A]

11. A transformer has a primary winding of 800 turns and a secondary winding of 200 turns. When the load current on the secondary is 80 A at 0.8 power factor lagging, the primary current is 25 A at 0.707 power factor lagging. Determine graphically or otherwise the no-load current of the transformer and its phase with respect to the voltage. **[Ans.** I_0 = 5.93 A ; ϕ_0 = 73.3°]

12. A 8 : 1 step down, single-phase transformer takes a no-load current of 0.6 A at a power factor of 0.8 lagging with l.v. winding kept open. If the secondary is connected to a load taking a current of 80 A at (*i*) 0.8 power factor lagging ; and (*ii*) 0.8 power factor leading find the primary current and power factor.
[Ans. 10.49 A, 0.7996 lagging ; 9.817 A, 0.8333 leading]

13. A 20 kVA, 2000/200 V, single-phase, 50-Hz transformer has a primary resistance of 2.5 Ω and reactance of 4.8 Ω. The secondary resistance and reactance are 0.01 Ω and 0.018 Ω respectively. Find :
(*i*) Equivalent resistance referred to primary,
(*ii*) Equivalent impedance referred to primary,
(*iii*) Equivalent resistance, reactance and impedance referred to secondary, and
(*iv*) Total copper loss of the transformer.
[Ans. (*i*) 3.5 Ω ; (*ii*) 7.47 Ω ; (*iii*) 0.035, 0.066 Ω, 0.0747 Ω ; (*iv*) 350 W]

14. A 50 kVA, 4400/220 V transformer has R_1 = 3.45 Ω, R_2 = 0.009 Ω. The values of reactances are X_1 = 5 2 Ω and X_2 = 0.015 Ω. Calculate for the transformer :
(*i*) Equivalent resistance as referred to primary,
(*ii*) Equivalent resistance as referred to secondary,
(*iii*) Equivalent reactance as referred to both primary and secondary,
(*iv*) Equivalent impedance as referred to both primary and secondary, and
(*v*) Total copper loss, first using individual resistances of the two windings and secondly, using equivalent resistances as referred to each side. **[Ans.** (*i*) 7.05 Ω ; (*ii*) 0.0176 Ω ; (*iii*) 11.2 Ω ; (*iv*) 13.23 Ω ; 0.0331 Ω ; (*iv*) 909 W]

15. A single-phase transformer has the following data :
Turn ratio = 19.5 : 1, R_1 = 25 Ω , X_1 = 100 Ω, R_2 = 0.06 Ω, X_2 = 0.25 Ω. No-load current = 1.25 A leading the flux by 30°.

The secondary delivers 200 A at a terminal voltage of 500 V and a power factor of 0.8 lagging. Determine by the aid of a vector diagram the primary applied voltage, the primary power factor and the efficiency.
[Ans. 12540 ∠186.7°, 0.698 (lag), 86.74%]

16. The high voltage and low voltage windings of a 1100/110 V single-phase 50 Hz transformer has resistances of 2.4 Ω and 0.02 Ω and reactances 1 Ω and 0.009 Ω respectively. The low voltage winding is connected to a load having an impedance of (3 + *j*2) Ω. Determine :
(*i*) Current in l.v. winding, (*ii*) Current in h.v. winding,
(*iii*) The load voltage, and (*iv*) Power consumed by the load.
[Ans. (*i*) 30.11 A ; (*ii*) 3.011 A ; 108.5 V (*iv*) 2720.7 W]

Equivalent Circuit and O.C. and S.C. Tests

17. The parameters of a 2300/230 V, 50 Hz transformer are given below :

R_1 = 0.286 Ω R_2' = 0.319 Ω
X_1 = 0.73 Ω X_2' = 0.73 Ω
R_0 = 250 Ω X_0 = 1250 Ω

The secondary load impedance Z_L = 0.387 + *j*0.29. Solve the exact equivalent circuit with normal voltage across the primary. **[Ans.** η = 78.8%, % Regulation = 2.7%]

18. A 230 V, 3 kVA single-phase transformer has an iron loss of 100 W at 40-Hz and 70 W at 30-Hz. Find the hysteresis and eddy current losses at 50-Hz. **[Ans.** 91.67 W, 41.67 W]

19. When a transformer is supplied at 400 V, 50-Hz the hysteresis loss is found to be 300 W and eddy current loss is found to be 250 W. Determine the hysteresis loss and eddy current loss when the transformer is supplied at 800 V, 100 Hz. **[Ans.** 600 W ; 1000 W]

20. When the primary of a transformer rated at 2200 V, 50-Hz is supplied at 2200 V, 50-Hz the wattmeter gives a reading of 1200 W on no-load. When it is supplied at 1100 V, 25-Hz, the wattmeter gives a reading

of 400 W on no-load. If the wattmeter is connected in the input circuit find the hysteresis loss and eddy current loss at normal voltage and frequency. **[Ans.** 400 V ; 800 W]

21. A 4400 V, 50-Hz transformer has a hysteresis loss of 1000 W, eddy current loss of 1500 W and full-load copper loss of 3500 W. If the transformer is supplied at 6600 V, 75-Hz, what will be the losses ? Assume that the full-load current remains the same. **[Ans.** P_h = 1500 W, P_e = 3375 W]

22. A 50 Hz, single-phase transformer has a turn ratio of 6. The resistances are 0.9 Ω, 0.03 Ω and reactances are 5 Ω and 0.13 Ω for high-voltage and low voltage windings respectively. Find :

(*i*) The voltage to be applied to the h.v. side to obtain full-load current of 200 A in the l.v. winding on short-circuit.

(*ii*) The power factor on short-circuit.

Draw the equivalent circuit and show therein all the values. **[Ans.** 329.3 V, 0.2]

23. A 200/2000 V transformer is fed from a 200 V supply. The total winding resistance and leakage reactance as referred to low voltage side are 0.16 Ω and 7.0 Ω respectively. The representing core loss is 400 Ω and the magnetising reactance is 231 Ω. A load of impedance 596 + j444 Ω is connected across the secondary terminals. Calculate :

(*i*) Input current ; (*ii*) The secondary terminal voltage ;

(*iii*) Primary power factor. **[Ans.** (*i*) 25.96 ∠– 40.78° A ; (*ii*) 1859 V ; (*iii*) 0.757 lagging]

24. Determine the approximate equivalent circuit of a given 200/2000 V single-phase 40 kVA transformer having the following test results :

O.C. test : 200 V, 6.4 A, 384 W .... on l.v. side.

S.C. test : 78 V, 20 A, 620 W ...... on h.v. side. **[Ans.** R_{01} = 0.0155 Ω, X_{01} = 0.0358 Ω, R_0 = 104.2 Ω, X_0 = 32.75 Ω]

25. Determine the equivalent circuit of a 200/400 V, 50-Hz, single-phase transformer from the following test data :

O.C. test (l.v. side) : 200 V, 0.7 A, 70 W

S.C. test (h.v. side) : 15 V, 10 A, 85 W

Calculate the secondary voltage when delivering 5 kW at 0.8 power factor, lagging, the primary voltage being 200 V. **[Ans.** R_0 = 571.4 Ω, X_0 = 330 Ω, R_{01} = 0.21 Ω, X_{01} = 0.31 Ω, V_2 = 377.8 V]

26. A single-phase, 10 kVA, 500/250 V, 50-Hz transformer has the following constants :

Resistance : Primary 0.2 Ω, Secondary 0.5 Ω

Reactance : Primary 0.4 Ω, Secondary 0.1 Ω

Resistance or equivalent exciting circuit referred to primary, R_0 = 1500 Ω, reactance of the equivalent exciting circuit referred to primary, X_0 = 750 Ω.

What would be the readings of the instruments when the transformer is connected for the open-circuit and short-circuit tests ? **[Ans.** O.C. test : 500 V, 0.745 A, 167 W ; S.C. test : 46.8 V, 20 A, 880 W]

27. Find the approximate equivalent circuit of a single-phase 400/4400 V transformer having the following test readings :

O.C. test (l.v. side) : 400 V, 5.2 A, 600 W

S.C. test (h.v. side) : 155 V. 50 A, 1850 W **[Ans.** R_0 = 266.7 Ω, X_0 = 80.34 Ω, Z_{01} = 0.02563 Ω, R_{01} = 0.006115 Ω, X_{01} = 0.02488 Ω]

28. The following readings were taken in open-circuit and short-circuit tests on a single-phase 20 kVA, 2000/200 V transformer :

O.C. test (l.v. side) : 200 V, 2 A, 120 W

S.C. test (h.v. side) : 30 V, 10 A, 140 W

Determine :

(*i*) Equivalent circuit referred to l.v. side

(*ii*) Secondary terminal voltage on full-load at 0.8 power factor leading.

[Ans. (*i*) R_0 = 333.3 Ω, X_0 = 104.8 Ω, R_{02} = 0.014 Ω, X_{02} = 0.02654 Ω ; (*ii*) 200.47 V]

Regulation and Efficiency of a Transformer

29. The corrected instrument readings obtained from open and short-circuit tests on 10 kVA, 450/120 V, 50-Hz transformer are :

O.C. Test :

V_1 = 120 V, I_1 = 4.2 A, W_1 = 80 W ; V_1, W_1 and I_1 were read on the low-voltage side.

S.C. Test :

V_1 = 9.65 V ; I_1 = 22.2 A ; W_1 = 120 W with low-voltage winding short-circuited.

Calculate :

(*i*) The equivalent circuit (approximate) constants.

(*ii*) Efficiency and voltage regulation for 0.8 lagging power.

(*iii*) The efficiency at half full-load and 0.8 lagging power factor load.

[**Ans.** (*i*) R_0 = 25.30 Ω, X_0 = 409 Ω ; (*ii*) 97.57%, 2.04% ; (*iii*) 97.34%]

30. The primary and secondary winding resistance of a 40 kVA, 6600/250 V single-phase transformer are 10 Ω and 0.02 Ω respectively. The equivalent leakage reactance as referred to the primary winding is 35 Ω. Find the full-load regulation for load power factors of (*i*) unity ; (*ii*) 0.8 lagging ; and (*iii*) 0.8 leading.

[**Ans.** (*i*) 2.202% ; (*ii*) 3.69% ; (*iii*) ; – 0.166 %]

31. The percentage resistance and reactance of a transformer are 2.5% and 4% respectively. Find the approximate regulation on full-load at

(*i*) unity power factor, (*ii*) 0.8 power factor lagging, and

(*iii*) 0.8 power factor leading. [**Ans.** (*i*) 2.5% ; (*ii*) 4.4% ; (*iii*) – 0.4%]

32. A single-phase 100 kVA, 2000/200 V, 50 Hz transformer has impedance drop of 10% and resistance drop of 5%.

(*i*) Find the regulation at full-load, 0.8 power factor lagging.

(*ii*) At what power factor is the regulation zero ? [**Ans.** 9.196% ; 0.866 leading]

33. The high voltage of a single-phase 200 kVA 4400/220 V transformer takes a current of 30 A and power of 1200 W at 75 V when the low voltage winding is short-circuited. Determine :

(*i*) The voltage to be applied to the high voltage winding on full-load 0.8 power factor lagging if the secondary terminal voltage is to be kept at 220 V, and

(*ii*) % regulation. [**Ans.** (*i*) 4506 V ; (*ii*) 2.355%]

34. A 20 kVA, 2200/220 V 50-Hz distribution transformer is tested for efficiency and regulation as follows ;

O.C. test (l.v. side) : 220 V, 4.2 A, 148 W.

S.C. test (h.v. side) : 86 V, 10.5 A, 360 W.

Determine :

(*i*) Core loss

(*ii*) Equivalent resistance referred to primary.

(*iii*) Equivalent resistance referred to secondary.

(*iv*) Equivalent reactance referred to primary.

(*v*) Equivalent reactance referred to secondary.

(*vi*) Regulation of transformer at 0.8 power factor lagging current.

(*vii*) Efficiency at full-load and half full-load at 0.8 power factor lagging current.

[**Ans.** 148 W, 3.26 Ω, 0.0326 Ω, 7.51 Ω, 0.0751 Ω, 2.95%, 97.4%, 97.3%]

35. A 100 kVA single-phase transformer has a full-load primary current of 400 A and total resistance referred to primary is 0.006 Ω. If the iron loss amounts to 500 W, find the efficiency on full-load and half-load at

(*i*) unity power factor, and (*ii*) 0.8 power factor.

[**Ans.** 98.58%, 98.56%, 98.22%, 98.19%]

Maximum Efficiency

36. A 50 kVA, single phase transformer has an iron loss of 400 W and full-load copper loss of 800 W.

(*i*) Find the load at which maximum efficiency occurs and the value of maximum efficiency at unity power factor.

(*ii*) If the maximum efficiency occurs at 80% of full-load, find the new core loss and full-load copper loss assuming that total full-load loss is constant.

[**Ans.** (*i*) 70.7% of full-load, 97.79% ; (*ii*) P_i = 468.36 W, P_c = 731.64 W]

37. A 1100/220 V transformer has a primary resistance of 0.25 Ω and secondary resistance of 0.03 Ω. If iron loss amounts to 250 W, determine the secondary current at which maximum efficiency occurs and find the maximum efficiency at 0.8 power factor. [**Ans.** 111.8 A ; 97.52%]

38. A single-phase 250 kVA transformer has an efficiency of 96% on full-load at 0.8 power factor and on half full-load at 0.8 power factor. Find :

(*i*) Iron loss, and (*ii*) full-load copper loss.

[**Ans.** P_i = 2.767·kW, P_c = 5.533 kW]

39. The efficiency of a 300 kVA, single-phase transformer is 97.8% when delivering full-load at 0.8 power factor lagging and 98.4% when delivering half full-load at unity power factor. Determine the efficiency at 80% of full-load at 0.8 power factor lagging. [**Ans.** 97.95%]

40. A 12 kVA, 400/200 V single-phase 50-Hz transformer has maximum efficiency of 95% at 85% of full-load at unity power factor. Determine the efficiency at full-load at 0.8 power factor lagging. [**Ans.** 93.76%]

41. A 50 kVA, 2000/250 V single-phase transformer has resistances of 1.1 Ω and 0.015 Ω and reactances of 4.4 Ω and 0.06 Ω for the high voltage and low voltage windings respectively. It has a maximum efficiency of 96.05% at 80% of full-load at unity p.f. The magnetising current for the h.v. side at 2000 V is 1.25 A. Find the readings of suitable instruments for open circuit test and short-circuit tests, supply being given to h.v. side in both cases.

[**Ans.** O.C. test : 2000 V, 1.316 A, 824 W
S.C. test : 212.3 V, 25 A, 1287 W]

42. A 100 kVA, 3300/300 V single-phase transformer has an efficiency of 97.5% both on full-load at unity p.f. and on half full-load at unity power factor. The power factor of the no-load current is 0.3 lagging and the regulation on full-load at 0.8 p.f. lagging is 3 per cent. Draw the equivalent circuit referred to the l.v. side.

[**Ans.** R_{01} = 105.3 Ω, X_0 = 33.11 Ω,
R_{02} = 0.01538 Ω, X_{02} = 0.02451 Ω]

43. The maximum efficiency of a single-phase 240 kVA, 2000/250 V transformer occurs at 70% of full-load and is equal to 98% at 0.8 p.f. lagging. Determine the efficiency and regulation on full-load at 0.8 p.f. lagging if the impedance of the transformer is 8 per cent. [**Ans.** 97.81%, 5.699%]

44. A 5 kVA, 230/115 V transformer takes 1.2 A and 75 W when 230 V is applied to h.v. winding and l.v. winding is kept open. It takes 21.75 A and 150 W when the l.v. winding is short-circuited and 17.4 V is applied to the h.v. winding. Find :

(*i*) The no-load current as a fraction of full-load input current at 0.8 p.f. lagging ;

(*ii*) Percentage regulation on full-load at 0.8 p.f. lagging ; and

(*iii*) The load at which maximum efficiency occurs and the maximum efficiency at unity p.f.

[**Ans.** 5.162% of full-load input current ; 6.566% ; 70.7% of full-load, 95.94%]

45. The efficiency of a 20 kVA, 2000/200 V transformer is 96.8% at full-load at unity p.f. and 96% at 60% of a load at 0.8 p.f. Find the regulation at full-load at :

(*i*) 0.8 p.f. lagging ; and (*ii*) 0.8 p.f. leading if the impedance is 7 per cent.

[**Ans.** (*i*) 5.469%, (*ii*) – 2.385%]

46. A 100 kVA, 2000/200 V single-phase transformer takes a current of 50 A and 2400 W at 100 V when the low voltage winding is short circuited. Determine the load-voltage and percentage regulation when delivering full-load current at 0.8 p.f. leading, the supply voltage being 2000 V.

[**Ans.** 200.142 V ; – 0.071%]

47. A single-phase, 25 kVA transformer has an iron loss of 240 W and full-load copper loss of 600 W.

(*i*) Find the load at which maximum efficiency occurs and maximum efficiency at 0.8 p.f.

(*ii*) If the maximum efficiency occurs at 70% of full-load, find the core loss and copper loss assuming the total loss to be the same as in the previous case.

[**Ans.** (*i*) 63.25% of full-load, 96.36% ;
(*ii*) 276.3 W, 563.7 W]

All-Day Efficiency

48. A 20 kVA, 2000/200 V transformer has an iron loss of 300 W and full-load copper loss of 400 W. During the day it is loaded as follows :

No. of hours	*Load*	*Power factor*
8	$\frac{1}{4}$ load	0.5
6	Full-load	0.8
6	$\frac{3}{4}$ load	unity
4	No-load	—

Find the all-day efficiency. **[Ans.** 94.84%]

49. A 30 kVA transformer has got a maximum efficiency of 97% at 80% of load at unity p.f. During the day it is loaded as follows :

No. of hours	*Load*	*Power factor*
10	4 kW	0.6 lag
9	20 kW	0.8 lag
5	24 kW	0.85 lag

Find the all-day efficiency. **[Ans.** 95.68%]

50. The all-day efficiency of 200 kVA transformer is 96% when it is loaded as follows :

No. of hours	*Load*	*Power factor*
12	120 kW	0.8
8	150 kW	unity
4	No-load	—

If maximum efficiency of this transformer occurs at 80% of full-load, find iron loss and full-load copper loss. **[Ans.** P_i = 2.645 kW, P_c = 4.133 kW]

51. An auto transformer supplies a load of 5 kW at 125 V and at unity power factor. If the primary voltage is 250 V, determine :

(*i*) Transformer ratio, (*ii*) Secondary current,
(*iii*) Primary current,
(*iv*) Number of turns across secondary if the total number of turns is 250,
(*v*) Power transformed, and
(*vi*) Power conducted directly from the supply mains to load. **[Ans.** $\frac{1}{2}$, 40 A, 20 A, 125, 9.5 kW, 2.5 kW]

52. The primary and secondary voltages of an auto-transformer are 500 V and 400 V respectively. Show with the aid of diagram the current distribution in the windings when the secondary current is 100 A and calculate the economy in copper. **[Ans.** 80 per cent]

53. A 200/250 V auto-transformer draws power from a 200 V line and supplies a 5 kW load with a power factor of 0.8 lagging. A second load of 2 kW is supplied at unity power factor from 100 V winding. Neglecting losses, calculate the current drawn by the transformer from the 200 V line and its power factor. **[Ans.** 42.64 A, 0.898]

54. A 5 kVA, single-phase 50-Hz transformer has full-load efficiency of 95 per cent and an iron loss of 50 W. The transformer is now connected as an auto-transformer to a 220 V supply. If it delivers a 5 kW load at unity power factor to a 110 V circuit, calculate the efficiency of the operation and the current drawn by the high-voltage side. **[Ans.** 79.27%, 23.2 A]

55. A 480/120 V, 5 kVA two winding transformer is to be used as an auto-transformer to supply power at 480 V from 600 V source. Draw the connection diagram and determine the kVA capacity as an auto-transformer. **[Ans.** 25 kVA]

Parallel Operation of Single-Phase Transformers

56. Two single-phase transformers with equal turns have impedance of (0.5 + j3) Ω and (0.6 + j10) Ω with respect to the secondary. If they operate in parallel, determine how they will share total load of 100 kW at power factor 0.8 lagging. **[Ans.** 78.2 kW, 21.8 kW]

57. Two single-phase transformers are connected in parallel on both primary and secondary. One transformer has full-load resistance drop of 1% and reactance drop of 6%. For the other transformer the corresponding values are 1.5% and 5% respectively. How will the following load be shared by the two transformers :

(*i*) 320 kW at 0.8 power factor lagging ; (*ii*) 400 kW at unity power factor ; and
(*iii*) 330 kW at 0.9 power factor leading.

[Ans. (*i*) 140.3 kW, 179.7 kW ; (*ii*) 184.8 kW, 215.2 kW ; (*iii*) 157.4 kW, 172.6 kW]

58. Two transformers of the same rating, connected in parallel, supply a load of 450 kW at 0.9 p.f. lagging. The percentage impedance of one transformer is $(1 + j4)\ \Omega$ and that of the other is $(1.5 + j5)\ \Omega$. Determine the load in kVA supplied by each transformer and its power factor.

[Ans. 279.4 kVA, 220.8 kVA, 0.9105 lagging]

59. Two 2200/1100 V transformers are operated in parallel to share a load of 125 kVA at 0.8 p.f. lagging.

Transformer 1	100 kVA	0.9% resistance and 10% reactance
Transformer 2	50 kVA	1% resistance and 5% reactance

Find the load shared by each transformer. **[Ans.** 48.4 kW at 0.767 p.f. lag, 51.6 kW at 0.83 p.f. lag]

60. Two 6600/250 V transformers have the following short-circuited characteristics :
Applied voltage 200 V, current 30 A, power input 1200 W for one of the transformers ; the corresponding data for other transformer being 120 V, 20 A and 1500 W. All values are measured on the h.v. side and with the l.v. terminals short-circuited. Find the approximate current and power factor of each transformer when working in parallel with each other on the high and low voltage sides and taking a total load of 300 kW at a p.f. of 0.8 lagging from the high voltage bus-bars.

[Ans. 27.7 A, 0.6266 (lag) ; 30.8 A, 0.915 (lag)]

61. Two transformers *A* and *B* are joined in parallel to the same load. Determine the current delivered by each transformer having given ; open circuit e.m.f. 6600 V for *A* and 6400 V for *B*. Equivalent leakage impedance in terms of the secondary = $0.3 + j3$ for *A* and $0.2 + j1$ for *B*. The load impedance is $8 + j6$.

[Ans. I_A = 195 A, I_B = 421 A]

62. Two transformers *A* and *B* of ratings 500 kVA find 250 kVA are supplying a load of 750 kVA at 0.8 power factor lagging. Their open-circuit voltages are 405 V and 415 V respectively. Transformer *A* has 1% resistance and 5% reactance and the transformer *B* has 1.5% resistance and 4% reactance. Find :
(*i*) Cross-current in the secondaries on no-load, and
(*ii*) Load shared by each transformer. **[Ans.** $-230\ \angle -72.9°$; $388\ \angle -35°$ kVA ; $350\ \angle -42.6°$ kVA]

Three-phase Transformers

63. A 3-phase, 50-Hz transformer has a delta-connected primary and star-connected secondary, the line voltages being 22,000 V and 400 V respectively. The secondary has a star-connected balanced load at 0.8 power factor lagging. The line current on the primary side is 5 A. Determine the current in each coil of the primary and in each secondary line. What is the output of the transformer in kW.

[Ans. $5/\sqrt{3}$ A ; 275 A ; 15.24 kW]

64. A 400 V, 3-phase supply is connected through a loss-free transformer 1 : 1 ratio, which has its primary connected in mesh and secondary in star to a load comprising the three 10 Ω resistors connected in mesh. Calculate the currents : (*i*) in the transformer windings ; (*ii*) in the resistors ; (*iii*) in the lines to the supply ; (*iv*) the load ; find also (*v*) the power supplied ; and (*vi*) power dissipated by each resistor.

[Ans. 120 A ; 120 A ; 69.3 A ; 207.8 A ; 120 A ; 144 kW ; 48 kW]

65. A 3-phase, 1200 kVA, 6600/1100 V transformer is delta-connected on the primary and star-connected on the secondary. The primary resistance/phase is 0.025 Ω. Determine the efficiency on full-load at :
(*i*) unity power factor ; and
(*ii*) 0.8 power factor if the iron loss is 20 kW. **[Ans.** 94.5%, 93.22%]

66. The resistance/phase and leakage reactance/phase of the primary winding of a 3-phase, 1500 kVA, 11000/2200 V, 50-Hz, step-down transformer are 0.8 Ω and 4.8 Ω and the corresponding values for the secondary winding are 0.01 Ω and 0.06 Ω respectively. If the primary winding is connected in delta and secondary in star. Calculate :
(*i*) The voltage to be applied at normal frequency to the primary to circulate full-load current through it when the secondary is short-circuited.
(*ii*) The power input under this condition. **[Ans.** 428.7 V, 9.612 kW]

Open delta Connection or V-V Connection

67. Three single-phase transformers are connected in mesh. If one transformer is found faulty and removed what will be the reduction in rating of each of the other transformers ? What must be the rating of each transformer in a *V-V* connection to supply a 3-phase balanced load of 100 kVA ? If a third similar transformer is included what will be the rated capacity of the set ?

[**Ans.** 13.3% of its rating ; 57.68 kVA ; 173 kVA]

68. Three single-phase transformers, connected in Δ-Δ, supply a balanced 3-phase load of 1200 kW at 4400 V at 0.8 power factor lagging. The transformers are supplied from 3-phase mains at 11000 V. Find the currents in the windings of each transformer. If one transformer is found faulty and the other two are connected in *V-V*, determine the currents in the windings of each transformer and its power factor.

[**Ans.** 113.2 A ; 45.48 A ; 196.9 A ; 78.76 ; 0.6928 lagging]

69. Two similar transformers each having a transformation ratio of $\frac{1}{5}$, are connected in *V-V* and supply a 3-phase load of 500 kW at 0.8 p.f. lagging at 440 V. Find :

(*i*) The line currents ; (*ii*) The currents in each transformer winding ; and

(*iii*) The power factor of the transformer.

If an identical transformer is added to the *V-V* set to convert it into a Δ-Δ set determine the additional load that can be supplied and the secondary line and phase currents if the temperature rise is same in both cases. [**Ans.** 820.4 A, 820.4 A, 164.1 A, 0.6928 lagging ; 364.8 kW, 1419 A, 820.4 A]

Scott or T-T Connection

70. Two single-phase electric furnaces *A* and *B* are supplied a 220 V from a 3-phase 1100 V supply by means of a Scott-connected transformer combination. If the total output is 400 kW at 0.8 power factor lagging determine transformation ratio and the currents in the windings of each transformer.

[**Ans.** $\frac{1}{5}$, 1136 A, 227.2 A ; 1/4.33, 1136 A, 262.4 A]

71. A Scott-connected transformer set supplies two electric furnaces *A* and *B* at 110 V each from a 3-phase 4400 V system. If the furnace connected to teaser transformer secondary takes 360 kW and the other one connected to main transformer secondary takes 600 kW ; (*i*) at unity p.f. ; and (*ii*) at 0.8 p.f. lagging ; find the line current taken from the 3-phase mains.

[**Ans.** 94.36 A, 144.3 A, 144.3 A ; 118 A, 186.4 A, 180.4 A]

72. A Scott-connected transformer set supplies two single-phase furnaces *A* and *B* at 120 V from a 3-phase 6600 V system. If the furnace connected to teaser transformer takes 400 kW at unity power factor and the other one connected to main transformer takes 660 kW at 0.8 power factor lagging determine the line currents from the 3-phase mains. Neglect losses. [**Ans.** I_R = 69.92 A ; I_Y = 148.7 A ; I_B = 107.7 A]

Parallel Operation of Three-Phase Transformers

73. A load of 800 kW at 0.8 lagging power factor is supplied by two 3-phase transformers rated 400 kVA and 600 kVA operating in parallel. The transformers have the same turn ratio and their per unit resistance and reactance are as follows :

400 kVA transformer — Resistance = 0.02; Reactance = 0.04

600 kVA transformer — Resistance = 0.01; Reactance = 0.05

Calculate kVA output and power factor of each transformer.

[**Ans.** 435.8 kVA at p.f. 0.882 (lag), 572.6 kVA at p.f. 0.725 (lag)]

74. Two 3-phase 4400/400 V transformers *A* and *B* have the following data :

Transformer A : % impedance/phase referred to secondary
= (1 + *j*3)

Rated kVA = 400

Transformer B : % impedance/phase referred to secondary
= (1 + *j*4)

Rated kVA = 800

Determine the load shared by each transformer and its power factor if the total load is 1000 kVA at 0.8 p.f. lagging. [**Ans.** 395 kVA at 0.8269 p.f. (lag), 605.9 kVA at 0.7814 p.f. (lag)]

3

Polyphase Induction Motors

3.1. General aspects 3.2. Classification of A.C. motors 3.3. Constructional details 3.4. Production of rotating magnetic field 3.5. Theory of operation of an induction motor 3.6. Slip 3.7. Frequency of rotor current 3.8. Rotor e.m.f. and rotor current 3.9. Torque and power 3.10. Effect of change in supply voltage on starting torque 3.11. Effect of change in supply voltage on torque and slip 3.12. Full-load torque and maximum torque 3.13. Starting torque and maximum torque 3.14. Torque-slip and torque-speed curves 3.15. Operating characteristics of a 3-phase squirrel-cage induction motor 3.16. Operating characteristics of a wound-rotor induction motor 3.17. Power stages in an induction motor 3.18. Measurement of slip 3.19. Induction motor as transformer 3.20. Equivalent circuit of an induction motor 3.21. Induction motor losses and efficiency 3.22. Circle diagram 3.23. Starting of induction motors 3.24. Resistance steps for 3-phase slip ring induction motor starter 3.25. Crawling 3.26. Cogging 3.27. Double squirrel cage motor 3.28. Speed control of induction motors 3.29. Electrical braking of polyphase induction motors 3.30. Factors governing performance of induction motors 3.31. Effects of operating conditions 3.32. Ratings of 3-phase induction motors 3.33. Squirrel cage motors—Advantages, disadvantages and applications 3.34. Wound rotor (or Slip ring) induction motors—Advantages, disadvantages and applications 3.35. Comparison of squirrel cage and phase wound induction motors 3.36. Comparison between induction and synchronous motors 3.37. Schrage motor 3.38. Induction generator 3.39. Linear induction motor—Highlights—Objective type questions—Theoretical questions—Unsolved problems.

3.1. GENERAL ASPECTS

Introduction. An induction motor is simply an *electric transformer* whose magnetic circuit is separated by an air gap into two relatively movable portions, one carrying the primary and the other the secondary winding. Alternating current supplied to the primary winding from an electric power system induces an opposing current in the secondary winding, when *latter is short-circuited* or *closed through an external impedance*. Relative motion between the primary and secondary structures is produced by the *electromagnetic forces corresponding to the power thus transferred across the air gap by induction*.

The *essential feature which distinguishes the induction machine from other types of electric motors is that the secondary currents are created solely by induction, as in a transformer instead of being supplied by a D.C. exciter or other external power source, as in synchronous and D.C. machines.*

Advantages. Three-phase induction motor is the *most commonly used motor* in industrial applications because of the *advantages* listed below :

1. Simple design
2. Rugged construction
3. Reliable operation
4. Low initial cost
5. Easy operation and simple maintenance
6. High efficiency
7. Simple control gear for starting and speed control.

Applications :

Induction motors are available with torque characteristics suitable for a *wide variety of loads* :

(*i*) The standard motor has a starting torque of about 120 to 150 per cent of full-load torque. Such motors are suitable for most applications.

(*ii*) For starting loads such as small refrigerating machines or plunger pumps operating against full pressure or belt conveyors, *high torque motors with a starting torque of twice normal full-load torque*, or more, are used.

(*iii*) For driving machines that use large flywheels to carry peak loads, such as punch presses and shears, a high-torque motor with a slip at full-load up to 10 per cent is available. *The high slip permits enough change in speed to make possible the proper functioning of the flywheel.*

(*iv*) By the use of a wound-rotor with suitable controller and external resistances connected in series with the rotor winding, it is possible to obtain any value of starting torque up to the maximum breakdown torque. *Such motors are well adapted as constant-speed drives for loads that have large friction loads to overcome at starting.*

3.2. CLASSIFICATION OF A.C. MOTORS

Different A.C. motors may be classified as follows :

1. **According to the 'type of current' :**
   (*i*) Single-phase (*ii*) Three-phase.
2. **According to 'speed' :**
   (*i*) Constant speed (*ii*) Variable speed
   (*iii*) Adjustable speed.
3. **According to 'principle of operation' :**
   (A) **Synchronous motors**
   (*i*) Plain (*ii*) Super.
   (B) **Asynchronous motors**
   (*a*) Induction motors :
   (*i*) Squirrel cage
   —single
   —double
   (*ii*) Slip-ring (external resistance)
   (*b*) Commutator motors :
   (*i*) Series
   —single-phase
   —universal
   (*ii*) Compensated
   —conductively
   —inductively.
4. **According to 'structural features' :**
   (*i*) Open (*ii*) Enclosed
   (*iii*) Semi-enclosed (*iv*) Ventilated
   (*v*) Pipe-ventilated (*vi*) Riveted frame eye
   (*vii*) Splash proof (*viii*) Totally enclosed fan-cooled
   (*ix*) Explosion proof (*x*) Water proof.

3.3. CONSTRUCTIONAL DETAILS

3.3.1. The Stator

- The stator frame consists of a symmetrical and substantial casting, having feet cast integral with it. The stator core, consisting of high grade, low loss electrical sheet-steel stampings, is assembled in the frame under hydraulic pressure. The thickness of stampings/laminations is usually from 0.35 to 0.6 mm. The stator laminations are punched in one piece for small induction motor (Fig. 3.1). In induction machines of large size the stator core is assembled from a large number of *segmental laminations*.

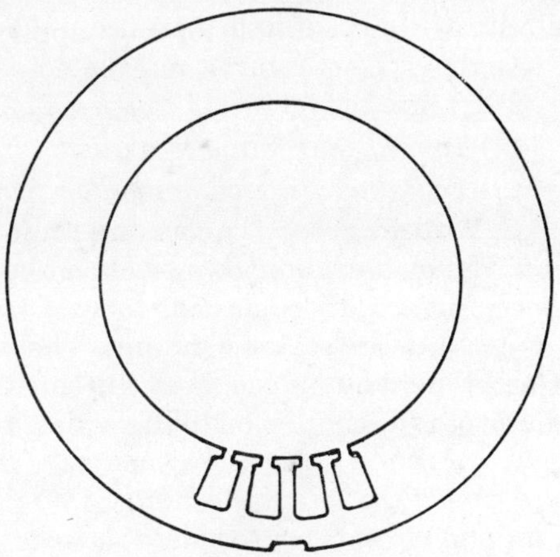

Fig. 3.1. Stator stamping.

The slots are *sometimes of the 'open type'* (*i.e.*, having parallel walls) for the accommodation of former wound coils. But the usual practice is to have practically *'enclosed slots' in order to reduce the effective* length of air-gap.

- The *stator windings* are given the utmost care to make them mechanically and electrically sound, so as to ensure long life and high efficiency. After the winding is in position it is thoroughly dried out whilst still hot and is completely immersed in a high grade synthetic resin varnish. It is then *acid, alkali, moisture* and *oil proof*.

  For *small motors* working at ordinary voltages, *single layer mush winding* is used. For *medium size machines double layer lap winding with diamond shaped coils* is used. *Single layer concentric windings are used for large motors working at high voltages.*
- *Frames* of electrical machines house the stator core. Frames of small and medium sizes of induction motors have hollow cylindrical form and that of larger motors have the shape of a circular box. *In small induction motors,* having a frame diameter of up to about 150 cm, *the frame also supports the end shields.* The frame should be *strong and rigid* as rigidity is very important in the case of induction motors of large dimensions. *This is because the length of the air gap is very small and if the frame is not rigid, it would create an irregular air gap around the machine resulting in production of unbalanced magnetic pull.* Frames for small machines are made as a *single unit* and are usually cast. The frames of *medium and large sized* machines are *fabricated from rolled steel plates.*

3.3.2. The Rotor. The rotors are of two types :

1. Squirrel-cage ;
2. Wound rotor.

Squirrel-cage. The squirrel-cage rotor is made up of stampings (Fig. 3.2) which are keyed directly to the shaft. The slots are partially closed and the winding consists of embedded copper bars to which the short-circuited rings are brazed. The squirrel cage rotor is so robust that it is almost indestructible.

The great majority of present day induction motors are manufactured with squirrel-cage rotors, a *common practice being to employ winding of cast aluminium.* In this construction the assembled rotor laminations are placed in a mould after which molten aluminium is forced in, under pressure, to form bars, end rings and cooling fans as extension of end rings. This

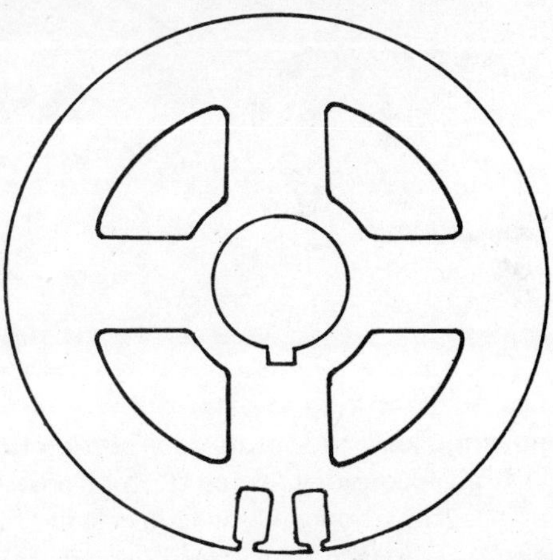

Fig. 3.2. Rotor stamping.

is known as *die cast rotor* and has become very popular as there are no joints and thus there is no possibility of high contact resistance.

In this type of rotor, it may be noted that *slots are not made parallel to the shaft but they are 'skewed'* to serve the following purposes :

(*i*) *To make the motor run quietly by reducing the magnetic hum.*

(*ii*) *To reduce the locking tendency of the rotor.*

2. **Wound rotor.** The wound rotor has also slotted stampings and the *windings are former wound.* The *wound rotor construction is employed for induction motors requiring speed control or extremely high values of starting torque.* The wound rotor has completely insulated copper windings very much like the stator windings. The windings can be connected in star or delta and the three ends are brought out at the three slip rings. The current is collected from these slip rings with carbon brushes from which it is led to the *resistances for starting purposes.* When the motor is *running, the slip rings are short-circuited by means of a collar, which is pushed along the shaft and connects all the slip rings together on the inside.* Usually the *brushes* are *provided with a device for lifting them from the slip rings when the motor has started up, thus reducing the wear and the frictional losses.*

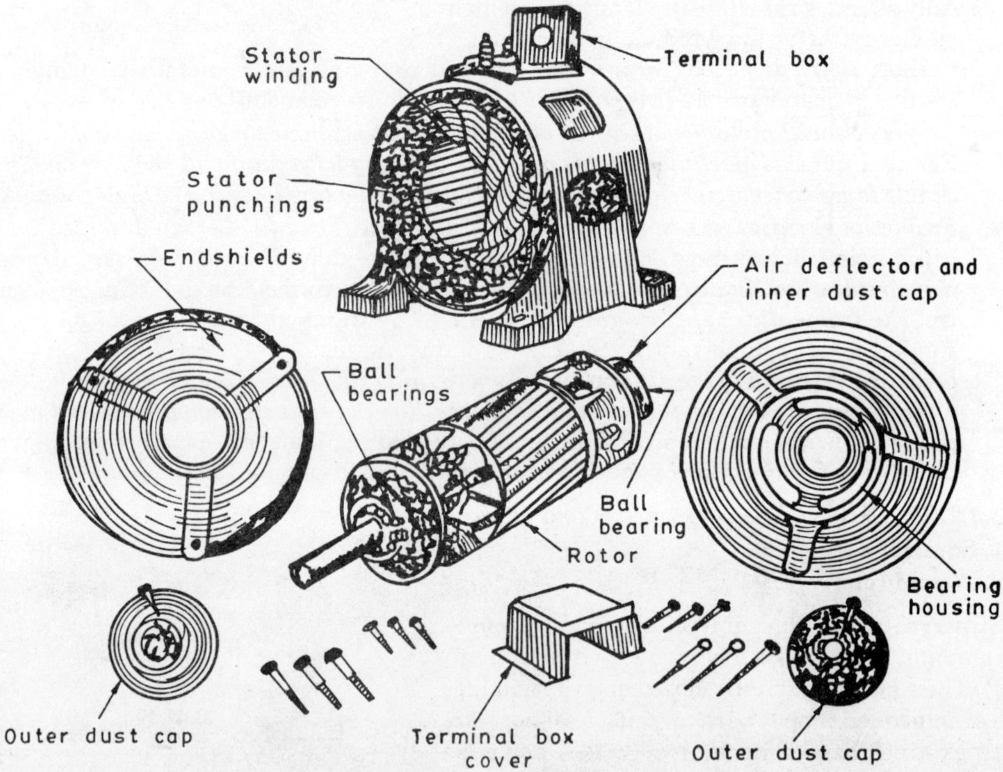

Fig. 3.3. Component parts of a small squirrel-cage induction motor.

The *number of slots in the rotor should never be equal to the number of slots in the stator.* If they are, there would be a *variation of reluctance of the magnetic path* from maximum, when teeth are opposite slots, to minimum when teeth are opposite teeth. The resulting flux pulsations would have a high frequency, since the periodic time would be the interval period for a tooth to occupy similar positions opposite two successive teeth. This will *not only cause extra iron loss but the rotor will tend to lock with the stator* if at the time of starting teeth are opposite teeth. The best plan is to make the number of the stator and the rotor teeth *prime to each other.*

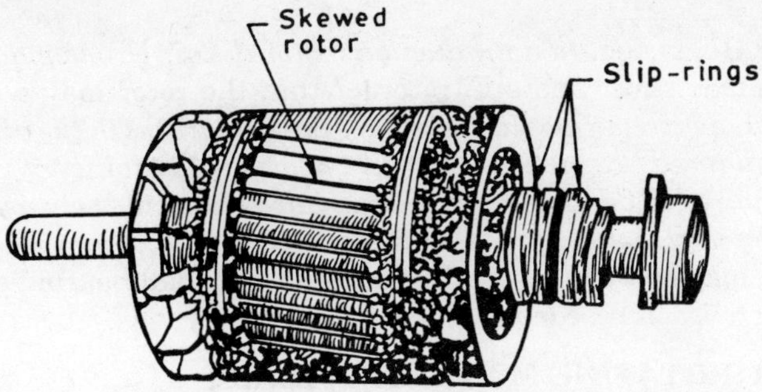

Fig. 3.4. Induction motor with phase-wound rotor, showing the three slip rings on the rotor shaft.

Figs. 3.5 and 3.6 show squirrel-cage and phase-wound induction motors respectively.

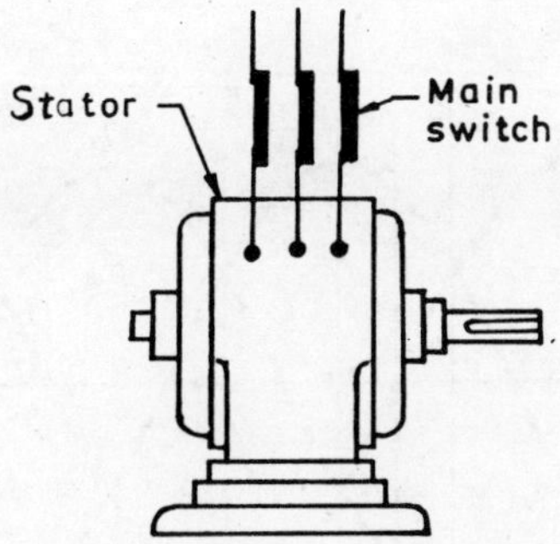

Fig. 3.5. Squirrel-cage motor.

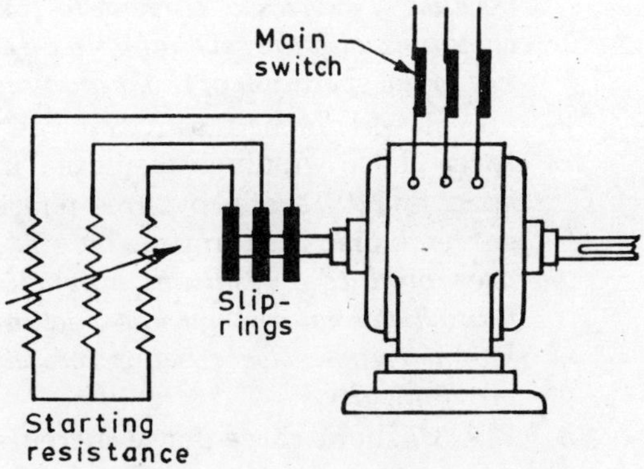

Fig. 3.6. Phase-wound motor connected to a three-phase star-connected starting resistance.

Advantages of a squirrel-cage motor over a phase-wound induction motor. As compared with a wound rotor a squirrel-cage induction motor entails the following *advantages* :

1. Slightly higher efficiency.
2. Cheaper and rugged in construction.
3. No slip rings, brush gear, short-circuiting devices, rotor terminals for starting rheostats are required. The star-delta starter is sufficient for starting.
4. It has better space factor for rotor slots, a shorter overhang and consequently a small copper loss.
5. It has a smaller rotor overhang leakage which gives a better power factor and a greater pull out torque and overload capacity.
6. It has bare end rings, a large space for fans and thus the cooling conditions are better.

The major *'disadvantage' of squirrel-cage motor* is that *it is not possible to insert resistance in the rotor circuit for the purpose of increasing the starting torque. The cage rotor has a smaller starting torque and large starting current as compared with wound rotor*.

3.3.3. Slip Rings. The slip rings for wound rotor machines are made of *either brass or phosphor bronze.* They are shrunk on to a cast iron sleeve with moulded silica insulation. This assembly is passed on to the rotor shaft. The slip rings are rotated either between the core and the bearing or on the shaft extension. In the latter case the shaft is made hollow to allow the three connections from rotor to slip rings to pass through bearings.

3.3.4. Shaft and Bearings. *In an induction motor the air gap is made as small as possible.* Therefore the shaft is made short and stiff in order that the rotor may not have any significant deflection, *as even a small deflection would create large irregularities in the air gap which would lead to production of an unbalanced magnetic pull. There is also a possibility of rotor and stator fouling with each other. Ball and roller bearings are generally used as with their use, accurate centering is much simpler than with journal bearings.* Also the *overall length of machine is reduced. For small motors,* a roller bearing may be used at the driving end and a ball bearing at the non-driving end. *For large and heavy rotors journal bearings are used.*

3.4. PRODUCTION OF ROTATING MAGNETIC FIELD

- *When a three-phase voltage is applied to the three-phase stator winding of the induction motor a rotating magnetic field is produced, which by transformer action induces a 'working' e.m.f. in the rotor winding. The rotor-induced e.m.f. is called a working e.m.f. because it causes a current to flow through the armature winding conductors. This combines with the revolving flux-density wave to produce torque. Thus revolving field is a key to the operation of the induction motor.*

  It will now be shown that when three-phase windings displaced in space by 120°, are fed by three-phase currents displaced in time by 120°, they *produce a resultant magnetic flux which rotates in space as if actual magnetic poles were being rotated mechanically.*

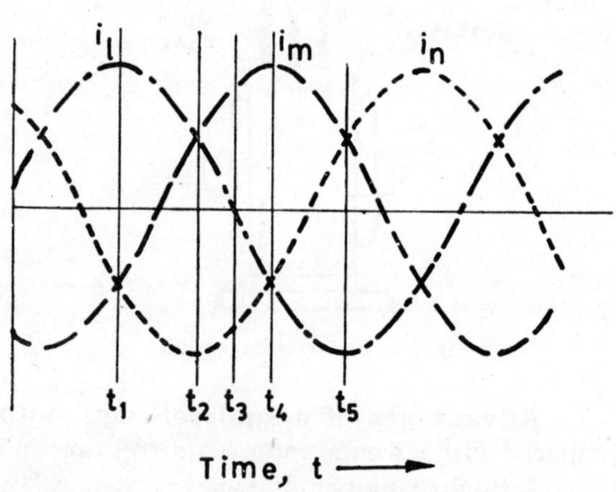

Fig. 3.7. Balanced three-phase alternating currents.

- Fig. 3.7 shows three-phase currents which are assumed to be flowing in phases '*l*', '*m*' and '*n*' respectively. These currents are time-displaced by the 120 electrical degrees.
- Fig. 3.8 shows the stator structure and the three-phase winding. Each phase (normally distributed over 60 electrical degrees) for convenience is represented by a single coil. Thus coil *l*-*l'* represents the entire phase '*l*' winding having its flux axis directed along the vertical. This means that whenever phase '*l*' carries current it produces a flux field directed along the vertical-up or down. The right-hand flux rule readily verifies this statement. Similarly, the flux axis of phase '*m*' is 120 electrical degrees displaced from phase '*l*' ; and that of phase '*n*' is 120 electrical degrees displaced from phase '*m*'. The unprimed letters refer to the beginning terminals of each phase.
- **Magnitude and direction of the resultant flux corresponding to time instant t_1.** (Fig. 3.7). At this instant the current into phase '*l*' is at its *positive maximum* value while currents in phases '*m*' and '*n*' are at *one-half their maximum values.* In Fig. 3.8 it is arbitrarily assumed that when current in a given phase is positive, it flows into the paper with respect to the unprimed conductors. Thus since, at time t_1, i_l is positive, a cross is used for conductor '*l*' [Fig. 3.8 (*i*)]. Of course a dot is used for *l'* because it refers to the return connection. Then by the right-hand rule it follows that phase '*l*' produces a flux contribution directed downward along the vertical. Moreover, the magnitude of this contribution is the maximum value because the current is at maximum. Hence, $\phi_l = \phi_{max}$, where ϕ_{max} is the maximum flux per pole of phase '*l*'. It is important to understand that

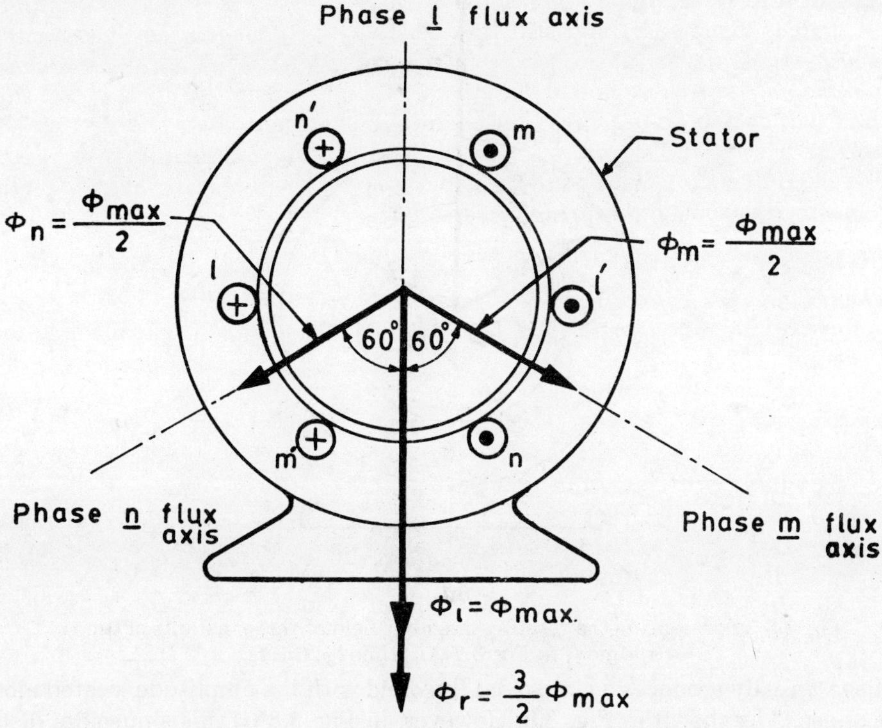
Phase l flux axis
n′
m
Stator
$\Phi_n = \frac{\Phi_{max}}{2}$
$\Phi_m = \frac{\Phi_{max}}{2}$
l
l′
60°
60°
m′
n
Phase n flux axis
Phase m flux axis
$\Phi_l = \Phi_{max.}$
$\Phi_r = \frac{3}{2}\Phi_{max}$

(i)

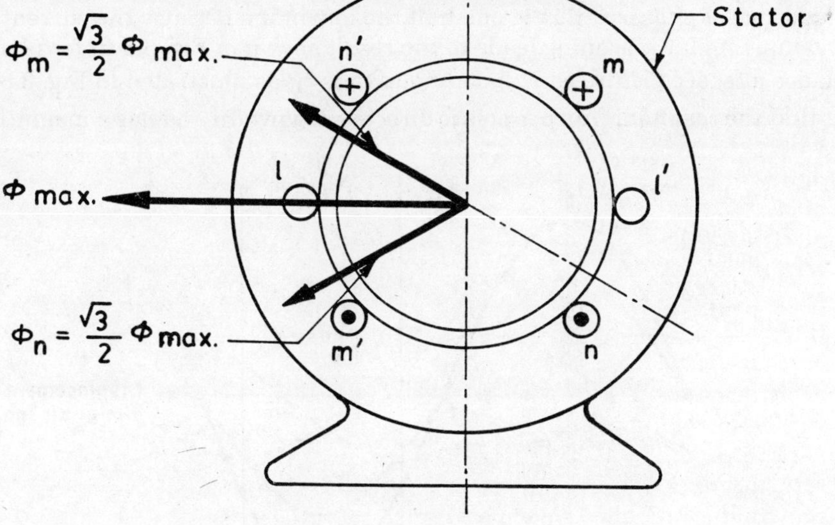
Stator
$\Phi_m = \frac{\sqrt{3}}{2}\Phi_{max.}$
n′
m
$\Phi_r = \frac{3}{2}\Phi_{max.}$
l
l′
$\Phi_n = \frac{\sqrt{3}}{2}\Phi_{max.}$
m′
n

(ii)

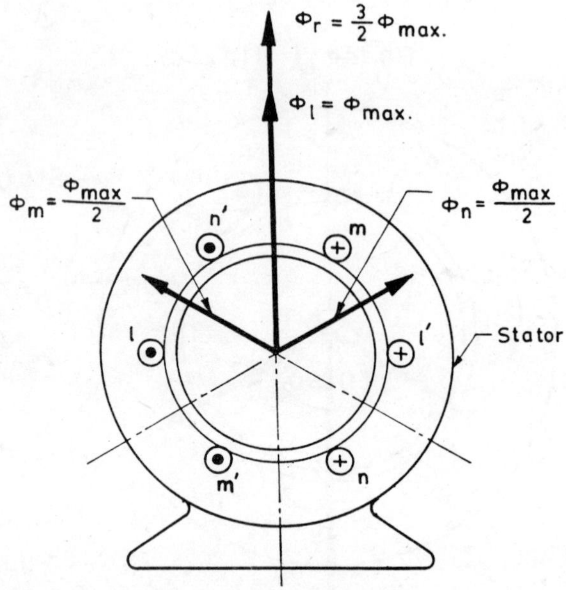

(iii)

Fig. 3.8. Representing the rotating magnetic field at three instants of time : (i) time t_1 in Fig. 3.7 ; (ii) time t_3 ; (iii) t_5.

phase '*l*' really produces a sinusoidal flux field with the amplitude located along the axis of phase '*l*' as shown in Fig. 3.9. However, in Fig. 3.8 (*i*) this sinusoidal distribution is conveniently represented by the vector ϕ_l.

For determining the magnitude and direction of the field contribution of phase '*m*' at time t_1, we first find that the current in phase '*m*' is negative with respect to that in phase '*l*'. Hence the conductor that stands for the beginning of phase '*m*' must be assigned a dot while '*m*' is assigned a cross. Hence the instantaneous flux contribution of phase '*m*' is directed downward along its flux axis and the magnitude of phase '*m*' flux is one-half the maximum because the current is at one-half its maximum value. Similar reasoning leads to the result shown in Fig. 3.8 (*i*) for phase '*n*'.

A glance at space picture corresponding to time t_1, as illustrated in Fig. 3.8 (*i*), should make it apparent that the resultant flux per pole is directed downward and has a magnitude of $\frac{1}{2}$ times the

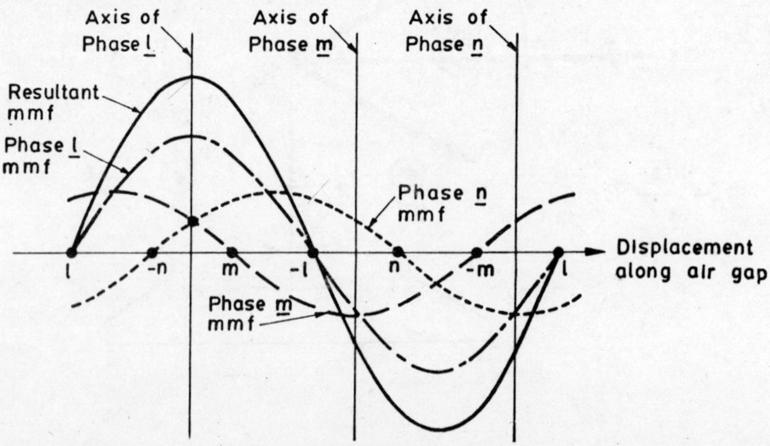

Fig. 3.9. Component and resultant field distributions corresponding to t_1 (Fig. 3.7).

maximum flux per pole of any one phase. Fig. 3.9 depicts the same results as Fig. 3.8 (*i*) but does so in terms of sinusoidal flux waves rather than flux vectors.

- **Magnitude and direction of the resultant flux corresponding to time instant t_3** [Fig. 3.8 (*ii*)]. Here phase '*l*' current is zero, yields no flux contribution. The current in phase '*m*' is positive and equal to $\frac{\sqrt{3}}{2}$ its maximum value. Phase '*n*' has the same current magnitude but is negative. Together phases '*m*' and '*n*' *combine to produce a resultant flux having the same magnitude as at time t_1*. See Fig. 3.8 (*ii*). It is important to note, too, that an *elapse of 90 electrical degrees in time results in rotation of the magnetic flux field of 90 electrical degrees.*
- **At time instant t_5.** A further elapse of time equivalent to an additional 90 electrical degrees leads to the situation depicted in Fig. 3.8 (*iii*).

From the above discussion it is apparent that the application of three-phase currents through a balanced three-phase winding gives rise to a rotating magnetic field that exhibits the following *two characteristics.*

1. It is of *constant amplitude.*

2. It is of *constant speed*. It follows from the fact that the resultant flux traverses through 2π electrical radians in space for every 2π electrical radians of variation in time for the phases currents. Hence for a 2-pole machines where electrical and mechanical degrees are identical, each cycle of variation of current produces one complete revolution of the flux field. Hence *this is a fixed relationship which is dependent upon the frequency of the currents* and number of poles for which the three-phase winding is designated. In the case where the winding is designed for *four* poles it requires *two cycles* of variation of the current to produce one revolution of the flux field. Therefore it follows that for a *p*-pole machine the relationship is :

$$N_s = \frac{120f}{p} \quad ...(3.1)$$

or

$$f = \frac{pN_s}{120} \quad ...[3.1\,(a)]$$

N_s (r.p.m.) is called the *synchronous speed* and all synchronous machines run at their respective synchronous speeds.

It may be noted that the speed of rotation of the field as described by eq. (3.1) is always given relative to the phase windings carrying the time-varying currents. Accordingly, if a situation arises where the *winding is itself revolving*, then the speed of rotation of the field relative to inertial force is different from it with respect to the winding.

3.5. THEORY OF OPERATION OF AN INDUCTION MOTOR

When a three-phase is given to the stator winding a rotating field is set-up. This field sweeps past the rotor (conductors) and by *virtue of relative motion,* an e.m.f. is induced in the conductors which form the rotor winding. Since this winding is in the form of a closed circuit, a current flows, the direction of which is, by Lenz's law, such as to oppose the change causing it.

Now, *the change is the relative motion of the rotating field and the rotor, so that, to oppose this, the rotor runs in the same direction as the field and* attempts to catch up with it. It is clear that torque must be produced to cause rotation, and *this torque is due to the fact that currents flow in the rotor conductors* which are situated in, and at right angles to, a magnetic field.

Fig. 3.10 shows the induction motor action.

- When the motor shaft is *not loaded*, the machine has only to rotate itself against the mechanical losses and the rotor speed is *very close to the synchronous speed.* However, the

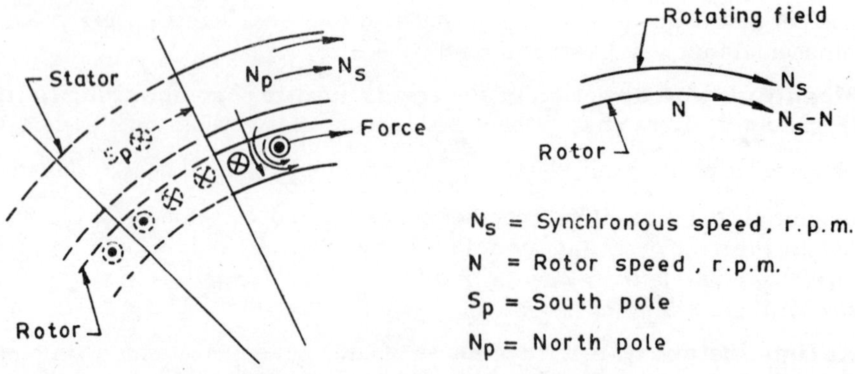

Fig. 3.10. Induction motor action.

rotor speed cannot become equal to the synchronous speed because if it does so, *the e.m.f. induced in the rotor winding would become zero and there will be no torque.* Hence the speed remains slightly less than the synchronous speed. If the motor shaft is *loaded,* the rotor will slow down and the relative speed of the rotor with respect to the stator rotating field will increase. The e.m.f. induced in the rotor winding will increase and will produce more rotor current which will increase the electromagnetic torque produced by the motor. Conditions of equilibrium are attained when the rotor speed has adjusted to a new value so that the electromagnetic torque is sufficient to balance the mechanical or load torque applied to the shaft. The speed of the motor *when running under full load conditions is somewhat less than the no-load speed.*

3.6. SLIP

- As earlier stated, *the rotor speed must always remain less than the synchronous speed. The difference between the synchronous speed and the rotor speed is known as* **'slip'.** It is usually expressed as a fraction of the synchronous speed. Thus slip s is

$$s = \frac{N_s - N}{N_s} \quad ...(3.2)$$

or $$N = N_s(1 - s) \quad ...[3.2\ (a)]$$

where N_s = synchronous speed (r.p.m.)

N = motor speed (r.p.m.).

In practice the value of slip is very small. At no-load, slip is around 1% or so and at full-load it is around 3%. For large efficient machines the slip at full-load may be around 1% only. The induction motor, is therefore, a motor with substantially constant speed and fills the same role as *D.C. shunt motor.*

- When the rotor is stationary (standstill) its speed is zero and $s = 1$. The rotor cannot run at synchronous speed because then there will be no rotor e.m.f. and no rotor current and torque. If the rotor is to run at synchronous speed an external torque is necessary. *If the rotor is driven such that $N > N_s$, the slip becomes negative, the rotor torque opposes the external driving torque and the* **machine acts as induction generator.**

- The induction motor derives its name from the fact that the *current in the rotor circuit is induced from the stator.* There is no external connection to the rotor except for some special purposes.

If the rotor reactance at standstill is X_2 its value at slip 's' becomes sX_2. *This is very desirable,* for at no-load the reactance becomes almost negligible and the rotor impedance is now all resistance.

Further if the rotor resistance is small the rotor current is large, so that motor works with a large torque which brings the speed near to synchronous speed, *i.e.* the slip is reduced.

3.7. FREQUENCY OF ROTOR CURRENT

At standstill (*i.e.* when the rotor is stationary), the *frequency of the rotor current is the same as the supply frequency (f)*. But when the rotor starts revolving, then the frequency depends upon the relative speed or slip-speed. If f_r is the frequency of the rotor current, then

$$N_s - N = \frac{120 f_r}{p} \qquad ...(i)$$

Also

$$N_s = \frac{120 f}{p} \qquad ...(ii)$$

Dividing (*i*) by (*ii*), we get

$$\frac{N_s - N}{N_s} = \frac{f_r}{f} \quad \text{or} \quad s = \frac{f_r}{f}$$

or

$$f_r = sf \qquad ...(3.3)$$

3.8. ROTOR E.M.F. AND ROTOR CURRENT

Rotor e.m.f :

When the rotor is stationary, an induction motor is equivalent to a 3-phase transformer with secondary short-circuited. Therefore, the induced e.m.f. per phase E_2 in the rotor at the instant of starting is given as :

$$E_2 = E_1 \times \frac{N_2}{N_1} \qquad ...(3.4)$$

where E_1 = applied voltage per phase to primary *i.e.*, stator winding,

N_1 = number of stator turns, and

N_2 = number of rotor turns.

When the rotor starts gaining speed, the relative speed of the rotor with respect to stator flux *i.e.* slip, is decreased. Hence induced e.m.f. in the rotor, which is *directly proportional to the relative speed i.e. slip*, is also decreased and is given by sE_2. Hence *for slip 's', the induced e.m.f. in the rotor is s times the induced e.m.f. in the rotor at standstill.*

Rotor current :

Let R_2 = rotor resistance/phase,

L_2 = rotor inductance/phase, and

E_2 = induced e.m.f. of rotor/phase at standstill.

At standstill :

Induced e.m.f. of rotor/phase $= E_2$

Rotor winding resistance/phase $= R_2$

Rotor winding reactance/phase, $X_2 = 2\pi f L_2$ where f is the supply frequency

Rotor impedance/phase, $Z_2 = \sqrt{R_2^2 + X_2^2}$

$\therefore$ Rotor current/phase $= \dfrac{E_2}{Z_2} = \dfrac{E_2}{\sqrt{R_2^2 + X_2^2}}$.

At slip 's' :

Induced e.m.f. of rotor/phase $= sE_2$

Rotor winding resistance $= R_2$

Rotor winding reactance $= 2\pi f_r L_2 = 2\pi s f L_2 = s(2\pi f L_2) = sX_2$

Rotor winding impedance/phase $= \sqrt{R_2^2 + (sX_2)^2}$

$\therefore$ Rotor current/phase,
$$I_2 = \frac{sE_2}{\sqrt{R_2^2 + (sX_2)^2}}$$
$$= \frac{sE_2}{\sqrt{R_2^2 + s^2X_2^2}} = \frac{E_2}{\sqrt{(R_2/s)^2 + X_2^2}} \quad \text{...(3.5)}$$

The rotor currrent I_2 lags the rotor voltage E_2 by rotor power factor angle ϕ_2 given by
$$\phi_2 = \tan^{-1}\left(\frac{sX_2}{R}\right)$$

Power factor of rotor current,
$$\cos\phi_2 = \frac{R_2}{\sqrt{R_2^2 + s^2X_2^2}} = \frac{R_2/s}{\sqrt{(R_2/s)^2 + X_2^2}}. \quad \text{...(3.6)}$$

3.9. TORQUE AND POWER

The torque of an induction motor (being due to interaction of a rotor and stator fields),
$$T \propto \phi I_2 \cos\phi_2$$
where ϕ = flux of rotating stator,

I_2 = rotor current/phase, and

$\cos\phi_2$ = rotor power factor.

Since rotor e.m.f./phase at standstill, $E_2 \propto \phi$
$$\therefore \quad T \propto E_2 I_2 \cos\phi_2$$
or
$$T = kE_2 I_2 \cos\phi_2 \text{ where } k \text{ is any constant} \quad \text{...(3.7)}$$

Substituting the value of I_2 and $\cos\phi_2$ from eqns. (3.5) and (3.6) in eqn. (3.7), we get
$$T = kE_2 \frac{sE_2}{\sqrt{R^2 + s^2X_2^2}} \times \frac{R_2}{\sqrt{R^2 + s^2X_2^2}}$$
i.e.,
$$T = \frac{ksR_2E_2^2}{R_2^2 + s^2X_2^2} \quad \text{...(3.8)}$$

3.9.1. Starting Torque. At start slip 's' = 1. Therefore, expression for starting torque may be obtained by putting $s = 1$ in eqn. 3.8.

Starting torque,
$$T_{st} = \frac{kR_2E_2^2}{R_2^2 + X_2^2}. \quad \text{...(3.9)}$$

3.9.2. Condition for Maximum Torque. The value of torque when motor is running is given by
$$T = \frac{ksR_2E_2^2}{R_2^2 + s^2X_2^2}$$

Torque will be maximum when,
$$\frac{sR_2}{R_2^2 + s^2X_2^2} \quad \text{or} \quad \frac{R_2}{\frac{R_2^2}{s} + sX_2^2}$$
or
$$\frac{R_2}{\left(\frac{R_2}{\sqrt{s}} - X_2\sqrt{s}\right)^2 + 2R_2X_2} \text{ is maximum, } viz., \ \frac{R_2}{\sqrt{s}} - X_2\sqrt{s} = 0$$

or $$s\ (= s_{mT}) = \frac{R_2}{X_2} \qquad ...(3.10)$$

(where s_{mT} = slip corresponding to maximum torque)

$\therefore$ Maximum torque, $$T_{max} = \frac{kE_2^2}{2X_2} \qquad ...(3.11)$$

From the above expression, the following conclusions can be drawn :

- Maximum torque is *independent* of rotor circuit resistance.
- Maximum torque varies *inversely as standstill reactance* of the rotor. Therefore to have maximum torque, standstill reactance (*i.e.* inductance) of the rotor should be kept as small as possible.
- The slip at which the maximum torque occurs depends upon the resistance of the rotor.

The condition for getting maximum torque at starting can be obtained by putting $s = 1$ in eqn. (3.10).

Thus, starting torque will be maximum if

$$\frac{R_2}{X_2} = s = 1 \quad \text{or} \quad R_2 = X_2.$$

3.9.3. Starting Torque of a Squirrel-Cage Motor. *The squirrel-cage rotor resistance is fixed* and *small* as compared to its reactance which is very large especially *at start* (because at standstill the frequency of rotor current is equal to that of supply frequency). Hence, the starting current I_2 of the rotor, though very large in magnitude, *lags by a very large angle* behind E_2 ; consequently the *starting torque per ampere is very poor.* It is roughly 1.5 times the full-load torque although the starting current is 5 to 7 times the full-load current. Thus such motors are *not suitable* for applications where these have to be *started against heavy loads.*

3.9.4. Starting Torque of a Slip Ring Motor. In a slip ring motor the torque is increased by *improving its power factor by adding external resistance in the rotor circuit from the star-connected rheostat* ; as the motor gains speed the rheostat resistance is gradually cut out. This additional resistance, however, increases the rotor impedance and so reduces the rotor current. At first, the effect of improved power factor *predominates the current-decreasing effect* of impedance, hence starting torque is increased. But after a certain point, the effect of increased impedance *predominates the effect of improved power factor and so the torque starts decreasing.*

3.9.5. Power. Eqn. (3.5) can be represented by a simple series circuit as shown in Fig. 3.11.

It is seen from this circuit that **per phase** power input (gross) to rotor,

$$P_g = E_2 I_2 \cos \phi_2$$

where $$\cos \phi_2 = \frac{R_2/s}{\sqrt{(R_2/s)^2 + X_2^2}}$$

$$\therefore \ P_g = \frac{E_2}{\sqrt{(R_2/s)^2 + X_2^2}} \cdot I_2 \frac{R_2}{s} = I_2^2 \frac{R_2}{s} \qquad ...(3.12)$$

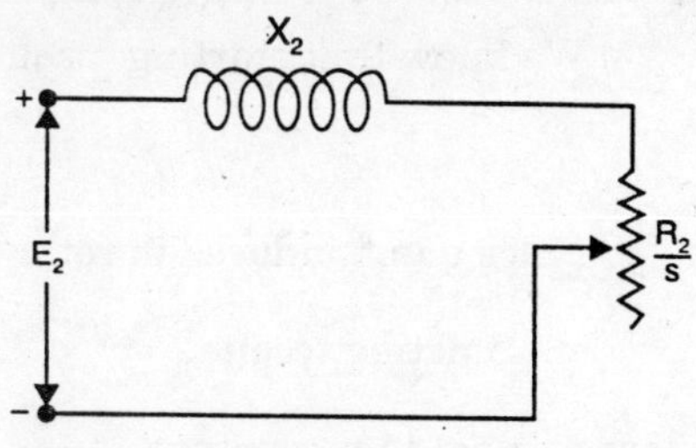

Fig. 3.11. Rotor equivalent circuit of an induction motor.

An examination of Fig. 3.11 also shows that per phase power input to rotor is equal to $I_2^2 \frac{R_2}{s}$ as the reactance is X_2 consumes no power.

P_g is the power transferred from stator to rotor across the air gap. In view of this, P_g is called the **air-gap power**. The expression for P_g may be written as

$$P_g = I_2^2 \frac{R_2}{s} = I_2^2 R_2 + I_2^2 R_2 \left(\frac{1-s}{s} \right)$$

= Rotor ohmic loss + internal mechanical power developed in rotor ($P_{mech.}$)

$$= sP_g + (1-s)P_g \quad ...(3.13\ a)$$

$$\therefore \quad P_{mech.} = (1-s)P_g = I_2^2 R_2 \left(\frac{1-s}{s} \right) \quad ...(3.13\ b)$$

$$\text{Rotor ohmic loss} = \left(\frac{s}{1-s} \right) P_{mech.} = sP_g \quad ...(3.13\ c)$$

Eqn. (3.12) reveals that ohmic loss

$$= I_2^2 R_2 = sP_g = s(\text{power input to rotor}) \quad ...(3.14)$$

Internal (or gross) torque developed per phase is given by

$$T_g = \frac{\text{internal mechanical power developed in rotor}}{\text{rotor speed in mechanical rad. per sec.}}$$

$$= \frac{P_{mech.}}{\omega_r} = \frac{(1-s)P_g}{(1-s)\omega_s} = \frac{P_g}{\omega_s} \quad ...(3.15)$$

Here $\omega_s \left(= \frac{2\pi N_s}{60} \right)$ is the *synchronous speed in mechanical radians per second.*

Also $$T = \frac{P_g}{\omega_s} = \frac{1}{\omega_s} \times \frac{I_2^2 R_2}{s} = \frac{\text{Rotor ohmic loss}}{(\omega_s)\text{ slip}} = \frac{1}{2\pi(N_s/60)} I_2^2 \frac{R_2}{s} \quad ...(3.16)$$

The power available at the shaft can be obtained from P_g as follows :

Output or shaft power, $P_{sh} = P_m$ – mechanical losses (friction, and windage losses)

or $P_{sh} = P_g$ – rotor ohmic loss – friction and windage losses.

Output or shaft torque,

$$T_{sh} = \frac{P_{sh}}{\text{rotor speed}} = \frac{P_{sh}}{(1-s)\omega_s}$$

If stator input is known, then air-gap power P_g is given by

P_g = stator power input – stator I^2R loss – stator core loss.

3.10. EFFECT OF CHANGE IN SUPPLY VOLTAGE ON STARTING TORQUE

We know that starting torque,

$$T_{st} = \frac{kR_2E_2^2}{R_2^2 + X_2^2}$$

Since e.m.f. induced in rotor (at standstill) $E_2 \propto \phi \propto V$

$\therefore$ Starting torque, $$T_{st} = \frac{k'R_2V_2^2}{R_2^2 + X_2^2}.$$

where k' is another constant

or $$T_{st} \propto V^2$$

i.e., Starting torque is proportional to the square of the applied voltage.

3.11. EFFECT OF CHANGE IN SUPPLY VOLTAGE ON TORQUE AND SLIP

The torque acting on the rotor when the motor is running with slip '*s*' is given by

$$T = \frac{ksR_2E_2^2}{R_2^2 + s^2X_2^2}$$

Since e.m.f. induced in rotor (at standstill), $E_2 \propto \phi \propto V$

$$\therefore \quad T = \frac{k' s R_2 V^2}{R_2^2 + s^2 X_2^2}, \quad \text{where } k' \text{ is another constant}$$

Since the slip '*s*' at full-load is *very low*, therefore $s^2 X_2^2$ can be *neglected* in comparison with R_2^2.

$$\therefore \quad T = \frac{k' s R_2 V^2}{R_2^2} = \frac{k' s V^2}{R_2} \quad \text{or} \quad T \propto sV^2$$

When the supply voltage is changed, it changes the torque under running condition also. With the decrease in supply voltage torque under running condition decreases, therefore, in order to *maintain the same torque, slip increases* or *speed decreases.*

3.12. FULL-LOAD TORQUE AND MAXIMUM TORQUE

Let, s_f = full-load slip of the motor, and

s_{mT} = slip corresponding to maximum torque = $\dfrac{R_2}{X_2}$

We know that,

Full-load torque, $$T_f = \frac{k s_f R_2 E_2^2}{R_2^2 + s_f^2 X^2} \quad \text{[From Art. 3.9]}$$

Maximum torque, $$T_m = \frac{k E_2^2}{2X_2} \quad \text{[From Art. 3.9.2]}$$

$$\therefore \quad \frac{T_f}{T_m} = \frac{k s_f R_2 E_2^2/(R_2^2 + s_f^2 X_2^2)}{k E_2^2/(2X_2)} = \frac{2 s_f R_2 X_2}{R_2^2 + s_f^2 X_2^2}$$

$$= \frac{2 s_f R_2 X_2/(X_2^2)}{(R_2^2 + s_f^2 X_2^2)/(X_2^2)} \quad \text{[Dividing numerator and denominator by } X_2^2\text{]}$$

$$= \frac{2 s_f (R_2/X_2)}{(R_2/X_2)^2 + s_f^2} = \frac{2 s_f s_{mT}}{s_{mT}^2 + s_f^2}$$

i.e., $$\frac{T_f}{T_m} = \frac{2 s_f s_{mT}}{s_f^2 + s_{mT}^2} \quad \text{...(3.17)}$$

$$= \frac{2}{\dfrac{s_f}{s_{mT}} + \dfrac{s_{mT}}{s_f}} \quad \text{...(3.17 a)}$$

[Dividing numerator and denominator by $s_f s_{mT}$]

3.13. STARTING TORQUE AND MAXIMUM TORQUE

$$\frac{T_{st}}{T_m} = \frac{k R_2 E_2^2/(R_2^2 + X_2^2)}{k E_2^2/(2X_2)} = \frac{2 R_2 X_2}{R_2^2 + X_2^2} = \frac{2(R_2/X_2)}{(R_2/X_2)^2 + 1}$$

(Dividing numerator and denominator by X_2^2)

or $$\frac{T_{st}}{T_m} = \frac{2 s_{mT}}{s^2_{mT} + 1} \quad \text{...(3.18)}$$

3.14. TORQUE-SLIP AND TORQUE-SPEED CURVES

The expression for torque is as follows :

$$T = \frac{ksR_2E_2^2}{R_2^2 + s^2X_2^2}$$

From the above expression, it is evident, that

- Torque is zero when slip $s = 0$ (*i.e.*, speed is synchronous).
- When slip 's' *is very low* the value of the term sX_2 is very small and is negligible in comparison with R_2, therefore torque T *is approximately proportional* to slip 's' if rotor resistance R_2 is constant. This means that at speeds near to synchronous speed the *torque-speed* and *torque-slip* curves are approximately straight lines (Figs. 3.12 and 3.13).
- When the slip 's' increases (*i.e.* as the speed decreases with increase in load) torque increases and reaches its maximum value when $s = \frac{R_2}{X_2}$. The *maximum torque is also known as* **'pull-out'** or **'break-down' torque.**

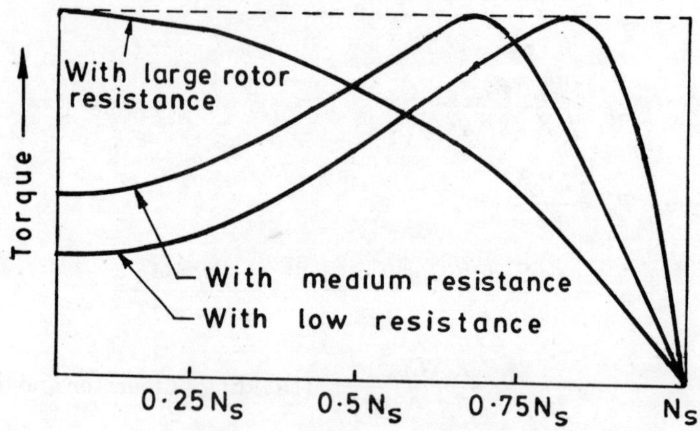

Fig. 3.12. Torque-speed curves.

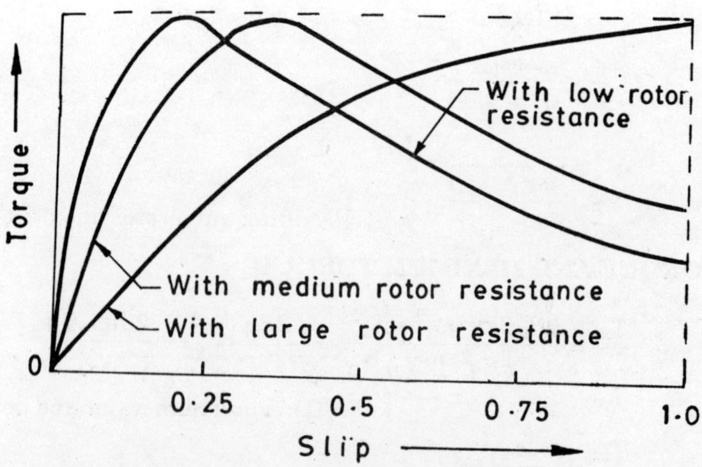

Fig. 3.13. Torque-slip curves.

- When the slip is further increased the torque decreases. The result is that *motor slows down* and *eventually stops. The motor operates for the value of slip between zero and that corresponding to maximum torque.*

With higher slip, R_2 becomes negligible as compared to sX_2 and torque varies as follows :

$$T \propto \frac{s}{s^2 X_2^2} \propto \frac{1}{s} \text{ if standstill reactance is constant.}$$

This means that *speed-torque* or *slip-torque* curves are rectangular hyperbola with the speed or slip beyond that corresponding to maximum torque. Figs. 3.12 and 3.13 show the *torque-speed* and *torque-slip* curves for different values of rotor resistance. It is observed that although maximum torque is independent of rotor resistance R_2, yet the exact location of T_m is dependent on it. *Greater the R_2, greater is the value of slip at which maximum torque occurs.*

3.15. OPERATING CHARACTERISTICS OF A 3-PHASE SQUIRREL-CAGE INDUCTION MOTOR

To start with, let us assume that the squirrel cage induction motor has been started without any load on it (*i.e.*, no electrical devices are connected to the generator, no metal is being cut by a machine tool, or the conveyor belt is completely disengaged). The motor will come to its no-load speed, which may be at a slip as low as 0.1 per cent. *Load is now placed on the motor* and the following reactions will be observed :

- *Speed decreases.*
- *Slip increases* (because the stator field continues to rotate at synchronous speed and therefore cuts the rotor bars more rapidly).
- As a consequence of increased cutting the *rotor induced e.m.f. increases.*
- *The current in the rotor bars increases.*
- *Torque developed by the rotor increases until it is sufficient to carry the load placed on the motor.*
- *The motor continues to rotate at that slip which will develop the required torque for that particular load.*

The motor may be loaded continuously till *pull-out (or break-down) torque* is developed, at which point the *motor will stop if more load is placed on it.* Pull-out occurs at a slip well beyond the

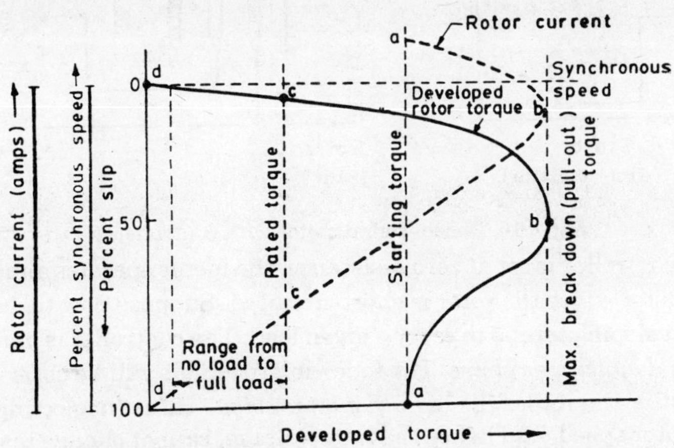

Fig. 3.14. Operating characteristics of a squirrel-cage induction motor.

normal range operation of the squirrel-cage induction motor. Even under normal operation, however, a suddenly applied load may require the temporary development of this maximum torque, after which the motor will speed up to its full-load value. Thus, in a manner similar to the shunt motor, the induction motor *inherently adjusts itself to the applied external torque*.

Fig. 3.14 shows typical operating characteristics of an induction motor.

Fig. 3.15 shows the effect of heavy loads on primary stator current and power factor.

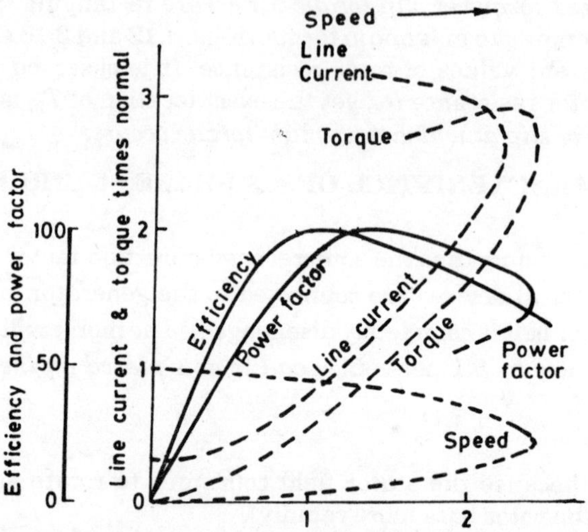

Fig. 3.15. Effect of heavy loads on primary stator current and power factor.

3.16. OPERATING CHARACTERISTICS OF A WOUND-ROTOR (SLIP RING) INDUCTION MOTOR

Fig. 3.16 shows a wound-rotor induction motor with controller (rheostat).

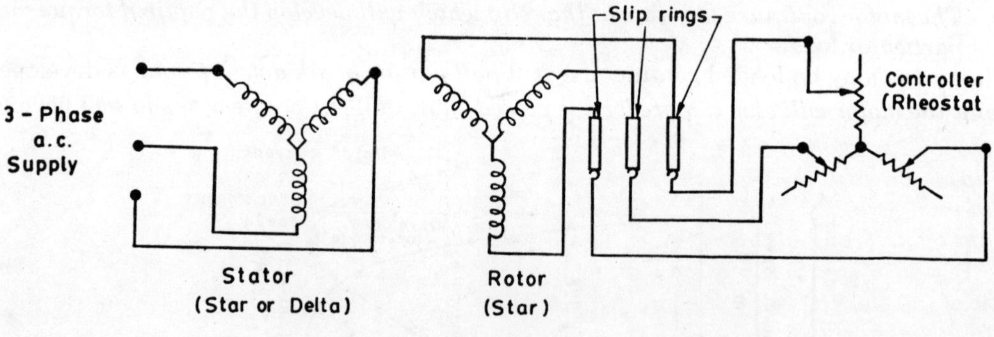

Fig. 3.16. Phase-wound motor with controller.

If the external controller is set at zero resistance, the motor operates almost exactly the same as the squirrel-cage motor, since the rotor is short-circuited. Suppose now the rotor is rotating at a constant speed and developing torque to carry a given load. The controller is moved so that the *rotor resistance per phase is doubled* per phase. The following reactions will be observed :

- The current in the rotor falls to *half of its previous value* (neglecting a transient effect), since the rotor speed, and hence the rotor voltage, cannot change instantaneously.
- The *developed torque decreases due to reduced rotor current.*

- The load on the motor has not changed, and with the external applied, or load, torque *now greater than the developed* torque, the *motor slows down.*
- The *slip increases* (increasing the rate at which the stator flux cuts the rotor conductors).
- The *rotor induced voltage increases.*
- The *rotor current now increases until it is sufficient to develop enough torque to again carry the load at a constant speed.*

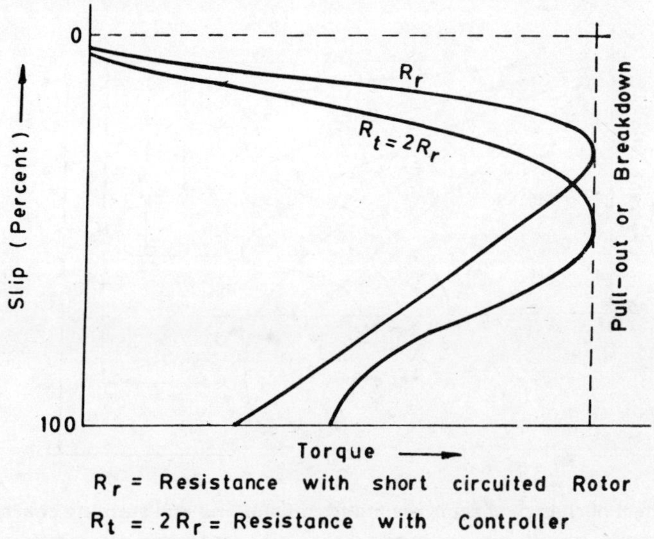

Fig. 3.17. Torque-slip characteristic of wound-rotor induction motor.

The induction motor now rotates at a lower speed, but with the same current as before, and develops the same torque as it did before the rotor circuit resistance was increased. The *frequency in the rotor is greater,* as is the rotor resistance. *Rotor resistance and reactance are both increased in same proportion,* however, and the *rotor power factor therefore remains the same.* If more load is now placed on the shaft, the rotor will develop the required torque, but always *at a lower speed i.e.* a *higher slip.* Eventually pull-out torque is attained. The *value of pull-out torque will be the same as with the short-circuited rotor, but it will occur at a slip which is twice as great as that with the shorted rotor, since the rotor circuit resistance is twice that of the rotor alone.*

Fig. 3.18 shows the torque-slip characteristic of the shorted rotor, together with that for a total resistance of, $R_t = 2R$.

It is worthnoting that the *torque developed at standstill is higher with increased rotor resistance.* This is explained by the fact that at standstill the rotor frequency is equal to the line frequency and rotor resistance is maximum. The addition of resistance in the rotor circuit obviously improves the power factor in that circuit while reducing the rotor current. Since standstill rotor resistance is much greater than

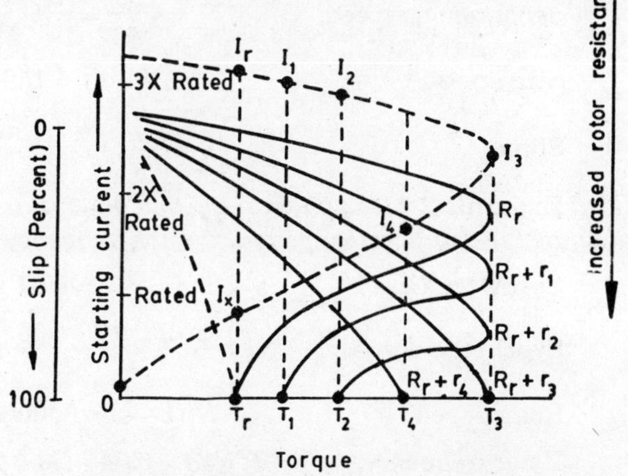

Fig. 3.18. Starting currents vs torque.

its resistance, the power factor increases more rapidly than the decrease of rotor current, and hence the developed torque is greater with added resistance.

Fig. 3.18 shows starting current vs torque.

Fig. 3.19 shows the effect of change of rotor resistance on starting and running characteristics.

In Figs. 3.18 and 3.19,

R_r = *basic rotor resistance*

r_1, r_2, r_3, r_4 = the resistance added to each phase of star-connected wound rotor respectively.

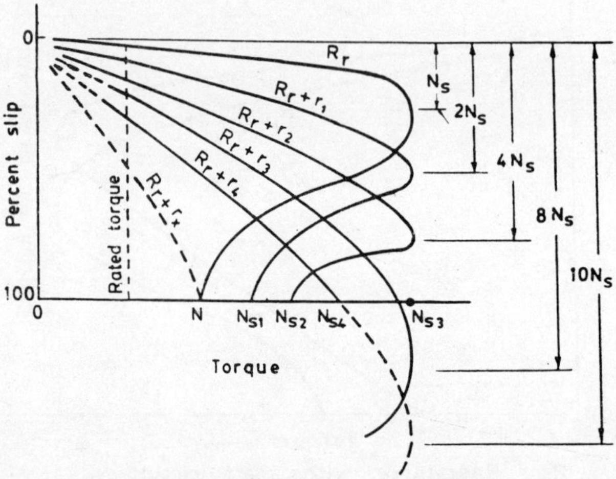

Fig. 3.19. Effect of change of rotor resistance on starting and running characteristics.

SLIP, STARTING TORQUE, MAXIMUM TORQUE

Example 3.1. *A 3-phase, 4-pole, 50-Hz induction motor is running at 1440 r.p.m. Determine the slip speed and slip.*

Solution. Number of poles, $p = 4$

Frequency, $f = 50$ Hz

Actual speed of rotor, $N = 1440$ r.p.m.

Slip speed = ?

Slip = ?

Synchronous speed, $N_s = \dfrac{120f}{p} = \dfrac{120 \times 50}{4} = 1500$ r.p.m.

Slip speed $= N_s - N = 1500 - 1440 = \mathbf{60\ r.p.m.}$ **(Ans.)**

Slip, $s = \dfrac{N_s - N}{N_s} = \dfrac{1500 - 1440}{1500} = \mathbf{0.04.}$ **(Ans.)**

Example 3.2. *A 3-phase induction motor has 2 poles and is connected to 400 V, 50 Hz supply. Calculate the actual rotor speed and rotor frequency when the slip is 4%.* **(B.T.E. U.P., 1998)**

Solution. *Given* : $p = 2$; $s = 4\% = 0.04$; $f = 50$ Hz.

Synchronous speed, $N_s = \dfrac{120f}{p} = \dfrac{120 \times 50}{2} = 3000$ r.p.m.

Rotor speed, $\mathbf{N} = N_s(1 - s) = 3000(1 - 0.04) = \mathbf{2880\ r.p.m.}$ **(Ans.)**

Rotor frequency, $f_r = s.f = 0.04 \times 50 = \mathbf{2\ Hz.}$ **(Ans.)**

Example 3.3. *A 3-phase alternator having 12-poles is driven at a speed of 500 r.p.m. It supplies power to an 8-pole, 3-phase induction motor. If the slip of the motor at full-load is 4%, calculate the full-load speed of the motor.*

Solution. Number of poles of the alternator = 12

Speed of alternator = 500 r.p.m.

Number of poles of induction motor = 8

Percentage slip = 4%

Full-load speed, N :

Supply frequency, $f = \frac{12 \times 500}{120} = 50$ Hz

Synchronous speed, $N_s = \frac{120 \times 50}{8} = 750$ r.p.m.

$\therefore$ % slip $= \frac{N_s - N}{N_s} \times 100$

$$4 = \frac{750 - N}{750} \times 100$$

$$\therefore \quad N = 750 - \frac{4 \times 750}{100} = 720 \text{ r.p.m.}$$

Hence, **full-load speed, N = 720 r.p.m. (Ans.)**

☞**Example 3.4.** *A 3-phase, 6-pole, 50-Hz induction motor has a slip of 1% at no-load and 3% at full-load. Find :*

(i) Synchronous speed, *(ii) No-load speed,*
(iii) Full-load speed, *(iv) Frequency of rotor current at standstill, and*
(v) Frequency of rotor current at full-load.

Solution. Number of poles, $p = 6$

No-load slip, $s_0 = 1\%$

Full-load slip, $s_f = 3\%$

(i) **Synchronous speed,**

$$N_s = \frac{120f}{p} = \frac{120 \times 50}{6} = \mathbf{1000 \text{ r.p.m.}} \text{ (Ans.)}$$

(ii) **No-load speed, N₀ :**

We know that, $s = \frac{N_s - N}{N_s}$ or $N = N_s(1 - s)$

or $$\mathbf{N_0} = N_s(1 - s_0) = 1000\left(1 - \frac{1}{100}\right) = \mathbf{990 \text{ r.p.m.}} \text{ (Ans.)}$$

(iii) **Full-load speed,**

$$N_f = N_s(1 - s_f) = 1000\left(1 - \frac{3}{100}\right) = \mathbf{970 \text{ r.p.m.}} \text{ (Ans.)}$$

(iv) **Frequency of rotor current at standstill, f_r :**

At standstill, $s = 1$

$$f_r = sf = 1 \times 50 = \mathbf{50 \text{ Hz.}} \text{ (Ans.)}$$

(v) **Frequency of rotor current at full-load, f_r = ?**

$$f_r = s \times f = \frac{3}{100} \times 50 = \mathbf{1.5 \text{ Hz.}} \text{ (Ans.)}$$

Example 3.5. *A 3-phase, 12-pole alternator is coupled to an engine running at 500 r.p.m. The alternator supplies an induction motor which has a full-load speed of 1455 r.p.m. Find the slip and number of poles of the motor.*

Solution. Number of poles of the alternator, $p_a = 12$

Speed of the engine, $N_e = 500$ r.p.m.

Full-load speed of the induction motor, $N_m = 1455$ r.p.m.

Slip, s = ?

Number of poles of the motor, $\mathbf{p_m}$ = ?

Supply frequency, $$f = \frac{N_a p_a}{120} = \frac{500 \times 12}{120} = 50 \text{ Hz.}$$

When the supply frequency is 50-Hz, the synchronous speed can be 3000, 1500, 1000, 750 r.p.m. etc. Since the full-load speed is 1455 r.p.m. and the full-load slip is *always* less than 5%, the synchronous is 1500 r.p.m.

Slip, $$s = \frac{N_s - N}{N_s} = \frac{1500 - 1455}{1500} = \mathbf{0.03 \text{ or } 3 \text{ per cent.}} \quad \textbf{(Ans.)}$$

Also, $$N_s = \frac{120f}{p_m}$$

$$\therefore \quad p_m = \frac{120f}{N_s} = \frac{120 \times 50}{1500} = 4 \text{ poles}$$

Hence, **number of motor poles = 4. (Ans.)**

Example 3.6. *A 3-phase induction motor runs at almost 1000 rpm at no load and 950 r.p.m. at full load when supplied with power from a 50 Hz 3-phase line.*

(i) How many poles has the motor ?

(ii) What is the percentage slip at full load ?

(iii) What is the corresponding frequency of rotor voltages ?

(iv) What is the corresponding speed of the rotor field with respect to rotor ?

(v) What is the corresponding speed of the rotor with respect to the stator ?

(vi) What is the corresponding speed of the rotor field with respect to the stator field ?

(vii) What is the rotor frequency at the slip of 10 per cent ? **(AMIE, 1997)**

Solution. *Given* : $N_0 = 1000$ r.p.m., $N_f = 950$ r.p.m., $f = 50$ Hz

Since no-load speed of motor is almost 1000 r.p.m., hence synchronous speed near to 1000 r.p.m. is 1000 r.p.m.

(*i*) Poles on motor, $$\mathbf{p} = \frac{120f}{N_s} = \frac{120 \times 50}{1000} = \mathbf{6.} \quad \textbf{(Ans.)}$$

(*ii*) Percentage slip at full load, $$\mathbf{s_f} = \frac{N_s - N_f}{N_s} \times 100 = \frac{1000 - 950}{1000} \times 100 = \mathbf{5\%.} \quad \textbf{(Ans.)}$$

(*iii*) Frequency of rotor voltage, $$f_r = s \times f = \frac{5}{100} \times 50 = \mathbf{2.5 \text{ Hz.}} \quad \textbf{(Ans.)}$$

(*iv*) Speed of rotor field with respect to rotor $$= \frac{120 \times f_r}{p} = \frac{120 \times 2.5}{6} = 50 \text{ r.p.m.} \quad \textbf{(Ans.)}$$

(*v*) Speed of rotor with respect to stator = **950 r.p.m. (since stator is stationary). (Ans.)**

(*vi*) Rotor field and stator field are revolving at the same speed of 1000 r.p.m., therefore, *speed of rotor field w.r.t. stator field is* **zero. (Ans.)**

(*vii*) Rotor frequency at slip of 10%, $$f_r = s' \times f = \frac{10}{100} \times 50 = \mathbf{5 \text{ Hz.}} \quad \textbf{(Ans.)}$$

Example 3.7. *In a 6-pole, 3-phase, 50 Hz motor with star-connected rotor, the rotor resistance per phase is 0.3 Ω, the reactance at standstill is 1.5 Ω per phase, and an e.m.f. between the slip rings on open circuit is 175 V. Find :*

(i) Slip at a speed of 950 r.p.m. ;

(ii) Rotor e.m.f. per phase ;

(iii) Rotor frequency and reactance at a speed of 950 r.p.m. **(B.T.E. Pb., 1997)**

Solution. *Given* : $p = 6$; $f = 50$ Hz ; $R_2 = 0.3\ \Omega$, $X_2 = 1.5\ \Omega$; $N = 950$ r.p.m.

Synchronous speed, $N_s = \dfrac{120f}{p} = \dfrac{120 \times 50}{6} = 1000$ r.p.m.

(*i*) **Slip, s :**

$$\mathbf{s} = \frac{N_s - N}{N_s} = \frac{1000 - 950}{1000} = 0.05 \text{ or } \mathbf{5\%}. \ \textbf{(Ans.)}$$

(*ii*) **Rotor e.m.f. per phase, E_2 :**

Rotor e.m.f. per phase at standstill $= \dfrac{E_2}{\sqrt{3}} = \dfrac{175}{\sqrt{3}} = 101$ V/phase

Rotor e.m.f. at 5% slip $= sE_2 = 0.05 \times 101 =$ **5.05 V/phase. (Ans.)**

(*iii*) **Rotor frequency and reactance at a speed of 950 r.p.m. :**

Rotor frequency, $\mathbf{f_r} = s \times f = 0.05 \times 50 =$ **2.5 Hz. (Ans.)**

Rotor reactance $= sX_2 = 0.05 \times 1.5 =$ **0.075 Ω/phase. (Ans.)**

Example 3.8. *A 50 Hz, 440 V, 3-phase, 4-pole induction motor develops half the rated torque at 1490 r.p.m. With the applied voltage magnitude remaining at the rated value, what should be its frequency if the motor has to develop the same torque at 1600 r.p.m. ? Neglect stator and rotor winding resistances, leakage reactances and iron losses.* **(GATE, 1995)**

Solution. *Given* : $f = 50$ Hz ; $p = 4$; $N = 1490$ r.p.m. ; New speed, $N_n = 1600$ r.p.m.

Synchronous speed, $N_s = \dfrac{120f}{p} = \dfrac{120 \times 50}{4} = 1500$ r.p.m.

Slip at a speed of 1490 r.p.m., $s = \dfrac{N_s - N}{N_s} = \dfrac{1500 - 1490}{1500} = 0.00667$

Since torque developed by an induction motor, $T \propto sV^2$ (Refer Art. 3.11)

Slip s for constant torque and constant applied voltage remains unchanged.

$\therefore$ New synchronous speed, $N_{sn} = \dfrac{N_n}{1 - s} = \dfrac{1600}{1 - 0.00667} = 1610.7$ r.p.m.

New frequency, $f_n = \dfrac{p \times N_{sn}}{120} = \dfrac{4 \times 1610.7}{120} =$ **53.7 Hz. (Ans.)**

☞**Example 3.9.** *A 3-phase slip ring induction motor gives a reading of 60 V across slip rings on open circuit when at rest with normal stator voltage applied. The motor is star-connected and has impedance of (0.6 + j6) Ω per phase. Find the rotor current when the machine is : (i) at standstill with slip rings joined to a star-connected starter with a phase impedance of (5 + j4) Ω ; and (ii) running normally with a 4 per cent slip.*

Solution. Impedance of motor per phase = $(0.6 + j6)\ \Omega$

Phase impedance of star-connected starter = $(5 + j4)\ \Omega$

(*i*) **At standstill** :

Induced e.m.f. in the rotor winding per phase,

$$E_2 = \frac{60}{\sqrt{3}} = 34.64 \text{ V}$$

Resistance per phase in the rotor circuit,

$$R_2 = 0.6 + 5 = 5.6\ \Omega$$

Reactance per phase in the rotor circuit,

$$X_2 = 6 + 4 = 10\ \Omega$$

Impedance per phase in rotor circuit,

$$Z_2 = \sqrt{R_2^2 + X_2^2} = \sqrt{(5.6)^2 + (10)^2} = 11.46\ \Omega$$

Rotor current per phase, $I_2 = \dfrac{E_2}{Z_2} = \dfrac{34.64}{11.46} = \mathbf{3.02\ A.}$ **(Ans.)**

Power factor, $\cos\phi_2 = \dfrac{R_2}{Z_2} = \dfrac{5.6}{11.46} = \mathbf{0.488\ (lag).}$ **(Ans.)**

(*ii*) **Running at 4 per cent slip :**

Induced e.m.f. on the rotor winding per phase

$$= sE_2 = 0.04 \times 34.64 = 1.38\ V$$

Rotor impedance per phase,

$$Z_2 = \sqrt{R_2^2 + (sX_2)^2} = \sqrt{(0.6)^2 + (0.04 \times 6)^2}$$

$$= 0.646\ \Omega$$ [Here $R_2 = 0.6\ \Omega$]

Rotor current per phase,

$$I_2 = \frac{sE_2}{Z_2} = \frac{1.38}{0.646} = \mathbf{2.136\ A.}\ \textbf{(Ans.)}$$

Power factor, $\cos\phi_2 = \dfrac{R_2}{Z_2} = \dfrac{0.6}{0.646} = \mathbf{0.928\ (lag).}$ **(Ans.)**

☞**Example 3.10.** *A 3-phase induction motor has a star-connected rotor. The rotor e.m.f. (between slip rings) at standstill is 50 V. The rotor resistance and standstill reactance are 0.5 Ω and 3 Ω respectively. Find :*

(*i*) *Rotor current per phase at starting and the slip rings short-circuited.*

(*ii*) *Rotor current per phase at starting if a star connected rheostat of resistance 6 Ω per phase is connected across the slip rings.*

(*iii*) *Full-load rotor current and rotor power factor if the full-load slip is 4 per cent.*

(*iv*) *Rotor e.m.f. per phase under full-load condition.*

Solution. Rotor e.m.f. at standstill (between slip rings)

$$= 50\ V$$

∴ Rotor e.m.f./phase at standstill,

$$E_2 = \frac{50}{\sqrt{3}} = 28.87\ V$$

Rotor resistance/phase, $R_2 = 0.5\ \Omega$

Rotor reactance/phase, $X_2 = 3\ \Omega$

(*i*) At starting $s = 1$

We know that rotor current/phase (I_2) is given by,

$$I_2 = \frac{sE_2}{\sqrt{R_2^2 + (sX_2)^2}} \quad ...(i)$$

∴ **Rotor current/phase at starting**

$$= \frac{E_2}{\sqrt{R_2^2 + X_2^2}} = \frac{28.87}{\sqrt{(0.5)^2 + (3)^2}} = \mathbf{9.49\ A.}\ \textbf{(Ans.)}$$

(*ii*) Total resistance in the rotor circuit

= rheostat resistance + rotor resistance = 6 + 0.5 = 6.5 Ω

∴ **Rotor current/phase with rheostat resistance,**

$$I_2 = \frac{28.87}{\sqrt{(6.5)^2 + (3)^2}} = \mathbf{4.03\ A.}\ \ \mathbf{(Ans.)}$$

(*iii*) Full-load slip, $s_f = 4\% = 0.04$

Full-load rotor current (as per eqn. (*i*)),

$$I_2 = \frac{0.04 \times 28.87}{\sqrt{(0.5)^2 + (0.04 \times 3)^2}} = \mathbf{2.246\ A.}\ \ \mathbf{(Ans.)}$$

Rotor power factor $= \dfrac{R_2}{\sqrt{R_2^2 + (s_f X_2)^2}} = \dfrac{0.5}{\sqrt{(0.5)^2 + (0.04 \times 3)^2}} = \mathbf{0.972.}\ \ \mathbf{(Ans.)}$

(*iv*) **Rotor e.m.f./phase under full-load condition**

$= sE_2 = 0.04 \times 28.87 = \mathbf{1.155\ V.}$ **(Ans.)**

Example 3.11. (*a*) *Explain why the rotor of a polyphase induction motor can never attain synchronous speed.*

(*b*) *A 1100 V, 50 Hz delta connected induction motor has a star-connected slip ring motor with a phase transformation ratio of 3.8. The rotor resistance and standstill leakage reactance are 0.012 Ω and 0.25 Ω per phase respectively. Neglecting stator impedance and magnetising current, determine :*

(*i*) *The rotor current at start with slip ring shorted ;*

(*ii*) *The rotor p.f. at start with slip ring shorted ;*

(*iii*) *The rotor current at 4% slip with slip ring shorted ;*

(*iv*) *The rotor power factor at 4% slip with slip ring shorted ;*

(*v*) *The external rotor resistance per phase required to obtain a starting current of 100 A in the stator supply lines.* **(AMIE Summer, 1997)**

Solution. (*a*) When the rotor speed equals the speed of rotating flux the relative speed between rotor and stator flux ceases and induced current in rotor becomes zero. There is no other agency to develop torque in rotor at $N_r = N_s$. Hence it is not possible for rotor to run at synchronous speed. As the rotor will approach synchronous speed its torque becomes zero and it falls back to lower speed. Therefore, *the only possibility for rotor is that it always rotates at a speed* N_r, *lower than synchronous speed.*

(*b*) Phase voltage of stator winding = 1100 V

Transformer ratio = 3.8

∴ Phase voltage induced in the rotor winding at standstill, $E_2 = \dfrac{1100}{3.8} = 289.5$ V

Rotor resistance per phase, $R_2 = 0.012\ \Omega$

Rotor reactance per phase, $X_2 = 0.25\ \Omega$

∴ Rotor impedance at standstill $= \sqrt{0.012^2 + 0.25^2} \simeq 0.25\ \Omega$

(*i*) Rotor starting current per phase at normal voltage with slip ring shorted

$$= \frac{289.5}{0.25} = \mathbf{1158\ A.}\ \ \mathbf{(Ans.)}$$

(*ii*) Rotor p.f. at start with slip ring shorted

$$= \cos\left(\tan^{-1}\frac{0.25}{0.012}\right) = \mathbf{0.048\ lag.}\ \ \mathbf{(Ans.)}$$

(*iii*) At 4% slip, rotor e.m.f. $= 0.04 \times E_2 = 0.04 \times 289.5 = 11.58$ V

Rotor reactance at 4% slip $= sX_2 = 0.04 \times 0.25 = 0.01\ \Omega$

Rotor impedance at 4% slip $= \sqrt{0.012^2 + 0.01^2} = 0.0156\ \Omega$

$\therefore$ Rotor current at 4% slip $= \dfrac{11.58}{0.0156} =$ **742.3 A. (Ans.)**

(*iv*) Rotor p.f. at 4% slip $= \dfrac{R_2}{Z_2} = \dfrac{0.012}{0.0156} =$ **0.769. (Ans.)**

(*v*) Rotor current corresponding to stator line current of 100 A

$$= \frac{100}{\sqrt{3}} \times 3.8 = 219.4 \text{ A}$$

$\therefore$ Rotor impedance $= \dfrac{289.5}{219.4} = 1.32\ \Omega$

$\therefore$ Rotor resistance required $= \sqrt{1.32^2 - 0.25^2} = 1.296\ \Omega$

$\therefore$ External rotor resistance per phase required $= 1.296 - 0.012 =$ **1.284 Ω. (Ans.)**

Example 3.12. *The slip ring rotor resistance and standstill reactance per phase of a 3300 V, 24-pole, 50-Hz, 3-phase star-connected induction motor are 0.02 Ω and 0.3 Ω respectively. Calculate* :

(*i*) *The speed at maximum torque.*

(*ii*) *Ratio of full-load torque to maximum torque if full-load torque is obtained at 246 r.p.m.*

Solution. Slip ring rotor resistance, $R_2 = 0.02\ \Omega$

Standstill reactance, $X_2 = 0.3\ \Omega$

We know that for maximum torque,

$$R_2 = s_{mT} X_2$$

or

$$s_{mT} = \frac{R_2}{X_2} = \frac{0.02}{0.3} = 0.0667 \text{ or } 6.67\%$$

(*i*) Rotor speed at maximum torque, $N = (1 - s_{mT})\, N_s$

But,
$$N_s = \frac{120 \times f}{p} = \frac{120 \times 50}{24} = 250 \text{ r.p.m.}$$

$\therefore$ Speed at maximum torque,

$$N = (1 - 0.0667) \times 250 = 233.3 \text{ r.p.m.}$$

(*ii*) Full-load slip, $s_f = \dfrac{250 - 246}{250} = 0.016$ or 1.6%

Now
$$\frac{T_f}{T_m} = \frac{2s_f \cdot s_{mT}}{s_f^{\,2} + s_{mT}^{\,2}} \qquad [\because \text{ Eqn. (3.17)}]$$

where $s_{mT} = \dfrac{R_2}{X_2} = \dfrac{0.02}{0.3} = 0.0667$

$$\therefore \quad \frac{T_f}{T_m} = \frac{2 \times 0.016 \times 0.0667}{(0.016)^2 + (0.0667)^2} = \frac{0.002134}{0.0047} = \mathbf{0.454.\ (Ans.)}$$

☞**Example 3.13.** *A 6-pole, 3-phase, 50-Hz induction motor runs on full-load with a slip of 5 per cent. If the rotor standstill impedance is (0.015 + j0.075) Ω, calculate the available maximum torque in terms of full-load torque. Also determine the speed at which the maximum torque occurs.*

Solution. Number of poles, $p = 6$

Rotor resistance per phase, $R_2 = 0.015\ \Omega$

Rotor standstill reactance per phase, $X_2 = 0.075\ \Omega$

Full-load slip, $s_f = 5\% = 0.05$

(*i*) $\mathbf{\dfrac{T_m}{T_f}}$ **:**

Ratio of maximum torque (T_m) to full-load torque (T_f) is given by :

$$\frac{T_m}{T_f} = \frac{s_f^2 + s_{mT}^2}{2 s_f s_{mT}}$$

where $s_{mT} = \dfrac{R_2}{X_2} = \dfrac{0.015}{0.075} = 0.2$

$$\therefore \quad \frac{T_m}{T_f} = \frac{(0.05)^2 + (0.2)^2}{2 \times 0.05 \times 0.2} = 2.125$$

∴ **Maximum torque,** $\mathbf{T_m = 2.125\ T_f}$. **(Ans.)**

(*ii*) **Speed at which maximum torque occurs, N :**

Slip corresponding to maximum torque,

$$s_{mT} = \frac{R_2}{X_2} = \frac{0.015}{0.075} = 0.2$$

Speed corresponding to maximum torque,

$$N = N_s (1 - s_{mT}) = \frac{120 \times 50}{6} (1 - 0.2) = \mathbf{800\ r.p.m.}\quad \textbf{(Ans.)}$$

Example 3.14. *A 6-pole, 3-phase, 50-Hz induction motor develops maximum torque of 300 Nm at a speed of 960 r.p.m. Determine the torque exerted by the motor at 5 per cent slip. The rotor resistance per phase is 0.6 Ω.*

Solution. Number of poles, $p = 6$

Maximum torque, $T_m = 300$ Nm

Speed, $N = 960$ r.p.m.

Rotor resistance per phase $= 0.6\ \Omega$

Synchronous speed, $N_s = \dfrac{120 \times f}{p} = \dfrac{120 \times 50}{6} = 1000$ r.p.m.

Percentage slip, $s_{mT} = \dfrac{N_s - N}{N_s} = \dfrac{1000 - 960}{1000} \times 100 = 4$ per cent.

For maximum torque, $R_2 = s_{mT} X_2$

$R_2 = 0.6\ \Omega$, $s_{mT} = 4\%$ or 0.04

$$\therefore \quad X_2 = \frac{R_2}{s_{mT}} = \frac{0.6}{0.04} = 15\ \Omega$$

Now, $T_m \propto \dfrac{s_{mT} R_2}{R_2^2 + s_{mT}^2 X_2^2}$

and $T \propto \dfrac{s R_2}{R_2^2 + s^2 X_2^2}$,

where T is the torque at 5% slip

$$\therefore \quad \frac{T_m}{T} = \frac{s_{mT} R_2}{s R_2} \times \frac{R_2^2 + s^2 X_2^2}{R_2^2 + s_{mT}^2 X_2^2} = \frac{s_{mT}}{s} \times \frac{R_2^2 + s^2 X_2^2}{R_2^2 + s_{mT}^2 X_2^2}$$

$$= \frac{0.04}{0.05} \times \frac{(0.6)^2 + (0.05 \times 15)^2}{(0.6)^2 + (0.04 \times 15)^2} = \frac{0.04}{0.05} \times \frac{0.36 + 0.5625}{0.36 + 0.36} = 1.025$$

$$\therefore \quad T = \frac{T_m}{1.025} = \frac{300}{1.025} = 292.68 \text{ N-m}$$

Hence, **torque developed by motor at 5% slip**
= 292.68 N-m. (Ans.)

Example 3.15. *An induction motor runs at a slip frequency of 2 Hz when supplied from a three-phase 400 V, 50 Hz supply. For the same developed torque, find the slip frequency at which motor will run when supplied from a three-phase 340 V, 40 Hz system. Slip at which the machine develops maximum torque using 50 Hz supply is 0.1. Neglect the stator impedance and assume linear torque-slip characteristic between zero torque and maximum torque in the working region.* **(GATE, 1998)**

Solution. Rating of induction motor = 400 V, 50 Hz

Slip frequency = 2 Hz

Slip at maximum torque, $s_{mT} = 0.1$

When the slip frequency is 2 Hz, the slip at this frequency, $s_1 = \frac{2}{50} = 0.04$

Let the slip at 340 V, 40 Hz be s_2.

$$\text{Torque } T \propto \frac{sE_2^2}{R_2^2 + (sX_2)^2} \quad ...(i)$$

Here stator impedance is neglected and as such $V = E_2$

Also slip at maximum torque, $s_{mT} = \frac{R_2}{X_2} = 0.1$

$$\therefore \quad R_2 = 0.1\, X_2$$

Substituting this value in (*i*), we get

$$T \propto \frac{sV^2}{(0.01 + s^2)X_2^2}$$

Since the developed torque for both the cases is same, therefore,

$$\frac{s_1 V_1^2}{(0.01 + s_1^2)X_2^2} = \frac{s_2 V_2^2}{(0.01 + s_2^2)X_2^2}$$

or $$\frac{0.04 \times (400)^2}{0.01 + (0.04)^2} = \frac{s_2 \times (340)^2}{0.01 + (s_2)^2}$$

or $$\frac{6400}{0.0116} = \frac{s_2 \times 115600}{0.01 + s_2^2}$$

or $$6400(0.01 + s_2^2) = 0.0116 \times s_2 \times 115600$$

or $$64 + 6400\, s_2^2 = 1340.96 s_2$$

or $$6400 s_2^2 - 1340.96 s_2 + 64 = 0$$

or $$s_2 = \frac{1340.96 \pm \sqrt{(1340.96)^2 - 4 \times 6400 \times 64}}{2 \times 6400}$$

$$= \frac{1340.96 \pm 399.7}{2 \times 6400} = 0.136 \text{ or } 0.0735$$

As the slip cannot be high, thus select the value of slip as 0.0735.

$\therefore$ Slip at 40 Hz = 0.0735

Hence, slip frequency = 0.0735 × 40 = **2.94 Hz. (Ans.)**

Example 3.16. *A 3-phase star-connected 6.6 kV, 20 pole 50 Hz induction motor has rotor resistance of 0.12 Ω and standstill reactance of 1.12 Ω.*

The motor has speed of 292.5 r.p.m. at full load.

Calculate slip at maximum torque and ratio of maximum torque to full load torque.

(AMIE Summer, 1996)

Solution. *Given* : $p = 20$; $f = 50$ Hz ; $R_2 = 0.12\ \Omega$; $X_2 = 1.12\ \Omega$; $N = 292.5$ r.p.m.

$\mathbf{s_{mT}}$; $\dfrac{\mathbf{T_m}}{\mathbf{T_f}}$:

Slip corresponding to maximum torque is given by

$$s_{mT} = \frac{R_2}{X_2} = \frac{0.12}{1.12} = 0.107 \text{ or } \mathbf{10.7\%.\ (Ans.)}$$

We know that,

$$\frac{T_f}{T_m} = \frac{2s_f s_{mT}}{s_f^{\,2} + s_{mT}^{\,2}} \qquad \text{[Eqn. (3.17)]}$$

where $s_{mT} = \dfrac{R_2}{X_2}$, and s_f = full-load slip of the motor.

Now

$$s_{mT} = \frac{0.12}{1.12} = 0.107$$

$$N_s = \frac{120f}{p} = \frac{120 \times 50}{20} = 300 \text{ r.p.m.}$$

$$\therefore \quad s_f = \frac{N_s - N}{N_s} = \frac{300 - 292.5}{300} = 0.025$$

Substituting the values in the above equation, we get

$$\therefore \quad \frac{T_f}{T_m} = \frac{2 \times 0.107 \times 0.025}{(0.107)^2 + (0.025)^2} = 0.443$$

Hence

$$\frac{T_m}{T_f} = \frac{1}{0.443} = \mathbf{2.257.\ (Ans.)}$$

Example 3.17. *A 3-phase induction motor has starting torque of 100% and a maximum torque of 200% of the full-load torque. Find slip at maximum torque.* **(UPSC, 1994)**

Solution. *Given* : Starting torque, T_{st} = 100% of T_f or = T_f

Maximum torque, T_m = 200% of T_f or = $2T_f$

Slip at maximum torque, $\mathbf{s_{mT}}$:

We know that,

$$\frac{T_{st}}{T_m} = \frac{2s_{mT}}{s_{mT}^{\,2} + 1} \qquad \text{[Eqn. (3.18)]}$$

$$\frac{T_f}{2T_f} = \frac{2s_{mT}}{s_{mT}^{\,2} + 1}$$

or

$$\frac{1}{2} = \frac{2s_{mT}}{s_{mT}^{\,2} + 1} \quad \text{or} \quad s_{mT}^{\,2} - 4s_{mT} + 1 = 0$$

or

$$s_{mT} = \frac{+4 \pm \sqrt{4^2 - 4 \times 1}}{2} = 0.268 \text{ (rejecting higher value)}$$

$\therefore$ Slip at maximum torque, s_{mT} = **26.8%. (Ans.)**

Example 3.18. *A delta-connected, 6-pole, 50 Hz, 3-phase induction motor has a rotor resistance of 0.15 Ω per phase and exerts maximum torque at 880 r.p.m. Calculate the percentage maximum torque that would be exerted*

(i) at standstill ;

(ii) at 940 r.p.m. **(AMIE Advanced Elect. M/Cs., Winter 1996)**

Solution. Given : $f = 50$ Hz ; $p = 6$; $R_2 = 0.15$ Ω/phase, $N = 880$ r.p.m.

Percentage maximum torque :

Synchronous speed, $N_s = \frac{120f}{p} = \frac{120 \times 50}{6} = 1000$ r.p.m.

Slip corresponding to maximum torque, $s_{mT} = \frac{N_s - N}{N_s} = \frac{1000 - 880}{1000} = 0.12.$

(*i*) **At standstill :**

$$\frac{T_{st}}{T_m} = \frac{2s_{mT}}{s_{mT}^2 + 1} = \frac{2 \times 0.12}{0.12^2 + 1} = 0.2366$$

or $T_{st} = 0.2366 T_m$ or **23.66 percent maximum torque. (Ans.)**

(*ii*) **At a speed of 940 r.p.m. :**

$$\text{Slip, } s = \frac{1000 - 940}{1000} = 0.06$$

$$\frac{T}{T_m} = \frac{2s\, s_{mT}}{s^2 + s_{mT}^2} = \frac{2 \times 0.06 \times 0.12}{0.06^2 + 0.12^2} = 0.8$$

or $T = 0.8 T_m$ or **80 percent maximum torque. (Ans.)**

Example 3.19. *In a 3-phase induction motor :*

Maximum torque = 2 × full-load torque

Starting torque = full-load torque.

Calculate :

(i) Full-load speed ; and *(ii) slip at which maximum torque occurs.*

Solution. Maximum torque, $T_m = 2 \times T_f$

Starting torque, $T_{st} = T_f$

(*i*) Now, ratio of starting torque to maximum torque,

$$\frac{T_{st}}{T_m} = \frac{T_f}{2T_f} = 0.5$$

But, $\frac{T_{st}}{T_m} = \frac{2s_{mT}}{s^2_{mT} + 1}$, where $s_{mT} = \frac{R_2}{X_2}$

$$\therefore \quad 0.5 = \frac{2s_{mT}}{s^2_{mT} + 1}$$

or $0.5\,(s^2_{mT} + 1) = 2s_{mT}$ or $s^2_{mT} - 4s_{mT} + 1 = 0$

or $$s_{mT} = \frac{4 \pm \sqrt{4^2 - 4}}{2} = \frac{4 \pm \sqrt{12}}{2} = \frac{4 \pm 3.464}{2}$$

$= 0.268$ neglecting higher value

Let s_f = full-load slip

Also, $$\frac{T_f}{T_m} = \frac{2s_f\, s_{mT}}{s_f^2 + s_{mT}^2}$$

$$\frac{T_f}{2T_f} = \frac{2 \times s_f \times 0.268}{s_f^2 + (0.268)^2}$$

$$0.5 = \frac{0.536 s_f}{s_f^2 + 0.0718}$$

or $\quad 0.5\,(s_f^2 + 0.0718) = 0.536 s_f \quad$ or $\quad s_f^2 + 0.0718 = 1.072 s_f$

or $\quad s_f^2 - 1.072 s_f + 0.0718 = 0$

or $$s_f = \frac{1.072 \pm \sqrt{(1.072)^2 - 4 \times 0.0718}}{2} = \frac{1.072 \pm 0.928}{2}$$

$= 0.072$, rejecting the higher value

Full-load speed, $\quad N_f = N_s\,(1 - s) = N_s\,(1 - 0.072) = \mathbf{0.928\ N_s}$.

i.e., 0.928 times the synchronous speed. **(Ans.)**

(*ii*) **Slip corresponding to maximum torque,**

$$s_{mT} = \left(= \frac{R_2}{X_2} \right) = \mathbf{0.268} \text{ (or } \mathbf{26.8\%}\text{).} \quad \textbf{(Ans.)}$$

Example 3.20. *A 3-phase, 50 Hz induction motor has a starting torque which is 1.25 times full-load torque and a maximum torque which is 2.5 times the full-load torque. Neglecting stator resistance and rotational losses and assuming constant rotor resistance, find :*

(i) Slip at maximum torque ; *(ii) The slip at full-load ;*

(iii) The current at starting in per unit of full-load current. **(Punjab University)**

Solution. *Given* : $f = 50$ Hz ; $T_{st} = 1.25\,T_f$; $T_m = 2.5 T_f$

(*i*) **The slip at maximum torque,** $\mathbf{s_{mT}}$ **:**

$$\frac{T_{st}}{T_m} = \frac{2 s_{mT}}{s^2_{mT} + 1} = \frac{1.25}{2.5} = 0.5 \quad \text{or} \quad s^2_{mT} - 4 s_{mT} + 1 = 0$$

or $$s_{mT} = \frac{4 \pm \sqrt{4^2 - 4}}{2} = 0.268, \text{ rejecting higher value.}$$

Hence slip at maximum torque = 0.268 or **26.8%. (Ans.)**

(*ii*) **The slip at full-load,** $\mathbf{s_f}$ **:**

$$\frac{T_f}{T_m} = \frac{2 s_f s_{mT}}{s_f^2 + s_{mT}^2} \quad \text{or} \quad \frac{1}{2.5} = \frac{2 s_f \times 0.268}{s_f^2 + 0.268^2}$$

or $\quad s_f^2 - 1.34 s_f + 0.0718 = 0$

or $$s_f = \frac{1.34 \pm \sqrt{(1.34)^2 - 4 \times 0.0718}}{2}$$

$$= \frac{1.34 \pm 1.228}{2} = 0.056, \text{ rejecting higher value}$$

$\therefore \quad \mathbf{s_f} = 0.056$ or **5.6%. (Ans.)**

(*iii*) $\dfrac{\mathbf{I_{st}}}{\mathbf{I_f}}$:

Rotor current at start, $I_{st} = \dfrac{E_2}{\sqrt{R_2^2 + X_2^2}}$

Rotor current at full load, $I_f = \dfrac{E_2}{\sqrt{\left(\dfrac{R_2}{s_f}\right)^2 + X_2^2}}$

$$\therefore \qquad \frac{I_{st}}{I_f} = \frac{\sqrt{\left(\dfrac{R_2}{s_f}\right)^2 + X_2^2}}{\sqrt{R_2^2 + X_2^2}} = \frac{\sqrt{\left(\dfrac{R_2/X_2}{s_f}\right)^2 + 1}}{\sqrt{\left(\dfrac{R_2}{X_2}\right)^2 + 1}}$$

$$= \frac{\sqrt{\left(\dfrac{0.268}{0.056}\right)^2 + 1}}{\sqrt{(0.268)^2 + 1}} = \frac{4.889}{1.0353} = 4.722$$

$$\left(\because \quad \frac{R_2}{X_2} = s_{mT} = 0.268, \text{ as above}\right)$$

∴ *Starting current* is **4.722 times full-load current. (Ans.)**

Example 3.21. *An induction motor runs at a slip frequency of 2 Hz when supplied from a 3-phase, 400 V, 50 Hz supply. For the same developed torque, find the slip frequency at which it will run when supplied from a 3-phase 340 V, 40 Hz system. Slip at which the machine develops maximum torque using 50 Hz supply is 0.1. Neglect the stator impedance and assume linear torque-slip characteristics between zero torque and maximum torque in the working region.* **(GATE, 1997)**

Solution. The maximum torque in a 3-phase induction motor, with stator impedance ignored, is given by :

$$T_m = \frac{3}{\omega_s} \cdot \frac{V^2}{2X_2}$$

$$=$$

For 400 V supply, $\qquad s_{mT} = 0.1$

When $f = 50$ Hz, $\qquad \dfrac{R_2}{X_2} = s_{mT} = 0.1 \propto \dfrac{R_2}{50} \qquad$ ...(*i*)

For $f = 40$ Hz, $\qquad s_{mT1} \propto \dfrac{R_2}{40} \qquad$ ...(*ii*)

From (*i*) and (*ii*), we have

$$s_{mT1} = 0.1 \times \frac{50}{40} = 0.125$$

Maximum torque, $\qquad T_m = \dfrac{3}{\omega_s} \cdot \dfrac{V^2}{2X_2} \quad \text{or} \quad T_m \propto \dfrac{V^2}{f^2} \qquad (\because \ \omega_s \propto f \text{ and } X_2 \propto f)$

As torque-slip characteristic is linear between zero torque and maximum torque, the developed torque (T_d) at a slip of 0.04 ($f_r = s.f.$ *i.e.* $2 = s \times 50$ or $s = 0.04$)

$$T_d = \frac{T_m}{s_{mT}} \times 0.04 \propto \left(\frac{400}{50}\right)^2 \times \frac{0.04}{0.1} \qquad ...(i)$$

For 340 V, 50 Hz, $\qquad T_d \propto \left(\dfrac{340}{40}\right)^2 \times \dfrac{s}{0.125} \qquad$ ...(*ii*)

For the same developed torque, from (*i*) and (*ii*), we get

$$\left(\frac{400}{50}\right)^2 \times \frac{0.04}{0.1} = \left(\frac{340}{40}\right)^2 \times \frac{s}{0.125}\text{, where } s \text{ is the new slip}$$

$$\therefore \qquad s = \left(\frac{400}{50}\right)^2 \times \frac{0.04}{0.1} \times 0.125 \times \left(\frac{40}{340}\right)^2 = 0.0443$$

$\therefore$ Slip frequency $= sf = 0.0443 \times 40 =$ **1.772 Hz. (Ans.)**

☞**Example 3.22.** *A 4-pole, 50-Hz, 3-phase, induction motor develops a maximum torque of 110 N-m at 1360 r.p.m. The resistance of the star-connected rotor is 0.25 Ω/phase. Calculate the value of resistance that must be inserted in series with each rotor phase to produce a starting torque equal to half maximum torque.* **(S.C.E.W.B. 1998)**

Solution. Number of poles, $p = 4$

Resistance of star-connected rotor, $R_2 = 0.25\ \Omega$/phase

Maximum torque developed at 1360 r.p.m. = 110 N-m

Synchronous speed, $N_s = \dfrac{120f}{p} = \dfrac{120 \times 50}{4} = 1500$ r.p.m.

Slip corresponding to maximum torque,

$$s_{mT} = \frac{N_s - N}{N_s} = \frac{1500 - 1360}{1500} = 0.093$$

Also $\qquad {}_{mT} = \dfrac{R_2}{X_2}$

$$\therefore \qquad X_2 = \frac{R_2}{s_{mT}} = \frac{0.25}{0.093} = 2.688\ \Omega$$

Now $\qquad T_m = \dfrac{kE_2^2}{2X_2} = \dfrac{K}{2X_2}$ $\qquad$ [where $K = kE_2^2$]

$$= \frac{K}{2 \times 2.688} = 0.186\text{ K}$$

If '*r*' is the external resistance inserted per phase in the rotor circuit, then

Starting torque, $\qquad T_{st} = \dfrac{kE_2^2 (R_2 + r)}{(R_2 + r)^2 + (X_2)^2} = \dfrac{K(0.25 + r)}{(0.25 + r)^2 + (2.688)^2}$

$\because \qquad T_{st} = \frac{1}{2} \times T_m$ $\qquad$ (*given*)

$$\therefore \qquad \frac{K(0.25 + r)}{(0.25 + r)^2 + (2.688)^2} = \frac{0.186K}{2}$$

or $\qquad 2(0.25 + r) = 0.186[(0.25 + r)^2 + (2.688)^2]$

$$\frac{2}{0.186}(0.25 + r) = (0.25 + r)^2 + (2.688)^2$$

$$2.688 + 10.75r = 0.0625 + r^2 + 0.5r + 7.225$$

$$r^2 - 10.25r + 4.6 = 0$$

$$r = \frac{10.25 \pm \sqrt{(10.25)^2 - 4 \times 4.6}}{2} = \frac{10.25 \pm 9.31}{2}$$

or $\qquad r =$ **9.78 Ω** or **0.47 Ω. (Ans.)**

Example 3.23. *A 4-pole, 50-Hz, 3-phase, induction motor has rotor resistance and reactance of 0.025 Ω and 0.1 Ω per phase respectively. Determine :*

(i) The value of speed at which maximum torque occurs.

(ii) The value of external rotor resistance per phase to be inserted to obtain 80% of maximum torque at staring.

Solution. Number of poles, $p = 4$

Rotor resistance/phase, $R_2 = 0.025\ \Omega$

Rotor reactance/phase, $X_2 = 0.1\ \Omega$

Synchronous speed, $N_s = \frac{120f}{p} = \frac{120 \times 50}{4} = 1500$ r.p.m.

Slip corresponding to maximum torque,

$$s_{mT} = \frac{R_2}{X_2} = \frac{0.025}{0.1} = 0.25.$$

(*i*) **Speed corresponding to maximum torque,**

$$N = N_s(1 - s_{mT}) = 1500\ (1 - 0.25) = \mathbf{1125\ r.p.m.\ \ (Ans.)}$$

(*ii*) Let '*r*' be the external resistance inserted in the rotor circuit.

Ratio of rotor circuit resistance to standstill reactance,

$$s_{mT} = \frac{R_2 + r}{X_2} = \frac{0.025 + r}{0.1} \qquad ...(i)$$

Ratio of starting torque to maximum torque,

$$\frac{T_{st}}{T_m} = \frac{2s_{mT}}{s^2_{mT} + 1}$$

or $$\frac{80}{100} = \frac{2s_{mT}}{s^2_{mT} + 1}$$

$$0.8(s^2_{mT} + 1) = 2s_{mT} \quad \text{or} \quad s^2_{mT} + 1 = 2.5\ s_{mT}$$

or $$s^2_{mT} - 2.5\ s_{mT} + 1 = 0$$

or $$s_{mT} = \frac{2.5 \pm \sqrt{(2.5)^2 - 4 \times 1}}{2} = \frac{2.5 \pm 1.5}{2}$$

$= 0.5$, rejecting the higher value

Substituting $s_{mT} = 0.5$ in expression (*i*), we get

$$0.5 = \frac{0.025 + r}{0.1}$$

$$r = 0.5 \times 0.1 - 0.025 = 0.025\ \Omega$$

Hence, **external rotor resistance per phase to be inserted**

$= \mathbf{0.025\ \Omega.}$ **(Ans.)**

Example 3.24. *A 3-phase induction motor has rotor (star-connected) resistance and standstill reactance of 0.025 Ω and 0.35 Ω respectively. The full-load slip is 5 per cent. Calculate the resistance per phase of a star-connected rheostat which when connected to the rotor, will give a pull out torque at one-half of the full-load speed. What is then the power factor ?*

Solution. Rotor resistance, $R_2 = 0.05\ \Omega$

Rotor standstill reactance, $X_2 = 0.35\ \Omega$

The full-load slip, $s_f = 5\% = 0.05$

Full-load speed, $N_f = N_s\,(1 - s_f) = N_s\,(1 - 0.05) = 0.95\,N_s$

One-half of the full-load speed,

$$N = \frac{1}{2} N_f = \frac{0.95}{2} N_s = 0.475\,N_s$$

Slip at one-half of the full-load speed,

$$s = \frac{N_s - 0.475\,N_s}{N_s} = 0.525$$

Rotor reactance per phase at one-half of full-load speed

$$= sX_2 = 0.525 \times 0.35\ \Omega = 0.184\ \Omega$$

Let 'r' be the resistance per phase of a star-connected rheostat connected to the motor.

Pull out torque or maximum torque will be at one-half of full-load speed, if resistance per phase in the rotor circuit is equal to reactance per phase of the rotor at one-half of full-load speed.

i.e., $0.05 + r = 0.184$

$\therefore$ $\mathbf{r = 0.134\ \Omega}$. **(Ans.)**

Power factor,
$$\cos\phi = \frac{R_2 + r}{\sqrt{(R_2 + r)^2 + (sX_2)^2}} = \frac{0.05 + 0.134}{\sqrt{(0.05 + 0.134)^2 + (0.184)^2}}$$
$$= \frac{0.184}{0.2602} = 0.707 \text{ (lag)}$$

Hence, **power factor = 0.707 (lag). (Ans.)**

Example 3.25. *A 746 kW, 3-phase, 50 Hz, 16-pole induction motor has a rotor impedance of (0.02 + j0.15) Ω at standstill. Full-load torque is obtained at 360 r.p.m. Calculate :*

(i) The ratio of maximum to full-load torque ;

(ii) The speed for maximum torque ;

(iii) The rotor resistance to be added to get maximum starting torque. **(Nagpur Univ., 1998)**

Solution. *Given* : $f = 50$ Hz ; $p = 16$, $R_2 = 0.02\ \Omega$; $X_2 = 0.15\ \Omega$; $N_f = 360$ r.p.m.

Synchronous speed,
$$N_s = \frac{120f}{p} = \frac{120 \times 50}{16} = 375 \text{ r.p.m.}$$

Full-load slip,
$$s_f = \frac{N_s - N_f}{N_s} = \frac{375 - 360}{375} = 0.04$$

Slip corresponding to maximum torque, $s_{mT} = \dfrac{R_2}{X_2} = \dfrac{0.02}{0.15} = 0.133$

(*i*) **The ratio of maximum torque to full-load torque,** $\dfrac{\mathbf{T_m}}{\mathbf{T_f}}$ **:**

We know that,
$$\frac{T_f}{T_m} = \frac{2 s_f s_{mT}}{s_f^2 + s_{mT}^2} \qquad \text{[Eqn. (3.17)]}$$
$$= \frac{2 \times 0.04 \times 0.133}{(0.04)^2 + (0.133)^2} = 0.5516$$

$\therefore$
$$\frac{T_m}{T_f} = \mathbf{1.813.}\ \textbf{(Ans.)}$$

(*ii*) **The speed for maximum torque, N :**

$$\mathbf{N} = N_s(1 - s_{mT}) = 375(1 - 0.133) = \mathbf{325\ r.p.m.}\ \textbf{(Ans.)}$$

***(iii)* The rotor resistance to be added to get maximum starting torque, r :**

For obtaining maximum starting torque,

$$s = \frac{(R_2 + r)}{X_2} = 1$$

or $R_2 + r = X_2$ or $0.02 + r = 0.15$

$\therefore$ **r = 0.13 Ω. (Ans.)**

Example 3.26. *The starting and maximum torques of a 3-phase induction motor are 1.5 times and 2.5 times its full-load torque. Determine the percentage change in rotor circuit resistance to obtain a full-load slip of 0.03. Neglect stator impedance.* **(Pb. Univ., 1998)**

Solution. *Given* : $T_{st} = 1.5\ T_f$; $T_m = 2.5\ T_f$; $s_f = 0.03$.

Percentage change in rotor circuit resistance :

$$\frac{T_{st}}{T_m} = \frac{2s_{mT}}{s^2_{mT} + 1} = \frac{1.5}{2.5} = 0.6$$

$$2s_{mT} = 0.6(s^2_{mT} + 1)$$

or $s^2_{mT} - 3.333s_{mT} + 1 = 0$

or $$s_{mT} = \frac{3.333 \pm \sqrt{3.333^2 - 4}}{2}$$

$$= \frac{3.333 \pm 2.666}{2} = 0.333 \text{ (rejecting higher value)}$$

or $$s_{mT} = \frac{R_2}{X_2} = 0.333$$

or Rotor resistance, $R_2 = 0.333\ X_2$

Also $$\frac{T_f}{T_m} = \frac{2s_f\,s_{mT}}{s_f^2 + s^2_{mT}}$$

or $$\frac{1}{2.5} = \frac{2 \times 0.03 \times s_{mT}}{0.03^2 + s^2_{mT}}$$

or $0.03^2 + s^2_{mT} = 2.5 \times 2 \times 0.03 \times s_{mT} = 0.15s_{mT}$

or $s^2_{mT} - 0.15s_{mT} + 0.0009 = 0$

or $$s_{mT} = \frac{0.15 \pm \sqrt{0.15^2 - 4 \times 0.0009}}{2} = \frac{0.15 \pm 0.1375}{2} = 0.1437$$

(other value is not feasible)

$\therefore$ New rotor resistance, $R_2' = s_{mT}\,X_2 = 0.1437X_2$

$\therefore$ Percentage reduction in rotor resistance

$$= \frac{R_2 - R_2'}{R_2} \times 100$$

$$= \frac{0.333X_2 - 0.1437X_2}{0.333X_2} = \textbf{56.85\%. (Ans.)}$$

Example 3.27. *In a 3-phase induction motor, maximum torque is twice the full-load torque and starting torque is 1.5 times the full-load torque. In order to get a full-load slip of 4 per cent, calculate the percentage reduction in rotor circuit resistance. Neglect stator impedance.*

Solution. *Given* : $T_m = 2T_f$; $T_{st} = 1.5T_f$; $s_f = 4\%$ or 0.04.

Percentage reduction in rotor-circuit resistance :

We know that, $$\frac{T_{st}}{T_m} = \frac{2s_{mT}}{s^2_{mT} + 1} \qquad \text{[Eqn. (3.18)]}$$

or $$\frac{1.5T_f}{2T_f} = \frac{2s_{mT}}{s^2_{mT} + 1} \quad \text{or} \quad 0.75(s^2_{mT} + 1) = 2s_{mT}$$

or $$s^2_{mT} - 2.667s_{mT} + 1 = 0$$

∴ $$s_{mT} = \frac{2.667 \pm \sqrt{(2.667)^2 - 4}}{2} = \frac{2.667 \pm 1.764}{2} = 0.45 \text{ or } 2.21$$

Thus, $s_{mT} = 0.45$ (rejecting the higher value)

The negligible stator impedance,

$$\frac{R_2}{X_2} = s_{mT} = 0.45 \quad \text{or} \quad R_2 = 0.45X_2$$

For a full-load slip of 0.04, we have

$$\frac{T_f}{T_m} = \frac{1}{2} = \frac{2s_f s_{mT}}{s_f^2 + s^2_{mT}} = \frac{2 \times 0.04 \times s_{mT1}}{(0.04)^2 + s^2_{mT1}}$$

or $$(0.04)^2 + s^2_{mT1} = 2 \times 2 \times 0.04\, s_{mT1}$$

or $$s^2_{mT1} - 0.16s_{mT1} + 0.0016 = 0$$

or $$s_{mT1} = \frac{0.16 \pm \sqrt{(0.16)^2 - 4 \times 0.0016}}{2} = \frac{0.16 \pm 0.139}{2} = 0.1495 \text{ or } 0.0105$$

∴ $s_{mT1} = 0.1495$ (rejecting the other value)

∴ New rotor resistance for obtaining a slip of 4%, $R_{2n} = 0.1495X_2$

Thus, reduction in rotor circuit resistance

$$= 0.45X_2 - 0.1495X_2 = 0.3X_2$$

∴ Percentage reduction in rotor-circuit resistance

$$= \frac{0.3X_2}{0.45X_2} \times 100 = \mathbf{66.67\%.} \quad \textbf{(Ans.)}$$

☞**Example 3.28.** *A 3-φ induction motor has a rotor resistance and reactance of 0.016 Ω and 0.1 Ω per phase respectively. At normal voltage, the full-load slip is 4 per cent. Calculate :*

(i) Percentage reduction in rotor voltage to develop full-load torque at half full-load speed.

(ii) Power factor.

Solution. Rotor resistance/phase, $R_2 = 0.016\ \Omega$

Rotor reactance/phase, $X_2 = 0.1\ \Omega$

Full-load slip, $s_f = 4\%$ (or 0.04)

Let N_s be the synchronous speed.

The full-load speed, $N_f = N_s(1 - s_f) = N_s(1 - 0.04) = 0.96\, N_s$

One-half of full-load speed $= \dfrac{0.96N_s}{2} = 0.48\, N_s$

Slip at one-half of full-load speed,

$$s_2 = \frac{N_2 - 0.48N_s}{N_s} = 0.52$$

We know that the value of torque when motor is running is given by,

$$T = \frac{ksR_2E_2^2}{R_2^2 + s^2X_2^2} = \frac{ksR_2V^2}{R_2^2 + s^2X_2^2} \qquad [\because\ E_2 \propto V]$$

As torque in both cases is the same,

$$\therefore \quad \frac{ks_1R_2V_1^2}{R_2^2 + s_1^2X_2^2} = \frac{ks_2R_2V_2^2}{R_2^2 + s_2^2X_2^2}$$

where V_2 is the voltage in the second case

$$\therefore \quad \left(\frac{V_1}{V_2}\right)^2 = \frac{s_2}{s_1} \times \frac{R_2^2 + s_1^2X_2^2}{R_2^2 + s_2^2X_2^2}$$

$$= \frac{0.52}{0.04} \times \frac{(0.016)^2 + (0.04)^2 \times (0.1)^2}{(0.016)^2 + (0.52)^2 \times (0.1)^2} = 13 \times \frac{0.000272}{0.00296} = 1.19$$

$$\therefore \quad \frac{V_1}{V_2} = 1.093$$

(*i*) **Percentage reduction in supply voltage**

$$= \frac{V_1 - V_2}{V_1} \times 100 = \left(1 - \frac{V_2}{V_1}\right) \times 100$$

$$= \left(1 - \frac{1}{1.093}\right) \times 100 = \mathbf{8.51\ per\ cent.}\ \textbf{(Ans.)}$$

(*ii*) **Power factor,** $\cos\phi = \frac{R_2}{R_2^2 + s_2^2X_2^2} = \frac{0.016}{\sqrt{(0.016)^2 + (0.52 \times 0.1)^2}} = \mathbf{0.294.}$ **(Ans.)**

3.17. POWER STAGES IN AN INDUCTION MOTOR

- **Stator iron loss** (consisting of eddy and hysteresis losses) depends on the supply frequency and the flux density in the iron core. It is practically *constant*.

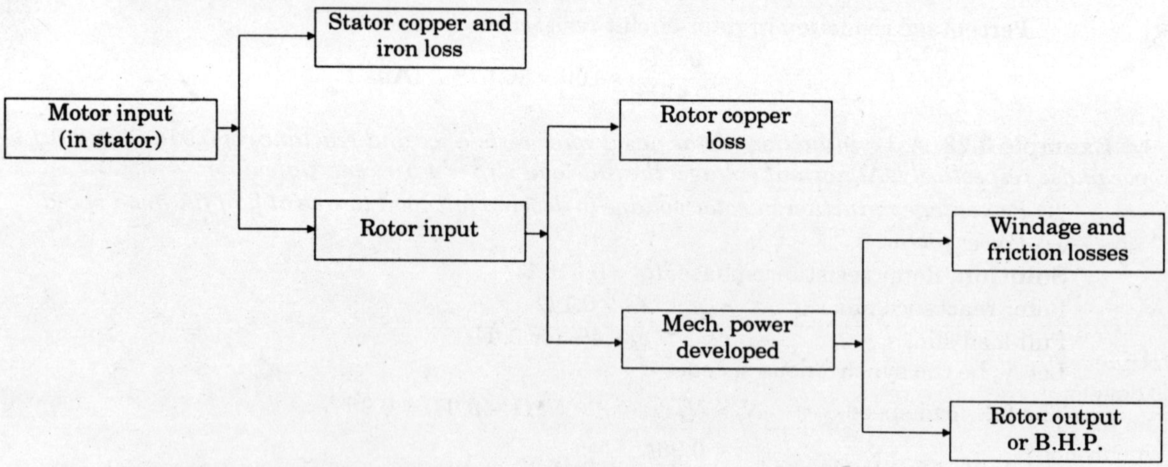

- Rotor iron loss is negligible because the frequency of rotor currents under normal running conditions is always small.
- Total rotor copper loss = $3I_2^2R_2$ (where I_2 and R_2 are current and resistance of the rotor respectively).

3.17.1. Rotor Output

Stator input = stator output + losses

Rotor input = stator output

[stator output is transferred entirely inductively to the rotor circuit]

Rotor gross output = rotor input – rotor copper losses.

This *rotor output is converted into mechanical* energy and gives rise to gross torque T_g. Out of this gross torque developed, some is lost due to windage and friction losses in the rotor and the rest appears as the *useful* or *shaft torque* T_{sh}. If T_g is in N-m and N is the actual speed of the rotor in r.p.m.

$$T_g = \frac{\text{rotor gross output in watts}}{(2\pi N/60)} \quad \text{...(3.19)}$$

If there were no copper losses in the rotor, then rotor output will equal rotor input and the rotor will run at synchronous speed,

$$T_g = \frac{\text{rotor input}}{(2\pi N_s/60)} \quad \text{...(3.20)}$$

From eqn. (3.19) and eqn. (3.20), we get

$$\text{Rotor output} = T_g \times 2\pi N/60$$

$$\text{Rotor input} = T_g \times 2\pi N_s/60 \quad \text{...(3.21)}$$

But rotor input = rotor output + rotor copper losses

$$\therefore \quad \text{Rotor copper losses} = \text{rotor input} - \text{rotor output}$$

$$= T_g \times 2\pi/60\,(N_s - N) \quad \text{...(3.22)}$$

From eqn. (3.21) and eqn. (3.22), we get

$$\frac{\text{Rotor copper loss}}{\text{Rotor input}} = \frac{N_s - N}{N_s} = s$$

$$\therefore \quad \text{Rotor copper loss} = s \times \text{rotor input}$$

$$= s \times \text{power across air-gap} \quad \text{...(3.23)}$$

$$\text{Rotor gross output} = \text{input} - \text{copper loss}$$

$$= \text{input} - s \times \text{rotor input}$$

$$= (1 - s)\ \text{input} \quad \text{...(3.24)}$$

$$\therefore \quad \frac{\text{Rotor gross output}}{\text{Rotor input}} = 1 - s = 1 - \frac{N_s - N}{N_s}$$

$$= \frac{N}{N_s} = \frac{\text{actual speed of rotor}}{\text{synchronous speed}}$$

$$\therefore \quad \text{Rotor efficiency} = \frac{N}{N_s}$$

Also,

$$\frac{\text{rotor copper loss}}{\text{rotor gross output}} = \frac{s}{1 - s}. \quad \text{...(3.25)}$$

3.17.2. Synchronous Watt. A synchronous watt is a new unit of torque and is defined as the *torque which develops a power of 1 watt at the synchronous speed of motor*. When we say that motor is developing a torque of 500 synchronous watts, we mean that the rotor input is 500 watts and that the torque is such that power developed would be 500 watts provided the rotor is running synchronously and developing the same torque.

$$\text{Rotor input} = T_{sw} \times 2\pi N_s$$

or

$$T_{sw} = \frac{\text{rotor input in watts}}{2\pi N_s}$$

Synchronous wattage of an induction motor equals the power transferred across the air-gap to the rotor.

3.17.3. Factors Determining Torque. When a 3-phase induction motor is operating on a certain load with slip s, then

Induced e.m.f. per phase in rotor $= sE_2$ volts (where E_2 = standstill rotor induced e.m.f.)

Rotor impedance per phase, $Z_2 = R_2 + jsX_2$

(where R_2 and X_2 are rotor resistance and standstill reactance per phase)

$$\text{Rotor current per phase} = \frac{sE_2}{Z_2} = \frac{sE_2}{\sqrt{R_2^2 + s^2X_2^2}}$$

$$\text{Rotor circuit power factor, } \cos\phi_2 = \frac{R_2}{\sqrt{R^2 + s^2X_2^2}}$$

Total rotor copper loss = number of phases × rotor induced e.m.f. per phase × rotor current per phase × rotor current power factor

$$= 3 \times sE_2 \times \frac{sE_2}{\sqrt{R_2^2 + s^2X_2^2}} \times \frac{R_2}{\sqrt{R_2^2 + s^2X_2^2}} = \frac{3R_2\, s^2E_2^2}{R_2^2 + s^2X_2^2} \quad ...(3.26)$$

$$\text{Also, total rotor copper loss} = s \times \text{rotor input} = \frac{s \times T \times 2\pi N_s}{60}$$

$$\therefore \quad s \times \frac{T \times 2\pi N_s}{60} = \frac{3R_2\, s^2E_2^2}{R_2^2 + s^2X_2^2}$$

or

$$\text{Torque developed by the rotor, } T = \frac{3}{(2\pi N_s/60)} \cdot \frac{R_2 sE_2^2}{R_2^2 + s^2E_2^2} \quad ...(3.27)$$

$$= \frac{3}{\omega_s} \cdot \frac{R_2\, sE_2^2}{R_2^2 + s_2{}^2E_2^2} \quad ...[3.27\ (a)]$$

or

$$T \propto \frac{R_2\, sE_2^2}{R_2^2 + s^2X_2^2} \propto \frac{R_2\, sV_1^2}{R_2^2 + s^2X_2^2}. \quad (\because\ E_2 \propto V_1, \text{ the supply voltage})$$

ROTOR OUTPUT, LOSSES AND EFFICIENCY

Example 3.29. *The power input to a 3-phase induction motor is 50 kW. The stator-losses total 1.2 kW. If the motor is running with a slip of 3 per cent, find the total mechanical power developed.*

Solution. Stator power input = 50 kW

Stator losses = 1.2 kW

Slip = 3% (= 0.03)

Mechanical power developed :

Rotor input = stator output = stator input – stator losses

= 50 – 1.2 = 48.8 kW

Now *mechanical power developed*

= (1 – s) rotor input = (1 – 0.03) × 48.8 = **47.34 kW. (Ans.)**

☞**Example 3.30.** *A 6-pole, 50-Hz, 3-phase induction motor running on full-load with 3% slip develops a torque of 160 N-m at its pulley rim. The friction and windage losses are 210 W and the stator copper and iron losses equal 1640 W. Calculate :*

(i) Rotor output, *(ii) Rotor copper loss, and*

(iii) Efficiency at full-load.

Solution. Number of poles, $p = 6$

Supply frequency, $f = 50$ Hz

Slip, $s = 3\%$ or 0.03

Torque developed at full-load, $T_f = 160$ N-m

Friction and windage losses = 210 W

Stator copper and iron losses = 1640 W

Synchronous speed, $N_s = \frac{120f}{p} = \frac{120 \times 50}{6} = 1000$ r.p.m.

Actual rotor speed, $N = (1 - s)\, N_s = (1 - 0.03) \times 1000 = 970$ r.p.m.

(*i*) **Rotor output**

$$= \frac{2\pi N T_f}{60} = \frac{2\pi \times 970 \times 160}{60} = \mathbf{16252.5\ W.}\ \textbf{(Ans.)}$$

(*ii*) **Rotor gross output**

= rotor output + friction and windage losses

= 16252.5 + 210 = 16462.5 W

Also, $\frac{\text{rotor gross output}}{\text{rotor input}} = \frac{N}{N_s}$

$\therefore$ Power input $= \frac{\text{rotor gross output} \times N_s}{N}$

$$= \frac{16462.5 \times 1000}{970} = 16971.6 \text{ W}$$

$\therefore$ **Rotor copper loss** = 16971.6 – 16462.5 = **509.1 W. (Ans.)**

(*iii*) **Stator output** = rotor input = 16971.6 W

$\therefore$ Stator input = stator output + stator copper and iron losses

= 16971.6 + 1640 = 18611.6 W

% Overall efficiency at full-load $= \frac{\text{rotor output}}{\text{stator input}}$

$$= \frac{16252.5}{18611.6} \times 100 = \mathbf{87.32\%.}\ \textbf{(Ans.)}$$

Example 3.31. *The power input to the rotor of a 440 V, 50-Hz, 3-phase, 6-pole induction motor is 50 kW. It is observed that the rotor e.m.f. makes 120 complete cycles per minute. Calculate :*

(*i*) *Slip,* (*ii*) *Rotor speed,*

(*iii*) *Rotor copper loss/phase,* (*iv*) *Mechanical power developed, and*

(*v*) *Rotor resistance/phase if rotor current is 50 A.*

Solution. Supply voltage = 440 V

Supply frequency, $f = 50$-Hz

Number of poles, $p = 6$

Rotor input = 50 kW or 50000 W

Rotor frequency, $f_r = \frac{120}{60} = 2$ Hz

(*i*) **Slip,** $s = \frac{f_r}{f} = \frac{2}{50} = \mathbf{0.04}$ **or 4%. (Ans.)**

(*ii*) **Rotor speed,** $N = (1 - s)N_s = (1 - s) \times \frac{120f}{p}$

$$= (1 - 0.04) \times \frac{120 \times 50}{6} = \textbf{960 r.p.m.} \quad \textbf{(Ans.)}$$

(*iii*) Rotor copper loss $= s \times$ rotor input $= 0.04 \times 50000 = 2000$ W

Rotor copper loss/phase $= \frac{2000}{3} = \textbf{666.67 W.}$ **(Ans.)**

(*iv*) **Mechanical power developed**

$= (1 - s) \times$ rotor input $= (1 - 0.04) \times 50000$

$=$ **48000 W or 48 kW. (Ans.)**

[or = rotor input – total copper loss]

(*v*) Let R_2 be the rotor resistance/phase

$$\therefore \quad 3I_2^2R_2 = 2000$$

$$R_2 = \frac{2000}{3 \times 50 \times 50} = 0.267\ \Omega$$

Hence, **rotor resistance/phase = 0.267 Ω. (Ans.)**

Example 3.32. *A 50 H.P., 6-pole, 50 Hz, slip-ring induction motor runs at 960 r.p.m. on full-load with a rotor current of 40 A. Allowing 300 W for copper loss in the short-circuiting gear and 1200 W for mechanical losses, find the resistance R_2 per phase of the 3-phase rotor winding.*

(GATE, 1998)

Solution. $P_{out} = 50$ H.P. $= 50 \times 735.5 = 36775$ W ; $p = 6$; $f = 50$ Hz ; $N = 960$ r.p.m. ; $I_2 = 40$ A ; copper loss in the short-circuiting gear = 300 W ; mechanical loss = 1200 W.

Resistance, R_2 :

Rotor output = motor output + short-circuiting gear loss + mechanical losses

$= 36775 + 300 + 1200 = 38275$ W

Synchronous speed, $$N_s = \frac{120f}{p} = \frac{120 \times 50}{60} = 1000 \text{ r.p.m.}$$

Slip, $$s = \frac{N_s - N}{N_s} = \frac{1000 - 960}{1000} = 0.04$$

Rotor copper losses $$= 3I_2^2R_2 = \frac{\text{Rotor output}}{1 - s} \times s = \frac{38275}{(1 - 0.04)} \times 0.04 = 1594.8 \text{ W}$$

or Rotor resistance, $$\mathbf{R_2} = \frac{1594.8}{3 \times 40^2} = \mathbf{0.332\ \Omega.} \quad \textbf{(Ans.)}$$

Example 3.33. (*a*) *Sketch the torque-speed characteristics of 3-phase induction motor. Show the regions of stable and brake modes of operation. What will be the effect of the following operations on the characteristics ?*

(*i*) *Applied voltage with normal frequency is reduced to half.*

(*ii*) *Both the applied voltage and the frequency are reduced to half.*

(*b*) *The star-connected starter winding of a 3-phase, 6-pole induction motor is supplied from a 240 V, 50 Hz source. The rotor resistance and standstill reactance are 0.12 Ω and 0.85 Ω per phase respectively. She stator to rotor turns ratio is 1.8. Find the pull out torque and the corresponding speed.*

(AMIE Winter, 1998)

Solution. (*a*) Fig. 3.20 shows the torque-speed characteristics of a 3-phase induction motor showing the regions of stable and brake modes of operation.

The torque equation is given by :

$$T = \frac{ksR_2E_2^2}{R_2^2 + s^2X_2^2}$$

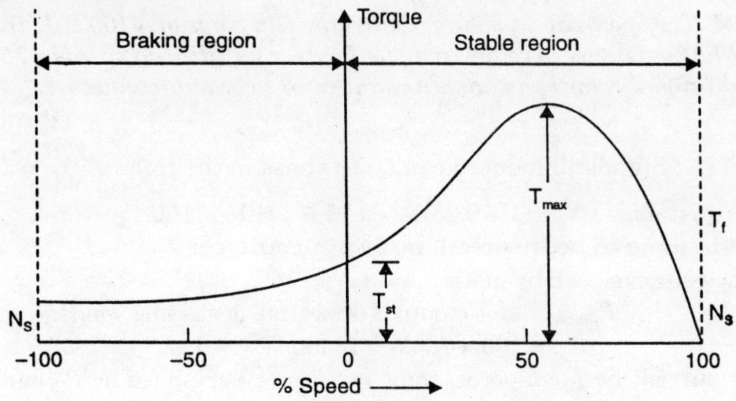

Fig. 3.20

(*i*) Thus with normal frequency, $T \propto E_2^2$ *i.e.* square of voltage, hence torque when applied voltage is reduced to half will be *reduced to* $\frac{1}{4}$ *th.* **(Ans.)**

(*ii*) Also $T \propto \frac{E^2}{N_s} \propto \frac{E_2^2}{f}$, hence when applied voltage and frequency both are reduced to half, the torque will be reduced to *one half of the normal value*. **(Ans.)**

(*b*) For pull-out torque or maximum torque,

$$R_2 = s_{mT} X_2$$

$$\therefore \quad s_{mT} = \frac{R_2}{X_2} = \frac{0.12}{0.85} = 0.141$$

Standstill e.m.f. of rotor per phase, $E_2 = KE_1$

$$= \frac{240}{1.8} = 133.33 \text{ V}$$

$\therefore$ Rotor e.m.f. (at maximum torque) $= s_{mT} E_2$

$$= 0.141 \times 133.33 = 18.8 \text{ V}$$

Rotor impedance, $= \sqrt{R_2^2 + (s_{mT} X_2)^2}$

$$= \sqrt{0.12^2 + (0.141 \times 0.85)^2} = 0.169 \ \Omega$$

$$\therefore \quad I_2 = \frac{18.8}{0.169} = 111.24 \text{ A}$$

Total rotor copper loss $= 3 \times (111.24)^2 \times 0.12 = 4454.76$ W

Rotor output $= \text{rotor copper loss} \times \left(\frac{1-s}{s}\right)$

$$= 4454.76 \times \left(\frac{1 - 0.141}{0.141}\right) = 27139.3 \text{ W}$$

Speed, $N = (1 - s)N_s = (1 - 0.141) \times \frac{120 \times 50}{6} =$ **859 r.p.m. (Ans.)**

Pull out or maximum torque, $T_m = \frac{\text{rotor gross output}}{(2\pi N/60)}$

$$= \frac{27139.3}{(2\pi \times 859/60)} = \textbf{301.7 N-m.}$$ **(Ans.)**

Example 3.34. *The open-circuit voltage across the slip rings of a 100 H.P. induction motor is 273 V at standstill. What resistance in rotor circuit will reduce its full-load speed by 25% ? The full-load slip is 2% with no additional rotor resistance. Assume rotor to be star-connected.*

(Punjab University)

Solution. *Given* : Standstill induced e.m.f. per phase in the rotor = $\frac{273}{\sqrt{3}}$ = 157.62 V ; s = 2% (with no additional resistance) ; N' = (1 – 0.25)N = 0.75 N ; H.P. = 100.

External resistance to be inserted in rotor circuit, r :

Mechanical power developed by motor,

$$P_{mech.} = \text{motor output (in watts), neglecting windage and frictional losses}$$
$$= 100 \times 735.5 = 73550 \text{ W}$$

Let the rotor current be I_2 amperes, rotor resistance per phase be R_2 and rotor reactance negligible.

Then,
$$I_2 = \frac{sE_2}{R_2} = \frac{0.02 \times 157.62}{R_2} = \frac{3.15}{R_2} \quad ...(i)$$

Mechanical power developed by the rotor = $\frac{\text{total copper loss}}{s_f}(1 - s_f)$

or
$$73550 = \frac{3I_2^2R_2}{0.02}(1 - 0.02) = 147I_2^2R_2 \quad ...(ii)$$

Substituting the value of I_2 from (*i*) into (*ii*), we get

$$147 \times \left(\frac{3.15}{R_2}\right)^2 \times R_2 = 73550 \quad \text{or} \quad R_2 = 0.02\ \Omega$$

New speed,
$$N' = (1 - 0.25)\ N = (1 - 0.25)\ [N_s(1 - s)]$$
$$= 0.75 \times N_s(1 - 0.02) = 0.735\ N_s$$

∴ Slip,
$$s' = \frac{N_s - N'}{N_s} = \frac{N_s - 0.735N_s}{N_s} = 0.265$$

Assuming load torque to be constant, $s \propto$ resistance

∴
$$\frac{s'}{s} = \frac{R_2 + r}{R_2}$$

or
$$\frac{0.265}{0.02} = \frac{0.02 + r}{0.02} \quad \text{or} \quad \mathbf{r = 0.245\ \Omega.} \quad \textbf{(Ans.)}$$

☞**Example 3.35.** *A 36 kW, 3-phase, 4-pole, 50-Hz induction motor has a full-load efficiency of 84 per cent. The friction and windage losses are one-third of the no-load losses and rotor copper losses equal the iron loss at full-load. Find the full-load speed. Neglect the stator resistance.*

Solution. Motor output = 36 kW or 36000 W

Full-load efficiency = 84 per cent

Friction and windage losses = $\frac{1}{3}$ × no-load losses

Rotor copper losses = iron loss (at full-load)

Full-load speed, N :

Motor input $= \frac{\text{motor output}}{\text{efficiency}} = \frac{36000}{0.84} = 42857$ W

Total losses = input – output = 42857 – 36000 = 6857 W

Also, Total losses = stator iron losses + stator copper losses + rotor copper losses + windage and friction losses

Since stator resistance is negligible stator losses can be neglected.

Since friction and windage losses are $\frac{1}{3}$rd of no-load losses, therefore, stator iron losses are $\frac{2}{3}$rd of no-load losses or double of friction and windage losses.

Total losses at full-load = stator iron losses + rotor copper losses + friction and windage losses

or 6857 = stator iron losses + stator iron losses + $\frac{1}{2}$ stator iron losses

[∵ Rotor copper losses equal iron losses and friction and windage are half of stator iron losses]

∴ Stator iron losses $= \frac{6857}{2.5} = 2743$ W

Rotor copper losses = stator iron losses = 2743 W

Friction and windage losses $= \frac{1}{2} \times$ stator iron losses $= \frac{1}{2} \times 2743 = 1371$ W

Mechanical power developed = motor output + friction and windage losses
= 36000 + 1371 = 37371 W

Rotor input = mechanical power developed + rotor copper losses
= 37371 + 2743 = 40114 W

Since, rotor output = $(1 - s) \times$ rotor input

$$\therefore \quad (1 - s) = \frac{\text{rotor output (gross)}}{\text{rotor input}} = \frac{37371}{40114} = 0.932$$

∴ **Motor speed at full-load**

$$= N_s (1 - s) = \frac{120f}{p} \times 0.932 = \frac{120 \times 50}{4} \times 0.932$$

= **1398 r.p.m. (Ans.)**

Example 3.36. *A 3-phase, 400 V, 4-pole, 50-Hz induction motor has a star-connected stator and rotor. The rotor resistance and standstill reactance/phase are 0.25 Ω and 1.2 Ω respectively. The ratio of stator to rotor turns is 1.3. The full-load slip is 4 per cent. Calculate :*

(i) The power and torque developed at full-load.

(ii) Maximum torque and the speed at which it occurs.

Solution. Rotor resistance/phase, $R_2 = 0.25\ \Omega$

Standstill reactance/phase, $X_2 = 1.2\ \Omega$

$$\frac{\text{Rotor turns}}{\text{Stator turns}} = \frac{1}{1.3} \ (= K)$$

Full-load slip = 4% or 0.04

Supply voltage = 400 V

Phase voltage $= \frac{400}{\sqrt{3}} = 230.9$ V

∴ Standstill e.m.f./rotor phase,

$$E_2 = 230.9 \times \frac{1}{1.3} = 177.6 \text{ V}$$

$$Z_2 = \sqrt{R_2^2 + (sX_2)^2} = \sqrt{(0.25)^2 + (0.04 \times 1.2)^2} = 0.254\ \Omega$$

$$I_2 = \frac{sE_2}{Z_2} = \frac{0.04 \times 177.6}{0.254} = 27.97 \text{ A}$$

Total copper losses $= 3I_2^2R_2 = 3 \times (27.97)^2 \times 0.25 = 586.7$ W

Also, $\dfrac{\text{Rotor copper loss}}{\text{Rotor gross output}} = \dfrac{s}{1-s}$

(i) $\therefore$ Rotor gross output $= \text{rotor copper loss} \times \left(\dfrac{1-s}{s}\right)$

$$= 586.7 \times \left(\frac{1-0.04}{0.04}\right) = 14081 \text{ W}$$

Hence, **power developed at full-load**

$$= \mathbf{14081 \text{ W or } 14.081 \text{ kW.}} \quad \textbf{(Ans.)}$$

Rotor speed $= (1-s)\,N_s = (1-0.04) \times \dfrac{120f}{p}$

$$= (1-0.04) \times \frac{120 \times 50}{4} = 1440 \text{ r.p.m.}$$

Gross torque developed,

$$T_g = \frac{\text{rotor gross output}}{(2\pi N/60)} = \frac{14081}{2\pi \times 1440/60} = \frac{14081 \times 60}{2\pi/1440}$$

$$= \mathbf{93.37 \text{ Nm.}} \quad \textbf{(Ans.)}$$

(ii) **Maximum torque and speed :**

We know that for *maximum torque* :

$$R_2 = sX_2$$

$$\therefore \quad s = \frac{R_2}{X_2} = \frac{0.25}{1.2} = 0.208$$

$$E_r = sE_2 = 0.208 \times 177.6 = 36.9 \text{ V}$$

$$Z_2 = \sqrt{R_2^2 + (sX_2)^2} = \sqrt{(0.25)^2 + (0.208 \times 1.2)^2} = 0.353 \ \Omega$$

$$I_2 = \frac{36.9}{0.353} = 104.5 \text{ A}$$

Rotor copper losses $= 3 \times (104.5)^2 \times 0.25 = 8190$ W

Rotor output $= \text{rotor copper loss} \times \left(\dfrac{1-s}{s}\right)$

$$= 8190 \times \left(\frac{1-0.208}{0.208}\right) = 31185 \text{ W or } 31.185 \text{ kW}$$

Speed, $N = (1-s)\,N_s = (1-0.208) \times \dfrac{120f}{p}$

$$= (1-0.208) \times \frac{120 \times 50}{4} = \mathbf{1188 \text{ r.p.m.}} \quad \textbf{(Ans.)}$$

$$\mathbf{T_m} = \frac{\text{rotor gross output}}{2\pi N/60} = \frac{31185}{2\pi \times 1188/60} = \mathbf{250.67 \text{ Nm.}} \quad \textbf{(Ans.)}$$

☞**Example 3.37.** *A 3-phase, 6-pole, 50-Hz induction motor has 160 N-m as its useful full-load torque. The rotor e.m.f. is observed to make 90 cycles per minute. Calculate :*

(i) Motor output in kW, *(ii) Copper losses in rotor,*

(iii) Motor input, and

(iv) Efficiency if mechanical torque lost in friction and windage is 20 N-m and stator losses are 800 W.

Solution. Number of poles, $p = 6$

Supply frequency, $f = 50$ Hz

Rotor e.m.f. frequency, $f_r = \frac{90}{60} = 1.5$ Hz

Useful full-load torque, $T_{sh} = 160$ N-m

Torque lost in friction and windage = 20 N-m

Stator losses = 800 W

Slip, $s = \frac{f_r}{f} = \frac{1.5}{50} = 0.03$

Rotor speed, $N = (1 - s) N_s$

$$= (1 - 0.03) \times \frac{120f}{p} = (1 - 0.03) \times \frac{120 \times 50}{6} = 970 \text{ r.p.m.}$$

(*i*) **Motor output** $= \frac{2\pi N T_{sh}}{60} = \frac{2\pi \times 970 \times 160}{60} = 16252$ W or **16.252 kW. (Ans.)**

(*ii*) **Rotor copper losses :**

Gross torque, $T_g = T_{sh} + T_{lost} = 160 + 20 = 180$ N-m

Rotor input $= \frac{2\pi N_s T_g}{60} = \frac{2\pi \times 1000 \times 180}{60} = 18850$ W

∴ **Rotor copper loss** $= s \times$ rotor input $= 0.03 \times 18850 =$ **565.5 W. (Ans.)**

(*iii*) **Motor input** = rotor input + stator losses = 18850 + 800 = **19650 W. (Ans.)**

(*iv*) **Efficiency :**

Efficiency $= \frac{\text{motor output}}{\text{motor input}} = \frac{16252}{19650} \times 100 =$ **82.7%. (Ans.)**

Example 3.38. *A 4 kW, 6-pole, 3-phase, 50-Hz induction motor operates with a slip of 2.2 per cent and requires 11.5 amperes and 3500 watts when delivering its usual load. When operating without load machine receives 4.4 amperes and 300 watts. When the rotor of the machine is locked, 445 watts at 50 volts are required to circulate a current of 15 amperes. Calculate output torque and efficiency of the motor when delivering its usual load.*

Solution. Rotor (or motor) output = 4 kW or 4000 W

Number of poles, $p = 6$

Supply frequency, $f = 50$ Hz

Slip, $s = 2.2\%$ or 0.022

Power input to machine at normal load = 3500 W

No load losses = 300 W

Power input to rotor at full-load = 3500 − 300 = 3200 W

We know that, $\frac{\text{rotor output}}{\text{rotor input}} = 1 - s$

∴ Rotor output $= (1 - s) \times$ rotor input

$= (1 - 0.022) \times 3200 = 3130$ W or 3.13 kW

Rotor speed, $N = (1 - s) N_s = (1 - 0.022) \frac{120f}{p}$

$$= (1 - 0.022) \times \frac{120 \times 50}{6} = 978 \text{ r.p.m.}$$

Also, rotor output $= \frac{2\pi N T_{sh}}{60}$

$$\therefore \quad 3130 = \frac{2\pi \times 978 \times T_{sh}}{60}$$

or
$$T_{sh} = \frac{3130 \times 60}{2\pi \times 978} = 30.56 \text{ N-m}$$

Hence, **output torque = 30.56 Nm. (Ans.)**

Efficiency $= \dfrac{\text{rotor output}}{\text{rotor input}} = \dfrac{3130}{3500} \times 100 =$ **89.42%. (Ans.)**

Example 3.39. *A 40 kW, 3-phase slip-ring induction motor of negligible stator impedance runs at a speed of 0.96 times synchronous speed at rated torque. The slip at maximum torque is 4 times the full-load value. If the rotor resistance of the motor is increased by 5 times, determine :*

(*i*) *The speed, power output and rotor copper losses at rated torque ;*

(*ii*) *The speed corresponding to maximum torque.* **(GATE, 1994)**

Solution. *Given* : P_{out} = 40 kW ; $N = 0.96\ N_s$, $s_{mT} = 4s_f$

Full-load slip,
$$s_f = \frac{N_s - N}{N_s} = \frac{N_s - 0.96N_s}{N_s} = 0.04$$

$$\therefore \quad s_{mT} = 4s_f = 4 \times 0.04 = 0.16$$

Now,
$$\frac{T_f}{T_m} = \frac{2s_f s_{mT}}{s_f^2 + s_{mT}^2} = \frac{2 \times 0.04 \times 0.16}{(0.04)^2 + (0.16)^2} = 0.4706$$

When the rotor circuit resistance is increased 5 times, the magnitude of maximum torque will remain unchanged because it is independent of load but the slip corresponding to the maximum torque will change. Let the new slip corresponding to maximum torque be s_{mTn}.

Since *slip corresponding to maximum torque is proportional to rotor resistance provided its standstill reactance is fixed*, so

$$s_{mTn} = s_{mT} \times \frac{R_{2n}}{R_2} = s_{mT} \times 5 = 0.16 \times 5 = 0.8$$

Now,
$$\frac{T_f}{T_m} = \frac{2s_{fn}\, s_{mTn}}{s_{fn}^2 + s_{mTn}^2}$$

or
$$0.4706 = \frac{2 \times s_{fn} \times 0.8}{s_{fn}^2 + (0.8)^2} \quad \text{or} \quad s_{fn}^2 + 0.64 = 3.4\, s_{fn}$$

or
$$s_{fn}^2 - 3.4s_{fn} + 0.64 = 0$$

or
$$s_{fn} = \frac{3.4 \pm \sqrt{(3.4)^2 - 4 \times 0.64}}{2} = \frac{3.4 \pm 3}{2} = 0.2, \text{ rejecting higher value.}$$

(*i*) **The speed, power output and rotor copper losses at rated torque :**

New speed at full load, **N′** $= (1 - s_{fn})N_s = (1 - 0.2)N_s =$ **0.8 N$_s$. (Ans.)**

Gross torque at full-load
$$= \frac{40 \times 1000}{(2\pi N/60)} = \frac{40 \times 1000 \times 60}{2\pi \times 0.96N_s}$$

$\therefore$ **Power output at full-load** = gross torque at full load $\times \dfrac{2\pi N'}{60}$

$$= \frac{40 \times 1000 \times 60}{2\pi \times 0.96N_s} \times \frac{2\pi \times 0.8N_s}{60} = 33333 \text{ W or } \mathbf{33.333\ kW.\ (Ans.)}$$

Rated copper losses at rated torque $= \dfrac{\text{power output}}{1 - s_{fn}} \times s_{fn}$

$$= \frac{33.333}{(1 - 0.2)} \times 0.2 = \mathbf{8.333\ kW.\ (Ans.)}$$

(*ii*) **The speed corresponding to maximum torque :**

Speed corresponding to maximum torque

$= N_s(1 - s_{mTn}) = N_s(1 - 0.8) = \mathbf{0.2\ N_s}$. **(Ans.)**

Example 3.40. (*a*) *'The operating speed of a 3-phase induction motor is very close to synchronous speed at no load and it reduces to meet load torque demand'. Justify the statement in brief.*

(*b*) *A 12-pole, 3-phase slip ring induction motor takes 270 kW from a 50 Hz supply and runs at 485 r.p.m. The total stator loss is 10 kW and efficiency is 93%. Determine the mechanical losses.*

If speed for the same input is reduced to 400 r.p.m. by adding resistance in the rotor circuit ; find the output power and overall efficiency. **(AMIE Winter, 1999)**

Solution. (*a*) The rotor of an induction motor always runs at a speed very close to synchronous speed at no load because the rotor starts rotating because of the revolving magnetic field which rotates at synchronous speed. The rotor tries to catch up the same speed as that of revolving field but in practice it will never be able to run at that speed *because of losses*.

When *load torque increases, the developed torque also has to increase.* At normal *speed close to synchronism, the term* sX_2 *is small and hence negligible with respect to* R_2 *and torque at this speed* $\propto \frac{s}{R_2} \propto s$ *as* R_2 *is constant.*

Hence with *increase in load torque slip increases* and *hence the operating speed will decrease.*

(*b*) *Given :* $p = 12$; $f = 50$ Hz ; stator input $= 270$ kW ; $N = 485$ r.p.m. ;

Total stator loss = 10 kW, $\eta = 93\%$.

Rotor input = stator output

= stator input – stator loss = 270 – 10 = 260 kW

Synchronous speed, $N_s = \frac{120f}{p} = \frac{120 \times 50}{12} = 500$ r.p.m.

Slip, $s = \frac{N_s - N}{N_s} = \frac{500 - 485}{500} = 0.03$

$$\frac{\text{Rotor copper loss}}{\text{Rotor input}} = s$$

∴ Rotor copper loss $= s \times$ rotor input

$= 0.03 \times 260 = 7.8$ kW

Rotor output $= \eta \times$ input

$= 0.93 \times 270 = 251.1$ kW

Mechanical losses = rotor input – rotor output – rotor copper loss

= 260 – 251.1 – 7.8 = 1.1 kW

New slip $= \frac{500 - 400}{500} = 0.2$

Rotor copper loss $= 0.2 \times 260 = 52$ kW

∴ Total losses = 52 + 10 + 1.1 = 63.1 kW

Output power = 270 – 63.1 = 206.9 kW

∴ $\eta = \frac{206.9}{270} = 0.7663$ or **76.63%. (Ans.)**

Example 3.41. *A 7.5 kW, 230 V, 3-phase, star-connected, 50 Hz, 4-pole, squirrel cage induction motor has its full-load internal torque at a slip of 0.04. The parameters of the motor are :*

$R_1 = 0.36\ \Omega$*/phase ;* $R_2 = 0.222\ \Omega$*/phase ;* $X_1 = X_2 = 0.47\ \Omega$*/phase ;* $X_m = 15.5\ \Omega$*/phase.*

Assume that the shunt branch is connected across the supply terminals. Find :

(i) Slip at maximum torque ;
(ii) Maximum internal torque at rated voltage and frequency ;
(iii) Internal starting torque at rated voltage and frequency. **(GATE, 1996)**

Solution. *Given* : $p = 4$; $f = 50$ Hz ; Per phase : $R_1 = 0.36\ \Omega$, $R_2 = 0.222\ \Omega$,
$X_1 = X_2 = 0.47\ \Omega$, $X_m = 15.5\ \Omega$

It is assumed that there is unity turn-ratio between stator and rotor (*i.e.* $K = 1$)

Phase voltage $= \dfrac{230}{\sqrt{3}} = 132.8$ V.

(*i*) **Slip at maximum torque, s_{mT} :**

Slip, $s_{mT} = \dfrac{R_2}{X_2} = \dfrac{0.222}{0.47} = \mathbf{0.4723.}$ **(Ans.)**

(*ii*) **Maximum internal torque at rated voltage and frequency :**

Equivalent impedance of motor at s_{mT},

$$Z = \left(R_1 + \frac{R_2}{s_{mT}}\right) + j(X_1 + X_2) = \left(0.36 + \frac{0.222}{0.4723}\right) + j(0.47 + 0.47)$$

$$= 0.83 + j0.94 = 1.254 \angle 48.56°$$

Rotor current per phase, $I_2 = \dfrac{E_2}{Z} = \dfrac{132.8}{1.254 \angle 48.56°} = 105.9$ A

Rotor copper loss $= 3I_2^2 R_2 = 3 \times 105.9^2 \times 0.222 = 7469$ W

Power input to rotor, $P_2 = \dfrac{\text{Rotor copper loss}}{s_{mT}} = \dfrac{7469}{0.4723} = 15814$ synchronous watts.

Synchronous speed, $N_s = \dfrac{120f}{p} = \dfrac{120 \times 50}{4} = 1500$ r.p.m.

Maximum internal torque $= \dfrac{P_2}{(2\pi N_s/60)} = \dfrac{15814 \times 60}{2\pi \times 1500} = \mathbf{100.7\ Nm.}$ **(Ans.)**

(*iii*) **Internal starting torque, T_{st} :**

Slip, $s = 1.0$ ...... (at starting)

Equivalent motor impedance, $Z = (R_1 + R_2) + j(X_1 + X_2)$
$= (0.36 + 0.222) + j(0.47 + 0.47)$
$= 0.582 + j0.94 = 1.106 \angle 58.24°$

Rotor current, $I_2 = \dfrac{E_2}{Z_2} = \dfrac{132.8}{1.106} = 120.1$ A

Power input to rotor, P_2 = total rotor copper losses
$= 3I_2^2 R_2 = 3 \times (120.1)^2 \times 0.222 = 9606$ W

Starting torque, $\mathbf{T_{st}} = \dfrac{P_2}{(2\pi N_s/60)} = \dfrac{9606 \times 60}{2\pi \times 1500} = \mathbf{61.15\ Nm.}$ **(Ans.)**

Example 3.42. *A 25 H.P., 6-pole, 50 Hz, 3-phase induction motor has stator/rotor phase voltage ratio of 6/5. The stator and rotor impedances per phase are (0.25 + j0.75) ohms and (0.173 + j0.52) ohms respectively. Find the starting torque exerted by the motor when an external resistance of 1.0 Ω is inserted in each phase, the motor being started directly on the 400 V supply system. Assume ⅄/⅄ connections.* **(S.C.E. & T.E. W.B. 1998)**

Solution. *Given* : H.P. = 25 ; $p = 6$; $f = 50$ Hz ; $K = \dfrac{5}{6}$; $R_1 = 0.25\ \Omega$/phase, $X_1 = 0.75\ \Omega$/phase ; $R_2 = 0.173\ \Omega$/phase, $X_2 = 0.52\ \Omega$/phase ; external resistance = 1 Ω, $V_L = 400$ volts ; connection : ⅄/⅄

Starting torque, T_{st} :

Supply voltage per phase $= \frac{400}{\sqrt{3}} = 230.9$ V

Equivalent resistance of motor as referred to rotor,

$$R_{02} = R_2 + K^2R_1 = 0.173 + \left(\frac{5}{6}\right)^2 \times 0.25 = 0.3466\ \Omega$$

Equivalent reactance of motor as referred to rotor,

$$X_{02} = X_2 + K^2X_1 = 0.52 + \left(\frac{5}{6}\right)^2 \times 0.75 = 1.041\ \Omega$$

When an external resistance of 1 Ω per phase is inserted in the rotor circuit, the equivalent motor impedance referred to rotor circuit,

$$Z_{02} = \sqrt{(R_{02} + 1)^2 + X_{02}^2} = \sqrt{(0.3466 + 1)^2 + 1.041^2} = 1.7\ \Omega$$

Induced e.m.f. in rotor at standstill, $E_2 = V_1 \times K = 230.9 \times \frac{5}{6} = 192.4$ V

Rotor current, $I_2 = \frac{E_2}{Z_{02}} = \frac{192.4}{1.7} = 113.18$ A

Total copper losses in rotor $= 3I_2^2(R_2 + 1) = 3 \times (113.18)^2 \times (0.173 + 1) = 45077$ W

Rotor power input at standstill, P_2 = total rotor copper losses = 45077 synchronous watts

Synchronous speed, $N_s = \frac{120f}{p} = \frac{120 \times 50}{6} = 1000$ r.p.m.

Starting torque, $T_{st} = \frac{P_2}{(2\pi N_s/60)} = \frac{45077 \times 60}{(2\pi \times 1000)} =$ **430.45 Nm. (Ans.)**

Example 3.43. *An induction motor has an efficiency of 0.9 when delivering an output of 37 kW. At this load, the stator copper loss and rotor copper loss each equals the iron loss. The mechanical losses are one-third of the no-load loss. Calculate the slip.*

(AMIE Advanced Elec. M/Cs Winter, 1998)

Solution. *Given* : $\eta = 0.9$; $P_{out} = 37$ kW ; stator copper loss = rotor copper loss = iron loss ; mechanical losses $= \frac{1}{3} \times$ no-load loss.

Slip, s :

Motor input, $P_{in} = \frac{P_{out}}{\eta} = \frac{37}{0.9} = 41.111$ kW

Total losses [including stator copper and iron losses, copper loss and rotor mechanical losses (rotor iron loss being negligible)]

$$= P_{in} - P_{out} = 41.111 - 37 = 4.111 \text{ kW} = 4111 \text{ W}$$

No-load loss of an induction motor consists of *stator iron loss* and *mechanical losses* (stator copper loss at no-load being negligible) ; moreover, these two losses are independent of load.

Let P_i = each of the stator copper loss, rotor copper loss and stator iron loss, and

P_{ml} = mechanical loss.

Then, no-load loss $= P_i + P_{ml}$

$$\therefore \quad P_{ml} = \frac{1}{3}(P_i + P_{ml}) \quad \text{or} \quad P_{ml} = \frac{P_i}{2}$$

Total losses = stator copper losses + stator iron losses + rotor copper losses + mechanical losses or $$4111 = 3P_i + \frac{P_i}{2} \quad \text{or} \quad P_i = 1174.6 \text{ W} \quad \text{or} \quad 1.1746 \text{ kW}$$

Now rotor output = motor output + mechanical losses

$$= 37 + \frac{1.1746}{2} = 37.587 \text{ kW}$$

Rotor input = rotor output + rotor copper losses

$$= 37.587 + 1.1746 = 38.762 \text{ kW}$$

Slip, $$s = \frac{\text{rotor copper losses}}{\text{rotor input}} = \frac{1.1746}{38.762} = 0.03 \text{ or } \mathbf{3\%}. \textbf{ (Ans.)}$$

Example 3.44. *A 3-phase, 50 Hz, 400 V induction motor has 4-pole star-connected stator winding. Rotor resistance and standstill reactance per phase are 0.1 Ω and 1.0 Ω respectively. The full-load slip is 4 per cent. Calculate :*

(i) The total torque developed ; *(ii) The horse power developed ;*

(iii) Maximum torque and the speed at maximum torque.

Assume that the stator to rotor turn-ratio is 2 : 1. **(S.C.E.W.B. 1999)**

Solution. *Given :* $f = 50$ Hz ; $V_1 = 400$ V ; $p = 4$; $R_2 = 0.1\ \Omega$; $X_2 = 1.0\ \Omega$; $s_1 = 4\%$ or 0.04.

Rotor standstill e.m.f. per phase, $$E_2 = \frac{V_1/\sqrt{3}}{\text{Turn-ratio}} = \frac{400/\sqrt{3}}{2} = 115.47 \text{ V}$$

Synchronous speed, $$N_s = \frac{120f}{p} = \frac{120 \times 50}{4} = 1500 \text{ r.p.m.}$$

Power input to rotor, $$P_2 = \frac{3R_2 s E_2^2}{R_2^2 + s^2 X_2^2} = \frac{3 \times 0.1 \times 0.04 \times 115.47^2}{0.1^2 + (0.04 \times 1.0)^2}$$

$$= 13793 \text{ synchronous watts.}$$

(*i*) **Total torque develop, T :**

$$T = \frac{P_2}{(2\pi N_s/60)} = \frac{13793 \times 60}{2\pi \times 1500} = \mathbf{87.81 \text{ Nm.}} \textbf{ (Ans.)}$$

(*ii*) **The horse power developed, H.P. :**

$$\text{H.P.} = \frac{P_2}{735.5} = \frac{13793}{735.5} = \mathbf{18.75 \text{ H.P. (metric).}} \textbf{ (Ans.)}$$

(*iii*) $\mathbf{T_m, s_{mT}}$ **:**

$$s_{mT} = \frac{R_2}{X_2} = \frac{0.1}{1.0} = 0.1$$

Speed at maximum torque, $\mathbf{N} = N_s(1 - s_{mT})$

$$= 1500(1 - 0.1) = \mathbf{1350 \text{ r.p.m.}} \textbf{ (Ans.)}$$

Power input to rotor $$= \frac{3R_2 s_{mT} E_2^2}{R_2^2 + s_{mT}^2 X_2^2} = \frac{3E_2^2}{2X_2} \qquad \text{(Putting } R_2 = s_{mT} X_2)$$

$$= \frac{3 \times (115.47)^2}{2 \times 1.0} = 20000 \text{ synchronous watts}$$

$$\therefore \quad \mathbf{T_m} = \frac{20000}{(2\pi N_s/60)} = \frac{20000 \times 60}{2\pi \times 1500} = \mathbf{127.32 \text{ Nm.}} \textbf{ (Ans.)}$$

Example 3.45. *A 3-phase, 4 pole, 50 Hz induction motor at standstill has 120 V induced across its star-connected rotor terminals. The rotor resistance and reactance per phase are 0.2 Ω and 1.0 Ω respectively. Calculate :*

(i) The speed when the rotor is drawing a current of 16 A at a particular load.

(ii) The speed at which the torque is maximum and the corresponding value of rotor input.

(AMIE Summer, 1999)

Solution. *Given* : $p = 4$; $f = 50$ Hz ; standstill voltage across star-connected terminals = 120 V, $R_2 = 0.2\ \Omega$, $X_2 = 1.0\ \Omega$.

(i) The speed when the rotor is drawing 16 A current, N :

Rotor impedance at any slip s,

$$Z_2 = R_2 + jsX_2$$

(where R_2 and X_2 are the rotor resistance and reactance at standstill)

At standstill, $E_2 = \dfrac{120}{\sqrt{3}} = 69.28$ V

∴ At slip s, induced voltage per phase, $E_r = sE_2 = 69.28s$

The current in the rotor circuit at slip $s = I_2 = 16$ A

∴ Rotor voltage, $E_r = I_2Z_2 = I_2(R_2 + jsX_2) = 16(0.2 + j1.0s)$

$$= 16\sqrt{0.2^2 + (1.0s)^2} \text{ volts/phase}$$

or $69.28s = 16\sqrt{0.2^2 + (1.0s)^2}$

or $(69.28)^2 s^2 = 256(0.04 + s^2)$

or $s^2 \times \dfrac{(69.28)^2}{256} = 0.04 + s^2$

or $18.75s^2 = 0.04 + s^2$

∴ $s = 0.04747$

∴ Motor speed, $\mathbf{N} = N_s(1 - s)$

$$= \frac{120 \times 50}{4}(1 - 0.04747) = \mathbf{1428.8\ r.p.m.}\ \textbf{(Ans.)}$$

(ii) The speed at maximum torque and value of rotor input :

For maximum torque, $R_2 = s_{mT}X_2$ or $s_{mT} = \dfrac{R_2}{X_2} = \dfrac{0.2}{1.0} = 0.2$

Speed at maximum torque $= N_s(1 - s)$

$$= \frac{120 \times 50}{4}(1 - 0.2) = \mathbf{1200\ r.p.m.}\ \textbf{(Ans.)}$$

At $s_{mT} = 0.2$, the rotor impedance,

$$Z_2 = \sqrt{R^2 + (s_{mT}X_2)^2} = \sqrt{0.2^2 + (0.2 \times 1.0)^2} = 0.2828\ \Omega$$

Induced e.m.f. in the rotor per phase,

$$s_{mT}E_2 = 0.2 \times \frac{120}{\sqrt{3}} = 13.86 \text{ V}$$

∴ Rotor current $= \dfrac{s_{mT}E_2}{Z_2} = \dfrac{13.86}{0.2828} = 49$ A.

Rotor impedance angle $= \cos^{-1}\left(\dfrac{R_2}{Z_2}\right) = \cos^{-1}\left(\dfrac{0.2}{0.2828}\right) = 45°$

Rotor input power $= \sqrt{3} \times$ rotor voltage × rotor current × power factor

$= \sqrt{3} \times 13.86 \times 49 \times \cos 45° = \mathbf{831.6\ W.}$ **(Ans.)**

Example 3.46. *(a) A balanced 3-phase induction motor has an efficiency of 0.85 when its output is 60 H.P. At this load the stator and rotor copper losses are equal to the core losses. The mechanical losses are one-fourth of the no-load loss. Calculate the slip.*

(b) The full-load efficiency and power factor of a 16 H.P., 440 V, 3-phase induction motor are 85% and 0.8 lag respectively. The blocked rotor line current is 45 A at 220 V. Calculate the ratio of starting to full-load current if the motor is provided with a star-delta starter. Neglect magnetising current.

(AMIE Winter, 1997)

Solution. (*a*) Motor output = 60 H.P. = 60 × 735.5 = 44130 W

Motor input $= \dfrac{\text{motor output}}{\text{efficiency}} = \dfrac{44130}{0.85} = 51917.6$ W

Let stator core loss = x watts

stator copper loss = x watts

rotor copper loss = x watts

No-load motor losses = stator core loss + mech. loss

But no-load losses = 4 × mech. loss (Given)

∴ 4 × mech. loss = stator core loss + mech. loss

or Mech. loss $= \dfrac{1}{3}$ stator core loss $= \dfrac{x}{3}$ watts

Total losses at 60 H.P. output = stator core loss + stator copper loss + rotor copper loss + mech. loss

or $x + x + x + \dfrac{x}{3}$ = motor input – motor output

= 51917.6 – 44130 = 7787.6 W

∴ x = 2336.3 W

Rotor input = stator input – stator core loss – stator copper loss

= 51917.6 – 2 × 2336.3 = 47245 W

∴ Slip, $s = \dfrac{\text{rotor copper loss}}{\text{rotor input}} = \dfrac{2336.3}{47245} = 0.0494$ or **4.94%. (Ans.)**

(*b*) The blocked rotor line current, when full voltage is applied

$$= I_{sc} = 45 \times \frac{440}{220} = 90 \text{ A}$$

Motor input $= \dfrac{16 \times 735.5}{0.85} = \sqrt{3} \times 440 \times I_f \times 0.8$

∴ Full-load current, $I_f = \dfrac{16 \times 735.5}{0.85 \times \sqrt{3} \times 440 \times 0.8} = 22.71$ A

When the motor is started with star-delta starter, the starting current,

$$I_{st} = \frac{I_{sc}}{\sqrt{3}} = \frac{90}{\sqrt{3}} = 51.26$$

∴ $\dfrac{I_{st}}{I_f} = \dfrac{51.96}{22.71} =$ **2.288. (Ans.)**

Example 3.47. *A 50 kW, 400 V, 3-phase, 6-pole, 50 Hz wound rotor induction motor has a full-load slip of 0.04 when operating at rated voltage and frequency with rotor winding short circuited at slip rings. The slip at maximum torque at is 0.2. Stator resistance and rotational losses are neglected. Determine :*

(i) The maximum torque ;

(ii) Full-load rotor ohmic losses.

Rotor resistance is now doubled by adding external series resistance in each rotor phase. For the rated power output, determine :

(iii) Slip at maximum torque ;

(iv) Full-load slip ;

(v) Full-load torque.

Solution. *Given* : Mechanical power developed, $P_{mech.}$ = 50 kW (since rotational losses are neglected) ; $p = 6$; $f = 50$ Hz ; $s_f = 0.04$, $s_{mT} = 0.2$.

Synchronous speed, $N_s = \frac{120f}{p} = \frac{120 \times 50}{6} = 1000$ r.p.m.

Rotor speed at full-load, $N = (1 - s_f)N_s = (1 - 0.04) \times 1000 = 960$ r.p.m.

Now, $P_{mech.} = \frac{2\pi N T_f}{60}$

$$\therefore \quad T_f = \frac{60 \times P_{mech.}}{2\pi N} = \frac{60 \times 50 \times 1000}{2\pi \times 960} = \mathbf{497.36\ N\text{-}m.}\ \textbf{(Ans.)}$$

(*i*) As stator resistance is neglected,

$$\frac{T_f}{T_m} = \frac{2s_f s_{mT}}{s_f^2 + s_{mT}^2}$$

$$= \frac{2 \times 0.04 \times 0.2}{0.04^2 + 0.2^2} = 0.385 \quad \text{...(Eqn. 13.7)}$$

$$\therefore \quad \mathbf{T_m} = \frac{T_f}{0.385} = \frac{497.36}{0.385} = \mathbf{1291.84\ N\text{-}m.}\ \textbf{(Ans.)}$$

(*ii*) **Full-load rotor ohmic losses**

$$= s_f P_g = \left(\frac{s_f}{1 - s_f}\right) P_{mech.}$$

$$= \left(\frac{0.04}{1 - 0.04}\right) \times 50000 = \mathbf{2083.33\ W.}\ \textbf{(Ans.)}$$

(*iii*) **Slip at maximum torque, s_{mT1}**

$$\frac{R_2}{X_2} = s_{mT} = 0.2$$

New slip at maximum torque,

$$s_{mT1} = \frac{2R_2}{X_2} = 2 \times 0.2 = \mathbf{0.4.}\ \textbf{(Ans.)}$$

(*iv*) **Full-load slip, s_{f1} :**

As stator resistance is neglected,

$$T_m = 1291.84 \text{ N-m}$$

Again, $\frac{T_f}{T_m} = \frac{2s_f s_{mT}}{s_f^2 + s_{mT}^2}$

$$\therefore \quad \frac{T_{f1}}{T_m} = \frac{2s_{f1}s_{mT1}}{s_{f1}^2 + s_{mT1}^2}$$

where s_{f1} and T_{f1} are new full-load slip and torque respectively.

or
$$T_{f1} = \frac{T_m \times 2s_{f1}s_{mT1}}{s_{f1}^2 + s_{mT1}^2} = \frac{1291.84 \times 2 \times s_{f1} \times 0.4}{s_{f1}^2 + (0.4)^2} = \frac{1033.47s_{f1}}{s_{f1}^2 + 0.16} \quad \text{...(i)}$$

Also
$$\frac{2\pi N T_{f1}}{60} = 50000 \quad \text{...(Given)}$$

or
$$T_{f1} = \frac{60 \times 50000}{2\pi N_s(1 - s_{f1})} \quad [\because N = (1 - s_{f1}) N_s]$$

or $$T_{fl} = \frac{60 \times 50000}{2\pi \times 1000(1 - s_{fl})} = \frac{50000}{104.72(1 - s_{fl})} \quad ...(ii)$$

From (*i*) and (*ii*), we have

$$\frac{1033.47}{(s_{fl}^2 + 0.16)} = \frac{50000}{104.72(1 - s_{fl})}$$

$$(s_{fl}^2 + 0.16) = \frac{1033.47}{50000} \times 104.72(1 - s_{fl})s_{fl}$$

$$= 2.16(1 - s_{fl})\, s_{fl} = 2.16s_{fl} - 2.16s_{fl}^2$$

or $$3.16s_{fl}^2 - 2.16\, s_{fl} + 0.16 = 0$$

or $$s_{fl} = \frac{2.16 \pm \sqrt{(2.16)^2 - 4 \times 3.16 \times 0.16}}{2 \times 3.16} = \frac{2.16 \pm 1.626}{6.32} = 0.084,\ 0.6$$

Higher value being redundant, the full-load slip, s_{fl} = **0.084 (or 8.4%). (Ans.)**

(*v*) **Full-load torque, T_{fl} :**

Substituting the value of s_{fl} = 0.084 in eqn. (*ii*), we get

$$T_{fl} = \frac{50000}{104.72(1 - 0.084)} = \mathbf{521.25\ N\text{-}m.}\ \textbf{(Ans.)}$$

3.18. MEASUREMENT OF SLIP

Following are some of the methods of measuring slip of an induction motor :

1. Electromechanical counter.
2. Mechanical differential counter.
3. Stroboscopic method.

3.18.1. Electromechanical Counter. Perhaps the most simple and direct method of measuring slip (*i.e.* difference between synchronous speed and rotor speed) is to compare the speed of induction motor with a small synchronous motor's speed as shown in Fig. 3.20.

The synchronous motor employed must have the same number of poles as the induction motor. At the end of each shaft is a cylinder of phenolic or other suitable insulating material into which has been fitted a circular slip ring to which a small contactor is connected as shown in Fig. 3.21. The slip rings are connected to a voltage source in series with an electric pulse counter, either of the electrochemical relay type or an electronic digital pulse counter. The synchronous motor always runs at synchronous speed, which is the speed of the rotating field of the induction motor. *Each time the induction motor slips, a revolution, the synchronous, and induction motor contactors close the circuit and register a pulse. The number of pulses registered per minute is the slip in r.p.m.*

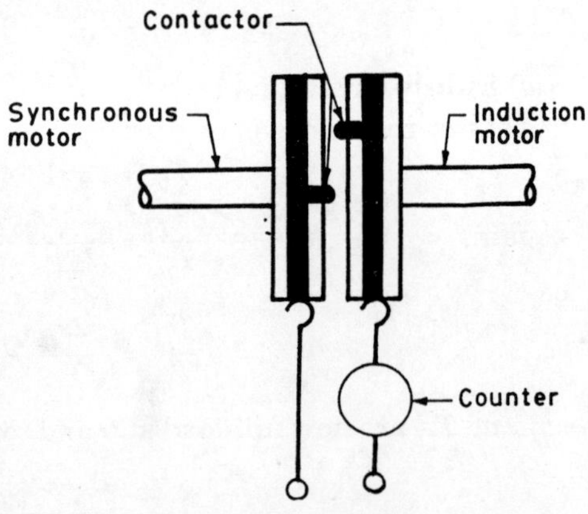

Fig. 3.21. Electromechanical counter.

3.18.2. Mechanical-differential Counter. Fig. 3.22 illustrates the use of a mechanical differential *whose output gear rotates at a speed equal to the difference of two input gears to which the synchronous and induction motor shafts are connected.*

A mechanical counter or a low-speed electric or mechanical tachometer is used to *record the slip speed* directly.

This method has a *disadvantage* in loading the induction motor slightly because of the friction and drag in the gearing of the differential, and it *should not be employed with small induction motors* for measurement of slip. An electric differential with less effective drag may be employed for small induction motors.

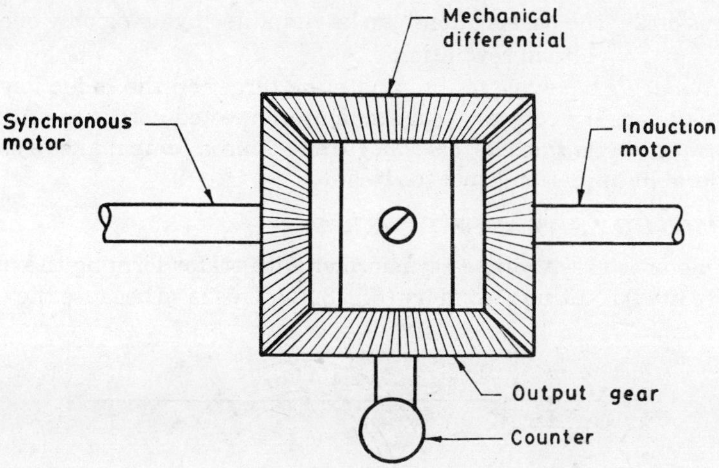

Fig. 3.22. Mechanical differential counter.

3.18.3. Stroboscopic Method. In Fig. 3.23 a black disc with white sectors, *equal in number to the number of poles of the induction motor*, is attached to the induction motor shaft. It is observed through another disc having an *equal number* of sector-shaped slits and carried on the shaft of *small self- starting synchronous motor*, in turn fitted with a revolution counter which can be thrown in and out of gear at will.

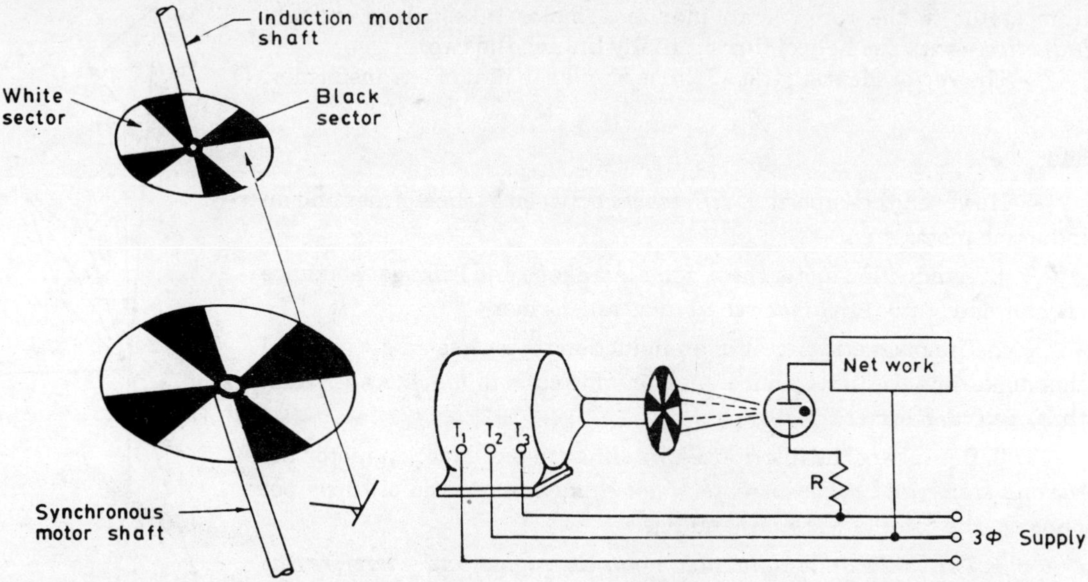

(*a*) (*b*)

Fig. 3.23. Slip measurement by stroboscopic method.

If n = number of passages of the sectors, then

$$\text{Slip in per cent} = \frac{100n}{n_s n_r}$$

where n_s = number of sectors, and

n_r = number of revolutions recorded by the counter during the interval of observation.

For large values of slip the observations can be simplified by using only one sector ; then

n = slip in revolutions.

With a *synchronous light source* to illuminate the target on the induction-motor shaft, the synchronous motor is no longer necessary. An arc lamp connected across the A.C. supply may be used, but the *carbons must be readjusted from time to time*. A *neon lamp* makes a satisfactory source of light when the general illumination is not too bright.

3.19. INDUCTION MOTOR AS TRANSFORMER

An induction motor is essentially a transformer with stator forming the primary and rotor forming (the short-circuited) rotating secondary (Fig. 3.24). This is so because the transfer of energy

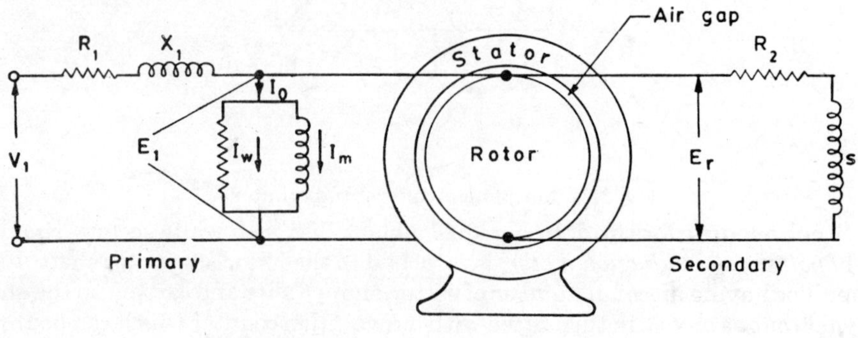

Fig. 3.24. Induction motor as transformer.

from stator to the rotor of an induction motor takes place entirely *inductively* with the help of flux mutually linking the two.

The vector diagram (Fig. 3.25) is similar to that of a transformer.

$$V_1 = E_1 + I_1 (R_1 + jX_1)$$

and

$$E_r = I_2 (R_2 + jsX_2)$$

However, the *important differences* between a transformer and an induction motor are :

1. In induction motor the magnetic leakage and leakage reactance of rotor and stator are *higher* than in a transformer.

2. The magnetic circuit of an induction motor has an air gap and this makes the per unit value of magnetising current *much higher* than that of a transformer.

3. Because of the distributed windings in an induction motor the *ratio of stator and rotor currents is not equal* to the ratio of turns per phase in the rotor and stator windings.

4. *The losses in an induction motor are higher and, therefore, the efficiency is lower than in a transformer.*

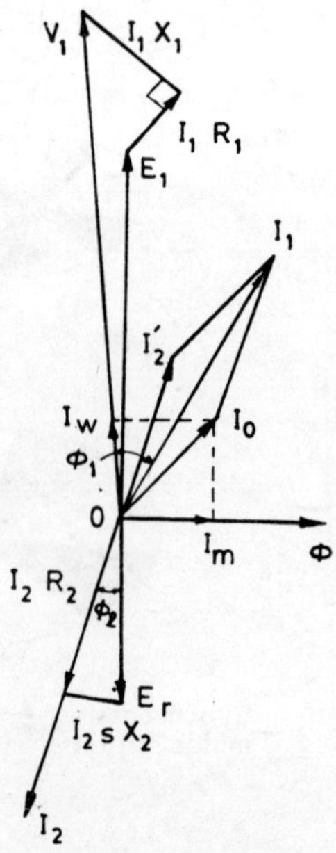

Fig. 3.25. Vector diagram of induction motor.

3.19.1. Rotor output. Primary current I_1 consists of two parts I_0 and I_2'. It is the latter which is transferred to the rotor, because I_0 is used in meeting the copper and iron losses in the stator itself. Out of the primary voltage V_1, some is absorbed in the primary itself ($= I_1Z_1$) and the remaining E_1 is transferred to the rotor. If the angle between E_2 and I_2' is ϕ, then

Rotor input/phase $= E_1I_1' \cos \phi$

Total rotor input $= 3E_1I_2' \cos \phi$

The electrical input to the rotor which is wasted in the form of the heat

$$= 3I_2E_r \cos \phi \text{ (or } 3I_2^2R_2)$$

Now $\quad I_2' = KI_2 \quad \text{or} \quad I_2 = \dfrac{I_2'}{K}$

$$E_r = sE_2 \text{ and } E_2 = KE_1$$

$\therefore \quad E_r = sKE_1$

$\therefore$ Electrical input wasted as heat

$$= 3 \times (I_2'/K) \times sKE_1 \cos \phi)$$
$$= 3E_1I_2' \cos \phi \times s = \text{rotor input} \times s$$

Now, $\quad$ rotor output = rotor input − losses

$$= 3E_1I_2' \cos \phi - 3E_1I_2' \cos \phi \times s$$
$$= (1 - s)\, E_1I_2' \cos \phi = (1 - s) \text{ rotor input}$$

$\therefore \quad \dfrac{\text{Rotor output}}{\text{Rotor input}} = 1 - s$

$\therefore \quad$ Rotor copper loss $= s \times$ rotor input

Rotor efficiency $\quad = 1 - s = \dfrac{N}{N_s} = \dfrac{\text{actual speed}}{\text{synchronous speed}}$.

3.19.2. Equivalent Circuit of the Rotor. The rotor current I_2, when motor is loaded, is given by

$$I_2 = \frac{sE_2}{\sqrt{R_2^2 + (sX_2)^2}} = \frac{E_2}{\sqrt{(R_2/s)^2 + X_2^2}}$$

From the above relation it appears that the rotor circuit which actually consists of a fixed resistance R_2 and a variable reactance sX_2 (proportional to slip) connected across $E_r = sE_2$ [Fig. 3.26 (*i*)] can be looked upon as equivalent to a rotor circuit having a fixed reactance X_2 connected in series with a variable resistance R_2/s (inversely proportional to slip) and supplied with constant voltage E_2 as shown in [Fig. 3.26 (*ii*)].

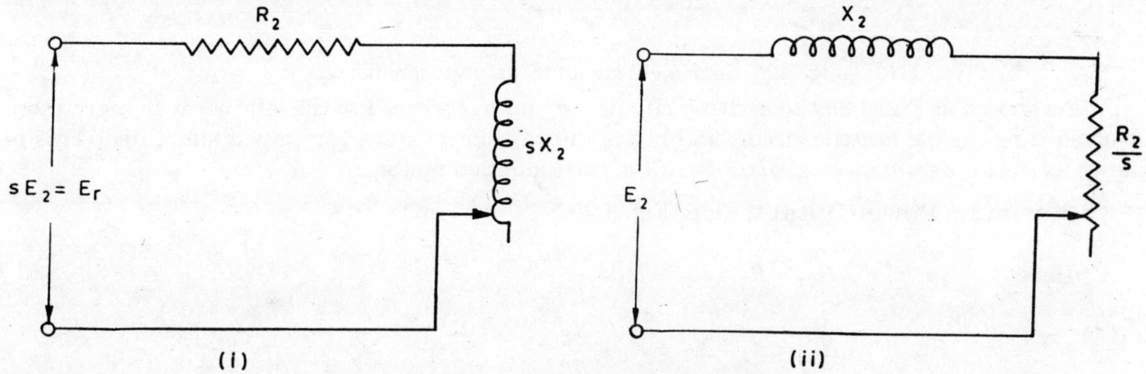

Fig. 3.26. Equivalent circuit of a rotor.

Also, the resistance R_2/s can be written as,

$$\frac{R_2}{s} = R_2 + R_2\left(\frac{1}{s} - 1\right).$$

It consists of two parts : (*i*) The part R_2 is the *rotor resistance* itself and *represents the rotor copper loss* (*ii*) the second part is $R_2\left(\frac{1}{s} - 1\right)$. This is known as *load resistance* R_L and is *the electrical equivalent of the mechanical load on the motor*. In other words the mechanical load on an induction motor can be represented by a *non-inductive resistance of the value* $R_2\left(\frac{1}{s} - 1\right)$.

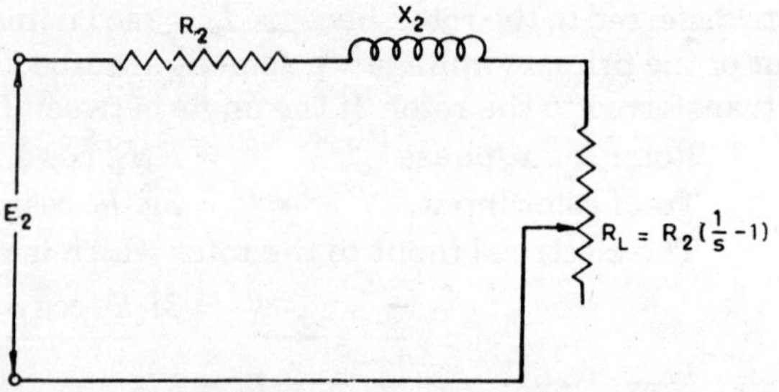

Fig. 3.27. Equivalent circuit of a rotor with load resistance R_L.

In Fig. 3.27 is shown the equivalent rotor circuit along with the load resistance R_L.

3.20. EQUIVALENT CIRCUIT OF AN INDUCTION MOTOR

The equivalent circuit for a polyphase induction motor is shown in Fig. 3.28, where

V_1 = applied voltage per phase
R_1 = stator resistance/phase
R_2 = rotor resistance/phase
X_1 = stator leakage reactance/phase
X_2 = rotor standstill leakage reactance/phase
K = turn-ratio of secondary to primary
R_0 = no-load resistance/phase ⎤ Rotor being driven at
X_0 = no-load reactance/phase ⎦ synchronous speed.

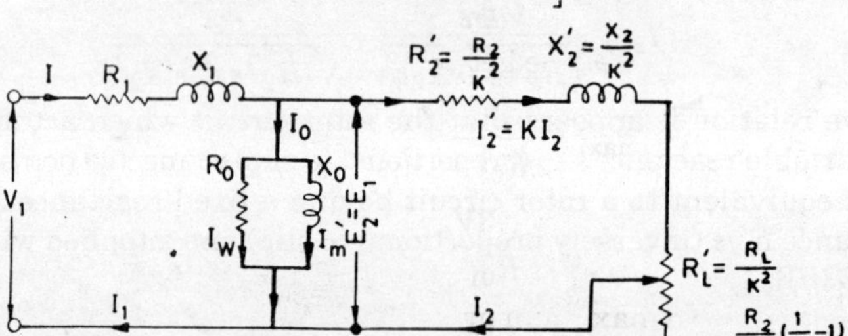

Fig. 3.28. Equivelent circuit of an induction motor.

As shown in Fig. 3.29 the exciting circuit may be transferred to the left, because inaccuracy involved is negligible but the circuit and hence the calculations are very much simplified. This is known as the *approximate equivalent* circuit of the induction motor.

Maximum Power Output. Refer Fig. 3.29.

Here $$R_{01} = R_1 + R_2' = R_1 + \frac{R_2}{K^2}$$

$$X_{01} = X_1 + X_2' = X_1 + \frac{X_2}{K^2}$$

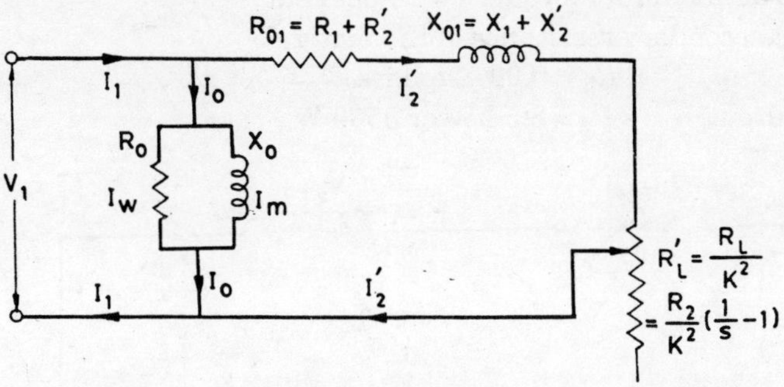

Fig. 3.29. Approximate equivalent circuit of an induction motor.

Load resistance, $$R_L' = R_L/K^2 = \frac{R_2}{K^2}\left(\frac{1}{s} - 1\right)$$

Gross mechanical power output,

$$P_g = 3I_2'^2 R_L'$$

But $$I_2' = \frac{V_1}{\sqrt{(R_{01} + R_L')^2 + (X_{01})^2}}$$

$\therefore$ $$P_g = \frac{3V_1^2 R_L'}{(R_{01} + R_L')^2 + (X_{01})^2} \qquad ...(3.27)$$

The condition for maximum power output can be obtained by differentiating the above equation and by equating, the first derivative to zero. If it is done, it will be found that

$$R_L' = \sqrt{(R_{01})^2 + (X_{01})^2} = Z_{01}$$

Hence the gross mechanical power output of the motor is maximum when the equivalent *load resistance R_L' is equal to the standstill leakage impedance of the motor Z_{01}*.

Putting $R_L' = Z_{01}$ in eqn. (3.27), we get

$$P_{g\,(max)} = \frac{3V_1^2 Z_{01}}{(R_{01} + Z_{01})^2 + (X_{01})^2} = \frac{3V_1^2 Z_{01}}{R_{01}^2 + X_{01}^2 + Z_{01}^2 + 2R_{01}Z_{01}}$$

$$= \frac{3V_1^2}{2(R_{01} + Z_{01})} \qquad ...(3.28)$$

The slip corresponding to maximum gross mechanical power output will be given as

$$Z_{01} = R_L' = \frac{R_2}{K^2}\left(\frac{1}{s} - 1\right)$$

or $$s = \frac{R_2/K^2}{R_2/K^2 + Z_{01}} \qquad ...(3.29)$$

EQUIVALENT CIRCUIT OF AN INDUCTION MOTOR

Example 3.48. *In a 3-phase induction motor maximum torque occurs at a slip of 10 per cent. The equivalent secondary resistance of the motor is 0.09 Ω/phase. If the gross power output is 8.5 kW. Calculate :*

(i) Equivalent load resistance, *(ii) Equivalent load voltage, and*

(iii) Current at this slip.

Solution. Slip at maximum torque, s = 10 per cent

The equivalent secondary resistance of the motor,

$$R_2 = 0.09\ \Omega/\text{phase}$$

Gross power output, P_g = 8.5 kW or 8500 W

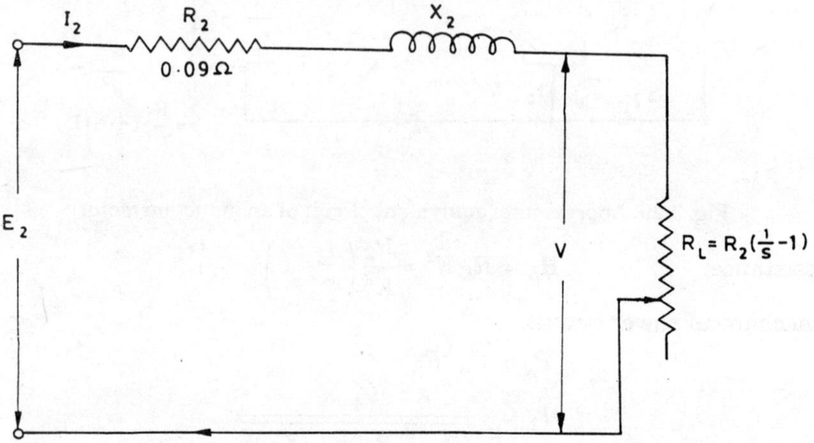

Fig. 3.30

(*i*) **Equivalent load resistance, R_L :**

We know that $R_L = R_2\left[\left(\frac{1}{s}\right) - 1\right] = 0.09\left(\frac{1}{0.1} - 1\right)$ = **0.81 Ω. (Ans.)**

(*ii*) **Equivalent load voltage, V :**

As shown in the equivalent circuit of the rotor in Fig. 3.30, V is a fictitious voltage drop equivalent to that consumed in the load connected to the secondary *i.e.*, rotor, $V = I_2R_L$

But, gross power, $P_g = \frac{3V^2}{R_L}$

∴ $V = \sqrt{\frac{P_g \times R_L}{3}} = \sqrt{\frac{8500 \times 0.81}{3}}$ = 47.9 V

Hence, **equivalent load voltage = 47.9 V. (Ans.)**

(*ii*) **Equivalent load current :**

Equivalent load current $= \frac{V}{R_L} = \frac{47.9}{0.81}$ = **59.1 A. (Ans.)**

Example 3.49. *The stator impedance and equivalent rotor resistance of a 400 V, 3-phase star-connected induction motor are (0.05 + j0.2) Ω and (0.05 + j0.24) Ω respectively. Neglecting exciting current determine the maximum gross power and the slip at which it occurs.*

Solution. Supply voltage V_1 = 400 V

Stator impedance = (0.05 + j0.2) Ω

Equivalent rotor resistance = (0.05 + j0.24) Ω

The equivalent circuit is shown in Fig. 3.31.

$$R_{01} = R_1 + R_2' = 0.05 + 0.05 = 0.1\ \Omega$$

$$X_{01} = X_1 + X_2' = 0.2 + 0.24 = 0.44\ \Omega$$

∴ $$Z_{01} = \sqrt{R_{01}^2 + X_{01}^2} = \sqrt{(0.1)^2 + (0.44)^2} = 0.45\ \Omega$$

Slip = ?

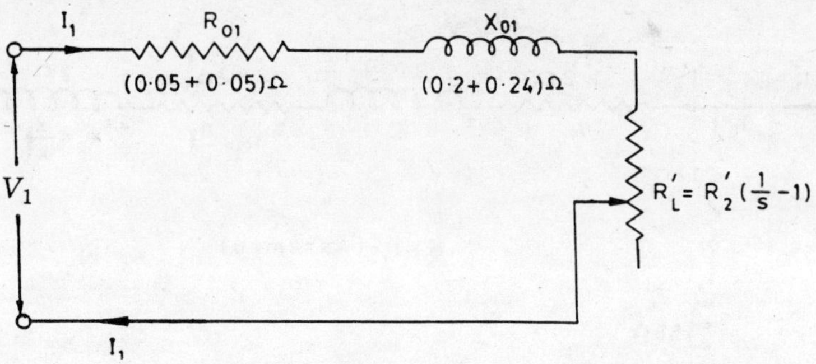

Fig. 3.31

We know that slip corresponding to maximum gross power output (Art. 3.20) is given by

$$s = \frac{\dfrac{R_2}{K^2}}{\dfrac{R_2}{K^2} + Z_{01}} = \frac{R_2'}{R_2' + Z_{01}} = \frac{0.05}{0.05 + 0.45} = 0.1 \text{ or } 10\%$$

Hence, **slip at which maximum gross power output occurs**

= **10%. (Ans.)**

Maximum gross power output

$$= \frac{3V_1^2}{2(R_{01} + Z_{01})} = \frac{3\left(\dfrac{400}{\sqrt{3}}\right)^2}{2(0.1 + 0.45)}$$

= **145454 W** or **145.454 kW. (Ans.)**

☞**Example 3.50.** *The following data pertain to an induction motor : stator impedance = (1.2 + j3.0) Ω ; rotor standstill impedance = (1.2 + j2.0) Ω ; no-load shunt impedance = (12 + j55) Ω ; volts/phase = 240. Determine :*

(i) Stator current, *(ii) Equivalent rotor current,*
(iii) Input power factor, *(iv) Mechanical power developed, and*
(v) Efficiency of the motor at a slip of 5%.

Solution. Stator impedance = (1.2 + j3.0) Ω
Rotor standstill impedance = (1.2 + j2.0) Ω
No-load shunt impedance = (12 + j55) Ω
Volts/phase = 240 V

The equivalent circuit per phase of motor referred to stator is shown in Fig. 3.32. It is *assumed* that the given impedance figures are *phase values* and there is *unity turn-ratio* between stator and rotor.

Equivalent load resistance,

$$R_L' = R_2'\left(\frac{1}{s} - 1\right) = \frac{R_2}{K^2}\left(\frac{1}{s} - 1\right) = 1.2\left(\frac{1}{0.05} - 1\right) = 22.8\ \Omega$$

Effective impedance/phase $= (R_1 + R_2' + R_L') + j(X_1 + X_2') = (1.2 + 1.2 + 22.8) + j(3 + 2)$
$= (25.2 + j5)\ \Omega$

(i) **Stator current, I_1 :**

Stator load current, $I_2' = \dfrac{240}{25.2 + j5} = \dfrac{240(25.2 - j5)}{(25.2 + j5)(25.2 - j5)}$

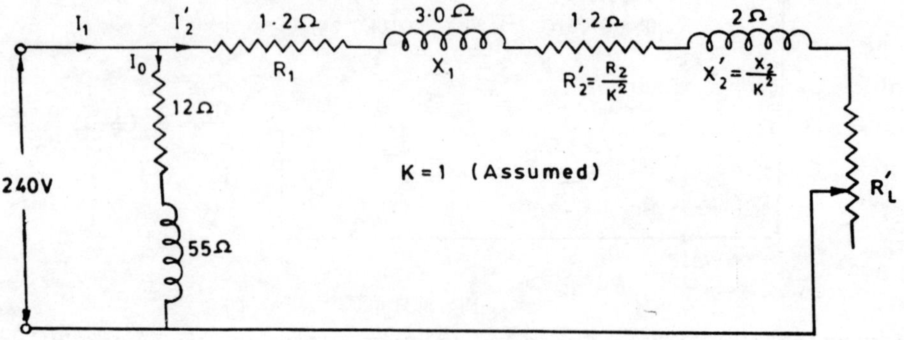

Fig. 3.32

$$= \frac{240(25.2 - j5)}{660} = (9.16 - j1.82) \text{ A}$$

Stator no-load current, $I_0 = \frac{240}{(12 + j55)} = \frac{240}{12 + j55} \times \frac{(12 - j55)}{(12 - j55)}$

$$= \frac{240(12 - j55)}{3169} = (0.91 - j4.16) \text{ A}$$

Total stator current, $I_1 = I_2' + I_0 = (9.16 - j1.82) + (0.91 - j4.16)$

$$= (10.07 - j5.98) \text{ A}$$

$\therefore$ $I_1 = \sqrt{(10.07)^2 + (5.98)^2} = \mathbf{11.71\ A.}$ **(Ans.)**

(*ii*) **Equivalent rotor current,**

$I_2' = \sqrt{(9.16)^2 + (1.82)^2} = \mathbf{9.34\ A.}$ **(Ans.)**

(*iii*) **Input power factor,**

$\cos \phi = \frac{10.07}{11.71} = \mathbf{0.86.}$ **(Ans.)**

(*iv*) **Mechanical power developed,**

$$= 3(I_2')^2 R_L' = 3 \times (9.34)^2 \times 22.8$$

$= \mathbf{5967\ W}$ or $\mathbf{5.967\ kW.}$ **(Ans.)**

(*v*) **Efficiency of the motor,**

$$\eta = \frac{\text{output}}{\text{input}} \times 100 = \frac{5967}{3 \times 240 \times 11.71 \times 0.86} \times 100$$

$= \mathbf{82.29\%.}$ **(Ans.)**

☞**Example 3.51.** *The following data refer to a 10-pole, 400 V, 50-Hz, 3-phase induction rotor :*

$R_1 = 1.75\ \Omega$, $X_1 = 5.5\ \Omega$, R_2' *(rotor resistance referred to stator)* $= 2.25\ \Omega$.

X_2' *(rotor standstill reactance referred to stator)* $= 6.6\ \Omega$.

When the motor is tested on no-load, it is observed that it takes 3.8 A (line current) and the total core loss is 310 W.

By using an approximate equivalent circuit at 4% slip, calculate :

(i) The rotor current, *(ii) Supply current and power factor,*

(iii) Mechanical power developed, and *(iv) Gross load torque.*

Solution. Stator resistance, $R_1 = 1.75\ \Omega$

Stator reactance, $X_1 = 5.5\ \Omega$

Rotor resistance referred to stator, $R_2' = 2.25\ \Omega$

Rotor standstill reactance referred to stator, $X_2' = 6.6\ \Omega$

Total core loss $= 310$ W

No-load line current, $I_0 = 3.8$ A

If $\cos \phi_0$ is the power factor at no-load, then

$$\sqrt{3}\ VI \cos \phi_0 = 310$$

$$\therefore \quad \cos \phi_0 = \frac{310}{\sqrt{3} \times 400 \times 3.8} = 0.1178$$

$$\phi_0 = 83.23°$$

$$\therefore \quad \sin \phi_0 = 0.993$$

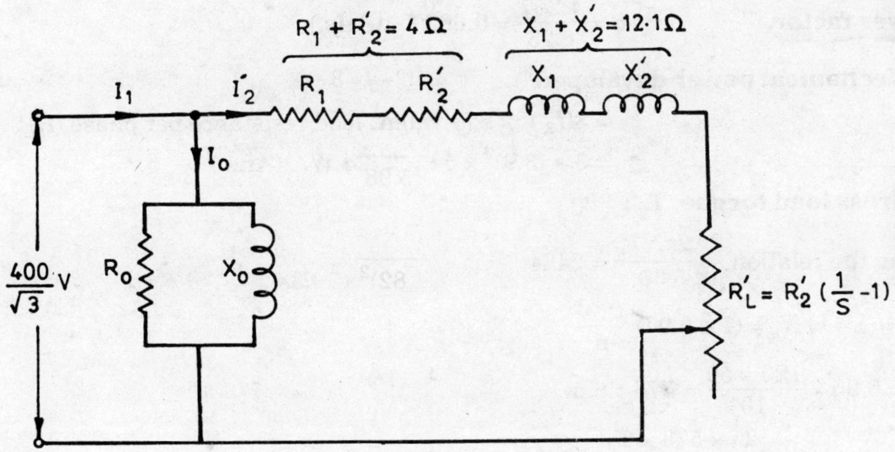

Fig. 3.33

So in the equivalent circuit,

$$R_0 = \frac{400}{\sqrt{3}\ I_0 \cos \phi_0} = \frac{400}{\sqrt{3} \times 3.8 \times 0.1178} = 515.9\ \Omega$$

$$X_0 = \frac{400}{\sqrt{3}\ I_0 \sin \phi_0} = \frac{400}{\sqrt{3} \times 3.8 \times 0.993} = 61.2\ \Omega$$

For a load slip of 4%

Equivalent load resistance,

$$R_L' = R_2'\left(\frac{1}{s} - 1\right) = 2.25\left(\frac{1}{0.04} - 1\right) = 54\ \Omega$$

$$R_1 + R_2' = 1.75 + 2.25 = 4\ \Omega$$

$$X_1 + X_2' = 5.5 + 6.6 = 12.1\ \Omega$$

Phase voltage $= \frac{400}{\sqrt{3}}$ V = 231 V

(*i*) **Rotor current referred to stator,**

$$\mathbf{I_2}' = \frac{231 + j0}{(4 + 54) + j12.1} = \frac{231 + j0}{58 + j12.1} = \frac{231 + j0}{(58 + j12.1)} \times \left(\frac{58 - j12.1}{58 - j12.1}\right)$$

$$= \frac{13398 - j2795}{3364 + 146.4} = 3.82 - j0.976$$

$\therefore$ $\mathbf{I_2}' = \sqrt{(3.82)^2 + (0.796)^2}$ = **3.9 A. (Ans.)**

Also $I_0 = 3.8 (\cos \phi_0 - j \sin \phi_0)$

$= 3.8 (0.1178 - j0.993) = 0.447 - j3.77$

(*ii*) **Supply current and power factor :**

Supply (or motor) current,

$$\mathbf{I_1} = I_0 + I_2' = (0.447 - j3.77) + (3.82 - j0.796)$$

$$= 4.267 - j4.566$$

$\therefore$ $\mathbf{I_1} = \sqrt{(4.267)^2 + (4.566)^2}$ = **6.25 A. (Ans.)**

Power factor, $\cos \phi = \frac{4.267}{6.25}$ = **0.6827. (Ans.)**

(*iii*) **Mechanical power developed**

$= 3(I_2')^2 \times$ equivalent load resistance per phase (R_L')

$= 3 \times (3.9)^2 \times 54$ = **2464 W. (Ans.)**

(*iv*) **Gross load torque, $\mathbf{T_g}$:**

Using the relation, $\frac{2\pi N T_g}{60} = 2464$

where $N = (1 - s) N_s = (1 - 0.04) \frac{120f}{p}$

$= 0.96 \times \frac{120 \times 50}{10} = 576$ r.p.m.

$\therefore$ $\frac{2\pi \times 576 \times T_g}{60} = 2464$

or $T_g = \frac{2464 \times 60}{2\pi \times 576} = 40.85$ N-m

Hence, **gross load torque = 40.85 N-m. (Ans.)**

Example 3.52. *Compute the value of maximum torque in N-m for a 3-phase 440 V star-connected 50-Hz, 4-pole induction motor for which $X_1 = 5.2\ \Omega$; $R_2' = 1.2\ \Omega$ and $X_2' = 4.5\ \Omega$. Also determine the slip corresponding to it.*

Solution. Stator resistance/phase, $R_1 = 0$

Stator leakage reactance/phase, $X_1 = 5.2\ \Omega$

Rotor resistance/phase, $R_2' = 1.2\ \Omega$

Rotor standstill reactance/phase, $X_2' = 4.5\ \Omega$

Supply voltage = 440 V

Maximum torque, T_m :
Slip, s :

We know that, $$P_{g(max)} = \frac{3V_1^2}{(R_{01} + Z_{01})}$$

where $R_{01} = R_1 + R_2'$
= stator resistance + rotor resistance referred to stator
$= 0 + 1.2 = 1.2\ \Omega$

$X_{01} = X_1 + X_2'$
= stator reactance + rotor reactance referred to stator
$= 5.2 + 4.5 = 9.7\ \Omega$

$$Z_{01} = \sqrt{R_{01}^2 + X_{01}^2} = \sqrt{(1.2)^2 + (9.7)^2} = 9.77\ \Omega$$

$$\therefore \quad P_{g(max)} = \frac{3 \times \left(\frac{440}{\sqrt{3}}\right)^2}{2(1.2 + 9.77)} = 8824 \text{ W}$$

Slip corresponding to the maximum power output,

$$s = \frac{R_2'}{R_2' + Z_{01}} = \frac{1.2}{1.2 + 9.77} = \mathbf{0.1094.} \quad \textbf{(Ans.)}$$

To find torque, T_g, using the relation

$$\frac{2\pi N T_g}{60} = 8824$$

where $N = (1 - s) N_s = (1 - 0.1094) \dfrac{120f}{p}$

$$= (1 - 0.1094) \times \frac{120 \times 50}{4} = 1336 \text{ r.p.m.}$$

$$\therefore \quad \frac{2\pi \times 1336 \times T_g}{60} = 8824 \quad \text{or} \quad T_g = \frac{8824 \times 60}{2\pi \times 1336} = \mathbf{63.07\ N\text{-}m.} \quad \textbf{(Ans.)}$$

☞**Example 3.53.** *The rotor resistance and reactance per phase of a 3-phase, 50-Hz, 400 V induction motor having 4-pole star connected stator winding are 0.15 Ω and 1.0 Ω respectively. If full-load slip is 5 per cent and stator to rotor ratio 2 : 1, calculate :*

(i) Total mechanical power developed in kW,
(ii) Total torque developed,
(iii) The speed at maximum torque, and *(iv) Maximum torque.*

Solution. Rotor resistance/phase $= 0.15\ \Omega$

Rotor reactance/phase $= 1.0\ \Omega$

Full-load slip, $s = 5\%$

Stator to rotor ratio, $= 2 : 1$ *i.e.*, $K = \frac{1}{2} = 0.5$

Supply frequency, $f = 50$ Hz

Voltage/phase $= \dfrac{400}{\sqrt{3}}$ V

Number of poles, $p = 4$

Synchronous speed, $N_s = \frac{120f}{p} = \frac{120 \times 50}{4} = 1500$ r.p.m.

Motor speed, $N = (1 - s) N_s = (1 - 0.05) \times 1500 = 1425$ r.p.m.

Rotor impedance/phase, $Z_2 = (0.15 + j1.0)\ \Omega$

Equivalent rotor resistance/phase,

$$\frac{Z_2}{K^2} = \frac{0.15 + j1.0}{(0.5)^2} = (0.6 + j4.0)\ \Omega$$

Equivalent load resistance,

$$R_L' = \frac{R_2}{K^2}\left(\frac{1}{s} - 1\right) = \frac{0.15}{(0.5)^2}\left(\frac{1}{0.05} - 1\right) = 11.4\ \Omega$$

Total effective impedance of rotor,

$$= Z_1 + \frac{Z_2}{K^2} + R_L' = 0 + (0.6 + j4.0) + (11.4 + j0)$$

$$= (12 + j4)\ \Omega, \text{ neglecting stator impedance}$$

Stator current, $I_1 = \frac{400/\sqrt{3}}{(12 + j4)} = \frac{231}{\sqrt{160}} = 18.26$ A

(*i*) **Total mechanical power developed**

$$= 3I_1^2R_L' = 3 \times (18.26)^2 \times 11.4 = \mathbf{11403\ W} \text{ or } \mathbf{11.403\ kW.} \quad \textbf{(Ans.)}$$

(*ii*) **Total torque developed, T :**

Using the relation, $= \frac{2\pi NT}{60} = 11403$

$\therefore$ $T = \frac{11403 \times 60}{2\pi \times 1425} = 76.41$ N-m

Hence, **total torque developed = 76.41 N-m. (Ans.)**

(*iii*) **Slip corresponding to maximum torque,**

$$s_{mT} = \frac{R_2}{X_2} = \frac{0.15}{1.0} = 0.15$$

$\therefore$ **Speed corresponding to maximum torque,**

$$N = (1 - s_{mT}) N_s = (1 - 0.15) \times 1500 = \mathbf{1275\ r.p.m.} \quad \textbf{(Ans.)}$$

(*iv*) **Maximum torque developed, T_m :**

Equivalent load resistance,

$$R_L' = \frac{R_2}{K^2}\left(\frac{1}{s_{mT}} - 1\right) = \frac{0.15}{(0.5)^2}\left(\frac{1}{0.15} - 1\right) = 3.4\ \Omega$$

Total effective impedance of motor

$$= (0.6 + j4.0) + (3.4 + j0) = (4 + j4)\ \Omega = \sqrt{4^2 + 4^2} = 5.66\ \Omega$$

Stator current/phase $= \frac{231}{5.66} = 40.8$ A

Mechanical power developed $= 3 \times (40.8)^2 \times 3.4 = 16979$ W

Using the relation, $\frac{2\pi NT_m}{60} = 16979$

$\therefore$ $T_m = \frac{16979 \times 60}{2\pi \times 1275} = 127.16$ N-m

Hence, **maximum torque developed = 127.16 N-m. (Ans.)**

Example 3.54. *A 3-phase star-connected, 220 V (line to line), 50-Hz, 4-pole induction motor has the following constants in Ω/phase : $r_1 = 0.3, r_2 = 0.15, X_1 = 0.5, X_2 = 0.2$ and magnetising reactance $X_0 = 14$ Ω. The total friction, windage and core loss may be assumed to be constant at 420 watts, independent of load. For a slip of 2 per cent, calculate :*

(i) Speed, *(ii) Output torque, and*

(iii) Stator current of the motor.

Solution. Stator impedance/phase, $Z_1 = (0.3 + j0.5)$ Ω

Equivalent rotor impedance/phase,

$$\frac{Z_2}{K^2} = (0.15 + j0.2)$$

Equivalent load resistance,

$$R_L' = \frac{R_2}{K^2}\left(\frac{1}{s} - 1\right) = 0.15\left(\frac{1}{0.02} - 1\right) = 7.35\ \Omega$$

Voltage/phase $= \dfrac{220}{\sqrt{3}} = 127$ V

Effective impedance/phase $= Z_1 + \dfrac{Z_2}{K^2} + R_L'$

$$= (0.3 + j0.5) + (0.15 + j0.2) + (7.35 + j0) = (7.8 + j0.7)\ \Omega$$

Stator current, I_2' (counter-balanced rotor current)

$$= \frac{127}{7.8 + j0.7} = \frac{127}{7.8 + j0.7} \times \frac{7.8 - j0.7}{7.8 - j0.7}$$

$$= \frac{127(7.8 - j0.7)}{61.33} = 16.15 - j1.45 = 16.21 \text{ A}$$

No-load losses/phase $= \dfrac{420}{3} = 140$ W

Wattful component of no-load current,

$$I_w = \frac{140}{127} = 1.1 \text{ A}$$

Magnetising component of no-load current,

$$I_m = \frac{V_{ph}}{X_0} = \frac{127}{14} = 9.07 \text{ A}$$

Hence no-load current, $I_0 = (1.1 - j9.07)$ A

Stator current, $I_1 = I_2' + I_0 = (16.15 - j1.45) + (1.1 - j9.07)$

$$= (17.25 - j10.52) = 20.2 \text{ A}$$

(i) **Speed** $= (1 - s)\, N_s = (1 - 0.02) \times \dfrac{120f}{p}$

$$= (1 - 0.02) \times \frac{120 \times 50}{4} = \mathbf{1470\ r.p.m.}\ \textbf{(Ans.)}$$

(ii) **Output torque, T :**

Using the relation, $\dfrac{2\pi NT}{60} = 3I_2'^2 R_L' = 3 \times (16.21)^2 \times 7.35$

$$\therefore \quad T = \frac{3 \times (16.21)^2 \times 7.35 \times 60}{2\pi \times 1470}$$

$$= \mathbf{37.64\ N\text{-}m.}\ \textbf{(Ans.)}$$

(iii) **Stator current of motor = 20.2 A. (Ans.)**

Example 3.55. *A 3 kW, 400 V/200 V, Delta/Star, 50 Hz, three-phase, 6 pole induction motor is found to draw a line current of 25 A at a power factor of 0.4, when a blocked rotor test is conducted at the rated voltage. Determine the stator and rotor winding resistances in ohms per phase, if the torque developed by the motor under the above conditions is 25 Nm.* **(GATE, 1999)**

Solution. *Given :* Rating : 3 kW, 400 V/200 V ; Delta/star, 50 Hz ; $p = 6$; $I_L = 25$ A ; p.f. = 0.4, Torque developed, $T = 25$ N-m.

Stator and rotor winding resistances, R_1, R_2 :

Analysing per phase basis :

Under blocked rotor condition, equivalent circuit is shown in Fig. 3.34.

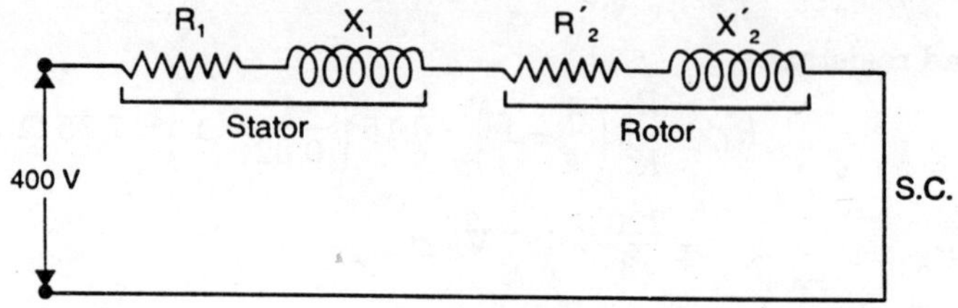

Fig. 3.34

$$Z_{01} = \frac{400}{(25/\sqrt{3})} = 27.7\ \Omega\text{/phase}$$

$$R_{01} = Z_{01} \times \text{p.f.} = 27.7 \times 0.4 = 11.0\ \Omega\text{/phase}$$

where $R_{01} = R_1 + R_2'$

Turn ratio, $$K = \frac{(200/\sqrt{3})}{400} = 0.2887$$

Rotor current, $$I_2 = \frac{(I_1/\sqrt{3})}{K} = \frac{(I_L/\sqrt{3})}{K} = \frac{25/\sqrt{3}}{0.2887} = 50 \text{ A}$$

Synchronous speed, $$N_s = \frac{120f}{p} = \frac{120 \times 50}{6} = 1000 \text{ r.p.m.}$$

We know that, $$\left(\frac{2\pi N_s}{60}\right) \times T = 3I_2^2 R_2$$

or $$\mathbf{R_2} = \frac{2\pi N_s T}{60 \times 3 \times I_2^2} = \frac{2\pi \times 1000 \times 25}{60 \times 3 \times 50^2}$$

$$= \mathbf{0.349\ \Omega/phase.}\ \textbf{(Ans.)}$$

$$R_2' = \frac{R_2}{K^2} = \frac{0.349}{(0.2887)^2} = 4.187\ \Omega$$

Also, $$R_{01} = R_1 + R_2'$$

$\therefore$ $$\mathbf{R_1} = R_{01} - R_2' = 11.08 - 4.187 = \mathbf{6.893\ \Omega/phase.}\ \textbf{(Ans.)}$$

Example 3.56. *A 25 H.P., 400 V, 50 Hz, 4-pole, star-connected induction motor has the following impedances per phase in ohms referred to the stator side :*

$$R_s = 0.641,\ R_r = 0.332\ ;\ X_s = 1.106,\ X_r = 0.464 \text{ and } X_{mag} = 26.30.$$

Rotational losses are assumed constant and are 1.1 kW and core losses are assumed negligible. If the slip is 2.2% at rated voltage and frequency, find :

(i) Speed ; (ii) Stator current ; (iii) Power factor ; (iv) Output and input-power ; (v) Efficiency of motor. **[U.P.S.C., 1997]**

Solution. Refer Fig. 3.35.

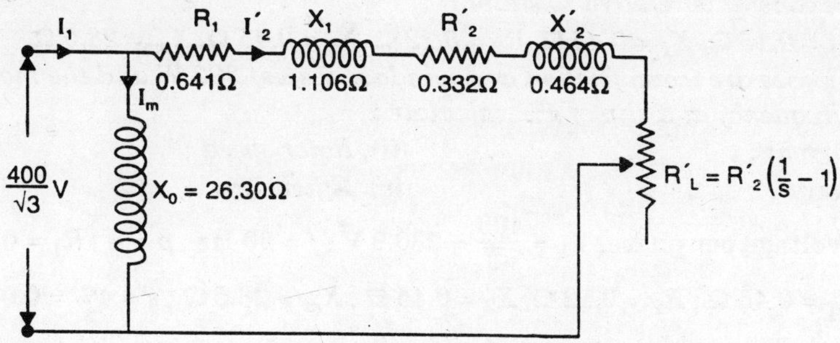

Fig. 3.35

Given : $R_s = R_1 = 0.641\ \Omega$; $R_2' = 0.332\ \Omega$; $X_1 = 1.106\ \Omega$; $X_2' = 0.464\ \Omega$; $X_{mag} = X_0 = 26.30\ \Omega$; rotational losses = 1.1 kW ; s = 2.2% or 0.022 ; p = 4 ; f = 50 Hz.

Stator impedance per phase, $Z_1 = R_1 + jX_1 = (0.641 + j1.106)\ \Omega$

Equivalent rotor impedance per phase, $Z_2' = R'_2 + jX_2' = (0.332 + j0.464)\ \Omega$

Equivalent load resistance, $R_L' = R_2'\left(\frac{1}{s} - 1\right) = 0.332\left(\frac{1}{0.022} - 1\right) = 14.759\ \Omega.$

Voltage applied per phase $= \frac{400}{\sqrt{3}} = 230.9$ V

Effective impedance per phase, $Z = Z_1 + Z_2' + R_L'$

$$= (0.641 + j1.106) + (0.332 + j0.464) + (14.759 + j0)$$
$$= (15.732 + j1.57) \text{ or } 15.81 \angle 5.7°\ \Omega$$

Counter-balancing rotor current per phase,

$$I_2' = \frac{230.9}{15.81 \angle 5.7°} = 14.6 \angle -5.7° = (14.53 - j1.45) \text{ A.}$$

Magnetising component of no-load current,

$$I_m = \frac{230.9}{X_0} = \frac{230.9}{26.30} = 8.78 \text{ A.}$$

(Energy component of no-load current, $I_w = 0$, because core losses are negligible.)

Hence no-load current, $I_0 = (0 - j8.78)$ A.

Stator current per phase, $I_1 = I_2' + I_0 = (14.53 - j1.45) + (0 - j8.78).$

$$= (14.53 - j10.23) \text{ or } 17.77 \angle -35.15° \text{ A.}$$

(*i*) **Speed,** $\mathbf{N} = N_s(1 - s) = \frac{120f}{p}(1 - s)$

$$= \frac{120 \times 50}{4}(1 - 0.022) = \mathbf{1467\ r.p.m.}\ \textbf{(Ans.)}$$

(*ii*) **Stator current** = **17.77 A. (Ans.)**

(*iii*) **Power factor,** $\cos\phi_1 = \cos(-35.15°) = \mathbf{0.8176}$ **lag. (Ans.)**

(*iv*) **Output power,** $\mathbf{P_{out}} = 3I_2^2R_L'$ – rotational losses

$$= [3 \times (14.6)^2 \times 14.759] - (1.1 \times 1000) = \mathbf{8338\ W.}\ \textbf{(Ans.)}$$

Power input, $\mathbf{P_{in}} = 3 \times 230.9 \times 17.77 \times 0.8176 = \mathbf{10064\ W.}$ **(Ans.)**

(*v*) **Efficiency of motor,** $\eta = \frac{P_{out}}{P_{in}} = \frac{8338}{10064} = 0.8285$ or **82.85%. Ans.**

Example 3.57. *A 3-phase, star-connected, 400 V, 50 Hz, 4-pole induction motor has the following per phase constants referred to stator :*

$$R_1 = 0.15\ \Omega, X_1 = 0.45\ \Omega, R_2 = 0.12\ \Omega, X_2 = 0.45\ \Omega, X_m = 28.5\ \Omega.$$

If the fixed losses are (core, friction and windage losses) 380 W and the motor is operated at rated voltage and frequency at a slip of 4%, calculate :

(i) Stator current ; *(ii) Rotor speed ;*
(iii) Output torque ; *(iv) Efficiency.*

Solution. Voltage per phase, $V_1 = \frac{400}{\sqrt{3}} = 230.9$ V ; $f = 50$ Hz ; $p = 4$; $R_1 = 0.15\ \Omega$;

$X_1 = 0.45\ \Omega$; $R_2 = 0.12\ \Omega$, $X_2 = 0.45\ \Omega$; $X_m = 28.5\ \Omega$; $s = 4\% = 0.04$.

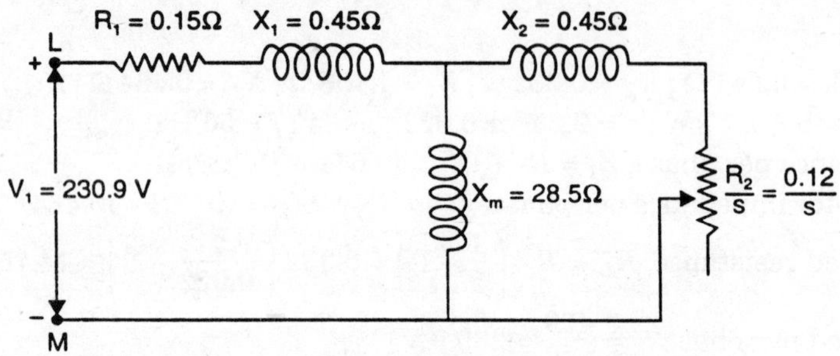

Fig. 3.36. Induction motor equivalent circuit.

The equivalent circuit of induction motor is shown in Fig. 3.36.

The impedance offered to the voltage source

$$\bar{Z}_{LM} = R_1 + jX_1 + \frac{\left(\frac{R_2}{s} + jX_2\right)(jX_m)}{\frac{R_2}{s} + j(X_2 + X_m)}$$

$$= R_1 + jX_1 + R_{rf} + jX_{rf}$$

$$\left[\text{where } Z_{rf} = R_{rf} + jX_{rf} = \frac{\left(\frac{R_2}{s} + jX_2\right)(jX_m)}{\frac{R_2}{s} + j(X_2 + X_m)}\right.$$

and is the *impedance offered to the stator by the rotating air-gap field*]

$$= 0.15 + j0.45 + \frac{\left(\frac{0.12}{0.04} + j0.45\right)(j28.5)}{\frac{0.12}{0.04} + j(0.45 + 28.5)}$$

$$= 0.15 + j0.45 + \frac{(3 + j0.45)(j28.5)}{3 + j28.95}$$

$$= (0.15 + j0.45) + \left[\frac{(3 + j0.45)(3 - j28.95)}{(3 + j28.95)(3 - j28.95)}\right](j28.5)$$

$$= 0.15 + j0.45 + \left[\frac{9 + j1.35 - j86.85 + 13.03}{9 + 838.1}\right](j28.5)$$

$$= 0.15 + j0.45 + \left(\frac{22.03 - j85.5}{847.1}\right) j28.5$$

$$= (0.15 + j0.45) + \left[\frac{j627.86 + 2436.75}{847.1}\right]$$

$$= (0.15 + j0.45) + (j0.741 + 2.87)$$

$$= 3.02 + j1.191 = 3.25 \angle 21.5°$$

From above, $R_{rf} = 2.87\ \Omega$

(*i*) **Stator current, I_1**

$$I_1 = \frac{230.9 \angle 0°}{3.25 \angle 21.5°} = \mathbf{71.05 \angle -21.5°}.\ \textbf{(Ans.)}$$

(*ii*) **Rotor speed, N :**

Power factor $= \cos(21.5°) = 0.93$ lag

Stator input $= \sqrt{3} \times 400 \times 71.05 \times 0.93 \times 10^{-3}$ kW = 45.779 kW

Total air-gap power, $P_g = 3 \times I_1^2 R_{rf} = 3 \times (71.05)^2 \times 2.87 \times 10^{-3}$ kW = 43.464 kW

Synchronous speed, $N_s = \dfrac{120 \times f}{p} = \dfrac{120 \times 50}{4} = 1500$ r.p.m.

∴ Rotor speed, $\mathbf{N} = N_s(1 - s) = 1500(1 - 0.04) =$ **1440 r.p.m. (Ans.)**

(*iii*) **Output torque, T_{sh} :**

$$\omega_s = \frac{2\pi N_s}{60} = \frac{2\pi \times 1500}{60} = 50\pi \text{ rad/s.}$$

Mechanical power developed, $P_{mech.} = (1 - s)P_g$

$= (1 - 0.04) \times 43.464 = 41.725$ kW

∴ Power output $= P_{sh} = P_{mech.}$ − fixed losses

$= 41.725 - 0.38 = 41.345$ kW

Also, $P_{sh} = T_{sh} \times \dfrac{2\pi N}{60}$

$$\therefore \quad \mathbf{T_{sh}} = \frac{P_{sh}}{(2\pi N/60)} = \frac{(41.345 \times 1000) \times 60}{2\pi \times 1440} = \mathbf{274.18\ N\text{-}m.}\ \textbf{(Ans.)}$$

(*iv*) **Efficiency :**

Total losses in induction motor

= stator ohmic losses + rotor ohmic losses + fixed losses

$= 3I_1^2 R_1 + sP_g + 380$ watts

$= 3 \times (71.05)^2 \times 0.15 + 0.04 \times 43464 + 380 = 4390.2$ W or 4.39 kW

∴ Power input $= P_{sh}$ + losses

$= 41.345 + 4.39 = 45.735$ kW

∴ Efficiency $= \dfrac{\text{output}}{\text{input}} = \dfrac{41.345}{45.735} = 0.904$ or **90.4%. (Ans.)**

3.21. INDUCTION MOTOR LOSSES AND EFFICIENCY

In an induction motor following losses occur :

1. **Copper losses** due to current in the conductors.
2. **Core loss in the iron.**
3. **Friction and windage** due to rotation.

There are copper losses and core losses in the *stator*, and copper losses and frictional losses in the *rotor*. Actually there is some core loss in the rotor. Under operating conditions, however, the rotor frequency is so low that it may logically be *assumed that all core losses occur in the stator only*.

The efficiency of induction motor can be determined (as with any rotating equipment) by loading the motor and measuring the input and output directly. This method, however, is not always desirable, because of the inherent inaccuracy, or may not even be feasible, because of the inability to simulate an actual load. The latter is especially true in the case of large machines. As with D.C. dynamos and the transformer, therefore, a method of *measuring the losses with electrical instruments only is used*. Since the circuit of the induction motor closely approximates that of the transformer, losses may be measured in a manner virtually identical with that used for transformers. The method is the following :

1. **No-load or Open-circuit Test.** This test is performed to determine the following :

(*i*) No-load current I_0

(*ii*) No-load power factor $\cos \phi_0$

(*iii*) Windage and friction losses

(*iv*) No-load core losses

(*v*) No-load input

(*vi*) No-load resistance R_0 and reactance X_0.

The test is conducted as follows :

The motor is uncoupled from its load, and rated voltage is applied to the stator. Since there is no power output, the power supplied to the stator furnishes its copper loss, its core loss, and the friction and windage loss in the rotor. Since slip at no-load is often within one-tenth of one per cent, the rotor current is practically zero, and hence the analogy to the no-load test of the transformer. However the no-load stator current of the induction motor is comparatively large, and therefore copper losses in the stator may not be neglected as they were for the transformer primary. It is thus necessary to *calculate this copper loss* and *subtract from the no-load input* in order to obtain the sum of the core and friction and windage losses of the motor.

Speed and flux vary but little from no-load to full-load, and these losses may be assumed constant for the operating range of the polyphase induction motor.

Knowing the total core losses, P_{core}, no-load current per phase I_0 and applied voltage per phase V (Fig. 3.37), the values of magnetising current, I_m, wattful current I_w, no-load resistance R_0 and no-load reactance X_0 are determined as follows :

$$\text{No-load power factor, } \cos \phi_0 = \frac{P_{core}}{3VI_0}$$

$$I_w = I_0 \cos \phi_0 = I_0 \times \frac{P_{core}}{3VI_0} = \frac{P_{core}}{3V}$$

$$I_m = \sqrt{I_0^2 - I_w^2}$$

$$R_0 = \frac{V}{I_w} \text{ and } X_0 = \frac{V}{I_m}.$$

2. Blocked Rotor or Short-circuit Test. This test is performed to determine :

(*i*) Short-circuit current, I_{sc} with normal voltage applied to stator.

(*ii*) Power factor on short-circuit.

(*iii*) Total equivalent resistance and reactance of the motor as referred to stator.

(*iv*) Full-load copper loss.

The test is conducted as follows (Fig. 3.37).

- The rotor is prevented from turning by blocking it, and a low voltage of less than 10 per cent of rated is applied to the stator. This voltage is best applied through a variable auto-transformer, or three-phase Variac. Suitable resistors in series with the motor-stator

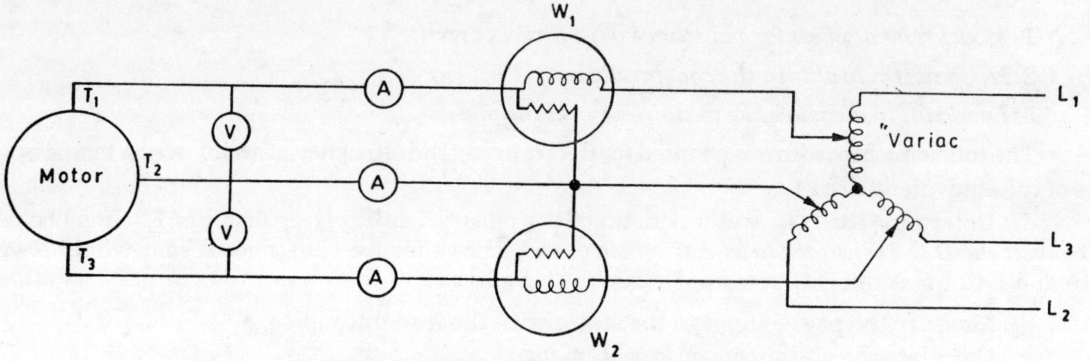

Fig. 3.37. Efficiency test of induction motor.

terminals may also be used, but care must be taken that the resistors are identical so that balanced voltage may be applied to the stator.

- The voltage is then gradually decreased until rated current is obtained in the stator. Since the voltage required is quite low, the power supplied to the stator will be dissipated as copper loss in both stator and rotor. It is thus assumed, as in the static transformer short-circuit test, that the core loss is negligible. This will enable us to calculate an equivalent resistance of the motor referred to the stator terminals and to use this value of resistance to correct our readings of the no-load test.

If at *short-circuit*,

V_s = applied voltage/phase

I_s = short current/phase in the stator winding

P_s = total input power, then

$$I_{sc} = I_s \frac{V}{V_s}$$

and power factor, $$\cos \phi_s = \frac{P_s}{3V_s I_s}$$

Since input on short-circuit meets with stator and rotor copper losses and core losses, core losses being very small can be neglected, therefore,

Input on short-circuit, $P_s = 3I_s^2 R_{01}$

or Equivalent resistance per phase referred to stator,

$$R_{01} = \frac{P_s}{3I_s^2}$$

Motor equivalent impedance per phase as referred to stator,

$$Z_{01} = \frac{V_s}{I_s}$$

Motor equivalent reactance per phase as referred to stator,

$$X_{01} = \sqrt{Z_{01}^2 - R_{01}^2}$$

The wiring diagram to be used for both tests is shown in Fig. 3.37.

Determining the efficiency of an induction motor by the use of the equivalent circuit.

It is seen that the efficiency calculation based on equivalent circuit of induction motor actually requires three tests :

1. *Measurement of stator resistance with direct current.*

2. *No-load test to obtain the constant loss.*

3. *Load run to measure the total input and the slip.*

The following procedure is adopted for determining the efficiency of an induction motor by the use of the equivalent circuit :

(*i*) Determine the D.C. stator resistance per phase. A multiplying factor of 1.25 may be used to *convert the D.C. resistance to 'effective resistance'*, since effective resistance is somewhat between 1.15 and 1.40 times the D.C. resistance (due to *skin effect*).

(*ii*) Measure the power input to the stator with the load uncoupled.

(*iii*) Calculate the stator copper loss at no-load.

(*iv*) Subtract the no-load copper loss from the no-load power input. This is the constant loss, which is the sum of the friction, windage, hysteresis, and eddy-current losses.

(*v*) Load the motor, and measure the power input, line current and slip.

(*vi*) Calculate the stator copper loss at the given load.

(*vii*) Subtract the stator copper loss from the motor power input to calculate the power input to the *rotor*.

(*viii*) Multiply the rotor power input by the slip *s* to calculate the rotor copper loss.

(*ix*) Calculate the total loss in the rotor by adding the rotor copper loss and the constant loss.

(*x*) Subtract the total rotor loss from the rotor power input to obtain the power output of the motor.

Although calculation of efficiency from the equivalent circuit is considered more accurate than the previous method, it does have *drawbacks. A load run must be performed, and slip must be accurately measured.* The second factor is not a serious drawback since a technique of direct simple measurement can be easily used. Nevertheless, the *efficiency test using the no-load run and blocked-rotor test is sometimes used, despite the fact the efficiency measured in this manner is slightly lower.*

3.22. CIRCLE DIAGRAM

For the purpose of studying the various characteristics of an induction motor, the **'circle-diagram'**, *which is a vector locus diagram of stator and rotor currents, is very useful and invaluable for estimating certain working conditions.*

Circle diagram of an induction motor can be drawn by using the data obtained from no-load test, short-circuit test and stator resistance test. The procedure is as follows :

Refer Fig. 3.38.

- Calculate I_0 and ϕ_0 from *no-load test* and lay off vector for I_0 lagging ϕ_0 behind the applied voltage V.

- Find from *blocked rotor test* or *short-circuit test*, short-circuit current I_{SC} *corresponding to normal voltage* and ϕ_s. The vector OL represents $I_{SN} = \left(I_s \times \frac{V}{V_s} \right)$ in magnitude and phase.

  Vector $O'L$ represents rotor current I_2' as referred to stator. Clearly, the two points O' and A lie on the required circle.

- To find centre G of this circle, bisect chord $O'L$ at right-angles, its bisector gives point G. Draw the diameter $O'H$ perpendicular to voltage vector.

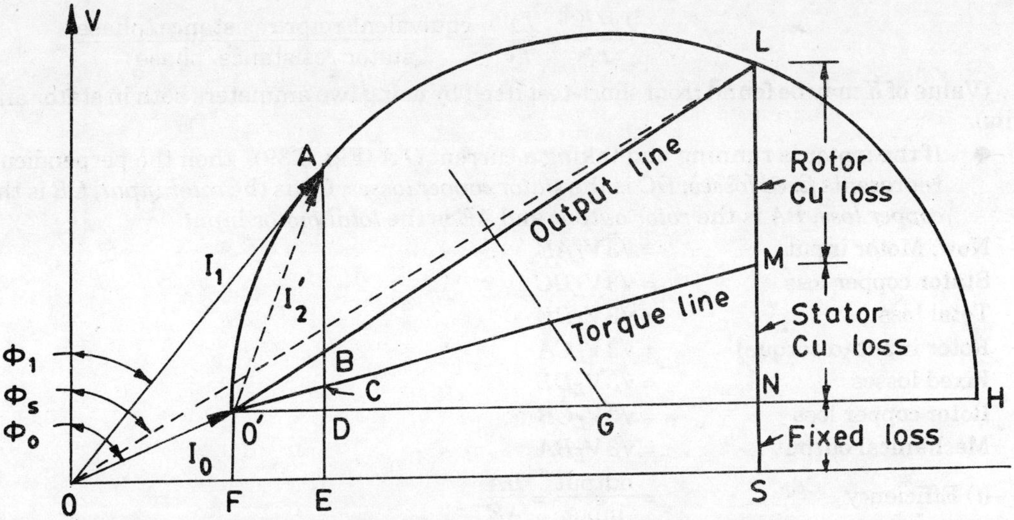

Fig. 3.38. Circle diagram for induction motor.

- With centre G and radius GO' draw the circle.

The line $O'L$ is known as *'output line'* or mechanical power developed line.

It may be noted that *as the voltage vector is drawn vertically, all vertical distances represent the active or power* or *energy components of currents.*

- **Torque line.** *This is the line which separates the stator and rotor copper losses.*

When the rotor is locked, then all the power supplied to the motor goes to meet core losses and copper losses in the stator and rotor windings. The power input is proportional to LS. Out of this, $NS = (O'F)$ represents the fixed loss *i.e.* stator core loss and rotor core losses. The point M is such that

$$\frac{LM}{MN} = \frac{\text{rotor copper loss}}{\text{stator copper loss}}$$

The line $O'M$ is known as *torque line* (or rotor input line).

The point M is located at follows :

(*i*) **Squirrel cage motor :**

Stator copper loss $= 3I_s^{\,2} R_1$

∴ Rotor copper loss $= P_s - 3I_s^{\,2}R_1$

$$\therefore \quad \frac{LM}{MN} = \frac{P_s - 3I_s^{\,2}R_1}{3I_s^{\,2}R_1}$$

Stator resistance/phase *i.e.*, R_1 is found from **stator resistance test.**

(*ii*) **Wound rotor.** In this case, rotor and stator resistances per phase r_2 and r_1 can be easily calculated. For any values of stator and rotor currents I_1 and I_2 respectively, we can write

$$\frac{LM}{MN} = \frac{I_2^{\,2} r_2}{I_1^{\,2} r_1} = \frac{r_2}{r_1}\left(\frac{I_2}{I_1}\right)^2$$

$$= \frac{r_2}{r_1} \times \frac{1}{K^2} \qquad \left[\because \frac{I_1}{I_2} = K = \text{transformation ratio}\right]$$

$$= \frac{r_2/K^2}{r_1} = \frac{r_2'}{r_1} = \frac{\text{equivalent rotor resistance/phase}}{\text{stator resistance/phase}}$$

(Value of K may be found from short-test itself by using two ammeters both in stator and rotor circuits).

- If the motor is running and taking a current $O'A$ (Fig. 3.39), then the perpendicular DE represents *fixed losses, DC* is the *stator copper losses, CA* is the *rotor input, CB* is the *rotor copper loss, BA* is the *rotor output* and *AE* is the *total motor input*.

Now, Motor input $= \sqrt{3}V_L AE$

Stator copper loss $= \sqrt{3}V_L DC$

Total loss $= \sqrt{3}V_L BE$

Rotor input (α torque) $= \sqrt{3}V_L CA$

Fixed losses $= \sqrt{3}V_L DE$

Rotor copper loss $= \sqrt{3}V_L CB$

Mechanical output $= \sqrt{3}V_L BA$

(*i*) Efficiency $= \frac{\text{output}}{\text{input}} = \frac{BA}{AE}$

(*ii*) $\frac{\text{Rotor copper loss}}{\text{Rotor input}} = \text{slip 's'} = \frac{CB}{CA}$

(*iii*) $\frac{\text{Rotor output}}{\text{Rotor input}} = 1 - s = \frac{BA}{CA}$

(*iv*) Power factor $= \frac{AE}{OA}$.

Maximum Quantities:

Refer Fig. 3.39.

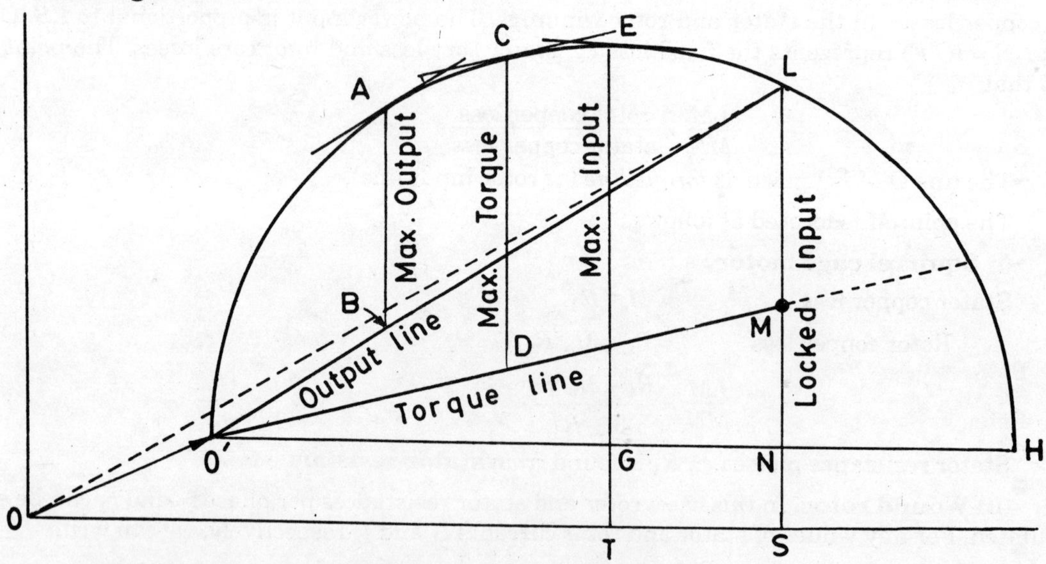

Fig. 3.39. Maximum quantities.

- **Maximum output.** Maximum output occurs at A where the tangent is parallel to the output line $O'L$. To locate the point A draw a line from point G such that it is perpendicular to the output line $O'L$.

Vertical line AB represents the *maximum output*.

- **Maximum Torque or Rotor Input.** Maximum torque or rotor input occurs at point C where the tangent is parallel to torque line $O'M$. Again, point C may be found by drawing GC perpendicular to the torque line.

  Vertical line CD represents the *maximum torque.* Maximum torque is also known as *pull out* or *stalling torque.*
- **Maximum Input Power.** It occurs at the highest point of the circle *i.e.* at point E where tangent to the circle is horizontal. It is proportional to ET. *As the point E is beyond the point of maximum torque, the induction motor will be unstable here.* However, *the maximum input is a measure of the size of the circle and is an indication of the ability of the motor to carry short-time overloads.* Generally ET is twice or thrice the motor input at rated load.

CIRCLE DIAGRAM

Example 3.58. *A 3-phase, 400 V induction motor gave the following test readings :*

No-load test : 400 V, 1250 W, 9 A

Short-circuit test : 150 V, 4 kW, 38 A.

Draw the circle diagram.

If the normal rating is 14.91 kW, find from the circle diagram, the full-load values of current, p.f. and slip.

Solution. No-load test : $I_0 = 9$ A, $V = 400$ V, $P_0 = 1250$ W

$$\therefore \quad \cos\phi_0 = \frac{1250}{\sqrt{3}\times 400\times 9} = 0.2004$$

$$\therefore \quad \phi_0 = 78.5°$$

Short-circuit test :

$$I_s = 38 \text{ A}, V_s = 150 \text{ V}, P_s = 4 \text{ kW}$$

$$\cos\phi_s = \frac{4000}{\sqrt{3}\times 150\times 38} = 0.405$$

$$\therefore \quad \phi_s = 66.1°$$

Short-circuit current with normal voltage,

$$I_{sc} = I_s \times \frac{V}{V_s}$$

where $I_s = 38$ A, $V = 400$ V, $V_s = 150$ V

$$\therefore \quad I_{sc} = 38 \times \frac{400}{150} = 101.3 \text{ A}$$

$$\text{Power taken} \quad = 4000 \times \left(\frac{400}{150}\right)^2 = 28444 \text{ W}$$

Refer Fig. 3.40.

Current scale 1 cm = 5 A

- OO' represents $I_0 = 9$ A. Vector $OO' = 9/5 = 1.8$ cm is drawn at an angle of $\phi_0 = 78.5°$ with the vertical OV which represents voltage.
- OL represents I_{sN} (short-circuit with normal voltage applied) equal to 101.3 A. It measures 101.3/10 = 10.13 cm and is drawn at an angle of $\phi_s = 66.1°$.
- Line $O'H$ is drawn parallel to OX. FG is the right-angle bisector of $O'L$.

  With G as the centre, the semi-circle $O'LH$ is drawn. *This semi-circle is the locus of the current vector for all load conditions from no-load to short-circuit.*
- LS represents 28444 W and measures 8.1 cm. Hence *power scale* becomes :

$$1 \text{ cm} = \frac{28444}{8.1} \simeq 3510 \text{ W}$$

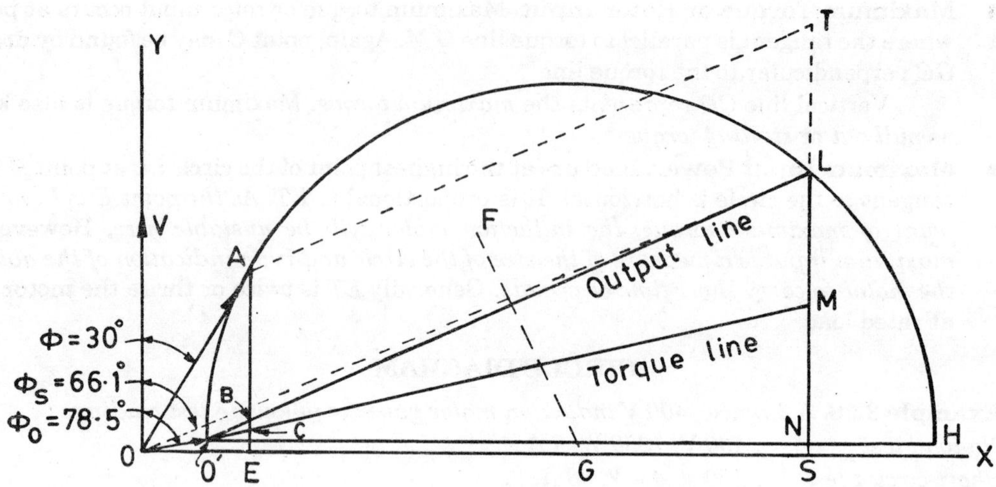

Fig. 3.40

Now full-load motor output = 14.91 kW = 14910 W. According to the above calculated power scale, the intercept between the semi-circle and output line $O'L$ should measure $= \frac{14910}{3510} = 4.25$ cm.

- For locating full-load point A, NL is extended. LT is made equal to 4.25 cm and TA is drawn ∥ to output line $O'L$. AE is ⊥ to OX.

From circle diagram :

(*i*) **Line current** $= OA = 6 \text{ cm} = 6 \times 5 =$ **30 A. (Ans.)**

$\phi = 30°$ (by measurement)

(*ii*) ∴ **p.f.,** $\cos \phi = \cos 30° =$ **0.866. (Ans.)**

(*iii*) **Slip** $= \dfrac{\text{rotor copper loss}}{\text{rotor input}}$

$= \dfrac{BC}{AC} = \dfrac{0.3}{4.5} = 0.0667$ or **6.67%. (Ans.)**

Note. The actual scale of the book diagram is different because it has been reduced during block making.

☞Example 3.59. *The following test data was obtained for a 3.73 kW, 200 V, 50-Hz, 4-pole, 3-phase star-connected induction motor.*

No-load test : Line voltage	*= 200 V*
Line current	*= 5 A*
Total input	*= 350 W*
Blocked rotor test : Line voltage	*= 100 V*
Line current	*= 26 A*
Total input	*= 1700 W*

Draw the circle diagram and estimate from the diagram for full-load condition the following :

(i) Line current, *(ii) Power factor, and*

(iii) Maximum torque in terms of the full-load torque.

The rotor copper loss at standstill is half the total copper loss. **(AMIE Winter, 1998)**

Solution. No-load test :

$I_0 = 5$ A, V = 200 V, $P_0 = 350$ W

$$\therefore \quad \cos \phi_0 = \frac{350}{\sqrt{3} \times 200 \times 5} = 0.202$$

$$\therefore \quad \phi_0 = 78° \ 20'.$$

Blocked rotor test :

$$I_s = 26 \text{ A}, V_s = 100 \text{ V}, P_s = 1700 \text{ W}$$

$$\therefore \quad \cos \phi_s = \frac{1700}{\sqrt{3} \times 100 \times 26} = 0.377$$

$$\phi_s = 67° \ 40'$$

Short-circuit current with normal voltage,

$$I_{sc} = I_s \times \frac{V}{V_s} = 26 \times \frac{200}{100} = 52 \text{ A}$$

Short-circuit/blocked rotor input with normal voltage,

$$P_{sc} = P_s \times \left(\frac{I_{sc}}{I_s}\right)^2 = 1700 \times \left(\frac{52}{26}\right)^2 = 6800 \text{ W}$$

Refer Fig. 3.41.

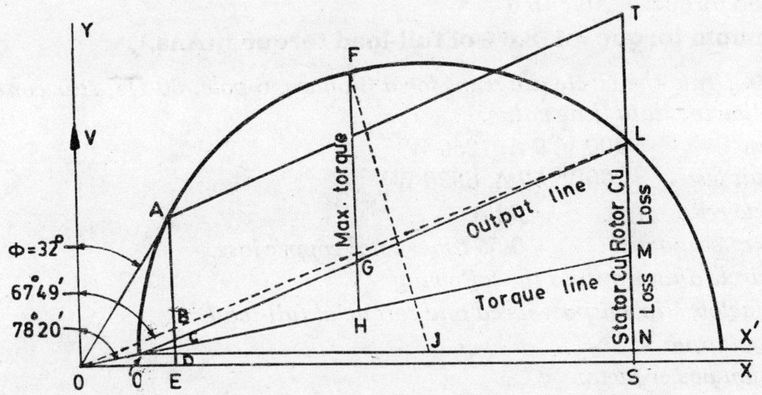

Fig. 3.41

Current scale 1 cm = 2 A

- Line OO' (I_0) = 5/2 = 2.5 cm and is drawn at an angle of ϕ_0 = 78° 20′.
- Line OL represents I_{sc} *i.e.* 52 A and measures = 52/2 = 26 cm.
- *O′L* represents *output line.*
- To locate the centre *J* of the circle, perpendicular bisector of *O′L* is drawn.
- With *J* as centre and radius *JO′*, a circle is drawn which passes through points *O′* and *L*.

  From *L*, a perpendicular is drawn to the base.
- *LS* represents total input of 6800 W for blocked rotor test. Out of this, *NS* represents no-load loss of 350 W and *LN* represents 6800 – 350 = 6450 W.

  Now *LS* = 9.8 cm and represents 6800 W

  ∴ Power scale = 6800/9.8

  ≃ 700 W/cm or 1 cm = 700 W.
- *LN* which represents total copper loss in rotor and stator, is bisected at point *M* to separate the two losses.

  O′M represents torque line.

- Motor output = 3.73 kW or 3730 W and is represented by the line = $\frac{3730}{700}$ = 5.33 cm.
- The output point A on the circle is located as follows :

  SL is extended and LT is cut = 5.33 cm.

  Line TA is drawn parallel to output line $O'L$ and cuts the circle at point A. Perpendicular AE is drawn and A is joined to origin O.
- The point F corresponding to maximum torque is obtained as follows :

  From centre J, a line JF is drawn such that it is perpendicular to torque line $O'M$. It cuts the circle at F which is the required point.

  FH is drawn vertical and it represents *maximum torque.*

From circle diagram (for full-load condition) :

(*i*) **Line current** = OA = 7.6 cm = 7.6 × 2 = **15.2 A. (Ans.)**

(*ii*) **Power factor** = $\frac{EA}{OA} = \frac{6.45}{7.6}$ = 0.848

[or by measurement ϕ = 32°, cos ϕ = cos 32° = 0.848].

(*iii*) $\frac{\text{Maximum torque}}{\text{Full-load torque}} = \frac{FH}{AC} = \frac{10}{5.6} = 1.786$

∴ **Maximum torque = 178.6% of full-load torque. (Ans.)**

☞**Example 3.60.** *Draw the circle diagram for a 3-phase, 6-pole, 50-Hz, star-connected induction motor from the following data (line values) :*

No-load test	*: 400 V, 9 A, 1250 W*
Short-circuit test	*: 200 V, 50 A, 6930 W*
Full-load current	*= 32 A*
Stator loss at standstill	*= 0.55 times total copper loss.*

From the circle diagram find the following :

(*i*) *Power factor, slip, output, speed and torque at full-load*

(*ii*) *Starting torque*

(*iii*) *Maximum power factor*

(*iv*) *Maximum power output*

(*v*) *Maximum power input.*

(*vi*) *Maximum torque in synchronous watts and slip for maximum torque.*

Solution. No-load test :

No-load input, $I_0 = 9$ A

Voltage applied, $V = 400$ V

No-load input, $P_0 = 1250$ W

∴ No-load power factor, $\cos \phi_0 = \frac{P_0}{\sqrt{3}VI_0} = \frac{1250}{\sqrt{3} \times 400 \times 9} = 0.2$

∴ $\phi_0 = 78° \, 28'$

Short-circuit test :

$$I_s = 50 \text{ A}, V_s = 200 \text{ V}, P_s = 6930 \text{ W}$$

Short-circuit current with normal voltage,

$$I_{sc} = 50 \times \frac{V}{V_s} = 50 \times \frac{400}{200} = 100 \text{ A}$$

Short-circuit input with normal voltage,

$$P_{sc} = 6930 \times \left(\frac{100}{50}\right)^2 = 27720 \text{ W}$$

Short-circuit power factor, $\cos \phi_s = \frac{6930}{\sqrt{3} \times 200 \times 50} = 0.4$

$\therefore \quad \phi_s = 66° \, 25'$

Assume a current scale of 1 cm = 5 A. The circle diagram of Fig. 3.42 is constructed as follows :

- No-load current vector *OO′* represents 9 A. Hence it measures 9/5 = 1.8 cm and is drawn at an angle of 78° 28′ with *OY*.

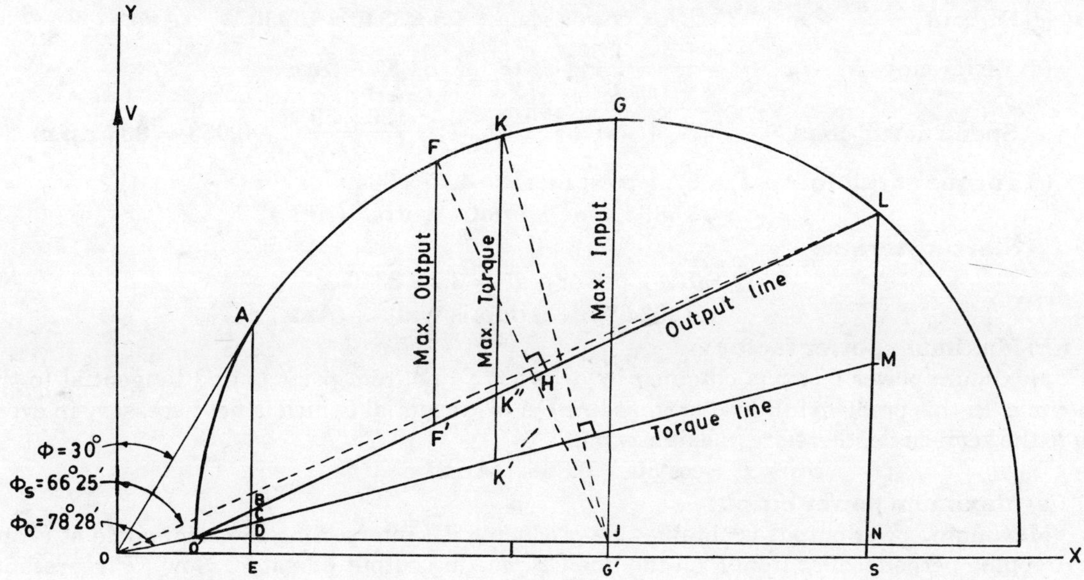

Fig. 3.42

- Vector *O′L* represents 100 A and measures 100/5 = 20 cm. It is drawn at an angle of 66° 25′ with *OY*.
- *O′L* represents output line. Perpendicular bisector of *O′L* is drawn to locate the centre *J* of the circle.
- With *J* as centre and radius = *O′J* a semi-circle is drawn which passes through *O′* and *L* as shown.
- From point *L*, a perpendicular is drawn to the base. *LS* represents total input on short-circuit with normal voltage which is 27720 W.

  Since $LS = 8.3$ cm

  Hence power scale is 1 cm $= \frac{27720}{8.3} = 3340$ W

  Since stator copper loss = 0.55 times total copper loss

  $\therefore$ Torque line *O′M* is obtained by taking,

  $$NM = 0.55 \, NL.$$

- Since full-load current is 32.0 A, hence it is represented by a length of 32/5 = 6.4 cm.

  With *O* as centre and 6.4 cm as radius an arc is drawn intersecting semi-circle at point *A*. This point represents *full-load conditions.*

  Point *O* and *A* are joined.

- From point *A*, perpendicular *AE* is drawn to line *OS* intersecting lines *O′L*, *O′M*, *O′N* and *OS* at *B, C, D* and *E* respectively.

From circle diagram :

(*i*) **At full-load :**

(*a*) **Power factor** $= \frac{EA}{OA} = \frac{5.5}{6.4} = \mathbf{0.86}$. **(Ans.)**

[Also power factor = cos ∠ *AOV* = cos 30° = 0.866]

(*b*) **Slip,** $s = \frac{BC}{AC} = \frac{0.30}{4.75} = 0.063$ or **6.3%. (Ans.)**

(*c*) **Output** = *BA* × power scale = 4.5 × 3340 = **15030 W. (Ans.)**

(*d*) **Efficiency** $= \frac{BA}{EA} = \frac{4.5}{5.5} = 0.818$ or **81.8%. (Ans.)**

(*e*) **Speed at full-load,** $= N_s(1 - s) = \frac{120f}{p}(1 - s) = \frac{120 \times 50}{6}(1 - 0.063) = \mathbf{937\ r.p.m.}$

(*f*) **Torque at full-load,** T_f = *CA* × power scale = 4.75 × 3340

= **15865 synchronous watts. (Ans.)**

(*ii*) **Starting torque,**

T_{st} = *LM* × power scale = 3.6 × 3340

= **12024 synchronous watts. (Ans.)**

(*iii*) **Maximum power factor :**

Maximum power factor is obtained by drawing a line from point *O* and tangential to the semi-circle. In this problem full-load current line *OA* is tangential (which is not necessary in every case) to the semi-circle, therefore, maximum p.f.,

$$\cos \phi_{max} = \cos 30° = \mathbf{0.866}. \textbf{ (Ans.)}$$

(*iv*) **Maximum power output :**

Maximum power output is obtained by producing *JH* intersecting the semi-circle at *F* and then dropping perpendicular from *F* on the line *OS* meeting output line at *F'*. Now *FF'* represents maximum power output.

Maximum power output = *FF'* × power scale

= 6.28 × 3340 = **20975 W. (Ans.)**

(*v*) **Maximum power input :**

It is obtained by drawing vertical line *GG'* from point *G*. *GG'* represents maximum power input.

∴ **Maximum power input** = *GG'* × power scale

= 10.3 × 3340 = **34400 W. (Ans.)**

(*vi*) **Maximum torque :**

It is obtained by drawing line *JK* perpendicular to *O'M* and dropping a vertical line from *K* intersecting torque line *O'M* at *K'*. Now *KK'* represents maximum torque.

∴ **Maximum torque** = *KK'* × power scale

= 7.7 × 3340 = **25718 synchronous watts. (Ans.)**

Slip for maximum torque $= \frac{K''K'}{KK'} = \frac{1.6}{7.7} = \mathbf{0.207}$ or **20.7%. (Ans.)**

Example 3.61. *The following test results refer to a 3-phase, 29.84 kW, 50-Hz, delta connected induction motor :*

No-load test : 415 V, 21 A, 1335 W.

Locked rotor test : 100 V, 45 A, 2730 W.

Draw the circle diagram of this induction motor and determine :

(i) Line current and power factor for rated output, and

(ii) Maximum torque.

Assume that at standstill, stator Cu loss = rotor Cu loss.

Solution. No-load test :

$$I_0 = 21 \text{ A}, V = 415 \text{ V}, P_0 = 1385 \text{ W}$$

No-load power factor,

$$\cos \phi_0 = \frac{P_0}{\sqrt{3}\, VI_0} = \frac{1385}{\sqrt{3} \times 415 \times 21} = 0.918$$

$\therefore$

$$\phi_0 = \cos^{-1} (0.0918) = 84° \, 44'$$

Locked rotor test :

$$I_s = 45 \text{ A}, V_s = 100 \text{ V}, P_s = 2730 \text{ W}$$

Power factor with locked rotor,

$$\cos \phi_s = \frac{2730}{\sqrt{3} \times 100 \times 45} = 0.3503$$

or

$$\phi_s = \cos^{-1} (0.3503) = 69° \, 30'$$

Short-circuit current (I_{sc}) if normal voltage were applied

$$= 45 \times \left(\frac{415}{100}\right) = 186.75 \text{ A}$$

Short-circuit power input (P_{sc}) with normal voltage,

$$= 2730 \times \left(\frac{186.75}{45}\right)^2 = 47017 \text{ W}$$

The circle diagram of Fig. 3.43 is constructed as follows :
Assume a current scale of 1 cm = 10 A

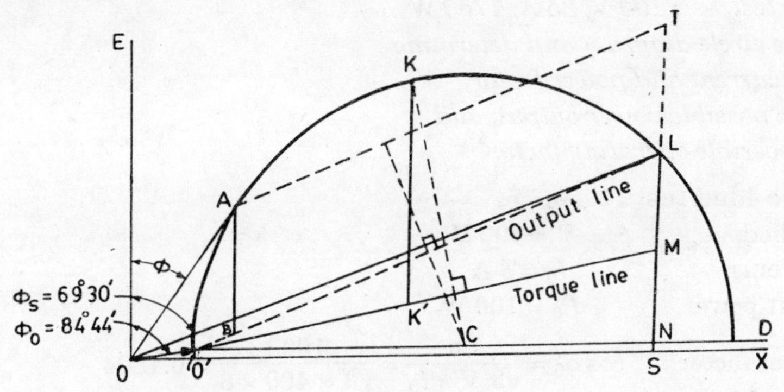

Fig. 3.43

- No-load current vector OO' represents 21 A. Hence it measures 21/10 = 2.1 cm and is drawn at an angle of 84° 44′ with OY.
- Vector OL represents 186.75 A and measures 186.75/10 = 18.675 cm. It is drawn at an angle of 69° 30′ with OY.
- $O'D$ is drawn parallel to OX. Perpendicular bisector of $O'L$ is drawn to locate the centre C of the circle.
- With C as centre and radius = CO', a semi-circle is drawn which passes through O' and L. *The semi-circle is the locus of the current vector for all load conditions from no-load to short-circuit.*
- The vertical line LS represents power input on short-circuit with normal voltage applied. LS measures 6.6 cm and represents 47017 W (as calculated above).

  Hence power scale becomes $= \dfrac{47017}{6.6} = 7124$ W.

From circle diagram :

(*i*) Full-load output = 29.84 kW or 29840 W. According to power scale, the intercept between the semi-circle and output line *O'L* should measure = 29840/7124 = 4.19 cm. Hence the line *AB* is found which measures 4.19 cm. Point *A* represents the *full-load operating point*.

(The *operating point* may also be found by making *LT* = 4.19 cm and drawing *TA* || to *O'L*).

Phase current $= OA = 6 \text{ cm} = 6 \times 10 = 60 \text{ A}$

Line current $= \sqrt{3} \times 60 =$ **103.9 A. (Ans.)**

Power factor $= \cos \angle AOE = \cos 35° =$ **0.8191. (Ans.)**

(*ii*) **Maximum torque :**

To obtain the maximum torque line *CK* is drawn ⊥ to the torque line *O'M*. Point *M* is such that

$$\frac{LM}{NM} = \frac{\text{rotor Cu loss}}{\text{stator Cu loss}}$$

Since the two Cu losses are equal, point *M* is the mid-point of *LN*.

Line *KK'* represents the maximum torque of the motor in synchronous watts.

Hence, **maximum torque** $= KK' = 7.3 \text{ cm} = 7.3 \times 7124$

= **52005 synchronous watts. (Ans.)**

☞**Example 3.62.** *A 400 V, 11 kW, 3-phase, 50-Hz, 4-pole delta-connected induction motor gave the following test data :*

No-load test : 400 V, 8 A, 1000 W

Locked rotor test : 100 V, 25 A, 1750 W

Construct the circle diagram and determine :

(*i*) *Full-load current and power factor,*

(*ii*) *Maximum possible power output, and*

(*iii*) *The best possible operating factor.*

Solution. No-load test :

Voltage applied $V = 400 \text{ V}$

No-load current, $I_0 = 8 \text{ A}$

No-load input power, $P_0 = 100 \text{ W}$

No-load power factor, $\cos \phi_0 = \frac{P_0}{\sqrt{3}\, V \times I_0} = \frac{1000}{\sqrt{3} \times 400 \times 8} = 0.1804$

∴ $\phi_0 = \cos^{-1} (0.1804) = 79.6°$

Locked rotor test :

$$V_s = 100 \text{ V}, I_s = 25 \text{ A}, P_s = 1750 \text{ W}$$

Short-circuit current with normal voltage,

$$I_{sc} = I_s \times \frac{V}{V_s} = 25 \times \frac{400}{100} = 100 \text{ A}$$

Short-circuit input power with normal voltage,

$$P_{sc} = P_s \times \left(\frac{I_{sc}}{I_s}\right)^2 = 1750 \times \left(\frac{100}{25}\right)^2 = 28000 \text{ W}$$

Short-circuit p.f., $\cos \phi_s = \frac{1750}{\sqrt{3} \times 100 \times 25} = 0.404$

∴ $\phi_s = \cos^{-1} (0.404) = 66.2°$

The circle diagram is constructed as described in Example 3.58.
Assume current scale : 1 cm = 10 A

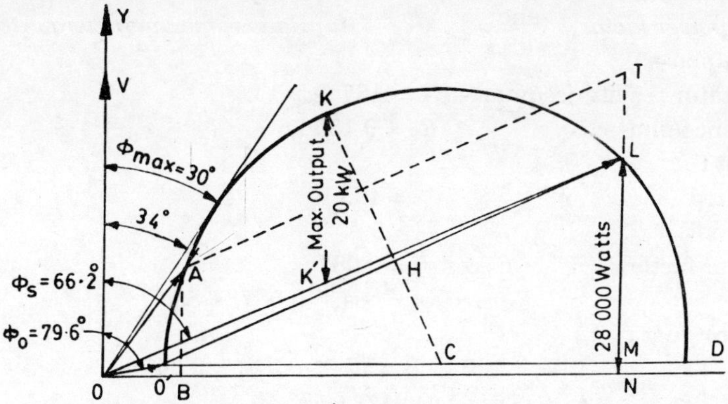

Fig. 3.44

- To determine power scale a perpendicular *LN* through point *L* is drawn.

LN represents total input on short-circuit with normal voltage applied *i.e.*, 28000 W.

But *LN* = 4 cm

$\therefore$ Power scale : 1 cm = $\frac{28000}{4}$ = 7000 W.

- Full-load motor output = 11 kW or 11000 W. According to above power scale, motor output will be represented by 11000/7000 = 1.57 cm on the circle diagram.
- Line *NL* is extended to point *T* so that *LT* = 1.57 cm. From point *T* line *TA* is drawn || to output line *O'L* intersecting the circle at point *A*.
- Point *A* is joined with origin *O* and perpendicular *AB* is drawn.

From circle diagram :

(*i*) **Full-load current** = *OA* = 2 cm = 2 × 10 = **20 A. (Ans.)**

Full-load power factor, $\cos \phi_f = \frac{BA}{OA} = \frac{1.62}{2}$ = **0.81. (Ans.)**

(*ii*) **Maximum possible power output :**

To find maximum possible power output *CH* is produced intersecting semicircle at *K* and then from point *K* a perpendicular is dropped on horizontal line *ON* meeting output line at *K'*. Now *KK'* represents maximum power output.

$\therefore$ Maximum power output = *KK'* × power scale
= 2.86 × 7000 = **20000 W** or **20 kW. (Ans.)**

(*iii*) The best *possible operating power factor* is obtained by drawing a line from point *O* and *tangential to the semicircle.*

By measurement, ϕ_{max} = 30°

$\therefore$ **Best possible operating power factor**
= cos 30° = **0.866. (Ans.)**

☞**Example 3.63.** *A 3-phase, 400 V, 5.6 kW, 4-pole, 50-Hz has a wound rotor, the ratio of primary to secondary turns on rotor is 2.62/1. When running light on 400 V, it took a current of 6 amperes at 0.087 power factor, and on standstill with the rotor locked, and an applied voltage of 100 V, the current* *taken was 12 A at a power factor of 0.347. The stator resistance/phase is 0.67 Ω and of rotor is 0.185 Ω. Determine :*

(i) Full-load current, *(ii) Full-load slip,*
(iii) Full-load power factor, *(iv) Ratio of maximum torque to full-load torque, and*
(v) Maximum power.

Solution. Stator resistance/phase, $R_1 = 0.67\ \Omega$

Rotor resistance/phase, $R_2 = 0.185\ \Omega$

No-load test :

Voltage applied $= 400$ V

No-load current, $I_0 = 6$ A

No-load power factor, $\cos \phi_0 = 0.087$

$\therefore$ $\phi_0 = \cos^{-1} (0.087) = 85°$

Locked rotor test :

$$V_s = 100 \text{ V}, I_s = 12 \text{ A}, \cos \phi_s = 0.347$$

$$\therefore \quad \phi_s = \cos^{-1} (0.347) = 69° \ 40'$$

Short-circuit current if normal voltage was applied,

$$I_{sc} = I_s \times \frac{V}{V_s} = 12 \times \frac{400}{100} = 48 \text{ A}$$

Short-circuit power input with this current,

$$P_{sc} = \sqrt{3} \times 400 \times 48 \times 0.347 = 11540 \text{ W}$$

Ratio of turns on stator to rotor,

$$K = 2.62$$

Assume a current scale of 1 cm = 2 A. Refer Fig. 3.45.

- No-load current vector OO' represents 6 A. Hence it measures 6/2 = 3 cm and is drawn at angle of 85° with OY.
- Vector OL represents 48 A and measures 48/2 = 24 cm. It is drawn at angle of 69° 40′.
- $O'G$ is drawn parallel to OX.
- Perpendicular bisector of $O'L$ is drawn to locate the centre J of the circle.
- With J as centre and radius = JO', a circle is drawn which passes through points O' and L.
- From point L, a perpendicular LS is drawn to the base, LS = 8.24 cm and represents 11540 W.

$$\therefore \text{ Power scale, 1 cm} = \frac{11540}{8.24} = 1400 \text{ W or } 1.4 \text{ kW}$$

- To draw the *torque line*, we proceed as follows :

  $K = 2.62$, $R_1 = 0.67\ \Omega$, $R_2 = 0.185\ \Omega$ (In practice an allowance of 10% is made for skin effect).

$$\therefore \quad \frac{\text{Rotor Cu loss}}{\text{Stator Cu loss}} = (2.62)^2 \times \frac{0.185}{0.67} = 1.895 \text{ say } 1.9$$

$$\therefore \quad \frac{\text{Rotor Cu loss}}{\text{Total Cu loss}} = \frac{1.9}{1.9 + 1} = \frac{1.9}{2.9} = 0.655$$

LN represents total copper loss and is divided at point M in the ratio of 1.9 : 1

$$\frac{LM}{LN} = \frac{1.9}{2.9}$$

or

$$LM = LN \times \frac{1.9}{2.9} = 8 \times 0.655 = 5.24 \text{ cm}$$

$O'M$ is the *torque line.*

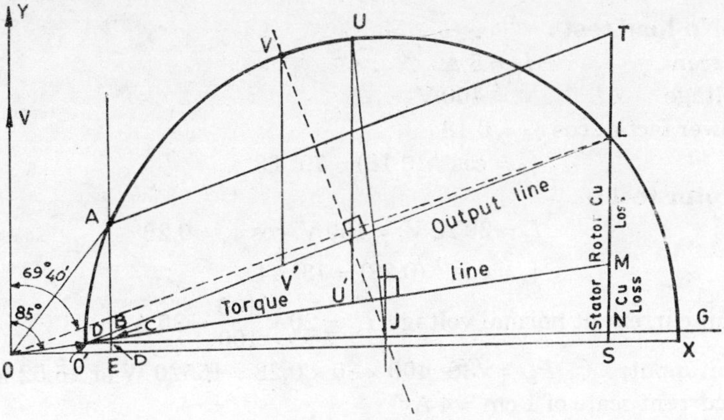

Fig. 3.45

- Full-load output = 5.6 kW or 5600 W. It is represented by a line = $\frac{5600}{1400}$ = 4 cm. *SL* is extended and *LT* = 4 cm. Line *TA* is drawn *parallel* to output line *O′L* and cuts the circle at point *A*.
  OA represents the full-load current.
- *AE* is drawn vertically.
- From centre *J*, a line is drawn ⊥ to torque line *O′M*. It cuts the circle at *U*. *UU′* is drawn vertical and its represents *maximum torque*.
- From centre *J*, a line is drawn ⊥ to outline *O′L*. It cuts the circle at *V*. *VV′* is drawn vertical and it represents *maximum output*.

From circle diagram :

(*i*) **Full-load current** = *OA* = 5.75 cm = 5.75 × 2 = **11.5 A. (Ans.)**

(*ii*) **Full-load slip** $= \frac{BC}{AC} = \frac{0.2}{4.25} = 0.047 =$ **4.7%. (Ans.)**

(*iii*) **Full-load power factor** $= \frac{EA}{OA} = \frac{4.6}{5.75} =$ **0.8. (Ans.)**

(*iv*) **Ratio of maximum torque to full-load torque,**

$$\frac{T_{max}}{T_f} = \frac{UU'}{AC} = \frac{10.06}{4.25} = \mathbf{2.367.}\ \textbf{(Ans.)}$$

(*v*) **Maximum power output** = *VV′* = 7.74 cm = 7.74 × 1.4 = **10.836 kW. (Ans.)**

Example 3.64. *Following are the test results of a 3-phase, 4- pole, 50-Hz slip ring induction motor.*

No-load test : *400 V, 8 A, p.f. = 0.18*
Blocked rotor test : *100 V, 20 A, p.f. = 0.28.*

The full-load current is 18 A. The ratio of stator to rotor turns per phase is 3 : 1. The stator and rotor copper losses are divided equally in the blocked rotor test.

Draw the circle diagram and determine the following :

(*i*) *Power factor, torque, output, efficiency and slip at full-load,*
(*ii*) *Standstill torque or starting torque, and*
(*iii*) *Resistance to be inserted in the rotor circuit for giving a starting torque 200% of the full-load torque. Also find the current and power factor under these conditions.*

Solution. No-load test :

No-load current, $I_0 = 8$ A

Applied voltage $= 400$ V

No-load power factor, $\cos \phi_0 = 0.18$

$$\phi_0 = \cos^{-1}(0.18) = 79° \ 38'$$

Blocked rotor test :

$$I_s = 20 \text{ A}, V_s = 100 \text{ V}, \cos \phi_s = 0.28$$

$$\therefore \quad \phi_s = \cos^{-1}(0.28) = 73° \ 44'$$

Short-circuit current at normal voltage, $I_{sc} = 20 \times \dfrac{400}{100} = 80$ A

Short-circuit input, $P_{sc} = \sqrt{3} \times 400 \times 80 \times 0.28 = 15520$ W or 15.52 kW

Assume a current scale of 1 cm ≑ 4 A

- In the circle diagram of Fig. 3.46, $OO' = 8/4 = 2$ cm drawn at an angle of 79° 38′ behind OV. Similarly, $OL = 80/4 = 20$ cm and is drawn an angle of 73° 44′ behind OV. The semicircle is drawn as usual.

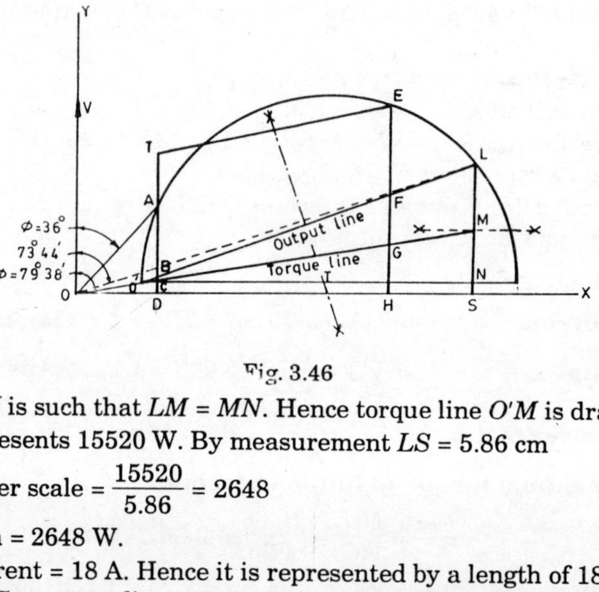

Fig. 3.46

Point M is such that $LM = MN$. Hence torque line $O'M$ is drawn.

LS represents 15520 W. By measurement $LS = 5.86$ cm

$$\therefore \quad \text{Power scale} = \frac{15520}{5.86} = 2648$$

$\therefore$ 1 cm = 2648 W.

- Full-load current = 18 A. Hence it is represented by a length of 18/4 = 4.5 cm. With O' as centre and 4.5 cm as radius, an arc is drawn which cuts the semicircle at point A. *This point represents full-load condition.* AD is drawn perpendicular to the base OS.

From circle diagram :

(*i*) (*a*) **Power factor, cos φ** $= \dfrac{AD}{OA} = \dfrac{3.7}{4.57} = $ **0.813. (Ans.)**

(*b*) **Torque** can be found by measuring the input.

Rotor input $= AC = 3.17$ cm $= 3.17 \times 2648 = 8394$ W

Now $$N_s = \frac{120f}{p} = \frac{120 \times 50}{4} = 1500 \text{ r.p.m.}$$

$$\therefore \quad T_s = \frac{\text{rotor input}}{2\pi\left(\dfrac{N_s}{60}\right)} = \frac{8394 \times 60}{2\pi \times 1500} = \mathbf{53.44 \text{ N-m.}} \ \textbf{(Ans.)}$$

(*c*) **Output** $= AB = 3.08 \text{ cm} = 3.08 \times 2648 =$ **8155.8 W. (Ans.)**

(*d*) **Efficiency** $= \dfrac{\text{output}}{\text{input}} = \dfrac{AB}{AD} = \dfrac{3.08}{3.7} = 0.832$ or **83.2%. (Ans.)**

(*e*) **Slip,** $s = \dfrac{\text{rotor Cu loss}}{\text{rotor input}} = \dfrac{BC}{AC} = \dfrac{0.09}{3.17} = 0.0284$ or **2.84% (Ans.)**

(*ii*) **Standstill torque** is represented by *LM*,

$$LM = 2.7 \text{ cm} = 2.7 \times 2648 = 7149.6 \text{ W}$$

∴ **Standstill torque,** $T_s = \dfrac{7149.6}{2\pi \times \left(\dfrac{1500}{60}\right)} =$ **45.5 N-m. (Ans.)**

(*iii*) We will now locate point *E* on the semicircle which corresponds to a starting torque twice the full-load torque.

Full-load torque = *AC*. Produce *CA* to point *T* such that *AT* = *AC*. From *T* draw a line parallel to torque line *O'M* cutting the semicircle at *E*. Draw *EG* perpendicular to the base.

At starting when rotor is stationary, *EG* represents total copper losses

FG = copper loss in rotor itself as before

EF = copper loss in external resistance

and $EF = 4.26 \text{ cm} = 4.26 \times 2648 = 11280 \text{ W}$

Copper loss/phase $= \dfrac{11280}{3} = 3760 \text{ W}$

Rotor current, $O'E = 15.9 \text{ cm} = 15.9 \times 4 = 63.6 \text{ A}$

Let *r* be the external resistance in the rotor circuit (as referred to stator), then

$$(63.6)^2 \times r = 3760$$

$$\therefore \quad r = \frac{3760}{(63.6)^2} = 0.929 \ \Omega$$

Now $K = \frac{1}{3}$

∴ **Rotor resistance/phase** (referred to rotor itself = $r \times K^2$)

$$= \frac{0.929}{3^2} = \mathbf{0.1032 \ \Omega.} \ \textbf{(Ans.)}$$

Stator current $= OE = 17.8 \text{ cm} = 17.8 \times 4 =$ **71.2 A. (Ans.)**

Power factor $= \dfrac{EH}{OE} = \dfrac{8.9}{17.8} =$ **0.5 (Ans.)**

Example 3.65. *Make use of the circle diagram or otherwise, prove that the maximum output of an induction motor per phase* $= \dfrac{V_1 (I_{sc} - I_0)}{2(1 + \cos \phi_{sc})}$ *where* V_1 *= source voltage per phase,* I_{sc} *= source current during blocked rotor test,* I_0 *= no-load current and* ϕ_{sc} *= angle of lag during blocked rotor test between source voltage and current.* **(AMIE Advance Elec. M/CS Summer, 1998)**

Solution. Maximum power output of an induction motor per phase is given as :

$$P_{g(max)} = \frac{V_1^2}{2(R_{01} + Z_{01})} \text{ watts} \qquad \text{[Eqn. (3.28)]}$$

where, V_1 = source voltage per phase,

R_{01} = motor equivalent resistance, ohms, and

Z_{01} = motor equivalent impedance, ohms.

(Referred to stator)

Neglecting the phase angle difference caused by the exciting current during blocked rotor test,

$$\cos \phi_{sc} = \frac{R_{01}}{Z_{01}}$$

and

$$I_{sc} - I_0 = \frac{V_1}{Z_{01}}$$

$$\therefore \quad P_{g(max)} = \frac{V_1 \times \dfrac{V_1}{Z_{01}}}{2\left[\dfrac{R_{01}}{Z_{01}} + 1\right]} = \frac{V_1 (I_{sc} - I_0)}{2(\cos \phi_{sc} + 1)}. \quad \textbf{(Proved.)}$$

3.23. STARTING OF INDUCTION MOTORS

Small induction motors (up to 2 kW) capacity may directly be switched on to the supply mains, but those of higher capacity must use some type of starting device, or starters as they are commonly called. *The function of these starters is to restrict the initial rush of current,* which, in the case of induction motors, is about 5 times the full-load current. This excessive current has two major upsetting effects, namely, a *large voltage drop in the distributing network* and *causing stoppage of machines which are already running on the supply mains.* Hence the Electrical Undertaking Authorities *forbid the users of large capacity induction motors to directly switch on their machines.*

The principle of all starting devices is to impress lower voltage on stator phases at the time of starting, or if the motor is slip ring or wound rotor, then to include external resistance in each rotor phase to keep the initial rotor current to a low value, this consequently means less current in the stator phases and therefore in the supply mains.

Direct-on-line starting of induction motors. This method means switching the motor directly on to the supply without using any device for reducing the starting current. The method is restricted to small motors up to about 2 kW. For these small motors, the *starting torque is about twice the full-load torque.* Hence the starting period lasts only a few seconds.

We know that,

Rotor input $= 2\pi N_s T = kT$

Also, rotor copper loss $= s \times$ rotor input

$$\therefore \quad 3I_2^2 R_2 = skT$$

$$\therefore \quad T \propto \frac{I_2^2}{s} \qquad \text{[if } R_2 \text{ is the same]}$$

Now $\quad I_2 \propto I_1$

$$\therefore \quad T \propto \frac{I_1^2}{s} \quad \text{or} \quad T = \frac{KI_1^2}{s}$$

At starting $\quad s = 1$

$$\therefore \quad T_{st} = KI_{st}^2 \quad \text{where } I_{st} = \text{starting current}$$

If $\quad I_f$ = normal full-load current

s_f = full-load slip

then

$$T_f = \frac{KI_f^2}{s_f}$$

$$\therefore \quad \frac{T_{st}}{T_f} = \left(\frac{I_{st}}{I_f}\right)^2 s_f \qquad \text{...[3.30 (a)]}$$

When the motor is direct-switched on to normal voltage, then starting current is the short-circuit current I_{sc}

$$\therefore \qquad \frac{T_{st}}{T_f} = \left(\frac{I_{sc}}{I_f}\right)^2 s_f \qquad \text{...[3.30 (b)]}$$

Some of the starting devices for starting induction motors are discussed below :

Squirrel-Cage Motors :

(*i*) Stator rheostat starter.
(*ii*) Auto-transformers (auto-starters).
(*iii*) Star-delta starter.

Slip Ring Motors :

(1) Rotor rheostat.

3.23.1. Squirrel-Cage Motors.

3.23.1.1. Stator Rheostat Starter. The connection diagram is shown in Fig. 3.47.

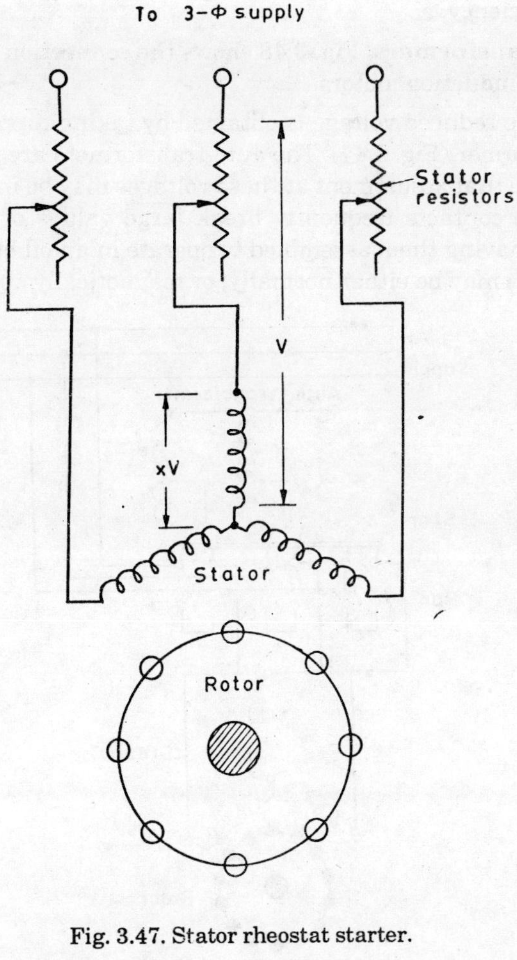

Fig. 3.47. Stator rheostat starter.

Reduced voltage is impressed on each stator phase due to the resistance R of the rheostat. Hence the initial current drawn from the supply mains will be *less* than if the machine were to be switched directly on to the supply mains.

If x = fraction of voltage (V) reduced by the stator resistors

Then $I_{st} = xI_{sc}$ and $T_{st} = x^2 T_{sc}$

$$\frac{T_{st}}{T_f} = \left(\frac{I_{st}}{I_f}\right)^2 s_f = \left(\frac{xI_{sc}}{I_f}\right)^2 s_f = x^2 \left(\frac{I_{sc}}{I_f}\right)^2 s_f \quad ...(3.31)$$

- This method is suitable for starting of *small machines only*.

Advantages : 1. High power factor during start.

2. Smooth acceleration.

3. Less expensive than auto-transformer starter in lower output ratings.

4. Closed transition starting.

Disadvantages : 1. Heat is given off by the resistors.

2. Expensive resistors are required because starting duration usually exceeds 5 seconds.

3. Low torque efficiency.

3.23.1.2. Auto-transformers. Fig. 3.48 shows the connection diagram for auto-transformer starting of squirrel-cage induction motors.

In this method the reduced voltage is obtained by taking tappings at suitable points from a three-phase auto-transformer (Fig. 3.47). The auto-transformers are generally tapped at the 50, 60 and 80 per cent points, so that adjustment at these voltages may be made for proper starting torque requirements. Since the contacts frequently break large values of current arcing is sometimes quenched effectively by having them assembled to operate in an oil bath.

Auto-transformers may be either normally, or magnetically operated.

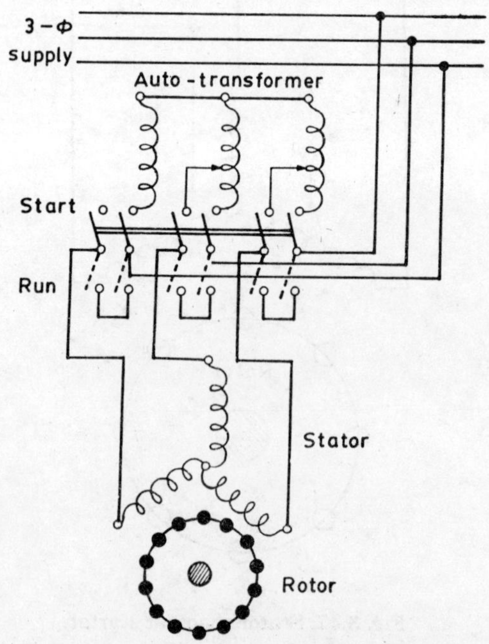

Fig. 3.48. Auto-transformer starter.

Relation between starting (T_{st}) and full-load (T_f) torques. Let the motor be started by an auto-transformer having transformation ratio K. If I_{sc} is the starting current when normal voltage is applied and applied voltage to stator winding at starting is KV then,

Motor input current, $I_{st} = KI_{sc}$

Supply current = Primary current of auto-transformer

$= K^2 I_{sc}$

$$\frac{T_{st}}{T_f} = \left(\frac{I_{st}}{I_f}\right)^2 \times s_f = K^2 \left(\frac{I_{sc}}{I_f}\right)^2 \times s_f \quad ...(3.32)$$

In may be noted that this expression is similar to eqn. (3.31) except that x has been replaced by the transformation ratio K.

Advantages :

1. Voltage is reduced by transformation and not by dropping the voltage in resistors, and therefore, the current and power drawn from the supply mains are also reduced in comparison to resistor starting.
2. Availability of highest torque per ampere of supply current.
3. Adjustment of starting voltage by selection of proper tap on the auto-transformer.
4. The method is suitable for long starting periods.
5. Motor current larger than supply current.
6. Closed transition starting.

Disadvantages :

1. Low power factor.
2. Higher cost in case of lower output rating motors.

- This method can be used for starting of star-connected as well as delta-connected motors.
- This method is often employed for starting of large cage motors (rating exceeding 20 kW).

3.23.1.3. Star-Delta Starter. Star-delta switching method is based upon the principle that with three windings connected in star, voltage across each winding is $\frac{1}{\sqrt{3}}$ *i.e.*, 57.7% of the line to line voltage whereas the same windings connected in delta will have full-line-to-line voltage across each.

The star-delta starter *connects the three stator windings in star* across the rated supply voltage *at the starting instant. After the motor attains speed the same windings through a change over switch are re-connected in delta across the same supply voltage.*

The basic connection diagram of a star-delta starter is shown in Fig. 3.49. An actual starter incorporates *under-voltage* and *over-load coils.* The starter is also provided with a mechanical inter-locking device to prevent the handle from being put in the 'Run' position first. Such starters are employed for starting 3-phase squirrel-cage induction motors of rating between 4 kW and 15 kW.

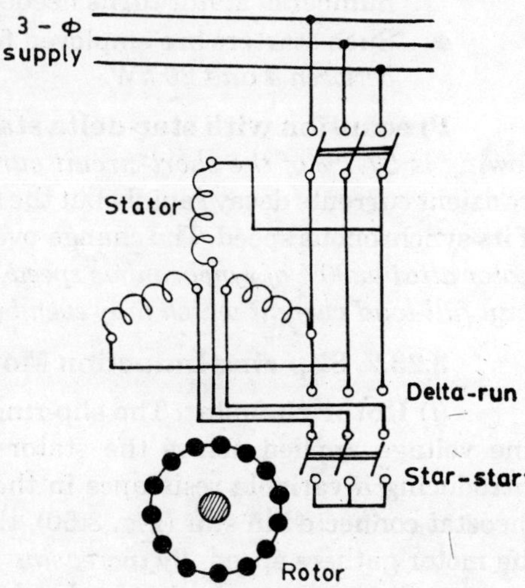

Fig. 3.49. Star-delta starter.

When *star-connected,* the applied voltage over each motor phase is reduced by a factor of $\frac{1}{\sqrt{3}}$ and hence the torque becomes $\frac{1}{3}$ of that which would have been developed if motor were directly connected in delta. The line current is reduced to $\frac{1}{3}$. Hence, during starting period when motor is star-connected, it takes $\frac{1}{3}$rd as much starting current and develops $\frac{1}{3}$rd as much torque as would have been developed were it directly connected in delta.

Relation between T_{st} and T_f :

$$I_{st} \text{ per phase} = \frac{1}{\sqrt{3}} I_{sc} \text{ per phase}$$

where I_{sc} is the current per phase which delta-connected motor would have taken if switched on to the supply directly ; however, line current at start is equal to $\frac{1}{3}$ of line I_{sc}.

Now, starting torque, $T_{st} \propto I_{st}^2 \quad (s = 1)$

Full-load torque, $T_f \propto I_f^2/s_f$

$$\therefore \quad \frac{T_{st}}{T_f} = \left(\frac{I_{st}}{I_f}\right)^2 s_f = \left(\frac{I_{sc}}{\sqrt{3} I_f}\right)^2 s_f = \frac{1}{3}\left(\frac{I_{sc}}{I_f}\right)^2 s_f$$

i.e.,

$$\frac{T_{st}}{T_f} = \frac{1}{3}\left(\frac{I_{sc}}{I_f}\right)^2 s_f \quad \text{...(3.33)}$$

Here, I_{st} and I_{sc} represent phase values.

- This method reduces the starting line current to one-third but the starting torque is also reduced by the same amount.
- *This method is cheap but limited to applications where high starting torque is not necessary e.g., machine tools, pumps, motor-generator sets etc.*
- The method is unsuitable for motors for voltage exceeding 3000 V because of the excessive number of stator turns needed for delta connection.
- Such starters are employed for starting *3-phase squirrel cage induction motors of rating between 4 and 20 kW.*

Precaution with star-delta starting. When the motor is started in star the initial current flowing is *57.7% of the short-circuit current in delta together with a transient in each phase*. The transient currents decay rapidly but the steady state is not reached until the motor has attained 70% of its synchronous speed. The change-over from star to delta connection *should not be made until the motor attains 90% of synchronous speed, otherwise there will be a current surge considerably greater than full-load current which may even be greater than the standstill current with star-connection.*

3.23.2. Slip-ring Induction Motors—Starting of

(*i*) **Rotor rheostat.** The slip-ring induction motors are practically always started with full line voltage applied across the stator terminals. The value of starting current is adjusted by introducing a variable resistance in the rotor circuit. The controlling resistance is in the form of rheostat connected in star (Fig. 3.50), the resistance being gradually cut out of the rotor circuit as the motor gathers speed. *By increasing the rotor resistance, not only is the rotor* (and hence stator) *current reduced at starting* but *at the same time torque is also increased due to improvement in power factor.*

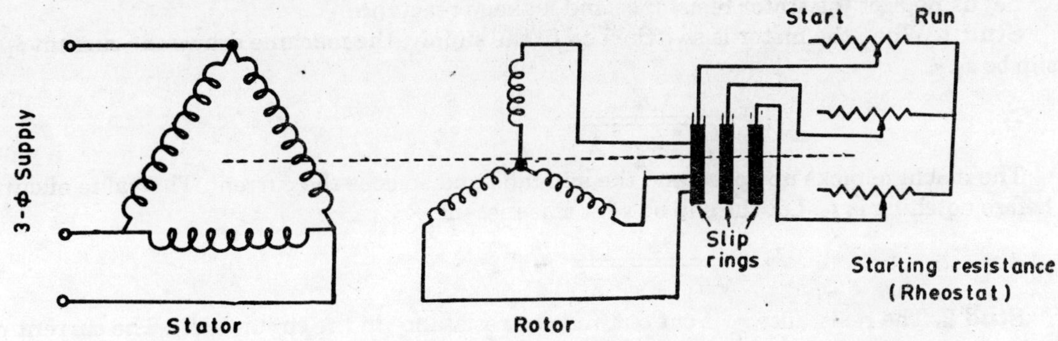

Fig. 3.50. Starting of slip-ring induction motor.

The rheostat is either of stud or contractor type and may be hand operated or automatic.

As discussed earlier, the introduction of additional external resistance in the rotor circuit enables slip-ring motor to develop a high starting torque with reasonably moderate starting current. Hence *such motors can be started under load.* When the motor runs under normal conditions the rings are short-circuited and brushes lifted from them.

3.24. RESISTANCE STEPS FOR 3-PHASE SLIP-RING INDUCTION MOTOR STARTER

Refer Fig. 3.51.

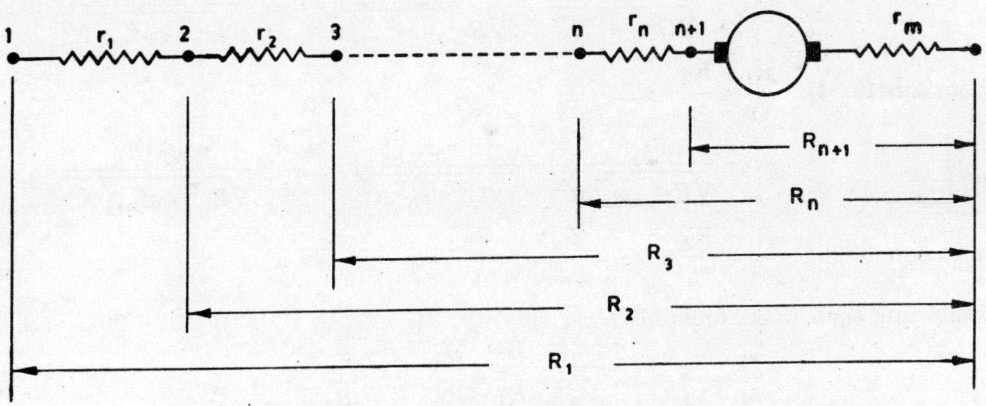

Fig. 3.51. Resistance steps for 3-phase slip-ring induction motor.

Let I_1 = upper current limit during starting,

I_2 = lower current limit during starting,

r_m = motor resistance,

V = motor line voltage,

E = rotor standstill e.m.f.,

s = slip,

X = rotor standstill e.m.f., and

n = number of sections/elements.

Let us neglect the stator resistance and leakage reactance.

Stud 1. When the motor is switched on to the supply, the machine draws the current I_1. Let the slip be s_1.

$$\therefore \qquad I_1 = \frac{s_1 E}{\sqrt{R_1^2 + s_1^2 X^2}}$$

The machine picks up speed and the slip falls and so does the current. The value of current just before notching is I_2. Let the slip be s_2 at this instant.

$$\therefore \qquad I_2 = \frac{s_2 E}{\sqrt{R_1^2 + s_2^2 X^2}}$$

Stud 2. The resistance r_1 is cut out and the resistance in the circuit is R_2. The current rises to its maximum value I_1. Let us consider that there is no change in speed during notching operation. Therefore the slip just after notching is s_2.

Hence,
$$I_1 = \frac{s_2 E}{\sqrt{R_2^2 + s_2^2 X^2}}$$

Just before notching at stud 2, the current falls to I_2 and slip s_3.

$$\therefore \qquad I_2 = \frac{s_3 E}{\sqrt{R_2^2 + s_3^2 X_2}}$$

Thus, from above, we have

$$I_1 = \frac{s_1 E}{\sqrt{R_1^2 + s_1^2 X^2}} = \frac{s_2 E}{\sqrt{R_2^2 + s_2^2 X^2}} = \ldots\ldots = \frac{s_{n+1} E}{\sqrt{r_m^2 + s_{n+1}^2 X^2}} \qquad \ldots(3.34)$$

From eqn. (3.34),
$$\frac{R_1}{s_1} = \frac{R_2}{s_2} = \ldots\ldots \frac{r_m}{s_{n+1}} \qquad \ldots(3.35)$$

and
$$I_2 = \frac{s_2 E}{\sqrt{R_1^2 + s_2^2 X^2}} = \frac{s_3 E}{\sqrt{R_2^2 + s_3^2 X^2}} = \ldots\ldots = \frac{s_{n+1} E}{\sqrt{R_n^2 + s_{n+1}^2 X^2}} \qquad \ldots(3.36)$$

From eqn. (3.36),
$$\frac{R_1}{s_2} = \frac{R_2}{s_3} = \ldots\ldots = \frac{R_n}{s_{n+1}} \qquad \ldots(3.37)$$

Combining eqns. (3.35) and (3.37)

$$\frac{s_2}{s_1} = \frac{s_3}{s_2} = \ldots\ldots \frac{s_{n+1}}{s_n} = \frac{R_2}{R_1} = \frac{R_3}{R_2} = \ldots\ldots \frac{r_m}{R_n} \qquad \ldots(3.38)$$

Now, slip at starting, $s_1 = 1$

and slip at $(n + 1)$th stud, s_{n+1} = slip at full-load = s_m

(This is true provided the upper limit of current is the full-load current).

$$\therefore \qquad \frac{s_2}{s_1} = \frac{s_3}{s_2} = \ldots\ldots = \frac{s_m}{s_n} = \frac{R_2}{R_1} = \frac{R_3}{R_2} = \ldots\ldots = \frac{r_m}{R_n} = \alpha$$

or
$$s_m = (\alpha)^n$$

$$\therefore \qquad \alpha = (s_m)^{1/n} \qquad \ldots(3.39)$$

Also
$$\frac{r_m}{R_1} = (\alpha)^n$$

or
$$r_m = R_1(\alpha)^n \qquad \ldots(3.40)$$

Further, $R_2 = \alpha R_1$

$R_3 = \alpha R_2 = \alpha^2 R_1$

$R_4 = \alpha R_3 = \alpha^3 R_1$

....

$R_n = \alpha R_{n-1} = \alpha^{n-1} R_1$

The sections are :

$r_1 = R_1 - R_2 = R_1(1 - \alpha)$

$r_2 = R_2 - R_3 = R_1(\alpha - \alpha^2) = \alpha r_1$

$r_3 = R_3 - R_4 = R_1(\alpha^2 - \alpha^3) = \alpha^2 r_1$

....

$r_n = R_n - R_{n+1} = R_1(\alpha^{n-1} - \alpha^n) = \alpha^{n-1} r_1$

From eqns. (3.34) and (3.36)

$$\left(\frac{I_2}{I_1}\right)^2 = \frac{s_2^2(R_1^2 + s_1^2X^2)}{s_1^2(R_1^2 + s_2^2X^2)} = \frac{s_2^2(R_1^2 + X^2)}{(R_1^2 + s_2^2X^2)}$$

$$= \frac{R_1^2 + X^2}{\dfrac{R_1^2}{s_2^2} + X^2} \cong \alpha^2 \text{ for normal motors} \quad ...(3.41)$$

The ratio of the rotor current limits, is therefore, approximately equal to α and this is also the approximate ratio of stator currents.

STARTING OF SQUIRREL-CAGE INDUCTION MOTORS

Example 3.66. *The short-circuit current of a small 3-phase induction motor is 3 times the full-load currents. Determine the starting torque as a percentage of full-load torque if full-load slip is 3 per cent.*

Solution. Short-circuit current, $I_{st} = 3I_f$

Full-load slip, $s_f = 3\%$ or 0.03

Starting torque, $T_{st} = T_f \left(\dfrac{I_{sc}}{I_f}\right)^2 s_f = T_f(3)^2 \times 0.03 = 0.27T_f.$

Hence, **starting torque is 27% of full-load torque. (Ans.)**

Example 3.67. *A 3-phase, squirrel-cage induction motor has a short-circuit current equal to 4 times the full-load current. Find the starting torque as a percentage of full-load torque if the motor is started by :*

(i) direct switching to the supply ; *(ii) a star delta starter ;*

(iii) an auto-transformer ; and *(iv) a resistance in the stator circuit.*

The stator current in (iii) and (iv) is limited to 2 times the full-load current and the full-load slip is 3%.

Solution. Short-circuit current, $I_{sc} = 4I_f$

Full-load slip, $s_f = 3\%$ or 0.03

(*i*) **Starting torque with direct switch,**

$$T_{st} = \left(\frac{I_{sc}}{I_f}\right)^2 s_f T_f = (4)^2 \times 0.03 \times T_f$$

$= 0.48T_f$ *i.e.* **48% of full-load torque. (Ans.)**

(*ii*) **Starting torque with star-delta starter,**

$$T_{st} = \frac{1}{3}\left(\frac{I_{sc}}{I_f}\right)^2 s_f T_f = \frac{1}{3} \times (4)^2 \times 0.03 \times T_f$$

$$= 0.16T_f \text{ i.e. } \mathbf{16\%\ of\ full\text{-}load\ torque.}\quad \textbf{(Ans.)}$$

(*iii*) **Supply line current**

$$= K^2 I_{sc} = 2I_f$$

or the transformation ratio,

$$K = \sqrt{\frac{2I_f}{I_{sc}}} = \sqrt{\frac{2I_f}{4I_f}} = \frac{1}{\sqrt{2}}$$

Starting torque with auto-transformer starter

$$= K^2\left(\frac{I_{sc}}{I_f}\right)^2 s_f T_f = \left(\frac{1}{\sqrt{2}}\right)^2 \times (4)^2 \times 0.03T_f$$

$$= 0.24T_f \text{ or } \mathbf{24\%\ of\ full\text{-}load\ torque.}\quad \textbf{(Ans.)}$$

(*iv*) **Starting torque with resistance in the stator circuit**

$$= \left(\frac{I_{sc}}{I_f}\right)^2 s_f T_f = (2)^2 \times 0.03 \times T_f$$

$$= 0.12T_f \text{ i.e. } \mathbf{12\%\ of\ full\text{-}load\ torque.}\quad \textbf{(Ans.)}$$

Example 3.68. *A 400 V, 50-Hz, induction motor, when started directly from the mains takes 4 times the full-load current and the torque produced is twice the full-load torque. Determine :*

(i) The motor current, the line current and the starting torque when started by means of an auto- transformer of ratio 2.5 : 1.

(ii) The voltage to be applied and the motor current if the full-load torque is to be obtained at starting.

Solution. Supply voltage $= 400$ V

Frequency $= 50$ Hz

Transformation ratio, $K = \frac{1}{2.5} = 0.4$

Let us assume that the winding is star-connected

Phase voltage, $E_{ph} = \frac{400}{\sqrt{3}} = 231$ V

Voltage across the motor terminal

$$= KE_L = 0.4 \times 400 = 160 \text{ V}$$

The phase voltage when started by an auto-transformer

$$= \frac{160}{\sqrt{3}} = 92.4 \text{ V}$$

As the current is proportional to the voltage per phase, therefore :

(*i*) **New starting phase current**

$$= \frac{92.4}{231} \times 4I_f = \mathbf{1.6\ I_f}.\quad \textbf{(Ans.)}$$

Line current :

In case of a transformer, $\frac{E_2}{E_1} = \frac{N_2}{N_1} = \frac{I_1}{I_2}$

But $\frac{E_2}{E_1} = \frac{1}{2.5} = 0.4\ (= K)$ and $I_2 = 1.6I_f$

$\therefore$ $\frac{I_1}{I_2} = K$ or $I_1 = KI_2 = K \times 1.6I_f = 0.4 \times 1.6I_f = 0.64\ I_f$

Hence, **line current = 0.64I$_f$. (Ans.)**

Let T_f = full-load torque

$$T \propto E^2$$

$$2T_f \propto (231)^2 \quad ...(i) \quad [\because\ T = 2T_f\ (given)]$$

When the applied voltage is 92.4 V let the torque = T

$$T \propto (92.4)^2 \quad ...(ii)$$

$$\frac{T}{2T_f} = \left(\frac{92.4}{231}\right)^2 = (0.4)^2$$

$\therefore$ **T = 0.32T$_f$. (Ans.)**

(*ii*) Let the voltage per phase be V to get full-load starting torque,

or $$T_f \propto V^2 \quad ...(iii)$$

So from eqns. (*i*) and (*iii*), we get

$$\frac{T_f}{2T_f} = \frac{V^2}{(231)^2}$$

$\therefore$ $$V^2 = \frac{(231)^2}{2}$$

or $$\mathbf{V} = \frac{231}{\sqrt{2}} = \mathbf{163.3\ V.\ (Ans.)}$$

$\therefore$ **Motor Voltage** $= 163.3 \times \sqrt{3} =$ **282.8 V. (Ans.)**

Motor current $= \frac{163.3}{231} \times 4I_f =$ **2.827 I$_f$. (Ans.)**

☞Example 3.69. *A squirrel-cage type induction motor when started by means of a star-delta starter takes 200 per cent of full-load line current and develops 45 per cent of full-load torque at starting. Calculate the starting current and current in terms of full-load values, if an auto-transformer with 80 per cent tapping were employed.* **(AMIE Summer, 2000)**

Solution. Short-circuit current, $I_{sc} = 2I_f$

Starting torque, $T_{st} = 0.45\ T_f$

Using the relation, $$\frac{T_{st}}{T_f} = \left(\frac{I_{sc}}{I_f}\right)^2 \times s_f \quad ...(i)$$

When the motor is supplying full normal load, it will be having delta connection and its line current is I_f *i.e.*, the full-load current.

Hence full-load phase current $= \frac{I_f}{\sqrt{3}}$

When started with a star/delta starter

$$T_{st} = 0.45T_f$$

and $I_{sc} = 2I_f$ (line current = phase current)

$$\frac{0.45T_f}{T_f} = \left(\frac{2I_f}{\frac{I_f}{\sqrt{3}}}\right)^2 \times s_f \quad \text{or} \quad 0.45 = (2\sqrt{3})^2 \times s_f$$

$$\therefore \quad s_f = \frac{0.45}{(2\sqrt{3})^2} = 0.0375 \text{ or } 3.75\%.$$

When started through an auto-transformer with 80% tapping and winding in delta connection, 80% of normal voltage is applied.

When the star-delta starter is used voltage applied is $\frac{1}{\sqrt{3}}$ times the normal voltage and the current is $2I_f$.

∴ Motor phase current with auto-transformer

$$= 2I_f \times \frac{0.8E}{\frac{E}{\sqrt{3}}} = 2.77I_f$$

∴ Motor line current $= \sqrt{3} \times 2.77I_f = 4.8I_f$

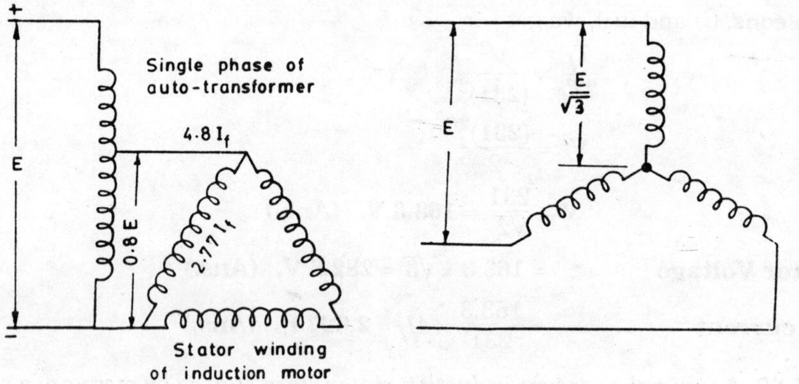

Fig. 3.52

Current (line value) **from mains,** I_1 **:**

$$\frac{E_{ph}}{0.8E_{ph}} = \frac{4.8I_f}{I_1}$$

$$\therefore \quad I_1 = 0.8 \times 4.8I_f = \mathbf{3.84I_f}. \quad \textbf{(Ans.)}$$

Starting torque with 80% tapping :

$$\frac{T_{st}}{T_f} = \left(\frac{2.77I_f}{\frac{I_f}{\sqrt{3}}}\right)^2 \times 0.0375$$

$$= 0.863 \text{ or } \mathbf{86.3\% \text{ full-load torque.}} \quad \textbf{(Ans.)}$$

Example 3.70. *A 15 kW, 400 V, 950 r.p.m., 3 phase, 50-Hz 6-pole cage motor with 400 V applied takes 6 times full-load current at standstill and develops 1.8 times the full-load running torque. The full-load current is 32 A.*

(i) What voltage must be applied to produce full-load torque at starting ?

(ii) What current will this voltage produce ?

(iii) If the voltage is obtained by an auto-transformer what will be the line current ?

(iv) If starting current is limited to full-load current by an auto-transformer, what will be the starting torque as a percentage of full-load torque ?

The magnetising current and stator impedance drops may be neglected.

(AMIE Summer, 1998)

Solution. Motor output = 15 kW

Supply voltage = 400 V

Motor speed, $N = 950$ r.p.m.

Current at standstill, $I_{sc} = 6$ times full-load current (I_f)

Full-load current, $I_f = 32$ A

Standstill torque = 1.8 times full-load running torque

(*i*) We know that, $T \propto V^2$

In the first case, $1.8T_f \propto 400^2$

$$T_f \propto V^2$$

$$\therefore \quad \left(\frac{V}{400}\right)^2 = \frac{T_f}{1.8T_f} \quad \text{or} \quad V = \mathbf{298\ V.\ (Ans.)}$$

(*ii*) Since currents are proportional to applied voltage

$$\therefore \quad 6I_f \propto 400$$

$$I \propto 298$$

$$\therefore \quad \frac{6I_f}{I} = \frac{400}{298}$$

$$\text{or} \quad I = 6I_f \times \frac{298}{400} = \frac{6 \times 32 \times 298}{400} = \mathbf{143\ A.\ (Ans.)}$$

(*ii*) Here $K = \frac{298}{400}$

$$\text{Line current} = K^2 I_{sc} = \left(\frac{298}{400}\right)^2 \times 6\, I_f$$

$$= \left(\frac{298}{400}\right)^2 \times 6 \times 32 = \mathbf{106.6\ A.\ (Ans.)}$$

(*iv*) We know that

Line current $= K^2 I_{sc}$

Now line current = full-load current, I_f

$$\therefore \quad 32 = K^2 \times 6 \times 32$$

$$\therefore \quad K^2 = \frac{1}{6}$$

Now using the relation, $\frac{T_{st}}{I_f} = K^2 \left(\frac{I_{sc}}{I_f}\right)^2 \times s_f$

$$\text{But} \quad s_f = \frac{N_s - N}{N_s} \times 100$$

$$\text{or} \quad s_f = \frac{1000 - 950}{1000} \qquad \left[\because\ N_s = \frac{120f}{p} = \frac{120 \times 50}{6} = 1000 \text{ r.p.m.}\right]$$

$$= 0.05 \text{ or } 5\%$$

$$\therefore \quad \frac{T_{st}}{T_f} = \frac{1}{6} \times \left(\frac{6I_f}{I_f}\right)^2 \times 0.05 = 0.3$$

$\therefore$ $T_{st} = 0.3T_f$ or **30% full-load torque. (Ans.)**

☞**Example 3.71.** *The ratio of maximum torque to full-load torque in a 3-phase squirrel-cage induction motor is 2.2 : 1. Determine the ratio of actual starting torque to full-load torque for the following cases :*

(i) Direct starting *(ii) Star-delta starting*

(iii) Auto-transformer starting with tapping of 70%.

The rotor resistance and standstill reactance per phase are 0.5 Ω and 5 Ω respectively.

Solution. $\frac{T_m}{T_f} = 2.2 : 1$

Rotor resistance/phase $R_2 = 0.5\ \Omega$

Rotor standstill reactance/phase,. $X_2 = 5\ \Omega$

(*i*) **Direct starting :**

For starting torque, $s = 1$

and $$T = \frac{KsE_2^2R_2}{R_2^2 + s^2X_2^2}$$

$$\therefore \quad T_{st} = \frac{KE_2^2R_2}{R_2^2 + X_2^2} \quad ...(i)$$

$$T_m = K\frac{\frac{R_2}{X_2}E_2^2R_2}{R_2^2 + \left(\frac{R_2}{X_2}\right)^2 X_2^2} = K\frac{E_2^2}{2X_2} \quad ...(ii)$$

But $$\frac{T_m}{T_f} = \frac{2.2}{1}$$

$$\therefore \quad T_f = \frac{T_m}{2.2} = \frac{1}{2.2}\frac{KE_2^2}{2X_2} = \frac{1}{2.2} \times \frac{KE_2^2}{2 \times 5} = \frac{KE_2^2}{22}$$

For direct switching,

$$\frac{\text{Starting torque } (T_{st})}{\text{Full-load torque } (T_f)} = \frac{KE_2^2R_2}{R_2^2 + X_2^2} \times \frac{22}{KE_2^2} = \frac{0.5 \times 22}{(0.5)^2 + (5)^2} = \mathbf{0.4356.} \ \textbf{(Ans.)}$$

(*ii*) **Star-delta starting :**

Starting voltage/phase $= \frac{E_2}{\sqrt{3}}$

$$\therefore \quad \frac{T_{st}}{T_f} = \frac{K\left(\frac{E_2}{\sqrt{3}}\right)^2 R_2}{R_2^2 + X_2^2} \times \frac{22}{KE_2^2} = \frac{1}{3} \times \frac{0.5 \times 22}{(0.5)^2 + (25)^2} = \mathbf{0.1452.} \ \textbf{(Ans.)}$$

(*iii*) **Auto-transformer starting :**

Star voltage/phase = 0.7 times the voltage on direct switching

or Rotor voltage at starting $= 0.7\,E_2$

$$\therefore \quad \frac{T_{st}}{T_f} = \frac{K(0.7E_2)^2 R_2}{R_2^2 + X_2^2} \times \frac{20}{KE_2^2} = (0.7)^2 \times \frac{0.5 \times 22}{(0.5)^2 + (5)^2} = \mathbf{0.2134.} \quad \textbf{(Ans.)}$$

Example 3.72. *A 12 kW, 3-phase, 6-pole, 50 Hz, 400 V, delta-connected induction motor runs at 960 r.p.m. on full-load. If it takes 85 A on direct starting, find the ratio of the starting torque to full-load torque with a star-delta starter. Full-load efficiency and power factor are 88% and 0.85 respectively.* **(Nagpur Univ. 1998)**

Solution. *Given* : Output = 12 kW ; $p = 6, f = 50$ Hz ; $V_L = 400$ V, $N_f = 960$ r.p.m.,

$\eta = 88\%$, full-load power factor = 0.85

$\mathbf{\frac{T_{st}}{T_f}}$:

Full-load line current drawn by a 3-phase delta-connected induction motor

$$= \frac{\text{Output}}{\sqrt{3}\, V_L \times \text{power factor} \times \eta} = \frac{12 \times 1000}{\sqrt{3} \times 400 \times 0.85 \times 0.88} = 23.16 \text{ A}$$

Full-load current per phase, $\quad I_f = \frac{23.16}{\sqrt{3}} = 13.37$ A

Short-circuit current per phase, $I_{sc} = \frac{85}{\sqrt{3}} = 49.07$ A

Synchronous speed, $\quad N_s = \frac{120f}{p} = \frac{120 \times 50}{6} = 1000$ r.p.m.

Full-load slip, $\quad s_f = \frac{N_s - N_f}{N_s} = \frac{1000 - 960}{1000} = 0.04$

Using the relation, $\quad \frac{T_{st}}{T_f} = \frac{1}{3}\left(\frac{I_{sc}}{I_f}\right)^2 \times s_f = \frac{1}{3}\left(\frac{49.07}{13.37}\right)^2 \times 0.04 = \mathbf{0.18.}$ **(Ans.)**

Example 3.73. *A squirrel cage type induction motor, when started by means of a star-delta starter takes 200% full-load current (line) and develops 44% of full-load torque at starting. Calculate the starting torque and current, if an auto-transformer with 75% tapping were employed.*

(S.C.E.W.B. 1999)

Solution. *Given* : $I_{st} = 200\%\ I_f$ or $2I_f$, $T_{st} = 44\%$ of $T_f = 0.44\ T_f$, $K = 75\%$ or 0.75

Supply line current at start, $I_{sc} = 3$ times, the current drawn by motor with star-delta starter

$$= 3 \times 2I_f = 6I_f$$

Starting torque with *star-delta starter*,

$$T_{st} = \frac{1}{3}\left(\frac{I_{sc}}{I_f}\right)^2 s_f = 0.44T_f$$

or $$\frac{T_{st}}{T_f} = \frac{1}{3}\left(\frac{I_{sc}}{I_f}\right)^2 s_f = 0.44$$

or $$\frac{1}{3} \times 6^2 \times s_f = 0.44$$

or Full-load slip, $\quad s_f = 0.0366$

Starting current (line value) with auto-transformer,

$$I_{st} = K^2 I_{sc} = (0.75)^2 \times 6I_f = 3.375I_f$$

or **337.5% of full-load current. (Ans.)**

Starting torque, $$T_{st} = K^2 \left(\frac{I_{sc}}{I_f}\right)^2 s_f T_f$$

$$= (0.75)^2 \times 6^2 \times 0.0366 T_f = 0.7411 T_f$$ or **74.11% of T_f. (Ans.)**

Example 3.74. *A 3-phase induction motor has a ratio of maximum torque to full-load torque as 2.5 : 1. Determine the ratio of actual starting torque to full-load torque for star-delta starting. Given $R_2 = 0.4\ \Omega$ and $X_2 = 4\ \Omega$.* **(S.C.E.T.E.W.B. 1998)**

Solution. *Given :* $\frac{T_m}{T_f} = 2.5$; $R_2 = 0.4\ \Omega$, $X_2 = 4\ \Omega$

$\mathbf{\frac{T_{st}}{T_f}}$:

We know that $$\frac{T_f}{T_m} = \frac{2 s_f s_{mT}}{s^2_f + s^2_{mT}}$$ [Eqn. (3.17)]

where s_f = full-load slip and s_{mT} (slip at maximum torque) $= \frac{R_2}{X_2} = \frac{0.4}{4} = 0.1$

or $$\frac{1}{2.5} = \frac{2s_f \times 0.1}{{s_f}^2 + 0.1^2} = \frac{0.2 s_f}{{s_f}^2 + 0.01} \quad \text{or} \quad {s_f}^2 - 0.5 s_f + 0.01 = 0$$

or $$s_f = \frac{0.5 \pm \sqrt{0.5^2 - 4 \times 0.01}}{2}$$

$$= \frac{0.5 \pm 0.458}{2} = 0.021, \text{ rejecting the higher value}$$

Full-load rotor current per phase, $$I_f = \frac{s_f E_2}{\sqrt{R^2 + s_f^2 {X_2}^2}}$$

Short-circuit rotor current per phase, $$I_{sc} = \frac{E_2}{\sqrt{{R_2}^2 + {X_2}^2}}$$ (∵ At standstill $s = 1$)

∴ $$\frac{I_{sc}}{I_f} = \frac{\sqrt{{R_2}^2 + s_f^2 {X_2}^2}}{s_f \sqrt{{R_2}^2 + {X_2}^2}} = \frac{\sqrt{0.4^2 + 0.021^2 \times 4^2}}{0.021 \sqrt{0.4^2 + 4^2}} = \frac{0.4087}{0.0844} = 4.84$$

Starting torque with star-delta starter, $$T_{st} = \frac{1}{3} T_f \left(\frac{I_{sc}}{I_f}\right)^2 s_f$$ [Eqn. (3.26)]

$$= \frac{1}{3} T_f \times (4.84)^2 \times 0.021 = 0.164 T_f$$

∴ $$\frac{T_{st}}{T_f} = \mathbf{0.164.} \ \textbf{(Ans.)}$$

Example 3.75. *A 50 kVA, 400 V 3-phase, 50 Hz squirrel cage induction motor has full-load slip of 5%. Its standstill impedance is 0.866 Ω/phase. It is started using a tapped auto-transformer. If the maximum allowable supply current at the time of starting is 100 A, calculate the tap position and the ratio of starting torque to full-load torque.* **(GATE, 1997)**

Solution. Given : Rating of induction motor = 50 kVA, 400 V ; s_f = 5% or 0.05 ; standstill impedance = 0.866 Ω per phase, I_{st} = 100 A.

Tap position, $\mathbf{\frac{T_{st}}{T_f}}$:

Full-load current $= \dfrac{\text{output in kVA} \times 1000}{\sqrt{3} \times \text{line voltage}}$, neglecting losses

$$= \frac{50 \times 1000}{\sqrt{3} \times 400} = 72.2 \text{ A}$$

Short-circuit current, $I_{sc} = \dfrac{(400/\sqrt{3})}{0.866} = 266.7$ A

Tap position of the transformer,

$$K = \sqrt{\frac{I_{st}}{I_{sc}}} = \sqrt{\frac{100}{266.7}} = 0.6123 \text{ or } \mathbf{61.23\%. \ (Ans.)}$$

Now,
$$\frac{T_{st}}{T_f} = K^2 \left(\frac{I_{sc}}{I_f}\right)^2 s_f = (0.6123)^2 \times \left(\frac{266.7}{72.2}\right)^2 \times 0.05 = \mathbf{0.256. \ (Ans.)}$$

Example 3.76. *A 3-phase induction motor takes a starting current which is 5 times full-load current at normal voltage. Its full-load slip is 4 percent. What auto-transformer ratio would enable the motor to be started with not more than twice the full-load current drawn from the supply ? What would be the starting torque under this condition ?* **(AMIE Summer, 1997)**

Solution. *Given :* $I_{st} = 5I_f$; $s_f = 4\%$ or 0.04

Let the transformation ratio of auto-transformer be K.

Now the starting current, $I_{st} = K^2 I_{sc}$

or
$$\frac{I_{st}}{I_f} = K^2 \frac{I_{sc}}{I_f} \quad \text{or} \quad 2 = K^2 \times 5$$

or
$$\mathbf{K} = \sqrt{\frac{2}{5}} = 0.6324 \text{ or } \mathbf{63.24\%. \ (Ans.)}$$

Starting torque,
$$\mathbf{T_{st}} = K^2 \left(\frac{I_{sc}}{I_f}\right)^2 s_f T_f = 0.6324^2 \times 5^2 \times 0.04 \times T_f = 0.4T_f$$

or **40% of full-load torque. (Ans.)**

Example 3.77. *Estimate approximately the starting torque of a 3-phase induction motor in terms of full-load torque when started by (i) star-delta starter (ii) an auto-transformer starter with 50 percent tapping. The short-circuit current of motor is 5 times the full-load current and the full-load slip is 5 percent.* **(Nagpur Univ. 1997)**

Solution. *Given :* $I_{sc} = 5I_f$; $s_f = 5\%$ or 0.05.

(*i*) Starting torque with *star-delta starter*,

$$T_{st} = \frac{1}{3}\left(\frac{I_{sc}}{I_f}\right)^2 s_f \times T_f = \frac{1}{3} \times 5^2 \times 0.05 \times T_f = 0.4167T_f$$

or *i.e.* **41.67% of full-load torque. (Ans.)**

(*ii*) Starting torque with an *auto-transformer starter* with 50% tapping (*i.e.* $K = 0.5$)

$$\mathbf{T_{st}} = K^2 \left(\frac{I_{sc}}{I_f}\right)^2 s_f T_f = 0.5^2 \times 5^2 \times 0.05 T_f = 0.3125 T_f$$

or *i.e.* **31.25% of full-load torque. (Ans.)**

Example 3.78. *A 20 H.P., 400 V, 950 r.p.m., 3-phase, 50 Hz, 6-pole cage induction motor draws 6 times the full-load current at standstill from a 400 V mains and develops 1.8 times full-load running torque. The full-load current is 30 A.*

(*i*) *What voltage must be applied to produce full-load torque at starting ?*

(*ii*) *What current will this voltage produce ?*

(*iii*) *If this voltage is obtained by an auto-transformer, what will be the line current ?*

(*iv*) *If the starting current in the line is limited to full-load current by means of an auto-transformer, what will be the starting torque as a percentage full-load torque ?*

Assume full-load slip as 5% and neglect magnetising current and the stator impedance drops.

(AMIE Summer, 1993)

Solution. *Given* : H.P. = 20 ; $V = 400$ volts ; $f = 50$ Hz ; $p = 6$; $I_{sc} = 6I_f$; $I_f = 30$ A ;
$T_{st} = 1.8T_f$; $s_f = 5\%$ or 0.05.

(*i*) Let the voltage applied to develop starting torque T_{st} equal to full-load torque be V' volts.

$$\because \quad T_{st} \propto (\text{applied voltage})^2$$

$$\therefore \quad \left(\frac{V'}{V}\right)^2 = \frac{T'_{st}}{T_{st}} = \frac{T_f}{1.8T_f} = \frac{1}{1.8}$$

or
$$V' = \sqrt{\frac{1}{1.8}} \times V = \frac{400}{\sqrt{1.8}} = \mathbf{298.14\ V.} \quad \textbf{(Ans.)}$$

(*ii*) **Current produced** at start by $V' = I_{sc} \times \frac{V'}{V} = 6I_f \times \frac{V'}{V}$

$$= (6 \times 30) \times \frac{298.14}{400} = \mathbf{134.16\ A.} \quad \textbf{(Ans.)}$$

Transformation ratio, $\quad K = \frac{V'}{V} = \frac{298.14}{400} = 0.745$

(*iii*) **Line current with auto-transformer** of ratio of 0.745

$$= K^2 I_{sc} = 0.745^2 \times 180 = \mathbf{99.9\ A.} \quad \textbf{(Ans.)}$$

(*iv*) Auto-transformer transformation ratio to produce starting line current equal to full-load current ;

$$K = \sqrt{\frac{I_{st}}{I_{sc}}} = \sqrt{\frac{I_f}{6I_f}} = 0.408$$

$$\therefore \quad \mathbf{T_{st}} = K^2 \left(\frac{I_{sc}}{I_f}\right)^2 sT_f = (0.408)^2 \times (6)^2 \times 0.05T_f = 0.3T_f$$

i.e. **30% of full-load torque. (Ans.)**

Example 3.79. *The starting torque and maximum torque of a 3-phase induction motor are 140 percent and 220 percent of the full-load torque. Neglecting stator resistance and assuming constant rotor resistance, calculate :*

(*i*) *The slip at maximum torque ;* (*ii*) *Full-load slip ;*

(*iii*) *Rotor current at starting in terms of full-load rotor current.* **(Allahabad University)**

Solution. *Given* : $T_{st} = 1.4T_f$; $T_m = 2.2T_f$

(*i*) **The slip at maximum torque, s_{mT} :**

We know that,
$$\frac{T_{st}}{T_m} = \frac{2s_{mT}}{s^2_{mT} + 1} \quad \text{[Eqn. 3.18]}$$

$$\frac{1.4}{2.2} = \frac{2s_{mT}}{s^2_{mT} + 1}$$

or $\quad 1.4(s^2_{mT} + 1) = 4.4s_{mT} \quad$ or $\quad s^2_{mT} - 3.143s_{mT} + 1 = 0$

or $$s_{mT} = \frac{3.143 \pm \sqrt{(3.143)^2 - 4}}{2} = \frac{3.143 \pm 2.424}{2} = 0.359, 2.78$$

$\therefore$ $\mathbf{s_{mT} = 0.359}$ (rejecting the other value). **(Ans.)**

(*ii*) **Full-load slip, T_f :**

Using the relation, $$\frac{T_f}{T_m} = \frac{2 s_f s_{mT}}{s_f^2 + s_{mT}^2}$$ [Eqn. (3.17)]

or $$\frac{1}{2.2} = \frac{2 s_f \times 0.359}{s_f^2 + (0.359)^2} = \frac{0.718 s_f}{s_f^2 + 0.1289}$$

$$= s_f^2 - 1.579 s_f + 0.1289 = 0$$

or $$s_f = \frac{1.579 \pm \sqrt{(1.579)^2 - 4 \times 0.1289}}{2} = \frac{1.579 \pm 1.406}{2} = 0.0865, 1.49.$$

Hence, $\mathbf{s_f = 0.0865}$, the other value rejected. **(Ans.)**

(*iii*) **Rotor current at starting in terms of full load current :**

We know that, $$T_f = \frac{1}{\omega_s} I_{2f}^2 \frac{R_2}{s_f}$$ ...(*i*) [Eqn. (3.16)]

Thus, $$T_{st} = \frac{1}{\omega_s} I_{2st}^2 R_2$$ ...(*ii*) ($\because$ $s = 1$ at starting)

Dividing (*ii*) by (*i*), we get

$$\frac{T_{st}}{T_f} = \left[\frac{I_{2st}}{I_{2f}}\right]^2 s_f$$

$\therefore$ $$\left[\frac{I_{2st}}{I_{2f}}\right]^2 = \frac{1.4}{0.0865} = 16.185 \text{ or } \mathbf{I_{2st} = 4.023\, I_{2f}}. \quad \textbf{(Ans.)}$$

Example 3.80. *In a 3-phase induction motor, leakage reactance is four times the resistance for both stator and rotor circuits. But stator leakage reactance is equal to rotor standstill leakage reactance referred to stator. The motor develops a torque of 220 Nm at a slip of 0.05. Calculate the starting and maximum torques.*

The exciting current may be neglected. **(Punjab University)**

Solution. *Given* : $X_1 = X_2 = 4R_1 = 4R_2$; $T = 220$ Nm ; $s = 0.05$.

Starting and maximum torques, T_{st}, T_m :

We know that, $$T = \frac{3}{\omega_s} \cdot I_2^2 \cdot \frac{R_2}{s}$$ (for 3-phases) ...[Refer Eqn. 3.16]

or $$T = \frac{3}{\omega_s}\left[\frac{V^2}{\left(R_2 + \frac{R_2}{s}\right)^2 + (X_1 + X_2)^2}\right] \cdot \frac{R_2}{s}$$

Substituting the values, we get

$$220 = \frac{3V^2}{\omega_s} \frac{1}{\left[\left(1 + \frac{1}{0.05}\right)^2 + (4+4)^2\right] R_2} \cdot \frac{R_2}{0.05}$$

or $$\frac{3V^2}{\omega_s} = 220\left[\left(1 + \frac{1}{0.05}\right)^2 + (4+4)^2\right] \times 0.05 = 5555$$

$$\therefore \quad \mathbf{T_{st}} = \frac{3V^2}{\omega_s} \frac{1}{(1+1)^2 + (4+4)^2} = 5555 \times \frac{1}{(4+64)} = \mathbf{81.69\ N\text{-}m.\ \ (Ans.)}$$

$$s_{mT} = \frac{R_2}{\sqrt{R_1^2 + (X_1 + X_2)^2}} = \frac{1}{\sqrt{1 + (4+4)^2}} = 0.124$$

$$\therefore \quad \mathbf{T_m} = \frac{3V^2}{\omega_s} \frac{1}{\left(1 + \frac{1}{0.124}\right)^2 + 8^2} \times \frac{1}{0.124}$$

$$= 5555 \times \frac{1}{146.16} \times \frac{1}{0.124} = \mathbf{306.5\ N\text{-}m.\ \ (Ans.)}$$

☞**Example 3.81.** *A squirrel-cage induction motor has a slip of 4% at full-load. Its starting current is five times the full-load current. The stator impedance and magnetising current may be neglected, the rotor resistance is assumed constant.*

(i) Calculate the maximum torque and the slip at which it would occur.

(ii) Calculate the starting torque.

Express torques in per unit (p.u.) of the full-load torque. **(ESE, 1995)**

Solution. *Given* : Slip at full load, $s_f = 4\% = 0.04$; $I_{st} = 5I_f$

(*i*) Maximum torque and the slip at which it would occur :

$$T_f = k_t \cdot \frac{s_f s_{mT}}{s_f^2 + s_{mT}^2}, \text{ where } s_{mT} = \frac{R_2'}{X_2'} \quad ...(i)$$

Stator impedance and no-load current are neglected ;

Hence, $$I_f = I'_2 = \frac{SE_1}{\sqrt{R_2'^2 + s^2 X_2'^2}} \quad ...(ii)$$

But $$I_{st} = \frac{E_1}{\sqrt{R_2'^2 + X_2'^2}} \quad (\because s = 1) \quad ...(iii)$$

Dividing (*iii*) by (*ii*), we get

$$\frac{I_{st}}{I_f} = \frac{1}{s}\left[\frac{R_2'^2 + s^2 X_2'^2}{R_2'^2 + X_2'^2}\right]^{1/2} = \frac{1}{s}\left[\frac{s_{mT}^2 + s_f^2}{s_{mT}^2 + 1}\right]^{1/2}$$

Substituting the values, we get

or $$5 = \frac{1}{0.04}\left[\frac{s_{mT}^2 + 0.04^2}{s_{mT}^2 + 1}\right]^{1/2}$$

Squaring both sides, we have

$$25 = \frac{1}{(0.04)^2}\left(\frac{s_{mT}^2 + 0.04^2}{s_{mT}^2 + 1}\right)$$

or $$25 \times (0.04)^2 \times (s_{mT}^2 + 1) = s_{mT}^2 + 0.04^2$$

or $$0.04(s_{mT}^2 + 1) = s_{mT}^2 + 0.0016$$

or $$0.04 s_{mT}^2 + 0.04 = s_{mT}^2 + 0.0016$$

or $$s_{mT}^2 (1 - 0.04) = 0.04 - 0.0016$$

$\therefore \qquad s_{mT} = 0.2$

Hence maximum torque occurs when $s = \mathbf{s_{mT} = 20\%}$. **(Ans.)**

But $\qquad T_m = \frac{1}{2} k_t \qquad$ [Putting $s_f = s_{mT}$ in eqn. (*i*)]

and $\qquad T_f = k_t \frac{0.04 \times 0.2}{(0.04)^2 + (0.2)^2} = \frac{k_t}{5.2} \qquad$ [Substituting the values in eqn. (*i*)]

Hence, $\qquad k_t = 5.2 T_f$

$$T_m = \frac{1}{2} \times 5.2 T_f = 2.6 T_f$$

$$\mathbf{T_{st}} = k_t \cdot \frac{s_{mT}}{1 + s_{mT}^2} \qquad \text{[Putting } s_f = 1 \text{ in eqn. (}i\text{)]}$$

$$= k_t \cdot \frac{0.2}{1 + 0.2^2} = 5.2 T_f \times \frac{0.2}{1.04} = \mathbf{T_f}. \quad \textbf{(Ans.)}$$

Thus, $\mathbf{T_m}$ is 2.6 p.u. and it occurs at a slip of 0.2 and $\mathbf{T_{st}}$ is 1 p.u. (Ans.)

Example 3.82. *A 3-phase, 4-pole, 1440 r.p.m., 50 Hz induction motor has star-connected rotor winding, having a resistance of 0.2 Ω per phase and a standstill leakage reactance of 1 Ω per phase. When the rotor is energised at rated voltage and frequency, the rotor induced e.m.f. at standstill is 125 V per phase.*

(i) Determine the rotor current, rotor power factor, and torque at starting and at full-load and compare these results.

(ii) If an external resistance of 1 Ω per phase is inserted in rotor circuit, calculate rotor current, rotor power factor and torque at the time of starting. **(U.P.S.C.)**

Solution. *Given* : $p = 4$; $N = 1440$ r.p.m. ; $f = 50$ Hz ; $R_2 = 0.2\ \Omega$; $X_2 = 1\ \Omega$; $E_2 = 125$ V.

(*i*) **Rotor current, rotor power factor and torque at starting and at full-load :**

Synchronous speed, $\qquad N_s = \frac{120f}{p} = \frac{120 \times 50}{4} = 1500$ r.p.m.

At starting :

Rotor current, $\qquad \mathbf{I_2} = \frac{E_2}{\sqrt{R_2^2 + X_2^2}} = \frac{125}{\sqrt{0.2^2 + 1^2}} = \mathbf{122.57\ A.}$ **(Ans.)**

Rotor power factor, $\qquad \cos \phi_2 = \frac{R_2}{\sqrt{R_2^2 + X_2^2}} = \frac{0.2}{\sqrt{0.2^2 + 1^2}} = \mathbf{0.196.}$ **(Ans.)**

Synchronous speed in radians per second,

$$\omega_s = \frac{2\pi N_s}{60} = \frac{2\pi \times 1500}{60} = 50\pi \text{ rad/s.}$$

Further, we know that, $\qquad T = \frac{3}{\omega_s} I_2^2 \frac{R_2}{s}$

$\therefore$ Rotor starting torque, $T_{st} = \frac{3}{\omega_s} I_2^2 \frac{R_2}{1} \qquad$ ($\because$ $s = 1$, at starting)

$$= \frac{3}{50\pi} \times 122.57^2 \times 0.2 = \mathbf{57.38\ N\text{-}m.} \quad \textbf{(Ans.)}$$

At full load :

Name plate speed of 1440 r.p.m. is the full-load (or rated) speed of the motor.

Full-load slip, $\qquad s_f = \frac{N_s - N}{N_s} = \frac{1500 - 1440}{1500} = 0.04$ (or 4%)

$\therefore$ Rotor current, $I_2 = \dfrac{s_f E_2}{\sqrt{R_2^2 + (s_f X_2)^2}} = \dfrac{0.04 \times 125}{\sqrt{0.2^2 + (0.04 \times 1)^2}}$ = **24.51 A. (Ans.)**

Power factor, $\cos \phi_2 = \dfrac{R_2}{\sqrt{R_2^2 + (s_f X_2)^2}} = \dfrac{0.2}{\sqrt{0.2^2 + (0.04 \times 1)^2}} = 0.98$

Torque, $T_f = \dfrac{3}{\omega_s} I_2^2 \dfrac{R_2}{s_f} = \dfrac{3}{50\pi} \times (24.51)^2 \times \dfrac{0.2}{0.04}$ = **57.37 N-m. (Ans.)**

$\therefore$ $\dfrac{(I_2)_{starting}}{(I_2)_{full\text{-}load}} = \dfrac{122.57}{24.51}$ = **5. (Ans.)**

and $\dfrac{(T)_{starting}}{(T)_{full\text{-}load}} = \dfrac{T_{st}}{T_f} = \dfrac{57.38}{57.37} \simeq 1.$

(*ii*) **Rotor current, rotor power factor and torque at starting, with external resistance :**

When external resistance is inserted in the rotor circuit, the total rotor circuit resistance becomes 0.2 + 1 = 1.2 Ω per phase.

$\therefore$ Rotor starting current, $I_2 = \dfrac{125}{\sqrt{(1.2)^2 + (1)^2}}$ = **80.02 A. (Ans.)**

Rotor power factor, $\cos \phi_2 = \dfrac{1.2}{\sqrt{(1.2)^2 + (1)^2}}$ = **0.768. (Ans.)**

Rotor torque, $T = \dfrac{3}{50\pi} \times (80.02)^2 \times \dfrac{1.2}{1}$ = **146.75 N-m. (Ans.)**

From the above solution we find that *with the insertion of external resistance in the circuit, starting current has decreased whereas starting torque has increased and rotor power factor at starting has improved.*

SLIP RING OR WOUND MOTORS STARTER

(Refer Art. 3.24)

Example 3.83. *Design the sections of a rotor starter for a 75 kW, 3-phase induction motor, using 7 notches. Rotor resistance per phase = 0.018 Ω. The upper current limit is to be full-load current, for which slip is 2 per cent.*

Solution. Motor output = 75 kW

Notches or studs = 7

$\therefore$ Number of elements, $n = 6$

Rotor resistance phase, $r_m = 0.018$ Ω

Upper current limit = full-load current

Slip, $s_m = 2\%$ or 0.02

Using the relation, $\alpha = (s_m)^{1/n} = (0.02)^{1/6} = 0.521$

Also, $r_m = R_1 (\alpha)^n = R_1 s_m$

$\therefore$ $R_1 = \dfrac{r_m}{s_m} = \dfrac{0.018}{0.02} = 0.9\ \Omega$

The resistances of elements are :

$$r_1 = R_1 (1 - \alpha) = 0.9\,(1 - 0.521) = 0.432\ \Omega$$
$$r_2 = \alpha r_1 = 0.521 \times 0.432 = 0.225\ \Omega$$
$$r_3 = \alpha^2 r_1 = (0.521)^2 \times 0.432 = 0.108\ \Omega$$

$r_4 = \alpha^3 r_1 \quad = (0.521)^3 \times 0.432 = 0.062\ \Omega$

$r_5 = \alpha^4 r_1 \quad = (0.521)^4 \times 0.432 = 0.032\ \Omega$

$r_6 = \alpha^5 r_1 \quad = (0.521)^5 \times 0.432 = 0.017\ \Omega$

$r_m \quad = 0.018\ \Omega$

Total $= 0.894\ \Omega$

This resistance is nearly equal to R_1.

Example 3.84. *Design the six sections of a seven stud rotor starter for a 3-phase wound induction motor. The slip at full-load current is 2 per cent and the maximum starting current is 1.5 times full-load current. The resistance of rotor is 0.02 Ω per phase.*

Solution. Number of sections, $n = 6$

Slip at full-load current $= 2$ per cent

Maximum starting current $= 1.5$ times full-load current

Rotor resistance/phase, $r_m = 0.02\ \Omega$

We know that

$$\text{Rotor current} = \frac{E}{\sqrt{\left(\frac{R}{s}\right)^2 + X^2}} \approx \frac{sE}{R} \text{ if rotor reactance is neglected.}$$

This means that slip changes directly with rotor current. The full-load slip is 2%.

∴ Slip at 1.5 times full-load current, s_m

$= 1.5 \times 2 = 3\%$

Using the relation, $\alpha = (s_m)^{1/n} = (0.03)^{1/6} = 0.557$

$$R_1 = \frac{r_m}{s_m} = \frac{0.02}{0.03} = 0.666\ \Omega$$

The resistances of the sections are :

$r_1 = (1 - \alpha)R_1 \quad = (1 - 0.557) \times 0.666 \quad = 0.295\ \Omega$

$r_2 = \alpha r_1 \quad = 0.557 \times 0.295 \quad = 0.164\ \Omega$

$r_3 = \alpha^2 r_1 \quad = (0.557)^2 \times 0.295 \quad = 0.092\ \Omega$

$r_4 = \alpha^3 r_1 \quad = (0.577)^3 \times 0.295 \quad = 0.051\ \Omega$

$r_5 = \alpha^4 r_1 \quad = (0.557)^4 \times 0.295 \quad = 0.028\ \Omega$

$r_6 = \alpha^5 r_1 \quad = (0.557)^5 \times 0.295 \quad = 0.016\ \Omega$

$r_m \quad = 0.020$

Total $= 0.666\ \Omega$

3.25. CRAWLING

- It is observed that induction motors, particularly the squirrel-cage type, sometimes *exhibit a tendency to run stably at speeds as low as one-seventh of their synchronous speed* N_s. This phenomenon is known as *crawling of an induction motor.*
- A 3-phase winding carrying sinusoidal currents produces harmonics of the order $n = 6N \pm 1$, where N is an integer. The movement of the harmonics is with or against the direction of rotation depending upon the sign (+ means with the rotation and—means against the rotation). *The synchronous speed of nth order harmonic is* $\frac{1}{n}$*th of the synchronous speed of fundamental.*

For $N = 1$, a three-phase winding would produce a forward rotating 7th harmonic and backward rotating 5th harmonic. Considering the 7th harmonic, the interaction between the fictitious stator and rotor 7th harmonic poles will produce a positive torque and *if the torque is sufficiently pronounced it may prevent the motor speed exceeding one seventh of normal*. Thus the motor crawls at about 1/7th the normal speed.

Fig. 3.53 shows the resultant slip torque characteristics due to fundamental and 7th harmonic. It is clear that there is a dip in the curve which represents the stable running region. The running speed is less than the synchronous speed corresponding to 7th harmonic and therefore torque produced is a harmonic induction torque. This is known as *crawling*.

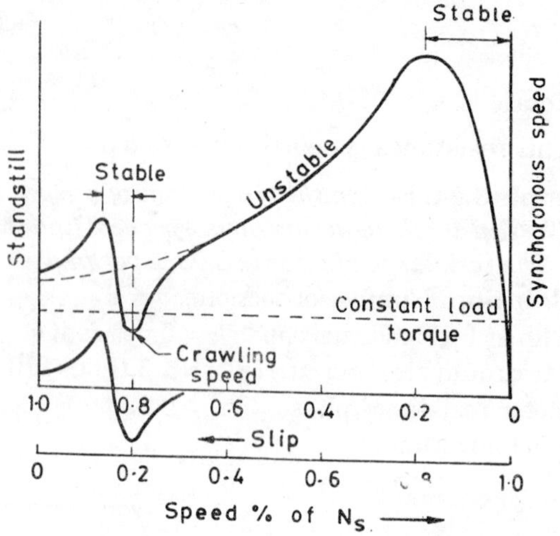

Fig. 3.53. Crawling of induction motor.

3.26. COGGING

When the number of rotor slots is equal to the number of stator slots, the speeds of all the harmonics produced by the stator slotting coincide with the speed of corresponding rotor harmonics. Thus harmonics of every order would try to exert synchronous torques at their synchronous speeds and the machine *would refuse to start*. This is known as *cogging* (or magnetic locking). Therefore the *number of stator slots should never be equal to the number of rotor slots.*

Cogging of squirrel cage motors can be easily overcome by *making the number of rotor slots prime to the number of stator slots.*

3.27. DOUBLE SQUIRREL CAGE MOTOR

- The ordinary squirrel-cage motor has excellent running characteristics, but the torque at starting is very poor. On the other hand the wound rotor induction motor can have a very high starting torque and good running characteristics. But the latter type is unsuitable for many types of service, because the starting rheostat has to be manipulated while the motor accelerates and extra cost of the external switch-gear is also another objection.

  It has been the aim of designing engineers to combine in one machine two characteristics : (*i*) good starting torque ; and (*ii*) excellent running performance. To realise (*i*), it is necessary to have a *high resistance* for the rotor phases during starting periods, and to realise (*ii*), the *rotor resistance must be low* to obtain excellent running performance.

  This ensures small fractional slip and high frequency.

  The desired characteristic is obtained by *housing two independent squirrel-cage windings on the rotor, each having its own set of slots.*

- Fig. 3.54 shows a double squirrel-cage rotor. There are two sets of squirrel-cage windings. The upper cage (starting cage) arranged nearer to the air-gap is made of *high resistivity material*—brass, aluminium, bronze etc. while the lower cage (running cage) is made of *copper*, the two cages are separated by narrow slits. As a result the lower cage has high permeance for leakage flux due to which the leakage reactance of lower cage is considerably higher than that of upper cage. The upper and lower cages may have either common short circuiting end rings or each of the cages has its own short circuiting ring.

The two cages, sometimes, may be made of cast aluminium. In this case a high active resistance for the upper cage is obtained decreasing its area of cross-section.

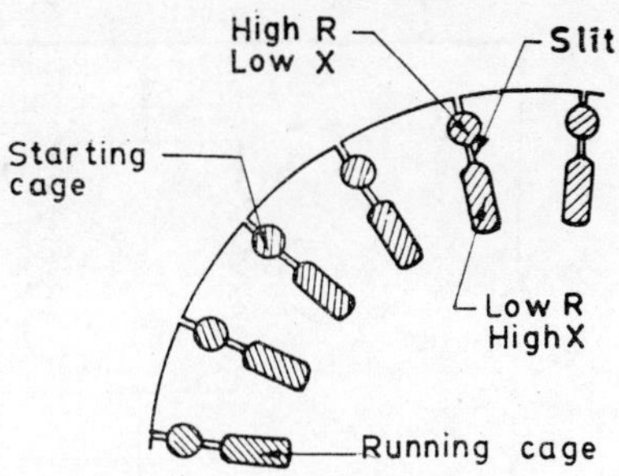

Fig. 3.54. Double squirrel-cage rotor.

At starting the frequency of rotor currents is high and is equal to supply frequency. The current is distributed between the upper and lower cages in the inverse proportion of their impedance and since the lower cage has a very high leakage reactance, its impedance is very large as compared with that of upper cage. Thus the current in the lower cage is small and therefore the current is confined to the upper cage which has a high resistance, this gives a *very good starting torque*.

As the motor speed increases, the frequency of rotor currents decreases and this results in reduction of leakage reactance of lower cage when the motor runs near about its full speed, the reactance of the cages becomes negligible and therefore the current divides itself in the inverse ratio of resistances. Since the resistance of upper cage is about 5 to 6 times that of lower cage, *most of the current is carried by the lower cage and the motor has the running characteristic of the low-slip general-purpose motor*.

The torque-slip (speed) characteristics of individual cages as well as resultant (or total) are shown in Fig. 3.55.

This motor, however, is *better adapted to adjust to suddenly applied loads, since at increased values of slip, the high-resistance winding again develops significant torque.* With a starting current not more than $5\frac{1}{2}$ times full-load current, the double-squirrel-cage motor is *known as a low-starting current, high-starting-torque, low-slip motor*.

The double-cage induction motor *is ideal for compressors*.

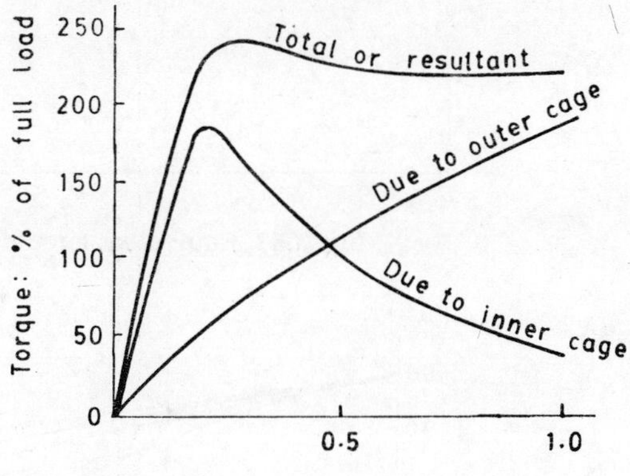

Fig. 3.55. Torque-slip curves of double-squirel-cage motor.

Equivalent Circuit of Double-squirrel-cage motor :

Let, R_1 = resistance/phase of stator,

X_1 = reactance/phase of stator,

R_0' = phase resistance of *outer cage* referred to the stator,

X_0' = phase reactance of *outer cage* referred to the stator,

R_i = phase resistance of *inner cage* referred to the stator, and

X_i' = phase reactance of *inner cage* referred to the stator.

Since both cages completely link the main flux so the impedances of the two cages can be considered in parallel.

The equivalent circuit for slip 's' for one of the phases and referred to primary (or stator) is shown in Fig. 3.56. If the magnetising current is neglected, then the figure is simplified to that shown in Fig. 3.57.

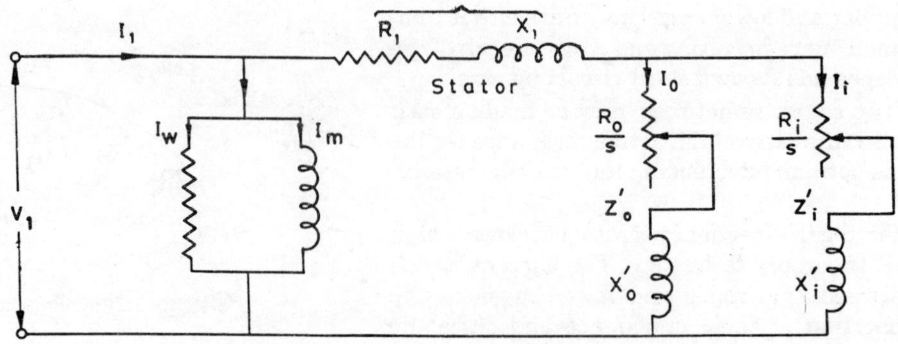

Fig. 3.56. Equivalent circuit of double-squirrel-cage motor.

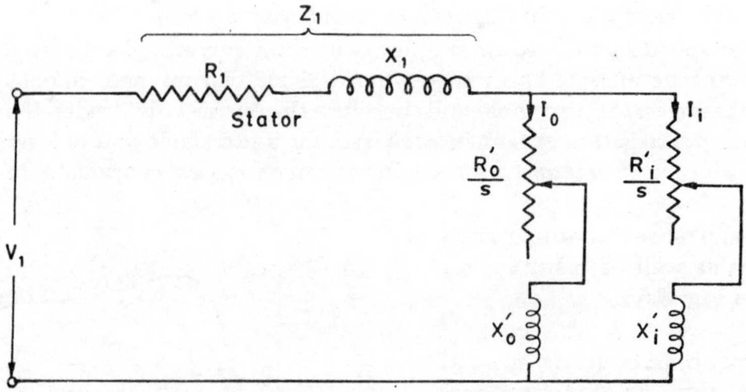

Fig. 3.57. Equivalent circuit of double-squirrel-cage motor.

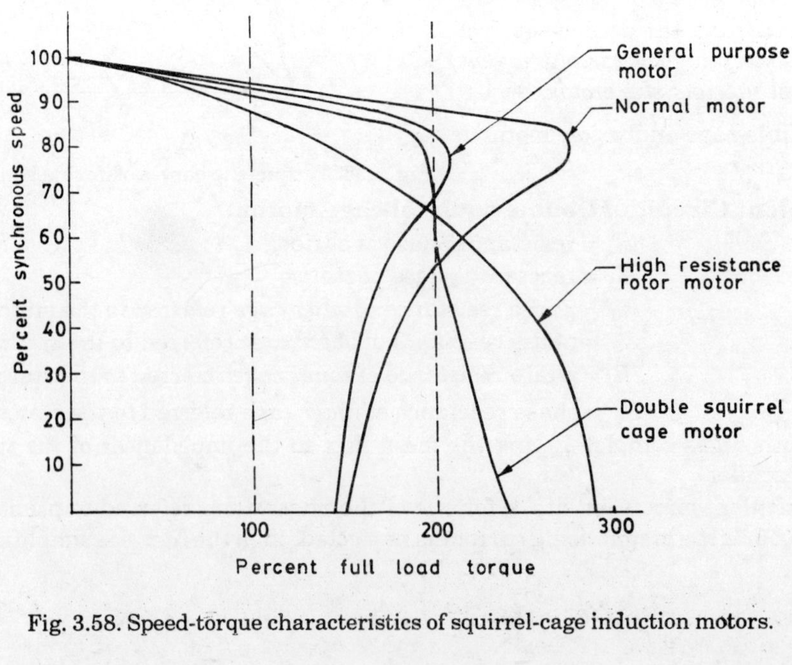

Fig. 3.58. Speed-torque characteristics of squirrel-cage induction motors.

Total impedance referred to primary is given by

$$Z_{01} = R_1 + jX_1 + \frac{1}{\frac{1}{Z_i'} + \frac{1}{Z_0'}} = R_1 + jX_1 + \frac{Z_i' Z_0'}{Z_0' + Z_i'}$$

Speed-torque characteristics of squirrel-cage induction motors. Fig. 3.58 shows the speed-torque characteristics of squirrel-cage induction motors.

DOUBLE SQUIRREL-CAGE MOTOR

☞**Example 3.85.** *The equivalent standstill impedance per phase of the first cage and of the second cage respectively of a double-cage induction motor are (4 + j1) and (1 + j4) Ω. Compare the torques of the two cages :*

(i) At standstill *(ii) At a slip of 4 per cent.*

Solution. Standstill impedance of first or outer cage,

$$Z_0 = (4 + j1)$$

Standstill impedance of second or inner cage,

$$Z_i = (1 + j4)$$

The equivalent circuit diagram of the double squirrel-cage motor is shown in Fig. 3.59.

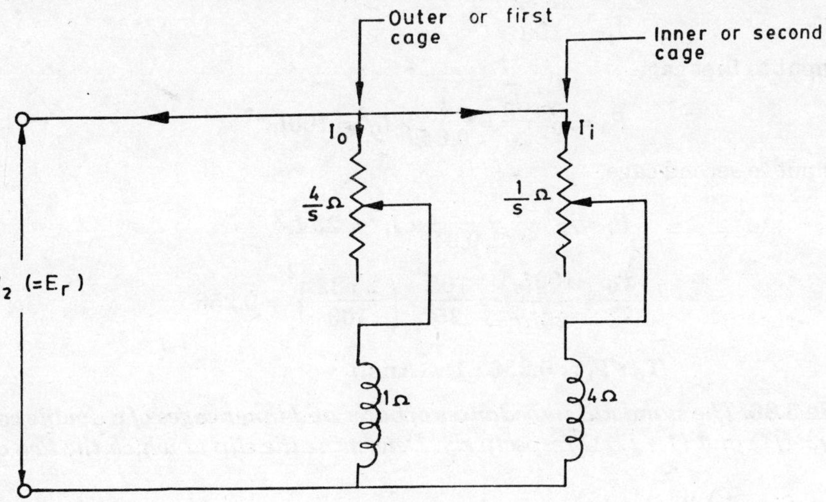

Fig. 3.59

(i) **At standstill :**

Slip, $s = 1$

$$Z_0 = \frac{4}{1} + j1 = 4 + j = \sqrt{4^2 + 1^2} = 4.123\ \Omega$$

Similarly, $$Z_i = \left(\frac{1}{1} + j4\right) = 1 + j4 = \sqrt{1^2 + 4^2} = 4.123\ \Omega$$

Since the impedances Z_0 and Z_i are equal, the currents will be divided equally between the two cages

$\therefore$ $$I_0 = I_i$$

Power input to first or outer cage,

$$P_0 = I_0^2 R_0 = 4I_0^2 \text{ W} \qquad (\because\ R_0 = 4\ \Omega)$$

Power input to second or inner cage

$$P_i = I_i^2 R_i = I_i^2 \text{ W} \qquad (\because\ R_i = 1\ \Omega)$$

$$\frac{\text{Torque of first cage}}{\text{Torque of second cage}} = \frac{P_0}{P_i} = \frac{4I_0^2}{I_i^2}$$

$$\therefore \qquad \frac{T_0}{T_i} = \frac{4}{1} = 4$$

$$\mathbf{T_0 : T_i :: 4 : 1.} \quad \textbf{(Ans.)}$$

(*ii*) **At a slip of 4 per cent :**

Slip, $s = 0.04$

$$Z_0 = \frac{4}{0.04} + j1 = 100 + j1 = \sqrt{100^2 + 1^2} = 100\ \Omega \text{ (app.)}$$

$$Z_i = \frac{1}{0.04} + j4 = 25 + j4 = \sqrt{25^2 + 4^2} = 25.32\ \Omega$$

Currents will be divided in the *inverse ratio of these impedances.*

$$\therefore \qquad \frac{I_0}{I_i} = \frac{25.32}{100}$$

Power input to first cage,

$$P_0 = I_0^2 \frac{R_0}{s} = \frac{4}{0.04} \times I_0^2 = 100I_0^2$$

Power input to second cage,

$$P_i = I_i^2 \frac{R_i}{s} = \frac{1}{0.04} \times I_i^2 = 25\, I_i^2$$

$$\therefore \qquad \frac{T_0}{T_i} = \frac{100I_0^2}{25I_i^2} = \frac{100}{25} \times \left(\frac{25.32}{100}\right)^2 = 0.256$$

Hence, $\mathbf{T_0 : T_i :: 0.256 : 1.}$ **(Ans.)**

Example 3.86. *The standstill impedances of outer and inner cages of a double cage induction motor are (4 + j2.4) Ω and (1 + j7) Ω respectively. Determine the slip at which the two cages develop equal torques.*

Solution. Standstill impedance of outer cage,

$$Z_0 = (4 + j2.4)\ \Omega$$

Standstill impedance of inner cage,

$$Z_i = (1 + j7)\ \Omega$$

Let s be the slip at which two cages develop equal torques.

Then
$$Z_0 = \sqrt{\left(\frac{4}{s}\right)^2 + (2.4)^2} = \sqrt{\frac{16}{s^2} + (2.4)^2}$$

$$Z_i = \sqrt{\left(\frac{1}{s}\right)^2 + (7)^2} = \sqrt{\frac{1}{s^2} + (7)^2}$$

Also
$$\frac{I_0}{I_i} = \frac{Z_i}{Z_0}$$

or
$$\left(\frac{I_0}{I_i}\right)^2 = \left(\frac{Z_i}{Z_0}\right)^2 = \frac{\left(\frac{1}{s}\right)^2 + 7^2}{\frac{16}{s^2} + (2.4)^2}$$

Power input to outer cage,
$$P_0 = I_0^2 R_0 = I_0^2 \times \frac{4}{s} \text{ W}$$

Similarly,
$$P_i = I_i^2 R_i = I_i^2 \times \frac{1}{s} \text{ W}$$

$$\therefore \quad \frac{T_0}{T_i} = \frac{P_0}{P_i} = \left(\frac{I_0}{I_i}\right)^2 \times 4 = 4 \times \frac{\left(\frac{1}{s^2} + 49\right)}{\left(\frac{16}{s^2} + 5.76\right)}$$

But $T_0 = T_i$ ...*(given)*

$$\therefore \quad \frac{16}{s^2} + 5.76 = \left(\frac{1}{s^2} + 49\right) \times 4$$

$$\frac{16}{s^2} + 5.76 = \frac{4}{s^2} + 196$$

or
$$\frac{12}{s^2} = 190.24 \quad \text{or} \quad s^2 = \frac{12}{190.24}$$

$$\therefore \quad s = 0.251 \text{ or } \mathbf{25.1\%.} \textbf{ (Ans.)}$$

Example 3.87. *The equivalent resistance and reactance values of a double cage induction motor for stator, outer cage and inner cages are 0.5, 2.0 and 0.3 Ω resistance and 7, zero and 6.0 Ω reactance respectively.*

If the phase voltage is 230 V and synchronous speed is 1000 r.p.m. find the starting torque.

Solution. Impedance of outer cage, $Z_0 = (2 + j0)$ Ω

Impedance of inner cage, $Z_i = (0.3 + j6)$ Ω

Stator impedance $= 0.5 + j7$

Supply phase voltage $= 230$ V

Synchronous speed, $N_s = 1000$ r.p.m.

Starting torque :

The equivalent circuit is shown in Fig. 3.60.

The two impedances, Z_0 and Z_i are in parallel. Therefore, their equivalent impedance,

$$Z_e = \frac{Z_0 Z_i}{Z_0 + Z_i} = \frac{(2 + j0)(0.3 + j6)}{(2 + j0) + (0.3 + j6)} = \frac{(0.6 + j12)}{2.3 + j6} = \frac{(0.6 + j12)(2.3 - j6)}{(2.3 + j6)(2.3 - j6)}$$
$$= \frac{1.38 - j3.6 + j27.6 + 72}{5.29 + 36} = \frac{73.38 + j24}{41.29} \quad \text{or} \quad Z_e = 1.777 + j0.58$$

Stator impedance $= 0.5 + j7$

∴ Total impedance, $Z_f = (1.777 + j0.58) + (0.5 + j7)$
$= (2.777 + j7.58)$ Ω $= (2.28 + j7.58)$ Ω (app.)

Current,
$$I = \frac{\text{phase voltage}}{\text{total phase impedance}} = \frac{230 + j0}{(2.28 + j7.58)}$$
$$= \frac{(230 + j0)(2.28 - j7.58)}{(2.28 + j7.58)(2.28 - j7.58)} = \frac{524.4 - j1743.4}{5.198 + 57.456}$$

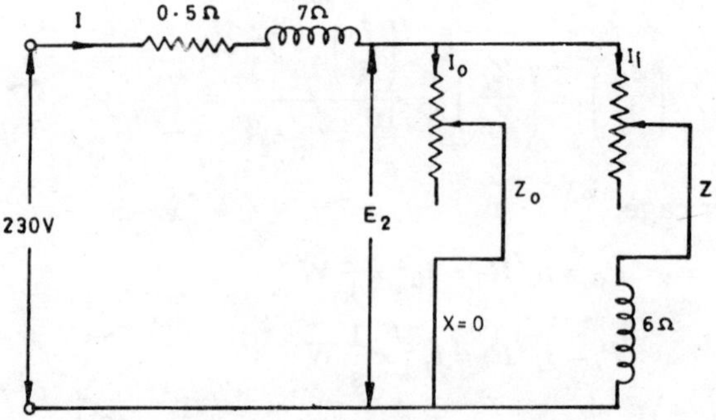

Fig. 3.60

$$= \frac{524.4 - j1743.4}{62.654} = (8.37 - j27.83) = 29.06 \text{ A}$$

Rotor copper loss/phase $= (\text{current})^2 \times$ total resistance/phase of two rotors

$= (29.06)^2 \times 1.777 \text{ W} = 1500 \text{ W}$

Now slip, $s = \frac{\text{rotor copper loss}}{\text{rotor input}}$

At starting $s = 1$

$\therefore$ Rotor input $= 1500$ W

Let, T = total torque developed at starting

Then, $\frac{T \times 2\pi N_s}{60} = 1500$

or $T = \frac{1500 \times 60}{2\pi N_s} = \frac{1500 \times 60}{2\pi \times 1000} = 14.32$ N-m

Hence, **torque developed at starting = 14.32 N-m. (Ans.)**

If it is required to find the torques developed by the two rotors separately, then find I_2, I_0 and I_i. Knowing these values T_0 and T_i can be found as discussed in previous examples.

3.28. SPEED CONTROL OF INDUCTION MOTORS

The rotor speed of an induction motor may be stated by the equation

$$N = \frac{120f}{p}(1 - s) \quad \text{...(3.42)}$$

where, N = rotor speed in r.p.m.,

f = frequency of supply,

p = number of poles, and

s = fractional slip.

This shows that a change in any one of these quantities will affect its speed, namely :

(*i*) change of frequency ; (*ii*) change of number of poles ; and

(*iii*) change of slip.

1. **Control by Changing Frequency.** This method is *impractical for most applications because frequency of the supply system must remain fixed.* In special cases where the motor load is the only load connected to generators, the speed of the prime-movers may be varied to change the supply frequency and thus change the motor speed. The range over which the speed may be varied in this way, however, is limited by the range of economical speeds of the drives.

This method of control has been applied to a limited extent in ship propulsion.

2. **Control by Changing Number of Poles.** Change in the number of poles is effected by making change in the stator winding connections with the help of suitable switching arrangement. When the speed ratio is 2 : 1, the '*Consequent Pole*' method is adopted and is shown diagrammatically in Fig. 3.61.

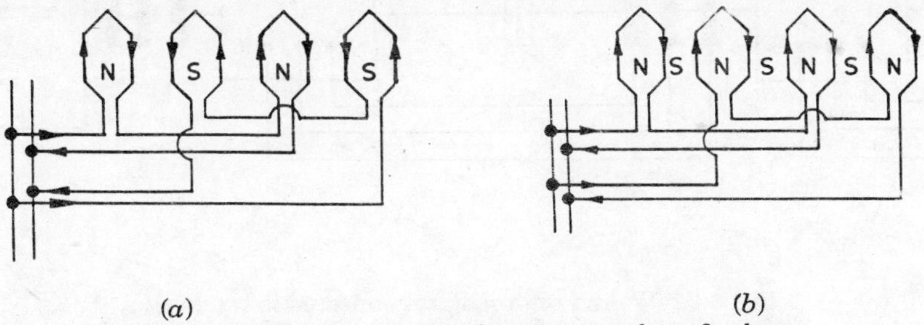

Fig. 3.61. Speed control by changing number of poles.

Two distinct windings are on the stator producing the same number of poles, one winding creates two *N-poles,* and the other two S-poles as shown in Fig. 3.61 (*a*). This gives *four* poles, and the synchronous speed at 50-Hz is 1500 r.p.m.

If the connection of one of the windings is reversed [Fig. 3.61 (*b*)], there will be in all four *N*-poles or four *S*-poles, depending upon which of the winding connection is reversed. But in between these poles other four poles of opposite polarity will be created. Thus the stator now has *eight* poles and its synchronous speed is 750 r.p.m.

The 2 : 1 change in speed is rather drastic. Hence motors are manufactured with *two windings* one is so wound that is created *8-poles* and the other *10-poles*, the synchronous speeds being 750 r.p.m. and 600 r.p.m. respectively. But the *disadvantage* is that at either speed only 50 per cent of stator copper is utilised, since only one winding is in use at a time.

In the case of squirrel-cage motors, change in the number of poles on the stator does not affect the existing arrangement on the rotor. But in case of wound rotors, the number of poles on the stator must equal to the number of poles on the rotor. Otherwise there will be a greater reduction in torque due to some rotor conductors developing a negative torque.

3. **Control by Changing Slip.** *Change in slip is effected by introducing* an *external resistance in rotor circuit of wound rotors.* But this is done at the *sacrifice of efficiency* and besides the *speed regulation is poor.*

These *disadvantages are more or less overcome by :*

- **Supplying counter e.m.fs. to the rotor at slip frequency.** This method requires *auxiliary commutating machine* which injects e.m.fs. into the rotor at rotor frequency through the slip rings. This method had the *advantage of providing a wide adjustment in speed without serious reduction in efficiency*. Its use in limited, however, to very special applications *that use large amounts of power in a single unit*.
- **Supplying counter e.m.fs. to the rotor at supply frequency.** This method requires a commutator for the rotor.

4. **Control by Cascade or Tandem Connection.** Where multiple speeds are desired, motors are sometimes operated in tandem or cascade. When so used, two motors are rigidly coupled to the same shaft or are otherwise mechanically linked, as by means of gears. The stator winding of the first is connected to the mains in the usual way, while that of the second stator is fed from the rotor winding of the first, as shown in Fig. 3.62.

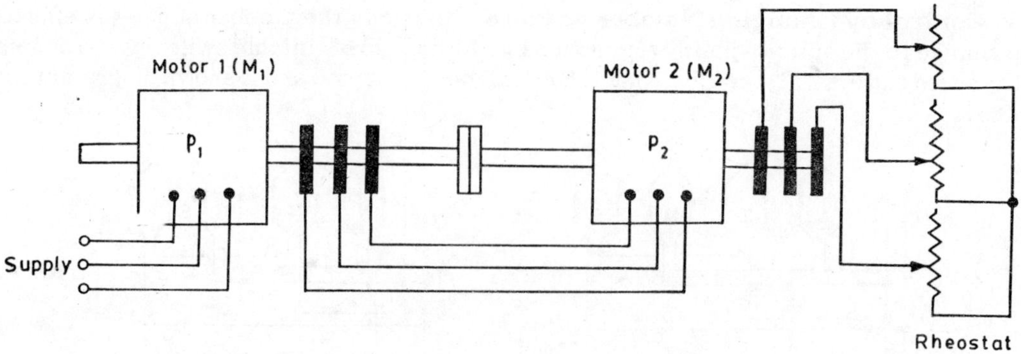

Fig. 3.62. Wound rotor motors connected in tandem.

If the two machines are designed for the same voltage, as is usually the case, the turn ratio of stator to rotor of the first machine should be *unity*. The second rotor may have a cage winding or a polyphase winding like its stator. In the latter case the rotor circuit of the second motor is connected to slip rings in the usual way, in order that resistance may be introduced while starting and for securing additional speed control when running.

The motors may be so connected that both tend to run in the same direction, or the phase rotation of one motor may be reversed, thus tending to make it rotate in the reverse direction. In either case the set will run after it is started, but in the latter case no starting torque is developed, and for this reason this connection is *little used*. If the first machine has p_1 poles and the second has p_2 poles, the synchronous speed of the set is that of a motor with *$p_1 + p_2$ poles for the first case* and *$p - p_2$ poles for the second*. If p_1 and p_2 are not equal, four synchronous speeds are possible, two with tandem operation and one for each motor separately. Some applications of this method of control are found in European railways.

The *expression for the speed of set* is derived as follows :

Let the frequency of the supply voltage be f_1 and let machines M_1 and M_2 have p_1 and p_2 number of poles respectively. Let the two machines, M_1 and M_2 run with slip of s_1 and s_2 respectively.

Speed in r.p.m. of $M_1 = N_1 = \dfrac{120 f_1}{p_1}(1 - s_1)$

Speed in r.p.m. of $M_2 = N_2 = \dfrac{120 f_2}{p_2}(1 - s_2)$

But the shafts are mechanically coupled, therefore

$$N_1 = N_2$$

Also $$f_2 = s_1 f_1$$

Hence substituting the value of f_2, we get

$$N_2 = \frac{120 s_1 f_1}{p_2}(1 - s_2)$$

Equating the expression for N_1 and N_2 and solving for s_1, we have

$$\frac{120 f_1}{N_1}(1 - s_1) = \frac{120 s_1 f_1}{p_2}(1 - s_2)$$

$$\therefore \qquad s_1 = \frac{p_2}{p_1 - p_1 s_2 + p_2} = \frac{p_2}{p_1(1 - p_2) + p_2} \qquad ...(3.43)$$

But when the rheostat is short-circuited, s_2 approaches zero, the above expression then reduces to

$$s_1 = \frac{p_2}{p_1 + p_2} \quad ...(3.44)$$

But $$s_1 = \frac{N_{s1} - N_1}{N_{s1}}$$

where N_{s1} is the synchronous speed of M_1

$$\therefore \quad \frac{N_{s1} - N_1}{N_{s1}} = \frac{p_2}{p_1 + p_2}$$

Solving for N_1, we have $$N_1 = N_{s1} \frac{p_1}{p_1 + p_2}$$

However, $N_{s1}p_1 = 120\ f_1$, so that substituting the value of $N_{s1}p_1$ in the above expression it becomes

$$N_1 = \frac{120 f_1}{p_1 + p_2} \text{ r.p.m.} \quad ...(3.45)$$

Equating (3.45) shows that the speed of the set is that of a single machine having the number of poles equal to the sum of the numbers of poles of the two machines. Hence the set can give four different speeds.

Example 3.88. *The full-load speed of an 8-pole, 50-Hz slip ring motor is 730 r.p.m. The rotor resistance per phase is 0.2 Ω. Calculate the external resistance per phase that must be added to lower the speed to 620 r.p.m. given that the torque is same in the two cases.*

Solution. Number of poles, $p = 8$

Supply frequency, $f = 50$ Hz

Full-load speed, $N = 730$ r.p.m.

Rotor resistance/phase, $R_2 = 0.2\ \Omega$

Final speed $= 620$ r.p.m.

External resistance/phase to be added per phase, r :

As a first approximation $$T \propto \frac{s}{R_2}$$

$$\therefore \quad T_1 \propto \frac{s_1}{R_2} \text{ and } T_2 \propto \frac{s}{(R_2 + r)}$$

or $$\frac{T_1}{T_2} = \frac{s_1}{s_2} \times \frac{(R_2 + r)}{R_2}$$

Now $$N_s = \frac{120 f}{p} = \frac{120 \times 50}{8} = 750 \text{ r.p.m.}$$

$$s_1 = \frac{750 - 730}{750} \times 100 = 2.66\%$$

$$s_2 = \frac{750 - 620}{750} \times 100 = 17.33\%$$

Also, $R_2 = 0.2\ \Omega$ and $T_1 = T_2$ *(given)*

$$\therefore \quad \frac{s_1}{R_2} = \frac{s_2}{(R_2 + r)}$$

or $$\frac{2.66}{0.2} = \frac{17.33}{(0.2 + r)} \quad \text{or} \quad 2.66(0.2 + r) = 0.2 \times 17.33$$

$$r = 1.1.\ \Omega$$

Hence, **external resistance to be added per phase = 1.1 Ω. (Ans.)**

Example 3.89. *A 4-pole, 50-Hz, 3-phase slip ring induction motor when fully loaded runs with a slip of 4 per cent. Determine the value of the resistance to be inserted in series per phase in the rotor circuit to reduce the speed by 12 per cent and the new slip.*

The rotor resistance per phase is 0.25 Ω.

Solution. Number of poles, $p = 4$

Supply frequency, $f = 50$

Slip at full-load $s_1 = 4\%$

Speed to be reduced $= 12$ per cent

Rotor resistance/phase, $R_2 = 0.25\ \Omega$

New slip :

Resistance to be inserted/phase, r :

Synchronous speed, $N_s = \dfrac{120f}{p} = \dfrac{120 \times 50}{4} = 1500$ r.p.m.

Full-load speed, $N_1 = N_s(1 - s_1) = 1500(1 - 0.04) = 1440$ r.p.m.

New speed, $N_2 = (1 - 0.12)N_1 = 0.88 \times 1440 = 1267$ r.p.m.

$\therefore$ **New slip,** $s_2 = \dfrac{N_s - N_2}{N_s} = \dfrac{1500 - 1267}{1500} = \mathbf{0.155}$ or **15.5% (Ans.)**

Since slip, $s = \dfrac{\text{rotor copper loss}}{\text{input power to rotor}} = \dfrac{3I_2^2 R_2}{\text{input power to rotor}}$

For constant input power to rotor and rotor current,

$$\text{slip } s \propto R_2$$

$\therefore$ If a resistance of r ohms be connected in series per phase, then

$$\frac{R_2 + r}{R_2} = \frac{s_2}{s_1}$$

$$\therefore \quad \frac{0.25 + r}{0.25} = \frac{15.5}{4}$$

$$4(0.25 + r) = 0.25 \times 15.5$$

$$\therefore \quad r = 0.719\ \Omega$$

Hence, **external resistance/phase to be added = 0.719 Ω. (Ans.)**

☞**Example 3.90.** *An 8-pole, 50-Hz, 3-phase induction motor is running at 4 per cent slip when delivering full-load torque. It has standstill rotor resistance of 0.1 Ω and reactance of 0.6 Ω per phase. Calculate the speed of the motor if an additional resistance of 0.5 Ω per phase is inserted in the rotor circuit. The full-load torque remains constant.*

Solution. Number of poles, $p = 8$

Supply frequency, $f = 50$ Hz

Slip at full-load, $s_1 = 4\%$

Standstill rotor resistance/phase, $R_2 = 0.1\ \Omega$

Standstill rotor reactance/phase, $X_2 = 0.6\ \Omega$

Additional resistance added/phase, $r = 0.5\ \Omega$

Speed of the motor, N_2 :

Synchronous speed, $N_s = \dfrac{120f}{p} = \dfrac{120 \times 50}{8} = 750$ r.p.m.

The motor torque is given by the expression,

$$T = \frac{KsR_2 E_2^2}{R_2^2 + s^2X_2^2}$$

Let the new slip be s_2.

Now $$T_1 = \frac{Ks_1R_2 E_2^2}{R_2^2 + s^2X_2^2} = \frac{K \times 0.04 \times 0.1 \times E_2^2}{(0.1)^2 + (0.04 \times 0.6)^2} = \frac{KE_2^2}{2.644}$$

and $$T_2 = \frac{Ks_2(R_2 + r) E_2^2}{(R_2 + r)^2 + (s_2X_2)^2}$$

$$= \frac{Ks_2 \times (0.1 + 0.5)E_2^2}{(0.1 + 0.5)^2 + (0.6s_2)^2} = \frac{0.6s_2 KE_2^2}{0.36 + 0.36s_2^2}$$

Also $T_1 = T_2$ *(given)*

$\therefore$ $$\frac{KE_2^2}{2.644} = \frac{0.6s_2 KE_2^2}{0.36 + 0.36s_2^2}$$

or $0.6(1 + s_2^2) = 2.644s_2$ or $s_2^2 - 4.407s_2 + 1 = 0$

or $s_2 = 0.24$ neglecting higher value, being impractical

$\therefore$ Speed of the motor, $N_2 = (1 - s_2)N_s = (1 - 0.24) \times 750 = 570$ r.p.m.

Hence, **speed of the motor = 570 r.p.m. (Ans.)**

Example 3.91. *The rotor of a 4-pole, 50 Hz, slip ring induction motor has a resistance of 0.25 Ω per phase and runs at 1440 r.p.m. at full-load. Calculate the external resistance per phase which must be added to lower the speed to 1200 r.p.m., the torque being the same as before.*

(AMIE Advanced Elec. M/Cs, 1998)

Solution. *Given* : $p = 4$; $f = 50$ Hz ; $R_2 = 0.25\ \Omega$; $N = 1440$ r.p.m. ; $N' = 1200$ r.p.m.

External resistance/phase to be added, r :

Synchronous speed, $$N_s = \frac{120f}{p} = \frac{120 \times 50}{4} = 1500 \text{ r.p.m.}$$

Full-load slip, $$s = \frac{N_s - N}{N_s} = \frac{1500 - 1440}{1500} = 0.04$$

After inserting resistance R ohms per phase in the rotor circuit,

$$N' = 1200 \text{ r.p.m.}$$

$\therefore$ Slip, $$s' = \frac{1500 - 1200}{1500} = 0.2$$

Since slip, $$s = \frac{\text{rotor copper loss}}{\text{input power to rotor}} = \frac{3I_2^2R_2}{\text{input power to rotor}}$$

For constant power input to rotor and rotor current, $s \propto R_2$.

Since for constant load torque power input to rotor and rotor current remains the same, therefore,

$$\frac{s'}{s} = \frac{R_2 + r}{R_2} \quad \text{or} \quad \frac{0.2}{0.04} = \frac{0.25 + r}{0.25}$$

or **r = 1 Ω. (Ans.)**

Example 3.92. *A 75 kW, 415 V, 4-pole, 3-phase, 50 Hz induction motor is delivering rated torque at 5% slip driving a pump load. The motor delivers maximum torque at 10% slip. A 3-phase a.c. regulator controls the r.m.s. supply voltage to effect limited speed control. Assume the torque-speed *characteristics of the motor in the stable region to be linear. Calculate the r.m.s. voltage at which motor runs at 1350 r.p.m.*

(AMIE Electric Drives, Summer, 1999)

Solution. *Given* : $V = 415$ volts ; $p = 4$; $f = 50$ Hz ; $s_f = 5\%$, 0.05 ; $s_{mT} = 10\% = 0.1$, $N' = 1350$ r.p.m.

R.M.S. voltage at which motor runs at 1350 r.p.m. :

Full-load speed, $N = N_s(1 - s_f) = \dfrac{120f}{p}(1 - s_f) = \dfrac{120 \times 50}{4}(1 - 0.05) = 1425$ r.p.m.

Also, $\dfrac{R_2}{X_2} = s_{mT} = 0.1$ or $X_2 = 10R_2$

When speed is $N' = 1350$ r.p.m.

$$s_f' = \frac{N_s - N'}{N_s} = \frac{(120f/p) - N'}{(120f/p)} = \frac{(120 \times 50/4) - 1350}{(120 \times 50/4)} = 0.1$$

For pump load *assuming that the torque varies as the speed so torque* corresponding to 1350 r.p.m.,

$$T' = T \times \frac{1350}{1425} \quad \text{or} \quad \frac{T'}{T} = \frac{1350}{1425}$$

Now, torque developed $= \dfrac{ksR_2E_2}{R_2^2 + s^2X_2^2}$ [Eqn. (3.8)]

$$\therefore \quad \frac{T'}{T} = \frac{(s_f' E_2'^2)/(R_2^2 + s_f'^2X_2^2)}{(s_fE_2^2)/(R_2^2 + s_f^2X_2^2)} = \frac{s_f'(V')^2(R_2^2 + s_f^2X_2)}{s_f(V)^2(R_2^2 + s_f'^2X_2^2)}$$

$\because$ $E_2 \propto V$, the supply voltage

or
$$V' = V \times \sqrt{\frac{T'}{T} \times \frac{s_f}{s_f'} \times \frac{(R_2^2 + s_f'^2X_2^2)}{(R_2^2 + s_f^2X_2^2)}}$$

$$= 415 \times \sqrt{\frac{1350}{1425} \times \frac{0.05}{0.1} \times \frac{(R_2^2 + 0.01 \times 100R_2^2)}{(R_2^2 + 0.0025 \times 100R_2^2)}}$$

$$= 415 \times \sqrt{\frac{1350}{1425} \times 0.5 \times 1.6} = \mathbf{361.3\ volts.} \quad \textbf{(Ans.)}$$

Example 3.93. *A 3-phase, 440 V, 1000 r.p.m. slip-ring induction motor is operating with 2% slip and taking a stator current of 50 A. Speed of the motor is reduced at constant torque to 500 r.p.m. using stator voltage control. Calculate the new value of stator current.*

(AMIE, M/Cs for Special Applications, Winter, 1996)

Solution. *Given* : Supply voltage, $V = 440$ V ; $N_s = 1000$ r.p.m., $s = 2\% = 0.02$; $I_1 = 50$ A ; $N' = 500$ r.p.m.

New stator current, I_1' :

Slip at reduced speed N', $s' = \dfrac{N_s - N'}{N_s} = \dfrac{1000 - 500}{1000} = 0.5$

Since $T \propto sV^2$, therefore, for constant torque

$$V \propto \sqrt{\frac{1}{s}} \quad \text{or} \quad V' = V \times \sqrt{\frac{s}{s'}} = 440\sqrt{\frac{0.02}{0.5}} = 88 \text{ V}$$

Also, stator current $I_1 \propto sV$

$\therefore$ New stator current, $\mathbf{I_1'} = I_1 \times \dfrac{s'V'}{sV} = 50 \times \dfrac{0.5 \times 88}{0.02 \times 440} = \mathbf{250\ A.}$ **(Ans.)**

Example 3.94. *A star-connected cage induction motor has the following data :*

440 V, 50 Hz, 4-pole, 1460 r.p.m., $R_s = 2\ \Omega$, $R_r' = 2\ \Omega$, $X_s = X_r' = 3\ \Omega$.

Calculate the starting torque and starting current of this motor at 50 Hz and 10 Hz for V/f control. **(AMIE Electric Drives Winter, 1999)**

Solution. *Given* : Supply voltage for phase, $V = \dfrac{440}{\sqrt{3}} = 254$ V ; $f = 250$ Hz ; $p = 4$,

$N = 1460$ r.p.m., $R_s = 2\ \Omega$, $R_r' = 2\ \Omega$, $X_s = X_r' = 3.0\ \Omega$.

At 50 Hz supply :

Equivalent resistance of motor referred to stator, $R_{01} = R_s + R_r' = 2 + 2 = 4\ \Omega$

Equivalent reactance of motor referred to stator, $X_{01} = X_s + X_r' = 3 + 3 = 6\ \Omega$

Starting current, $\mathbf{I_{st}} = \dfrac{V}{\sqrt{R_{01}^2 + X_{01}^2}} = \dfrac{254}{\sqrt{4^2 + 6^2}} = \mathbf{35.22\ A.}$ **(Ans.)**

Total copper losses in rotor $= 3I_{st}^2 \times R_r' = 3 \times 35.22^2 \times 2 = 7442.7$ W

Rotor power input at standstill, P_2 = total copper losses = 7442.7 W

Synchronous speed, $N_s = \dfrac{120f}{p} = \dfrac{120 \times 50}{4} = 1500$ r.p.m.

Starting torque, $\mathbf{T_{st}} = \dfrac{P_2}{(2\pi N_s/60)} = \dfrac{7442.7 \times 60}{2\pi \times 1500} = \mathbf{47.38\ Nm.}$ **(Ans.)**

At 10 Hz supply :

Voltage supply per phase, $V' = V \times \dfrac{f'}{f} = 254 \times \dfrac{10}{50} = 50.8$ V

Equivalent resistance of motor referred to stator,

$R_{01}' = 4\ \Omega$, the same as before

Equivalent reactance of motor referred to stator,

$$X_{01}' = X_{01} \times \frac{f'}{f} = 6 \times \frac{10}{50} = 1.2\ \Omega$$

Starting current, $\mathbf{I_{st}}' = \dfrac{V'}{\sqrt{R'^2_{01} + X'^2_{01}}} = \dfrac{50.8}{\sqrt{4^2 + 1.2^2}} = \mathbf{12.16\ A.}$ **(Ans.)**

Power input to rotor at standstill,

P_2' = total copper losses $= 3(I'_{st})^2 R'_r$

$= 3 \times 12.16^2 \times 2 = 887.2$ W

Synchronous speed, $N_s' = \dfrac{120f'}{p} = \dfrac{120 \times 10}{4} = 300$ r.p.m.

Starting torque, $\mathbf{T'_{st}} = \dfrac{P_2'}{(2\pi N_s'/60)} = \dfrac{887.2 \times 60}{2\pi \times 300} = \mathbf{28.24\ Nm.}$ **(Ans.)**

Example 3.95. *Two induction motors with 4-poles and 6-poles respectively are connected in cumulative cascade. The frequency in the secondary circuit of the 6-pole motor is observed to be 2.0 Hz. Determine the combined speed of the set and the slip in each machine.*

Assume supply frequency as 50 Hz. **(AMIE Advanced Elec. M/Cs, Summer, 1999)**

Solution. Number of poles, $p_1 = 4$

Number of poles, $p_2 = 6$

Frequency in the secondary circuit of 6-poles motor,

$f_2 = 2.0$ Hz

Supply frequency, $f_1 = 50$ Hz

Concatenated speed of the set, $N_{sc} = \frac{120 f_1}{p_1 + p_2} = \frac{120 \times 50}{4 + 6} = 600$ r.p.m.

Also slip $= \frac{2}{f_1} = \frac{2}{50} = 0.04$ or 4%

Actual speed of the set,

$N = (1 - 0.04) \times 600 =$ **576 r.p.m. (Ans.)**

Synchronous speed of 4-pole induction motor,

$$N_{s1} = \frac{120 f_1}{p_1} = \frac{120 \times 50}{4} = 1500 \text{ r.p.m.}$$

Slip for 4-pole induction motor,

$$s_1 = \frac{N_{s1} - N}{N_{s1}} = \frac{1500 - 576}{1500} = \mathbf{0.616 \text{ or } 61.6\%.} \text{ (Ans.)}$$

The impressed frequency on the second motor,

$$f_2 = s_1 f_1 = 0.616 \times 50 = 30.8 \text{ Hz}$$

Synchronous speed of 6-pole induction motor with frequency f_2,

$$N_{s2} = \frac{120 f_2}{p_2} = \frac{120 \times 30.8}{6} = 616 \text{ r.p.m.}$$

Slip for 6-pole induction motor,

$$s_2 = \frac{N_{s2} - N}{N_{s2}} = \frac{616 - 576}{616} = \mathbf{0.065 \text{ or } 6.5\%.} \text{ (Ans.)}$$

☞**Example 3.96.** *Two 50 Hz, 3-phase induction motors having six and four poles respectively are cumulatively cascaded, the 6-pole motor being connected to the main supply. Determine the frequencies of the rotor currents and the slips referred to each stator field if the set has a slip of 3 per cent.*

Solution. Supply frequency, $f_1 = 50$ Hz

$p_1 = 6, p_2 = 4$

Slip of the set, $s = 3\%$

Synchronous speed of the set,

$$N_{sc} = \frac{120 \times 50}{6 + 4} = 600 \text{ r.p.m.}$$

Actual rotor speed $= (1 - s)N_{sc} = (1 - 0.03) \times 600 = 582$ r.p.m.

Synchronous speed of the stator field of 6-pole motor,

$$N_{s1} = \frac{120 f_1}{p_1} = \frac{120 \times 50}{6} = 1000 \text{ r.p.m.}$$

Slip referred to this stator field,

$$s_1 = \frac{N_{s1} - N}{N_{s1}} = \frac{1000 - 582}{1000} = 0.418 \text{ or } \mathbf{41.8\%} \text{ (Ans.)}$$

Frequency of the rotor currents of 6-pole motor,

$$f_2 = s_1 f_1 = 0.418 \times 50 = \mathbf{20.9 \text{ Hz.}} \text{ (Ans.)}$$

This is also the frequency of the stator currents of the 4-pole motor.

The synchronous speed of the stator of 4-pole motor

$$N_{s2} = \frac{120 f_2}{p_2} = \frac{120 \times 20.9}{4} = 627 \text{ r.p.m.}$$

The slip as referred to the 4-pole motor is

$$s_2 = \frac{N_{s2} - N}{N_{s2}} = \frac{627 - 582}{627} = 0.0717 \text{ or } \mathbf{7.17\%. \ (Ans.)}$$

The frequency of rotor current of 4-pole motor,

$$f_2' = s_2 f_2 = 0.0717 \times 20.9 = \mathbf{1.5 \ Hz \ (app.) \ (Ans.)}$$

$$[\textit{Check}: f_2' = sf_1 = 0.03 \times 50 = 1.5 \text{ Hz.}]$$

Example 3.97. *Two 50 Hz, 3-phase induction motors having 6 and 4 poles respectively are cumulatively cascaded, the 6-pole motor being connected to the main supply. Determine the frequency of the rotor currents and slips referred to each stator field if the set has a slip of 2%.*

(Madras Univ., 1999)

Solution. *Given* : $f = 50$ Hz ; $p_1 = 6$; $p_2 = 4$; $s = 2\% = 0.02$.

s_1 ; f_1 ; s_2 ; f_2 :

Synchronous speed of the set $N_{sc} = \dfrac{120f}{p_1 + p_2} = \dfrac{120 \times 50}{6 + 4} = 600$ r.p.m.

Actual speed of the set, $N = N_{sc}(1 - s) = 600(1 - 0.02) = 588$ r.p.m.

6-pole motor :

Synchronous speed of 6-pole motor stator field,

$$N_{s1} = \frac{120f}{p_1} = \frac{120 \times 50}{6} = 1000 \text{ r.p.m.}$$

Slip referred to the stator field,

$$\mathbf{s_1} = \frac{N_{s1} - N}{N_{s1}} = \frac{1000 - 588}{1000} = 0.412 \text{ or } \mathbf{41.2\%. \ (Ans.)}$$

The 6-pole motor current frequency (or the impressed frequency on 4-pole motor, stator field)

$$\mathbf{f_1} = s_1 f = 0.412 \times 50 = \mathbf{20.6 \ Hz. \ (Ans.)}$$

4-pole motor :

The synchronous speed of 4-pole induction motor with frequency f_1,

$$N_{s2} = \frac{120 \times f_1}{p_2} = \frac{120 \times 20.6}{4} = 618 \text{ r.p.m.}$$

The slip as referred to the 4-pole induction motor,

$$\mathbf{s_2} = \frac{N_{s2} - N}{N_{s2}} = \frac{618 - 588}{618} = 0.0485 \text{ or } \mathbf{4.85\%. \ (Ans.)}$$

Frequency of rotor current of 4-pole induction motor,

$$\mathbf{f_2} = s_2 f_1 = 0.0485 \times 20.6 = \mathbf{1.0 \ Hz. \ (Ans.)}$$

Example 3.98. *A cascade set comprises two motors 1 and 2 with four poles and six poles respectively. The set is connected to a 50 Hz supply. Find :*

(i) The synchronous speed of the set ;

(ii) The electric power transferred to motor **2** *when the input to motor* **1** *is 30 kW. Neglect losses.*

(Utkal Univ., 1998)

Solution. *Given* : $p_1 = 4$; $p_2 = 6$, Input to motor **1** = 30 kW.

(i) **The speed of the set, N_{sc} :**

Synchronous speed of the set, $N_{sc} = \dfrac{120f}{p_1 + p_2} = \dfrac{120 \times 50}{4 + 6} =$ **600 r.p.m. (Ans.)**

(*ii*) **The electric power transferred to motor 2 :**

Since the outputs of the two motors are proportional to the number of their poles, therefore, power transferred to motor **2** with 6-poles

= power output of motor **2**, neglecting losses

$$= 30 \times \frac{p_2}{p_1 + p_2} = 30 \times \frac{6}{4+6} = \textbf{18 kW.} \quad \textbf{(Ans.)}$$

3.29. ELECTRICAL BRAKING OF POLYPHASE INDUCTION MOTORS

The most simple way to stop an induction motor (or any other type of motor) is to *disconnect the terminals from the supply*. Torque is no longer developed, and the combined effect of the rotor and external load brings the motor to rest.

When *rapid and more positive action is required*, mechanical or electrical braking may be employed, but the *latter* has many advantages, particularly where *precise control of the stopping moment and smoothness of operation are required.*

Following are the *three* main methods of electrical braking of induction motors :

1. Plugging (or counter-current braking) ;
2. Dynamic (or rheostatic) braking ;
3. Regenerative braking.

1. Plugging :

- It is known that the rotor of a polyphase induction motor develops torque in the same direction as the rotating magnetic field set up by the stator winding. Also if any two stator leads are reversed, the rotating magnetic field is also reversed. *If, therefore, a pair of stator leads are reversed while a motor is rotating, torque is suddenly produced opposite to the original direction of rotation.* This reverse torque causes rotation in the opposite direction as soon as the motor stops, therefore provision must be made *to disconnect stator completely from the supply lines when the motor stops. A plugging controller must therefore be used in conjunction with a switch.* The plugging switch is coupled to the motor and opens when the motor stops. It is connected in the control circuit so that it permits operation of controller's reversing current when the motor rotates in a given direction only. *When the motor stops, the reversing circuit is opened by the plugging switch, and the motor is completely disconnected from the supply lines.*
- The moment the stator field reverses, the slip suddenly increases from a *small per cent to* **two hundred per cent,** since the rotor and field are now rotating in opposite directions. The rotor induced voltage also increases by a great amount, and hence there is a large in rush of stator current. Moreover, *the high slip causes a high rotor reactance,* and *hence a very low power factor in the stator.*

  After plugging has started, *the best possible conditions occur when the motor stops,* slip is unity, and the current and power factor are the same as when the motor is started with full voltage. Although the above conditions occur during a very brief period of time, they may *cause disturbances in the supply* and therefore, this, as well as how frequently plugging is required, must be considered before a decision can be made to as whether or not plugging is the best way of bringing a motor to quick halt.
- During the period of plugging, since the induced e.m.f. in the rotor is very high, the rotor current and thereby stator current are very high. However, braking current can be reduced by inserting external resistance in the rotor circuit and as such wound rotor motors are beneficial as compared to squirrel cage rotor motors.

— Squirrel cage motors of about 20 kW output are plugged direct, using the star connection if a star-delta switch is provided. *Larger motors need stator resistors.*

— Wound rotor motors employ rotor resistors for current limitation as well as developing higher braking torques.

- Plugging is *advantageously employed* in the case of *reversing devices* where braking and starting up of induction motor in reverse direction comprises stages of the same continuous process.

2. Dynamic (or rheostatic) braking :

- The dynamic or rheostatic braking in case of a polyphase induction motor can be realised by disconnecting the stator winding from the A.C. supply and exciting it from (*a*) D.C. source *to produce a stationary D.C. field.*
- In dynamic or rheostatic braking, the stator winding is employed on a D.C. field winding and the rotor winding as an armature winding.

— With a wound rotor machine, external resistors can be inserted into the rotor circuit to provide a load.

— With squirrel cage machines, however, the rotor winding itself has to form the load.

The source of excitation may be provided by :

(*i*) an independent D.C. source, or

(*ii*) from the A.C. mains through a transformer-rectifier set.

- The magnitude of the braking torque developed by the motor depends upon

(*i*) The excitation (strength of the field developed by the stator winding) ;

(*ii*) The rotor circuit resistance ;

(*iii*) The speed of the motor.

Braking torque can be controlled by any or both of the following methods :

1. By controlling the D.C. excitation.

2. By varying the rotor resistance.

- A.C. dynamic braking is *not popular* owing to relatively high cost of capacitor banks.
- D.C. dynamic braking *takes little power from the supply and provides smooth braking torque, useful for mine winders and high inertia loads.*

It entails the following *advantages over plugging* :

(*i*) The absence of the reverse-rotation air-gap field (and, therefore, no tendency for the machine to run backwards).

(*ii*) Lower rotor I^2R loss.

- The advent of automatic control of dynamic braking of induction motors employing closed loop systems has made induction motors more popular than D.C. motors, especially for the drives employed in mine hoists.

3. Regenerative braking :

- When an induction motor runs at speed above synchronous speed, it operates as an induction generator and feeds power back to the supply line ; thus *regenerative braking is an inherent characteristic of an induction motor.*

The following processes may be employed to operate 3-phase induction motor at speed above synchronous speed :

(*i*) Downward motion of a loaded hoisting mechanism such as crane hoists, excavators etc.

(*ii*) Switching over to a low frequency supply in frequency controlled induction motors in order to reduce the speed of operation of the drive.

(*iii*) Switching over to a large number of pole operation from a smaller one in multi-speed squirrel cage motors.

In all the processes mentioned above the slip and torque developed become *negative* and thus the *machine acts as a generator*, receiving mechanical energy and giving it back to the supply system in the form of electrical energy.

- In case of a squirrel cage induction motor, stable speed is obtained at a speed considerably in excess of the synchronous speed and the regenerative braking cannot be applied unless the motor is specially designed to withstand the excessive speed.
- The main *disadvantage* of this method of braking is the possibility of braking only at super-synchronous speeds and, therefore, *seldom used for braking*.
- This method *can be used only in hoisting type of mechanism or with a multi-speed squirrel cage motor*. It returns about 20% of the total energy on certain railway runs and saves a great deal of brake shoe wear.

Example 3.99. *The following parameters relate to a 3-phase, 400 V, 50 Hz, 8-pole star-connected induction motor :*

$R_1 = 0.12\ \Omega$, $R_2 = 0.6\ \Omega$, $X_1 = 0.72\ \Omega$, $X_2 = 2.88\ \Omega$, $s_f = 0.05$ *and the ratio of effective stator to rotor turns* = 1 : 2.

The motor is to be braked at rated speed and an external resistance of 2.4 Ω per phase (referred to stator) has been inserted into the rotor circuit. Determine the initial braking torque in (i) plugging and (ii) d.c. dynamic braking.

Solution. *Given* : Supply voltage = 400 V ; $f = 50$ Hz ; $p = 8$; $R_1 = 0.12\ \Omega$; $R_2 = 0.6\ \Omega$; $X_1 = 0.72\ \Omega$; $X_2 = 2.88\ \Omega$; $s_f = 0.05$, $K = 2$; external resistance, $R' = 2.4\ \Omega$.

Rotor resistance referred to stator, $R_2' = \dfrac{R_2}{K^2} = \dfrac{0.6}{2^2} = 0.15\ \Omega$

Rotor leakage reactance referred to stator, $X_2' = \dfrac{X_2}{K^2} = \dfrac{2.88}{2^2} = 0.72\ \Omega$

(*i*) **During plugging :**

Slip, $s = 2 - s_f = 2 - 0.05 = 1.95$

Synchronous speed, $N_s = \dfrac{120f}{p} = \dfrac{120 \times 50}{8} = 750$ r.p.m.

Initial rotor current during plugging,

$$I_2' = \frac{V}{\left[R_1 + \dfrac{R_2' + R'}{s}\right]^2 + (X_1 + X_2')^2}$$

$$= \frac{(400/\sqrt{3})}{\sqrt{\left[0.12 + \dfrac{(0.15 + 2.4)}{1.95}\right]^2 + (0.72 + 0.72)^2}} = 113.9 \text{ A}$$

Initial braking torque, $\mathbf{T_b} = \dfrac{3}{(2\pi N_s/60)} I_2'^2 \times \dfrac{R_2' + R'}{s}$

$$= \frac{3 \times 60}{2\pi \times 750} \times (113.9)^2 \times \frac{(0.15 + 2.4)}{1.95} = \mathbf{648\ Nm.}\ \ \textbf{(Ans.)}$$

(*ii*) **During D.C. dynamic braking :**

Figs. 3.63 and 3.64 show respectively the equivalent circuit of the motor corresponding to slip of 0.05 and equivalent circuit of the machine during dynamic braking at rated speed.

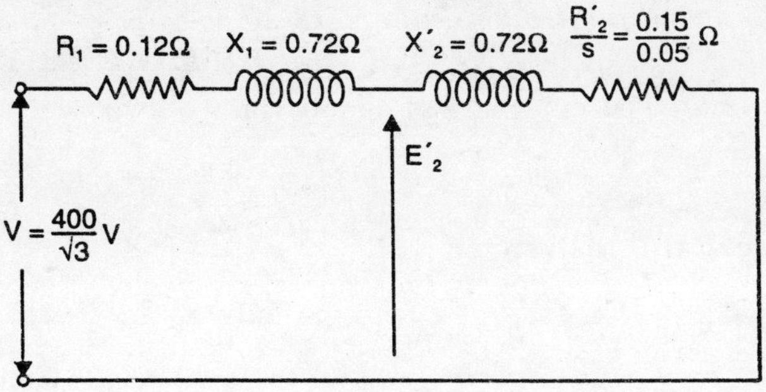

Fig. 3.63

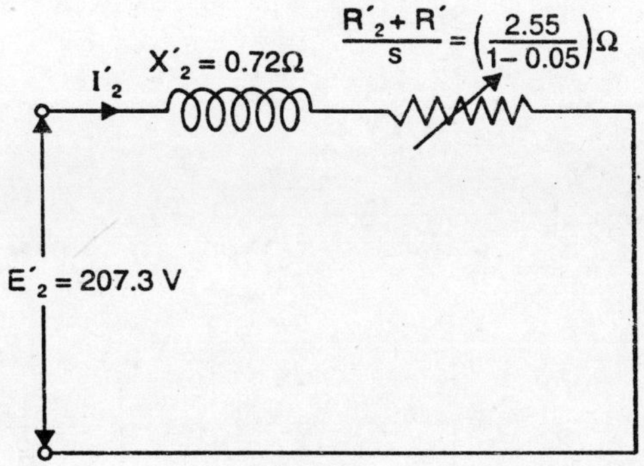

Fig. 3.64

$$E_2' = \frac{400}{\sqrt{3}}\sqrt{\frac{\left(\frac{0.15}{0.05}\right)^2 + (0.72)^2}{\left(\frac{0.15}{0.05} + 0.12\right)^2 + (0.72 + 0.72)^2}}$$

$$= 230.9\sqrt{\frac{9.518}{9.734 + 2.074}} = 207.3 \text{ V}$$

During D.C. dynamic braking,

$$s = 1 - 0.05 = 0.95$$

$$I_2' = \frac{207.3}{\sqrt{\left(\frac{2.55}{0.95}\right)^2 + (0.72)^2}} = 74.6 \text{ A}$$

Initial braking torque, $\mathbf{T_b} = \dfrac{3}{(2\pi N_s/60)} \times I_2'^2 \times \dfrac{R_2' + R'}{s}$

$$= \frac{3 \times 60}{2\pi \times 750} \times (74.6)^2 \times \frac{2.55}{0.95} = \mathbf{570.6\ N\text{-}m.}\ \ \textbf{(Ans.)}$$

Example 3.100. *A 30 kW rated output, 400 V, 3-phase delta-connected, 4-pole, 50 Hz induction motor has full-load slip of 5%. If the ratio of standstill reactance to resistance per rotor phase is 4,* *estimate the plugging torque at full-load speed. Ignore stator leakage impedance and magnetizing reactance.*

(AMIE Elect. Drives, Summer, 1997)

Solution. *Given* : Rated output = 30 kW, supply voltage = 400 V ;

$$p = 4 \,;\, f = 50 \text{ Hz} \,;\, s_f = 5\% \text{ or } 0.05 \,;\, \frac{X_2}{R_2} = 4$$

Plugging torque at full-load speed :

Synchronous speed, $N_s = \frac{120f}{p} = \frac{120 \times 50}{4} = 1500$ r.p.m.

Full-load speed, $N_f = N_s(1 - s_f) = 1500(1 - 0.05) = 1425$ r.p.m.

Full-load torque, $T_f = \frac{\text{rated output in kW} \times 1000}{(2\pi N/60)}$

$$= \frac{30 \times 1000 \times 60}{2\pi \times 1425} = 201.04 \text{ N-m}$$

Now, $T \propto \frac{sR_2E_2^2}{R_2^2 + (sX_2)^2}$

$$\therefore \quad \frac{T_p}{T_f} = \frac{(s_pR_2E_2^2)/(R_2^2 + s_p^2X_2^2)}{(s_fR_2E_2^2)/(R_2^2 + s_f^2X_2^2)} = \frac{s_p\,(R_2^2 + s_f^2X_2^2)}{s_f\,(R_2^2 + s_p^2X_2^2)}$$

or
$$\frac{T_p}{T_1} = \frac{s_p}{s_f}\left[\frac{1 + s_f^2\left(\frac{X_2}{R_2}\right)^2}{1 + s_f^2\left(\frac{X_2}{R_2}\right)^2}\right] = \frac{s_p}{s_f}\left[\frac{1 + 16s_f^2}{1 + 16s_p^2}\right] \qquad \left(\because \frac{X_2}{R_2} = 4\right)$$

Putting $s_f = 0.05$ and s_p = slip at plugging = $2 - s_1 = 2 - 0.05 = 1.95$, we get

Plugging torque, $\mathbf{T_p} = \frac{1.95}{0.05}\left[\frac{1 + 16 \times (0.05)^2}{1 + 16 \times (1.95)^2}\right] T_f$

$$= 39 \times \frac{1.04}{61.84} \times 201.04 = \mathbf{131.86 \text{ N-m.}} \quad \textbf{(Ans.)}$$

3.30. FACTORS GOVERNING PERFORMANCE OF INDUCTION MOTORS

The operating or performance characteristics of an induction motor are governed by the following factors/parameters :

1. Rotor resistance ; 2. Air gap length ;
3. Shapes of both stator and rotor slots and teeth.

1. Rotor resistance :

- The rotor resistance is the most important parameter and is determined primarily by the material used in construction and the area of cross-section of the end rings in case of a squirrel case induction motor. The commonly used materials are copper and aluminium.
- An increase in rotor resistance affects the performance of the motor as mentioned below :
  - — With the *increase in starting resistance the starting torque increases until it attains the maximum value, after which it diminishes.*
  - — An *increase in rotor resistance reduces the starting current and operating efficiency and causes increase in per cent regulation.* The maximum torque remains unaffected because it depends upon the rotor reactance only.

2. Air gap length :

- The air gap is one of the two main sources of the low power factor at which the induction motor operates.
- Too much reduction of air-gap may cause
  — *increase in noise, pulsation losses and magnetic pull ;*
  — *reduction in over-load capacity.*

3. Shapes of both stator and rotor slots and teeth :

- The shapes of rotor teeth and slots affect the reactance, which in turn, affects the starting current and maximum torque. The induction motors with deeper slots have more leakage reactance and so less value of starting current, maximum torque and slip corresponding to maximum torque.
- The slots in the rotor are *not always* made parallel to the shaft, but are given a slight twist or skew. The *motor noise and vibration, cagging and synchronous cups can be reduced or even eliminated by skewing either the stator or rotor slots* (in India, practice is to skew the rotor slots).

  Harmonic torques can be *reduced* by :
  —chording ;
  —integral slot winding ;
  —increasing air-gap length ;
  —skewing of rotor slots.

Note. Space harmonic fields are developed by the **windings, slotting, magnetic saturation, gap length irregularity**. These harmonic fields induce e.m.fs. and circulate harmonic currents in the rotor windings. These harmonic currents in the rotor interact with the harmonic fields and develop *harmonic torques, vibration* and *noise*.

3.31. EFFECTS OF OPERATING CONDITIONS

Effects of operating conditions are briefly described below :

1. Effect of unbalance supply voltage :

- An unbalance supply voltage produces a rotating field which will move at a non-uniform rate and in a non-uniform strength. It provides *unbalanced currents in the stator winding which produce unequal heating.*

2. Effect of loading :

- As the *load increases, speed falls and consequently torque increases, till a point is reached when the maximum torque will be developed.*
- If the load is *increased further*, there is further *fall in the speed and the driving torque decreases and ultimately the rotor comes to standstill.*

3. Effects of variation in line frequency and line voltage :

- The variation in supply frequency and supply voltage not only affects the speed and torque of the motor but also the starting current, full-load current, starting torque, maximum torque and operating temperature of various parts of the machine and thus make their *operation unsatisfactory*.

4. Effects of break in one phase :

- When a polyphase induction motor is in operation and if its one phase breaks, may be due to any reason, the motor will continue to operate single phase, provided that load does not exceed 57.7 percent of the normal rating, with about the same temperature rise as when carrying rated load as a 3-phase induction motor. It will not, however, start single phase but if brought upto speed by some external means, such as putting on the belt, it will continue to operate satisfactorily on load below 57.7 percent of rated one.

- A break in one phase of the rotor of a wound-rotor induction motor may prevent it from starting but if it is brought upto speed before the circuit is opened, it will operate under reduced load but usually with considerable vibrations.

3.32. RATINGS OF 3-PHASE INDUCTION MOTORS

- Three-phase induction motors are rated in terms of the following :

(*i*) Power output in kW ; (*ii*) Speed ;
(*iii*) Voltage ; (*iv*) Frequency ;
(*v*) Phase (single or three) ; (*vi*) Line current ;
(*vii*) Temperature rise in specified time.

— If a motor is designed to be operated on more than one voltage or at more than one speed by reconnecting the winding, a connection diagram is also given on the name plate.

— Class of motors *A, B, C, D, E* or *F* is also to be mentioned on the name plate.

- *Standard types of squirrel cage motors are* :
  — Class *A* motors—Low-impedance squirrel cage rotor motors.
  — Class *B* motors—High-reactance squirrel cage rotor motors.
  — Class *C* motors—High starting torque, low starting current-double squirrel cage motors.
  — Class *D* motors—High resistance squirrel cage rotor motors.
  — Class *E* motors.
  — Class *F* motors.

3.33. SQUIRREL-CAGE MOTORS—ADVANTAGES, DISADVANTAGES AND APPLICATIONS

Advantages :

1. Cheaper in cost.
2. Simple and rugged in construction.
3. Maintenance cost is low.
4. Explosion proof.
5. Can be cooled better because of base end rings.
6. More pull-out torque and greater maximum power output.
7. Simple starting arrangement.
8. Nearly constant speed.
9. High overload capacity.
10. High power factor.

Disadvantages :

1. *Poor starting torque* due to low rotor resistance.
2. At the instant of switching on, *draws a large current* from the line.
3. Very *sensitive to change in supply voltage.*
4. *At light load the power factor is very low.*
5. Speed regulation *not possible.*

Applications. Suitable for industrial drives of small power where *speed control is not required* such as :

- Flour mills
- Printing machinery
- Other shaft drives of small power.

Note. When the squirrel cage motors are used for crane or hoist work where large starting torque is more important than high efficiency owing to the intermittent nature of load, the rotor resistance is increased by employing the end rings, and sometimes the bars as well, of high resistance metal such as German silver.

3.34. WOUND ROTOR (OR SLIP RING) INDUCTION MOTORS-ADVANTAGES, DISADVANTAGES AND APPLICATIONS

Advantages :

1. High starting torque.
2. High over-load capacity.
3. Nearly constant speed.
4. Low starting current (in comparison with squirrel cage motor).

Disadvantages :

1. Low power factor at light loads.
2. Lower efficiency and lower power factor (in comparison to squirrel cage motor).
3. Sensitivity to fluctuations in supply voltage.
4. Higher initial and maintenance costs.
5. Speed regulation is poor when operated with external resistances in the rotor circuit.

Applications. Suitable for most industrial drives of high power where high starting torque is required such as for driving :

- Line shafts
- Pumps
- Lifts
- Generators
- Winding machines
- Mills etc.

3.35. COMPARISON OF A SQUIRREL CAGE AND A SLIP RING (OR PHASE WOUND) INDUCTION MOTORS

	Property	*Squirrel cage induction motor*	*Phase wound induction motor*
1.	**Starting**	The motor can be started with a star-delta starter and it eliminates the hazards of slip rings, brush gear etc.	The motor requires slip rings, brush gear, short-circuiting device and starting resistance etc.
2.	**Overhang**	Less	Large
3.	**Space factor in slots**	Better	Poor
4.	**Cooling**	The losses can be dissipated much efficiently since the end rings are bars and there is more space for providing an efficient fan.	Most of the space is taken up by the overhang and less space is left for the provision of a good cooling fan ; thus the cooling is not quite efficient.
5.	**Efficiency**	High efficiency (only for machines, which are not redesigned for high starting torque).	Low efficiency.
6.	**Starting torque**	For plain squirrel cage motor starting torque is poor which cannot be increased in an ordinary way.	Starting torque can be increased by the insertion of external resistance in each phase of the rotor.

7.	**Copper losses**	Since overhang is less and space factor is better the copper losses are *small*.	Due to large overhang and poor space factor, the copper losses are *more*.
8.	**Maintenance**	Less	More
9.	**Cost**	*Cheaper* (since there are less components and less labour required for winding).	Costlier

3.36. COMPARISON BETWEEN INDUCTION AND SYNCHRONOUS MOTORS

The comparison between induction and synchronous motors is given below :

S. No.	*Induction motor*	*Synchronous motor*
1.	It has got self-starting torque and no special means are required for starting.	It is inherently not self-starting and some external means are required for its starting.
2.	Does not require D.C. excitation.	Requires D.C. excitation.
3.	Speed can be controlled but to small extent.	Speed control not possible.
4.	Its speeds falls with the increase in load and is always less than synchronous speed.	Its average speed is constant and independent of load.
5.	It operates at only lagging power factor, which becomes very poor at light loads.	It can be operated under a wide range of power factor, both lagging and leading.
6.	Its torque is more sensitive to change in supply voltage.	Its torque is less sensitive to change in supply voltage.
7.	Breakdown torque is proportional to the square of the supply voltage.	Breakdown torque is proportional to the supply voltage.
8.	More simple and less costly comparatively.	More complicated and more costly comparatively.
9.	Employed for supplying mechanical load only.	Employed for supplying mechanical load as well as for power factor improvement.

3.37. SCHRAGE MOTOR

The *schrage motor is a useful motor having shunt characteristics and an adjustable speed.*

Fig. 3.65 shows connection diagram of the schrage motor, showing the primary winding and regulating winding which are mounted on the rotor, and the secondary or stator winding with its starting resistance.

Fig. 3.66 shows the general arrangement of schrage motor.

The *primary winding* unlike that of the induction motor is *located on the rotor* and it is connected to the supply by means of slip rings and the brushes ; the *secondary winding is located on the stator.*

Another winding called the *regulating winding*, is *housed in the rotor slots* in addition to the primary winding, and this is connected to a commutator like the armature winding of a D.C. motor.

Two brush rockers each carrying three brush sets, are provided, and these are arranged so that they can be rotated in opposite directions by means of a handwheel or a remote-control device. The three brushes of one rocker are connected to the respective 'starts' of the three-phase of the stator or secondary winding, while the 'finishes' of these windings are connected to the corresponding

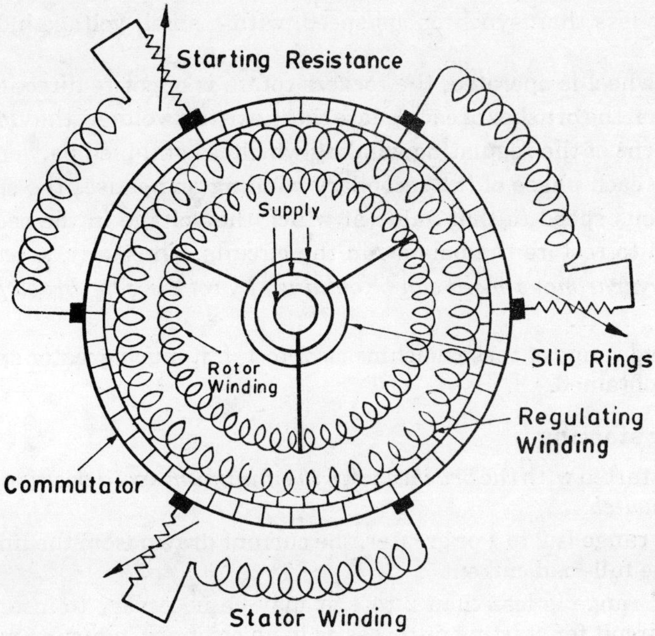

Fig. 3.65. Connection diagram of the schrage motor.

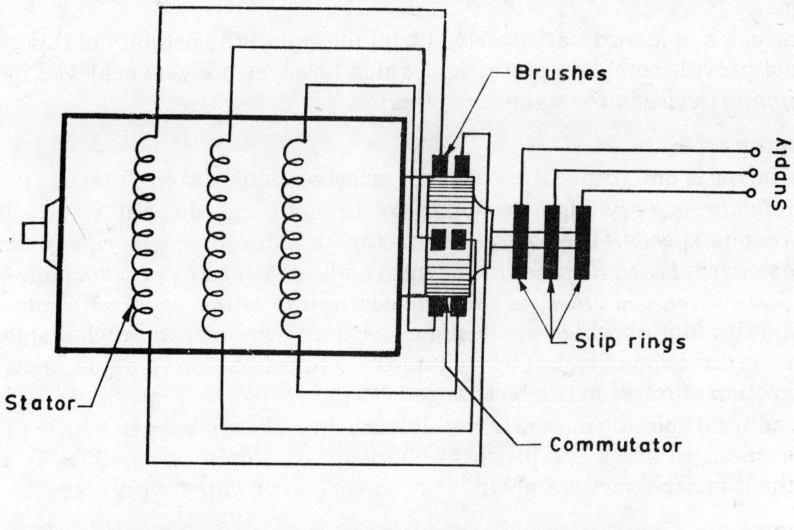

Fig. 3.66. General arrangement of the schrage motor.

brushes on the second rocker. In this way, the *greater the separation of the brushes, the greater is the amount of regulating winding which is connected in series with each phase of the secondary winding.*

Control of speed :

When the brush rockers are 'in line', so that the brushes connected to the ends of the same phase of the secondary winding are in contact with the same commutator segments, the secondary winding is short-circuited through these segments. The machine is then virtually an induction motor, so that it runs slightly less than synchronous speed, with a small voltage induced in each of the secondary windings.

When the handwheel is operated, the rockers rotate is opposite directions and the brushes separate, so that between the brushes of each phase there exists a voltage, the value of which depends upon the number of turns of the regulating winding which is contained between the brushes. This voltage is injected into each phase of the secondary winding and causes the speed of the motor to change from synchronous speed to that value at which the voltage in the secondary windings is increased or decreased to restore the balance in the circuits. Thus, *any, speed within reasonable limits, above or below synchronous speed can be obtained by rotating the brush gear in one direction or the other.*

The normal speed range of these machines is 3 to 1, but much greater speed ranges up to 15 to 1, or higher may be obtained.

Conditions for starting :

The machine is started with the *brushes set in the minimum speed position, when it is necessary only to close the main switch.*

- If the speed range is 2 to 1 or greater, the current drawn from the line is limited to about 1.5 times the full-load current.
- If the speed range is less than 2 to 1, it may be necessary to insert resistances in the secondary circuit for starting purposes as in an ordinary induction motor ; otherwise, the starting torque may be inadequate and the starting current excessive.
- When starting, to *ensure that the brushes are in the minimum speed position, it is usual to provide a small switch operated by the brush gear and interlocked with the starting switch.*
- The inclusion of an ordinary overload tripping coil in the primary of this type of machine does not provide complete protection, which however, may be achieved by *incorporating the tripping device in the secondary circuit.*

Power factor :

- Power factor is one respect in which the schrage motor differs from the induction motor.
- Power factor is very high at hyper-synchronous speeds, but relatively low at sub-synchronous speeds. The low power factor at reduced speeds *can be improved by an unsymmetrical brush displacement* ; this can be *achieved by allowing one brush rocker to move faster than the other, so that the centre line of the brushes is rotated around the commutator.* Motors which are designed with unsymmetrical brush displacement can be run in one direction only, but by a simple re-connection and re-adjustment of the brushes the direction of rotation can be changed.
- The crane and hoisting motors (reversible machines) are designed with symmetrical brush displacement to make the brush mechanism as simple as possible. To reverse such a machine it is necessary simply *to interchange two primary supply leads.*

Advantages :

1. High average efficiency.
2. High power factors.
3. It has a continuous speed regulation within the required range, so that the driven machinery may be run at the best possible speeds.

Important industrial applications :

1. Printing industry, in which variable speed is essential.

2. Suitable for driving the smallest presses as for driving the large rotary presses used in newspaper production.

3.38. INDUCTION GENERATOR

- *Any induction motor, if driven above its synchronous speed when connected to an A.C. power source, will deliver power to the external circuit.*

  This generator operation is easily visualised for motor-circle diagram (Fig. 3.38) corresponding to the lower half of the circle in which the current vector is described below the *OS* line. A unique feature is that the power factor of the output is fixed in value by the generator characteristics and is always leading, independent of the external circuit. The explanation is that the generator draws all its excitation from the system and so must receive a definite amount of lagging kVA for a given voltage and load currents. For this reason, induction generators alone cannot supply a power system but *must always operate in parallel with synchronous machines or with capacitors*. They are, therefore, no more helpful in system stability than the addition of parallel reactors with a rating equal to the generator magnetizing reactance.

  An induction generator delivers an instantaneous 3-phase short-circuit current equal to the terminal voltage divided by its standstill reactance, but *its rate of decay is much faster than that of a synchronous generator of the same rating,* and its sustained short-circuit current is zero.

  Such an induction generator *must have a laminated rotor,* to *provide for the slip frequency rotor magnetic field,* its construction is not adapted to high speeds as synchronous machines employing solid steel rotors. For these various reasons, induction generators have found few *practical applications,* their chief use being perhaps in *variable-ratio frequency converter sets,* where the induction end of the set operates as a motor or a generator depending on the direction of power flow through the set.

- The principle of the induction generator is important, however, in the *dynamic braking of induction motors* ; the machine acts as an overspeed device and produces braking action when the motor speed exceeds synchronous speed, such an induction motor will automatically become an induction generator in that event. It is also important in computing short-circuit system protection when induction motors are used.

Advantages :

1. It does not hunt or drop out of synchronism.
2. Simple and rugged in construction.
3. Cheaper in cost.
4. Easy maintenance.
5. Induction regulators provide a constant voltage adjustment depending on the loading of the lines.

Disadvantages :

1. Cannot be operated independently.
2. Can deliver only leading current (whereas most commercial loads require lagging current).
3. Dangerously high voltages may occur over long transmission lines if the synchronous machines at the far end become disconnected and the line capacitance excites the induction machines.
4. The induction generators are not helpful in system stability.

Applications : The induction generator is very useful in the following *cases / applications* :

1. When the prime-mover does not run at constant speed (such as in hydro-electric stations having a variable low-head water supply.

2. For installation in small power stations where it can be operated in parallel and feeding into a common mains without attendant.
3. For braking purposes in railway work.

3.39. LINEAR INDUCTION MOTOR

3.39.1. Introduction. A linear induction motor is a special type of induction motor which *provides linear motion* and works on the following principle (same as that of conventional induction motor).

"*Whenever a relative motion occurs between the field and short-circuited conductors, currents are induced in them which results in electro-magnetic forces and under the influence of these forces, according to Lenz's law, the conductors try to move in such a way as to eliminate the induced currents*". In this case the field's movement is linear and so is the conductors' movement.

3.39.2. Construction and Working

Construction. A linear induction motor, in its simplest form, consists of field system having a 3-ϕ distributed winding placed in slots as shown in Fig. 3.67 (short-single primary).

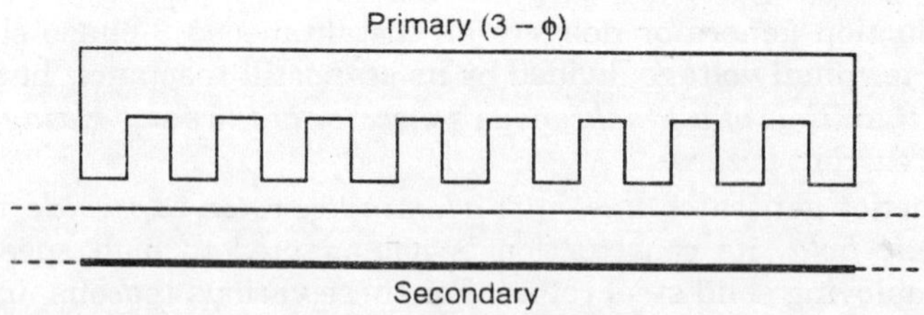

Fig. 3.67. Short-single primary.

The field system may be single or double primary system. The *secondary* of this type of induction motor is normally a *conducting plate made of either copper or aluminium in which interaction currents are induced*.

Depending upon the particular requirements either member can be the *stator*, the other being the *rotor*.

— The ferromagnetic plate, in a single primary system, is usually placed on the other side of the conducting plate to provide a path of low reluctance to the main flux. The ferromagnetic plate, however, gets attracted towards the primary when the field is energised ; consequently *unequal gap length results on the two sides of the plate. Double primary system* can be used to overcome this problem.

— The use of the motor decides which of the two primary and secondary will be shorter in length compared to other. The *primary is made shorter than secondary when the operating distance is large* (since winding a very long 3-phase primary is costly proposition) and the short secondary is used when the operating distance is limited.

Working :

When the 3-ϕ primary winding of the motor is energised from a balanced three phase source, a magnetic field moving in a straight line from one end to other at a *linear synchronous* speed V_s is produced. The linear synchronous speed V_s is given as :

$$V_s = 2\,\tau f \text{ m/s} \qquad \text{...(3.45)}$$

where, τ = pole pitch in metres, and

f = supply frequency in Hz.

As the flux moves linearly, it drags the rotor plate alongwith it in the same direction with speed V. Consequently the relative speed of travel of the flux w.r.t. rotor plate decreases. (In case rotor plate speed equals that of magnetic field, the latter would be stationary when viewed from rotor plate. When the rotor plate moves faster than synchronous speed of magnetic field, the direction of force is reversed and a form of regenerative braking based on the principle of induction generator, will come into existence).

Slip of the motor(s) is given as :

$$s = \frac{V_s - V}{V_s} \quad ...(3.46)$$

where V_s = linear synchronous speed, and
V = actual speed of the rotor plate.

Thrust for force or tractive effort (F) is given as :

$$F = \frac{P_2}{V_s} \quad ...(3.47)$$

where P_2 = actual power supplied to the rotor.

Also, copper losses in rotor $= sP_2$, ...(3.48)

and mechanical power developed, $P_{mech.} = (1 - s)P_2$ ...(3.49)

Fig. 3.68 shows the thrust or tractive effort-speed characteristics. In a linear induction motor the following peculiar effects are encountered :

(*i*) Transverse edge effect ;

(*ii*) End effect.

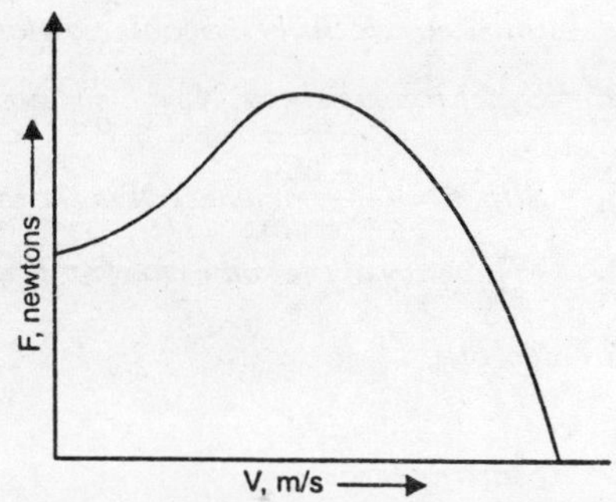

Fig. 3.68. Thrust-speed characteristics.

Due to the secondary of this motor being a solid conducting state, the paths of the induced currents in the secondary are not well defined. The portion of the current paths parallel to the direction of motion of the secondary does not contribute anything towards the production of useful thrust but only contributes towards losses. This effect causes reduction in thrust and increases the losses and is known as **transverse edge effect** (since the current paths parallel to the direction of motion are more towards the edges of the conducting plate).

Advantages :

1. Simpler in construction.
2. Better power to weight ratio.
3. Low initial cost.
4. No over-heating of rotor (since the motor moves continuously over cool rotor plate leaving behind heated rotor portion).
5. Owing to absence of rotating parts the maintenance cost is low.
6. No limitation of tractive effort due to adhesion between the wheel and the rail.
7. No limitation of maximum speed due to centrifugal forces.

Disadvantages :

1. Owing to transverse edge and end effects utilisation of motor is poor.
2. Capital cost of reaction rail fixed along the centre line of the track is very high.
3. Provision of three phase collector system along the track involves complications and high cost.
4. Maintaining adequate clearance at points and crossings entails a lot of difficulties.

5. Requirement of larger air-gap and non-magnetic reaction rail (rotor plate) necessitates more magnetising current resulting in poor efficiency and low-power factor.

Applications :

Following are the fields of application of linear induction motor :

- Electromagnetic pumps
- Conveyors
- High-speed rail traction
- Trolley cars (for internal transport in workshops)
- As booster accelerator for moving heavy trains from rest or up the planes or on curves.
- Metallic belt conveyors etc.

Owing to design difficulties and economic considerations the use of linear induction motor is limited to only a few applications.

HIGHLIGHTS

1. *An induction motor is simply an electric transformer whose magnetic circuit is separated by an air gap into two relatively movable portions, one carrying the primary and the other the secondary winding.*
2. *Synchronous speed,* $N_s = \dfrac{120f}{p}$, *where f is the frequency and p the number of poles.*
3. *Slip,* $s = \dfrac{N_s - N}{N_s}$, *where N is the motor speed (r.p.m).*
4. *Frequency of the rotor current,* $f_r = sf$, *where s is the slip and f is the supply frequency.*
5. *Torque,* $T = \dfrac{ksR_2 E_2^2}{R_2^2 + s^2X_2^2}$

   Starting torque, $T_{st} = \dfrac{kR_2 E_2^2}{R_2^2 + X_2^2}$

   Condition for maximum torque, $s = \dfrac{R_2}{X_2}$

   Maximum torque, $T_{max} = \dfrac{kE_2^2}{2X_2}$
6. **Starting torque** *is proportional to the* **square of applied voltage.**
7. $\dfrac{T_f}{T_m} = \dfrac{2s_f s_{mT}}{s_f^2 + s_{mT}^2}$, *where* $s_f = \dfrac{\text{rotor resistance}}{\text{rotor standstill reactance}}$.

   $\dfrac{T_{st}}{T_{max}} = \dfrac{2s_{mT}}{s_{mT}^2 + 1}$.
8. $\dfrac{\text{Rotor copper loss}}{\text{Rotor gross output}} = \dfrac{s}{1 - s}$.
9. $\dfrac{\text{Rotor output}}{\text{Rotor input}} = 1 - s$.

OBJECTIVE TYPE QUESTIONS

(A) Choose the Correct Answer :

1. Which of the following components is usually fabricated out of silicon steel ?

   (*a*) Bearings (*b*) Shaft

   (*c*) Stator core (*d*) None of the above.

2. The frame of an induction motor is usually made of
(*a*) silicon steel (*b*) cast iron
(*c*) aluminium (*d*) bronze.
3. The shaft of an induction motor must be
(*a*) stiff (*b*) flexible
(*c*) hollow (*d*) any of the above.
4. The shaft of an induction motor is made of
(*a*) high speed steel (*b*) stainless steel
(*c*) carbon steel (*d*) cast iron.
5. In an induction motor, on no-load the slip is generally
(*a*) less than 1% (*b*) 1.5%
(*c*) 2% (*d*) 4%.
6. In medium sized induction motors, the slip is generally around
(*a*) 0.04% (*b*) 0.4%
(*c*) 4% (*d*) 14%.
7. In squirrel cage induction motors, the rotor slots are usually given slight skew in order to
(*a*) reduce windage losses (*b*) reduce eddy currents
(*c*) reduce accumulation of dirt and dust (*d*) reduce magnetic hum.
8. In case the air gap in an induction motor is increased
(*a*) the magnetising current of the rotor will decrease
(*b*) the power factor will decrease
(*c*) speed of motor will increase (*d*) the windage losses will increase.
9. Slip rings are usually made of
(*a*) copper (*b*) carbon
(*c*) phosphor bronze (*d*) aluminium.
10. A 3-phase 440 V, 50 Hz induction motor has 4% slip. The frequency of rotor e.m.f. will be
(*a*) 200 Hz (*b*) 50 Hz
(*c*) 2 Hz (*d*) 0.2 Hz.
11. If N_s is the synchronous speed and s the slip, then actual running speed of an induction motor will be
(*a*) N_s (*b*) $s \cdot N_s$
(*c*) $(1 - s)N_s$ (*d*) $(N_s - 1)s$.
12. The efficiency of an induction motor can be expected to be nearly
(*a*) 60 to 90% (*b*) 80 to 90%
(*c*) 95 to 98% (*d*) 99%.
13. The number of slip rings on a squirrel cage induction motor is usually
(*a*) two (*b*) three
(*c*) four (*d*) none.
14. The starting torque of a squirrel-cage induction motor is
(*a*) low (*b*) negligible
(*c*) same as full-load torque (*d*) slightly more than full-load torque.
15. A double squirrel-cage induction motor has
(*a*) two rotors moving in opposite direction (*b*) two parallel windings in stator
(*c*) two parallel windings in rotor (*d*) two series windings in stator.
16. The ratio of starting torque to normal torque in case of a star-delta starter will be
(*a*) 0.35 (*b*) 0.67
(*c*) 1.07 (*d*) 1.35.
17. Star-delta starting of motors is not possible in case of
(*a*) single phase motors (*b*) variable speed motors
(*c*) low horse power motors (*d*) high speed motors.

18. The terms 'cogging' is associated with
(*a*) three phase transformers (*b*) compound generators
(*c*) D.C. series motors (*d*) induction motors.

19. In case of the induction motors the torque is
(*a*) inversely proportional to ($\sqrt{\text{slip}}$) (*b*) directly proportional to $(\text{slip})^2$
(*c*) inversely proportional to slip (*d*) directly proportional to slip.

20. Stepless speed control of induction motor is possible by which of the following methods ?
(*a*) E.m.f. injection in rotor circuit (*b*) Changing the number of poles
(*c*) Cascade operation (*d*) None of the above.

21. Rotor rheostat control method of speed control is used for
(*a*) squirrel-cage induction motors only (*b*) slip ring induction motors only
(*c*) both (*a*) and (*b*) (*d*) none of the above.

22. In the circle diagram for induction motor, the diameter of the circle represents
(*a*) slip (*b*) rotor current
(*c*) running torque (*d*) line voltage.

23. For which motor the speed can be controlled from rotor side ?
(*a*) Squirrel-cage induction motor (*b*) Slip-ring induction motor
(*c*) Both (*a*) and (*b*) (*d*) None of the above.

24. If any two phases for an induction motor are interchanged
(*a*) the motor will run in reverse direction (*b*) the motor will run at reduced speed
(*c*) the motor will not run (*d*) the motor will burn.

25. An induction motor is
(*a*) self-starting with zero torque (*b*) self-starting with high torque
(*c*) self-starting with low torque (*d*) non-self starting.

26. The maximum torque in an induction motor depends on
(*a*) frequency (*b*) rotor inductive reactance
(*c*) square of supply voltage (*d*) all of the above.

27. In three-phase squirrel-cage induction motors
(*a*) rotor conductor ends are short circuited through slip rings
(*b*) rotor conductors are short-circuited through end rings
(*c*) rotor conductors are kept open
(*d*) rotor conductors are connected to insulation.

28. In a three-phase induction motor, the number of poles in the rotor winding is always
(*a*) zero (*b*) more than the number of poles in stator
(*c*) less than number of poles in stator (*d*) equal to number of poles in stator.

29. *DOL* starting of induction motors is usually restricted to
(*a*) low horsepower motors (*b*) variable speed motors
(*c*) high horsepower motors (*d*) high speed motors.

30. No-load current in an induction motor is nearly
(*a*) 0.1 per cent of full load current (*b*) 1 to 5 per cent of full-load current
(*c*) 15 to 25 per cent of full-load current (*d*) 50 to 60 per cent of full-load current.

31. The speed of a squirrel-cage induction motor can be controlled by all of the following except
(*a*) changing supply frequency (*b*) changing number of poles
(*c*) changing winding resistance (*d*) reducing supply voltage.

32. The 'crawling' in an induction motor is caused by
(*a*) high loads (*b*) low voltage supply
(*c*) improper design of machine (*d*) harmonics developed in the motor.

33. The power factor of an induction motor under no-load conditions will be closer to
(*a*) 0.2 lagging (*b*) 0.2 leading
(*c*) 0.5 leading (*d*) unity.

34. The 'cogging' of an induction motor can be avoided by

(*a*) proper ventilation (*b*) using *DOL* starter

(*c*) auto-transformer starter

(*d*) having number of rotor slots more or less than the number of stator slots (not equal).

35. One phase of a three-phase partly loaded induction motor power supply got opened up while the motor was in motion. This is likely to result in

(*a*) stalling of the motor

(*b*) burning of one phase winding of the motor

(*c*) burning of all three phases of winding of the motor

(*d*) undisturbed running of motor.

36. If an induction motor with certain ratio of rotor to stator slots, runs at $\frac{1}{7}$ of the normal speed, the phenomenon will be termed as

(*a*) humming (*b*) hunting

(*c*) crawling (*d*) cogging.

37. It is advisable to avoid line starting of induction motor and use starter because

(*a*) it will run in reverse direction

(*b*) it will pick up very high speed and may go out of step

(*c*) motor takes five to seven times its full-load current

(*d*) starting torque is very high.

38. The speed characteristics of an induction motor closely resemble the speed-load characteristics of which of the following machines ?

(*a*) D.C. series motor (*b*) D.C. shunt motor

(*c*) Universal motor (*d*) None of the above.

39. In a slip-ring induction motor, resistance is connected in rotor phases

(*a*) to increase starting torque (*b*) to limit starting current

(*c*) to limit the starting current and increase the starting torque

(*d*) none of the above.

40. Which type of bearing is provided in small induction motors to support the rotor shaft ?

(*a*) Ball bearings (*b*) Cast iron bearings

(*c*) Bush bearings (*d*) None of the above.

41. A pump induction motor is switched on to a supply 30% lower than its rated voltage. The pump runs. What will eventually happen ? It will

(*a*) stall after sometime (*b*) stall immediately

(*c*) continue to run at lower speed without damage

(*d*) get heated and subsequently get damaged.

42. 5 H.P., 50-Hz, 3-phase, 440 V, induction motors are available for the following r.p.m. Which motor will be the costliest ?

(*a*) 730 r.p.m. (*b*) 960 r.p.m.

(*c*) 1440 r.p.m. (*d*) 2880 r.p.m.

43. A 3-phase slip ring motor has

(*a*) double cage rotor (*b*) wound rotor

(*c*) short-circuited rotor (*d*) any of the above.

44. The torque developed in an induction motor is nearly proportional to

(*a*) $\frac{1}{V}$ (*b*) V

(*c*) V^2 (*d*) $\frac{1}{V^2}$

where V = applied voltage.

45. Short-circuit test on an induction motor cannot be used to determine
(*a*) windage losses (*b*) copper losses
(*c*) transformation ratio (*d*) power scale of circle diagram.

46. In a three-phase induction motor
(*a*) iron losses in stator will be negligible as compared to that in rotor
(*b*) iron losses in motor will be negligible as compared to that in rotor
(*c*) iron losses in stator will be less than that in rotor
(*d*) iron losses in stator will be more than that in rotor.

47. In case of 3-phase induction motors, plugging means
(*a*) putting the motor directly on line without a starter
(*b*) locking of rotor due to harmonics
(*c*) starting the motor on load which is more than the rated load
(*d*) interchanging two supply phases for quick stopping.

48. Which is of the following data is required to draw the circle diagram for an induction motor ?
(*a*) Block rotor test only (*b*) No load test only
(*c*) Block rotor test, and no-load test
(*d*) Block rotor test, no-load test and stator resistance test.

49. In three-phase induction motors sometimes copper bars are placed deep in the rotor to
(*a*) improve starting torque (*b*) reduce copper losses
(*c*) improve efficiency (*d*) improve power factor.

50. In a three-phase induction motor
(*a*) power factor at starting is high as compared to that while running
(*b*) power factor at starting is low as compared to that while running
(*c*) power factor at starting is the same as that while running.

51. The circle diagram for an induction motor cannot be used to determine
(*a*) efficiency (*b*) power factor
(*c*) frequency (*d*) output.

52. The current taken by a three-phase induction motor during plugging will be
(*a*) normal full-load current (*b*) more than blocked rotor current
(*c*) equal to blocked rotor current (*d*) less than blocked rotor current.

53. Blocked rotor test on induction motors is used to find out
(*a*) leakage reactance (*b*) power factor on short circuit
(*c*) short-circuit current under rated voltage (*d*) all of the above.

54. Lubricant used for ball bearing is usually
(*a*) graphite (*b*) grease
(*c*) mineral oil (*d*) molasses.

55. An induction motor can run at synchronous speed when
(*a*) it is run on load (*b*) it is run in reverse direction
(*c*) it is run on voltage higher than the rated voltage
(*d*) e.m.f. is injected in the rotor circuit.

56. Which of the following can be reduced by skewing in induction motor ?
(*a*) Noise (*b*) Vibration
(*c*) Synchronous cusps (*d*) All of the above.

57. Which motor is preferred for use in mines where explosive gases exist ?
(*a*) Air motor (*b*) Induction motor
(*c*) D.C. shunt motor (*d*) Synchronous motor.

58. The torque developed by a 3-phase induction motor least depends on
(*a*) rotor current (*b*) rotor power factor
(*c*) rotor e.m.f. (*d*) shaft diameter.

59. In an induction motor if air-gap is increased

(*a*) the power factor will be low (*b*) windage losses will be more

(*c*) bearing friction will reduce (*d*) copper loss will reduce.

60. In an induction motor

(*a*) $s = \frac{N - N_s}{N}$ (*b*) $s = \frac{N_s - N}{N_s}$

(*c*) $s = \frac{N_s - N}{N}$ (*d*) $s = \frac{N - N_s}{N_s}$.

61. When R_2 is the rotor resistance, X_2 the rotor reactance at supply frequency and s the slip, then the condition for maximum torque under running conditions will be

(*a*) $sR_2X_2 = 1$ (*b*) $sR_2 = X_2$

(*c*) $R_2 = sX_2$ (*d*) $R_2 = s^2X_2$.

62. How many different speeds can be obtained in cascade control of two induction motors ?

(*a*) 2 (*b*) 4

(*c*) 8 (*d*) 16.

63. In case of a double cage induction motor, the inner cage has

(*a*) high inductance and low resistance (*b*) low inductance and high resistance

(*c*) low inductance and low resistance (*d*) high inductance and high resistance.

64. Open circuit test is usually conducted on

(*a*) slip ring motors only (*b*) wound rotor motors only

(*c*) either of (*a*) and (*b*) above (*d*) none of the above.

65. When two motors of 12 and 8 poles respectively are connected to a 50-Hz supply, which of the following no-load speed cannot be obtained ?

(*a*) 300 r.p.m. (*b*) 450 r.p.m.

(*c*) 500 r.p.m. (*d*) 750 r.p.m.

(*e*) 1500 r.p.m.

66. The low power factor of induction motor is due to

(*a*) rotor leakage reactance (*b*) stator reactance

(*c*) the reactive lagging magnetizing current necessary to generate the magnetic flux

(*d*) all of the above.

67. During back to back test

(*a*) both machines run as induction motors (*b*) both machines run as induction generators

(*c*) one machine runs as induction motor and the other as induction generator

(*d*) one machine runs as induction motor and the other as synchronous generator.

68. Phase advancers are used with induction motors to

(*a*) reduce noise (*b*) reduce vibrations

(*c*) reduce copper losses (*d*) improve power factor.

69. During dynamic braking of induction motor

(*a*) stator connections are reversed (*b*) motor is used as induction generator

(*c*) stator winding is excited by direct current (*d*) fictitious load is applied on motor.

70. The operation of repeated starting and stopping of an induction motor for exact positioning of the driven equipment is called

(*a*) crawling (*b*) balancing

(*c*) plugging (*d*) jogging.

71. In moist surroundings which frame of motor will be preferred ?

(*a*) Open type (*b*) Totally enclosed fan cooled

(*c*) Drip proof (*d*) Splash proof.

72. The principle of operation of a three-phase induction motor closely resembles to that of which of the following ?

(*a*) Synchronous motor (*b*) D.C. series motor

(*c*) Transformer with short-circuited secondary (*d*) Transformer with open-circuited secondary.

73. Under which of the following conditions a squirrel-cage induction motor would be preferred over a wound rotor induction motor ?
(*a*) When external voltage is to be necessarily introduced into the rotor
(*b*) When wide range of speed control is desired
(*c*) When cost is the major consideration
(*d*) When higher starting torque is required.

74. Which of the following statement is not valid when a squirrel-cage induction motor operates under no-load ?
(*a*) Rotor induced e.m.f. is low (*b*) Rotor current is low
(*c*) Power factor is low (*d*) Slip is low.

ANSWERS

1. (*c*)	**2.** (*b*)	**3.** (*a*)	**4.** (*c*)	**5.** (*a*)	**6.** (*c*)	**7.** (*d*)
8. (*b*)	**9.** (*c*)	**10.** (*c*)	**11.** (*c*)	**12.** (*b*)	**13.** (*d*)	**14.** (*a*)
15. (*c*)	**16.** (*a*)	**17.** (*a*)	**18.** (*d*)	**19.** (*d*)	**20.** (*b*)	**21.** (*b*)
22. (*b*)	**23.** (*b*)	**24.** (*a*)	**25.** (*c*)	**26.** (*d*)	**27.** (*b*)	**28.** (*d*)
29. (*a*)	**30.** (*d*)	**31.** (*c*)	**32.** (*d*)	**33.** (*a*)	**34.** (*d*)	**35.** (*d*)
36. (*c*)	**37.** (*c*)	**38.** (*b*)	**39.** (*b*)	**40.** (*a*)	**41.** (*d*)	**42.** (*a*)
43. (*b*)	**44.** (*c*)	**45.** (*a*)	**46.** (*d*)	**47.** (*d*)	**48.** (*d*)	**49.** (*a*)
50. (*b*)	**51.** (*a*)	**52.** (*b*)	**53.** (*d*)	**54.** (*b*)	**55.** (*d*)	**56.** (*d*)
57. (*a*)	**58.** (*d*)	**59.** (*a*)	**60.** (*b*)	**61.** (*c*)	**62.** (*b*)	**63.** (*a*)
64. (*a*)	**65.** (*b*)	**66.** (*d*)	**67.** (*c*)	**68.** (*d*)	**69.** (*c*)	**70.** (*d*)
71. (*b*)	**72.** (*c*)	**73.** (*c*)	**74.** (*d*).			

(B) Say 'Yes' or 'No' :

1. Three-phase induction motor has a low efficiency.
2. An induction motor is simply an electric transformer whose magnetic circuit is separated by an air gap into two relatively movable portions, one carrying the primary and the other the secondary winding.
3. Frames of electrical machines house the stator core.
4. The number of slots in the rotor should always be equal to the number of slots in the stator.
5. For large and heavy rotors journal bearings are used.
6. The slips rings are made of aluminium.
7. The difference between the synchronous speed and rotor speed is known as slip.
8. Starting torque, $T_{st} = \dfrac{kR_2E_2^{\,2}}{R_2^{\,2} + X_2^{\,2}}$.
9. Starting torque is inversely proportional to the square of the applied voltage.
10. Stator iron loss is practically constant.
11. $\dfrac{\text{Rotor copper loss}}{\text{Rotor gross output}} = \dfrac{1-s}{s}$.
12. Circle diagram of an induction motor can be drawn by using the data obtained from no-load test, short-circuit test and stator resistance test.
13. The synchronous speed of the nth order harmonic is $\dfrac{1}{n}$th of the synchronous speed of fundamental.
14. 'Cogging' of squirrel-cage motors can be easily overcome by making the number of rotor slots prime to the number of stator slots.
15. A double squirrel-cage motor has two independent squirrel-cage windings on the rotor, each having its own set of slots.

ANSWERS

1. No	2. Yes	3. Yes	4. No	5. Yes	6. No	7. Yes
8. Yes	9. No	10. Yes	11. No	12. Yes	13. Yes	14. Yes
15. Yes.						

THEORETICAL QUESTIONS

1. Explain the principle of operation of the polyphase induction motor.
2. Show that a rotating magnetic field can be produced by the use of 3-phase currents of equal magnitude.
3. What is meant by slip in an induction motor ? Develop an expression for the frequency of rotor currents in it.
4. Derive an expression for the torque of an induction motor and obtain the condition for maximum torque.
5. Show that in a 3-phase induction motor with negligible stator impedance maximum torque is developed at slip $s = \frac{R_2}{X_2}$, where R_2 and X_2 are rotor resistance and standstill reactance respectively.
6. State the effects of increasing rotor resistance on starting current, starting torque, maximum torque and full-load slip of an induction motor.
7. Show that in an induction motor the rotor input : power developed : rotor copper losses : : 1 : (1 – s) : s, where s is the fractional slip.
8. Draw equivalent circuit of a 3-phase induction motor.
9. Describe and explain how to perform a locked rotor test. What data can be obtained by the test ?
10. What are no-load and blocked rotor tests ? What sort of losses can be measured by these tests ?
11. How would you determine circle diagram of a 3-phase induction motor experimentally ?
12. Describe the constructional details of a 3-phase squirrel-cage and phase wound induction motors. Also discuss the applications of various types of starters used for starting these motors.
13. Why at all starters are necessary for starting the induction motors ? What are the various types of starters used for squirrel-cage motors ? Discuss them.
14. Explain with the aid of diagrams the principle of operation of double cage induction motor. Sketch the torque-slip curves of such a motor.
15. Explain briefly the different methods of speed control of 3-phase induction motors.
16. Discuss briefly the following :
    1. Crawling
    2. Cogging.
17. Write a short note on induction generator.

UNSOLVED PROBLEMS

1. A 3-phase, 4-pole, 50-Hz induction motor is running at 1455 r.p.m. Find the slip speed and slip.
   [Ans. 45 r.p.m., 3 per cent]
2. An induction motor having 8-poles runs at 50-Hz supply. If it operates at full-load at 720 r.p.m., calculate the slip. **[Ans.** 4 per cent]
3. A 6-pole alternator running at 1000 r.p.m. supplies an 8-pole induction motor. What is the actual speed of the motor if the slip is 2.5%. **[Ans.** 731.25 r.p.m.]
4. A slip ring motor runs at 290 r.p.m. on full-load when connected to 50-Hz supply. Calculate :
   (*i*) Number of poles
   (*ii*) Absolute slip
   (*iii*) Fractional slip
   (*iv*) Percentage slip
   (*v*) Frequency of rotor currents. **[Ans.** 20, 10 r.p.m., 0.0333, 3.33%, 1.66 Hz]
5. A 12-pole, 3-phase alternator is coupled to an engine running at 500 r.p.m. It supplies a 3-phase induction motor having a full- load speed of 1440 r.p.m. Find the percentage slip and number of poles of motor.
   [Ans. 4 per cent, 4 poles]

6. A 4-pole induction motor is run at 1450 r.p.m. from a 50-Hz supply. Find the percentage slip and frequency of rotor current. **[Ans.** 3.33%, 1.67 Hz]

7. If the e.m.f. in the stator of an 8-pole induction motor has a frequency of 50-Hz and that in the rotor 1.5 Hz, at what speed is the motor running and what is the slip ? **[Ans.** 727.5 r.p.m., 3 per cent]

8. A three-phase slip ring induction motor gives a reading of 55 V across slip rings on open circuit when at rest with normal stator voltage applied. The rotor is star-connected and has impedance (0.7 + *j*5) Ω per phase. Find the rotor current when the machine is
   (*i*) at standstill with the slip rings joined to a star-connected starter with a phase impedance of (4 + *j*3) Ω ; and
   (*ii*) running normally with a 5 per cent slip.

   [Ans. (*i*) 3.425 A at 0.506 p.f. (lag) ; (*ii*) 2.14 A at 0.942 p.f. (lag)]

9. A 3-phase, 400 V, star-connected induction motor has a star-connected rotor with a stator to rotor turn ratio of 6.5. The rotor resistance and stand-still reactance per phase are 0.05 and 0.25 Ω respectively. What should be the value of external resistance per phase to be inserted in the rotor circuit to obtain maximum torque at starting and what will be the rotor starting current with this resistance ?

   [Ans. 0.2 Ω, 100 A (app.)]

10. If the motor has a rotor resistance of 0.02 Ω and a standstill reactance of 0.1 Ω what must be the value of the total resistance of a starter for the rotor circuit for maximum torque to be exerted at starting ?

    [Ans. 0.08 Ω]

11. The rotor of a 3-phase 440 V, 50 Hz, 4-pole induction motor has a resistance of 0.3 Ω per phase and an inductance of 0.008 H per phase. The ratio of stator to rotor turns is 2 : 1. The stator is delta connected and rotor is star-connected. Determine the voltage between slip rings when the motor is at standstill. Calculate the standstill rotor current and rotor current when the motor is operating with 3 per cent slip.

    [Ans. 381 V, 87 A, 22 A]

12. A 3300 V, 24-phase, 50-Hz, 3-phase star-connected induction motor has a slip-ring rotor resistance of 0.016 Ω and standstill reactance of 0.265 Ω per phase. Calculate :
    (*i*) The speed at maximum torque.
    (*ii*) Ratio of full-load torque to maximum torque if full-load torque is obtained at 247 r.p.m.

    [Ans. 235 r.p.m., 0.382]

13. A 6-pole, 3-phase, 50 Hz induction motor runs on full-load with a slip of 4%. If the rotor standstill impedance per phase is (0.01 + *j*0.05) Ω, calculate the available maximum torque in terms of full-load torque. Also determine the speed at which the maximum torque occurs. $\left[\textbf{Ans. } \frac{T_m}{T_f} = 2.6, \text{ 800 r.p.m.}\right]$

14. A 6-pole, 3-phase, 50 Hz induction motor develops maximum torque of 200 Nm at a speed of 960 r.p.m. Determine the torque exerted by the motor at 5 per cent slip. The rotor resistance per phase is 0.5 Ω.

    [Ans. [illegible] Nm]

15. The rotor resistance and standstill reactance per phase of a 3-phase slip ring induction motor are 0.02 Ω and 0.1 Ω respectively. What should be the value of the external resistance per phase to be inserted in the rotor circuit to give maximum torque at starting ? **[Ans.** 0.08 Ω]

16. The star-connected rotor of an induction motor has a standstill impedance of (0.4 + *j*4) Ω per phase, and the rheostat impedance per phase is (6 + *j*2) Ω. The motor has an induced e.m.f. of 80 V between slip rings at standstill when connected to its normal supply voltage. Find :
    (*i*) The rotor current at standstill with the rheostat in the circuit.
    (*ii*) When running short-circuited with slip of 0.03. **[Ans.** 5.27 A, 3.3 A]

17. A 4-pole, 50 Hz, 3-phase, induction motor develops a maximum torque of 115 N-m at 1365 r.p.m. The resistance of the star-connected rotor is 0.2 Ω/phase. Calculate the value of resistance that must be inserted in series with each rotor phase to produce a starting torque equal to half the maximum torque.

    [Ans. 0.4 Ω or 8.1 Ω]

18. A 4-pole, 50-Hz, 3-phase induction motor has rotor resistance and reactance of 0.03 Ω and 0.12 Ω per phase respectively. Determine :
    (*i*) The value of speed at which maximum torque occurs.

(*ii*) The value of external rotor resistance per phase to be inserted to obtain 80% of maximum torque at starting. [**Ans.** 1125 r.p.m., 0.02418 Ω]

19. The resistance and standstill reactance of each phase of a 3-phase induction motor with star-connected rotor are 0.06 Ω and 0.4 Ω respectively. The full-load slip is 4%. Calculate the resistance per phase of a star-connected rheostat, which when connected to the rotor, will give a pull out torque at one half of the full-load speed. What is then the power factor ? [**Ans.** 0.148 Ω, 0.707 (lag)]

20. The rotor resistance and standstill reactance of a 3-phase induction motor are respectively 0.015 Ω and 0.09 Ω per phase. At normal voltage, the full-load slip is 3 per cent. Estimate the percentage reduction in stator voltage to develop full-load torque at half full-load speed. Also calculate the power factor. [**Ans.** 22.9%, 0.31 (app.)]

Rotor Output Losses, Efficiency

21. The power input to a 3-phase induction motor is 60 kW. The stator-losses total 1.5 kW. Find the total mechanical power developed if the motor is running with a slip of 4 per cent. [**Ans.** 56.16 kW]

22. The rotor e.m.f. of a 3-phase, 6-pole, 400 V, 50-Hz induction motor alternates at 3-Hz. Compute the speed and percentage slip of the motor. Find the rotor copper loss per phase if the full input to the rotor is 111.9 kW. [**Ans.** 940 r.p.m., 6%, 2.238 kW]

23. A 3-phase, 50-Hz, induction motor draws 50 kW from the mains. If the stator losses are 2 kW and the rotor e.m.f. is observed to make 100 complete oscillations/alterations per minute, determine :

(*i*) Rotor copper loss. (*ii*) Gross mechanical output. [**Ans.** 1.6 kW, 46.4 kW]

24. The power input to the rotor of a 3-phase, 50-Hz, 6-pole, slip ring induction motor is 40 kW and the motor runs at 960 r.p.m. The rotor resistance per phase is 0.25 Ω. Determine the value of the rotor current per phase. [**Ans.** 46.2 A]

25. The power input to the rotor of a 440 V, 50-Hz, 3-phase, 6-pole induction motor is 60 kW. It is observed that the rotor e.m.f. makes 90 complete cycles per minute. Calculate :

(*i*) Slip, (*ii*) Rotor speed,

(*iii*) Rotor copper loss per phase, (*iv*) Mechanical power developed, and

(*v*) Rotor resistance/phase if the rotor current is 60 A.

[**Ans.** 0.03, 970 r.p.m., 600 W, 58.2 kW, 0.167 Ω]

Equivalent Circuit of an Induction Motor

26. The maximum torque of a 3-phase induction motor occurs at a slip of 12 per cent. The motor has an equivalent secondary resistance of 0.08 Ω/phase. Calculate the equivalent load resistance R_L, the equivalent load voltage V_L and the current at this slip if the gross power output is 9 kW. [**Ans.** 0.587 Ω/phase, 42 V, 71.6 A]

27. The stator impedance and equivalent rotor resistance of a 400 V, 3-phase star-connected induction motor are $(0.06 + j0.2)$ Ω and equivalent rotor impedance of $(0.06 + j0.22)$ Ω. Neglecting exciting current, find the maximum gross power and the slip at which it occurs. [**Ans.** 12%, 142.9 kW]

28. Estimate the stator current, equivalent rotor current, efficiency, output, and its power factor at a slip of 5% for a motor having the following data :

Stator impedance = $1.0 + j3.0$ Ω, rotor standstill impedance = $1.0 + j2.0$ Ω, no-load shunt impedance = $10 + j50$ Ω, volts/phase = 250 volts. [**Ans.** 14.34 A, 11.6 A, 83.6%, 7,670 W, 0.853]

29. The primary leakage impedance and equivalent secondary impedance referred to stator at standstill of a 3-phase, 220 V, 50-Hz, 4-pole induction motor are $(0.4 + 3j)$ Ω and $(1.2 + j1)$ Ω per phase respectively. If the magnetizing current is neglected, calculate the maximum torque the motor can develop and slip at which it occurs. [**Ans.** s = 0.2985, 1820 W]

30. A 3-phase, star-connected, 220 V (line to line), 50-Hz, 4-pole induction motor has the following constants in ohm per phase referred to stator. $r_1 = 0.29$, $r_2 = 0.14$, $X_1 = 0.5$, $X_2 = 0.21$ and magnetising reactance $X_0 = 13.25$ Ω. The total friction, windage and core loss may be assumed to be constant at 400 watts. For a slip of 2 per cent calculate :

(*i*) Speed, (*ii*) Output, and

(*iii*) Stator current of the motor. [**Ans.** 1470 r.p.m., 40 N-m, 21.5 A]

Circle Diagram

31. Draw the circle diagram for a 400 V, 5 H.P. (3.6775 kW) delta-connected 3-phase induction motor from the following test results (line values) :

No-load test : 400 V, 3.0 A, cos ϕ_0 = 0.2

Short-circuit test : 200 V, 12.0 A, cos ϕ_s = 0.4

From the circle diagram determine :

(*i*) Full-load current, (*ii*) Full-load power factor, and

(*iii*) Starting torque in terms of full-load torque at the normal voltage. Assume the copper losses to be equally divided between the stator and the rotor.

[**Ans.** 6.8 A, 0.815, 0.836 times the full-load torque]

32. Draw the circle diagram from no-load and short-circuit tests of a 3-phase, 14.29 kW, 400 W, 6-pole induction motor from the following test results (line values).

No-load test : 400 V, 11 A, p.f. = 0.2

Short-circuit test : 100 V, 25 A, p.f. = 0.4

Rotor copper loss at standstill is half the total copper loss.

From the circle diagram find :

(*i*) Line current, slip, efficiency and p.f. at full-load

(*ii*) Maximum torque. [**Ans.** (*i*) 32.5 A, 5.6%, 80%, 0.84, (*ii*) 27030 syn. watts]

33. Draw the circle diagram for a 3.677 kW, 200 V, 50-Hz, 4-pole, 3-phase, star-connected induction motor from the following data :

No-load test : 200 V, 5 A, 350 W

Locked rotor test : 100 V, 26 A, 1700 W

Rotor copper loss at standstill = 0.5 times the total copper loss.

Determine :

(*i*) Full-load current, (*ii*) Power factor, and

(*iii*) Speed and torque. [**Ans.** 15 A, 0.866, 1435 r.p.m., 4010 syn. watts]

34. A 440 V, 3-phase, 4-pole, 50-Hz slip ring induction motor gave the following test results :

No-load test : 440 V, 9 A, p.f. = 0.2

Blocked rotor test : 110 V, 22 A, p.f. = 0.3

The ratio of stator to rotor turns per phase is 3.5/1. The stator and rotor copper losses are divided equally in the blocked rotor test. The full-load current = 20 A. Draw the circle diagram and obtain the following :

(*i*) Power factor, torque, output, efficiency and slip at full-load.

(*ii*) Standstill torque (or starting torque).

(*iii*) Resistance to be inserted in the rotor circuit for giving a starting toque 200% of the full-load torque. Also find the current and power factor under these conditions.

[**Ans.** (*i*) 0.81, 61 N-m, 10.21 kW, 83%, 3% ; (*ii*) 60.25 N-m, (*iii*) 0.076 Ω, 78.4 A, 0.498]

35. Draw the circle diagram of a 7.46 kW, 200 V, 50-Hz, 3-phase slip-ring induction motor with a star-connected stator and rotor, a winding ratio of unity, a stator resistance of 0.38 Ω/phase and a rotor resistance of 0.24 Ω/phase. The following are the test readings :

No-load test : 200 V, 7.7 A, cos ϕ_0 = 0.195

Short-circuit test : 100 V, 47.6 A, cos ϕ_s = 0.454

Find :

(*i*) Starting torque, (*ii*) Maximum torque, both in synchronous watts,

(*iii*) Maximum power factor, (*iv*) Slip for maximum torque, and

(*v*) Maximum output. [**Ans.** 5570 syn. watts, 12440 syn. watts, 0.879, 0.195, 10270 W]

Starting of Induction Motor

36. A small 3-phase induction motor has a short-circuit current equal to 4 times the full-load current. Determine the starting torque as a percentage of full-load torque if full-load slip is 2.5 per cent.

[**Ans.** 40% of full-load torque]

37. A 3-phase, squirrel-cage induction motor has a short-circuit current equal to 5 times the full-load current. Find the starting torque as a percentage of full-load torque if the motor is started by

(*i*) direct switching to the supply ; (*ii*) a star-delta starter ;
(*iii*) an auto-transformer ; and (*iv*) a resistance in the stator circuit.

The starting current in (*iii*) and (*iv*) is limited to 2.5 times the full-load current and the full-load slip is 4 per cent. [**Ans.** T_f, 33.3% T_f, 50%T_f, 25% T_f]

38. Determine a suitable auto-transformer ratio for starting an induction motor with a supply current not exceeding twice full-load current. Use the following data :
Short-circuit current = 5 times the full-load current, full-load slip = 3%. Estimate the torque in terms of full-load torque. Ignore magnetising current. [**Ans.** 30%]

39. Find the percentage tapping required on an auto-transformer required for a cage motor to start the motor against $\frac{1}{4}$ of full-load torque. The short-circuit current at normal voltage is 4 times the full-load current and the full-load slip is 3 per cent. [**Ans.** K = 72.2%]

40. A squirrel-cage type induction motor when started by means of a star-delta starter takes 20 per cent of full-load line current and develops 40 per cent of full-load torque at starting. Calculate the starting torque and current in terms of full-load values, if an auto-transformer with 75 per cent tapping were employed. [**Ans.** 3.375I_f, 67% full-load torque]

41. A 10 kW, 200 V, 3-phase induction motor has at full-load an efficiency of 87% and power factor of 0.85. At standstill the motor draws 5 times the full-load current and develops 1.5 times full-load torque. An auto-transformer is installed to reduce the starting current to give full-load torque at starting. Neglecting magnetising current, determine :

(*i*) The voltage to be applied at starting. (*ii*) The full-load line current.
(*iii*) The current drawn by the motor at start.
(*iv*) The current supplied to the primary of the autotransformer. [**Ans.** 163.3 V, 39 A, 159.2 A, 130 A]

42. A 3-phase squirrel-cage induction motor has a ratio of maximum torque to full-load torque as 2.5 : 1. Determine the ratio of actual starting torque to full-load torque for :

(*i*) direct starting ; (*ii*) star-delta starting ;
(*iii*) auto-transformer starting with tapping of 60%.

The rotor resistance and standstill reactance per phase are 0.4 Ω and 4 Ω respectively. [**Ans.** 0.495, 0.165, 0.179]

Wound Rotor Induction Motor Starter Steps

43. Calculate the steps in a 5-step rotor resistance starter for a 3-phase induction motor. The slip at the maximum starting current is 2 per cent with slip ring short-circuited and the resistance per rotor phase is 0.015 Ω. [**Ans.** 0.407 Ω, 0.186 Ω, 0.0853 Ω, 0.0389 Ω, 0.0178 Ω]

44. Calculate the steps in 5-section rotor starter for a 3-phase induction motor from the following data :

Maximum starting current	= full-load current
Full-load slip	= 0.018
Rotor resistance/phase	= 0.015 Ω

[**Ans.** 0.460 Ω, 0.206 Ω, 0.092 Ω, 0.0415 Ω, 0.0185 Ω]

Double Squirrel-cage Motor

45. The equivalent standstill impedances per phase of the first cage and of the second cage respectively of a double-cage induction motor are (5 + j1) and (1 + 5j) Ω. Compare the torques of the two cages :

(*i*) At standstill,
(*ii*) At a slip of 4 per cent. [**Ans.** 5 : 1, 1 : 4.8]

46. If the standstill impedance of the outer cage machine is 0.3 + j0.4 Ω and of the inner cage is 0.1 + j1.5 Ω, compare the torques of the two cages :

(*i*) At standstill,
(*ii*) At a slip of 5 per cent. [**Ans.** 27.1 : 1, 0.52 : 1]

47. **In a double-cage induction motor, if the outer cage has an impedance at standstill of $(2 + j1.2)\ \Omega$, determine the slip at which the two cages develop equal torques if the inner cage has an impedance of $(0.5 + j3.5)\ \Omega$ at standstill.** **[Ans. 25.1 per cent]**

Speed Control of Induction Motors

48. **The rotor resistance per phase of an 8-pole, 50 Hz slip-ring motor is $0.25\ \Omega$ and its full-load speed is 720 r.p.m. Calculate the external resistance per phase that must be added to lower the speed to 600 r.p.m. given that the torque is the same in the two cases.** **[Ans. $1\ \Omega$]**

49. **A 4-pole, 50-Hz, 3-phase slip-ring induction motor when fully loaded runs with a slip of 3 per cent. Determine the value of resistance to be inserted in series per phase in the rotor circuit to reduce the speed by 10 per cent and the new slip. The rotor resistance per phase is $0.2\ \Omega$.** **[Ans. $0.6467\ \Omega$]**

50. **A 4-pole induction motor and a 6-pole induction motor are connected in cumulative cascade. The frequency in the secondary circuit of the 6-pole motor is observed to be 1.0 Hz. Determine the slip in each machine and the combined speed of the set.**
Assume supply frequency as 50-Hz. **[Ans. 60.8%, 3.3%, 588 r.p.m.]**

51. **Two 50-Hz, 3-phase induction motors having six and four-poles respectively are cumulatively cascaded, the 6-pole motor being connected to the main supply. Determine the frequencies of the rotor currents and the slips referred to each stator field if the set has a slip of 2 per cent.**
[Ans. 41.2%, 20.6 Hz, 4.85%, 1.0 Hz]

4

Single Phase Motors

4.1. GENERAL

- The number of machines operating from single-phase supplies is greater than all other types taken in total. For the most part, however, they are only used in the smaller sizes, less than 5 kW and mostly in the fractional H.P. range. They operate at *lower power-factors* and are relatively *inefficient when compared with polyphase motors.* Though simplicity might be expected in view of the two-line supply, the *analysis is quite complicated.*
- *Single phase motors* perform a great variety of useful services in the home, the office, the factory, in business establishments, on the farm, and many other places where electricity is available. Since the requirements of the numerous applications differ so widely, the motor-manufacturing industry has developed several types of such machines, each type having operating characteristics that meet definite demands. For example, one type operates satisfactorily on direct current or any frequency up to 60 cycles ; another rotates at absolutely constant speed, regardless of load ; another develops considerable starting torque and still another, although not capable of developing much starting torque, is nevertheless extremely cheap to make and very rugged.

4.2. TYPES OF SINGLE-PHASE MOTOR

The single-phase motors may be of the following types :

1. **Single-phase Induction Motors :**

A. Split-phase motors

(*i*) Resistance-start motor (*ii*) Capacitor-start motor

(*iii*) Permanent-split (single-value) capacitor motor

(*iv*) Two-value capacitor motor.

B. Shaded-pole induction motor.

C. Reluctance-start induction motor.

D. Repulsion-start induction motor.

2. **Commutator-Type, Single-Phase Motors :**

A. Repulsion motor. B. Repulsion-induction motor.

C. A.C. series motor. D. Universal motor.

3. Single-phase Synchronous Motors :

A. Reluctance motor. B. Hysteresis motor.

C. Sub-synchronous motor.

4.3. SINGLE-PHASE INDUCTION MOTORS

4.3.1. Applications and Disadvantages

Applications :

- Single phase induction motors are in very wide use in industry especially in *fractional horse-power field.*

  They are extensively used for electric drive for *low power constant speed apparatus* such as *machine tools, domestic apparatus* and *agricultural machinery* in circumstances where a three-phase supply is *not* readily available.
- There is a large demand for single-phase induction motors in sizes ranging from a *fraction of horse-power upto about 5 H.P.*

Disadvantages :

Though these machines are useful for small outputs, they are not used for large powers as they suffer from many disadvantages and are never used in cases where three-phase machines can be adopted.

The main *disadvantages* of single-phase induction motors are :

1. Their output is only 50% of the three-phase motor, for a given frame size and temperature rise.
2. They have lower power factor.
3. Lower-efficiency.
4. These motors do not have inherent starting torque.
5. More expensive than three-phase motors of the same output.

4.3.2. Construction and Working

Construction :

- A single phase induction motor is similar to a 3-ϕ squirrel cage induction motor in physical appearance. Its rotor is essentially the same as that used in 3-ϕ induction motors. Except for shaded pole motors, the stator is also very similar. There is a uniform air-gap between the stator and rotor but no electrical connection between them. It can be wound for any even number of poles, two, four and six being most common. Adjacent poles have opposite magnetic property and synchronous speed equation, $N_s = \frac{120f}{p}$ also applies.
- The stator windings differ in the following two aspects :
  - **Firstly** single phase motors are usually provided with concentric coils.
  - **Secondly,** these motors normally have two stator windings. In motors that operate with both windings energised, the winding with the *heaviest wire* is known as the **main winding** and the other is called the **auxiliary winding.** If the motor runs with auxiliary winding open, these windings are usually referred as *running* and *starting.*
  - In *most of motors the main winding is placed at the bottom of the slots and the starting winding on top but shifted 90° from the running winding.*

Working :

When the stator winding of a single phase induction motor is connected to single phase A.C. supply, a *magnetic field* is developed, *whose axis is always along the axis of stator coils*. The magnetic field produced by the stator coils is pulsating, varying sinusoidally with time. Currents are induced in the rotor conductors by transformer action, these currents being in such a direction as to oppose the stator m.m.f. Then the axis of the rotor m.m.f. wave coincides with that of the stator field, the *torque angle* is, therefore, *zero* and *no torque is developed on starting*. However, if the rotor is given a push by hand or by other means in either direction, it will pick-up the speed and continue to rotate in the same direction developing operating torque. Thus a single phase induction motor is not inherently self starting and requires some special means for starting.

The above mentioned behaviour of this type of motor can be explained by any one of the following theories :

1. Double revolving field theory
2. Cross-field theory.

The results given by both the theories are approximately same.

Double revolving field theory is described below :

The magnetic field produced by the stator coils is *pulsating*, varying sinusoidally with time. *Ferrari* pointed out that *such a field can be resolved into two equal fields but rotating in opposite directions with equal angular velocities. The maximum value of each component is equal to half the maximum of the pulsating field.*

If the initial time is such that the rotating vectors of the two component fields are along the *Y*-axis in the positive direction, the two component waves ϕ_1 and ϕ_2 coincide. The resultant of these

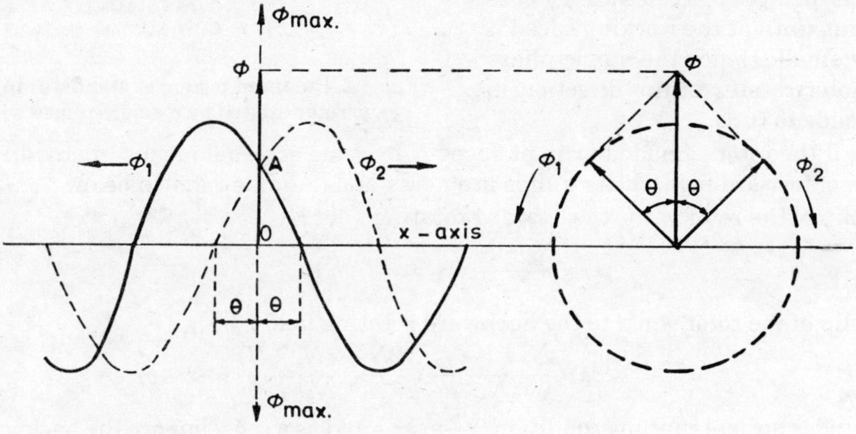

Fig. 4.1

two is ϕ_{max}. After a short interval of time the two vectors rotate, through an angle θ in their respective directions and the waves are shown to occupy the positions in Fig. 4.1. These waves intersect at *A* on the *Y*-axis and as the waves travel *A* moves along the *Y*-axis. Hence the resultant of these two component waves at any instant is equal to 2*OA*.

$$\phi_1 = OA = \phi_{1(max)} \cos(\omega t - \theta) \quad ...(i)$$

$$\phi_2 = OA = \phi_{2(max)} \cos(\omega t + \theta) \quad ...(ii)$$

and

$$\phi_{1(max)} = \phi_{2(max)}$$

By expanding and adding (*i*) and (*ii*),

$$\phi_1 + \phi_2 = 2\phi_{1(max)} \cos\theta \cos\omega t$$

$$2OA = \phi_{(max)} \cos\theta \cos\omega t \quad ...(iii)$$

which is the equation of the pulsating field and **proves Ferrari's statement.** Thus a single-phase induction motor is *not inherently self-starting*.

The existence of these two fluxes (*forward and backward*) rotating in opposite directions can be verified by supplying a fractional horsepower single-phase induction motor with rated voltage. The motor does not start, but if the shaft is turned by hand, say in clockwise direction, the rotor picks up speed. This means that the rotor conductors are rotating in the direction of that field which rotates in clockwise direction. When the motor is braked and stopped without switching off the supply, the rotor remains at rest. If now the shaft is turned by hand in anti-clockwise direction, the motor picks up speed in that direction. This means that the rotor conductors are now rotating in the direction of the other field.

This behaviour of the motor is *due to the presence of two opposing torques due to the two field.* When the rotor is at rest, (*i.e.* slip = 1) the two torques are equal but opposite in direction. Hence the net torque is zero and therefore the rotor remains at rest. Fig. 4.2. shows the torque variations due to the two fields. If the rotor is made to speed up in one direction, say in that in which T_1 increases, T_1 exceeds the opposing torque T_2 and the motor beings to accelerate. T_2 goes on diminishing until at the working speed it is negligibly small. Hence the single-phase induction motor rotates in the direction in which it is made to run.

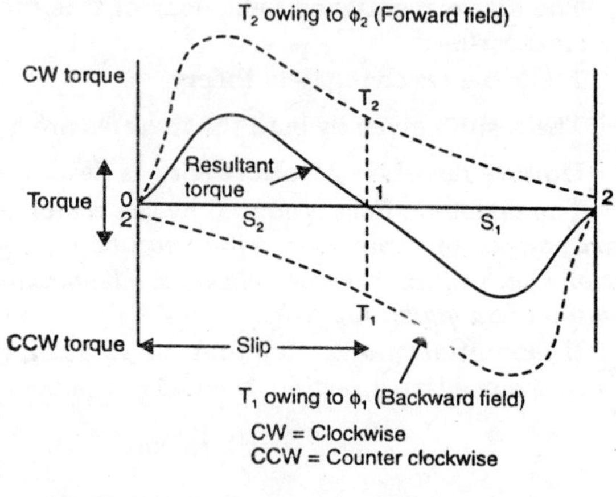

Fig. 4.2. Balanced torque at standstill in squirrel cage rotor excited by a single-phase winding.

Thus, if the rotor is made to run at speed N by some external means in any direction, say in the direction of forward field, the two slips are now s and $(2 - s)$, as shown below :

The slip of the rotor w.r.t. the forward rotating field F_f,

$$s_f = \frac{N_s - N}{N_s} = s \qquad \text{...(4.1)}$$

The slip of the rotor w.r.t. to the backward rotating field F_b,

$$s_b = \frac{N_s - (-N)}{N_s} = \frac{2N_s - (N_s - N)}{N_s} = (2 - s) \qquad \text{...(4.2)}$$

- Under normal running condition $(2 - s) >> s$ and as a consequence the backward field rotor currents are much larger than at standstill and have a low power factor. The corresponding opposing rotor m.m.f., owing to stator impedance, causes the backward field to be greatly reduced in strength. On the other hand, the low-slip forward rotating field induces smaller currents of a high power factor in the rotor than at standstill. This leads to greatly strengthening of forward field. Thus weakening of backward field and strengthening of forward field depends upon the slip or speed of rotor and the difference increases with the decrease in slip w.r.t. the forward field or with the increase in rotor speed in forward direction.
- In a single phase induction motor, *the increase in rotor resistance increases the effectiveness of the backward field, which reduces the breakdown torque, lowers the efficiency and increases the slip corresponding to maximum torque.*
- The **performance characteristics** of a single phase induction motor are somewhat *inferior* to that of a 3-phase induction motor due to the presence of backward rotating field.
  — A single phase induction motor has a *lower breakdown torque at larger slip and increased power losses.*

— Greater power input.
— The *speed regulation* tends to be *poorer* than that for a polyphase motor.
— The *power factor tends to be lower* (since the normal slip of a single phase induction motor under load conditions is rather greater than that of the corresponding 3-phase motor).

- In view of the above factors, a single phase induction motor has a *larger frame size* than that of 3-phase motor.
- Single phase motors *tend to be somewhat noiser than 3-phase motors which have no such pulsating torque.*

4.3.3. Equivalent Circuit and Performance Characteristics

Equivalent Circuit :

- The equivalent circuit of a single phase induction motor can be developed on the basis of two revolving field theory. For developing the equivalent circuit it is first imperative to consider standstill or blocked rotor conditions. The motor with a blocked rotor merely acts like a transformer with its secondary short circuited and its equivalent circuit will be as shown in Fig. 4.3, E_m being e.m.f. induced in the stator.

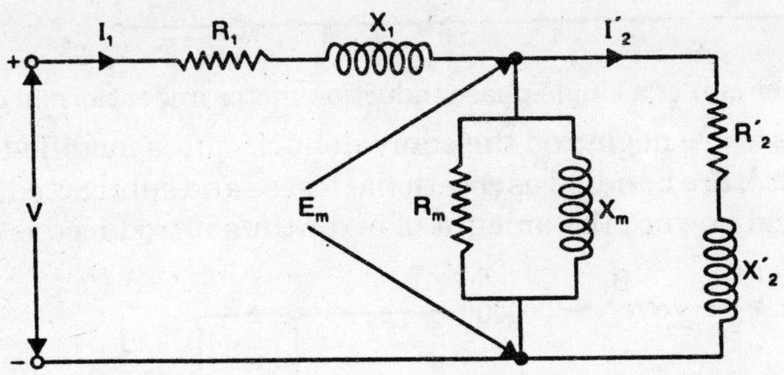

Fig. 4.3. Equivalent circuit of a single phase induction motor at standstill.

- Fig. 4.4 shows the equivalent circuit of a single phase induction motor at standstill on the basis of two revolving field theory. The phasor sum of E_{mf} and E_{mb} equals the applied voltage V (less the voltage drops in stator resistance R_1 and leakage reactance X_1).

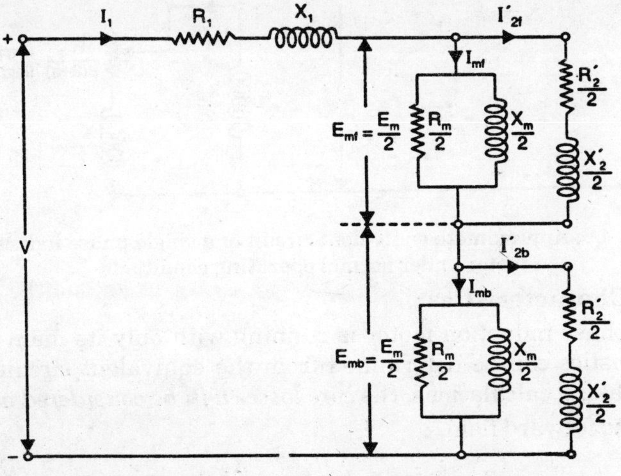

Fig. 4.4. Equivalent circuit of a single phase induction motor at standstill on the basis of two revolving field theory.

- When the rotor runs at speed N with respect to forward field, the slip is s w.r.t. forward field and $(2 - s)$ w.r.t. backward field and the equivalent circuit becomes as shown in Fig. 4.5.

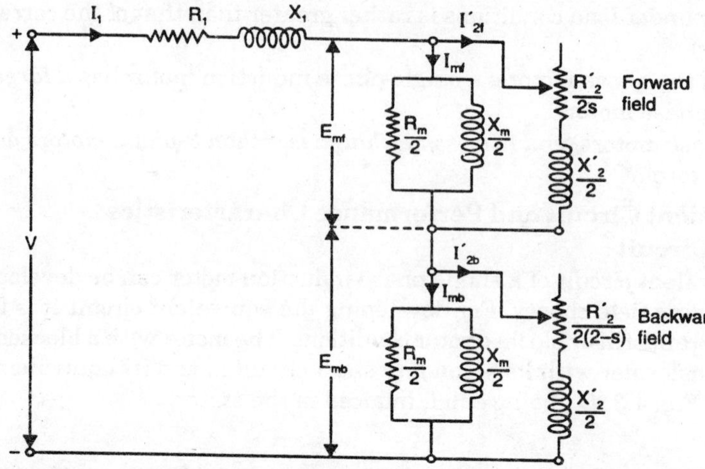

Fig. 4.5. Equivalent circuit of a single phase induction motor under normal operating conditions.

- If the core losses are neglected the equivalent circuit is modified as shown in Fig. 4.6. The core losses, here, are handled as rotational losses and subtracted from the power converted into mechanical power ; the amount of error thus introduced is relatively small.

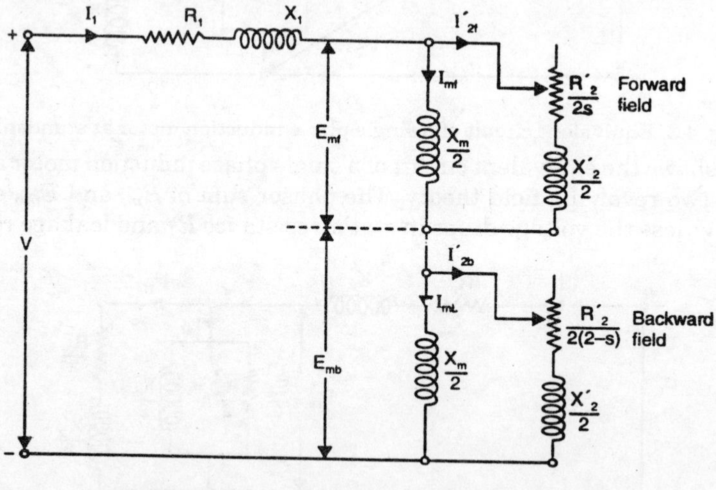

Fig. 4.6. Approximate equivalent circuit of a single phase induction motor under normal operating conditions.

Performance Characteristics :

When a single phase induction motor is running with only its main winding energised, its performance characteristics can be determined from the equivalent circuit diagram *for different values of slip* ; while making calculations, the *core losses will be considered as rotational losses.*

Impedance due to forward field,

$$Z_f = R_f + jX_f = \frac{R_2'}{2s} + j\frac{X_2'}{2} \text{ in parallel with } \frac{jX_m}{2} \qquad ...(4.3)$$

Impedance due to backward field,

$$Z_b = R_b + jX_b = \frac{R_2'}{2(2-s)} + j\frac{X_2'}{2} \text{ in parallel with } \frac{jX_m}{2} \quad ...(4.4)$$

Total equivalent impedance, $Z_{eq} = Z_1 + Z_f + Z_b$ ...(4.5)

Motor current, $$I_1 = \frac{V}{Z_{eq}} = \frac{V}{Z_1 + Z_f + Z_b} \quad ...(4.6)$$

Power factor $$= \frac{R_{eq}}{Z_{eq}} = \frac{R_1 + R_f + R_b}{Z_1 + Z_f + Z_b} \quad ...(4.7)$$

$$E_{mf} = I_1 Z_f \quad ...(4.8)$$

$$E_{mb} = I_1 Z_b \quad ...(4.9)$$

$$I'_{2f} = \frac{E_{mf}}{\frac{R'_2}{2s} + j\frac{X'_2}{2}} = \frac{I_1 Z_f}{\sqrt{\left(\frac{R'_2}{2s}\right)^2 + \left(\frac{X'_2}{2}\right)^2}} \quad ...(4.10)$$

$$I'_{2b} = \frac{E_{mb}}{\frac{R'_2}{2(2-s)} + j\frac{X'_2}{2}} = \frac{E_{mb}}{\sqrt{\left(\frac{R'_2}{2(s-s)}\right)^2 + \left(\frac{X'_2}{2}\right)^2}} \quad ...(4.11)$$

Air-gap power for forward field,

$$P_{\text{air gap }(f)} = (I'_{2f})^2 \frac{R'_2}{2s} \text{ watts} \quad ...(4.12)$$

Air-gap power for backward field,

$$P_{\text{air gap }(b)} = (I'_{2b})^2 \frac{R'_2}{2(2-s)} \text{ watts} \quad ...(4.13)$$

Mechanical power output for the forward field,

$$P_{\text{mech }(f)} = (1-s)\, P_{\text{air gap }(f)} = (I'_{2f})^2 \frac{R'_2}{2} \cdot \left(\frac{1-s}{s}\right) \quad ...(4.14)$$

Mechanical power output for the backward field,

$$P_{\text{mech }(b)} = [1-(2-s)]\, P_{\text{air gap }(b)} = -(I'_{2b})^2 \frac{R'_2}{2} \cdot \frac{1-s}{2-s} \quad ...(4.15)$$

Mechanical power output (net),

$$\begin{aligned} P_{\text{mech (net)}} &= P_{\text{mech }(f)} + P_{\text{mech }(b)} \\ &= (1-s)\, P_{\text{air gap }(f)} + [1-(2-s)]\, P_{\text{air gap }(b)} \\ &= (1-s)\, [P_{\text{air gap }(f)} - P_{\text{air gap }(b)}] \end{aligned} \quad ...(4.16)$$

Torque developed for the forward field,

$$T_f = \frac{P_{\text{air gap }(f)}}{(2\pi N_s/60)} \quad ...(4.17)$$

Torque developed for the backward field,

$$T_b = \frac{P_{\text{air gap }(b)}}{(2\pi N_s/60)} \quad ...(4.18)$$

Net torque developed, $$T_{\text{net}} = \frac{[P_{\text{air gap }(f)} - P_{\text{air gap }(b)}]}{(2\pi N_s/60)} \quad ...(4.19)$$

Net power output = $P_{\text{mech (net)}}$ **– friction and windage losses – core losses** ...(4.20)

Motor efficiency, $$\eta = \frac{\text{output power}}{\text{input power}} \times 100 \qquad ...(4.21)$$

Power delievered to the backward field

$$= (2 - s)\, P_{\text{air gap }(b)} = P_{\text{air gap }(b)} + (1 - s)\, P_{\text{air gap }(b)}$$

$$= P_{\text{air gap }(b)} - P_{\text{mech }(b)} \qquad ...(4.22)$$

All of the power delivered to the backward field is converted into heat through the rotor copper losses caused by the backward field.

Tests for determining equivalent circuit parameters :

The most common tests performed on single phase induction motors to determine various constants are as follows :

(*i*) *Measurement of stator winding resistance.* The resistance of each stator winding is measured separately ; in case of capacitor start motor the condenser is omitted.

(*ii*) *Blocked rotor test.* This test is conducted with main winding and starting winding separately and input voltage, amperes, and watts are measured in each case.

This test is sometimes performed at reduced voltage (about 40% of rated voltage) to avoid excessive short-circuit current and heating.

(*iii*) *No-load test.* No-load test is performed at rated voltage, exciting the main winding only and input current, voltage and power, as well as speed, are measured.

(*iv*) *Determination of friction and windage losses.* In this test the voltage applied to the motor is reduced until the motor will just run and power input is measured. The power input, just sufficient to keep the motor running on no-load, gives approximately the friction and windage losses.

4.4. SPLIT-PHASE MOTORS

Since the single-phase induction is not self-starting, means must be provided to create an initial torque. But the initial torque is only possible if a rotating flux is created in the stator. It is known that *a rotating flux is produced when there is a difference of 90° between the currents of two stationary coils. Or if the stator possesses two fluxes having a large phase difference, the result is a rotating flux.*

4.4.1. Split-phase Resistance—Start Induction Motor

- In a split-phase induction motor the stator is provided with two parallel windings displaced 90 electrical degrees in space and somewhat less than 90° in time. Fig. 4.7 (*a*) shows the winding diagram of the *two windings* of a split-phase induction motor.

  The *starting winding* has fewer turns and is wound of smaller diameter copper than the running winding. The starting winding, therefore, *has a high resistance and low reactance.*

  The *running or main, winding* (heavier wire of more turns) has a *low resistance* and *high reactance.* Because of its lower impedance, the current in the running winding I_r is *higher* than the current in the starting winding, I_s.

  The phase relations of the lock-motor currents at the instant of starting are shown in Fig. 4.7 (*b*). The starting winding I_s lags the supply voltage by about 15°, while the greater running winding current lags the single-phase voltage by about 40°. Despite the fact that the current in the two space-quadrature windings are *not* equal, the quadrature components are practically equal.

  If the windings are displaced by 90° in space, and if their quadrature current components, which are displaced by 90° in time, are practically equal, an equivalent two-phase rotating field is produced at starting which develops sufficient starting torque to accelerate the rotor, in the direction of the rotating field produced by the currents.

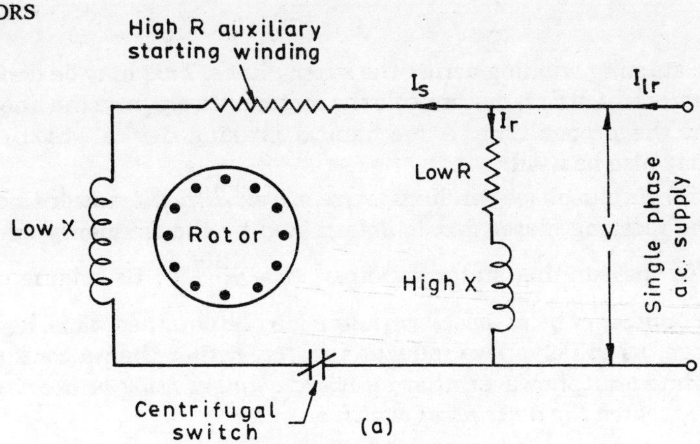

(*a*) Connection diagram.

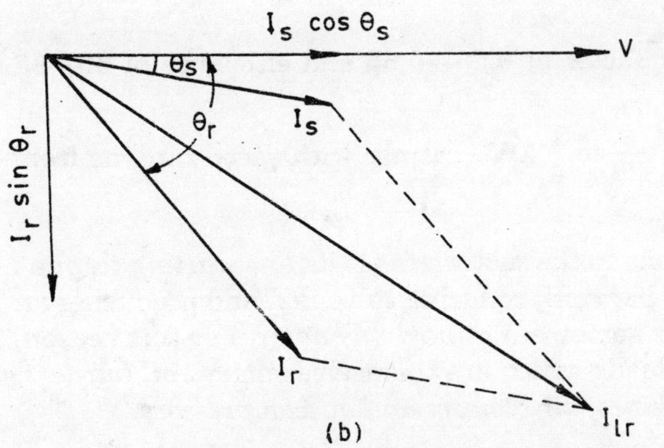

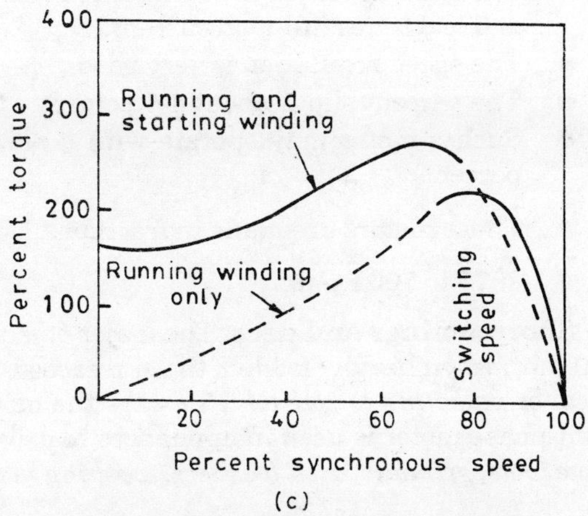

(*b*) Phase relations. (*c*) Typical torque speed characteristic.

Fig. 4.7. Split phase resistance-start induction motor.

- As the *motor speeds up,* the *torque developed increases.* Above 85 per cent of synchronous speed, the torque developed by the running winding (main winding) alone is actually greater than that developed by both windings, and it might be advantageous to *open the auxiliary circuit at this cross over point.* To allow for individual variations among motors and switches, however, the contacts are usually designed to open at 75 per cent of synchronous speed. This does not seriously affect the operation, because the running (or main) winding alone usually develops approximately 200 per cent of full-load torque at this speed.
- The *starting winding is not designed for continuous operation, and care should be exercised that it does not remain connected to the supply after it should have been disconnected by the switch.* This series switch is usually *centrifugally operated, and is rather inexpensive.* In case of a hermatically sealed motor the switch is magnetically operated, and is opened in the de-energized condition.
- Split-phase induction motors may be *reversed by reversing the line connections of either the main or the auxiliary winding.* If however, reversal is attempted under normal running condition, nothing will happen.

If it is necessary to *reverse the motor while it is rotating,* then some means must be incorporated to slow the motor down to the speed where the starting-switch contacts close, placing the starting winding across the supply lines. This may be done by *incorporating a timing device* which first disconnects the motor entirely from the line and then reverses one field at the proper time. A mechanical braking device which can be electrically operated may also be used.

- Speed control of split-phase windings is a *relatively difficult* matter since the synchronous speed of the rotating stator flux is determined by the frequency and number of poles developed in the running stator winding $\left(N_s = \frac{120f}{p}\right)$. By adding stator windings to change the number of poles, speed variation may be obtained. This, however, is a stepped speed change, as in polyphase induction motor, rather than a continuous variation. It must be pointed out, however, that *all speed changes must be accomplished in a range above that at which the centrifugal switch operates.*

Fig. 4.8 (*c*) shows the typical torque speed characteristics :

- The starting torque is 1.5 times to twice the full-load starting torque and starting current is 6 to 8 times full-load current.
- The speed regulation is very good.
- The percent slip is about 4-6 percent.
- Such a motor may operate with a power factor of 0.55—0.65 and efficiency of 60—65 percent.
- These motors are made in fractional kW $\left(\frac{1}{20} \text{ to } \frac{1}{4} \text{ kW}\right)$ ratings with speed ranging from 2875 to 700 r.p.m.

Shortcomings and uses. The major objections to the motor are (1) its *low starting torque* ; and (2) that, when heavily loaded, the slip exceeds 5 per cent, reducing the e.m.f. and producing an elliptical or *pulsating torque which makes* the motor somewhat *annoyingly noisy*. For this reason, the split-phase motor is used in appliances to drive loads which are themselves noisy : *oil burners, machine tools, grinders, dish washers, washing machines, air blowers and air compressors.*

- *Because of low starting torque these motors are seldom employed in sizes larger than $\frac{1}{4}$ kW.*

4.4.2. Split-phase Capacitor—Start Induction Motor

Another method of splitting the single-phase supply into two phases to be applied to the stator windings is *placing a capacitor in series with the starting auxiliary winding*. In this manner, the current in the starting winding may be made to lead the line voltage. Since the running winding current lags the line voltage, the phase displacement between the two currents be made to approximately 90° on starting. The circuit of capacitor-start motor is shown in Fig. 4.8 (*a*), while the vector diagram of the currents and voltage is shown in Fig. 4.8 (*b*). The values of the angles shown are fairly representative, and are rounded off for convenience. *One of the factors upon which the starting torque depends is the sine of the angle between the currents in the two windings.* The value of series capacitor may therefore be reduced, while maintaining a phase-shift angle of about 90°.

- The increase in phase angle between starting and running winding currents is not the only difference between the split-phase and capacitor start-motors. *The split phase motor must keep the number of starting-winding turns low, so that the current may be nearly in phase with the line voltage.* This, *however, is unnecessary in a capacitor-start motor, since the capacitor can overcome the inductance of the winding while still providing the proper phase shift.* There are thus *more auxiliary starting turns* in the capacitor-start motor than in the comparable split-phase motor. This provides a greater number of ampere-turns, hence a *larger rotating flux,* and therefore a *further increase in the starting torque.*

Also it is seen that for the same magnitudes of field currents, the *current I_{lr} is less in capacitor-start motor,* because of the greater angle between the two field currents. In addition, the *starting power factor is also better*. For a given line current, the starting torque is thus much higher for a capacitor-start motor than for a split-phase induction motor. *The starting torque of capacitor-start motor is from 3 to 4.5 times the full-load torque, while that of split-phase resistance start induction motor rarely exceeds twice the full-load torque.*

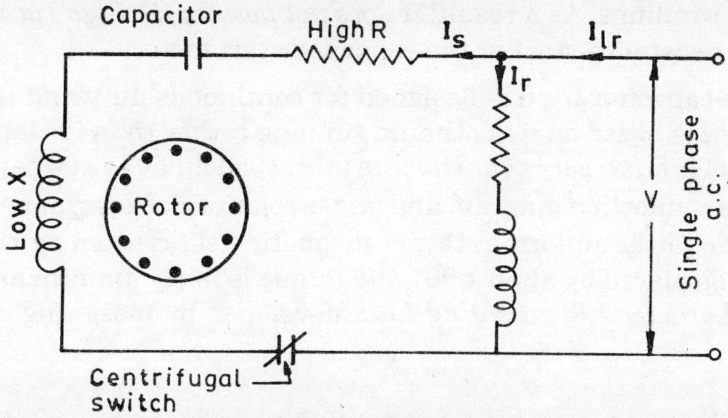

(*a*) Circuit diagram

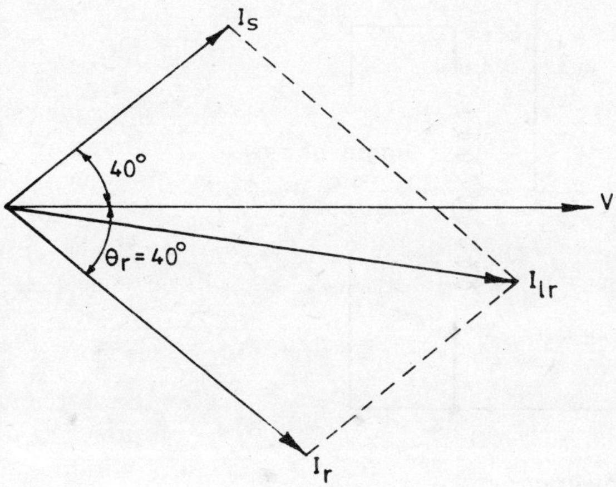

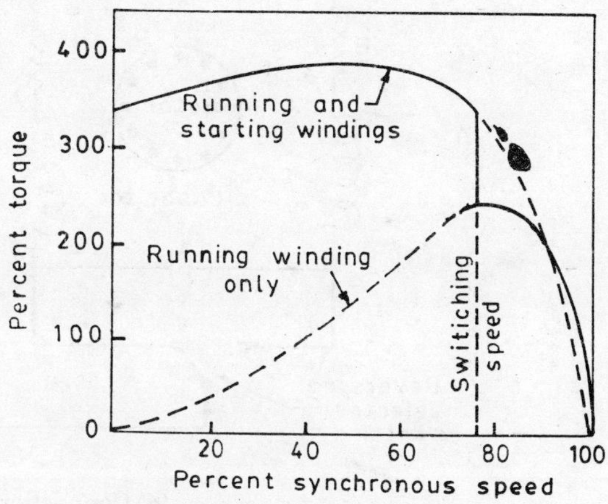

(*b*) Vector diagram (*c*) Typical torque-speed characteristic

Fig. 4.8. Capacitor-start induction motor.

- These motors are manufactured in *ratings ranging from $\frac{1}{10}$ kW to $\frac{3}{4}$ kW, but larger sizes are also available.*
- The capacitor-start motor may be reversed by changing the connections of one of the windings, but it is subject to the same limitations as the resistance-start induction motor.

Uses. By virtue of their higher starting torque, capacitor-start split-phase motors are used for *pumps, compressors, refrigeration units, air-conditioners,* and *large washing machines,* where a split-phase motor is required that will develop high starting torque under load.

4.4.3. Permanent-split Capacitor Induction Motor

- A permanent-split capacitor motor (also called single-value capacitor motor) is a single-phase induction motor which has the *same capacitor* in series with the starting (or auxiliary) winding for both starting and running. Because it runs continuously as a permanent split-phase motor, no centrifugal switch is required.
- The motor starts and runs by virtue of the quadrature phase-splitting produced by the two identical windings. As a result, *it does not possess the high running torque* produced by either the resistance-start or the capacitor-start motor.

Furthermore, the capacitor used is designed for continuous duty and is of an oil-filled type. The value of the capacitor is based on its optimum running rather than its starting characteristics. The result is that this motor *has a very poor starting torque,* about 50 to 100 per cent of rated torque.

Fig. 4.9 shows the connection diagram and phase relations of a permanent split-phase motor.

- Because of the fairly uniform rotating magnetic field created by equal windings whose currents are displaced by almost 90°, the torque is fairly uniform and the motor *does not exhibit the characteristic pulsating hum* developed by most single-phase motors when loaded.

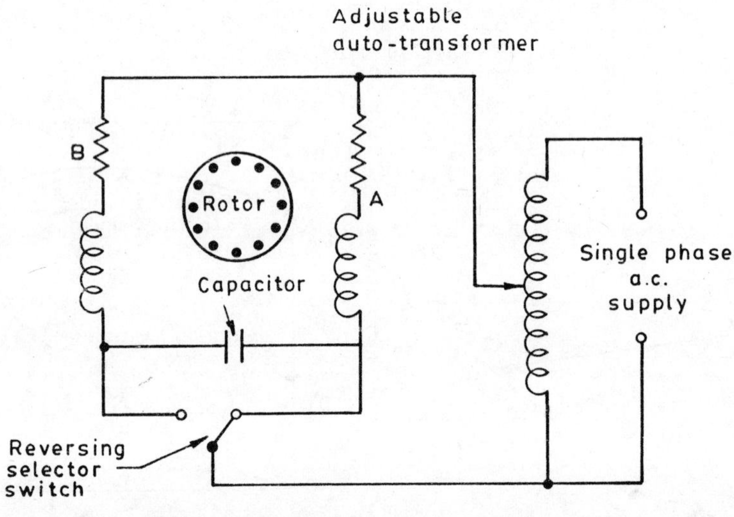

(*a*) Connection diagram.

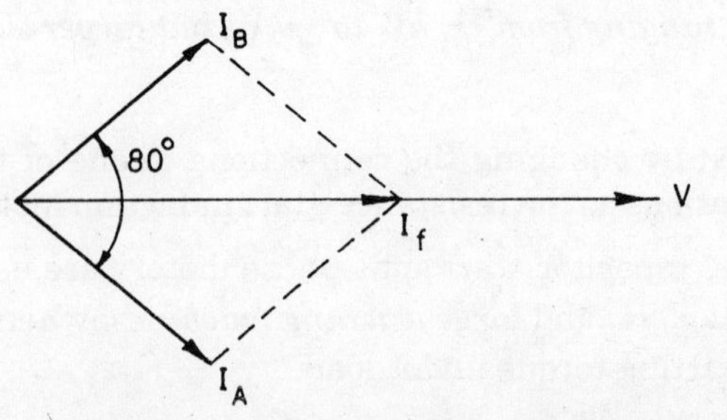

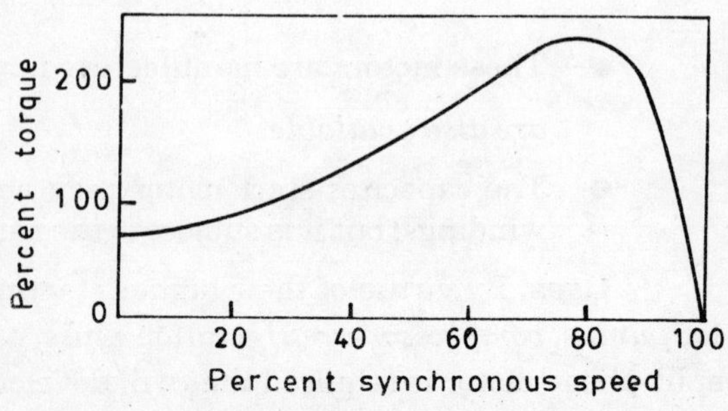

(*b*) Vector diagram or phase relations. (*c*) Typical torque-speed characteristic.

Fig. 4.9. Permanent split-phase motor.

This motor possesses the following *merits* :

(*i*) Higher power factor at full-load

(*ii*) Lower full-load line current

(*iii*) Higher full-load efficiency

(*iv*) Increased pull-out torque.

- The permanent-split capacitor motor is *more expensive* than the equivalent split-phase or capacitor-start induction motor. This is primarily due to the fact that the auxiliary (or starting) winding is now also a running winding. It must therefore have a continuous duty rating and as such is heavier than if it were were short-time-rated.

Uses. Because of its *instant response as a reversing motor,* its *quieter operation,* and the *possibility of speed control*, the permanent-split capacitor motor is used for *exhaust and intake fans* and *blowers, unit heaters* and *office machines.*

4.4.4. Two-value Capacitor Induction Motor

- Refer Figs. 4.10 and 4.11. *The two-value capacitor motor combines the quiet operation and limited speed control advantages of a running permanent-split capacitor with high starting torque of the capacitor-start motor.*

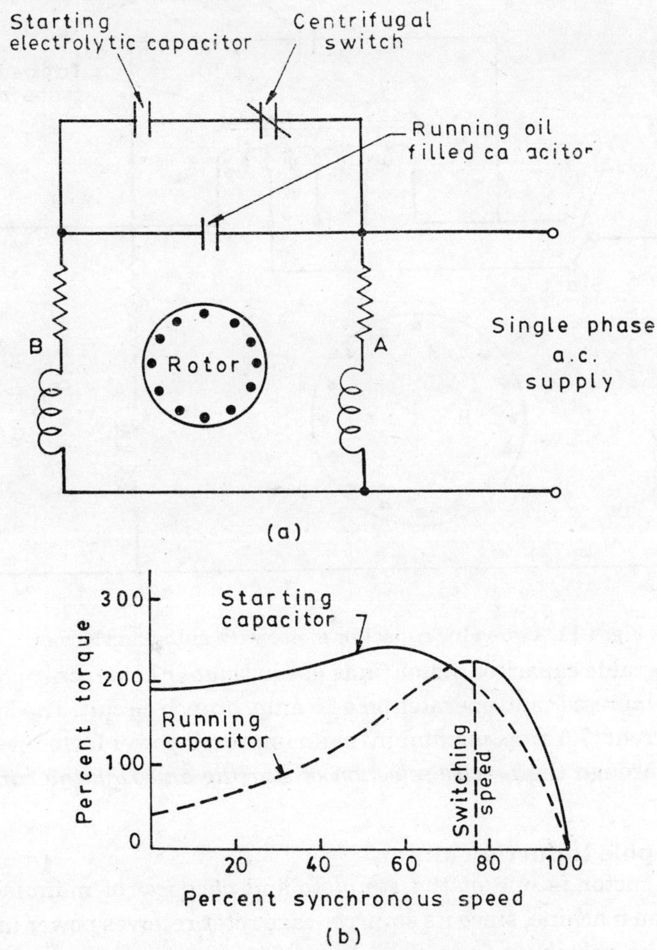

Fig. 4.10. Connection diagram for two-value capacitor motor and typical torque-speed characteristic.

- Two capacitors (Fig. 4.10) are employed during the starting period. One of these, an electrolytic starting capacitor, similar to that used for the intermittent duty of the capacitor-start split-phase motor, is of fairly high capacity (about 10 to 15 times the value of the running capacitor) and is cut out of the circuit by a centrifugal switch when slip reaches about 25 per cent.
- The *major advantage of the two value capacitor motor is its high starting torque, coupled with quiet operation and good running torque.* Still classed as a reversing motor, when the line leads of one winding are reversed, it is reversed in the usual manner. When the speed drops below 25% slip during reversing, the centrifugal switch closes its starting contact, providing the maximum torque as the motor slows down and reverses. The contacts open again when the motor is up to 75% synchronous speed in the reverse direction. Frequent reversals, therefore will reduce the life of the centrifugal switch. For this reason, where frequent reversals are accomplished, permanent-split, single-value capacitor, using no centrifugal switch whatever, is preferable.

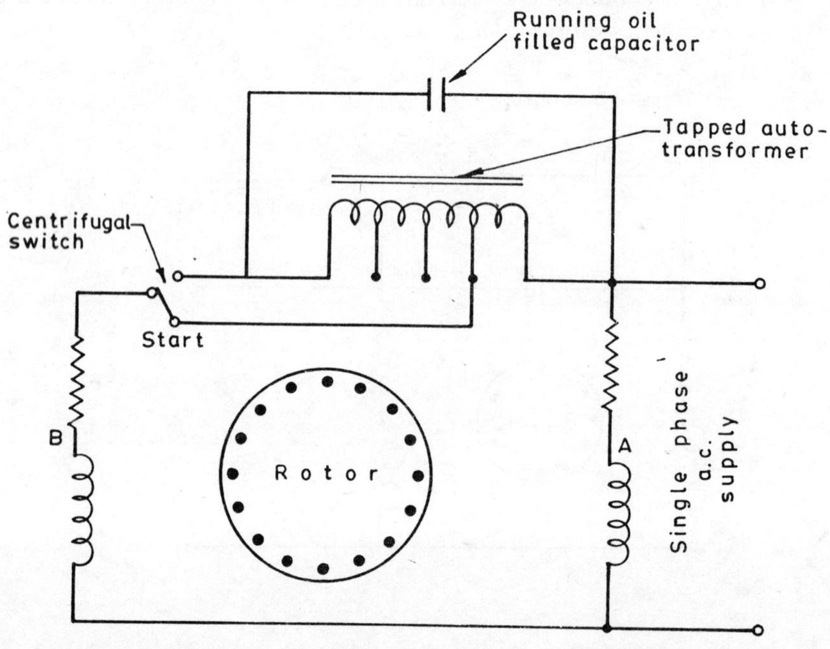

Fig. 4.11. Two-value capacitor motor with auto-transformer.

Uses. This two-value capacitor motor finds use in *smaller home air-conditioning units* which use this motor in its *compressor* and operates on a 15 amp. branch circuit. The lower starting current and lower running current (7.5 amp. maximum) at an improved power factor over the capacitor-start motor, are obtained through the *precise selection of starting and running capacitors* for the fixed compressor load.

4.4.5. Shaded-pole Induction Motor

A shaded-pole motor is one of the *simplest* and *cheapest* of manufactured motors. It is essentially an induction machine, since its squirrel-cage rotor receives power in much the same way as does the rotor of the polyphase induction motor. There is however, one extremely important difference between the two. Whereas the poly-phase induction motor creates a true revolving field, in the sense that it is constant in magnitude and rotates at synchronous speed *completely round the entire core*, the field of the shaded-pole motor is not constant in magnitude but *merely shifts from one side of the pole to the other.* Because the shaded-pole motor does not create a true revolving field, the *torque is not uniform but varies from instant to instant.*

Fig. 4.12 shows the *general construction and principle of shaded pole motor.*

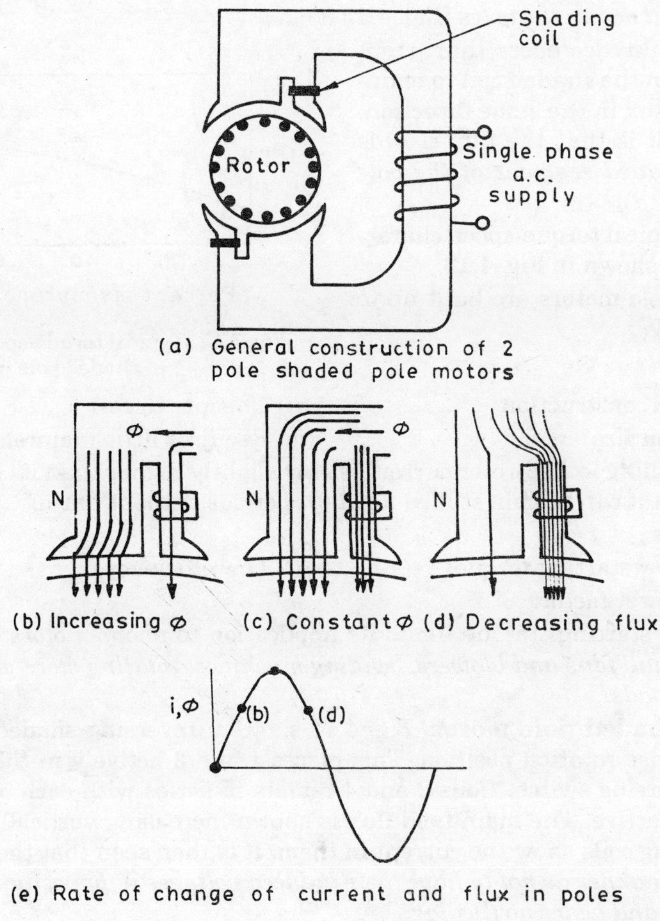

Fig. 4.12. General construction and principle of shaded pole motor.

Construction. Each of the laminated poles of the stator has a slot cut *across* the laminations about one-third the distance from one edge. Around the smaller of the two areas formed by this slot is placed a heavy *copper short-circuited coil,* called a *shading coil* ; the iron around which the shading coil is placed is called the *shaded* part of the pole, while the free portion of the pole is the *unshaded* part. The exciting coil surrounds the entire coil.

Principle of operation. When the exciting winding is connected to an A.C. source of supply, the *magnetic axis will shift from the unshaded part of the pole to the shaded part of the pole.* This shift in the magnetic axis is, in effect *equivalent to an actual physical motion of the pole* ; the result is that the squirrel cage rotor will rotate in a direction from the unshaded part to the shaded part. The shifting of flux is explained below.

- Refer Fig. 4.12 (*b*). When the flux in the field poles tend to increase, a short-circuit current is induced in the shading coil, which by Lenz's law opposes the force and the flux producing it. Thus, as the flux increases in each field pole, there is a *concentration of flux in the main segment of each pole*, while the shaded segment opposes the main field flux.
- At point (*c*) shown in Fig. 4.12 (*e*), the rate of change of flux and of current is zero, and there is no voltage induced in the shaded coil. Consequently, the *flux is uniformly distributed across the poles* [Fig. 4.12 (*c*)].
- When the flux decreases, the current reverses in the shaded coil to maintain the flux in the same direction. The result is that the flux crowds in the *shaded segment of the pole* [Fig. 4.12 (*d*)]

  A typical torque-speed characteristic is shown in Fig. 4.13.
- Shaded pole motors are built up to about 40 W.

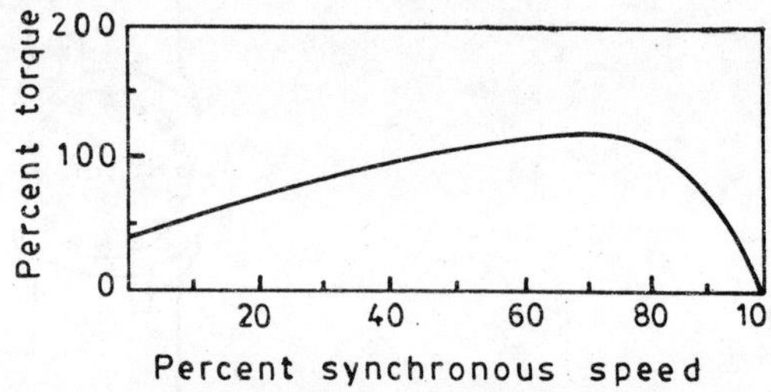

Fig. 4.13. Typical torque-speed characteristic of shaded pole motor.

Merits :

(*i*) Rugged construction (*ii*) Cheaper in cost

(*iii*) Small in size (*iv*) Requires little maintenance

(*v*) Its stalling locked-rotor current is only slightly higher than its normal rated current, so that it can remain stalled for short periods without harm.

Demerits :

(*i*) Very low starting torque (*ii*) Low efficiency

(*iii*) Low power factor.

Uses. Its low starting torque limits its application to *phono-motors* or *turntables, motion picture projectors, small fans and blowers, bending machines, rotating store-window display tables, and relatively light loads.*

Reversing shaded pole motor. Fig. 4.14 shows a reversing shaded-pole motor with the switch in the clockwise rotation position. This places 1 and 3 actively in the circuit, being short-circuited by the reversing switch. Coils 2 and 4 remain in series with each other, but are on open circuit and so are inactive. The main-field flux is shown increasing vertically downward, and the arrows on the shading coils show the current in them. It is then seen that the *shading coil must be connected in such a manner as not to have their induced voltages in opposition, or there may not be any current in them, and hence no flux lag.*

4.4.6. Reluctance-start Induction Motor

A reluctance-start induction motor is shown in Fig. 4.15. Its characteristics are similar to that of shaded pole motor. In this motor too the magnetic field shifts across the pole, but the effect is obtained by the *non-uniform air gap of salient poles. Where there is a greater air gap, the flux in that portion is more nearly in phase with the current. There is a greater lag between flux and current where there is a lower reluctance or where the air gap is smaller.* Since both fluxes are produced by the same current, the flux across the larger air gap *leads* the flux across the smaller one. The two fluxes are obviously displaced in time, and so the *magnetic field shifts across the poles from larger air gap to the shorter gap*. Thus the *direction of rotation is firmly fixed by the construction, and the motor cannot be reversed at all.*

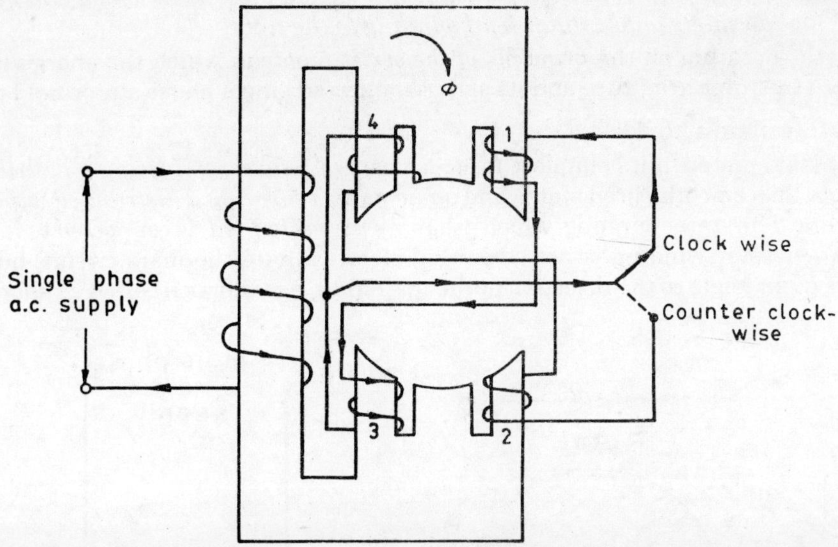

Fig. 4.14. Reversing shaded pole motor.

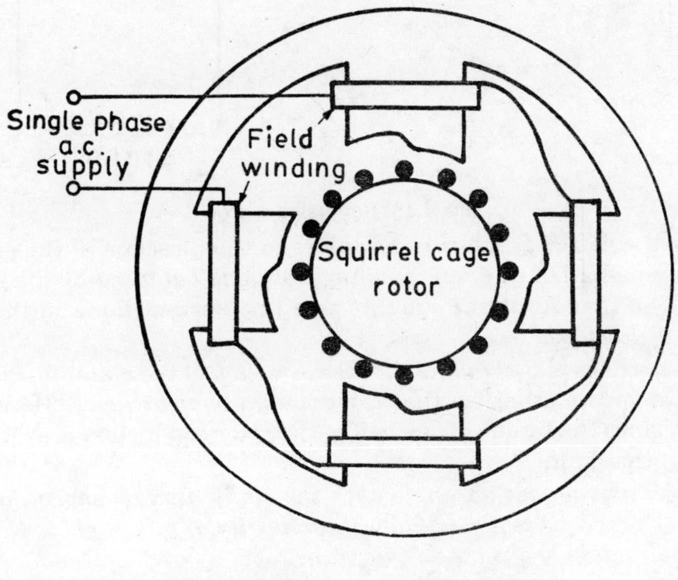

Fig. 4.15. Reluctance start motor.

Uses. For most small power applications, the shaded pole motor is preferred, and the reluctance-start motor has limited use, usually only where *starting torque requirements are low.*

Note. This motor is an induction motor and should not be confused with *reluctance motor* which is actually a non-excited synchronous motor.

4.5. SINGLE-PHASE COMMUTATOR MOTORS

The commutator motors are so called because the wound rotor of this kind of motor is equipped with a *commutator and brushes.* This group consists of the following two classes :

1. Those operating on *'repulsion principle'* (repulsion motors) *in which energy is inductively transferred from the single phase stator field winding to the rotor.*

2. Those operating on the *principle of the series motor* in which the energy is *conductively* carried both to the rotor armature and its series-connected single phase stator field.

4.5.1. Repulsion Motor

A repulsion motor in its simplest form consists of a *field comprising a distributed winding,* housed in slots, in a smooth-cored stator and *an armature carrying a distributed winding connected to a commutator.* The stator winding, which produces the *main field,* is connected to the main supply. The armature or rotor winding is not connected electrically to the main circuit, but the brushes, which are set at an angle to the direction of the main flux, are short-circuited as shown in Fig. 4.16.

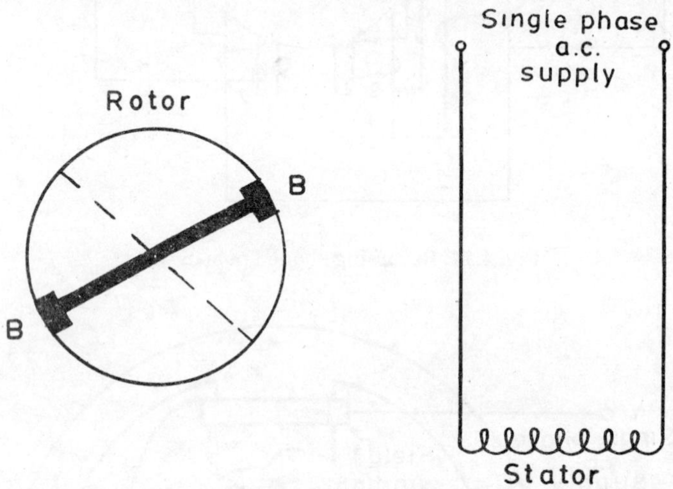

Fig. 4.16. Repulsion motor.

- If the brush axis *BB* is set *at right angles* to the direction of the stator flux, the e.m.f. induced in one-half of the rotor winding is *exactly balanced* by the e.m.f. induced in the other half, so that resultant e.m.f. is *zero* ; no current flows in the rotor winding and *no-torque is developed.*
- If the brush axis is placed *in line with the direction* of the stator flux, the e.m.f. in one half assists that in the other, so that *a maximum current flows.* Here again *no torque is developed,* since the torque of one half of the rotor conductors is exactly balanced by that due to the other half.
- If *the brush axis makes an angle with the stator flux as shown, a resultant torque is produced.* The torque is a *maximum theoretically, if this angle is 45°, but in practice the angle of inclination is about half this value.*

It is clear from above that the *speed of repulsion motor depends upon the brush position. Speed control of* such a machine can be provided by *mounting the brushes on a rocker* which can be rotated by a lever handle mounted on the motor end-shield. If remote control is required, the lever handle may be manipulated by a simple system of rods and cranks.

Alternatively, if the motor is to be totally enclosed, or remote control from a considerable distance is required, speed control may be obtained by the *use of an external series resistance with fixed brush gear.*

- The direction of rotation of a simple repulsion motor may be reversed by swinging the brushes into the position shown *dotted* in Fig. 4.16.

Atkinson repulsion motor. A modification of the simple repulsion motor is the Atkinson repulsion motor, in which the stator winding comprises two windings at right-angles to each other and connected in series, as shown in Fig. 4.17. One advantage obtained by this method is that the *direction of rotation can be reversed by reversing the connections to one of the stator windings.* Instead of moving the brush rocker, it is necessary only to throw the reversing switch, shown in Fig. 4.17.

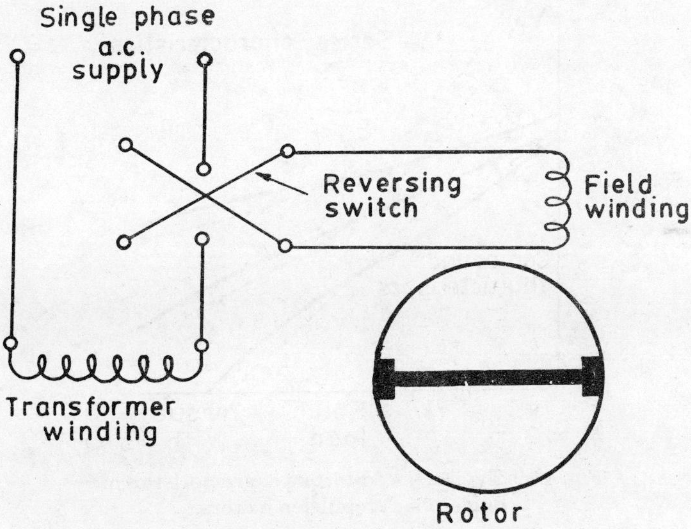

Fig. 4.17. Atkinson repulsion motor with reversing switch.

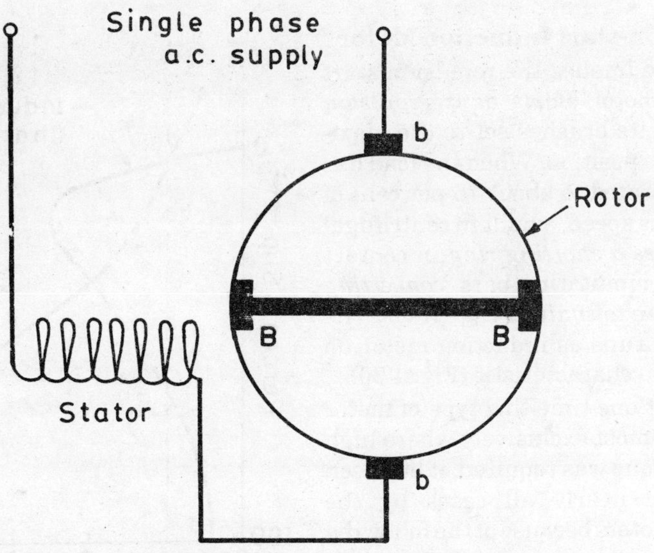

Fig. 4.18. Compensated repulsion motor.

It will be observed that, as the rotor is electrically connected to the stator, the compensated repulsion motor is not able to operate directly from a high-voltage supply, as was the case with the simple repulsion motor.

Fig. 4.19 shows the typical speed-torque characteristics of single-phase repulsion motor.

Compensated repulsion motor. The *power factor of the machine may be improved by compensation,* and this is done by providing *an additional set of brushes* as shown in Fig. 4.18. The axis of one pair, *BB*, of brushes coincides with the stator winding, these brushes are *short-circuited.*

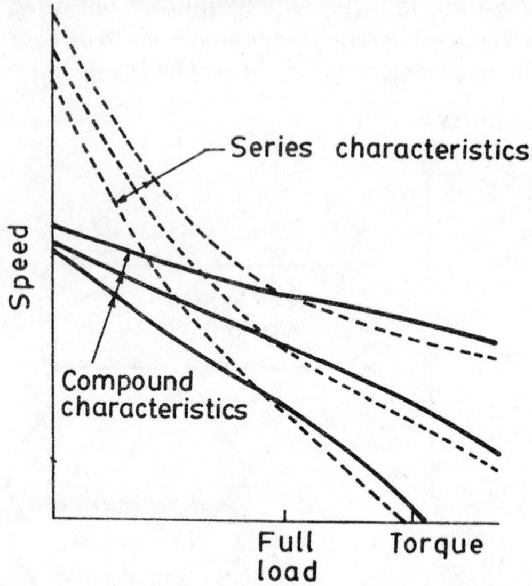

Fig. 4.19. Typical speed-torque characteristics of single-phase repulsion motors.

The other pair, *bb*, of brushes is set at right-angles to the former and is connected in *series with the stator winding.*

4.5.2. Repulsion-start Induction Motor

- As its name implies the repulsion-start induction motor *starts as a repulsion motor* with its brushes set to the maximum torque position. When the load has been accelerated to about 75 per cent of synchronous speed, a built in centrifugal device *places a shorting ring* in contact with the commutating bars, *converting the armature to squirrel-cage rotor.* The motor then runs as induction motor on its induction characteristic (Fig. 4.20).
- Although at one time this type of motor was used almost exclusively where high starting torque was required, it has been replaced in nearly all cases by the capacitor motors because of the following reasons :

  (*i*) Requires more maintenance.

  (*ii*) More expensive.

  (*iii*) Makes quite a bit of noise on starting.

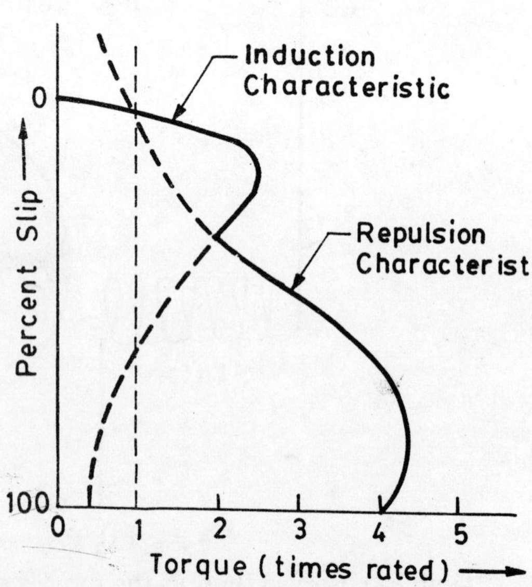

Fig. 4.20. Speed-torque characteristic of repulsion-start induction motor.

(*iv*) Causes radio interference when starting

(*v*) Cannot be reversed easily.

Repulsion-start motors, despite these disadvantages, are still used in *integral-horsepower sizes* because of the following reasons :

(*i*) High starting torque. (*ii*) Low starting current.

(*iii*) Ability to accelerate a heavy load more rapidly than high-capacitance dual-capacitor motors.

4.5.3. Repulsion-induction Motor

- A single-phase repulsion-induction motor combines the constant-speed characteristics of the single-phase induction motor with the good starting characteristics of the repulsion motor.
- The *stator of this machine has a simple single-phase winding like that of single-phase induction motor.* The rotor, however, is built up of *laminations,* each of which has *two concentric sets of slots. These slots contain two distinct windings ; in the outer slots is wound a commutator winding similar to that of a D.C. armature, while in the inner slots is a cast aluminium squirrel-cage winding which clamps the laminations.*

Fig. 4.21 (*a*) illustrates an armature for a repulsion induction motor complete with a squirrel-cage winding.

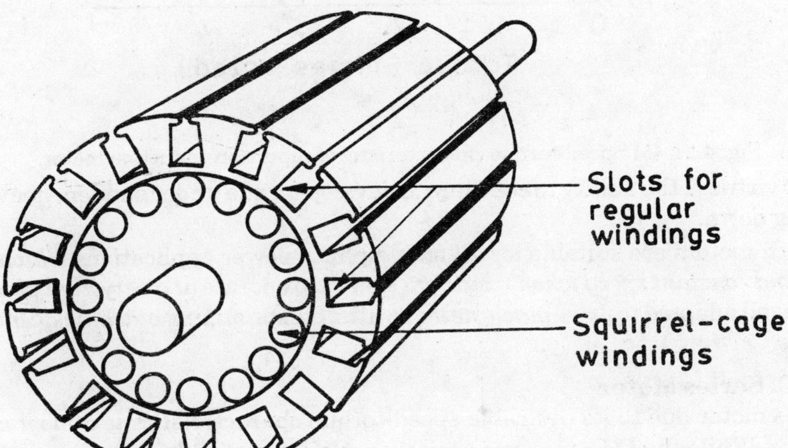

Fig. 4.21. (*a*) Repulsion induction motor.

At starting and during the acceleration period, the magnetic flux produced by the stator *embraces only the commutator winding in the outer slots owing to the high reactance of the squirrel-cage.* The *motor starts up virtually as a repulsion motor and develops a high starting torque. As the motor speeds up the reactance of the squirrel-cage decreases, so that this winding assists the commutator winding to supply the running torque.*

Fig. 4.21 (*b*) shows the speed-torque characteristic of repulsion-induction motor.

Merits. The repulsion-induction motor has the following merits :

(*i*) High starting torque. (*ii*) Fairly good speed regulation.

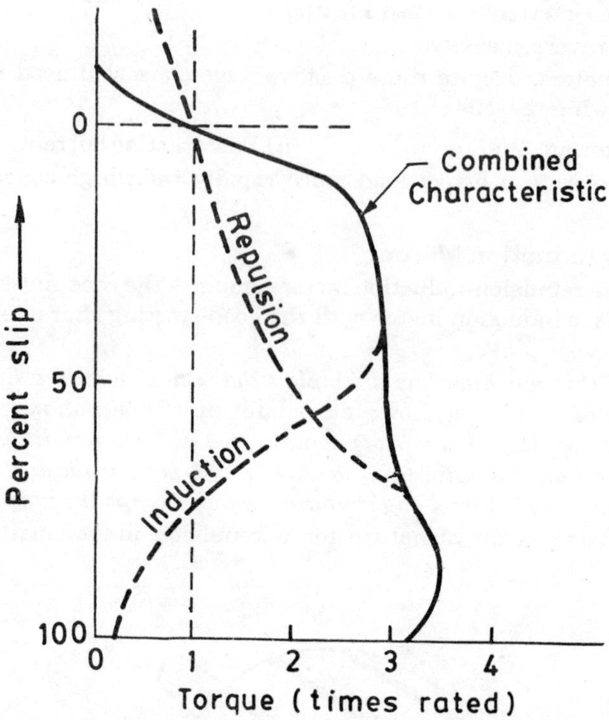

Fig. 4.21. (*b*) Speed-torque characteristic of repulsion-induction motor.

(*iii*) Major virtue is the ability to continue to develop torque under sudden, heavy applied loads without breaking down.

Uses. Such motors are suitable for all single-phase power applications *which require a high starting torque and constant speed when running* ; they also operate *at a very high power factor.* They are particularly well adapted to drive *machine tools,* lifts, hoists, *mixing machines, centrifugal pumps, fans and blowers.*

4.5.4. A.C. Series Motor

The series motor due to its desirable speed-torque characteristics *is almost exclusively used in railway service.* While the D.C. motor is entirely satisfactory for this class of work service and is generally used on street railway cars and trolley coaches, the fact that it is more convenient and more economical to transmit power and to transform voltages in A.C. systems than with direct currents has lead to the development of the A.C. series motor for use on some of the important steam-road electrifications.

Working principle. The working principle of an A.C. series motor is the same as that of the D.C. series motor. The armature and field are wound and interconnected in the same manner as the D.C. series motor.

When an alternating e.m.f. is applied to the terminals, since field and armature windings are connected in series, the field flux and armature current reverse simultaneously every half cycle, but the direction of the torque remains unchanged. The torque is pulsating, but its average value is equal to that which a D.C. motor will develop if it had the same r.m.s. value of flux and current. Motor connections, direction of torque, etc. for two successive half cycles are shown in Fig. 4.22. If the field and armature core are run at low saturation, the air-gap flux is approximately proportional to the current and the torque is approximately proportional to current squared.

Although it is theoretically possible to operate a D.C. series motor from an A.C. circuit, the following structural changes must be made in the motor to make it a practical and reasonable efficient machine :

- The entire magnetic circuit must be *laminated,* and materials with low iron-loss co-efficients should be used as in transformers.

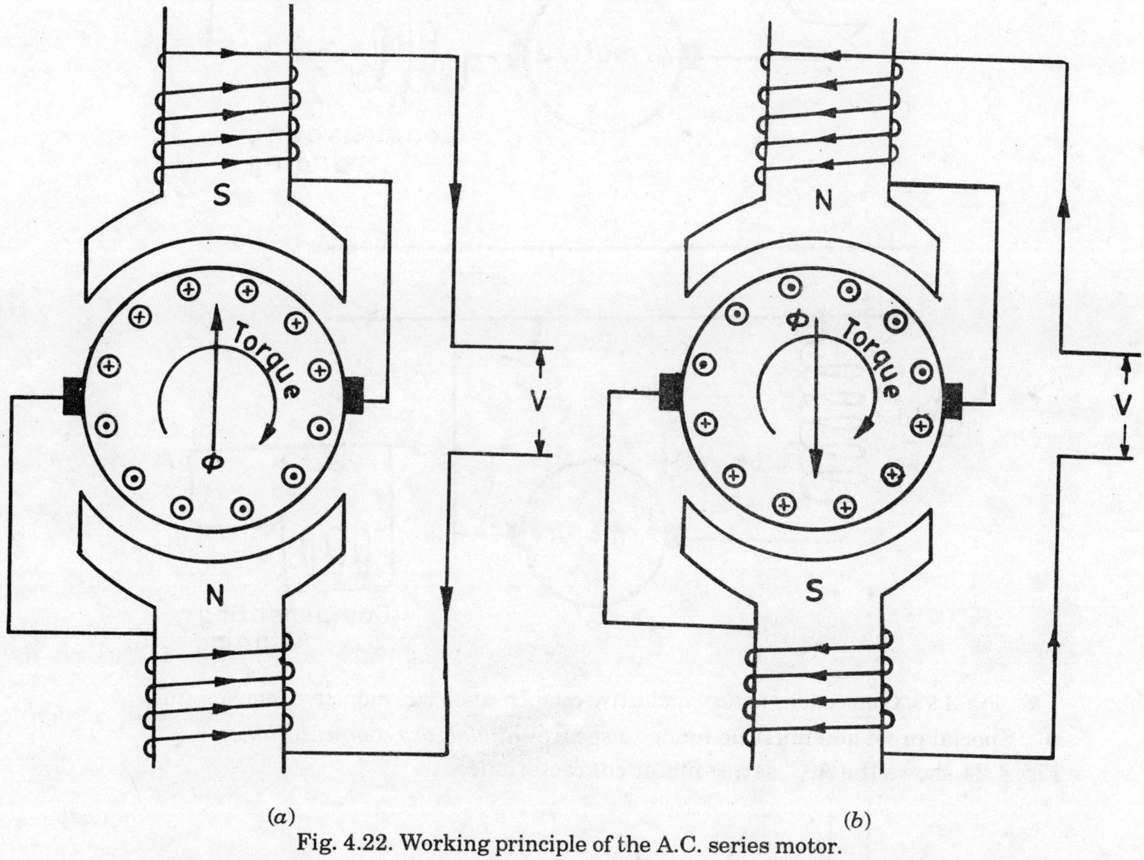

Fig. 4.22. Working principle of the A.C. series motor.

- The field circuit must be designed for a *much lower reactance* than the corresponding D.C. motor field in order to reduce the reactance voltage drop of the field to a minimum and to improve the power factor of the motor.
- A *distributed compensating winding* is required to reduce the reactance of the armature winding by reducing the leakage flux and to neutralize the cross-magnetising effect of the armature ampere turns.

The compensating winding may be connected in series with the series-field and armature windings, or it may be short-circuited upon itself and receive its excitation voltage by transformer action, since it is inductively coupled with the armature cross-field (Fig. 4.23). In the first case, the motor is said to be *conductively* compensated, while in the second it is *inductively* compensated. *Conductive compensation* is required on motors which are intended for operation in *D.C. as well as A.C. circuits.*

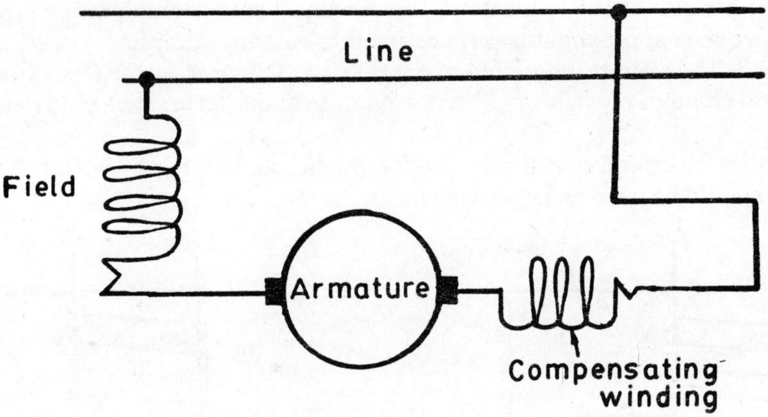

(a)

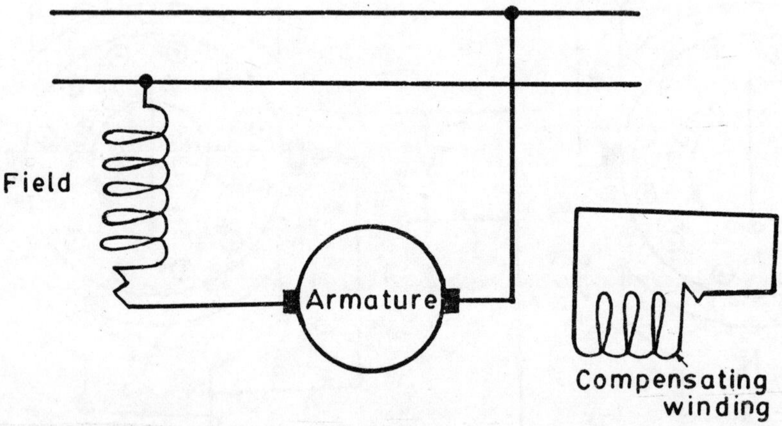

(b)

Fig. 4.23. Connections for (a) conductive compensation ; (b) inductive compensation.

- Special provision must be made to secure *satisfactory commutation*.

Fig. 4.24 shows the A.C. series motor characteristic.

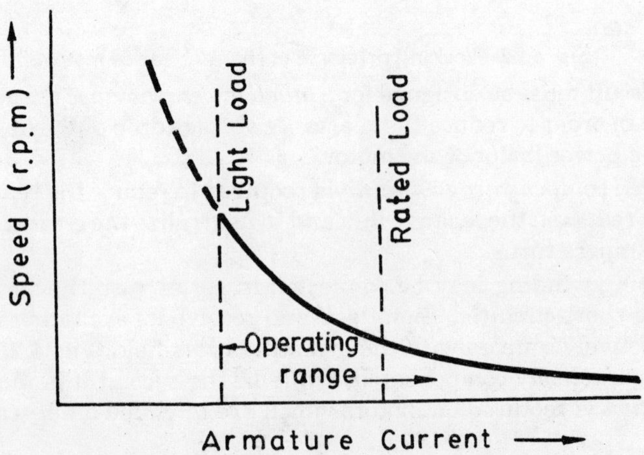

Fig. 4.24. A.C. series motor characteristics.

4.5.5. Universal Motor

- Fractional-horsepower series motors that are adapted for use on either D.C. or A.C. circuits of a given voltage are called *universal motors*.
- The universal motor is designed for commercial frequencies from 60 cycles down to D.C. (zero frequency), and for voltage from 250 V to 1.5 V. A commercial universal motor may have a somewhat weaker series field and more armature conductors than a D.C. series motor of equivalent horsepower. *It is manufactured in ratings up to 3 / 4 H.P., particularly for vacuum cleaners and industrial sewing machines.* In smaller sizes of $\frac{1}{4}$ H.P. or less, it is used in *electric hand drills*.

Like all series motors, the *no-load speed of the universal motor is universally high.* Quite frequently, *gears trains* are built into the motor housing of some universal motors to provide exceedingly high torque at low speeds.

When these motors are used in commercial appliances such as *electric shavers, sewing machines, office machines, and small hand hair dryers or vacuum cleaners*, they are always *directly loaded* with little danger of motor runaway.

Advantages of a universal motor :

1. High speed from above 3600 r.p.m. to around 25000 r.p.m.
2. High power output in small physical sizes for use in portable tools.
3. High torque at low and intermediate speeds to carry a particularly severe load.
4. Variable speed by adjustable governor, by line voltage or especially by modern pulse techniques.

Disadvantages :

1. Increased service requirement due to use of brushes and commutators. The life of these parts is limited in severe service.
2. Relatively high noise level at high speeds.
3. Moderate to severe radio and television interference due to brush sparking.
4. Requirement for careful balancing to avoid vibration.
5. Requirement for reduction gearing in most portable tools.

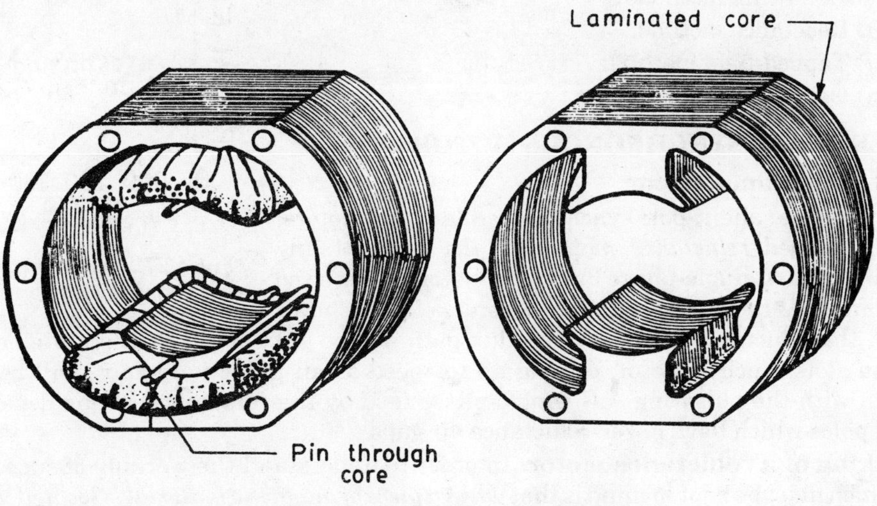

Fig. 4.25. Field core of a two pole universal motor.

Universal motors are manufactured in two types :

1. Concentrated-pole, non-compensated type (low H.P. rating).
2. Distributed field compensated type (high H.P. rating).

Fig. 4.25 shows the laminated field structure of a typical concentrated field universal motor.

Operation of a Universal Motor. As explained in Art. 4.5.4, such motors develop unidirectional torque regardless of whether they operate on D.C. or A.C. supply. The production of unidirectional torque when the motor runs on A.C. supply can be easily understood from Fig. 4.26. The motor works on the same principle as a D.C. motor *i.e.*, the force between the main pole flux and the current carrying armature conductors. This is true regardless of whether current is alternating or direct.

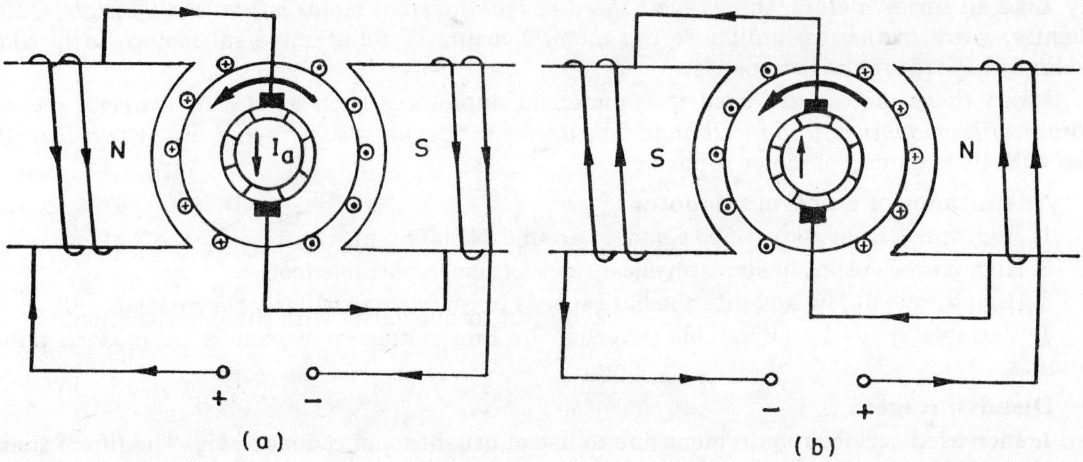

Fig. 4.26. Universal motor operation.

- Fig. 4.27 shows the typical torque characteristics of a universal motor both for D.C. and A.C. supply.
- The speed of a universal motor may be controlled by the following methods :
  (*i*) Reactance method.
  (*ii*) Tapped-field method.
  (*iii*) Centrifugal mechanism.

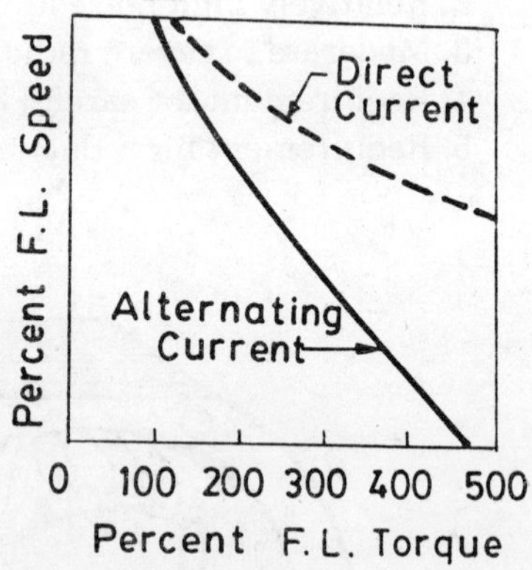

Fig. 4.27. Typical torque characteristics of a universal motor.

4.6. SINGLE PHASE SYNCHRONOUS MOTORS

4.6.1. Reluctance Motor

Single-phase salient-pole synchronous-induction motors, are generally called *reluctance motors*. If the rotor of any uniformly distributed single-phase induction motor is altered so that the laminations tend to produce *salient rotor poles*, as shown in Fig. 4.28, the reluctance of the air-gap flux path will be greater where there are no conductors embedded in slots. Such a motor, coming up to speed as an induction motor, will be pulled into synchronism with the pulsating A.C. single-phase field by the reluctance torque developed at the salient iron poles which have lower-reluctance air gaps.

Working of a reluctance motor. In order to understand the working of such a motor the basic fact which must be kept in mind is that *when a piece of magnetic material is located in a magnetic field, a force acts on the material, tending to bring it into the densest portion of the field. The force* *tends to align the specimen of material in such a way that the reluctance of the magnetic path that passes through the material will be minimum.*

When supply is given to the stator winding, the revolving magnetic field will exert reluctance torque on the unsymmetrical rotor tending to align the salient pole axis of the rotor with the axis of the revolving magnetic field (because in this position, the reluctance of the magnetic path would be minimum). If the reluctance torque is sufficient to start the motor and its load, the rotor will pull into step with the revolving field and continue to run at the speed of the revolving field. (Actually the motor starts as an induction motor and after it has reached its maximum speed as an induction motor, the reluctance torque pulls its rotor into step with the revolving field so that the motor now runs as synchronous motor by virtue of its saliency).

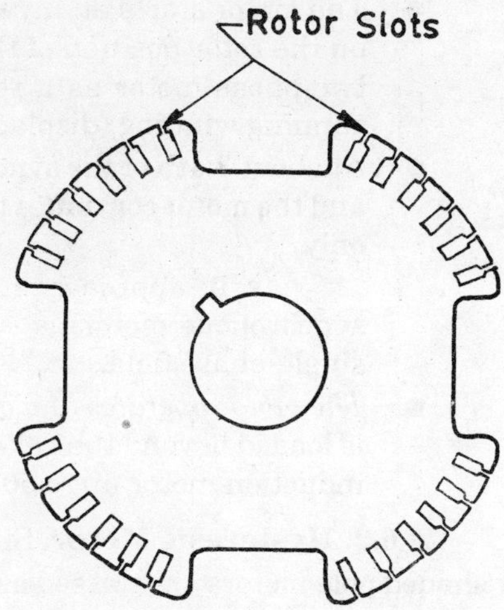

Fig. 4.28. Reluctance-motor lamination.

Reluctance motors have approximately *one-third* the horsepower rating they would have as induction motors with cylindrical rotors, although the ratio may be increased to one-half by proper design of the field windings. *Power factor and efficiency are poorer than for the equivalent induction motor*. Reluctance motors are subject to '*cogging*', since, the locked-rotor torque varies with the rotor position, but the effect may be *minimized* by *skewing the rotor bars* and by *not* having the number of rotor slots exactly equal to an exact multiple of the number of poles.

Uses. Despite its short-comings, the reluctance motor is widely used for many *constant speed speed* applications such as *recording instruments, time devices, control apparatus, regulators*, and *phonograph turntables*.

- *Reversing* is obtained as in any single-phase induction motor.

Speed-torque characteristics. Fig. 4.29 shows speed-torque characteristics of a typical single-phase reluctance motor.

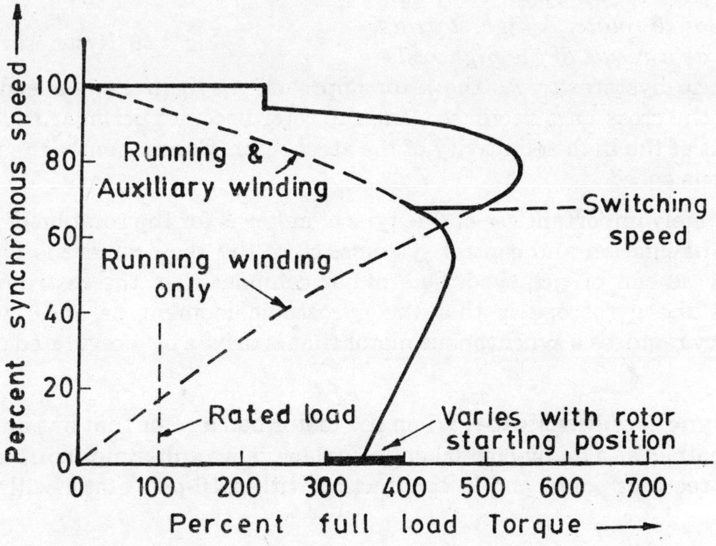

Fig. 4.29. Speed-torque characteristics of a single-phase reluctance motor.

- The motor starts at anywhere from 300 to 400 per cent of its full-load torque (depending on the rotor position of the unsymmetrical rotor with respect to the field windings) as a two-phase motor as a result of the magnetic rotating field created by a starting and running winding (displaced) 90° in both space and time.
- At about 3/4th of the synchronous speed, a centrifugal switch opens the starting winding, and the motor continues to develop a single-phase torque produced by its running winding only.

  As it approaches synchronous speed, the reluctance torque (developed as a synchronous motor) is sufficient to pull the rotor into synchronism with the pulsating single-phase field.
- *The motor operates at a constant speed up to a little over 200% of its full-load torque.* If it is loaded beyond the value of pull-out torque, it will continue to operate as a single-phase induction motor up to 500% of its rated output.

4.6.2. Hysteresis Motor. Single-phase cylindrical (non-salient pole) synchronous-induction or shaded pole motors are classed as *hysteresis motors*. A hysteresis motor has neither a salient pole rotor nor direct excitation, but nevertheless it rotates at synchronous speed. This type of motor *runs into synchronism and runs on hysteresis torque*.

Hysteresis-type lamination, shown in Fig. 4.30, are usually made of *hardened, high retentivity steel* rather than commercial, low retentivity dynamo steel.

Working. As a result of a rotating magnetic field produced by phase splitting or a shaded-pole stator, *eddy currents are induced in the steel of the rotor which travel across the two bar paths of the rotor* as shown in Fig. 4.30. A high-retentivity steel produces a high hysteresis loss, and an appreciable amount of energy is consumed from the rotating field in reversing the current direction of the rotor. At the same time the *rotor magnetic field set up by the eddy currents causes the rotor to rotate. A high starting torque is produced as a result of the high resistance* (proportional to hysteresis). As the rotor approaches synchronous speed, the frequency of current reversal in the cross-bars decreases, and the rotor becomes *permanently magnetized* in one direction as a result of the high retentivity of the steel rotor. Consequently the motor continues to rotate at synchronous speed.

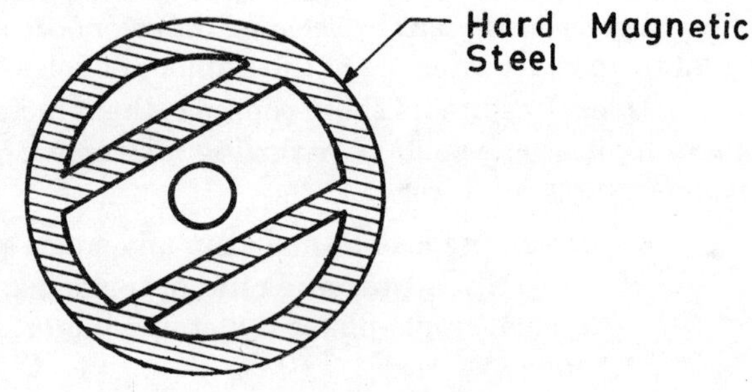

Fig. 4.30. Hysteresis rotor.

- An extremely important *use* of this type of motor is for the rotation of *gyroscope rotors* in inertial navigation and control systems. Here the requirement is for as near absolute accuracy as can be achieved. One major component of the instrument accuracy that contains the gyroscope is that the gyroscopic moment be *absolutely constant.* This constancy requires a synchronous motor that is driven by a regulated constant-frequency source.

4.6.3. Sub-synchronous Motor. When the motor has a rotor that has an overall cylindrical outline and yet is toothed as a many-pole salient-pole rotor, it is a **sub-synchronous motor.** A typical rotor may have 16 teeth or poles, and in conjunction with a 16-pole stator will normally rotate at synchronous 450 r.p.m. when operated on 60 Hz. If this motor were temporarily overloaded, it would drop out of synchronism. Then the speed drops down toward the maximum torque point, and the motor will *again lock into synchronism* at a sub-multiple speed of 225 r.p.m. Hence the name of sub-synchronous motor.

This type of motor starts and accelerates with hysteresis torque just as the hysteresis synchronous motor does. There is no equivalent of induction-motor torque as in the reluctance motors.

This type of motor in any given size will develop a higher starting torque but a lesser synchronous speed torque than a reluctance motor.

Fig. 4.31 shows a sub-synchronous rotor.

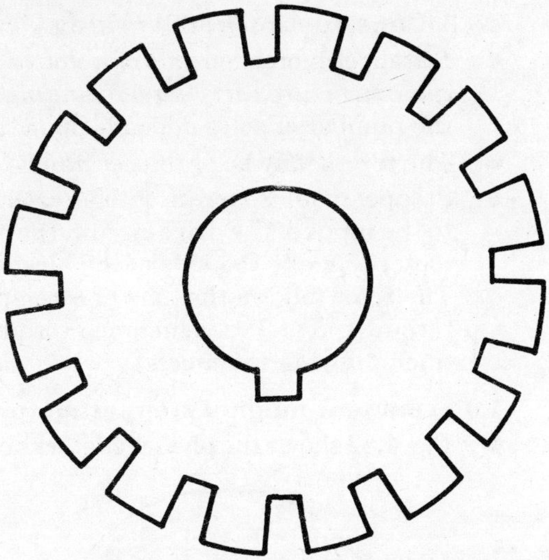

Fig. 4.31. Sub-synchronous rotor.

4.7. STEPPER MOTORS

4.7.1. Introduction

- A **stepper motor** is an *incremental motion machine* (*i.e.* the motor which turns in discrete movement called the steps). *It does not rotate continuously as a conventional motor does.*
- The stepper motor is a special type *of synchronous motor which is designed to rotate through a specific number of degrees for each electrical pulse received by its control unit.* Typical steps are 2°, 2.5°, 5°, 7.5° and 15° per pulse. These motors are built to follow signals as rapid as 1200 pulses per second and with equivalent power ratings upto several kW.
- The *stepper motor is used in digitally controlled position control system in open loop mode.* The input command is in the form of a train of pulses to turn a shaft through a specified angle.

4.7.2. Advantages and Applications

Advantages : The stepper motor (a position control device) entails the following advantages :

1. Compatibility with digital systems.
2. The angular displacement can be precisely controlled without any feedback arrangement.
3. No sensors are needed for position and speed sensing.
4. It can be readily interfaced with microprocessor (or computer based controller).

Applications : Stepper motors have a wide range of applications, mentioned below :

- Paper feed motors in typewriters and printers.
- Positioning of print heads.
- Pens in XY-plotters.
- Recording heads in computer disc drives.
- Positioning of workables and tools in numerically controlled machining equipment.
- Also employed to perform many other functions such as *metering, mixing, cutting, blending, stirring* etc. in several commercial, military and medical applications.

4.7.3. Construction and Working

- A stepper motor consists of a *slotted stator having multi-pole, multi-phase winding and a rotor structure carrying no winding*. They typically use three and four phase windings, the number of poles depends upon the required angular change per input pulse.
- The rotors may be of the *permanent magnet* or *variable reluctance type*.
- Stepper motors operate with an *external drive logic circuit*. When a train of pulse is applied to the input of the drive circuit, the circuit supplies currents to the stator windings of the motor to make the axis of the air-gap field around in coincidence with the input pulses. The rotor follows the axis of air-gap magnetic field by virtue of the permanent magnet torque and/or the reluctance torque, depending upon the pulse rate and load torque (including inertia effects).

1. Permanent-magnet stepper motor :

- Fig. 4.32 shows the phases or stacks of a 2-phase, 4-pole permanent-magnet stepper motor.

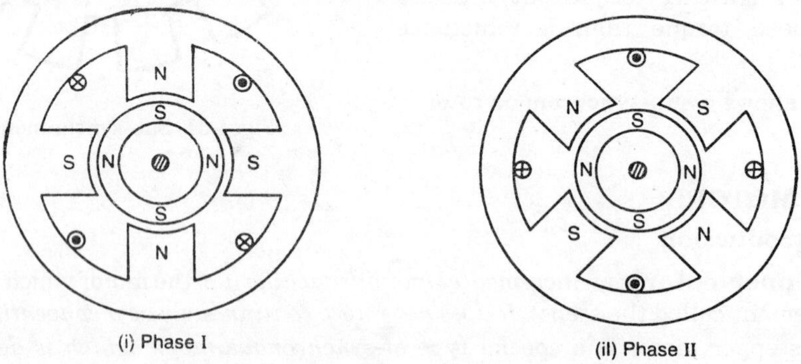

Fig. 4.32. Permanent-magnet stepper motor.

- The rotor is made of *ferrite or rare-earth material* which is *permanently magnetised*.
- The stator stack of phase II is staggered from that of phase 'I' by an angle of 90°.
- When the phase 'I' is excited, the rotor is aligned as shown in Fig. 4.32 (*i*). If now the phase 'II' is also excited, the effective stator poles shift anti-clockwise by 22.5° [Fig. 4.32 (*ii*)] causing the rotor to move accordingly. Now, keeping the phase 'II' still energised, if the phase 'I' is now de-energised, the rotor will move another step of 22.5°. The reversal of phase 'I' winding current will produce a further forward movement of 22.5°, and so on. It can be a easily observed/visualised as to how the direction of movement can be reversed.
- Each phase is provided with double coils to simplify the switching arrangement (which is electronically accomplished).

2. Variable-reluctance stepper motor :

- A variable-reluctance stepper motor has no permanent magnet on the rotor and the rotor employed is a ferro-magnetic multi-toothed one.
- *The large differences in magnetic reluctances that exist between the direct and quadrature axes develop the torque*. The stationary field developed by the direct current in some stator coils tends to develop a torque which causes the rotor to move to the position where the reluctance of the flux path is minimum.

Stepping angle, irrespective of the type of stepper motor is given as :

$$\alpha = \frac{360°}{\text{Number of phases} \times \text{number of poles}} = \frac{360}{np} \qquad ...(4.23)$$

3. Hybrid stepped motor :

- This is *infact in permanent-magnet stepper motor with constructional features of toothed and stacked rotor adopted from the variable-reluctance motor.*
- The stator has only one set of winding-excited poles which interact with the two rotor stacks.
- The permanent magnet is placed *axially along the rotor in the form of an annular cylinder over the motor shaft.*
- The stacks at each end of the rotor are toothed. So all the teeth on the stack at one end of the rotor acquire the *same polarity* while the teeth of the stack at the other end of the rotor acquire the *opposite polarity*. The two sets of the teeth are displaced from each other by *one half of the tooth pitch* (also called pole pitch).
- The primary **advantage** of the hybrid motor is that if *stator excitation is removed, the rotor continues to remain locked into the same position, as before removal of excitation.* This is due to the reason that the rotor is prevented to move in either direction by torque because of the permanent magnet excitation.
- Typical step angles for stepper motors are 15°, 7.5°, 2° and 0.72°. The choice of the angle depends upon the angular resolution required for application.

WORKED EXAMPLES

Example 4.1. *The following data pertains to a single-phase induction motor :*

Number of poles	= 4
Supply voltage	= 110 V
Rated output	= 140 W
Slip at rated output	= 5 per cent
Total copper loss at full-load	= 28 W
Rotational losses	= 28 W

Calculate the full-load efficiency and the rotor copper loss caused by the backward field. Neglect stator copper loss.

Solution. Number of poles, $p = 4$

Motor output $= 140$ W

Slip, $s = 5\%$

Rotational losses $= 28$ W

Mechanical power developed by rotor,

$$P_{mech.} = \text{output} + \text{rotational losses} = 140 + 28 = 168 \text{ W}$$

Power input to motor

$$= P_{air\ gap\ (f)} - P_{air\ gap\ (b)}$$

$$= \frac{\text{mechanical power developed by rotor}}{(1 - s)}$$

$$= \frac{168}{1 - 0.05} = 176.84 \text{ W}$$

$\left[\begin{array}{l}\text{where } P_{air\ gap\ (f)} \text{ means power delivered to the forward field rotor} \\ \quad P_{air\ gap\ (b)} \text{ means power delivered to the backward field rotor}\end{array}\right]$

i.e., $$P_{air\ gap\ (f)} - P_{air\ gap\ (b)} = 176.84 \text{ W} \quad ...(i)$$

and $$sP_{air\ gap\ (f)} + (2 - s)\, P_{air\ gap\ (b)} = \text{total rotor copper loss}$$

$$0.05P_{air\ gap\ (f)} + 1.95P_{air\ gap\ (b)} = 28$$

or $P_{\text{air gap}\,(f)} + 39P_{\text{air gap}\,(b)} = 560$ ...(*ii*)

Solving (*i*) and (*ii*), we get

$$P_{\text{air gap}\,(b)} = 9.58 \text{ W}$$

Rotor copper losses caused by backward field

$$= (2 - s)P_{\text{air gap}\,(b)}$$
$$= (2 - 0.05) \times 9.58 = \mathbf{18.68\ W.}\ \ \textbf{(Ans.)}$$

Full-load efficiency :

Input power to motor = power input to rotor + stator copper loss

= 176.84 W, neglecting stator copper loss

$$\therefore \quad \eta_{\text{full-load}} = \frac{\text{power output}}{\text{power input}} = \frac{140}{176.84} = 0.79 \text{ or } \mathbf{79\ per\ cent.}\ \ \textbf{(Ans.)}$$

Example 4.2. *A 230 V, 50 Hz, 4-pole, class A, single phase induction motor has the following parameters at an operating temperature of 63°C :*

$$r_{1m} = 2.51\ \Omega,\ r_2' = 7.81\ \Omega,\ x_m = 150.88\ \Omega,\ x_m = 4.62\ \Omega,\ x_2' = 4.62\ \Omega$$

Determine the main winding current and power factor when the motor is running at a slit of 0.05 at a specified temperature of 63°C. **(AMIE Summer, 1997)**

Solution. *Given* : Resistance of stator main winding, $R_1 = 2.51\ \Omega$

Leakage reactance of stator main winding, $X_1 = 4.62\ \Omega$

Magnetizing reactance of stator main winding, $X_m = 150.88\ \Omega$

Rotor resistance, $R_2' = 7.81\ \Omega$

Standstill rotor reactance, $X'_2 = 4.62\ \Omega$

Slip, $s = 0.05$

$$(2 - s) = 2 - 0.05 = 1.95$$

Main winding current (I₁), power factor (cos φ) :

Stator impedance, $Z_1 = (2.51 + j4.62)\ \Omega$

$$\text{Forward field impedance, } Z_f = \frac{\dfrac{jX_m}{2}\left[\dfrac{R_2'}{2s} + \dfrac{jX_2'}{2}\right]}{\dfrac{R_2'}{2s} + j\left(\dfrac{X_m + X_2'}{2}\right)} = \frac{j\,\dfrac{150.88}{2}\left[\dfrac{7.81}{2 \times 0.05} + j\,\dfrac{4.62}{2}\right]}{\dfrac{7.81}{2 \times 0.05} + j\left(\dfrac{150.88 + 4.62}{2}\right)}$$

or
$$Z_f = \frac{j75.44(78.1 + j2.31)}{78.1 + j77.75} = \frac{75.44\ \angle 90^\circ \times 78.134\ \angle\ 1.694^\circ}{110.203\ \angle 44.87^\circ}$$
$$= 53.487\ \angle 46.824^\circ\ \Omega$$

Similarly backward field impedance,

$$Z_b = \frac{\dfrac{jX_m}{2}\left[\dfrac{R_2'}{2(2 - s)} + j\,\dfrac{X_2'}{2}\right]}{\dfrac{R_2'}{2(2 - s)} + j\left(\dfrac{X_m + X_2'}{2}\right)} = \frac{j75.44\left[\dfrac{7.81}{2 \times 1.95} + j2.31\right]}{\dfrac{7.81}{2 \times 1.95} + j\,\dfrac{150.88 + 4.62}{2}}$$

$$= \frac{j75.44(2.003 + j2.31)}{2.003 + j77.75}$$

or
$$Z_b = \frac{75.44\ \angle 90^\circ \times 3.0575\ \angle 49.1^\circ}{77.776\ \angle 88.52^\circ} = 2.965\ \angle 50.58^\circ\ \Omega$$

Total equivalent impedance, $Z_{eq} = Z_1 + Z_f + Z_b$

$$= (2.51 + j4.62) + 53.487 \angle 46.824° + 2.965 \angle 50.58°$$
$$= (2.51 + j4.62) + 36.598 + j39 + 1.883 + j2.29$$
$$= 40.991 + j45.91 = 61.547 \angle 48.24° \ \Omega$$

Stator main winding current, $\mathbf{I_1} = \dfrac{V}{Z_{eq}} = \dfrac{230}{61.547 \angle 48.24°} = $ **3.737 A. (Ans.)**

Power factor, $\cos\phi = \cos 48.24° =$ **0.666 lag. (Ans.)**

Example 4.3. *The constants of a one-quarter H.P., 230 V, 4-pole, 60 Hz, single phase induction motor are as follows :*

Stator resistance, $R_1 = 10.0\ \Omega$; stator reactance, $X_1 = 12.8\ \Omega$; rotor resistance referred to stator, $R_2' = 11.65\ \Omega$; rotor reactance referred to stator, $X_2' = 12.8\ \Omega$; magnetising reactance, $X_m = 258.0\ \Omega$.

The total load is such that the machine runs at 3% slip, when the voltage is at 210 V. The iron losses are 35.5 W at 210 V. Calculate :

(i) Input current ; *(ii) Power developed ;*
(iii) Shaft power (if mechanical losses are 7 W) ;
(iv) Efficiency. **(AMIE Advance Elec. M/Cs. Summer, 1994)**

Solution. *Given :* $R_1 = 10.0\ \Omega$; $X_1 = 12.8\ \Omega$; $R_2' = 11.65\ \Omega$; $X_2' = 12.8\ \Omega$; $X_m = 258.0\ \Omega$; $s = 3\% = 0.03$; $V = 210$ volts ; $P_i = 35.5$ W ; mechanical losses = 7 W.

Core loss current, $I_w = \dfrac{35.5}{210} = 0.169$ A.

Since under running condition V_f is very high (almost 90% of the applied voltage) and V_s is very low, so most of the core loss takes place in the 'forward motor' consisting of the common stator and forward running motor. So half value of core loss equivalent resistance, $\dfrac{R_m}{2} = \dfrac{210}{0.169} = 1242\ \Omega$.

For forward field : Refer Fig. 4.4.

Conductance of core-loss branch $= \dfrac{1}{(R_m/2)} = \dfrac{1}{1242} = 0.0008$ S

Susceptance of magnetising branch $= \dfrac{-j}{(X_m/2)} = -\dfrac{j}{(258/2)} = -j0.00775$ S

Rotor admittance
$$= \frac{1}{\dfrac{R_2'}{2s} + \dfrac{jX_2'}{2}} = \frac{1}{\dfrac{11.65}{2 \times 0.03} + \dfrac{j12.8}{2}} = \frac{1}{194.17 + j6.4}$$
$$= \frac{1}{194.17 + j6.4} \times \frac{194.17 - j6.4}{194.17 - j6.4}$$
$$= \frac{194.17 - j6.4}{37743} = (0.005145 - j0.00017) \text{ S}$$

Admittance
$$Y_f = (0.0008 - j0.00775 \text{ S} + 0.005145 - j0.00017) \text{ S}$$
$$= (0.005945 - j0.00792) \text{ S}$$

Impedance,
$$Z_f = \frac{1}{Y_f} = \frac{1}{0.005945 - j0.00792}$$
$$= 60.62 + j80.759 = 100.979 \angle 53.1° \ \Omega$$

For backward field :

Rotor admittance
$$= \frac{1}{\dfrac{R_2'}{2(2-s)} + \dfrac{jX_2'}{2}} = \frac{1}{\dfrac{11.65}{2(2-0.03)} + \dfrac{j12.8}{2}}$$

$$= \frac{1}{2.957 + j6.4} = (0.0595 - j0.12876) \text{ S}$$

Admittance, $Y_b = (0.0008 - j0.00775 + 0.0595 - j0.12876)$ S
$= (0.0603 - j0.1365)$ S

Impedance, $Z_b = \frac{1}{Y_b} = \frac{1}{0.0603 - j0.1365} = 2.708 + j6.13 = 6.70 \angle 66.17° \ \Omega$

Equivalent impedance, $Z_{eq} = Z_1 + Z_f + Z_b$
$= (10 + j12.8) + (60.62 + j80.759) + j(2.708 + j6.13)$
$= 73.323 + j99.689 = 123.75 \angle 53.66° \ \Omega$

(*i*) **Input current,** $\mathbf{I_1} = \frac{V}{Z_{eq}} = \frac{230}{123.75} = \mathbf{1.8586\ A.}$ **(Ans.)**

Power input $= VI_1 \cos\phi = 230 \times 1.8586 \times \cos 53.66° = 253.3$ W

$$E_{mf} = I_1 \times Z_f = 1.8586 \times 100.979 = 187.68 \text{ V}$$
$$E_{mb} = I_1 \times Z_b = 1.8586 \times 6.70 = 12.45 \text{ V}$$

$$I'_{2f} = \frac{E_{mf}}{\sqrt{\left(\frac{R'_2}{2s}\right)^2 + \left(\frac{X'_2}{2}\right)^2}} = \frac{187.68}{\sqrt{(194.17)^2 + (6.4)^2}} = 0.966 \text{ A}$$

$$I'_{2b} = \frac{E_{mb}}{\sqrt{\left(\frac{R'_2}{2(2-s)}\right)^2 + \left(\frac{X'_2}{2}\right)^2}} = \frac{12.45}{\sqrt{(2.2957)^2 + (6.4)^2}} = 1.831 \text{ A}$$

Air-gap power for forward field, $P_{\text{air gap}(f)} = (I'_{2f})^2 \times \frac{R_2'}{2s} = (0.966)^2 \times \frac{11.65}{2 \times 0.03} = 181.2$ W

Air-gap power for backward field, $P_{\text{air gap}(b)} = (I'_{2b})^2 \times \frac{R_2'}{2(2-s)} = (1.831)^2 \times \frac{11.65}{2 \times 1.97} = 9.9$ W

(*ii*) **Power developed** $\mathbf{P_{mech\ (net)}} = (1 - s)\, P_{\text{air gap}(f)} - P_{\text{air gap}(b)}$
$= (1 - 0.03)(181.2 - 9.9) =$ **166.2 W. (Ans.)**

(*iii*) **Shaft power** $= P_{\text{mech (net)}}$ − mechanical losses = 166.2 − 7 = **159.2 W. (Ans.)**

(*iv*) **Efficiency,** $\eta = \frac{\text{Power output}}{\text{Power input}} \times 100 = \frac{159.2}{253.3} \times 100 =$ **62.85%. (Ans.)**

Example 4.4. *The following test results were obtained in case of a 220 V single-phase induction motor :*

No-load test : 220 V, 5.8 A ; 310 W
Locked rotor test : 120 V, 13.8 A ; 530 W
Stator winding resistance = 1.4 Ω
Determine the approximate equivalent circuit of the motor.

Solution. Stator winding resistance $R_1 = 1.4 \ \Omega$

Locked rotor test :

Voltage applied, $V = 120$ V
Current drawn, $I = 13.8$ A
Power consumed, $P = 530$ W

Equivalent impedance of motor windings,

$$Z_{eq} = \frac{\text{voltage applied}}{\text{current drawn}} = \frac{120}{13.8} = 8.69 \ \Omega$$

Equivalent resistance of motor winding,

$$R_{eq} = \frac{P}{I^2} = \frac{330}{(10.8)^2} = 2.78\ \Omega$$

Rotor winding resistance referred to stator,

$$R_2' = R_{eq} - R_1 = 2.78 - 1.4 = \mathbf{1.38\ \Omega.}\ \textbf{(Ans.)}$$

Equivalent reactance of the motor windings,

$$X_{eq} = \sqrt{Z_{eq}^{\ 2} - R_{eq}^{\ 2}} = \sqrt{(8.69)^2 - (2.78)^2} = 8.23\ \Omega$$

Rotor winding reactance referred to stator,

X_2' = stator winding reactance,

$$X_1 = \frac{X_{eq}}{2} = \frac{8.23}{2} = \mathbf{4.11\ \Omega.}\ \textbf{(Ans.)}$$

Assuming that magnetising reactance is the same in both the cross and main axes,

$$X_m = \frac{2 \times \text{no-load voltage}}{\text{no-load current}} = \frac{2 \times 220}{5.8} = \mathbf{75.86\ \Omega.}\ \textbf{(Ans.)}$$

The approximate equivalent circuit is shown in Fig. 4.33.

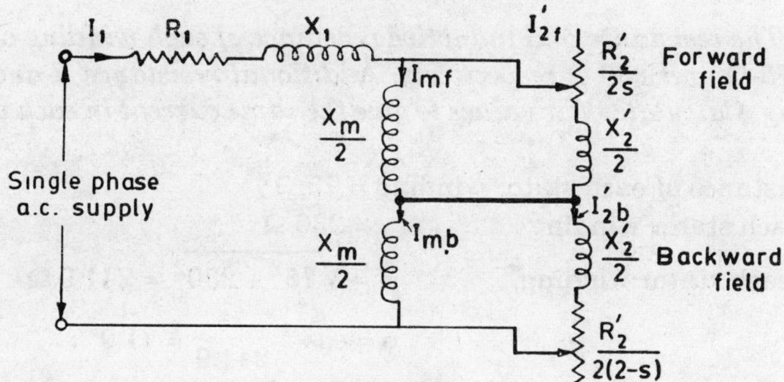

Fig. 4.33

Example 4.5. *The following test results were obtained in case of a 230 V, single-phase, 4-pole, 50-Hz induction motor :*

No-load test : 230 V, 0.5 A, p.f. 0.18 lagging.

Locked rotor test : 58 V, 1.48 A, p.f. 0.58 lagging.

Determine the approximate equivalent circuit.

Assume : (i) Stator and rotor I^2R losses at standstill are equal.

(ii) Rotor leakage reactance referred to the stator and stator leakage reactance are equal.

Solution. Locked rotor test :

Voltage applied, $V = 58$ V

Current drawn, $I = 1.48$ A

Power factor $= 0.58$

∴ Equivalent impedance of motor windings,

$$Z_{eq} = \frac{V}{I} = \frac{58}{1.48} = 39.19\ \Omega$$

Equivalent resistance of motor windings,

$$R_{eq} = Z_{eq} \cos \phi = 39.19 \times 0.58 = 22.73\ \Omega$$

Equivalent reactance of motor windings,

$$X_{eq} = \sqrt{Z_{eq}^2 - R_{eq}^2} = \sqrt{(39.19)^2 - (22.73)^2} = 31.92\ \Omega$$

Rotor winding resistance referred to stator

$$= \text{stator winding resistance}$$

$$= \frac{R_{eq}}{2} = \frac{22.73}{2} = \mathbf{11.36\ \Omega.\ (Ans.)}$$

(Since the stator and rotor I^2R losses at standstill are equal)

Rotor winding reactance referred to stator

$$= \text{stator winding reactance}$$

$$= \frac{X_{eq}}{2} = \frac{31.92}{2} = \mathbf{15.96\ \Omega.\ (Ans.)}$$

Assuming that the magnetising reactance is same in both cross and main axes,

$$X_m = \frac{2 \times \text{no-load voltage}}{\text{no-load current}} = \frac{2 \times 230}{0.5} = \mathbf{920\ \Omega.\ (Ans.)}$$

For approximate equivalent circuit refer Fig. 4.33.

Example 4.6. *The resistance and inductive reactance of each winding of a 50-Hz split-phase induction motor are 75 Ω and 230 Ω respectively. Additional resistance R and condenser C are in series with one winding. Calculate their values to give the same current in each winding with a phase difference of 90°.*

Solution. Resistance of each stator winding = 75 Ω

Reactance of each stator winding = 230 Ω

Impedance of each stator winding, $Z = \sqrt{75^2 + 230^2} = 241.9\ \Omega$

Phase angle, $\phi = \cos^{-1} \dfrac{75}{241.9} = 71.9°$

The impedance of starting, after connecting resistance R and capacitance C in series with any of the winding should be equal to its old impedance *i.e.*, 241.9 Ω and phase angle 18.1° (90° – 71.9°) so that the current in each winding is same and have phase difference of 90°,

i.e., $Z_s = 241.9\ \Omega,\ \phi_s = 18.1°$

Resistance of starting winding,

$$R_s = Z_s \cos \phi_s = 241.9 \cos 18.1° = 229.93\ \Omega$$

Hence, **additional resistance required,**

$$R = 229.93 - 75 = \mathbf{154.93\ \Omega.\ (Ans.)}$$

Capacitive reactance of starting winding,

$$X_s = Z_s \sin \phi_s = 241.9 \sin 18.1° = 75.15\ \Omega$$

Hence, additional capacitive reactance required,

$$X_c = 75.15 - (-230) = 305.15\ \Omega$$

∴ **Additional capacitance required,**

$$C = \frac{10^6}{2\pi f X_c} = \frac{10^6}{2\pi \times 50 \times 305.15} = \mathbf{10.43\ \mu F.\ (Ans.)}$$

Fig. 4.34 shows the voltage-current relationship. Suffix '*s*' stands for 'starting' winding and '*m*' stands for main or running winding.

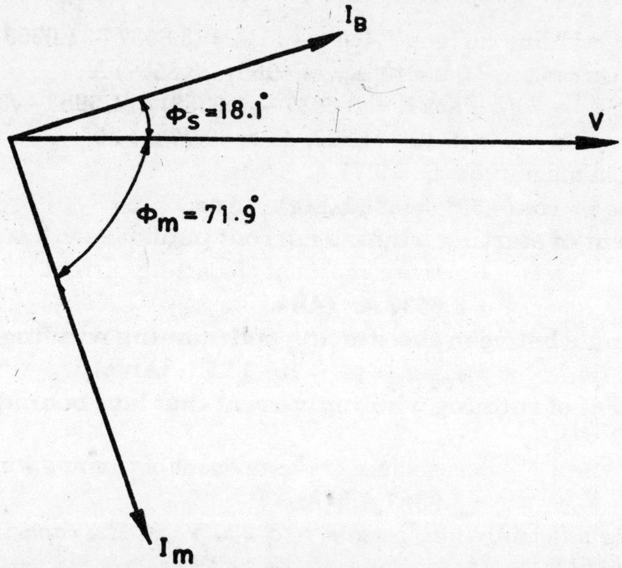

Fig. 4.34

Example 4.7. *The name plate of a 1-phase, 4-pole induction motor gives the following data : Output = 410 W ; Supply voltage = 230 V ; Frequency = 50 Hz ; Input current = 3.2 A ; Power factor = 0.7 ; Speed = 1410 r.p.m.*

Calculate :

(i) The efficiency of the motor, and

(ii) The slip of the motor when delivering the rated output. **(Bharathiyar University)**

Solution. *Given :* $p = 1$; $P_{out} = 410$ W ; $V = 230$ volts ; $f = 50$ Hz ;

$I = 3.2$ A ; $\cos \phi = 0.7$; $N = 1410$ r.p.m.

(*i*) **Motor efficiency, η :**

Power in, $P_{in} = VI \cos \phi = 230 \times 3.2 \times 0.7 = 515.2$ W

$\therefore$ $\eta = \dfrac{P_{out}}{P_{in}} = \dfrac{410}{515.2} = 0.7958$ or **79.58%. (Ans.)**

(*ii*) **The slip of the motor, s :**

Synchronous speed, $N_s = \dfrac{120f}{p} = \dfrac{120 \times 50}{4} = 1500$ r.p.m.

$\therefore$ Slip, $s = \dfrac{N_s - N}{N_s} = \dfrac{1500 - 1410}{1500} = 0.06$ or **6%. (Ans.)**

Example 4.8. *A $\frac{1}{4}$ H.P., split-phase motor draws its starting winding current of 4 A lagging the supply voltage by 15° elec. and its running winding current is 6A lagging by 40° elec. Calculate :*

(i) The total current and power factor (at steady state) ;

(ii) The component of starting winding current in phase with voltage ;

(iii) The phase angle between the starting and running winding currents ;

(iv) The component of running winding current that lags the supply voltage by 90°.

(AMIE Advance Elect. M/Cs. Summer, 1996)

Solution. Starting winding current, $I_s = 4\ \angle -15°\text{A} = (3.8637 - j1.0353)$ A

Running winding current, $I_r = 6\ \angle -40°\text{A} = (4.5963 - j3.8567)$ A

Total current, $I_{lr} = I_s + I_r = (3.8637 - j1.0353) + (4.5963 - j3.8567)$

$= (8.46 - j4.892)$ A or $9.77\ \angle -30°$

(*i*) **Total current** in magnitude, I_{lr} = **9.77 A. (Ans.)**

Power factor, $\cos\phi = \cos(-30°)$ = **0.866 (lag). (Ans.)**

(*ii*) **The component of starting winding current in phase with supply voltage**

= active component of starting current

= **3.8637 A. (Ans.)**

(*iii*) **The phase angle between the starting and running winding currents,**

$\theta = \phi_r - \phi_s = 40° - 15° =$ **25°. (Ans.)**

(*iv*) **The component of running winding current that lags behind the supply voltage by 90°**

= reactance or *j*-component of running winding current

= **3.8567 A. (Ans.)**

Example 4.9. *The following data pertain to a 230 V, 50 Hz, capacitor-start single phase induction motor at standstill :*

Main winding excited alone = *100 V, 2 A, 40 W*

Auxiliary winding excited alone = *80 V, 1 A, 50 W*

Determine the value of capacitance for determining the maximum starting torque.

(AMIE Advance Elect. M/Cs. Winter, 1996)

Solution. *Given* : $V_m = 100$ V, $I_m = 2$ A ; $P_m = 40$ W ; $V_a = 80$ V ; $I_a = 1$ A ; $P_a = 50$ W

Value of capacitance for maximum torque :

Main winding :

Equivalent impedance with main winding, $Z_{01m} = \dfrac{V_m}{I_m} = \dfrac{100}{2} = 50\ \Omega$

Equivalent resistance with main winding, $R_{01m} = \dfrac{P_m}{I_m^2} = \dfrac{40}{2^2} = 10\ \Omega$

Assuming rotor resistance and reactance referred to stator, R_2' and X_2' equal to stator resistance R_1 and reactance X_1 respectively

$$R_{1m} = R'_2 = \frac{R_{01m}}{2} = \frac{10}{2} = 5\ \Omega$$

and
$$X_{1m} = X_2' = \frac{X_{01m}}{2} = \frac{\sqrt{50^2 - 10^2}}{2} = 24.5\ \Omega$$

Angle of lag of main winding current, $\phi_m = \tan^{-1}\dfrac{X_{1m}}{R_{1m}} = \tan^{-1}\dfrac{24.5}{5} = 78.46°$

Auxiliary winding :

Equivalent impedance with auxiliary winding, $Z_{01a} = \dfrac{V_a}{I_a} = \dfrac{80}{1} = 80\ \Omega$

Equivalent resistance with auxiliary winding, $R_{01a} = \dfrac{P_a}{I_a^2} = \dfrac{50}{1^2} = 50\ \Omega$

Equivalent reactance with auxiliary winding, $X_{01a} = \sqrt{80^2 - 50^2} = 62.45\ \Omega$

Auxiliary winding resistance $R_a = R_{01a} - R'_2 = 50 - 5 = 45\ \Omega$

Auxiliary winding reactance, $X_a = X_{01a} - X'_2 = 62.45 - 24.5 = 37.95\ \Omega$

Torque developed at start will be maximum, if the phase angle between main winding current I_m and auxiliary winding current I_a is 90° or the auxiliary winding current leads the applied voltage by (90° – 78.46°) or 11.54°.

If X_c is the capacitive reactance of the capacitor connected in series with the auxiliary winding, then impedance of the auxiliary winding will be given as

$$Z_a = R_a + X_a - X_c = 45 + j37.95 - jX_c = [45 + j(37.95 - X_c)] \text{ ohms}$$

For auxiliary winding, $\tan \phi_a = \dfrac{37.95 - X_c}{45}$

or $X_c = 37.95 - 45 \tan \phi_a = 37.95 - 45 \tan(-11.54°)$

$= 37.95 + 9.19 = 47.14\ \Omega$

Capacitance of capacitor, $C = \dfrac{1}{2\pi f X_c} = \dfrac{1}{2\pi \times 50 \times 47.14} \times 10^6 = \mathbf{67.52\ \mu F.}$ **(Ans.)**

Example 4.10. *A 250 W, 230 V, 50 Hz single-phase capacitor-start induction motor has the following constants for the main and auxiliary windings. Main winding Z_m = (4.5 + j3.7) Ω, auxiliary winding Z_a = (9.5 + j3.5) Ω. Determine the value of starting capacitor that will place the main and auxiliary winding currents in quadrature at starting.* **(Bharathiyar Univ., 1997)**

Solution. Main (or running) winding impedance, $Z_m = (4.5 + j3.7)\ \Omega$ or $5.826 \angle 39.43°\ \Omega$ (obviously, main winding current I_m (or I_r) lags behind the applied voltage by 39.43°).

Auxiliary starting winding impedance, $Z_a = (9.5 + j3.5)\ \Omega$

Since, time phase angle between auxiliary winding current I_a (or I_s) and main winding current I_m is 90°, so auxiliary current I_a must lead the applied voltage by (90° – 39.43°) or 50.57°.

If X_c is the capacitive reactance of the capacitor connected in series with the auxiliary winding, then impedance of the auxiliary winding will be given as

$$Z_a = (9.5 + j3.5 - jX_c) \text{ or } [9.5 + j(3.5 - X_c)]\ \Omega.$$

For auxiliary winding, $\tan \phi_c = \dfrac{3.5 - X_c}{9.5}$

or $X_c = 3.5 - 9.5 \tan \phi_a = 3.5 - 9.5 \tan(-50.57°) = 15.053\ \Omega$

$\therefore$ Capacitance of capacitor, $\mathbf{C} = \dfrac{1}{2\pi f X_c} = \dfrac{1}{2\pi \times 50 \times 15.053} \times 10^6\ \mu F = \mathbf{211.5\ \mu F.}$ **(Ans.)**

Example 4.11. (*a*) *Explain the following in case of single phase induction motor :*

(*i*) *The forward flux wave is several times greater than the backward flux wave at normal rotor speed but are equal at standstill condition.*

(*ii*) *The direction of rotation is always from auxiliary winding to main winding whether a resistance or capacitance is connected in series with auxiliary winding.*

(*b*) *A universal motor (a.c. operated) has two-pole armature with 690 conductors. At a certain load the motor speed is 5000 r.p.m. and armature current is 4.6 amp, the armature terminal voltage and input are 100 V and 300 watts respectively. If the armature resistance is 3.5 ohms, compute effective armature reactance and maximum value of useful flux/pole.* **(UPSC Exams.)**

Solution. (*a*) (*i*) From double field revolving theory the alternating mmf in the main axis of a single-phase induction motor produced by the stator winding may be considered to be resolved into two rotating mmf's, each having half the amplitude of the simple alternating mmf and rotating in opposite directions, but at synchronous speed. On resolving the original alternating stator mmf into two equal rotating components, the corresponding fluxes in the actual machine will be equal only when the rotor is stationary. *Under running conditions the slip is approaching zero with respect to the positively rotating field, the operating point on one of the circle diagrams will be close to the no-load point, while on the other circle diagram it will be beyond the standstill point. The circles cannot then* *be the same in size, because the line representing the primary current must have the same length in both cases, in order to agree with the fact that the primary mmf was resolved into two equal components. It follows then that the circle corresponding to the negatively rotating field is much smaller than the other one and this in turn means that the actual magnitude of negatively rotating field is small in comparison with that of the forwardly rotating field.*

(*ii*) A single-phase induction motor is temporarily converted into a 2 phase motor by providing an extra winding on the stator in addition to the main or running winding. The *starting winding or the auxiliary winding is made highly resistive* and the *main winding highly reactive*. The phase difference between the currents drawn by them can be made sufficiently large. The motor behaves like a two-phase motor for starting purposes. The starting torque $T_s = kI_sI_m \sin \alpha$ where k is a constant governed by motor design parameters.

As the angle of lag of *main winding is very near to 90°* and that of *auxiliary winding near to 0° lagging in case of resistive phase splitting and leading in case of capacitive phase spliting, the torque produced will always be more due to auxiliary winding as compared to that produced by main winding, hence direction of rotation is always from auxiliary winding to main winding.*

(*b*) *Given* : $p = 2$; $Z = 690$; $N = 5000$ r.p.m. ; $I_a = 4.6$ A ; $V = 100$ volts ;
$P_{in} = 300$ W ; $R_a = 3.5\ \Omega$.

Effective armature reactance, X :

Input power, $P_{in} = VI \cos \phi$

$$\cos \phi = \frac{P_{in}}{VI} = \frac{300}{100 \times 4.6} = 0.652$$

Now, $$\cos \phi = \frac{R_a}{Z}$$

$\therefore$ $$Z = \frac{R_a}{\cos \phi} = \frac{3.5}{0.652} = 5.37\ \Omega$$

Now $$\mathbf{X} = \sqrt{Z^2 - R_a^2} = \sqrt{5.37^2 - 3.5^2} = \mathbf{4.07\ \Omega}. \quad \textbf{(Ans.)}$$

Maximum useful flux/pole, φ :

For universal motor, $$E = \frac{p\phi ZN}{60a}$$

$$E = V + I_aR_a = 100 + 4.6 \times 3.5 = 116.1 \text{ V}$$

Universal motors are always 2-pole motors,

$\therefore$ $$116.1 = \frac{2 \times \phi \times 690 \times 5000}{60 \times 2}$$

or $$\phi = \frac{116.1 \times 60 \times 2}{2 \times 690 \times 5000} = 2.02 \times 10^{-3} \text{ Wb} = \mathbf{2.02\ m\ Wb}. \quad \textbf{(Ans.)}$$

Example 4.12. *The resistance and total inductance of a single-phase fractional horsepower series motor are 35 Ω and 0.55 H respectively. It draws 0.85 A current and runs at 1800 r.p.m. when connected to 250 V D.C. supply. Calculate the speed and power factor when connected to 250 V, 50-Hz supply and loaded to take the same current.*

Solution. Resistance of motor $= 35\ \Omega$

Total inductance of motor $= 0.55$ H

$\therefore$ Total inductive reactance, $X = 2\pi fL = 2\pi \times 50 \times 0.55 = 172.78\ \Omega$

When motor is connected to D.C. supply :

Back e.m.f. $E_b = V - IR = 250 - 0.85 \times 35 = 220.25$ V

Speed, $N = 1800$ r.p.m.

When motor is connected to A.C. supply :

Voltage applied, $V = 250$ V

Voltage drop across resistance $= IR = 0.85 \times 35 = 29.75$ V

Voltage drop across inductance $= IX = 0.85 \times 172.78 = 146.86$ V

Refer Fig. 4.35 (vector diagram)

$$\sin \phi = \frac{IX}{V} = \frac{146.86}{250}$$

$$\therefore \quad \phi = \sin^{-1}\left(\frac{146.86}{250}\right) = 36°$$

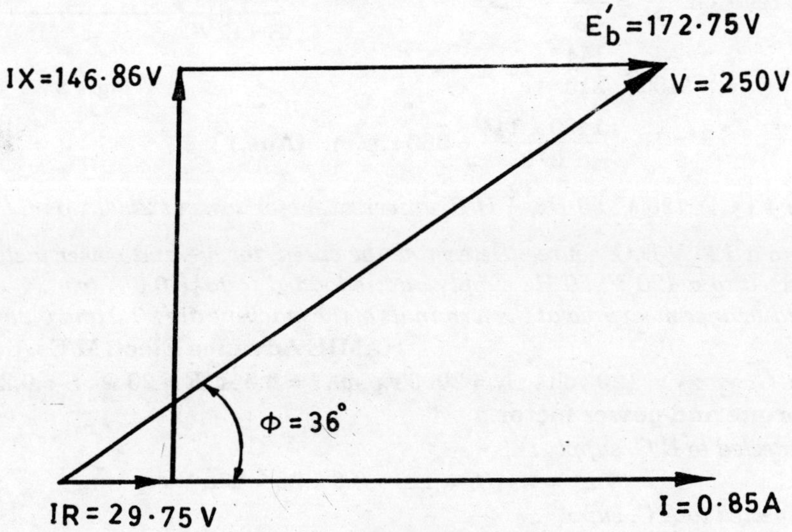

Fig. 4.35

Power factor, $\cos \phi = \cos 36° =$ **0.81. (Ans.)**

Back e.m.f. developed, $E_b' = V \cos \phi - IR = 250 \times 0.81 - 29.75 = 172.75$ V

Using the relation, $\dfrac{N'}{N} = \dfrac{E_b'}{E_b}$

$\therefore$ **Speed,** $N' = \dfrac{E_b'}{E_b} \times N = \dfrac{172.75 \times 1800}{220.25} =$ **1411.8 r.p.m. (Ans.)**

Example 4.13. *A single phase series motor has total resistance of 1 Ω and total inductance of 0.05 H. When connected to 220 V D.C., the motor draws 10 A and runs at 1400 r.p.m. Calculate the speed and power factor when the motor draws the same current from 220 V, 50 Hz source.*

(AMIE Summer, 2000)

Solution. *Given* : $R = 1\ \Omega$; $L = 0.05$ H ; $V = 220$ volts ; $I = 10$ A ; $N = 1400$ r.p.m. ; $f = 50$ Hz.

Speed (N′) and power factor (cos φ) :

When motor is connected to D.C. supply :

Back e.m.f., $E_b = V - IR = 220 - 10 \times 1 = 210$ V

Speed, $N = 1400$ r.p.m.

When motor is connected to A.C. supply :

Voltage drop across resistance $= IR = 10 \times 1 = 10$ V

Voltage drop across reactance $= IX = 10 \times (2\pi \times 50 \times 0.05) = 157$ V

Refer Fig. 4.36 (Vector diagram)

$$\sin \phi = \frac{IX}{V} = \frac{157}{220} = 0.714$$

or $$\phi = \sin^{-1}(0.714) = 45.6°$$

∴ **Power factor,**

$$\cos \phi = \cos(45.6°) = \mathbf{0.7.}\ \textbf{(Ans.)}$$

Back e.m.f. developed,

$$E'_b = V \cos \phi - IR$$
$$= 220 \times 0.7 - 10 = 144 \text{ V}$$

Using the relation : $\frac{N'}{N} = \frac{E'_b}{E_b}$

or $$\frac{N'}{1400} = \frac{144}{210}$$

∴ $$N' = \frac{1400 \times 144}{210} = \textbf{960 r.p.m. (Ans.)}$$

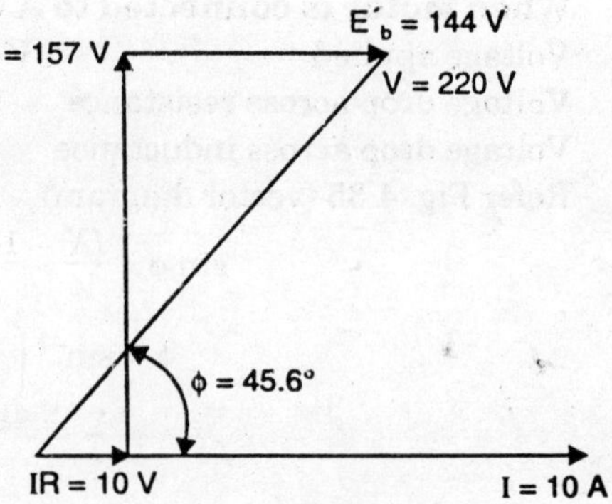

Fig. 4.36

Example 4.14. *A 120 V, 60 Hz, $\frac{1}{4}$ H.P. universal motor runs at 2000 r.p.m. and takes 0.6 A when connected to a 120 V D.C. source. Determine the speed, torque and power factor of the motor when it is connected to a 120 V, 50 Hz supply and is loaded to take 0.6 A (r.m.s.) of current. The resistance and inductance measured at the terminals of the machine are 20 Ω and 0.25 H respectively.*

(AMIE Advance Elect. M/Cs. Summer, 1997)

Solution. *Given* : $V = 120$ volts ; $N = 2000$ r.p.m., $I = 0.6$ A, $R = 20\ \Omega$; $L = 0.25$ H ; $f = 50$ Hz

Speed torque and power factor :

When connected to D.C. supply :

Back e.m.f. $E_b = V - IR = 120 - 0.6 \times 20 = 108$ V

When connected to A.C. supply :

Resistance drop $= IR = 0.6 \times 20 = 12$ V

Reactance drop $= IX = 0.6 \times (2\pi \times 50 \times 0.25) = 47.12$ V

$$\sin \phi = \frac{IX}{V} = \frac{47.12}{120} = 0.393$$

or $$\phi = \sin^{-1}(0.393) = 23.14°$$

∴ **Power factor,** $\cos \phi = \cos(23.14°) = \mathbf{0.92.}$ **(Ans.)**

$$E'_b = V \cos \phi - IR = 120 \times 0.92 - (0.6 \times 20) = 98.4 \text{ V}$$

Using the relation, $\frac{N'}{N} = \frac{E_b'}{E_b}$

or $$\frac{N'}{2000} = \frac{98.4}{108} \text{ or } N' = \frac{2000 \times 98.4}{108} = \textbf{1822 r.p.m. (Ans.)}$$

Torque developed, $$T = \frac{E'_b I}{(2\pi N'/60)} = \frac{98.4 \times 0.6 \times 60}{2\pi \times 1822} = \textbf{0.309 N-m. (Ans.)}$$

Example 4.15. *The resistance and total inductance of a single-phase fractional horse power series motor are 30 Ω and 0.5 H respectively. It draws 0.8 A current and runs at 2000 r.p.m. when connected to a 250 V D.C. supply. Calculate the speed and power factor when connected to a 250 V, 50 Hz supply and takes the same load current. How much voltage is required for getting 2000 r.p.m. with A.C. supply ? Assume resistance and reactance remain constant.* **(AMIE Winter, 1996)**

Solution. *Given* : $R = 30\ \Omega$; $L = 0.5$ H ; $I = 0.8$ A ; $N = 2000$ r.p.m., $V = 250$ volts.

(*i*) **Speed, N′** :

When motor is connected to D.C. supply :

Back or counter e.m.f., $E_b = V - IR = 250 - 0.8 \times 30 = 226$ V

Speed, $N = 2000$ r.p.m.

When motor is connected to A.C. supply :

Voltage drop across resistance $= IR = 0.8 \times 30 = 24$ V

Voltage drop across reactance $= IX = 0.8 \times (2\pi \times 50 \times 0.5) = 125.66$ V

Refer Fig. 4.37 (Vector diagram)

$$\sin \phi = \frac{IX}{V} = \frac{125.66}{250} = 0.5026$$

$$\therefore \quad \phi = \sin^{-1} (0.5026) = 30.17°$$

Power factor = cos 30.17° = **0.864 (lag). (Ans.)**

Back e.m.f. developed,

$$E_b' = V \cos \phi - IR$$
$$= 250 \times 0.864 - 24 = 192 \text{ V}$$

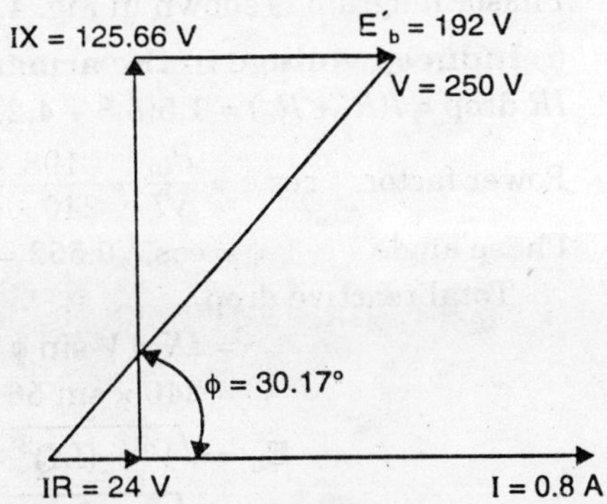

Fig. 4.37

Using the relation, $\frac{N'}{N} = \frac{E_b'}{E_b}$

or $$\frac{N'}{2000} = \frac{192}{226}$$

or $$N' = \frac{192 \times 2000}{226}$$

$$\simeq \mathbf{1700\ r.p.m.}\ \textbf{(Ans.)}$$

(*ii*) **Voltage required for getting 2000 r.p.m. with A.C. supplies, V′** :

Let V' be the A.C. supply voltage to give a speed of 2000 r.p.m.

The motor will operate at a speed of 2000 r.p.m. When $E_b' = E_b = 226$ V

$$E_b' = V' \cos \phi' - IR \text{ (where } \phi' \text{ is the phase angle in the second case)}$$

$$226 = V' \times \sqrt{1 - (\sin \phi')^2} - 24$$

or $$250 = V' \times \sqrt{1 - \left(\frac{IX}{V'}\right)^2}$$

$$= \sqrt{V'^2 - (IX)^2} = \sqrt{V'^2 - (125.66)^2} \qquad \left(\because \sin \phi' = \frac{IX}{V'}\right)$$

Squaring both sides, we get

$$(250)^2 = V'^2 - (125.66)^2$$

or $$\mathbf{V'} = \sqrt{(250)^2 + (125.66)^2} = \mathbf{279.8\ V.}\ \textbf{(Ans.)}$$

Example 4.16. *The following data refer to a 240 V, 2-pole uncompensated series motor when run on a 50 Hz single phase supply :*

Applied voltage	*= 240 V (r.m.s.) ;*
Input current	*= 1.5 A ;*
Input watts	*= 198.75 W ;*
Field winding resistance	*= 8.8 Ω ;*
Armature resistance	*= 4.2 Ω ;*
Speed	*= 4800 r.p.m.*

Number of turns of the field winding = 300 ;
Number of armature conductors = 1200 ;
Leakage reactance drop = 20 *percent of total reactive drop in the motor. Draw the phasor diagram and compute :*
(i) *Induced voltage in the armature ;*
(ii) *Field flux per pole ;*
(iii) *Armature (quadrature axis) flux per pole.* **(AMIE Advance Elect. M/Cs. Summer, 1996)**

Solution. *Given* : V = 240 volts (r.m.s.) ; I = 1.5 A ; P_{in} = 198.75 W ; R_f = 8.8 Ω ;
R_a = 4.2 Ω ; N = 4800 r.p.m. ; T_f = 300 ; Z = 1200,
Leakage reactance drop = 20% of total reactive drop.

Phasor diagram is shown in Fig. 4.38.

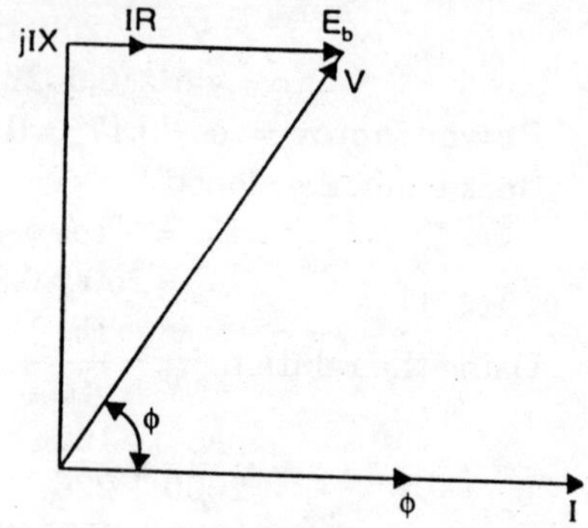

Fig. 4.38

(i) **Induced voltage in the armature :**

IR drop = $I(R_f + R_a)$ = 1.5(8.8 + 4.2) = 19.5 V

Power factor, $\cos \phi = \frac{P_{in}}{VI} = \frac{198.75}{240 \times 1.5} = 0.552$

Phase angle $\phi = \cos^{-1} 0.552 = 56.5°$

∴ Total reactive drop

$= IX = V \sin \phi$

$= 240 \times \sin 56.5° = 200$ V

$\mathbf{E_b} = \sqrt{V^2 - (IX)^2} - IR$

$= \sqrt{(240)^2 - (200)^2} - 19.5$

= **113 V. (Ans.)**

(ii) **Field flux per pole, ϕ_{max} :**

The rotational e.m.f. E_r is given as :

$$E_r = E_b = \frac{\phi_{max}}{\sqrt{2}} \times \frac{ZN}{60} \times \frac{p}{a}$$

∴
$$113 = \frac{\phi_{max}}{\sqrt{2}} \times \frac{1200 \times 4800}{60} \times \frac{2}{2}$$

or
$$\phi_{max} = \frac{113 \times \sqrt{2} \times 60}{1200 \times 4800} = 1.665 \times 10^{-3} \text{ Wb or } \mathbf{1.665 \text{ mWb. (Ans.)}}$$

(iii) **Armature (quadrature axis) flux per pole, ϕ'_{max} :**

Drop due to leakage reactance, $IX_L = \frac{20}{100} IX = \frac{20}{100} \times 200 = 40$ V

Voltage drop in field turns due to air-gap flux from the field,

$$V_{mf} = 4.44 \phi T_f f = 4.44 \times 1.665 \times 10^{-3} \times 300 \times 50 = 110.89 \text{ V}$$

Also
$$IX = V_{ma} + IX_L + V_{mf}$$
(where V_{ma} = drop in armature due to air-gap flux in the brush-axis)

or
$$V_{ma} = IX - IX_L - V_{mf} = 200 - 40 - 110.89 = 49.11 \text{ V}$$

Also,
$$V_{ma} = E_t \text{ (transformer e.m.f.)} = \sqrt{2}\, \phi'_{max} \times \frac{Z}{a} \times f$$

or
$$49.11 = \sqrt{2} \times \phi'_{max} \times \frac{1200}{2} \times 50$$

∴
$$\phi'_{max} = \frac{49.11 \times 2}{\sqrt{2} \times 1200 \times 50} = 1.16 \times 10^{-3} \text{ Wb} = \mathbf{1.16 \text{ mWb. (Ans.)}}$$

Example 4.17. *A 75 W, 2-pole, 50 Hz, 230 V, 1-phase series motor having salient poles has the following data :*

Total resistance = 25 Ω ;
Total reactance = 261 Ω ;
Mutual reactance = 121 Ω *on the direct axis ;*
Stray power loss = 15 W.

Calculate current, power factor and output when running at 6000 r.p.m.

(AMIE Summer, 1995)

Solution. Let X_{md} = mutual reactance in the field (or direct axis)
X_{mq} = mutual reactance in the brush axes, and
X_L = total leakage reactance of all the windings.

Then, $X_{md}I$ = induced voltage in the field due to air-gap flux in the field or direct axis, and
$X_{mq}I$ = induced voltage in the armature due to air-gap flux in the brush or quadrature axis.

In the case of *fully compensated motor* X_{mq} is neglected and the total reactance is, therefore,

$$X = X_L + X_{md},$$

and total mutual reactance, $X_m = X_{md}$.

However, in case of *uncompensated series motor*,

$$X = X_L + X_{md} + X_{mq}$$

Rotational or counter e.m.f. can be expressed in terms of direct-axis mutual reactance X_{md}. If the brushes were shifted to the pole axis, its induced e.m.f. due to transformer action of the pole flux would have been $X_{md}I$. The same pole flux induces rotational e.m.f. E_r in the armature with the brushes in the normal position *i.e.*, in the interpolar axis and its magnitude is $\frac{f_r}{f}$ times X_{md}.

Thus, counter or rotational e.m.f. $E_r = X_{md}\, I \times \frac{f_r}{f}$

Rotational frequency, $$f_r = \frac{Np}{120} = \frac{6000 \times 2}{120} = 100$$

∴ $$E_r = X_{md}\, I \times \frac{f_r}{f} = 121 \times I \times \frac{100}{50} = 242I \text{ volts}$$

∴ Supply volts, $$V = IR + jIX + E_r$$

$$= IR + jIX + X_{md}\, I \times \frac{f_r}{f} = I\left(R + jX + X_{md} \times \frac{f_r}{f}\right)$$

$$= I(25 + j261 + 242) = 373.4I \angle 44.35° \text{ V}$$

or, **Current,** $$\mathbf{I} = \frac{V}{373.4} = \frac{230}{373.4} = \mathbf{0.616\ A.}\ \textbf{(Ans.)}$$

Power factor = cos φ = cos 44.35° = **0.715 (lag). (Ans.)**

Output power = E_rI – stray power losses
= (242 × 0.616) × 0.616 – 15 = **76.83 W. (Ans.)**

Example 4.18. (*a*) *Mention the problems usually encountered when a D.C. series motor is operated on A.C. What design modifications are to be incorporated for its satisfactory operation on A.C. ? Compare the speed for D.C. and A.C. operations.*

(*b*) *Sketch and explain the nature of the torque-slip characteristics of a single phase induction motor based on double revolving field theory. Hence explain why its full load slip is higher than that of a corresponding 3-phase induction motor.*

(AMIE Summer, 1994)

Solution. (*a*) If an alternating current is passed through an ordinary D.C. series motor, the field and armature currents reverse simultaneously every half cycle, so that a unidirectional torque is produced and the armature rotates. The motor has to be considerably modified for running it with alternating currents. The various operational difficulties and the required design modifications are given below :

1. With operation on alternating currents the *eddy current loss increases considerably*. Hence in *addition to the armature, the field structure has to be laminated.*
2. The D.C. series motor when connected to A.C. supply has a *much greater tendency to produce sparking*. Also its *power factor is very low. Both excessive sparking and poor factor are prevented by* **compensating** *the motor, the aim of which is to neutralize the effect of self induction.* The armature is compensated by employing coils carrying the armature current and placed in such a way as to neutralize the armature flux. The compensating coil can be connected in series with the main field or can be short circuited on itself as shown in Fig. 4.39 (*a*) and (*b*) respectively.

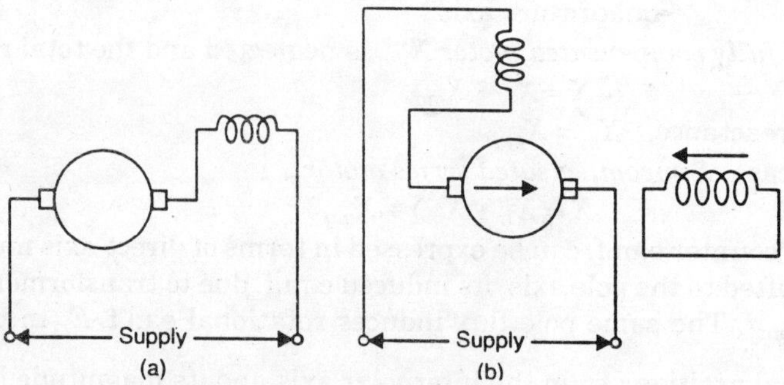

Fig. 4.39

3. Sparking at the brushes is further reduced by *using leads of high resistance between the windings and the commutator segments* as shown in Fig. 4.40.

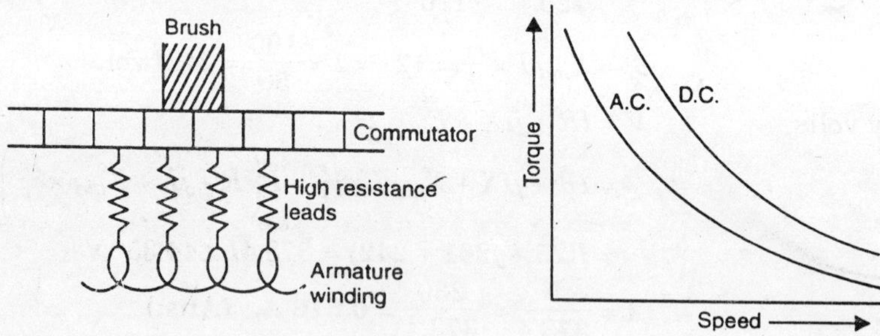

Fig. 4.40

4. *The air gap is kept small because of the weak main field.* The number of turns in the field winding has to be *kept low to reduce the reactance of the field winding.*
5. Because of commutation difficulties the operation of large A.C. motors is confined to *low frequencies* as $16\frac{2}{3}$ and 25 Hz.

Torque-speed characteristics for the series motor when run on A.C. and D.C. supply are shown in Fig. 4.40. The torque obtained with D.C. is somewhat higher than that obtained with A.C. supply.

(*b*) The magnetic field produced by a single phase current is an alternating field. This alternating field can be resolved into *two rotating fields travelling in opposite directions*. The magnitude of each flux phasor is *half the maximum flux value*. Based on this two revolving field theory the alternating field produced by the stator current in a single phase induction motor can be resolved into two rotating components and if the motor is running at nearly synchronous speed under the influence of the forward component, then its *velocity relative to the backward component will be nearly twice the synchronous speed*. The total torque is the sum of the torques developed by each of these fields. *The torque-speed curve for the forward field is exactly similar to that of a polyphase motor. That for the backward field is below the horizontal axis and in a reversed manner*. The two component curves are shown dotted in Fig. 4.40 and the total (or resultant) torque is indicated by the full-time curve. It is clear from the figure that the torque is *zero at synchronous speed and it is zero again at zero speed*. It is also clear from the upper dotted curve which is similar to the curve for a 3-phase induction motor and the full line curve for single phase induction motor, the *slips at full load are different. The slip for a single motor is larger as compared to the slip for a 3-phase motor*.

Example 4.19. *Explain with reasons why :*

(*a*) *Single phase induction motors are not self-starting.*

(*b*) *Single phase series motors are preferred for fraction service rather than induction or synchronous motors.*

(*c*) *Compensating winding is provided in a universal motor for A.C. operation.*

(*d*) *Reluctance motor has low efficiency.*

(*e*) *In a repulsion motor, the armature current is very high when the brush axis is aligned in the direction of stator field axis.* **(AMIE Winter, 1994)**

Solution. (*a*) Constructionally a single phase motor is similar to a 3-phase induction motor *except that the stator is provided with a single phase winding*. When this winding is fed from a single phase supply, its stator produces a flux which is only alternating along the space axis. It is not a synchronously revolving or rotating flux as in the case of two or three phase stator winding fed from a 2-phase or 3-phase supply. *An alternating flux acting on a stationary squirrel cage rotor cannot produce relation*. This is *why a single phase motor is not self-starting*.

(*b*) Now-a-days single phase motors are manufactured as fractional horse power motors and mostly they are used with domestic appliances, such as tape recorder, record player, hair drier, electric clock, refrigerator, mixi, fans, pumps etc. As these are *light duty machines, are less costly and hence preferred.*

(*c*) The *compensating winding* is provided in a universal motor *to reduce the reactance voltage* present in the armature when motor runs on A.C. supply. This reactance voltage is caused by the alternating flux by transformer action because the flux produced by the field winding is alternating.

(*d*) The construction of stator of single phase reluctance motor is similar to single phase induction motor. To make the motor self starting the starting winding is provided with the main winding. The rotor of the motor is a modified cage rotor with *some of the teeth removed*. This *actually provides variation of air gap reluctance*. When motor is connected to single phase supply it starts like an induction motor. When motor speed reaches near to synchronous speed it will be pulled into synchronous speed with the stator field *due to the reluctance torque, because of the variation of the air gap reluctance*, now the motor continues to run at synchronous speed, as a non-excited synchronous motor.

The *efficiency and power factor of this type of motor is poor because of the varying reluctance.*

(*e*) When the brush axis is aligned in the direction of the stator field axis the direction of induced currents in the armature will be such that the armature will become an electromagnet with a N-pole on its top directly under the main N-pole and with a S-pole at the bottom directly over the main S-pole, *because of this face to face position the armature current becomes very high*. However, because of this face to face position of the main and induced magnetic pole *no torque is developed*.

Example 4.20. *Calculate the stepping angle for a 3-stack, 16-tooth variable reluctance motor.*

(AMIE Advance Elect. M/Cs. Summer, 1996)

Solution. Number of stacks or phases, $n = 3$

Number of teeth or poles, $p = 16$

Stepping angle, $\alpha = \frac{360°}{np} = \frac{360}{3 \times 16} = 7.5°$. **(Ans.)**

HIGHLIGHTS

1. *There are two basic types of single-phase induction motors which start on the split-phase principle, namely resistance-start induction-run motor and the capacitor-start induction-run motor.*
2. *The resistance-start induction-run motor has a relatively poor starting torque because the phase angle between the running winding current and the starting winding current is only from 30 to 50 electrical degrees.*
3. *The capacitor-start induction-run motor has a relatively good starting torque because the phase angle between the running winding current and the starting current is practically 90 electrical degrees.*
4. *The capacitor-start, capacitor-run induction motor has relatively good starting torque and operates at a comparatively high power factor. The main winding and auxiliary windings are both energized at all times the motor is in operation.*
5. *A resistance-start induction-run motor or a capacitor-start induction-run motor may be reversed by interchanging the leads of the starting winding circuit. The direction of rotation of either type of motor can also be reversed by interchanging the leads of the running winding circuit. However,* **never attempt to change the rotation of either type of motor by reversing the line wires.**
6. *The* **repulsion motor** *operates on the repulsion principles at all times. It has* **excellent starting torque** *and* **a relatively wide range of speeds.**
7. *The repulsion-start induction-run motor starts as a repulsion motor and after it reaches 75 per cent of rated speed, operates as an induction motor. It has excellent starting torque and very good speed regulation.*
8. *There are two types of brush mechanism used with repulsion start induction-run motors : The* **brush-lifting type** *and* **brush-riding** *type.*
9. *To change the direction of rotation of either a repulsion motor or a repulsion-start induction-run motor, move the brushes to the other side of stator field poles so that there is an angle of 15 electrical degrees in the new position between the brushes and the stator field pole centres.*
10. *The A.C. series motor has excellent starting torque and can be used in those applications requiring different speeds. It has characteristics comparable to those of the D.C. series motor.*
11. *The direction of rotation of an A.C. series motor can be reversed by interchanging the connections to the armature.*
12. *The shaded pole induction motor is usually made in fractional horsepower sizes in excess of 1/10 horsepower. This type of motor has no centrifugal starting device or mechanism which requires maintenance. However, the* **starting torque** *for this type of motor is* **small.**

OBJECTIVE TYPE QUESTIONS

(A) Choose the Correct Answer :

1. In a split phase motor, the running winding should have
   (*a*) high resistance and low inductance (*b*) low resistance and high inductance
   (*c*) high resistance as well as high inductance (*d*) low resistance as well as low inductance.

2. If the capacitor of a single-phase motor is short-circuited
(*a*) the motor will not start (*b*) the motor will burn
(*c*) the motor will run in reverse direction
(*d*) the motor will run in the same direction at reduced r.p.m.
3. In capacitor start single-phase motors
(*a*) current in the starting winding leads the voltage
(*b*) current in the starting winding lags the voltage
(*c*) current in the starting winding is in phase with voltage in running winding.
(*d*) none of the above.
4. In a capacitor start and run motors the function of the running capacitor in series with the auxiliary winding is to
(*a*) improve power factor (*b*) increase overload capacity
(*c*) reduce fluctuations in torque (*d*) to improve torque.
5. In a capacitor start motor, the phase displacement between starting and running winding can be nearly
(*a*) 10° (*b*) 30°
(*c*) 60° (*d*) 90°.
6. In a split phase motor
(*a*) the starting winding is connected through a centrifugal switch
(*b*) the running winding is connected through a centrifugal switch
(*c*) both starting and running windings are connected through a centrifugal switch
(*d*) centrifugal switch is used to control supply voltage.
7. The torque developed by a single-phase motor at starting is
(*a*) more than the rated torque (*b*) rated torque
(*c*) less than the rated torque (*d*) zero.
8. Which of the following motors will give relatively high starting torque ?
(*a*) Capacitor start motor (*b*) Capacitor run motor
(*c*) Split phase motor (*d*) Shaded pole motor.
9. Which type of capacitor is preferred for capacitor start and run motor ?
(*a*) Electrolyte capacitor (*b*) Paper capacitor (oil filled)
(*c*) Ceramic capacitor (*d*) Air capacitor.
10. Which of the following motors will have relatively higher power factor ?
(*a*) Capacitor run motor (*b*) Shaded pole motor
(*c*) Capacitor start motor (*d*) Split phase motor.
11. In a shaded pole motor, the shading coil usually consist of
(*a*) a single turn of heavy wire which is in parallel with running winding
(*b*) a single turn of heavy copper wire which is short-circuited and carries only induced current
(*c*) a multilayer fine gauge copper wire in parallel with running winding
(*d*) none of the above.
12. In a shaded pole single-phase motor, the revolving field is produced by the use of
(*a*) inductor (*b*) capacitor
(*c*) resistor (*d*) shading coils.
13. A centrifugal switch is used to disconnect starting winding when motor has
(*a*) run for about 1 minute (*b*) run for about 5 minutes
(*c*) picked up about 50 to 70 per cent of rated speed
(*d*) picked up about 10 to 25 per cent of rated speed.
14. If a particular application needs high speed and high starting torque, then which of the following motors will be preferred ?
(*a*) Universal motor (*b*) Shaded pole type motor
(*c*) Capacitor start motor (*d*) Capacitor start and run motor.

15. Which motor has the highest power to weight ratio ?
(*a*) Capacitor run motor (*b*) Capacitor start motor
(*c*) Universal motor (*d*) None of the above.

16. In a universal motor the width of the brush is nearly
(*a*) half the width of commutator segment (*b*) equal to width of the commutator segment
(*c*) equal to width of two commutator segment (*d*) equal to width of four commutator segment.

17. The range of efficiency for shaded pole motors is
(*a*) 95% to 99% (*b*) 80% or 90%
(*c*) 50% to 75% (*d*) 5% to 35%.

18. In a capacitor start single-phase motor, when capacitor is replaced by a resistance
(*a*) torque will increase (*b*) the motor will consume less power
(*c*) motor will run in reverse direction (*d*) motor will continue to run in same direction.

19. The power factor of a single-phase induction motor is usually
(*a*) lagging (*b*) always leading
(*c*) unity (*d*) unity to 0.8 leading.

20. A shaded pole motor can be used for
(*a*) toys (*b*) hair dryers
(*c*) circulators (*d*) any of the above.

21. A hysteresis motor works on the principle of
(*a*) hysteresis loss (*b*) magnetisation of rotor
(*c*) eddy current loss (*d*) electromagnetic induction.

22. Which of the following motors will give the highest starting torque ?
(*a*) D.C. shunt motor (*b*) Schrage motor
(*c*) Repulsion start and induction run motor (*d*) Universal motor.

23. In repulsion motors by shifting the position of brushes
(*a*) only speed can be controlled (*b*) speed and direction of rotation can be controlled
(*c*) speed, direction of rotation and torque can be controlled
(*d*) none of the above.

24. Which of the following single-phase motors is reversible ?
(*a*) Capacitor start split-phase motor (*b*) Resistance start split-phase motor
(*c*) Reluctance motor (*d*) None of the above.

25. For which of the applications a reluctance motor is preferred ?
(*a*) Electric shavers (*b*) Refrigerators
(*c*) Signalling and timing devices (*d*) Lifts and hoists.

26. In repulsion motors
(*a*) speed remains constant (*b*) speed variation is within 5%
(*c*) speed varies with load (*d*) speed varies only with change in applied voltage.

27. In single-phase repulsion motors power factor is
(*a*) always leading (*b*) high at low speeds
(*c*) always unity (*d*) high at high speeds.

28. The motor used on small lathes is usually
(*a*) universal motor (*b*) d.c. shunt motor
(*c*) single-phase capacitor run motor (*d*) 3-phase synchronous motor.

29. Which of the following motors is preferred for tape recorders ?
(*a*) Shaded pole motor (*b*) Hysteresis motor
(*c*) Two value capacitor motor (*d*) Universal motor.

30. A single-phase induction motor is
(*a*) inherently self-starting with high torque (*b*) inherently self-starting with low torque
(*c*) inherently non-self-starting with low torque (*d*) inherently non-self-starting with high torque.

31. A schrage motor can run on
(*a*) zero slip (*b*) negative slip
(*c*) positive slip (*d*) all of the above.

32. A universal motor can run on
(*a*) A.C. only (*b*) D.C. only
(*c*) either A.C. or D.C. (*d*) none of the above.

33. Which of the following single-phase motors is suitable for timing and control purposes ?
(*a*) Reluctance motor (*b*) Series motor
(*c*) Repulsion motor (*d*) Universal motor.

34. In case of a shaded pole motor the direction of rotation of the motor is
(*a*) from main pole to shaded pole (*b*) from shaded pole to main pole
(*c*) either of the above depending on voltage (*d*) either of the above depending on power factor.

35. In case of high speed universal motor which of the following needs more attention ?
(*a*) End play (*b*) Air gap
(*c*) Insulation in rotor (*d*) Balancing of rotor.

36. The wattage rating for a ceiling fan motor will be in the range
(*a*) 50 to 250 W (*b*) 250 to 500 W
(*c*) 50 to 150 W (*d*) 10 to 20 W.

37. The wattage of motor for driving domestic sewing machine will be around
(*a*) 100 to 150 W (*b*) 40 to 75 W
(*c*) 10 to 30 W (*d*) 5 to 10 W.

38. Which of the following single-phase motors has relatively poor starting torque ?
(*a*) Universal motor (*b*) Repulsion motor
(*c*) Capacitor motor (*d*) All single phase motors have zero starting torque.

39. Which type of load is offered by cranes and hoists ?
(*a*) gradually varying load (*b*) non-reversing, no-load start
(*c*) reversing, light start (*d*) reversing, heavy start.

40. The speed of a universal motor is generally reduced by using
(*a*) gear trains (*b*) V-belts
(*c*) brakes (*d*) chains.

41. Which of the following motors can be used for unity power factor ?
(*a*) Capacitor run motor (*b*) Shaded pole motor
(*c*) Hysteresis motor (*d*) Schrage motor.

42. When a D.C. series motor is connected to A.C. supply, the power factor will be low because of
(*a*) high inductance of field and armature circuits
(*b*) induced current in rotor due to variations of flux
(*c*) fine copper wire winding (*d*) none of the above.

43. The direction of rotation of universal motor can be reversed by reversing the flow of current through
(*a*) armature winding (*b*) field winding
(*c*) either armature winding or field winding (*d*) none of the above.

44. In which single-phase motor, the rotor has no teeth or winding ?
(*a*) Split phase motor (*b*) Reluctance motor
(*c*) Hysteresis motor (*d*) Universal motor.

45. Which motor is normally free from mechanical and magnetic vibrations ?
(*a*) Split phase motor (*b*) Universal motor
(*c*) Hysteresis motor (*d*) Shaded pole motor.

46. As hysteresis motors are free from mechanical and magnetic vibrations therefore these are considered as suitable for

(*a*) fans (*b*) blowers

(*c*) sound equipment (*d*) mixer grinders.

47. A reluctance motor

(*a*) is self-starting (*b*) is constant speed motor

(*c*) needs no D.C. excitation (*d*) all of the above.

48. In a hysteresis motor, the rotor must have

(*a*) retentivity (*b*) resistivity

(*c*) susceptibility (*d*) none of the above.

49. The rotor of a hysteresis motor is made of

(*a*) aluminium (*b*) cast iron

(*c*) chrome steel (*d*) copper.

50. The electric motor used in portable drills is

(*a*) capacitor run motor (*b*) hysteresis motor

(*c*) universal motor (*d*) repulsion motor.

51. Which of the following applications always has some load whenever switched on ?

(*a*) Vacuum cleaners (*b*) Fan motors

(*c*) Pistol drills (*d*) All of the above.

52. The speed control of universal motor used for sewing machines is by

(*a*) friction (*b*) varying the resistance

(*c*) tapping the field (*d*) centrifugal mechanism.

ANSWERS

1. (*b*)	**2.** (*a*)	**3.** (*a*)	**4.** (*a*)	**5.** (*d*)	**6.** (*a*)	**7.** (*d*)
8. (*a*)	**9.** (*b*)	**10.** (*a*)	**11.** (*b*)	**12.** (*d*)	**13.** (*c*)	**14.** (*a*)
15. (*c*)	**16.** (*c*)	**17.** (*d*)	**18.** (*d*)	**19.** (*a*)	**20.** (*d*)	**21.** (*a*)
22. (*b*)	**23.** (*b*)	**24.** (*a*)	**25.** (*c*)	**26.** (*c*)	**27.** (*d*)	**28.** (*c*)
29. (*b*)	**30.** (*c*)	**31.** (*d*)	**32.** (*c*)	**33.** (*a*)	**34.** (*a*)	**35.** (*d*)
36. (*c*)	**37.** (*a*)	**38.** (*c*)	**39.** (*d*)	**40.** (*a*)	**41.** (*d*)	**42.** (*a*)
43. (*c*)	**44.** (*c*)	**45.** (*c*)	**46.** (*c*)	**47.** (*d*)	**48.** (*a*)	**49.** (*c*)
50. (*c*)	**51.** (*c*)	**52.** (*b*).				

(B) Say 'Yes' or 'No' :

1. In a capacitor start motor, the phase displacement between starting and running winding can be nearly 90°
2. In a split-phase motor the staring winding is connected through a centrifugal switch.
3. In a shaded pole single-phase motor, the revolving field is produced by the use of capacitor.
4. A universal motor has the highest power to weight ratio.
5. The range of efficiency for shaded pole motors is 80 to 90%.
6. The power factor of a single-phase induction motor is usually lagging.
7. A hysteresis motor works on the principle of eddy current loss.
8. Capacitor-start split-phase motor is reversible.
9. A reluctance motor is preferred for signalling and timing devices.
10. The wattage rating for a ceiling fan motor varies from 50 to 150 W.

ANSWERS

1. Yes	**2.** Yes	**3.** No	**4.** Yes	**5.** No	**6.** Yes	**7.** No
8. Yes	**9.** Yes	**10.** Yes.				

THEORETICAL QUESTIONS

1. Explain what is meant by the split-phase method of motor starting.
2. Explain why the starting torque of a capacitor-start induction run motor is better than that of a resistance-start induction run-motor.
3. How is the direction of rotation reversed for each of the following ?
   (*i*) Resistance-start induction-run motor, and
   (*ii*) Capacitor-start induction-run motor.
4. (*a*) Compare operating characteristics of a resistance-start induction-run motor with those of a capacitor-start induction-run motor.
   (*b*) List three applications for :
   (*i*) A resistance-start induction-run motor
   (*ii*) A capacitor-start induction-run motor.
5. Explain how a repulsion motor operates.
6. Explain how a repulsion-start induction-run motor operates.
7. How is the direction of rotation of each of the following motors reversed ?
   (*i*) The repulsion motor, and (*ii*) The repulsion-start induction-run motor.
8. What is the difference between A.C. and D.C. series motor ?
9. Why are small fractional horsepower A.C. series motors called universal motors ?
10. What is the difference between a conductively compensated series motor and an inductively compensated series motor ?
11. Describe the construction and operation of a shaded pole motor.
12. Write a short note on a sub-synchronous motor and a hysteresis motor.
13. What is a stepper motor ? Enumerate its advantages and applications.
14. Explain briefly permanent-magnet stepper motor briefly. How does it differ from variable reluctance motor ?

UNSOLVED PROBLEMS

1. The following data pertains to a single-phase motor :

Number of poles	= 4
Supply voltage	= 110 V
Rated output	= 125 W
Slip at rated output	= 6 per cent
Total copper loss at full-load	= 25 W
Rotational losses	= 25 W

Calculate the full-load efficiency and the rotor copper loss caused by the backward field.
Neglect stator copper loss. **[Ans.** 14.94 W, 78.12%]

2. The following test results were obtained in case of a 220 V, single-phase induction motor :
   No-load Test : 220 V, 6.15 A, 348 W
   Locked rotor test : 126 V, 15.0 A, 577 W.
   Stator winding resistance = 1.5 Ω.
   Determine the approximate equivalent circuit of the motor.
   [Ans. $R_1 = 1.5\ \Omega$, $R_2' = 1.064\ \Omega$, $X_1 = 4.0\ \Omega$, $X_m = 71.5\ \Omega$]
3. The following test results were obtained in case of a 230 V, single-phase, 4-pole 50-Hz induction motor :
   No-load test : 30 V, 0.535 A, p.f. 0.174 lagging
   Locked rotor test : 50 V, 1.50 A, p.f. 0.6 lagging.

Determine the approximate equivalent circuit.

Assume : (*i*) Stator and rotor I^2R losses at standstill are equal.

(*ii*) Rotor leakage reactance referred to the stator leakage reactance are equal.

[**Ans.** 12 Ω, 16 Ω, 860 Ω]

4. The resistance and inductive reactance of each winding of a 50-Hz split-phase induction motor are 80 Ω and 237.5 Ω respectively. Additional resistance R and condenser C are in series with one winding. Calculate their values to give the same current in each winding with a phase difference of 90°.

[**Ans.** 157.5 Ω, 10 μF]

5. The resistance and total inductance of a single-phase fractional horsepower series motor are 30 Ω and 0.5 H respectively. It draws 0.8 A current and runs at 2000 r.p.m., when connected to 250 V D.C. supply. Calculate the speed and power factor when connected to 250 V, 50-Hz supply and loaded to take the same current. [**Ans.** 1700 r.p.m., p.f. = 0.8648]

5

Synchronous Generators or Alternators

5.1. INTRODUCTION

- *A machine for generating alternating currents is referred to as an alternator.*
- The term 'A.C. generator' is also frequently used, in place of alternator and this is often contracted to just 'generator' when it is obvious that an A.C. machine is meant. In older literature, the term 'alternating current dynamo' will also be found, but the present tendency is to reserve the use of the word dynamo for D.C. generators.
- *High-speed* alternators driven by steam turbines differ considerably in their construction from the *slow speed* types and are distinguished by the use of the terms *'turbo-alternator'* or *'turbo-generator'* whilst the slow engine-driven machines are often described as being of the *'flywheel-type'*.

5.2. CLASSIFICATION AND OPERATING PRINCIPLE

- In *D.C. generators,* the field poles are stationary and the armature conductors rotate. The *alternating* voltage induced in armature conductors is converted to a *direct* voltage at the brushes by means of the commutator.
- *A.C. generators* commonly called alternators, have *no commutators* as they are required to supply electrical energy with an *alternating* voltage. Therefore, it is not necessary that armature be the rotating member.

Alternators, according to their construction, are divided into the following two classifications :

1. Revolving-armature type.
2. Revolving-field type.

1. **Revolving-armature type alternator**

- It has *stationary field poles* and *revolving armature.*
- It is usually of relatively *small* kVA *capacity* and *low-voltage rating.* It resembles a D.C. generator in general appearance except that it *has slip-rings instead of a commutator.* The field excitation must be direct current and therefore, must be supplied from an external direct current source.

2. **Revolving-field type alternator :**

- It has a *stationary* armature or *stator,* inside of which the *field poles rotate.*
- Most alternators are of the revolving-field type, in which the *'revolving-field structure'* or *'rotor'* has slip rings and brushes to *supply the excitation current from an outside D.C. source.* The armature coils are placed in slots in a laminated core, called the *'stator'*, which is made up of thin steel punchings or laminations securely clamped and held in place in the steel frame of the generator. Usually the field voltage is between 100 and 250 volts and the amount of *power delivered to the field circuit is relatively small.*

The following are the principal **advantages** of the revolving-field type alternators :

1. The armature windings are more easily braced to prevent deformation under the mechanical stresses due to short-circuit currents and centrifugal forces.

2. The armature (stator) winding must be insulated for a high voltage, while the voltage of field circuit is low (100 to 250 volts). *It is much easier to insulate the high-voltage winding when it is mounted on the stationary structure.*

3. Only a small amount of power at low voltage is handled by the slip ring contacts.

4. It is easier to build and properly balance high-speed rotors when they carry the field structure.

5. The armature winding is cooled more readily because the stator core can be made large enough and with many air passages or cooling ducts for forced air circulation.

Operating principle (*Revolving-field type*). When the rotor rotates, the stator conductors (being stationary) are cut by the magnetic flux, hence they have induced e.m.f. produced in them. Because the magnetic poles are alternately *N* and *S*, they induce an e.m.f. and hence current in armature conductors, which first flows in one direction and then in the other. Hence, an alternating e.m.f. is produced in the stator conductors whose frequency depends on the number of *N* and *S* poles moving past a conductor in one second and whose direction is given by Fleming's right-hand rule.

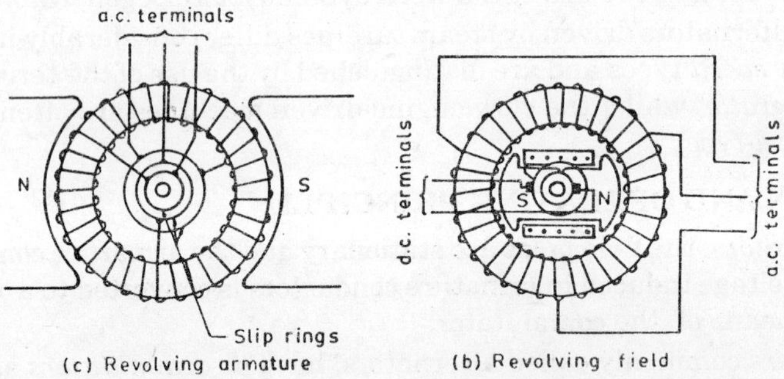

Fig. 5.1. Operating principle of a three-phase alternator.

Fig. 5.1 shows the operating principle of a three-phase alternator.

Note. All synchronous A.C. generators and motors require direct current for excitation. Excitation is supplied by a D.C. generator called an *exciter*. The capacity of the exciter is only a *small percentage* of the rated capacity of the alternator. The exciter may be directly connected to the shaft of the alternator, or it may be driven by a separate electric motor, water wheel, or small turbine. Large power stations usually have several exciters employing different methods of drive as insurance against the failure of excitation.

5.3. CONSTRUCTIONAL DETAILS

Refer Fig. 5.2.

5.3.1. Stator. The *stator* of an alternator consists essentially of a cast iron or a welded-steel frame supporting a slotted ring made of soft laminated sheet-steel punchings (Fig. 5.3) in the slots of which the armature coils are assembled.

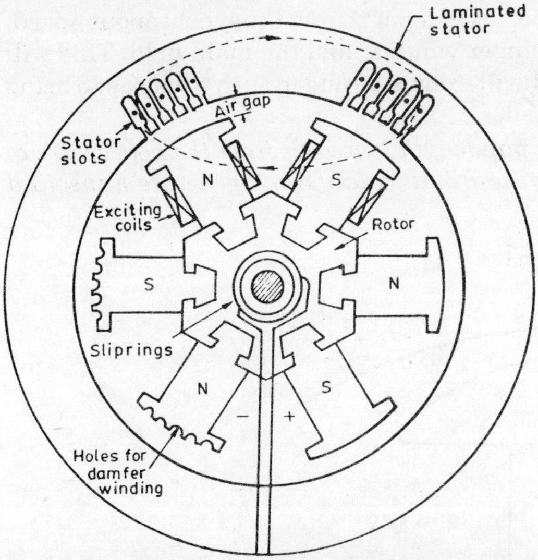

Fig. 5.2. Alternator.

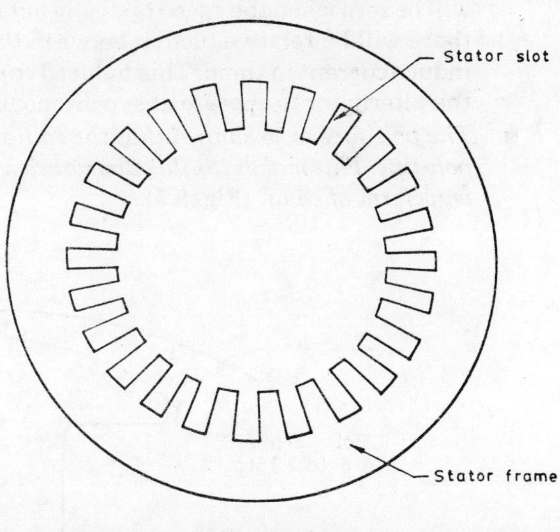

Fig. 5.3. Alternator stator.

- The *laminations* are annealed and are insulated from each other by a thin coating of oxide and an enamel (as in D.C. machines, transformers etc.)
- *Open slots* are used, permitting easy installation of stator coils and easy removal in case of repair. Suitable spacing blocks are inserted at intervals between laminations to leave radial *air ducts,* open at both ends, through which cooling air may circulate.
- The *coils* are shaped much like the coils of a D.C. generator, the two sides of the coil being approximately a pole pitch apart. All coils are alike, and therefore, interchangeable. They are insulated before being inserted in the slots and are further protected by a horn-fibre slot lining. When in place on the stator, the coils are connected together in groups to form a winding of the required number of phases, three phase star-connected windings being common.
- A fractional rather an *integral number of slots per pole is often used in order to eliminate harmonics in the waveform.*

5.3.2. Rotor. The revolving field structure is usually called the rotor. There are two types of rotors :

1. Salient pole type rotor.
2. Smooth cylindrical type rotor.

Salient pole type rotor :

This type of rotor is used for *slow-speed machines which have large diameters and small axial lengths.*

- The poles are made of thick steel laminations rivetted together and attached to a rotor by a devetail joint as shown in Fig. 5.4. The overhang of the pole gives mechanical support to the field coil.
- In most of the alternators, where the oscillation or the hunting effect is very high, the *damper winding* in the pole faces is provided. The copper bars short circuited at both ends are placed in the specially provided holes. The relative velocity of the damping winding with respect to main field

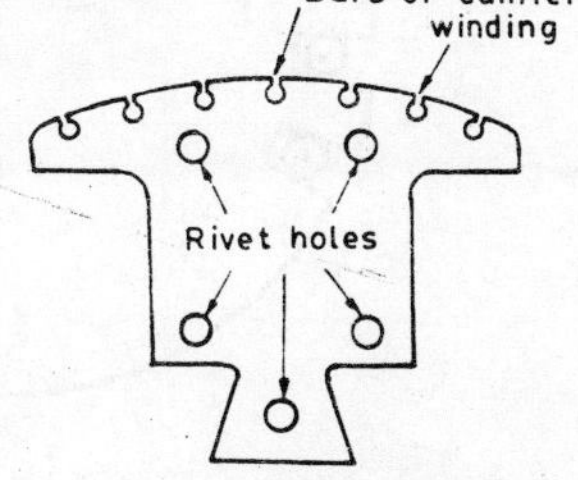

Fig. 5.4. Typical lamination of a salient pole rotor.

will be zero when the speed is steady but as soon as it departs from the synchronous speed, there will be relative motion between the damper winding and the main field. This will induce current in them. This induced current will exert a torque in such a way as to bring the alternator to operate at synchronous speed.

- *The pole face is so shaped that the radial air gap length increases from the pole centre to pole tips. This makes the flux distribution over the armature uniform to generate sinusoidal waveform of e.m.f.* (Fig. 5.5).

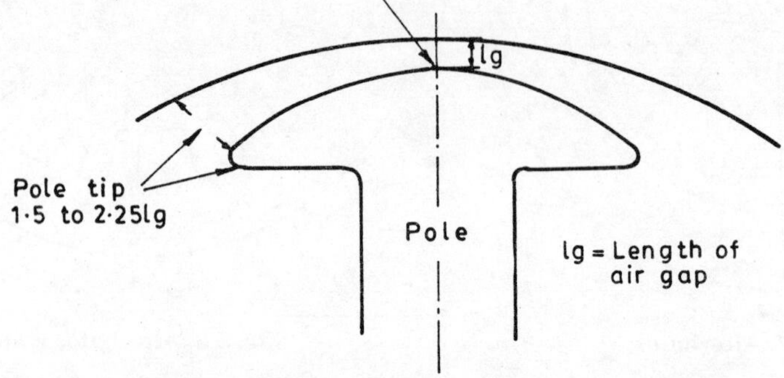

Fig. 5.5

The salient pole field structure has the following **special features :**

(*i*) They have *large diameter* and *short axial length*.

(*ii*) The pole shoe covers about $\frac{2}{3}$ rd of pole pitch.

(*iii*) Poles are laminated to reduce eddy current losses.

(*iv*) These are employed with hydraulic turbines or diesel engines. The speed is 100 to 375 r.p.m.

Smooth cylindrical rotor :

- This type of rotor is used for alternators which are coupled to *steam turbines* which run at very high speeds. To reduce the peripheral speed of the alternator the *diameter of the*

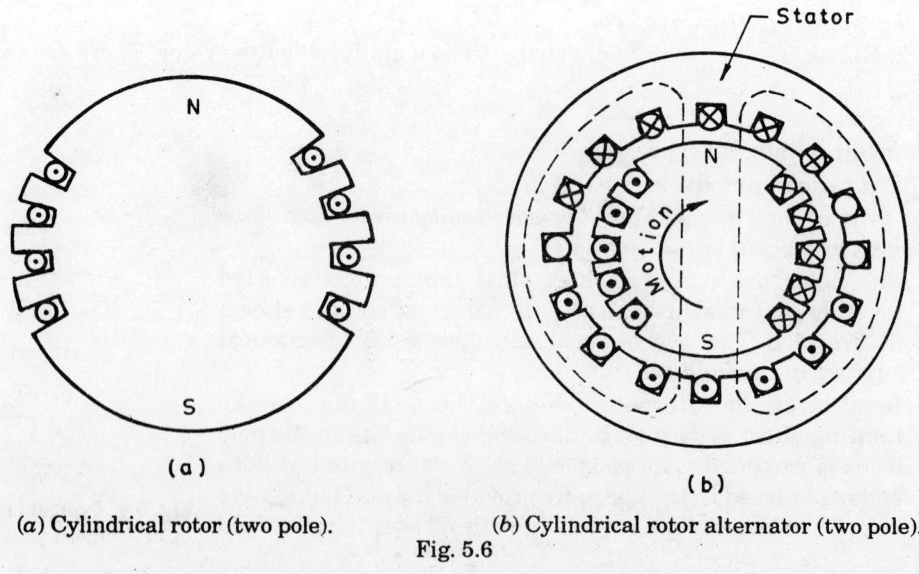

(*a*) Cylindrical rotor (two pole). (*b*) Cylindrical rotor alternator (two pole).

Fig. 5.6

rotor is reduced and axial length is increased. The number of poles of the rotor are *two* or *four.*

Figs. 5.5 and 5.6 show a cylindrical rotor and cylindrical rotor alternator respectively.

- *These rotors are made from solid forgings of alloy steel.* The outer periphery of rotor has slots in which the field winding is placed. About 2/3rd of rotor pole pitch is slotted, leaving the 1/3rd unslotted for the pole centre. Heavy wedges of non-magnetic steel are forced into the grooves in the teeth outside the field coils to keep the field coils in position.
- *Cylindrical rotor machines have always horizontal configuration.*
- Since these rotors have large lengths of core forced ventilation is necessary for proper cooling. *Forced air cooling is used up to about 50 MVA sizes and for bigger sizes hydrogen cooling is invariably employed because the conductivity of hydrogen is about 7 times that of air.*

The non-salient field structure has the following **special features** :

(*i*) They are of small diameter and of very long axial length.
(*ii*) Robust construction.
(*iii*) High operating speed (3000 r.p.m.).
(*iv*) Noiseless operation.
(*v*) Dynamic balancing is better.
(*vi*) Less windage (air-resistance) loss.
(*vii*) No need to provide damper windings, except in special cases to assist in synchronising.
(*viii*) Better e.m.f. waveform.

Note. It may be worth mentioning that *cylindrical rotors* will most likely be located on alternators where *steam power* is readily available. *Salient-pole rotors* will be found where *water power* is the prime-mover source of energy. Diesel engine, gas engine, and gas turbine prime-movers are considered medium-speed machines, and their alternators will also have salient poles. Where alternators are driven by other electrical machines, either A.C. or D.C. motors, there are no such restrictions on the rotor construction. A design is developed that is compatible with the space limitations, speed considerations, and heat dissipation for both electrical machines. If may be noted here that the terms high-speed and low-speed rotors are sometimes used synonymously with cylindrical and salient-pole rotors, respectively.

5.3.3. Bearings

- Although antifriction bearings are occasionally used on alternators of the smaller ratings, the great majority are furnished with oil-lubricated babbitted bearings. For *horizontal shafts* these will be self-contained ring-oiled bearings wherever design conditions permit. At higher shaft peripheral speeds and higher bearing loadings, ring oiling is supplemented with recirculation of externally cooled coil. The rings may be eliminated, or they may be retained to afford some degree of emergency oil supply in the event of a failure of the external system. *Lead-base babbits are commonly used for journal bearings,* although tin-base babbits may be employed for some heavy-duty application.
- Two principal types of *thrust bearings* are used on vertical alternators : the *pivoted shoe type* and the *spring type*.

5.4. FREQUENCY

In case of a generator which has two poles, the induced e.m.f. passes through one complete revolution in one revolution of the machine. In a multipolar machine one cycle of e.m.f. would be generated when the field structure rotates through an angle subtended by a double pole pitch. Therefore, in a machine with *p*-poles, the number of cycles of e.m.f. in one revolution will be *p*/2. If a machine has a speed of N_s revolutions per minute, the frequency will be

$$\left(\frac{p}{2}\right) \times \frac{N_s}{60} \text{ per second.}$$

Thus,
$$f = \frac{p}{2} \times \frac{N_s}{60} = \frac{N_s p}{120} \text{ Hz} \qquad ...(5.1)$$

In order to keep the frequency constant, the speed N_s must remain unchanged. Therefore, a synchronous generator (*i.e.*, alternator) runs at a constant speed known as synchronous speed.

The following table gives the number of poles and speeds for three frequencies : 60-Hz, 50-Hz and 25-Hz.

Higher frequencies are today being used for specific purposes, especially when it is necessary to use high-speed induction motors, but these frequencies are not for general distribution. A great deal of aircraft equipment operates with voltage having a frequency of 400 Hz.

Table 5.1

Number of poles	*Speed, r.p.m.*		
	60 Hz	*50 Hz*	*25 Hz*
2	3600	3000	1500
4	1800	1500	750
6	1200	1000	500
8	900	750	375
10	720	600	300
12	600	500	250

Note. *The standard frequency in India is 50 Hz.*

5.5. ARMATURE WINDINGS

5.5.1. Winding Types. A wide variety of winding types are possible to produce a desired voltage in the proper number of phases and with a suitable waveshape. Practical considerations, *mainly economic,* limit the usual alternator winding to a *double-layer 3-phase lap winding, arranged in 60° phase belts in open slots*. The number of coils, the number of turns per coil, the coil pitch, the number of circuits, and the connection of the phases are selected to give the desired voltage and waveform.

Double-layer windings in open slots permit the use of form-wound coils which are all alike in a given machine. These coils have the characteristic diamond shape in the end area. Three-phase windings may be either Δ-or *Y*-connected ; Y-connected machines are much more common, particularly in the larger sizes. The winding may be arranged to be connected either *Y* or Δ, with leads brought out from both ends of each phase to make this possible.

- **Double-layer windings** in open-slots have the following *advantages* over single layer windings in semi-enclosed slots :
  (*i*) Ease in manufacture of coils and lower cost of winding.
  (*ii*) Less number of coils are required as spare in the case of winding repairs.
  (*iii*) Fractional slot windings can be employed.
  (*iv*) Fractional pitch coils can be used.
- **Single-layer windings** have the following *advantages* :
  (*i*) Higher efficiency and quieter operation because of narrow slot openings.
  (*ii*) Space factor for slots is higher owing to absence of inter layer separator.

Modern practice all over the world favours use of double layer windings. Single layer windings are popular only in the continent.

5.5.2. Number of Coils. For integral slot windings, where the number of slots per phase per pole is a whole number, all coil groups are identical, and the phase voltages are balanced with respect to magnitude and angle. Fractional slot windings, where the number of slots per phase per pole is not an integer, have unequal coil groups. These can be arranged to produce balanced voltages if the number of phases is a factor of the number of slots per phase at least as many times as it is a factor of the number of the poles. For example, a balanced 3-phase winding for a 36-pole alternator will require a number of slots per phase which is a multiple of 3-times 3, since 3 appears twice as a factor of 36. Similarly, a balanced 2-phase winding for a 16-pole alternator will require a number of slots per phase which is a multiple of 16 (2 times, 2 times, 2 times).

Other numbers of slots than those which satisfy the above criterion may be used by making coils which do not have the same number of turns. Special groupings of the coils may also permit a balanced wiring when the number of slots does not conform to this criterion. These expedients may be used because of limitations in available punching dies.

5.5.3. Conductor Design. *Conductors are stranded* except in the smallest ratings to *enable easier shaping of the coil and to limit eddy-current losses* which result from the flux that crosses the field. The effect of this flux is to produce a voltage within the strand which results in circulating currents. The thinner the strand, the less the voltage and the resulting eddy currents. Since the strands ordinarily are all connected together at the joints between coils, there will also be eddy currents circulating between strands because of the difference in flux linked by the various strands. This source of loss can often be reduced sufficiently by the *use of more and thinner conductors per coil, with a corresponding increase in the number of parallel circuits.* As an alternative, some kind of *transposition* is to arrange the strands in one part of the coil with respect to another so that the induced voltages cancel each other.

5.5.4. Skewing. Sometimes *slots are skewed to minimize the effects of voltages resulting from the ripple in the air-gap flux produced by the stator slots. Skewing a slot, essentially a full slot pitch in length of the core, will eliminate voltages due to slot ripple.* The effect of a number of slots which produces undesirable ripple voltages can be limited by some degree of skew.

5.5.5. Dead Coils. Coils can be left unconnected in the armature circuit during manufacture *to obtain operating characteristics not readily obtainable from adjustment of number of coils and turns.* They may also be cut out of machines in service in *order to permit operation after one coil or more has been damaged.* Where the winding has parallel circuits, the number of coils cut out must be the same in each circuit in order to avoid damaging circulating currents. If more than a few per cent of the coils are cut out, it may also be necessary to cut out a like number of coils in the other phases of the machine. The controlling consideration is the voltage balance and the circulating ground currents. This situation would be most critical in a Δ-connected winding.

5.5.6. Coil Pitch. Most economical use of the flux in the magnetic circuit results when the stator coil is full pitch, or as near to full pitch as a fractional slot winding will permit. In a fractional slot winding, the maximum pitch ordinarily used is full pitch for the integer, *i.e.*, for $7\frac{5}{7}$ slots per pole winding, a coil pitch of 1 to 8 would be the maximum ordinarily used. *Less than full pitch coils are used to obtain adjustments in the voltage generated or to limit harmonics.*

5.5.7. Single Phase Windings. These are usually 3-phase windings with 1-phase not used. Sometimes the coils for the third phase are omitted, but most often they are wound in the machine and might be considered as spares. In small sizes a special concentric winding may be used. A *low-resistance damper winding* is generally necessary on single phase machines to *reduce the flux pulsations* that are set up by the single-phase armature reaction and to reduce the effective armature reactance. Single-phase machines have an inherent torque pulsation at twice rated frequency. The resulting noise and vibration are noticeable on small machines and may require special construction on large machines.

5.5.8. Two-phase Windings. These windings differ from 3-phase windings only in the grouping of the stator coils. Ninety-degree phase belts are ordinarily used.

5.5.9. Double Windings. These windings are sometimes *used to reduce short-circuit currents and to simplify switch gear and bus structure problems*. The windings have standard coil design with special end connections. The electrical designs must allow for the effects of unbalanced armature reaction if the two windings are not equally loaded and for the mutual reactance between the windings.

5.5.10. Multispeed Windings. For some applications a winding is required which will permit operation at more than one speed, but at the same frequency. An example is a hydraulic pump-turbine unit, which may generate at one speed and pump at another. Such machines may have two windings in the same slot or in adjacent slots or a single armature winding may be reconnected by means of suitable switches to serve this purpose. Special rotor construction and field-winding reconnections are also necessary.

5.6. CHORDING OF WINDINGS

Chording of the pole windings is one of the design factors. If, on the 36-slot, four pole machine, an individual coil enters slot 1 and comes back in slot 10, it will have spanned 90 mechanical degrees of the stator circular structure. Since there are four poles, 90 mechanical degrees is 180 electrical degrees (because *total electrical degrees in one revolution = 180 p*, where p, is the number of poles).

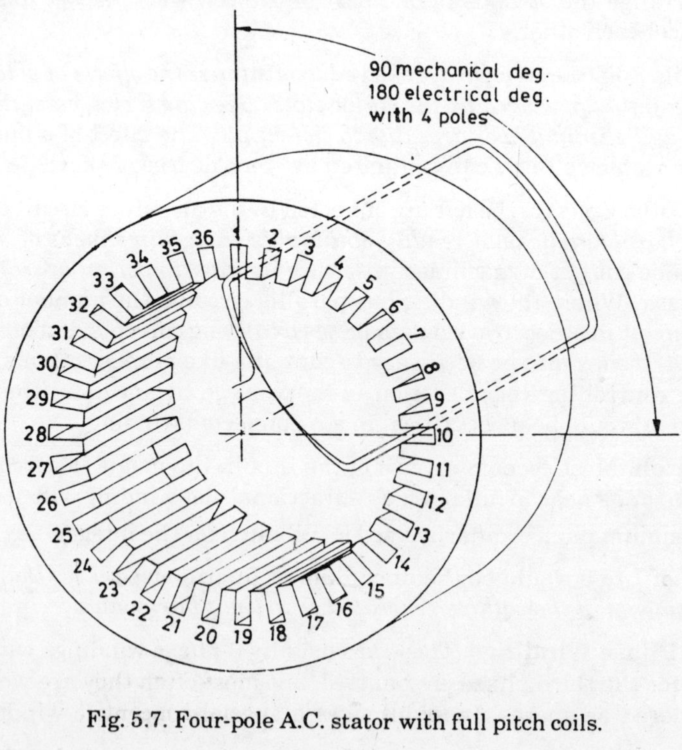

Fig. 5.7. Four-pole A.C. stator with full pitch coils.

Thus, the two sides of the coil are in the same relative position on the adjacent north and south pole positions. This is a full-pitch coil construction (Fig. 5.7).

The more usual A.C. machine coil will cover less of the periphery of the machine and is then said to be *fractional pitch.* A typical situation might have a coil enter slot 1 and leave slot 7. This then covers six out of a possible nine slot pitches, and is a 6/9 or 66.7% pitch. *The majority of A.C. machine coils are of fractional-pitch type.*

Advantages of Fractional Pitch Winding :

(*i*) The ends of the coils are shorter, which means *less copper loss due to less total length.*

(*ii*) The end coils can be formed more compactly. The end belts will need less winding space resulting in a shorter unit.

(*iii*) Improved waveform of the induced e.m.f.

(*iv*) Fractional number of slots per pole can be used which in turn reduces tooth ripples.

(*v*) Mechanical strength of the coils is increased.

(*vi*) There is distinct *reduction in machine harmonics* due to cancellation of higher harmonics. Since all A.C. equipment is designed to operate on a pure sine wave, the generation of harmonics is to be avoided. This is especially so when the factor that achieves it is otherwise desirable.

5.7. PITCH FACTOR

In a full pitch coil, the e.m.fs. in the two coil sides are in phase and therefore the coil e.m.f. is twice the e.m.f. of each coil side. In a *short pitch coil* the e.m.fs. of the two coil sides are not in phase and must be *added vectorially* to give the coil e.m.f. The *factor by which the e.m.f. per coil is reduced, because of the pitch being less, is known as pitch factor* (or coil span factor) k_p. Thus,

$$k_p = \frac{\text{vector or phasor sum of induced e.m.fs. per coil}}{\text{arithmetic sum of the induced e.m.fs per coil}} \quad ...(5.2)$$

It is *always less than unity.*

Let E_s be induced e.m.f. in each side of the coil. If the coil were full-pitched, then the total induced e.m.f. in the coil would have been $2E_s$ [Fig. 5.8 (*a*)].

If it is short-pitched by angle α° (electrical) then their resultant [Fig. 5.8 (*b*)] is E which is vector sum of two voltages, α° (electrical) apart.

$$\therefore \quad E = 2E_s \cos\frac{\alpha}{2}$$

$$\therefore \quad k_p = \frac{\text{vector sum}}{\text{arithmetic sum}}$$

$$= \frac{E}{2E_s} = \frac{2E_s\cos\frac{\alpha}{2}}{2E_s} = \cos\frac{\alpha}{2}$$

$$\text{Hence,} \quad k_p = \cos\frac{\alpha}{2} \quad ...(5.3)$$

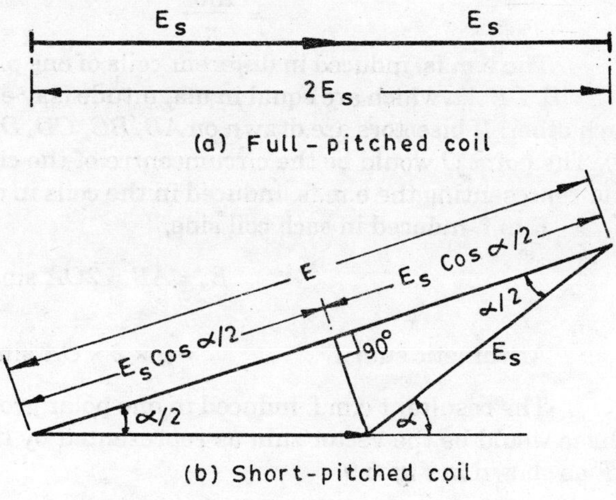

Fig. 5.8

The pitch factor given by eqn. (5.3) is for the *fundamental component of e.m.f.* If the flux density distribution contains space harmonics, the pitch factor for the *n*th harmonic is given by,

$$k_{pn} = \cos\frac{n\alpha}{2} \quad ...(5.4)$$

The nth harmonic e.m.f. is reduced to zero if the angle α is such that

$$\cos \frac{n\alpha}{2} = 0 \quad \text{or} \quad \frac{n\alpha}{2} = 90° \qquad ...(5.5)$$

This enables the windings to be designed such that specified harmonics will not be generated *e.g.* if $\alpha = 60°$, there can be no third harmonic generation. *Thus fractional pitch windings result in a voltage wave form which resembles a sinusoidal to a better degree than would a full-pitch winding.*

Note. If the value of α is not given in question, then assume $k_p = 1$.

5.8. DISTRIBUTION OR BREADTH OR WINDING FACTOR

When the coils comprising a phase of the winding are distributed in two or more slots per pole, the e.m.fs. in the adjacent coils will be out of phase with respect to one another and their resultant will be less than their algebraic sum. *The ratio of the vector sum of the e.m.fs. induced in all the coils distributed in a number of slots under one pole to the arithmetic sum of the e.m.fs. induced (or to the resultant of the e.m.fs. induced in all the coils concentrated in one slot under one pole) is known as distributed factor* k_d

or $$k_d = \frac{\text{e.m.f. induced in a distributed winding}}{\text{e.m.f. induced if the winding would have been concentrated}}$$

$$= \frac{\text{vector sum}}{\text{arithmetic sum}}.$$

The distribution factor is always less than unity.

Let n = number of slots/pole

q = number of slots/pole/phase

E_s = induced e.m.f. in each coil side

β = angular displacement between the slots

$= \frac{180°}{n}$

The e.m.fs. induced in different coils of one phase under one pole are represented by side AB, BC, CD, DE ..... which are equal in magnitude (say each equal to E_s) and differ in phase (say $\beta°$) from each other. If bisectors are drawn on AB, BC, CD, DE ...... they would meet at common point (say at O). The point O would be the circumcentre of the circle having AB, BC, CD, DE, ... as the chords and representing the e.m.fs. induced in the coils in different slots.

E.m.f. induced in each coil side,

$$E_s = AB = 2OA \sin \frac{\beta}{2}$$

Arithmetic sum $$= q \times 2 \times OA \sin \frac{\beta}{2}$$

The resultant e.m.f. induced in one polar group of one phase would be the vector sum as represented by the vector AF as shown in Fig. 5.9.

∴ The resultant e.m.f.,

$$E_r = AF$$

$$= 2 \times OA \sin \frac{AOF}{2} = 2 \times OA \sin \left(\frac{q\beta}{2} \right)$$

Distribution factor,

$$k_d = \frac{\text{vector sum}}{\text{arithmetic sum}}$$

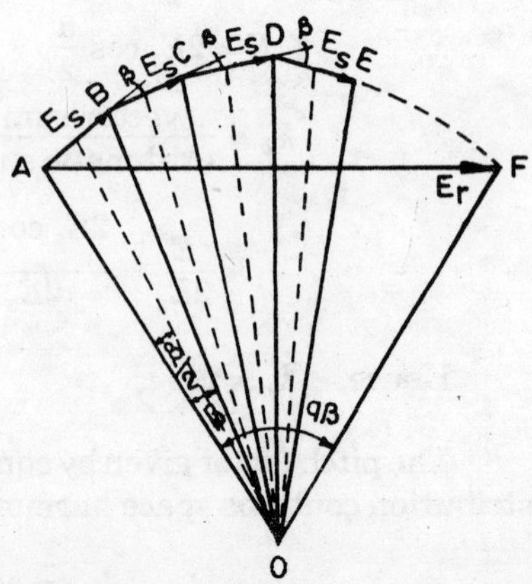

Fig. 5.9. Calculation of distribution factor.

$$= \frac{2 \times OA \sin\left(\frac{q\beta}{2}\right)}{q \times 2 \times OA \sin\frac{\beta}{2}} = \frac{\sin\left(\frac{q\beta}{2}\right)}{q \sin\frac{\beta}{2}}$$

Hence,
$$k_d = \frac{\sin\left(\frac{q\beta}{2}\right)}{q \sin\frac{\beta}{2}} \quad ...(5.6)$$

$q\beta$ is also known as the *phase spread* and is expressed in electrical radians.

Value of the distribution factor for *nth harmonic* is,

$$k_d = \frac{\sin\left(\frac{qn\beta}{2}\right)}{q \sin\left(\frac{n\beta}{2}\right)}.$$

5.9. E.M.F. EQUATION

Let Z_{ph} = number of conductors or coil sides in series *per phase*
$= 2T_{ph}$, where T_{ph} is the number of coils or turns per phase (one turn coil has two sides),
p = number of poles,
ϕ = useful flux per pole, webers,
N = rotational speed of rotor, r.p.m.,
f = frequency, Hz,
k_d = distribution factor, and
k_p = pitch factor.

Consider a conductor on the stator of the alternator. Let the alternator rotor move through one revolution in $t\left(=\frac{60}{N}\right)$ seconds, the flux cut by the conductor

$$= p\phi \text{ webers}$$

The average e.m.f. induced in the conductor,

$$E_{av} = \frac{d}{dt}(\text{flux}) \text{ volts} = p \cdot \frac{\phi}{t} = \frac{p\phi}{\frac{60}{N}} = \frac{p\phi N}{60} \quad ...(5.7)$$

We know that, $f = \frac{Np}{120}$ or $N = \frac{120f}{p}$

Substituting the value of N in eqn. (5.7), we get

$$E_{av} = \frac{p\phi \times 120f}{60 \times p} = 2f\phi \text{ volts/conductor}$$

Average e.m.f. per phase,

$$E_{av}/\text{phase} = 2f\phi \times 2T_{ph} = 4f\phi T_{ph}$$
$$E_{r.m.s.}/\text{phase} = E_{av} \times \text{form factor} = 4f\phi T_{ph} \times 1.11$$
$$= 4.44 f\phi T_{ph} \text{ volts} \quad ...(5.8)$$

The above equation of e.m.f. induced per phase is true only, *if the winding is concentrated in one slot but practically it is not true*, as the winding for each phase under each pole is *distributed* and for such cases k_p and k_d must be considered.

Thus, e.m.f. induced (for sinusoidal wave) per phase will be

$$E_{r.m.s.}/\text{phase} = 4.44 f \phi T_{ph} k_p k_d \text{ volts} \quad ...(5.9)$$

For full-pitched and concentrated windings $k_p = 1$, $k_d = 1$.

If the alternator is star-connected (as is usually the case) then the line voltage,

$$E_L = \sqrt{3}\, E_{r.m.s.} \text{ phase} = \sqrt{3} \times 4.44 f \phi T_{ph} k_p k_d \text{ volts.}$$

Example 5.1. *Determine the pitch (or coil span) factors for the following windings :*

(i) 36 stator slots, 4 poles, coil span 1 to 8. (ii) 96 stator slots, 6 poles, coil span 1 to 12. Sketch the two coil spans.

Solution. (*i*) 36 stator slots, 4 poles, coil span 1 to 8 : [Fig. 5.10 (*a*)]

In this case coil span falls short by,

$$\left(\frac{2}{9}\right) \times 180° = 40° \quad \textit{i.e.,} \quad \alpha = 40°$$

∴ Pitch factor, $k_p = \cos \frac{40°}{2} = \cos 20° = 0.94.$

(*ii*) **96 stator slots, 6 poles, coil span 1 to 12 :** [Fig. 5.10 (*b*)]

Here, coil span falls short by,

$$\left(\frac{5}{16}\right) \times 180° = 56° \quad \textit{i.e.,} \quad \alpha = 56°$$

∴ **Pitch factor,** $k_p = \cos \frac{\alpha}{2} = \cos \frac{56°}{2} = \cos 28° = \mathbf{0.883.}$ **(Ans.)**

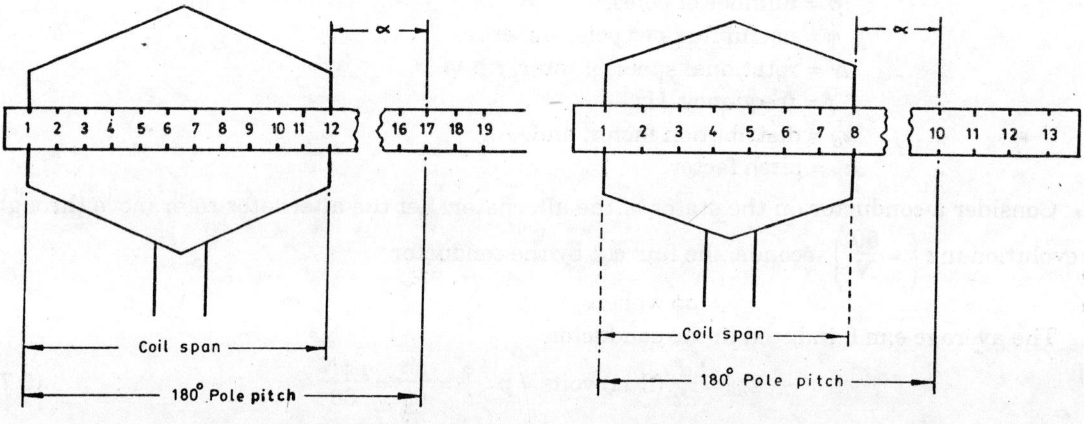

(*a*) 36 stator slots, 4 poles, coil span 1 to 8. (*b*) 96 stator slots, 6 poles, coil span 1 to 12.

Fig. 5.10

☞**Example 5.2.** *What are the effects of the coil and distribution span factors on the output and waveform of an alternator ? Determine their values for a 3-phase winding with 4 slots per pole per phase, the coil span being 10 slot pitches. Calculate the percentage increase in R.M.S. value of the phase voltage due to a 25 per cent third harmonic.*

Solution. Effects of the coil and distribution span factors on the output and waveform of alternator :

Coil span factor. At the fundamental frequency this factor is $k_p = \cos \frac{\alpha}{2}$ where α is the angle in electrical degrees by which the span of the coil is less than a pole-pitch. *The output at the fundamental frequency is reduced in the same ratio as this factor.* The *n*th harmonic is reduced in the ratio $k_{pn} = \cos \frac{n\alpha}{2}$. *Short-chording can thus be used to reduce or eliminate troublesome harmonics.*

Distribution factor. The effect of the distribution factor on the output of an alternator is to reduce it by an amount depending on the spread of the winding. The vector sum of the coil e.m.fs. is less than their arithmetic sum which would be given if all the coils were located in the same slot. The distribution factor is not always the same for harmonics as for the fundamental. For a phase spread of 120° $k_d = 0$ for the third harmonic and multiples of three, which are thus *eliminated from the waveform.*

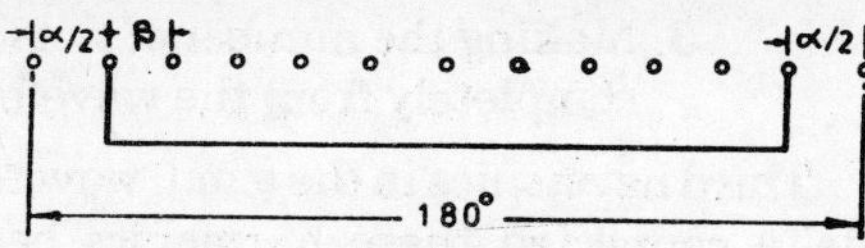

Fig. 5.11

In the problem given, we see from Fig. 5.11 that

$$\beta = \frac{180°}{4 \times 3} = 15°, \quad \alpha = \frac{2}{12} \times 180 = 30°$$

∴ Coil span factor, $$k_p = \cos\frac{\alpha}{2} = \cos\frac{30°}{2} = \cos 15° = 0.966$$

Distribution factor, $$k_d = \frac{\sin\left(\frac{q\beta}{2}\right)}{q \sin\frac{\beta}{2}}$$

[where q = no. of slots/pole/phase = 4]

$$= \frac{\sin\left(\frac{4 \times 15°}{2}\right)}{4 \times \sin\frac{15°}{2}} = \frac{0.5}{4 \times 0.1305} = 0.958$$

R.M.S. value of a complex wave = $E = \sqrt{E_1^2 + E_h^2}$

where E_1 = R.M.S. value of the fundamental, and

E_h = R.M.S. value of the harmonic.

$$\therefore \quad E = E_1\sqrt{1 + (E_h/E_1)^2} = E_1\sqrt{1 + (0.25)^2} = 1.03\, E_1$$

i.e. **a 25% harmonic increases the R.M.S. value of the phase e.m.f. by 3%. (Ans.)**

☞**Example 5.3.** *What are the causes of harmonics in the voltage and current waves of electrical machinery and what means are taken in design to reduce them ?*

The phase voltage of a 750 kW, 2200 V, 3-phase, 50 Hz alternator has a 5 per cent third harmonic. What is the circulating current on normal voltage if the machine is delta connected ? The resistance and reactance per phase are 0.25 Ω and 0.7 Ω respectively. Express the loss due to the circulating current as a percentage of full-load output.

Solution. Causes of harmonics in the wave form :

1. Non-sinusoidal waveform of the field flux.
2. Variation in the reluctance of air-gap due to the slotting of the stator core.

Design measures to reduce the harmonics :

(A) 1. Shaping the poles by chamfering the pole tips.
2. Skewing the poles.
3. Short-chording the armature winding by making the coil-span less than a full pole-pitch.
4. Distributing the armature winding.

(B) 1. Increasing the number of slots per pole.
2. Skewing the poles through one slot-pitch.
3. Making the number of slots per pole pair an odd number. This eliminates tooth ripples completely from the waveform.

Third harmonics in the e.m.f. waveform are in time-phase in all three phases. When the phases are delta connected these harmonics become *additive* around the closed circuit and *produce a circulating current*.

Given : Phase e.m.f., $E_{ph} = 2200$ V

Hence, third harmonic/phase $= 0.05 \times 2200 = 110$ V

Output $= 750$ kW

Resistance/phase, $R_{ph} = 0.25\ \Omega$

Reactance/phase at third harmonic frequency,

$$X_{ph} = 3 \times 0.7 = 2.1\ \Omega$$

Impedance per phase at this frequency,

$$Z_{ph} = \sqrt{(0.25)^2 + (2.1)^2} = 2.11\ \Omega$$

Circulating current $= \dfrac{110}{2.11} = 52.1$ A

Copper loss in three phases $= 3 \times (52.1)^2 \times 0.25 = 2035.8$ W

Percentage of full load output

$$= \frac{2035.8}{750 \times 1000} \times 100 = \mathbf{0.271\ per\ cent.}\ \textbf{(Ans.)}$$

☞**Example 5.4.** *A 3-phase, 16-pole alternator has the following data :*

Number of slots = 192 ; Conductors / slot = 8 (conductors of each phase are connected in series) ; coil span = 160 electrical degrees ; speed of the alternator = 375 r.p.m. ; flux / pole = 55 m Wb. Calculate the phase and line voltages.

Solution. Number of poles, $p = 16$

Number of slots $= 192$

Conductors/slot $= 8$

Coil span $= 160°$ (electrical)

Speed of the alternator $= 375$ r.p.m.

Flux/pole, $\phi = 55$ mWb $= 0.055$ Wb

Here, $\alpha = 180° - 160° = 20°$

$\therefore$ Pitch factor, $k_p = \cos\dfrac{\alpha}{2} = \cos\dfrac{20°}{2} = \cos 10° = 0.9848$

Number of slots/pole, $n = \dfrac{192}{16} = 12$

$$\therefore \quad \beta = \frac{180°}{n} = \frac{180°}{12} = 15°$$

$$q = \text{number of slots/pole/phase} = \frac{192}{16 \times 3} = 4$$

$\therefore$ Distribution factor, $k_d = \dfrac{\sin \dfrac{q\beta}{2}}{q \sin \dfrac{\beta}{2}} = \dfrac{\sin\left(\dfrac{4 \times 15°}{2}\right)}{4 \times \sin\left(\dfrac{15°}{2}\right)} = \dfrac{\sin 30°}{4 \sin 7.5°} = 0.9577$

Now, total number of slots/phase $= \dfrac{192}{3} = 64$

Number of conductors/slot $= 8$ *(given)*

$\therefore$ Total number of conductors/phase

$= 64 \times 8 = 512$

$\therefore$ Turns/phase, $T_{ph} = \dfrac{512}{2} = 256$

Also, frequency, $f = \dfrac{Np}{120} = \dfrac{375 \times 16}{120}$

$f = 50\text{-Hz}$

Now, Using the relation, $E_{ph} = 4.44 f \phi T_{ph} k_p k_d$ volts

$= 4.44 \times 50 \times 0.55 \times 256 \times 0.9848 \times 0.9577 = 2948$ V

$\therefore$ **Line voltage,** $E_L = \sqrt{3}\, E_{ph} = \sqrt{3} \times 2948 =$ **5106 V. (Ans.)**

☞**Example 5.5.** *A 3-phase, star-connected alternator has the following data : voltage required to be generated on open-circuit = 4000 V (at 50-Hz) ; speed = 500 r.p.m. ; stator slots/pole/phase = 3 ; conductors/slot = 12. Calculate :*

(i) Number of poles, and *(ii) Useful flux pole.*

Assume all conductors per phase to be connected in series and coil to be full pitch.

Solution. Line voltage, $E_L = 4000$ V

$\therefore$ $E_{r.m.s.}$/phase, $E_{ph} = \dfrac{4000}{\sqrt{3}} = 2309$ V

Speed of the alternator, $N = 500$ r.p.m.

Number of slots/pole/phase,

$q = 3$

Number of conductors/slot $= 12$

Frequency, $f = 50\text{-Hz}.$

(*i*) **Number of poles, p :**

Using the relation, $f = \dfrac{Np}{120}$

or $p = \dfrac{120f}{N} = \dfrac{120 \times 50}{500} =$ **12 poles. (Ans.)**

(*ii*) **Useful flux/pole, ϕ :**

Pitch factor, $k_p = 1$ *(given)* [$\because$ Coils are full pitched]

Number of slots/pole, $n = 3 \times 3 = 9$

Angular displacement between slots,

$\beta = \dfrac{180}{n} = \dfrac{180}{9} = 20°$

$\therefore$ Distribution factor, $k_d = \dfrac{\sin \dfrac{q\beta}{2}}{q \sin \dfrac{\beta}{2}} = \dfrac{\sin\left(\dfrac{3 \times 20°}{2}\right)}{3 \times \sin\left(\dfrac{20°}{2}\right)} = \dfrac{\sin 30°}{3 \sin 10°} = 0.96$

Number of turns/phase, $T_{ph} = \frac{Z_{ph}}{2} = \frac{12 \times 9 \times 12}{2 \times 3} = 216$

Now, using the relation,

$$E_{r.m.s.}/\text{phase} = 4.44 f \phi T_{ph} k_p k_d \text{ volts}$$

$$2309 = 4.44 \times 50 \times \phi \times 216 \times 1 \times 0.96$$

$$\therefore \quad \phi = \frac{2309}{4.44 \times 50 \times 216 \times 0.96} = 0.05 \text{ Wb}$$

Hence, **useful flux/pole = 0.05 Wb. (Ans.)**

Example 5.6. *A 3-phase, 10-pole alternator has 2 slots per pole per phase on its stator with 10 conductors per slot. The air gap flux is sinusoidally distributed and equals to 0.05 Wb. The stator has double layer winding with a coil span of 150° electrical degrees. If the alternator is running at 600 r.p.m., calculate the e.m.f. generated per phase at no load.* **(AMIE Winter, 1997)**

Solution. *Given* : $p = 10$; q number of slots/pole/phase = 2, conductors/slot = 10, $\phi = 0.05$ Wb, coil span = 150° (electrical), $N = 600$ r.p.m.

E.m.f. generated per phase, E_{ph} :

Number of slots/pole, $n = 2 \times 3 = 6$

Angular displacement between the slots, $\beta = \frac{180°}{n} = \frac{180°}{6} = 30°$

Angle of chording $= 180° - 150° = 30°$

$\therefore$ Pitch factor, $k_p = \cos\left(\frac{30°}{2}\right) = 0.9659$

Distribution factor, $k_d = \dfrac{\sin\left(\frac{q\beta}{2}\right)}{q \sin\left(\frac{\beta}{2}\right)} = \dfrac{\sin\left(\frac{2 \times 30°}{2}\right)}{2 \times \sin\left(\frac{30°}{2}\right)} = 0.9659$

Number of conductors in series/phase, $Z_{ph} = 10 \times \text{slots/phase} = 10 \times (2 \times 10) = 200$

$\therefore$ Number of turns/phase, $T_{ph} = \frac{Z_{ph}}{2} = \frac{200}{2} = 100$

$$\therefore \quad E_{ph} = 4.44 f \phi\, T_{ph}\, k_p k_d \text{ volts}$$

$$= 4.44 \times 50 \times 0.05 \times 100 \times 0.9659 \times 0.9659$$

$$= \mathbf{1035.6\ V.\ (Ans.)}$$

Example 5.7. *Calculate the speed and open-circuit line and phase voltages of a 4-pole, 3 phase, 50 Hz star-connected alternator with 36 slots and 30 conductors per slot. The flux per pole is 0.05 Wb sinusoidally distributed.* **(Madras Univ. April, 1997)**

Solution. *Given* : $p = 4$; $f = 50$ Hz ; number of slots = 36 ;
number of conductors per slot = 30 ; $\phi = 0.05$ Wb.

Speed, phase and line voltages :

Speed of the alternator, $\mathbf{N} = \frac{120f}{p} = \frac{120 \times 50}{4} = \mathbf{1500\ r.p.m.\ (Ans.)}$

Number of conductors connected in series per phase,

$$Z_{ph} = \frac{\text{No. of slots} \times \text{no. of conductors per slot}}{\text{No. of phases}} = \frac{36 \times 30}{3} = 360$$

Number of turns per phase, $T_{ph} = \frac{Z_{ph}}{2} = \frac{360}{2} = 180$

Number of slots per pole, $n = \frac{36}{4} = 9$

Number of slots per pole per phase, $q = \frac{9}{3} = 3$

Angular displacement between the slots, $p = \frac{180°}{n} = \frac{180°}{9} = 20°$

Distribution factor, $$k_d = \frac{\sin\left(\frac{q\beta}{2}\right)}{q\sin\left(\frac{\beta}{2}\right)} = \frac{\sin\left(\frac{3 \times 20°}{2}\right)}{3 \times \sin\left(\frac{20°}{2}\right)} = 0.96$$

Assuming coils to be full-pitched,

Coil span factor, $k_p = 1$

Open-circuit phase voltage, $E_{ph} = 4.44\, f\phi\, T_{ph}\, k_p k_d$ volts

$= 4.44 \times 50 \times 0.05 \times 180 \times 1 \times 0.96 = 1918.1$ V

Open-circuit line voltage, $E_L = \sqrt{3}\, E_{ph} = \sqrt{3} \times 1918.1 =$ **3322.2 V. (Ans.)**

Example 5.8. *A 4-pole, 50 Hz star-connected alternator has a flux per pole of 0.12 Wb. It has 4 slots per pole per phase, conductors per slot being 4. If the winding coil span is 150°, find the e.m.f.*

(AMIE Summer, 1995)

Solution. *Given* : $p = 4$; $f = 50$ Hz ; $\phi = 0.12$ Wb ; $q = 4$;
conductors per slot = 4 ; coil span = 150°.

E.m.f. induced :

Number of slots per pole, $n = q \times$ number of phases $= 4 \times 3 = 12$

Number of slots per phase $= q \times$ number of poles $= 4 \times 4 = 16$

Number of conductors connected in series per phase,

Z_{ph} = Number of conductor/slot × number of slots/phase

$= 4 \times 16 = 64$

∴ Numbers of turns per phase, $T_{ph} = \frac{Z_{ph}}{2} = \frac{64}{2} = 32$

Angular displacement between the slots, $\beta = \frac{180°}{n} = \frac{180°}{12} = 15°$

Distribution factor, $$k_d = \frac{\sin\left(\frac{q\beta}{2}\right)}{q\sin\left(\frac{\beta}{2}\right)} = \frac{\sin\left(\frac{4 \times 15°}{4}\right)}{4 \times \sin\left(\frac{15°}{2}\right)} = 0.958$$

Chording angle, $\alpha = 180° -$ coil span $= 180° - 150° = 30°$

∴ Pitch factor, $k_p = \cos\frac{\alpha}{2} = \cos\left(\frac{30°}{2}\right) = 0.966$

$E_{ph} = 4.44 f\, \phi\, T_{ph}\, k_p\, k_d$ volts

$= 4.44 \times 50 \times 0.12 \times 32 \times 0.966 \times 0.958$ V $= 788.91$ V

∴ $E_L = \sqrt{3}\, E_{ph} = \sqrt{3} \times 788.91 =$ **1366.4 V. (Ans.)**

Example 5.9. *A 3-phase, 4-pole, 50 Hz star-connected alternator has 60 slots with 2 conductors per slot and having armature winding of the double layer type. Coils are short-pitched, that is, if one coil side lies in slot number 1, the other coil side lies in slot number 13. Find the useful flux per pole required to induce a line voltage of 6.6 kV.* **(Bharathiyar Univ. April, 1997)**

Solution. *Given* : $p = 4$, number of slots = 60 ; conductors per slot = 2, $E_L = 6.6$ kV.

Useful flux per pole, ϕ :

Number of slots/pole, $n = \dfrac{60}{4} = 15$

Number of slots/pole/phase, $q = \dfrac{15}{3} = 5$

Number of conductors connected in series/phase, $Z_{ph} = \dfrac{60 \times 2}{3} = 40$

Number of series turns/phase, $T_{ph} = \dfrac{Z_{ph}}{2} = \dfrac{40}{2} = 20$

Angular displacement between the slots, $\beta = \dfrac{180°}{n} = \dfrac{180°}{15} = 12°$

Distribution factor, $$k_d = \frac{\sin\left(\dfrac{q\beta}{2}\right)}{q \sin\left(\dfrac{\beta}{2}\right)} = \frac{\sin\left(\dfrac{5 \times 12°}{2}\right)}{5 \sin\left(\dfrac{12°}{2}\right)} = 0.957$$

Coil span $$= \frac{\text{coil span in terms of slots}}{\text{number of slots/pole}} \times 180° = \left(\frac{13-1}{15}\right) \times 180° = 144°$$

Coil span factor, $$k_p = \cos \alpha = \cos\left(\frac{180° - 144}{2}\right) = 0.951$$

Also, $$E_L = \sqrt{3} \times 4.44\, f\phi\, T_{ph}\, k_p k_d \text{ volts}$$

$\therefore$ $$6.6 \times 10^3 = \sqrt{3} \times 4.44 \times 50 \times \phi \times 20 \times 0.951 \times 0.957$$

or $$\phi = \frac{6.6 \times 10^3}{\sqrt{3} \times 4.44 \times 50 \times 20 \times 0.951 \times 0.957} = \mathbf{0.943\ Wb.}\ \textbf{(Ans.)}$$

Example 5.10. *An 8-pole, 3-phase, 60° spread, double layer winding has 72 coils in 72 slots. The coils are short-pitched by two slots. Calculate the winding factor for the fundamental and third harmonic.* **(Madurai Univ. Nov., 1996)**

Solution. *Given* : $p = 8$; Number of coils = 72 ; Number of slots = 72.

Winding factor, k_w :

Number of slots per pole, $n = \dfrac{72}{8} = 9$

Number of slots per pole per phase, $q = \dfrac{72}{8 \times 3} = 3$

Angular displacement between the slots, $\beta = \dfrac{180°}{n} = \dfrac{180°}{9} = 20°$ (electrical)

Coil span $$= \frac{180° \times \text{coil span in terms of slots}}{\text{number of slots/pole}} = \frac{180°\,(9-2)}{9} = 140°$$

Chording angle, $\alpha = 180° - \text{coil span} = 180° - 140° = 40°$

For the fundamental component :

Distribution factor, $$k_d = \frac{\sin\left(\dfrac{q\beta}{2}\right)}{q \sin\left(\dfrac{\beta}{2}\right)} = \frac{\sin\left(\dfrac{3 \times 20°}{2}\right)}{3 \times \sin\left(\dfrac{20°}{2}\right)} = 0.96$$

Pitch factor, $k_p = \cos\left(\frac{\alpha}{2}\right) = \cos\left(\frac{40°}{2}\right) = 0.94$

Winding factor, $\mathbf{k_w} = k_d \times k_p = 0.96 \times 0.94 = \mathbf{0.9}$. **(Ans.)**

For the third harmonic component :

Value of the distribution factor for *nth* harmonic is given as :

Distribution factor, $$k_n = \frac{\sin\left(\frac{qn\beta}{2}\right)}{q \sin\left(\frac{n\beta}{2}\right)} \text{ or } k_{d_3} = \frac{\sin\left(\frac{3 \times 3 \times 20°}{2}\right)}{3 \sin\left(\frac{3 \times 20°}{2}\right)} = 0.666$$

Pitch factor, $$k_{p_3} = \cos\left(\frac{n\alpha}{2}\right) = \cos\left(\frac{3 \times 40°}{2}\right) = 0.5$$

$\therefore$ Winding factor, $$\mathbf{k_{w_3}} = k_{d_3} \times k_{p_3} = 0.666 \times 0.5 = \mathbf{0.333}. \textbf{ (Ans.)}$$

☞Example 5.11. *Calculate the r.m.s. value of the induced e.m.f. per phase of a 10-pole, 3-φ, 50 Hz alternator with 2 slots per pole per phase and 4 conductors per slot in two layers. The coil span is 150°. The flux per pole has a fundamental component of 0.12 Wb and a 20% third harmonic component.* **(AMIE Winter, 1999)**

Solution. *Given* : $p = 10$, $f = 50$ Hz ; $q = 2$; number of conductors/slot = 4 ; coil span = 150°.

Induced emf/phase, $\mathbf{E_{ph}}$:

Number of slots/pole, $n = q \times$ no. of phases $= 2 \times 3 = 6$

Number of slots/phase $= q \times$ no. of poles $= 2 \times 10 = 20$

Number of conductor connected in series, $Z_{ph} = 20 \times 4 = 80$

Number of series turns/phase, $T_{ph} = \frac{Z_{ph}}{2} = \frac{80}{2} = 40$

Angular displacement between adjacent slots, $\beta = \frac{180°}{n} = \frac{180°}{6} = 30°$

Distribution factor, $$k_d = \frac{\sin\left(\frac{q\beta}{2}\right)}{q \sin\left(\frac{\beta}{2}\right)} = \frac{\sin\left(\frac{2 \times 30°}{2}\right)}{2 \sin\left(\frac{30°}{2}\right)} = 0.966$$

Coil span/pitch factor, $$k_p = \cos\left(\frac{\alpha}{2}\right) = \cos\left(\frac{180° - 150°}{2}\right) = 0.966$$

Induced e.m.f. per phase (*fundamental component*),

$$E_{ph-1} = 4.44 f \phi T_{ph} k_p k_d \text{ volts}$$
$$= 4.44 \times 50 \times 0.12 \times 40 \times 0.966 \times 0.966 = 994.4 \text{ V}$$

For third harmonic component of flux :

Distribution factor, $$k_{d_3} = \frac{\sin\left(\frac{qn\beta}{2}\right)}{q \sin\left(\frac{n\beta}{2}\right)} = \frac{\sin\left(\frac{2 \times 3 \times 30°}{2}\right)}{2 \sin\left(\frac{3 \times 30°}{2}\right)} = 0.707$$

Coil span/pitch factor, $$k_{p_3} = \cos\left(\frac{n\alpha}{2}\right) = \cos\left[\frac{3 \times (180° - 150°)}{2}\right] = 0.707$$

(where n stands for nth harmonic)

Frequency $= 3 \times f = 3 \times 50 = 150$ Hz

Flux/pole, $\phi_3 = \frac{1}{3} \times 0.12 \times \frac{20}{100} = 0.008$ Wb

Induced e.m.f. the phase (*Third harmonic component*),

$$E_{ph-3} = 4.44\, f_3\, \phi_3\, T_{ph}\, k_{p_3}\, k_{d_3} \text{ volts}$$

$$= 4.44 \times 150 \times 0.008 \times 40 \times 0.707 \times 0.707 = 106.53 \text{ V}$$

Induced e.m.f. per phase, $\mathbf{E_{ph}} = \sqrt{(E_{ph-1})^2 + (E_{ph-3})^2}$

$$= \sqrt{(994.4)^2 + (106.53)^2} = \mathbf{1000\ V.} \quad \textbf{(Ans.)}$$

Example 5.12. *The flux distribution curve of a smooth-core, 50 Hz generator is B = 10 sin θ + 2 sin 3θ + 2 sin 5θ + 2 sin 7θ kilo-gauss when θ is measured from the neutral axis. The pole pitch is 35 cm, the core length 32 cm, and stator coil span four-fifths of the pole pitch. Determine the equation for the e.m.f. induced in one turn, and its r.m.s. value.* **(Bharathiyar Univ. Nov., 1996)**

Solution. *Given* : $f = 50$ Hz ; Flux distribution curve : $B = 10 \sin\theta + 2\sin 3\theta + 2\sin 5\theta + 2\sin 7\theta$ kilo-gauss (θ measured from neutral axis) ; pole pitch = 32 cm = 0.32 m ; core length = 32 cm = 0.32 cm ; stator coil span = $\frac{4}{5}$ th of the pole pitch.

Equation for e.m.f. induced and its r.m.s. value :

Area of pole pitch $= 0.35 \times 0.32 = 0.112 \text{ m}^2$

Fundamental component of flux per pole,

$$\phi_1 = \text{Average flux density} \times \text{area}$$

$$= \frac{2}{\pi} \times 10 \times 1{,}000 \times 10^{-4} \times 0.112 = 0.0713 \text{ Wb}$$

$\left(\because \text{ Average value of flux density} = \frac{2}{\pi} \text{ peak value of flux density, and 1 kilo-gauss} = 10^3 \times 10^{-4} \text{ Wb/m}^2 \right)$

RMS value of fundamental voltage per conductor

$$= 1.11 \times 2f \times \phi_1 = 1.11 \times 2 \times 50 \times 0.0713 = 7.914 \text{ V}$$

Peak value of fundamental voltage per conductor

$$= \sqrt{2} \times 7.914 = 11.192 \text{ V}$$

Since harmonic conductor voltages are in proportion to their flux-densities, so

Third harmonic voltage (maximum value) $= 11.192 \times \frac{2}{10} = 2.24$ V

Fifth harmonic voltage (maximum value) $= 11.192 \times \frac{2}{10} = 2.24$ V

Seventh harmonic voltage (maximum value) $= 11.192 \times \frac{2}{10} = 2.24$ V

Coil span $= \frac{4}{5} \times 180° = 144°$

So chording angle, $\alpha = 180° - 144° = 36°$

Pitch factor for fundamental component, $k_{p_1} = \cos\frac{\alpha}{2} = \cos\frac{36°}{2} = 0.951$

Pitch factor for third harmonic component, $k_{p_3} = \cos\frac{3\alpha}{2} = \cos\left(\frac{3 \times 36°}{2}\right) = 0.588$

Pitch factor for fifth harmonic component, $K_{p_5} = \frac{\cos 5\alpha}{2} = \cos\left(\frac{5 \times 36°}{2}\right) = 0$

Pitch factor for seventh harmonic component, $K_{p_7} = \frac{\cos 7\alpha}{2} = \cos\left(\frac{7 \times 36°}{2}\right) = -0.588$

Since peak values of induced e.m.fs. in one turn = $2 \times K_{p_n}$ × peak value of induced e.m.f. per conductor,

Therefore, *peak values of induced e.m.fs. in one turn are* :

(*i*) For fundamental component, $E_{1\,max} = 2 \times 0.951 \times 11.192 = 21.3$ V

(*ii*) For third harmonic component, $E_{3\,max} = 2 \times 0.588 \times 2.24 = 2.63$ V

(*iii*) For 5th harmonic component, $E_{5\,max} = 2 \times 0 \times 2.24$ = zero

(*iv*) For 7th harmonic component, $E_{7\,max} = 2 \times (-0.588) \times 2.24 = 2.63$ V

Hence the equation for the e.m.f. induced in one turn is given as

$$\mathbf{e = 21.3 \sin \theta + 2.63 \sin 3\theta - 2.63 \sin 7\theta \text{ volts.}} \quad \textbf{(Ans.)}$$

RMS value of e.m.f. induced in one turn,

$$E_{rms} = \sqrt{\left(\frac{21.3}{\sqrt{2}}\right)^2 + \left(\frac{2.63}{\sqrt{2}}\right)^2 + \left(\frac{2.63}{\sqrt{2}}\right)^2} = \mathbf{15.29\ V.} \quad \textbf{(Ans.)}$$

5.10. SHAPE OF THE E.M.F. WAVE

The instantaneous value of the induced e.m.f. in a conductor moving across a magnetic field is proportional to the rate of cutting the flux at that instant. Hence the waveform of the induced e.m.f. is similar to the wave of the magnetic flux in the air-gap. When the distribution of the flux in the air-gap is sinusoidal, the induced e.m.f. waveform will also be sinusoidal. But the flux wave in practice is never *exactly* sinusoidal. Further harmonics do creep up due to various causes. The *pulsating flux tufts at the armature teeth also cause harmonics.*

The following methods are employed to obtain e.m.f. wave whose shape is very close approximation to a sine wave :

1. The pole shoes are *skewed.* The field form due to a well designed skewed pole-shoe is shown in Fig. 5.12 (*b*), while Fig. 5.12 (*a*) shows a skewed pole. This method of obtaining a very good shape for the e.m.f. wave is simple and interesting. Since the value of e.m.f. induced in a conductor is given by the expression,

$$e = Blv \text{ volts}$$

and since v (velocity) is assumed to be constant, either B (flux density) or l (active length of conductors) must vary according to a sine law. How the active length l varies is shown in Fig. 5.12 (*a*). As the conductor A moves relative to the poles, the length of conductor under the influence of a pole varies over a certain distance from instant to instant. Hence the pole-shoes should be given such a shape that the variation of the length l is according to a sine law.

2. The flux density under the gap can be made to vary according to a sine law if the gap length varies from the centre of the pole to the tips of the pole-shoes. The increase in the gap length is made almost *twice* and the ratio $\frac{\textit{pole arc}}{\textit{pole pitch}}$ is kept round about 0.67.

3. In turbo-alternators, where rotor is cylindrical in form (*i.e.*, there are no salient poles), the unslotted portion should be 0.3 times the pole pitch (The main poles are formed at unslotted portions).

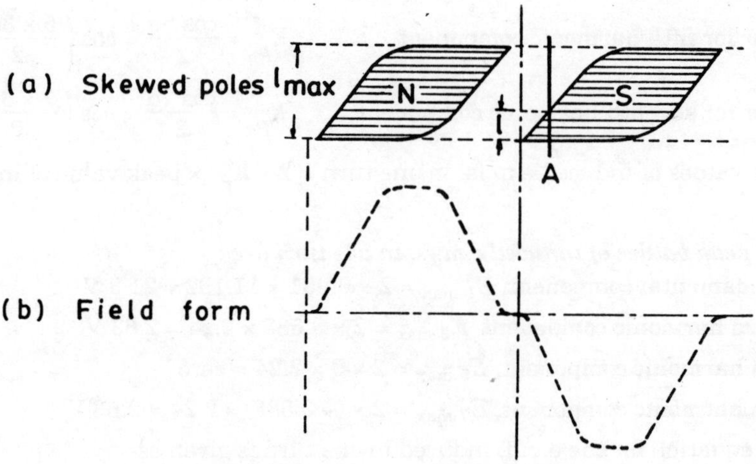

Fig. 5.12. (*a*) Skewed poles, (*b*) Field form.

The field windings are accommodated in several pairs of slots, the windings of each pair contribute towards the total m.m.f. The *resultant gap density variation approaches very near to a sine wave law*.

5.11. RATING OF ALTERNATORS

- The alternators are rated in *kVA* or *kW at specified power factor*.
- Other name plate datas include :
  (*i*) Voltage ; (*ii*) Current ;
  (*iii*) Frequency ; (*iv*) Speed ;
  (*v*) Number of phases ; (*vi*) Field ampere and voltage ;
  (*vii*) Maximum temperature rise.
- The prime-movers which drive the alternators, have a rating independent of power factor. It depends on kW output.

Example 5.13. *A 3-phase water-wheel generator is rated at 100 MVA, unity power factor, 11 kV star-connected 50 Hz, 120 r.p.m. Determine :*

(*i*) *The number of poles ;* (*ii*) *The kW rating ;*
(*iii*) *The current rating ;*
(*iv*) *The input at rated kW load if the efficiency is 97% (excluding the field loss) ;*
(*v*) *Prime mover torque applied to the generator shaft.*

Solution. *Given* : Rating : 100 MVA, unity p.f., 11 kV (star-connected) ;

N = 120 r.p.m., η = 97%.

(*i*) The number of poles, p :

$$\mathbf{p} = \frac{120f}{N} = \frac{120 \times 50}{120} = \mathbf{50.}\ \textbf{(Ans.)}$$

(*ii*) The kW rating :

$$\text{kW rating} = \text{kVA} \times \cos\phi = (100 \times 1000) \times 1 = \mathbf{10^5\ kW.}\ \textbf{(Ans.)}$$

(*iii*) The current rating, I :

$$\mathbf{I} = \frac{\text{kVA}}{\sqrt{3}\ E_L\ (\text{in kV})} = \frac{100 \times 10^3}{\sqrt{3} \times 11} = \mathbf{5248.6\ A.}\ \textbf{(Ans.)}$$

(*iv*) The input :

$$\text{Input} = \frac{\text{output}}{\eta} = \frac{100 \times 1}{0.97} = \mathbf{103.1\ MW.}\ \textbf{(Ans.)}$$

(*v*) **Prime-mover torque, T :**

$$T = \frac{\text{Input in watts}}{(2\pi N/60)} = \frac{103.1 \times 10^6 \times 60}{2\pi \times 120} = \mathbf{8.2 \times 10^6\ Nm.} \quad \textbf{(Ans.)}$$

5.12. ALTERNATOR ON LOAD

When load on an alternator varies, its terminal voltage also varies. This variation in terminal voltage is due to the following reasons :

(*i*) Voltage drop due to armature resistance,
(*ii*) Voltage drop due to armature leakage reactance X_L, and
(*iii*) Voltage drop due to armature reaction.

(*i*) **Armature Resistance.** The voltage drop caused by armature resistance per phase R_a is IR_a and is in phase with current I. This drop, however, is *practically negligible.*

Effective Resistance. The *effective resistance* of the armature winding is greater than the conductor resistance as measured by direct current. This is because additional energy over purely I^2R value is expended inside and sometimes outside of the conductor, owing to the alternating current.

The *chief sources of this additional energy loss are* :
(*i*) eddy currents in the surrounding material ;
(*ii*) magnetic hysteresis in the surrounding material ; and
(*iii*) eddy currents or unequal currents distribution in the conductor itself.

In many cases it is sufficiently accurate to measure the resistance of an armature by direct current and increase it to a fictitious value, called the *effective resistance*, which is large enough to take care of these extra losses. The exact value can vary widely from 1.25. to 1.75, or more × D.C. resistance, depending upon design. *Extreme accuracy is not necessary in this factor,* since the voltage change due to resistance is small in comparison to the changes produced by leakage reactance and armature reaction.

(*ii*) **Armature Leakage Reactance.** The load current, flowing through the armature winding, builds up local flux which on cutting the winding *generates a counter* (or *reactance*) *e.m.f.* This effect gives the armature a *reactance* which is numerically equal to $2\pi fL$. L, in henries, is the leakage inductance of the winding. This armature reactance is called the *leakage reactance* X_L (also known as Potier reactance X_P) since the flux *which causes it is around the armature turns only and does not effect the field flux directly. This leakage flux* (Fig. 5.13) is proportional to the armature current, since the magnetic path it covers is not normally saturated. X_L varies somewhat with the position of the armature and field poles. It is usually assumed to be constant.

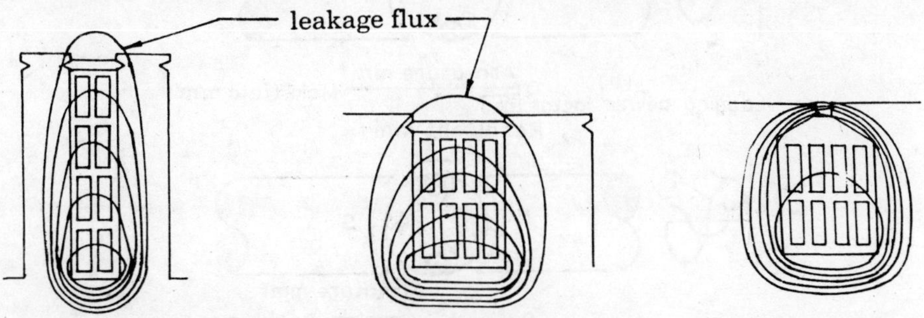

(*a*) Deep, narrow, open slot. (*b*) Shallow, wide, open slot. (*c*) Shallow, wide, partially closed slot.

Fig. 5.13. Slot-leakage flux.

The voltage drop due to X_L is IX_L. A part of generated e.m.f. is used up in overcoming this voltage drop (reactance e.m.f.).

$$\therefore \qquad E = V + I(R_a + jX_L) \qquad ...(5.9)$$

The fact is illustrated in Fig. 5.14.

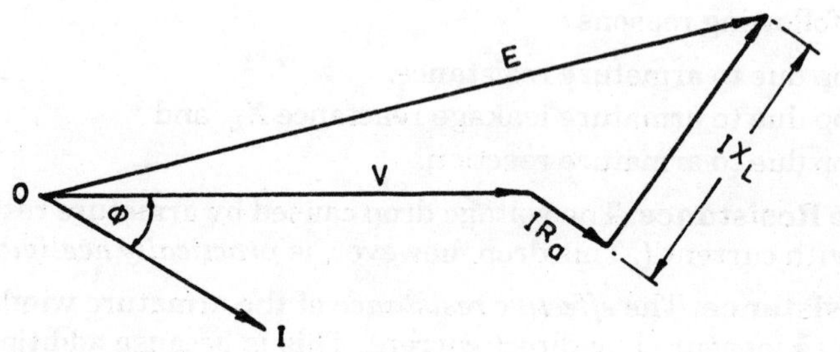

Fig. 5.14

(iii) **Armature Reaction.** Armature reaction is the magneto-motive force produced by the armature currents in the armature conductors.

Unity Power Factor. At unity power factor, armature reaction consists of distortion of the main field flux.

Lagging Power Factor. When an alternator operates with a lagging power factor a m.m.f. is set up by the current in the armature conductors which opposes the m.m.f of the main field and

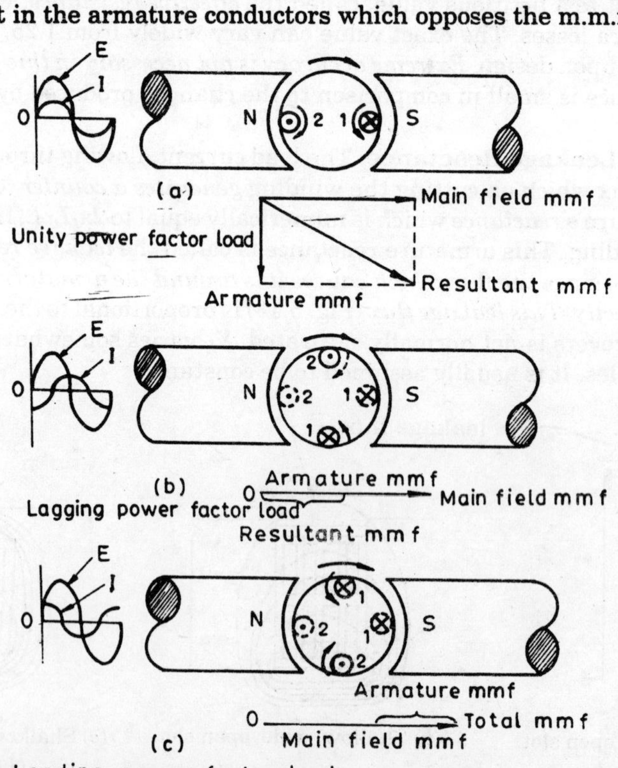

Fig. 5.15. Effects of armature reaction.

causes a *decrease in the main flux*. This, in turn, results in decrease in the induced voltage. The *lower the value of lagging power factor*, the *greater will be the armature m.m.f. which will oppose and weaken the field*.

Leading Power Factor. In those limited applications where an alternator supplies a load with a leading power factor, the armature current sets up m.m.f. in the armature which aids the m.m.f. of the main field and *causes the main field flux to increase*. Therefore, the voltage of the alternator will increase with an increase in load current. The *lower the value of the leading power factor, the greater will be the armature m.m.f. aiding the m.m.f. of the main field to cause an increase in the main field flux*.

The following illustration in Fig. 5.15 shows how the main field flux is affected by the armature m.m.f. for different load power factors. For ease of illustration, a two-pole rotating type alternator is used. However, the same conditions take place in the revolving field type alternator.

A study of illustrations in Fig. 5.15 and vector diagrams for different power factor loads, show that the *voltage drop due to inductive reactance and armature reaction are the same in their effect on the terminal voltage*. Both of these effects are proportional to the armature current.

Synchronous Reactance. The leakage reactance X_L (or X_p) and the armature reactance X_a may be combined to give *'synchronous reactance'* X_s.

Hence, $X_s = X_L + X_a$

[The ohmic value of X_a varies with the power factor of the load because armature reaction depends on load power factor].

From the above discussion, it is clear that for the same field excitation, terminal voltage is decreased from its no-load value E_0 to V (for a lagging power factor). This is because of :

(*i*) Drop due to armature resistance IR_a.

(*ii*) Drop due to synchronous reactance IX_s.

Therefore, *total voltage drop* in an alternator under load is

$$= IR_a + jIX_s = I(R_a + jX_s) = IZ_s$$

where Z_s is known as *'synchronous impedance'*. (The word *synchronous simply refers to the working conditions*)

Fig. 5.16 shows the vector diagram for the above fact.

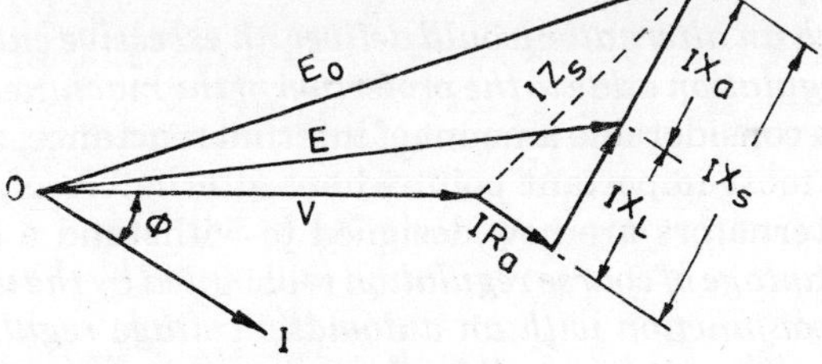

Fig. 5.16

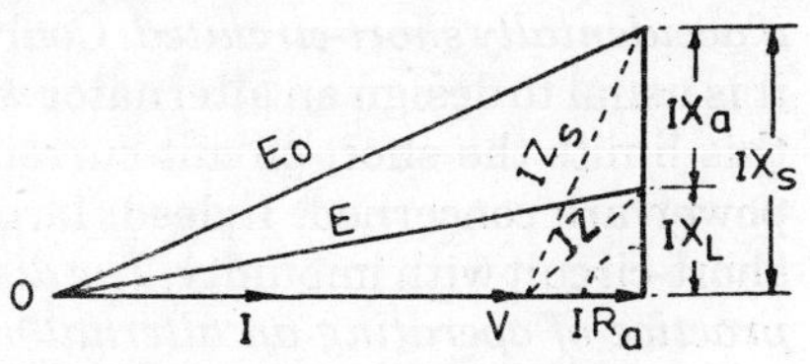

Fig. 5.17. Unity power factor.

Vector Diagrams of a Loaded Alternator. The vector diagrams for unity *power factor, for lagging power factor* and for *leading power factor* have been shown in Figs. 5.17, 5.18 and 5.19, respectively. All these diagrams apply to one phase of a 3-phase machine. Diagrams for the other phases can also be drawn in the same way.

In all these diagrams,

E_0 = no-load e.m.f. This being the voltage induced in armature in the absence of IR_a, IX_L and IX_a. Hence it represents the maximum value of induced e.m.f.

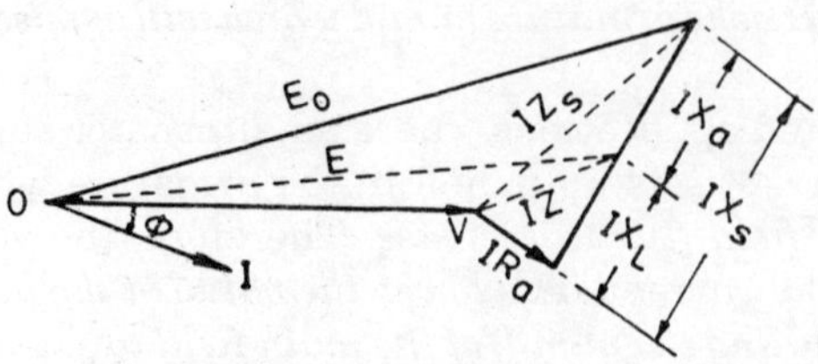

Fig. 5.18. Lagging power factor.

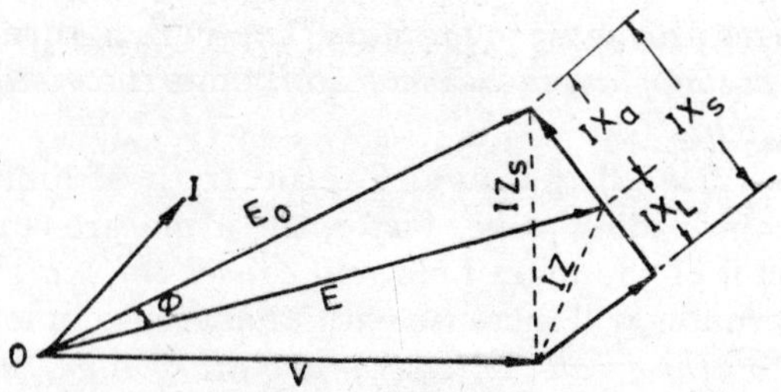

Fig. 5.19. Leading power factor.

E = it is the induced e.m.f. after allowing for armature reaction. E is vectorially less than E_0 by IX_a

V = terminal voltage. It is vectorially less than E by IZ where $Z = \sqrt{R^2 + X_L^2}$

I = armature current/phase and

ϕ = load power factor angle.

5.13. VOLTAGE REGULATION

- When an alternator is subjected to a varying load, the voltage at the armature terminals varies to a certain extent, the amount of this variation determines the *regulation* of the machine. The numerical value of *regulation* is defined as *the percentage rise in voltage when full-load at the specified power-factor is switched off, the excitation being adjusted initially to give normal voltage.* Thus,

$$\% \text{ regulation 'up'} = \frac{E_0 - V}{V} \times 100 \quad \text{...(5.10)}$$

This gives a lower figure than the percentage drop in voltage when full-load is *switched on,* because in the latter case the excitation is only that required for normal voltage on open circuit. As actually defined, the excitation is greater, since it is that required to give normal voltage on full-load, and owing to the greater saturation of magnetic circuit in these circumstances, the rise in voltage when the load is switched off is less than the drop in voltage when the load is switched on.

- A normal alternator has a regulation of about 8 to 10 percent at unity power factor, but the *voltage rise is considerably increased at lagging power-factors.* At 0.8 lagging power-factor the value lies between 25 and 35 per cent, or even more.
- *Close regulation is not desired, since such an alternator would deliver an excessive current if accidentally short-circuited. Coarse regulation adds to the protection of the machine,* and it is usual to design an alternator with a considerable amount of internal reactance, since this limits the short-circuit current, a most important point where alternators of high power are concerned. Indeed, large alternators are now designed to withstand a dead short-circuit with impunity. *The disadvantage of coarse regulation is obviated by the usual practice of operating an alternator in conjunction with an automatic voltage regulator, which maintains an approximately constant voltage at all loads.*

5.13.1. Determination of Voltage Regulation. It is not usually possible or desirable to measure the regulation by direct experiment, indirect methods being adopted which do not necessitate the loading of the alternator. For this purpose, open circuit and short-circuit characteristics are required.

The following *indirect methods* are used to determine voltage regulation :

1. Synchronous impedance or E.M.F. method.
2. The Ampere-turn or M.M.F. method.
3. Zero power-factor or Potier method.

All these methods require :

(*i*) Armature (or stator) resistance R_a.

(*ii*) Open-circuit/no-load characteristic.

(*iii*) Short-circuit characteristic (but zero power factor lagging characteristic for Potier method).

(*i*) **Value of R_a.** Armature resistance R_a per phase can be measured directly by voltmeter and ammeter method or by using wheatstone bridge. The effective value of R_a, however, under working conditions is increased due to skin effect. The value of R_a so obtained is increased by sixty per cent or so to allow this effect. *Generally, a value 1.6 times D.C. value is taken.*

(*ii*) **Open-circuit Characteristic.** This type of characteristic (as in D.C. machines) is plotted by running the machine on no-load and noting the values of induced voltage and field excitation current.

(*iii*) **Short-circuit Characteristic.** Short-circuit characteristic is obtained by short-circuiting the armature (*i.e.*, stator) windings through a low resistance ammeter. The excitation is so adjusted as to give 1.5 to 2 times the value of full-load current. *The speed* (not necessarily synchronous) *is kept constant during the test.*

5.13.1.1. Synchronous Impedance Method. This method involves the following **steps** :

(*i*) Plot the open-circuit characteristic (O.C.C.) from a given data as shown in Fig. 5.20.

(*ii*) Plot short-circuit characteristic (S.C.C.) from the data given by short-circuit test.

Both these curves are drawn on a *common field-current base.*

Consider a field current I_f. Corresponding to this field current the O.C. voltage is E_1. When winding is short-circuited, the terminal voltage is zero. Hence, it may be assumed that the whole of this voltage E_1 is being used to circulate the armature short-circuit current I_1 against the synchronous impedance Z_s.

$$\therefore \quad E_1 = I_1 Z_s$$

or
$$Z_s = \frac{E_1 \text{ (open-circuit)}}{I_1 \text{ (short-circuit)}}$$

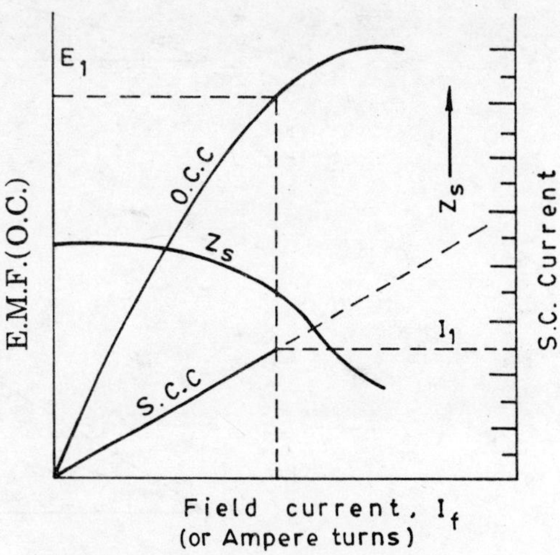

Fig. 5.20. O.C. and S.C. test curves of an alternator.

(*iii*) Find synchronous reactance X_s as follows :

$$X_s = \sqrt{Z_s^2 - R_a^2}$$

(R_a can be found as discussed earlier)

(*iv*) After finding R_a and X_s vector diagrams for any load any power factor may be drawn. Three cases are considered in Fig. 5.21, (*a*) *unity power factor*, (*b*) *lagging power factor*, and (*c*) *leading power factor*.

- **Unity power factor.** Refer Fig. 5.21 (*a*).

$$OC^2 = (OA + AB)^2 + BC^2$$

$$E_0^2 = (V + IR_a)^2 + (IX_s)^2$$

$$\therefore \quad E_0 = \sqrt{(V + IR_a)^2 + (IX_s)^2} \quad \ldots(5.11)$$

- **Lagging power factor.** Refer Fig. 5.21 (*b*).

$$OC^2 = (OA + AB)^2 + (BD + DC)^2$$

$$E_0^2 = (V \cos \phi + IR_a)^2 + (V \sin \phi + IX_s)^2$$

$$\therefore \quad E_0 = \sqrt{(V \cos \phi + IR_a)^2 + (V \sin \phi + IX_s)^2} \qquad ...(5.12)$$

- **Leading power factor.** Refer Fig. 5.21 (*c*).

$$OC^2 = (OF + FD)^2 + (BD - BC)^2$$

$$E_0^2 = (V \cos \phi + IR_a)^2 + (V \sin \phi - IX_s)^2$$

$$\therefore \quad E_0 = \sqrt{(V \cos \phi + IR_a)^2 + (V \sin \phi - IX_s)^2} \qquad ...(5.13)$$

The procedure is similar to that adopted in the case of transformers. The *approximate method* of calculating the total drop as was used for transformers *must not be used for alternators* where the *magnitudes involved are much greater in percentages.*

The results obtained by this method are *too high,* owing to the fact that the synchronous impedance determined at short-circuit condition is too *large due to a very low degree of saturation.* Hence this method is called the **'pessimistic'** method.

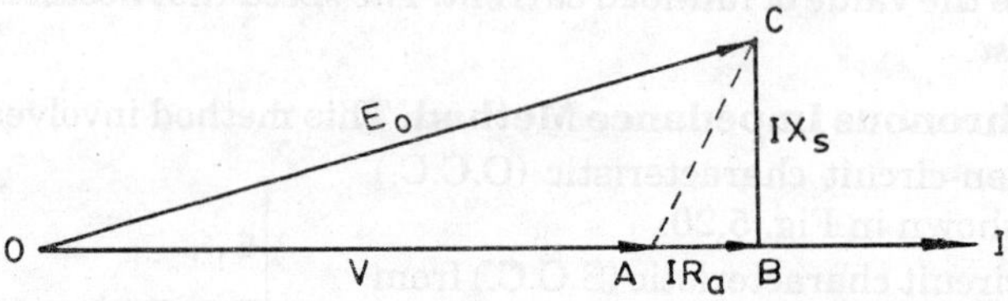

(*a*) Unity power factor.

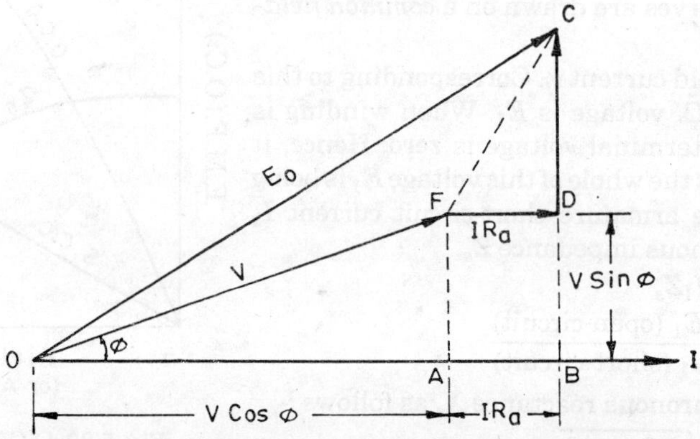

(*b*) Lagging power factor.

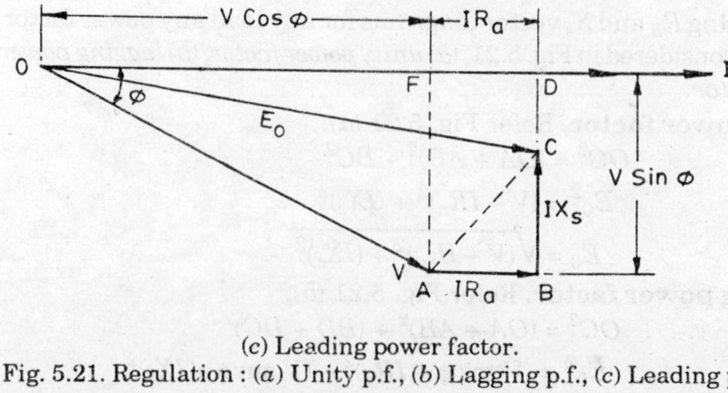

(*c*) Leading power factor.

Fig. 5.21. Regulation : (*a*) Unity p.f., (*b*) Lagging p.f., (*c*) Leading p.f.

Assumptions inherent in the synchronous impedance method. The assumptions are as follows :

(*i*) *The effect of the armature-reaction flux can be replaced by a voltage drop proportional to the armature current*. The substitution of voltage for flux is the reason that the synchronous-impedance method is also called *e.m.f. method*.

(*ii*) *Since the voltage caused by the main-field flux is added vectorially to that caused by the armature flux, it is also assumed that both fluxes have sinusoidal distribution*. Little error is introduced because of this with non-salient-pole machines, but the error is much greater in the case of the salient-pole ones which have concentrated field windings.

(*iii*) *The magnetic reluctance to the armature flux is constant regardless of the power factor*. This is substantially true for a non-salient, or round-rotor machine, whose air gap is almost constant but introduces considerable error with salient poles, since the position of the armature flux relative to the field poles is determined by the power factor.

(*iv*) There is no saturation effect.

(*v*) The flux under test conditions is the same as that under load conditions.

Example 5.14. (*a*) *Compare the constructional features of hydel and turbo-generators.*

(*b*) *A 200 kVA, 480 V, 50 Hz Y-connected synchronous generator with a rated field current of 5 A was tested, and the following data were obtained :*

— *Open circuit terminal voltage 540 V (line-to-line)*

— *Short circuit current 300 A.*

When a d.c. voltage of 10 V was applied to two of its terminals, a current of 25 A was measured. Find the value of the synchronous reactance. **(AMIE Summer, 1998)**

Solution. (*a*) A comparison of constructional features of hydel and turbo-generators in a tabular form is given below :

Constructional features of hydel and turbo-generators :

S. No.	*Aspects*	*Hydel generators*	*Turbo-generators*
1.	*Numbers of poles*	A large numbers of poles	Only two poles
2.	*Speed*	Low	High (normally 3000 r.p.m.)
3.	*Pole construction*	Salient	Smooth rotor construction
4.	*Length, diameter*	Less, more	Large, small
5.	*Cooling*	Indirect	Direct
6.	*Rating*	Less, lower rated comparatively	11 kV, upto 550 MW

(*b*) *Given* : Rating : 200 kVA, 480 V, I_f = 5 A, O.C. terminal voltage = 540 V (line-to-line) ;
S.C. current = 300 A ; D.C. voltage = 10 V, current = 25 A.

Synchronous reactance, $\mathbf{X_s}$:

For an alternator, $$Z_s = \frac{\text{open-circuit voltage/phase}}{\text{short-circuit current}} = \frac{540/\sqrt{3}}{300} = 1.04\ \Omega$$

Resistance, $$R_a = \frac{\text{D.C. voltage}}{\text{current}} = \frac{10}{25} = 0.4\ \Omega$$

$$\therefore \quad \mathbf{X_s} = \sqrt{Z_s^2 - R_a^2} = \sqrt{1.04^2 - 0.4^2} = \mathbf{0.96\ \Omega}. \quad \textbf{(Ans.)}$$

Example 5.15. *A 600 V, 60 kVA, single-phase alternator has an effective resistance of 0.2 Ω. A field current of 10 A produces an armature current of 210 A on short-circuit and an e.m.f. of 480 V on open-circuit. Calculate :*

(*i*) *Synchronous impedance and reactance.*

(*ii*) *Full-load regulation with 0.8 p.f. lagging.*

Solution. Terminal voltage, $V = 600$ V
Alternate output in kVA $= 60$ kVA
Armature effective resistance, $R_a = 0.2\ \Omega$
On short-circuit : Field current $= 10$ A
Armature current $= 210$ A
Open-circuit voltage $= 480$ V.

(*i*) **Synchronous impedance and reactance :**

Synchronous impedance,

$$Z_s = \frac{\text{open-circuit voltage on short-circuit}}{\text{short-circuit current}}$$

$$= \frac{480}{210} = \mathbf{2.28\ \Omega.} \quad \textbf{(Ans.)}$$

∴ **Synchronous reactance,**

$$X_s = \sqrt{Z_s^2 - R_a^2} = \sqrt{(2.28)^2 - (0.2)^2} = \mathbf{2.27\ \Omega.} \quad \textbf{(Ans.)}$$

(*ii*) **Full-load regulation with 0.8 p.f. lagging :**

Refer Fig. 5.22.

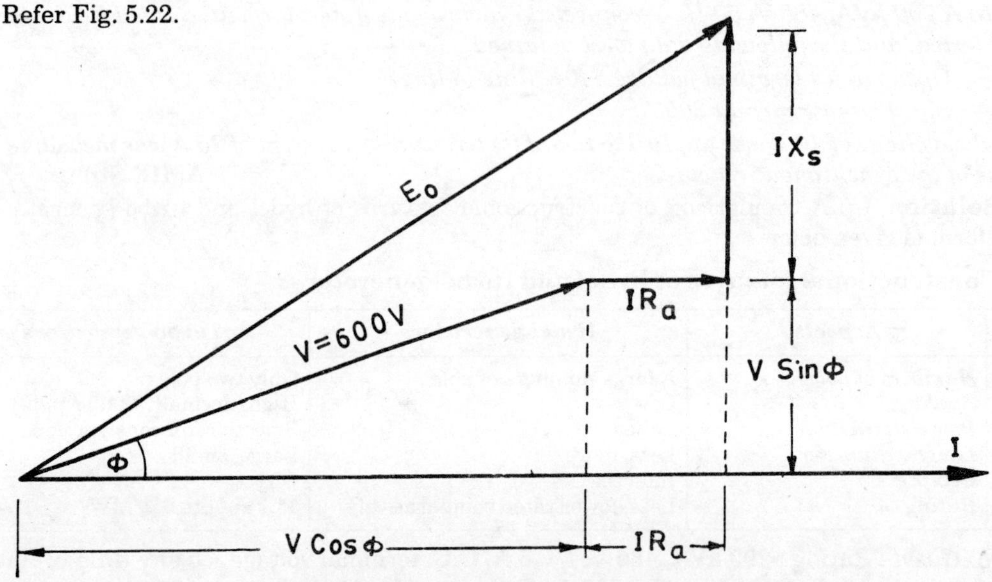

Fig. 5.22

$$E_0 = \sqrt{(V \cos \phi + IR_a)^2 + (V \sin \phi + IX_s)^2}$$

Now $\cos \phi = 0.8$ and $\sin \phi = 0.6$

Full-load current $= \dfrac{60 \times 1000}{600} = 100$ A

∴ $$E_0 = \sqrt{(600 \times 0.8 \times + 100 \times 0.2)^2 + (600 \times 0.6 + 100 \times 2.27)^2}$$

$$= \sqrt{(480 + 20)^2 + (360 + 227)^2} = 771 \text{ V}$$

∴ Percentage regulation $= \dfrac{E_0 - V}{V} = \dfrac{771 - 600}{600} \times 100 = \mathbf{28.5\%.}$ **(Ans.)**

Example 5.16. *A 3-phase star-connected alternator is rated at 1500 kVA, 12000 V. The armature effective resistance and synchronous reactance are 2 Ω and 35 Ω respectively per phase. Calculate the percentage regulation for a load of 1200 kW at power factors of :*

(i) 0.8 lagging *(ii) 0.8 leading.*

Solution. Terminal voltage, $E_L = 12000$ V

Alternator rating $= 1500$ kVA

Load $= 1200 \text{ kW} = \dfrac{1200}{0.8} = 1500$ kVA

Armature effective resistance/phase, $R_a = 2\ \Omega$

Synchronous reactance/phase, $X_s = 35\ \Omega$

Phase voltage, $E_{ph}\ (= V) = \dfrac{12000}{\sqrt{3}} = 6928$ V.

(*i*) **When power factor is 0.8 lagging :**

$$\cos \phi = 0.8 \text{ and } \sin \phi = 0.6$$

∴ Open-circuit voltage/phase,

$$E_0 = \sqrt{(V \cos \phi + IR_a)^2 + (V \sin \phi + IX_s)^2}$$

Now load current at a load of 1200 kW, 0.8 p.f.

$$= \frac{1200 \times 1000}{0.8 \times \sqrt{3} \times 12000} = 72.2 \text{ A}$$

$$\therefore \quad E_0 = \sqrt{(6928 \times 0.8 + 72.2 \times 2)^2 + (6928 \times 0.6 + 72.2 \times 35)^2}$$

$$= \sqrt{(5686.8)^2 + (6683.8)^2} = 8775.7 \text{ V}$$

∴ **Percentage regulation**

$$= \frac{E_0 - V}{V} \times 100 = \frac{8775.7 - 6928}{6928} \times 100 = \mathbf{26.67\%}. \quad \textbf{(Ans.)}$$

(*ii*) **When power factor is 0.8 leading :**

$$\cos \phi = 0.8 \text{ and } \sin \phi = 0.6$$

Open-circuit voltage/phase,

$$E_0 = \sqrt{(V \cos \phi + IR_a)^2 + (V \sin \phi - IX_s)^2}$$

$$= \sqrt{(6928 \times 0.8 + 72.2 \times 2)^2 + [6928 \times 0.6 - 72.2 \times 35]^2}$$

$$= \sqrt{(5686.8)^2 + (1629.8)^2} = 5915.7 \text{ V}$$

∴ **Percentage regulation**

$$= \frac{E_0 - V}{V} \times 100 = \frac{5915.7 - 6928}{5915.7} \times 100 = -\mathbf{17.11\%}. \quad \textbf{(Ans.)}$$

Example 5.17. *A 3300 V, 3-phase star-connected alternator has a full load current of 100 A. On short-circuit a field current of 5 amperes was necessary to produce full-load current. The e.m.f. on open-circuit for the same excitation was 900 volts. The armature resistance was 0.8 Ω/phase. Determine the full-load voltage regulation for :*

(i) 0.8 p.f. lagging *(ii) 0.8 p.f. leading.*

Solution. Rated voltage (line) = 3300 V

Rated voltage (phase) $= \dfrac{3300}{\sqrt{3}} = 1905$ V

E.m.f. on open-circuit corresponding to a 5 A excitation = 900 V

Short-circuit current for the same excitation = 100 A

∴ Synchronous impedance/phase,

$$Z_s = \frac{900}{100} = 9\ \Omega$$

Resistance/phase, $R_a = 0.8\ \Omega$

Full-load current, $I = 100$ A

∴ Synchronous reactance, $X_s = \sqrt{Z_s^2 - R_a^2} = \sqrt{9^2 - (0.8)^2} = 8.96\ \Omega$

(*i*) **At 0.8 p.f. lagging.** Refer Fig. 5.23.

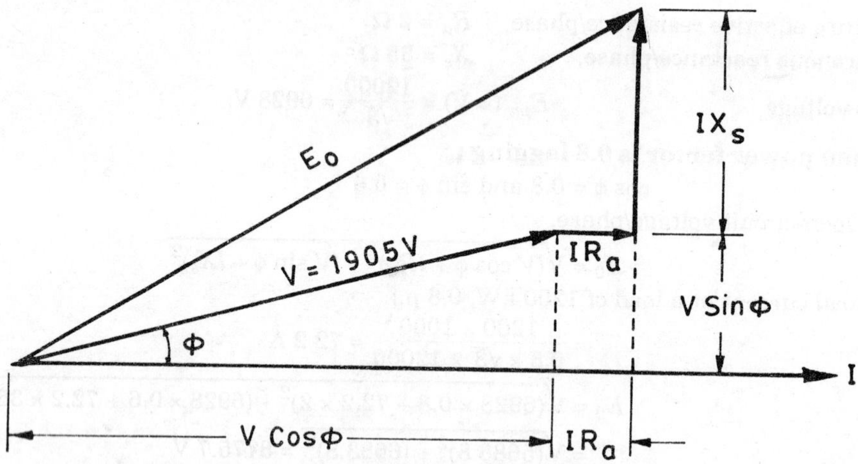

Fig. 5.23

$$E_0 = \sqrt{(V \cos \phi + IR_a)^2 + (V \sin \phi + IX_s)^2}$$

$$= \sqrt{(1905 \times 0.8 + 100 \times 0.8)^2 + (1905 \times 0.6 + 100 \times 8.96)^2}$$

$$= \sqrt{(1604)^2 + (2039)^2} = 2594.3 \text{ V.}$$

∴ **Percentage regulation** $= \frac{E_0 - V}{V} = \frac{2594.3 - 1905}{1905} \times 100 =$ **36.18%. (Ans.)**

(*ii*) **At 0.8 p.f. leading.** Refer Fig. 5.24.

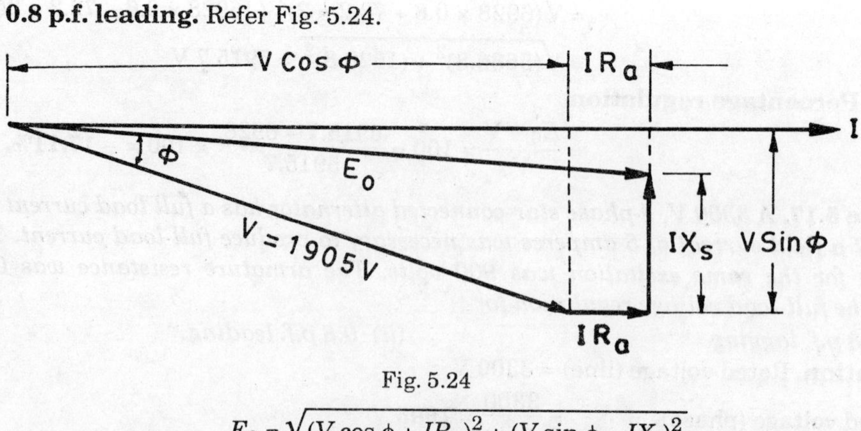

Fig. 5.24

$$E_0 = \sqrt{(V \cos \phi + IR_a)^2 + (V \sin \phi - IX_s)^2}$$

$$= \sqrt{(1905 \times 0.8 + 100 \times 0.8)^2 + (1905 \times 0.6 - 100 \times 8.96)^2}$$

$$= \sqrt{1604)^2 + (247)^2} = 1622.9 \text{ V}$$

$\therefore$ **% Regulation** $= \dfrac{E_0 - V}{V} \times 100 = \dfrac{1622.9 - 1905}{1905} \times 100 = -$ **14.8%. (Ans.)**

Example 5.18. *A 3.3 kV, 3-phase star-connected alternator has a full-load current of 100 A. Under short-circuit condition it takes 5 A field current to produce full-load short-circuit current. The e.m.f. on open circuit for the same excitation is 900 V (line-to-line). The armature resistance is 0.9 Ω/phase. Determine synchronous reactance per phase and regulation for (i) 0.8 p.f. lagging, (ii) 0.8 p.f. leading.* **(AMIE Winter, 1996)**

Solution. Terminal voltage/phase on O.C. $= \dfrac{900}{\sqrt{3}} = 519.6$ V

Now, synchronous impedance, $Z_s = \dfrac{\text{O.C. voltage}}{\text{S.C. current}}$ = for the same field excitation

$$= \frac{519.6}{100} = 5.196\ \Omega$$

$$\therefore \quad X_s = \sqrt{Z_s^2 - R_a^2} = \sqrt{5.196^2 - 0.9^2} = 5.117\ \Omega$$

Percentage regulation :

(i) At 0.8 p.f. lagging :

Given rated phase voltage $= \dfrac{3.3 \times 1000}{\sqrt{3}} = 1905.3$ V

$$E_0 = \sqrt{(V\cos\phi + IR_a)^2 + (V\sin\phi + IX_s)^2}$$

$$= \sqrt{(1905.3 \times 0.8 + 100 \times 0.9)^2 + (1905.3 \times 0.6 + 100 \times 5.117)^2}$$

$$= \sqrt{(1614.24)^2 + (1654.88)^2} = 2311.8 \text{ V}$$

$\therefore$ **Percentage regulation** $= \dfrac{E_0 - V}{V} = \dfrac{2311.8 - 1905.3}{1905.3} \times 100 =$ **21.33%. (Ans.)**

(ii) At 0.8 p.f. leading :

$$E_0 = \sqrt{(V\cos\phi + IR_a)^2 + (V\sin\phi - IX_s)^2}$$

$$= \sqrt{(1905.3 \times 0.8 + 100 \times 0.9)^2 + (1905.3 \times 0.6 - 100 \times 5.117)^2}$$

$$= \sqrt{(1614.24)^2 + (631.48)^2} = 1733.4 \text{ V}$$

$\therefore$ **Percentage regulation** $= \dfrac{E_0 - V}{V} = \dfrac{1733.4 - 1905.3}{1905.3} \times 100 = -$ **9.02%. (Ans.)**

Example 5.19. *A 3-phase, star-connected, 1000 kVA, 11000 V alternator has rated current of 52.5 A. The A.C. resistance of the winding per phase is 0.45 Ω. The test results are given below :*

O.C. test : Field current = 12.5 A, voltage between lines = 422 V

S.C. test : Field current = 12.5 A, line current = 52.5 A.

Determine the full-load voltage regulation of the alternator (a) 0.8 p.f. lagging and (b) 0.8 p.f. leading. **(Madras Univ. April, 1998)**

Solution. *Given* : Rating : 1000 kVA, 11000 V ; $I = 52.5$ A ; $R_a = 0.45\ \Omega$;

$I_{ph} = I_L = 52.5$ A, ($\because$ alternator is star-connected) ; O.C. line voltage = 422 V.

Full-load voltage regulation :

Synchronous impedance, $Z_s = \dfrac{\text{O.C. phase voltage}}{\text{S.C. current per phase}} = \dfrac{(422/\sqrt{3})}{52.5} = 4.64\ \Omega$

$\therefore$ Synchronous reactance, $X_s = \sqrt{Z_s^2 - R_a^2} = \sqrt{4.64^2 - 0.45^2} = 4.62\ \Omega$

$$E_{ph}(= V) = \frac{11000}{\sqrt{3}} = 6351 \text{ V}$$

(*a*) *When power factor is 0.8 lagging :*

$$E_0 = \sqrt{(V\cos\phi + IR_a)^2 + (V\sin\phi + IX_s)^2}$$
$$= \sqrt{(6351 \times 0.8 + 52.5 \times 0.45)^2 + (6351 \times 0.6 + 52.5 \times 4.62)^2}$$
$$= \sqrt{(5104.4)^2 + (4053.15)^2} = 6518 \text{ V}$$

$\therefore$ **Percentage regulation** $= \dfrac{E_0 - V}{V} \times 100 = \dfrac{6518 - 6351}{6351} \times 100 =$ **2.63%. (Ans.)**

(*b*) *When power factor is 0.8 leading :*

$$E_0 = \sqrt{(V\cos\phi + IR_a)^2 + (V\sin\phi - IX_s)^2}$$
$$= \sqrt{(6351 \times 0.8 + 52.5 \times 0.45)^2 + (6351 \times 0.6 - 52.5 \times 4.62)^2}$$
$$= \sqrt{(5104.4)^2 + (3568.05)^2} = 6228 \text{ V}$$

$\therefore$ **Percentage regulation** $= \dfrac{E_0 - V}{V} \times 100 = \dfrac{6228 - 6351}{6351} \times 100 =$ **– 1.94%. (Ans.)**

Example 5.20. (*a*) *Develop the terms generally used for reducing harmonic voltages from the induced e.m.f.*

(*b*) *A 3-phase, 50 Hz, 6000 kVA, 6000 V star-connected alternator has an effective resistance of 0.2 Ω. A field current of 10 A produces 480 V on open circuit and a field current of 5 A gives armature current of 105 A.*

Calculate the voltage regulation of this alternator at 0.8 power for lagging.

(AMIE Summer, 1996)

Solution. (*a*) For reducing the harmonics from the induced e.m.f. the winding is to be designed properly. The different methods of reducing harmonics are :

1. **Distribution.** The magnitude of harmonic e.m.fs. depends upon their distribution factors. *The distribution factor is small for higher order harmonics.*

2. **Chording.** The e.m.f. generated is proportional to $\cos\left(\dfrac{n\alpha}{2}\right)$ where α is the angle of chording and n is the order of harmonic. *The harmonic e.m.f. can therefore be considerably reduced by choosing a proper value of* α.

3. **Skewing.** The *slot harmonics can be eliminated by skewing the pole face.*

4. **Fractional slot winding.** The *harmonic e.m.fs. can be considerably reduced by using fractional slot windings on account of the fact that these windings give a much smaller distribution factor.*

5. **Large length of air gap.** By *using large air gap length, the reluctance is increased and therefore the magnitude of slot harmonics is reduced.*

(*b*) *Given* : Rating : 6000 kVA, 6000 V ; $R_a = 0.2\ \Omega$; $\cos\phi = 0.8$

Full-load current, $I = \dfrac{6000 \times 1000}{\sqrt{3} \times 6000} = 577.35$ A

As the O.C. voltage of 480 V appears at a field current of 10 A, hence the field current of 10 A under S.C. test will give an armature current of 105 × 2 = 210 A.

Phase voltage $= \dfrac{6000}{\sqrt{3}} = 3464$ V

Synchronous impedance, $Z_s = \dfrac{\text{O.C. voltage/phase}}{\text{S.C. current/phase}} = \dfrac{(480/\sqrt{3})}{210} = 1.32\ \Omega$

$\therefore$ $X_s = \sqrt{{Z_s}^2 - {R_a}^2} = \sqrt{1.32^2 - 0.2^2} = 1.305\ \Omega$

Now, $E_o = \sqrt{(V\cos\phi + IR_a)^2 + (V\sin\phi + IX_s)^2}$

$= \sqrt{(3464 \times 0.8 + 577.35 \times 0.2)^2 + (3464 \times 0.6 + 577.35 \times 1.305)^2}$

$= \sqrt{(2886.67)^2 + (2831.84)^2} = 4043.8$ V

$\therefore$ Percentage regulation $= \dfrac{4043.8 - 3464}{3464} \times 100 =$ **16.74%. (Ans.)**

Example 5.21. (*a*) *Why the modern alternators are designed with a high ratio of armature synchronous reactance to resistance ratio ? Why in addition external reactors are provided in series with each phase of an alternator ?*

(*b*) *A 3-phase, Y-connected, 11 kV, 165 MVA alternator has a synchronous reactance of 1 ohm and armature resistance of 0.1 ohm per phase. Calculate short circuit current at the instant of short circuit and sustained short circuit current.*

(*c*) *The data obtained on 100 kVA, 1100 V, 3-phase alternator in d.c. resistance test,*

E between lines = *6 V d.c.*
I in lines = *10 A d.c.*
Open circuit test, Field current = *12 A d.c.*
Line voltage = *420 V a.c.*
Short circuit test, Field current = *12.5 A d.c.*
Line current = *rated value*

Calculate the voltage regulation of alternator at 0.8 p.f. lagging. **(AMIE Winter, 1993)**

Solution. (*a*) Modern alternators are designed with larger values of reactance so as to protect them in the event of a short circuit near the alternator terminals. As a consequence their voltage regulation *is inherently poor*. But this problem can be *overcome by providing automatic voltage regulators*.

When large number of alternators operate in parallel, the fault current increases considerably when a short circuit occurs in the system. In order to limit the fault current *external reactors are connected in series with the generators*. Again due to expansion of the existing power system the fault current may become so large that the existing circuit breakers may not be able to clear this fault current. *By adding reactors it may become possible to use the existing circuit breakers.*

(*b*) The value of the short circuit current at the instant of fault *depends upon the value of voltage at the instant and the value of subtransient reactance*. Since nothing is stated, it is assumed that *sub-transient is same as synchronous reactance.*

Synchronous impedance $= \sqrt{{R_a}^2 + {X_s}^2} = \sqrt{(1)^2 + (0.1)^2}$

$Z_s = 1.005\ \Omega$

S.C. current at the instant of fault

$$= 2\sqrt{2}\,\frac{E_{ph}}{Z_s} = \frac{2\sqrt{2} \times \frac{1000}{\sqrt{3}}}{1.005} = \textbf{17873.56 A. (Ans.)}$$

Sustained S.C. current $= \dfrac{E_{ph}}{Z_s} = \dfrac{\frac{11000}{\sqrt{3}}}{1.005} =$ **6319.26 A. (Ans.)**

(*c*) Assuming the alternator to be star-connected.

Rated current $= \dfrac{100 \times 1000}{\sqrt{3} \times 1100} = 52.48$ A

Armature resistance per phase $= \dfrac{6}{2 \times 10} = 0.3\ \Omega$ (The D.C. voltage is across the two phases)

Synchronous impedance, $Z_s = \dfrac{\text{O.C. voltage/phase}}{\text{S.C. current}}$, for same excitation

$$= \frac{(420/\sqrt{3})}{52.48} = 4.62\ \Omega$$

Synchronous reactance, $X_s = \sqrt{Z_s^2 - R_a^2} = \sqrt{4.62^2 - 0.3^2} = 4.61\ \Omega$

$$E_0 = \sqrt{(V\cos\phi + IR_a)^2 + (V\sin\phi + IX_s)^2}$$

$$= \sqrt{\left(\frac{1100}{\sqrt{3}} \times 0.8 + 52.48 \times 0.3\right)^2 + \left(\frac{1100}{\sqrt{3}} \times 0.6 + 52.48 \times 4.61\right)^2}$$

$$= \sqrt{(523.81)^2 + (622.98)^2} = 813.93\ \text{V}$$

$\therefore$ Percentage regulation $= \dfrac{813.93 - 1100(\sqrt{3})}{(1100/\sqrt{3})} \times 100 =$ **28.16. (Ans.)**

Example 5.22. *A 1500 kVA, 6600-V, 3-phase star-connected alternator with a resistance of 0.4 ohm and reactance of 6 ohms per phase, delivers full-load current at power factor 0.8 lagging, and normal rated voltage. Estimate the terminal voltage for the same excitation and load current at 0.8 power factor leading.* **(AMIE Summer, 1999)**

Solution. *Given* : Rating : 1500 kVA, 6600 V ; $R_a = 0.4\ \Omega$/phase ; $X_s = 6\ \Omega$/phase ; $\cos\phi = 0.8$.

Terminal voltage, V :

F.L. output current $= \dfrac{1500 \times 10^3}{\sqrt{3} \times 6600 \times 0.8} = 164$ A

IR_a drop $= 164 \times 0.4 = 65.6$ V

IX_a drop $= 164 \times 6 = 984$ V

Terminal voltage/phase $= \dfrac{6600}{\sqrt{3}} = 3810$ V.

$$\phi = \cos(0.8) = 36.87°,\ \sin\phi = 0.6$$

$\therefore$ Generated e.m.f., $E = \sqrt{(V\cos\phi + I_aR_a)^2 + (V\sin\phi + I_aX_s)^2}$ (p.f. lagging)

$$= \sqrt{(3810 \times 0.8 + 65.6)^2 + (3810 \times 0.6 + 984)^2}$$

$$= \sqrt{(3113.6)^2 + (3270)^2} = 4515\ \text{V}.$$

Now for the same excitation and load current at 0.8 p.f. *leading*, the voltage equation is

$$E = \sqrt{(V\cos\phi + I_aR_a)^2 + (V\sin\phi - I_aX_s)^2}$$

$\therefore$ $4515 = \sqrt{(V \times 0.8 + 65.6)^2 + (V \times 0.6 - 984)^2}$

or $4515 = \sqrt{0.64V^2 + 4303.4 + 104.96V + 0.36V^2 + 968256 - 1180.8V}$

or $4515 = \sqrt{V^2 - 1075.84V + 972559}$

or $V^2 - 1075.84\ V + 972559 = 20385225$

$V^2 - 1075.84\ V - 19412666 = 0$

$\therefore$ $V = \dfrac{1075.84 \pm \sqrt{(1075.84)^2 + 4 \times 19412666}}{2} = \dfrac{1075.84 \pm 8877}{2}$

$\therefore$ **V = 4976.4 volts. (Ans.)**

Example 5.23. (*a*) *A 3-phase, star-connected, 5 kVA, 400 V, 50 Hz, 4-pole alternator has the following test data at rated speed :*

(i) *Exciting current (in amperes)*	*0.5*	*1.0*	*1.5*	*2.0*	*2.5*	*3.0*	*3.5*	*4.0*	*4.5*	*5.0*	*6.0*	*8.0*
Per phase O.C. volts	*75*	*140*	*173*	*202*	*224*	*240*	*250*	*257*	*260*	*263*	*266*	*271*

(ii) Exciting current (amperes)	*1.0*	*2.0*	*3.0*
S.C. line current	*3.6*	*7.2*	*10.8*

(iii) Armature resistance per phase 2.0 Ω.

Draw O.C. and S.C. characteristics on a squared graph paper and then determine unsaturated value of synchronous reactance per phase in ohms and in per unit.

(b) For the synchronous machine in 5.23 (a) determine per cent voltage regulation at rated load at 0.8 p.f. lag and lead by synchronous impedance method under unsaturated condition. Draw relevant phasor diagrams. **(A.M.I.E. Summer, 1997)**

Solution. *Given* : Rating : 5 kVA, 400 V, 50 Hz, 4-pole ; R_a = 2.0 Ω

(*a*) **Unsaturated value of synchronous reactance per phase :**

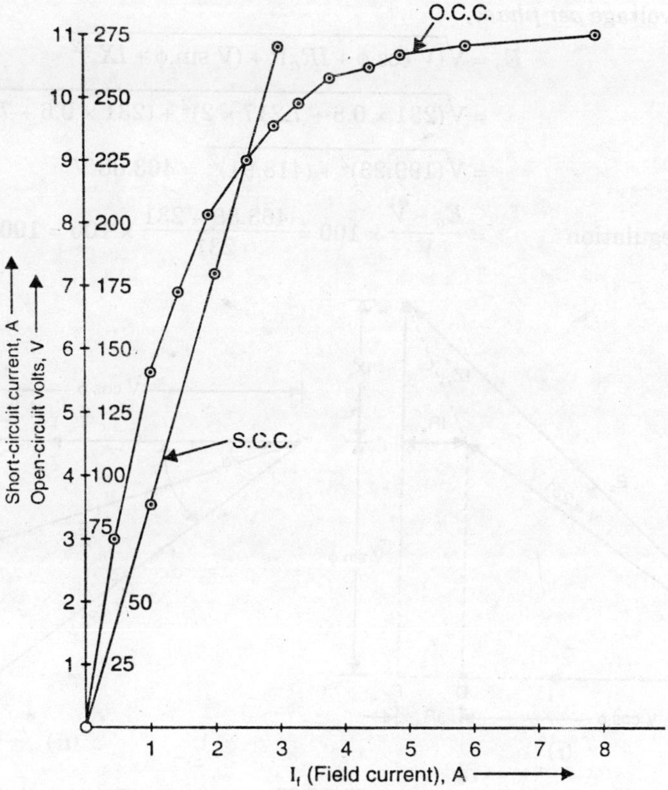

Fig. 5.25

Fig. 5.25 shows the O.C.C. and S.C.C. of the machine alongwith the approximate scales.

Unsaturated value of synchronous reactance is the value of reactance which is obtained for a linear portion of O.C.C.

For a field current of 1 A,

Open-circuit line e.m.f. = 140 V

Short-circuit current, I_{sc} = 3.6 A

Synchronous impedance per phase, $Z_s = \dfrac{\text{O.C. voltage per phase}}{\text{S.C. phase current}} = \dfrac{140}{3.6} = 38.89\ \Omega$

Armature resistance per phase, $R_a = 2.0\ \Omega$

Synchronous reactance per phase, $\mathbf{X_s} = \sqrt{(Z_s)^2 - R_e^2} = \sqrt{(38.89)^2 - (2)^2} = \mathbf{38.84\ \Omega}$. **(Ans.)**

Rated phase voltage, $V_{ph} = \dfrac{V_L}{\sqrt{3}} = \dfrac{400}{\sqrt{3}} = 231$ V

Phase current, I_{ph} = Line current, $I_L = \dfrac{5 \times 1,000}{\sqrt{3} \times 400} = 7.217$ A

Per unit value of synchronous reactance,

$$\mathbf{X_s} = \frac{IX_s}{V_{ph}} = \frac{7.217 \times 38.84}{231} = \mathbf{1.213\ p.u.}\ \ \textbf{(Ans.)}$$

(*b*) **Percentage regulation :**

(*i*) *For 0.8 lagging power factor :*

Open-circuit voltage *per phase*,

$$E_o = \sqrt{(V\cos\phi + IR_a)^2 + (V\sin\phi + IX_s)^2} \qquad \text{(Here, } V = V_{ph})$$

$$= \sqrt{(231 \times 0.8 + 7.217 \times 2)^2 + (231 \times 0.6 + 7.217 \times 38.84)^2}$$

$$= \sqrt{(199.23)^2 + (418.91)^2} = 463.88 \text{ V}$$

Percentage regulation $= \dfrac{E_o - V}{V} \times 100 = \dfrac{463.88 - 231}{231} \times 100 = \mathbf{100.8\%}$. **(Ans.)**

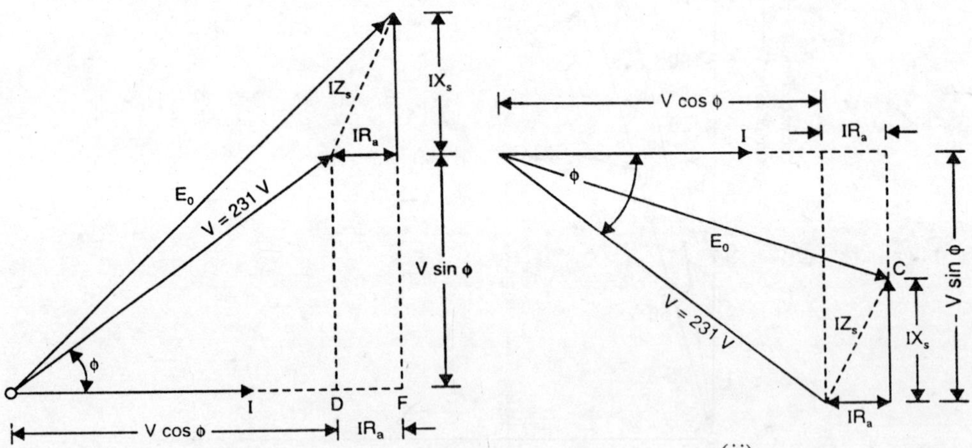

Fig. 5.26

(*ii*) *For 0.8 leading power factor :*

Open-circuit voltage per phase, $E_o = \sqrt{(V\cos\phi + IR_a)^2 + (V\sin\phi - IX_s)^2}$

$$= \sqrt{(231 \times 0.8 + 7.217 \times 2)^2 + (231 \times 0.6 - 7.217 \times 38.84)^2}$$

$$= \sqrt{(199.23)^2 + (-141.71)^2} = 244.5 \text{ V.}$$

Percentage regulation $= \dfrac{E_o - V}{V} \times 100$

$= \dfrac{244.5 - 231}{231} \times 100 =$ **5.84%. (Ans.)**

Phasor diagrams for 0.8 power factor lagging and leading are shown in **Fig.** 5.26 (*i*) and (*ii*) respectively.

☞Example 5.24. *A 3-phase star-connected synchronous generator is rated at 1.5 MVA, 11 kV. The armature effective resistance and synchronous reactance are 1.2 Ω and 25 Ω respectively per phase. Calculate the percentage regulation for a load of 1.4375 MVA at (i) 0.8 p.f. lagging and (ii) 0.8 p.f. leading.*

Also find out the power factor at which the regulation becomes zero. **(U.P.S.C., 1992)**

Solution. *Given* : Rating : 1.5 MVA, 11 kV ; $R_a = 1.2\ \Omega$; $X_s = 25\ \Omega$, Load = 1.4375 MVA

Percentage regulation :

Phase voltage, $E_{ph} = \dfrac{E_L}{\sqrt{3}} = \dfrac{11 \times 1000}{\sqrt{3}} = 6351$ V

Load current, $I = \dfrac{1.4375 \times 10}{\sqrt{3} \times 11000} = 75.45$ A

(*i*) *When power factor is 0.8 lagging :*

Open-circuit voltage/phase, $E_0 = \sqrt{(V\cos\phi + IR_a)^2 + (V\sin\phi + IX_s)^2}$

$= \sqrt{(6351 \times 0.8 + 75.45 \times 1.2)^2 + (6351 \times 0.6 + 75.45 \times 25)^2}$

$= \sqrt{26742757 + 32454099.9} = 7694$ V

∴ **Percentage regulation** $= \dfrac{E_0 - V}{V} \times 100 = \dfrac{7694 - 6351}{6351} \times 100 =$ **21.15%. (Ans.)**

(*ii*) *When power factor is 0.8 leading :*

Open-circuit voltage/phase, $E_0 = \sqrt{(V\cos\phi + IR_a)^2 + (V\sin\phi - IX_s)^2}$

$= \sqrt{(6351 \times 0.8 + 75.45 \times 1.2)^2 + (6351 \times 0.6 - 75.45 \times 25)^2}$

$= \sqrt{26742757 + 3703123} = 5518$ V

Percentage regulation $= \dfrac{E_0 - V}{V} \times 100 = \dfrac{5518 - 6351}{6351} \times 100 =$ **13.12%. (Ans.)**

Power factor at which the regulation becomes zero :

In the phasor diagram shown in Fig. 5.27.

$\angle OAC = 180 - [(90 - \alpha) + \phi] = 90 + \alpha - \phi = [90 - (\phi - \alpha)]$

$\phi = \cos^{-1}(0.8) = 36.87°$

$\alpha = \tan^{-1}\left(\dfrac{IR_a}{IX_s}\right) = \tan^{-1}\left(\dfrac{1.2}{25}\right) = 2.75°$

$OC = E_0,\ OA = V$

$AC = \sqrt{AB^2 + BC^2} = \sqrt{(IR_a)^2 + (IX_s)^2}$

$= \sqrt{(75.45 \times 1.2)^2 + (75.45 \times 25)^2}$

$= 1888.4$ V

Now, $OC^2 = OA^2 + AC^2 - 2 \times OA \times AC \cdot \cos[90 - (\phi - \alpha)]$

or $OC^2 = OA^2 + AC^2 - 2 \times OA \times AC \times \sin(\phi - \alpha)$

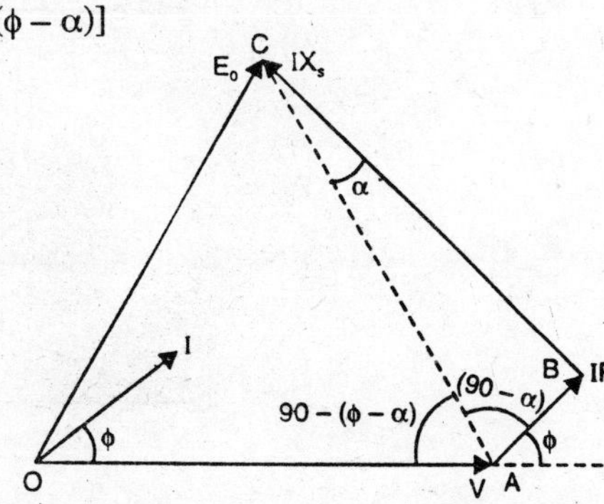

Fig. 5.27. Phasor diagram.

or $$\sin(\phi - \alpha) = \frac{AC^2}{2 \times OA \times AC}$$

$$= \frac{AC}{2 \times OA} \quad (\because \text{ For zero regulation, } E_0 = V \text{ or } OC = OA)$$

$$= \frac{1888.4}{2 \times 6351} = 0.1487 \quad (\because OA = 6351 \text{ V})$$

$\therefore$ $$\phi - \alpha = \sin^{-1}(0.1487) = 8.55°$$

or $$\phi = 8.55 + \alpha = 8.55 + 2.75 = 11.3°$$

$\therefore$ Power factor, $\cos\phi = \cos(11.3°) =$ **0.981 (leading). (Ans.)**

5.13.1.2. The Ampere Turn or M.M.F. Method. The graphs of open and short circuit tests are used in this method. The method is converse of the 'e.m.f. method' in the sense that *armature leakage reactance is treated as an additional armature reaction.*

- From O.C. test field current I_f' is determined to give rated voltage V on no-load, neglecting armature resistance drop, and I_f'' is determined to cause short-circuit current, equal to full-load current, on short-circuit.
- The field excitation I_f'', on short-circuit, *balances the impedance drop in addition to armature reaction on full-load.* But since R_a is usually very small and X_L is also small for low voltage on short circuit, so impedance drop can be neglected. Hence *power factor on short-circuit is almost zero lagging and field ampere turns are used entirely to overcome the armature reaction. Therefore, I_f'' gives demagnetising ampere-turns on full-load.*

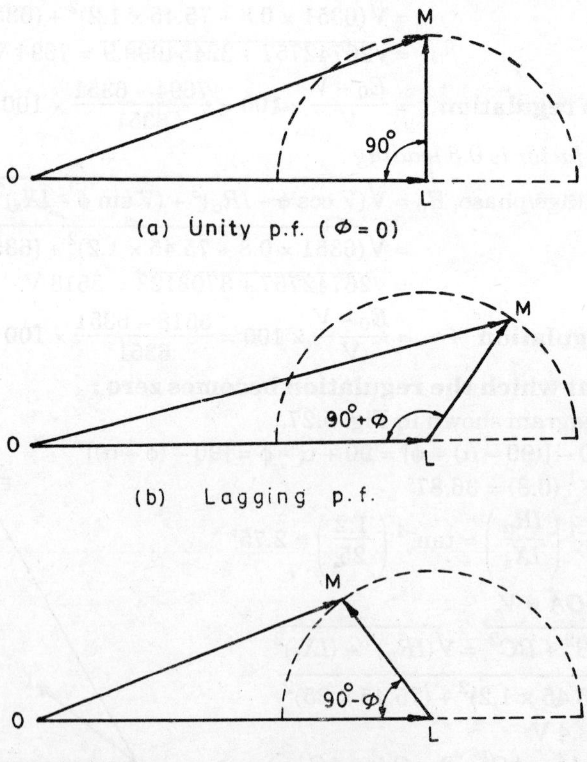

Fig. 5.28

- Now let us consider a general case when the alternator supplies full-load current at a power factor of cos ϕ.
  (*i*) Draw *OL* representing I_f' to give full-load rated voltage, *V* (or more exactly $V + IR_a \cos \phi$).
  (*ii*) Draw *LM* at an angle $(90° \pm \phi)$ representing I_f'' to give full-load current on short circuit (+ ve sign for lagging p.f. and – ve sign for leading p.f.).
  (*iii*) Find field current I_f, measuring *OM*, which will give open circuit e.m.f. E_0 which can be determined from open circuit characteristic (O.C.C.).

The percentage regulation then, can be obtained from the following relation :

$$\% \text{ Regulation} = \frac{E_0 - V}{V} \times 100.$$

- *Regulation given by this method is much lower than that given by the synchronous impedance method, but it is nearer the correct value. This method is called the* **'Optimistic'** *method.*

Example 5.25. *The no-load excitation of an alternator required to give rated voltage is 90 A. In a short-circuit test with full-load current flowing in the armature., the field excitation was 70 A Determine the approximate excitation that will be required to give full-load current at 0.8 p.f. lag and at the rated terminal voltage.*

Solution. Field current required to give full-load rated voltage on open circuit,

$$I_f' = 90 \text{ A}$$

Field current required to give full-load current on short circuit in armature,

$$I_f'' = 70 \text{ A}$$

To find the field current required to give full-load current at 0.8 p.f. lag and at the rated terminal voltage proceed as follows :

(*i*) Draw *OL* equal to 90 A (as shown in Fig. 5.29)

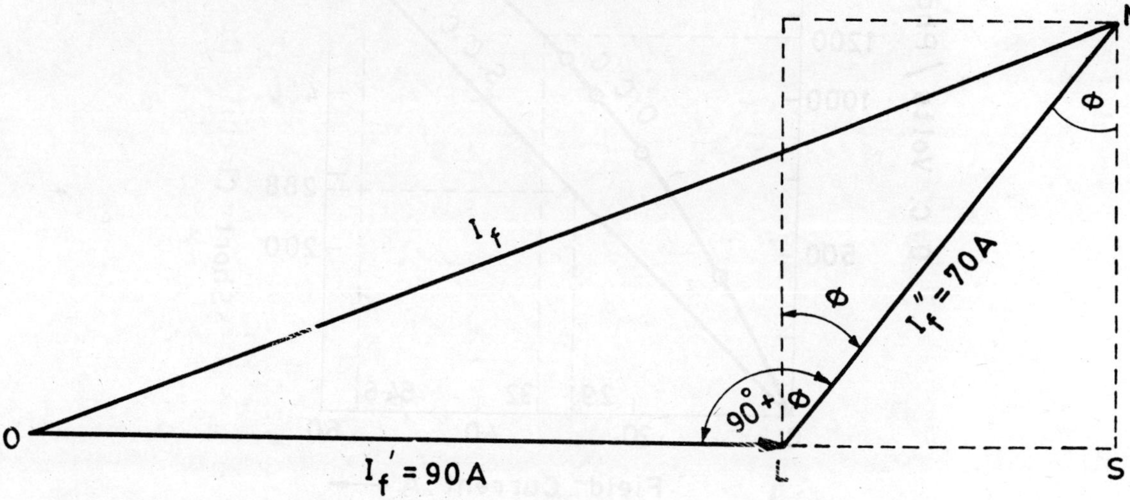

Fig. 5.29

(*ii*) Draw *LM* at an angle $(90 + \phi)$ with *OL*, where $\phi = \cos^{-1} 0.8$.

(*iii*) The required field current will be *OM*, the vector sum of *OL* and *LM*.

∴ The required field current I_f to give full-load current at 0.8 p.f. lag and at the rated terminal voltage

$$= OM = \sqrt{(OL + LS)^2 + MS^2} = \sqrt{(OL + LM \sin \phi)^2 + (LM \cos \phi)^2}$$

$$= \sqrt{(90 + 70 \times 0.6)^2 + (70 \times 0.8)^2} = \mathbf{143.39\ A.}\ \mathbf{(Ans.)}$$

☞**Example 5.26.** *A 3-phase star-connected, 1000 kVA, 2000 V, 50-Hz, alternator gave the following open-circuit and short-circuit test readings :*

Field current,	*A :*	*10*	*20*	*25*	*30*	*40*	*50*
Open-circuit voltage,	*V :*	*800*	*1500*	*1760*	*2000*	*2350*	*2600*
Short-circuit armature current, A :		—	*200*	*250*	*300*	—	—

Draw the characteristic curves and estimate the full-load percentage regulation at :

(i) 0.8 power factor lagging ; and *(ii) 0.8 power factor leading.*

The armature effective resistance/phase may be taken as 0.2 Ω.

Solution. The phase voltages are :

462 V, 866 V, 1016 V, 1154 V, 1357 V, 1502 V

Full-load phase voltage $= 1154$ V $\left[i.e., \frac{2000}{\sqrt{3}} \right]$

Plot the O.C.C and S.C.C. as shown in Fig. 5.30.

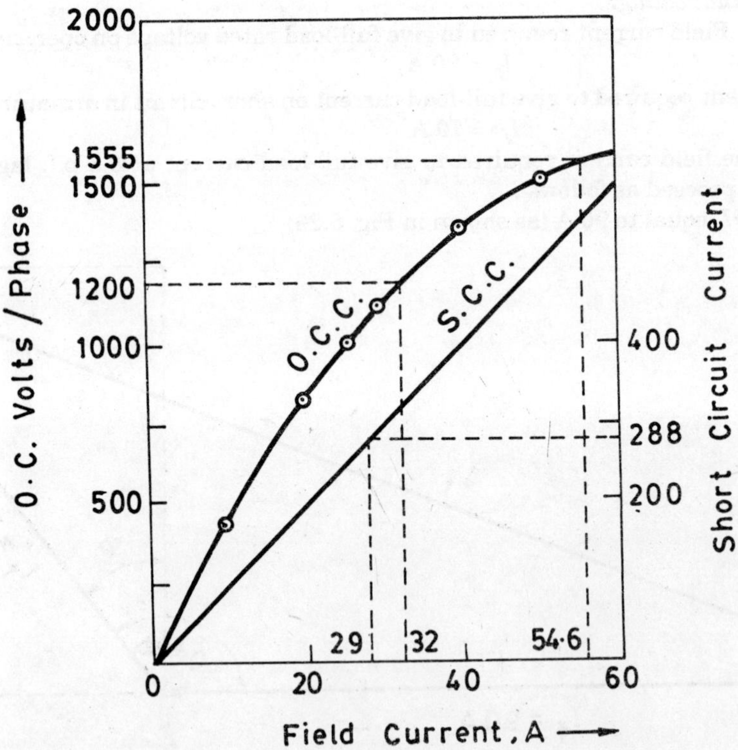

Fig. 5.30

Full-load current $= \frac{1000 \times 1000}{\sqrt{3} \times 2000} = 288.7$ A

Voltage/phase at full-load at 0.8 power factor

$$= V + IR_a \cos \phi = 1154 + (288.7 \times 0.2 \times 0.8) = 1200\text{ V.}$$

— From O.C.C., the field current required to produce this voltage (*i.e.*, 1200 V) = 32 A.

— From S.C.C., the field current required to produce full-load current of 288.7 A = 29 A

(*i*) **0.8 power factor lagging :**

$$\cos \phi = 0.8 \quad \text{or} \quad \phi = \cos^{-1} 0.8 = 36° \, 52' \text{ (lag)}.$$

Refer Fig. 5.31 (Graphical method).

Draw to some scale ;

$$OL = 32 \text{ A}, LM = 29 \text{ A at an angle of } (90 + \phi)$$

or $(90 + 36° \, 52') = 126° \, 52'$

Then OM = total field current at full-load 0.8 p.f. lagging = 54.6 A.

From O.C.C. the open circuit volts corresponding to a field current of 54.6 A = 1555 V

$$\therefore \ \% \textbf{ Regulation} = \frac{E_0 - V}{V} \times 100 = \frac{1555 - 1154}{1154} \times 100 = \mathbf{34.75\%. \ (Ans.)}$$

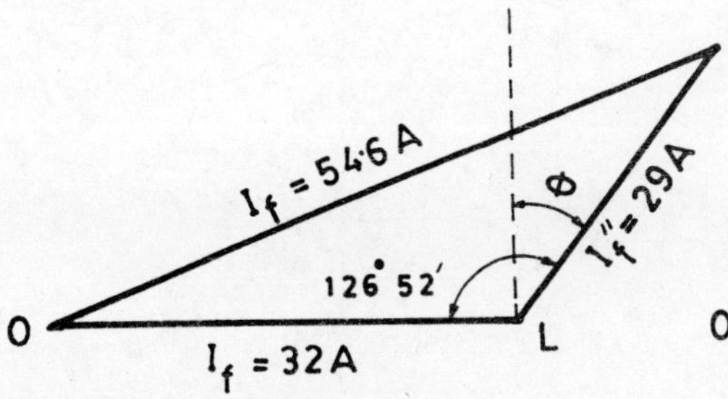

Fig. 5.31

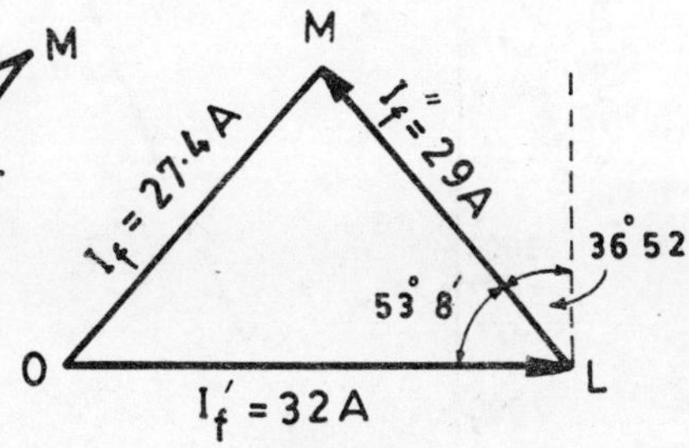

Fig. 5.32

(*ii*) **0.8 power factor leading :**

Refer Fig. 5.32.

In this case since power factor is leading, draw LM with OL at an angle of (90° – 36° 52') = 53° 8'.

$$OM = 27.4 \text{ A}$$

From O.C.C. the open circuit volts corresponding to a field current of 27.4 A = 1080 V

$$\therefore \ \% \textbf{ Regulation} = \frac{1080 - 1154}{1154} \times 100 = \mathbf{-6.41\%. \ (Ans.)}$$

☞**Example 5.27.** *A 6600 V alternator gave the following test results :*

O.C. voltage, V :	*3100*	*4900*	*6600*	*7500*	*8300*
Field current, A :	*16*	*25*	*37.5*	*50*	*70*

A field current of 22 A is found necessary to circulate full-load current on short-circuit of the armature. Calculate the full-load regulation at 0.8 power factor lagging by :

(*i*) *The ampere-turn method.*

(*ii*) *The synchronous impedance method.*

Comment on the two values of regulation obtained by the above methods.

(Madras University, Nov. 1998)

Solution. (*i*) **Ampere turn Method :**

Draw the O.C.C. (Fig. 5.33) from the data given.

From open circuit curve the field current required to give normal voltage of 6600 V = 37.5 A (*i.e.* I_f').

Field current required to give full-load current is 22 A (*i.e.* I_f'')

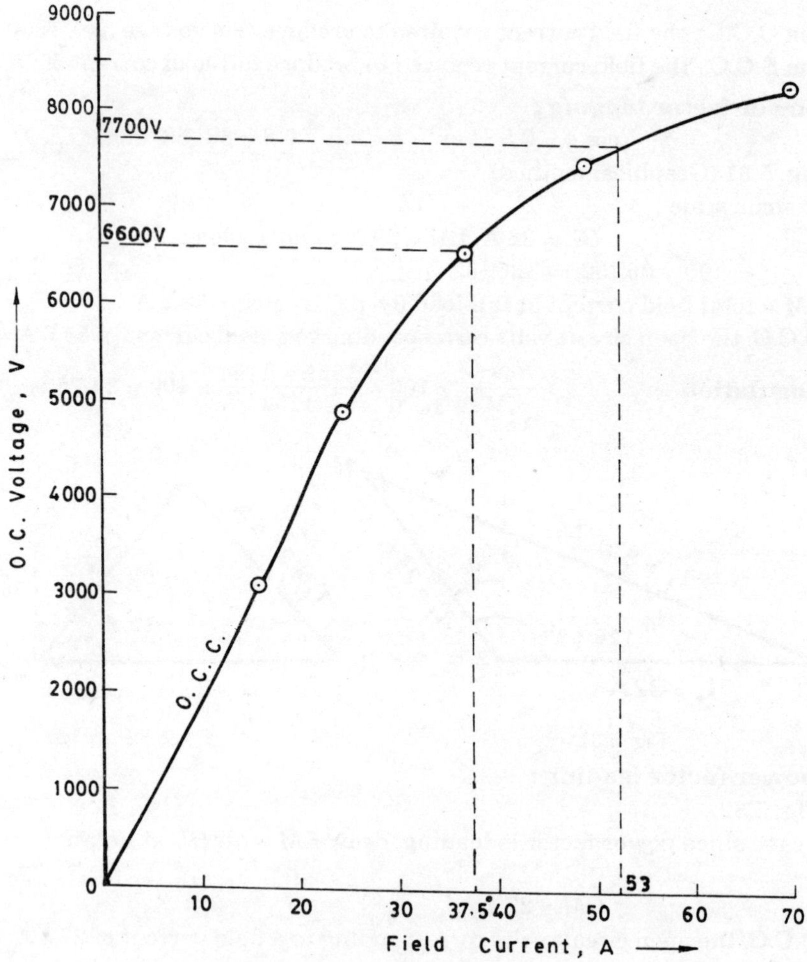

Fig. 5.33

To find the field current required to give a terminal potential difference of 6600 V on full-load the field excitation of 22 A must be added vectorially with the normal value of 37.5 A as follows :

Graphical method (Fig. 5.34) :

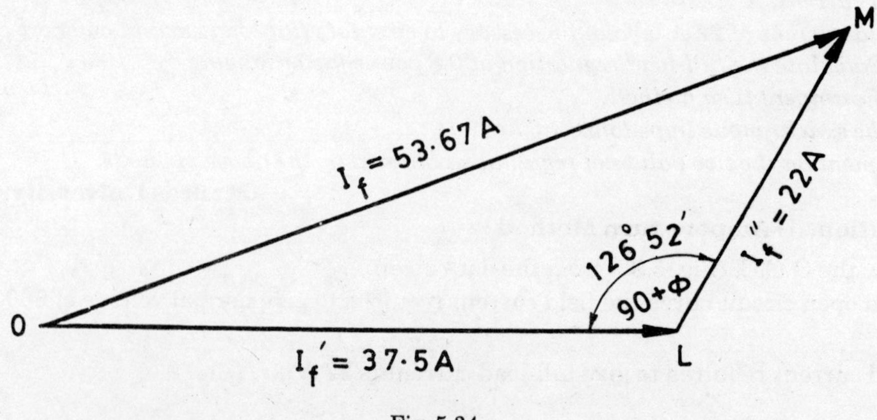

Fig. 5.34

(*i*) Draw OL (or I_f') = 37.5 A (according to some suitable scale).

(*ii*) Draw LM (or I_f'') = 22 A at an angle of $(90 + \phi)$

or $(90 + 36° 52') = 126° 52'$ $[\because \cos\phi = 0.8 \text{ or } \phi = \cos^{-1} 0.8 = 36° 52']$

(*iii*) Measure OM and multiply with scale.

Then OM (or I_f) = 53.67 A

Analytical method :

$$I_f = \sqrt{(I_f')^2 + (I_f'')^2 - 2I'I_f'' \times \cos(90 + \phi)}$$

$$= \sqrt{(37.5)^2 + (22)^2 - 2 \times 37.5 \times 22 \times \cos(126° 52')}$$

$$= \sqrt{1406.25 + 484 - 1650 \times (-0.6)} = \sqrt{2880.25} = 53.67 \text{ A}$$

From O.C.C., corresponding to a field current of 53.67 A the generated e.m.f., E_0 = 7700 V.

$$\therefore \ \% \textbf{ Regulation} = \frac{E_0 - V}{V} \times 100 = \frac{7700 - 6600}{6600} \times 100 = \mathbf{16.67\%}. \ \textbf{(Ans.)}$$

(*iii*) **Synchronous Impedance Method :**

Let the voltage of 6600 V be taken as 100%.

Let the excitation of 37.5 A be required to give 6600 V on open-circuit = 100%.

Full-load or 100% armature current is produced on short-circuit by field current of 22 A. If 100% *i.e.*, 37.5 A of field current were applied on short circuit, then short-circuit current would be :

$$100 \times \frac{37.5}{22} = 170.45\%$$

Percentage impedance, $Z_s = \dfrac{\text{open-circuit voltage}}{\text{short-circuit current}}$ for given excitation

$$= \frac{100}{170.45} \times 100 = 58.67\%.$$

Since armature resistance is negligible, therefore, impedance drop IZ_s = 58.67% of the normal voltage will lead the load current by 90°.

Hence, the open circuit voltage, E_0 (Fig. 5.35) which is the vector sum of normal voltage and impedance drop is given by :

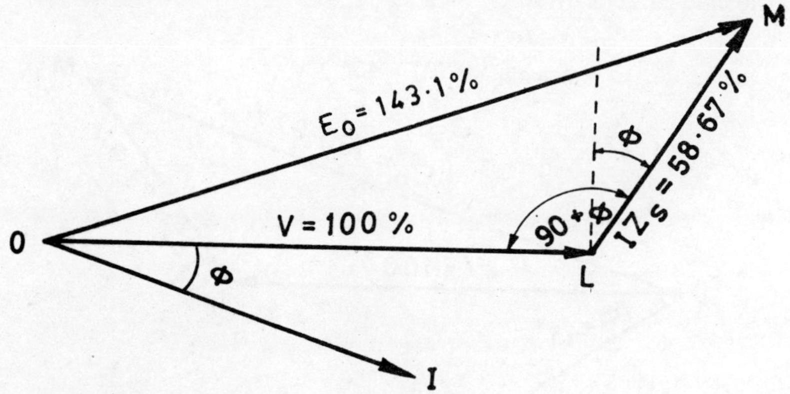

Fig. 5.35

$$E_0 = \sqrt{V^2 + (IZ_s)^2 - 2 \times V \times IZ_s \cos (90° + \phi)}$$

$$= \sqrt{(100)^2 + (58.67)^2 - 2 \times 100 \times 58.67 \times \cos (90° + 36° \, 52')}$$

$$= \sqrt{10000 + 3442.2 + 7040.4} = 143.1\%$$

% Regulation $= \dfrac{E_0 - V}{V} \times 100 = \dfrac{143.1 - 100}{100} \times 100 =$ **43.1%. (Ans.)**

Comments on the two values of regulation. It is found that the two values of the regulation calculated above differ widely from each other. The first method gives a little lower value as compared to the actual value while the other gives somewhat higher value. However, *the first value is more likely to be nearer to the actual value because the second method employs synchronous impedance Z_s which is not constant but depends upon field excitation.*

Example 5.28. *The following O.C.C. was obtained in case of a 3-phase, 6000 V alternator running at normal speed.*

Field current, A	*: 14*	*18*	*23*	*30*	*43*
Terminal voltage, V	*: 4000*	*5000*	*6000*	*7000*	*8000*

The field current is 17 A when armature is short-circuited and the full-load current is flowing. Using synchronous impedance method, determine the voltage regulation of the alternator supplying the full-load of 200 kVA at 0.707 lagging.

Neglect armature resistance.

Solution. Let the voltage of 6000 V be taken as = 100%.

Let the excitation of 23 A (from O.C.C.) to give 6000 V on open circuit be = 100%

Full-load or 100% armature current is produced on short-circuit by field current of 17 A. If 100% *i.e.*, 23 A of field current were applied on short-circuit then short-circuit current would be :

$$100 \times \frac{23}{17} = 135.3\%$$

Percentage impedance, $Z_s = \dfrac{\text{open-circuit voltage}}{\text{short-circuit current}}$ for given excitation

$$= \frac{100}{135.3} \times 100 = 73.9\%.$$

Since armature resistance is negligible, therefore, impedance drop IZ_s = 73.9% of the normal voltage will lead the load current by 90°.

Now, $\cos \phi = 0.707$ *(given)*

$\therefore$ $\phi = \cos^{-1} 0.707 = 45°$

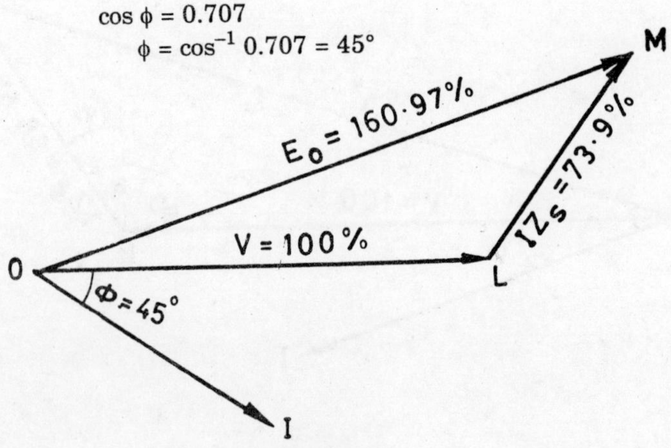

Fig. 5.36

The open-circuit voltage, E_0 (Fig. 5.36) which is the vector sum of normal voltage and impedance drop is given by :

$$E_0 = \sqrt{V^2 + (IZ_s)^2 - 2 \times V \times (IZ_s) \times \cos (90 + \phi)}$$
$$= \sqrt{(100)^2 + (73.9)^2 - 2 \times 100 \times 73.9 \times \cos (90 + 45°)}$$
$$= \sqrt{10000 + 5461 + 10449.5} = 160.97 \text{ \%}.$$

$$\therefore \quad \text{\% Regulation} = \frac{E_0 - V}{V} \times 100 = \frac{160.97 - 100}{100} \times 100 = \mathbf{60.97 \text{ \%}}. \quad \textbf{(Ans.)}$$

☞**Example 5.29.** *A 275 kW, 3-phase, 6600 V non-salient pole type alternator gave the following test results :*

O.C.C. :

Exciting amperes, A :	*46.5*	*58*	*67.5*	*96*
Volts, V :	*5600*	*6600*	*7240*	*8100*

S.C.C. :

Stator current 35 amperes with an exciting current of 50 amperes. Leakage reactance on full-load : 8 per cent.

Calculate as accurately as possible the exciting current (for full-load) at 0.8 p.f. lagging. Neglect armature resistance.

Solution. The phase voltages are : 3233, 3810, 4180, 4676

Plot O.C.C. as shown in Fig. 5.37.

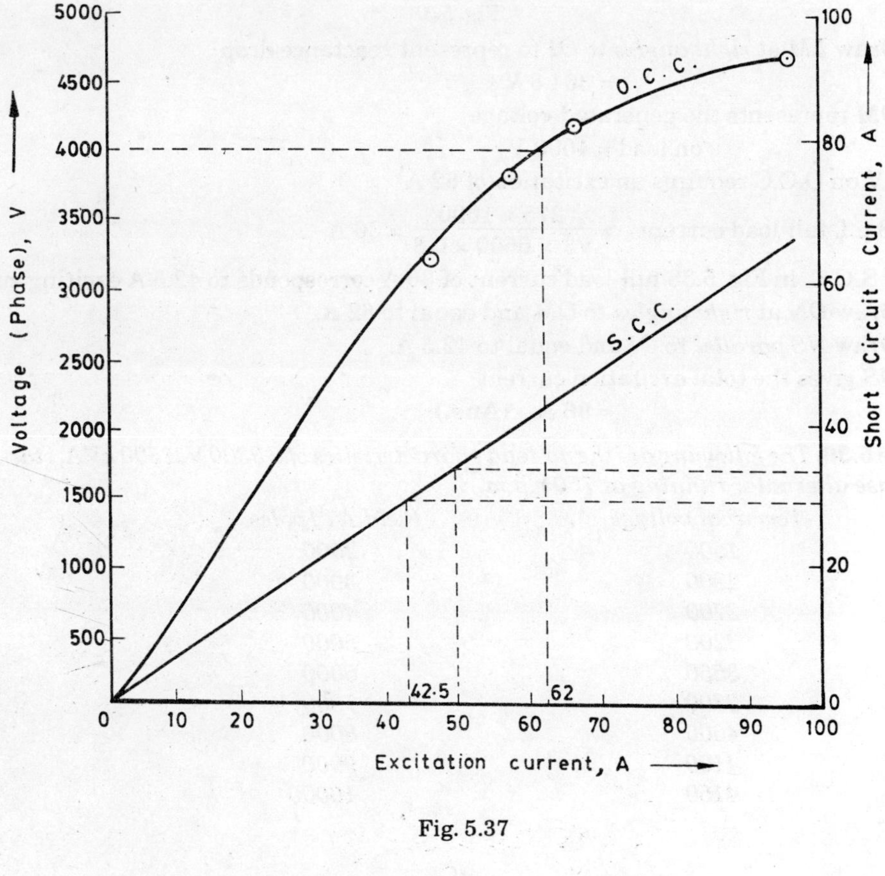

Fig. 5.37

Leakage reactance drop $= 8\% = \dfrac{8}{100} \times 3810 = 304.8$ V

Load phase voltage $= OL = \dfrac{6600}{\sqrt{3}} = 3810$ V

Now consider Fig. 5.38.

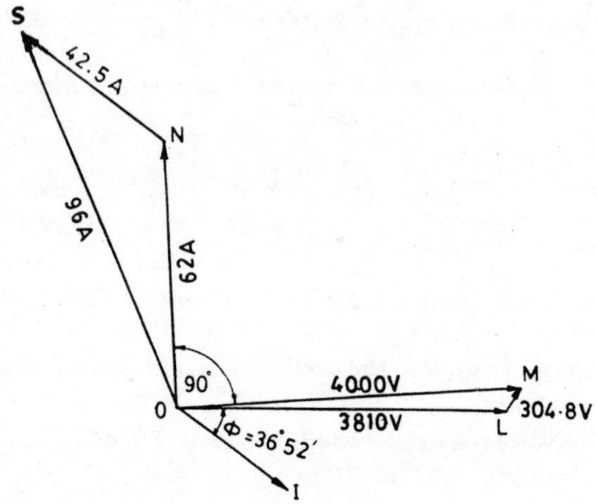

Fig. 5.38

- Draw *LM* at *right angles* to *OI* to represent reactance drop

  = 304.8 V.
- *OM* represents the generated voltage

  on load = 4000 V

4000 V on O.C.C requires an excitation of 62 A

At 0.8 p.f. full-load current $= \dfrac{275 \times 1000}{\sqrt{3} \times 6600 \times 0.8} = 30$ A

From S.C.C. in Fig. 5.38 full-load current of 30 A corresponds to 42.5 A exciting current.

- Draw *ON* at *right angles* to *OM* and equal to 62 A.
- Draw *NS parallel* to *OI* and equal to 42.5 A.
- *OS* gives the total excitation current

  = **96 A. (Ans.)**

☞**Example 5.30.** *The following are the no-load characteristics of a 3300 V, 1500 kVA, star-connected, 8-pole, 3-phase alternator running at 750 r.p.m. :*

Terminal voltage	*Field AT/poles*
1500	*2000*
2200	*3000*
2700	*4000*
3200	*5000*
3550	*6000*
3800	*7000*
4000	*8000*
4100	*9000*
4150	*10000*

The resistance measured between terminals is 0.5 Ω and its true reactance per phase is 1.4 Ω. The number of turns per phase is 60. Assuming that the armature reaction ampere turns per pole are equal to 1.6 times amperes turns/pole/phase and the ratio of effective to the measured ohmic resistance is 1.5, determine the percentage regulation when full-load at normal voltage and at p.f. 0.8 lag is thrown off and the speed of the alternator increases to 790 r.p.m.

Solution. Number of poles, $p = 8$

Measured resistance between terminals = 0.5 Ω

True reactance/phase = 1.4 Ω

Number of turns/phase = 60

Power factor, $\cos \phi = 0.8$

Percentage regulation :

Phase volts corresponding to terminal volts are :

866, 1270, 1559, 1848, 2050, 2194, 2309, 2367, 2396

Plot the O.C.C. as shown in Fig. 5.39.

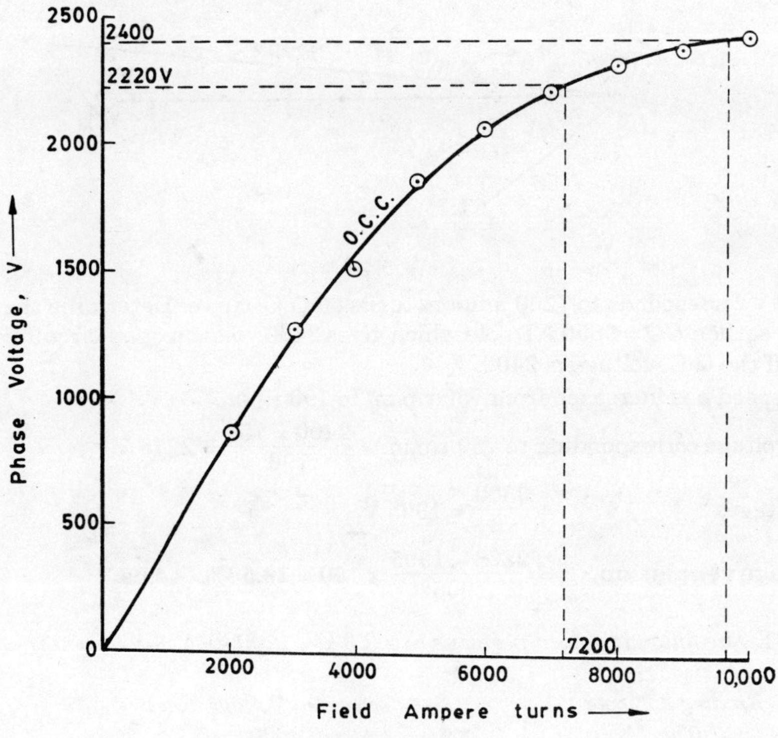

Fig. 5.39

Full-load current $= \dfrac{1500 \times 1000}{\sqrt{3} \times 3300} = 262$ A

Measured resistance/phase $= \dfrac{\text{measured resistance between terminals}}{2} = \dfrac{0.5}{2} = 0.25\ \Omega$

Also, effective resistance = 1.5 × measured resistance *(given)*

∴ Effective resistance = 1.5 × 0.25 = 0.375 Ω

Full-load resistance drop = 0.375 × 262 = 98.25 V

Leakage reactance drop = 1.4 × 262 = 366.8 V

Armature reaction ampere turns/pole/phase

$$= \frac{262 \times 60}{8} \times 1.6 = 3141.$$

In Fig. 5.40, *ON* represents the generated e.m.f. on load

$$= 2220 \text{ V}$$

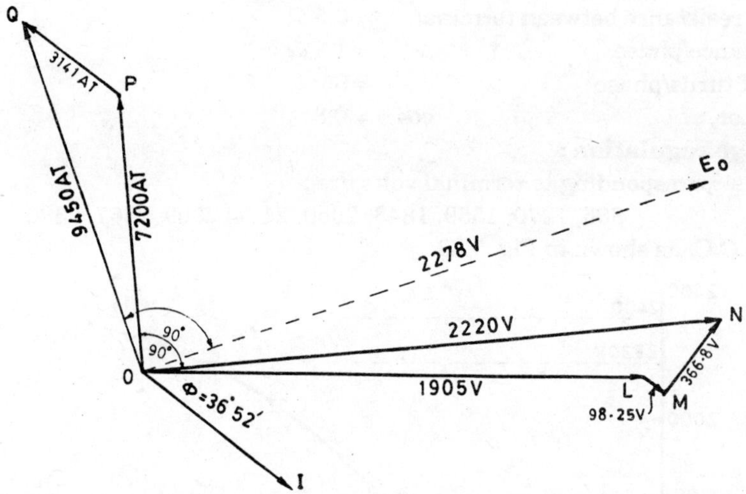

Fig. 5.40

And 2220 V corresponds to 7200 ampere turns on O.C. curve. Determine the total excitation *OQ* as explained earlier, *OQ* = 9650 AT/pole which gives 2400 volts on open-circuit. Hence, when the load is thrown off the O.C. voltage is 2400 V.

Now the speed has increased from 750 r.p.m. to 790 r.p.m.

$\therefore$ O.C. voltage corresponding to 750 r.p.m. $= \frac{2400 \times 750}{790} = 2278$ V

Phase voltage, $= \frac{3300}{\sqrt{3}} = 1905$ V

Percentage regulation, $= \frac{2278 - 1905}{1905} \times 100 =$ **19.58%. (Ans.)**

☞**Example 5.31.** *The following data pertains to a 3.3 kV, 1200 kVA, 3-phase star-connected salient pole alternator :*

Exciting AT / pole	*Generated Voltage (line)*
2000	*1500*
3600	*2500*
5000	*3150*
6000	*3450*
7000	*3680*
8000	*3850*
9000	*4000*
10000	*4050*

Resistance per phase $= 0.5\ \Omega$

Leakage reactance per phase $= 1.3\ \Omega$

Equivalent armature reaction (AT/pole)
= 15 times the line current
Reluctance to cross flux = 2 times direct flux
Determine the regulation at full-load at 0.8 power factor lag.

Solution. Resistance/phase = 0.5 Ω
Leakage reactance/phase = 1.3 Ω
Equivalent armature reaction (AT/pole) = 15 × line current
Reluctance to cross flux = 2 × direct flux

Regulation at full-load at 0.8 p.f. lag :

Converting line voltages to phase voltages.

Phase voltages are : 866 ; 1443 ; 1819 ; 1992 ; 2125 ; 2223 ; 2309 ; 2338

Plot the O.C.C. as shown in Fig. 5.41.

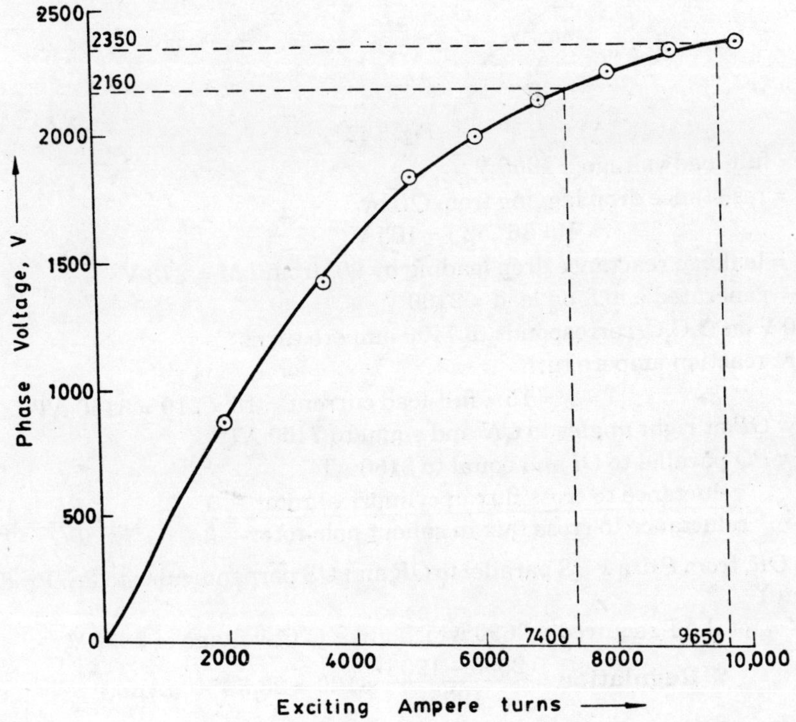

Fig. 5.41

Full-load current of the alternator

$$= \frac{1200 \times 1000}{\sqrt{3} \times 3.3 \times 1000} = 210 \text{ A}$$

Resistive drop/phase $= 210 \times 0.5 = 105$ V

Leakage reactance drop $= 210 \times 1.3 = 273$ V

Full-load phase voltage $= \dfrac{3330}{\sqrt{3}} = 1905$ V

In Fig. 5.42,

$$\cos \phi = 0.8, \ \phi = \cos^{-1} 0.8 = 36° \ 52'$$

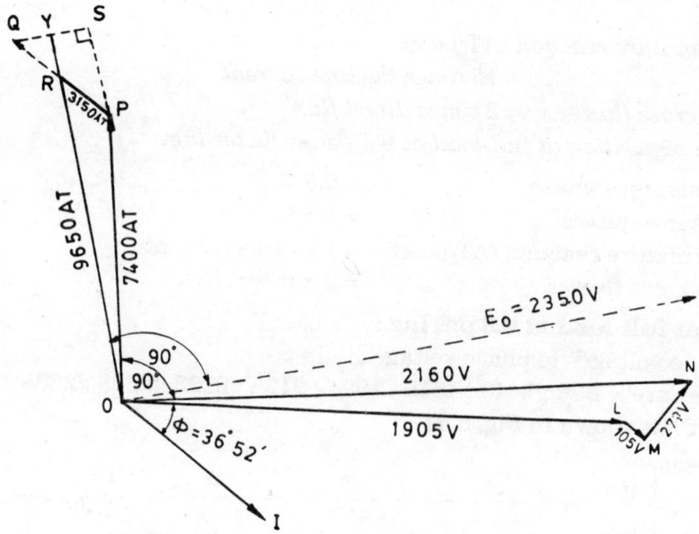

Fig. 5.42

- OL = full-load voltage = 1905 V
- LM = resistance drop lagging from OL by
$$\phi(= 36° 52') = 105 \text{ V}$$
- MN = leakage reactance drop leading by 90° from LM = 273 V
- ON = generated e.m.f. on load = 2160 V
- 2160 V on O.C.C. corresponds to 7400 ampere turns

Armature reaction ampere turns

$$= 15 \times \text{full-load current} = 15 \times 210 = 3150 \text{ AT}$$

- Draw OP at right angles to ON and equal to 7400 AT
- Draw PQ parallel to OI and equal to 3150 AT

Now, $\dfrac{\text{reluctance to cross flux in cylindrical rotor}}{\text{reluctance to cross flux in salient pole rotor}} = \dfrac{1}{3}$ *i.e.*, $\dfrac{PR}{PQ} = \dfrac{1}{2}$

- Join OR, from P draw PS parallel to OR and QS perpendicular to PS. Produce OR to meet QS in Y.

Then OY = total AT required = 9650 which gives O.C. voltage of 2350 V

$$\% \textbf{ Regulation} = \frac{2350 - 1905}{1905} \times 100 = \mathbf{23.36\%}. \textbf{ (Ans.)}$$

5.13.1.3. Zero Power Factor or Potier Method. This method *gives more accurate results* since it is based on the *separation of armature leakage reactance drop and the armature reaction effects*. The following experimental data is required in this method :

(*i*) No-load or open circuit curve.

(*ii*) Full-load zero power factor curve (not S.C.C.).

From (*ii*) the reduction in voltage due to armature reaction is found out and voltage drop due to armature leakage reactance (also called Potier reactance) X_L is found from both (*i*) and (*ii*). By combining the two, E_0 can be calculated.

The above two curves are similar and displaced horizontally by the m.m.f. due to armature reaction in terms of the field current.

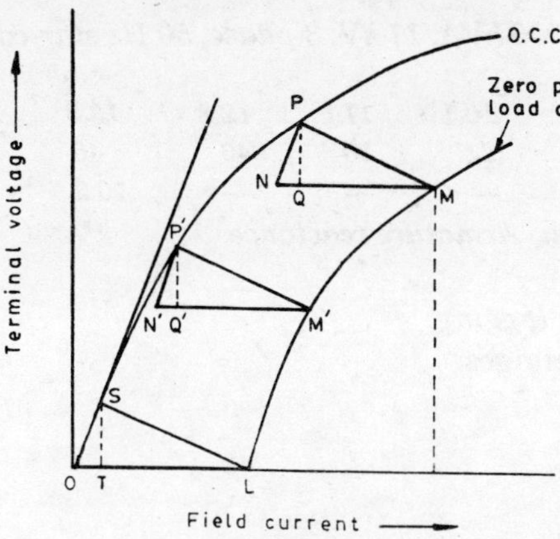

Fig. 5.43

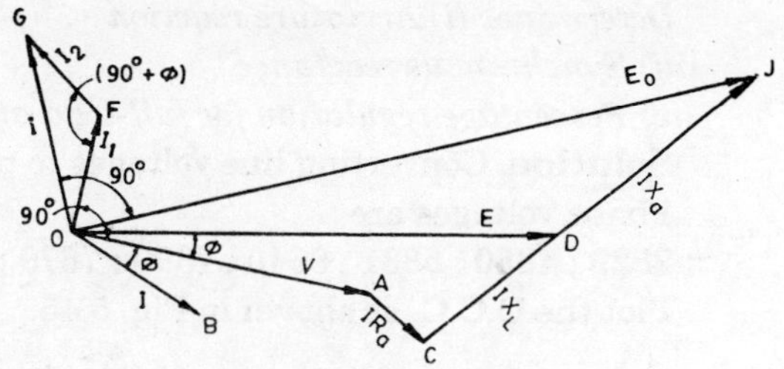

Fig. 5.44

- Zero power factor, full-load voltage excitation characteristic can be drawn by knowing two points L and M. Point L is obtained from a short circuit test with full-load armature current. Hence OL represents field current (excitation) required to *overcome demagnetising effect of armature reaction and to balance leakage reactance drop at full-load.* Point M is obtained when full-load current flows through the armature and wattmeter reading is zero. *Zero power factor curve may be drawn as follows* :
  (*i*) From M draw line MN equal and parallel to OL.
  (*ii*) Through point N draw a line parallel to initial straight part of O.C.C. (parallel to OS), cutting the O.C.C. at P.
  (*iii*) Join PN and drop a perpendicular PQ on MN.
  (*iv*) Impose the triangle MPQ at various-points of O.C.C. to obtain corresponding points on the zero power factor curve.

  In triangle MPQ :
  Length PQ represents *leakage reactance drop* (IX_L)
  Length MQ represents *armature reaction excitation.*

Potier Regulation Diagram. Following is the procedure to draw Potier regulation diagram :

(*i*) Draw OA horizontally to represent terminal voltage V on full load and OB to represent full load current (I) at a given power factor.

(*ii*) Draw AC (= IR_a), voltage drop due to resistance R_a (if resistance is given) parallel to OB.

(*iii*) Draw CD perpendicular to AC and equal to reactance drop IX_L.

Now OD represents generated e.m.f. E.

(*iv*) From O.C.C., find the field current I_1 corresponding to this generated e.m.f. E and draw OF (equal to I_1) perpendicular to OD. Draw FG parallel to load current OB (*i.e.*, I) to represent excitation (field current) *equivalent to full load armature reaction.*

OG gives *total field current* required.

(*v*) If the load is thrown off, then terminal voltage will be equal to generated e.m.f. corresponding to field excitation OG. Hence e.m.f. E_0 may be obtained from O.C.C. corresponding to field excitation OG. Vector OJ will lag behind vector OG by 90°. DJ represents *voltage drop due to armature reaction.*

Now regulation may be obtained from the following relation :

$$\% \text{ Regulation} = \frac{E_0 - V}{V} \times 100.$$

☞**Example 5.32.** *The following data pertains to a 15000 kVA, 11 kV, 3-phase, 50 Hz star-connected turbo-alternator :*

O.C. line kV	:	*4.9*	*8.4*	*10.1*	*11.5*	*12.8*	*13.3*	*13.65*
Field AT in 10^3	:	*10*	*18*	*24*	*30*	*40*	*45*	*50*
Zero p.f. full-load line kV	:	—	*0*	—	—	—	*10.2*	—

Determine : (i) Armature reaction *(ii) Armature reactance*

(iii) Synchronous reactance

(iv) Percentage regulation for full-load at 0.8 p.f. lagging.

Solution. Converting line voltages to phase voltages.

Phase voltages are :

2829 ; 4850 ; 5831 ; 6640 ; 7390 ; 7679 ; 7880

Plot the O.C.C. as shown in Fig. 5.45.

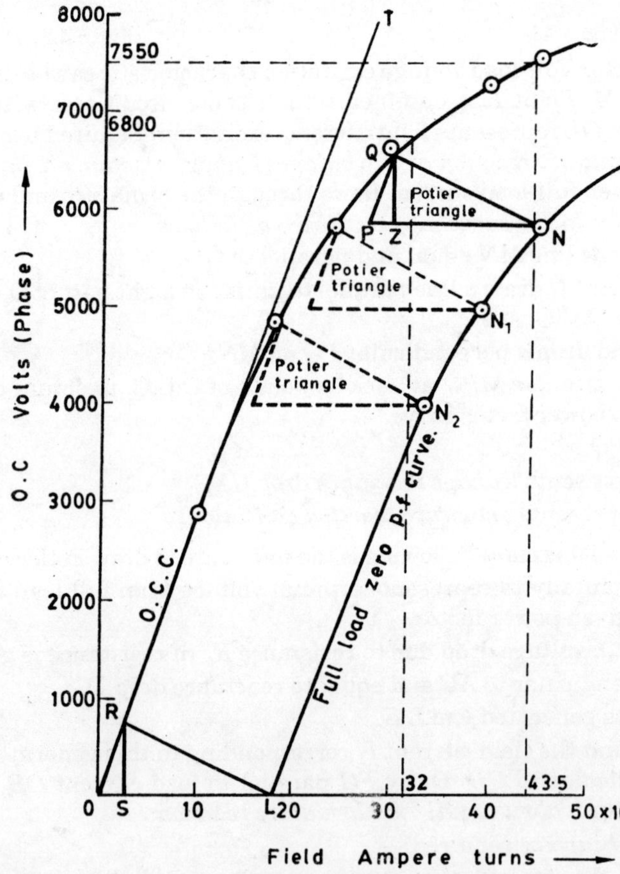

Fig. 5.45

Full-load zero p.f. line is drawn because two points are known *i.e.*, (18, 0) and $\left[45, \frac{10200}{\sqrt{3}}\right.$ *i.e.*, 5890 $\left.\right]$ and other points on this curve can be found by transferring the Potier triangle (as shown in Fig. 5.45) as follows :

- At N, draw $NP \parallel$ to and equal to OL and from P draw $PQ \parallel$ to OR. Join QN. Hence NPQ is *Potier triangle*.
- Draw line $QZ \perp$ to PN

(*i*) NZ = *field ampere-turns for balancing armature reaction only*

= **15500. (Ans.)**

$QZ = RS = 670$ V

= leakage reactance drop/phase

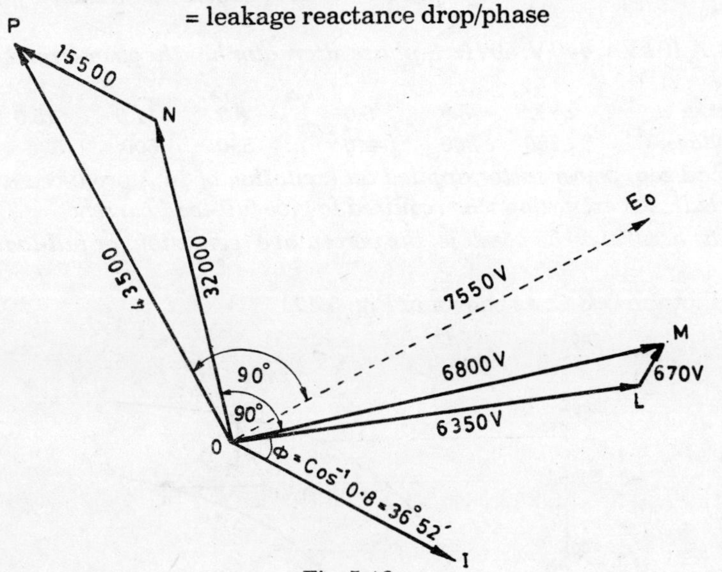

Fig. 5.46

Short-circuit AT required $= OL = 18000$

Full-load current, $I = \dfrac{15000 \times 1000}{\sqrt{3} \times 11000} = 787$ A

$\because$ $IX_L = 670$

(*ii*) $\therefore$ $X_L = \dfrac{670}{I} = \dfrac{670}{787} = \mathbf{0.85\ \Omega}$. **(Ans.)**

(*iii*) From O.C.C. we find that 18000 AT corresponds to an O.C. voltage of $\dfrac{8400}{\sqrt{3}}$ = **4850 V**

$\therefore$ Synchronous impedance,

$$Z_s = \frac{\text{O.C. volts}}{\text{short-circuit current}} = \frac{4850}{787} = 6.16\ \Omega$$

Since R is negligible, hence

$$Z_s = X_s = 6.16\ \Omega$$

Hence, **synchronous reactance, X_s = 6.16 Ω. (Ans.)**

(*iv*) **Regulation at 0.8 p.f. lag :** Refer Fig. 5.46.

$$OL = \text{phase voltage} = \frac{11000}{\sqrt{3}} = 6350 \text{ V}$$

$LM = 670$ V is drawn at right angles to OI or $(90 + \phi)$ to OL. Resultant is $OM = 6800$ V

From O.C.C., field ampere-turns corresponding to 6800 V is

$ON = 32000$ and is drawn $\perp OM$

NP = armature reaction at full-load = 15500 and is drawn || to OI at (90 + ϕ) to ON

Hence, OP = 43500

From O.C.C. e.m.f. corresponding to 43500 AT of rotor = 7550 V

$\therefore$ **% Regulation** $= \dfrac{7550 - 6350}{6350} \times 100 =$ **18.89 %. (Ans.)**

☞**Example 5.33.** *A 10 kVA, 440 V, 50 Hz, 3-phase alternator has the open-circuit characteristic given below :*

Field current, A	*: 1.5*	*3.0*	*5.0*	*8.0*	*11.0*	*15.0*
Terminal voltage, V	*: 150*	*300*	*440*	*550*	*600*	*635*

With full-load zero power factor applied an excitation of 14 A produced a terminal voltage of 500 V on short-circuit, 4 A excitation was required to give full-load current.

Determine as accurately as possible, the percentage regulation for full-load 0.8 p.f. (i) lagging (ii) leading.

Solution. Plot the O.C.C. as shown in Fig. 5.47.

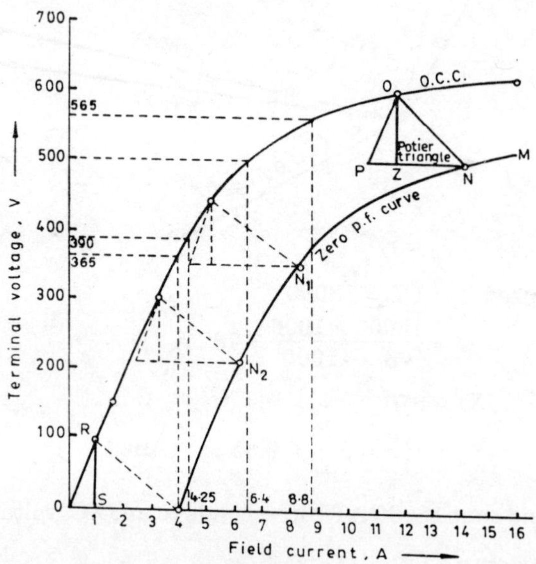

Fig. 5.47

The points L (4 A, 0 V) and M (14 A, 500 V) are on the curve of full-load zero p.f. The zero power factor curve may be obtained as follows :

- Draw NP horizontal, parallel and equal to OL.
- Draw PQ || to OT.
- Impose the Potier triangle at various points on O.C. curve to obtain points, N_1, N_2, ... on zero power factor curve.

In ORL excitation OL is required to generate voltage RS which equalizes leakage reactance. Excitation SL = (NZ) balances armature reaction.

(*i*) **Regulation at 0.8 p.f. lagging :**

Leakage reactance drop RS(= QZ) measured = 92 V

In Fig. 5.48, OL = 440 V ; LM = 92 V

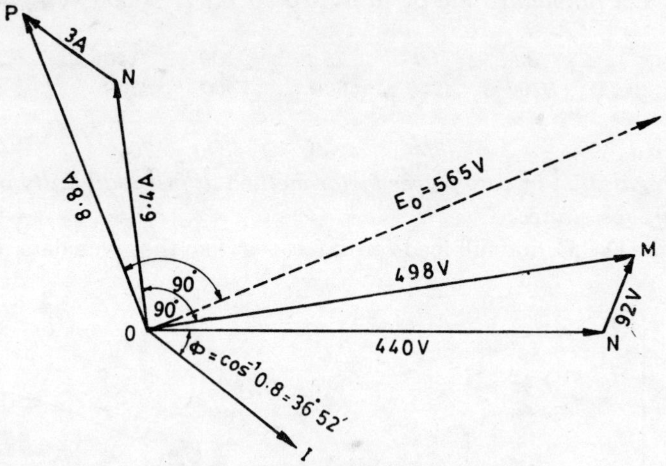

Fig. 5.48

Then OM = induced voltage on load = 498 V

498 V corresponds on O.C.C to an excitation current of 6.4 A

Draw $ON \perp$ to OM and equal to 6.4 A

Draw $NP \parallel$ to OI and equal to $SL (= NZ = 3\ A)$

Then OP = 8.8 A and equals the total excitation which corresponds to 565 V on O.C.C.

$$\therefore \ \% \textbf{ Regulation} = \frac{565 - 440}{440} \times 100 = \textbf{28.4 \%}. \ \textbf{(Ans.)}$$

(ii) **Regulation at 0.8 p.f. leading :**

Refer Fig. 5.49.

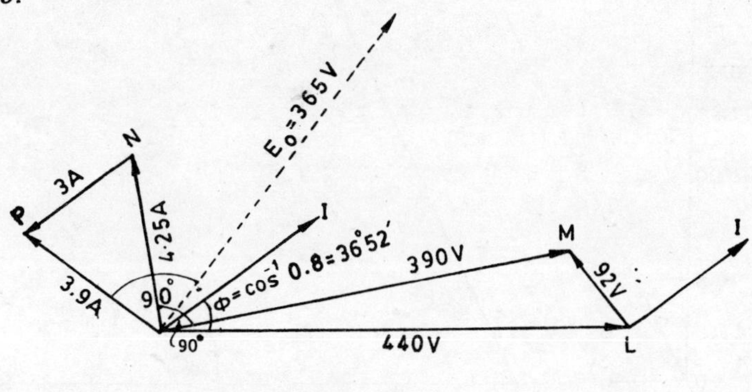

Fig. 5.49

The induced voltage on load = OM = 390 V

390 V requires an excitation of 4.25 A.

- Draw $ON \perp$ to OM and equal to 4.25 A
- Draw $NP \parallel$ to OI and equal to $SL\ (= NZ) = 3$ A

OP = total excitation required = 3.9 A

3.9 A corresponds to 365 V (from O.C.C.)

$$\therefore \ \% \textbf{ Regulation} = \frac{365 - 440}{440} \times 100 = -\textbf{17.05\%}. \ \textbf{(Ans.)}$$

Example 5.34. *The following data pertains to a 5000 kVA, 6600 V, 3-phase, 50-Hz star-connected alternator :*

Field current, A	: 32	50	75	100	140
O.C. terminal voltage, V	: 3100	4900	6600	7500	8300
Full-load current					
Zero p.f. test line p.d., V	: 0	1850	4250	5800	7000

Determine the regulation by zero power factor method at full-load unity power factor. Neglect armature resistance.

Solution. Plot the O.C.C. and full-load zero p.f. curve from the given data as shown in Fig. 5.50.

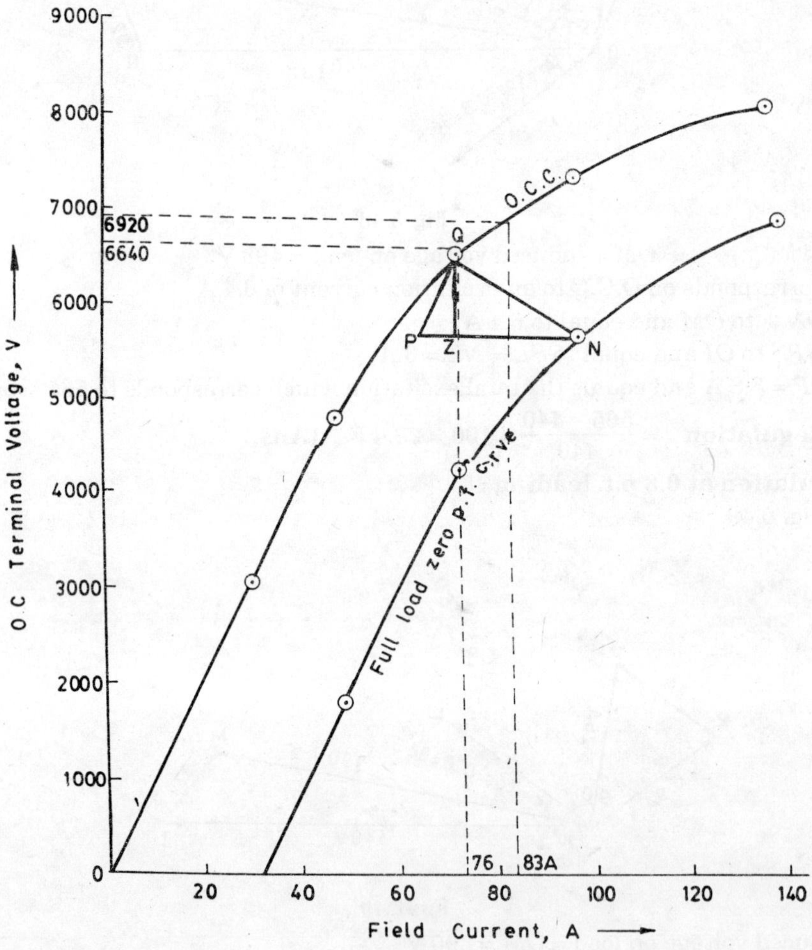

Fig. 5.50

From the Potier triangle *NPQ*, field current required to overcome *demagnetising effect of armature reaction on full-load NZ* = 25 A.

Full-load reactance drop, *QZ* = 800 V

- Draw *OL* = 6600 V.
- Draw *LM* = *QZ* = 800 V (full-load reactance drop) ⊥ to *OI* (or *OL* since p.f. is unity *i.e.*, $\phi = 0$)

OM = induced voltage on load = 6640 V

From O.C.C., the field current required to induce 6640 V = 76 A

- Draw $ON \perp$ to OM and equal to 76 A, and $NP \parallel$ to OI and equal to 25 A.

Total field current = OP = 83 A

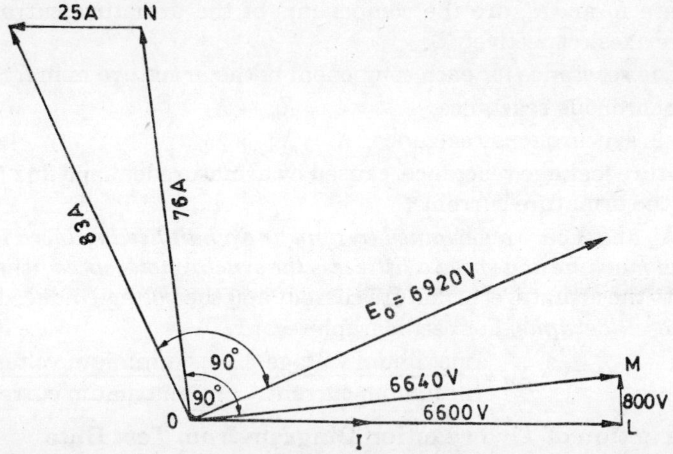

Fig. 5.51

From O.C.C., the line e.m.f. corresponding to this field current of 83 A = 6920 V

$$\therefore \ \textbf{\% Regulation} = \frac{6920 - 6600}{6600} \times 100 = \textbf{4.85 \%}. \ \textbf{(Ans.)}$$

5.14. TWO-REACTANCE CONCEPT FOR SALIENT POLE SYNCHRONOUS MACHINES

5.14.1. General Aspects

- In case of a *multi-polar machine having* **cylindrical rotor**, the air gap is uniform and as a consequence its reactance remains the same, irrespective of the position occupied by the rotor. Thus the effect of armature reaction, fluxes and voltage induced can be treated in a single way with the concept of a synchronous reactance and taking it as constant for all positions of field poles w.r.t. the armature.
- Refer Fig. 5.52. In case of **salient pole machine**, the reluctance of the magnetic path on which the e.m.f. acts is **different** *along the "direct axis" and "quadrature axis"*.
  - — The *reluctance of the* **direct axis** magnetic circuit is *due to yoke and teeth of the stator, air gap pole and core of the rotor*.
  - — The *reluctance, in* **quadrature axis**, is *mainly due to large air-gap in the inter-polar space*.
- When there is a *phase difference* between armature and excitation voltage (some angle between 0° and 90°), the armature m.m.f. will have a *direct acting* as well as *a quadrature component*.
  - — The *'direct-acting component' is proportional to the sine of the phase angle* between the armature current and excitation voltage.

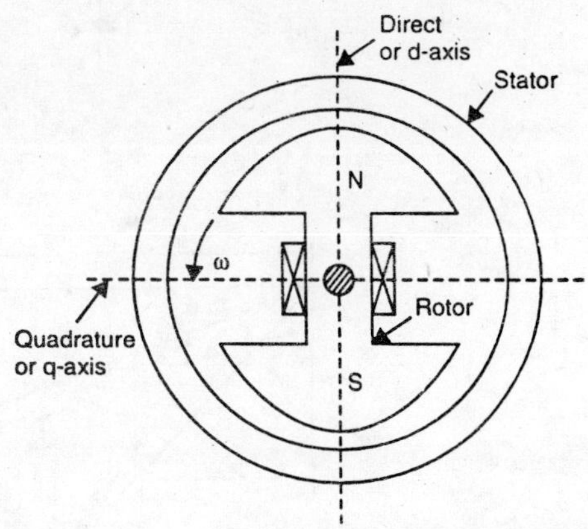

Fig. 5.52

— The quadrature (or cross) component is proportional to the *cosine of the angle*, between the armature current and excitation voltage.

- The two-reactance concept for salient pole machines replaces the effect of armature reaction by two fictitious voltage ; these reactance voltages are respectively, $I_d X_{ad}$ and $I_q X_{aq}$, where I_d and I_q are the components of the armature current along direct and quadrature axes respectively.

The synchronous reactance for each component of the armature m.m.f. is as follows :

Direct axis synchronous reactance, $X_d = X_{ad} + X_L$

Quadrature axis synchronous reactance, $X_q = X_{aq} + X_L$

where X_L = The armature leakage reactance, caused by armature leakage flux (same value for each of the components of the armature current).

The values of X_d and X_q are *determined by applying a balanced reduced external voltage, say, V volts, to an unexcited machine at a speed a little less the synchronous speed*, (the slip being less than 1%). Applied voltage to the armature, armature current and the voltage induced in the field winding are measured by the *oscillographs*. For oscillograph record :

$$X_d = \frac{\text{maximum voltage}}{\text{minimum current}} \; ; X_q = \frac{\text{minimum voltage}}{\text{maximum current}} .$$

5.14.2. Construction of Two-reaction Diagram from Test Data

After finding the values of X_d and X_q the two reaction may be constructed as follows : Refer Fig. 5.53.

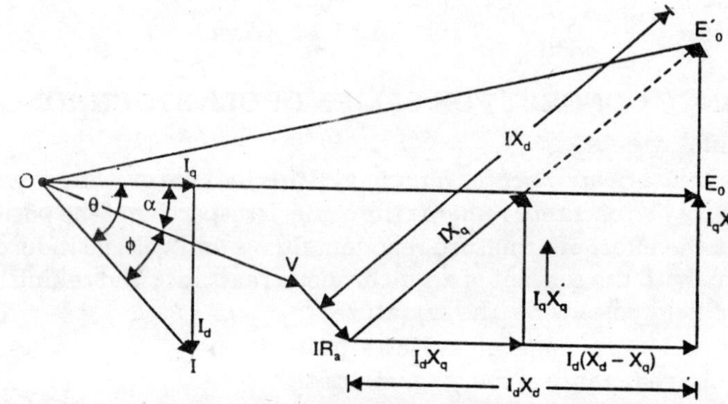

Fig. 5.53

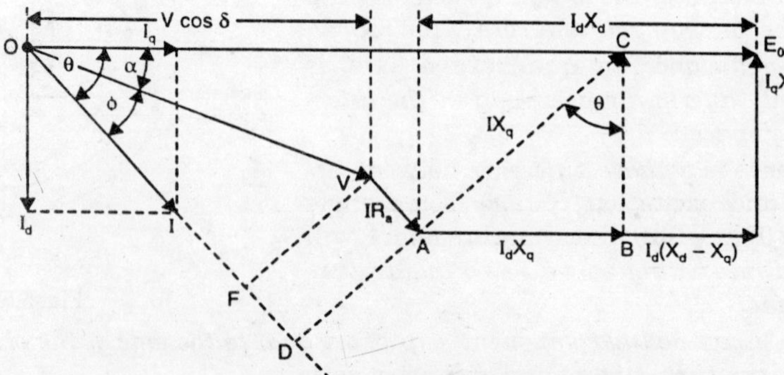

Fig. 5.54

- **From the load specified, draw the voltage phasor and current phasors at the load power factor angle.**
- **From the extremity of the voltage phasor draw a line parallel to current phasor OI and equal to IR_a.**
- **From the extremity of the phasor IR_a draw a line perpendicular to current phasor OI and equal to IX_q. Then draw a line from the origin O, passing through the extremity of IX_q phasor ; this line gives the direction of excitation voltage E_0.**
- **After knowing the direction of E_0 determine the values of I_d and I_q. Thereafter, determine the value of E_0 by adding vectorially I_dX_d and I_qX_q to the extremity of resistive drop phasor IR_a.**

  It may be noted that the *triangle involving the voltages IX_q and I_qX_q is similar to the triangle of the currents*. Therefore, the left hand portion of the phase I_dX_d must be equal to I_dX_q.
- **Produce the perpendicular drawn from the extremity of the phasor IR_a in such a way that its length equals ($IX_d - IX_q$) and obtain the phasor E_0' (which is normally obtained from the synchronous impedance method). In Fig. 5.54 the extremity of E_0' is shown on the line perpendicular to I_dX_d and it is slightly greater than E_0, making the regulation as obtained by the synchronous impedance method greater than that obtained by the two reaction method. The *difference depends upon the saliency (ratio of pole pitch to pole arc) and increases with the increase in the value of saliency*. The major *difference between the two methods is the magnitude of the torque angle, which is much larger when determined from the syncrhonous impedance phasors than when found from the two reaction diagram.***
- **The angle θ between E_0 and I is called the '*internal power factor angle*', while α, the angle between E_0 and V is called the '*load or power angle*'.**
- **Fig. 5.54 shows the phasor diagram redrawn, with E_0 drawn in horizontal direction.**

Here, $I_d = I \sin \theta \; ; I_q = I \cos \theta$

From ΔODC,

$$\tan \theta = \frac{CD}{OD} = \frac{DA + AC}{OF + FD}$$

$$= \frac{V \sin \phi + IX_q}{V \cos \phi + IR_a} \quad \text{...(5.13)} \quad (Generating)$$

$$= \frac{V \sin \phi - IX_q}{V \cos \phi - IR_a} \quad \text{...(5.14)} \quad (Motoring)$$

The angle θ can be found out from the above equations.

Load angle $\alpha = \theta - \phi$ *(Generating)*

$= \phi - \theta$ *(Motoring)*

Angle φ is taken + ve and – ve for lagging and leading power factors respectively.

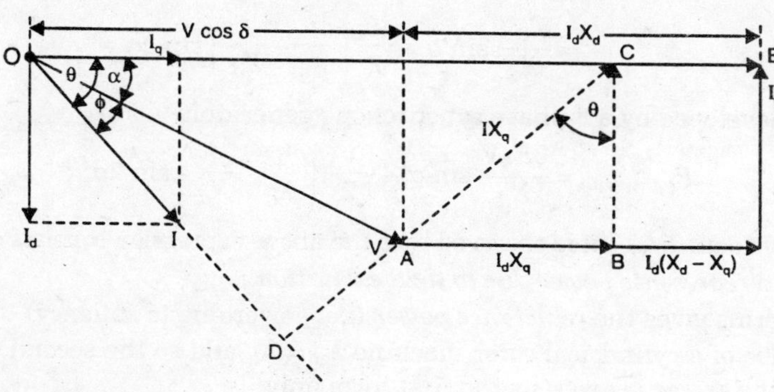

Fig. 5.55

Fig. 5.55 shows the phasor diagram when armature resistance is neglected.

The load angle, $\alpha = \theta - \phi$ for *generating* ; $\alpha = (\phi - \theta)$ *for motoring*.

or $$\theta = \phi \pm \alpha \qquad \textit{General case} \qquad ...(5.15)$$

Direct axis component, $I_d = I \sin \theta = I \sin (\phi \pm \alpha)$ ...(5.16)

Quadrature axis component, $I_q = I \cos \theta = I \cos (\phi \pm \alpha)$ ...(5.17)

$$V \sin \delta = I_q X_q = I \cos \theta \,.\, X_q = IX_q \cos (\phi \pm \alpha) = IX_q (\cos \phi \cos \alpha + \sin \phi \sin \alpha)$$

or $$V \sin \delta = IX_q \cos \phi \cos \alpha \mp IX_q \sin \phi \sin \alpha$$

or $$V = IX_q \cos \phi \cot \alpha \mp IX_q \sin \phi$$

or $$IX_q \cos \phi \cot \alpha = V \pm IX_q \sin \phi$$

or $$\tan \alpha = \frac{IX_q \cos \phi}{V \pm IX_q \sin \phi} \qquad ...(5.18)$$

(+ ve and – ve signs correspond to synchronous generator and synchronous motor respectively.)

When R_a is neglected, we have

$$E_0 = V \cos \alpha \pm I_d X_d$$

When R_a is considered, we have

$$E_0 = V \cos \alpha + IR_a \cos \theta + I_d X_d$$

or $$E_0 = V \cos \alpha + I_q R_a + I_d X_d \qquad \textit{(Generating)} \quad ...(5.19)$$

and $$E_0 = V \cos \alpha - I_q R_a - I_d X_d \qquad \textit{(Motoring)} \quad ...(5.20)$$

5.14.3. Power Developed by a Synchronous Generator

Power developer per/phase, P_{phase} = power output (P_{out}) per phase

$= VI \cos \phi$ (when R_a or copper loss is negligible)

From Fig. 5.55, we have

$$I_q X_q = V \sin \alpha \qquad ...(i)$$

$$I_d X_d = E_0 - V \cos \alpha \qquad ...(ii)$$

Also, $$I \cos \phi = I_d \sin \alpha + I_q \cos \alpha$$

$$P_{phase} = VI_d \sin \alpha + VI_q \cos \alpha$$

Substituting the values of I_d and I_q from (*i*) and (*ii*), we get

$$P_{phase(dev.)} = V \,.\, \left(\frac{E_b - V \cos \alpha}{X_d} \right) \sin \alpha + V \,.\, \frac{V \sin \alpha}{X_q} \,.\, \cos \alpha$$

$$= \frac{E_0 V}{X_d} \sin \alpha - \frac{V^2 \sin \alpha \cos \alpha}{X_d} + \frac{V^2 \sin \alpha \,.\, \cos \alpha}{X_q}$$

$$= \frac{E_0 V}{X_d} \sin \alpha + V^2 \left(\frac{1}{X_q} - \frac{1}{X_d} \right) \sin \alpha \cos \alpha$$

$$= \frac{E_0 V}{X_d} \sin \alpha + \frac{V^2}{2} \left(\frac{1}{X_q} - \frac{1}{X_d} \right) \sin 2\alpha \qquad ...(5.21)$$

Total power developed by a 3-phase synchronous generator,

$$P_{total(dev.)} = \frac{3E_0 V}{X_d} \sin \alpha + \frac{3V^2}{2} \left(\frac{1}{X_q} - \frac{1}{X_d} \right) \sin 2\alpha \qquad ...(5.22)$$

From the above eqn. (5.21), it is observed that the above expression consists of the two terms :

- First term *represents power due to field excitation* ;
- Second terms gives the *reluctance power* (*i.e.* power due to saliency)
  - — In case of a cylindrical rotor machine $X_d = X_q$ and so the second term becomes zero and the power is given by the first term only.

— In the absence of field excitation (*i.e.* E_0) the first term becomes zero and the power is given by the second term.

Note. In case of an *alternator α is positive, whereas in case of a motor it is negative.*

Example 5.35. *A 2200 V, 50 Hz, 3-phase, star-connected alternator has an effective resistance of 0.5 Ω per phase. A field current of 30 A produced the full-load current of 200 A on short-circuit and a line-to-line e.m.f. of 1100 V on open-circuit. Determine :*

(i) The power angle of the alternator when it delivers full-load at 0.8 p.f. (lag).

(ii) The SCR of the alternator. **(AMIE Winter, 1994)**

Solution. *Given* : $V = 2200$ volts, $R_a (= R_e) = 0.5$ Ω/phase ; $I_f = 30$ A ; $I = 200$ A,

Open circuit voltage (line) = 1100 V ; $\cos \phi = 0.8$

(*i*) **Power angle of the alternator,** α :

Phase voltage, $$V = \frac{V_L}{\sqrt{3}} = \frac{2200}{\sqrt{3}} = 1270.2 \text{ V}$$

Synchronous impedance per phase,

$$Z_s = \frac{\text{open-circuit phase voltage}}{\text{short-circuit current per phase}} \quad \text{(for the same excitation)}$$

$$= \frac{1100/\sqrt{3}}{200} = 3.175 \ \Omega$$

Synchronous reactance per phase,

$$X_s = \sqrt{Z_s^2 - R_a^2} = \sqrt{3.175^2 - 0.5^2} = 3.136 \ \Omega$$

Open-circuit voltage per phase,

$$E_0 = \sqrt{(V \cos \phi + IR_a)^2 + (V \sin \phi + IX_s)^2}$$

$$= \sqrt{(1270.2 \times 0.8 + 200 \times 0.5)^2 + (1270.2 \times 0.6 + 200 \times 3.136)^2}$$

$$= \sqrt{(1116.16)^2 + (1389.32)^2} = 1782 \text{ V}$$

Power output per phase $$= V_{ph} I_{ph} \cos \phi = \frac{2200}{\sqrt{3}} \times 200 \times 0.8 = 203227 \text{ W}$$

Power developed per phase due to field excitation, *neglecting losses*,

$$P_{phase(dev.)} = \frac{E_0 V}{X_s} \sin \alpha$$

or $$203227 = \frac{1782 \times 1270.2}{3.136} \times \sin \alpha$$

or $$\sin \alpha = \frac{203227 \times 3.136}{1782 \times 1270.2} = 0.2816$$

∴ Power angle $\alpha = \sin^{-1} (0.2816) = \mathbf{16.36°}$. **(Ans.)**

(*ii*) **SCR of the alternator :**

Short-circuit ratio (SCR) of a synchronous machine is *defined as the ratio of field current to produce rated voltage on open circuit to field current required to circulate rated current on short-circuit while the machine is mechanically driven at synchronous speed.*

$$SCR = \frac{1}{\left[\dfrac{\text{Per unit voltage on open circuit}}{\text{Corresponding per unit current on short circuit}}\right]} = \frac{1}{Z_s} = \frac{1}{X_s} \quad (R_a << X_s)$$

∴ $$SCR = \frac{1}{3.136} = \mathbf{0.319}. \quad \textbf{(Ans.)}$$

[The SCR is reciprocal of per unit synchronous reactance X_s. The *value of X_s depends upon saturated conditions of the machine but SCR is specific and defined at rated voltage.*]

Example 5.36. *A 3-phase synchronous generator has per phase a direct axis synchronous reactance of 1.0 per unit and a quadrature axis synchronous reactance of 0.65 per unit. Draw a phasor diagram of the machine when operating at full-load at a power factor of 0.8 lagging and estimate from there :*

(*i*) *The load angle* ; (*ii*) *Per unit no-load e.m.f.*

Neglect armature resistance. **(AMIE Advanced Elect. M/cs Winter, 1996)**

Solution. *Given* : X_d = 1.0 p.u. ; X_q = 0.65 p.u. ; cos ϕ = 0.8,

or $\phi = \cos^{-1}(0.8) = 36.87°$ and $\sin\phi = \sin 36.87° =$ **0.6.**

(*i*) **Load angle,** α **:**

Terminal voltage, V = 1.0 p.u.

Armature current, I = 1.0 p.u.

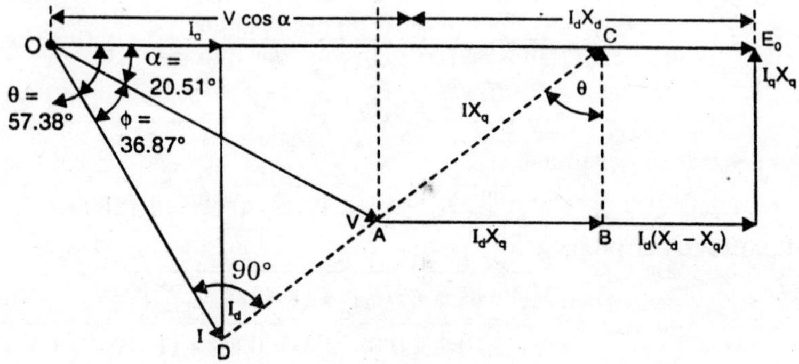

Fig. 5.56

From phasor diagram, Fig. 5.56, we have

$$\tan\theta = \frac{DC}{OD} = \frac{DA + AC}{OD} = \frac{V\sin\phi + IX_q}{V\cos\phi} = \frac{1.0 \times 0.6 + 1.0 \times 0.65}{1 \times 0.8} = 1.5625$$

and $\theta = \tan^{-1}(1.5625) = 57.38°$

$\therefore$ Load angle, $\alpha = \theta - \phi = 57.38° - 36.87° =$ **20.51°. (Ans.)**

(*ii*) **Per unit no-load e.m.f., E_0 :**

Direct axis component of current, $I_d = I\sin\theta = 1.0 \times \sin 57.38° = 0.8423$

$\therefore$ No-load e.m.f., $E_0 = V\cos\alpha + I_dX_d = 1.0 \times \cos(20.51°) + 0.8423 \times 1.0$

= **1.779 per unit. (Ans.)**

Example 5.37. *A 5 kVA, 220 V, star-connected 3-phase salient pole alternator with direct and quadrature axes reactances of 12 Ω and 7 Ω respectively, delivers full-load current at unity power factor. Calculate the excitation voltage, neglecting resistance.* **(Bharathiyar Univ. April, 1997)**

Solution. *Given* : Rating : 5 kVA, 220 V ; X_d = 12 Ω, X_q = 7 Ω, cos ϕ = 1

i.e. $\phi = \cos^{-1}(1) = 0°$ and $\sin 0° = 0$

Armature current, $$I = \frac{5 \times 1000}{\sqrt{3} \times 220 \times 1} = 13.122 \text{ A}$$

Also, $$\tan\alpha = \frac{IX_q\cos\phi}{V + IX_q\sin\phi}$$ [Eqn. (5.18)]

$$= \frac{13.122 \times 7 \times 1}{(220/\sqrt{3}) + 13.122 \times 7 \times 0} = 0.723$$

∴ Load angle, $\alpha = \tan^{-1} (0.723) = 35.87°$

Angle, $\theta = \phi + \alpha = 0° + 35.87° = 35.87°$

Now, $I_d = I \sin \theta = 13.122 \sin 35.87° = 7.689$ A

Excitation voltage, $E_0 = V \cos \alpha + I_d X_d$

$$= \left(\frac{220}{\sqrt{3}}\right) \times \cos (35.87°) + 7.689 \times 12 = 195.2 \text{ V per phase}$$

Excitation line voltage, $\mathbf{E_{OL}} = \sqrt{3} \times 195.2 =$ **338.1 V. (Ans.)**

Example 5.38. *A 10 kVA, 380 V, 50 Hz, 3-phase, star-connected salient pole alternator has direct axis and quadrature axis reactances of 12 Ω and 8 Ω respectively. The armature has a resistance of 1 Ω per phase. The generator delivers rated load at 0.8 power factor lagging with the terminal voltage being maintained at rated value. If the load angle is 16.15° determine :*

(i) The direct axis and quadrature axis components of armature current.

(ii) Excitation voltage of the generator. **(GATE, 1993)**

Solution. *Given* : Rating : 10 kVA, 380 V ; $X_d = 12$ Ω, $X_q = 8$ Ω, $R_a = 1$ Ω/phase,

$$\cos \phi = 0.8 \text{ or } \phi = \cos^{-1} (0.8) = 36.87° ; \alpha = 16.15°$$

(*i*) $\mathbf{I_d, I_q}$ **:**

Refer Fig. 5.57.

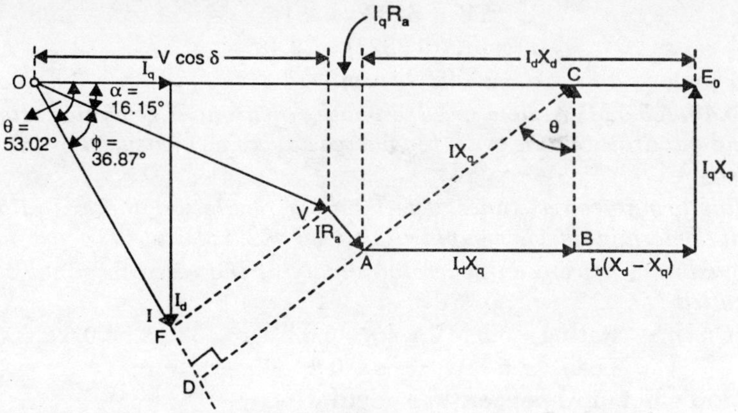

Fig. 5.57

$$\theta = \phi + \alpha = 36.87° + 16.15° = 53.02°$$

Terminal voltage per phase, $V = \dfrac{380}{\sqrt{3}} = 219.4$ V

Armature current, $I = \dfrac{10 \times 1000}{\sqrt{3} \times 380} = 15.2$ A

The phasor diagram is shown in Fig. 5.57.

Now, $\mathbf{I_d} = I \sin \theta = 15.2 \times \sin (53.02°) =$ **12.14 A. (Ans.)**

$\mathbf{I_q} = I \cos \theta = 15.2 \times \cos (53.02°) =$ **9.14 A. (Ans.)**

(*ii*) **Excitation voltage of generator, $\mathbf{E_0}$:**

$$E_0 = V \cos \alpha + I_q R_a + I_d X_d$$

$$= 219.4 \times \cos (16.15°) + 9.14 \times 1 + 12.14 \times 12 = 365.56 \text{ V/phase}$$

∴ E_0 (line-to-line) $= \sqrt{3} \times 365.56 =$ **633.2 V. (Ans.)**

Example 5.39. *A 3-phase star-connected alternator is delivering 20 MW and 8 MVA to an infinite bus at 11 kV. The alternator has synchronous impedance of 0 + j3Ω. Determine the load angle and the excitation emf of the alternator.* **(AMIE Summer, 2000)**

Solution. For an alternator, $E = V + IZ_s$

Taking V as reference ; $VI \cos \phi = 20 \text{ MW} = 20 \times 10^6 \text{ W}$

$VI \sin \phi = 8 \text{ MVA} = 8 \times 10^6 \text{ VARs}$

$\therefore \quad \tan \phi = \frac{8}{20} = 0.4 \text{ or } \phi = \tan^{-1} (0.4) = 21.8°$

Z_s/phase $= 0 + j3$ ...(Given)

Load current, $I = \frac{20 \times 10^6}{\sqrt{3} \times 11000 \times \cos (21.8°)} = 1130.6 \text{ A}$

$\therefore$ Excitation e.m.f. of the alternator (per phase),

$$\mathbf{E} = \frac{11 \times 1000}{\sqrt{3}} \angle 0° + (1130.6 \angle -21.8°) \times 3\angle 90°$$

$$= 6351 + 3392\angle 68.2° = 6351 + 1260 + j3149$$

$$= 7611 + j3149 = \mathbf{8237 \angle 22.48°}. \ \textbf{(Ans.)}$$

Now, $P = \frac{EV}{X_s} \sin \alpha$, neglecting losses

or $\sin \alpha = \frac{P.X_s}{EV} = \frac{\frac{20}{3} \times 10^6 \times 3}{8237 \times 6351} = 0.3823$

or $\alpha = \sin^{-1} (0.3823) = 22.48°$

Hence load angle $\alpha = \mathbf{22.48°}$. **(Ans.)**

Example 5.40. *A 3.5 MVA, slow speed, 3-phase synchronous generator rated at 6.6 kV has 32 poles. Its direct and quadrature axes synchronous reactances as measured by the slip test are 9.6 Ω and 6 Ω respectively,*

(i) Neglecting armature resistance, determine the regulation and excitation e.m.f. needed to maintain 6.6 kV at the terminals when supplying a load of 2.5 MW at 0.8 power factor lagging.

(ii) What maximum power can generator supply at the rated terminal voltage, if the field becomes open-circuited ? **(U.P.S.C., 1994)**

Solution. *Given* : Rating = 3.5 MVA, $V_L = 6.6$ kV, $p = 32$; $X_d = 0.6\ \Omega$; $X_q = 6\ \Omega$;
Load = 2.5 MW, $\cos \phi = 0.8$.

(*i*) Excitation e.m.f. and percentage regulation :

Terminal voltage per phase, $V = \frac{6.6 \times 1000}{\sqrt{3}} = 3810.5 \text{ V}$

Armature current, $I = \frac{2.5 \times 10^6}{\sqrt{3} \times 6600 \times 0.8} = 273.37 \text{ A}$

Load phase angle, $\phi = \cos^{-1} (0.8) = 36.87°$, $\sin \phi = \sin 36.87° = 0.6$

We know that, $\tan \alpha = \frac{IX_q \cos \phi}{V + IX_q \sin \phi} = \frac{273.37 \times 6 \times 0.8}{3810.5 + 273.37 \times 6 \times 0.6} = 0.2737$

or $\alpha = \tan^{-1} (0.2737) = 15.3°$

$\therefore$ Angle $\theta = \phi + \alpha = 36.87° + 15.3 = 52.17°$

Now, $I_d = I \sin \theta = 273.37 \times \sin (52.17°) = 215.92 \text{ A}$

Excitation e.m.f. per phase, $E_0 = V \cos \alpha + I_d X_d$

$$= 3810.5 \times \cos (15.3°) + 215.92 \times 9.6 = 5748.3 \text{ V}$$

Excitation voltage (line-to-line) $= \sqrt{3} \times 5748.3 = \mathbf{9956.3 \ V.}$ **(Ans.)**

Percentage regulation $= \frac{9956.3 - 6600}{6600} \times 100 = \mathbf{50.85\%}.$ **(Ans.)**

(ii) **Maximum power the generator can supply, with field open-circuited :**

The *power developed will be maximum* for *sin* 2α = 1 and so the *maximum power, that the generator can supply at the rated terminal voltage, per phase, with field open-circuited*

$$= \frac{V_L^2}{2}\left(\frac{1}{X_q} - \frac{1}{X_d}\right) \times 1 = \frac{(6600)^2}{2}\left(\frac{1}{6} - \frac{1}{9.6}\right)$$

$$= 1361250 \text{ W or } \mathbf{1.361 \text{ MW.}} \quad \textbf{(Ans.)}$$

Example 5.41. *A 3-phase, 17.32 kVA, 400 V, star-connected alternator is delivering rated load at 400 V and at p.f. 0.8 lag. Its synchronous impedance is 0.2 + j2 Ω per phase. Find the load angle at which it is operating.*

Now, with the magnitude of excitation voltage held constant, the nature of rated load p.f. is altered to p.f. 0.8 leading. Determine the new value of terminal voltage and the load angle.

(AMIE Winter, 1998)

Solution. *Given :* Rating = 17.32 kVA, 400 V, star-connected ;

cos φ = 0.8, Z_s = (0.2 + *j*2 Ω) per phase

Load angle, α (between *E* and *V*) :

(i) When cos φ = 0.8 (lagging) :

Load current, $I = \frac{17.32 \times 1000}{\sqrt{3} \times 400} = 25 \text{ A}$

Phase voltage, $V = \frac{400}{\sqrt{3}} = 231 \text{ V}$

Now,

$$E = V + IZ_s$$
$$= 231 + 25(0.8 - j0.6)(0.2 + j2)$$
$$= 231 + 25(0.16 + j1.6 - j0.12 + 1.2)$$
$$= 231 + 25(1.36 + j1.48) = 231 + 34 + j37$$
$$= 265 + j37 = 267.57\angle 7.95°$$

Hence load angle α between *E* and *V* = **7.95°. (Ans.)**

Load angle and new terminal voltage :

(ii) When cos φ = 0.8 (leading), excitation voltage held constant.

In this case,

$$E = V + 25(0.8 + j0.6)(0.2 + j2)$$
$$267.57 = V + 25(0.16 + j1.6 + j0.12 - 1.2)$$

or $$267.57 = V + 25(-1.04 + j1.72) = V - 26 + j43$$

or $$(V - 26)^2 + (43)^2 = (267.57)^2$$

or $$(V - 26) = \sqrt{(267.57)^2 - (43)^2} = 264$$

or $$V = 264 + 26 = \mathbf{290 \text{ V.}} \quad \textbf{(Ans.)}$$

∴ **Load angle,** $\alpha = \frac{43}{(V - 26)} = \frac{43}{290 - 26} = \frac{43}{264} = \mathbf{9.25°}.$ **(Ans.)**

Example 5.42. *A 10000 kVA, 3-phase, star-connected 11000 V, 2-pole turbo-generator has a synchronous impedance of (0.0145 + j0.5) ohms per phase. The various losses in this generator are as follows :*

Open-circuit core loss at 11000 V	*= 90 kW ;*
Windage and friction loss	*= 50 kW ;*
Short-circuit load loss at 525 A	*= 220 kW ;*
Field winding resistance	*= 3 Ω ;*
Field current	*= 175 A.*

Ignoring the change in field current, compute the efficiency at :

(i) rated load, 0.8 power factor leading ;

(ii) half rated load, 0.9 power factor lagging. **(GATE, 1996)**

Solution. Phase current, $I_{ph} = \frac{10000 \times 1000}{\sqrt{3} \times 11000} \simeq 525$ A

Field copper loss $= \frac{I_f^2 R_f}{1000} = \frac{(175)^2 \times 3}{1000} = 91.875$ kW

Efficiency :

(i) At full load, 0.8 power factor (lead) :

Output $= 10000 \times 0.8 = 8000$ kW

Total losses $= 90 + 50 + 220 + 91.875 = 451.875$ kW

Efficiency, $\eta = \frac{\text{output}}{\text{output + losses}} = \frac{8000}{8000 + 451.875} = 0.9465$ or **94.65%. (Ans.)**

(ii) At half load, 0.9 power factor (lag) :

Output $= \frac{1}{2} \times 10000 \times 0.9 = 4500$ kW

Armature copper loss $= \left(\frac{1}{2}\right)^2 \times 220 = 55$ kW

Total losses $= 90 + 50 + 55 + 91.875 = 286.875$ kW

$\therefore$ Efficiency $\eta = \frac{4500}{4500 + 286.875} = 0.94$ or **94%. (Ans.)**

5.15. LOSSES AND EFFICIENCY

Losses. The following losses occur in an alternator :

1. **Copper Losses.** These losses occur in the armature winding and in the field coils.

2. **Core Loss.** The core loss consists of eddy-current and hysteresis loss in the pole faces, teeth, and stator core due to the flux resulting from the combined rotor and armature fields.

3. **Friction and Windage Loss.** This loss is due to the bearing and brush friction and to the power required to circulate the cooling air.

4. **Load Loss.** This is due to the *armature leakage flux* which causes eddy-currents and hysteresis in the iron surrounding the armature conductors. If, however, effective resistance is used to calculate the armature copper loss, then these load losses are also included in the calculation.

Efficiency. The efficiency of an alternator is calculated as follows :

$$\text{Alternator efficiency} = \frac{\text{output}}{\text{output + losses}} = \frac{\text{kVA (p.f.)}}{\text{kVA(p.f.) + losses}}$$

Maximum efficiency occurs at that load point where the constant losses (friction, windage, core loss and field copper loss) *are equal to variable losses* (armature copper and load loss). The maximum efficiency usually occurs at about 80% of full load.

Determination of Losses. Losses may be determined by the following methods :

Measurement of losses method. As in most electric machines, efficiency measurement of an alternator by direct loading is rather impractical. It may also be a physical impossibility to obtain the required load, and even if it could be obtained, the cost may be prohibitive. Also, the measurement of the mechanical power input is somewhat difficult, and any inaccuracy with measurement is *reflected directly in the final efficiency calculation*. Efficiency is therefore calculated by '*measurement of losses*' method which entails the following *advantages* :

(*i*) At any one time, only part of the losses need be provided. Therefore the source of power required for testing has a capacity of less than 5% of the rating of the alternator.

(*ii*) There is no need to put an electric load on the alternator.

(*iii*) Greater accuracy can be obtained since electrical instruments can be used for all measurements.

(*iv*) An inaccuracy that occurs in the test *is not directly reflected in the final efficiency calculation*, since the error occurs on only a small portion of the name plate rating.

Use of a calibrated D.C. motor to drive the alternator. A *quite accurate and simple method* of determining the losses is to use a calibrated D.C. motor to drive the alternator. The motor is calibrated in the sense that all its losses have been determined for varying conditions of operation, so that its output is then readily obtained.

This method involves the following procedural *steps* :

Step 1. Drive the alternator at synchronous speed, but *without field excitation*. The D.C. motor output is the alternator input, and hence its *friction and windage loss*.

Step 2. Repeat step 1, but this time *with field excitation*. The excitation should be that at which the alternator normally operates. If this is unknown, then the open-circuit voltage is adjusted to be equal to the rated voltage plus the internal voltage drops as determined by one of the methods used to calculate voltage regulation.

The difference between the motor output of step 2 and that of step 1 is the *core loss*.

Step 3. Short-circuit the armature and adjust the field current to obtain rated line current. The difference between this motor output and that of step 1 is the *armature copper loss plus load loss*. It is assumed that the flux density under short-circuit conditions is so low as to make the core loss negligible.

Step 4. The *field copper loss* is measured by simple D.C. measurement.

Example 5.43. *A 2 MVA, 3.3 kV, 3-phase, star-connected alternator delivers rated load at a p.f. 0.8 lagging. The effective resistance between armature winding terminals is 0.12 Ω. The field takes 70 A at 130 V. Friction and windage loss is 15 kW and the core losses are 32 kW.*

Determine the full-load efficiency of the alternator.

Solution. Load p.f. = 0.8

Line voltage = 3300 V

Armature effective resistance, R_a = 0.12 Ω

Field current = 70 A at 130 V

Friction and windage loss = 15 kW

Core losses = 32 kW

Full-load efficiency :

Output of 2 MVA alternator at 0.8 p.f. (lag)

$$= 2 \times 1000 \times 0.8 = 1600 \text{ kW}$$

Load current on full-load, $I = \dfrac{2000 \times 1000}{\sqrt{3} \times 3300} = 350$ A

Field copper loss $= \dfrac{70 \times 130}{1000} = 9.1$ kW

Armature copper loss $= \dfrac{3 \times (350)^2 \times 0.12}{1000} = 44.1$ kW

Total losses = 15 + 32 + 9.1 + 44.1 = 100.2 kW

Power input = 1600 + 100.2 = 1700.2 kW

Full-load efficiency $= \dfrac{\text{output}}{\text{input}} \times 100 = \dfrac{1600}{1700.2} \times 100 =$ **94.1%. (Ans.)**

5.16. PARALLEL OPERATION OF ALTERNATORS

5.16.1. Necessity. Alternators may be put in parallel because of the following *reasons* :

1. Local or regional power use may *exceed* the power of a *single* available generator.

2. Parallel alternators allow one or more units to be *shut down* for scheduled or emergency *maintenance* while the load is being supplied with power.

3. Generators are *inefficient at part load,* so shutting down one or more generators allows the remaining load to be carried with less machines that are efficiently loaded.

4. Load growth can be handled by *added* machines without disturbing the original installation.

5. Available machine primemovers and generators can be *matched* for economic first cost and flexible use.

5.16.2. Requirements for Paralleling. The *requirements* for paralleling include the requirements for D.C. machines plus a few others.

1. The *voltages* must be the *same* at the *paralleling point* or junction even though not the same at the alternators.

2. The *phase sequence* for three phase (or any multiple phase) must be the *same* at the paralleling point.

3. The incoming machine must be *in phase* at the moment of paralleling. It will stay in phase under normal conditions after paralleling. It is important to recognize that *phase sequence* and *in phase* are *not the same thing*.

4. The line frequencies must be *identical* at the paralleling point. In the vast majority of cases, this means the same frequency at the generator because frequency changing is not economic. Mixed frequencies must be paralleled through some frequency conversion means for compatibility at the point of interconnection.

5. The primemovers must have relatively similar and *drooping speed-load characteristics*. This is to prevent a machine with a rising speed load characteristic from taking more and more of the load until it fails from overload.

Violation of these requirements for paralleling would result in circulating currents between the machines varying from uneconomic, to serious, to disastrous.

5.16.3. Alternator Synchronising Procedure. It may logically be assumed that one alternator is placed in parallel with one or more other alternators only when additional load requires it. Those alternators already carrying load are known as the *running machines,* while that which is to be placed in the system is known as the *incoming machine*. At the time of synchronizing, the following *conditions* must be met.

1. The effective voltage of the incoming alternator *must be exactly equal to that of the others, or of the bus-bars connecting them.*

2. The *phase rotation,* or sequence of the running and incoming alternators, *must be the same.*

3. The individual phase voltages which are to be connected to each other must be in *exact phase opposition*. This is the same as saying that D.C. generators must be connected + to + and – to –.

4. The frequencies should be the same, although it is more desirable that the frequencies at the *instant* of paralleling be almost, but not quite, identical.

Synchronizing lamps :

- To satisfy the first condition of paralleling (*i.e.*, effective voltages be the same) a voltmeter can be used as shown in (Fig. 5.58).
- Satisfaction of the other conditions of *phase sequence, voltage opposition,* and *frequency* may be *determined by the use of the incandescent lamps connected between the two machines* (Fig. 5.58). The auxiliary equipment is omitted to simplify the sketch.

At any instant it is seen that the voltage across the lamp is the sum of the individual phase voltages.

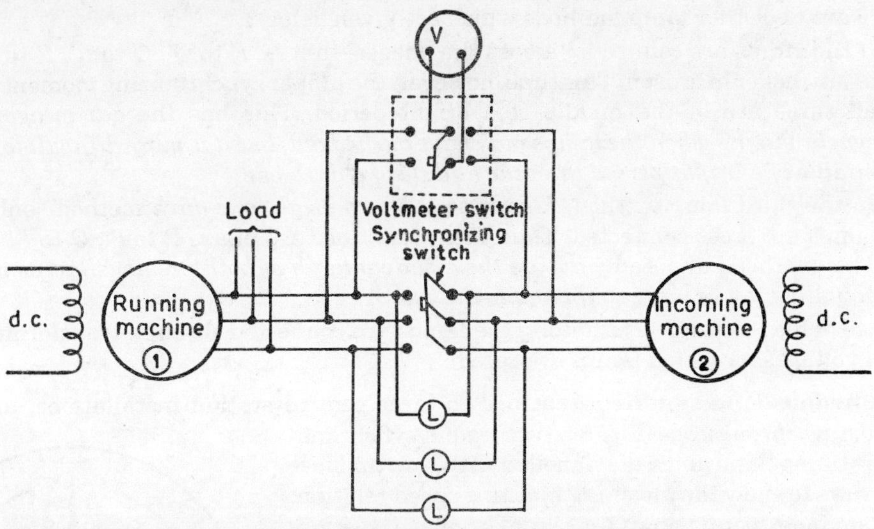

Fig. 5.58. Parallel operation of alternators : Synchronizing lamps and voltmeter.

The procedural steps for putting incoming alternator in parallel with the running machine are as follows :

Step 1. The primemover of the incoming machine is started, and the alternator is brought up to near its rated speed.

Step 2. By adjusting the field current, the terminal voltage of incoming machine is made the same as that of the running alternator.

The lamp in the circuit will now flicker at a rate equal to the difference in frequency of the two alternators. *If the phases are properly connected, all lamps will be bright and dark at the same time.* If this is not the case, then it means that the phase sequences are not correct, and it is merely necessary to interchange two of the line leads of the incoming machine.

Step 3. Further adjustment of the incoming primemover is now necessary, until the lamps flicker at a very low rate, *usually less than one dark period per second.*

Step 4. Final adjustment of the incoming voltage again made and the synchronizing switch is thrown in the middle of a dark period. The voltage across the lamps varies from zero to twice the phase voltage, and therefore, the lamps must be rated for this higher voltage. It is not convenient, however, to stock special lamps for this purpose, and two lamps of standard voltage ratings may be connected in series in each line.

The *lamp method* has the following *advantages* and *disadvantages :*

Advantages :

(*i*) Equipment is inexpensive.

(*ii*) Proper phase sequence is readily obtained.

Disadvantages :

(*i*) Lamps go dark at somewhat less than half their rated voltage, and so the synchronizing switch might be closed when there is a considerable phase difference between the machines, with a high circulating current resulting in possible damage to the machines. An experienced operator, however, can minimize this danger, since he can quickly learn to estimate the middle of the dark period.

(*ii*) The lamp filament can burn out.

(*iii*) The flicker of the lamps does not indicate which machine has the higher frequency.

There are two other lamp methods which are given below :

- If the lamps are connected across the phases, that is, P to Q', Q to R', R to P', they will again flicker in unison. This time, however, the proper synchronizing moment occurs when all three are in the middle of a bright period. This has the *advantage of avoiding synchronizing when the lamps may have burned out, but it is more difficult to estimate the middle of a bright period than the middle of a dark one.*
- In the third lamp method, known as the '*two-bright one-dark* method', only two of the lamps are cross-connected. Thus the connections are phase P to P', Q to R' and R to Q'. This method supposedly *avoids the disadvantages of both previous lamps methods, but actually may be more confusing to the eye.*

In case of high-voltage alternators, the lamps are connected through transformers to obtain the nominal voltages for which lamps are rated.

Synchronoscope synchronization. In large central station installations, an additional device called a *synchronoscope* (Fig. 5.59) is used. A synchronoscope has a rotating hand and a dial labelled with slow and fast direction arrows to show the incoming machine speed relation. In addition, an index point shows the actual in-phase position. During synchronization, as the incoming machine rotational speed approaches near synchronism, the speed of the synchronoscope hand drops enough to become visible. The *hand speed is proportional to the difference in speed.* The slow indication is accompanied with an arrow showing that *counter-clockwise* hand rotation means *below* synchronous speed. Similarly *clockwise* rotation means above speed. When the speeds are matched as that the hand speed is very slow, the hand is matched until it points to the index mark, where upon the paralleling switch may be closed.

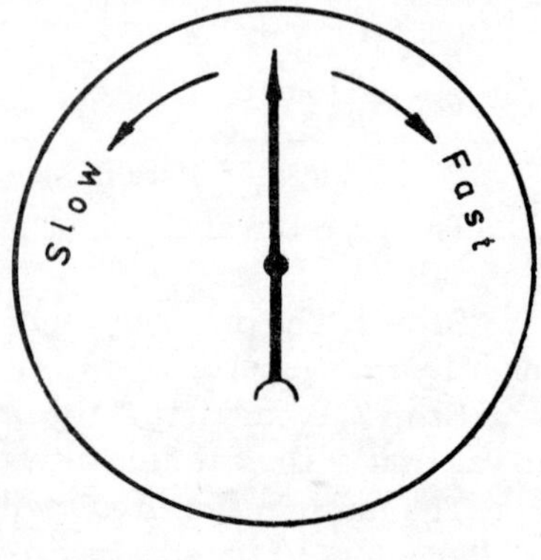

Fig. 5.59. Synchronoscope.

There are following three principal types of synchronoscopes :

(*i*) Polarized vane. (*ii*) Moving iron.

(*iii*) Crossed coil.

All are used similarly. However, *none* of the types will detect an *out-of-phase sequence condition.* They are used as a convenient and accurate means of routinely achieving synchronous speed and in-phase indications. The phase sequence problem is really an original test problem since, once sorted out, the phase sequence will remain correct until repair or other circuit changes are needed.

5.16.4. Synchronising Current, Power and Torque

Synchronising current :

- When in exact synchronism, the two alternators have equal inducted e.m.fs. and are in exact phase opposition as shown in Fig. 5.60.

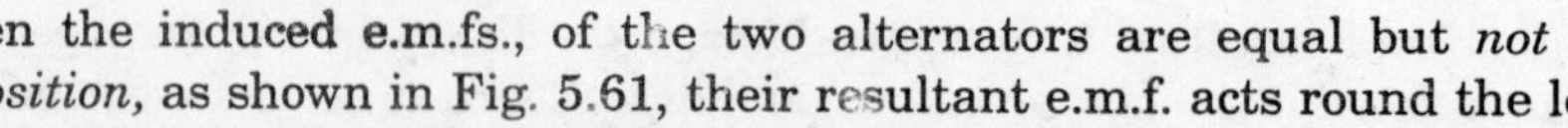

Fig. 5.60

- When the induced e.m.fs., of the two alternators are equal but *not in exact phase opposition,* as shown in Fig. 5.61, their resultant e.m.f. acts round the local circuit and causes flow of current called the '*synchronising current*'.

Let $(180° - \alpha)$ = angular phase difference between the two induced e.m.fs. E_1 and E_2 ;

E = each e.m.f. (phase e.m.f.)

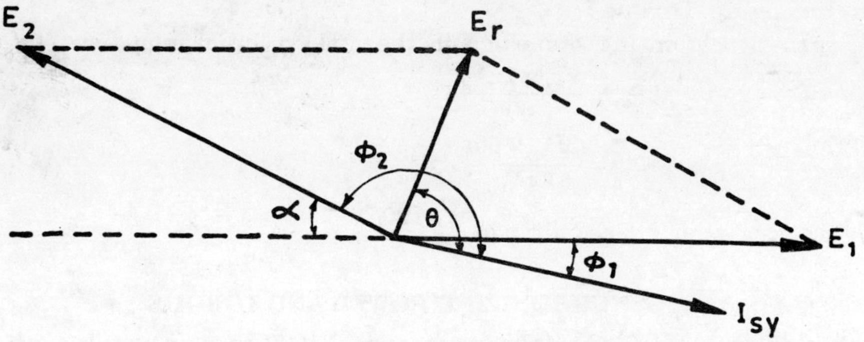

Fig. 5.61

Resultant e.m.f., $E_r = 2E\cos\left(\frac{180° - \alpha}{2}\right) = 2E\cos\left(90° - \frac{\alpha}{2}\right) = 2E\sin\frac{\alpha}{2}$

$= 2E \times \frac{\alpha}{2}$ [∵ α is very small]

$= E\alpha$

Synchronising current, $I_{sy} = \frac{E_r}{Z} = \frac{E\alpha}{Z}$ lagging behind the resultant e.m.f. E_r by θ, where $\theta = \tan^{-1}\frac{X_s}{R}$, where Z is the *combined* impedance per phase of the two alternators or of one alternator only if, it is connected to infinite bus bars. If resistance is very small as compared to synchronous reactance of the alternator, then

Synchronising current, $I_{sy} = \frac{E\alpha}{X_s}$ lagging behind E_r by 90°.

Synchronising power : Refer Fig. 5.61.

Power supplied of by machine No. 1 $= E_1 I_{sy} \cos\phi_1$

Power supplied by machine No. 2 $= E_2 I_{sy} \cos\phi_2$

Also, Power supplied by machine No. 1 = power supplied by machine No. 2 + copper losses

or $E_1 I_{sy} \cos\phi_1 = E_2 I_{sy} \cos\phi_2$ + copper losses

The power supplied by machine No. 1 is called *synchronising power* and is given by the expression

$$P_{sy} = E_1 I_{sy} \cos\phi_1 = E_1 I_{sy}$$

$$= E \times \frac{E\alpha}{X_s} \quad [\because E_1 = E \text{ and } \phi_1 \text{ is very small}]$$

$$= \frac{\alpha E^2}{X_s} \quad ...(5.23)$$

Total synchronising power for 3-phases

$$= 3P_{sy} = \frac{3\alpha E^2}{X_s} \quad ...(5.24)$$

Synchronising Torque :

If T_{sy} is the synchronising torque in N-m, the total synchronising power,

$$3P_{sy} = \frac{T_{sy} \times 2\pi N_s}{60}$$

or $$T_{sy} = \frac{3P_{sy} \times 60}{2\pi N_s} \qquad ...(5.25)$$

where N_s is the synchronous speed in r.p.m. $\left(= \frac{120f}{P}\right)$.

SYNCHRONIZING POWER AND TORQUE

Example 5.44. *A 3 MVA, 6-pole alternator runs at 1000 r.p.m. in parallel with other machines on 3.3 kV bus-bars. The synchronous reactance is 20%. Calculate the synchronizing power per one mechanical degree of displacement and the corresponding synchronizing torque.*

Solution. Phase voltage $E_{ph} = \frac{3.3 \times 1000}{\sqrt{3}} = 1905$ V

Speed of the alternator, $N_s = 1000$ r.p.m.

Synchronous reactance, $X_s = 20\%$

Rotor displacement, $\alpha = 1°$ (mech.)

Number of poles, $p = 6$

Synchronizing power :

Synchronizing torque :

$\alpha = 1°$ (mech.) $= 1° \times$ number of pair of poles (degrees elect.)

$$= 1° \times \left(\frac{6}{2}\right) = 3° \text{ (electrical)} = 3 \times \frac{\pi}{180} = \frac{\pi}{60} \text{ radian}$$

Full-load current $$= \frac{3 \times 10^6}{\sqrt{3} \times 3.3 \times 1000} = 524.9 \text{ A}$$

Synchronous impedance, $$Z_s = \frac{\text{phase voltage}}{\text{full-load current}} = \frac{1905}{524.9} = 3.63\ \Omega$$

Also, $$X_s = 0.2Z_s = 0.2 \times 3.63 = 0.726\ \Omega$$

Synchronizing power, P_{sy} :

Using the relation, $$3P_{sy} = \frac{3\alpha E_{ph}^2}{X_s} = \frac{3 \times \frac{\pi}{60} \times (1905)^2}{0.726 \times 1000} \text{ kW}$$

$$= \textbf{785.19 kW (for three-phase).} \quad \textbf{(Ans.)}$$

Synchronizing torque, T_{sy} :

Using the relation, $$T_{sy} = \frac{3P_{sy} \times 60}{2\pi N_s} = \frac{785.19 \times 1000 \times 60}{2\pi \times 1000} = \textbf{7498 N-m.} \quad \textbf{(Ans.)}$$

☞**Example 5.45.** *A 4 MVA, 10 kV, 1500 r.p.m., 50 Hz alternator runs in parallel with other machines. Its synchronous reactance is 25%. Find for (i) no-load, (ii) full-load, at p.f. 0.8 lagging, synchronizing power per unit mechanical angle of phase displacement, and calculate the synchronizing torque if the mechanical displacement is 0.7.*

Solution. Phase voltage, $E_{ph} = \frac{10 \times 1000}{\sqrt{3}} = 5774$ V

Speed of the alternator, $N_s = 1500$ r.p.m.

Frequency, $f = 50$ Hz

Synchronous reactance, $X_s = 25\%$

Full-load current $= \dfrac{4 \times 10^6}{\sqrt{3} \times 10000} = 231$ A

Synchronous impedance, $Z_s = \dfrac{E_{ph}}{\text{full-load current}} = \dfrac{5774}{231} = 25\ \Omega$

$\therefore \quad X_s = 0.25 Z_s = 0.25 \times 25 = 6.25\ \Omega$

Also $\quad f = \dfrac{Np}{120}$

$\therefore \quad p = \dfrac{120f}{N} = \dfrac{120 \times 50}{1500} = 4$

$\therefore$ No. of pair of poles $= \dfrac{p}{2} = \dfrac{4}{2} = 2$

$\therefore \quad \alpha = 1° \times 2 = 2°$ (elect). $= 2 \times \dfrac{\pi}{180} = \dfrac{\pi}{90}$ rad.

(i) **At No-load :**

Synchronizing power (for 3-phases)

$$3P_{sy} = \frac{3\alpha E_{ph}^2}{X_s} = 3 \times \frac{\pi}{90} \times \frac{(5774)^2}{6.25 \times 1000} \text{ kW} = \mathbf{558.6\ kW.}\ \ \textbf{(Ans.)}$$

$\therefore$ Synchronizing torque $T_{sy} = \dfrac{3P_{sy} \times 60}{2\pi N_s} = \dfrac{558.6 \times 1000 \times 60}{2 \times \pi \times 1500} = 3556.2$ N-m.

$\therefore \quad$ **T_{sy} for 0.7°** $= 3556.2 \times 0.7 =$ **2489.3 N-m. (Ans.)**

(ii) **At full-load 0.8 lagging :**

Let us first find E_0.

Refer Fig. 5.62.

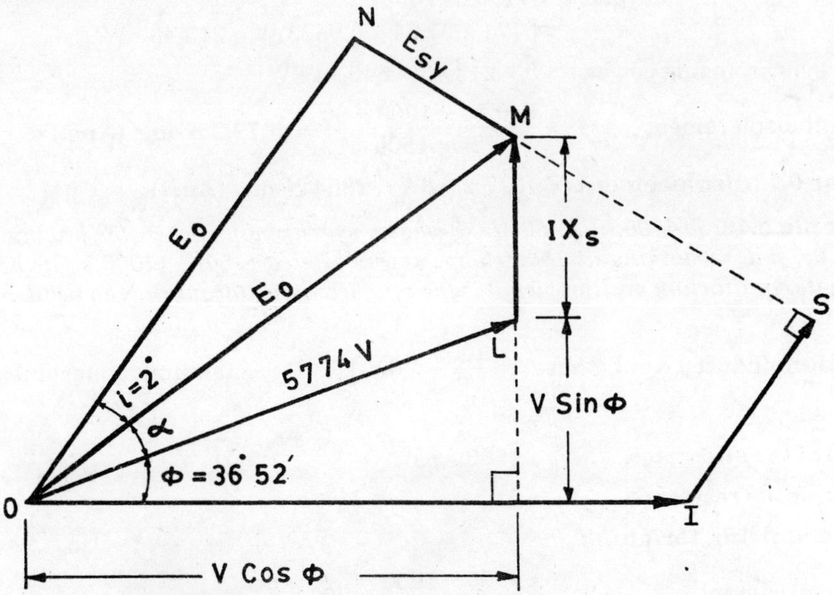

Fig. 5.62

- Draw OL = 5774 V, $\phi°$ (*i.e.* 36° 52′) ahead of current vector OI.
- LM represents reactance drop IX_s and is at right angles to OI.

Hence, OM is ahead of V by an angle α.

$$\therefore \quad E_0 = \sqrt{(V\cos\phi)^2 + (V\sin\phi + IX_s)^2}$$

Now, $\cos\phi = 0.8$ and $\sin\phi = 0.6$

$$X_s = 6.25\ \Omega$$

$$\therefore \quad E_0 = \sqrt{(5774 \times 0.8)^2 + (5774 \times 0.6 + 231 \times 6.25)^2}$$

$$= \sqrt{(4619.2)^2 + (4908.15)^2} \simeq 6740 \text{ V}$$

In ΔOMQ, $$\tan(\alpha + \phi) = \frac{MQ}{OQ} = \frac{IX_s + V\sin\phi}{V\cos\phi} = \frac{231 \times 6.25 + 5774 \times 0.6}{5774 \times 0.8} = 1.0625$$

$$\therefore \quad \alpha + \phi = 46.7° \text{ or } 46° 40'$$

or $$\alpha = 46° 40' - \phi = 46° 40' - 36° 52' = 9° 48'$$

In Fig. 5.62 if E_0 is further shifted through an angle of 2° (elect.) then an additional e.m.f., $E_{sy} = MN$ is created which produces a synchronizing current I_{sy}. This current (shown by vector IS) lags behind E_{sy} by 90° as shown.

Obviously $$E_{sy} = 2E_0 \sin\left(\frac{2}{2}\right) = 2 \times 6740 \sin 1° = 235.26 \text{ V}$$

$$I_{sy} = \frac{E_{sy}}{X_s} = \frac{235.26}{6.25} = 37.64 \text{ A}$$

It can be shown from the diagram that E_{sy} leads E_0 by an angle $\left(90 + \frac{2}{2}\right)$ and I_{sy} is lagging E_{sy} by 90°, hence I_{sy} is leading V by an angle $(\alpha + 1) = 9° 48' + 1° = 10° 48'$.

$$\therefore \quad P_{sy}\text{/phase} = VI_{sy} \cos 10° 48'$$

$$= 5774 \times 37.64 \times 0.9823 \text{ W} = 213.48 \text{ kW}$$

Total synchronizing power $= 3 \times 213.48 = 640.44$ kW

T_{sy}/unit displacement $= \dfrac{640.44 \times 1000 \times 60}{2\pi \times 1500}$ = **4077.2 N-m. (Ans.)**

T_{sy} for 0.7° displacement = 4077.2 × 0.7 = **2854 N-m. (Ans.)**

Example 5.46. *A 4500 kVA, 50 Hz, 3-phase, synchronous generator having a synchronous reactance of 0.3 p.u. is running at 1500 r.p.m. and is excited to give 11000 V. If the rotor deviates slightly from its equilibrium position, what is the synchronizing torque in N-m per degree mechanical displacement.*

Solution. Induced e.m.f./phase $= \dfrac{11000}{\sqrt{3}} = 6351$ V (assuming machine star-connected)

Frequency, $f = 50$ Hz

Speed of the alternator, $N_s = 1500$ r.p.m.

Synchronous reactance = 0.25 p.u.

Synchronising torque, T_{sy} :

Full-load current $= \dfrac{4500 \times 1000}{\sqrt{3} \times 11000} = 236.2$ A

Now, $IX_s = 0.3 \times 6351$

$$\therefore \qquad X_s = \frac{0.3 \times 6351}{236.2} = 8.06\ \Omega$$

Number of poles, $p = \frac{120f}{N_s} = \frac{120 \times 50}{1500} = 4$

Rotor displacement, $\alpha = 1 \times \left(\frac{p}{2}\right) = 1 \times \frac{4}{2} = 2°$ (electrical) $= 2 \times \frac{\pi}{80} = \frac{\pi}{90}$ rad.

Total synchronizing power,

$$3P_{sy} = \frac{3\alpha E_{ph}^2}{X_s} = \frac{3 \times \pi \times (6351)^2}{90 \times 8.06} = 524056 \text{ W}$$

Synchronizing torque, $T_{sy} = \frac{3P_{sy} \times 60}{2\pi N_s} = \frac{524056 \times 60}{2 \times \pi \times 1500} \simeq$ **3336 N-m. (Ans.)**

Example 5.47. *A 2-pole, 50-Hz, 3-phase, turbo-alternator is excited to generate the bus-bar voltage of 11 kV on no-load. The machine is star-connected and the short-circuit current for this excitation is 1000 A.*

Calculate the synchronizing power per degree of mechanical displacement of the rotor and the corresponding synchronizing torque.

Solution. Number of poles, $p = 2$

Frequency, $f = 50$ Hz

E.m.f. per phase, $E_{ph} = \frac{11000}{\sqrt{3}} = 6351$ V

Short-circuit current, $I_{sc} = 1000$ A

Rotor displacement $= 1°$ mech.

$$= 1 \times \frac{p}{2} = 1 \times \frac{2}{2} = 1° \text{ (electrical)} = 1 \times \frac{\pi}{180} = \frac{\pi}{180} \text{ rad.}$$

Synchronous speed, $N_s = \frac{120f}{p} = \frac{120 \times 50}{2} = 3000$ r.p.m.

Synchronous reactance, $X_s = \frac{E_{\text{phase}}}{I_{sc}} = \frac{6351}{1000} = 6.35\ \Omega$

Total synchronizing power for three phases,

$$\mathbf{3P}_{sy} = \frac{3\alpha E_{ph}^2}{X_s} = \frac{3 \times \pi \times (6351)^2}{180 \times 6.35} = \mathbf{332590\ W \text{ or } 332.6\ kW.\ (Ans.)}$$

Synchronizing torque, $T_{sy} = \frac{3P_{sy} \times 60}{2\pi N_s} = \frac{332590 \times 60}{2 \times \pi \times 3000} =$ **1058.67 N-m. (Ans.)**

Example 5.48. *3 MVA, 6-pole alternator runs at 1000 r.p.m. on 3.3 kV bus-bars. The synchronous reactance is 25%. Calculate the synchronising power and torque per mechanical degree of displacement when alternator is supplying full-load at 0.8 power factor lagging.*

(Madras Univ., Nov. 1996)

Solution. *Given* : Rating : 3 MVA ; $p = 6$; $N_s = 1000$ r.p.m. ;
Bus-bar voltage = 3.3 kV ; $X_s = 25\%$; $\cos \phi = 0.8$

Synchronising power and torque :

Voltage per phase, $V = \frac{3.3 \times 10^3}{\sqrt{3}} = 1905$ V

Full-load current, $I = \frac{3 \times 10^6}{\sqrt{3} \times 3.3 \times 10^3} = 524.9$ A

Now, $$IX_s = \frac{25}{100} \times 1905 = 476.25 \text{ V}$$

∴ **Synchronous reactance,** $$X_s = \frac{476.25}{524.9} = 0.9074 \ \Omega$$

Power factor, $\cos \phi = 0.8$; $\phi = \cos^{-1}(0.8) = 36.87°$ and $\sin 36.87° = 0.6$.

Rotor displacement in electrical degrees,

$$\alpha = \text{rotor displacement in mechanical degrees} \times \frac{1}{2}$$

$$= 1 \times \frac{6}{2} = 3° \quad \text{or} \quad 3 \times \frac{\pi}{180} \text{ rad.} = 0.05236 \text{ rad.}$$

Induced e.m.f.,

$$E = V + jIX_s$$
$$= 1905 + j524.9(0.8 - j0.6) \times 0.9074$$
$$= 1905 + j381 + 285.8 = 2190.8 + j381 = 2224\angle 9.87°$$

Total synchronising power for 3-phases,

$$3P_{sy} = \frac{3\alpha EV}{X_s} = \frac{3 \times 0.05236 \times 2224 \times 1905}{0.9074} = \mathbf{733418 \ W.} \quad \textbf{(Ans.)}$$

Synchronising torque, $$T_{sy} = \frac{3P_{sy}}{(2\pi N_s/60)}$$

$$= \frac{3 \times 733418}{(2\pi \times 1000/60)} = \mathbf{21010.8 \ Nm.} \quad \textbf{(Ans.)}$$

5.16.5. Effect of Reactance. It has been observed that *without reactance, the synchronous machines cannot develop motor and generator power, respectively to restore synchronism and consequently parallel operation is not possible.*

5.16.6. Effect of increasing the Excitation of one of the Alternators

- If the *excitation of an alternator* operating in parallel with other alternators is *increased above its normal value of excitation, its power factor changes in the lagging direction,* and *its current output increases with no appreciable change in its kW load.*

  When the alternator is *under-excited* its power factor becomes *more leading* and its *current output increases with no change in kW output.* This increase in current in either case is *not applied to the load but circulates between the alternators connected to the system, thereby increasing their losses and reducing their useful capacity.* It is desirable in most cases, therefore, to *operate at the same factor (at or near the load power factor), keeping the circulating currents to the minimum.*
- The proper amount of excitation for alternators operating in parallel, in general, is the amount of excitation each alternator would need if it were carrying its load alone at the same frequency and voltage.
- The voltage of the system in which several alternators are operating in parallel may be raised or lowered by varying the excitation of all the machines up or down as needed ; this is accomplished automatically through the voltage regulators.

5.16.7. Effect of increasing the driving Torque of one of the Alternators

- When two alternators are running in parallel and if the driving torque of one of the alternators is increased, it is further loaded and other is relieved of its load. If the output of the alternator, whose driving torque has been increased, becomes more than the total load supplied, then the other alternator runs as a synchronous motor.

- Governor controls are adjusted to make kW load division between the alternators.

5.16.8. Effect of Changes in Speed of one of the Alternators

- When in exact synchronism, the two alternators have equal induced e.m.fs and are in exact phase opposition as shown in Fig. 5.60. *No circulating current flows round the local circuit.*
- When the speed of alternator 2 is *reduced*, E_2 falls back by a phase angle α electrical degrees as shown in Fig. 5.61, the magnitude of induced e.m.fs., E_1 and E_2 remaining the same. It will cause resultant e.m.f. E_r acting in the local circuit ; which will *cause a flow of synchronising current I_{sy} in the local circuit. This synchronising current sets up a synchronising torque tending to retard the generating alternator 1* and accelerating the motoring alternator 2. Thus the *two alternators come again in synchronism.*
- Thus by increasing or reducing the speed of all the alternators simultaneously, the frequency of the system can be raised or lowered.

5.16.9. Effect of Unequal Voltages. Let there be two alternators having e.m.fs. E_1 and E_2 ($E_1 > E_2$) exactly in phase opposition relative to external load circuit (Fig. 5.63). The resultant e.m.f (E_r) equal to E_1–E_2 acts in the local circuit and causes synchronising current I_{sy} around the local

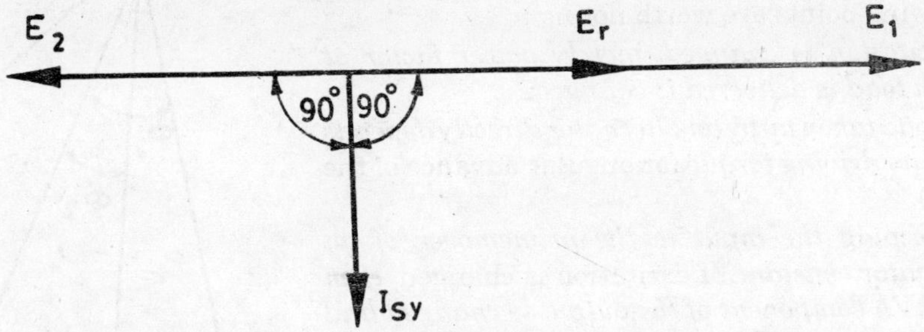

Fig. 5.63

circuit. The current I_{sy} lags behind E_r or E_1 by 90° and leads E_2 by 90°. Thus current I_{sy} produces *demagnetising effect on machine No. 1, resulting thereby reduction in* E_1 and *magnetising effect on machine No. 2, resulting thereby increase in* E_2. Thus difference of E_1 and E_2 is reduced and stable condition is established.

5.16.10. Distribution of Load. Under this heading we shall discuss the following :

(*i*) Effect of change in driving torque.

(*ii*) Effect of change in excitation.

It will be shown that :

(*i*) Amount of load taken up by an alternator running in parallel with other machines is solely determined by its *driving torque i.e.,* by the *power input* to its primemover.

(*ii*) *Any change in its excitation merely changes its kVA output but not its kW output.* In other words, it *merely changes the power factor* at which load is delivered.

- Let us consider two single-phase alternators running in parallel with generated e.m.fs. E_1 and E_2 respectively. If they are in phase with external circuit, then they will be in direct opposition with respect to their local circuit.
- Let their excitations and power inputs to their primemovers are changed such that they deliver currents of I_1 and I_2 at respective power factors of cos ϕ_1 and cos ϕ_2. Then, total current supplied to the load = vector sum of I_1 and I_2.

Consider that the *power input to the primemover of the second machine is increased*. Then its e.m.f. vector swings ahead by a certain angle as shown in Fig. 5.64. Now, a resultant voltage E_r or E_{sy} is produced which, acting on the local circuit, sets flowing a current I_{sy} which lags almost by 90° behind E_r. The new armature current I_2' of second machine is vector sum of I_{sy} and I_2. It will be noted that *I_2' is greater than I_2 and is less lagging than it.* Hence, the second alternator takes up increased share of the load. This *synchronising current is taken up by the first machine.* Its *armature current also changes from I_1 to I_1'*. It may be noted that *I_1' is lesser and is at a greater angle from E_1*. Hence, this alternator is taking up a *lesser share of its load.*

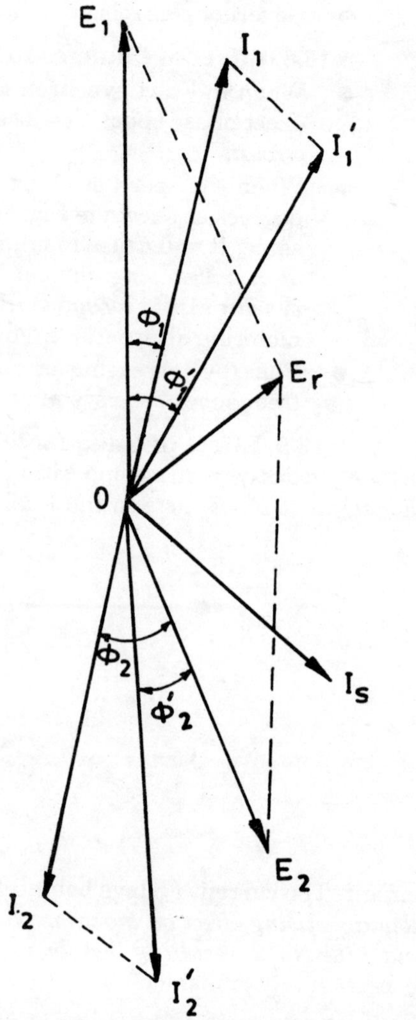

Fig. 5.64

Thus we find that by *increasing the input to its primemover, an alternator can be made to take a greater share of load, though at different power factor.*

The following points are worth noting :

- *If excitation is changed, merely power factor at which load is delivered is changed.*
- *The load taken up by an alternator directly depends upon its driving torque* (or angular advance of the rotor).
- By *keeping the input to the primemover of an alternator constant,* if excitation is changed, *then only kVA component of the output is changed and not kW.*

5.16.11. Load Sharing between Two Alternators. Let us consider two machines with identical speed characteristics running in parallel with a common terminal voltage V and load impedance Z.

$$I = I_1 + I_2 \quad ...(i)$$

$$I = \frac{V}{Z} \quad ...(ii)$$

From Fig. 5.64,

$$I_1 = \frac{E_1 - V}{Z_1} \quad ...(iii)$$

$$I_2 = \frac{E_2 - V}{Z_2} \quad ...(iv)$$

From eqns. (*i*), (*ii*), (*iii*) and (*iv*), we have

$$V = (I_1 + I_2)Z = \left(\frac{E_1 - V}{Z_1} + \frac{E_2 - V}{Z_2}\right)Z$$

$$= \left(\frac{E_1}{Z_1} + \frac{E_2}{Z_2}\right)Z - V\left(\frac{1}{Z_1} + \frac{1}{Z_2}\right)Z \quad ...(v)$$

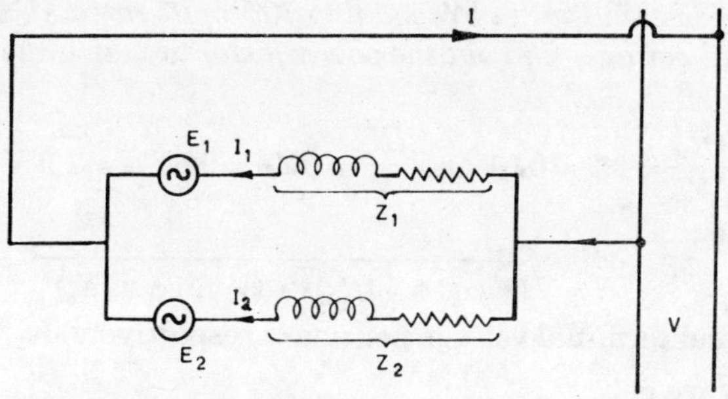

Fig. 5.65. Equivalent circuit for two alternators in parallel.

Rearranging eqn: (*v*), we get

$$V\left(\frac{1}{Z}+\frac{1}{Z_1}+\frac{1}{Z_2}\right)=\frac{E_1}{Z_1}+\frac{E_2}{Z_2}$$

or
$$V=\frac{\left(\dfrac{E_1}{Z_1}+\dfrac{E_2}{Z_2}\right)}{\left(\dfrac{1}{Z}+\dfrac{1}{Z_1}+\dfrac{1}{Z_2}\right)} \qquad ...(5.26)$$

5.17. HUNTING OF ALTERNATORS

- When two alternators are operating in parallel, any *instantaneous reduction in the angular velocity of one machine* causes :

  (*i*) *a change in load division between them* ; and

  (*ii*) *a circulating current*.

The *circulating current acts as additional load on one machine and lightens the load on the other*. This *retards the former* and permits the *latter to accelerate* until the two are once more in the proper relative phase positions where no circulating current flows if the excitations have been equal. The change to correct phase position cannot be accomplished without some *'overshooting'* on the part of the rotors, accompanied by a retardation, with a repetition of the entire cycle. That is, the alternators hunt, their actions being exactly equivalent to those of the synchronous motors under similar conditions. The period of the swing agrees with the natural oscillating period of the rotor as a torsional pendulum.

- The torque output will *pulsate* if the primemover of one of the alternators is a reciprocating engine. If this pulsation has a forced frequency within 20 per cent of that of the natural oscillating frequency of the alternator rotor, the oscillation following any load change will be *cumulative* and the machines may be thrown out of synchronism.
- It is customary to specify, for alternators to be operated in parallel, the allowable *torque-angle variation*.
- *Machines driven by internal combustion engines must have large flywheels or heavy damping windings to prevent excessive oscillation.*

PARALLEL OPERATION OF ALTERNATORS

Example 5.49. *A number of alternators are working in parallel with their terminal voltage equal to the rated value. One of the machines, which has a synchronous reactance of 50% and a* *resistance of 1%, delivers a power output in kW equal to 70% of its rated kVA. If the e.m.f. of this unit equals 1.2 times the terminal voltage, find out the power factor at which the machine is operating.*

(I.A.S.)

Solution. *Given* : $\frac{IR_a}{V} = 1\% = 0.01$ and $\frac{IX_s}{V} = 50\% = 0.5$; $E = 1.2$ V

Power factor, cos ϕ :

We know that, $$E = \sqrt{(V\cos\phi + IR_a)^2 + (V\sin\phi + IX_s)^2}$$

(where E and V are e.m.f. and terminal voltage per phase respectively)

Squaring both sides, we get

$$E^2 = (V\cos\phi + IR_a)^2 + (V\sin\phi + IX_s)^2$$

$$= V^2\left[\left(\cos\phi + \frac{IR_a}{V}\right)^2 + \left(\sin\phi + \frac{IX_s}{V}\right)^2\right]$$

Here $I = 0.7\, I_{rated}$

$\therefore$ $$(1.2V)^2 = V^2[(\cos\phi + 0.01 \times 0.7)^2 + (\sin\phi + 0.5 \times 0.7)^2]$$

$$1.44 = [(\cos^2\phi + 0.007^2 + 2 \times 0.007 \times \cos\phi) + (\sin^2\phi + 0.35^2 + 2 \times 0.35\sin\phi)]$$

$$= 0.014\cos\phi + 0.7\sin\phi + 1.12255$$

or $$0.014\cos\phi + 0.7\sin\phi = 0.31745 \quad ...(i)$$

We know that, $$A\cos\phi + B\sin\phi = \sqrt{(A^2 + B^2)}\,\sin\left(\phi + \tan^{-1}\frac{A}{B}\right)$$

$\therefore$ Eqn. (*i*) can be written as

$$\sqrt{[(0.014)^2 + (0.7)^2]}\,\sin\left\{\phi + \tan^{-1}\left(\frac{0.014}{0.7}\right)\right\} = 0.31745$$

$$0.70014\sin(\phi + 1.14577°) = 0.31745$$

or $$\sin(\phi + 1.14577°) = 0.4534$$

or $$\phi + 1.14577° = \sin^{-1}(0.4534) = 26.962°$$

or $$\phi = 25.82°$$

$\therefore$ Power factor, $\cos\phi = \cos(25.82°) =$ **0.9 (lagging). (Ans.)**

☞**Example 5.50.** *Two 1-phase alternators operate in parallel and supply a load impedance of (3 + j4) Ω. If the impedance of each machine is (0.2 + j2) and e.m.fs. are (200 + j0) and (220 + j0) volts respectively, determine for each machine :*

(*i*) *Terminal voltage,* (*ii*) *Power factor, and* (*iii*) *Output.*

Solution. Load impedance, $Z_L = (3 + j4)\ \Omega$

Impedance of each machine $= (0.2 + j2)\ \Omega$

E.m.f. of machine 1, $E_1 = (200 + j0)$

E.m.f. of machine 2, $E_2 = (220 + j0)$

Let V = terminal voltage,

I_1 = current supplied by machine 1, and

I_2 = current supplied by machine 2.

Now $V = E_1 - I_1Z_1$

or $$V = (200 - j0) - I_1(0.2 + j2) \quad ...(i)$$

Similarly $$V = (E_2 - I_2Z_2) = (220 + j0) - I_2(0.2 + j2) \quad ...(ii)$$

Also $V = IZ_L = (I_1 + I_2)Z_L = (I_1 + I_2)(3 + j4)$ ...(*iii*)

Adding, (*i*) and (*ii*), we get

$$2V = (420 + j0) - (I_1 + I_2)(0.2 + j2)$$

or $$V = (210 + j0) - (I_1 + I_2)(0.1 + j1) \quad ...(iv)$$

Multiply (*iii*) by (0.1 + *j*1) and (*iv*) by (3 + *j*4), we get

$$(0.1 + j1)V = (I_1 + I_2)(3 + j4)(0.1 + j1) \quad ...(v)$$

$$(3 + j4)V = 210(3 + j4) - (I_1 + I_2)(3 + j4)(0.1 + j1) \quad ...(vi)$$

Adding (*v*) and (*vi*), we get

$$(3.1 + j5)V = 210(3 + j4) = 630 + j840$$

or $$V = \frac{630 + j840}{3.1 + j5} = \frac{630 + j840}{3.1 + j5} \times \frac{(3.1 - j5)}{(3.1 - j5)}$$

$$= \frac{630 \times 3.1 + j840 \times 3.1 - 630 \times j5 - j^2(840 \times 5)}{(3.1)^2 + (5)^2}$$

$$= \frac{1953 + j2604 - j3150 + 4200}{34.61} = \frac{6153 - j546}{34.61}$$

$$= 177.8 - j15.77 = \mathbf{178.5 \angle -5.07°} \quad \textbf{(Ans.)}$$

Machine 1 :

(*i*) ∴ **Terminal voltage, V = 178.5 ∠– 5.07°. (Ans.)**

Current, $$I_1 = \frac{E_1 - V}{Z_1}$$

$$= \frac{(200 + j0) - (177.8 - j15.77)}{(0.2 + j2)} = \frac{22.2 + j15.77}{(0.2 + j2)}$$

$$= \frac{22.2 + j15.77}{(0.2 + j2)} \times \frac{(0.2 - j2)}{(0.2 - j2)}$$

$$= \frac{22.2 \times 0.2 + j15.77 \times 0.2 - 22.2 \times j2 - j^2(2 \times 15.77)}{(0.2)^2 + (2)^2}$$

$$= \frac{4.44 + j3.154 - j44.4 + 31.54}{4.04} = \frac{35.98 - j41.25}{4.04}$$

$$= 8.9 - j10.27 = 13.58 \angle -49.08° \text{ A}$$

(*ii*) **Power factor** $= \cos \phi_1$

$= \cos [-49.08 - (-5.07)] = \mathbf{0.7192}$ **(lag). (Ans.)**

(*iii*) **Output** $= VI_1 \cos \phi_1 = 178.5 \times 13.58 \times 0.7192 = \mathbf{1743.4\ W}.$ **(Ans.)**

Machine 2 :

(*i*) **Terminal voltage,** $V = 178.5 \angle -5.07°$ (same as machine 1)

Current, $$I_2 = \frac{E_2 - V}{Z_2} = \frac{(220 + j0) - (177.8 - j15.77)}{(0.2 + j2)}$$

$$= \frac{42.2 + j15.77}{(0.2 + j2)} = \frac{(42.2 + j15.77)(0.2 - j2)}{(0.2 + j2)(0.2 - j2)}$$

$$= \frac{42.2 \times 0.2 + j15.77 \times 0.2 - 42.2 \times j2 - j^2(15.77 \times 2)}{(0.2)^2 + (2)^2}$$

$$= \frac{8.44 + j3.154 - j84.4 + 31.54}{4.04} = \frac{39.98 - j81.246}{4.04}$$

$$= 9.89 - j20.1 = 22.4 \angle -63.8° \text{ A}$$

(*ii*) **Power factor,**

$$\cos \phi_2 = \cos [-63.8° - (-5.07°)] = \cos 58.73° = \mathbf{0.519\ (lag).\ \ (Ans.)}$$

(*iii*) **Output** $= VI_2 \cos \phi_2 = 178.5 \times 22.4 \times 0.519 =$ **2075 W. (Ans.)**

Example 5.51. *Two alternators working in parallel supply a lighting load of 3000 kW and a motor load aggregating to 5000 kW at a p.f. 0.72. One machine is loaded up to 5000 kW at 0.8 p.f. lagging. What is the load and power factor of the other machine ?*

Solution. Lighting load = 3000 kW at unity p.f.

Motor load = 5000 kW at 0.72 p.f. (lag)

Load of machine 1 = 5000 kW at 0.8 p.f. (lag)

Load and p.f. of the machine 2 :

Phase angle of light load, $\phi_l = \cos^{-1} 1.0 = 0°$

Phase angle of motor load, $\phi_m = \cos^{-1} 0.7 = 43.94°$

Reactive kVA of load = load in kW × tan of phase angle

Reactive kVA of lighting load = 0

Reactive kVA of motor load = 5000 tan 43.94°

= 5000 × 0.964 = 4820 kVAR

Total reactive kVA = 0 + 4820 = 4820 kVAR

Total load = 3000 + 5000 = 8000 kW

Load supplied by machine 1 = 5000 kW

Reactive kVA supplied by machine, 1

= 5000 tan ($\cos^{-1}$ 0.8) = 3750 kVAR

Load supplied by machine 2 = 8000 – 5000 = **3000 kW. (Ans.)**

kVAR supplied by other machine = 4820 – 3750 = 1070 kVAR

Phase angle of machine 2, $\phi_2 = \tan^{-1} \dfrac{\text{kVAR}}{\text{kW}} = \tan^{-1} \dfrac{1070}{3000} = 19.63°$

∴ **Power factor of machine 2,** cos 19.63° = **0.942. (Ans.)**

Example 5.52. *The following loads are supplied by two alternators running in parallel :*

(*i*) *1400 kW at p.f. 0.86 lagging,* (*ii*) *900 kW at p.f. 0.8 lagging.*

(*iii*) *800 kW at p.f. unity.* (*iv*) *500 kW at p.f. 0.8 leading.*

If the load on one machine is adjusted to 2100 kW at p.f. 0.92, find the load and p.f. of the other machine.

Solution. Phase angle of **first** load, $\phi_1 = \cos^{-1} 0.86 = 30°\ 41'$

Phase angle of **second** load, $\phi_2 = \cos^{-1} 0.8 = 36°\ 52'$

Phase angle of **third** load, $\phi_3 = \cos^{-1} = 0°$

Phase angle of **fourth** load, $\phi_4 = \cos^{-1} 0.8 = (-36°\ 52')$

Reactive kVA of each load, kVAR

= load in kW × tan. of phase angle

∴ Reactive kVA of **first** load, kVAR

= 1400 × tan 30° 41′ = 830.7

Reactive kVA of **second** load, $kVAR_2$

= 900 × tan 36° 52′ = 675

Reactive kVA of **third** load, $kVAR_3$

= 800 × tan 0° = 0

Reactive kVA of **fourth** load, $kVAR_4$

= 500 × tan (– 36° 52′) = – 375

Reactive kVA of total load, kVAR

= 830.7 + 675 + 0 + (– 375) = 1130.7

Total load = 1400 + 900 + 800 + 500 = 3600 kW.

Load supplied by machine 1 = 2100 kW

Reactive kVA supplied by machine

$= 2100 \tan (\cos^{-1} 0.92) = 894.6$ kVAR

Load supplied by machine 2

= 3600 – 2100 = **1500 kW. (Ans.)**

kVAR supplied by machine 2 = 1130.7 – 894.6 = 236.1 kVAR

∴ Phase angle of machine 2,

$$\phi = \tan^{-1}\left(\frac{\text{kVAR}}{\text{kW}}\right) = \tan^{-1}\left(\frac{236.1}{1500}\right) = 8.94°$$

∴ **Power factor of machine 2.**

$\cos \phi = \cos 8.94° =$ **0.9878. (Ans.)**

Example 5.53. *The following data pertains to two similar alternators : 3-phase, 6600 V, 60-Hz, 1200 kVA, star-connected.*

Resistance/phase = *1.05 Ω*

Synchronous reactance/phase = *5 Ω*

Saturation curve :

Field current, A :	*150*	*200*	*250*	*300*	*350*	*500*
Terminal voltage :	*5600*	*6490*	*7000*	*7400*	*7750*	*8500*

When operating in parallel with a terminal voltage of 6600 V the first machine supplies 90 A at p.f. 0.8 lagging. If the load power factor is 0.707 lagging and total load is 1600 kW, determine the excitation of second alternator.

Solution. Terminal voltage of each machine = 6600 V

Phase angle of the load, $\phi = \cos^{-1}(0.707) = 45°$

Reactive power of the load = load in kW × tan ϕ

= 1600 × tan 45° = 1600 kVAR

kW supplied by machine 1 $= \dfrac{\sqrt{3} V I_1 \cos \phi_1}{1000} = \dfrac{\sqrt{3} \times 6600 \times 90 \times 0.8}{1000} = 823$ kW

Reactive power supplied by machine 1

= load in kW × tan ϕ_1

$= 823 \times \tan (\cos^{-1} 0.8) = 617.2$ kVAR

Power supplied by machine 2

= 1600 – 823 = 777 kW

Reactive power supplied by machine 2

= 1600 – 617.2 = 982.8 kVAR

Since power supplied $= \dfrac{\sqrt{3} V I_2 \cos \phi_2}{1000}$

$$\therefore \quad I_2 \cos \phi_2 = \frac{\text{power supplied (kW)} \times 1000}{\sqrt{3} V} = \frac{777 \times 1000}{\sqrt{3} \times 6600} = 67.97 \text{ A}$$

Since reactive power supplied

$$= \frac{\sqrt{3} V I_2 \sin \phi_2}{1000}$$

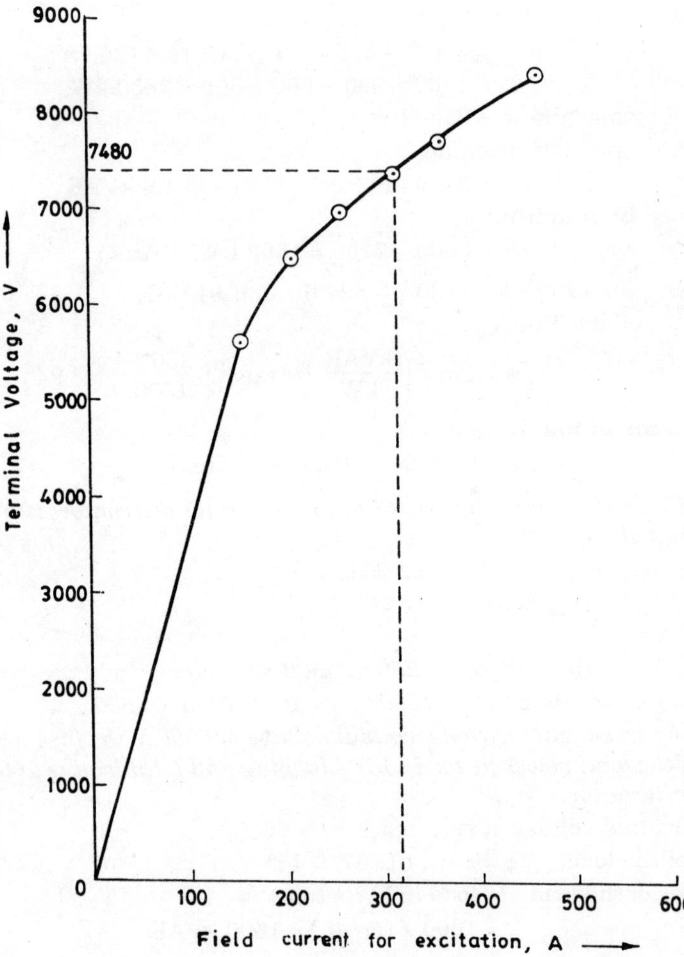

Fig. 5.66

$$\therefore \qquad I_2 \sin \phi_2 = \frac{\text{reactive power (kVAR)} \times 1000}{\sqrt{3}V} = \frac{982.8 \times 1000}{\sqrt{3} \times 6600} = 85.97 \text{ A.}$$

Phase voltage $= \dfrac{6600}{\sqrt{3}} = 3810$ V

Now,
$$E_0 = [(V + I_2R \cos \phi_2 + I_2X \sin \phi_2)^2 + (I_2X \cos \phi_2 - I_2R \sin \phi_2)^2]^{1/2}$$
$$= [(3810 + 67.97 \times 1.05 + 85.97 \times 5)^2 + (67.97 \times 5 - 85.97 \times 1.05)^2]^{1/2}$$
$$= [(4311.2)^2 + (249.6)^2]^{1/2} = 4318.4 \text{ V}$$

E_0 between lines $= 4318.4 \times \sqrt{3} = 7480$ V

From saturation curve (Fig. 5.66) excitation required to induce an e.m.f. of 7480 V = **308 A. (Ans.)**

☞**Example 5.54.** *Two identical 3-phase alternators work in parallel and supply a total load of 1600 kW at 11000 V at a power factor of 0.92. Each machine supplies half the total power. The synchronous reactance of each is 50 Ω/phase and resistance is 2.5 Ω/phase. The field excitation of the* *first machine is adjusted so that armature current is 50 A lagging. Determine the armature current of the second alternator, the power factor at which each machine is working and generated voltage of the first alternator.* **(AMIE Advanced Elect. M/Cs, Winter, 1998)**

Solution. Total load supplied = 1600 kW (at 11000 V and 0.92 p.f.)

Synchronous reactance of each machine/phase = 50 Ω

Resistance/phase = 2.5 Ω

Load current at 0.92 p.f. $= \dfrac{1600 \times 1000}{\sqrt{3} \times 11000 \times 0.92} = 91.3$ A

$\cos \phi = 0.92, \sin \phi = 0.392$

∴ Working component $= 91.3 \times \cos \phi = 91.3 \times 0.92 = 84$ A.

Wattless component $= 91.3 \times \sin \phi = 91.3 \times 0.392 = 35.8$ A.

Each alternator supplies half of each component when conditions are identical as shown in current diagram of Fig. 5.67.

Total current supplied per machine $= \dfrac{91.3}{2} = 45.65$ A.

Since the steam supply of machine 1 is not changed hence the working components of both machines would remain the same at 42 A. But the *reactive / wattless components will be redivided due to change in excitation.*

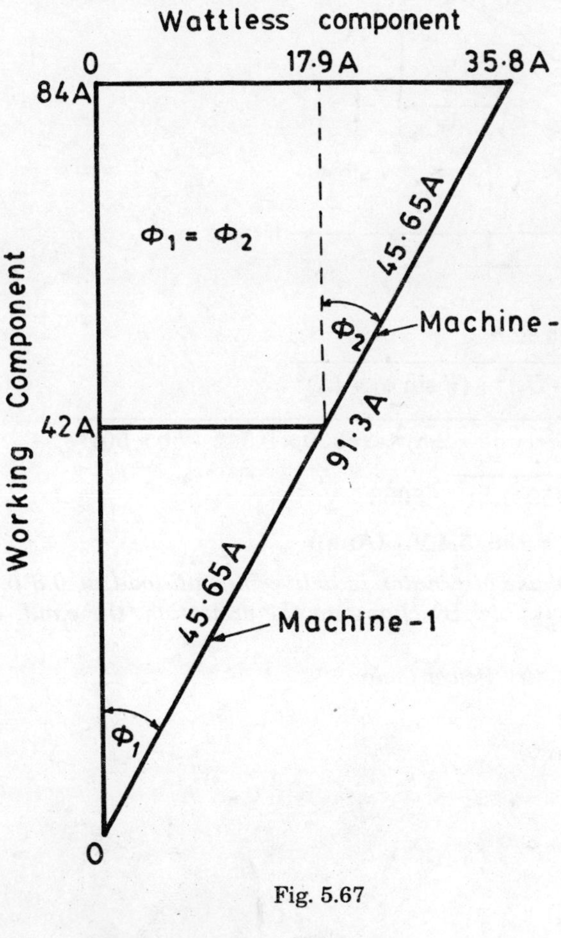

Fig. 5.67

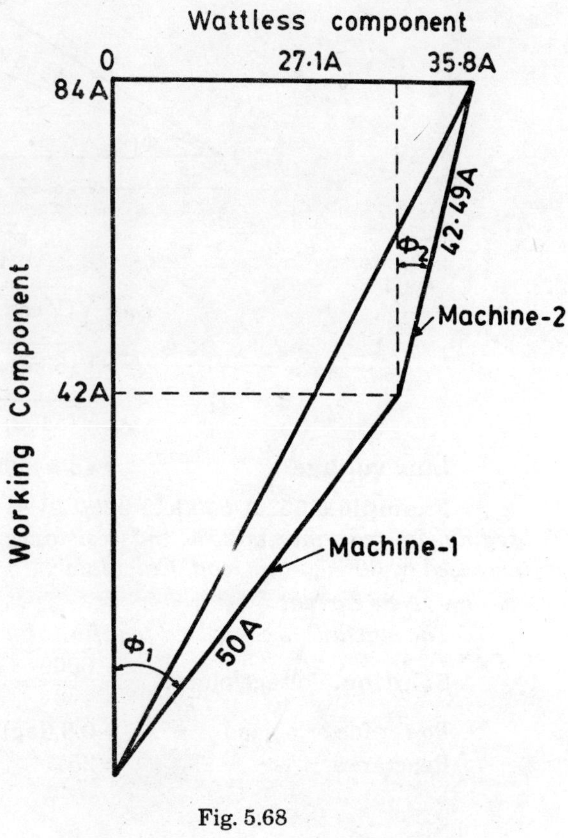

Fig. 5.68

The total current taken by machine 1 is changed from 45.65 A to 50 A.

∴ Wattless component for machine 1

$$= \sqrt{50^2 - 42^2} = 27.1 \text{ A}$$

∴ Wattless current delivered by machine 2

$$= 35.8 - 27.1 = 8.7 \text{ A}$$

New current diagram is shown in Fig 5.68.

(*i*) **New total current for machine 2**

$$I_2 = \sqrt{(42)^2 + (8.7)^2} = \mathbf{42.89\ A.\ (Ans.)}$$

(*ii*)
$$\cos \phi_1 = \frac{42}{50} = \mathbf{0.84.\ (Ans.)}$$

$$\cos \phi_2 = \frac{42}{42.89} = \mathbf{0.979.\ (Ans.)}$$

(*iii*) **Generated voltage of machine 1 :**

Terminal voltage/phase $= \dfrac{11000}{\sqrt{3}} = 6350$ V

$$\cos \phi_1 = 0.84, \sin \phi_1 = 0.542$$

Refer Fig. 5.69.

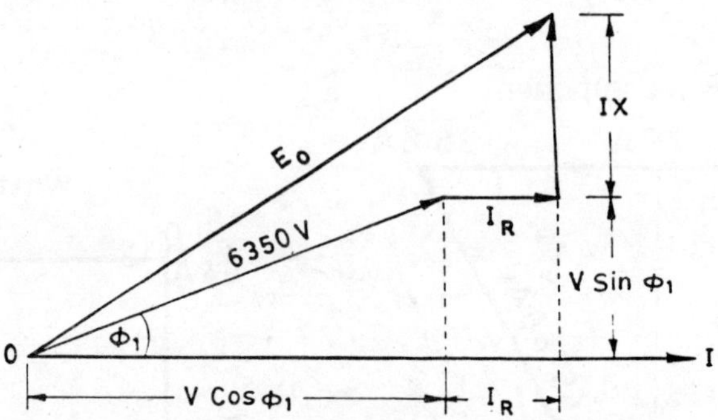

Fig. 5.69

$$E_0 = \sqrt{(V \cos \phi_1 + IR)^2 + (V \sin \phi_1 + IX)^2}$$

$$= \sqrt{(6350 \times 0.84 + 50 \times 2.5)^2 + (6350 \times 0.542 + 50 \times 50)^2}$$

$$= \sqrt{(5459)^2 + (5941.2)^2} = 8068.7 \text{ V}$$

Line voltage $= \sqrt{3} \times 8068.7 =$ **13975.4 V. (Ans.)**

Example 5.55. *A 6600 V, 1200 kVA, 3-phase alternator is delivering full-load at 0.8 p.f. lagging. Its reactance is 25% and resistance negligible. By changing the excitation, the e.m.f. is increased by 30%, at this load. Calculate :*

(*i*) *New current* (*ii*) *Power factor.*

The machine is connected to infinite bus-bars.

Solution. Voltage/phase $= \dfrac{6600}{\sqrt{3}} = 3810$ V

Power factor of load = 0.8 (lag)

Reactance = 25%

New current and power factor :

Full-load current $= \dfrac{1200 \times 1000}{\sqrt{3} \times 6600} = 105$ A (app.)

Reactance $= \dfrac{3810 \times 25}{105 \times 100} = 9.07\ \Omega$

In Fig. 5.70 current vector OI is taken along X-axis. OV represents bus-bar or terminal voltage and is hence constant.

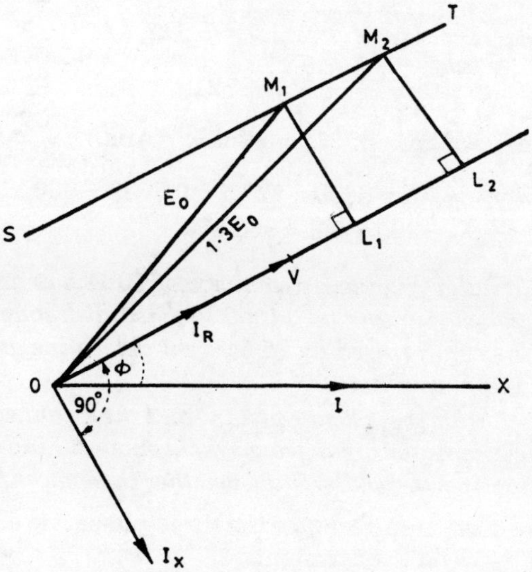

Fig. 5.70

$$I_R = \text{active component of current } I$$
$$= 105 \times 0.8 = 84 \text{ A}$$
$$I_X = \text{reactive component of current } I$$
$$= 105 \times 0.6 = 63 \text{ A}$$
$$VL_1 = I_X \,.\, X = 63 \times 9.07 = 571.4 \text{ V}$$
$$L_1M_1 = I_R \,.\, X = 84 \times 9.07 = 761.9 \text{ V}$$
$$E_0 = OM_1 = \sqrt{(V + I_X \,.\, X)^2 + (I_R \,.\, X)^2}$$
$$= \sqrt{(3810 + 571.4)^2 + (761.9)^2} = 4447 \text{ V}$$

When e.m.f. is increased by 30% e.m.f. becomes equal to

$$4447 \times 1.3 = 5781 \text{ V}$$

The locus of the extremity of E_0 lies on the line ST, which is parallel to OV. Since the kW is unchanged, I_R and hence I_RX will remain the same. It is the only I_X. X component which will be changed. Let OM_2 be the new value of E_0. Then $L_2M_2 = L_1M_1 = I_RX$ as shown before. But the $I_X \,.\, X$ component will change. Let I' be the new line current having active component I_R (the same as before) and the new reactive component I_X'. Then

$$I_X'X = VL_2$$

From right angled triangle OL_2M_2,

$$OM_2^2 = OL_2^2 + L_2M_2^2$$

$$(5781)^2 = (3810 + VL_2)^2 + (761.9)^2$$

or $$(3810 + VI_2)^2 = (5781)^2 - (761.9)^2$$

or $$3810 + VL_2 = 5730$$

$\therefore$ $$VL_2 = 1920 \text{ V}$$

or $$I_X'X = 1920 \quad \text{or} \quad I_X' = \frac{1920}{9.07} = 211.7 \text{ A.}$$

(i) $\therefore$ **New line current,** $I' = \sqrt{(84)^2 + (211.7)^2} = 227.7$ A.

(ii) Now angle of lag, $\phi' = \tan^{-1} \dfrac{211.7}{84} = 68° \, 21'$

$\therefore$ **Power factor,** $\cos \phi' = \cos 68° \, 21' =$ **0.3689. (Ans.)**

[As a check, the new power $= \sqrt{3} \times 6600 \times 227.7 \times 0.3689 = 960$ kW
which is same as before $= 1200 \times 0.8 = 960$ kW]

☞**Example 5.56.** *A turbo-alternator having a reactance of 10 Ω has an armature current of 220 amperes at unity power factor when running on 11000 V constant frequency bus bars. If the steam admission be unchanged and the e.m.f. raised by 25 per cent determine graphically or otherwise the new value of machine current and power factor.*

If this higher value of excitation were kept constant and the steam supply gradually increased, at what power output would the alternator break from synchronism ? Find also the current and power factor to which this maximum load corresponds. State whether this power factor is lagging or leading.

Solution. Let us assume that the alternator is a three-phase one and that the reactance given is per phase. Hence reactance between the terminals,

$$X = \sqrt{3} \times 10 = 17.32 \; \Omega.$$

At unity power factor the reactance drop $IX = 220 \times 17.32 = 3810$ V.

The vector diagram for this condition is shown in Fig. 5.71 (*a*) and the e.m.f. of the alternator is

$$E = \sqrt{V^2 + (IX)^2} = \sqrt{11000^2 + 3810^2} = 11640 \text{ V.}$$

If *the excitation is increased with an unchanged steam supply, i.e. with the same power input, the power output will not change. The result is to give the armature current a lagging reactive component which exerts a demagnetising effect on the main field and neutralizes the increase in excitation.* The vector diagram is Fig. 5.71 (*b*), which shows that the locus of the e.m.f. vector is dotted line perpendicular to I_RX . I_R is the power component of the total current = 220 A as before and I_X is the reactive demagnetising component.

With the increased excitation the e.m.f. is

$$E_1 = 1.25 \, E = 1.25 \times 11640 = 14550 \text{ V}$$

From Fig. 5.71 (*b*), we have

$$V + I_XX = \sqrt{E_1^2 - (I_RX)^2}$$

$$11000 + I_XX = \sqrt{14550^2 - 3810^2} = 14040 \text{ V}$$

$\therefore$ $$I_XX = 14040 - 11000 = 3040 \text{ V} \quad \text{or} \quad I_X = \frac{3040}{17.32} = 175.5 \text{ A}$$

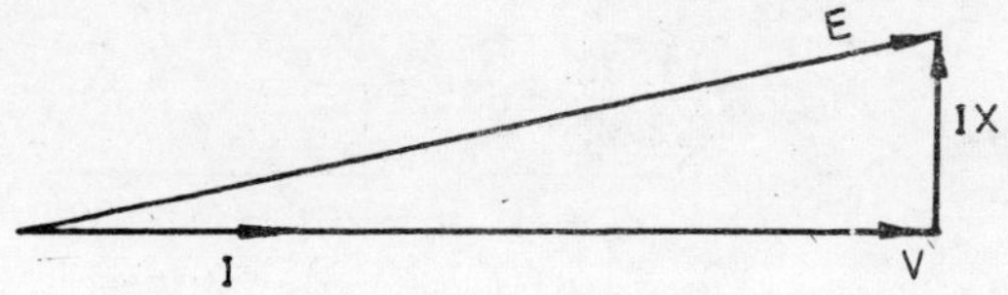

(a)

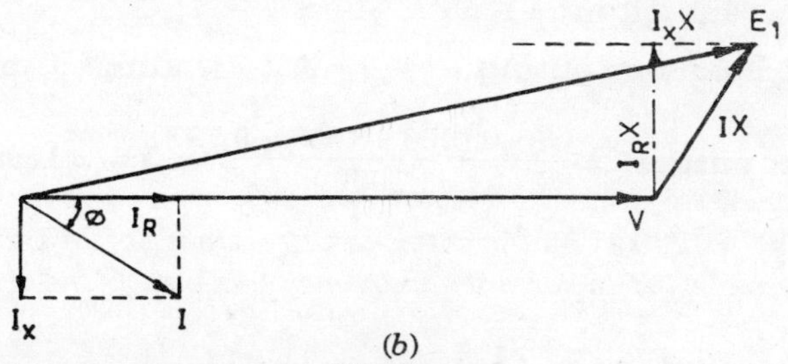

(b)

Fig. 5.71

New value of the machine current,

$$I = \sqrt{I_R^2 + I_X^2} = \sqrt{220^2 + 175.5^2} = \mathbf{281.4\ A.}\ \ \textbf{(Ans.)}$$

New power factor, $\cos \phi = \dfrac{I_R}{I} = \dfrac{220}{281.4} = \mathbf{0.78\ lagging.}$ **(Ans.)**

When the excitation is kept constant and steam supply is increased the locus of the e.m.f. vector is a circle (Fig. 5.72). A series of dotted lines may be drawn as shown, each of which represents the locus of the e.m.f. vector for a certain constant power output at varying excitation. *Maximum power output conditions are reached when the circular e.m.f. locus is tangential to this family of loci, and if the steam supply is increased still further the alternator will break from synchronism.*

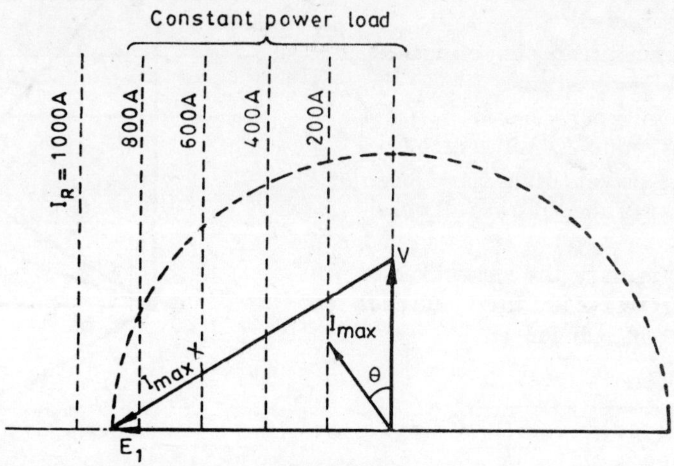

Fig. 5.72

Therefore maximum output is being given when

$$I_RX = E_1 \quad \text{and} \quad I_XX = V$$

Hence $$I_R = \frac{14550}{17.32} = 840\ \text{A}$$

$$I_X = \frac{11000}{17.32} = 635.1 \text{ A}$$

$$I_{max} = \sqrt{I_R^2 + I_X^2} = \sqrt{840^2 + (635.1)^2} = 1053 \text{ A}$$

Phase angle for maximum output, $\theta = \tan^{-1}\frac{I_X}{I_R} = \tan^{-1}\frac{635.1}{840} = 37.1°$

$\therefore$ **Maximum current output = 1053 A. (Ans.)**

Power factor at maximum output = cos θ = **0.797 leading. (Ans.)**

Maximum power output $= \frac{\sqrt{3} \times 11000 \times 1053 \times 0.797}{1000}$ kW = **15989 kW. (Ans.)**

☞**Example 5.57.** *A 6600 V, 1200 kVA alternator has a reactance of 25% and is delivering full-load at 0.8 p.f. lagging. It is connected to constant frequency bus-bar. If the steam supply is gradually increased, calculate :*

(i) At what output will the power factor become unity.

(ii) The maximum load which it can supply without dropping out of synchronism, and the corresponding power factor.

Solution. Phase voltage $= \frac{6600}{\sqrt{3}} = 3810$ V

Reactance = 25%

In example 5.54, we have found that,

$I = 105$ A, $X = 9.07\ \Omega$

$E_0 = 4447$ V, $I_R = 84$ A, $I_X = 63$ A

and $IX = 105 \times 9.07 = 952.4$ V

Using this data, vector diagram of Fig. 5.73 can be constructed.

- **Since excitation is constant, hence E_0 is constant.**
- The extremity of E_0 lies on the arc of a circle of radius E_0 and centre O.
- Constant power lines have been shown dotted and they are all parallel to OV.
- When p.f. is unity, the current vector lies along OV, I_1Z is ⊥ to OV and cuts the arc at M_1. Obviously,

$$VM_1 = \sqrt{OM_1^2 - OV^2}$$
$$= \sqrt{(4447)^2 - (3810)^2}$$
$$= 2293.4 \text{ V}$$

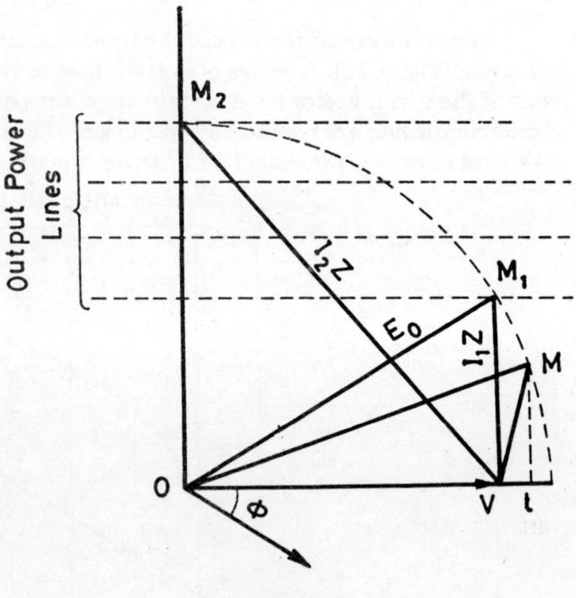

Fig. 5.73

Now, $Z = X$

$\therefore \quad I_1X = 2293.4$

$\therefore \quad I_1 = \frac{2293.4}{9.07} = 252.8$ A

(*i*) $\therefore$ **Power output at unity p.f.**

$$= \frac{\sqrt{3} \times 6600 \times 252.8}{1000} = \mathbf{2889.9\ kW.}\ \textbf{(Ans.)}$$

(*ii*) A vector OM moves *upwards* along the arc, *output power goes on increasing i.e.*, point M shifts on to a higher output power line. Maximum output power is reached when OM reaches the position OM_2, where it is vertical to OV. The output power line passing through M_2 represents the maximum output for that excitation. If OM is further rotated, then point M_2 shifts down to a lower point line *i.e.*, power is decreased. Hence $M_2V = I_2Z$ where I_2 is the new current corresponding to maximum output.

From triangle OM_2V, we have

$$M_2V = \sqrt{(OV)^2 + (OM_2)^2} = \sqrt{(3810)^2 + (4447)^2} = 5856 \text{ V}$$

$\therefore$ $I_2Z = 5856$

$$I_2 = \frac{5856}{9.07} = 645.6 \text{ A}$$

Let I_{2R} = power component of I_2

I_{2X} = wattless component of I_2

Then $I_{2R}\,X = OM_2 = 4447$ V

$\therefore$ $$I_{2R} = \frac{4447}{9.07} = 490.3 \text{ A}$$

Similarly, $$I_{2X} = \frac{3810}{9.07} = 420 \text{ A}$$

$$\tan \phi_2 = \frac{420}{490.3} = 0.8566$$

$\therefore$ $\phi_2 = 40.58°$

Power factor $\cos \phi_2$ = **0.759. (Ans.)**

$\therefore$ **Maximum power output** $= \frac{\sqrt{3} \times 6600 \times 645.6 \times 0.759}{1000}$ = **5601.57 kW. (Ans.)**

☞**Example 5.58.** *Two 3-phase synchronous mechanically coupled generators operate in parallel on the same load. Determine the kW output and power factor of each machine under the following conditions. Synchronous impedance of each generator : 0.2 + j2 Ω/phase. Equivalent impedance of the load : 3 + j4 Ω/phase. Induced e.m.f. / phase, 2000 + j0 volts for machine 1 and 2200 + j100 for machine 2.*

Solution. Synchronous impedance of each generator

$= 0.2 + j2$ Ω/phase

Equivalent impedance of load

$= 3 + j4$ Ω/phase

Induced e.m.f./phase for machine 1,

$E_1 = 2000 + j0$ volts

Induced e.m.f./phase for machine 2,

$E_2 = 2200 + j100$ volts

kW output and p.f. of each machine :

Current of machine 1,

$$I_1 = \frac{E_1 - V}{0.2 + j2}$$

or $E_1 - V = I_1(0.2 + j2)$

or $2000 - V = I_1(0.2 + j2)$ ...(*i*)

Similarly $E_2 - V = I_2(0.2 + j2)$

or $(2200 + j100) - V = I_2\,(0.2 + j2)$ ...(*ii*)

and $V = (I_1 + I_2)(3 + j4)$ ...(*iii*)

Solving from above, we get

$$I_1 = 68.2 - j102.5$$

and $$I_2 = 127 - j196.4$$

and $$I = I_1 + I_2 = 195.2 - j299$$

Now $$V = IZ = (195.2 - j299)(3 + j4)$$

$$= [195.2 \times 3 - j299 \times 3 + 195.2 \times j4 - j^2(299 \times 4)]$$

$$= 585.6 - j897 + j780.8 + 1196 = 1781 - j116$$

Output of machine 1 : output

$$= VI_1 = \text{dot product of } V \text{ and } I_1 = (1781 - j116)(68.2 - j102.5)$$

$$= 1781 \times 68.2 + 116 \times 102.5$$

$$= 133.35 \text{ kW per phase}$$

$$= 3 \times 133.35 = \mathbf{400\ kW} \text{ (for 3 phases).}$$ **(Ans.)**

Now $$\tan^{-1}\frac{102.5}{68.2} = 56.36° \text{ or } 56° \ 22'$$

$$\tan^{-1}\frac{116}{1781} = 3.72° \text{ or } 3° \ 43'$$

∴ **For machine 1, power factor**

$$= \cos\phi = \cos(56° \ 22' - 3°43') = \mathbf{0.6068}.$$ **(Ans.)**

For machine 2 :

$$\text{Output} = \text{dot product of } V \text{ and } I_2$$

$$= (1781 - j116)(127 - j196.4) = 1781 \times 127 + 116 \times 196.4$$

$$= 248.97 \text{ kW per phase} = 3 \times 248.97$$

$$= \mathbf{746.91\ kW} \text{ (for 3 phases).}$$ **(Ans.)**

$$\tan^{-1}\frac{196.4}{127} = 57.11° \text{ or } 57° \ 6'$$

Power factor, $\cos\phi = \cos(57° \ 6' - 3° \ 43') = \mathbf{0.596}.$ **(Ans.)**

Example 5.59. *Two exactly similar turbo-alternators are rated at 25 MW each. They are running in parallel. The speed-load characteristics of the driving turbines are such that the frequency of alternator 1 drops uniformly from 50-Hz on no-load to 48-Hz on full-load, and that of alternator 2 from 50-Hz to 48.5-Hz. How will the two machines share a load of 30 MW ?*

Solution. Rating of each alternator = 25 MW

Change of frequency of alternator 1 = 50 Hz to 48 Hz

Change of frequency of alternator 2 = 50 Hz to 48.5 Hz

Total load to be shared = 30 MW

Load shared by each alternator :

Fig. 5.74 shows the speed-load characteristics for driving turbines of alternators 1 and 2.

- *NL* represents the speed-load characteristic for turbine driving alternator 1.
- *NM* represents the speed-load characteristic for turbine driving alternator 2.

Since the machines are running in parallel, therefore their frequency *must* be same.

- Let *PQ* be drawn through *U* so that total load supplied is 30 MW (or 30000 kW).

Load supplied by alternator 1

$$= UP = OL - LR$$

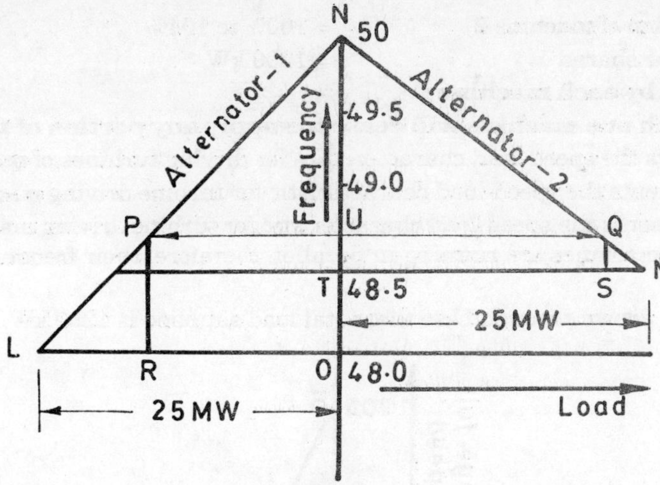

Fig. 5.74

$$= 25 - \frac{PR}{ON} \times OL$$

$$= 25 - \frac{OU}{50 - 48} \times 25$$

From similar ∆s PLR and NLO,

$$\frac{PR}{LR} = \frac{ON}{OL}$$

$$\therefore \quad LR = \frac{PR}{ON} \times OL$$

i.e., $UP = 25 - 12.5\ OU$ ...(*i*)

Load supplied by alternator 2

$$= UQ = TM - MS = 25 - \frac{QS}{NT} \times MT = 25 - \frac{OU - 0.5}{50 - 48.5} \times 25$$

i.e., $UQ = 25 - \frac{25}{1.5}(OU - 0.5)$ ...(*ii*)

Total load supplied $= UP + UQ = 30$

i.e., $(25 - 12.5OU) + \left[25 - \frac{25}{1.5}(OU - 0.5)\right] = 30$

$$25 - 12.5OU + 25 - 16.67OU + 8.33 = 30$$

$$29.17OU = 28.33$$

$\therefore$ $OU = 0.9712$

Load supplied/shared by alternator 1

$= 25 - 12.5 \times 0.9712 =$ **12.86 MW. (Ans.)**

Load supplied/shared by alternator 2

$= 25 - \frac{25}{1.5}(0.9712 - 0.5) =$ **17.14 MW. (Ans.)**

☞**Example 5.60.** *Two 800 kW alternators operate in parallel. The speed regulation of one set is 100% to 103% from full-load to no-load and that of the other is 100% to 104%. How will the two alternators share a load of 1200 kW and at what load will one machine cease to supply any portion of the load ?*

Solution. Rating of each alternator = 800 kW

Speed regulation of machine 1 = 100% to 103%

Speed regulation of machine 2 = 100% to 104%
Total load to be shared = 1200 kW

Load shared by each machine :

Load at which one machine will cease to supply any portion of the load :

Fig. 5.75 shows the speed-load characteristics for driving turbines of machines 1 and 2.

- *NL* represents the speed-load characteristic for turbine driving machine 1.
- *NM* represents the speed-load characteristic for turbine driving machine 2.
  Since the machines are running in parallel, therefore, their frequency or speed *must* be same.
- Let *PQ* be drawn through *U* so that total load supplied is 1200 kW.

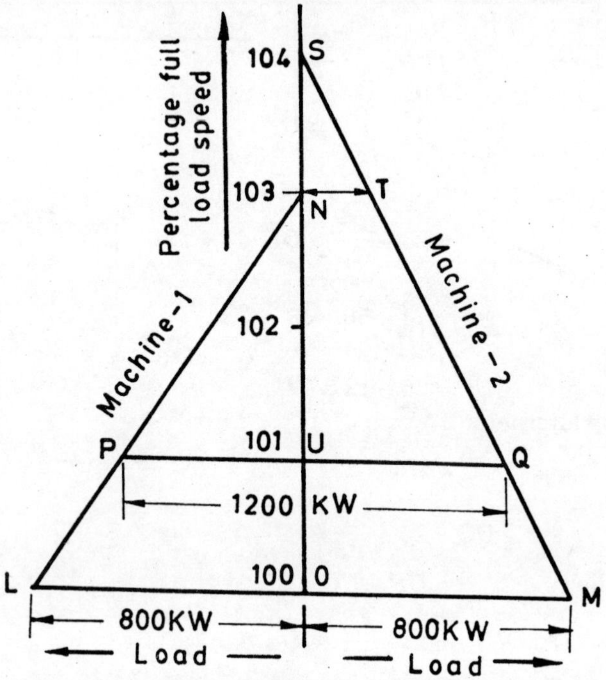

Fig. 5.75

Machine 1 :

Load supplied by machine, 1 = *UP*

From Δs *NPU* and *NLO*, we get

$$\frac{UP}{OL} = \frac{NU}{NO}$$

$$\therefore \qquad UP = \frac{NU}{NO} \times OL = \frac{3 - OU}{103 - 100} \times 800 = \frac{800}{3}(3 - OU)$$

$$\therefore \qquad UP = \frac{800}{3}(3 - OU) \qquad ...(i)$$

Machine 2 :

Load supplied by machine 2 = *UQ*

From Δs *SUQ* and *SOM*, we get

$$\frac{UQ}{OM} = \frac{SU}{SO}$$

$$\therefore \quad UQ = \frac{SU}{SO} \times OM = \frac{(4 - OU)}{104 - 100} \times 800$$

$$= \frac{800}{4}(4 - OU) = 200(4 - OU)$$

$$\therefore \quad UQ = 200(4 - OU) \quad ...(ii)$$

Since $UP + UQ = PQ$, total load supplied

$$= 1200 \text{ kW}$$

$$\therefore \quad \frac{800}{3}(3 - OU) + 200(4 - OU) = 1200$$

$$2400 - 800OU + 2400 - 600U = 3600$$

$$1400OU = 1200$$

$$\therefore \quad OU = \frac{1200}{1400} = 0.857$$

Load shared/supplied by machine 1

$$= \frac{800}{3}(3 - 0.857) = \textbf{571.4 kW. (Ans.)}$$

Load shared/supplied by machine 2

$$= 200(4 - 0.857) = \textbf{628.6 kW. (Ans.)}$$

Machine 1 will cease to supply any load when the line PQ is shifted to point N. In this case only machine 2 will supply load equal to *NT* which is found as follows :

From Δs *SNT* and *SOM*, we get

$$\frac{NT}{OM} = \frac{SN}{SO}$$

$$\therefore \quad NT = \frac{SN}{SO} \times OM = \frac{104 - 103}{104 - 100} \times 800 = \frac{1}{4} \times 800 = \textbf{200 kW. (Ans.)}$$

Example 5.61. *Two 60 MVA, 3-phase alternators operate in parallel. The settings of the governors are such that the rise in speed from full-load to no-load is 2% in one machine and 3% in the other, the characteristics being straight lines in both cases. If each machine is fully loaded when the total load is 120 MW, what will be the load on each machine when the total load is reduced to 80 MW?*

Solution. Rating of each machine = 60 MVA

Rise in speed from full-load to no-load :

For machine 1 = 2 per cent

For machine 2 = 3 per cent.

Fig. 5.76 shows the speed-load characteristics for machines 1 and 2, represented by lines *NL* and *SM* respectively.

Since the machines are running in parallel, therefore their frequency or speed must be same.

- Let *PQ* be drawn through *U* so that total load supplied is 80 MW.

Machine 1 :

Load supplied/shared by machine 1 = *UP*

From Δs *NPU* and *NLO*, we have

$$\frac{UP}{OL} = \frac{NU}{NO}$$

$$\therefore \quad UP = \frac{NU}{NO} \times OL = \frac{ON - OU}{NO} \times OL = \frac{2 - OU}{2} \times 60$$

$$\therefore \quad UP = 30(2 - OU) \quad ...(i)$$

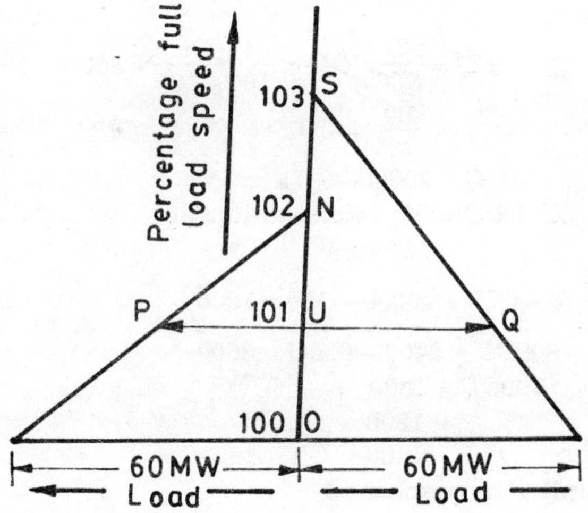

Fig. 5.76

Machine 2 :

Load supplied/shared by machine 2 = UQ

From Δs SUQ and SOM, we have

$$\frac{UQ}{OM} = \frac{SU}{SO}$$

or
$$UQ = \frac{SO - OU}{SO} \times OM = \frac{3 - OU}{3} \times 60$$

$\therefore$
$$UQ = 20(3 - OU) \quad ...(ii)$$

Since $UP + UQ = PQ$, total load supplied

$= 80$ MW

$\therefore \quad 30(2 - OU) + 20(3 - OU) = 80$

or $\quad 60 - 30OU + 60 - 20OU = 80$

$$50OU = 40$$

$$OU = \frac{40}{50} = 0.8$$

$\therefore$ **Load supplied/shared by machine 1**

$= 30(2 - 0.8) =$ **36 MW. (Ans.)**

Load supplied/shared by machine 2

$= 20(3 - 0.8) =$ **44 MW. (Ans.)**

Example 5.62. *The following two synchronous machines are operating in parallel :*

Machine A : 50 MW 6% speed regulation

Machine B : 50 MW 3% speed regulation

(i) Determine the load taken by each machine for a total load of 80 MW when the speed changes are set to give rated speed at 100% rated output.

(ii) The speed changer of machine A is so adjusted that 80 MW is equally shared. Find the output of machine A for the rated speed and also its percentage speed, at no-load. **(U.P.S.C. 1992)**

Solution. *Given* : Rating of each machine = 50 MW

Speed regulation of machine $A = 6\%$

Speed regulation of machine $B = 3\%$

Total load to be shared $= 80$ MW

(i) Load shared by each machine :

Speed-load characteristics for machine, A and B are represented by lines SL and NM respectively in Fig. 5.77.

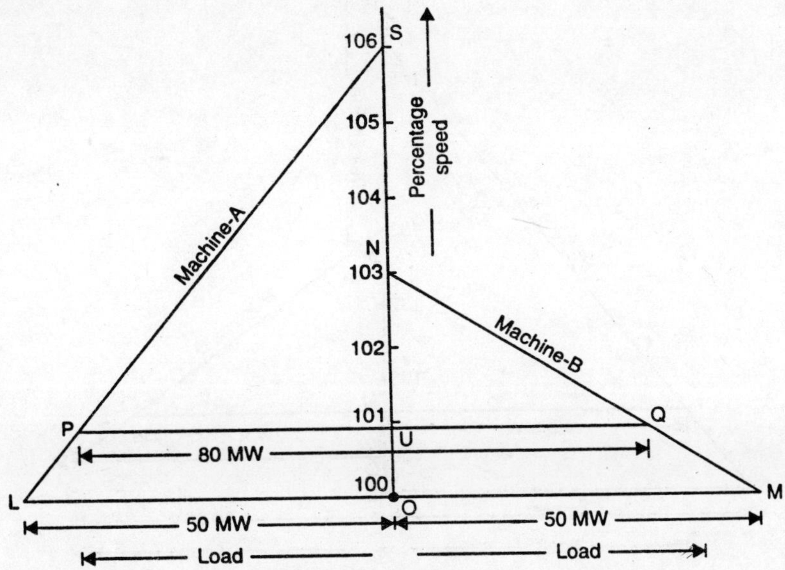

Fig. 5.77

Since the machines are operating in parallel, therefore, their frequency or percentage speed must be the same. Let PQ be drawn through U so that the total load supplied is 80 MW.

Load supplied by machine A, $PU = LO \times \frac{SU}{SO}$ (From similar Δs PSU and LSO)

$$= 50 \times \frac{SO - OU}{SO} = 50 \times \frac{6 - OU}{6} = \frac{25}{3}(6 - OU)$$

Load supplied by machine B, $QU = MO \times \frac{NU}{NO}$ (From similar Δs UNQ and ONM)

$$= 50 \times \frac{NO - OU}{NO} = 50 \times \frac{3 - OU}{3} = \frac{50}{3}(3 - OU).$$

Also $PU + UQ = PQ =$ total load supplied

or $$\frac{25}{3}(6 - OU) + \frac{50}{3}(3 - OU) = 80$$

or $$50 - \frac{25}{3} \times OU + 50 - \frac{50}{3} \times OU = 80$$

or $$25\,OU = 20 \quad \text{or} \quad OU = 0.8.$$

$\therefore$ **Load supplied by machine A** $= \frac{25}{3}(6 - OU)$

$= \frac{25}{3}(6 - 0.8) =$ **43.333 MW. (Ans.)**

Load supplied by machine B $= \frac{50}{3}(3 - OU)$

$= \frac{50}{3}(3 - 0.8) =$ **36.667 MW. (Ans.)**

(*ii*) **Percentage no-load speed of machine A and its output at the rated speed :** Refer Fig. 5.78.

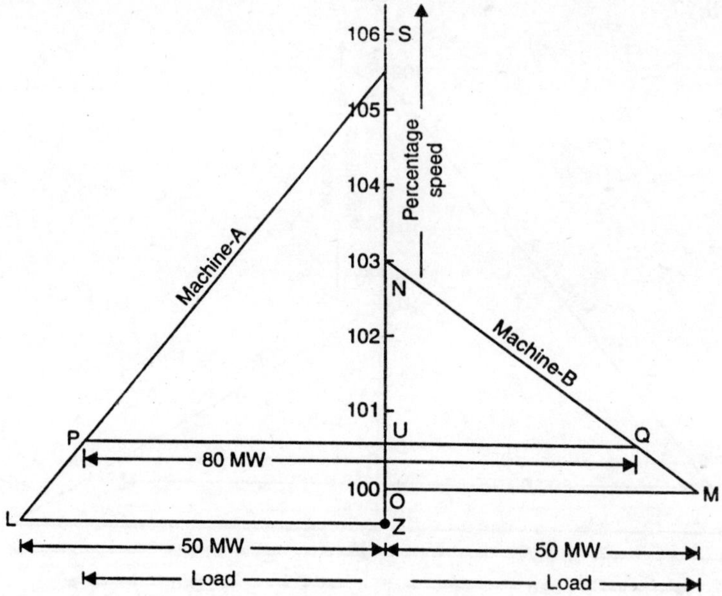

Fig. 5.78

Now speed changer is so adjusted that a total head of 80 MW is shared equally *i.e.* load shared by machine $B = \frac{1}{2} \times 80$ MW = 40 MW

$$\therefore \quad UQ = \frac{50}{3}(3 - OU) = 40 \quad \text{or} \quad OU = 3 - \frac{40 \times 3}{50} = 0.6$$

i.e. speed of operation will be 100.6%.

From Fig. 5.78, we have

$$SU = \frac{PU}{LZ} \times SZ \qquad \text{(From similar } \Delta\text{s } PSU \text{ and } LSZ)$$

$$= \frac{40}{50}(106 - 100) = 4.8$$

$$\therefore \quad OS = OU + SU = 100.6 + 4.8 = 105.4$$

Hence, percentage no-load speed of machine *A* is **105.4%. (Ans.)**

Load at rated speed $= LZ \times \frac{SO}{SZ}$

$$= 50 \times \frac{(105.4 - 100)}{(106 - 100)} = \textbf{45 MW. (Ans.)}$$

Example 5.63. *Two synchronous generators are supplying a common load. Generator 1 has no-load frequency of 51.5 Hz and regulation of 1 MW/Hz. Generator 2 has no-load frequency of 51.0 Hz and regulation of 1 MW/Hz. The total load is 2.5 MW at 0.8 p.f. lagging.*

(*i*) *At what frequency, are the generators supplying this load and how much power is supplied by each generator ?*

(*ii*) *An additional load of 1 MW is attached to the system. What will be the new frequency ? Also calculate the power supplied by each generator in this case.*

(iii) How much is governor set point of generator 2 to be adjusted to bring the system frequency at 50 Hz for 3.5 MW system load. **(U.P.S.C., 1997)**

Solution. *(i)* Let the frequency of supply be x Hz at a load of 2.5 MW

Load supplied by generator 1 $= (51.5 - x) \times 1 = (51.5 - x)$ MW

Load supplied by generator 2 $= (51 - x) \times 1 = (51 - x)$ MW

Total load supplied $= (51.5 - x) + (51 - x) = 2.5$ or $x =$ **50 Hz. (Ans.)**

Load supplied by generator 1 $= (51.5 - 50) \times 1 =$ **1.5 MW. (Ans.)**

Load supplied by generator 2 $= (51 - 50) \times 1 =$ **1.0 MW. (Ans.)**

(ii) Let the new frequency of supply be x_1 Hz

Now total load supplied $= (51.5 - x_1) \times 1 + (51 - x_1) \times 1 = (2.5 + 1)$

or $x_1 =$ **49.5 Hz. (Ans.)**

Load supplied by generator 1 $= (51.5 - 49.5) \times 1 =$ **2 MW. (Ans.)**

Load supplied by generator 2 $= (51.0 - 49.5) \times 1 =$ **1.5 MW. (Ans.)**

(iii) Load supplied by generator 1 at system frequency of 50 Hz

$= (51.5 - 50) \times 1 = 1.5$ MW

Load to be supplied by generator 2 at system frequency of 50 Hz

$= 3.5 - 1.5 = 2$ MW

$$\text{No-load frequency of generator 2} = 50 + \frac{\text{load to be supplied in MW}}{\text{generator regulation in MW/Hz}}$$

$$= 50 + \frac{2}{1} = \textbf{52 Hz. (Ans.)}$$

5.18. TIME PERIOD OF OSCILLATION

There is a natural time period of oscillation for every synchronous machine. Many causes, such as *variation in load* or *variation in mechanical driving torque create phase swinging of the machine.* If the time period of these oscillations coincides with that of the natural time period, then the amplitude of the oscillations may become so greatly developed *as to swing the machine out of synchronism.*

For a synchronous machine, the expression for the natural time period of oscillations may be derived as under :

Let T = torque per mechanical radian, N-m/mech. radian,

$J = \Sigma mr^2$ – moment of inertia, kg-m^2,

E = generated voltage/phase,

P_{sy} = synchronous power,

T_{sy} = synchronous torque,

I_{sc} = short-circuit current,

n_{sy} = synchronous speed, r.p.s.,

Z_s = synchronous impedance, and

p = number of poles.

If the rotor is displaced by α electrical radians the total synchronising power developed by all three phases $= \dfrac{3\alpha E^2}{Z_{sy}}$.

Synchronous power *per electrical radian* displacement $= \dfrac{3E^2}{Z_s}$

But, displacement in electrical radians

$$= \frac{p}{2} \times \text{displacement in mechanical radians}$$

∴ Synchronous power *per mechanical radian* displacement,

$$3P_{sy} = 3\frac{E^2}{Z_s} \times \frac{p}{2} = \frac{3E^2p}{2Z_s}$$

Synchronous torque, $T_{sy} = \frac{\text{synchronous power (in watts)}}{2\pi n_s}$

$$= \frac{3E^2p}{2Z_s \times 2\pi n_s} = \frac{3Ep}{4\pi n_s} \times \frac{E}{Z_s}$$

$$= \frac{3pEI_{sc}}{4\pi n_s} \quad \left[\because I_{sc} = \frac{E}{Z_s}\right]$$

$$= \frac{3EI_{sc}}{4\pi} \times \frac{2f}{(n_s)^2} \quad \left[\because p = \frac{2f}{n_s}\right]$$

Natural time period is given by,

$$T = 2\pi\sqrt{\frac{J}{T_{sy}}} = 2\pi\sqrt{\frac{J(n_s)^2}{0.478\,EI_{sc}f}} = 9.11n_s\sqrt{\frac{J}{EI_{sc}f}}$$

Hence, $$T = 9.11n_s\sqrt{\frac{J}{EI_{sc}f}} \quad ...(5.27)$$

Example 5.64. *A 12 MVA, 5000 V, 3-phase, 4-pole, 50-Hz alternator is connected to infinite bus-bars. The short-circuit current is 4.0 time the normal full-load current and the moment of inertia of the rotating system is 22000 kg m². Determine the normal period of oscillation.*

Solution. Rating of the alternator = 12 MVA

Number of poles, $p = 4$

Frequency, $f = 50$ Hz

Short-circuit current, $I_{sc} = 4 \times$ full-load current (I_f)

Phase voltage, $E = \frac{5000}{\sqrt{3}} = 2886.8$ V

Moment of inertia of the rotating system, $J = 22000$ kg m²

Normal period of oscillation, T :

Synchronous speed, $N_s = \frac{120f}{p} = \frac{120 \times 50}{4} = 1500$ r.p.m. = 25 r.p.s.

Full-load current, $I = \frac{12 \times 10^6}{\sqrt{3} \times 5000} = 1385.7$ A

∴ Short-circuit current, $I_{sc} = 4 \times I_f = 4 \times 1385.7 = 5542.8$ A

Normal time period,

$$T = 9.11n_s\sqrt{\frac{J}{EI_{sc}f}}$$

$$= 9.11 \times 25\sqrt{\frac{22000}{2886.8 \times 5542.8 \times 50}} = \textbf{1.192 s. (Ans.)}$$

Example 5.65. *The following figures pertain to a 2-pole, 50-Hz, 3-phase turbo-alternator connected to the infinite bus.*

Rating = 120 MVA, supply voltage (line) = 33000 V, synchronous reactance = 0.6 p.u., moment of inertia = 1.2 × 10^6 kg m^2.

Calculate its natural frequency of oscillation.

Solution. Rating of the alternator = 120 MVA

Number of poles, $p = 2$

Frequency, $f = 50$ Hz

Phase voltage, $E = \dfrac{33000}{\sqrt{3}} = 19053$ V

Synchronous reactance $= 0.6$ p.u.

Moment of inertia of the rotating system, $J = 1.2 \times 10^6$ kg m^2

Synchronous speed, $N_s = \dfrac{120f}{p} = \dfrac{120 \times 50}{2} = 3000$ r.p.m. $= 50$ r.p.s.

Per unit reactance $= \dfrac{IX}{V}$

or $0.6 = \dfrac{IX}{V}$ or $X = \dfrac{0.6\,V}{I}$

Short-circuit current, $I_{sc} = \dfrac{V}{X} = \dfrac{V}{\dfrac{0.6\,V}{I}} = \dfrac{I}{0.6} = 1.667\,I$

Full-load current, $I = \dfrac{120 \times 10^6}{\sqrt{3} \times 19053} = 3636$ A

Short-circuit current, $I_{sc} = 1.667 \times 3636 = 6061$ A

Natural time period, $T = 9.11 n_s \sqrt{\dfrac{J}{EI_{sc}f}}$

$$= 9.11 \times 50 \times \sqrt{\frac{1.2 \times 10^6}{19053 \times 6061 \times 50}} = 6.566 \text{ s}$$

∴ **Natural frequency of oscillation**

$$= \frac{1}{T} = \frac{1}{6.566} = \mathbf{0.1523\ Hz.}\ \textbf{(Ans.)}$$

Example 5.66. *A 10000 kVA, 4-pole, 6600 V, 50 Hz, 3-phase star-connected alternator has a synchronous reactance of 25% and operates on constant voltage, constant frequency bus-bars. If the natural period of oscillation while operating at full-load and unity power factor is to be limited to 1.5 seconds, calculate the moment of inertia of the rotating system.* **(Andhra University)**

Solution. *Given* : Rating = 10000 kVA ; $E_L = 6600$ V ; $f = 50$ Hz ; $p = 4$; $X_s = 25\%$; $T = 1.5$ s.

Moment of inertia of the rotating system, J :

Phase voltage, $E = \dfrac{6600}{\sqrt{3}} = 3810.5$ V

Synchronous speed, $n_s = \dfrac{2f}{p} = \dfrac{2 \times 50}{4} = 25$ r.p.s.

Full-load current, $I = \dfrac{10000 \times 1000}{\sqrt{3} \times 6600} = 874.77$ A

Short-circuit current, $I_{sc} = I \times \dfrac{100}{25} = 874.77 \times \dfrac{100}{25} = 3499.08$ A

Using the relation, $T = 9.1\, n_s \sqrt{\dfrac{J}{EI_{sc}f}}$ [Eqn. (5.27)]

$$1.5 = 9.1 \times 25 \sqrt{\frac{J}{3810.5 \times 3499.08 \times 50}}$$

or $$\left(\frac{1.5}{9.1 \times 25}\right)^2 = \frac{J}{3810.5 \times 3499.08 \times 50}$$ (Squaring both sides)

or $$\mathbf{J} = \left(\frac{1.5}{9.1 \times 25}\right)^2 \times 3810.5 \times 3499.08 \times 50$$

$$= \mathbf{28981.8\ kg\ m^2}.\ \textbf{(Ans.)}$$

Example 5.67. *A 100 MVA synchronous generator operates on full-load at frequency of 50 Hz. The load is suddenly reduced to 50 MW. Due to time lag in governor system the steam valve begins to close after 0.4 second. Determine the change in frequency that occurs in this time. Given H = 5 kW-s / kVA of generating capacity.* **(UPSC, 1995)**

Solution. *Given* : Generating capacity, G = 100 MVA = 100000 kVA ;

Inertia constant, H = 5 kW-s/kVA ; f = 50 Hz

Frequency at the end of 0.4 s, f′ :

Kinetic energy stored in rotating parts of the synchronous generator and prime-mover at synchronous speed of 50 Hz

$$= H \times G = 5 \times 100000 = 5 \times 10^5 \text{ kW-s}$$

Excess power input to the generator shaft before the steam valve begins to close = 50 MW

Excess energy transferred to the rotating parts is 0.4 second

$$= 50 \times 1000 \times 0.4 = 20000 \text{ kW-s}$$

Since kinetic energy stored is proportional to square of the speed, therefore,

$$\frac{5 \times 10^5 + 0.2 \times 10^5}{5 \times 10^5} = \left(\frac{f'}{f}\right)^2$$

$$\therefore \quad \mathbf{f'} = f \times \sqrt{\frac{5.2}{5}} = 50 \times \sqrt{\frac{5.2}{5}} \simeq \mathbf{51\ Hz.}\ \textbf{(Ans.)}$$

5.19. MAXIMUM POWER OUTPUT

There is a maximum power output that the alternator is capable of delivering for *given values of voltage, frequency and excitation.* Full-load conditions for a cylindrical rotor (IR_a drop neglected) is shown in Fig. 5.79.

Power output/phase, $$P = VI \cos \phi = \frac{VIX_s \cos \phi}{X_s} \quad ...(i)$$

From ΔOMN, we get $$\frac{IX_s}{\sin \theta} = \frac{E}{\sin (90 + \phi)} = \frac{E}{\cos \phi}$$

or $$IX_s \cos \phi = E \sin \theta \quad ...(ii)$$

From (*i*) and (*ii*), we get $$P = \frac{EV \sin \theta}{X_s}$$

If V, E and X_s are regarded as constant (of course E is fixed by excitation), power becomes *maximum* when $\theta = 90°$.

$$\therefore \quad P_{max} = \frac{EV}{X_s} \quad ...(5.28)$$

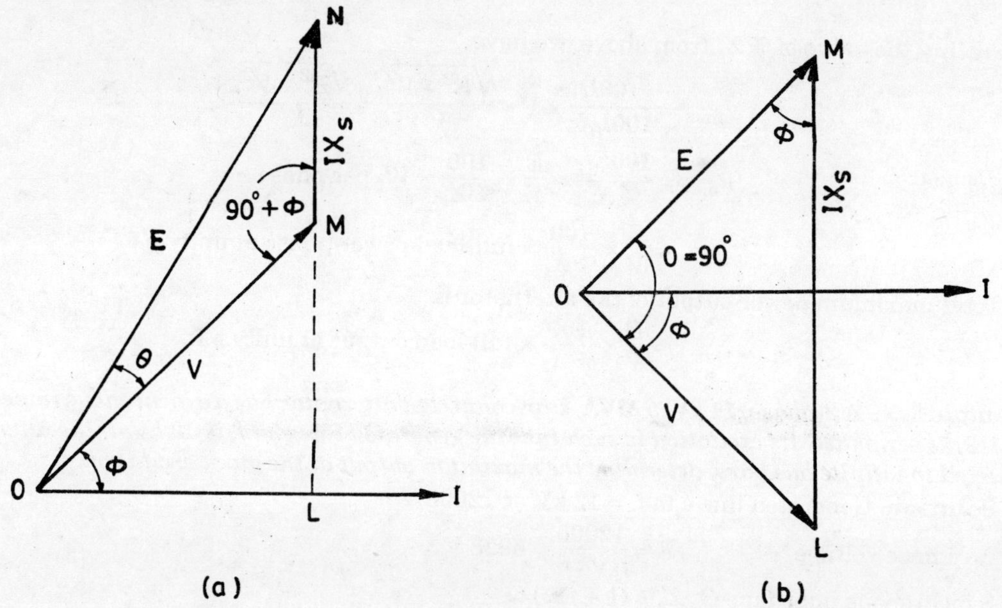

Fig. 5.79

From Fig. 5.79 (*b*), it is evident that under conditions for maximum power output, I leads V and since IX_s leads I by 90°, angle ϕ and hence $\cos\phi$ is fixed $= \dfrac{E}{IX_s}$

Let us now consider ΔOLM (right angled)

$$(IX_s)^2 = E^2 + V^2 \qquad \text{or} \qquad IX_s = \sqrt{E^2 + V^2}$$

∴ Power factor corresponding to maximum power output is

$$\cos\phi = \frac{E}{\sqrt{E^2 + V^2}} \qquad ...(5.29)$$

The maximum power output/phase may also be expressed as :

$$P_{max} = VI_{max} \cos\phi$$

or

$$P_{max} = V_{max} \frac{E}{\sqrt{E^2 + V^2}} \qquad ...(5.30)$$

where I_{max} represents the current/phase for maximum power output.

If I_f = full-load current

$\% X_s$ = percentage synchronous reactance

Then, $$\% X_s = \frac{I_f X_s}{V} \times 100$$

∴ $$\frac{V}{X_s} = \frac{I_f \times 100}{\% X_s}$$

Now, $$P_{max} = VI_{max} \frac{E}{\sqrt{E^2 + V^2}} = \frac{EV}{X_s} = \frac{EI_f \times 100}{\% X_s}$$

From above equations, we have

(*i*) $$I_{max} = \frac{100 I_f}{\% X_s} \times \frac{\sqrt{E^2 + V^2}}{V}$$

Putting the value of $\% X_s$ from above, we have

$$I_{max} = \frac{100 I_f}{100 I_f X_s} \times V \times \frac{\sqrt{E^2 + V^2}}{V} = \frac{\sqrt{E^2 + V^2}}{X_s}$$

(ii)
$$P_{max} = \frac{100 E I_f}{\% X_s} = \frac{E}{V} \times \frac{100}{\% X_s} \times V I_f \text{ per phase}$$

$$= \frac{E}{V} \times \frac{100}{\% X_s} \times \text{full-load power/phase at unity p.f.}$$

Total maximum power output of the alternator is

$$= \frac{E}{V} \times \frac{100}{\% X_s} \times \text{full-load output at unity p.f.}$$

Example 5.68. *A 3-phase, 11 kV, 5 MVA, star-connected alternator has a synchronous impedance of (1 + j12) Ω per phase. Its excitation is such that the generated line e.m.f. is 12 kV. If the alternator is connected to infinite bus bars, determine the maximum output at the given excitation.*

Solution. Generated line e.m.f. = 12 kV or 12000 V

∴ Phase voltage, $E = \frac{12000}{\sqrt{3}} = 6928$ V

Synchronous impedance, $Z_s = (1 + j12)$ Ω

Maximum output :

We know that, $P_{max}/\text{phase} = \frac{EV}{X_s}$ if R_a is neglected

$$= \frac{V}{Z_s} (E - V \cos \alpha) \text{ if } R_a \text{ is considered}$$

Now, $E = 6928$ V

Terminal voltage/phase, $V = \frac{11000}{\sqrt{3}} = 6351$ V

$$\cos \alpha = \frac{R_a}{Z_s} = \frac{1}{\sqrt{1 + (12)^2}} = 0.083$$

∴
$$P_{max}/\text{phase} = \frac{V}{Z_s} (E - V \cos \alpha)$$

$$= \frac{6351}{\sqrt{1 + (12)^2}} (6928 - 6351 \times 0.083)$$

$$= 527.4 \times \frac{6401}{1000} \text{ kW} = \mathbf{3375.88 \text{ kW}}. \text{ (Ans.)}$$

Total $P_{max} = 3 \times 3375.88 = $ **10127.64 kW. (Ans.)**

Example 5.69. *A 3-phase, 11 kV, 10 MW, star-connected synchronous generator has a synchronous impedance of (0.6 + j10.0) Ω per phase. If the excitation is such that the open-circuit voltage is 12 kV, determine :*

(i) Maximum output of the generator.

(ii) Current and p.f. at the maximum output.

Solution. (i) If we neglect R_a, then

$$P_{max}/\text{phase} = \frac{EV}{X_s}$$

where, V = terminal voltage (or bus-bar voltage in general), and

E = e.m.f. of the machine.

Here, $E\text{/phase} = \frac{12000}{\sqrt{3}} = 6928 \text{ V}$

$V\text{/phase} = \frac{11000}{\sqrt{3}} = 6351 \text{ V}$

$P_{max}\text{/phase} = \frac{6928 \times 6351}{10} = 4400 \text{ kW/phase}$

Total $P_{max} = 4400 \times 3 = 13200 \text{ kW} = \mathbf{13.2\ MW.}$ **(Ans.)**

(*ii*) $\mathbf{I}_{max} = \frac{\sqrt{E^2 + V^2}}{X_s} = \frac{\sqrt{(6928)^2 + (6351)^2}}{10} = \mathbf{939.8\ A.}$ **(Ans.)**

Power factor, $\text{p.f.} = \frac{E}{\sqrt{E^2 + V^2}}$

$= \frac{6928}{\sqrt{(6928)^2 + (6351)^2}} = \mathbf{0.737\ (lead).}$ **(Ans.)**

HIGHLIGHTS

1. *An alternator consists of a* **stator** *and* **rotor.** *The stator provides the armature windings whereas rotor provides the rotating magneting field.*
2. *There are two types of rotors :*
   (*i*) **Salient pole type.** *Used for low and medium-speed (engine-driven) alternators.*
   (*ii*) **Smooth cylindrical type.** *Used for turbo-alternators i.e. for turbine-driven alternators.*
3. *The frequency of the alternating current produced is,* $f = \frac{N_s P}{120}$ *Hz.*
4. *The e.m.f. induced (for sinusoidal wave) per phase will be,* $E_{r.m.s.}/\text{phase} = 4.44 f \phi T_{ph} k_p k_d$ *volts where* k_p *= pitch factor,* k_d *= distribution or breadth factor*

$$= \frac{\sin\left(\frac{q\beta}{2}\right)}{q \sin \frac{\beta}{2}}$$

Value of k_d *for nth harmonic is,* $k_d = \frac{\sin\left(\frac{qn\beta}{2}\right)}{q \sin\left(\frac{n\beta}{2}\right)}$.

5. *The terminal voltage V is less than* E_0 *because of*
   (*i*) *Armature drop* IR_a. (*ii*) *Synchronous reactance drop* IX_s.
   (*iii*) *Armature reaction.*
6. *Voltage regulation is given by,*
   $\% \text{ regulation} = \frac{E_0 - V}{E_0} \times 100$
   It can be found by the following methods :
   (*i*) *Synchronous impedance or E.M.F. method.* (*ii*) *The Ampere-turn or M.M.F. method.*
   (*iii*) *Zero power factor or Potier method.*
7. *The following requirements should be met with, for paralleling of alternators :*
   (*i*) *The voltages must be the same at the paralleling point or junction even though not the same at the alternators.*

(ii) The phase sequence for three-phase (or any multiple phase) must be the same at the paralleling point.
(iii) The incoming machine must be in phase at the moment of paralleling.
(iv) The line frequencies must be identical at the paralleling point.
(v) The primemovers must have relatively similar and dropping speea-load characteristics.

8. *(i) Synchronizing current,* $I_{sy} = \frac{E\alpha}{Z}$

(ii) Synchronizing power, $P_{sy} = \frac{3\alpha E^2}{X_s}$ *; and*

(iii) Synchronizing torque, $T_{sy} = \frac{3P_{sy} \times 60}{2\pi N_s}$.

9. Time period of oscillation, $T = 9.11 n_s \sqrt{\frac{J}{EI_{sc}f}}$

10. *In the case of alternator with cylindrical rotor, under conditions of maximum power output (if R_a is neglected) :*

(i) Power factor $= \frac{E}{\sqrt{E^2 + V^2}}$

(ii) $P_{max}/phase = VI_{max} \frac{E}{\sqrt{E^2+V^2}} = \frac{EV}{X_s} = \frac{EI_f \times 100}{\% X_s} = \frac{E}{V} \cdot \frac{100}{\% X_s} \times$ *full-load power per phase at unity p.f.*

where E = e.m.f. of the machine ; V = terminal or bus-bar voltage ;
I_f *= full-load current.*

OBJECTIVE TYPE QUESTIONS

(A) Choose the Correct Answer :

1. The speed of a 4-pole 60 Hz synchronous machine will be
   (*a*) 1800 r.p.m. (*b*) 2400 r.p.m.
   (*c*) 3000 r.p.m. (*d*) 3600 r.p.m.
2. The speed of a *p*-pole synchronous machine in r.p.m. is given by
   (*a*) $\frac{120f}{p}$ (*b*) $\frac{120p}{f}$
   (*c*) $\sqrt{120fp}$ (*d*) $120fp$.
3. What is the largest size of alternator being manufactured in India ?
   (*a*) 500 MW (*b*) 250 MW
   (*c*) 210 MW (*d*) 110 MW.
4. Which of the following organisations is engaged in the manufacture of large size alternators for power plants in India ?
   (*a*) Department of Science and Technology (*b*) Electricity Authority of India
   (*c*) National Thermal Power Corporation Ltd. (*d*) Bharat Heavy Electricals Ltd.
5. An exciter is a nothing but a
   (*a*) D.C. series motor (*b*) D.C. shunt motor
   (*c*) D.C. shunt generator (*d*) D.C. series generator.
6. Hydrogen is used in large alternators mainly to
   (*a*) reduce distortion of waveform (*b*) cool the machine
   (*c*) strengthen the magnetic field (*d*) reduce eddy current losses.
7. An alternator coupled to which primemover will usually have the highest rotating speed ?
   (*a*) Steam engine (*b*) Reciprocating diesel engine
   (*c*) Francis turbine (*d*) Steam turbine.

8. In an alternator the voltage generated per phase is proportional to
(*a*) number of turns in coil (*b*) flux per pole
(*c*) frequency of waveform (*d*) all of the above.

9. Salient pole type alternators are generally used on
(*a*) low voltage alternators (*b*) hydrogen cooled primemovers
(*c*) high speed primemovers (*d*) low and medium speed primemovers.

10. Turbo-alternators are generally used to run at
(*a*) 1500 r.p.m. (*b*) 3000 r.p.m.
(*c*) 5000 r.p.m. (*d*) 15000 r.p.m.

11. The rotor preferred for alternators applied to hydraulic turbines are
(*a*) salient pole type (*b*) cylindrical rotor type
(*c*) solid rotor type (*d*) any of the above.

12. The frequency of voltage generated in large alternators is
(*a*) 50 Hz (*b*) 60 Hz
(*c*) in kilo cycles (*d*) in mega cycles.

13. Which of the following primemovers is least efficient ?
(*a*) Gas turbine (*b*) Petrol engine
(*c*) Diesel engine (*d*) Steam engine.

14. In which coil the harmonic component of the generated e.m.f. will be more ?
(*a*) Full pitch coil (*b*) Short pitch coil
(*c*) Long pitch coil (*d*) Same in all coils.

15. In case of turbo-alternators the rotor is usually made of
(*a*) cast iron (*b*) forged steel
(*c*) laminated stainless steel (*d*) manganese steel.

16. In case a three-phase alternator supplies resistive load
(*a*) the armature reaction flux will be almost zero
(*b*) the armature reaction flux will be along the axis of the field
(*c*) the armature reaction flux will be at 90° with the field axis
(*d*) the armature reaction flux will be at 180° with the field axis.

17. A 12-pole alternator will pass through how many electrical degrees in one complete revolution
(*a*) 60° (*b*) 360°
(*c*) 1080° (*d*) 2160°.

18. The number of poles in turbo-alternators is usually
(*a*) 2 (*b*) 4
(*c*) 12 (*d*) 50.

19. Which harmonic will be totally eliminated in an alternator by using a fractional pitch of 4/5 ?
(*a*) Third (*b*) Fifth
(*c*) Seventh (*d*) Ninth.

20. Dirt accumulation in generators can cause all of the following *except*
(*a*) low power factor (*b*) poor voltage regulation
(*c*) flashovers (*d*) overheating.

21. Three-phase alternators are usually star-connected
(*a*) to save on copper (*b*) to reduce conductor size
(*c*) to obtain higher terminal voltage (*d*) to reduced windage losses.

22. The torque angle for an alternator is usually
(*a*) 30° to 60° (*b*) 10° to 30°
(*c*) 1° to 3° (*d*) 0.1° to 0.3°.

23. If an alternator is operating at leading power factor, then it can be concluded that
(*a*) the alternator is under-excited (*b*) the alternator is over-excited
(*c*) the torque angle of the alternator has negative value
(*d*) the residual magnetism of the poles is zero.

24. Which fractional pitch will eliminate the seventh harmonic from the voltage waveform of an alternators ?
(*a*) 6/7 (*b*) 7/8
(*c*) 5/6 (*d*) none of the above.

25. Damper winding on alternator results in all of the following *except*
(*a*) increases instability of machine
(*b*) elimination of harmonic effects
(*c*) absorption of energy of oscillations when operating in parallel with other alternators
(*d*) suppression of energy of spontaneous hunting when supplying power to transmission line with high resistance to reactance ratio.

26. Short-circuit ratio for turbo-alternators is usually
(*a*) 0.5 (*b*) between 0.5 and 0.7
(*c*) between 0.1 and 1.5 (*d*) between 2 and 3.

27. In a cylindrical rotor how much portion of the rotor is wound ?
(*a*) One third (*b*) Half
(*c*) Two third (*d*) Full.

28. Dampers in large generators
(*a*) reduce voltage fluctuations (*b*) increase stability
(*c*) reduce frequency fluctuations (*d*) none of the above.

29. The permissible voltage variation for domestic power supply is
(*a*) 0.1% (*b*) 0.5%
(*c*) 1% (*d*) 10%.

30. The speed of an alternator is changed from 3000 r.p.m. to 1500 r.p.m. The generated e.m.f./phase will become
(*a*) one fourth (*b*) half
(*c*) double (*d*) unchanged.

31. Zero power factor method for an alternator is generally used to determine
(*a*) synchronous impedance of alternator (*b*) efficiency of alternator
(*c*) voltage regulation of the alternator (*d*) none of the above.

32. In an alternator the armature reaction is mainly influenced by
(*a*) the power factor of the load (*b*) short-circuit ratio
(*c*) speed of the alternator (*d*) total current drawn.

33. The effect of cross-magnetising in an alternator can be reduced by
(*a*) using interpoles (*b*) improving cooling
(*c*) shifting brush position (*d*) reducing rotor speed.

34. When two alternators are to be operated in parallel, the blank lamp test is performed to ensure
(*a*) proper voltage matching (*b*) proper phase sequence
(*c*) proper frequency matching (*d*) none of the above.

35. The Potier's triangle is used to separate
(*a*) copper losses and winding losses (*b*) iron losses and copper losses
(*c*) armature leakage reactance and reaction m.m.f.
(*d*) field m.m.f. and armature m.m.f.

36. If the input to the primemover of an alternator is constant but the excitation is changed, then the
(*a*) power factor of the load remains constant (*b*) reactive component of the output is changed
(*c*) active component of the output is changed (*d*) all of the above.

37. Which of the following has significant influence on the power factor of an alternator ?
(*a*) Excitation (*b*) Speed of alternator
(*c*) Horsepower of primemover (*d*) Connected load.

38. An alternator is supplying a load alone. The driving torque of the alternator is increased while excitation remain constant. It will result in
(*a*) unity power factor (*b*) leading power factor
(*c*) power factor less lagging (*d*) none of the above.

39. An alternator is connected to infinite bus-bars. Its excitation is increased while its driving power remains constant. This is likely to result in
(*a*) less lagging power factor (*b*) more lagging power factor
(*c*) terminal voltage drop (*d*) none of the above.

40. In which case the terminal voltage rise of an alternator will be more ?
(*a*) When unity power factor load is thrown off (*b*) When leading power factor load is thrown off
(*c*) When lagging power factor load is thrown off
(*d*) None of the above.

41. The speed of the primemover for an alternator is maintained constant by
(*a*) coupling (*b*) filters
(*c*) brakes (*d*) generator.

42. If two alternators are running in proper synchronism and the voltage of one machine is suddenly increased
(*a*) the machine will burn (*b*) both machines will stop
(*c*) synchronizing torque will be produced to restore further synchronism
(*d*) none of the above.

43. Two alternators 1 and 2 are running in parallel. What will happen when the excitation of 1 is increased ?
(*a*) Alternator 2 will burn-out (*b*) The wattless component will change
(*c*) Power output will reduce (*d*) The machine will stop.

44. Two alternators 1 and 2 have the same percentage impedance. They will share any additional load
(*a*) equally (*b*) in proportion to their power factor
(*c*) in proportion to their kVA ratings (*d*) in proportion to their short-circuit ratio.

45. Which of the following methods is preferred for determination of voltage regulation of an alternator ?
(*a*) M.m.f. method (*b*) Synchronous impedance method
(*c*) E.m.f. method (*d*) Potier triangle method.

46. The armature reaction of an alternator influences
(*a*) operating speed (*b*) waveform
(*c*) e.m.f. phase (*d*) windage loss.

47. Alternators used for electric power supply system on air-craft have frequency of
(*a*) 20-Hz (*b*) 25-Hz
(*c*) 50-Hz (*d*) 400-Hz.

48. Negative regulation can be expected in case of
(*a*) lagging power factor load (*b*) leading power factor load
(*c*) slow speed alternators (*d*) high speed alternators.

ANSWERS

1. (*a*)	**2.** (*a*)	**3.** (*a*)	**4.** (*d*)	**5.** (*c*)	**6.** (*b*)	**7.** (*d*)
8. (*d*)	**9.** (*d*)	**10.** (*b*)	**11.** (*a*)	**12.** (*a*)	**13.** (*d*)	**14.** (*a*)
15. (*b*)	**16.** (*c*)	**17.** (*d*)	**18.** (*a*)	**19.** (*b*)	**20.** (*a*)	**21.** (*a*)
22. (*c*)	**23.** (*b*)	**24.** (*a*)	**25.** (*a*)	**26.** (*b*)	**27.** (*c*)	**28.** (*a*)
29. (*d*)	**30.** (*b*)	**31.** (*c*)	**32.** (*a*)	**33.** (*c*)	**34.** (*c*)	**35.** (*d*)
36. (*b*)	**37.** (*d*)	**38.** (*c*)	**39.** (*b*)	**40.** (*c*)	**41.** (*d*)	**42.** (*c*)
43. (*b*)	**44.** (*a*)	**45.** (*d*)	**46.** (*c*)	**47.** (*d*)	**48.** (*b*)	

(B) Say 'Yes' or 'No' :

1. The short-circuit characteristic of an alternator is always linear.
2. Under short-circuit conditions the power factor of an alternator is almost zero lagging.
3. The synchronous impedance of a large size modern alternator is about 0.1 p.u.
4. The ratio of armature leakage reactance to synchronous reactance of a large size modern alternator is about 0.2.

5. The rated voltage of alternators used in power stations is usually 400 kV.
6. High speed alternators usually have cylindrical rotors.
7. Cylindrical rotor alternators have small length to diameter ratio.
8. Salient pole machines have small number of poles.
9. To ensure proper cooling, cylindrical rotor alternators use both radial and axial ducts.
10. The stators of modern alternators are wound for 60° phase groups.
11. The value of voltage regulation as determined by synchronous impedance method is greater than that found by actual loading.
12. Alternators are usually operated at a power angle of about 40°.
13. An alternator is connected to the bus-bars and is supplying load. Its primemover is suddenly shut down. The alternator will continue to work as alternator.
14. The biggest size of alternators used in India is 500 MW.
15. Time period of oscillation is given by $T = 3.11\, n_s \sqrt{\dfrac{J}{EI_{sc} f^2}}$.

ANSWERS

1.	Yes	**2.** Yes	**3.** No	**4.** Yes	**5.** No	**6.** Yes	**7.** No
8.	No	**9.** Yes	**10.** Yes	**11.** Yes	**12.** Yes	**13.** No	**14.** Yes
15.	No.						

THEORETICAL QUESTIONS

1. What is an alternator ? What is its operating principle ?
2. How are alternators classified ?
3. Give reasons :
   (*i*) The speed of a synchronous generator should be kept constant at synchronous speed.
   (*ii*) Generally cylindrical rotor construction is used for high speed machines and salient pole construction for low speed machines.
   (*iii*) Cylindrical rotor alternators have small diameters and large lengths of core.
   (*iv*) Salient pole alternators have large diameters and small core lengths.
   (*v*) Both axial and radial ventilating ducts are necessary in cylindrical rotor machines where only radial ducts are usually sufficient in salient pole machines.
4. Describe the main constructional features of cylindrical rotor and salient pole alternators.
5. What is a distribution factor ? What is its effect ? Derive an expression for distribution factor of a winding having q slots per pole per phase and a slot angle of β.
6. What is the meaning of 'electrical angle' ? How is it different from mechanical angle ? Can they be equal ?
7. What is a fractional pitch winding ?
8. Derive an expression for the voltage induced in an alternator. Discuss the role of different factors which appear in the expression.
9. Draw the phasor diagrams of an alternator at unity, lagging and leading power factors.
10. What is voltage regulation ? Discuss the synchronous impedance and m.m.f. method of calculating voltage regulation ?
11. (*a*) What is the necessity of parallel operation of alternators ?
    (*b*) State the requirements for paralleling alternators.
12. Describe briefly alternator synchronising procedure.
13. Explain briefly 'hunting of alternators'.
14. Derive an expression of 'time period of oscillation'.

UNSOLVED PROBLEMS

Pitch, Distribution Factor and E.m.f. Equation

1. For a 3-phase winding with 4 slots/pole/phase and with the coil span of 10 slots pitch, calculate the values of the pitch factor and distribution factor. [**Ans.** 0.966, 0.9576]

2. A 2200 volt, 3-phase alternator is running at 300 r.p.m. and has 24 poles. Find the number of conductors in the stator winding, if the magnetic flux is 5×10^{-2} Wb/pole.
Assume distribution factor as 0.96. [**Ans.** 600]

3. A 4-pole alternator has an armature with 25 slots and 8 conductors/slot and rotates at 1500 r.p.m. and the flux/pole is 0.05 Wb. Calculate the e.m.f. generated, if winding factor is 0.96 and all conductors are in series. [**Ans.** 1065.6 V]

4. The stator of a 3-phase, 20-pole alternator has 120 slots, and there are 4 conductors per slot accommodated in two layers. If the speed of the alternator is 300 r.p.m. calculate the e.m..f. induced per phase. Resultant flux in the air-gap is 0.055 Wb/ pole.
Assume the coil span as 150° electrical. [**Ans.** 905 V]

5. A 16-pole, 3-phase alternator has a star-connected winding with 144 slots and 10 conductors per slot. The flux per pole is 0.03 Wb distributed sinusoidally and the speed is 375 r.p.m. Find the line voltage. [**Ans.** 2658 V]

6. Calculate the e.m.f. of a 4-pole, 3-phase star-connected alternator running at 1500 r.p.m., from the following data :

Flux per pole	= 0.1 Wb
Number of slots	= 48
Conductors/slot (in two layers)	= 4
Coil span	= 150°.

[**Ans.** 1138 V]

7. A 3-phase, 16-pole, star-connected alternator has 192 stator slots with eight conductors/slot and the conductors of each phase are connected in series. The coil span is 150 electrical degrees. Determine the phase and line voltage if the machine runs at 375 r.p.m. and the flux per pole is 64 m Wb distributed sinusoidally over pole. [**Ans.** 3367 V, 5830 V]

8. A 3-phase, 4-pole star-connected alternator has 60 slots with 2 conductors/slot. The pitch of the coil is 3 slot less than the pole pitch. The flux/pole is 12.5×10^{-2} Wb sinusoidally distributed. Calculate the no-load terminal voltage for a frequency of 50-Hz.

9. A 3-phase, star-connected alternator on open-circuit is required to generate a line voltage of 3600 V, 50-Hz, when driven at 500 r.p.m. The stator has 3 slots per pole, and 10 conductors per slot. Calculate :
(*i*) number of poles (*ii*) useful flux/pole.
Assume all conductors per phase to be connected in series and coils to be full pitch.
[**Ans.** (*i*) 12, (*ii*) 0.054 Wb]

Voltage Regulation

10. A 500 V, 50 kVA, single-phase alternator has an effective resistance of 0.2 Ω. A field current of 10 A produces an armature current of 200 A on short-circuit and an e.m.f. of 450 V on open-circuit. Calculate :
(*i*) Synchronous impedance and reactance.
(*ii*) Full-load regulation with 0.8 p.f. lagging. [**Ans.** 2.25 Ω, 2.24 Ω, 34.4%]

11. A 3-phase star-connected alternator is rated at 1600 kVA, 13500 V. The armature effective resistance and synchronous reactance are 1.5 Ω and 30 Ω respectively per phase.
Calculate the percentage regulation for a load of 1280 kW at power factors of :
(*i*) 0.8 lagging (*ii*) 0.8 leading. [**Ans.** 18.59%, – 11.98%]

12. A 1000 kVA, 3000 V, 50-Hz, 3-phase star-connected alternator has an armature effective resistance of 0.2 Ω. A field current of 40 A produces on short-circuit current of 200 A and an open-circuit e.m.f. of 1040 V (line value). Calculate the full-load percentage regulation at a power factor of 0.8 lagging. How will the regulation be affected if the alternator delivers its full-load output at a power factor of 0.8 leading
[**Ans.** 24.4% – 13.5%]

13. A 3-phase, 4-pole, star-connected turbo-alternator has a cylindrical rotor. The reactance per phase of winding is 2.5 Ω and resistance per phase is 0.15 Ω. The alternator has a terminal potential difference of 6600 V when delivering a current of 250 amperes. Calculate :

(*i*) The generated e.m.f. at 0.6 power factor lagging, and

(*ii*) The regulation at 0.8 p.f. lagging. [**Ans.** 7350 V (line), 11.36%]

14. A 3300 V, 3-phase star-connected alternator has a full-load current of 100 amperes. On short-circuit a field current of 5 amperes was necessary to produce full-load current. The e.m.f. on open-circuit for the same excitation was 900 volts. The armature resistance was 0.9 Ω/phase. Determine the full-load voltage regulation for :

(*i*) 0.8 p.f. lagging (*ii*) 0.8 p.f. leading [**Ans.** 36.5%, – 14.3%]

15. A full-load current of 100 amperes requires an excitation of 3 amperes when a single-phase 1200 V alternator is short circuited by an ammeter of resistance 0.01 Ω (it also includes the resistance of leads). On open-circuit the same excitation produced 350 V. The resistance of the armature is 0.5 Ω. Calculate the regulation of the alternator at a p.f. of

(*i*) 0.8 lag (*ii*) 0.8 lead. [**Ans.** 22.5%, – 10.25%]

[**Hint** : Effective resistance = 0.5 + 0.01 = 0.51 Ω]

16. The no-load excitation of an alternator required to give rated voltage is 100 A. In a short-circuit test with full-load current flowing in the armature, the field excitation was 80 A. Determine the approximate excitation that will be required to give full-load current at 0.8 p.f. lag and at the rated terminal voltage. [**Ans.** 161.2 A]

17. The following test results are obtained on a 6600 V alternator :

O.C. voltage, V	:	3100	4900	6600	7500	8300
Field current, A	:	16	25	37.5	50	70

A field current of 20 A is found necessary to circulate full-load current on short-circuit of armature. Calculate by :

(*i*) the ampere-turn method, and

(*ii*) the synchronous impedance method the full-load regulation at 0.8 p.f. (lag).

Neglect resistance and leakage reactance. [**Ans.** 15.15%, 38.7%]

18. The O.C.C. of a 1800 kVA, 6928 volts, 3-phase star-connected synchronous generator is as follows :

Field Current, A :									
0	5	10	15	20	25	30	35	40	45
O.C. e.m.f./phase :									
0	1120	2160	3000	3600	4000	4320	4600	4800	4960

Field current of 25 A produces an excitation of 5000 AT/pole. The armature leakage reactance/phase is 2.5 Ω, the resistance being negligible. The joint phase armature reaction when carrying (balanced) currents of 150 A is 2000 AT/pole. Calculate the field current required :

(*i*) to produce rated armature current on short-circuit.

(*ii*) to supply rated current at normal voltage at a p.f. of 0.866 lagging and corresponding regulation [**Ans.** (*i*) 11.7 A ; (*ii*) 35.15 A, 15.5%]

19. Estimate the percentage regulation at full-load and p.f. 0.8 lagging of a 1000 kVA, 6600 V, 3-phase, 50 Hz star-connected salient pole synchronous generator. The O.C.C. is as follows :

Terminal voltage, V	:	4000	6000	6600	7200	7800
Field/AT	:	5200	8500	10000	12500	17500

Leakage reactance is 10% and resistance is 2%. Short-circuit characteristics give full-load current with a field excitation of 500 AT. Take the permanence to cross armature reaction as 35% of that to direct reaction. [**Ans.** 20%]

20. A 3-phase, 6000 V alternator gave the following open-circuit characteristics at normal speed :

Terminal voltage, V	:	4000	5000	6000	7000	8000
Field current, A	:	14	18	23	30	43.

With the armature short-circuited and full-load current flowing the field current is 16 A, and when the machine is supplying the full-load kVA at zero power factor the field current is 42.5 A when the terminal voltage is 6000 V. Determine the field current required when the machine is supplying the full-load of 2000 kVA at 0.8 p.f. lagging. Determine also the leakage reactance in Ω/phase and the field current required to overcome the armature reaction. [**Ans.** 41.6 A, 3.3 Ω, 12.5 A]

21. An 11 kV, 3-phase cylindrical rotor-type alternator has the following open-circuit characteristic at rated speed :

Line voltage, V	:	7300	10300	12400	14000
Field current, A	:	40	60	80	100

The excitation to produce full-load current on short-circuit is 34 A, and when the machine supplies full-load output at 11 kV and zero power factor, the excitation is 106 A. Determine :

(*i*) % synchronous reactance drop. (*ii*) % leakage reactance drop.

(*iii*) The armature reaction in equivalent field amperes at full-load.

Neglect the armature resistance. [**Ans.** 57%, 12.7%, 26 A]

22. The O.C.C. of a 10 MVA, 6.6 kV, 3-phase star-connected alternator is given by the following data :

Exciting current, A	:	0	35.5	60	82	100	118	142
Line voltage, (kV)	:	0	3	4.8	6.0	6.8	7.2	7.8

and the full-load zero power factor characteristic is given by :

Exciting current, A	:	40	92	128	156	170
Line voltage, (kV)	:	0	3	4.8	6.0	6.6

Assuming the leakage reactance to be 15% and ignoring resistance, determine the excitation required for full-load, normal voltage at a power factor of 0.8 lagging. [**Ans.** 147.5 A]

23. An 11 kV, 1000 kVA, 3-phase, star-connected alternator has a resistance of 2 Ω/phase. The open circuit curve and the characteristic with rated full-load current at zero power factor are given in the following table :

Field current, A	: 40	50	110	140	180
Line volts O.C.C. V	: 5800	7000	12500	13750	15000
Line volts zero p.f., V	: 0	1500	8500	10500	12500

[**Ans.** 21.5%]

24. The no-load and full-load zero factor characteristics for a 23.5 MVA, 13.8 kV, 3-phase, star-connected, turbo-generator are given below in per unit values :

No-load Characteristic :

I_f : 0.10	0.20	0.40	0.60	0.80	1.0	1.2	1.4	1.6	2.0	2.5
V : 0.13	0.23	0.45	0.69	0.87	1.0	1.09	1.15	1.21	1.28	1.36

Zero p.f. Characteristic :

I_f :	1.2	1.3	1.4	1.6	1.8	2.0	2.2	2.4	2.6
V :	0.015	0.13	0.25	0.49	0.69	0.83	0.92	0.99	1.04

Determine the regulation at full-load, 0.8 p.f. lag by the zero p.f. method.

Neglect armature resistance. [**Ans.** 28%]

Synchronizing Torque and Power

25. A 3 MVA, 6 pole alternator runs at 1000 r.p.m. in parallel with other machines on 3300 V bus-bars. The synchronous reactance is 25%. Calculate the synchronizing power per one mechanical degree of displacement and the corresponding synchronizing torque. [**Ans.** P_{sy} = 628.4 kW, T_{sy} = 6000 N-m]

26. A 5 MVA, 10 kV, 1500 r.p.m., 50 Hz alternator runs in parallel with other machines. The synchronous reactance is 20%. Find for :

(*i*) no-load

(*ii*) full-load, at power factor 0.8 lagging, synchronizing power per unit mechanical angle of phase displacement, and calculate the synchronizing torque if the mechanical displacement is 0.5.

[**Ans.** 873.4 kW, 2782 N-m, 3118.5 N-m]

27. A 5500 kVA, 50 Hz, 3-phase synchronous generator having a synchronous reactance of 0.3 p.u. is running at 1500 r.p.m. and is excited to give 11 kV. If the rotor deviates 1° (mech.) from its equilibrium position, what is the synchronizing torque ? [**Ans.** 4074 N-m]

28. A 2-pole, 50 Hz, 3-phase, turbo-alternator is expected to generate the bus-bar voltage of 11000 V on no-load. Calculate the synchronizing power per degree of mechanical displacement of the rotor and the corresponding synchronizing torque. The machine is star-connected and the short-circuit current for this excitation is 1200 A. [**Ans.** 399 kW ; 1270 N-m]

Parallel Operation of Alternators

29. Two alternators working in parallel supply a lighting load of 300 kW and a motor load aggregating to 5000 kW at a p.f. of 0.71. One machine is loaded up to 5000 kW at 0.8 p.f. lagging. What is the load and power factor of the other machine ? [**Ans.** 3000 kW ; 0.928]

30. Two alternators running in parallel supply the following loads :

(*i*) 1500 kW at p.f. 0.9 lagging. (*ii*) 1000 kW at p.f. 0.8 lagging.

(*iii*) 800 kW at p.f. unity. (*iv*) 600 kW at p.f. 0.8 leading.

The load on one machine is adjusted to 2000 kW at p.f. 0.95 lagging. Find the load and p.f. of the other machine. [**Ans.** 1900 kW ; 0.98]

30A. Two similar alternators display the following data : 3-phase 6600 V, 60 Hz, 1000 kVA, resistance per phase of each of the star connected alternator = 1.02 Ω.

Saturation Curve :

Field current	:	150	200	250	300	350	500
Terminal voltage	:	5000	6490	7000	7400	7750	8500

Synchronous reactance per phase = 4 Ω.

When operating in parallel with a terminal voltage of 6600 V, the first alternator supplies 85 A at a p.f. of 0.85 lagging. If the load power factor is 0.78 lagging and total load is 1500 kW, determine the excitation of the second alternator. [**Ans.** 265 A]

31. A 6.6 kV, 1 MVA, 3-phase alternator is delivering full-load at 0.8 p.f. lagging. Its reactance is 20% and resistance negligible. By changing the excitation, the e.m.f. is increased by 25% at this load. Calculate the new current and the power factor.

The machine is connected to infinite bus-bars. [**Ans.** 191 A ; 0.3665]

32. A 6.6 kV, 1 MVA alternator has a reactance of 20% and is delivering full-load at 0.8 p.f. lagging. It is connected to constant frequency bus-bar. If the steam supply is gradually increased, calculate :

(*i*) At what output will the power factor become unity ?

(*ii*) The maximum load which it can supply without dropping out of synchronism, and the corresponding power factor. [**Ans.** (*i*) 2652 kW ; (*ii*) 5658 kW ; 0.749]

33. Two exactly similar turbo-alternators are rated 20 MW each. They are running in parallel. The speed-load characteristics of the driving turbines are such that the frequency of alternator 1 drops uniformly from 50 Hz on no-load to 48 Hz on full-load, and that of alternator 2 from 50 Hz to 48.5 Hz. How will the two machines share a load of 30 MW ? [**Ans.** 12.8572 MW ; 17.1428 MW]

34. Two 750 kW alternators operate in parallel. The speed regulation of one set is 100% to 103% from full-load to no-load and that of the other is 100% to 104%. How will the two alternators share a load of 1000 kW and at what load will one machine cease to supply any portion of the load ?

[**Ans.** 464.3 kW ; 535.7 kW ; 187.5 kW]

35. Two 50 MVA, 3-phase alternators operate in parallel. The settings of the governors are such that the rise in speed from full-load to no-load is 2% in one machine and 3% in the other, the characteristics being straight lines in both cases. If each machine is fully loaded when the total load is 100 MW, what will be the load on each machine when the total load is reduced 60 MW ? [**Ans.** 26 MW, 34 MW]

36. Two 30 MVA, 3-phase alternators operate in parallel. The settings of governors are such that the rise in speed from full-load to no-load is one per cent in one machine and two per cent in the other, the characteristics being straight lines in both cases. If each machine is fully loaded when the total load is 60 MW, what will be the load on each machine when the total load is reduced to 40 MW ?

[**Ans.** 16.7 MW (app.), 33.3 MW (app.)]

37. Two identical 2000 kVA alternators operate in parallel. The governor of first machine is such that the frequency drops uniformly for 50 Hz on no-load to 48 Hz on full-load. The corresponding uniform speed drop of the second machine is 50 to 47.5 Hz.

(*i*) How will the two machines share a load of 3000 kW ?

(*ii*) What is the maximum load at unity p.f. that can be delivered without over loading either machine ?

[**Ans.** 1333 kW assuming u.p.f. ; 1667 kW assuming u.p.f. ; 3600 kW]

Time Period of Oscillations

38. A 10 MVA, 5000 V, 3-phase 4-pole. 50 Hz alternator is connected to infinite bus-bars. The short-circuit current is 3.5 times the normal full-load current and the moment of inertia of the rotating system is 21000 kg m^2. Determine its normal period of oscillation. [**Ans.** 1.365 seconds]

39. A 500 kVA, 3-phase, 10000 V, 50 Hz, alternators runs at 1500 r.p.m. connected to constant frequency, constant voltage bus- bars. If the moment of the inertia of the entire system is 1.5×10^4 kg-m^2 and the steady short-circuit current is 5 times the normal full-load current, find the natural time period of oscillations [**Ans.** 1.364 second]

40. A 2-pole, 50 Hz, 3-phase, 100 MVA, 33000 V turbo-alternator connected to the infinite bus has a moment of inertia of 10^6 kgm^2 in its rotating parts. It has a synchronous reactance of 0.5 p.u. Calculate is natural frequency of oscillation. [**Ans.** 0.127 Hz]

41. A 10000 kVA, 4-pole, 6600 V, 50 Hz, 3-phase star-connected alternator has a synchronous reactance of 25% and operates at constant-voltage, constant-frequency bus-bars. If the natural period of oscillation while operating at full-load and unity p.f. is to be limited to 1.5 seconds, calculate the moment of inertia of the rotating system [**Ans.** 2.89×10^4 kg m^2]

Maximum Power Output

42. A 3-phase, 11 kV, 5 MVA, star-connected alternator has a synchronous impedance of $(1 + j10)\ \Omega$ per phase. Its excitation is such that the generated line e.m.f. is 14 kV. If the alternator is connected to infinite bus-bars, determine the maximum output at the given excitation. [**Ans.** 14133 kW]

43. A 3-phase, 11 kV, 10 MW, star-connected synchronous generator has a synchronous impedance of $(0.8 + j8.0)\ \Omega$ per phase. If the excitation is such that the open circuit voltage is 14 kV determine :

(*i*) Maximum output of the generator.

(*ii*) Current and p.f. at the maximum output [**Ans.** 19.25 MW ; 1287 A, 0.78 (lead)]

6

Synchronous Motors

6.1. INTRODUCTION

- The synchronous motor is the one type of 3-phase A.C. motors *which operates at a constant speed from no-load to full-load.* It is similar in construction to 3-phase A.C. generator in that *it has a revolving field which must be separately excited from a D.C. source. By changing the D.C. field excitation current, the power factor of this type of motor can be varied over a wide range of lagging and leading values.*
- This motor is used in many individual applications because of its *fixed speed from no-load to full-load, its high efficiency* and *low initial cost.* It is also used *to improve the power factor of 3-phase A.C. industrial circuits.*

6.2. CHARACTERISTIC FEATURES, ADVANTAGES AND DISADVANTAGES

Characteristic Features :

The following **characteristic features** of a synchronous motor are worth noting :

1. *It runs either at synchronous speed or not at all.* The speed can be changed by *changing the frequency only* (since $N_s = 120f/p$).

2. *It is not inherently self-starting.* It has to be *run up to synchronous or near synchronous speed* by some means before it can be synchronized to the supply.

3. *It can operate under a wide range of power factors both lagging and leading.*

4. On no-load the motor draws very little current from the supply to meet the internal losses. With fixed excitation the input current increases with the increase in load. After the input current reaches maximum no further increase in load is possible. If the motor is further loaded, the motor will stop.

Advantages. Synchronous motors entail the following *advantages* :

1. These motors can be used for *power factor correction* in addition to supplying torque to drive loads.

2. They are more efficient (when operated at unity power factor) than induction motors of corresponding output (kW) and voltages rating.

3. The field pole rotors of synchronous motors *can permit the use of wider air-gaps* than the squirrel-cage designs used on induction motors, requiring *less bearing tolerance and permitting greater bearing wear.*

4. They may be less costly for the same output, speed, and voltage ratings as compared to induction motors.

5. They give constant speed from no-load to full-load.

6. Electro-magnetic power varies linearly with the voltage.

Disadvantages. The *disadvantages* of synchronous motors are :

1. They require D.C. excitation which must be supplied from external source.
2. They have a tendency to *hunt.*
3. They cannot be used for variable speed jobs as speed adjustment cannot be done.
4. They require collector rings and brushes.
5. They cannot be started under load. Their starting torque is zero.
6. They may fall out of synchronism and stop when overloaded.

6.3. APPLICATIONS

The synchronous motors have the following *fields of application* :

1. **Power houses and sub-stations.** Used in power houses and sub-stations in parallel to the bus-bars to improve the power factor.

2. **Factories.** Used in factories having large number of induction motors or other power apparatus, operating at *lagging power factor, to improve the power factor.*

3. **Mills-industries etc.** Used in textile mills, rubber mills, mining and other big industries, cement factories for power applications.

4. **Constant speed equipments.** Used to drive continuously operating and constant speed equipment such as :

- Fans.
- Blowers.
- Centrifugal pumps.
- Motor generator sets.
- Ammonia and air compressors etc.

6.4. CONSTRUCTION

A three-phase synchronous motor consists of the following *essential parts* :

1. **Laminated stator core** with three-phase armature winding.
2. **Revolving field complete with amortisseur winding and slip rings.**
3. **Brushes and brush holders.**
4. **Two end shields** to house the bearings that support the shaft.

- The stator core and windings of a synchronous motor are similar to those of a 3-phase squirrel-cage induction motor or a wound-rotor induction motor. The leads for the stator winding, marked T_1, T_2 and T_3, terminate in a terminal box usually mounted on the side of the motor frame.
- The rotor is generally a salient pole rotor. *The number of rotor field poles must equal the number of stator field poles.* In order *to eliminate hunting* and *to develop the necessary starting torque* when A.C. voltage is applied to the stator, the rotor poles contain pole-face conductors which are short-circuited at their ends as shown in Fig. 6.1. This amortisseur or damper winding consists of *solid copper bars* embedded at the surface of the pole face and short-circuited at each end by means of a shorting strip as shown in Fig. 6.2.
- The field circuit leads are brought out to two *slip rings* mounted on the rotor shaft. Carbon brushes mounted in brush holders make contact with the two slip rings. The terminals of the field circuit are brought out from the brush holders to a second terminal box mounted on the motor frame. The two leads for the field circuit are marked F_1 and F_2.

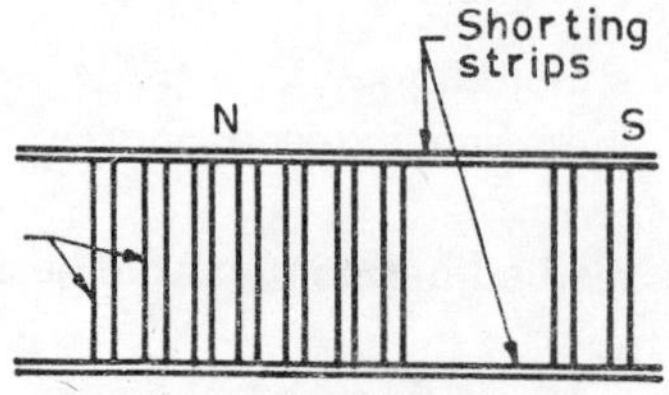

Fig. 6.1. Pole of an A.C. synchronous motor.

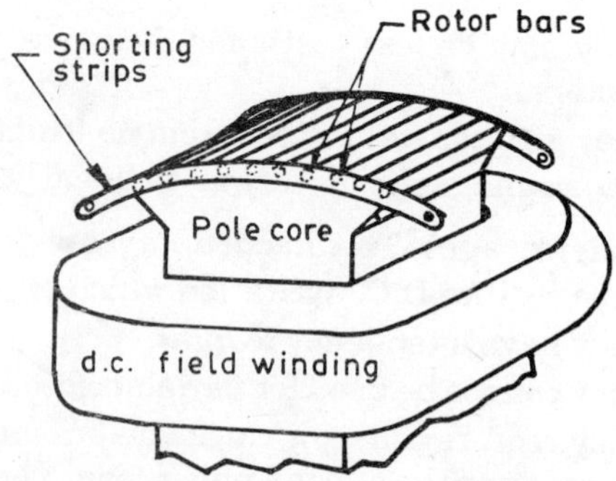

Fig. 6.2. Damper winding.

Fig. 6.3 shows a typical arrangement of synchronous motor, with overhung exciter for providing the D.C. supply to the rotor of the machine.

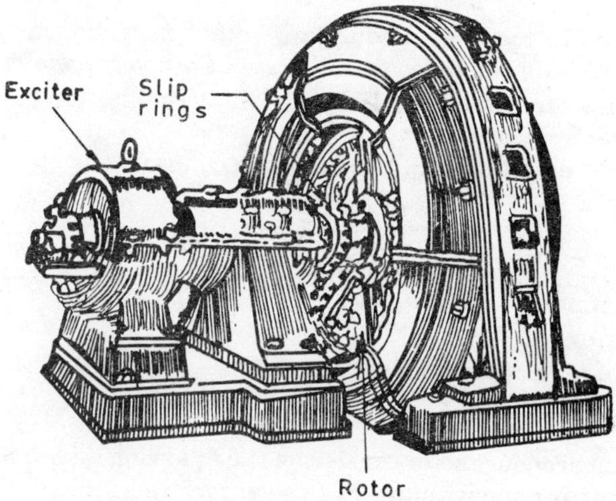

Fig. 6.3. Synchronous motor.

6.5. PRINCIPLE OF OPERATION

When the stator windings of a 3-phase synchronous motor are supplied with rated 3-phase voltage, a rotating field travelling at synchronous speed is set up. The synchronous speed is found from the relation $N_s = \frac{120f}{p}$ where N_s (in r.p.m.), f and p are synchronous speed, frequency and number of poles respectively. This rotating magnetic field cuts across the amortisseur or squirrel-cage winding of the rotor and induces voltages and currents in the bars of this winding. *The resultant magnetic field of squirrel-cage winding embedded in the rotor field poles reacts with the stator field in*

Fig. 6.4. Operating principle of a synchronous motor.

such a manner as to cause the rotation of the motor. The rotor will increase its speed to a point *slightly below* the synchronous speed of the stator field. The rotor of the typical synchronous motor accelerates to about 95 to 97% of synchronous speed when started as an induction motor with amortisseur windings. *The field circuit is now excited from an outside source of D.C. and magnetic poles of fixed polarity are set up in the rotor field cores. The fixed magnetic poles of the rotor are attracted to unlike poles of the rotating magnetic field set up by the stator windings.*

Fig. 6.4 shows the locking of rotor field poles with unlike poles of stator field. The rotor then runs at the same speed as that of the stator field *i.e.*, N_s.

6.6. SYNCHRONOUS MOTORS—STARTING

As earlier stated, the synchronous motor must be brought to a speed sufficiently *close to synchronous speed* in order to lock into synchronism with the rotating field. The means by which it is brought up to speed are :

1. **A D.C. motor coupled to the synchronous motor shaft.** This method is sometimes used in laboratories with synchronous motors not equipped with damper windings. Generally, the synchronous motor is intended as the constant speed primemover for the D.C. generator. But in order to bring the motor upto synchronism, the D.C. generator is operated as motor, and the A.C. synchronous dynamo is synchronized to the A.C. supply as an alternator. Once in parallel with the supply, the synchronous dynamo is operated as a motor. The D.C. *'motor' will now act as a generator if its field current is increased so that its generated e.m.f. exceeds the D.C. bus.*

2. **Using the field excited generator as a D.C. motor.** This method is the same as the first, except that the *exciter* (D.C. shunt generator) *is operated as a motor,* and the A.C. synchronous dynamo is synchronized to the A.C. supply.

3. **A small induction motor of at least one pair of poles less than the synchronous motor.** This method involves the same synchronizing procedure for A.C. synchronous motor as an alternator . At least one pair of poles fewer is required on the induction motor to compensate for the loss in induction motor speed due to slip.

In the above three methods the following *conditions* should be met with :

- There should be *little or no-load* on the synchronous motor.
- The *capacity of the starting motor* (D.C. or A.C.) should be between *5 and 10% of the rating of the synchronous motor* coupled to it.

4. **Using the damper windings as a squirrel-cage induction motor.**

Note. It is practically impossible to start a synchronous motor with its D.C. field energized. Even when left de-energized, the rapidly rotating magnetic field of the stator will induce extremely high voltages in the many turns of the field winding. It is customary, therefore, *to short-circuit the D.C. field winding during the starting period* ; whatever voltage and current are induced in it may then aid in producing induction motor action. In very large synchronous motors, field sectionalising or field-splitting switches are used which short-circuit individual field windings to prevent cumulative addition of induced voltages from pole to pole.

6.7. EFFECT OF LOAD ON A SYNCHRONOUS MOTOR

When mechanical load on a D.C. motor or an A.C. motor is increased, the speed decreases. This, is turn, decreases the back or counter e.m.f. (E_b) so that the source is able to supply more current to meet the increased load demands. However, this action *cannot take place in the synchronous motor for the rotor must turn at synchronous speed at all loads.*

Fig. 6.5 (*a*) shows the relative position of a stator and rotor pole at *no-load*, poles centres are directly *in line* with each other.

Fig. 6.5 (*b*) represents the relative position of the stator and the rotor poles after mechanical load has been added to the motor. Now there has been a *shift* of the rotor pole in a direction opposite to that of the stator field flux and the direction of the rotor. It may be kept in mind that there has

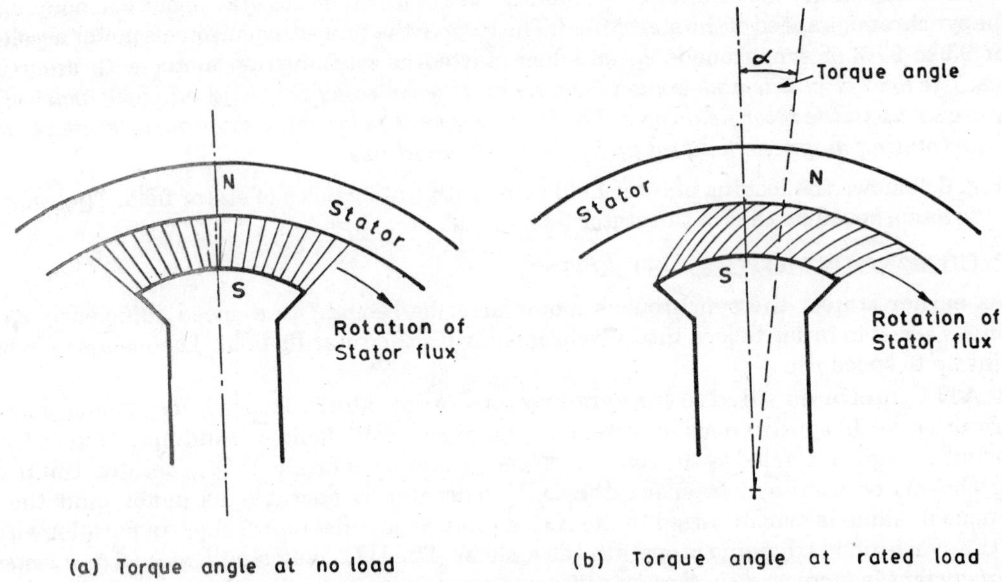

Fig. 6.5. Relative positions of stator and rotor poles.

been no change in speed as the rotor will continue to rotate at synchronous speed. There is *only an angular displacement between the centres of the stator and rotor field poles*. The angular displacement shown in Fig. 6.5 (*b*) is called the **'torque angle'.**

No-load condition-vector diagrams. Fig. 6.6 shows the conditions when the motor (properly synchronised to the supply) is running on *no-load* and is having no losses. It is seen that $V = E_b$, hence their vector difference is zero and so is the armature current. Motor intake is zero, as there is neither load nor losses to be met by it. In other words, the motor just *floats*.

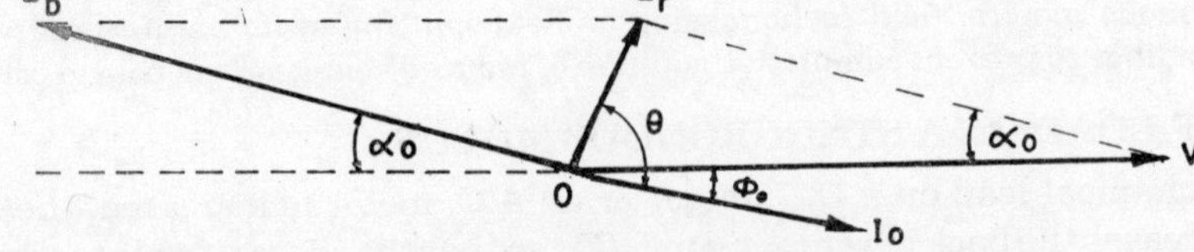

Fig. 6.6. No-load (no losses).

Fig. 6.7 shows the vector diagram when the motor is on no-load but has losses. The vector for E_b falls back by a certain angle α_0, so that a resultant voltage E_r and hence current I_0 is brought into existence which supplies losses.

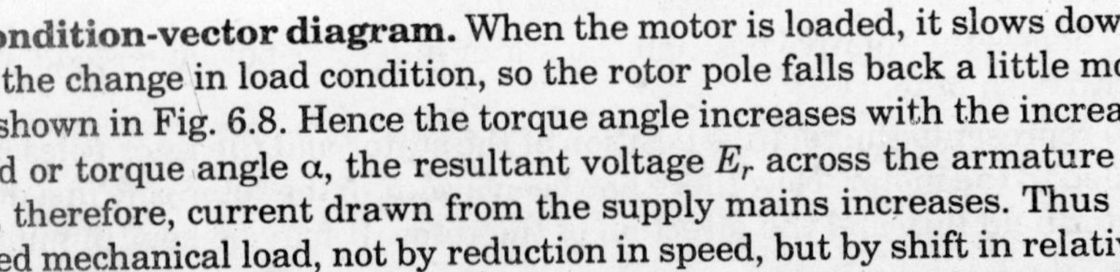

Fig. 6.7. No-load (with losses).

Load condition-vector diagram. When the motor is loaded, it slows down momentarily to adjust itself to the change in load condition, so the rotor pole falls back a little more relative to the stator pole, as shown in Fig. 6.8. Hence the torque angle increases with the increase in load. Due to increase in load or torque angle α, the resultant voltage E_r across the armature (or stator) circuit increases, and, therefore, current drawn from the supply mains increases. Thus a motor is able to supply increased mechanical load, not by reduction in speed, but by shift in relative positions of the rotor and rotating magnetic field (or stator flux). From Fig. 6.8 it is obvious that *for increasing load with a constant value of back e.m.f. E_b the phase angle ϕ Increases in lagging direction.*

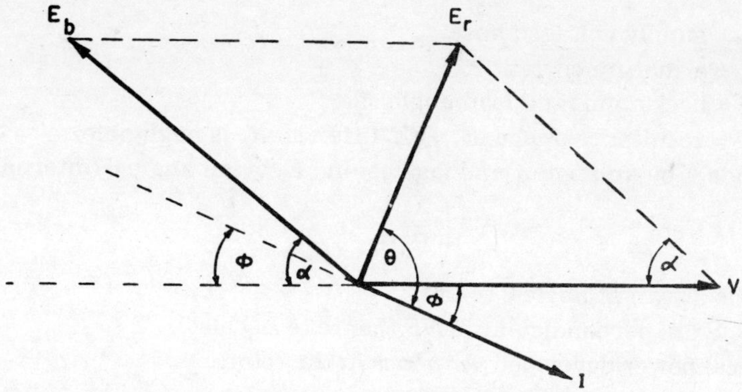

Fig. 6.8. Synchronous motor no-load vector diagram.

If *the angle between stator and rotor pole centres become too great, due to a serious over-load then the rotor will pull out of synchronism and operate as an induction motor with the aid of the amortisseur winding.* The maximum value of torque which a synchronous motor can develop without dropping out of synchronism is called the **'pull-out torque'.** *In most synchronous motors this is 150 to 200 per cent of rated torque output.*

6.8. TORQUE DEVELOPED BY THE MOTOR

Refer Fig. 6.9.

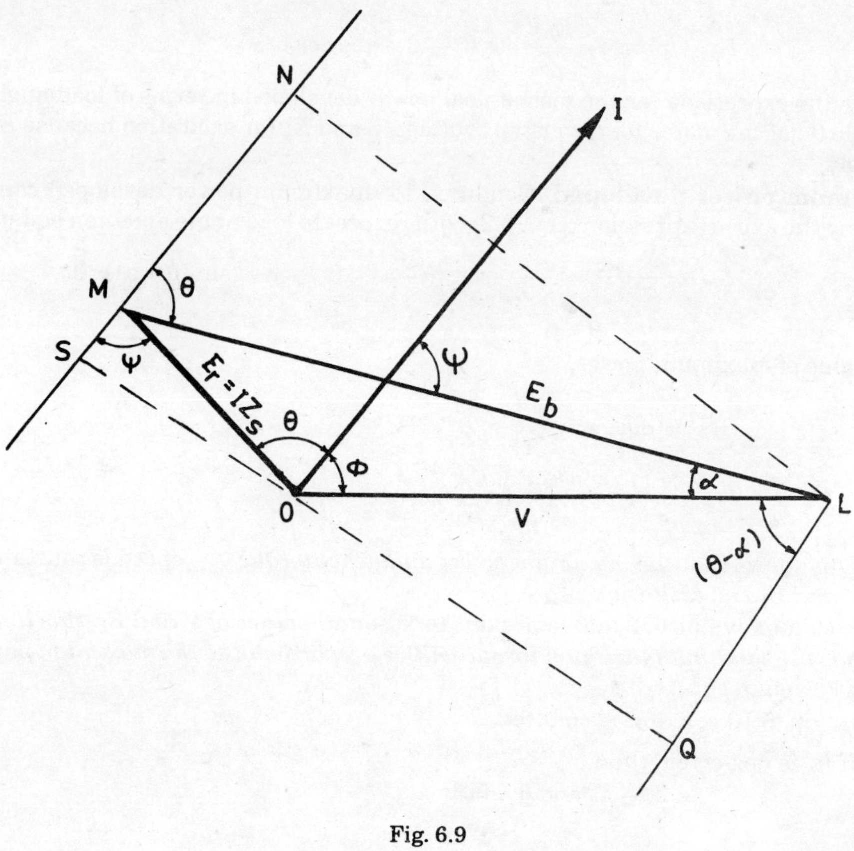

Fig. 6.9

OL = supply voltage/phase
I = armature current
LM = back e.m.f. at a load angle of α
OM = resultant voltage, $E_r = IZ_s$ (or IX_s if R_a is negligible)

- I lags/leads V by an angle ϕ and lags behind E_r by an angle θ (internal angle)

$$= \tan^{-1}\left(\frac{X_s}{R_a}\right).$$

- Line NS is drawn at angle θ to LM.
- LN and QS are perpendicular to NS (hence to LQ also).

Mechanical power developed *per phase* in the rotor,

$$P_{mech} = E_b I \cos \psi \qquad ...(6.1)$$

In ΔOMS,

$$MS = IZ_s \cos \psi$$

Now,

$$MS = NS - NM = LQ - NM$$

$$\therefore \quad IZ_s \cos \psi = V \cos (\theta - \alpha) - E_b \cos \theta$$

$$\therefore \quad I \cos \psi = \frac{V}{Z_s} \cos (\theta - \alpha) - \frac{E_b}{Z_s} \cos \theta$$

Putting this value in (6.1), we get

$$P_{mech}/\text{phase} = E_b \left[\frac{V}{Z_s} \cos (\theta - \alpha) - \frac{E_b}{Z_s} \cos \theta \right]$$

or

$$P_{mech}/\text{phase} = \frac{E_b V}{Z_s} \cos (\theta - \alpha) - \frac{E_b^2}{Z_s} \cos \theta \qquad ...(6.2)$$

This is the expression for the mechanical power developed in terms of load angle (α) and the internal angle (θ) of the motor for a constant voltage V and E_b (or excitation because E_b depends on excitation only).

Maximum power developed. Condition for maximum power developed can be found by differentiating the above expression (eqn. 6.2) with respect to load angle and then equating it to zero.

$$\therefore \quad \frac{dP_{mech}}{d\alpha} = -\frac{E_b V}{Z_s} \sin (\theta - \alpha) = 0 \quad \text{or} \quad \sin (\theta - \alpha) = 0$$

$$\therefore \quad \theta = \alpha$$

$\therefore$ Value of maximum power,

$$(P_{mech})_{max} = \frac{E_b V}{Z_s} - \frac{E_b^2}{Z_s} \cos \alpha$$

or

$$\frac{E_b V}{Z_s} - \frac{E_b^2}{Z_s} \cos \theta \qquad ...(6.3)$$

- This shows that the *maximum power and hence torque* ($\because$ speed is constant) *depends on V and E_b i.e., excitation.*
- Maximum value of θ and hence α is 90°. For *all values of V and E_b, this limiting value of α is the same but maximum torque will be proportional to the maximum power developed* as given in eqn. (6.3).
- In Fig. 6.10 eqn. (6.2) is plotted.
- If R_a is neglected, then

$$Z_s \simeq X_s \text{ and } \theta = 90°$$

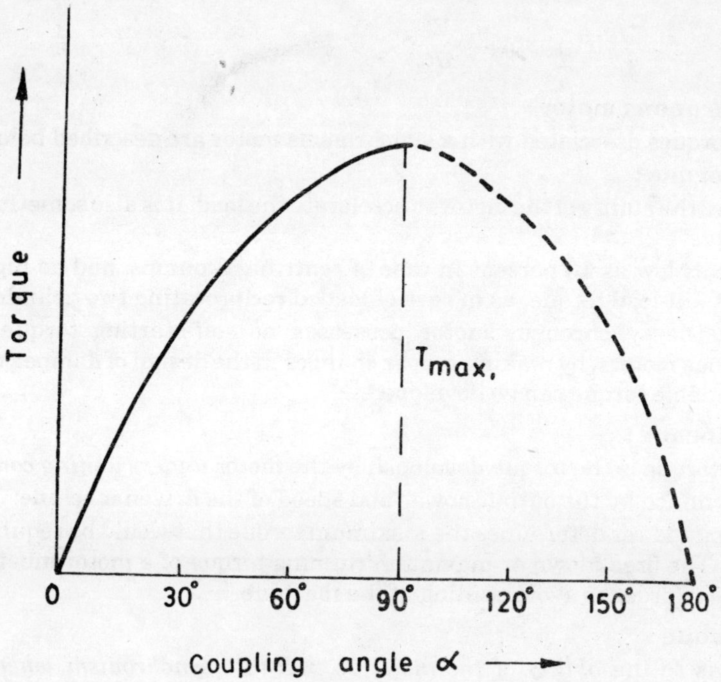

Fig. 6.10

$\therefore$ $\cos \theta = 0$

$$P_{mech} = \frac{E_b V}{X_s} \cos (90° - \alpha) \quad \text{[from eq. (6.2)]}$$

i.e.,

$$P_{mech} = \frac{E_b V}{X_s} \sin \alpha \quad ...(6.4)$$

This gives the value of mechanical power developed in terms of α—the basic variable of a synchronous machine.

$$\therefore \quad (P_{mech})_{max} = \frac{E_b V}{X_s} \quad \text{[from eqn. (6.3)]}$$

This corresponds to the *'pull-out'* torque.

The above value can be obtained by putting α = 90° in eqn. (6.4).

- **To determine the value of excitation or induced e.m.f. E_b to give maximum power developed possible, differentiate eqn. (6.3) with respect to E_b and equate to zero**

$$\therefore \quad \frac{d(P_{mech})_{max}}{dE_b} = \frac{V}{Z_s} - \frac{2E_b}{Z_s} \cos \theta = 0 \quad \text{or} \quad E_b = \frac{V}{2 \cos \theta}$$

Substituting $E_b = \frac{V}{2 \cos \theta}$ in eqn. (6.3), we get

$$(P_{mech})_{max} = \frac{V^2}{2Z_s \cos \theta} - \frac{V^2}{4Z_s \cos \theta}$$

$$= \frac{V^2}{4Z_s \cos \theta} = \frac{V^2}{4R_a} \quad [\because Z \cos \theta = R_a]$$

where R_a = effective resistance of the motor.

Hence $$(P_{mech})_{max} = \frac{V^2}{4R_a}$$ ...(6.5)

Torques of a synchronous motor :

The various torques associated with a synchronous motor are described below :

1. Starting torque :

- It indicates the ability of the motor to accelerate the load. It is also sometimes called *"break away torque"*.
- It may be as low as 10 percent in case of centrifugal pumps, and as high as 200 or 250 percent of full-load torque, as in case of loaded reciprocating two-cylinder compressors.
- Although the synchronous motor possesses no self-starting torque yet in modern synchronous motors, by making proper changes in the design of damper windings, almost any reasonable torque can be developed.

2. Running torque :

- Running torque is the torque developed by the motor *under running conditions.*
- It is determined by the output power and speed of the driven machine.
- Peak output power determines the maximum torque that would be required by the driven machine. The breakdown or maximum running torque of a motor must be greater than this value in order to avoid stalling of the machine.

3. Pull-in torque :

- It pertains to the *ability of the motor to pull-into synchronism when changing from induction to synchronous motor operation.*

4. Pull-out torque :

- It *is the maximum torque which the motor will develop without pulling out of step or synchronism.*
- Its value varies from 1.25 to 3.5 times the full-load torque.

6.9. TWO REACTANCE CONCEPT FOR SALIENT POLE SYNCHRONOUS MOTORS

- The steady-state performance of salient pole synchronous motors can be predicted quite accurately by the synchronous reactance concept, however, when high degree of accuracy is required or when problems concerning transients or power stability are to be handled, the *two-reactance theory* is required. This theory is considered below :
- The salient pole synchronous motor has the following two axes :
  - — Field pole axis, called the *direct axis* or *d-axis* ; and
  - — The axis passing through the centre of interpolar space, called *quadrature axis* or *q-axis* (as in case of an alternator).

I_d and I_q are the components of the armature current resolved along d-axis and q-axis respectively.

In the Fig. 6.11 and 6.12 are shown the complete phasor diagrams of a salient pole synchronous motor, for a leading power factor, considering and neglecting armature resistance respectively.

From Fig. 6.12, using E_b as reference phasor, we have

$$V \cos \alpha = E_b - I_d X_d$$ ...(6.6)

and $$V \sin \alpha = I_q X_q$$ ...(6.7)

∴ $$I_d = \frac{E_b - V \cos \alpha}{X_d}$$ ...(6.8)

and $$I_q = \frac{V \sin \alpha}{X_q}$$ ...(6.9)

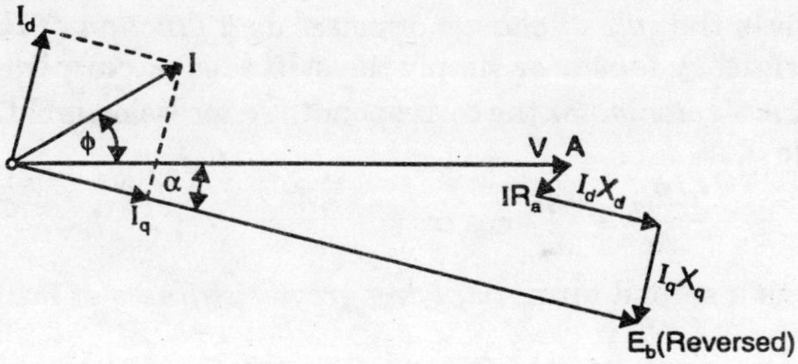

Fig. 6.11. Phasor diagram for synchronous motor-leading p.f. (Considering armature resistance).

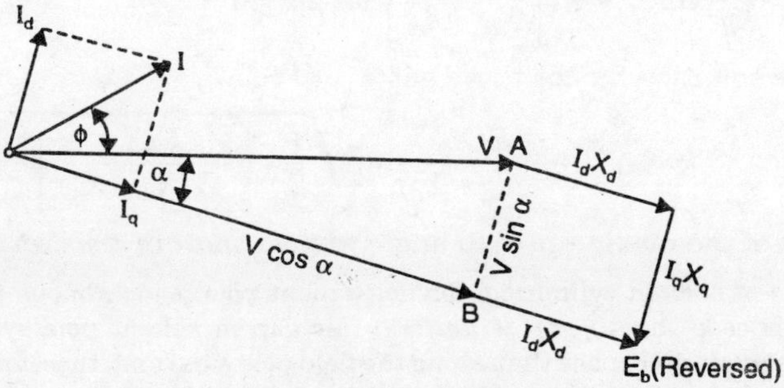

Fig. 6.12. Phasor diagram for synchronous motor-leading p.f. (Neglecting armature resistance).

Regardless of the axis of reference, power input is given by the product of the in-phase components of the current and voltage plus the product of the quadrature component.

$$\therefore \qquad P_{in} = I_q V \cos\alpha + I_d V \sin\alpha$$

Substituting the values of I_d and I_q from eqns. (6.8) and (6.9), we get

$$P_{in(per\ phase)} = \frac{V\sin\alpha}{X_q} \times V\cos\alpha + \frac{E_b - V\cos\alpha}{X_d} \times V\sin\alpha$$

$$= \frac{V^2 \sin\alpha\cos\alpha}{X_q} + \left[\frac{E_b V\sin\alpha - V^2\sin\alpha\cos\alpha}{X_d}\right]$$

$$= \frac{E_b V}{X_d}\sin\alpha + \frac{V^2}{2}\left[\frac{1}{X_q} - \frac{1}{X_d}\right]\sin 2\alpha \qquad ...(6.10)$$

Total power input (3 times of the above)

$$= \frac{3E_b V}{X_d}\sin\alpha + \frac{3V^2}{2}\left[\frac{1}{X_q} - \frac{1}{X_d}\right]\sin 2\alpha$$

$$= \frac{E_{bL} V_L}{X_d}\sin\alpha + \frac{{V_L}^2}{2}\left[\frac{1}{X_q} - \frac{1}{X_d}\right]\sin 2\alpha \qquad ...(6.11)$$

(Power developed will be equal to power input *minus* copper losses)

Stability and maximum load angle :

Differentiating eqn. (6.10) w.r.t. load angle α, we get

$$\frac{dP_{in}}{d\alpha} = \frac{E_b V}{X_d}\cos\alpha + V^2\left[\frac{1}{X_q} - \frac{1}{X_d}\right]\cos 2\alpha \qquad ...(6.12)$$

This eqn. (6.12) gives the *rate of change of power as a function of the load angle* α. This is called **stability factor, rigidity factor** or simply the **stiffness of coupling.**

In a *smooth cylindrical rotor motor* the corresponding expression is obtained by differentiating eqn. (6.4) w.r.t. load angle α, so

$$\frac{dP_{in}}{d\alpha} = \frac{E_b V}{X_s} \cos \alpha$$

Eqn. (6.12) contains a second term, *implying greater stiffness of the coupling for the salient pole motor.*

Maximum load angle can be determined by equating the R.H.S. of the eqn. (6.12) equal to zero,

i.e.
$$\frac{E_b V}{X_d} \cos \alpha + V^2 \left[\frac{1}{X_q} - \frac{1}{X_d} \right] \cos 2\alpha = 0$$

Solving the above equation for cos α, we get

$$\cos \alpha \simeq \frac{-E_b X_q}{4V(X_d - X_q)} + \sqrt{\frac{1}{2} + \left[\frac{E_b X_q}{4V(X_d - X_q)} \right]^2} \quad ...(6.13)$$

This is the cosine of the maximum load angle when *armature resistance is neglected.*

Note. The analysis of smooth cylindrical (or non-salient pole) synchronous machine is much easier comparatively since in this case the *air-gap is uniform* (air gap in salient pole synchronous motor is not uniform-much larger in the interpolar space than along the field pole axis) and, therefore, the effect of armature reaction, fluxes and voltages induced can be treated in a simple way with synchronous reactance concept.

6.10. EFFECT OF EXCITATION ON ARMATURE CURRENT AND POWER FACTOR (V-CURVES)

Consider a synchronous motor in which the *mechanical load is constant* and hence output is also constant if losses are neglected.

Case 1. 100% Excitation :

The case for 100% excitation *i.e.*, when $E_b = V$ is shown in Fig. 6.13 (*i*). Here the armature current I lags behind V by a small angle ϕ. Its angle with E_R is fixed by stator constants *i.e.*,

$$\tan \theta = \frac{X_s}{R_a}.$$

Case 2. Excitation less than 100% :

Fig. 6.13 (*ii*) represents the condition for *under-excited* motor *i.e.*, $E_b < V$. Here E_r is advanced clockwise and so is the armature current (because it lags behind E_r by a fixed angle θ). We find that :

The magnitude of I is increased, but *its power factor is decreased* ($\because$ ϕ has increased) :

— Since input as well as V are constant, hence the power component of I *i.e.*, $I \cos \phi$ remains the same as before, but wattless component $I \sin \phi$ is increased. Hence, as excitation is decreased, I will increase, but power factor will decrease so that power component *i.e.*, $I \cos \phi = OL$ will remain constant. *The locus of the extremity of current vector would be a horizontal straight line.*

Case 3. Excitation greater than 100% :

In Fig. 6.13 (*iii*) excitation is greater than 100% *i.e.*, $E_b > V$ (*i.e.*, motor is over-excited). Here the resultant voltage vector E_r is pulled-anticlockwise and so is I. It may be noted that now motor is drawing a *leading current*. It may also happen for some value of excitation, that I may be in phase

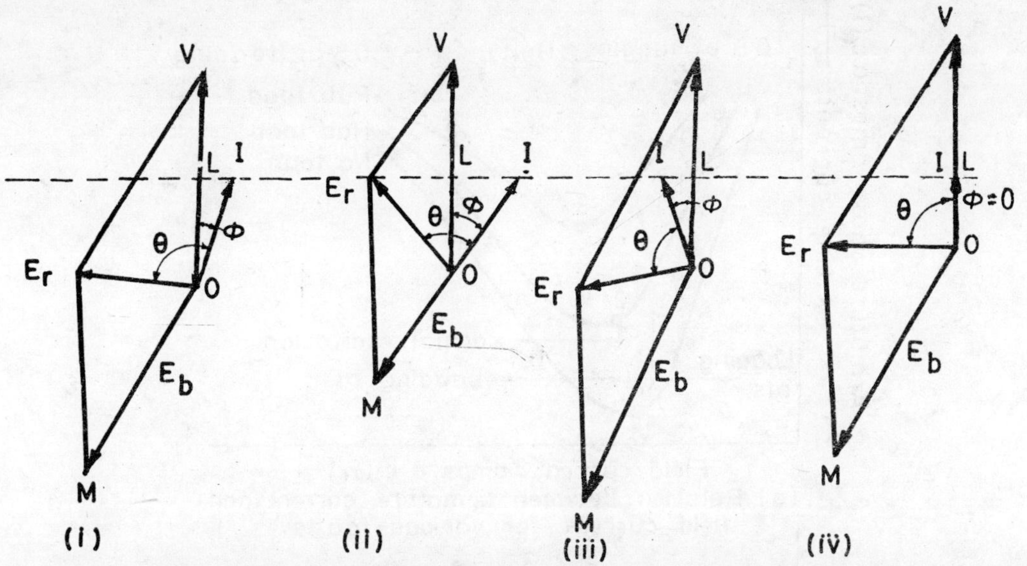

Fig. 6.13. Effect of excitation on armature current and power factor.

with *V i.e.,* power factor *is unity* [Fig. 6.13 (*iv*)]. At that time the *current drawn by motor would* be *minimum.*

V-curves of a Synchronous Motor :

- It has been stated above that, when the field current (*i.e.*, excitation) of a synchronous motor is reduced, a lagging armature current [Fig. 6.13 (*ii*)] *I* is produced which exceeds the minimum current at unity power or at normal excitation.
- Similarly, when motor is *over-excited,* the armature current also rises [Fig. 6.13 (*iii*)] and exceeds the current required at normal excitation to develop the necessary torque, at any given load.
- By applying a given *constant load* to the shaft of a synchronous motor and varying the field current from under-excitation to *over-excitation,* recording the armature current at each step, the curves of Fig. 6.14 (*a*) are obtained. The A.C. armature current is plotted against the D.C. field current for *no-load, half-load*, and *full-load* values, respectively.
- The wattmeter connections shown in Fig. 6.15 will also yield the power factor for each value of armature and field current at any given load condition. (The method employed in Fig. 6.15 is the two-wattmeters method. Since the synchronous motor is a balanced three-phase load, the one-wattmeter method, three-wattmeters method, industrial analyzer, or polyphase wattmeter would do as well in providing the power factor of the load).

Thus, as shown in Fig. 6.14. (*b*), the power factor as determined from the wattmeter readings, is plotted against the field current for the various given loads. It is worth noting that both curves show that a *slightly increased field current is required to produce normal excitation as the load is increased* (points 1, 2 and 3, respectively).

— At *no-load, the armature current at unity power factor* (normal excitation) *is not zero* but *some small value of A.C. armature current per phase, necessary to produce torque to counter balance rotational losses.* As load is applied (neglecting armature reaction) not only does the armature current rise but it is also necessary to increase the excitation to bring the armature current back in phase with the bus phase voltage. Each of the curves in the

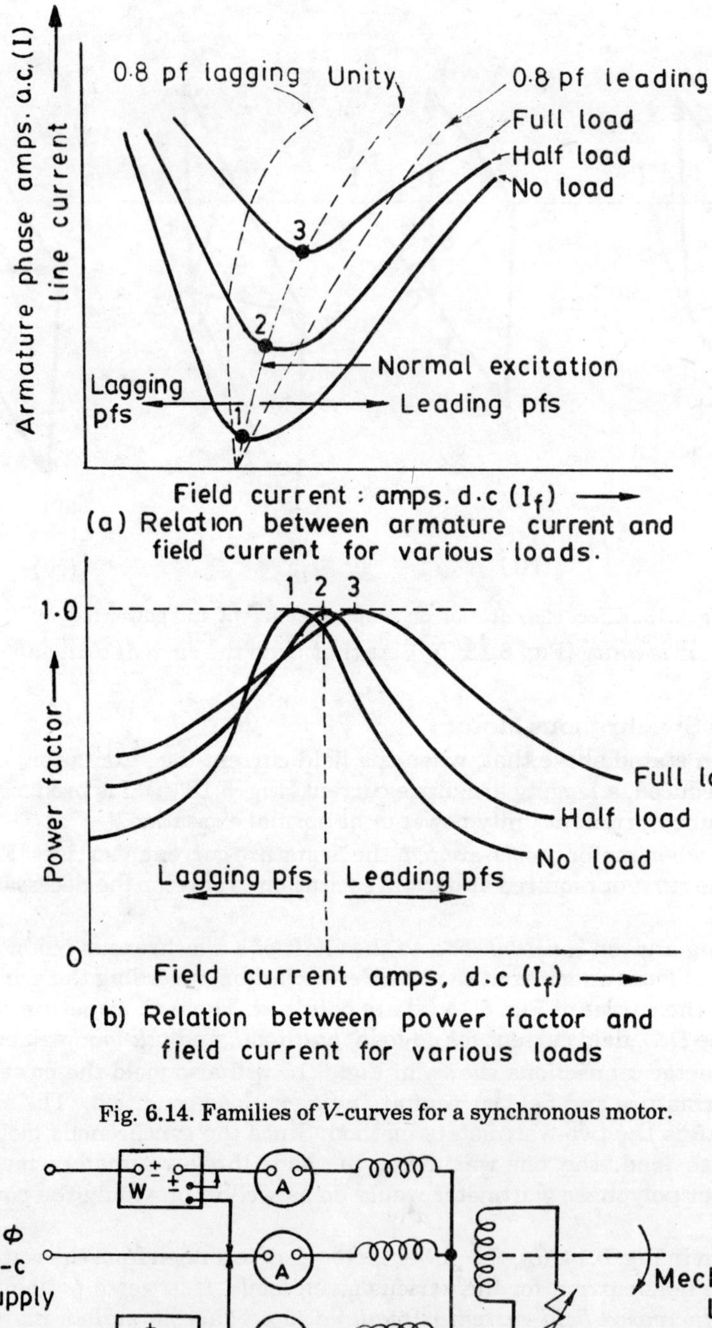

Fig. 6.14. Families of *V*-curves for a synchronous motor.

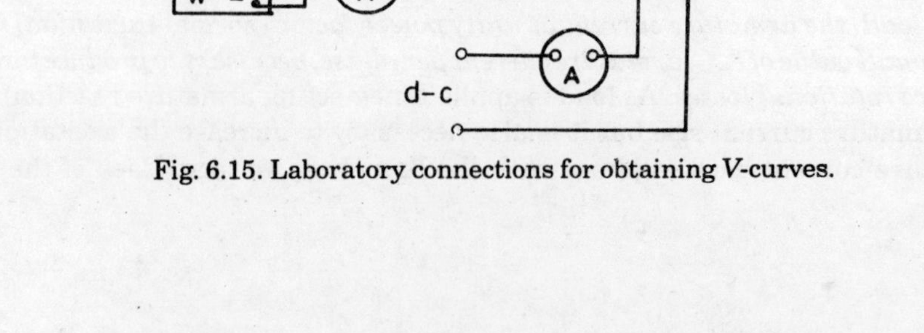

Fig. 6.15. Laboratory connections for obtaining *V*-curves.

family, therefore, will have a shift to the right as the load is increased, as shown in Figs. 6.14 (*a*) and (*b*), to provide the excitation required to obtain the same power factor (0.8 lagging, unity, or 0.8 leading) at an increased load. Thus, the *V-curves represent the phasor diagrams*, and *vice versa* for various conditions of load and power factor.

Constant power lines :

Refer Fig. 6.16.

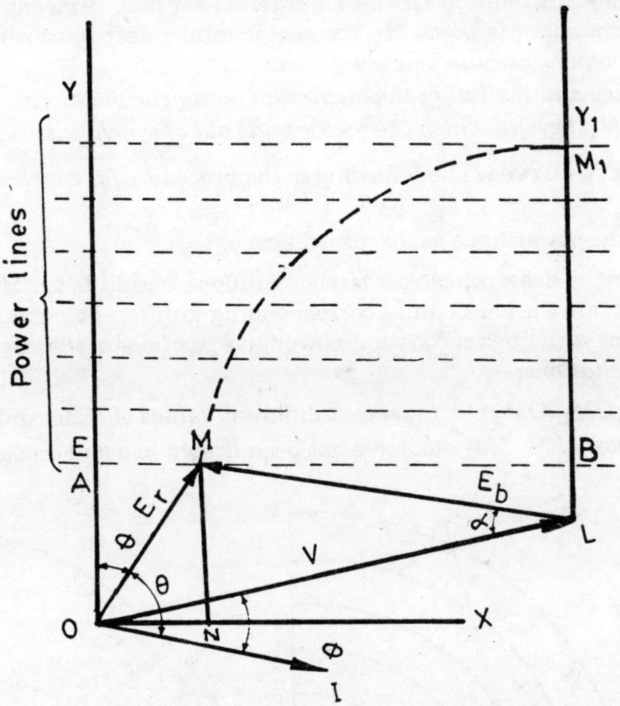

Fig. 6.16

OL = applied voltage/phase of the motor, V

LM = back e.m.f./phase, E_b

OM = resultant voltage, E_r.

OI = armature current lagging behind E_r by an angle

$$\theta = \tan^{-1} \frac{X_s}{R_a}, \text{ value of } I = \frac{E_r}{Z_s}$$

- OY is drawn at an angle ϕ with OM (or at an angle θ with OL).
- MN is drawn perpendicular to OX which is at right angles to OY. Vector OM when referred to OY also represents, on a different scale, the current both in magnitude and phase.
- The power input/phase of the motor

$$= VI \cos \phi = V \times MN$$

As V is constant, power input is dependent on MN. If the motor is working with a constant intake then locus of M is a straight line $\parallel$ to OX and $\perp$ OY *i.e.*, line AY for which MN is constant. Hence AB represents a *constant power input line for a given voltage but varying excitation.* Similarly, a series of such parallel lines can be drawn each representing a definite power intake of the motor. Regarding these power lines the following points are worth noting :

(*i*) For equal increase in intake, the power lines are parallel and equally spaced.

(*ii*) The line *OX* represents zero power line.

(*iii*) The perpendicular distance from working point *M* to *OX* represents the motor intake.

(*iv*) If the excitation is fixed *i.e.*, *LM* is constant in length, then, as the load on the motor is increased, α increases. In other words, locus of *M* is a *circle* with radius = *LM* and centre at *L*. With increasing load, *M* goes on to the lines of higher power till point M_1 is reached. Any further increase in load on the motor will bring point *M* down to a lower power line. It means that as load increases beyond the value corresponding to point M_1, the motor intake decreases which is impossible. The area to the right of LY_1 *represents unstable condition.*

For a given voltage and excitation the maximum power the motor can develop, is determined by the location of point M_1, beyond which the motor pulls out of synchronism.

Construction of V-curves. The following is the *procedure for construction of V-curves* : Refer Fig. 6.17.

- Draw constant-power lines as discussed above.
- With *L* as centre, draw concentric circles of different radii *LM*, LM_1, LM_2 etc. where *LM*, LM_1, LM_2 etc. are the back e.m.fs. corresponding to different excitations. The intersection of these circles with lines of constant power give *positions of the working points for specific loads and excitations.*

The vectors *OM*, OM_1, OM_2 etc. represent different values of E_r for different excitations. (Back e.m.fs. vectors *LM*, LM_1 etc. have not been drawn to avoid confusion).

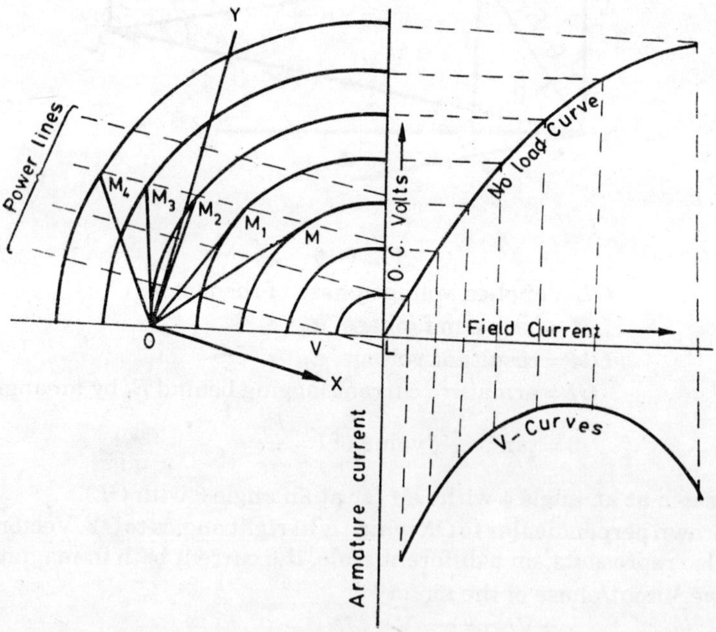

Fig. 6.17

- Project the different values of back e.m.f. on the magnification curve and read the corresponding values of the field amperes from it.
- Plot the field amperes against the corresponding armature currents to get *V*-curves.

WORKED EXAMPLES

Example 6.1. *A 400 V, single-phase synchronous motor gives a net output mechanical power of 7.35 kW and operates at 0.92 p.f. lagging. Its effective resistance is 0.7 Ω. If the iron and mechanical losses are 550 W and excitation losses are 750 W, calculate :*

(i) armature current ; and *(ii) commercial efficiency.*

Solution. Supply voltage = 400 V

Net output mechanical power = 7.35 kW or 7350 W

Power factor = 0.92 (lag)

Effective resistance = 0.7 Ω

Iron and mechanical losses = 550 W

Excitation losses = 750 W

Armature current :

Commerical efficiency :

(*i*) Let I be the input armature current

Power input $= VI \cos \phi = 400 \times I \times 0.92 = 368I$ watts

Power output = power input – mechanical losses – copper losses

$$7350 = 368I - 550 - I^2 \times 0.7$$

or

$$0.7I^2 - 368I + 7900 = 0$$

$$\therefore \quad I = \frac{368 \pm \sqrt{(368)^2 - 4 \times 0.7 \times 7900}}{2 \times 0.7} = \frac{368 \pm 336.6}{1.4} = 22.43 \text{ A}$$

Hence, armature current **= 22.43 A. (Ans.)**

(*ii*) Power input including excitation losses

$$= (400 \times 22.43 \times 0.92 + 750) = 9004 \text{ W}$$

Commerical efficiency, $\eta = \frac{\text{output}}{\text{input}} \times 100 = \frac{7350}{9004} \times 100 = $ **81.6%. (Ans.)**

Example 6.2. *A 2200 V, 3-phase star-connected synchronous motor has a resistance of 0.22 Ω/phase and a synchronous reactance of 2.4 Ω/phase. The motor is operating at 0.6 power factor leading with a line current of 180 A.*

Determine the value of the generated e.m.f./phase. **(Bangalore University Nov., 1996)**

Solution. Phase voltage, $V = \frac{2200}{\sqrt{3}} = 1270$ V

Resistance/phase, $R = 0.22$ Ω

Synchronous reactance/phase, $X_s = 2.4$ Ω

Power factor, $\cos \phi = 0.6$ (leading)

Line current, $I = 180$ A

Generated e.m.f./phase, E_b :

Here, $\cos \phi = 0.6$

$\therefore \quad \phi = \cos^{-1} = (0.6) = 53.13°$ (lead)

Also $\tan \theta = \frac{X_s}{R} = \frac{2.4}{0.22} = 10.91$

$\therefore \quad \theta = 84.8°$

$\therefore \quad (\theta + \phi) = 53.13 + 84.8° = 137.93°$

$\cos(\theta + \phi) = \cos 137.93° = -0.7423$

Synchronous impedance, $Z_s = \sqrt{(0.22)^2 + (2.4)^2} = 2.41\ \Omega$/phase

Impedance drop/phase, $E_r = IZ_s = 180 \times 2.41 = 433.8$ V

The vector diagram is shown in Fig. **6.18.**

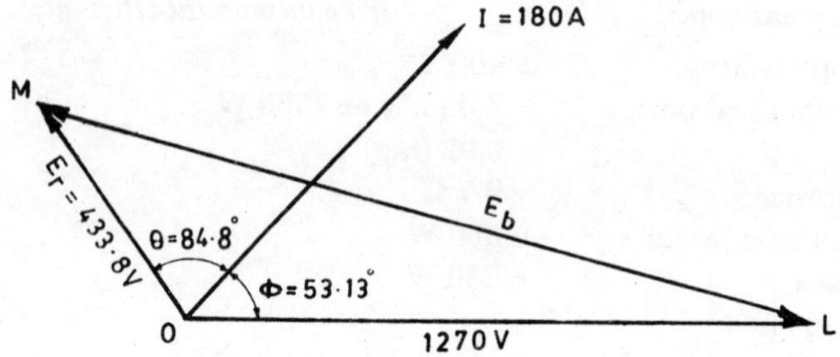

Fig. 6.18

$\therefore$ **Generated e.m.f./phase,** $E_b = \sqrt{V^2 + (E_r)^2 - 2V \times E_r \cos(\theta + \phi)}$

$$= \sqrt{(1270)^2 + (433.8)^2 - 2 \times 1270 \times 433.8 \cos(53.31° + 84.8°)}$$

$$= 1618.3 \text{ V}$$

Hence, **synchronous e.m.f. (line) = $\sqrt{3}$ × 1618.3 = 2803 V. (Ans.)**

☞**Example 6.3.** *A 6600 V, 3-phase, star-connected synchronous motor draws a full-load current of 80 A at 0.8 p.f. leading. The armature resistance is 2.2 Ω and reactance 22 Ω per phase. If the stray losses of the machine are 3200 W, find :*

(i) E.m.f. induced, *(ii) Output power, and*

(iii) Efficiency of the machine.

Solution. Phase voltage, $V = \dfrac{6600}{\sqrt{3}} = 3810$ V

Full-load current, $I = 80$ A at 0.8 p.f. leading

Armature resistance per phase, $R_a = 2.2\ \Omega$

Synchronous reactance/phase, $X_s = 22\ \Omega$

Stray loss = 3200 W

(i) E.m.f. induced, E_b :

Power factor, $\cos \phi = 0.8$

$\therefore \quad \phi = \cos^{-1}(0.8) = 36.9°$

Synchronous impedance/phase,

$$Z_s = \sqrt{R_a^2 + X_s^2} = \sqrt{(2.2)^2 + (22)^2} = 22.11\ \Omega$$

$$\tan \theta = \frac{X_s}{R_a} = \frac{22}{2.2} = 10 \quad \text{or} \quad \theta = \tan^{-1} 10 = 84.3°$$

Impedance drop/phase, $E_r = IZ_s = 80 \times 22.11 = 1768.8$ V

The vector diagram is shown in Fig. 6.19.

Induced e.m.f, E_b/phase, $E_b = \sqrt{V^2 + E_r^2 - 2VE_r \cos(\theta + \phi)}$

$$= \sqrt{(3810)^2 + (1768.8)^2 - 2 \times 3810 \times 1768.8 \cos(84.3° + 36.9°)}$$

$$= 4962.5 \text{ V}$$

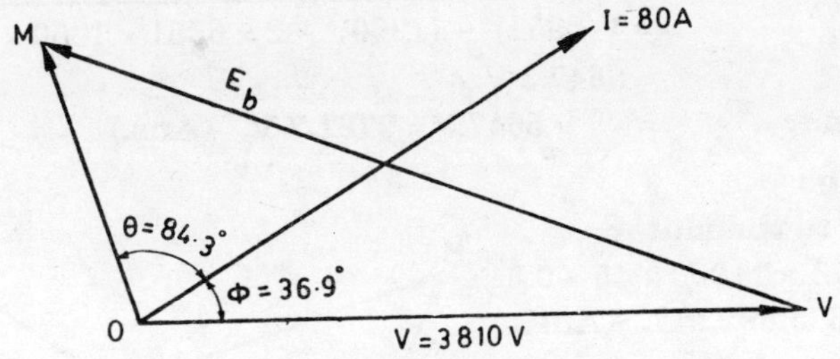

Fig. 6.19

Induced line e.m.f. $= \sqrt{3} \times 4962.5 =$ **8595 V. (Ans.)**

(*ii*) **Power output :**

Total input $= \sqrt{3} E_L I_L \cos \phi = \sqrt{3} \times 6600 \times 80 \times 0.8 = 731618$ W

Total copper losses $= 3I^2 R_a = 3 \times (80)^2 \times 2.2 = 42240$ W

Total stray losses = 3200 W

∴ **Power output** = power input – copper losses – stray losses

= 731618 – 42240 – 3200 = **686178 W. (Ans.)**

(*iii*) **Efficiency :**

Efficiency, $\eta = \frac{\text{output}}{\text{input}} = \frac{686178}{731618} \times 100 =$ **93.79%. (Ans.)**

Example 6.4. *A 11 kV, 3-phase star-connected synchronous motor draws a current of 45 A. The effective resistance and synchronous reactance per phase are 0.9 Ω and 28 Ω respectively. Calculate the power supplied to the motor and induced e.m.f. for a power factor of :*

(i) 0.8 lagging *(ii) 0.8 leading.*

Solution. Supply voltage/phase $= \frac{E_L}{\sqrt{3}} = \frac{11 \times 1000}{\sqrt{3}} = 6351$ V

Current drawn, $I = 45$ A

Effective resistance/phase, $R_a = 0.9\ \Omega$

Synchronous resistance/phase, $X_s = 28\ \Omega$

Synchronous impedance/phase,

$$Z_s = \sqrt{R_a^2 + X_s^2} = \sqrt{(0.9)^2 + (28)^2} = 28\ \Omega \text{ (app.)}$$

$$\tan \theta = \frac{X_s}{R_a} = \frac{28}{0.9} = 31.111$$

∴ $\theta = 88.1°$

Impedance drop/phase, $E_r = IZ_s = 45 \times 28 = 1260$ V.

(*i*) **0.8 p.f. lagging :**

$\cos \phi = 0.8$

or $\phi = \cos^{-1} 0.8 = 36.9°$

Power supplied to the motor

$= \sqrt{3} E_L I_L \cos \phi$

$= \sqrt{3} \times 11000 \times 45 \times 0.8$

= 685892 or **685.892 kW. (Ans.)**

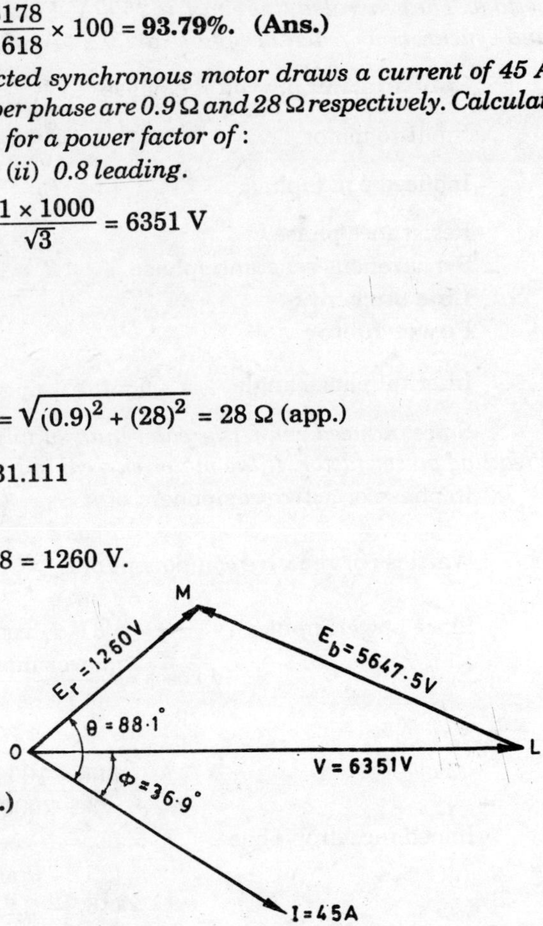

Fig. 6.20

Refer Fig. 6.20.

Induced e.m.f./phase,

$$E_b = \sqrt{V^2 + E_r^2 - 2VE_r \cos(\theta - \phi)}$$

$= \sqrt{(6351)^2 + (1260)^2 - 2 \times 6351 \times 1260 \times \cos(88.1° - 36.9°)}$

$= 5647.5$ V

Induced line e.m.f. $= \sqrt{3} \times 5647.5 =$ **9781.7 V. (Ans.)**

(*ii*) **0.8 p.f. leading :**

Power supplied to the motor

$= \sqrt{3} E_L I_L \cos \phi = \sqrt{3} \times 11000 \times 45 \times 0.8$

$= 685892$ W or **685.892 kW. (Ans.)**

Refer Fig. 6.21.

Induced e.m.f./phase,

$$E_b = \sqrt{V^2 + E_r^2 - 2VE_r \cos(\theta + \phi)}$$

$$= \sqrt{(6351)^2 + (1260)^2 - 2 \times 6351 \times 1260 \cos(88.1° + 36.9°)}$$

$= 7148.6$ V

Induced line e.m.f. $= \sqrt{3} \times 7148.6 =$ **12381.7 V. (Ans.)**

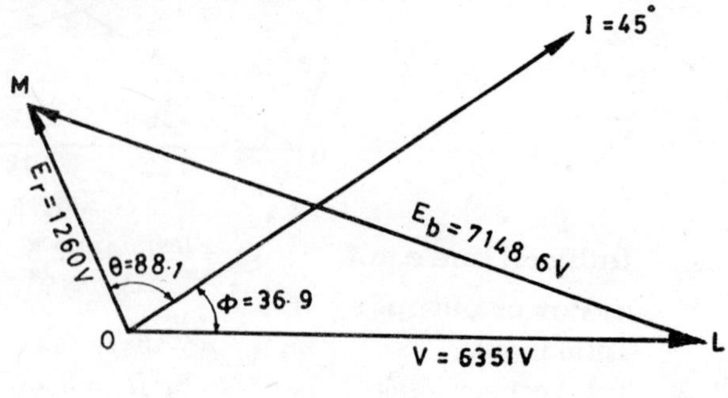

Fig. 6.21

☞**Example 6.5.** *The input to 2000 V, 3-phase star-connected synchronous motor is 800 kW at normal voltage. The generated line e.m.f. is 2500 V. Find the line current and power factor when the resistance and synchronous reactance/phase are 0.2 Ω and 2.2 Ω respectively.*

Solution. Supply voltage/phase, $V = \frac{2000}{\sqrt{3}} = 1155$ V

Input to motor $= 800$ kW

Induced e.m.f./phase, $E_b = \frac{2500}{\sqrt{3}} = 1443$ V

Resistance/phase, $R_a = 0.2\ \Omega$

Synchronous reactance/phase, $X_s = 2.2\ \Omega$

Line current :

Power factor :

Internal phase angle, $\theta = \tan^{-1} \frac{X_s}{R} = \tan^{-1} \frac{2.2}{0.2} = 84.8°$.

Since induced e.m.f. is greater than supply voltage, therefore the motor must be operating with a leading power factor. If the motor current is I, then :

In phase or active component of I

$= I \cos \phi$

Wattless or reactive component of I

$= I \sin \phi$

Since power input $= \sqrt{3} V_L I_L \cos \phi$

$\therefore \quad I \cos \phi = \frac{\text{power input}}{\sqrt{3} V_L}$

[∵ In star-connection, phase current = line current]

$= \frac{800 \times 1000}{\sqrt{3} \times 2000} = 231$ A

Impedance drop/phase, $E_r = IZ_s = (I \cos \phi + jI \sin \phi)(0.2 + j2.2)$

$= (231 + jI \sin \phi)(0.2 + j2.22)$

$= (46.2 - 2.2I \sin \phi) + j(508.2 + 0.2I \sin \phi)$

Refer Fig. 6.22.

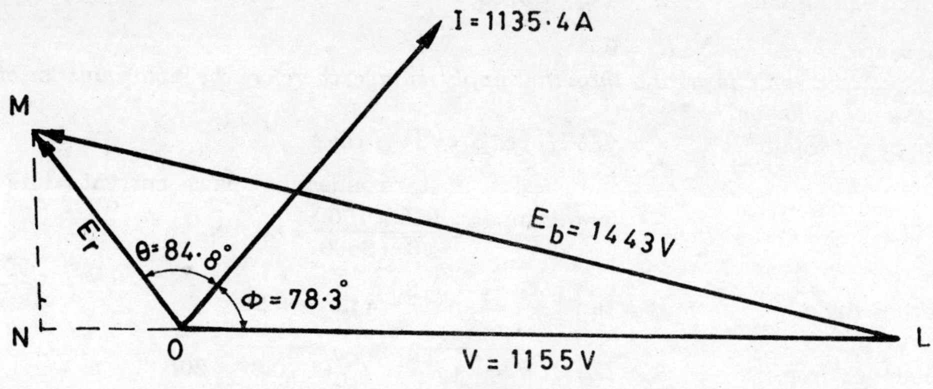

Fig. 6.22

$$OL = 1155 \text{ V}$$
$$ON = 46.2 - 2.2I \sin \phi$$
$$MN = 508.2 + 0.2I \sin \phi$$

From right angle ΔLMN, we have

$$LM^2 = LN^2 + MN^2$$
$$(1443)^2 = (OL + ON)^2 + MN^2$$
$$= (1155 + 46.2 - 2.2I \sin \phi)^2 + (508.2 + 0.2I \sin \phi)^2$$
$$= (1201.2 - 2.2I \sin \phi)^2 + (508.2 + 0.2I \sin \phi)^2$$

$\therefore$ $$2082249 = (1442881 + 4.84I^2 \sin^2 \phi - 5285.3I \sin \phi) + (258267 + 0.04I^2 \sin^2 \phi + 203.3I \sin \phi)$$
$$= 1701148 + 4.88I^2 \sin^2 \phi - 5082I \sin \phi$$

$\therefore$ $$4.88I^2 \sin^2 \phi - 5082I \sin \phi - 381101 = 0$$

or $$I^2 \sin^2 \phi - 1041.4I \sin \phi - 78094 = 0$$

$\therefore$ $$I \sin \phi = \frac{1041.4 \pm \sqrt{(1041.4)^2 + 4 \times 1 \times 78094}}{2}$$
$$= \frac{1041.4 \pm 1181.9}{2} = 1111.65 \text{ A}$$

$\therefore$ $$I = \sqrt{(I \cos \phi)^2 + (I \sin \phi)^2} = \sqrt{(231)^2 + (1111.65)^2} = 1135.4 \text{ A}$$

Hence current $= 1135.4$ A

Power factor $= \frac{I \cos \phi}{I} = \frac{231}{1135.4} =$ **0.203 (leading). (Ans.)**

$[\phi = \cos^{-1} 0.203 = 78.3°]$.

☞**Example 6.6.** *The synchronous reactance per phase of a 3-phase star-connected 6600 V synchronous motor is 20 Ω. For a certain load input, the input is 915 kW at normal voltage and the induced line e.m.f. is 8942 V. Neglecting resistance, determine :*

(i) line current, and *(ii) power factor.* **(S.C.E.T.E. W.B., 1996)**

Solution. Synchronous reactance/phase, $X_s = 20\ \Omega$

Input to motor $= 915$ kW

Supply phase voltage, $V = \frac{6600}{\sqrt{3}} = 3810$ V

Induced e.m.f./phase, $E_b = \dfrac{8942}{\sqrt{3}} = 5163$ V

Resistance, $R_a = 0.$

Since induced e.m.f. is greater than the supply voltage, therefore the motor must be operating with a leading power factor.

Since power input $= \sqrt{3}V_L I_L \cos\phi = \sqrt{3}V_L I \cos\phi$

[∵ In star-connection : phase current = line current]

∴ $I\cos\phi = \dfrac{\text{power input}}{\sqrt{3}V_L} = \dfrac{915 \times 1000}{\sqrt{3} \times 6600} = 80$ A

Internal angle, $\theta = \tan^{-1}\dfrac{X_s}{R_a} = \tan^{-1}\dfrac{20}{0} = \tan\infty = 90°$

Impedance drop, $E_r = IZ_s = I\sqrt{R_a^2 + X_s^2} = I\sqrt{0^2 + 20^2} = 20I$

Refer Fig. 6.23.

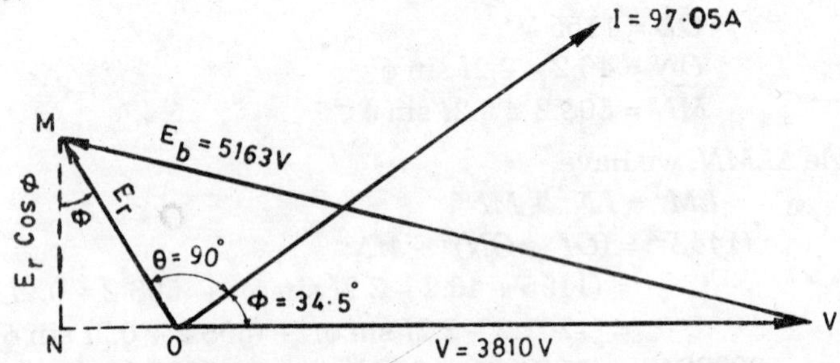

Fig. 6.23

From right angle ΔLMN, we have

$$LM^2 = LN^2 + MN^2$$

$$(5163)^2 = LN^2 + (E_r\cos\phi)^2 = LN^2 + (20I\cos\phi)^2$$

$$= LN^2 + (20 \times 80)^2 \quad [\because I\cos\phi = 80 \text{ A ..... calculated above}]$$

∴ $LN^2 = (5163)^2 - (1600)^2$

∴ $LN = 4909$ V

But $ON = LN - OL = 4909 - 3810 = 1099$ V

Now $E_r = (OM) = \sqrt{(ON)^2 + MN^2} = \sqrt{(1099)^2 + (20 \times 80)^2} = 1941$ V

(*i*) **Line current,** I_L = phase current, I

$$= \frac{E_r}{Z_s} = \frac{1941}{20} = \mathbf{97.05\ A.}\ \textbf{(Ans.)}$$

(*ii*) **Power factor,** $\cos\phi = \dfrac{I\cos\phi}{I} = \dfrac{80}{97.05} = \mathbf{0.8243\ (leading).}\ \textbf{(Ans.)}$

$[\phi = \cos^{-1} 0.8243 = 34.5°]$.

Example 6.7. *A 6.6 kV star-connected, 3-phase synchronous motor works at constant voltage and constant excitation. Its synchronous reactance is 20 Ω per phase, neglect resistance. When the input is 1000 kW, the power factor is 0.8 leading. Find the power factor when the input is increased to 1500 kW.* **(AMIE Advance Elec. M/cs Summer, 1998)**

Solution. *Given* : Supply voltage per phase, $V = \dfrac{6.6 \times 1000}{\sqrt{3}} = 3810$ V ; $X_s = 20\ \Omega$;

$$\text{Internal angle, } \theta = \tan^{-1}\left(\frac{20}{0}\right) = \tan^{-1}(\infty) = 90°$$

When the input is 1000 kW at 0.8 p.f. leading : Refer Fig. 6.24.

Armature current/phase, $$I = \frac{1000 \times 1000}{\sqrt{3} \times 6600 \times 0.8} = 109.35 \text{ A}$$

$$\cos \phi = 0.8, \quad \therefore \quad \phi = \cos^{-1}(0.8) = 36.87°$$

Impedance drop per phase, $E_r = IZ_s = 109.35 \times 20 = 2187$ V $\quad (\because Z_s = X_s = 20\ \Omega)$

Now, $$E_b = \sqrt{V^2 + E_r^2 - 2VE_r \cos(\theta + \phi)}$$

$$= \sqrt{(3810)^2 + (2187)^2 - 2 \times 3810 \times 2187 \times \cos(90° + 36.87°)}$$

$$= 5413 \text{ V}$$

When the input is increased to 1500 kW : Refer Fig. 6.25.

Now when the load on the machine is increased, the angle of retardation α will increase. The phasor diagram is shown in Fig. 6.24.

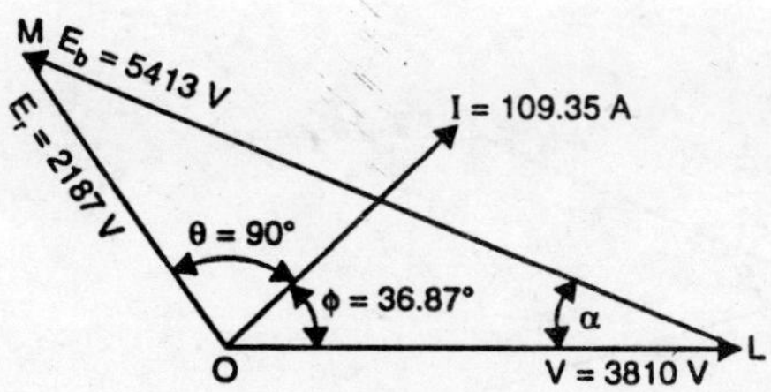

Fig. 6.24

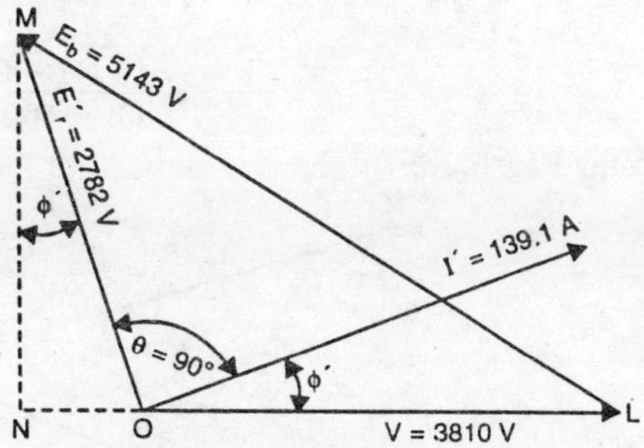

Fig. 6.25

The excitation (5413 V per phase) as well as supply voltage (per phase) remain constant in this case.

Now, $$1500 \times 1000 = \sqrt{3} V_L I' \cos \phi' \quad \text{(where } I' = \text{new armature current)}$$

or $$I' \cos \phi' = \frac{1500 \times 1000}{\sqrt{3} \times 6600} = 131.2 \text{ A}$$

Impedance drop per phase, $E_r' = I'X_s = 20I'$

In ΔLMN of phasor diagram shown in Fig. 6.25, we have

$$LM^2 = MN^2 + LN^2$$

or $$LN = \sqrt{LM^2 - MN^2} = \sqrt{E_b^2 - (E_r' \cos \phi')^2}$$

$$= \sqrt{(5413)^2 - (20 \times 131.2)^2} = 4734.5 \text{ V}$$

$$ON = LN - LO = 4734.5 - 3810 = 924.5 \text{ V}$$

$$E_r' = \sqrt{MN^2 + ON^2} = \sqrt{(20 \times 131.2)^2 + (924.5)^2} = 2782 \text{ V}$$

$$\therefore \quad I' = \frac{E_r'}{Z_s} = \frac{2782}{20} = 139.1 \text{ A}$$

Power factor, $$\cos \phi' = \frac{I' \cos \phi'}{I'} = \frac{131.2}{139.1} = \mathbf{0.943 \text{ (leading)}}.$$ **(Ans.)**

Example 6.8. *A 3-phase, 400 V, star-connected synchronous motor has an equivalent reactance of 1 Ω/phase. The exciting current is adjusted to such a value that the generated e.m.f./phase is 270 V. Find the power factor at which the motor would operate when taking 40 kW from the supply.*

Solution. Supply phase voltage, $V = \frac{400}{\sqrt{3}} = 231$ V

Equivalent reactance, $X_s = 1\ \Omega$/phase

Generated e.m.f./phase, $E_b = 270$ V

Input to motor $= 40$ kW

Power factor :

Since $E_b > V$, motor is running with **a leading** p.f.

Internal angle, $\theta = \tan^{-1} = \frac{X_s}{R_a} = \frac{1}{0} = \infty$ [Since $R_a = 0$]

$\therefore$ $\theta = 90°$

Since power input $= \sqrt{3} V_L I_L \cos\phi$

$\therefore$ $I \cos\phi = \frac{\text{power input}}{\sqrt{3}\, V_L}$ [$\because$ In star-connection : phase current = line current]

$= \frac{40 \times 1000}{\sqrt{3} \times 400} = 57.7$ A

Refer Fig. 6.26.

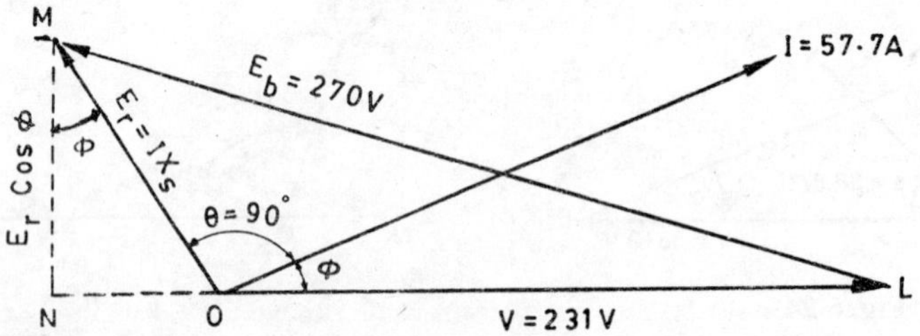

Fig. 6.26

From right-angled ΔLMN,

$$LM^2 = LN^2 + MN^2$$

$$(270)^2 = LN^2 + (E_r \cos\phi)^2 = LN^2 + (IX_s \cos\phi)^2$$

$$= LN^2 + (I \cos\phi .\ X_s)^2 \quad [\because\ Z_s = X_s]$$

$$= LN^2 + (57.7 \times 1)^2$$

$\therefore$ $LN^2 = (270)^2 - (57.7)^2$ or $LN = 263.7$ V

But $LN = OL + ON$

$\therefore$ $ON = LN - OL = 263.7 - 231 = 32.7$ V

$\therefore$ $\tan\phi = \frac{ON}{MN} = \frac{32.7}{57.7}$

$\therefore$ $\phi = 29.54°$

$\therefore$ **Power factor, $\cos\phi$** $= \cos 29.54° =$ **0.87 (lead). (Ans.)**

Example 6.9. *A 2000 V, 3-phase, 4-pole star-connected synchronous motor runs at 1500 r.p.m. The excitation is constant and corresponding to an open-circuit voltage of 2000 V. The resistance is negligible in comparison with synchronous reactance of 3.5 Ω/phase. For an armature current of 200 A, determine :*

(i) Power factor, *(ii) Power input, and*

(iii) Torque developed.

Solution. Supply voltage/phase, $V = \dfrac{2000}{\sqrt{3}} = 1155$ V

Armature current $= 200$ A

Synchronous speed, $N_s = 1500$ r.p.m.

Induced e.m.f./phase, $E_b = \dfrac{2000}{\sqrt{3}} = 1155$ V

Synchronous reactance/phase, $X_s = 3.5\ \Omega$/phase

Resistance/phase = negligible

$\therefore$ Synchronous impedance, $Z_s = X_s = 3.5\ \Omega$

$\therefore$ Internal angle, $\theta = 90°$

Impedance drop, $E_r = IZ_s = 200 \times 3.5 = 700$ V

Assuming armature current lagging behind the supply voltage, as shown in Fig. 6.27.

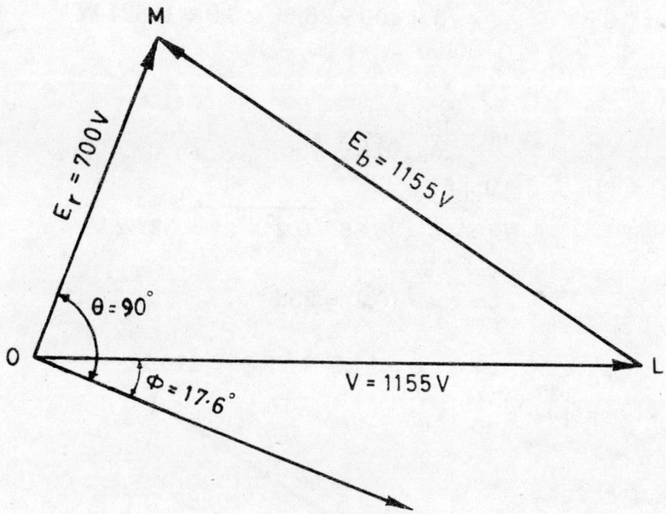

Fig. 6.27

In ΔLOM, we have $E_b^2 = V^2 + E_r^2 - 2VE_r \cos(\theta - \phi)$

$$(1155)^2 = (1155)^2 + (700)^2 - 2 \times 1155 \times 700 \cos(90 - \phi)$$

$\therefore$ $2 \times 1155 \times 700 \sin\phi = 700^2$

$\therefore$ $\sin\phi = \dfrac{(700)^2}{2 \times 1155 \times 700} = 0.303$

$\therefore$ $\phi = \sin^{-1}(0.303) = 17.6°$.

(*i*) **Power factor,** $\cos\phi = \cos 17.6° = \mathbf{0.9532}$. **(Ans.)**

(*ii*) **Power input,** $= \sqrt{3} V_L I_L \cos\phi = \sqrt{3} \times 2000 \times 200 \times 0.9532$

$= 660396$ W or **660.396 kW. (Ans.)**

(*iii*) Torque developed $= \dfrac{\text{power input} - \text{copper losses}}{2\pi N_s/60}$

$= \dfrac{660396 \times 60}{2\pi \times 1500}$

[$\because$ Copper losses are negligible as resistance is negligible]

= **4204.2 N-m. (Ans.)**

Example 6.10. *The efficiency of a 3-phase, 400 volts, star-connected synchronous motor is 95% and it takes 24 A at full-load and unity p.f. What will be the back e.m.f. generated and total mechanical power developed in kW for full-load and 0.9 p.f. leading ? The impedance/phase is (0.2 + j2) Ω.*

(AMIE Winter, 1997)

Solution. Efficiency of synchronous motor = 95%

Full-load current, $I = 24$ A

Power factor = unity

Impedance/phase = $(0.2 + j2)\ \Omega$

Mechanical power developed for full-load and 0.9 p.f. leading :

The current at 0.9 p.f. $= \dfrac{24}{0.9} = 26.66$ A

Power input at 0.9 p.f. $= \sqrt{3} \times 400 \times 26.66 \times 0.9 = 16624$ W

Phase voltage, $V = \dfrac{400}{\sqrt{3}} = 231$ V

$$\tan \theta = \frac{X_s}{R_a} = \frac{2}{0.2} = 1\dot{0}$$

$\therefore$ $\theta = 84.3°$

Impedance drop, $E_r = IZ = 26.66 \sqrt{0.2^2 + 2^2} = 53.32$ V

$$\cos \phi = 0.9$$

$$\phi = \cos^{-1} (0.9) = 25.8°$$

Refer Fig. 6.28.

$$\angle MOL = \theta + \phi = 84.3° + 25.8° = 110.1°$$

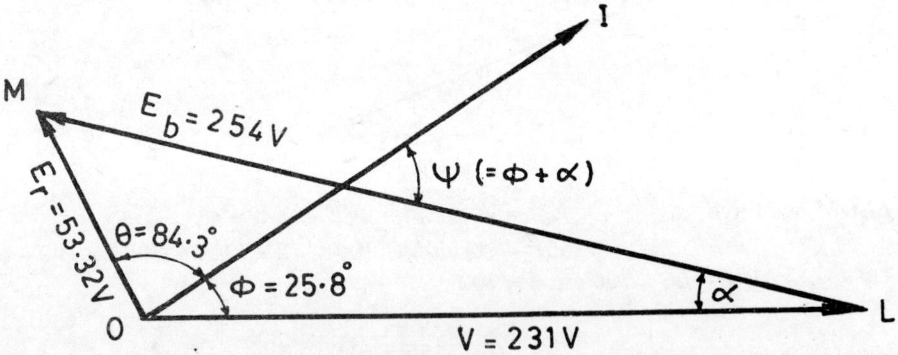

Fig. 6.28

In ΔLOM, we have

$$LM^2 = OL^2 + OM^2 - 2OL \times OM \times \cos (\theta + \phi)$$

i.e., $$E_b^2 = V^2 + E_r^2 - 2 \times V \times E_r \times \cos (\theta + \phi)$$

$$E_b = \sqrt{(231)^2 + (53.32)^2 - 2 \times 231 \times 53.32 \times \cos 110.1°} = 254 \text{ V}$$

Induced line e.m.f. $= \sqrt{3} \times 254 = 440$ V

In ΔLOM, $$\frac{OM}{\sin \alpha} = \frac{LM}{\sin (\theta + \phi)}$$

$$\frac{53.32}{\sin \alpha} = \frac{254}{\sin 110.1°}$$

$\therefore$ $\sin \alpha = \frac{53.32}{254} \times \sin 110.1° = 0.1972$

$\therefore$ $\alpha = \sin^{-1} (0.1972) = 11.4°$

Mechanical power developed = 3 (e.m.f. induced per phase × armature current × cosine of the angle between the armature current and the e.m.f. induced)

Let ψ = angle between the armature current and the e.m.f. induced

$= \phi + \alpha = 25.8° + 11.4° = 37.2°$

$\therefore$ **Mechanical power developed**

$= 3 \times 254 \times 26.66 \times \cos 37.2° \times 10^{-3}$ kW = **16.18 kW. (Ans.)**

Example 6.11. *A 75 kW, 400 V, 4-pole, 3-phase, star-connected synchronous motor has a resistance and synchronous reactance per phase of 0.04 Ω and 0.4 Ω respectively. Compute for full load 0.8 p.f. lead the open-circuit e.m.f. per phase and gross mechanical power developed. Assume an efficiency of 92.5%.* **(AMIE Summer, 1997)**

Solution. *Given* : Motor output = 75 kW ; V_L = 400 volts,

$R_a = 0.04\ \Omega$, $X_s = 0.4\ \Omega$, $\cos \phi = 0.8$ (lead) ; $\eta = 92.5\%$.

Open-circuit e.m.f., E_b :

Phase voltage, $V = \frac{400}{\sqrt{3}} = 230.9$ V

$\cos \phi = 0.8$, $\phi = \cos^{-1} (0.8) = 36.87°$ and $\sin \phi = \sin (36.87°) = 0.6$.

Motor input $= \frac{\text{motor output}}{\eta}$

$= \frac{75}{0.925} = 81.081$ kW or 81081 W

Armature current, $I = \frac{81081}{\sqrt{3} \times 400 \times 0.8} = 146.3$ A

Resultant voltage, $E_r = IZ_s = 146.3 \sqrt{R_a^2 + X_s^2}$

$= 146.3 \sqrt{(0.04)^2 + (0.4)^2}$

$= 58.81$ V

$\tan \theta = \frac{X_s}{R_a} = \frac{0.4}{0.04} = 10$

$\therefore$ Internal angle $\theta = \tan^{-1} (10) = 84.3°$

The vector diagram is shown in Fig. 6.29

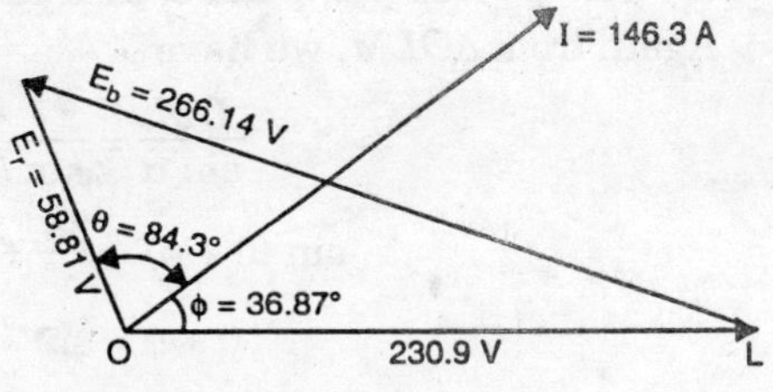

Fig. 6.29

$$\mathbf{E_b} = \sqrt{V^2 + E_r^2 - 2VE_r \cos (\theta + \phi)}$$

$$= \sqrt{(230.9)^2 + (58.81)^2 - 2 \times 230.9 \times 58.81 \cos (84.3° + 36.87°)}$$

$$= \sqrt{53314.81 + 3458.62 + 14056.65} = \mathbf{266.14\ V.\ (Ans.)}$$

Gross mechanical power developed :

Gross mechanical power developed for all 3 phases,

$(\mathbf{P_{mech}})_{\mathbf{gross}} = P_{in} - 3I^2R_a$

$= 81081 - 3 \times (146.3)^2 \times 0.04$

= 78512 W or **78.512 kW. (Ans.)**

Example 6.12. *A 20-pole, 30 kW, 660 V, 50 Hz, 3-φ star-connected synchronous motor is operating with its generated voltage per phase exactly equal to the phase voltage applied to its armature. At loaded condition the motor is retarded by 5 mechanical degrees from its synchronous* *position. The synchronous reactance and armature resistance are 10 Ω and 1 Ω per phase. Calculate :*

(i) Armature current per phase ;

(ii) The power per phase, and the total power drawn by the motor from the bus ;

(iii) The developed power. **(AMIE Summer, 1995)**

Solution. *Given :* $p = 20$; Power output = 30 kW, $V_L = 660$ volts, $R_a = 1\ \Omega$; $X_s = 10\ \Omega$.

(i) Armature current/phase, I :

Induced e.m.f. per phase, $E_b = V = \frac{660}{\sqrt{3}} = 381$ volts

Load angle, $\alpha = \frac{p}{2} \times$ angle of retardation in mechanical degrees

$$= \frac{20}{2} \times 5 = 50° \text{ electrical}$$

Impedance/phase, $Z_s = \sqrt{R_a^2 + X_s^2}$

$$= \sqrt{1^2 + 10^2} = 10.05\ \Omega$$

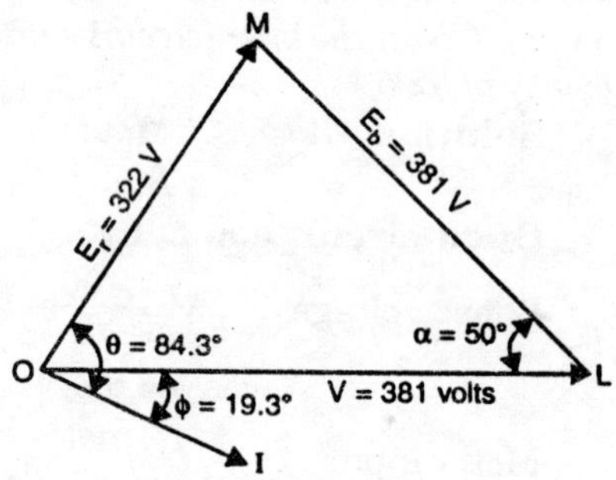

Fig. 6.30

Now from ΔOLM of vector diagram in Fig. 6.30, we have

$$E_r = \sqrt{E_b^2 + V^2 - 2E_bV\cos\alpha}$$

$$= \sqrt{381^2 + 381^2 - 2 \times 381 \times 381 \times \cos 50°}$$

$$= 322 \text{ V}$$

$$\therefore \quad \mathbf{I} = \frac{E_r}{Z_s} = \frac{322}{10.05} = \mathbf{32\ A.\ (Ans.)}$$

(ii) The power per phase, and total power drawn :

Again from ΔOLM, we have

$$\frac{OM}{\sin\alpha} = \frac{E_r}{\sin(\theta + \phi)} \quad \text{or} \quad \frac{322}{\sin 50°} = \frac{381}{\sin(\theta + \phi)}$$

or $$\sin(\theta + \phi) = \frac{381 \times \sin 50°}{322} = 0.9064$$

or $$\theta + \phi = \sin^{-1}(0.9064) = 65°$$

and $$\theta = \tan^{-1}\left(\frac{X_s}{R_a}\right) = \tan^{-1}\left(\frac{10}{1}\right) = 84.3°$$

$$\therefore \quad \phi = 65° - \theta = 65° - 84.3° = -19.3° \text{ i.e. } 19.3° \text{ lagging}$$

∴ **Power per phase** $= VI\cos\phi = 381 \times 32 \times \cos 19.3°$

$= 11507$ W or **11.5 kW. (Ans.)**

Total power drawn $= 3 \times 11.5 =$ **34.5 kW. (Ans.)**

(iii) Power developed :

Power developed = Total power drawn $- 3I^2R_a$

$$= 34.5 - \frac{3 \times 32^2 \times 1}{1000} = \mathbf{31.43\ kW.\ (Ans.)}$$

☞**Example 6.13.** *A 9 kW, 400 V, 3-phase star-connected synchronous motor has synchronous impedance per phase of (0.4 + j3) Ω. Find the angle of retard and the voltage to which the motor must be excited to give a full-load output at 0.8 leading power factor.*

Assume an efficiency of 90%. **(Manamanian Sundarnar Univ., 1996)**

Solution. Motor output $= 9$ kW or 9000 W

Efficiency $= 90\%$

Synchronous impedance/phase,

$$Z_s = (0.4 + j3)\ \Omega$$

Power factor $= 0.8$ (leading)

Induced e.m.f., E_b :

Angle of retard, α :

Motor input $= \dfrac{\text{motor output}}{\text{efficiency}} = \dfrac{9000}{0.9} = 10000$ W

Motor input current, $I = \dfrac{\text{motor input}}{\sqrt{3} V_L \cos\phi} = \dfrac{10000}{\sqrt{3} \times 400 \times 0.8} = 18$ A

Synchronous impedance/phase,

$$Z_s = \sqrt{(0.4)^2 + 3^2} = 3.03\ \Omega$$

Internal angle, $\theta = \tan^{-1} \dfrac{X_s}{R_a} = \tan^{-1} \dfrac{3}{0.4} = 82.4°$

Power factor angle, $\phi = \cos^{-1}(0.8) = 36.9°$

Supply voltage/phase $= \dfrac{400}{\sqrt{3}} = 231$ V

Impedance drop/phase, $E_r = IZ_s = 18 \times 3.03 = 54.54$ V.

Refer Fig. 6.31.

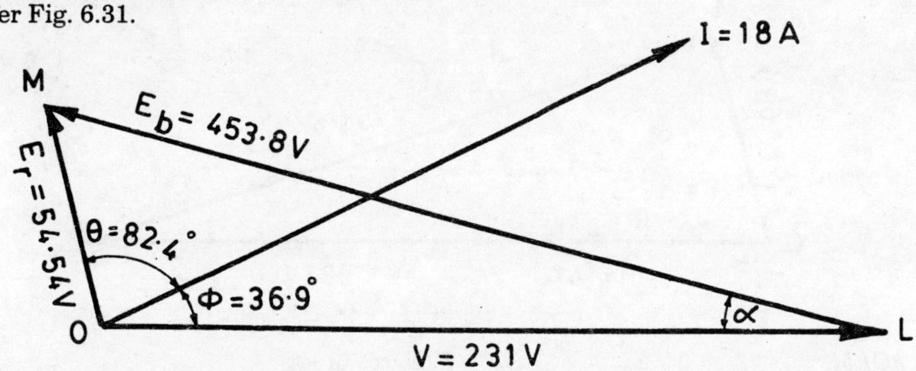

Fig. 6.31

In ΔOLM, $E_b^2 = V^2 + E_r^2 - 2VE_r \cos(\theta + \phi)$

$$= (231)^2 + (54.54)^2 - 2 \times 231 \times 54.54 \cos(82.4° + 36.9°)$$

$$= 53361 + 2974.6 + 12331$$

$\therefore$ $E_b = 262$ V

Induced e.m.f. E_b (line) $= \sqrt{3} \times 262 =$ **453.8 V. (Ans.)**

Also in ΔOLM, $\dfrac{E_r}{\sin\alpha} = \dfrac{E_b}{\sin(\theta + \phi)}$

$\therefore$ $\dfrac{54.54}{\sin\alpha} = \dfrac{262}{\sin(82.40° + 36.9°)}$

$\therefore$ $\sin\alpha = \dfrac{54.54 \sin(82.4° + 36.9°)}{262}$ or $\alpha = 10.44°$

Hence, **angle of retard** $=$ **10.44°. (Ans.)**

Example 6.14. *A 400 V, 6-pole, 3-phase, 50-Hz, star-connected synchronous motor has a resistance and synchronous impedance of 0.6 Ω and 4.5 Ω per phase respectively. It takes a current of 14 A at unity p.f. when operating with a certain field current. If the load torque is increased until the line current is increased to 50 A, the field current remaining unchanged, calculate :*

(*i*) *New power factor, and* (*ii*) *Gross-torque developed.*

Solution. Supply voltage/phase $= \dfrac{400}{\sqrt{3}} = 231$ V

Resistance/phase, $R_a = 0.6\ \Omega$.

Synchronous impedance/phase,

$Z_s = 4.5\ \Omega$.

Synchronous reactance, $X_s = \sqrt{Z_s^2 - R_a^2} = \sqrt{(4.5)^2 - (0.6)^2} = 4.46\ \Omega$

Internal angle, $\theta = \tan^{-1}\dfrac{X_s}{R_a} = \tan^{-1}\dfrac{4.46}{0.6} = 82.3°$

1st case. Current, $I = 14$ A

Power factor, $\cos\phi$ = unity *i.e.* 1

$\therefore$ $\phi = \cos^{-1} 1 = 0°$

Impedance drop/phase, $E_r = IZ_s = 14 \times 4.5 = 63$ V.

Refer Fig. 6.32.

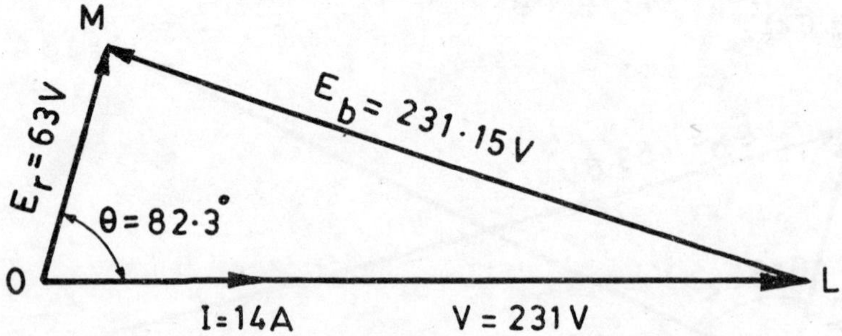

Fig. 6.32

In ΔOLM, $E_b^2 = V^2 + E_r^2 - 2VE_r \cos(\theta + \phi)$

$= (231)^2 + (63)^2 - 2 \times 23 \times 63 \cos(82.3° + 0°)$

$= 53361 + 3969 - 3900$

$\therefore$ $E_b = 231.15$ V

2nd case. Current, $I = 50$ A.

When motor load is increased, the angle α between the applied voltage and the induced (or back) e.m.f. is increased. The vector diagram is shown in Fig. 6.33.

Let ϕ = new phase angle

Impedance drop/phase, $E_r = IZ_s = 50 \times 4.5 = 225$ V

Since the field current remains constant, the value of E_b *remains the same.* Refer Fig. 6.33.

$\therefore$ $E_b^2 = V^2 + E_r^2 - 2VE_r \cos(\theta - \phi)$

$(231.15)^2 = (231)^2 + (225)^2 - 2 \times 231 \times 225 \cos(82.3° - \phi)$

$53430 = 53361 + 50625 - 103950 \cos(82.3° - \phi)$

$\therefore$ $\cos(82.3° - \phi) = 0.486$

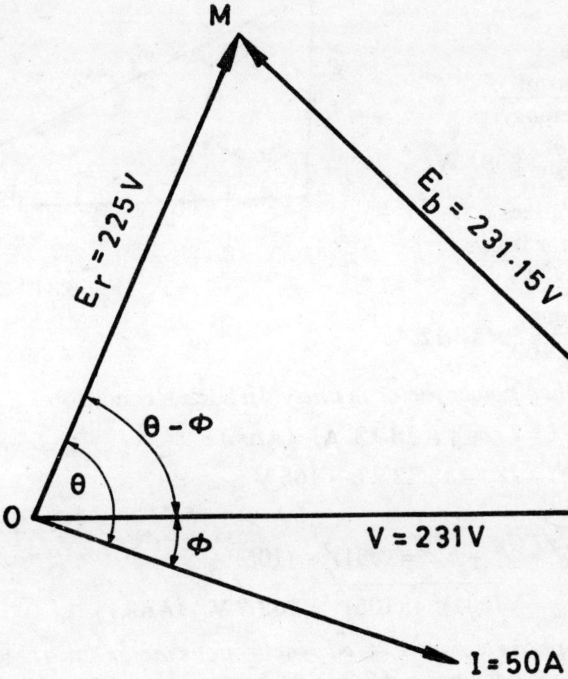

Fig. 6.33

∴ $(82.3° - \phi) = \cos^{-1} 0.486 = 60.9°$

∴ $\phi = 21.4°$ (lag).

(*i*) **New power factor,**

$$\cos \phi = \cos 21.4° = \mathbf{0.931 \text{ (lag)}}. \quad \textbf{(Ans.)}$$

(*ii*) **Gross torque developed :**

Motor input $= \sqrt{3} V_L I_L \cos \phi = \sqrt{3} \times 400 \times 50 \times 0.931 = 32250$ W

Total armature copper loss $= 3 \times 50^2 \times 0.6 = 4500$ W

Electrical power converted into mechanical power

$$= 32250 - 4500 = 27750 \text{ W}$$

Synchronous speed, $N_s = \frac{120f}{p} = \frac{120 \times 50}{6} = 1000$ r.p.m.

∴ **Gross torque developed,**

$$T_g = \frac{27750}{2\pi N_s/60} = \frac{27750 \times 60}{2\pi \times 1000} = \mathbf{265 \text{ N-m.}} \quad \textbf{(Ans.)}$$

Example 6.15. *A 400 V, 8 kW 3-phase synchronous motor has negligible resistance and synchronous reactance of 8 Ω/phase. Determine the minimum current and the corresponding induced e.m.f. for full-load condition.*

Assume an efficiency of 88%.

Solution. Supply voltage/phase,

$$V = \frac{400}{\sqrt{3}} = 231 \text{ V}$$

Motor output = 8 kW or 8000 W

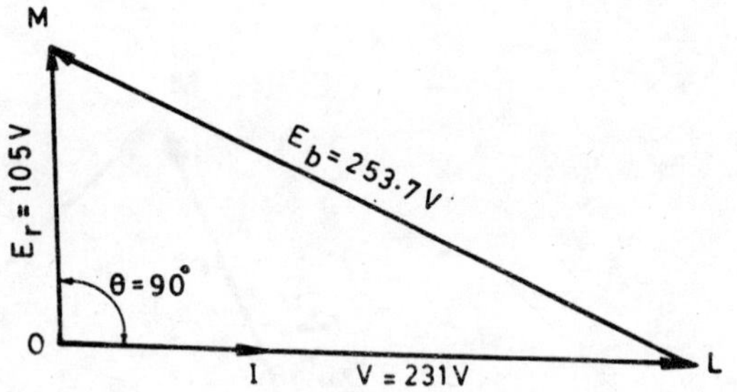

Fig. 6.34

Synchronous reactance/phase,

$$X_s = 8\ \Omega$$

Efficiency $= 88\%$

Motor input $= \dfrac{\text{output}}{\text{efficiency}}$

$= \dfrac{8000}{0.88} = 9091$ W

Since motor input $= \sqrt{3} V_L I \cos \phi$

$\therefore \quad I \cos \phi = \dfrac{\text{motor input}}{\sqrt{3} V_L}$

$= \dfrac{9091}{\sqrt{3} \times 400} = 13.12$ A.

The *current is minimum when power factor is unity*. In such a condition,

$$I = I \cos \phi = \mathbf{13.12\ A.}\ \textbf{(Ans.)}$$

Impedance drop, $\quad E_r = IZ_s = 13.12 \times 8 = 105$ V

Refer Fig. 6.34.

$$E_b^{\ 2} = V^2 + E_r^{\ 2} = (231)^2 + (105)^2$$

$\therefore \quad E_b = \sqrt{(231)^2 + (105)^2} = \mathbf{253.7\ V.}$ **(Ans.)**

Example 6.16. *A 2000 V, 3-phase star-connected synchronous motor has a resistance per phase of 0.2 Ω and a synchronous reactance per phase of 2.2 Ω. Determine the value of the generated e.m.f. and angular retardation of the rotor when the input is 800 kW at :*

(i) 0.8 p.f. lagging load. *(ii) 0.8 p.f. leading load.*

Solution. Supply phase voltage, $V = \dfrac{2000}{\sqrt{3}} = 1155$ V

Resistance/phase, $\quad R_a = 0.2\ \Omega$

Synchronous reactance/phase, $X_s = 2.2\ \Omega$

Motor input $= 800$ kW

Synchronous impedance, $\quad Z_s = \sqrt{R_a^{\ 2} + X_s^{\ 2}} = \sqrt{(0.2)^2 + (2.2)^2} = 2.21\ \Omega$

Internal angle, $\quad \theta = \tan^{-1} \dfrac{X_s}{R_a} = \tan^{-1} \dfrac{2.2}{0.2} = 84.8°$

Motor input $= \sqrt{3} V_L I \cos \phi$

$\therefore \quad I = \dfrac{800 \times 1000}{\sqrt{3} \times 2000 \times 0.8} = 288.6$ A

Impedance drop/phase, $\quad E_r = IZ_s$ $\quad \left[\because \text{ In star-connection, phase current = line current}\right]$

$= 288.6 \times 2.21 = 637.8$ V.

(i) **At 0.8 p.f. lagging :**

Refer 6.35.

$$E_b^{\ 2} = V^2 + E_r^{\ 2} - 2VE_r \cos (\theta - \phi)$$

$= (1155)^2 + (637.8)^2 - 2 \times 1155 \times 637.8 \times \cos (84.8° - 36.9°)$ $\quad \left[\because \cos \phi = 0.8,\ \therefore \phi = 36.9°\right]$

$= 1334025 + 406789 - 987752$

$\therefore$ $\quad$ $\mathbf{E_b = 867.8\ V.}$ **(Ans.)**

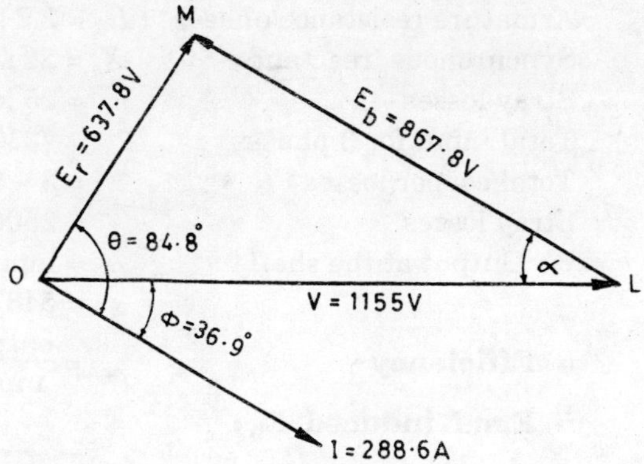

Fig. 6.35

In ΔOLM, $\dfrac{E_r}{\sin \alpha} = \dfrac{E_b}{\sin (\theta - \phi)}$

$$\frac{637.8}{\sin \alpha} = \frac{867.8}{\sin (84.8° - 36.9°)}$$

$\therefore$ $$\sin \alpha = \frac{637.8 \times \sin (84.8° - 36.9°)}{867.8}$$

$$= 0.545$$

$\therefore$ $\quad$ $\alpha = \mathbf{33°}$. **(Ans.)**

(ii) At 0.8 p.f. leading :

Refer Fig. 6.36.

$$E_b^{\,2} = V^2 + E_r^{\,2} - 2VE_r \cos (\theta + \phi)$$
$$= (1155)^2 + (637.8)^2 - 2 \times 1155 \times 637.8 \cos (84.8° + 36.9°)$$
$$= 1334025 + 406789 + 774187$$

$\therefore$ $\quad$ $\mathbf{E_b = 1585.8\ V.}$ **(Ans.)**

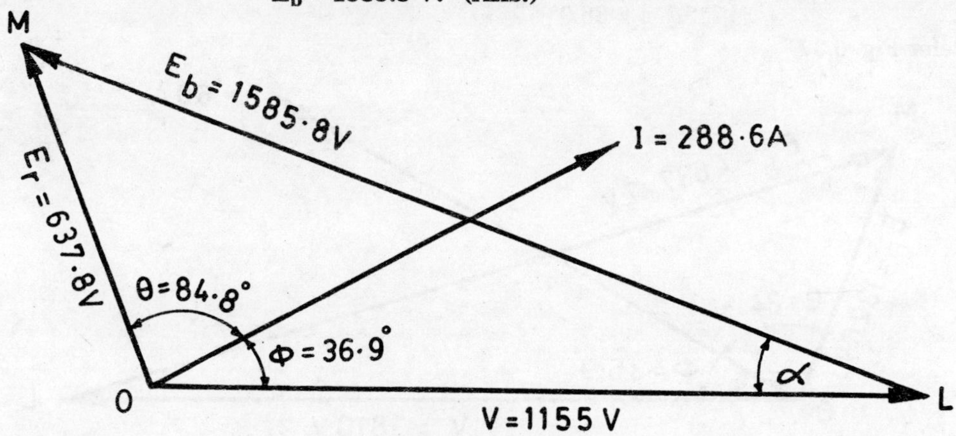

Fig. 6.36

In ΔOLM, $$\frac{E_r}{\sin \alpha} = \frac{E_b}{\sin (\theta + \phi)}$$

$\therefore$ $$\frac{637.8}{\sin \alpha} = \frac{1585.8}{\sin (84.8° + 36.9°)}$$

$\therefore$ $$\sin \alpha = \frac{637.8 \times \sin (84.8° + 36.9°)}{1585.8} = 0.3422$$

$\therefore$ $\quad$ $\alpha = \sin^{-1} (0.3422) = \mathbf{20°}$. **(Ans.)**

Example 6.17. *A 6600 V, 3-phase, star-connected synchronous motor draws a full-load current of 60 A at 0.8 p.f. leading. The armature resistance and reactance per phase are 2.2 Ω and 22 Ω respectively. If the stray losses total 25 kW, determine :*

(i) Efficiency of the machine, and. $\qquad$ *(ii) E.m.f. induced.*

Solution. Supply phase voltage, $V = \dfrac{6600}{\sqrt{3}} = 3810$ V

Full-load current, $\quad I = 60$ A

Power factor = 0.8 (leading)

Armature resistance/phase, $R_a = 2.2\ \Omega$

Synchronous reactance, $X_s = 22\ \Omega$

Stray losses = 25 kW

Total input for 3-phases $= \sqrt{3}V_L I_L \cos\phi = \sqrt{3} \times 6600 \times 60 \times 0.8 = 548714$ W

Total copper losses $= 3 \times 60^2 \times 2.2 = 23760$ W

Stray losses = 25000 W *(given)*

∴ Output at the shaft = input – copper losses – stray losses

= 548714 – 23760 – 25000 = 499954 W

(*i*) **Efficiency** $= \dfrac{\text{output}}{\text{input}} = \dfrac{499954}{548714} = 0.911$ or **91.1%. (Ans.)**

(*ii*) **E.m.f. induced, E_b :**

Synchronous impedance/phase, $Z_s = \sqrt{R_a^2 + X_s^2} = \sqrt{(2.2)^2 + (22)^2} = 22.1\ \Omega$

Internal angle, $\theta = \tan^{-1}\dfrac{X_s}{R_a} = \tan^{-1}\dfrac{22}{2.2} = 84.3°$

Impedance drop/phase, $E_r = IZ_s = 60 \times 22.1 = 1326$ V

Power factor, $\cos\phi = 0.8$

∴ $\phi = 36.9°$

Refer Fig. 6.37.

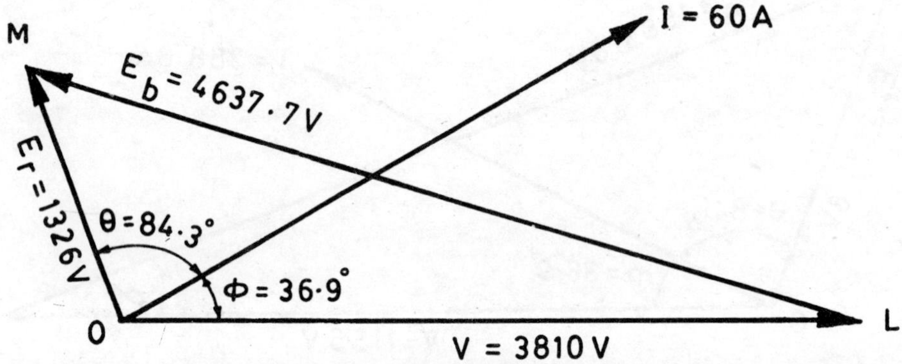

Fig. 6.37

In ΔOLM, $E_b^2 = V^2 + E_r^2 - 2VE_r\cos(\theta + \phi)$

$= (3810)^2 + (1326)^2 - 2 \times 3810 \times 1326 \times \cos(84.3° + 36.9°)$

= 14516100 + 1758276 + 5234207

∴ $E_b = 4637.7$ V

Induced e.m.f., E_b (line value)

$= \sqrt{3} \times 4637.7 =$ **8032.7 V. (Ans.)**

☞**Example 6.18.** *A 6-pole 2200 V, 50 Hz, 3-phase, star-connected synchronous motor has armature resistance of 0.4 Ω per phase and synchronous reactance of 4 Ω/phase. While running on no-load, the excitation has been adjusted so as to make the e.m.f. numerically equal to and antiphase with the terminal voltage. With a certain load torque applied, if the rotor gets retarded by 3 mechanical degrees, calculate :*

(*i*) *armature current, and* (*ii*) *power factor of the motor.*

Solution. Supply voltage/phase $= \frac{2200}{\sqrt{3}} = 1270$ V

Number of poles, $p = 4$

Armature resistance/phase, $R_a = 0.4\ \Omega$

Synchronous reactance/phase, $X_s = 4\ \Omega$

Load angle, $\alpha = 3$ mechanical degrees

Induced e.m.f., E_b/phase $= 1270$ V

Load angle $\alpha = \frac{p}{2} \times \left(\begin{array}{l}\text{angle of retardation in} \\ \text{mechanical degrees}\end{array}\right) = \frac{6}{2} \times 3 = 9°$

Synchronous impedance/phase,

$$Z_s = \sqrt{R_a^2 + X_s^2} = \sqrt{(0.4)^2 + (4)^2} = 4.02\ \Omega$$

Internal angle, $\theta = \tan^{-1} \frac{X_s}{R_a} = \tan^{-1} \frac{4}{0.4} = 84.3°$

Refer Fig. 6.38.

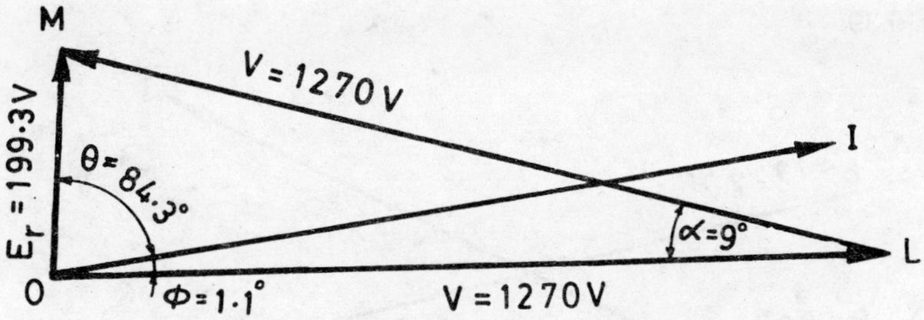

Fig. 6.38

In ΔOLM, $E_r = \sqrt{E_b^2 + V^2 - 2VE_b \cos \alpha}$

$$= \sqrt{(1270)^2 + (1270)^2 - 2 \times 1270 \times 1270 \cos 9°}$$

$$= \sqrt{1612900 + 1612900 - 3186085} = 199.3 \text{ V}$$

Input current, $I = \frac{E_r}{Z_s} = \frac{199.3}{4.02} = 49.6$ A

Again in ΔOLM, $\frac{E_b}{\sin(\theta - \phi)} = \frac{E_r}{\sin 9°}$

∴ $\frac{1270}{\sin(84.3° + \phi)} = \frac{199.3}{\sin 9°}$

or $\sin(84.3° + \phi) = \frac{1270}{199.3} \times \sin 9° = 0.9968$

∴ $84.3° + \phi = 85.4$

∴ $\phi = 1.1°$ (lead)

Power factor, $\cos \phi = \cos 1.1° =$ **0.9998 (lead). (Ans.)**

Example 6.19. *A 2200 V, 3-phase, 50 Hz, star-connected synchronous motor has a synchronous reactance of 10 Ω/phase. When it delivers 150 kW, the efficiency is found to be 88% and power angle is 15 electrical degrees. Determine :*

(*i*) *Induced excitation voltage ; and* (*ii*) *Line current.*
Neglect armature resistance.

Solution. Supply phase voltage, $V = \frac{2200}{\sqrt{3}} = 1270$ V

Synchronous reactance/phase, $X_s = 10\ \Omega$

Motor output $= 150$ kW

Efficiency $= 88\%$

Load or power angle, $\alpha = 15$ electrical degrees

Motor input $= \frac{\text{output}}{\text{efficiency}} = \frac{150 \times 1000}{0.88} = 170454$ W

Since motor input $= \sqrt{3} V_L I \cos \phi$

$\therefore \quad I \cos \phi = \frac{170454}{\sqrt{3} \times 2200} = 44.7$ A

Synchronous impedance/phase, $Z_s = X_s = 10\ \Omega$ [$\because$ armature resistance is negligible]

Impedance/phase $= IX_s = 10I$

Internal angle, $\theta = 90°$.

Refer Fig. 6.39.

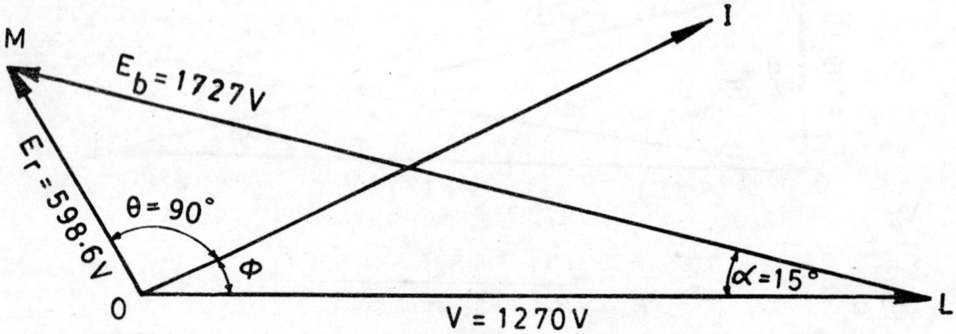

Fig. 6.39

From ΔOLM,
$$\frac{OM}{\sin \alpha} = \frac{LM}{\sin (\theta + \phi)}$$
$$\frac{E_r}{\sin 15°} = \frac{E_b}{\sin (90° + \phi)}$$

or
$$E_b = \frac{E_r \sin (90° + \phi)}{\sin 15°} = \frac{E_r \cos \phi}{\sin 15°}$$
$$= \frac{10 I \cos \phi}{\sin 15°} = \frac{10 \times 44.7}{\sin 15°} \quad [\because I \cos \phi = 44.7 \text{ A}]$$
$$= 1727 \text{ V}.$$

(*i*) **Induced e.m.f.**, E_b (line value)
$$= \sqrt{3} \times 1727 = \mathbf{2991\ V.} \quad \textbf{(Ans.)}$$

(*ii*) **Line current :**

Again from ΔOLM, we have
$$E_r = \sqrt{E_b^2 + V^2 - 2E_b V \cos \alpha}$$
$$= \sqrt{(1727)^2 + (1270)^2 - 2 \times 1727 \times 1270 \times \cos 15°}$$

$$= \sqrt{2982529 + 1612900 - 4237111} = 598.6 \text{ V}$$

∴ **Line current,** $I = \dfrac{E_r}{X_s} = \dfrac{598.6}{10}$ = **59.86 A. (Ans.)**

☞**Example 6.20.** *The excitation of a 400 V, 3-phase, mesh connected synchronous motor is such that the induced e.m.f. is 510 V. The impedance per phase is (0.6 + j 4.5) Ω. If the friction and iron losses are constant at 800 W, calculate :*

(i) Line current, *(ii) Power output,*
(iii) Power factor, and *(iv) Efficiency for maximum power output.*

Solution. Supply voltage/phase, $V = 400$ V [∵ motor is mesh connected]

Induced e.m.f./phase, $E_b = 510$ V

Iron and friction losses = 800 W

Synchronous impedance/phase, $Z_s = \sqrt{(0.6)^2 + (4.5)^2} = 4.54\ \Omega$

Internal angle, $\theta = \tan^{-1} \dfrac{4.5}{0.6} = 82.4°$.

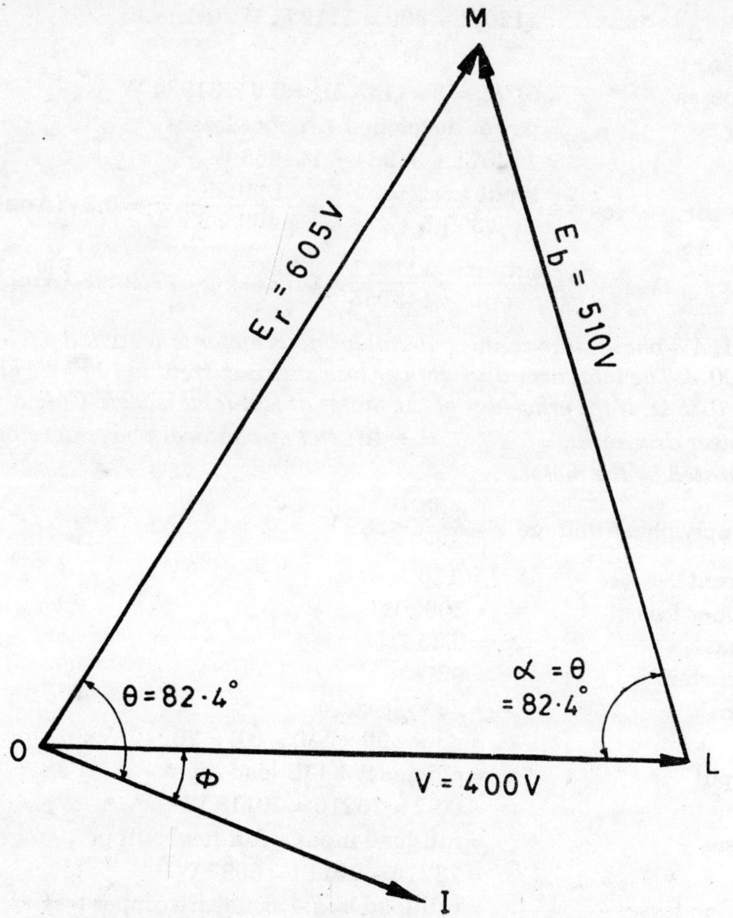

Fig. 6.40

For a constant value of supply voltage, excitation (or induced e.m.f.), armature resistance R_a and synchronous reactance X_s maximum power will be developed when lead angle,

$$\alpha = \theta = 82.4°$$

Refer Fig. 6.40.

$$E_r = \sqrt{(E_b)^2 + (V)^2 - 2E_b \times V \times \cos 82.4°}$$

$$= \sqrt{(510)^2 + (400)^2 - 2 \times 510 \times 400 \cos 82.4°}$$

$$= \sqrt{260100 + 160000 - 53961} = 605 \text{ V/phase}$$

Load current, $I = \dfrac{E_r}{Z_s} = \dfrac{605}{4.54} = 133.3$ A/phase

$\therefore$ (*i*) **Line current** $I_L = \sqrt{3} \times I = \sqrt{3} \times 133.3 = 230.9$ A [$\because$ motor is mesh-connected]

Maximum power developed,

$$(P_{mech})_{max} = \frac{E_b V - E_b^2 \cos \theta}{Z_s} = \frac{510 \times 400 - (510)^2 \times \cos 82.4°}{4.54} = 37357 \text{ W}$$

Maximum power for 3-phases = 3 × 37357 = 112071 W.

(*ii*) **Output power** = power developed – iron and friction losses
= 112071 – 800 = **111271 W. (Ans.)**

(*iii*) **Power factor :**

Total copper losses $= 3I^2R_a = 3 \times (133.3)^2 \times 0.6 = 31984$ W

Input to motor = power developed + copper losses
= 112071 + 31984 + 144055 W.

$\therefore$ **Power factor,** $\cos \phi = \dfrac{\text{input to motor}}{\sqrt{3} V_L I_L} = \dfrac{144055}{\sqrt{3} \times 400 \times 230.9} = \mathbf{0.9}$ **(Ans.)**

(*iv*) **Efficiency** $= \dfrac{\text{output}}{\text{input}} = \dfrac{111271}{144055} = 0.7724$ or **77.24%. (Ans.)**

Example 6.21. *A 3-phase, star-connected synchronous motor takes at 0.8 p.f. lagging full-load current of 110 A at 500 V. The total armature copper loss at this current is 1000 W. The reactance per phase of the motor is 0.35 Ω. If the efficiency of the motor at full-load is 92%. Calculate :*

(*i*) *Internal power developed,* (*ii*) *Net mechanical power developed, and*

(*iii*) *E.m.f. generated by the motor.*

Solution. Supply phase voltage $V = \dfrac{500}{\sqrt{3}} = 289$ V

Full-load current, $I = 110$ A

Armature copper loss = 1000 W

Reactance/phase, $X_s = 0.35\ \Omega$

Efficiency of motor = 92%

Full-load input $= \sqrt{3} V_L I_L \cos \phi$
$= \sqrt{3} \times 500 \times 110 \times 0.8 = 76210$ W

Full-load output = efficiency × full-load input
= 0.92 × 76210 = 70113 W

Full-load losses = full-load input – full-load output
= 76210 – 70113 = 6097 W

Iron and friction losses = full-load loss – armature copper loss
= 6097 – 1000 = 5097 W.

(*i*) **Internal power developed** = electrical power converted into mechanical power in the armature

= full-load input—armature copper loss

= 72610 – 1000 = **71610 W. (Ans.)**

(*ii*) **Net mechanical power developed,**

P_{mech} = net useful power available at the shaft

= full-load out = **70113 W. (Ans.)**

[Also, P_{mech} = 75210 – iron and friction loss

= 75210 – 5097 = 70113 W]

(*iii*) **E.m.f. generated, E_b :**

$\cos \phi = 0.8$ (lag)

$\therefore \quad \phi = \cos^{-1} (0.8) = 36.9°$

Internal angle, $\theta = 90°$ $[\because R_a = 0]$

Synchronous impedance, $Z_s = X_s = 0.35\ \Omega$

Impedance drop/phase $= IZ_s = 110 \times 0.35 = 38.5$ V.

Refer Fig. 6.41.

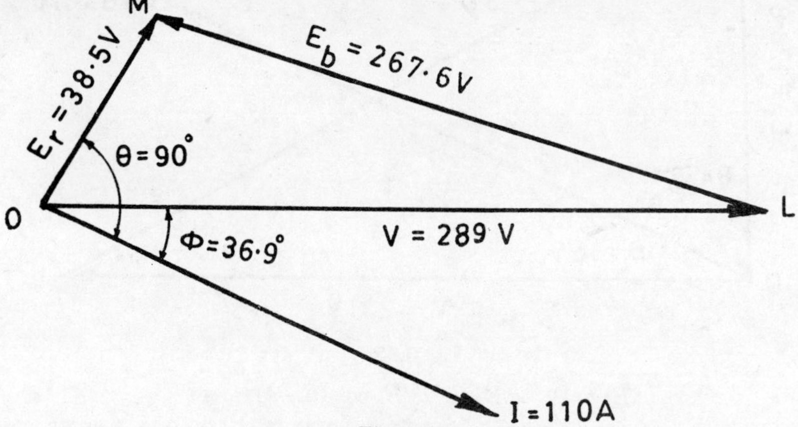

Fig. 6.41

$$E_b^2 = V^2 + E_r^2 - 2VE_r \cos (\theta - \phi)$$

$$= (289)^2 + (38.5)^2 - 2 \times 289 \times 38.5 \cos (90° - 36.9°)$$

$$= 83521 + 1482 - 13361$$

$$\therefore \quad E_b = 267.6 \text{ V}$$

Hence, **e.m.f. generated,** E_b (line value)

$= \sqrt{3} \times 267.6 =$ **463.5 V. (Ans.)**

☞**Example 6.22.** *A 3-phase, 32 kW, 400 V, 50 Hz star-connected synchronous motor has a full-load efficiency of 90%. The synchronous impedance of the motor is (0.25 + j2) Ω per phase. If the excitation of the motor is adjusted to give a leading p.f. of 0.8, calculate for full-load :*

(*i*) *Induced e.m.f., and* (*ii*) *total mechanical power developed.*

Solution. Motor output = 32 kW or 32000 W

Supply voltage/phase $= \dfrac{400}{\sqrt{3}} = 231$ V

Full-load efficiency = 90%

Synchronous impedance/phase, $Z_s = (0.25 + j2)$

i.e., $Z_s = \sqrt{(0.25)^2 + (2)^2} = 2.01\ \Omega$

Full-load current $= \dfrac{\text{motor input}}{\sqrt{3}V_L \cos\phi} = \dfrac{(32000/0.9)}{\sqrt{3} \times 400 \times 0.8} = 64.1$ A

$\therefore$ Impedance drop/phase,

$$E_r = IZ_s = 64.1 \times 2.01 = 128.8 \text{ V}$$

Internal angle, $\theta = \tan^{-1} \dfrac{2}{0.25} = 82.9°$

$$\cos\phi = 0.8,\ \phi = \cos^{-1} 0.8 = 36.9°.$$

(*i*) **Induced e.m.f., E_b :**

Refer Fig. 6.42.

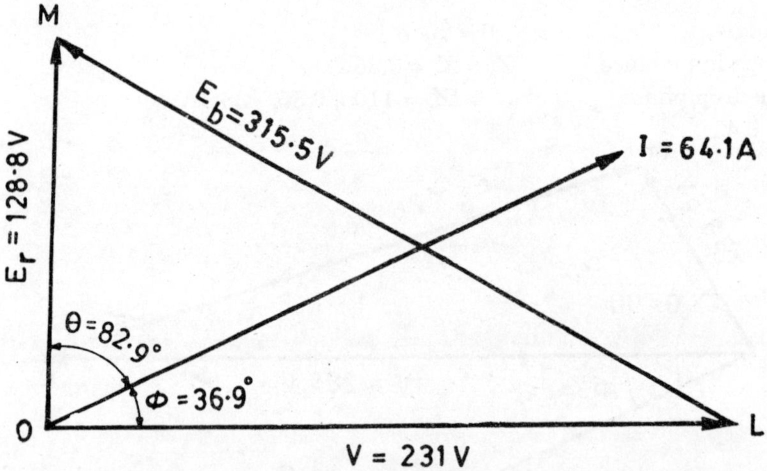

Fig. 6.42

$$E_b^2 = V^2 + E_r^2 - 2VE_r \cos(\theta + \phi)$$
$$= (231)^2 + (128.8)^2 - 2 \times 231 \times 128.8 \cos(82.9° + 36.9°)$$
$$= 53361 + 16589 + 29573$$

$\therefore$ $E_b = 315.5$ V

$\therefore$ **Induced e.m.f., E_b (line value)**

$= \sqrt{3} \times 315.5 =$ **546.5 V. (Ans.)**

(*ii*) **Total mechanical power developed** = ?

Total motor input $= \dfrac{32000}{0.9} = 35555.5$ W

Total copper losses $= 3 \times I^2 \times R_a = 3 \times (64.1)^2 \times 0.25 = 3081.6$ W

$\therefore$ **Total mechanical power developed**

= Total motor input – Total copper loss

= 35555.5 – 3081.6 = 32473.9 or **32.47 kW. (Ans.)**

☞**Example 6.23.** *A 15 kW, 3-phase, 400 V, star connected synchronous motor operating on full-load from infinite bus-bar, has its excitation so adjusted that power factor is 0.8 lagging. Load being kept constant, excitation is now increased by 25%. Synchronous reactance is 1.0 per unit. Find the new power factor.*

Solution. Motor output = 15 kW or 15000 W

Supply voltage/phase = $400/\sqrt{3}$ = 231 V

Synchronous reactance = 1.0 per unit

Since per unit reactance $= \frac{IX}{V}$

∴ Synchronous reactance drop,

$$IX = V \times \text{per unit reactance} = 231 \times 1 = 231 \text{ V}$$

Neglecting armature resistance

Impedance drop/phase, $E_r = IZ_s = IX = 231$ V

Internal angle, $\theta = 90°$

Power factor, $\cos\phi = 0.8$

∴ $\phi = \cos^{-1}(0.8) = 36.9°$.

Refer Fig. 6.43. $E_b^2 = V^2 + E_r^2 - 2VE_r \cos(\theta - \phi)$

$= (231)^2 + (231)^2 - 2 \times 231 \times 231 \cos(90 - 36.9°)$

$= 53361 + 53361 - 64078$

∴ $E_b = 206.5$ V

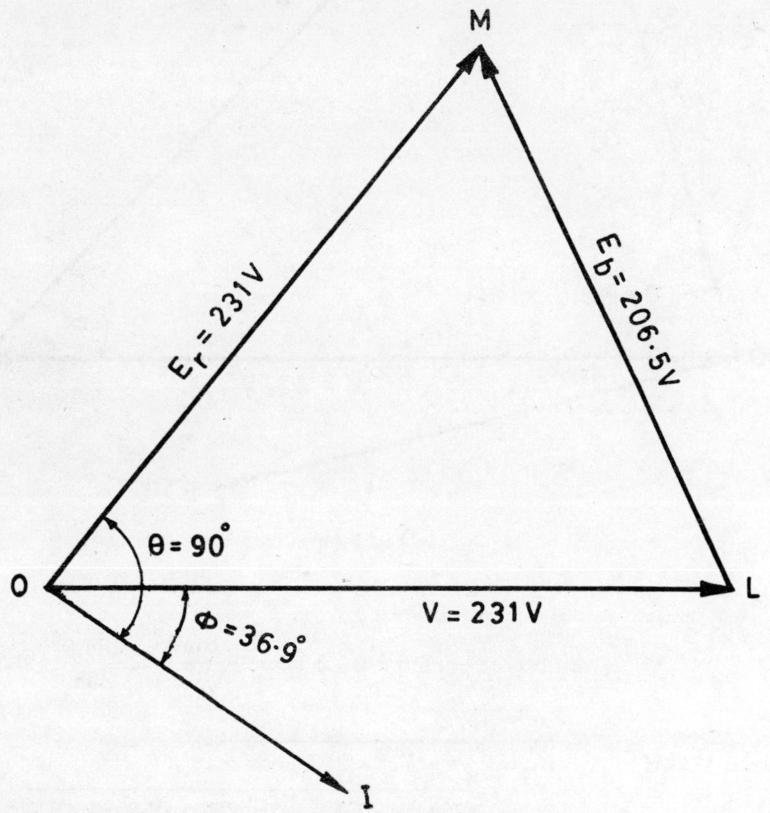

Fig. 6.43

When excitation is increased by 25%

$$E_b' = 1.25 \times 206.5 = 258 \text{ V}$$

Let the phase angle between phase voltage V and the current I be ϕ'. Let the new current be I'.

As load remains the same and since supply voltage is same so active component of current drawn from the 3-phase A.C. supply would remain the same.

$$\therefore \quad I' \cos \phi' = I \cos \phi = 0.8I$$

and impedance drop/phase,. $E_r' = I'Z = I'X$

and $E_r' \cos \phi' = I'X \cos \phi' = I \cos \phi \times X = 0.8IX = 0.8 \times 231 = 184.8$ V.

Refer Fig. 6.44.

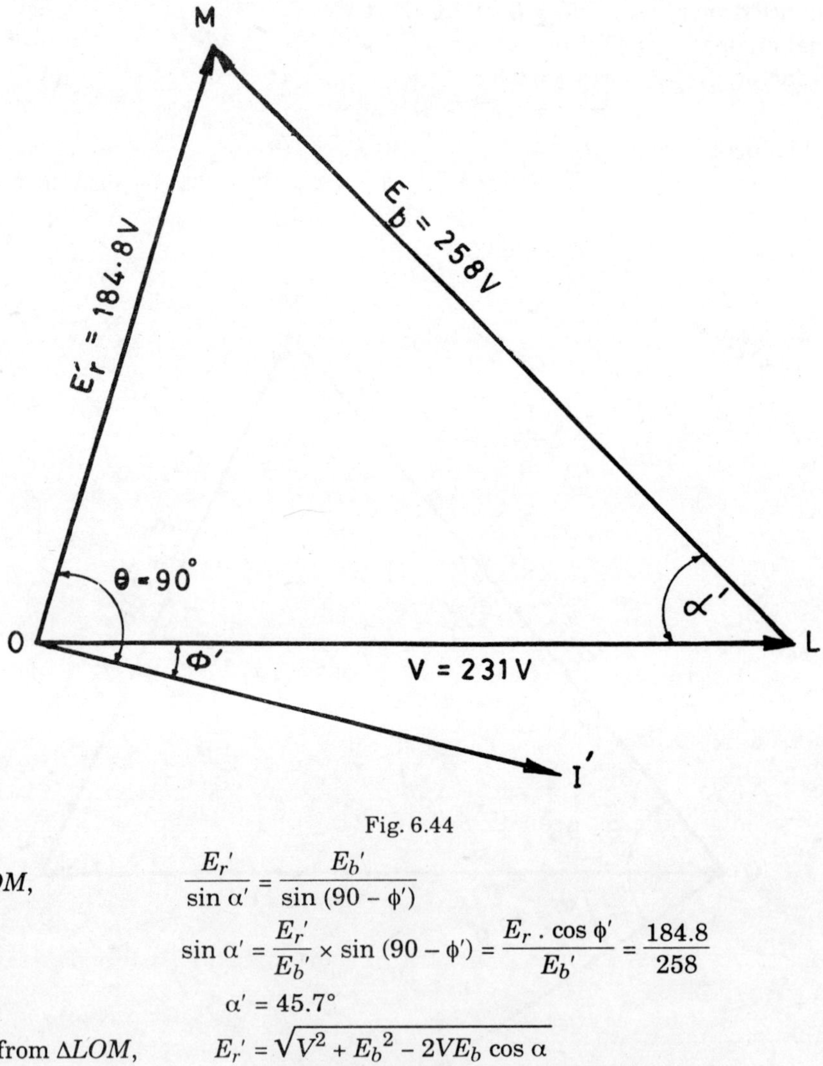

Fig. 6.44

In ΔLOM, $$\frac{E_r'}{\sin \alpha'} = \frac{E_b'}{\sin (90 - \phi')}$$

$$\therefore \quad \sin \alpha' = \frac{E_r'}{E_b'} \times \sin (90 - \phi') = \frac{E_r \cdot \cos \phi'}{E_b'} = \frac{184.8}{258}$$

$$\therefore \quad \alpha' = 45.7°$$

Again from ΔLOM, $$E_r' = \sqrt{V^2 + E_b^2 - 2VE_b \cos \alpha}$$

$$= \sqrt{(231)^2 + (258)^2 - 2 \times 231 \times 258 \cos 45.7°}$$

$$= \sqrt{53361 + 66564 - 83248} = 191.5 \text{ V}$$

$\therefore$ **New power factor,** $\cos \phi' = \dfrac{E_r' \cos \phi'}{E_r'} = \dfrac{184.8}{191.5} =$ **0.965 (lagging). (Ans.)**

Example 6.24. *A 3 phase, star-connected synchronous motor takes 48 kW at 693 V, the power factor being 0.8 lagging. The induced e.m.f. is increased by 30 percent, the power taken remaining the same. The machine has a synchronous reactance of 2 Ω/phase and negligible resistance. Find :*

(i) New power factor, and *(ii) New current.*

Solution. Supply voltage/phase, $V = 693/\sqrt{3} = 400$ V

Synchronous reactance, $X_s = 2\ \Omega$

Since, resistance is negligible

$\therefore$ $Z_s = X_s$

Full-load current, $I = \dfrac{48 \times 1000}{\sqrt{3} \times 693 \times 0.8} = 50$ A

Impedance drop/phase $= IZ_s = 50 \times 2 = 100$ V

Internal angle, $\theta = \tan^{-1} \dfrac{2}{0} = \tan^{-1} \infty$

$\therefore$ $\theta = 90°$

$\cos \phi = 0.8$

$\therefore$ $\phi = \cos^{-1} (0.8) = 36.9°$.

Refer Fig. 6.45.

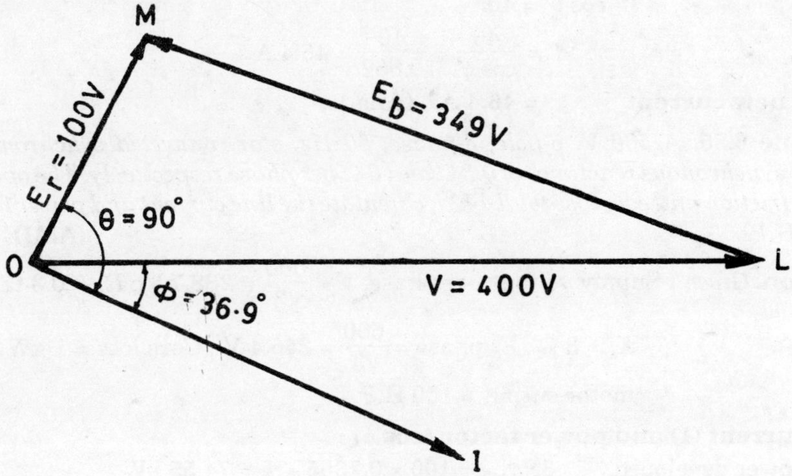

Fig. 6.45

From ΔOLM, $E_b^2 = V^2 + E_r^2 - 2VE_r \cos (\theta - \phi)$

$= (400)^2 + (100)^2 - 2 \times 400 \times 100 \times \cos (90 - 36.9°)$

$= 160000 + 10000 - 48000$

$\therefore$ $E_b = 349$ V.

The vector diagram for increased e.m.f. is shown in Fig. 6.46. Now, $E_b' = 1.3 \times 349 = 454$ V. It can be safely assumed that in the second case, the current is leading voltage V by some angle ϕ'.

Let I' = new current, and

ϕ' = leading angle of current I'

As power input remains the same and V is also constant, $I \cos \phi$ should be the same for the same input.

$\therefore$ $I \cos \phi = 50 \times 0.8 = 40 = I' \cos \phi'$

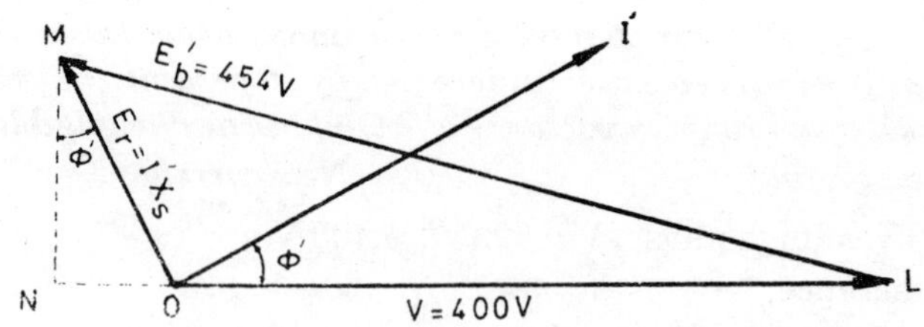

Fig. 6.46

In ΔLMN, $LM^2 = LN^2 + MN^2$

Now $MN = E_r' \cos \phi' = I' X_s \cos \phi'$ $[\because OM = E_r = I' X_s]$

$= I' \cos \phi X_s = 40 \times 2 = 80$ V

$\therefore$ $(454)^2 = LN^2 + (80)^2$

or $LN^2 = (454)^2 - (80)^2$ or $LN = 447$ V

$\therefore$ $ON = LN - OL = 447 - 400 = 47$ V

$\therefore$ $\tan \phi' = \dfrac{NO}{MN} = \dfrac{47}{80}$

$\therefore$ $\phi' = 30.4°$.

(*i*) **New power factor,** $\cos \phi' = \cos 30.4° =$ **0.862 (leading). (Ans.)**

(*ii*) Also $I' \cos \phi' = 40$

$\therefore$ $I' = \dfrac{40}{\cos \phi'} = \dfrac{40}{0.862} = 46.4$ A

Hence, **new current** **= 46.4 A. (Ans.)**

Example 6.25. *A 500 V, 6-pole, 3-phase, 50 Hz, star-connected synchronous motor has a resistance and synchronous reactance of 0.3 Ω and 3 Ω per phase respectively. The open-circuit voltage is 600 V. If the friction and core loss total 1 kW, calculate the line current and power factor when motor output is 100 H.P.* **(AMIE Summer, 1994)**

Solution. *Given* : Supply voltage per phase, $V = \dfrac{500}{\sqrt{3}} = 288.7$ V ; $R_a = 0.3\ \Omega$;

$X_s = 3\ \Omega$; E_b/phase $= \dfrac{600}{\sqrt{3}} = 346.4$ V ; Core loss = 1 kW ;

motor output = 100 H.P.

Line current (I) and power factor (cos φ) :

Total power developed, $3P_{mech} = 100 \times 0.7355 + 1 = 74.55$ kW

$\therefore$ Power developed/phase $= \dfrac{74.55}{3} = 24.85$ kW or 24850 W

Internal angle, $\theta = \tan^{-1} \dfrac{X_s}{R_a} = \tan^{-1}\left(\dfrac{3}{0.3}\right) = 84.3°$

Synchronous impedance/phase, $Z_s = \sqrt{0.3^2 + 3^2} = 3.015\ \Omega$

Power developed per phase is given as

$$P_{mech} = \frac{E_b V}{Z_s} \cos(\theta - \alpha) - \frac{{E_b}^2}{Z_s} \cos \theta \quad \text{[Eqn. (6.2)]}$$

or $$24850 = \frac{346.4 \times 288.7}{3.015} \cos(84.3° - \alpha) - \frac{(346.4)^2}{3.015} \cos 84.3°$$

or $$24850 = 33169.4 \cos(84.3° - \alpha) - 3952.8$$

or $\cos(84.3° - \alpha) = 0.8683$

or $84.3° - \alpha = \cos^{-1}(0.8683) = 29.74°$

$\therefore$ $\alpha = 54.56°$

From phasor diagram shown in Fig. 6.47, we have

$$E_r = \sqrt{E_b^2 + V^2 - 2E_bV\cos\alpha}$$

$$= \sqrt{(346.4)^2 + (288.7)^2 - 2 \times 346.4 \times 288.7 \times \cos(54.56°)}$$

$$= 295.8 \text{ V}$$

Line current = phase current,

$$\mathbf{I} = \frac{E_r}{Z_s} = \frac{295.8}{3.015} = \mathbf{98.1\ A.\ (Ans.)}$$

Again, from phasor diagram, we have

$$\frac{E_b}{\sin(\theta + \phi)} = \frac{E_r}{\sin\alpha}$$

or $$\frac{346.4}{\sin(\theta + \phi)} = \frac{295.8}{\sin 54.56°}$$

or $$\sin(\theta + \phi) = \frac{346.4 \times \sin 54.56°}{295.8} = 0.9541$$

or $\theta + \phi = \sin^{-1}(0.9541) = 72.57°$

or $\phi = 72.57 - \theta = 72.57 - 84.3°$

$= -11.73°$ in $11.73°$ (lag)

$\therefore$ Power factor, **cos ϕ** $= \cos(-11.73°) =$ **0.98 (lag). (Ans.)**

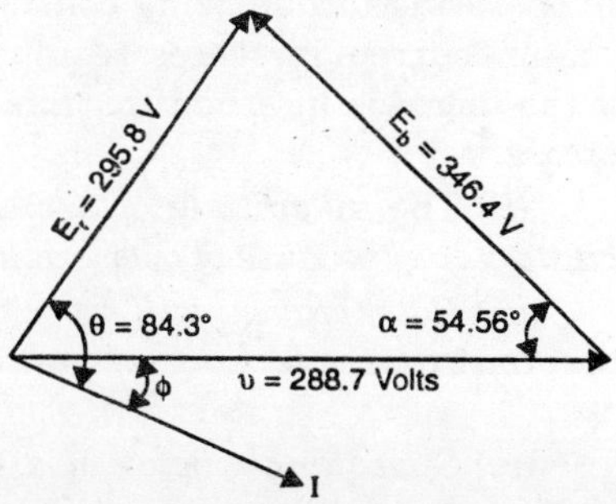

Fig. 6.47

Example 6.26. (*a*) *A 220 V, delta connected 50 Hz synchronous motor has a synchronous reactance of 2.5 ohms and negligible armature resistance. Its friction and windage losses are 1.5 kW, and its core losses are 1.0 kW. The shaft is supplying a 22.5 kW load. Find the load angle. Induced e.m.f. in the stator winding is 255 V.*

(*b*) *A synchronous motor is connected to an infinite bus and supplying constant torque and operates at unity power factor. If the field current is increased, explain what happens to the following :*

(*i*) *The magnitude of resultant flux wave ;*

(*ii*) *The magnitude of armature current and its power factor ;*

(*iii*) *Rotor current will move away from resultant m.m.f. wave or towards it ;*

(*iv*) *Space angle between armature m.m.f. and resultant air gap m.m.f.*

(AMIE Summer, 1998)

Solution. (*a*) *Given* : Supply phase voltage = 220 V ; $X_s = 2.5\ \Omega$;
friction and windage losses = 1.5 kW ; core losses = 1.0 kW ;
shaft load = 22.5 kW ; $E_b = 255$ V

Load angle α :

For a synchronous motor expression for power developed (R_a neglected) is given by

$$P_{mech}/\text{phase} = \frac{E_bV}{X_s}\sin\alpha \qquad \text{[Eqn. (6.4)]}$$

$$P_{mech} = \text{shaft load + core losses + friction and windage losses}$$

$$= 22.5 + 1.0 + 1.5 = 25 \text{ kW}$$

$$P_{mech}/\text{phase} = \frac{25 \times 1000}{3} = \frac{25000}{3} \text{ W}$$

Substituting the values in the above eqn., we get

$$\frac{25000}{3} = \frac{255 \times 220}{2.5} \times (\sin \alpha)$$

or $$\sin \alpha = \frac{25000 \times 2.5}{3 \times 255 \times 220} = 0.3714$$

$\therefore$ α (*i.e.* load angle) $= \sin^{-1}(0.3714) =$ **21.8°. (Ans.)**

(*b*) (*i*) For a synchronous motor connected to infinite bus and supplying constant torque operating at unity power factor, *when field current is increased, the machine becomes over-excited*. The operating torque being constant, the reactive component of armature current and hence the armature current increases, resulting into increase in armature m.m.f. The field m.m.f. also increases but the increase in armature m.m.f. is more and hence the *magnitude of resultant flux wave will increase.*

(*ii*) The magnitude of *armature current* as discussed above *will increase with increase in excitation and because of over excitation the power factor will become leading p.f.*

(*iii*) *Rotor current will move towards the resultant m.m.f. wave because the resultant flux wave moves towards the leading p.f. side resulting into reduction of angle between rotor current and m.m.f. wave.*

(*iv*) *Space angle between armature m.m.f. and resultant air gap m.m.f. will also decrease because resultant flux wave moves towards the air gap m.m.f.*

Example 6.27. *A 3-phase, 415 V, 6-pole, 50 Hz star-connected synchronous motor has e.m.f. of 520 (L-L). The stator winding has a synchronous reactance of 2 Ω per phase, and motor develops a torque of 220 N-m. The motor is operating at 415 V, 50 Hz bus.*

(*i*) *Calculate the current drawn from the supply and its power factor ;*

(*ii*) *Draw the phasor diagram showing all the relevant quantities.*

(GATE, 1992 ; AMIE Summer, 2000)

Solution. *Given* : V (phase voltage) $= \frac{415}{\sqrt{3}} = 239.6$ V ; $p = 6$;

$f = 50$ Hz ; E_b (per phase) $= \frac{520}{\sqrt{3}} = 300$ V ;

$X_s = 2$ Ω/phase, torque developed = 220 N-m.

(*i*) **I and cos ϕ :**

Synchronous speed of motor, $N_s = \frac{120f}{p} = \frac{120 \times 50}{6} = 1000$ r.p.m.

Total power developed, $3P_{mech} = \frac{T \times 2\pi N_s}{60} = \frac{220 \times 2\pi \times 1000}{60} = 23038$ W

Power developed per phase $= \frac{23038}{3} = 7679$ W

Synchronous impedance per phase, $Z_s = 2\angle 90°$ Ω

Power developed per phase is given as,

$$P_{mech} = \frac{E_b V}{Z_s} \cos(\theta - \alpha) - \frac{E_b^2}{Z_s} \cos \theta \qquad \text{[Eqn. (6.2)]}$$

$$= \frac{E_b V}{Z_s} \sin \alpha \qquad (\because \theta = 90°)$$

$$\therefore \quad \frac{E_b V}{Z_s} \sin \alpha = 7679$$

or $$\sin \alpha = \frac{7679 \times Z_s}{E_b V} = \frac{7679 \times 2}{300 \times 239.6} = 0.2137$$

or $$\alpha = \sin^{-1}(0.2137) = 12.34°$$

From phasor diagram shown in Fig. 6.48, we have

$$E_r = \sqrt{E_b^2 + V^2 - 2E_b V \cos \alpha}$$
$$= \sqrt{(300)^2 + (239.6)^2 - 2 \times 300 \times 239.6 \times \cos 12.34°}$$
$$= 83.48 \text{ V}$$

∴ Current drawn per phase,

$$I = \frac{E_r}{Z_s} = \frac{83.48}{2} = \textbf{41.74 A.} \quad \textbf{(Ans.)}$$

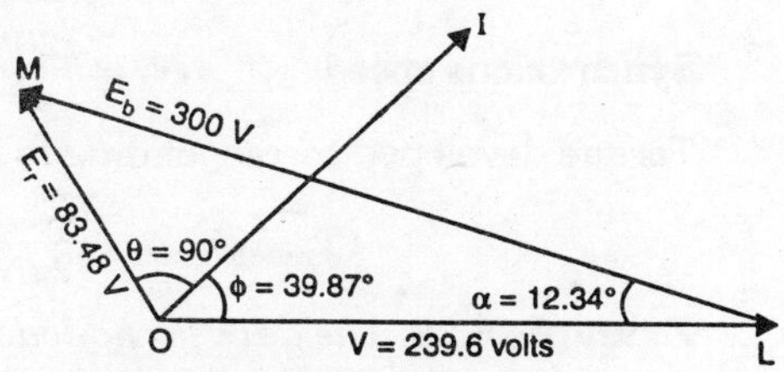

Fig. 6.48. Phasor diagram.

Again from phasor diagram, we have

$$\frac{E_b}{\sin(\theta + \phi)} = \frac{E_r}{\sin \alpha}$$

or $$\sin(\theta + \phi) = \frac{E_b \sin \alpha}{E_r} = \frac{300 \sin 12.34°}{83.48} = 0.768$$

∴ $\theta + \phi = \sin^{-1}(0.768) = 129.87°$ (since $E_b > V$, power factor will be leading one)

or $\phi = 129.87° - \theta = 129.87 - 90° = 39.87°$ (leading)

∴ Power factor, **cos ϕ** = cos (39.87°) = **0.768 leading. (Ans.)**

(*ii*) **Phasor diagram :**

Phasor diagram assimilating all the relevant quantities is shown in Fig. 6.48.

Example 6.28. *A 50 Hz, 4-pole, 3-phase, star-connected synchronous motor has a synchronous reactance of 12 Ω/phase and negligible armature resistance. The excitation is such as to give an open-circuit voltage of 13.2 kV. The motor is connected to 11.5 kV, 50 Hz supply.*

What maximum load can the motor supply before loosing synchronism ? What is the corresponding motor torque, line current and power factor ? **(AMIE Summer, 1996)**

Solution. *Given :* $X_s = 12\ \Omega$; E_b/phase $= \frac{13.2 \times 1000}{\sqrt{3}} = 7621$ V,

$$V/\text{phase} = \frac{11.5 \times 1000}{\sqrt{3}} = 6640 \text{ V}$$

Internal angle θ, $\tan^{-1}\left(\frac{X_s}{R_a}\right) = \tan^{-1}\left(\frac{12}{0}\right) = \tan^{-1}(\infty) = 90°$

Synchronous impedance/phase, $Z_s = X_s = 12\ \Omega$

Power developed will be maximum when,

Load angle $\alpha = \theta = 90°$

Impedance drop per phase, $E_r = \sqrt{E_b^2 + V^2 - 2E_b V \cos \alpha}$

$$= \sqrt{(7621)^2 + (6640)^2 - 2 \times 7621 \times 6640 \times \cos 90°} = 10108 \text{ V}$$

Line current = phase current, $\mathbf{I} = \frac{E_r}{Z_s} = \frac{10108}{12} = \textbf{842.3 A.}$ **(Ans.)**

The maximum load the motor can deliver,

$$(P_{mech})_{max} = 3\left[\frac{E_b V}{Z_s} - \frac{E_b^2}{Z_s} \cos \theta\right] \qquad \text{[Eqn. (6.3)]}$$

$$= 3\left[\frac{7621 \times 6640}{12} - \frac{(7621)^2}{12} \times \cos 90°\right]$$

$$= 12650860 \text{ W or } \mathbf{12.65 \text{ MW. (Ans.)}}$$

Power supplied to motor, $P_{in} = P_{mech}$ + armature copper loss

$= P_{mech} = 12.65$ MW (∵ armature resistance is negligible)

$$\therefore \quad \sqrt{3}\, V_L I_L \cos\phi = 12.65 \times 10^6$$

or

$$\cos\phi = \frac{12.65 \times 10^6}{\sqrt{3} \times 11500 \times 842.3} = \mathbf{0.754.\ (Ans.)}$$

Synchronous speed, $N_s = \frac{120f}{p} = \frac{120 \times 50}{4} = 1500$ r.p.m.

Torque developed corresponding to maximum power developed,

$$(\mathbf{T_{mech}})_{\mathbf{max}} = \frac{(P_{mech})_{max}}{(2\pi N_s/60)} = \frac{12.65 \times 10^6 \times 60}{2\pi \times 1500} = \mathbf{80532 \text{ N-m. (Ans.)}}$$

Example 6.29. *The data for no-load saturation curve of a 6600 V, 1000 kVA, 3-phase, 50-Hz, star-connected synchronous motor is given below :*

O.C. voltage, V :	*3400*	*5700*	*7200*	*7900*	*8400*
Field current (I_f), A :	*50*	*100*	*150*	*175*	*200*

The resistance and synchronous reactance per phase are 0.5 Ω and 12 Ω respectively.
Plot the V-curve for this machine when the input is maintained constant at 250 kW.

Solution. Rating of motor = 1000 kVA

Input which is maintained constant = 250 kW

Armature resistance/phase, $R_a = 0.5\ \Omega$

Synchronous reactance/phase $X_s = 12\ \Omega$

Supply line voltage, $V_L = 6600$ V

Supply current,

$$I = \frac{\text{motor input}}{\sqrt{3} V_L \cos\phi}$$

$$= \frac{250 \times 1000}{\sqrt{3} \times 6600 \times \cos\phi} = \frac{21.9}{\cos\phi} \text{ A}$$

$$\therefore \quad I\cos\phi = 21.9$$

Impedance, $Z_s = \sqrt{R_a^2 + X_s^2} = \sqrt{(0.5)^2 + (12)^2} = 12.01\ \Omega$

Internal angle, $\theta = \tan^{-1}\frac{X_s}{R_a} = \tan^{-1}\frac{12}{0.5} = 87.6°$

Internal drop (line), $E_r = \sqrt{3} IZ = \sqrt{3} \times \frac{21.9}{\cos\phi} \times 12.01 \simeq \frac{455}{\cos\phi}$

Refer Fig. 6.49.

- Draw OL to represent the applied potential difference of 6000 V.
- Draw line OY at an angle $\theta = 87.6°$ with OL.

Make OM to represent 455 V. Draw a line ⊥ to OY and passing through M. This line represent the power of 250 kW, and is the locus of the end of the impedance drop vector since any line joining point O to any point on this line has a length representing 455/cos ϕ volts where ϕ is the angle between this line and OY. ϕ is also the angle between OL and the current vector drawn at an angle of 87.6° from impedance drop vector *e.g.* when $\phi = \angle YON$, the impedance drop = ON then OI_N lagging LN by 87.6° the current and $\angle LOI_N = \phi$.

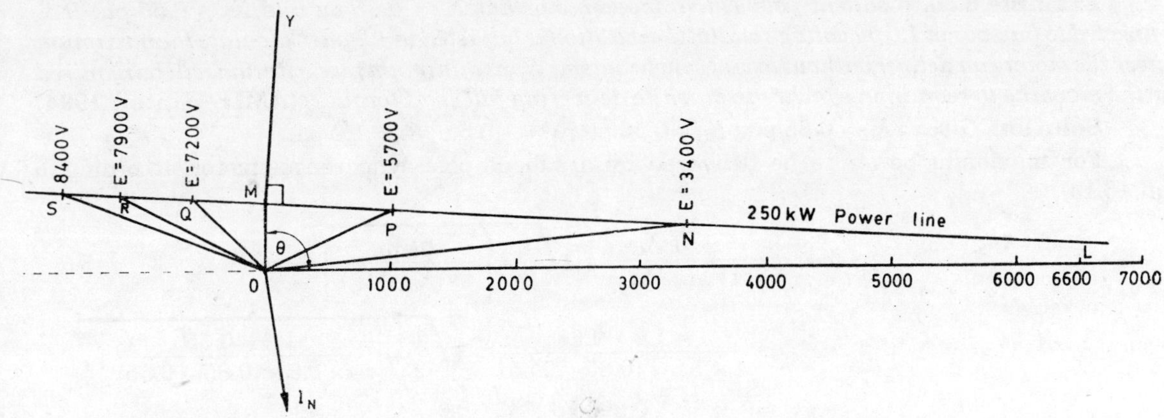

Fig. 6.49

- With centre L and radii representing 3400, 5700, 7200, 7900 and 8400 volts respectively draw arcs to cut power line of 250 kW at N, P, Q, R and S. Join these points to O. Then ON, OP, OQ, OR and OS represent the impedance drops for the condition when the induced e.m.fs. (line) are LN, LP, LQ, LR and LS *i.e.* 3400, 5700, 7200, 7900, and 8400 volts respectively.
- Now armature current, $I = \dfrac{\text{impedance drop}}{\sqrt{3} Z_s}$

$$= \frac{\text{impedance drop}}{\sqrt{3} \times 12.01} \qquad [\because \; Z_s = 12.01 \; \Omega]$$

$$= 0.04807 \text{ impedance drop}$$

- The results from the diagram (Fig. 6.49) are tabulated below :

O.C. voltage,	V:	3400	5700	7200	7900	8400
Field current (I_f),	A:	50	100	150	175	200
Impedance drop (line),	V:	3400	1100	700	1300	1750
Armature current (I),	A:	164.4	52.9	33.6	62.5	84.1

From the above figures draw the V-curve as shown in Fig. 6.50.

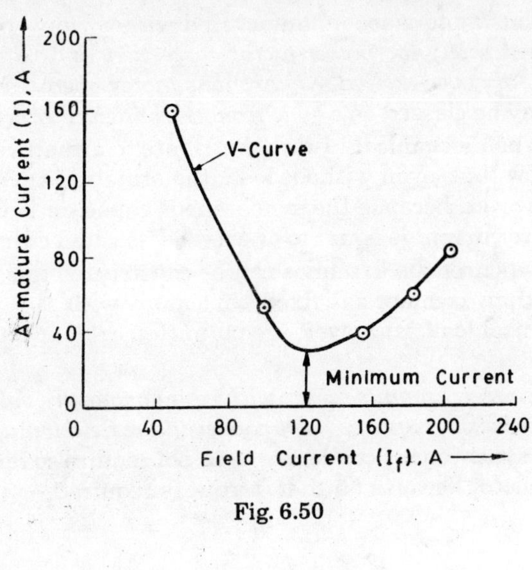

Fig. 6.50

Example 6.30. *A salient pole synchronous motor has X_d = 0.85 pu and X_q = 0.55 pu. It is connected to bus-bar of 1.0 pu voltage while its excitation is adjusted to 1.2 pu. Calculate the maximum power the motor can deliver without loss of synchronism. What minimum pu excitation will be required for the machine to remain in synchronism while delivering full-load torque.* **(AMIE Winter, 1994)**

Solution. *Given* : $X_d = 0.85$ pu ; $X_q = 0.55$ pu ; $V = 1.0$ pu ; $E_b = 1.2$ pu.

For maximum power to be delivered by a salient pole synchronous motor, according to eqn. (6.13),

$$\cos \alpha = \frac{-E_b X_q}{4V(X_d - X_q)} + \sqrt{\frac{1}{2} + \left[\frac{E_b X_q}{4V(X_d - X_q)}\right]^2}$$

$$= \frac{-1.2 \times 0.55}{4 \times 1.0 \times (0.85 - 0.55)} + \sqrt{\frac{1}{2} + \left[\frac{1.2 \times 0.55}{4 \times 1.0 \times (0.85 - 0.55)}\right]^2}$$

$$= -0.55 + 0.896 = 0.346$$

or Load angle, $\alpha_{(max)} = \cos^{-1} 0.346 = 69.8°$

Now maximum power is given by

$$\mathbf{P_{max}} = \frac{E_b V}{X_d} \sin \alpha + \frac{V^2}{2}\left[\frac{1}{X_q} - \frac{1}{X_d}\right] \sin 2\alpha$$

$$= \frac{1.2 \times 1.0}{0.85} \sin 69.8° + \frac{1}{2}\left[\frac{1}{0.55} - \frac{1}{0.85}\right] \sin (2 \times 69.8°)$$

$$= 1.325 + 0.321 \times 0.648 = \mathbf{1.533 \text{ pu.}} \quad \textbf{(Ans.)}$$

The power delivered due to excitation is given as

$$P = \frac{E_b V}{X_d} \sin \alpha$$

Substituting $V = 1.0$, $X_d = 0.85$, $\alpha = 69.8°$ and $P = 1.533$ in the above equation, we get

$$1.533 = \frac{E_b \times 1.0}{0.85} \times \sin 69.8°$$

or

$$\mathbf{E_b} = \frac{1.533 \times 0.85}{\sin 69.8°} = \mathbf{1.388 \text{ pu.}} \quad \textbf{(Ans.)}$$

6.11. SYNCHRONOUS CONDENSERS

- Several synchronous motors are manufactured without any shaft extensions whatever, ostensibly designed *solely for power-factor correction* and to be operated without any mechanical load. *Any over-excited synchronous motor operated without any mechanical load,* however, may be classed as a *synchronous condenser or synchronous capacitor.*
- Although there is no mechanical load to contribute to armature current, the *V*-curves of Fig. 6.14. will show that, even without load, the armature current is high. This is not a disadvantage, however, because the synchronous capacitor is over-excited to that point where its armature current is equal to or exceeds its rated current.

Alternating-current synchronous dynamos can be constructed *less expensively* in extremely high ratings and voltages than commercial fixed capacitors with the result that synchronous capacitors driving no mechanical load, whatever, are permitted to *'float'* on the line for the purposes of power factor improvement..

- *As compared with a synchronous motor with equal armature voltage and current ratings, a 'synchronous condenser' requires more copper in the field winding to carry the increased field current.* The synchronous condenser does not require so large shaft and bearings as the synchronous motor because no shaft torque is required.

- Synchronous condensers are sometimes operated at power factors ranging from lagging through unity to leading for voltage control. When applied in this manner a synchronous condenser is called a *synchronous reactor*.
- It is customary not to attempt correction of the power factor of the system all the way to a unity power factor. There is an *economic reason* for this, despite the fact that large, high capacity synchronous capacitors are available and for the same kVA rating are usually *less expensive* than synchronous motors. The capacitors are less expensive because (*i*) they are started and run without load, requiring no special heavy cage windings, and (*ii*) they require smaller shaft diameters and lighter bearings, although their field windings are some-what heavier.

Example 6.31. *A 440 V, 50-Hz, 3-phase circuit takes 18 A at a lagging power factor of 0.8. A synchronous motor is used to raise the power factor to unity. Calculate the kVA input to the motor, and its power factor when driving a mechanical load of 6 kW. The motor has an efficiency of 88%.*

Solution. Supply voltage, $V_L = 440$ V

Current drawn, $I_L = 18$ A

Power factor, $\cos \phi = 0.8$

$\therefore$ $\sin \phi = 0.6$

Mechanical load $= 6$ kW

Efficiency of the motor $= 88\%$

kVAR drawn by the 3-phase circuit

$$= \sqrt{3} V_L I_L \sin \phi \times 10^{-3} = \sqrt{3} \times 440 \times 18 \times 0.6 \times 10^{-3} = 8.23$$

Power supplied by the motor,

$$P = \frac{\text{output in kW}}{\text{efficiency}} = \frac{6}{0.88} = 6.82 \text{ kW}$$

Power factor will be raised to unity when kVAR (leading) drawn by 3-phase synchronous motor will become equal to kVAR (lagging) drawn by 3-phase A.C. circuit.

i.e., kVAR drawn by synchronous motor = 8.23 (lead)

kW draw by the synchronous motor = 6.82

$\therefore$ **kVA input to motor** $= \sqrt{(\text{kW})^2 + (\text{kVAR})^2}$

$$= \sqrt{(6.82)^2 + (8.23)^2} = \mathbf{10.69 \text{ kVA.}} \quad \textbf{(Ans.)}$$

Example 6.32. *An alternator supplying 600 kW at 0.8 power factor lagging has its power factor raised to unity by means of an over-excited synchronous motor. With the armature current unchanged, how much power can the alternator thus supply to the synchronous motor and what power can the latter develop when operating at an efficiency of 86% ? Find the power factor at which synchronous motor runs.*

Solution. Before installation of synchronous motor :

Load supplied by the alternator $kW_1 = 600$ kW

Power factor of the load, $\cos \phi = 0.8$

$\therefore$ $\sin \phi_1 = \sqrt{1 - (0.8)^2} = 0.6$

kVA supplied by the alternator,

$$\text{kVA}_1 = \frac{\text{kW}}{\cos \phi_1} = \frac{600}{0.8} = 750 \text{ kVA}$$

Load $\text{kVAR} = \text{kVA} \sin \phi = 750 \times 0.6 = 450$ kVAR

After installation of synchronous motor :

Power factor of alternator,

$$\cos \phi_2 = 1.0$$

kVA supplied by alternator,

$$kVA_2 = kVA_1 = 750 \qquad [\because \text{ armature current remains the same}]$$

Load supplied by the alternator,

$$kW_2 = kVA_2 \times \cos \phi_2 = 750 \times 1 = 750 \text{ kW}$$

Load supplied to the synchronous motor

$$= kW_2 - kW_1 = 750 - 600 = \mathbf{150\ kW.}\ \textbf{(Ans.)}$$

Leading kVAR drawn by synchronous motor

$$= \text{load kVAR} = 450$$

kVA drawn by synchronous motor

$$= \sqrt{(150)^2 + (450)^2} = 474.3 \text{ kVA}$$

Power factor at which the synchronous motor operates

$$= \frac{kW}{kVA} = \frac{150}{474.3} = \mathbf{0.3162.}\ \textbf{(Ans.)}$$

Power developed by the synchronous motor

$$= 150 \times 0.86 = \mathbf{129\ kW.}\ \textbf{(Ans.)}$$

Example 6.33. *A substation operating at full-load of 1200 kVA supplies a load at 0.7 power factor lagging. Calculate the permissible additional load at this power factor and the rating of synchronous condenser to raise the substation power factor to 0.9 lagging.*

Solution. Rating of power plant = 1200 kVA

Power factor, $\cos \phi_1 = 0.7$

New power factor, $\cos \phi_2 = 0.9$

Load supplied at 0.7 p.f. $= 1200 \times 0.7 = 840$ kW

Load supplied at 0.9 p.f. $= 1200 \times 0.9 = 1080$ kW

Increase in load supplied $= 1080 - 840 = 240$ kW

Additional load supplied at 0.7 p.f. lagging

$$= \frac{240}{0.7} = \mathbf{342.86\ kVA.}\ \textbf{(Ans.)}$$

Reactive power drawn by load $= \text{new load in kW} \times \tan \phi_1$

$$= 1080 \times \tan (\cos^{-1} 0.7) = 1080 \times 1.0202 = 1101.8 \text{ kVAR}$$

Reactive power supplied by the plant

$$= P \sin \phi_2 = 1200 \times \sin (\cos^{-1} 0.9) = 523 \text{ kVAR}$$

Rating of ideal synchronous condenser

= reactive power drawn by load

– reactive power supplied by the plant

$$= 1101.8 - 523 = \mathbf{578.8\ kVAR.}\ \textbf{(Ans.)}$$

Example 6.34. *An industrial plant has a load of 800 kW at a power factor of 0.8 lagging. It is desired to purchase a synchronous motor of sufficient capacity to deliver a load of 200 kW and also serve to correct the overall plant power factor to 0.92. Assuming that the synchronous motor has an efficiency of 92%, determine its kVA input rating and power factor at which it will operate.*

Solution. Industrial plant load = 800 kW

Power factor, $\cos \phi_1 = 0.8$

Reactive kVAR of load $= 800 \tan (\cos^{-1} 0.8) = 600$ kVAR

Synchronous motor output	$= 200$ kW
Synchronous motor input	$= \dfrac{200}{0.92} = 271.4$ kW
Net total industrial load	$= 800 + 217.4 = 1017.4$ kW
New power factor,	$\cos \phi_2 = 0.92$
∴	$\phi_2 = \cos^{-1}(0.92) = 23°$
	$\tan \phi_2 = \tan 23° = 0.4245$
New	kVAR $= 1017.4 \times 0.4245 = 423$ kVAR
kVAR neutralized	$= 600 - 432 = 168$

∴ **Power factor at which synchronous motor is working,**

$$\cos\left(\tan^{-1}\frac{168}{217.4}\right) = \mathbf{0.7912.}\ \textbf{(Ans.)}$$

Rating of motor $= \dfrac{217.4}{0.7912} =$ **274.8 kVA. (Ans.)**

Example 6.35. *A 3-phase synchronous motor absorbing 60 kW is connected in parallel with a factory load of 240 kW having a lagging power factor of 0.8. If the combined load has a p.f. of 0.9, what is the value of the leading kVAR supplied by the motor and at what power factor is it working ?*

(Madras University, April, 1997)

Solution. *Given* : Factory load, $P_L = 240$ kW ; $\cos \phi_L = 0.8$;
synchronous motor load, $P_M = 60$ kW ;
combined power factor, $\cos \phi = 0.9$ (or $\phi = \cos^{-1} 0.9 = 25.84°$).

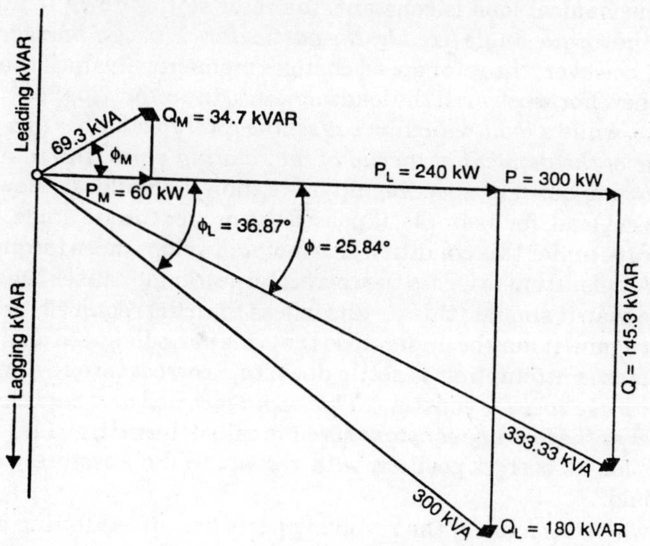

Fig. 6.51

Value of leading kVAR :

Factory load kVAR,	$Q_L = 240 \tan \phi = 240 \tan(\cos^{-1} 0.8) = 180$ kVAR
Total load	$P = P_L + P_M = 240 + 60 = 300$ kW
Combined kVAR,	$Q = P \tan \phi = 300 \tan(\cos^{-1} 0.9) = 145.3$ kVAR (lag)

Leading kVAR supplied by motor, $\mathbf{Q_M} = Q_L - Q = 180 - 145.3 =$ **34.7 kVAR. (Ans.)**

∴ kVA supplied by the motor $= \sqrt{P_M^2 + Q_M^2}$

$$= \sqrt{(60)^2 + (34.7)^2} = \mathbf{69.3\ kVAR.}\ \textbf{(Ans.)}$$

Power factor of motor, $\cos \phi_m = \dfrac{60}{69.3} = \mathbf{0.866}$ **(leading). (Ans.)**

6.12. SYNCHRONOUS MOTOR RATINGS

- Synchronous motors may be purchased in three standard ratings namely :
  (*i*) unity power factor ;
  (*ii*) 90% leading power factor ; and
  (*iii*) 80% leading power factor.

Other ratings are obtainable from motor manufacturers by special quotation.

- If a motor is rated at unity power factor it can be operated with a leading power factor. However, the mechanical load must be decreased sufficiently so that no more than rated A.C. stator current will flow at the reduced power factor. In other words, the mechanical load in horsepower output and the electrical load in leading kVARs must not exceed the rating of the motor. When a synchronous motor is rated at 80 or 90 per cent lead power factor it will have a large current capacity for a given horsepower output. This is necessary in order to supply rated horsepower output with the larger current at the reduced power factor.

6.13. HUNTING OF SYNCHRONOUS MOTORS

- When the mechanical load is constant, the rotor settles down to an absolutely constant speed with the *torque angle fixed by the particular delivered horsepower.* Should the load be changed, however, the rotor speed changes momentarily until the torque angle adjusts itself to the new horsepower, if the load *increases,* the *rotor slips backwards* to an increased torque angle, while a *load reduction causes the rotor to advance to a smaller torque angle.* But *because of the moment of inertia of the rotating parts,* the rotor *overshoots the final position,* slowing down or speeding up more than it should. In slowing down as a result of an increased load, for example, it passes the proper torque angle, giving up some of the kinetic energy ; under this condition, the motor develops more torque than it requires and speeds up. Acceleration to locate the correct torque angle causes the rotor to move forward a little more than it should ; this results in less than the required torque, so that the motor slows-down again. It may be understood that this periodic speed change is only *momentary* while the rotor is attempting to settle down to a correct torque angle and that it goes on while the *average speed* is constant. *This rapid forward and backward motion of the rotor as it revolves at the average constant speed* is called **'hunting'** (*i.e.,* the rotor may be said to '*hunting*' for its correct position with respect to the absolutely constant speed of the revolving field.

In some cases, when the mass of the revolving parts has an oscillating period that is same as, a some multiple of, the hunting period, it is possible for swings to grow progressively greater ; under this condition, the torque angle might even *exceed the pullout* value, *causing the motor to drop out of synchronous and stall.*

- **'Hunting'** *is an objectionable characteristic of all synchronous motors, since it produces severe mechanical stresses as well as great variation in current and power taken by the motor.*
- Fortunately, the very same pole-face of squirrel-cage that provides the machine with its starting torque is instrumental in *damping the oscillation*. Its effectiveness in doing this,

however, depends upon the resistance of the squirrel-cage ; *the lower the resistance the stronger is the damping action.* But since a high squirrel-cage resistance is necessary, if the synchronous motor is to have a good starting torque, it is customary to employ a *compromise value* for a cage. The term *amortisseur* is generally applied to the squirrel-cage in connection with its damping action. *This damping action results because by Lenz's law any change in flux linking the amortisseur as it attempts to oscillate back and forth causes a current to flow in the cage ; this current then flows in such a direction as to oppose a change in the flux that normally links the armature and the rotor.*

6.14. COMPARISON BETWEEN SYNCHRONOUS AND INDUCTION MOTORS

S. No.	Synchronous motor	Induction motor
1.	It is inherently not self-starting and some external means are required for its starting.	It has got self-starting torque and no special means are required for starting.
2.	Requires D.C. excitation.	Does not require D.C. excitation.
3.	Speed control not possible.	Speed can be controlled but to small extent.
4.	Its average speed is constant and independent of load.	Its speeds falls with the increase in load and is always less than synchronous speed.
5.	It can be operated under a wide range of power factor, both lagging and leading.	It operates at only lagging power factor, which becomes very poor at light loads.
6.	Its torque is less sensitive to change in supply voltage.	Its torque is more sensitive to change in supply voltage.
7.	Breakdown torque is proportional to the supply voltage.	Breakdown torque is proportional to the square of the supply voltage.
8.	More complicated and more costly comparatively.	More simple and less costly comparatively.
9.	Employed for supplying mechanical load as well as for power factor improvement.	Employed for supplying mechanical load only.

HIGHLIGHTS

1. *A synchronous motor is electrically identical with an alternator or A.C. generator.*
2. *Some characteristic features of a synchronous motor are :*
   — *It is not inherently self-starting.*
   — *It runs at synchronous speed,* $N_s = \frac{120f}{p}$.
   — *It can be operated under a wide range of power factors both lagging and leading.*
3. *Mechanical power developed/phase,*

$$(P_{mech})/phase = \frac{E_b V}{Z_s} \cos(\theta - \alpha) - \frac{E_b^2}{Z_s} \cos\theta$$

where α = *load angle,* θ = *internal angle.*

4. *Maximum power developed,* $(P_{mech})_{max} = \frac{E_bV}{Z_s} - \frac{E_b^2}{Z_s} \cos \theta$.

5. *The magnitude of armature current varies with excitation. The current has large values both for low and high values of excitation (though it is lagging for low excitation and leading for higher excitation).*
   — *For the same input, armature current varies between a wide range and so causes the power factor also to vary accordingly. When over-excited, motor runs with leading p.f. and with lagging p.f. under-excited.*

6. *A condition knows as* ***'hunting'*** *is produced when a synchronous motor is used for driving a* **varying load.** *Hunting can be reduced by the used of dampers or damping grids.*

OBJECTIVE TYPE QUESTIONS

(A) Choose the Correct Answer :

1. Synchronous motors are generally not self-starting because
   (*a*) the direction of rotation is not fixed.
   (*b*) the direction of instantaneous torque reverses after half cycle
   (*c*) starters cannot be used on these machines
   (*d*) starting winding is not provided on these machines.
2. Zero power factor test is generally conducted on synchronous machines of
   (*a*) large size (*b*) less than 10 kW
   (*c*) miniature size (*d*) multiple speed type.
3. Zero power factor test can be used to ascertain
   (*a*) pull in torque (*b*) the temperature rise
   (*c*) the synchronous reactance (*d*) all of the above.
4. In case one phase of a three-phase synchronous motor is short-circuited, the motor will
   (*a*) not start (*b*) run at $\frac{2}{3}$ of synchronous speed
   (*c*) run with excessive vibrations (*d*) take less than the rated load.
5. A pony motor is basically a
   (*a*) small induction motor (*b*) D.C. series motor
   (*c*) D.C. shunt motor (*d*) double winding A.C./D.C. motor.
6. A synchronous motor can develop synchronous torque
   (*a*) when under loaded (*b*) while over-excited
   (*c*) only at synchronous speed (*d*) below or above synchronous speed.
7. A synchronous motor can be started by
   (*a*) pony motor (*b*) D.C. compound motor
   (*c*) providing damper winding (*d*) any of the above.
8. As compared to an induction motor of the same size the air-gap in a synchronous motor is
   (*a*) three to five times (*b*) same
   (*c*) half to three half (*d*) less than half.
9. A three-phase synchronous motor will have
   (*a*) no slip-rings (*b*) one slip-ring
   (*c*) two slip-rings (*d*) three slip-rings.
10. Under which of the following conditions hunting of synchronous motor is likely to occur ?
    (*a*) Periodic variation of load (*b*) Over-excitation
    (*c*) Over-loading for long periods (*d*) Small and constant load.
11. When the excitation of an unloaded salient pole synchronous motor suddenly gets disconnected
    (*a*) the motor stops (*b*) it runs as a reluctance motor at the same speed
    (*c*) it runs as a reluctance motor at a lower speed
    (*d*) none of the above.

12. When V is the applied voltage, then the breakdown torque of a synchronous motor varies as

(*a*) V (*b*) $V^{3/2}$

(*c*) V^2 (*d*) $1/V$.

13. The power developed by a synchronous motor will be maximum when the load angle is

(*a*) zero (*b*) 45°

(*c*) 90° (*d*) 120°.

14. A synchronous motor can be used as a synchronous capacitor when it is

(*a*) under-loaded (*b*) over-loaded

(*c*) under-excited (*d*) over-excited.

15. A synchronous motor is running on a load with normal excitation. Now if the load on the motor is increased

(*a*) power factor as well as armature current will decrease

(*b*) power factor as well as armature current will increase

(*c*) power factor will increase but armature current will decrease

(*d*) power factor will decrease and armature current will increase.

16. For *V*-curves for a synchronous motor the graph is drawn between

(*a*) field current and armature current (*b*) terminal voltage and load factor

(*c*) power factor and field current (*d*) armature current and power factor.

17. The back e.m.f. of a synchronous motor depends on

(*a*) speed (*b*) load

(*c*) load angle (*d*) all of the above.

18. When V is the supply voltage and R is the rotor resistance per phase, the mechanical power developed by the synchronous motor will be proportional to

(*a*) $\frac{V}{R}$ (*b*) $\frac{V^2}{R}$

(*c*) $\frac{R}{V^2}$ (*d*) $\frac{V}{R^2}$.

19. In a synchronous motor which loss varies with load ?

(*a*) Windage loss (*b*) Bearing friction loss

(*c*) Copper loss (*d*) Core loss.

20. A synchronous motor can be made self-starting by providing

(*a*) damper winding on rotor poles (*b*) damper winding on stator

(*c*) damper winding on stator as well as rotor poles

(*d*) none of the above.

21. In a synchronous motor pull out torque represents

(*a*) the ability of the motor to maintain operation during the change over from induction to synchronous running

(*b*) the torque when motor is running on full-load and then all of a sudden the load is switched off

(*c*) the torque on which the synchronous running of the motor depends

(*d*) none of the above.

22. The oscillation in a synchronous motor can be damped out by

(*a*) maintaining constant excitation (*b*) running the motor on leading power factors

(*c*) providing damper bars in the rotor pole faces (*d*) oscillations cannot be damped.

23. The permissible maximum temperature for *B* class insulation is

(*a*) 75°C (*b*) 90°C

(*c*) 105°C (*d*) 155°C.

24. The shaft of synchronous motor is made of

(*a*) mild steel (*b*) chrome steel

(*c*) alnico (*d*) stainless steel.

25. When the field of a synchronous motor is under-excited, the power factor will be
(*a*) leading (*b*) lagging
(*c*) unity (*d*) zero.

26. The speed regulation of a synchronous motor is always
(*a*) 1% (*b*) 0.5%
(*c*) positive (*d*) zero.

27. The percentage slip in case of a synchronous motor is
(*a*) 1% (*b*) 100%
(*c*) 0.5% (*d*) zero,

28. The operating speed of a synchronous motor can be changed to new fixed value by
(*a*) changing the load (*b*) changing the supply voltage
(*c*) changing frequency (*d*) using brakes.

29. A synchronous motor will always stop when
(*a*) supply voltage fluctuates (*b*) load in motor varies
(*c*) excitation winding gets disconnected (*d*) supply voltage frequency changes.

30. An unexcited single-phase synchronous motor is
(*a*) A.C. series (*b*) reluctance motor
(*c*) universal motor (*d*) repulsion motor.

31. Hunting in a synchronous motor takes place
(*a*) when supply voltage fluctuates (*b*) when load varies
(*c*) when power factor is unity (*d*) motor is under loaded.

32. A synchronous motor is provided with damper winding against mechanical load. The number of slip ring will be
(*a*) 5 (*b*) 3
(*c*) 2 (*d*) nil.

33. A synchronous motor has better power factor as compared to that of an equivalent induction motor. This is mainly because
(*a*) synchronous motor has no slip
(*b*) stator supply is not required to produce magnetic field
(*c*) mechanical load on the rotor remains constant
(*d*) synchronous motor has large air-gap.

34. A synchronous motor working at leading power factor can be used as
(*a*) voltage booster (*b*) phase advancer
(*c*) noise generator (*d*) mechanical synchronizer.

35. Slip rings are usually made of
(*a*) carbon or graphite (*b*) brass or steel
(*c*) silver or gold (*d*) copper or aluminium.

36. An over-excited synchronous motor is used for
(*a*) fluctuating loads (*b*) variable speed load
(*c*) low torque loads (*d*) power factor corrections.

37. When the voltage applied to a synchronous motor is increased, which of the following will reduce ?
(*a*) Stator flux (*b*) Pull in torque.
(*c*) Both (*a*) and (*b*) (*d*) None of the above.

38. The speed of the synchronous motor
(*a*) increases with increase in load (*b*) decreases as load increases
(*c*) varies with power factor (*d*) always remains constant.

39. Dust and dirt accumulation on motor winding will result in
(*a*) rise in winding temperature (*b*) increased core losses
(*c*) shorting of coils (*d*) eddy current flow.

40. The efficiency of a properly designed synchronous motor will usually fall in range
(*a*) 60 to 70% (*b*) 75 to 80%
(*c*) 85 to 95% (*d*) 99 to 99.5%.

41. To limit the operating temperature an electrical machine should have proper
(*a*) voltage rating (*b*) current rating
(*c*) power factor (*d*) speed.

42. Slip-rings in a synchronous motor carry
(*a*) direct current (*b*) alternating current
(*c*) no current (*d*) all of the above.

43. The clutch method of starting an induction motor is used
(*a*) to save energy in starting (*b*) when starting torque is high
(*c*) for high speed synchronous motors (*d*) for low voltage synchronous motors.

ANSWERS

1. (*b*)	**2.** (*a*)	**3.** (*b*)	**4.** (*a*)	**5.** (*a*)	**6.** (*c*)	**7.** (*d*)
8. (*a*)	**9.** (*b*)	**10.** (*a*)	**11.** (*a*)	**12.** (*a*)	**13.** (*c*)	**14.** (*d*)
15. (*d*)	**16.** (*a*)	**17.** (*c*)	**18.** (*b*)	**19.** (*c*)	**20.** (*d*)	**21.** (*a*)
22. (*c*)	**23.** (*c*)	**24.** (*a*)	**25.** (*b*)	**26.** (*d*)	**27.** (*d*)	**28.** (*c*)
29. (*c*)	**30.** (*b*)	**31.** (*b*)	**32.** (*a*)	**33.** (*b*)	**34.** (*b*)	**35.** (*b*)
36. (*d*)	**37.** (*d*)	**38.** (*d*)	**39.** (*a*)	**40.** (*c*)	**41.** (*b*)	**42.** (*a*)
43. (*b*).						

(B) 'Yes' or 'No' :

1. An over-excited synchronous motor operates at unity p.f.
2. The amortisseur winding in a synchronous motor provides starting torque and eliminates hunting.
3. Under full-load running condition the slip of a synchronous motor is zero.
4. When the synchronous motor is to be started the field system is energised in the very beginning.
5. When a synchronous motor is started, the field winding is initially short-circuited.
6. V-curves of a synchronous motor show the relation between applied voltage and field current.
7. For a given load, the armature current of a synchronous motor is minimum when the power factor is unity.
8. Synchronous motors generally have salient pole rotor.
9. Synchronous motors are usually operated at leading power factor.
10. A salient pole synchronous motor is running at no-load. Its field current is switched off. The motor will come to stop.
11. As the load on a synchronous motor is increased the torque angle increases.
12. Synchronous motors are used in all sizes.
13. An under-excited synchronous motor operates at unity p.f.
14. A synchronous motor always runs at synchronous speed because of magnetic locking between the stator and rotor.
15. In a synchronous motor the generated e.m.f. can be less than or greater than the terminal voltage.

ANSWERS

1. No.	**2.** Yes	**3.** Yes	**4.** No	**5.** Yes.	**6.** No	**7.** Yes
8. Yes	**9.** Yes.	**10.** No	**11.** Yes	**12.** No	**13.** No.	**14.** Yes
15. Yes.						

THEORETICAL QUESTIONS

1. Is the synchronous motor self-starting ? If not, what are the methods adopted for starting it ?
2. What is the difference between a synchronous motor and an induction motor ?

3. State the characteristic features of a synchronous motor.
4. What are the advantages and disadvantages of a synchronous motor ?
5. What are the applications of a synchronous motor ?
6. Does the speed of the synchronous motor change when loaded ? If not explain how the motor takes load ?
7. What is meant by *V*-curves of a synchronous motor ?
8. What happens when the D.C. excitation of the D.C. machine is changed ? Is the effect same in a synchronous motor ?
9. What is meant by '*hunting*' in a synchronous motor ?
10. Draw and explain the phasor diagram of a synchronous motor operating at :
    (*i*) lagging power factor (*ii*) leading power factor.
11. Explain the purpose of using damper windings in synchronous machines.
12. Describe the method of starting a synchronous motor. Also discuss the effect of varying excitation of synchronous motor under constant load. Explain the circumstances leading a synchronous motor to work as an ideal synchronous condenser.
13. Explain the construction and principle of operation of a synchronous motor.
14. Show that mechanical power developed/phase of a synchronous motor is given by

$$P_{mech}/\text{phase} = \frac{E_b V \cos(\theta - \alpha)}{Z_s} - \frac{E_b^2 \cos\theta}{Z_s}.$$

where V, E_b, Z_s and α are the applied voltage, excitation voltage, synchronous impedance and torque angle respectively, and $\theta = \sin^{-1}\frac{X_s}{Z_s}$, where X_s is the synchronous reactance.

15. What is the significance of torque angle with reference to synchronous motor ? What torque angle gives maximum torque ?
16. What are *V*-curves ? How are they determined experimentally ?
17. Show that the current locus of a synchronous motor developing constant power is a circle. Determine its centre and radius.
18. Write short notes on the following :
    (*i*) *V*-curves of synchronous motor.
    (*ii*) Synchronous condenser.
    (*iii*) Power factor improvement with synchronous motors.
    (*iv*) Hunting of synchronous motors.

UNSOLVED PROBLEMS

1. A 500 V, single-phase synchronous motor gives a net output mechanical power of 7.46 kW and operates at 0.9 p.f. lagging. Its effective resistance is 0.8 Ω. If the iron and mechanical losses are 500 W and excitation losses are 800 W, estimate the armature current. Calculate the commercial efficiency. **[Ans.** 18.3 A, 82.56%**]**
2. A 2300 V, 3-phase, star-connected synchronous motor has a resistance of 0.2 Ω/phase and a synchronous reactance of 2.2 Ω/phase. The motor is operating at 0.5 p.f. leading with a line current of 200 A. Determine the value of the generated e.m.f./phase. **[Ans.** 1708 V/phase**]**
3. A 6600 V, 3-phase, star-connected synchronous motor draws a full- load current of 70 A at 0.8 p.f. leading. Find the e.m.f. induced, output power and efficiency of the machine if the stray losses total to 30 kW ; and the armature resistance is 2 Ω and reactance 20 Ω/phase. **[Ans.** 4696 V/phase, 580.6 kW, 90.72%**]**

4. The input to a 11000 V, 3-phase, star-connected synchronous motor is 50 A. The effective resistance and synchronous reactance per phase are 0.95 Ω and 29 Ω respectively. Calculate the power supplied to the motor and the induced e.m.f. for a power factor of :
(*i*) 0.8 lagging (*ii*) 0.8 leading.
[**Ans.** (*i*) 9635 V, 762.1 kW, (*ii*) 12610 V 762.1 kW]

5. A 200 V, 3-phase star-connected synchronous motor has an effective resistance and synchronous reactance of 0.2 Ω and 2.2 Ω respectively. The input is 800 kW at normal voltage and the induced line e.m.f. is 2500 V. Calculate the line current and power factor. [**Ans.** 254.15 A, 0.909 (lead)]

6. A 3-phase star-connected synchronous motor has a synchronous reactance of 4 Ω per phase and is working on 1100 volts. Calculate the power factor of the machine when taking 90 kW from the mains, the excitation being adjusted to a value corresponding to an induced e.m.f. of 1200 volts.
Neglect armature resistance. [**Ans.** 0.986 A]

7. A synchronous motor takes 25 kW from 400 V supply mains. The synchronous reactance of the motor is 4 Ω. Find the power factor at which the motor would operate when the exciting current is so adjusted that the generated e.m.f. is 500 volts. [**Ans.** 0.9914 leading]

8. A 2500 V, 3-phase, 4-pole star-connected synchronous motor runs at 1500 r.p.m. The excitation is constant and corresponding to an open-circuit voltage of 2500 V. The resistance is negligible in comparison with synchronous reactance of 3 Ω/phase. Determine the power input, power factor and torque developed for an armature current of 250 A. [**Ans.** 1045.3 kW, 0.9656 (lag), 6655 N-m]

9. A 7.355 kW, 400 V, 3-phase star-connected synchronous motor has synchronous impedance per phase of (0.33 + *j*2.8) Ω. Find the angle of retard and the voltage to which motor must be excited to give a full-load output at 0.866 leading power factor.
Assume an efficiency of 88%. [**Ans.** 431 V (line), 8.36°]

10. A 400 V, 6-pole, 3-phase, 50 Hz, star-connected synchronous motor has a resistance and synchronous impedance of 0.5 Ω and 4 Ω per phase respectively. It takes a current of 15 A at unity p.f. when operating with a certain field current. If the load torque is increased until the line current is increased to 60 A, the field current remaining unchanged, calculate the gross torque developed and the new power factor.
[**Ans.** 327 N-m ; 0.954 (lead)]

11. A 6600 V, 3-phase, star-connected synchronous motor draws a full-load current of 70 A at 0.8 p.f. leading. Find the e.m.f. induced and efficiency of the machine if the stray losses total 30 kW and the armature resistance is 2 Ω and the reactance is 20 Ω per phase. [**Ans.** 8130 V (line), 90.7%]

12. An 8-pole, 3300 V, 50-Hz, 3-phase, start-connected synchronous motor has armature resistance of 0.5 Ω per phase and synchronous reactance of 5 Ω/phase. While running on no-load, the excitation has been adjusted so as to make the e.m.f. numerically equal to and antiphase with the terminal voltage. With a certain load torque applied if the rotor gets retarded by 3 mechanical degrees calculate the armature current and the power factor of the motor. [**Ans.** 79.6 A, 0.9992 (lag)]

13. A 2300 V, 3-phase, 50 Hz, star-connected synchronous motor has a synchronous reactance of 11 Ω per phase. When it delivers 147.1 kW the efficiency is found to be 90% and the power angle is 15 electrical degrees. Neglecting ohmic resistance determine :
(*i*) Inducted excitation voltage/phase, and (*ii*) Line current.
[**Ans.** (*i*) 3018 V (line value), 52.18 A]

14. The excitation of a 415 V, 3-phase, mesh connected synchronous motor is such that the induced e.m.f. is 520 V. The impedance per phase is (0.5 + *j*4.0) Ω. If the friction and iron losses are constant at 1000 W, calculate line current, power factor and efficiency for maximum power output.
[**Ans.** 268.2 A, 0.89, 78.46%]

15. A 3-phase, star-connected synchronous motor takes at 0.8 p.f. (lag) full-load current of 116 A at 500 V. The total armature copper loss at this current is 1200 W. The reactance per phase of the motor is 0.3 Ω. Calculate the internal power developed, net mechanical power developed and e.m.f. generated by the motor. Assume 95% efficiency at full-load. [**Ans.** 79170 W, 76351 W, 269 V]

16. A 400 V, 50 Hz, 3-phase, 37.3 kW star-connected synchronous motor has a full-load efficiency of 88%. The synchronous impedance of the motor is $(0.2 + j1.6)\ \Omega$ per phase. If the excitation of the motor is adjusted to give a leading p.f. of 0.9 calculate for full-load :

(*i*) The inducted e.m.f, and (*ii*) the total mechanical power developed.

[**Ans.** (*i*) 495 V (line value). (*ii*) 39.3 kW]

17. The input to a 1100 V, 3-phase star-connected synchronous motor is 60 A. The effective resistance and synchronous reactance per phase are respectively 1 Ω and 30 Ω. Find the power supplied to the motor and the induced e.m.f. for p.f. of 0.8 leading. [**Ans.** 913 kW, 13090 V]

18. A 3-phase, 2200 V synchronous motor has a synchronous reactance of 10 Ω/phase and negligible resistance when the input is 90 kW and p.f. is 0.8 leading. Estimate the current and power factor when the input to the motor is increased to 495 kW. Assume that the motor works at constant voltage and constant current. [**Ans.** 135.6 A, 0.957 (lead)]

Synchronous Condensers

19. A 500 V, 50-Hz, 3-phase circuit takes 20 at a lagging p.f. of 0.8. A synchronous motor is used to raise the power factor to unity. Calculate the kVA input to the motor, and its power factor when driving a mechanical load of 7.355 kW (10 h.p.). The motor has an efficiency of 85%.

[**Ans.** 13.523 kVA, 0.64 (lead)]

20. An alternator supplying 800 kW at 0.8 power factor lagging has its power factor raised to unity by means of an overexcited synchronous motor. With the armature current unchanged, how much power can the alternator thus supply to the synchronous motor and what power in kW can the latter develop when operating at an efficiency of 90%. Find at what power factor synchronous motor runs.

[**Ans.** 200 kW, 180 kW]

21. A substation operating at its full-load of 1000 kVA supplies a load at 0.71 power factor lagging. Calculate the permissible additional load at this power factor and the rating of synchronous condenser to raise the substation power factor to 0.87 lagging. [**Ans.** 225.35 kVA, 370 kVAR]

22. An industrial load of 4000 kW is supplied at 11000 volts, the power factor being 0.8 lagging. A synchronous motor is required to meet an additional load of 1103.25 kW (1500 h.p.) and at the same time to raise the resultant power factor to 0.95 lagging. Determine the kVA capacity of the motor and the power factor at which it must operate. The efficiency of the motor is 80%. [**Ans.** 1850 kVA, 0.746 (leading)]

7

Converters and Rectifiers

7.1. GENERAL

Although A.C. system has been universally adopted for generation, transmission and distribution yet there are several cases where the use of D.C. current is either essential or else is advantageous to a degree sufficient to make it preferable to A.C. current. A few *applications of D.C. system* are given below :

- D.C. motors (because of their excellent and over-load characteristics) are most suitable for *traction purposes in tramways and railways.*
- Arc welding.
- Battery charging work.
- Arc lamps for searchlights and cinema projectors.
- For operating relays, telephones, switches, and circuit breakers etc.
- In rolling mills, paper mills, colliery winding, etc., where fine speed control or speeds in both directions are required.
- For electrolytic and electro-chemical processes.

Hence, it becomes necessary/essential sometimes to convert A.C. current into D.C. current by some suitable arrangement. The following methods may be used to convert A.C. system to D.C. :

1. Motor-generator sets.
2. Motor-converters.
3. Rotary converters.
4. Rectifiers.

7.2. MOTOR-GENERATOR SET

A motor-generator set consists of two machines *viz.* an A.C. motor and a D.C. generator which are mechanically coupled. A.C. motor may be synchronous or an induction motor. In the case of larger

units, the A.C. motor is invariably synchronous and the D.C. generator is usually compound. The synchronous motor if employed should have damper winding on its pole faces to avoid hunting.

- **For starting synchronous motor D.C. supply is used. If D.C. supply is not available then auxiliary starting methods are employed.**
- **In case of an induction motor any one of the starting methods (already discussed) can be employed. Fig. 7.1 shows a slip-ring induction motor coupled with a D.C. generator.**

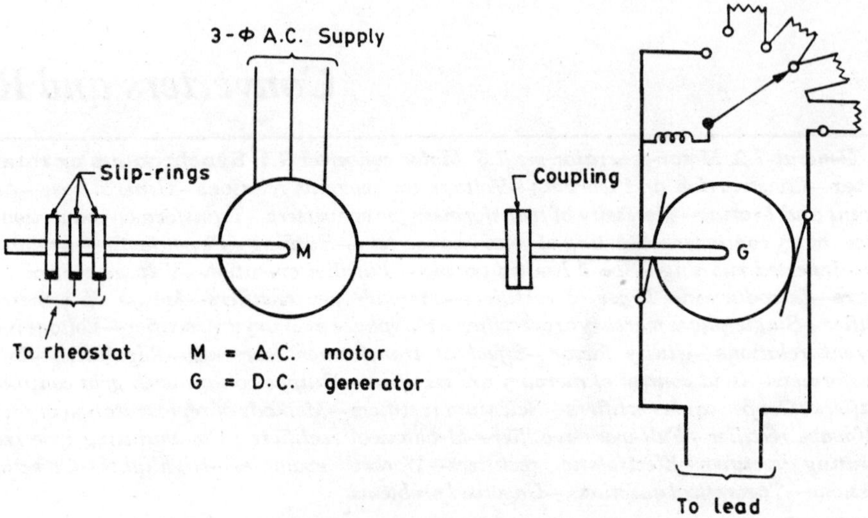

Fig. 7.1. Motor generator set.

- **Because of its better operation, synchronous motor generator set is usually preferred than the induction motor generator set.**

Following are the *advantages and disadvantages of a motor generator set*.

Advantages :

1. The motor-generator set is *simple* and *reliable*.

2. D.C. voltage at nearly constant magnitude can be obtained from synchronous motor generator set.

3. D.C. voltage can be obtained between wide ranges.

4. The synchronous motor generator set is reversible.

5. The synchronous motor generator set can be employed for improving the power factor.

6. The set can be operated on any frequency.

7. The set is free from reversal of polarity and flashover.

Disadvantages :

1. Since a motor-generator set requires two machines, therefore, it occupies more space, costs more and has poor efficiency.

2. The set is very heavy and cannot be easily moved from one place to another place.

3. Induction motor generator set is not reversible.

4. Induction motor-generator set operates at low power factor unless it is compensated.

5. At very high load, the synchronous motor generator set falls out of synchronism.

6. The synchronous motor set is liable to hunting and requires a damping winding on its poles faces.

CONVERTERS

7.3. MOTOR CONVERTER

A motor converter essentially consists of an ordinary slip-ring induction motor coupled both mechanically and electrically to a D.C. generator. The rotor of the induction motor is usually wound for 12 phases to obtain better performance. A.C. voltage up to 11 kV can be directly fed to the stator. If A.C. voltage available is more than 11 kV then it is economical to step down the voltage before feeding it to the stator winding. The rotor of the slip-ring induction motor is mechanically as well as electrically coupled to D.C. generator. D.C. generator is also known as *armature of the converter*. The motor input energy is partly transmitted mechanically (through the shaft) and partly electrically *i.e.*, as slip ring energy from the rotor of the induction motor. Fig. 7.2 shows the motor-converter connections.

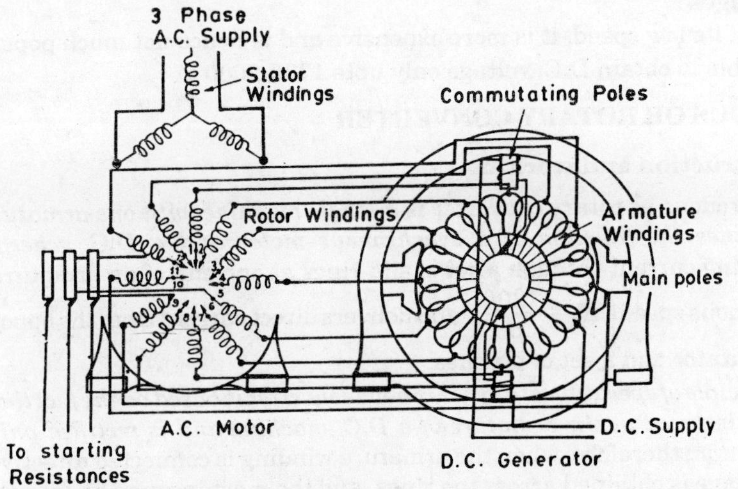

Fig. 7.2. Motor converter connections.

Since the secondary output of the induction motor is fed directly, or cascaded, into the armature of the rotary converter, therefore, this machine is sometimes also known as *cascade converter*. If p_a and p_d represent the number of poles on the A.C. and D.C. ends then it can be shown that :

- The set runs at cascade speed *i.e.*, the synchronous speed corresponding to the sum of the poles of both machines.

$$N = \frac{120f}{(p_a + p_d)}.$$

- The frequency of currents in D.C. generator is

$$= f \times \frac{p_d}{p_a + p_d}.$$

- Mechanical power transmitted through shaft

$$= \text{input} \times \frac{p_a}{p_a + p_d}.$$

- Power transmitted electrically

$$= \text{input} \times \frac{p_d}{p_a + p_d}.$$

The motor converter possesses characteristics which lie in between those of motor-generator set on one side and rotary converter on the other.

Following are the advantages and disadvantages of the motor converter set :

Advantages :

1. It is self-starting.
2. The set is simple to operate and reliable.
3. It has sufficiently high power factor.
4. It permits of wide regulation in D.C. voltage.
5. It can be operated at high voltages.
6. It can be operated on any frequency.

Disadvantages :

1. Because of its low speed, it is more expensive and is hence not much popular.
2. It is possible to obtain D.C. voltage only upto 1700-2000 V.

7.4. SYNCHRONOUS OR ROTARY CONVERTER

7.4.1. Construction and working

- A synchronous or rotary converter is *a single machine with one armature and one field. It combines the function of a synchronous motor and a D.C. generator*. It receives alternating current through a set of slip-rings at one end of an armature that rotates at synchronous speed $\left(N_s = \frac{120f}{p}\right)$ and delivers direct current from the opposite end through a commutator and a set of brushes.
- *The principle of operation of a synchronous converter is based on the fact that an alternating e.m.f. is induced in the armature of a D.C. machine and is rectified only by means of a commutator*; therefore, when the armature winding is connected directly to slip-rings, an A.C. voltage is obtained across the rings, and the machine may be connected through the latter to an A.C. circuit.
- Normally, a synchronous or rotary converter is used when a *large-scale conversion from A.C. to D.C. power is required.*

In addition to performing the above function, this machine may be :

(*i*) driven as a D.C. or A.C. generator ;

(*ii*) run as a synchronous or D.C. motor ; and

(*iii*) run as an inverter *i.e.*, for changing D.C. into A.C.

- The rotary converter is reversible in action. It can receive power from a D.C. supply mains and convert it into A.C. power. When thus used, it is called an *inverted rotary converter*.

Construction. Fig. 7.3 shows the main parts of a synchronous converter. A synchronous or rotary converter, in general construction and design is more or less like any D.C. machine. It consists of :

- Stationary field magnet system.
- A rotating armature.
- A commutator and a set of brushes (which is larger than that of a D.C. generator of the same size because the D.C. output of converter is greater than for a corresponding D.C. generator).
- A shunt field, usually a series field and interpoles to help the commutation.

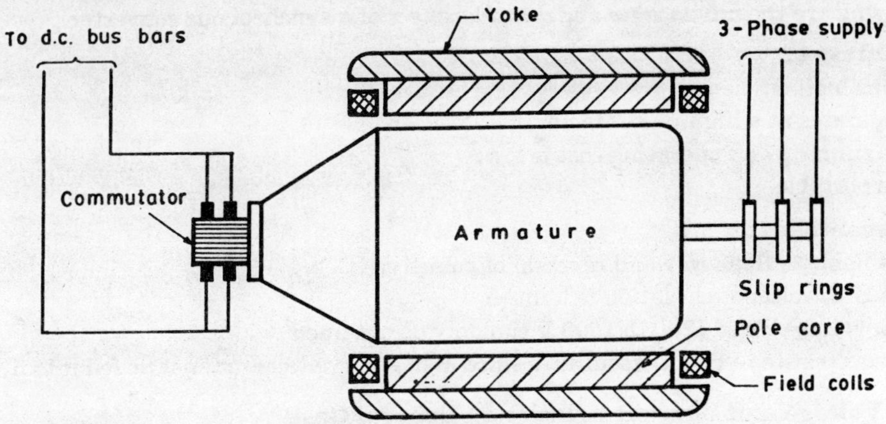

Fig. 7.3. Synchronous converter.

- Slip-rings mounted on the shaft on the opposite side of the commutator. The tappings from the armature winding are taken to slip-rings as well as to commutator.

A synchronous converter can be operated as a synchronous motor of the rotating armature type by connecting slip-rings to A.C. supply mains and keeping the D.C. brushes open circuited. On the other hand this machine can be operated as D.C. shunt or compound motor by connecting commutator and brushes to D.C. supply mains and disconnecting the slip-rings.

When a synchronous converter is driven mechanically and current drawn from commutator only, from slip-rings and from both simultaneously, then the machine is known as *D.C. generator, alternator* and *double current generator* respectively.

Rules for tapping the armature winding :

Following are the *rules* for tapping the armature windings :

1. Wave-wound armature. In a wave-wound armature, irrespective of number of poles, there are only two parallel paths. There is one tapping taken to each slip-ring. For a single-phase supply, the number of slip rings is two and there will be an equal number of armature conductors-half the total number of armature conductors, in each path from one tapping point to the other. In case of a 3- phase supply there will be three slip-rings and three equidistant tappings dividing the armature into three equal parts and so on.

Wave winding is suitable for small machines and is *rarely* used.

2. Lap-wound armature. In a lap-wound armature, the number of parallel paths in armature is equal to the number of poles. Therefore, the number of equi-potential points on the armature is equal to the number of pair of poles. The number of tappings taken to each slip-ring, is therefore, equal to the number of pair of poles. *The total number of tappings is equal to the product of number of slip-rings and the number of pair of poles. Example :* In case of a 3-phase, 6-pole synchronous converter there will be 3 slip-rings and 3 taps for each slip-ring, making a total of 9 taps to the winding.

Note. For machines whose output is less than 100 kW it is the usual practice to adopt a wave winding. When wave winding is used, only one connection is required per slip-ring. But when lap, or multiple circuit, winding is adopted, then each slip-ring will have connections equal to the number of pole pairs. *A simple rule to remember is that if a converter has N number of phases, there must be N slip-ring tapping point per pole pair.* This rule does not apply to single-phase machine.

Following are the *advantages* and *disadvantages* of a synchronous converter :

Advantages :

1. It has high efficiency.
2. It operates at a high power factor.
3. The running and operating cost is low.
4. It is portable.

Disadvantages :

1. It is liable to flashover and reversal of polarity.
2. The D.C. voltage regulation is limited.
3. D.C. voltage up to 1200 to 1500 V can only be obtained.
4. It requires step-down transformer since high A.C. voltages cannot be fed into it.

7.4.2. Voltage and Current Relations—General Case

Let us derive expressions for voltage and current relations for an m-phase, 2-pole converter having m-rings. By giving desired values to m, values for any number of phases can be found. In Fig. 7.4, circle representing the armature also represents the voltage vector diagram.

(*i*) **Voltage Relations :**

Refer Fig. 7.4.

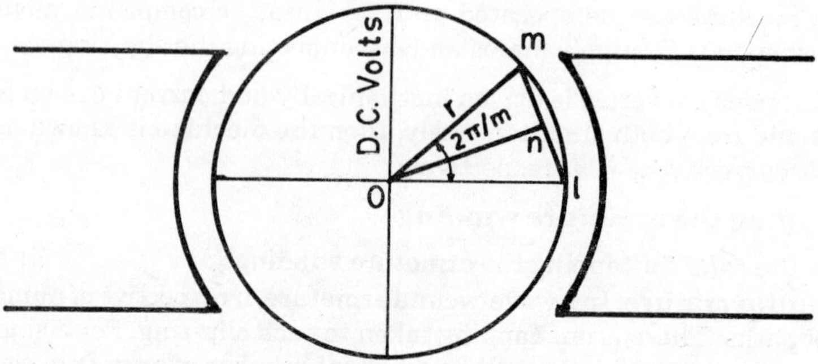

Fig. 7.4

Angle between two adjacent tappings $= \frac{2\pi}{m}$ radians.

Voltage between adjacent tappings l and m,

$$E_{p(\max)} = \text{chord } lm \text{ (= maximum A.C. voltage between the adjacent rings)}$$

D.C. brush voltage, $E_{dc} = 2r$ (diameter)

$$E_{p(\max)} = lm = 2ln = 2r \sin\left(\frac{2\pi}{2m}\right) = 2r \sin\frac{\pi}{m} = E_{dc} \sin\frac{\pi}{m}$$

$$\therefore \quad E_p = \frac{E_{dc}\sin\frac{\pi}{m}}{\sqrt{2}} \quad \text{...[7.1 (a)]}$$

$$\therefore \quad \frac{E_p}{E_{dc}} = \frac{\text{r.m.s. value of armature phase voltage}}{\text{D.C. output voltage}}$$

i.e.,

$$\frac{E_p}{E_{dc}} = \frac{\sin\frac{\pi}{m}}{\sqrt{2}} \quad \text{...[7.1 (b)]}$$

***(ii)* Current Relations :**

Let E_p = r.m.s. phase voltage *i.e.*, potential difference, across two armature tappings or E_{sr}

I_p = r.m.s. phase current *i.e.*, A.C. current flowing through the armature between adjacent tappings. It is *not* the same as I_{sr} *i.e.*, load current.

A.C. power input = $mE_pI_p \cos \phi$

Now A.C. line current at slip-ring, I_{sr} is vector difference of phase currents. The phase difference between these two phase currents is $\frac{2\pi}{m}$ radians.

As shown in Fig. 7.5.

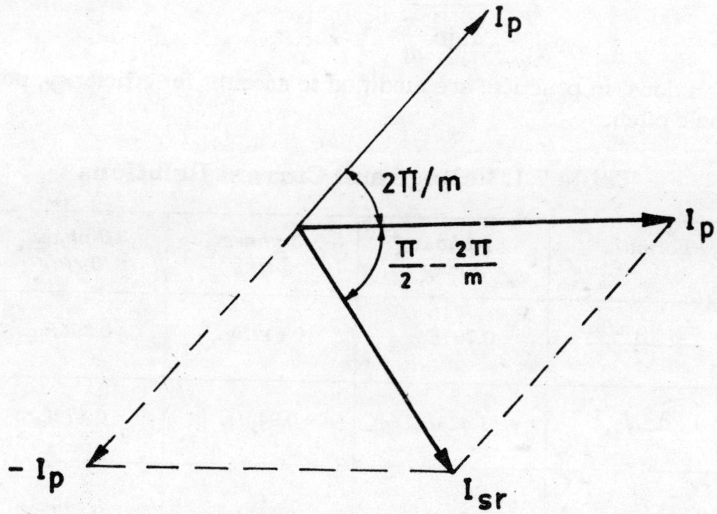

Fig. 7.5

$$I_{sr} = 2I_p \cos\left(\frac{\pi}{2} - \frac{\pi}{m}\right) = 2I_p \sin\frac{\pi}{m}$$

$$\therefore \quad I_p = \frac{I_{sr}}{2 \sin \frac{\pi}{m}} \qquad \text{...(7.2)}$$

Now A.C. input $= mE_pI_p \cos \phi$

Substituting the values of E_p and I_p as found above, we have

$$\text{A.C. input} = m \times \frac{E_{dc} \sin \frac{\pi}{m}}{\sqrt{2}} \times \frac{I_{sr}}{2 \sin \frac{\pi}{m}} \times \cos \phi$$

$$= \frac{m}{2\sqrt{2}} E_{dc}\, I_{sr} \cos \phi \qquad \text{...(7.3)}$$

D.C. output $= E_{dc}\, I_{dc}$

Assuming 100% efficiency,

$$\frac{m}{2\sqrt{2}} E_{dc}\, I_{sr} \cos \phi = E_{dc} \cdot I_{dc}$$

$$\therefore \quad I_{sr} = \frac{2\sqrt{2}}{m} I_{dc} \sec \phi$$

$$= \frac{2\sqrt{2}}{m} I_{dc} \text{ (assuming unity power factor)}$$

$$= \frac{2.83}{m} I_{dc}$$

The relation between I_p and I_{dc} can be found to be,

$$I_p = \frac{\sqrt{2} I_{dc}}{m \sin \frac{\pi}{m}} \qquad ...(7.4)$$

In general, it is seen from above that

$$I_p = \frac{I_{sr}}{2 \sin \frac{\pi}{m}}$$

The above relations, in practice, are modified to account for efficiency, power factor and the ratio, the pole arc/pole pitch.

Table 7.1. Voltage and Current Relations

	Formula	*1-phase 2-ring*	*3-phase 3-ring*	*6-phase 6-ring*	*12-phase 12-ring*
E_p or E_{sr}	$\frac{E_{dc} \sin \pi/m}{\sqrt{2}}$	$0.707E_{dc}$	$0.612E_{dc}$	$0.354E_{dc}$	$0.182E_{dc}$
I_{sr} current between leads	$\frac{2\sqrt{2}}{m} I_{dc}$	$1.414I_{dc}$	$0.943I_{dc}$	$0.472I_{dc}$	$0.236I_{dc}$
I_p current between two armature taps	$\frac{\sqrt{2} I_{dc}}{m \sin \pi/m}$	$0.707I_{dc}$	$0.544I_{dc}$	$0.472I_{dc}$	$0.455I_{dc}$

7.4.3. Armature Current and Heating

- *The resultant current in the armature winding of a rotary converter is the algebraic sum of direct and alternating currents flowing at the instant under consideration. This current varies from instant to instant and also differs in different conductors, depending upon their proximity to the slip-ring tappings.*
- Let us consider a single-phase, bipolar case. The A.C. component of the armature current reaches a maximum when the conductors connected to the slip-ring tappings *come under the commutator brushes*, neglecting losses and assuming unity power-factor, and at this same instant the D.C. component of current reverses. These two components are shown graphically in Fig. 7.6 (*a*), together with their resultant. The r.m.s. value of the A.C. component is $\sqrt{2}$ times and the maximum value double that of the D.C. component, so that the resultant current momentarily reaches a peak value equal to *three times* that of the D.C. component. The r.m.s. value of the resultant current is considerably in excess of that of D.C. component above, but this does not apply to all other conductors. For example, the conductors situated midway between two slip-ring tappings also have maximum A.C. flowing through them at the same instant that maximum current flows through all other conductors, but in this case the armature has to rotate through a further 90° (electrical), or half a pole pitch, before the D.C. component reverses. These conditions are

represented in Fig. 7.7, which also shows the graph of the resultant current. It is obvious, from inspection, that the r.m.s. value of this current is *much smaller* than r.m.s. value of the current shown in Fig. 7.6. The most complete neutralization occurs at those conductors situated midway between the slip-ring tappings, and these conductors carry the minimum resultant current. As the slip-ring tappings are approaches the r.m.s. value of the resultant current increases until it attains a maximum at the conductors nearest to the tappings themselves. *The heating of the armature conductors is thus different at different points relative to the slip-ring tappings. Those conductors nearest to the tappings generate the most heat, and those conductors midway between tappings generate the least heat.*

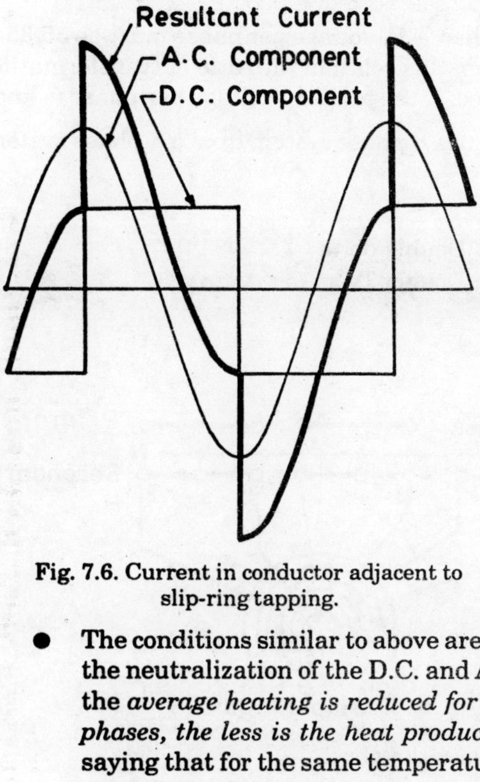

Fig. 7.6. Current in conductor adjacent to slip-ring tapping.

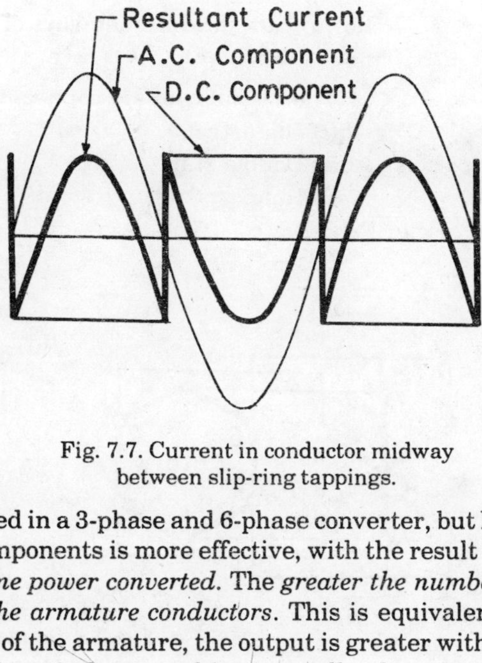

Fig. 7.7. Current in conductor midway between slip-ring tappings.

- **The conditions similar to above are obtained in a 3-phase and 6-phase converter, but here the neutralization of the D.C. and A.C. components is more effective, with the result that the *average heating is reduced for the same power converted.* The *greater the number of phases, the less is the heat produced in the armature conductors.* This is equivalent to saying that for the same temperature rise of the armature, the output is greater with the greater number of slip-rings, or, conversely, a 6-phase machine is *smaller* than one with fewer phases. With a *still larger number of phases the additional benefits derived are outweighted by the complications due to extra number of slip-rings.***

7.4.4. Necessity of Transformers for Converters. The use of transformers between the A.C. source and slip-rings is essential for the operations of polyphase synchronous converters due to reasons given below :

- A definite theoretical relation between D.C. output voltage and the input A.C. voltage—the latter usually having an odd value which in most cases does not match any standard distribution network.
- The A.C. source voltage is usually much higher than the A.C. voltage required at slip-rings.
- In the case of 6- and 12-phase converters it is necessary to change 3-phase supply into 6- and 12-phase respectively. This can be accomplished by properly connected transformers.

7.4.5. Transformer Connections for Six-phase Rotary Converter. Normally the rotary converters are 6-phase type, owing to their improved performance. Hence, the various types of transformer connections which convert a 3-phase A.C. system into a 6-phase A.C. system, are discussed in this section.

- The most popular and common connection is the *diametral,* which is shown in Fig. 7.8. In this method of conversion, from 3-phase to 6-phase, ordinary 3 single-phase transformers or one 3-phase, transformer can be used. This is not the case with other methods of transformation from 3-phase to 6-phase, wherein the secondary has two similar windings per phase.

  If D.C. voltage is assumed as 100 V, then A.C. voltage per phase must be 35.35 V in the case of a 6-phase rotary converter. From this relation the ratio of transformation or the turns ratio per phase is easily calculated, if the primary voltage per phase is known.

- Following are the different ways of obtaining a 6-phase system from a 3-phase system by using transformers :

  The secondaries are connected in :

  (*i*) Diametral (*ii*) Double-delta
  (*iii*) Double-star (*iv*) Double-*T*
  (*v*) Ring.

Refer Figs. 7.8 to 7.10.

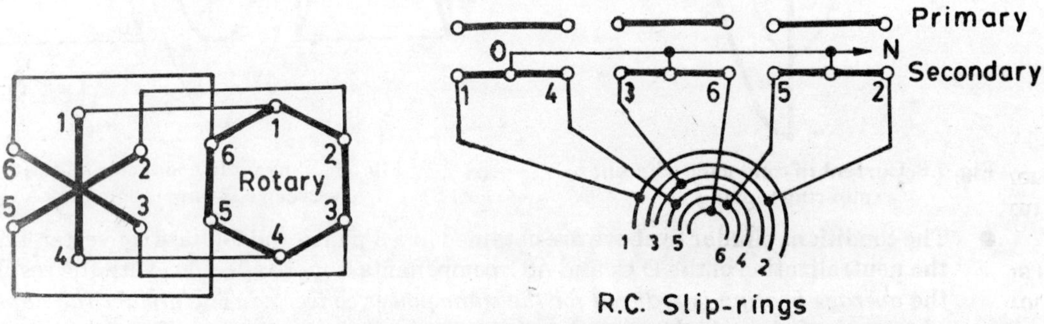

Fig. 7.8. Diametral, 3-phase to 6-phase.

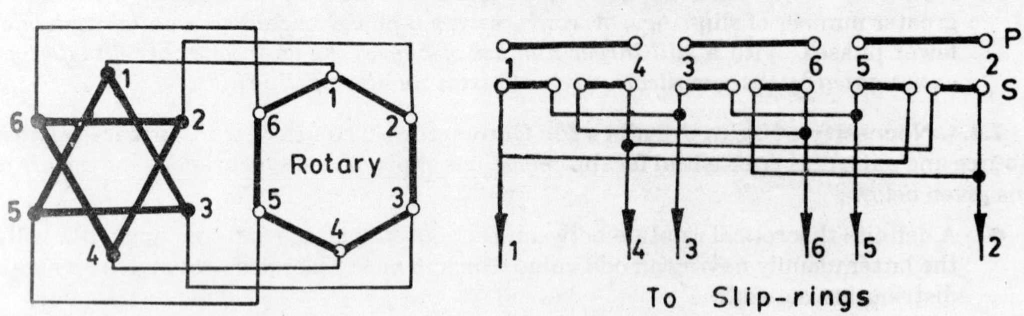

Fig. 7.9. Double-delta, 3-phase to 6-phase.

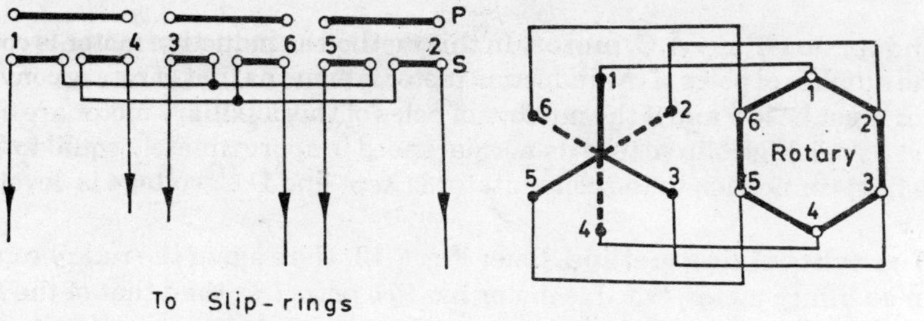

Fig. 7.10. Double-star, 3-phase to 6-phase.

Transformation from 2-phase to 6-phase can be accomplished by the double-T connection as shown in Fig. 7.11.

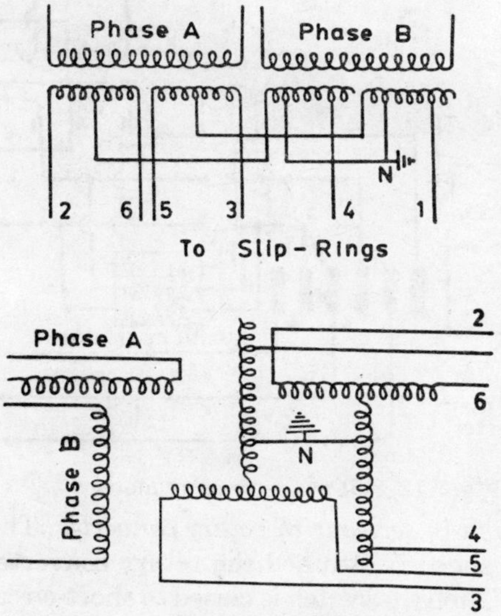

Fig. 7.11. Double-***T***, 2-phase to 6-phase.

The following relations are worth noting :

(*i*) **Diametral connection :**

(*a*) Voltage across each secondary = 0.707 E_{dc} $\left(= \dfrac{E_{dc}}{\sqrt{2}} \right)$.

(*b*) Voltage across two armature tappings = 0.354 E_{dc}.

(*ii*) **Double-delta connection :**

(*a*) P.d. across each secondary = 0.612E_{dc}.

(*b*) Phase e.m.f. = 0.354 E_{dc}.

(*iii*) **Double-star connection :**

Transformer secondary voltage per phase = 0.354 E_{dc}.

7.4.6. Starting of Rotary Converters. Rotary converters can be started by two methods :

1. From A.C. side
2. From D.C. side.

7.4.6.1. Starting of Rotary Converter from A.C. Side. Following are the methods of starting rotary converter from A.C. side :

(*i*) **Using an auxiliary A.C. motor.** In this method as induction motor is coupled to a rotary converter. If the number of poles of the induction motor is same as that of rotary converter the normal slip of the motor must be low and if the number of poles of this auxiliary motor are equal or less than this motor must have a high slip so that its normal speed is approximately equal to that of the rotary converter. The field connection to the commutator is kept and D.C. voltage is developed just as in a shunt generator.

(*ii*) **Self-synchronizing method.** Refer Fig. 7.12. Here again the rotary converter is started by means of an auxiliary motor, but this motor has *two poles less* than that of the rotary converter.

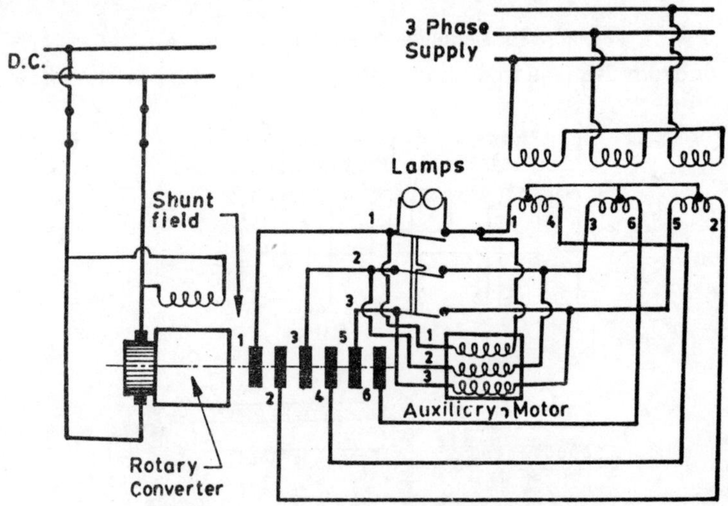

Fig. 7.12. Self-synchronizing method.

The *stator of the motor is connected in series with rotary converter*. This motor brings the speed to synchronism and as soon as the speed is attained the rotary converter jumps in step as both the machines have the same current. Then the switch is closed to short-circuit the stator of the induction motor.

In this method the brushes are not lifted and there is no possibility of developing the D.C. voltage in opposite direction since the rotating field produced due to current in the rotary armature is not so strong to neuralize the residual magnetism of the salient poles and due to the same reason there is no need of field break up switch.

(*iii*) **Started as an inverted induction motor :**

- This method of starting rotary converter is also called as *changing taps on the main transformer method.*
- In order to reduce the hunting of the rotary converter the main poles are slotted, into these slots are placed thick conductors shorted as both ends, such a winding is called *damper winding.*
- For starting, the bushes on the D.C. side are lifted and very reduced voltage is applied to the slip-ring. The three-phase current when it flows in the armature produces a rotating field which cuts the damper winding of the synchronous machine and thus the *rotary converter starts as an inverted induction motor*. Hence, at starting the rotating field cuts the shunt field conductors in the main poles with synchronous speed, so an enormous e.m.f. is induced in this winding, hence to limit this generated e.m.f. the winding must be sectionalized.

The current on the commutator side is not of constant nature but pulsates at slip-frequency ; that is why the brushes are lifted off the commutator, otherwise the excitation will be a pulsating one. When the speed of the rotary converter reaches the synchronous speed the relative speed of the rotor with respect to the rotating field becomes zero and appears that the rotor field comes from the salient poles and the current on the commutator side becomes of direct nature and the rotary converter is said to have *pulled in step*. The polarity of the brushes attained is quite arbitrary and depends upon the relative position of the field and armature at the instant of pulling into step. So it is possible that the D.C. voltage may develop in a wrong direction which is indicated by the centre zero voltmeter. In order that the voltage is developed in the correct direction the field for a moment is reversed and again it is brought in the proper direction, which will cause the D.C. voltage to develop in correct direction.

7.4.6.2. Starting of Rotary Converter from D.C. Side

- In power stations, generally, an alternating current is generated and for special purposes where D.C. supply is a necessity a rotary converter is used for the conversion. So the question of starting the rotary converter from D.C. side does not arise, but when an arrangement for auxiliary D.C. supply is there this method is useful.
- Before starting the rotary converter the switch on the A.C. side is put off. Then the D.C. supply is switched and the rotary converter is started similar to that of a D.C. compound motor.
- Before switching on the rotary converter to the A.C. side generated e.m.f. available at the slip rings *must* be synchronised with the supply mains with the help of a synchronoscope. When the conversion for synchronism is indicated by the meter the oil circuit breaker is closed, then the alternating current starts flowing through armature and the rotary converter starts converting from alternating current (A.C.) to direct current (D.C.).

7.4.7. Hunting. The rotary converters, like synchronous motors, are liable to give hunting trouble. But whereas in the synchronous motor there is no commutator, and the armature poles rotate at synchronous speed, in the rotary converter the armature poles, are stationary in space under perfect conditions. If hunting sets in, these armature poles oscillate to and for about their ideal position. This movement of poles, during hunting, sets up e.m.f. in the coils undergoing commutation and causes heavy sparking at the brushes.

Hunting can be avoided by providing damper winding on the pole shoes.

7.4.8. Power Factor. The *power factor* of rotary converters, like synchronous motors, *depends upon excitation*. However, in the case of rotary converters there is no freedom of running it at any desired power factor, because the *heating in the armature conductors is minimum at unity power factor, and the heat loss increases very much as the power factor varies from unity. The conductors near the tapping points are affected most.* Hence a rotary converter may be used to improve the power factor of a system but the improvement is restricted by the limitations mentioned above. In fact, whether the rotary converter improves the power factor of the system, or not, it is essential that it works at as near unity p.f. as is possible.

7.4.9. Armature Reaction. In a D.C. generator the armature reaction acts in such a direction as to require a forward lead of the brushes, while in a motor the armature reaction is in the opposite direction. In the rotary converter the D.C. and A.C. currents may be considered as being superimposed on each other in the armature so that there are two armature reactions tending to *neutralize each other*. In a three-phase machine A.C. and D.C. currents are roughly of the same magnitude, so that the two reactions are more or less equal. Successful commutation may therefore be obtained for all loads up to full-load, and even for over-loads, by fixing the brushes in the no-load neutral position. *This neutralisation of the armature reaction has the effect of considerably raising the limiting load*

from the sparking point of view, this being much higher than would be the case if the machine were used as a D.C. generator.

7.4.10. Inverted Running. When converting D.C. to A.C. the machine is called an *inverted rotary converter*, in which case its speed depends upon its field strength if running alone. If it is running in parallel with synchronous alternators of considerably greater capacity it must, however, run at the same frequency. When running above a weakened field causes a rise in speed, and if the load be inductive, the lagging current weakens the field and the speed may rise to a dangerous degree. To overcome this difficulty the excitation is obtained, not directly from the commutator, but from a *separate exciter coupled directly to the main armature*. An increase in the speed now brings about a greater increase in the exciter voltage, thus strengthening the field and checking the speed variation. Inverted rotary converters are usually compound wound as well, so that any increase in reactive current will bring about a counteracting increase in the field strength.

7.4.11. Speed-limiting Device. The speed of an inverted rotary converter may become excessive if the field is unduly weakened, and as a protection against this a *speed-limiting device* may be fitted.

A speed-limiting device consists of an arm mounted on the shaft and revolving with it. Due to centrifugal force this arm tends to fly outwards against the resistance of a spring, and at a predetermined speed it locks a trigger which closes the tripping circuit of the D.C. circuit breaker.

7.4.12. Parallel Operation

- Normally rotary converters can be operated in parallel without any difficulty. For successful operation, drooping external characteristics are required on D.C. side. On the A.C. side, it is preferable to have separate transformer banks for different converters for providing flexibility in voltage control and load division.
- For guarding against accidental interruption of A.C. supply and for preventing the converter from racing to dangerously high speeds by operating as a D.C. motor with reduced field excitation, the *D.C. circuit breakers must be provided with reverse-current relays and speed limiting devices should be fitted*. It is advantageous to *interlock A.C. and D.C. circuit breakers* so that wherever the A.C. side is open, the D.C. side will also become open.

7.4.13. Voltage Control. The voltage of D.C. generators can be controlled by changing the excitation but when the rotary converter is converting from A.C. to D.C. it normally works as a synchronous motor. The change in excitation will cause the change in power factor but there will not be any change in D.C. volts. The D.C. volts can be controlled by :

(*i*) controlling the D.C. volts of slip-rings ; and

(*ii*) by having some device on the D.C. side.

Generally the voltage is regulated by the earlier method. The A.C. applied volts may be controlled by the following methods :

1. By using tap-changing transformer.
2. Reactance control method.
3. Induction regulator method.
4. Synchronous booster control method.

7.4.13.1. By Using Tap-changing Transformer. The D.C. voltage can be controlled by changing the tappings on the main transformer interposed in between the slip-rings and the mains.

Advantages :

(*i*) Cheap method.

(*ii*) Causes no interference with power factor or commutation.

Disadvantages :

(*i*) Excessive wear and tear at the tappings from which heavy currents are taken out.

(*ii*) Fine control of D.C. volts cannot be obtained as voltage can be changed only in jumps.

7.4.13.2. Reactance Control Method. Fig. 7.13 shows the wiring diagram for this method.

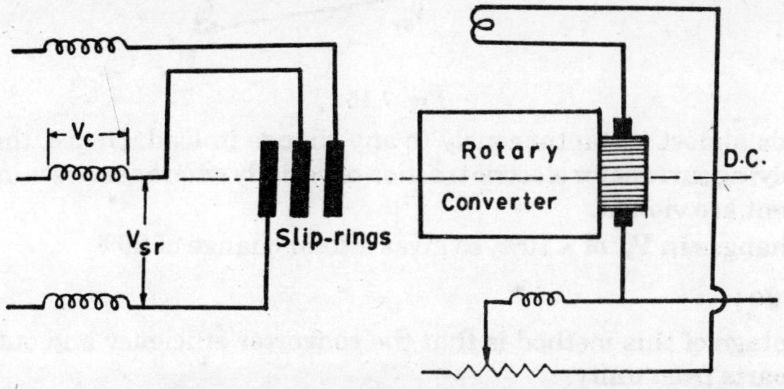

Fig. 7.13. Reactance control method.

The converter is compound wound. In this method a reactor or a choking coil of pure inductance is inserted between the slip-rings and the transformer, so the voltage applied to the slip-ring will be the vector difference of applied voltage per phase and the reactance drop. The power factor at which the rotary converter works can be varied with excitation. When it is over-excited it works at leading power factor.

Let V_t = transformer applied voltage per phase

X_c = reactance

I = load current.

- If I is in phase with voltage V_t

  The voltage drop in the reactance

$$V_c = IX_c \quad \text{...(7.5)}$$

  V_c will lead I by an angle of 90°, represented by LM (Fig. 7.14).

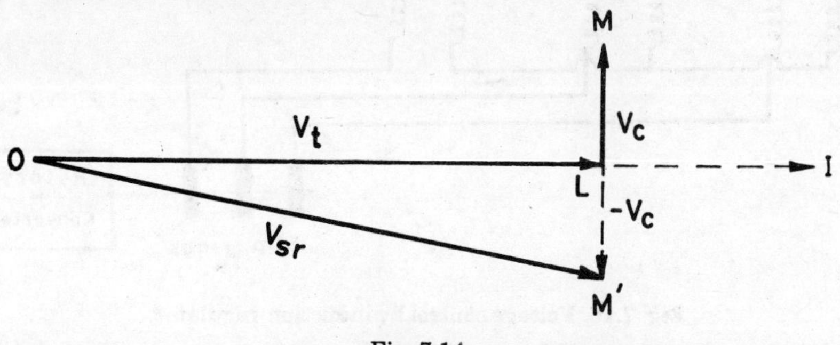

Fig. 7.14

The slip-ring voltage, V_{sr} will be $V_t - V_c = OM'$

- If the current I is leading by angle ϕ, the slip-ring voltage will be more than V_t as represented in Fig. 7.15.

Advantages :

(*i*) The method is relatively cheaper.

(*ii*) Extremely simple.

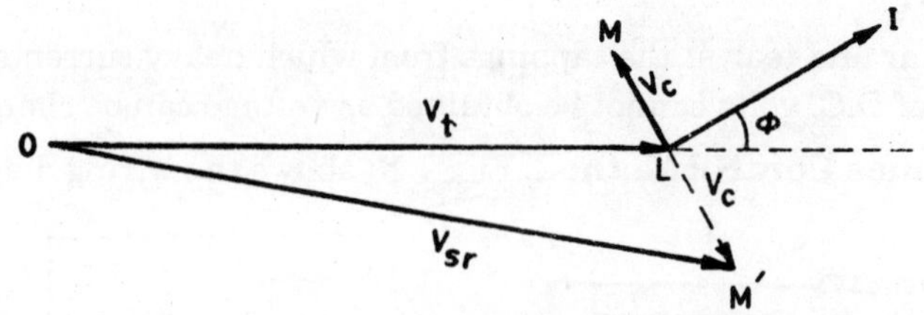

Fig. 7.15

(*iii*) It responds almost instantaneously to any change in load. Hence, this method has been widely used for supplying current for electric traction especially where service is infrequent and hence fluctuations in current are violent.

(*iv*) It gives changes in V_{sr} of ± 10%, so gives a total change of 20%.

Disadvantage :

The disadvantage of this method is that the converter efficiency and output are reduced as the power factor departs from unity.

7.4.13.3. Induction Regulator Method. In this method the *voltage is injected into the slip-rings with the help of an induction regulator*. The power factor at which the voltage is injected can also be controlled by *changing the rotor position of the induction-regulator*.

The method of connecting an induction-regulator is shown in Fig. 7.16.

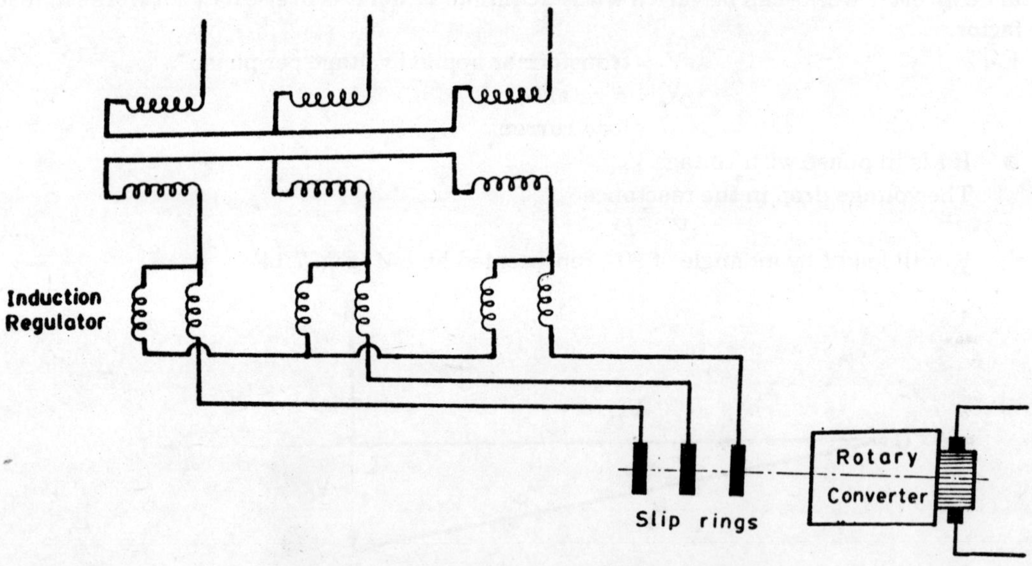

Fig. 7.16. Voltage control by induction-regulator.

The regulator can be arranged for hand or automatic operation and variation of voltage on D.C. side, as high as 30%, can be achieved with its use. In this method *voltage control is independent of power factor and the voltage may be varied smoothly without interruption of current*.

Advantages :

(*i*) The voltage variations follow a smooth curve over the operating range.

(*ii*) No additional switches are necessary in the main circuit and constructionally the regulator is simple and robust.

(*iii*) Can be used for converters running inverted (*i.e.*, inverters).

(*iv*) Variations of up to 30 per cent in D.C. voltage can be obtained.

7.4.13.4. Synchronous Booster Control Method :

- This method is employed in cases where control of *D.C. voltage output is extremely important.*
- The synchronous booster consists of an A.C. generator connected electrically in series with the rotary converter armature and is mounted on the shaft of the rotary converter. If the booster has a series as well as shunt field its voltage will increase with the increase of D.C. load.

  Fig. 7.17 shows the connections of the booster.

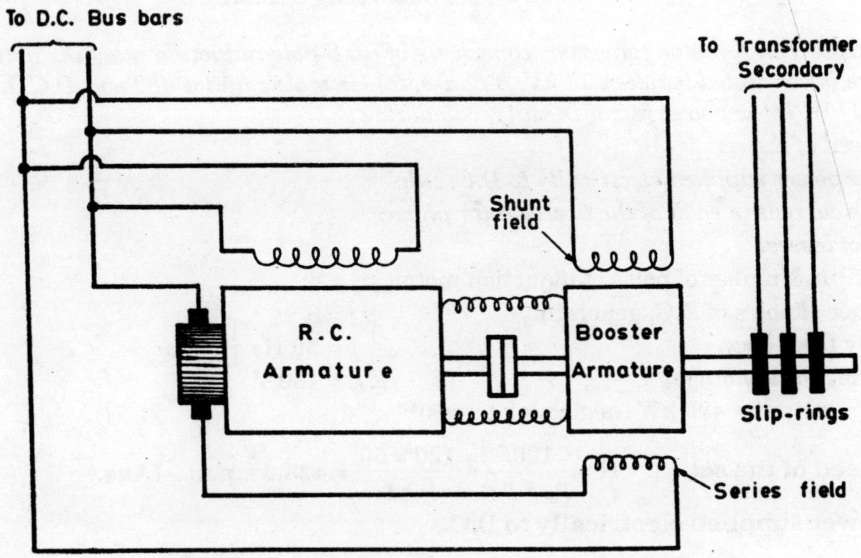

Fig. 7.17. Voltage control by synchronous booster.

Advantages :

(*i*) By simple adjustment of the booster field rheostat a wide range of voltage variations can be obtained.

(*ii*) Automatic boosting can be achieved by compound winding of booster.

(*iii*) Power factor is independent of load and can be made unity or even leading.

Disadvantages :

(*i*) There is some commutation trouble in this method.

(*ii*) Synchronous booster converters are somewhat costly.

Uses. This method is employed where wide range of voltage on D.C. side is required and where the *fluctuations of load are very rapid as in traction work*. This method is quite generally used *for large converters* such as used in the electro-chemical industries.

WORKED EXAMPLES

Converters

Example 7.1. *The induction motor and D.C. generator of a 800 kW motor-generator set have 6-poles and 8-poles respectively. If supply frequency is 50-Hz, find :*

(*i*) *Speed of the set.* (*ii*) *Frequency of currents in the generator.*

(*iii*) *Power transmitted mechanically through the shaft.*

Solution. Number of poles of induction motor, $p_a = 6$

Number of poles of D.C. generator, $p_d = 8$

Supply frequency, $f = 50$ Hz

(*i*) **Speed of the set, N :**

Cascade speed, $N = \frac{120f}{p_a + p_d} = \frac{120 \times 50}{6 + 8} = $ **428.6 r.p.m. (Ans.)**

(*ii*) **Frequency of current in the generator**

$= 50 \times \frac{8}{14} = $ **28.6 Hz. (Ans.)**

(*iii*) **Power transmitted through shaft**

$= 800 \times \frac{6}{14} = $ **342.8 kW. (Ans.)**

Example 7.2. *A motor converter, consisting of an 8-pole induction machine driving a 6-pole D.C. machine is connected between 11 kV, 3-phase, 50 Hz mains and a 480 volt D.C. net work. The output is 490 kW. If the power factor is unity, calculate :*

(*i*) *Speed of the set.*

(*ii*) *The power supplied electrically to D.C. side.*

(*iii*) *The current in each of the twelve rotor phases.*

Neglect losses.

Solution. Number of poles of induction motor, $p_a = 8$

Number of poles of D.C. generator, $p_d = 6$

Supply frequency, $f = 50$ Hz

D.C. net work voltage, $E_{dc} = 480$ V

Input = output = 480 kW (neglecting losses)

(*i*) **Speed of the set,** $\mathbf{N} = \frac{120f}{p_a + p_d} = \frac{120 \times 50}{8 + 6} = $ **428.6 r.p.m. (Ans.)**

(*ii*) **Power supplied electrically to D.C.,**

$$P_2 = \text{power input} \times \frac{p_d}{p_a + p_d} = 480 \times \frac{6}{8 + 6} = \mathbf{205.7\ kW.\ (Ans.)}$$

Rotor phase voltage, $E_2 = \frac{E_{dc}}{2\sqrt{2}} = \frac{480}{2\sqrt{2}} = 169.7$ V

(*iii*) **Current in each phase of rotor I_2**

$$= \frac{P_2}{\text{rotor phases} \times E_2} = \frac{205.7 \times 1000}{12 \times 169.7} = \mathbf{101\ A.\ (Ans.)}$$

☞**Example 7.3.** *A 3 kW single-phase rotary converter operates at full-load from 230 volt A.C. source. Assuming unity p.f. and 100% efficiency, calculate :*

(*i*) *D.C. voltage and current.* (*ii*) *A.C. input current at slip-rings.*

Solution. Power output = 3 kW

A.C. source voltage, $E_p = 230$ V (r.m.s. value)

Power factor = unity

Efficiency = 100%

(*i*) **D.C. voltage,** $E_{dc} = \dfrac{E_p \times \sqrt{2}}{\sin\left(\dfrac{\pi}{m}\right)}$

$$= \frac{230 \times \sqrt{2}}{\sin\left(\frac{\pi}{2}\right)} \qquad \left[\because \text{ For single-phase converter } m = 2\right]$$

$$= \mathbf{325\ V.\ (Ans.)}$$

Current on D.C. side, $I_{dc} = \frac{\text{power output}}{E_{dc}} = \frac{3 \times 1000}{325} = \mathbf{9.23\ A.\ (Ans.)}$

(*ii*) **A.C. input current at slip-rings,**

$$I_{sr} = \frac{I_{dc} \times 2\sqrt{2}}{\eta \times m \times \cos\phi}$$

$$= \frac{9.23 \times 2\sqrt{2}}{1 \times 2 \times 1} \qquad [\because \eta = 1, \cos\phi = 1 \ (given)]$$

$$= \mathbf{13.05\ A.\ (Ans.)}$$

Example 7.4. *A 3-ring rotary converter supplies direct current of 80 A at 240 volts. If the power factor is 0.86 and efficiency is 100%, calculate :*

(*i*) *Slip-ring voltage.* (*ii*) *Slip-ring current.*

Solution. Output D.C. voltage, $E_{dc} = 240$ V
Output direct current, $I_{dc} = 80$ A
Power factor, $\cos\phi = 0.86$
Efficiency of converter, $\eta = 100\%$ *i.e.*, $\eta = 1$
Number of phases, $m = 3$

(*i*) **Slip-ring voltage,** $E_{sr} = E_p = \frac{E_{dc} \sin\frac{\pi}{m}}{\sqrt{2}} = \frac{240 \sin\frac{\pi}{3}}{\sqrt{2}} = \mathbf{147\ V.\ (Ans.)}$

(*ii*) **Slip-ring current,** $I_{sr} = \frac{2\sqrt{2} I_{dc}}{\eta \times m \times \cos\phi} = \frac{2\sqrt{2} \times 80}{1 \times 3 \times 0.86} = \mathbf{87.7\ A.\ (Ans.)}$

☞**Example 7.5.** *A 120 kW, 3-phase rotary converter has a D.C. voltage of 500 V, calculate :*

(*i*) *Full-load D.C. current.* (*ii*) *A.C. phase current.*
(*iii*) *Slip-ring voltage and current.*
Assume 100% efficiency and unity p.f.

Solution. D.C. output voltage, $E_{dc} = 500$ V
Power factor, $\cos\phi = 1$
Efficiency, $\eta = 100\%$ or 1
Number of phases, $m = 3$
Power output on D.C. side = 120 kW (since $\eta = 1$)

(*i*) **Full-load D.C. current,**

$$I_{dc} = \frac{120 \times 1000}{500} = \mathbf{240\ A.\ (Ans.)}$$

(*ii*) **A.C. phase current,** $I_p = \frac{\sqrt{2} I_{dc}}{m\eta \cos\phi \sin\frac{\pi}{m}} = \frac{\sqrt{2} \times 240}{3 \times 1 \times 1 \times \sin\left(\frac{\pi}{3}\right)} = \mathbf{130.6\ A.\ (Ans.)}$

(*iii*) **Slip-ring voltage,** $E_{sr} = E_p = \frac{E_{dc} \times \sin\frac{\pi}{m}}{\sqrt{2}} = \frac{500 \times \sin\left(\frac{\pi}{3}\right)}{\sqrt{2}} = \mathbf{306\ V.\ (Ans.)}$

Slip-ring current, $I_{sr} = \dfrac{2\sqrt{2}\, I_{dc}}{\eta m \cos\phi} = \dfrac{2\sqrt{2} \times 240}{1 \times 3 \times 1}$ **= 226.3 A. (Ans.)**

Example 7.6. *A 1600 kW, 6-phase synchronous converter delivers a full-load D.C. voltage of 500 volts. Find :*

(i) A.C. voltage between slip-rings. *(ii) A.C. input current at slip-rings.*

(iii) A.C. phase current.

Assume power factor to be unity and efficiency 100%.

Solution. Power input = power output = 1600 kW ($\because$ $\eta = 1$)

Full-load D.C. voltage, $E_{dc} = 500$ V

Number of phases, $m = 6$

(*i*) **A.C. voltage between slip-rings,**

$$E_{sr} = E_p = \frac{E_{dc} \times \sin\left(\frac{\pi}{m}\right)}{\sqrt{2}} = \frac{500 \times \sin\left(\frac{\pi}{6}\right)}{\sqrt{2}} = \mathbf{176.8\ V.\ (Ans.)}$$

(*ii*) **A.C. input current at slip-rings, I_{sr} :**

Full-load D.C. current, $I_{dc} = \dfrac{1600 \times 1000}{500} = 3200$ A

$$\therefore \quad I_{sr} = \frac{2\sqrt{2} I_{dc}}{\eta m \cos\phi} = \frac{2\sqrt{2} \times 3200}{1 \times 6 \times 1} = \mathbf{1508.5\ A}$$

Now $I_{sr} = \mathbf{I_p}$ ($\because$ converter is 6-phase)

$\therefore$ $I_{sr} = $ **1508.5 A. (Ans.)**

Example 7.7. *A 4500 kW, 1200 V, 12-phase rotary converter has a full-load efficiency of 95% and power factor of 0.9. Calculate :*

(i) The A.C. slip-ring voltage. *(ii) Full-load D.C. output current.*

(iii) A.C. current at slip-rings. *(iv) A.C. phase current.*

Solution. Load in kW = 4500 kW

D.C. voltage, $E_{dc} = 1200$ V

Number of phases, $m = 12$

Power factor, $\cos\phi = 0.9$

Efficiency of converter, $\eta = 95\%$

(*i*) **A.C. slip-ring voltage,**

$$E_{sr} = E_p = \frac{E_{dc} \times \sin\frac{\pi}{m}}{\sqrt{2}} = \frac{1200 \times \sin\left(\frac{\pi}{12}\right)}{\sqrt{2}} = \mathbf{219.6.\ V.\ (Ans.)}$$

(*ii*) **Full-load D.C. output current,**

$$I_{dc} = \frac{\text{load in kW} \times 1000}{E_{dc}} = \frac{4500 \times 1000}{1200} = \mathbf{3750\ A.\ (Ans.)}$$

(*iii*) **A.C. current at slip-rings,**

$$I_{sr} = \frac{2\sqrt{2} I_{dc}}{\eta m \cos\phi} = \frac{2\sqrt{2} \times 3750}{0.95 \times 12 \times 0.9} = \mathbf{1033.8\ A.\ (Ans.)}$$

(*iv*) **A.C. phase current,** $I_p = \dfrac{\sqrt{2} I_{dc}}{m\eta \cos\phi \sin\frac{\pi}{m}} = \dfrac{\sqrt{2} \times 3750}{12 \times 0.95 \times 0.9 \times \sin\left(\frac{\pi}{12}\right)}$ **= 1997.1 A. (Ans.)**

Example 7.8. *A 3-phase, 220 kW synchronous converter develops 240 V on the commutator and is supplied from a 2200 V, 3-phase network through three single-phase transformers Δ-connected on the h.v. side and Y-connected on the l.v. side. The converter operates on full-load at p.f. 0.86 lag and with an efficiency of 92%. Determine the voltage and current rating of both sides of the transformers neglecting its losses.*

Solution. Load supplied = 220 kW

D.C. output voltage = 240 V

Power factor = 0.86 (lagging)

Efficiency of the converter = 92% or 0.92

Number of phases, $m = 3$

Output D.C. current, $$I_{dc} = \frac{\text{load in kW} \times 1000}{E_{dc}} = \frac{220 \times 1000}{240} = 916.7 \text{ A}$$

Slip-ring voltage, $$E_{sr} = E_p = \frac{E_{dc} \times \sin\left(\frac{\pi}{m}\right)}{\sqrt{2}} = \frac{240 \times \sin\left(\frac{\pi}{3}\right)}{\sqrt{2}} = \mathbf{147 \text{ V.}} \quad \textbf{(Ans.)}$$

Slip-ring current, $$I_{sr} = \frac{2\sqrt{2} I_{dc}}{\eta \times m \times \cos\phi} = \frac{2\sqrt{2} \times 916.7}{0.92 \times 3 \times 0.86} = \mathbf{1092.3 \text{ A.}} \quad \textbf{(Ans.)}$$

Voltage/phase across transformer secondary,

$$E_p = \frac{147}{\sqrt{3}} = 84.9 \text{ V} \qquad (\because \text{ secondary is star-connected})$$

Voltage/phase across transformer primary

$$= 2200 \text{ V} \qquad (\because \text{ Primary is } \Delta\text{-connected})$$

Turn-ratio, $$K = \frac{84.9}{2200} = \frac{1}{25.9}$$

∴ Phase current on primary side

$$= KI_{sr} = \frac{1}{25.9} \times 1092.3 = 42.2 \text{ A}$$

Line current on primary side,

$$I_L = 42.2 \times \sqrt{3} = \mathbf{73.1 \text{ A.}} \quad \textbf{(Ans.)}$$

☞**Example 7.9.** *A 250 kW, 3-phase, rotary converter delivers D.C. output voltage of 250 V. It receives its A.C. power from 2200 V source through a bank of 3 transformers connected delta/star. The machine works with a full-load efficiency of 96% and p.f. of 0.9. Calculate :*

(i) D.C. output current. *(ii) Three-phase A.C. voltage at slip-rings.*

(iii) A.C. current delivered by transformer secondaries to the slip-rings.

(iv) Ratio of transformation *(v) Current in transformer coils.*

(vi) Line current on the primary side.

Solution. Load supplied = 250 kW

Number of phases, $m = 3$

D.C. output voltage, $E_{dc} = 250$ V

Full-load efficiency of the machine, $\eta = 96\%$ or 0.96

Power factor, $\cos\phi = 0.9$

A.C. voltage on primary side = 2200 V

(*i*) **D.C. output current,**

$$I_{dc} = \frac{\text{load in kW} \times 1000}{E_{dc}} = \frac{250 \times 1000}{250} = \mathbf{1000 \text{ A.}} \quad \textbf{(Ans.)}$$

(*ii*) **Three-phase A.C. voltage at slip-rings,**

$$E_{sr} = E_p = \frac{E_{dc} \times \sin \frac{\pi}{m}}{\sqrt{2}} = \frac{250 \times \sin\left(\frac{\pi}{3}\right)}{\sqrt{2}} = \mathbf{153\ V.\ \ (Ans.)}$$

(*iii*) **A.C. current delivered by transformer secondaries to the slip-rings,**

$$I_{sr} = \frac{2\sqrt{2} I_{dc}}{\eta m \cos \phi} = \frac{2\sqrt{2} \times 1000}{0.96 \times 3 \times 0.9} = \mathbf{1091.2\ A.\ \ (Ans.)}$$

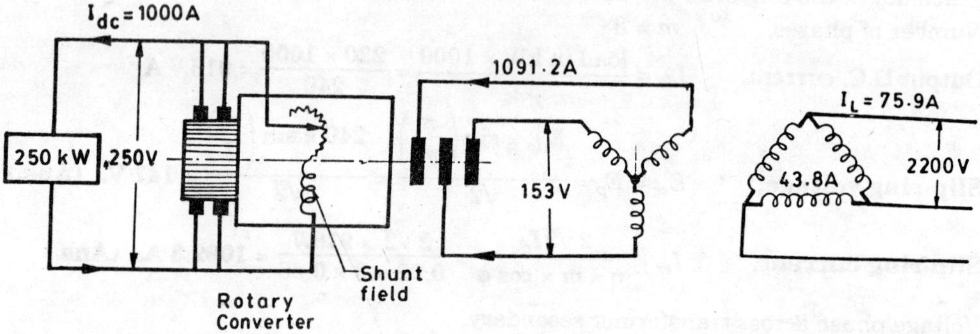

Fig. 7.18

[This current can also be calculated as :

$$I_{sr} = \frac{\text{input power}}{\sqrt{3} \times E_p \times \cos \phi} = \frac{250 \times 1000}{0.96 \times \sqrt{3} \times 153 \times 0.9} \simeq 1091.2 \text{ A}]$$

(*iv*) Voltage/phase across transformer secondary

$$= \frac{153}{\sqrt{3}} = 88.34 \text{ V} \qquad (\because \text{ secondary is star-connected})$$

$\therefore$ **Ratio of transformation,**

$$K = \frac{88.34}{2200} = \frac{1}{24.9}. \ \textbf{(Ans.)}$$

(*v*) **Current in each coil of transformer primary**

$$= 1091.2 \times \frac{1}{24.9} = \mathbf{43.8\ A.\ \ (Ans.)}$$

(*vi*) **Line current on primary side,**

$$I_L = \sqrt{3} \times 43.8 = \mathbf{75.9\ A.\ \ (Ans.)}$$

[This current can also be found as :

$$I_L = \frac{250 \times 1000}{\sqrt{3} \times 2200 \times 0.96 \times 0.9} = 75.9 \text{ A}]$$

Diametral Connection

Example 7.10. *A synchronous converter is delivering direct current at 400 V. The converter has 6 slip-rings. Determine :*

(i) Diametral voltage of the armature.

(ii) Voltage between adjacent slip-rings taps.

(iii) Voltage between alternate slip-rings.

Solution. D.C. voltage, $E_{dc} = 400$ V

Number of phases, $m = 6$.

(*i*) **Diametral voltage of the armature**

$$= 0.707E_{dc} = 0.707 \times 400 = \mathbf{282.8\ V.}\ \ \textbf{(Ans.)}$$

(*ii*) **Voltage between adjacent slip-ring taps,**

$$E_{sr} = E_p = \frac{E_{dc} \sin \frac{\pi}{m}}{\sqrt{2}} = \frac{400 \times \sin\left(\frac{\pi}{6}\right)}{\sqrt{2}} = \mathbf{141.4\ V.}\ \ \textbf{(Ans.)}$$

(*iii*) **Voltage between alternate slip-ring taps,**

$$= \sqrt{3}\ E_p = \sqrt{3} \times 141.4 = \mathbf{244.9\ V.}\ \ \textbf{(Ans.)}$$

Example 7.11. *A synchronous converter supplies 400 kW at 400 V from the commutator side. Assume diametral tappings. If the efficiency of the machine is 92% and power factor 0.9 find the voltage and current of one secondary phase of the 3-phase transformer.*

Solution. Load supplied = 400 kW

Output D.C. voltage, $E_{dc} = 400$ V

Efficiency of the machine, $\eta = 92\%$ or 0.92

Power factor, $\cos \phi = 0.9$

Number of phases, $m = 6$

Output D.C. current, $I_{dc} = \frac{\text{load in kW} \times 1000}{E_{dc}} = \frac{400 \times 1000}{400} = 1000$ A

R.m.s. value of secondary phase voltage,

E_2 = voltage across diametral slip-rings

$$= 0.707\ E_{dc} = 0.707 \times 400 = \mathbf{282.8\ V.}\ \ \textbf{(Ans.)}$$

Secondary phase current = slip-ring current,

$$I_{sr} = \frac{2\sqrt{2} I_{dc}}{\eta m \cos \phi} = \frac{2\sqrt{2} \times 1000}{0.92 \times 6 \times 0.9} = \mathbf{569.3\ A.}\ \ \textbf{(Ans.)}$$

☞**Example 7.12.** *A six-ring rotary converter supplies a load of 200 kW to A.C. net work. It has 10-poles, 900 lap-wound armature conductors and flux per pole is 45 mWb. The transformer is supplied from 3300 V, 50-Hz mains and the primary is delta connected. Find :*

(*i*) *Turn-ratio of the transformer if the machine has diametral tappings.*

(*ii*) *Current per slip-ring if the efficiency is 92% and load power factor unity.*

Solution. Load supplied = 200 kW

Number of poles, $p = 10$

Armature conductors, $Z = 900$

(Lap wound ; no. of parallel paths, $a = 10$)

Flux, $\phi = 45$ mWb = 0.045 Wb

Frequency, $f = 50$ Hz

A.C. supply voltage = 3300 V

Efficiency of converter, $\eta = 92\%$ or 0.92

Load p.f., $\cos \phi = 1$

Speed, $N = \frac{120f}{p} = \frac{120 \times 50}{10} = 600$ r.p.m.

E.m.f. induced, $E_{dc} = \frac{p\phi ZN}{60a} = \frac{10 \times 0.045 \times 900 \times 600}{60 \times 10} = 405$ V

Output D.C. current, $I_{dc} = \frac{\text{load in kW} \times 1000}{E_{dc}} = \frac{200 \times 1000}{405} = 493.8$ A

Transformer secondary phase voltage,

$E_2 = 0.707 E_{dc} = 0.707 \times 405 = 286.3$ V

Transformer primary phase voltage,

E_1 = primary line voltage

= 3300 V (∵ primary is Δ-connected)

(*i*) **Turn-ratio per phase** $= \frac{286.3}{3300} = \mathbf{1 : 11.53.}$ **(Ans.)**

(*ii*) **Current per slip-ring,**

$$I_{sr} = \frac{2\sqrt{2} I_{dc}}{\eta m \cos\phi} = \frac{2\sqrt{2} \times 493.8}{0.92 \times 6 \times 1} = \mathbf{253\ A.}\ \textbf{(Ans.)}$$

Example 7.13. *A six-ring, 8-pole, lap-wound rotary converter with diametrical tapping runs at 750 r.p.m. and supplies direct current of 1000 A. There are 800 armature conductors and the flux per pole is 52.6 mWb. The transformer is delta connected to 6600 V, 3-phase supply. Assuming unity p.f. and an efficiency of 95%, calculate :*

(*i*) *Turn-ratio of the transformer.*

(*ii*) *Current per ring.*

(*iii*) *Current in the armature connection between a slip-ring and a tapping point.*

Solution. Number of poles, $p = 8$

Output D.C. current, $I_{dc} = 1000$ A

Armature conductors = 800

No. of parallel paths, $a = 8$ (being lap-wound)

Speed, $N = 750$ r.p.m.

Flux/pole, $\phi = 52.6$ mWb or 0.0526 Wb

Power factor, $\cos\phi = ?$

Efficiency of the converter, η = 95% or 0.95

Now, $E_{dc} = \frac{p\phi ZN}{60a} = \frac{10 \times 0.0526 \times 760 \times 750}{60 \times 10} = 500$ V

R.m.s. value of secondary voltage/phase

$= 0.707 \times 500 = 353.5$ V

Transformer primary phase voltage,

E_1 = primary line voltage

= 6600 V (∵ primary is delta connected)

(*i*) **Turn-ratio of the transformer**

$$= \frac{\text{secondary voltage/phase}}{\text{primary voltage/phase}} = \frac{353.5}{6600}$$

= **1/18.67** *i.e.*, **1 : 18.67. (Ans.)**

(*ii*) **Current per ring,** $I_{sr} = \frac{2\sqrt{2} I_{dc}}{\eta m \cos\phi} = \frac{2\sqrt{2} \times 1000}{0.95 \times 6 \times 1} = \mathbf{496.2\ A.}$ **(Ans.)**

(iii) **Current in the armature connection between a slip-ring and a tapping point**

$$= \frac{496.2}{4} = \mathbf{124.05\ A.\ (Ans.)}$$

Example 7.14. *A 250 kW, 500 volts 4-pole synchronous converter is supplied through a transformer from a 33 kV, 50-Hz mains. The transformer has delta connections on the primary side and diametral connections on the secondary side. Calculate the number of turns and cross-sectional area required on each side of the transformer winding. The synchronous converter has a 92 per cent efficiency and is operating at unity power factor. Assume a maximum core flux of 100 m Wb and a current density of 2.4 A per mm^2.*

Neglect magnetising current and losses of the transformer.

Solution. Load supplied = 250 kW

Output D.C. voltage, E_{dc} = 500 V

Number of pole, $p = 4$

Efficiency of converter, η = 92% or 0.92

Power factor, $\cos \phi = 1$

Maximum core flux, ϕ_{max} = 100 mWb or 0.1 Wb

Current density, δ = 2.4 A/mm^2

Number of turns :

Cross-sectional areas :

Full-load D.C. current, $I_{dc} = \frac{\text{load in kW} \times 1000}{E_{dc}} = \frac{250 \times 1000}{500} = 500$ A.

Maximum voltage per phase in diametral connection

$$= E_{dc} = 500 \text{ V}$$

R.m.s. voltage/phase, $E_2 = \frac{500}{\sqrt{2}} = 353.5$ V.

(i) **Number of turns, $N_1 : N_2$:**

Using the relation, $E_2 = 4.44 f\phi_{max} N_2$ volts

or $353.5 = 4.44 \times 50 \times 0.1 \times N_2$

$\therefore$ $N_2 = \frac{353.5}{4.44 \times 50 \times 0.1} = \mathbf{15.92\ say.\ (Ans.)}$

As $\frac{E_2}{E_1} = \frac{N_2}{N_1}$

$\therefore$ $\frac{353.5}{33000} = \frac{16}{N_1}$

From which, $\mathbf{N_1} = \frac{16 \times 33000}{353.5} = \mathbf{1494.\ (Ans.)}$

(ii) **Cross-sectional areas :**

Using the relation, Output = slip-ring current × number of phases × phase voltage × p.f. × η

$$250 \times 1000 = I_{sr} \times 3 \times 353.5 \times 1 \times 0.92$$

$\therefore$ $I_{sr} = \frac{250 \times 1000}{3 \times 353.5 \times 1 \times 0.92} = 256.2$ A

Secondary phase current, $I_p = I_{sr} = 256.2$ A

Cross-sectional area of the secondary winding conductor

$$= \frac{256.2}{2.4} = \mathbf{106.75\ mm^2.\ (Ans.)}$$

Primary phase current $= \frac{16}{1494} \times 256.2$ $\left[\because \frac{N_2}{N_1} = \frac{I_1}{I_2} \text{ or } I_1 = \frac{N_2 I_2}{N_1} = \frac{16}{1494} \times 256.2 \text{ A} \right]$

$= 2.74$ A

∴ **Cross-sectional area of primary conductor**

$= \frac{2.74}{2.4} = \mathbf{1.142\ mm^2}$. **(Ans.)**

Example 7.15. *A six-ring converter delivers 400 kW, at 400 V on the D.C. side. The efficiency and power factor at this load are 95% and 0.9 respectively. Calculate :*

(*i*) *slip-ring current*

(*ii*) *voltage across :*

(*a*) *consecutive* (*b*) *alternate* (*c*) *diametral slip-rings on the input side.*

Solution. Load supplied = 400 kW

D.C. output voltage, $E_{dc} = 400$ V

Efficiency of converter, $\eta = 95\%$ or 0.95

Power factor, $\cos \phi = 0.9$

Output D.C. current, $I_{dc} = \frac{\text{load in kW} \times 1000}{E_{dc}} = \frac{400 \times 1000}{400} = 1000$ A.

(*i*) ∴ **Slip-ring current** $= \frac{2\sqrt{2} I_{dc}}{\eta m \cos \phi} = \frac{2\sqrt{2} \times 1000}{0.95 \times 6 \times 0.9} = \mathbf{551.3\ A.}$ **(Ans.)**

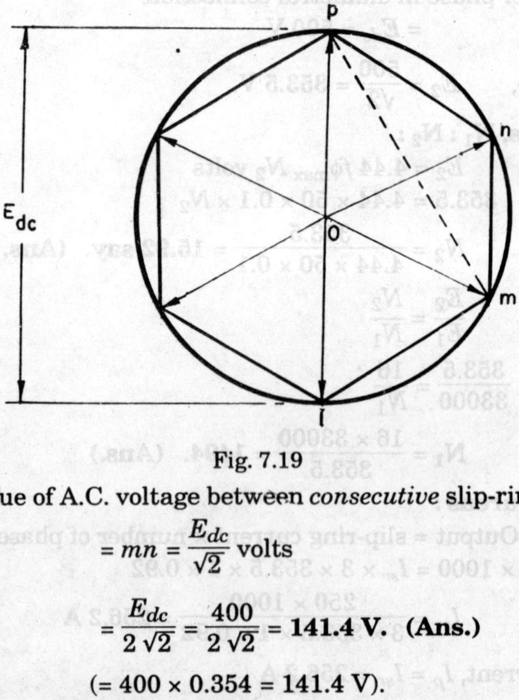

Fig. 7.19

(*ii*) (*a*) Maximum value of A.C. voltage between *consecutive* slip-rings

$= mn = \frac{E_{dc}}{\sqrt{2}}$ volts

R.m.s. value $= \frac{E_{dc}}{2\sqrt{2}} = \frac{400}{2\sqrt{2}} = \mathbf{141.4\ V.}$ **(Ans.)**

$(= 400 \times 0.354 = 141.4 \text{ V}).$

(*b*) **Voltage between alternate rings** = *mp*

Now slip-ring voltages have a mutual phase difference of 60°.

∴ R.m.s. value of voltage between alternate rings

$$= 2 \times 141.4 \cos\left(\frac{60°}{2}\right) = \mathbf{245\ V.\ \ (Ans.)}$$

(c) **Maximum voltage between diametral rings** (which is also equal to secondary voltage/phase)

$$= lp = E_{dc} = 400\ \text{V}$$

∴ R.m.s. value $= \frac{400}{\sqrt{2}} = \mathbf{282.8\ V.\ \ (Ans.)}$

Example 7.16. *If the voltage between the centre of a D.C. mains is 400 V and that of 3-phase system is 6.6 kV, calculate the turn-ratio of the transformer windings and the voltage between opposite and between adjacent slip-rings. The machine has diametral tappings and the transformer is delta connected on the high voltage side.*

Solution. D.C. output voltage, E_{dc} = 400 V

A.C. voltage on h.v. side = 6.6 kV or 6600 V

Number of phases, $m = 6$

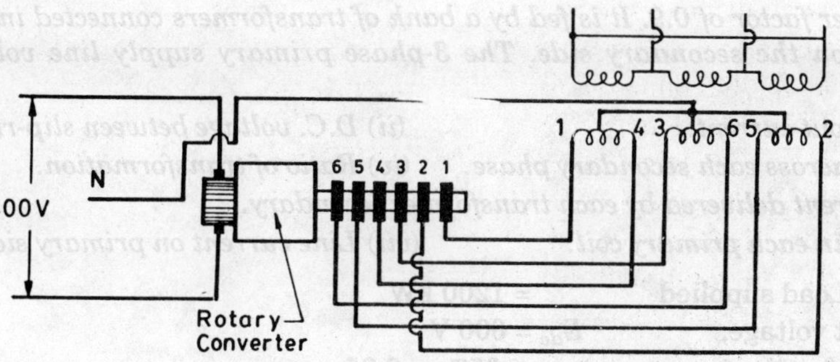

Fig. 7.20

$$E_p = \frac{E_{dc} \times \sin\frac{\pi}{m}}{\sqrt{2}} = \frac{400 \times \sin\left(\frac{\pi}{6}\right)}{\sqrt{2}} = 141.4\ \text{V}$$

E.m.f. induced between two adjacent slip-rings,

$$E_{sr} = E_p = 141.4\ \text{V}$$

E.m.f. induced per phase on the secondary side

$$= 141.4 \times 2 = 282.8\ \text{V}$$

∴ **Turn-ratio** $= \frac{\text{secondary voltage/phase}}{\text{primary voltage/phase}} = \frac{282.8}{6600} = \mathbf{1 : 23.34.\ \ (Ans.)}$

Example 7.17. *The transformers feeding a 2200 kW, 800 V, D.C. 6-phase converter are connected diametrical on the secondary side. Assuming a full-load efficiency of 95% and a power factor of 0.91, calculate :*

(i) 6-phase voltage between adjacent slip-rings.

(ii) Voltage of each transformer secondary.

(iii) Alternating current delivered by each of the transformer secondaries.

Solution. Load supplied = 2200 kW

D.C. output voltage, E_{dc} = 800 V

Number of phases, $m = 6$

Efficiency of converter, η = 95% or 0.95

Power factor, $\cos\phi = 0.91$

Output D.C. current, $I_{dc} = \frac{\text{load in kW} \times 1000}{E_{dc}} = \frac{2200 \times 1000}{800} = 2750 \text{ A}$

(*i*) **Voltage between adjacent slip-rings,**

$$E_{sr} = \frac{E_{dc} \sin \frac{\pi}{m}}{\sqrt{2}} = \frac{800 \times \sin\left(\frac{\pi}{6}\right)}{\sqrt{2}} = \mathbf{282.9\ V.\ \ Ans.}$$

(*ii*) **Voltage of each transformer secondary,**

$$E_2 = \frac{E_{dc}}{\sqrt{2}} = \frac{800}{\sqrt{2}} = \mathbf{565.7\ V.\ \ (Ans.)}$$

(*iii*) **Alternating current delivered by each of the transformer secondaries** = slip-ring current,

$$I_{sr} = \frac{2\sqrt{2} I_{dc}}{\eta m \cos\phi} = \frac{2\sqrt{2} \times 2750}{0.95 \times 6 \times 0.91} = \mathbf{1499.5\ A.\ \ (Ans.)}$$

☞**Example 7.18.** *A 1200 kW, 600 V, D.C. 6-phase synchronous converter has a full-load efficiency of 95% and a power factor of 0.9. It is fed by a bank of transformers connected in delta on primary and diametrical on the secondary side. The 3-phase primary supply line voltage is 13200 V. Calculate :*

(*i*) *D.C. output current.* (*ii*) *D.C. voltage between slip-rings.*
(*iii*) *Voltage across each secondary phase.* (*iv*) *Ratio of transformation.*
(*v*) *A.C. current delivered by each transformer secondary.*
(*vi*) *Current in each primary coil.* (*vii*) *Line current on primary side.*

Solution. Load supplied = 1200 kW
D.C. output voltage, E_{dc} = 600 V
Efficiency of converter, η = 95% or 0.95
Power factor, $\cos\phi$ = 0.9

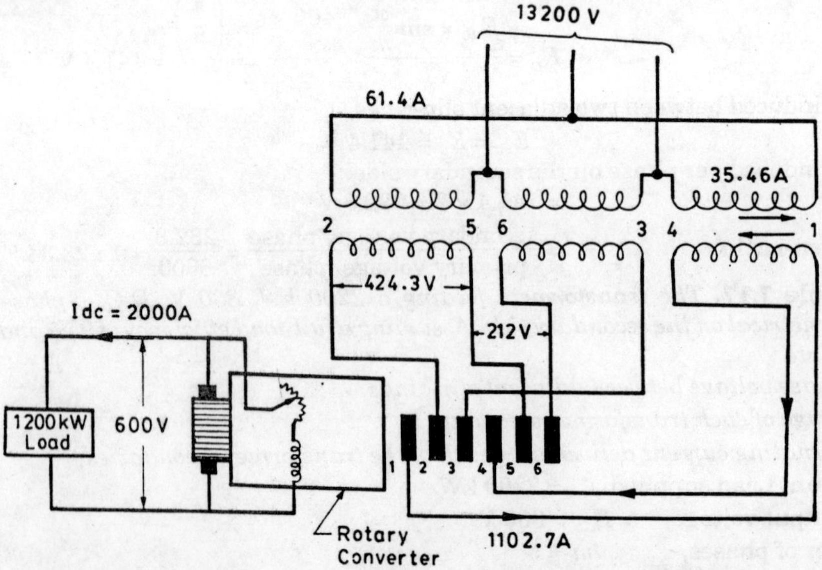

Fig. 7.21

Number of phases, $m = 6$

Primary supply line voltage $= 13200$ V

(*i*) **D.C. output current,** $I_{dc} = \dfrac{\text{load in kW} \times 1000}{E_{dc}} = \dfrac{1200 \times 1000}{600} = $ **2000 A. (Ans.)**

(*ii*) **Voltage between slip rings,**

$$E_{sr} = \frac{E_{dc}}{2\sqrt{2}} = \frac{600}{2\sqrt{2}} = \mathbf{212\ V.\ (Ans.)}$$

$(= 0.354 \times 600 = 212 \text{ V})$

(*iii*) Maximum value of voltage across each secondary phase

$= 600$ V

∴ **R.m.s. value** $= \dfrac{600}{\sqrt{2}} = $ **424.3 V. (Ans.)**

(*iv*) Transformation ratio, $K = \dfrac{424.3}{13200} = $ **1 : 31.1. (Ans.)**

(*v*) **Slip ring current,** $I_{sr} = \dfrac{2\sqrt{2} I_{dc}}{\eta m \cos \phi} = \dfrac{2\sqrt{2} \times 2000}{0.95 \times 6 \times 0.9} = $ **1102.7 A. (Ans.)**

(*vi*) **Primary phase current** $= \dfrac{1102.7}{31.1} = $ **35.46 A. (Ans.)**

(*vii*) **Primary line current,** $I_L = 35.46 \times \sqrt{3} = $ **61.4 A. (Ans.)**

[Primary line current can also be found as :

$$I_L = \frac{1200 \times 1000}{\sqrt{3} \times 13200 \times 0.95 \times 0.9} = 61.4 \text{ A} \Big]$$

Double-Delta Connection

☞**Example 7.19.** *A 500 kW, 440 V D.C., 6-phase rotary converter is fed by a bank of transformers connected in star on primary side and double-delta on the secondary side. The three-phase primary line voltage is 11 kV. Assuming a full-load efficiency of 90 per cent at full-load and power factor of 0.9, calculate* :

(*i*) *Secondary voltage per phase.* (*ii*) *A.C. voltage between adjacent rings.*

(*iii*) *Slip-ring current.* (*iv*) *Transformation ratio.*

(*v*) *Primary line current.*

Solution. Load supplied $= 500$ kW

D.C. output voltage, $E_{dc} = 440$ V

Primary line voltage $= 11$ kV or 11000 V

Efficiency of converter $= 90\%$ or 0.9

Power factor, $\cos \phi = 0.9$

Refer Fig. 7.54.

(*i*) Maximum voltage per phase

$$= lm = 2 \times \frac{E_{dc}}{2} \times \sin 60° = E_{dc} \times \frac{\sqrt{3}}{2} = 440 \times \frac{\sqrt{3}}{2} = 381 \text{ V}$$

∴ **R.m.s. value** $= \dfrac{381}{\sqrt{2}} = $ **269 V. (Ans.)**

(or voltage across each secondary $= 0.612 \times 440 = 269$ V)

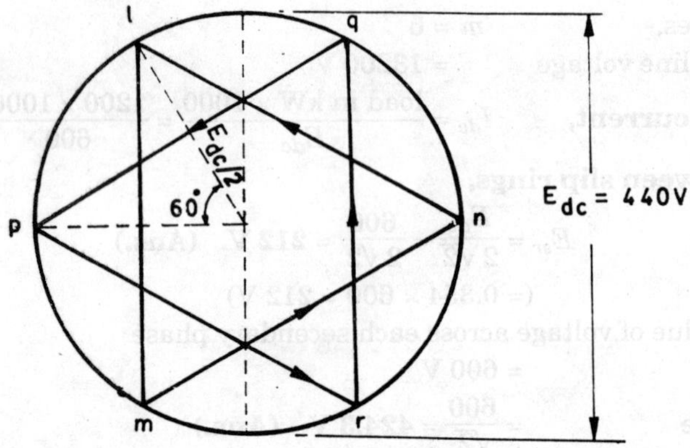

Fig. 7.22

(*ii*) **Maximum A.C. voltage between slip-rings**

$$I_p = \frac{E_{dc}}{2} = \frac{440}{2} = 220 \text{ V}$$

R.m.s. value $= \frac{220}{\sqrt{2}} = \mathbf{155.6\ V.\ \ (Ans.)}$

(or $E_{sr} = 0.354 \times 440 = 155.6$ V).

(*iii*) **Output** $= 500 \times 1000 = 500000$ W

Input $= \frac{500000}{0.9}$ W

A.C. input $= 6 \times E_p I_p \cos \phi$

$\therefore \quad 6 \times 155.6 \times I_p \times 0.9 = \frac{500000}{0.9}$.

$\therefore \quad I_p = \frac{500000}{6 \times 155.6 \times 0.9 \times 0.9} = 661$ A

$\therefore \quad I_{sr} = 661 \times \sqrt{3} = \mathbf{1144.9\ A.\ \ (Ans.)}$

(*iv*) **Since each secondary phase of the transformer has two parts the r.m.s. value of the secondary**

$$= 2 \times 269 = 538 \text{ V}$$

$\therefore$ **Transformation ratio,**

$$K = \frac{528}{\frac{11000}{\sqrt{3}}} \simeq \frac{1}{12.03}. \quad \textbf{(Ans.)}$$

(*v*) **Primary phase current** $= K \times$ secondary current/phase

$$= \frac{1}{12.03} \times 661 = \mathbf{54.9\ A.\ \ (Ans.)}$$

Double-star Connection

Example 7.20. *A six-ring 250 kW, 500 V, D.C. synchronous converter has an efficiency of 95%. The primary of the transformer is delta and secondary double-star. The primary is supplied by 3-phase, 3.3 kV line. Assuming the power factor as unity and neglecting magnetising current and losses in transformer calculate :*

(*i*) *Secondary phase voltage and current.* (*ii*) *Primary phase current.*

Solution. Load supplied = 250 kW

D.C. output voltage, E_{dc} = 500 V

Efficiency of the converter, η = 95% or 0.95

Primary supply voltage = 3.3 kV or 3300 V

The maximum value of voltage induced in each secondary phase is given by the radius of the circle (Fig. 7.23).

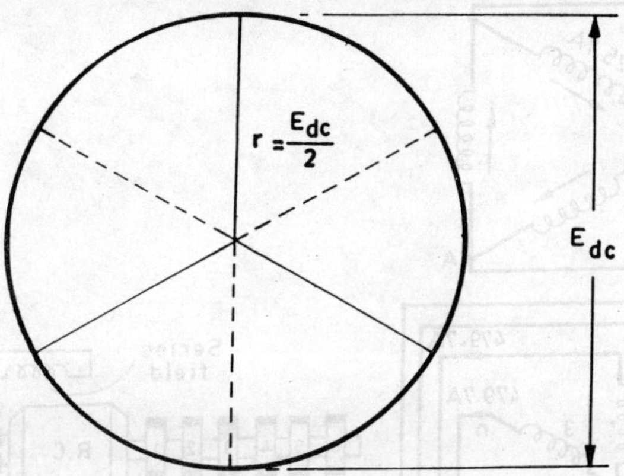

Fig. 7.23

$$\therefore \quad E_{s(\max)} = \frac{E_{dc}}{2} = \frac{500}{2} = 250 \text{ V.}$$

(*i*) **R.m.s. value,** $E_s = \frac{250}{\sqrt{2}} =$ **177 V. (Ans.)**

(or $E_s = 0.354 \times 500 = 177$ V)

D.C. output current, $I_{dc} = \frac{\text{load in kW} \times 1000}{E_{dc}} = \frac{250 \times 1000}{500} = 500 \text{ A}$

$$I_{sr} = \frac{2\sqrt{2} I_{sc}}{\eta m \cos\phi} = \frac{2\sqrt{2} \times 500}{0.95 \times 6 \times 1} = 248 \text{ A}$$

Hence, **secondary phase current,**

$$I_p = I_{sr} = \mathbf{248\ A.\ \ (Ans.)}$$

(*ii*) Primary line current, $I_L = \frac{250 \times 1000}{\sqrt{3} \times 3300 \times 0.95 \times 1} = 46 \text{ A}$

$\therefore$ **Primary phase current,**

$$= \frac{46}{\sqrt{3}} = \mathbf{26.56\ A.\ \ (Ans.)}$$

☞**Example 7.21.** *A 450 kW, 6 ring converter is supplied from 6600 V, 50-Hz, 3-phase system through transformer the primaries of which are connected in delta and secondaries in a 6-phase star system. The converter is working at a power factor of 0.92 and at an efficiency of 96%. The D.C. output voltage is 500 V. Neglecting no-load current and transformer losses calculate :*

(*i*) *Full-load current.* (*ii*) *Voltage between two adjacent slip-rings.*

(*iii*) *slip ring current*

(*iv*) *Secondary transformer voltage for diametral connections.*

(*v*) *Transformer primary current.* (*vi*) *A.C. line current.*

(*vii*) *kVA rating of each transformer phase.*

Solution. Load supplied $= 450$ **kW**

D.C. output voltage, $E_{dc} = 500$ **V**

Efficiency of converter, $\eta = 96\%$ **or 0.96**

Power factor, $\cos \phi = 0.92$

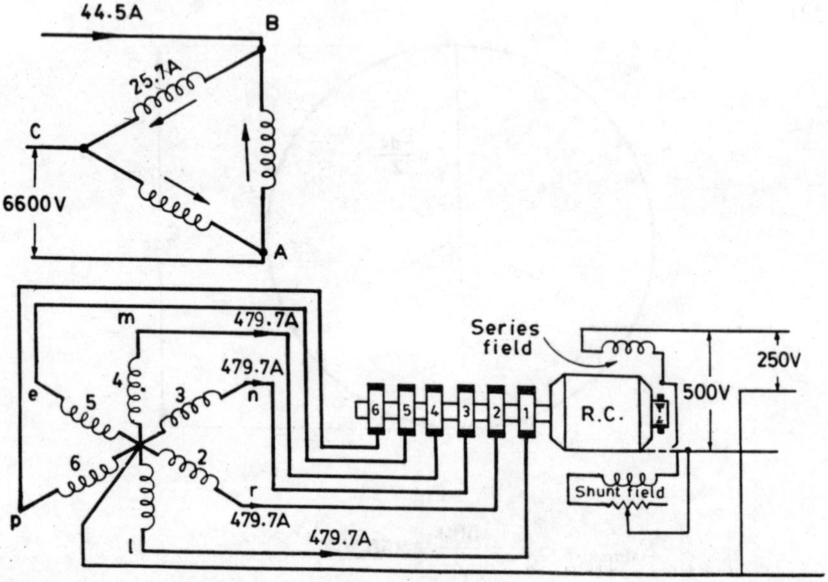

Fig. 7.24

(*i*) **Full-load output current,**

$$I_{dc} = \frac{\text{load in kW} \times 1000}{E_{dc}} = \frac{450 \times 1000}{500} = \mathbf{900\ A.}\ \ \textbf{(Ans.)}$$

(*ii*) **Maximum voltage between two adjacent slip-rings**

$$= \frac{E_{dc}}{2} = \frac{500}{2} = 250 \text{ V}$$

R.m.s. value, $$E_{sr} = \frac{250}{\sqrt{2}} = \mathbf{177\ V.}\ \ \textbf{(Ans.)}$$

(*iii*) **Slip-ring current, I_{sr} = ?**

$$\text{Output} = 450 \times 1000 = 450000 \text{ W}$$

$$\text{Input} = 6 \times E_{sr} \times I_{sr} \times \text{p.f.}$$

But $$\text{Output} = \text{input} \times \eta$$

$\therefore$ $$450000 = 6 \times E_{sr} \times I_{sr} \times \text{p.f.} \times \eta = 6 \times 177 \times I_{sr} \times 0.92 \times 0.96$$

$\therefore$ $$\mathbf{I_{sr}} = \frac{450000}{6 \times 177 \times 0.92 \times 0.96} = \mathbf{479.7\ A.}\ \ \textbf{(Ans.)}$$

(*iv*) Since in diametral connections, the voltage between adjacent slip-rings is half of the e.m.f. induced per phase,

$\therefore$ **Phase voltage of each of the secondary (r.m.s. value)**

$$= 2 \times 177 = \mathbf{354\ V.}\ \ \textbf{(Ans.)}$$

(*v*) Transformation ratio, $K = \dfrac{354}{6600}$.

$\therefore$ **Transformer primary phase current**

$$= K \times 479.7 = \frac{354}{6600} \times 479.7 = \mathbf{25.7\ A.}\ \textbf{(Ans.)}$$

(*vi*) **A.C. line current** $= \sqrt{3} \times 25.7 = \mathbf{44.5\ A.}$ **(Ans.)**

(*vii*) **kVA rating of each transformer phase**

$= \text{phase voltage} \times \text{phase current} \times 10^{-3}$

$= 6600 \times 25.7 = \mathbf{169.6\ kVA.}$ **(Ans.)**

Voltage Control—Reactance Control Method

Example 7.22. *A reactor of reactance 0.6 Ω and negligible resistance put in each line of supply of a 3-ring synchronous converter takes a leading current of 70 A at a p.f. of 0.86. If the supply voltage is kept constant at 440 V, calculate the D.C. output voltage.*

Neglect all losses.

Solution. Reactance of the reactor = 0.6 Ω

Current drawn by the converter = 70 A

Power factor, $\cos\phi = 0.86$ (leading)

$\therefore$ $\phi = \cos^{-1} 0.86 = 30.7°$

Supply voltage = 440 V

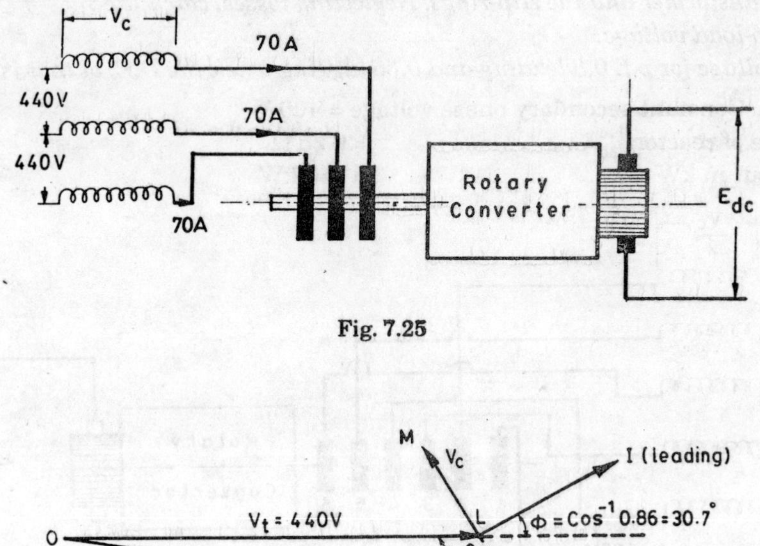

Fig. 7.25

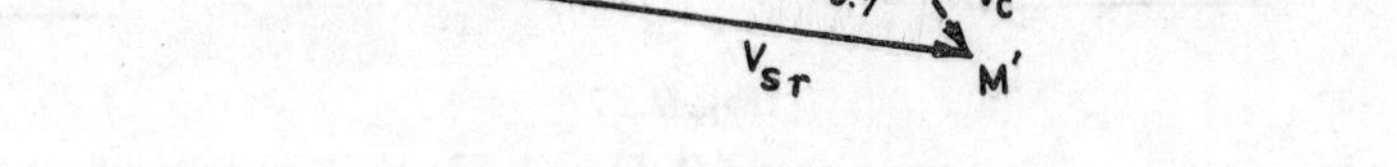

Fig. 7.26

D.C. output voltage, E_{dc} = ?

The reactance voltage V_c in the vector diagram is at right angles to I as shown in Fig. 7.26 and its value is

$$V_c = 70 \times 0.6 = 42 \text{ V (phase value)}$$

$$= 42 \times \sqrt{3} = 72.7 \text{ V (line value)}$$

(V_c has been changed to line value since V_c and V_{sr} are line values)
Voltage between two adjacent slip-rings

$$= \text{terminal voltage } (V_t) - \text{voltage drop in the reactance } (V_c)$$

Refer Fig. 7.58.

$$\phi = \cos^{-1} 0.86 = 30.7°$$

$$\theta = 90 + \phi = 90 + 30.7 = 120.7°$$

From $\Delta OLM'$, we have $V_{sr}^{\ 2} = V_t^{\ 2} + V_c^{\ 2} - 2V_t \times V_c \times \cos 120.7°$

$$= (440)^2 + (72.7)^2 - 2 \times 440 \times 72.7 \times \cos 120.7°$$

$$= 193600 + 5285 + 32662$$

$\therefore \quad V_{sr} = 481 \text{ V}$

Phase voltage, $E_p = \dfrac{481}{\sqrt{3}} = 277.7 \text{ V}$

D.C. output voltage, $E_{dc} = \dfrac{\sqrt{2}E_p}{\sin \dfrac{\pi}{m}} = \dfrac{\sqrt{2} \times 277.7}{\sin\left(\dfrac{\pi}{3}\right)} = \mathbf{453\ V.}$ **(Ans.)**

Example 7.23. *A six-ring converter with diametral tappings is supplied from a transformer having a constant secondary phase voltage of 400 volts. A reactor of 0.25 Ω is included in each line connecting the transformer and the slip-rings. Neglecting losses, calculate :*

(i) D.C. no-load voltage.

(ii) D.C. voltage for p.f. 0.8 leading and 0.86 lagging when the D.C. output is 240 kW.

Solution. Constant secondary phase voltage = 400 V

Reactance of reactor = 0.25 Ω

D.C. output in kW = 240 kW

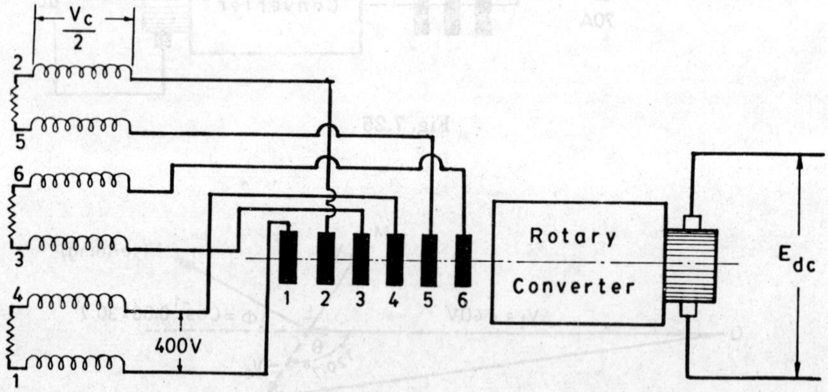

Fig. 7.27

(*i*) D.C. voltage = maximum A.C. volts between diametral tappings on no-load

$= \sqrt{2} \times 400 =$ **565.7 V. (Ans.)**

The power supplied by each phase in diametral connection

$$= \frac{240}{3} = 80 \text{ kW}$$

Assuming 100% efficiency

$$V_{ac} \times I_{sr} \times \cos \phi = 800 \times 1000$$

(ii) (a) **At 0.8 p.f. leading** $I_{sr} = \frac{80 \times 1000}{400 \times 0.8} = 250$ A

$\therefore$ Reactance drop to each line, $= 250 \times 0.25 = 62.5$ V

Since each phase comprises *two* reactors

$\therefore$ $V_c = 2 \times 62.5 = 125$ V

Let V_{sr} be the voltage between diametral tappings

V_t the terminal voltage per phase = 400 V

Refer Fig. 7.28.

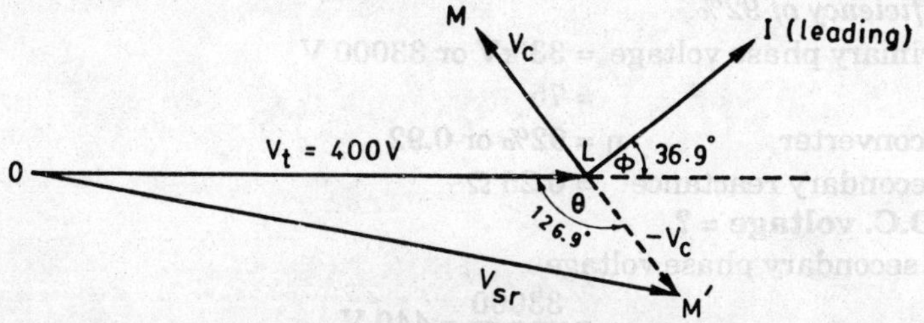

Fig. 7.28

$$\phi = \cos^{-1} 0.8 = 36.9°$$

$$\theta = 90 + \phi = 90 + 36.9 = 126.9°$$

From $\Delta OLM'$, we have $V_{sr}^2 = V_t^2 + V_c^2 - 2 \times V_t \times V_c \times \cos\theta$

$$= (400)^2 + (125)^2 - 2 \times 400 \times 125 \times \cos 126.9°$$

$$= 160000 + 15625 + 60042$$

or $V_{sr} = 485.4$ V

$\therefore$ **D.C. voltage,** $E_{dc} = \sqrt{2} \times 485.4 =$ **686.4 V. (Ans.)**

(b) **At 0.86 p.f. lagging,** $I_{sr} = \frac{80 \times 1000}{400 \times 0.86} = 232.5$ A

$\therefore$ Reactance drop to each line $= 232.5 \times 0.25 = 58$ V

Since each phase comprises *two* reactors

$\therefore$ $V_c = 2 \times 58 = 116$ V

Refer Fig. 7.29.

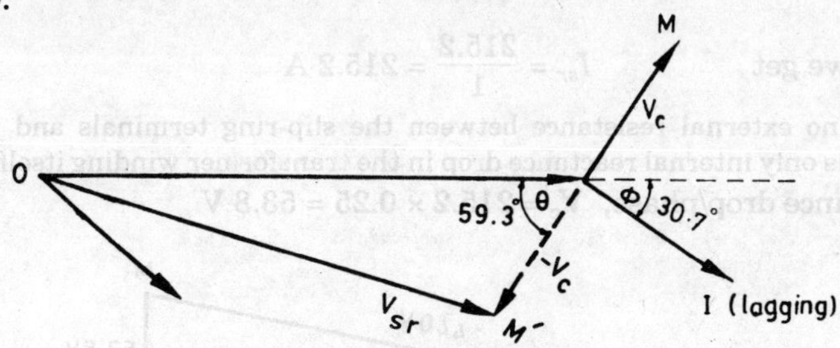

Fig. 7.29

$$\phi = \cos^{-1} 0.86 = 30.7°$$

$$\theta = (90 - \phi)° = 90 - 30.7 = 59.3°$$

From $\Delta LOM'$, we have $V_{sr}^2 = V_t^2 + V_c^2 - 2 \times V_t \times V_c \times \cos\theta$

$$= (400)^2 + (116)^2 - 2 \times 400 \times 116 \times \cos 59.3°$$

$$= 160000 + 13456 - 47378$$

or $V_{sr} = 355$ V

∴ **D.C. voltage,** $E_{dc} = \sqrt{2} \times 355 = $ **502 V. (Ans.)**

☞**Example 7.24.** *A six-ring converter supplies a direct load current of 420 amps and the power to it is supplied from a 3-phase 33 kV transformer with Δ-connection on the high voltage side and diametral connections on the secondary side. The transformer has a turn-ratio of 75 and equivalent secondary reactance of 0.25 Ω. Calculate the change in D.C. voltage when the power factor at slip-ring changes from unity to 0.86 leading.*

Assume an efficiency of 92%.

Solution. Primary phase voltage = 33 kV or 33000 V

Turn ratio = 75

Efficiency of converter η = 92% or 0.92

Equivalent secondary reactance = 0.25 Ω

Change in D.C. voltage = ?

Transformer secondary phase voltage

$$= \frac{33000}{75} = 440 \text{ V}$$

Let E_{dc} be the D.C. voltage

Power supplied $= E_{dc} \times 420$ W

Voltage between diametral slip-rings,

$$V_{sr} = \frac{E_{dc}}{\sqrt{2}}$$

Power supplied by each phase $= \dfrac{E_{dc} \times 420}{3} = 140\, E_{dc}$ W

$$\therefore \quad \frac{E_{dc}}{\sqrt{2}} \times I_{sr} \times \eta \times \cos\phi = 140\, E_{dc}$$

or $$I_{sr} = \frac{140 E_{dc} \times \sqrt{2}}{E_{dc} \times \eta \times \cos\phi} = \frac{140 \times \sqrt{2}}{0.92 \cos\phi}$$

or $$I_{sr} = \frac{215.2}{\cos\phi} \qquad \text{...}(i)$$

At unity p.f. :

From eqn. (*i*), we get $I_{sr} = \dfrac{215.2}{1} = 215.2$ A

Note. There is no external resistance between the slip-ring terminals and transformer secondary terminals, and so there is only internal reactance drop in the transformer winding itself.

Internal reactance drop/phase, $V_c = 215.2 \times 0.25 = 53.8$ V

Refer Fig. 7.30.

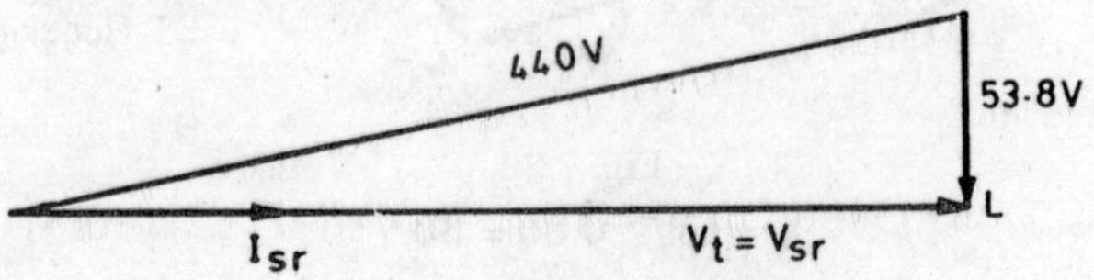

Fig. 7.30

From Δ*OLM*, $(440)^2 = V_{sr}^2 + (53.8)^2$

$$V_{sr}^2 = (440)^2 - (53.8)^2$$

$\therefore \quad V_{sr} = 436.7$ V

$\therefore \quad E_{dc} = \sqrt{2} \times 436.7 =$ **617.6 V. (Ans.)**

At 0.86 p.f. loading, $\quad \phi = \cos^{-1} 0.86 = 30.7°$

From eqn. (*i*), we get $\quad I_{sr} = \dfrac{215.2}{0.86} = 250.2$ A

Internal reactance drop/phase, $V_c = 250.2 \times 0.25 = 62.5$ V.

Refer Fig. 7.31.

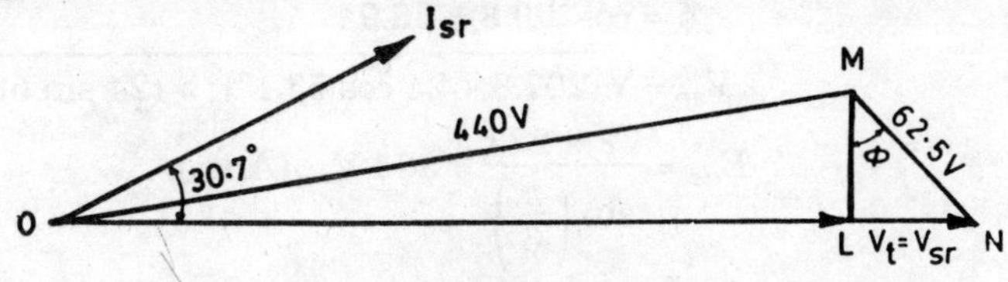

Fig. 7.31

$$V_{sr} = ON$$
$$OL^2 = OM^2 - ML^2 = (440)^2 - (62.5 \cos \phi)^2$$
$$= (440)^2 - (62.5 \times 0.86)^2$$

or $\quad OL = 436.7$ V

$$LN = 62.5 \sin \phi = 62.5 \times \sin 30.7° = 31.9 \text{ V}$$
$$V_{sr} = ON = OL + LN = 436.7 + 31.9 = 468.6 \text{ V}$$
$$E_{dc} = \sqrt{2} \times 468.6 = 662.7 \text{ V}$$

Hence, **change in D.C. volts** $= 662.7 - 617.6 =$ **45.1 V. (Ans.)**

Example 7.25. *A reactance of 0.8 Ω is included in each lead of a 3-ring converter which is supplied from a transformer giving a line voltage of 360 V. Find the commutator brush voltage, when the slip-ring current is 30 A at power factor of (i) unity, (ii) 0.8 leading.*

Neglect all losses in converter.

Solution. Value of reactance $= 0.8\ \Omega$

Transformer line voltage $= 360$ V

Slip-ring current $= 30$ A

Number of phases, $\quad m = 3$

Commutator brush voltage = ?

Transformer phase voltage $= \dfrac{360}{\sqrt{3}} = 207.8$ V

Reactance drop $= 30 \times 0.8 = 24$ V

(*i*) **At unity p.f.**

Refer Fig. 7.32.

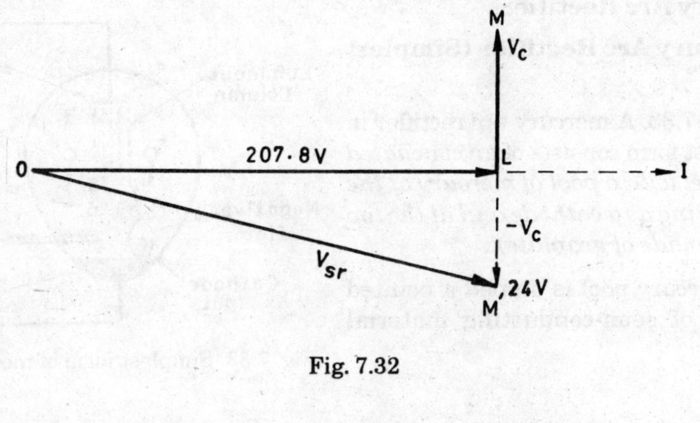

Fig. 7.32

$$V_{sr}^2 = (207.8)^2 + (24)^2$$

$$\therefore \quad V_{sr} = 209.2 \text{ V}$$

$$E_{dc} = \frac{\sqrt{2} \times 209.2}{\sin \frac{\pi}{m}} = \frac{\sqrt{2} \times 209.2}{\sin\left(\frac{\pi}{m}\right)} = \textbf{341.6 V. (Ans.)}$$

(*ii*) **At 0.8 p.f. leading**

$$\cos \phi = 0.8$$

$$\phi = \cos^{-1} 0.8 = 36.9°$$

$$V_{sr} = \sqrt{(207.8 + 24 \cos 53.1°)^2 + (24 \sin 53.1°)^2} = 223 \text{ V}$$

$$E_{dc} = \frac{\sqrt{2} \times 223}{\sin\left(\frac{\pi}{3}\right)} = \textbf{364 V. (Ans.)}$$

7.5. RECTIFIERS

7.5.1. Introduction. A rectifier may be defined as "*a device which converts alternating current into unidirectional current by virtue of a characteristic permitting appreciable flow of current in only one direction*". There are several types of rectifiers, but all exhibit the important and useful property indicated because the resistance to the flow of current *in the transmitted direction is extremely low* compared with that in the blocked direction.

Rectifiers find wide applications as converters to link together existing A.C. sources of supply and devices or power equipment that must operate from D.C. sources.

7.5.2. Types of Rectifiers. Some of the common types of rectifiers used are as given below :

1. Mercury are rectifiers.
2. Metal rectifiers (such as copper oxide or selenium rectifiers).
3. Mechanical rectifiers.
4. Electrolytic rectifiers.

- Where the power requirements are comparatively *small*, it is customary to employ solid materials types such as *copper-oxide or selenium rectifiers*, or the glass-enclosed hot-cathode type (such as Tungar or Rectigon). Mechanical and electrolytic types of rectifiers have been used on occasion, but they are generally subject to *operating difficulties and are of low efficiency*.
- For installations requiring *large amounts of power* such as traction motors and the electro-mechanical industries, the *mercury-arc type of rectifier* is the accepted standard. The commutator and its brushes on a D.C. machine are, of course, a *mechanical rectifier* that performs an important and reliable function, especially when used with interpoles and compensating windings.

7.5.3. Mercury Arc Rectifiers

7.5.3.1. Mercury Arc Rectifier (Simplest Form)

- Refer Fig. 7.33. A mercury arc rectifier in its simplest form consists of an *evacuated glass vessel with a pool of mercury at the bottom acting as a cathode and at the top is anode (made of graphite)*.
- In the mercury pool is dipped a pointed electrode of semi-conducting material

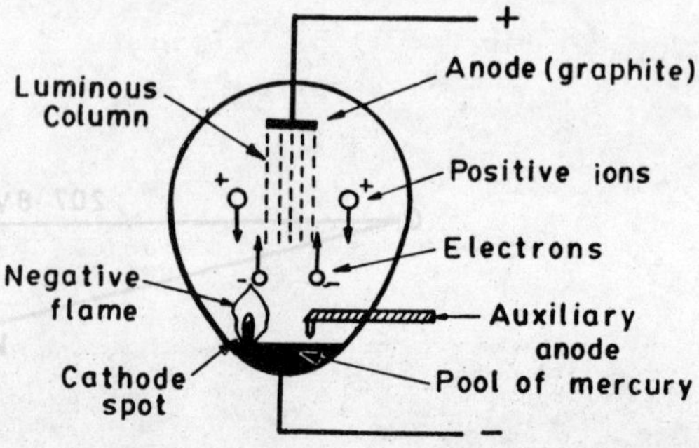

Fig. 7.33. Simplest form of mercury arc rectifier.

Boron carbide called the *ignited* or *auxiliary anode*. When the rectifier is fired, for a short duration current pulse is applied between the auxiliary anode and cathode which heats the surface of mercury and produces certain initial electrons. The high positive potential applied to anode attracts these electrons. The electrons reaching anode takes the form of an arc which starts at the anode and it is heated up subsequently.

- There is a luminous column next to anode which gives light. After the column there is a dark gap followed by a white hot cathode spot which travels irregularly over the mercury pool with high speed.
- The rectifier's cathode can only emit these electrons and is unable to receive them. Thus in Fig. 7.34, if the battery terminals are reversed making the anode negative and cathode positive, the electrons will be repelled and no current will flow. Thus it has a valve action which allows current to flow only in one direction *i.e.* from anode to cathode. If the battery is replaced by an A.C. supply there will be current flowing only during the + ve half of the cycle.

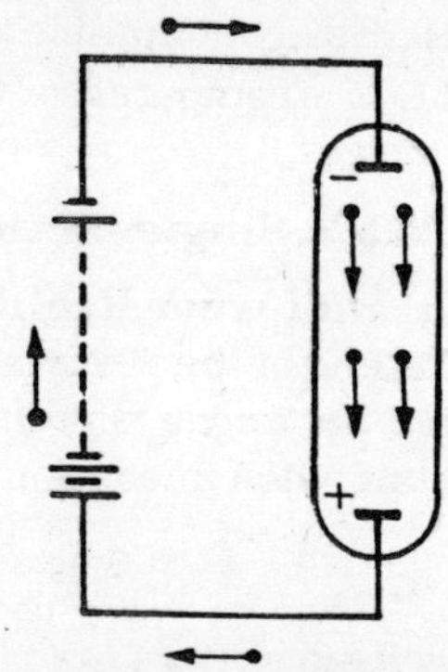

Fig. 7.34. Glass bulb containing gas and two electrodes against which potential difference is applied.

Arc volt drop. The voltage drop between the anode and cathode comprises the following :

(*i*) Anode drop. (*ii*) Cathode drop.
(*iii*) Arc drop.

- **Anode drop.** The anode collects the electrons emitted by the cathode and the waiting electrons here form a negative space and the coming electrons are repelled by this space charge. The repelled electrons overcome a certain potential drop which is of tune of about 5 V.
- **Cathode drop.** The cathode is bombarded by the positive ions, consequently a high potential gradient is caused due to which a large voltage drop takes place. The cathode drop is about 9 V.
- **Arc drop.** The potential drop in the arc varies from 0.05 V to 0.2 V per cm length of the arc. The average value of this drop can be assumed to be $0.1\ V \times l$, where l is the arc length.

Thus total drop equals $(5 + 9 + 0.1\ l)$V.

Choice of mercury as cathode. The mercury has been chosen for cathode (in preference to any other solid) due to the following **reasons** :

(*i*) Mercury possesses a self-restoring characteristic. The mercury vapours after being condensed, trickle back to the mercury pool.
(*ii*) Its ionization potential is relatively low (10.4 V approx.).
(*iii*) Its atomic weight is quite high.
(*iv*) Its boiling point and specific heat are low.
(*v*) Its latent heat of evaporation is small.
(*vi*) At ordinary temperature it remains in liquid state.

7.5.3.2. Merits of a Mercury Arc Rectifier. The mercury arc rectifier has made large inroads in the railway and manufacturing sub-station field. It entails the following *merits/advantages* :

1. Light weight and small floor space.
2. Comparatively simple operation.
3. Readily made automatic in operation.
4. Comparatively noiseless.

5. High efficiency.
6. Quick response to load demands.
7. Maintenance and attention small.
8. High over-load capacity.
9. Rectifiers of higher rating are economical.
10. Easily adaptable to variable frequency operation.
11. It is comparatively less effected by A.C. system disturbances especially on account of high voltage surges.

7.5.3.3. Single-phase Mercury Arc Rectifier

(*a*) **Half-wave Rectification.** Fig. 7.35 shows a half-wave rectifier having one anode only to which one end of the secondary of the transformer is connected. The *current in the external load flows only when the anode is positive, i.e.,* during the first-half of the cycle, no current flows during the second-half when anode is negative.

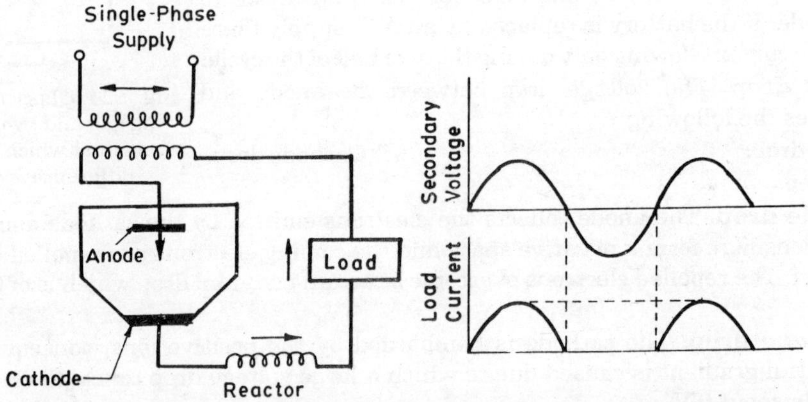

Fig. 7.35. Single-phase half-wave glass bulb mercury arc rectifier.

(*b*) **Full-wave rectification.** Fig. 7.36 shows a full wave rectifier. It has two anodes A_1 and A_2 and the load is connected between the cathode C and the centre of the secondary of the transformer.

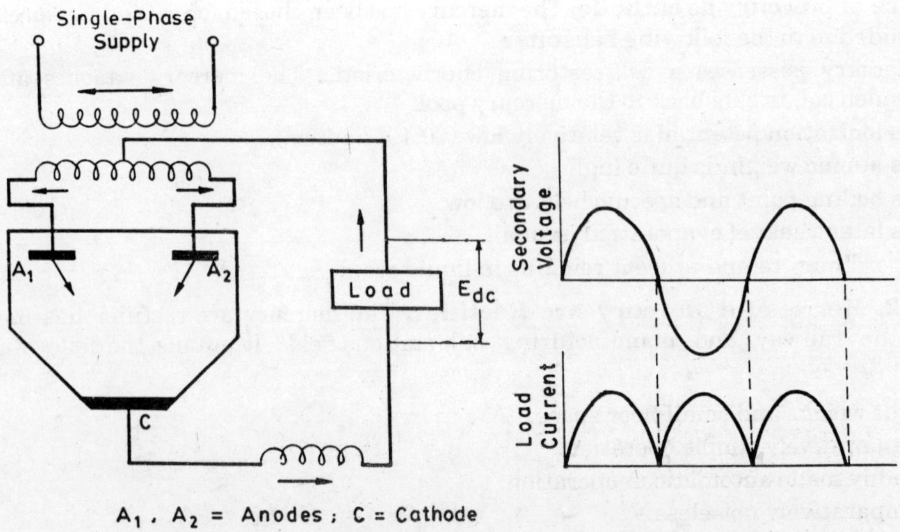

Fig. 7.36. Single-phase full wave glass bulb mercury arc rectifier.

When the current in the secondary is + ve, say from right to left, the anode A_1 will conduct. During the second half of the cycle the current in the secondary of the winding is from left to right. So now anode A_2 will conduct while the direction of the current in the load remains same. The period of conduction of each anode is $2\pi/2 = \pi$. Such a rectifier which utilises both halves of the alternating wave is called a full wave rectifier.

7.5.3.4. Polyphase Mercury-arc Rectifiers

(i) **Three-phase Rectifier.** A three-phase rectifier has three anodes and is connected to the 3-phase supply through three-phase transformers connected star-star or delta-star fashion (Fig. 7.37). In this case, each anode conducts for one-third of a cycle *i.e.*, over an angular distance of $2\pi/3$.

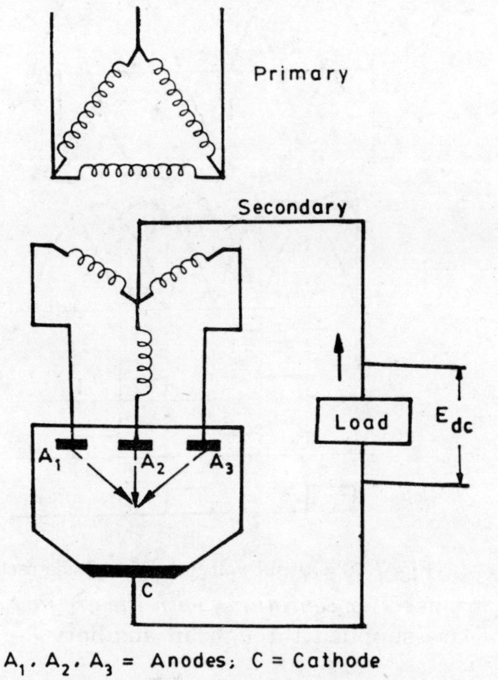

Fig. 7.37. Three-phase three-anodes half wave rectifier.

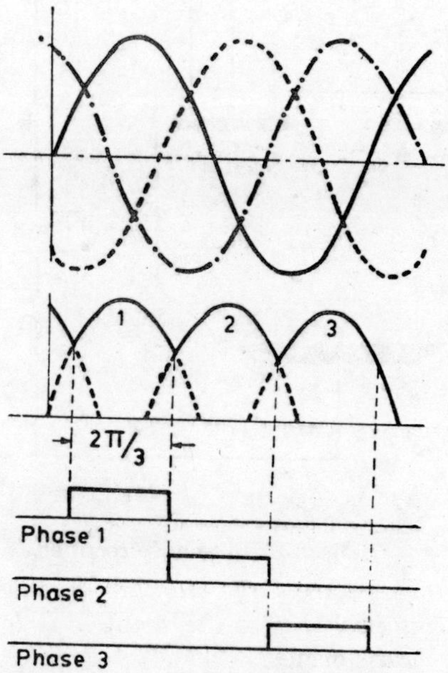

Fig. 7.38. Output voltage of three-phase three-anodes rectifier.

The output voltage wave of three-phase three-anode rectifier is shown in Fig. 7.38.

(ii) **Six-phase rectifier.** Fig. 7.39 shows simplest form of six-phase connection for a mercury-arc rectifier. This is known as a delta-double star connection. The primary supply is 3-phase but 6-phase secondary supply is obtained by having two secondaries for each primary. Each anode conducts for one-sixth of a cycle in each cycle *i.e.*, corresponding to angular distance of $2\pi/6$. In general if there are m phases each phase will be delivering current for a period of $2\pi/m$ radians.

The output voltage wave of six-phase rectifier is shown in Fig. 7.40. On comparing Figs. 7.38 and 7.40 we find that the ripple of the rectified wave is smoothened as the number of phases are increased.

Working of six-phase rectifier. Fig. 7.41 shows the detailed connections for a 6-phase mercury are rectifier.

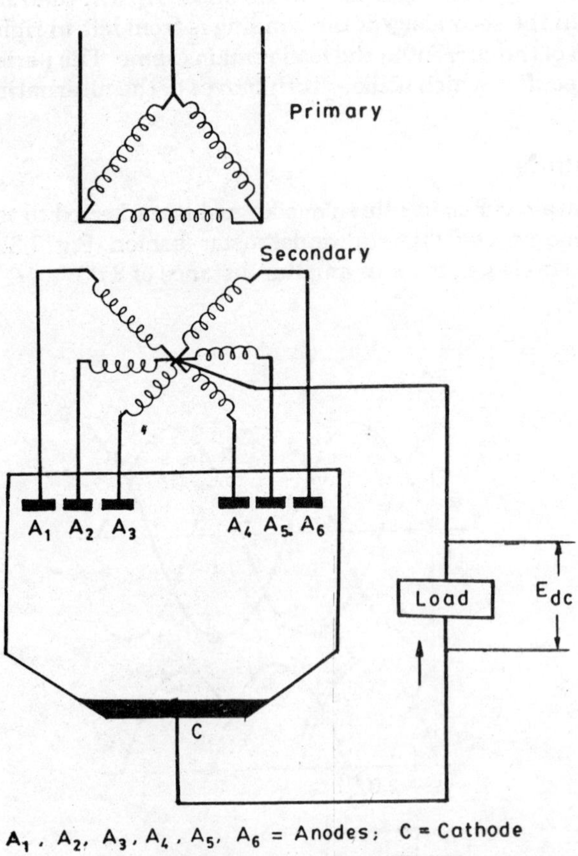

Fig. 7.39. Six-phase rectifier.

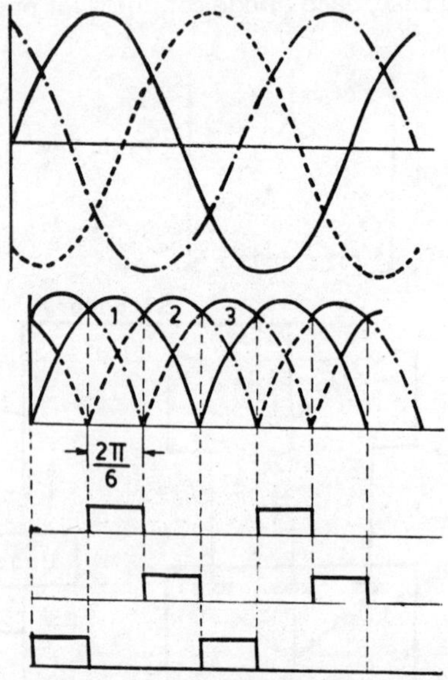

Fig. 7.40. Output voltage of six-phase rectifier.

- There are three *auxiliary anodes.* These are used for *continuing with the arc just after ignition before the rectifier is loaded* and are suppiied through an auxiliary 3-phase transformer.
- On switching the transformer the relay contact is closed and subsequently the magnet is energised. The magnet then pulls the flexible ignition electrode causing it to dip into mercury, a current limited by the reactor flows into the circuit.
- As the current flows through the cathode, the electromagnet is short-circuited and the ignition anode is released, at the same time the relay is opened.
- After ignition the *cathode spot* is formed and the *arc* is struck between the auxiliary anode turn by turn.
- When the rectifier is *loaded* the arc is transferred to main electrodes.
- When the load is *switched off*, then the arc is transferred to auxiliary anodes.

Thus the *auxiliary anodes keep the rectifier active when it is not loaded.*

The anode current supplied by each anode will be *rectangular waveform*, which can be resolved into *mean D.C. component and harmonic components*. The primary current will be having purely A.C, components alone. The D.C. component of the secondary will be confined to secondary side alone and produces a constant flux which will have its path completed from core to the tank body. This magnetization of the core due to D.C. components should be avoided since it leads to *saturation of*

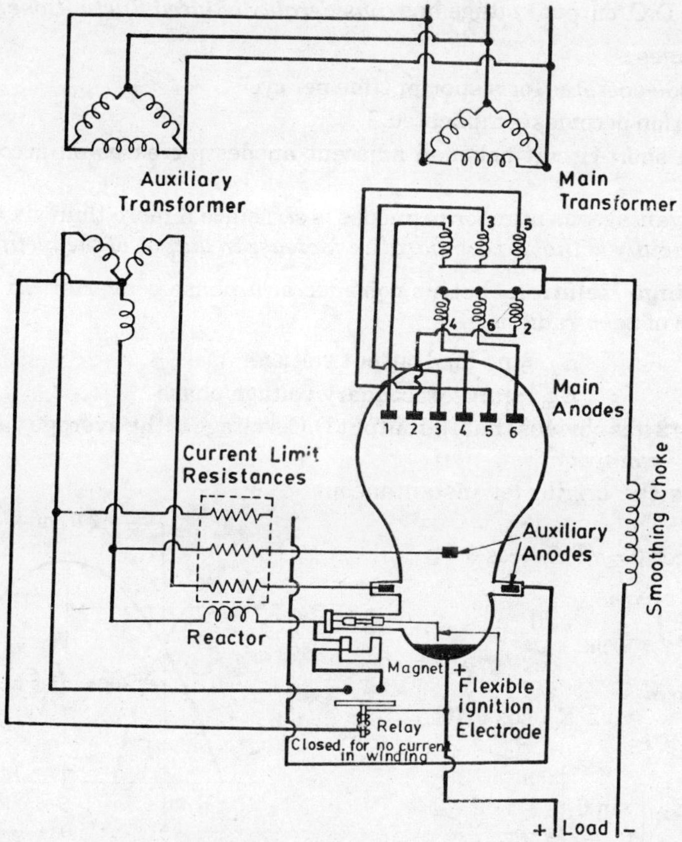

Fig. 7.41. Connection diagram of 6-phase rectifier.

the cores, the production of harmonics causes high magnetizing current. To avoid such defects '*zig-zag transformer* may be used. In such a transformer each phase has two winding sections in series wound in opposite direction on two different cores. As the D.C. component current in each phase is same, so that D.C. ampere turn on each core is zero.

Note 1. *Glass rectifiers,* because of inherent mechanical weakness of glass, cannot be manufactured for very large outputs. These are usually made into units capable of D.C. output of 500 A at 500 V (maximum continuous rating).

Note 2. *Steel rectifiers* are mounted in a vacuum chamber of steel, are *dismountable, continuously evacuated* and are *water cooled.* Because water-cooling is employed, hence the condensing chamber is much smaller than that of the glass bulb type rectifier. The arc operates in a large vacuum tank which is made out of a special quality of sheet iron that *resists the action of mercury vapours and helps in maintenance of consistently good vacuum.* In between the condensing chamber and the vacuum tank is the anode plate which carries the main anodes in the ring. Necessary high vacuum is maintained by a pumping system consisting of a mercury vapour diffusion pump backed by a rotary pump working under oil. Ignition and excitation of steel tank rectifiers is *different* from that of the glass-bulb type rectifiers. Such units have been constructed for as much as 16000 A at comparatively low voltages, 2500 kW at 3.3 kV and for voltages as high as 20 kV at 600 kW. Voltages up to 30 kV can be obtained only by derating the current.

Choice of number of anodes. The use of a greater number of anodes has the following *advantages* and *disadvantages* :

Advantage. D.C. output voltage has *considerably reduced fluctuations*.

Disadvantages :

(*i*) Each anode operates for a shorter time per cycle.

(*ii*) Construction becomes complicated.

(*iii*) Risk of a short-circuit between adjacent anodes increases on account of their relative nearness.

The most advantageous number of anodes is *six*, since if more than six anodes are used then the *cost increases rapidly without a comparative increase in output of the rectifier*.

7.5.3.5. Voltage Relation. Let us consider a m-phase converter. In this case anode will conduct for a period of $2\pi/m$ radians.

Let E_{dc} = no-load output voltage

E_{ac} = r.m.s. secondary voltage/phase

From Fig. 7.42 it is obvious that the output D.C. voltage is the average value of applied voltage between ordinates $-\pi/m$ to π/m.

Taking O as the origin, let instantaneous voltage be given by

$$e = E_{max} \cos \theta = \sqrt{2}\, E_{ac} \cos \theta$$

$$E_{dc} = \frac{1}{2\pi/m} \int_{-\pi/m}^{\pi/m} e d\theta$$

$$= \frac{m}{2\pi} \int_{-\pi/m}^{\pi/m} \sqrt{2}\, E_{ac} \cos \theta \, d\theta$$

$$= \frac{m}{2\pi} \sqrt{2} E_{ac} \Big[\sin \theta \Big]_{-\pi/m}^{\pi/m}$$

$$= \frac{m}{2\pi} \sqrt{2} E_{ac} \times 2 \sin \frac{\pi}{m} = \sqrt{2}\, \frac{m}{\pi} E_{ac} \sin \frac{\pi}{m}$$

$$\therefore \quad E_{dc} = \sqrt{2}\, E_{ac} \frac{m}{\pi} \sin \frac{\pi}{m} \qquad ...(7.6)$$

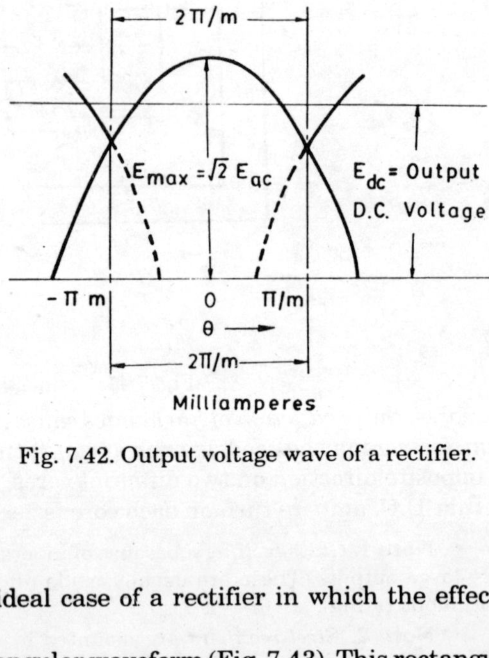

Fig. 7.42. Output voltage wave of a rectifier.

Terminal D.C. voltage,

$$V_{dc} = E_{ac} - E_a$$

where E_a = voltage drop in arc.

7.5.3.6. Current Relations. Let us take an ideal case of a rectifier in which the effect of transformer reactance is neglected.

Let I_{dc} be the pure output current having a rectangular waveform (Fig. 7.43). This rectangular wave of current shown in Fig. 7.43 (*a*) can be resolved into fundamental and harmonic components of cosine function of the form $a_n \cos n\theta$ between limits from $-\pi/m$ to π/m. The amplitude of nth harmonic is given as

$$a_n = I_{dc} \times \int_{-\pi/m}^{\pi/m} \cos n\theta \, \theta d\theta = \frac{2I_{dc}}{\pi} \sin \frac{n\pi}{m}$$

Since each phase conducts only for a period of $2\pi/m$ radian so the current supplied by each phase is as in Fig. 7.43 (*b*).

$\therefore$ Average current supplied by each phase $= \dfrac{I_{dc}}{m}$

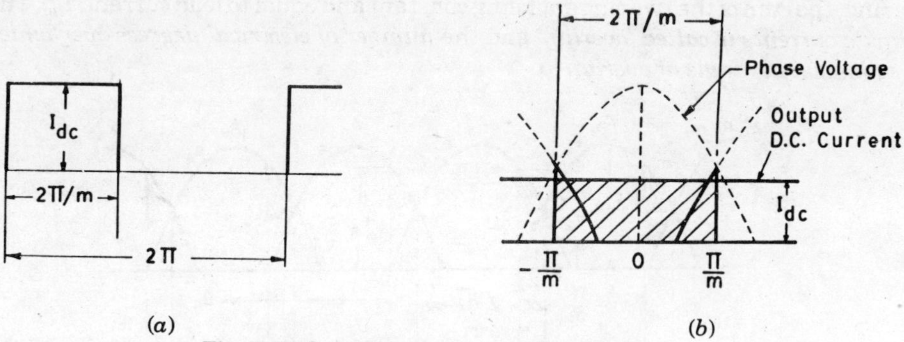

(a) (b)

Fig. 7.43. Output current wave of a rectifier.

If r.m.s. value of current/phase is $I_{r.m.s.}$, then

$$I^2_{r.m.s.} = \frac{1}{2\pi}\int_0^{2\pi/m} I_{dc}^2\, d\theta = \frac{1}{2\pi} I_{dc}^2 \Big[\theta\Big]_0^{2\pi/m} = I_{dc}^2 \times \frac{2\pi}{2\pi \times m} = \frac{I_{dc}^2}{m}$$

or

$$I_{r.m.s.} = \frac{I_{dc}}{\sqrt{m}} \qquad ...(7.7)$$

Now

$$I_{av}/\text{phase} = \frac{1}{2\pi}\int_0^{2\pi/m} I_{dc}\, d\theta = \frac{1}{2\pi} \times I_{dc}\Big[\theta\Big]_0^{2\pi/m} = I_{dc} \times \frac{2\pi}{2\pi \times m} = \frac{I_{dc}}{m}$$

$$\therefore \quad I_{ac}/\text{phase} = \frac{I_{dc}}{m}. \qquad ...(7.8)$$

7.5.3.7. Utility Factor. Utility factor is defined as the *ratio of the volt-amperes obtained from the rectifier on the output side to volt-amperes supplied from A.C. input side.*

$$\therefore \text{ Utility factor } = \frac{E_{dc}\, I_{dc}}{mE_{ac}\, I_{r.m.s.}}$$

$$= \frac{\sqrt{2}E_{ac}\left(\frac{m}{\pi}\right)\sin\frac{\pi}{m} \times \sqrt{m} \times I_{r.m.s.}}{mE_{ac}\, I_{r.m.s.}}$$

$$= \frac{(\sqrt{2m})}{\pi}\sin\frac{\pi}{m} \qquad ...(7.9)$$

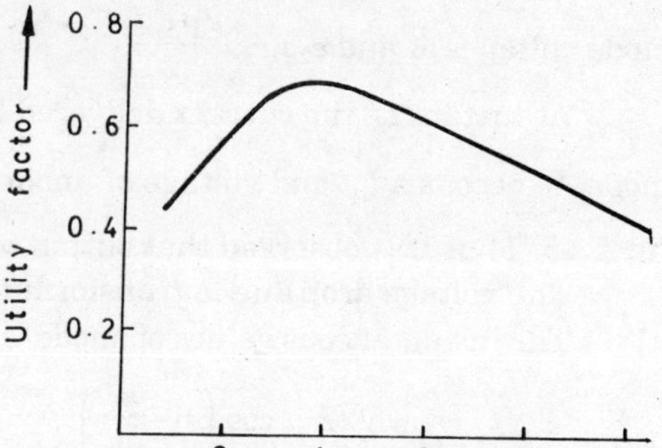

Fig. 7.44. Relation between utility factor and m.

Fig. 7.44 shows the relation between utility factor and m and it is obvious that utility factor is maximum for m = 2.7. Since m has integral values, hence utility factor is maximum for a rectifier which has *three* phases.

7.5.3.8. Effect of Transformer Reactance. In the Article 7.5.3.6 it was assumed that the current supplied by each phase has rectangular blocks, *i.e.*, the current supplied by the conducting anodes subsides immediately and the other anode begins to conduct instantaneously. But due to the leakage reactance of the transformer there can never be an infinite rise of current at the time when the anode begins to conduct and there can never be an infinite rate of decrease of current when the conduction of anode sub-sides. In practice the anode current gradually decreases towards the end of the conducting period of one anode, similarly, increases gradually when conduction begins at the next anode. In other words, it means that while the current of one anode is decreasing, that of another is increasing, the sum of the two current being constant and equal to load current, I_{dc}. This *co-existence of two anode currents is called 'overlap'* and the *number of electrical degrees over which co-existence persists is known as 'angle of overlap'* α.

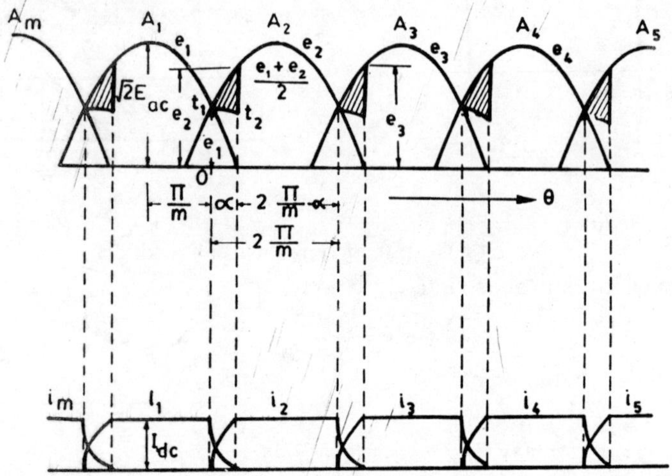

Fig. 7.45. Current and voltage waveforms in a rectifier.

Let us consider m-anode rectifier.

Consider an instant t_1, the instantaneous voltage e_1 of anode A_1 becomes equal to the instantaneous voltage e_2 of anode A_2. At this instant the anode A_2 starts conducting and current i_1 supplied by A_1 would have extinguished immediately in case of no transformer leakage reactance, but on account of transformer reactance i_1 falls gradually. The resultant voltage during the period of overlap of currents supplied by the anodes A_1 and A_2 is equal to mean of instantaneous values of anode voltages e_1 and e_2 *i.e.*, $\frac{e_1 + e_2}{2}$.

At instant t_2, the current delivered by anode A_1 becomes zero, therefore current delivered by anode A_2 becomes I_{dc} and voltage of anode A_2 *i.e.*, e_2 rises sharply from $\frac{e_1 + e_2}{2}$ to e_2 as shown in Fig. 7.45. Thus it is observed that output voltage is reduced due to transformer reactance.

The voltage drop due to transformer reactance is calculated as follows :

The instantaneous values of anode voltages, taking the origin at 0, are given by the relations :

$$e_1 = \sqrt{2} E_{ac} \cos\left(\theta + \frac{\pi}{m}\right)$$

and $$e_2 = \sqrt{2} E_{ac} \cos\left(\theta - \frac{\pi}{m}\right)$$

D.C. output voltage,

$$E_{dc} = \frac{\int_0^{\alpha} \left(\frac{e_1 + e_2}{2}\right) d\theta + \int_{\alpha}^{2\pi/m} e_2 \, d\theta}{2\pi/m}$$

$$= \frac{m}{2\pi} \sqrt{2} \, E_{ac} \left[\int_0^{\alpha} \frac{1}{2}\left\{\cos\left(\theta + \frac{\pi}{m}\right) + \cos\left(\theta - \frac{\pi}{m}\right)\right\} d\theta + \int_{\alpha}^{2\pi/m} \cos\left(\theta - \frac{\pi}{m}\right) d\theta\right]$$

$$= \frac{m}{2\pi} \sqrt{2}\, E_{ac} \left[\int_0^{\alpha} \cos\theta \,.\, \cos\frac{\pi}{m}\, d\theta + \int_{\alpha}^{2\pi/m} \cos\left(\theta - \frac{\pi}{m}\right) d\theta \right]$$

$$= \frac{m}{2\pi} \sqrt{2}\, E_{ac} \left[\cos\frac{\pi}{m} \left\{ \sin\theta \right\}_0^{\alpha} + \left\{ \sin\left(\theta - \frac{\pi}{m}\right) \right\}_{\alpha}^{2\pi/m} \right]$$

$$= \frac{m}{2\pi} \sqrt{2}\, E_{ac} \left[\cos\frac{\pi}{m} \sin\alpha + \sin\frac{\pi}{m} - \sin\left(\alpha - \frac{\pi}{m}\right) \right]$$

$$= \frac{m}{2\pi} \sqrt{2}\, E_{ac} \left[\cos\frac{\pi}{m} \sin\alpha + \sin\frac{\pi}{m} - \sin\alpha \cos\frac{\pi}{m} + \cos\alpha \sin\frac{\pi}{m} \right]$$

$$= \frac{m}{2\pi} \sqrt{2}\, E_{ac} \sin\frac{\pi}{m} (1 + \cos\alpha) = \frac{m}{\pi} \sqrt{2}\, E_{ac} \sin\frac{\pi}{m} \left(\frac{1 + \cos\alpha}{2} \right)$$

$$= \frac{m}{\pi} \sqrt{2}\, E_{ac} \sin\frac{\pi}{m} \cos^2\frac{\alpha}{2} \qquad \text{...}(i)$$

Neglecting reactance drop, D.C. output voltage is given by,

$$E_{dc} = \frac{m}{\pi} \sqrt{2}\, E_{ac} \sin\frac{\pi}{m} \qquad \text{...}(ii)$$

From (*i*) and (*ii*),

D.C. output voltage taking *reactance drop into account*

$$= \text{D.C. output voltage with } \textit{no reactance drop} \times \cos^2\frac{\alpha}{2}$$

$$= E_{dc} \times \cos^2\frac{\alpha}{2}. \qquad \text{...[7.10 }(a)]$$

Voltage drop due to reactance,

$$= \frac{m}{\pi} \sqrt{2}\, E_{ac} \sin\frac{\pi}{m} - \frac{m}{\pi} \sqrt{2}\, E_{ac} \sin\frac{\pi}{m} \cos^2\frac{\alpha}{2}$$

$$= \frac{m}{\pi} \sqrt{2}\, E_{ac} \sin\frac{\pi}{m} \left(1 - \cos^2\frac{\alpha}{2} \right) = E_{dc} \sin^2\frac{\alpha}{2}.$$

Thus, **voltage drop due to reactance**

$$= E_{dc} \sin^2\frac{\alpha}{2}. \qquad \text{...[7.10 }(b)]$$

Angle of Overlap. It is the leakage reactance of the transformer on which the magnitude of angle of overlap depends. If L is the equivalent leakage reactance per phase of transformer referred to secondary, then resultant D.C. output voltage during the period of overlap (neglecting resistance drop) is given by

$$e = e_1 - L\,.\,\frac{di_1}{dt} \qquad \text{...(7.11)}$$

Also,

$$e = e_2 - L\,\frac{di_2}{dt} \qquad \text{...(7.12)}$$

where e_1, e_2 = phase voltage

and i_1, i_2 = phase currents

Adding eqns. (7.11) and (7.12), we get

$$2e = e_1 + e_2 - L\left(\frac{di_1}{dt} + \frac{di_2}{dt}\right) = e_1 + e_2 - L\left[\frac{d}{dt}(i_1 + i_2)\right]$$

$$= e_1 + e_2 - L \,.\, \frac{dI_{dc}}{dt} \qquad (\because\ i_1 + i_2 = I_{dc})$$

$$= e_1 + e_2 - L \times 0 \qquad (\because\ I_{dc} \text{ is constant})$$

$$= e_1 + e_2$$

$$\therefore \qquad e = \frac{e_1 + e_2}{2}$$

Putting the value of e in eqn. (7.11), we get

$$\frac{e_1 + e_2}{2} = e_1 - L \,.\, \frac{di_1}{dt}$$

or $$L\frac{di_1}{dt} = e_1 - \frac{e_1 + e_2}{2} = \frac{e_1 - e_2}{2} \quad \text{or} \quad \frac{di_1}{dt} = \frac{e_1 - e_2}{2L}$$

Since $$e_1 - e_2 = \sqrt{2}E_{ac}\left[\cos\left(\theta + \frac{\pi}{m}\right) - \cos\left(\theta - \frac{\pi}{m}\right)\right]$$

$$= \sqrt{2}\, E_{ac}\left[-2 \sin\theta \sin\frac{\pi}{m}\right] = -2\sqrt{2}E_{ac} \sin\frac{\pi}{m} \sin\theta$$

$$\therefore \qquad \frac{di_1}{dt} = \frac{1}{2L}\left(-2\sqrt{2}E_{ac} \sin\frac{\pi}{m} \sin\theta\right)$$

$$= -\frac{\sqrt{2}}{L} E_{ac} \sin\frac{\pi}{m} \sin\omega t \qquad (\because\ \theta = \omega t)$$

or $$i_1 = \frac{-\sqrt{2}}{L} E_{ac} \sin\frac{\pi}{m}\int \sin\omega t dt$$

or $$i_1 = \frac{\sqrt{2}E_{ac}}{\omega L} \sin\frac{\pi}{m} \cos\omega t + k$$

where k is a constant of integration.

Now when $t = 0$, $i_1 = I_{dc}$, therefore to get the value of k, substituting $t = 0$ and $i_1 = I_{dc}$ in the above expression, we get

$$I_{dc} = \frac{\sqrt{2}E_{ac}}{\omega L} \sin\frac{\pi}{m} + k$$

or $$k = I_{dc} - \frac{\sqrt{2}E_{ac}}{\omega L} \sin\frac{\pi}{m}$$

Hence, $$i_1 = \frac{\sqrt{2}E_{ac}}{\omega L} \sin\frac{\pi}{m} \cos\omega t + I_{dc} - \frac{\sqrt{2}E_{ac}}{\omega L} \sin\frac{\pi}{m}$$

or $$i_1 = I_{dc} - \frac{\sqrt{2}E_{ac}}{\omega L} \sin\frac{\pi}{m}(1 - \cos\omega t)$$

or $$i_1 = I_{dc} - \frac{\sqrt{2}E_{ac}}{\omega L} \sin\frac{\pi}{m}(1 - \cos\theta) \qquad (\because\ \omega t = \theta)$$

or $$i_1 = I_{dc} - \frac{\sqrt{2}E_{ac}}{X} \sin\frac{\pi}{m}(1 - \cos\theta) \qquad (\because\ X = \omega L)$$

Since when $\theta = \alpha$, $i_1 = 0$, therefore substituting $i_1 = 0$ and $\theta = \alpha$ in above expression, we get

$$0 = I_{dc} - \frac{\sqrt{2}E_{ac}}{X} \sin\frac{\pi}{m}(1 - \cos\alpha)$$

or
$$1 - \cos\alpha = \frac{I_{dc} X}{\sqrt{2}E_{ac} \sin\frac{\pi}{m}}$$

or
$$\cos\alpha = 1 - \frac{I_{dc} X}{\sqrt{2}E_{ac} \sin\frac{\pi}{m}} \quad ...[7.13\ (a)]$$

$$\alpha = \cos^{-1}\left(1 - \frac{I_{dc} X}{\sqrt{2}E_{ac} \sin\frac{\pi}{m}}\right) \quad ...[7.13\ (b)]$$

From the above expression we find that the angle of overlap can be determined, if D.C. output current I_{dc}, r.m.s. value of A.C. voltage between each anode and neutral, E_{ac}, the number of phases and leakage reactance per phase, X are known.

Now
$$\cos\alpha = 1 - 2\sin^2\frac{\alpha}{2}$$

Substituting this value in eqn. [7.13 (*a*)], we get

$$1 - 2\sin^2\frac{\alpha}{2} = 1 - \frac{I_{dc} X}{\sqrt{2}E_{ac} \sin\frac{\pi}{m}}$$

or
$$\sin^2\frac{\alpha}{2} = \frac{I_{dc} X}{2\sqrt{2}E_{ac} \sin\frac{\pi}{m}} \quad ...(7.14)$$

or
$$\sin\frac{\alpha}{2} = \left(\frac{I_{dc} X}{2\sqrt{2}E_{ac} \sin\frac{\pi}{m}}\right)^{1/2} \quad ...(7.15)$$

Hence, angle of overlap depends directly on I_{dc} and X.

Now, multiplying both sides of eqn. (7.14) by E_{dc}, we get

$$E_{dc}\sin^2\frac{\alpha}{2} = \frac{E_{dc} I_{dc} X}{2\sqrt{2}E_{ac} \sin\frac{\pi}{m}} \quad ...(7.16)$$

It *shows that reduction in output voltage due to transformer reactance is a linear function of the load current I_{dc}. That is why a rectifier possesses a straight line regulation characteristics.*

7.5.3.9. Regulation. When a rectifier is loaded the following three types of voltage drops take place :

1. Voltage drop in transformer reactance.
2. Voltage drop in resistance of transformer and smoothing chokes.
3. Arc voltage drop = E_a. It may be assumed constant for all practical purposes.

Voltage drop in transformer reactance

$$= \frac{m}{\pi}\sqrt{2}E_{ac} \sin\frac{\pi}{m}\sin^2\frac{\alpha}{2}$$

But $$\sin^2 \alpha = \frac{1 - \cos \alpha}{2} = \frac{1 - \left[1 - \dfrac{I_{dc} X}{\sqrt{2} E_{ac} \sin \dfrac{\pi}{m}}\right]}{2} = \frac{I_{dc} X}{2\sqrt{2} E_{ac} \sin \dfrac{\pi}{m}}.$$

∴ Voltage drop in transformer reactance

$$= \frac{m}{\pi} \sqrt{2} E_{ac} \sin \frac{\pi}{m} \times \frac{I_{dc} X}{2\sqrt{2} E_{ac} \sin \dfrac{\pi}{m}}$$

$$= \frac{m}{2\pi} I_{dc} X \quad \text{...[7.17 (a)]}$$

Now, total voltage drop $= \frac{m}{2\pi} I_{dc} X + I_{ac} R + E_a$

Terminal voltage at no-load,

$$V_{dco} = E_{dc} - E_a$$

Terminal voltage on load,

$$V_{acl} = E_{dc} - \left(\frac{m}{2\pi} I_{dc} X + I_{dc} R + E_a\right)$$

% Regulation $= \frac{V_{dco} - V_{dcl}}{V_{dco}} \times 100$ ...[7.17 (b)]

7.5.3.10. Interphase Transformer. The following are the factors on which the number of phases for which a rectifier should be designed depend :

1. The voltage regulation of the rectifier should be low.
2. In the output circuit there should be no harmonics.
3. The power factor of the system should be high.
4. The rectifier supply transformer should be utilized to the best advantage.

The condition 2 requires that the number of phases should be high but the conditions 1, 3 and 4 demand that the number of phases should be kept small/low. The merits of a *three-phase rectifier* such as *better transformer utilisation factor, high power factor* and *low voltage regulation* and merits of *six* or *twelve* phase rectifier of low harmonic percentage can be had together by using *interphase transformer* with six anodes (or twelve anodes) rectifiers as shown in Fig. 7.46.

In this arrangement/scheme the m anodes of the rectifier are split into a number of star-connected groups of three *e.g.* two groups for a six-anode and four groups for a twelve anode rectifier. Each group has its own individual secondary winding and the star points of the previous groups, instead of being directly connected together are connected *through interphase transformer.* The common neutral point, forms the – ve pole of the system.

Fig. 7.46 shows the connection diagram/scheme for a six-anode rectifier. The interphase transformer performs *the function of equilising the potential of two anodes at a time so that at any given instant the load may be shared by those two anodes which are effectively working in parallel.* Each phase of transformer operates for one-third of the cycle instead of one-sixth and the mean output voltage is the same as that of a rectifier with three anodes instead of six.

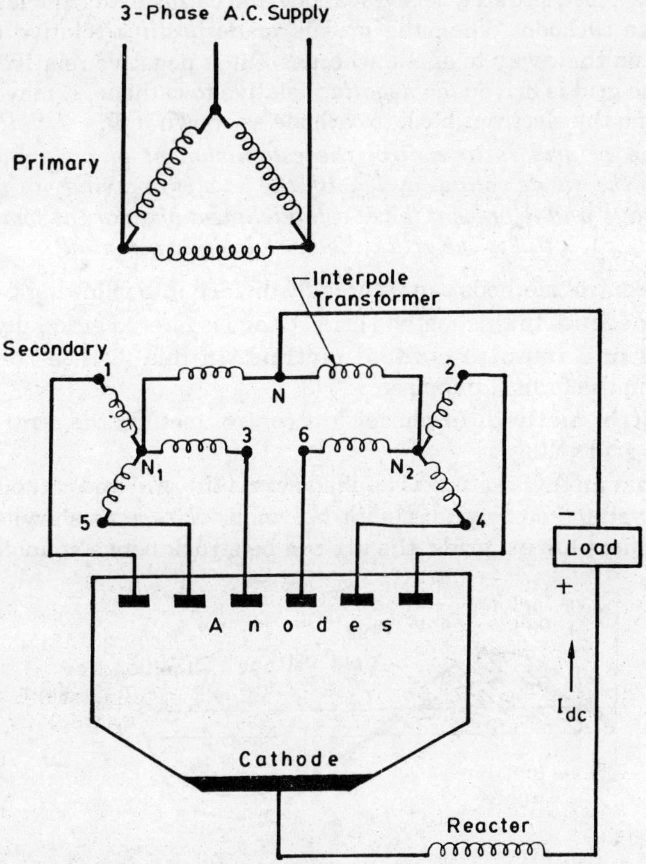

Fig. 7.46. Connection of interphase transformer on six-phase rectifier.

7.5.3.11. Grid Control of Mercury Arc Rectifiers. If a perforated *grid* is interposed between the anode and cathode the movement of the electrons towards the anode can be controlled by means of assigning proper voltages to it.

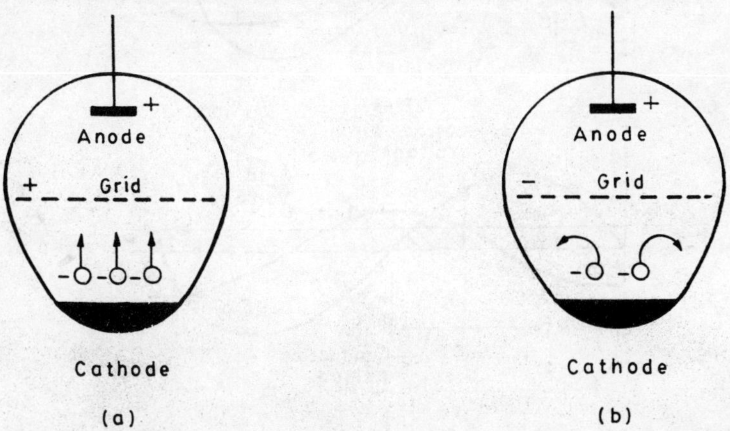

Fig. 7.47. (*a*) Grid having + ve potential, (*b*) Grid having – ve potential.

Fig. 7.47 (*a*) shows a grid situated between anode and cathode, the anode being kept at a + ve potential with respect to cathode. When the grid is made *positive* relative to cathode, then it *accelerates the electrons* on their way to anode whereas if it is negative relative to cathode, then it *deaccelerates* them. If the grid is driven *too negative* relative to cathode, it may altogether stop the anode current by repelling the electrons back to cathode as shown in Fig. 7.47 (*b*).

Thus the *function of grid is to control the establishment of an arc on the anode at a predetermined instant of the anode voltage cycle with the help of applying +ve potential to the grid known as 'liberating voltage' and to prevent the re-establishment of an arc at some other instant with the application of negative potential to the grid called as 'blocking potential'.*

The various grid control methods can be broadly divided into following two classes :

1. **Soft control method.** In this method grid voltage is altered gradually.

2. **Hard control and impulse control method.** In this method the variations of grid voltages are sudden *i.e.* in the form of impulses.

Phase-shift control method. In phase-shift control method the control is carried out by varying the phase of the grid voltage.

Let us consider that an A.C. voltage is applied across the grid and cathode.

- Let the grid voltage be *'in phase' with the anode voltage* as shown in Fig. 7.48 (*a*). In mercury rectifiers having grids, the arc can be struck between anode and cathode only

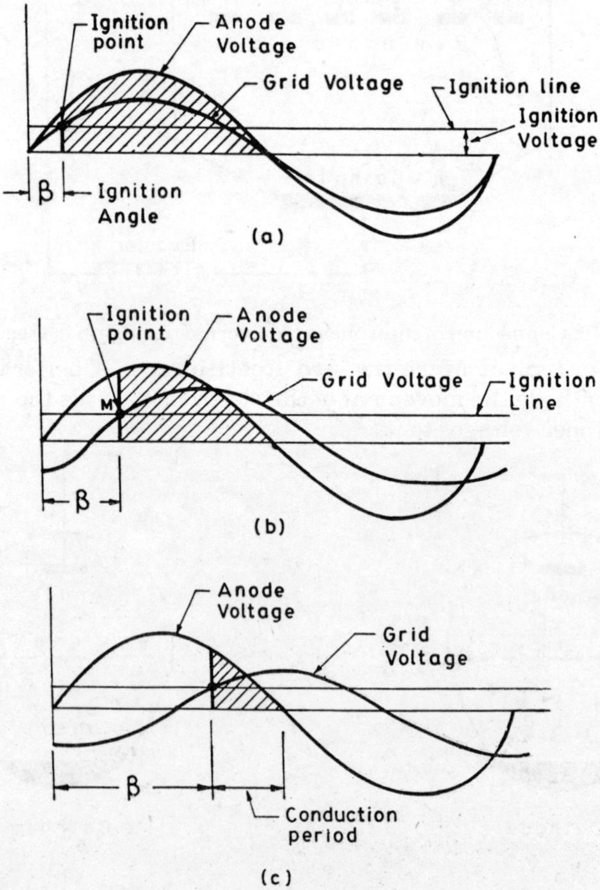

Fig. 7.48. Phase-shift control method.

when the grid attains a certain minimum potential, this potential being known as *'ignition voltage'* or *'critical grid voltage'*. The ordinate of the ignition line represents the ignition voltage or critical grid voltage. The ignition point is given by the intersection of ignition line and the grid voltage curve (*i.e.*, point L in Fig. 7.48 (*a*)). Hence the arc is established at L and once it is struck, the grid has no further effect on the performance of the rectifier and the arc continues till anode voltage falls to zero. The angle β, measured from zero of the anode voltage, by which *ignition is delayed* is known as *'ignition angle'*. In this case ignition angle is small, hence, rectifier conducts practically for one-half cycle.

- Let the grid voltage be *retarded in phases* with respect to the anode voltage [Fig. 7.48 (*b*)] so that now grid voltage curve intersects the ignition line at M. Now it will be observed that ignition angle β increases and consequently ignition is delayed by a longer time. It means that the time of delivery of current by anode is reduced considerably thereby decreasing D.C. output.
- Fig. 7.48 (*c*) represents the case when the ignition angle β becomes too large and the ignition is delayed by too large time. The grid voltage is almost in phase opposition to the anode voltage and hence time of delivery of anode current is drastically reduced. Hence, practically, no D.C. output is available.

From above discussion we arrive at the conclusion that the D.C. output can be controlled by varying the phase difference between anode and grid voltage. *This phase shift between anode and grid voltages can be achieved by means of an 'induction regulator'*. Induction regulator is an induction motor having a wound rotor which is capable of rotation through 180 electrical degrees. The primary (usually rotor) is connected to the supply line and the phase difference between anode and grid voltage is created and controlled by changing the angular position of rotor with respect to the secondary (stator) as shown in Fig. 7.49.

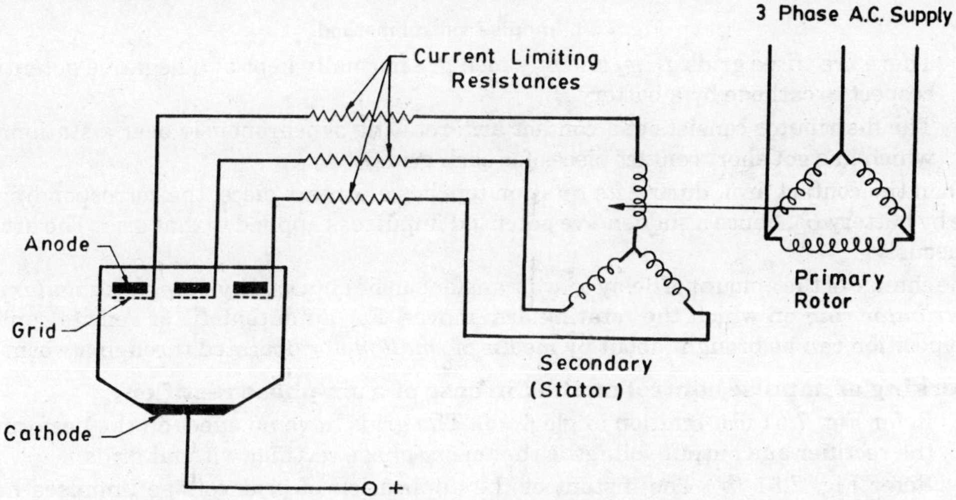

Fig. 7.49. Use of an induction regulator for phase shift between anode and grid voltages.

The major *shortcomings* of soft-control methods are :

1. It is difficult to determine precisely the moment of arc ignition.

2. Due to gradualness of phase-shift, *regulation of the rectifier becomes poor*.

Impulse control method. In phase-control method we have seen that there is a gradualness in phase-shifts which affects the active performance of the rectifier. This gradualness can be removed by giving sudden impulses of voltages to the grid instead of making their potential sinusoidal. This will in no way interfere with the performance of grids because otherwise also they are in-operative once the arc is established. These instantaneous voltage impulses *intersect the ignition line vertically and not obliquely as in case of sinusoidal grid voltage*. The sudden positive voltage impulses can be assigned to the grid at the desired instant of arc transfer and during the remaining period the grid can be kept at a negative potential, which is sufficient to prevent arc transfer.

One of the methods of applying these sudden impulses is to use a synchronously rotating distributor with a segment for each grid driven by a synchronous motor Fig. 7.50.

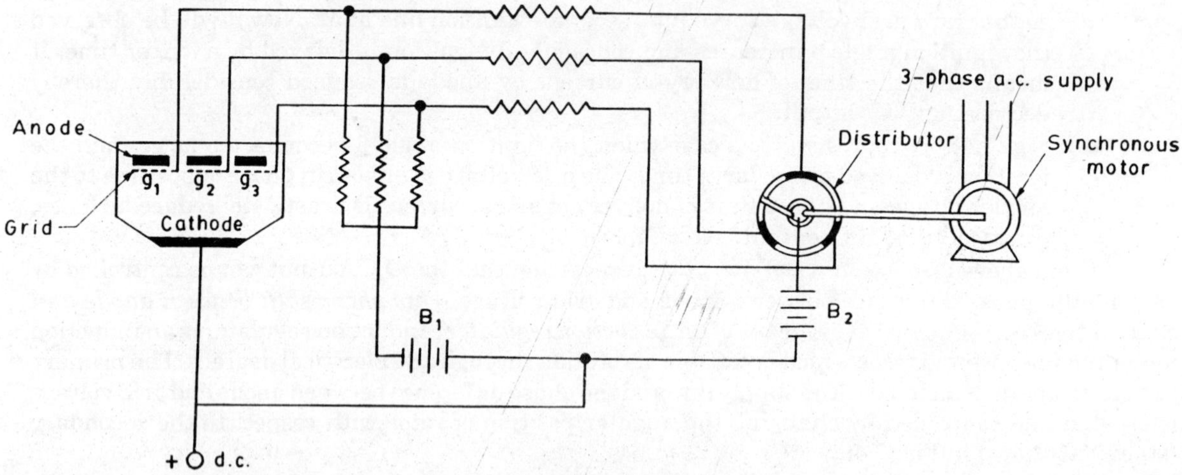

Fig. 7.50. Impulse control method.

- There are three grids g_1, g_2 and g_3 which are normally kept at a negative potential with respect to cathode by a battery B_1.
- The distributor consists of a contact arm rotating synchronously over a stationary ring which has got short contact pieces for each grid.

When the contact arm, during its rotation touches a contact piece, the corresponding grid is made +ve by battery B_2, hence a sudden +ve potential impulse is applied to that grid. The arc is fixed instantaneously.

The change of the amount of delay of arc transfer can be obtained by *changing angular position of the distributor ring* on which the rotating arm moves. For an automatic or remote control this change in position can be brought about by means of *small motor* operated through a worm gear.

Working of impulse control method in case of a six-phase rectifier :

- Refer Fig. 7.51 (*a*). Ignition angle $\beta = 0$. The grids have no effect on the performance of the rectifier and output voltage is that of six-phase rectifier without grids.
- Refer Fig. 7.51 (*b*). The instant of the application of grid voltage impulses has been advanced by an angle $\beta = 30°$.
  The arc starts suddenly at anode whose grid has been given the impulse and it persists on that anode until the impulse has been applied to the grid of the next anode in sequence.

The average output voltage, given by shaded area is 86.6% of that $\beta = 0$.

- Refer Fig. 7.51 (*c*). Ignition angle $\beta = 60°$. The average output becomes 50% of with that $\beta = 0$.
- Refer Fig. 7.51 (*d*). Ignition angle $\beta = 90°$. The average output becomes 13.5% of that $\beta = 0$.
- Refer Fig. 7.51 (*e*). Ignition angle $\beta = 120°$. The output voltage becomes zero.

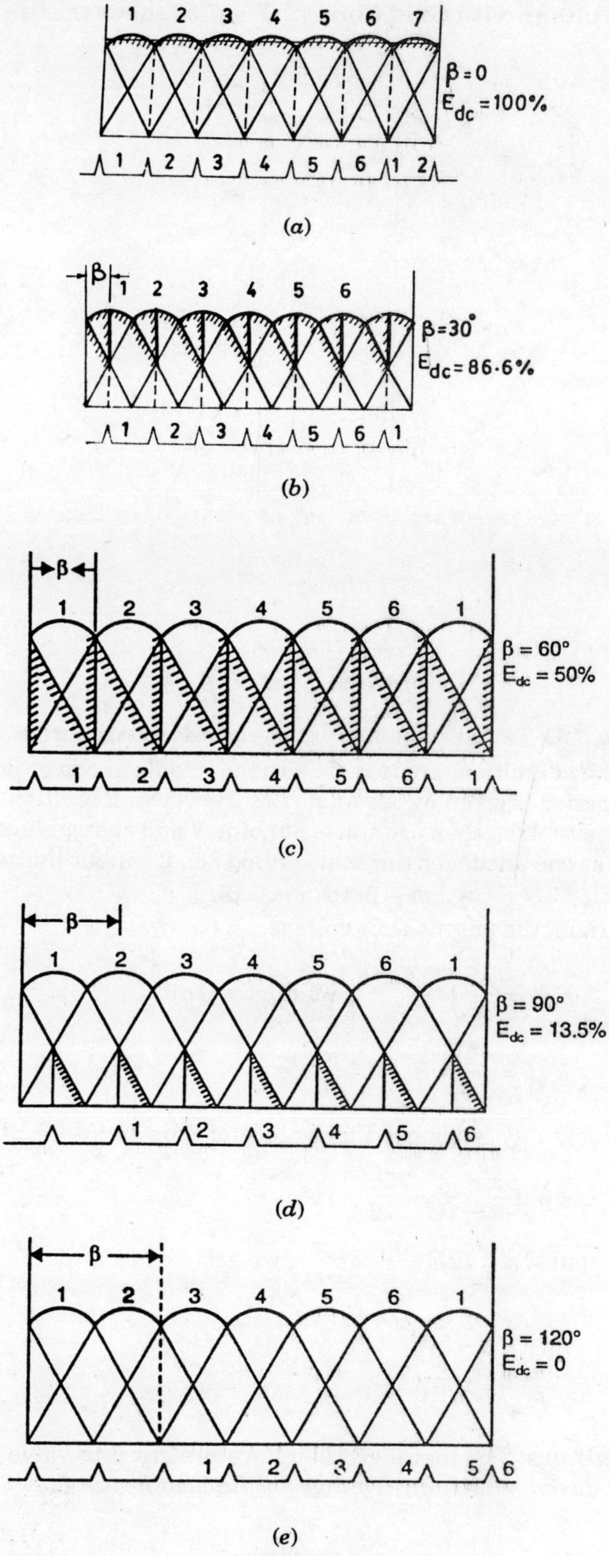

Fig. 7.51

7.5.3.12. Output Voltage with Grid Control. Fig. 7.52 shows the two adjacent intersection of anode voltage curves.

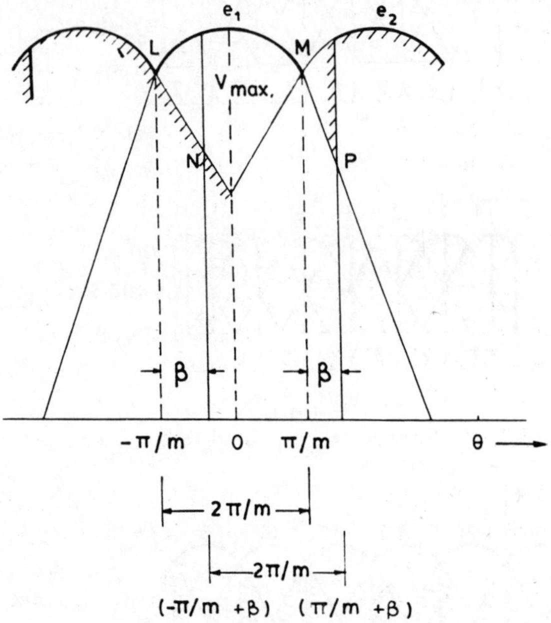

Fig. 7.52. Two adjacent intersection of anode voltage curves.

In case of uncontrolled ignition, arc transfers to the middle anode at point L and persists till point M *i.e.*, the ignition period is given by the angle LM. However, if ignition is retarded or delayed by angle β by means of grid control, then arc starts at point N and changes over to the next anode at point P. The arc persists at one anode for the same period *i.e.*, $2\pi/m$ but limits now become

$$(-\pi/m + \beta) \text{ to } (\pi/m + \beta).$$

Taking O as the origin, the output D.C. voltage,

$$E_{dc\beta} = \frac{m}{2\pi}\int_{-\pi/m+\beta}^{\pi/m+\beta} \sqrt{2}\, E_{ac} \cos\theta \, d\theta$$

or

$$E_{dc\beta} = \frac{\sqrt{2}m}{\pi} E_{ac} \sin\frac{\pi}{m} \cos\beta \quad \text{...(7.18)}$$

The above expression only holds good for values of β up to $\frac{\pi}{2} - \frac{\pi}{m}$. Thus with a 12-anode rectifier the limiting value of β is $\frac{\pi}{2} - \frac{\pi}{12} = \frac{5\pi}{12} = 75°$.

In eqn. (7.18) if we put $E_{dc} = \sqrt{2}E_{ac}\frac{m}{\pi}\sin\frac{\pi}{m}$, we get

$$E_{dc\beta} = E_{dc} \times \cos\beta$$

$$\therefore \quad \frac{E_{dc\beta}}{E_{dc}} = \cos\beta \quad \text{...(7.19)}$$

7.5.4. Metal Rectifiers. The metal rectifiers are preferred to valve rectifiers as these are *mechanically strong* and *do not require any voltage for filament heating.*

The metal rectifiers are of the following two types :

1. Copper oxide rectifiers. 2. Selenium rectifiers.

7.5.4.1. Copper Oxide Rectifiers

- Copper oxide rectifiers are made from copper discs which are usually 25 mm in diameter or smaller. These discs are heated to form a copper oxide on their surface. After the heat treatment they are quenched and the oxide film is removed from one side. The remaining *oxide film provides rectifying action*. Contact is made to the film through the copper on which it is formed and a lead disc pressed on to the upper surface (Fig. 7.53). The rectifier thus formed *will permit a current flow from the film to the copper and will block current from flowing in the opposite direction*. This device is *not a perfect rectifier*.

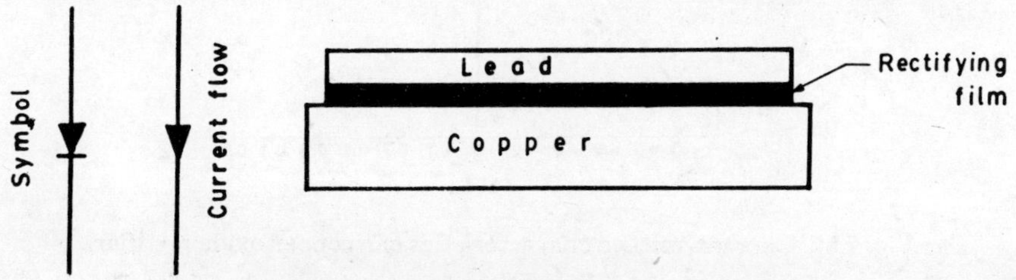

Fig. 7.53. Elements of the copper oxide rectifier.

Fig. 7.54 shows the construction of copper oxide rectifier.

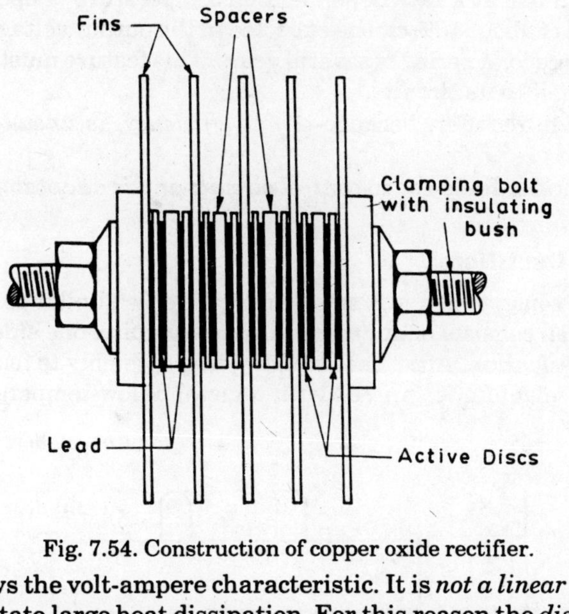

Fig. 7.54. Construction of copper oxide rectifier.

- Fig. 7.55 shows the volt-ampere characteristic. It is *not a linear device* and its resistances losses necessitate large heat dissipation. For this reason the *discs are limited in size*, and radiating fins are provided when the units are assembled in series (Fig. 7.54) in order to provide cooling. The temperature of the rectifier has considerable effect on the rectifying action as shown in the curves of Fig. 7.55.

- This type of rectifier is usually designed to operate at 45°C or less. The voltage impressed on each cell is usually limited to a range of 5 to 11 volts, depending on whether the rectifier is used on continuous or intermittent duty.

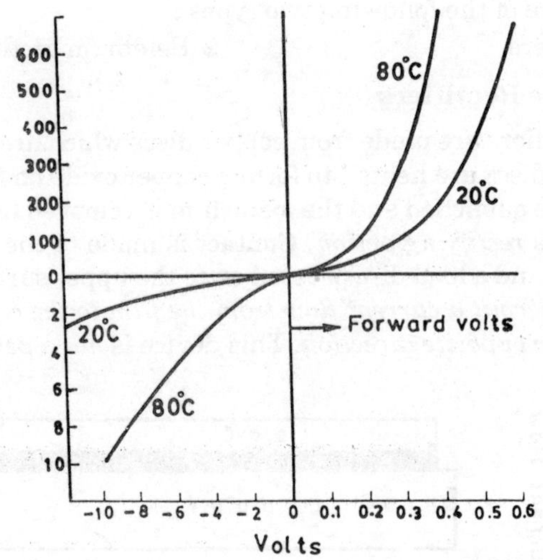

Fig. 7.55. Current-voltage characteristics of a copper-oxide rectifier.

- The *efficiency* of the copper oxide rectifier seldom exceeds 65% to 75%. This indicates a voltage *regulation* ranging between 25% and 35%.
- This rectifier is *not stable during early life*. The resistance in the forward direction tends to increase with use at a rate dependent on temperature of operation and its load. The over-all change of about 15% can be expected in the output voltage. The change is gradual and may continue for a period of several years. This feature must be taken into consideration in adapting it to its circuits.
- The copper-oxide rectifier, because of *low efficiency, is unsuitable for rectifying large amounts of power*.
- This type of rectifier finds use in *control circuits and is adaptable to various voltages and currents.*

7.5.4.2. Selenium Rectifiers

- The method of construction of a selenium rectifier is similar to that of the copper oxide rectifier. The cell consists of an *iron* or *aluminium* disc, one side of which is coated with a *thin layer of selenium*. After heat treating this assembly to form the copper crystalline structure, the selenium is sprayed with a metal of low-temperature melting point. The

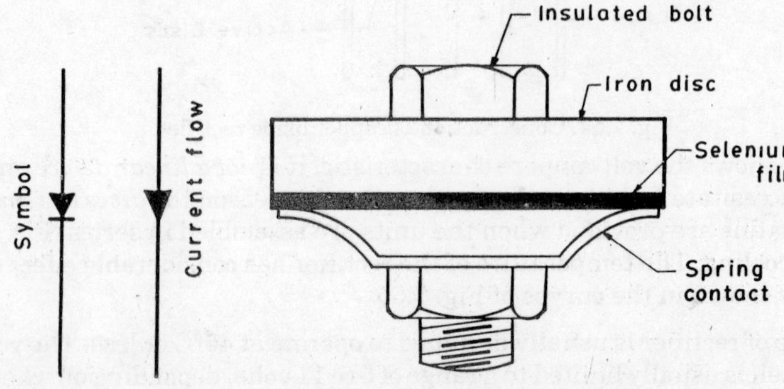

Fig. 7.56. Elements of the selenium rectifier.

cell is then 'formed' by supplying it with a direct current in its reverse direction for several hours. The *rectifying action takes place at the selenium film.*

Fig. 7.57 shows the details of construction of the selenium rectifier.

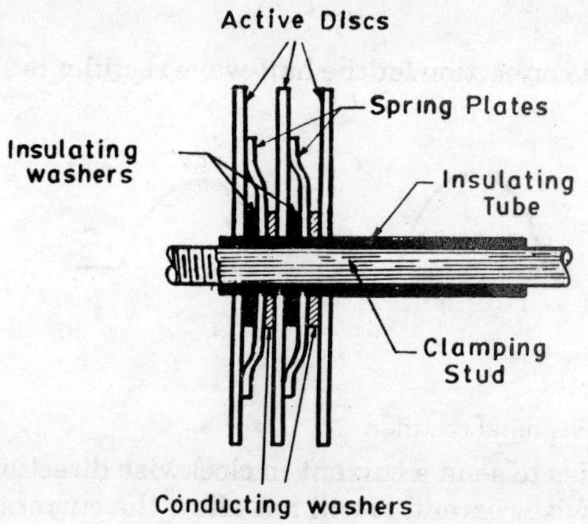

Fig. 7.57. Construction of the selenium rectifier.

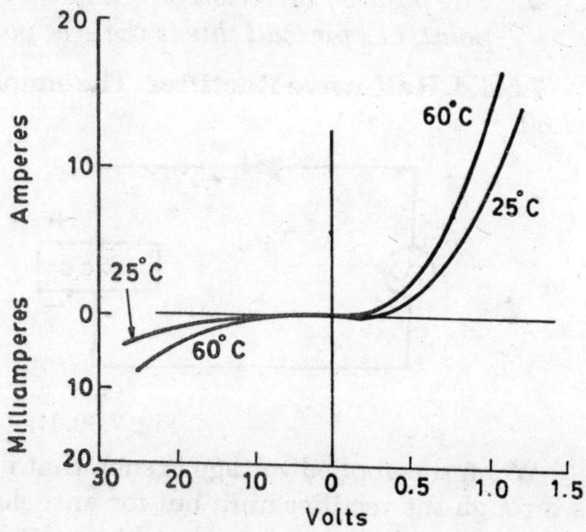

Fig. 7.58. Current voltage characteristic of the selenium rectifier.

- Fig. 7.58 shows the volt-ampere characteristics of the selenium cell. From these characteristics it is evident that the selenium cell has the following advantages over the copper oxide cell :
  (*i*) Changes in temperature have less effect on the selenium cell than on the copper oxide unit.
  (*ii*) The selenium cell can withstand larger reverse voltages.
  (*iii*) It can be operated at temperatures as high as 75°C.
  (*iv*) *Efficiencies* ranging from 75% to 85% are attainable.
- If the selenium cell is idle for a long period of time, its reverse resistance tends to decrease and current that it takes when reconnected to the circuit is abnormally high. Ageing of this rectifier may change the output voltage by 5% to 10%.

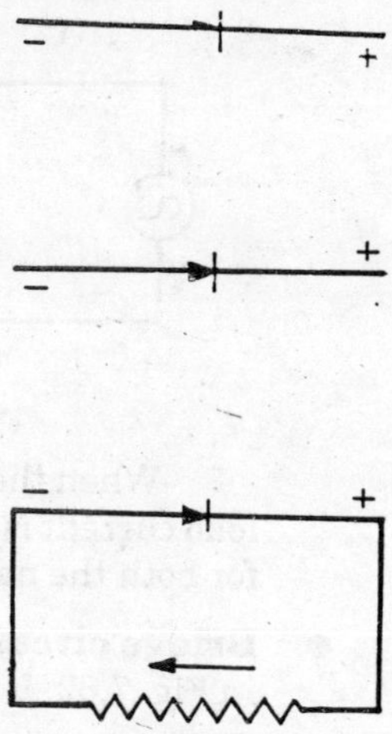

Fig. 7.59. Conventional method for representation of rectifiers.

The following points of selenium rectifiers are worth noting :

(*i*) They can be used on ployphase circuits to supply large values of current.

(*ii*) They are easily adapted to various voltages and currents by selecting proper series and parallel connection of the cells.

(*iii*) They are *less efficient than the mercury-arc rectifier* when used on high-voltage systems.

(*iv*) They are available in capacities up to *50* or *100 kW*, and their range can be extended.

(*v*) Their applications is usually limited to potential of 100 V or less.

These rectifiers have been used extensively in supplying direct current for *electroplating*.

7.5.4.3. Methods of Representation of Rectifiers. Fig. 7.59 shows the conventional method for representation of rectifiers.

- The *full arrow or half arrow shows the forward direction or the direction of current flow.*
- The *position direction or red mark represents that the current leaves the rectifier at this point, i.e., for load this is the + ve point.*

7.5.4.4. Half-wave Rectifier. The simplest connection for the half-wave rectifier is shown in Fig. 7.60.

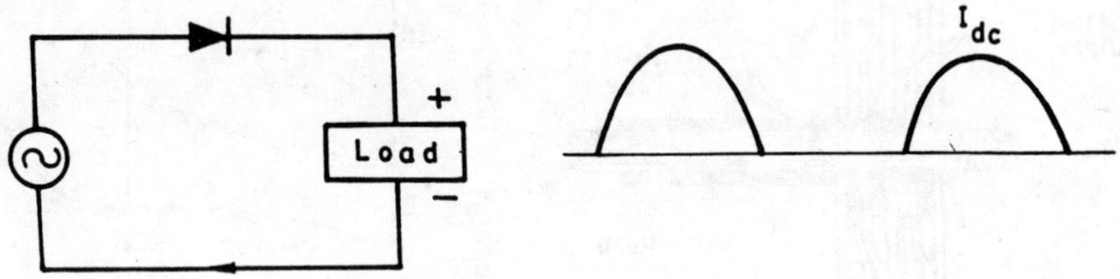

Fig. 7.60. Half-wave metal rectifier.

When the applied voltage is such that it tries to send a current in clockwise direction, it will pass through the rectifier unit, but for anti-clockwise current, it will not allow the current to flow, thus the current in through load will be in loops as shown in Fig. 7.60.

7.5.4.5. Full-wave Rectifier

- The circuit and output wave for a full-wave rectifier is shown in Fig. 7.61.

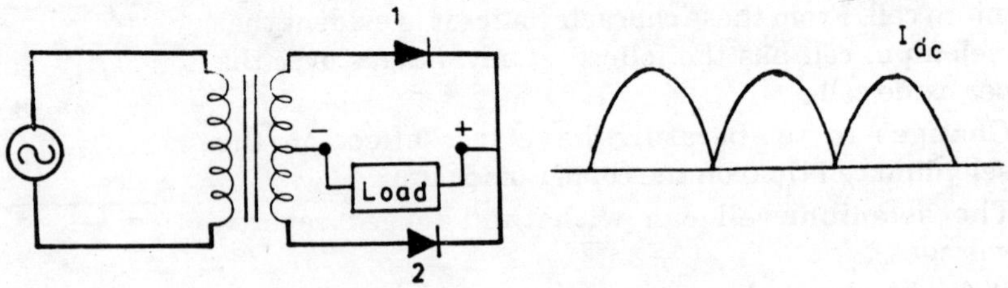

Fig. 7.61. Circuit and output wave for full-wave rectifier.

When the current is clockwise the rectifier unit '1' will be conducting and when the load current is anticlockwise unit '2' will conduct. Thus current will flow through the load for both the negative and positive loops of the supply voltage.

- **Bridge circuit.** Another circuit for the full-wave rectifier called *bridge circuit* is shown in Fig. 7.62. It consists of *four* rectifying units. When the current is clockwise the current

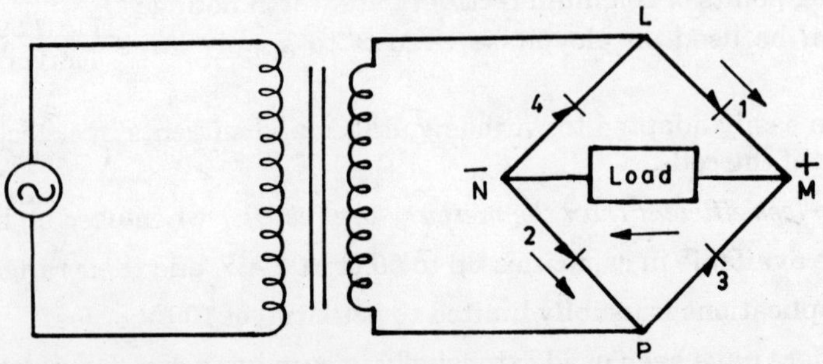

Fig. 7.62. Bridge circuit for full-wave rectification by metal rectifier.

is through anodes 1 and 2 and path of current is *LMNP*, for the reversed direction of current the anodes 4 and 3 conduct *i.e.*, the path of current is *PMNL*. For both negative and positive loops the current through the load resistance is from *M* to *N*.

7.5.5. Mechanical Rectifiers. Mechanical rectifiers are of the following two types :

1. Commutating type rectifier.
2. Vibrating rectifier.

7.5.5.1. Commutating Type Rectifier. A commutating type rectifier consists of commutator driven by a synchronous motor. The connections of segments are so made that when alternating current reverses, the connections to the D.C. load circuit are reversed simultaneously as shown in Fig. 7.63. Thus it gives unidirectional current.

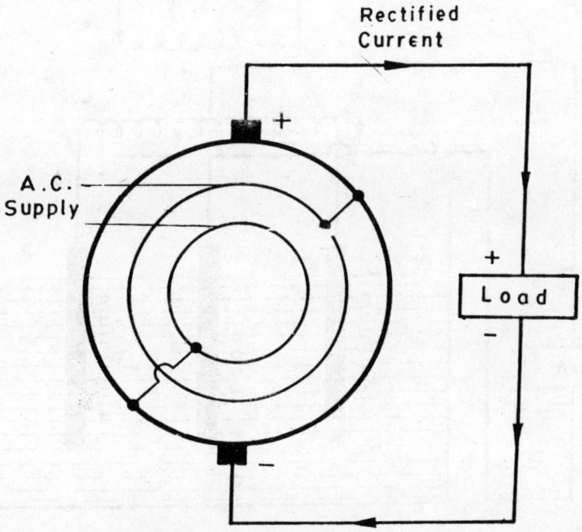

Fig. 7.63. Commutating type rectifier.

Since the brushes cannot be of zero width, therefore it is difficult to commutate at the point of zero current and the current and voltage are *rarely zero* at the same time. Hence, such devices are subject to sparking, and, therefore, its use is limited to the applications where small currents and voltages are to be dealt with.

7.5.5.2. Vibrating Rectifier

- Its working principle is same as of commutator type but the connections on A.C. side are reversed each half cycle by a polarised armature, which is caused to operate synchronously by means of an A.C. magnet.
- These rectifiers had been used for charging of small capacity storage batteries but for this purpose have been superseded by rectifiers of other types such as redox and selenium types. However this type of vibrator is *widely used for inverting direct current into alternating current, although the resulting wave is far from sinusoidal.* Such inverters are used for radio set where 6 volts D.C. supply from battery is inverted to an alternating voltage stepped up to 125 to 400 volts by a transformer, and is then rectified by a tube supply, the plate and grid circuit voltages.

7.5.6. Electrolytic Rectifiers

- The working principle of electrolytic rectifiers is as follows :

  "*If two plates of different metals* (*e.g.*, lead and aluminium) *are placed in an electrolyte* (*e.g.*, sodium bicarbonate or aluminium phosphate solution) *and connected to an A.C. voltage then current can pass only in one direction.*"

In case of an electrolytic rectifier consisting of lead and aluminium plates immersed in a solution of sodium bicarbonate or ammonium phosphate, current can pass from the solution to the aluminium plate, but as soon as the current attempts to flow in reverse direction a thin insulating film of aluminium is *instantly formed* over the aluminium plate and acts as an insulator up to about 150 volts, due to which current cannot flow from aluminium to solution. Fig. 7.64 shows an electrolytic rectifier giving a full-wave rectification.

- **These rectifiers are *primarily used for charging of low voltage batteries from A.C. supply.* Their efficiency is low, nearly 60%.**

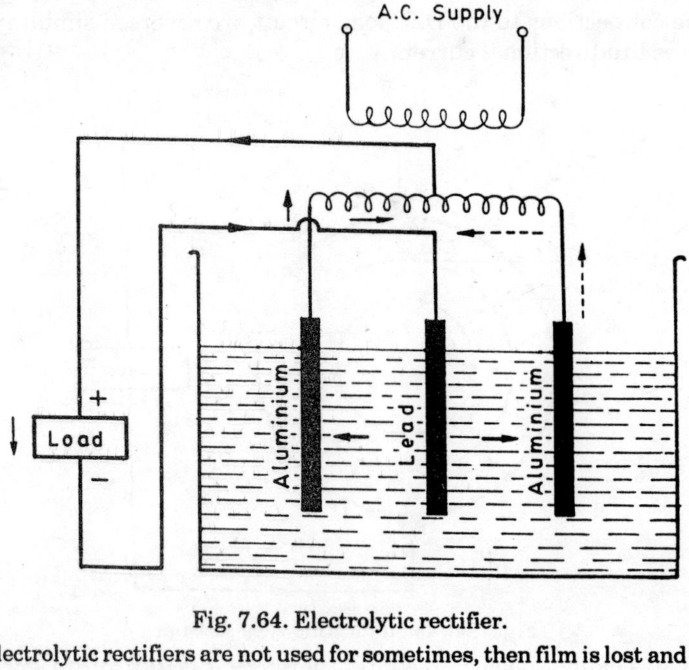

Fig. 7.64. Electrolytic rectifier.

Notes. (*i*) If electrolytic rectifiers are not used for sometimes, then film is lost and film has to be reformed if they are to be reused. The loss of film can be avoided by taking out the plates from the electrolyte and drying them.

(*ii*) In order to keep the density of the electrolyte to a proper value, only pure water must be added from time to time to make up loss on account of evaporation. ***Buzzing sound indicates the weakness of electrolyte and heating of electrolyte indicates that alternating current is being allowed to pass. If the heating of the electrolyte is excessive it indicates that rectifier needs to be recharged.***

RECTIFIERS

☞**Example 7.26.** *A 3-anode mercury arc rectifier is supplied through the star-connected secondary of a 3-phase transformer. It delivers an output of 40 kW at 400 V on D.C. side. Calculate the transformer secondary phase voltage and current. Allow an arc drop of 20 V.*

Solution. Number of phases, $m = 3$

D.C. output power $= 40$ kW

D.C. terminal voltage, $V_{dc} = 400$ V

Arc drop $= 20$ V

Secondary phase voltage, E_{ac} :

R.m.s. current/phase, $I_{r.m.s.}$:

We know that no-load output voltage,

$$E_{dc} = V_{dc} + \text{arc drop} = 400 + 20 = 420 \text{ V}$$

Now, using the relation, $E_{dc} = \sqrt{2} \, . \, E_{ac} \times \frac{m}{\pi} \times \sin\frac{\pi}{m}$

$$420 = \sqrt{2} \times E_{ac} \times \frac{3}{\pi} \sin\frac{\pi}{3}$$

$$\therefore \quad E_{ac} = \frac{420 \times \pi}{\sqrt{2} \times 3 \sin\frac{\pi}{m}} = 359 \text{ V}$$

Hence, **secondary phase voltage = 359 V. (Ans.)**

D.C. output current, $I_{dc} = \frac{\text{D.C. power output}}{\text{D.C. terminal voltage}} = \frac{40 \times 1000}{400} = 100 \text{ A}$

∴ R.m.s. current/phase,

$$I_{r.m.s.} = \frac{I_{dc}}{\sqrt{m}} = \frac{100}{\sqrt{3}} = 57.7 \text{ A}$$

Hence, $\mathbf{I_{r.m.s.}}$ **= 57.7 A. (Ans.)**

Example 7.27. *Find the values of rectified voltage and current when a 3-phase rectifier is working on 500 volts, 50-Hz supply. Current drawn from main is 8 A/phase.*

Solution. A.C. voltage, $E_{ac} = 500$ V

Number of phases, $m = 3$

R.m.s. current/phase, $I_{r.m.s.} = 8$ A

Rectified voltage, $\mathbf{E_{dc}}$:

Rectified current, $\mathbf{I_{dc}}$:

Rectified voltage is given by the relation,

$$\mathbf{E_{dc}} = \sqrt{2} E_{ac} \times \frac{m}{\pi} \times \sin\frac{\pi}{m} = \sqrt{2} \times 500 \times \frac{3}{\pi} \times \sin\frac{\pi}{3} = \mathbf{584.7 \text{ V.}} \ \textbf{(Ans.)}$$

Rectified current, $\mathbf{I_{dc}} = I_{r.m.s.} \times \sqrt{m} = 8 \times \sqrt{3} = \mathbf{13.86.}$ **(Ans.)**

Example 7.28. *In a six-phase rectifier the full-load terminal voltage on the output side is to be 1400 V and the sum of all drops 20 V. Calculate the transformer secondary voltage.*

Solution. Number of phases, $m = 6$

D.C. terminal voltage, $V_{dc} = 1400$ V

Sum of all drops $= 20$ V

Transformer secondary voltage :

No-load D.C. output voltage, $E_{dc} = 1400 + 20 = 1420$ V

R.m.s. value of A.C. voltage between each anode and neutral,

E_{ac} is found by the relation, $E_{dc} = \sqrt{2} E_{ac} \times \frac{m}{\pi} \sin\frac{\pi}{m}$

or

$$E_{ac} = \frac{E_{dc}}{\sqrt{2} \times \frac{m}{\pi} \times \sin\frac{\pi}{m}} = \frac{1420}{\sqrt{2} \times \frac{6}{\pi} \times \sin\frac{\pi}{6}} = 1051.5 \text{ V}$$

Secondary phase voltage $= 2E_{ac} = 2 \times 1051.5 =$ **2103 V. (Ans.)**

Example 7.29. *A 6600 V star-connected transformer is connected to a 6-anode mercury arc rectifier which supplies load at 400 volts. Calculate the turn-ratio of the transformer. Allow an arc drop of 20 V for each anode.*

Solution. D.C. terminal voltage, $V_{dc} = 400$ V

Arc drop/anode $= 20$ V

Primary phase voltage of the transformer

$$= \frac{6600}{\sqrt{3}} = 3810 \text{ V}$$

Turn-ratio of transformer :

No-load output D.C. voltage,

$$E_{dc} = 400 + 20 = 420 \text{ V}$$

R.m.s. value of A.C. voltage between each anode and neutral, E_{ac} is given by

$$E_{dc} = \sqrt{2} E_{ac} \frac{m}{\pi} \sin \frac{\pi}{m}$$

or

$$E_{ac} = \frac{E_{dc}}{\sqrt{2} \times \frac{m}{\pi} \sin \frac{\pi}{m}} = \frac{420}{\sqrt{2} \times \frac{6}{\pi} \times \sin \frac{\pi}{6}} = 311 \text{ V}$$

Secondary phase voltage $= 2 \times 311 = 622$ V

∴ **Primary to secondary turn-ratio**

$$= \frac{3810}{622} = \textbf{6.12 : 1.} \quad \textbf{(Ans.)}$$

Example 7.30. *A 3-anode rectifier supplies a load of 20 kW at 500 volts. If arc drop is 25 volts, calculate the kVA rating of the transformer. Neglect utility factor.*

Solution. Number of phases, $m = 3$

D.C. load supplied $= 20$ kW or 20000 W

Arc drop $= 25$ V

kVA rating of the transformer :

No-load output D.C. voltage, $E_{dc} = 500 + 25 = 525$ V

Using the relation, $E_{dc} = \sqrt{2} E_{ac} \times \frac{m}{\pi} \sin \frac{\pi}{m}$

(where E_{ac} = voltage between each anode and neutral of the secondary of the transformer)

or

$$E_{ac} = \frac{E_{dc}}{\sqrt{2} \times \frac{m}{\pi} \times \sin \frac{\pi}{m}} = \frac{525}{\sqrt{2} \times \frac{3}{\pi} \times \sin \frac{\pi}{3}} \simeq 449 \text{ V}$$

Now, D.C. load current, $I_{dc} = \frac{20000}{500} = 40$ A

R.m.s. value of the secondary phase current,

$$I_{\text{r.m.s.}} = \frac{I_{dc}}{\sqrt{3}} = \frac{40}{\sqrt{3}} = 23.1 \text{ A}$$

Transformer rating/phase $= E_{ac} \times I_{\text{r.m.s.}}$

$$= 449 \times \frac{23.1}{1000} \text{ kVA} = 10.37 \text{ kVA}$$

∴ **Transformer rating (kVA)** $= 10.37 \times 3 =$ **31.11 kVA. (Ans.)**

☞**Example 7.31.** *The following test results were obtained when a transformer was connected to a 3-anode mercury arc rectifier.*

Open-circuit test : 500 W consumed at normal voltage.

Short-circuit test : 900 W for 25 A current on secondary.

If the rectifier is supplying a load of 25 kW at 450 volts, determine :
(i) The bulb efficiency, and *(ii) The overall efficiency.*
Assume an arc drop of 20 volts and a loss of 120 watts in smoothing reactance.

Solution. D.C. output voltage = 450 volts
Arc drop = 20 volts
D.C. load output = 25 kW or 25000 W
No-load output D.C. voltage = 450 + 20 = 470 V

(*i*) **The bulb efficiency of the rectifier**

$$= \frac{\text{D.C. output voltage}}{\text{no-load output D.C. voltage}}$$

$$= \frac{450}{470} = 0.957 \text{ or } \mathbf{95.7\%}.$$

(*ii*) D.C. load current, $I_{dc} = \dfrac{25 \times 1000}{450} = 55.55$ A **(Ans.)**

R.m.s. value of the secondary phase current,

$$I_{r.m.s.} = \frac{I_{dc}}{\sqrt{3}} = \frac{55.55}{\sqrt{3}} = 32.07 \text{ A}$$

For 25 A secondary current, copper loss

= 900 W

When secondary current is 32.07 A, copper loss

$$= 900 \times \left(\frac{32.07}{25}\right)^2 = 1481 \text{ W}$$

Arc loss = 20 × 55.55 = 1111 W

Total losses of the rectifier including transformer are :

No-load transformer loss = 500 W
Load copper losses of transformer
= 1481 W
Arc loss = 1111 W
Reactance loss = 120 W
Total = 3212 W

Now, **overall efficiency,** $\eta = \dfrac{\text{output}}{\text{output + losses}} = \dfrac{25000}{25000 + 3212}$

= 0.886 or **88.6%. (Ans.)**

Example 7.32. *A 40 volt battery is charged by a half-wave rectifier from an A.C. supply of 230 volts. The resistance of the rectifier in one direction is 200 Ω and in the other direction very high. Find the average charging current per cycle.*

Assume the resistance of the battery to be negligible.

Solution. Battery voltage, $E_b = 40$ V
A.C. supply voltage = 230 V
Resistance of rectifier in one direction = 200 Ω

Average charging current/cycle :

It is evident that battery will get charged during one-half per cycle. During the charging half cycle, the battery e.m.f. E_b opposes the supply voltage $V_{max} \sin \theta$, hence the equation of the resultant voltage across battery is

$$e = V_{max} \sin \theta - E_b = \sqrt{2} \times 230 \times \sin \theta - 40 = 325 \sin \theta - 40.$$

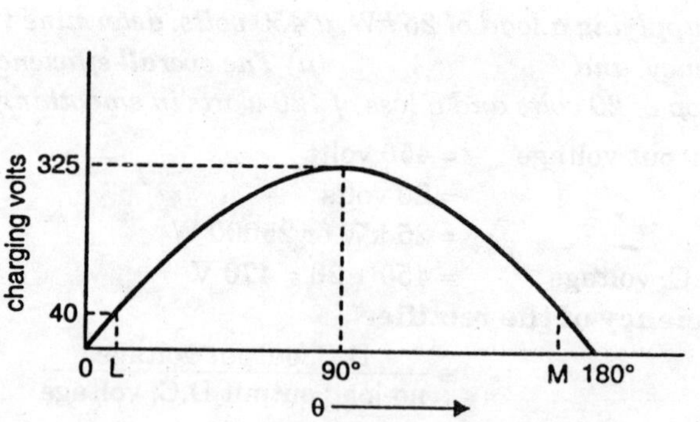

Fig. 7.65

Refer Fig. 7.65. Charging beings at point L when supply voltage has risen to 40 V and finishes when it is fallen to 40 V as at point M. The starting angle OL is found by putting e = 40 V

$$\therefore \quad 40 = 325 \sin\theta \quad \text{or} \quad \sin\theta = \frac{40}{325} = 0.123$$

$$\therefore \quad \theta = \sin^{-1} 0.123 = 7.07° \text{ or } 7° \, 4'$$

Charging stops at M

$\therefore$ Angle OM = 172° 56′

$\therefore$ Average value of charging voltage over the half-cycle is

$$= \int_{7°\,4'}^{172°\,56'} \left(\frac{325 \sin\theta \, d\theta - 40}{\pi} \right) = \frac{325}{\pi} \left| -\cos\theta \right|_{7°\,4'}^{172°\,56'} - 40$$

$$= \frac{325}{\pi} [-\cos(90° + 82° \, 56') + \cos(90° - 82° \, 56')] - 40$$

$$= \frac{325}{\pi} \times 2 \sin 82° \, 56' - 40 = 205.3 - 40 = 165.3 \text{ V}$$

$\therefore$ Charging current/half cycle

$$= \frac{165.3}{200} = 0.826 \text{ A}$$

$\therefore$ **Charging current over the whole cycle**

$$= \frac{0.826}{2} = \mathbf{0.413 \text{ A.}} \quad \textbf{(Ans.)}$$

Example 7.33. *In a 3-phase rectifier, the phase volts on the secondary side of the transformer are 1500 (r.m.s.) volts. The arc drop is 20 volts, and the full-load overlap is 30°.*

Calculate the full-load output voltage.

Solution. Phase volts on secondary side of the transformer.

E_{ac} = 1500 V

Number of phases, m = 3

Arc drop = 20 volts

Full-load overlap, α = 30°

Full-load output D.C. voltage :

D.C. output voltage (taking reactance 'drop into account')

$$= E_{dc} \times \cos^2 \frac{\alpha}{2} = \sqrt{2} \, E_{ac} \times \frac{m}{\pi} \sin \frac{\pi}{m} \times \cos^2 \frac{\alpha}{2}$$

$$= \sqrt{2} \times 1500 \times \frac{3}{\pi} \times \sin \frac{\pi}{3} \times \cos^2 \left(\frac{30°}{2} \right)$$

$$= \sqrt{2} \times 1500 \times \frac{3}{\pi} \times \sin\frac{\pi}{3} \times \cos^2 15° = 1636.5 \text{ V}$$

∴ **Full-load output D.C. voltage**

$$= 1636.5 - 20 = \mathbf{1616.5\ V.\ \ (Ans.)}$$

Example 7.34. *A 3-phase rectifier is supplied by a delta-star connected transformer with an equivalent leakage reactance per phase of 0.25 Ω. The effective secondary phase voltage is 220 volts. If the load current is 110 amperes, calculate :*

(i) D.C. no-load voltage.

(ii) Drop due to the effect of overlap on the wave shape.

(iii) Angle of overlap.

Solution. Number of phases, $m = 3$

Leakage reactance/phase $= 0.25\ \Omega$

Secondary phase voltage, $E_{ac} = 220$ V

Load current, $I_{dc} = 110$ A.

(*i*) **D.C. no-load voltage is given by,**

$$E_{dc} = \sqrt{2}E_{ac} \times \frac{m}{\pi}\sin\frac{\pi}{m} = \sqrt{2} \times 220 \times \frac{3}{\pi}\sin\frac{\pi}{3} = \mathbf{257\ V.\ \ (Ans.)}$$

(*ii*) Voltage drop due to transformer reactance is

$$= E_{dc}\sin^2\frac{\alpha}{2} = E_{dc}\left(\frac{1-\cos\alpha}{2}\right) \qquad ...(i)$$

where α is the angle of overlap.

Also, $$\cos\alpha = 1 - \frac{I_{dc}X}{\sqrt{2}E_{ac}\sin\frac{\pi}{m}} \qquad ...[\text{Eqn. 7.13 }(a)]$$

or $$1 - \cos\alpha = \frac{I_{dc}X}{\sqrt{2}E_{ac}\sin\frac{\pi}{m}} \qquad ...(ii)$$

Putting the value of (1 – cos α) in eqn. (*i*), we get

Drop in transformer reactance

$$= \frac{E_{dc}}{2} \times \frac{I_{dc}X}{\sqrt{2}E_{ac}\sin\frac{\pi}{m}} = \frac{\sqrt{2} \times E_{ac} \times \frac{m}{\pi} \times \sin\frac{\pi}{m}}{2} \times \frac{I_{dc}X}{\sqrt{2}E_{ac}\sin\frac{\pi}{m}}$$

$$= \frac{m}{2\pi} \times I_{dc}X = \frac{3}{2\pi} \times 110 \times 0.25 = 13.13 \text{ V.}$$

(*iii*) If α is the angle of overlap, then from eqn. (*ii*), we have

$$1 - \cos\alpha = \frac{110 \times 0.25}{\sqrt{3} \times 220 \times \sin\frac{\pi}{3}} = 0.1021$$

or $$\cos\alpha = 1 - 0.1021 = 0.8979$$

∴ $$\alpha = \cos^{-1} 0.8979 = 26.1°$$

Hence, **angle of overlap = 26.1°. (Ans.)**

☞**Example 7.35.** *A 6-phase rectifier is to deliver a load current of 420 amperes at 500 volts. The rectifier is supplied from a delta-star transformer with diametral connections. The arc drop is 20 volts. Calculate :*

(i) The voltage to neutral on the transformer secondaries.

(ii) The rating of the primary and secondary winding of each transformer for which these should be designed.

Take utility factor for primary and secondary as 0.80 and 0.6 respectively.

Solution. Number of phases, $m = 6$

D.C. output load current, $I_{dc} = 420$ A

D.C. terminal voltage $= 500$ V

Arc drop $= 20$ V

Utility factor for primary $= 0.8$

Utility factor for secondary $= 0.6.$

(*i*) No-load D.C. voltage, $E_{dc} = 500 + 20 = 520$ V

But
$$E_{dc} = \sqrt{2} \times E_{ac} \times \frac{m}{\pi} \sin \frac{\pi}{m}$$

$$520 = \sqrt{2} \times E_{ac} \times \frac{6}{\pi} \times \sin \frac{\pi}{6}$$

$$\therefore \quad E_{ac} = \frac{520}{\sqrt{2} \times \frac{6}{\pi} \times \sin \frac{\pi}{6}} = 385 \text{ V}$$

Hence, **voltage to the neutral on the transformer secondaries**

= 385 V. (Ans.)

(*ii*) Power supplied by each phase $= \dfrac{520 \times 420}{6} = 36400$ W

Secondary utility factor $= \dfrac{\text{power output of the rectifier/phase}}{\text{volt ampere on secondary side of transformer phase}}$

$$0.6 = \frac{36400}{\text{secondary volt-ampere}}$$

∴ **Secondary volt-ampere** $= \dfrac{36400}{0.6} =$ **60666.7. (Ans.)**

Primary utility factor $= 0.8$

∴ Power output/phase referred to primary

$$= \frac{520 \times 420}{3}$$

∴ **Primary volt-ampere** $= \dfrac{520 \times 420}{3 \times 0.8} =$ **91000. (Ans.)**

Example 7.36. *A three-anode mercury-arc rectifier has D.C. full-load current of 100 ampere. Determine the equivalent inductance per transformer secondary phase if the overlap angle on full-load is not to exceed 45°. The transformer secondary phase voltage is 475 V and the frequency 50-Hz. The load inductance may be considered such as to maintain a sensible constant output current.*

Solution. Number of phases, $m = 3$

D.C. full-load current, $I_{dc} = 100$ A

Overlap angle, $\alpha = 45°$

Transformer secondary phase voltage,

$E_{ac} = 475$ V

Frequency, $f = 50$-Hz

Equivalent inductance per transformer secondary phase, L :

Using the relation, $\cos \alpha = 1 - \dfrac{I_{dc} X}{\sqrt{2} E_{ac} \sin \dfrac{\pi}{m}}$

$$\cos 45° = 1 - \frac{100X}{\sqrt{2} \times 475 \times \sin \frac{\pi}{3}}$$

$$0.707 = 1 - \frac{100X}{581.66}$$

$$\therefore \quad \frac{100X}{581.66} = 0.293$$

or $$X = \frac{0.293 \times 581.66}{100} = 1.704$$

or $$2\pi f L = 1.704 \quad \text{or} \quad 2\pi \times 50 \times L = 1.704$$

$$\therefore \quad L = \frac{1.704}{2\pi \times 50} = 0.005424 \text{ or } 5.424 \text{ mH}$$

Hence, **equivalent inductance, L = 5.424 mH. (Ans.)**

Example 7.37. (Regulation). *A 3-anode mercury arc rectifier has an anode current of overlap 30°. What is the percentage regulation, neglecting the arc drop.*

Solution. Number of phases, $m = 3$

Angle of overlap, $\alpha = 30°$

%age regulation = ?

Voltage drop due to reactance is

$$= E_{ac} \sin^2 \frac{\alpha}{2} = E_{dc} \times \sin^2 \left(\frac{30°}{2} \right) = 0.067 \, E_{dc}$$

But $$E_{dc} = \sqrt{2} E_{ac} \times \frac{m}{\pi} \times \sin \frac{\pi}{m} = \sqrt{2} \times E_{ac} \times \frac{3}{\pi} \times \sin \frac{\pi}{3} = 1.17 \, E_{ac}$$

$\therefore$ Voltage drop due to reactance

$$= 0.067 \times 1.17 E_{ac} = 0.0784 E_{ac}$$

% Regulation $$= \frac{\text{voltage drop on load}}{\text{voltage on no-load}} = \frac{0.0784 E_{ac}}{1.17 \, E_{ac}} \times 100$$

$$= 0.067 \text{ or } \mathbf{6.7}. \text{ (Ans.)}$$

☞**Example 7.38.** *A three-phase rectifier is supplied by a delta-star connected transformer with an equivalent leakage reactance per leg of 0.2 Ω. The effective voltage from one secondary leg is 120 volts. The full-load D.C. output voltage is 130 volts. Determine :*

(i) Overlap angle. *(ii) Load current.*

Solution. Number of phase, $m = 3$

Equivalent leakage reactance/leg, $X = 0.2 \ \Omega$

R.m.s. value of the transformer secondary phase e.m.f.,

$$= 120 \text{ V}$$

Full-load D.C. output voltage, $V_{dc} = 130$ V

Overlap angle :

Load current, I_{ac} :

(*i*) The D.C. no-load voltage,

$$E_{dc} = \sqrt{2}E_{ac} \times \frac{m}{\pi} \times \sin\frac{\pi}{m} = \sqrt{2} \times 120 \times \frac{3}{\pi}\sin\frac{\pi}{3} = 140.3 \text{ V}$$

Full-load D.C. output voltage,

$$V_{dc} = E_{dc}\cos^2\frac{\alpha}{2} \quad \text{or} \quad 130 = 140.3\cos^2\frac{\alpha}{2}$$

$$\therefore \quad \cos^2\frac{\alpha}{2} = \frac{130}{140.3} = 0.926, \quad \cos\frac{\alpha}{2} = 0.9623$$

$$\therefore \quad \frac{\alpha}{2} = 15.78° \quad \text{or} \quad \alpha = 31.56°$$

Hence, **angle of overlap = 31.56°. (Ans.)**

(*ii*) Voltage drop due to transformer reactance

$$= \frac{m}{2\pi} I_{dc} X \quad \text{[(Eqn. 7.17 (a)]}$$

$$(140.3 - 130) = \frac{3}{2\pi} I_{dc} \times 0.2$$

or

$$I_{dc} = \frac{10.3 \times 2\pi}{3 \times 0.2} = 107.8 \text{ A}$$

Hence, **load current = 107.8 A. (Ans.)**

Example 7.39. *A 3300 V, 3-phase transformer has a phase turn-ratio of 162 / 11 ; the primary is mesh connected while the secondary is arranged for 6-phase delta-star connections. Such a transformer is connected to a 6-anode rectifier the grid of which is biased to give a delay of 30°. Calculate the voltage of the rectifier.*

Solution. Primary phase voltage of the transformer
= 3300 V

R.m.s. A.C. voltage between neutral and anode,

$$E_{ac} = \frac{3300 \times 11}{162} = 224 \text{ V}$$

Consider m phase rectifier and neglecting the effect of overlap, the angle of delay is 30°. If there is no biasing of the grid and $E_{dc\beta}$ is the average D.C. voltage of the rectifier, then

$$E_{dc\beta} = \sqrt{2}\frac{m}{\pi}E_{ac}\sin\frac{\pi}{m}\cos\beta$$

$$= \sqrt{2} \times \frac{6}{\pi} \times 224 \times \sin\frac{\pi}{6} \times \cos 30° = 261.9 \text{ V}$$

Hence, **voltage of the rectifier = 261.9 V. (Ans.)**

HIGHLIGHTS

1. *A.C. to D.C. can be converted by the following methods :*
   (*i*) *Motor generator set.*
   (*ii*) *Motor converters.*
   (*iii*) *Rotary converters.*
   (*iv*) *Mercury arc rectifiers.*
   (*v*) *Metal rectifiers.*

2. *In general, for m-phase rotary converter,*

(i) $\dfrac{E_p}{E_{dc}} = \dfrac{\sin \frac{\pi}{m}}{\sqrt{2}}$ (ii) $I_p = \dfrac{\sqrt{2} I_{dc}}{\eta m \sin \frac{\pi}{m} \cos \phi}$

$I_p = \dfrac{\sqrt{2} I_{dc}}{m \sin \frac{\pi}{m}}$ *assuming unity p.f. and 100% efficiency*

(iii) $I_{sr} = \dfrac{2.83 I_{dc}}{m}$ *assuming unity p.f. and 100% efficiency.*

3. *Conversion from 3-phase to 6-phase can be easily achieved by having two similar secondary windings for each of the primary of the three-phase transformers. There are many ways of connecting the six secondaries for this purpose such as :*
   (i) *Diametral.* (ii) *Double-delta.*
   (iii) *Double-star.*
4. *The output D.C. voltage of rotary converter can be changed by the following methods :*
   (i) *By using tap-changing transformer.* (ii) *By reactance control method.*
   (iii) *By induction regulator method.* (iv) *By synchronous booster control method.*
5. *A 'rectifier' is a device which converts alternating current into unidirectional current by virtue of a characteristic permitting appreciable flow of current in only one direction.*
   Some of the common rectifiers used are :
   — *Mercury arc rectifiers.*
   — *Metal rectifiers.*
   — *Mechanical rectifiers.*
   — *Electrolytic rectifiers.*
6. *In a mercury arc rectifier the voltage drop between anode and cathode comprises of :*
   (i) *Anode drop* (ii) *Cathode drop*
   (iii) *Arc drop.*
7. **Glass bulb mercury arc rectifiers,** *because of inherent mechanical weakness of glass, cannot be manufactured for very large outputs. These are usually made into units capable of D.C. outputs of 500 A at 500 V (maximum continuous rating).*

   Steel rectifiers *are made of special quality of sheet iron that resists the action of mercury vapours and helps in maintenance of consistently good vacuum. These units have been constructed for as much as 16000 A at comparatively low voltages, 2500 kW at 3.3 kV and for voltages as high as 20 kV at 600 kW.*

8. **Voltage relation,** $E_{dc} = \sqrt{2} E_{ac} \dfrac{m}{\pi} \sin \dfrac{\pi}{m}$

   where E_{ac} = *average value of no-load D.C. voltage*
   and E_{ac} = *r.m.s. value of secondary phase neutral voltage*

   Current relations, $I_{r.m.s.} = \dfrac{I_{dc}}{\sqrt{m}} = \dfrac{\sin \frac{\pi}{m}}{\sqrt{2}}$

   where $I_{r.m.s.}$ = *value of secondary current*

   $I_{av}/phase = \dfrac{I_{dc}}{m}$.

9. **'Utility factor'** *is the ratio of the volt-amperes obtained from the rectifier on the output side to volt-amperes supplied from A.C. input side.*
10. *The current in a mercury-arc rectifier can be controlled with the help of perforated grids placed between the anode and the cathode. Various grid-control methods can be broadly divided into two classes.*
    (i) *Soft control method.* (ii) *Hard control or impulse control method.*

11. *By increasing the relative phase difference between the anode and grid voltages, the establishment of the arc between the anode and cathode can be delayed by any amount and so D.C. output can be controlled. Such method is known 'phase-shift control method'. The phase-shift can be achieved with the help of an induction regulator or by bias-shift control method.*

OBJECTIVE TYPE QUESTIONS

Say 'Yes' or 'No' :

1. The motor-generator set is simple and reliable.
2. Induction motor-generator set is reversible.
3. The synchronous motor set is liable to hunting and requires a damping winding on its pole faces.
4. A motor-converter is self-starting.
5. A motor-converter can be operated on any frequency.
6. A rotary converter is a single machine with one armature and one field.
7. A motor converter has a low power factor.
8. With a motor-converter it is possible to obtain D.C. voltage only up to 1000 V.
9. A synchronous converter has a low efficiency.
10. The running and operating cost of a synchronous converter is low.
11. A synchronous converter is liable to flash over and reversal of polarity.
12. The rotary converters are liable to give hunting trouble.
13. Reactance control method of voltage control is extremely simple.
14. In induction regulator method variations up to 30% in D.C. voltage can be obtained.
15. Synchronous booster control method is employed in cases where control of D.C. voltage output is not of much importance.
16. A rectifier is a device which converts alternating current into unidirectional current.
17. Glass rectifiers cannot be manufactured for very large outputs.
18. Utility factor is the ratio of the volt-amperes obtained from the rectifier on the output side to volt-amperes supplies from A.C. input side.
19. In 'soft control method' grid voltage is altered gradually.
20. Buzzing sound in an electrolytic rectifier indicates the weakness of electrolyte.

ANSWERS

1. Yes	**2.** No	**3.** Yes	**4.** Yes	**5.** Yes	**6.** Yes	**7.** No
8. No	**9.** No	**10.** Yes	**11.** Yes	**12.** Yes	**13.** Yes	**14.** Yes
15. No	**16.** Yes	**17.** Yes	**18.** Yes	**19.** Yes	**20.** Yes	

THEORETICAL QUESTIONS

Converters

1. Why is it essential to convert A.C. into D.C. ? Give different methods of obtaining D.C. from A.C.
2. Name the common types of converting apparatus used and explain the working principle of any one of them, giving full details together with its merits and demerits over others.
3. What is a synchronous converter ?
4. Explain the construction of a rotary converter.
5. Name the types of rotating machinery used for converting A.C. into D.C. on large scale. What factors are favourable for use of rotary converters ?

6. Explain the advantages of a rotary converter over the motor-generator for conversion from A.C. to D.C. Describe the method of obtaining D.C. from A.C. by a synchronous converter.
7. Explain the action of synchronous converter and explain why it is more economical to use a rotary converter with large number of phases.
8. How many taps must be made to the armature winding of a 3-phase (*i*) six-pole converter, (*ii*) eight-pole converter.
9. Why a transformer is generally interposed between the slip-rings and the A.C. mains.
10. Derive an expression for voltage and current relationship for a m-phase rotary converter.
11. How does the power factor at which a converter operates affect the heating and output of a synchronous converter ?
12. Discuss the armature heating in rotary converters. Indicate the waveform of the armature current in a conductor situated at the centre of a phase.
13. Explain how a 6-phase supply may be obtained from a 3-phase supply using transformers. Discuss the different methods.
14. Discuss any two types of transformer connections used in conjunction with 6-ring rotary converter and comment on their applications.
15. Explain with vector diagram the reactance method of voltage control of synchronous converter.
16. Enumerate the various means of controlling D.C. output voltage of a rotary converter.
17. Explain how induction regulator can be used as a means of controlling the D.C. voltage in a converter. List the advantages of this method. How does this method compare with the tap changing transformer method of voltage control.
18. Explain clearly the underlying principle of the synchronous booster in connection with the D.C. output voltage of rotary converters.
19. Discuss the parallel operation of synchronous converters. Why is it preferable to have separate transformer banks for different converters ?
20. Enumerate the various methods available for the voltage control of a synchronous converter and explain with the vector diagram how the D.C. voltage can be controlled with a choke coil.

Rectifiers

1. What is a rectifier ? How does it work ?
2. Enumerate different types of rectifiers.
3. Draw a circuit diagram of a three anode mercury arc rectifier having the axcitation and ignition circuit.
4. What special properties of mercury make it very desirable rectifier cathode ?
5. What is the arc voltage drop made up of ?
6. Describe with sketches and a diagram of connections a method of ignition and excitation for a 6-anode mercury arc rectifier.
7. Explain the terms 'angle of delay' and 'angle of overlap' as applied to mercury arc rectifiers. How are the delay and overlap brought about ? Drive the expression for voltage ratio, for rectifier working with overlap angle β.
8. What is the necessity of an interphase transformer ? Give a diagram of connections of the transformer and rectifier with the interphase transformer.
9. What are the advantages of using an interphase transformer in mercury arc rectifiers.
10. Explain the action of control grids in a mercury arc rectifier.
11. Describe the construction and working of a 3-phase grid controlled mercury arc rectifier.
12. Distinguish between half-wave and a full-wave rectification each working on a single-phase source.
13. Describe the construction, operation and characteristics of the copper oxide rectifier.
14. Describe the construction and operation of selenium rectifier. Why is this preferred to copper oxide ?
15. What is the working principle of an electrolytic rectifier ?
16. Write a short note on 'mechanical rectifiers'.

UNSOLVED PROBLEMS

Converters

1. In a 700 kW motor generator set the induction motor has six poles and the D.C. generator has eight poles. If the supply frequency is 50-Hz, calculate :
   (*i*) Speed of the set. (*ii*) Frequency of currents in the generator.
   (*iii*) Power transmitted mechanically through the shaft. [**Ans.** 428.6 r..p.m., 28.6 Hz, 300 kW]
2. A motor-converter set consisting of 6-pole induction motor driving a 4-pole D.C. machine is fed from a 3-phase, 50-Hz, 6.6 kV supply and supplies a D.C. network at 380 V. If the machine is loaded to 300 kW, calculate :
   (*i*) Speed of the set. (*ii*) input current.
   (*iii*) Power supplied electrically to the D.C. side from A.C. machine.
   (*iv*) Output current. [**Ans.** 600 r.p.m., 26.25 A, 120 kW, 789.9 A]
3. A 2.5 kW single-phase rotary conveter operates at full-load from 230 volts A.C. supply. Assuming unity power factor and 100% efficiency, calculate :
   (*i*) D.C. voltage and current. (*ii*) A.C. input current at slip-rings.
   [**Ans.** 325 V, 7.7 A, 10.9 A]
4. A 2 kW single-phase rotary converter operates at full-load from 230 volt A.C. source. Assuming unity p.f. and 100% efficiency, calculate :
   (*i*) D.C. voltage. (*ii*) A.C. input current at slip-rings.
   [**Ans.** 325 V, 6.15 A, 8.7 A]
5. A 3-ring rotary converter supplies a direct current of 100 A at 250 volts. If power factor is 0.8 and efficiency is 100% calculate the slip-ring voltage and current. [**Ans.** 153 V, 117.8 A]
6. A 25 kW 3-phase, rotary converter delivers a D.C. output voltage of 230 V. Assuming unity p.f. and 100% efficiency, calculate :
   (*i*) Full-load D.C. output current at brushes. (*ii*) A.C. input voltage between slip-rings.
   (*iii*) A.C. input current per line at full-load. [**Ans.** 108.7 A, 141 V, 102.3 A]
7. A 3-phase synchronous converter delivers a current of 100 A on the D.C. side. If the power factor and efficiency of the machine are each 90 per cent, determine the alternating line current. [**Ans.** 106.3 A]
8. A 3-phase rotary converter supplies a load of 100 kW at 550 V on D.C. side. Determine the current supplied from the A.C. side and the slip-ring voltage if the converter operates of 0.9 p.f. and 0.92 efficiency.
   [**Ans.** 207.5 A, 336.5 V]
9. A 5000 kW, 1200 V, 12-phase rotary converter has a full-load efficiency of 96% and power factor of 0.9. Calculate.
   (*i*) A.C. slip-ring voltage. (*ii*) Full-load D.C. output current.
   (*iii*) A.C. current at slip-rings. (*iv*) A.C. phase current.
   [**Ans.** 218.4 V, 4167 A, 1283 A, 2195 A]
10. A 3-phase, 250 kW synchronous converter develops 250 V on the commutator and is supplied from a 2200 V, 3-phase net work through single-phase transformers Δ-connected on the h.v. side and *Y*-connected on the l.v. side. The converter operates on full-load at p.f. 0.9 lag and with efficiency of 91%. Determine the voltage and current rating of both sides of the transformers, neglecting its losses.
    [**Ans.** 153 V, 1151 A, 80 A]
11. A 3-ring rotary converter is required to give 480 volts at the commutator. The supply is 6600 V, 3-phase. The transformer primaries are connected in delta and secondaries in star. Determine the ratio of primary to secondary turns in the transformer. Neglect drops. [**Ans.** 39 : 1]
12. A synchronous converter is delivering direct current at 600 V. The converter has 6 slip-rings. Determine :
    (*i*) The diametral voltage of the armature
    (*ii*) The voltage between adjacent slip-ring taps, and
    (*iii*) The voltage between alternate slip-rings. [**Ans.** 424.2 V, 212.1 V, 367.4 V]

13. A six-rings synchronous converter supplies a load of 250 kW to a D.C. network. It has 10 poles 1000 lap wound armature conductors and the flux per pole is 50 mWb. The transformer is supplied from 3300 V, 50-Hz mains and the primary is delta connected. Find :

(*i*) The turn-ratio of the transformer if the machine has diametral tappings.

(*ii*) The current per slip-ring if the efficiency is 95 per cent. [**Ans.** 1 : 9.33, 248 A (at unity p.f.)]

14. A six-ring, 8-pole, lap-wound synchronous converter with diametrical tappings runs at 750 r.p.m. and supplies a direct current of 1200 A. There are 800 armature conductors and the flux per pole is 50 mWb. The transformer is delta connected to a 6600 V, 3-phase supply. Assuming unity power factor and an efficiency of 94%, calculate :

(*i*) The turn-ratio of the transformer. (*ii*) The current per ring.

(*iii*) The current in the armature connection between a slip-ring and a tapping point.

[**Ans.** 1 : 18.66, 603 A, 150.8 A]

15. A 200 kW, 400 volts, 4-pole synchronous converter is supplied through a transformer from a 33 kV, 50-Hz mains. The transformer has delta connections on the primary side and diametral connections on the secondary side. Calculate the number of turns and cross-sectional area required on each side of the transformer winding. The synchronous converter has a 95 per cent efficiency and is operating at unity p.f. Assume a maximum core flux of 100 m Wb and a current density of 2.5 A/mm^2. Neglect magnetizing current and losses in the transformer. [**Ans.** 13, 1520, 99.2 mm^2, 0.848 mm^2]

16. A six-ring converter delivers 500 kW, at 500 V on D.C. side, the efficiency and power factor at this load being 92% and 0.95 respectively. Calculate :

(*i*) slip-ring current (*ii*) voltage across.

(*a*) consecutive

(*b*) alternate

(*c*) diametral slip-rings on the input side. [**Ans.** (*i*) 540 A ; (*ii*) 177 V, 306 V, 354 V]

17. A 1500 kW, 8-pole, 50-Hz, 6-ring diametrically connected synchronous converter is required to supply a 500 V net work from a 11 kV, 3-phase system. If the transformers between A.C. lines and the slip-rings are mesh connected on the high voltage side, determine the number of turns on each winding and the cross- section of conductors. Take the current density as 3 A/mm^2, the total flux as 0.085 Wb and neglect losses. Assume a p.f. of 0.9. [**Ans.** 19, 594, 524.3 mm^2, 16.8 mm^2]

18. The transformer feeding a 2000 kW, 750 V, A.C. six-phase synchronous converter are connected diametrical on the secondary side. On the assumption of a full-load efficiency of 96% and a power factor of 0.95, calculate :

(*i*) The six-phase voltage between adjacent slip-rings.

(*ii*) The voltage of each transformer secondary.

(*iii*) The alternating current delivered by each of the transformer secondaries.

[**Ans.** 265 V, 530 V, 1375 A]

19. A 1500 kW, 750 V, D.C. 6-phase synchronous converter has a full-load efficiency of 96% and a power factor of 0.92. It is fed by a bank of transformers connected in delta on primary and diametrical on the secondary side. The 3-phase primary supply line voltage is 13200 V. Calculate :

(*i*) D.C. output current. (*ii*) A.C. voltage between slip-rings.

(*iii*) Voltage across each secondary phase. (*iv*) Ratio of transformation.

(*v*) A.C. current delivered by each transformer secondary.

(*vi*) Current in each primary coil. (*vii*) Line current on primary side.

[**Ans.** 2000 A, 265 V, 530 V, 1 : 24.9, 1069 A, 42.9 A, 74.4 A]

20. A 1500 kW, 750 V, D.C. 6-phase rotary converter is fed by a bank of transformers connected in star on primary side and double-delta on the secondary side. The three phase primary line voltage is 23000 V. Assuming a full-load efficiency of 96% and power factor of 0.92, calculate :

(*i*) Secondary voltage/phase. (*ii*) A.C. voltage between adjacent rings.

(*iii*) Slip-ring current. (*iv*) Transformation ratio.

(*v*) Primary line current. [**Ans.** 459.4 V, 265 V, 1849 A, 1/14.45, 73.9 A]

21. A six-ring 200 kW, 500 V, D.C. synchronous converter has an efficiency of 96 per cent. The primary of the transformer is delta and secondary double star. If the primary is supplied by 3-phase, 3300 volt line, calculate :

(*i*) Primary phase current. (*ii*) Secondary phase voltage and current.

Assume power factor as unity. Neglect magnetizing current and losses in transformer.

[**Ans.** 21 A, 177 V, 196.7 A]

22. A reactor of reactance 0.5 Ω and negligible resistance is put in each line of supply of a 3-ring synchronous converter. The converter takes a leading current of 80 A at a p.f. of 0.8. Calculate D.C. voltage output, if the supply voltage is kept constant at 400 V. Neglecting all losses. [**Ans.** 420 V]

23. A six-ring converter with diametral tappings is supplied from a transformer having a constant secondary phase voltage of 420 volts. A reactor of 0.2 Ω is included in each line connecting the transformer and the slip-rings. Neglecting losses, calculate :

(*i*) D.C. no-load voltage

(*ii*) D.C. voltage for p.f. 0.71 leading and 0.8 lagging when the D.C. output is 270 kW.

[**Ans.** 724 V, 516 V]

24. A 6-ring rotary converter supplies a direct load current of 400 A and the power to it is supplied from a 3-phase 33 kV transformer with Δ-connections on the high voltage side and diametral connections on the secondary side. The transformer has a turn-ratio of 60 and equivalent secondary reactance of 0.2 Ω. Calculate the change in D.C. voltage when the power factor at slip-ring changes from unity to 0.8. Assume an efficiency of 94%. [**Ans.** 45 V]

25. A 6-ring converter has diametral connections, the voltage on low voltage side of the 3-phase transformer being 300 V. Calculate the reactance in each of the slip-ring leads if the output is 500 kW at 500 V and the power factor at the transformer terminals is unity. Neglect losses. [**Ans.** 0.199 Ω]

Rectifiers

26. A 3-anode mercury arc rectifier is supplied through the star-connected secondary of a 3-phase transformer. It delivers an output of 50 kW at 500 V on D.C. side. Calculate the transformer secondary phase voltage and current. Allow an arc drop of 25 V. [**Ans.** 449 V, 57.8 A]

27. Determine the values of rectifier voltage and current when a 3-phase rectifier is working on 400 volts, 50-Hz supply. Current drawn from the mains is 5 A/phase. [**Ans.** 468 V, 8.65 A]

28. In a 6-phase rectifier the full-load terminal voltage on the output side is to be 1500 V and the sum of all drops is 25 V. Calculate the transformer secondary voltage. [**Ans.** 2260 V]

29. A D.C. load is supplied at 550 V (mean) from a 6-anode mercury arc rectifier. The high voltage supply is 11 kV and the transformer is star-connected on the primary side. Calculate the turn-ratio of the transformer. Allow an arc drop of 30 V for each anode. [**Ans.** 7.4]

30. A 3-anode rectifier supplies a load of 15 kW at 500 volts. If the arc drop is 25 volts calculate the kVA rating of the transformer. Neglect utility factor. [**Ans.** 23.25 kVA]

31. A 6-anode mercury-arc rectifier has a D.C. output voltage of 230 V. The transformer is star-connected to a 3300 V, 3-phase line. Calculate the turn-ratio of the transformer assuming that the sum of all the voltage drops in the rectifier is 25 V for each anode. [**Ans.** 5.04 : 1]

32. The turn-ratio of a 3-phase delta-star transformer operating from 6600 V mains is 15 : 1. Find the mean value of D.C. output voltage of 3-anode mercury-arc rectifier allowing 25 volts for arc drop.

[**Ans.** 490 V]

33. Calculate the bulb and over all efficiencies of a 3-anode mercury arc rectifier supplying a load of 40 A at 500 V. The arc drop is 20 V and transformer no-load loss is 560 W. The short-circuit loss of the transformer at 25 A is 900 W and loss due to reactance cores is 120 W. [**Ans.** 96.2%, 89.9%]

34. A transformer connected to a 3-anode mercury arc rectifier has the following test results :

Open-circuit test. At normal voltage consumes 450 W.

Short circuit test. 800 W for 20 A current on secondary. If the rectifier is supplying a load of 20 kW at 450 volts determine :

(*i*) the bulb efficiency (*ii*) the overall efficiency.

Assume an arc drop of 15 volts and a loss of 100 watts in smoothing reactance. [**Ans.** 96.8%, 88.6%]

35. The secondary phase voltage of a 3-anode rectifier is $1300/\sqrt{2}$ volts (r.m.s.). Full-load overlap is 30°. If arc voltage drop is 25 volts, determine the no-load and full-load output voltages. [**Ans.** 1075 V, 977 V]

36. A 3-phase rectifier is supplied by a delta-star connected transformer with an equivalent leakage reactance per phase of 0.2 Ω. The effective secondary phase voltage is 200 volts. If the load current is 100 amperes, calculate :'

(*i*) D.C. on-load voltage ; and

(*ii*) drop due to the effect of overlap on the wave shape and also the angle of overlap.

[**Ans.** 234 V, 9.54 V, 23.3°]

37. A 6-phase rectifier is to deliver a load current of 400 amperes at 500 volts. The rectifier is supplied from a delta-star transformer with diametral connections. The arc drop is 15 volts. Calculate :

(*i*) The voltage to neutral on the transformer secondaries.

(*ii*) The rating of the primary and secondary winding of each transformer for which these should be designed.

Assume utility factor for primary and secondary as 0.78 and 0.55 respectively.

[**Ans.** 382 V, secondary VA = 62400, Primary VA = 88000]

38. A three-anode rectifier has D.C. full-load current of 120 A. Estimate the equivalent inductance per transformer secondary phase if the overlap angle on full-load is not to exceed 45°. The transformer secondary phase voltage is 480 V and the frequency 50-Hz. The load inductance may be considered such as to maintain a sensible constant output current. [**Ans.** 4.566 mH]

39. A 3-phase rectifier is supplied by a delta-star connected transformer with an equivalent leakage reactance per leg of 0.25 Ω. The effective voltage from one secondary leg is 100 volts. The full-load D.C. output voltage is 105 volts. Determine the overlap angle and load current.

40. A 6-anode grid controlled mercury arc rectifier is supplied from a 3-phase 440 V, 50-Hz system. The transformer primary is mesh connected and central points of the secondary phases are joined together. The phase turns-ratio is 11/10. Determine the mean output voltage with a highly inductive load when the grids are biased to delay the ignition of the anode by :

(*i*) 0, (*ii*) $\pi/4$, (*iii*) $\pi/3$. [**Ans.** 217 V, 153.5 V, 108.5 V]

ADDITIONAL TOPICS
(COMMON TO D.C. AND A.C. MACHINES)

1. **Special Types of D.C. Motors**
2. **Electronic Control of D.C. Motors**
3. **Eddy Current Drives**
4. **Electronic Control of A.C. (Induction) Motors**
5. **Synchronous Motors – Types, Starting, Speed Control and Braking**
6. **Digital Control of Electrical Motors**
7. **Transient Conditions in Electrical Drives**
8. **Size and Rating of Motors**
9. **Enclosures for Rotating Electrical Machines**
10. **Selection of Electric Motor for any Application**
11. **Motors for Particular Applications**
12. **Energy Conservation in Electrical Drives.**

1. SPECIAL TYPES OF D.C. MOTORS

1.1. PERMANENT MAGNET D.C. MOTORS

In these motors, *field excitation* is obtained by suitably mounting permanent magnets on the stator. Magnets made from ferrites or rare earth (cobalt samarium) are used.

— *Ferrites* are commonly used because of *lower cost*, but machine becomes *bulky due to low retentivity*.

— Rare earths because of their high retentivity allow a large *reduction in weight and size, but they are very expensive*.

Use of permanent magnets for excitation *eliminates field copper loss and need for field supply.*

Advantages :

As compared to field wound motors, these motor possess the following *advantages :*

(*i*) More efficient.

(*ii*) More reliable.

(*iii*) More study and compact.

(*iv*) The field flux remains constant for all loads giving a more linear speed-torque characteristic.

(*v*) In a separately excited motor, failure of field can lead to runaway condition. This does not happen in permanent magnet motors.

Limitation. As the flux is constant in these motors, speed *cannot be controlled above base speed.*

Applications :

- These motors are mainly employed in Fractional H.P. range, but they are available upto 5 kW rating,
- They find applications in electric vehicles like *wheel chairs, mopeds, forklift trucks* etc.

1.2. D.C. SERVOMOTORS

The term *servo* or servo-mechanism refers to a *feedback control system* in which the controlled variable is :

— Mechanical position, or

— Time derivatives *e.g.* velocity and acceleration.

Following *characteristics* are usually required for a feedback control system :

(*i*) High accuracy ; (*ii*) Remote operation ;

(*iii*) Fast-response (*iv*) Unattended control.

Following are the *essentials* of a feedback control system :

1. **An error detecting device.** It determines when the regulated quantity is *different* from the reference quantity and sends out the error signal to the *amplifier*.

2. **An amplifier.** The amplifier receives the error signal and then supplies power to the error-correcting device, which in turn changes the regulated quantity so that it matches the reference input.

A *servo-motor* should entail the following *characteristics* :

1. *The output torque of the motor should be proportional to the voltage applied (i.e.,* the control voltage which is developed by the amplifier in response to an error signal).

2. *The direction of the torque developed by the servo-motor should depend upon the instantaneous polarity of the control voltage.*

Types of servo-motors

The servo-motors are of the following *two* types :

1. *D.C. servo-motors.*

2. *A.C. servo-motors.*

D.C. servo-motors :

These motors are preferred for very high power systems since they operate more efficiently (as compared to A.C. servo-motors).

These motors may be of the following types :

— Series motors ;

— Split series motors ;

— Shunt control motors ;

— Permanent magnet (*fixed excitation*) shunt motor.

(*i*) ***Series motors :***

- This motor has a high starting torque.
- It draws large current.
- The speed regulation is poor.
- Reversal can be obtained by reversing field voltage polarity with split series field winding.

(*ii*) ***Split series motor :***

- The D.C. series motor with split field (small fractional kW) may be operated as a separately excited field-controlled motor (Fig. 1.1).

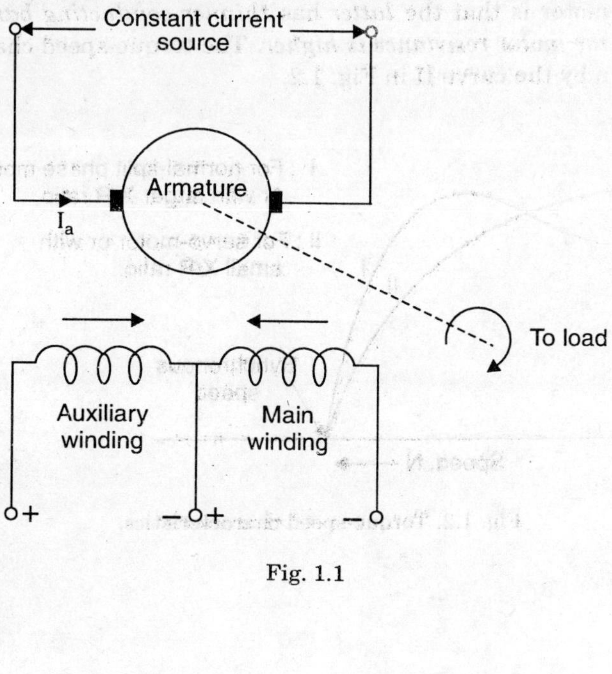

Fig. 1.1

The armature may be supplied from a constant current source.

- A typical torque curve shows the following :
  — High stall torque ;
  — Rapid reduction in torque with increase in speed.

(*iii*) ***Shunt control motor :***

- This type of motor has *two separate windings : Field winding placed on the stator and the armature winding placed on the rotor of the machine.* Both the windings are connected to a D.C. supply source.
- Whereas in a conventional D.C. shunt motor, the two windings are connected in parallel across the D.C. supply mains, but in a *servo-application* the windings are driven by *separate D.C. supplies.*

(*iv*) ***Permanent magnet shunt motor :***

- It is a fixed excitation shunt motor where the field is actually supplied by a permanent magnet.
- Its performance is similar to that of armature controlled fixed field motor.

A.C. servo-motors

Applications :

- These motors are best suited for *low power applications.*
- Precision servo-motors are used in :
  — Instrument servos ;
  — Computers ;
  — Inertial guidance systems etc.
- The mechanical output power of A.C. servo-motor varies from 2 watts to a few hundred watts.
- An A.C. servo-motor is basically a two-phase induction motor except for certain special design features. The *main important difference* between a standard split-phase motor and an A.C. servo-motor is that the *latter* has *thinner conducting bars in the squirrel cage motor, so that the motor resistance is higher.* The torque-speed characteristics should be *linear* as shown by the curve II in Fig. 1.2.

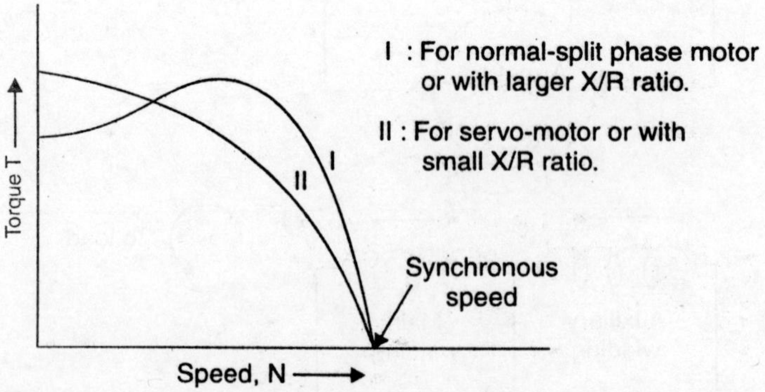

Fig. 1.2. Torque-speed characteristics.

Description of A.C. servo-motors :

1. **Drag-cup rotor servo-motor.** Refer Fig. 1.3.

- Drag-cup construction is used for *very low inertia applications.*
- In this type of motor the rotor construction is usually of squirrel cage or drag-cup type ; here only a *light cup rotates while the rotor core is stationary* (thus inertia is quite small).

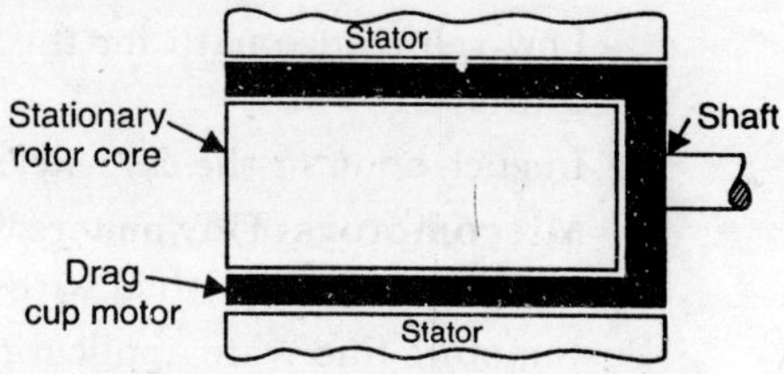

Fig. 1.3. Drag-cup rotor servo-motor.

- The servo-motors contains *two* windings namely, *main winding* (sometimes called fixed or reference winding) and *control winding.* The voltages applied to the windings are at *right angles* to one another. Usually one winding is excited with a fixed voltage while the other one is excited by the control voltage (which is the output from servo-amplifier).
- *While in operation, the output torque of the motor is roughly proportional to the applied control voltage, and the direction of torque is determined by the polarity of the control voltage.*

2. **Shaded-pole type servo-motor :**

- This type of motor employs a *phase-sensitive relay to actuate those contacts which produce a short-circuit of the shaded-pole winding to produce rotation in the desired direction.*
- *The main shortcoming of this motor is that it responds only when the amplifier error signal is of adequate magnitude to cause the relay to operate.*

1.3. MOVING COIL MOTORS

There are certain applications which require acceleration much higher than what can be achieved in a conventional D.C. servo-motor. The armatures of moving coil D.C. motors have special constructions which allow a substantial reduction in armature inertia and inductance, permitting very high accelerations.

Moving coil motors are of the following two types :

1. Shell type
2. Disc or Pancake type.

1. Shell type moving coil motor. Refer to Fig. 1.4.

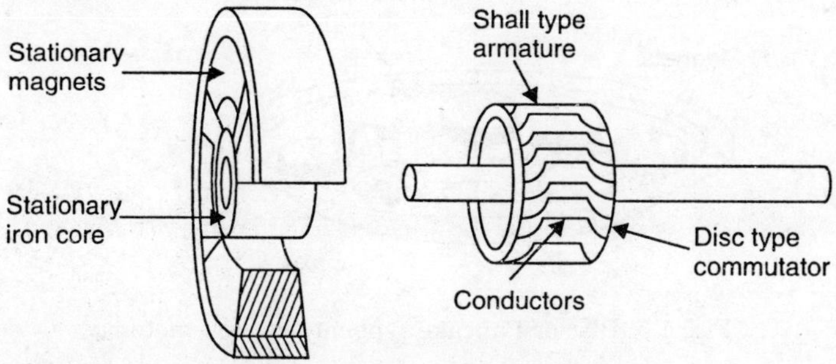

Fig. 1.4. Shell type moving coil motor.

- In this type of motor, the rotor consists of *only armature winding* due to which it has very low inertia ; consequently high acceleration is obtained. Armature winding consists of conductors assembled to form a thin walled cylinder. The *commutator* may have a cylindrical construction as in conventional D.C. motors or disc type construction.

  Low reluctance path for the stator field is provided by a stationary magnetic material cylinder.

  In such a motor the *current is axial and flux is radial.*
- **Micromotors** (Tiny motors with diameters around 1 cm) have armature winding consisting of simply varnished wires arranged in cylindrical form and a disc type commutator.

Such motors find wide applications in *card readers, video systems, cameras* etc.

In *bigger size motors* the armature winding is made by bonding conductors together using polymer resins and fibreglass to provide adequate mechanical strength.

2. Disc or Pancake type moving coil motor. Refer Fig. 1.5.

In this motor armature is made in disc or pancake form, and armature conductors resemble spokes on a wheel. The armature winding is formed by stamping conductors from a sheet of copper, welding them together and placing them on a light weight disc. Conductor segments are then joined with a *commutator* at the centre of the disc.

Here the direction of *flux is axial* and *armature current is radial* (just opposite to shell type conventional motors).

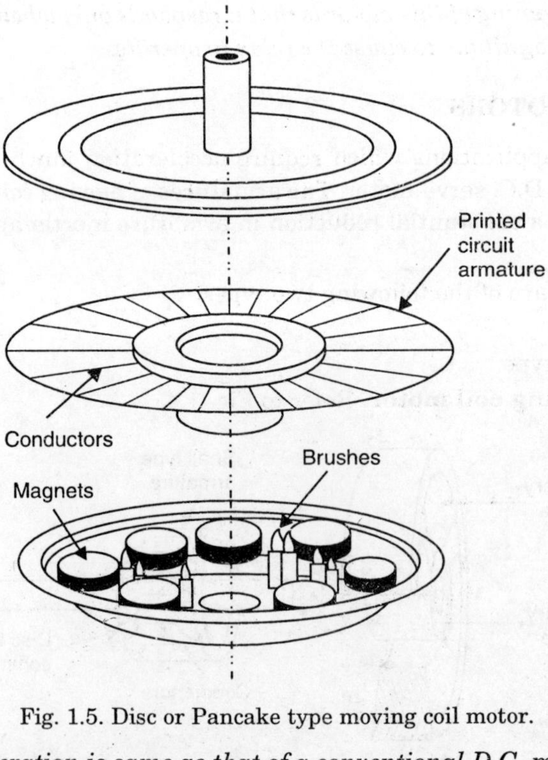

Fig. 1.5. Disc or Pancake type moving coil motor.

The *principle of operation is same as that of a conventional D.C. motor.*

- These motors are more robust and available in sizes *upto few kilowatts.*

- They find *applications* where *axial space is at a premium* such as *machine tools, disc drives* etc.

1.4. TORQUE MOTORS

"Torque motors" are *the D.C. motors designed to run for long periods in a stalled or a low speed condition.* Some torque motors are designed to operate at low speeds intermittently.

The torque motor *applications* can be divided into the following three types :

(*i*) Motor is required to *operate in stalled condition.*

— The purpose of the motor is to develop the required tension or pressure on a material, similar to spring.

(*ii*) Motor is required to *move through only a few revolutions* or *degrees of revolution.*

Examples. Opening of valves, switches, clamping devices etc.

(*iii*) This category involves *continuous movement of the motor at low speed.*

Example. Reel drive.

1.5. BRUSHLESS D.C. (OR TRAPEZOIDAL PMAC) MOTORS

Fig. 1.6 shows the cross-section of a 3-phase 2-pole trapezoidal PMAC motor.

The stator has three concentrated phase windings (AA', BB' and CC') which are displaced by 120° and each phase winding spans 60° on each side. The voltages induced in three phases are shown in Fig. 1.6. The reason for getting trapezoidal waveforms is explain below :

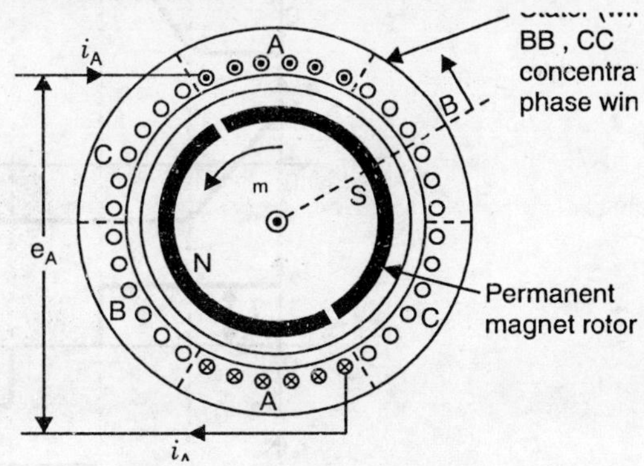

Fig. 1.6. Cross-section of a trapezoidal PMAC motor.

When revolving in the counter-clockwise direction, upto 120° rotation from the position shown in Fig. 1.3, all top conductors of phase *A* will be linking the *S*-pole and all bottom conductors of phase A will be linking the *N*-pole. Hence the voltage induced in phase *A* will be the same during 120° rotation (Fig. 1.6). Beyond 120°, some conductors in the top link *N*-pole and others the *S*-pole. Same happens with bottom conductors. Hence, the voltage induced in phase *A* linearly reverses in next 60° rotation. Rest of the waveform of phase *A* and waveforms of *B* and *C* can be explained on the same lines.

Fig. 1.7 shows the induced voltage, phase current and torque waveforms of a brushless D.C. motor.

An *inverter fed trapezoidal PMAC motor drive operating in self-controlled mode is called a* **brushless D.C. motor.** This motor is also conceived as *electronically commutated D.C. motor*, because inverter here performs the same function as the brushes and commutator in a D.C. motor *i.e.*, to shift currents between armature conductors to keep the stator and rotor fields stationary, and in quadrature with respect to each other.

Advantages :

Owing to the absence of brushes and commutator, brushless D.C. motors claim the following *advantages over the conventional D.C. motors :*

(*i*) Long life.

(*ii*) Require practically no maintenance.

(*iii*) High reliability.

(*iv*) Low inertia and friction

(*v*) Low radio frequency interference and noise.

(*vi*) Because armature windings are on the stator, cooling is much better, *i.e.*, *higher specific outputs* can be obtained.

(*vii*) They have a faster acceleration (due to low inertia and friction) and can be run at much higher speeds-upto 100 000 r.p.m. and higher are common.

(*viii*) High efficiency, exceeding 75 percent (whereas wound field motors of low power ratings have much lower efficiency).

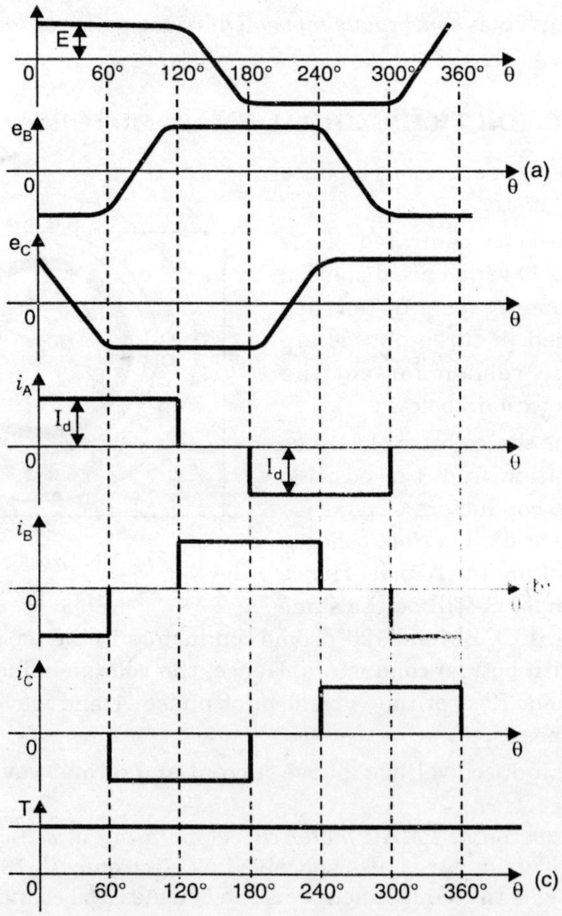

Fig. 1.7. Induced voltage, phase current and torque waveforms of a brushless D.C. motor.

Disadvantages :

(*i*) High cost.

(*ii*) Low stalling torque.

The size of a brushless D.C. motor is nearly the same as that of the conventional D.C. motor.

Applications :

The brushless D.C. motor finds applications in :

(*i*) Tape drive for video recorders ;

(*ii*) Turn table drives in record players ;

(*iii*) Spindle drives in hard disk drives for computers ;

(*iv*) Low cost and low power drives in computer peripherals, instruments and control systems.

(*v*) Gyroscope motors ;

(*vi*) Cryogenic coolers ;

(*vii*) Artificial heart pumps ;

(*viii*) Cooling fans for electronic circuits and heat sinks.

2. ELECTRONIC CONTROL OF D.C. MOTORS

2.1. INTRODUCTION

Normally, it is essential to vary the speed of electrical drives in different fields of application. Usually, in all process industries, it is desired that the system be set at slow speed in the beginning and then gradually increased to meet the maximum production rate, *e.g. newspaper printing press.*

One of the major achievements of *thyristor technology* in the field of control is the control of D.C. and A.C. motor drives. *Thyristor controlled schemes* have totally dominated the field of control of D.C. as well as A.C. motors because of the following *advantages* :

(*i*) *Compactness* ; (*ii*) *Fast response* ;
(*iii*) *More efficiency* ; (*iv*) *More control capabilities* ;
(*v*) *More reliability* ; (*vi*) *Less cost etc.*

2.2. ADVANTAGES OF ELECTRONIC CONTROL SYSTEMS

The electronic control system claim the following *advantages over conventional methods :*

1. Very compact and small in size.
2. Consumes very less power.
3. Very fast in response.
4. Much more accurate and efficient than a conventional system.
5. Control ranges are much more than any other systems.
6. High reliability comparatively.
7. Economical, since maintenance cost is minimum.
8. Highly protective.
9. In electronic systems more automation, as required for highly sophisticated machines, is possible.

2.3. D.C. MOTOR SPEED CONTROL

There are several methods by which the speed of a D.C. shunt motor can be controlled *using thyristors*, some of the commonly used methods are discussed below.

1. Armature voltage control method :

This is also called the *phase control method of speed control*. The complete diagram for this scheme is shown in Fig. 2.1.

- The field of motor is excited by a constant D.C. obtained from the *full-wave rectifier*.
- The armature voltage is *varied by varying the firing angle of the SCRs of the thyristor bridge*. Voltage across the armature terminals will be variable D.C. obtained from the full-wave half-controlled thyristor bridge. In the positive half-cycle SCR_1 and diode D_1 will conduct whereas in the negative half cycle SCR_2 and diode D_2 will conduct. Gates of *SCRs* will be given signal from the triggering circuit (not shown in the Fig. 2.1).

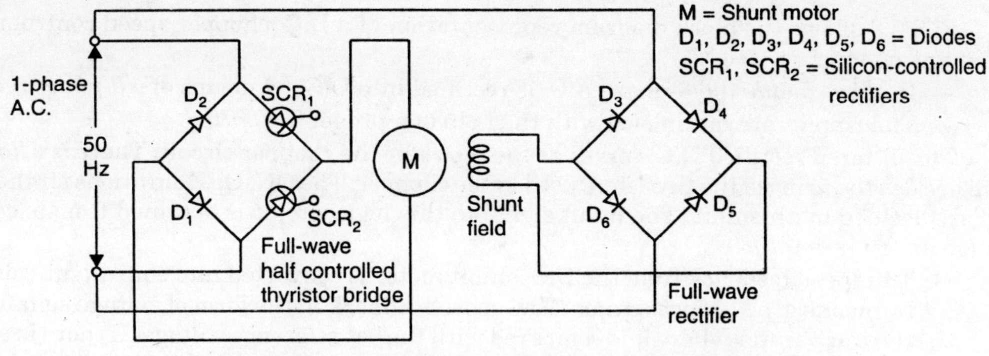

Fig. 2.1. Complete circuit diagram for the armature voltage control method for speed control of D.C. shunt motor.

- The wave shapes for the A.C. input voltage and controlled D.C. armature voltage are shown in Fig. 2.2.

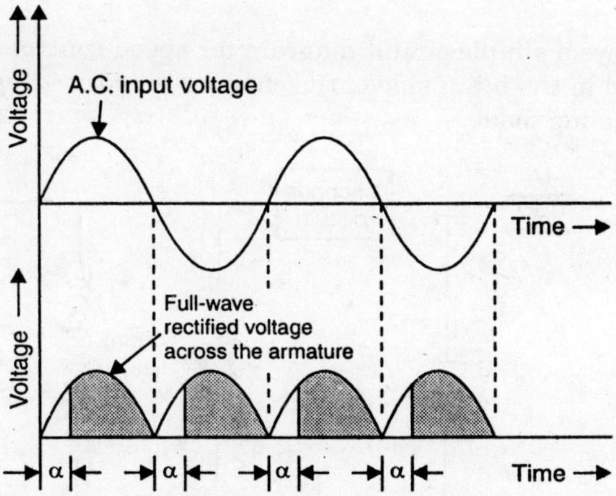

Fig. 2.2. Wave shapes for A.C. input voltage and controlled D.C. armature voltage.

2. D.C. chopper speed control :

A *D.C. chopper can give variable D.C. at its output.* This variable can be utilised for the purpose of speed control of D.C. shunt motors. This method of speed control has gained popularity since the introduction of semiconductor devices.

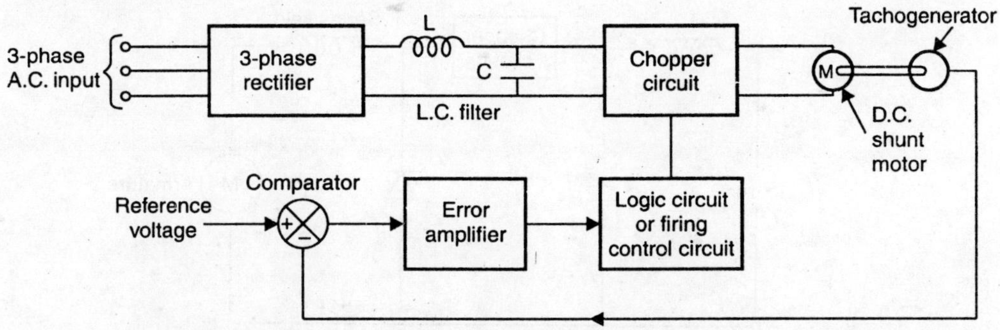

Fig. 2.3. Block diagram representation of a D.C. chopper speed control scheme for D.C. shunt motors.

Fig. 2.3 shows the block diagram representation of a D.C. chopper speed control scheme for D.C. shunt motors.

— In this scheme the 3-phase A.C. is rectified into D.C. by means of a *3-phase rectifier*.

— The ripples are minimised with the help of a proper *'L-C filter'*.

This filtered rectified D.C. serves as the *input* for the chopper circuit. There is a *'logic circuit'* which decides the firing of the thyristors used in the chopper. The ON, OFF durations for the thyristors used are decided by this unit. The input signal to this logic circuit is obtained from a *'comparator'* through an *'error amplifier'*.

— The speed feedback from the D.C. shunt motor is converted into equivalent voltage signal by means of a *'tachogenerator'*. The speed feedback in the form of voltage signal is given to the comparator where it is compared with the set reference voltage. If there is a difference between the two, it will generate an error signal which is amplified by the error amplifier and sent to the logic circuit to decide the ON, OFF duration of the thyristors connected in the chopper.

— Choppers are built by using one or two SCRs depending upon the type and circuitry used. *This is a very efficient method and is widely used in industries these days because of its fast response.*

- Fig. 2.4 shows a simple circuit diagram for speed control of a D.C. shunt motor. An *L-C* filter is used in the input side of the chopper to *reduce ripples* in the D.C. input. Diode is the freewheeling diode.

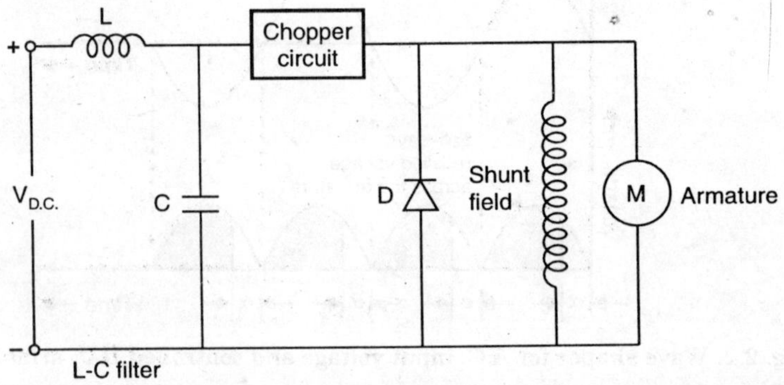

Fig. 2.4. Circuit diagram of a D.C. chopper for speed control of a D.C. shunt motor.

- Fig. 2.5 shows a simple chopper circuit which may be used for controlling the speed of a D.C. series motor.

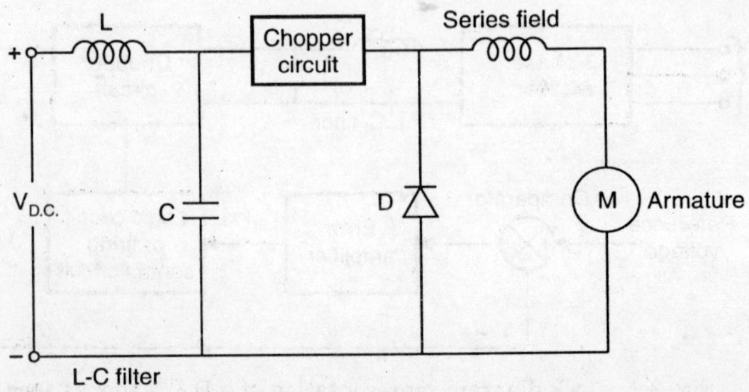

Fig. 2.5. D.C. chopper application for speed control of a D.C. series motor.

— Variation of T_{ON} and T_{OFF} will vary the load voltage at the output of the chopper which will change the speed of the motor accordingly.

— Diode *D* has been used as a *freewheeling diode to provide low resistance path for the current which will flow even at the OFF period of the thyristors.* This current flows for a little time due to the stored energy in the winding which is inductive in nature.

— An *L-C* filter has been used in the input side of the chopper to reduce the ripples in the D.C. input voltage.

3. Speed control by using a dual converter :

A dual converter, as the name indicates, uses two converters a *rectifier* and an *inverter*. Both the bridges are built by using *SCRs*. A dual converter may be used to obtain the following controls of a D.C. motor :

- Reversible speed control ;
- Plugging ;
- Regenerative braking.

The above controls are discussed below.

Fig. 2.6 shows the circuit diagram for speed control of a D.C. shunt motor using a *dual converter.*

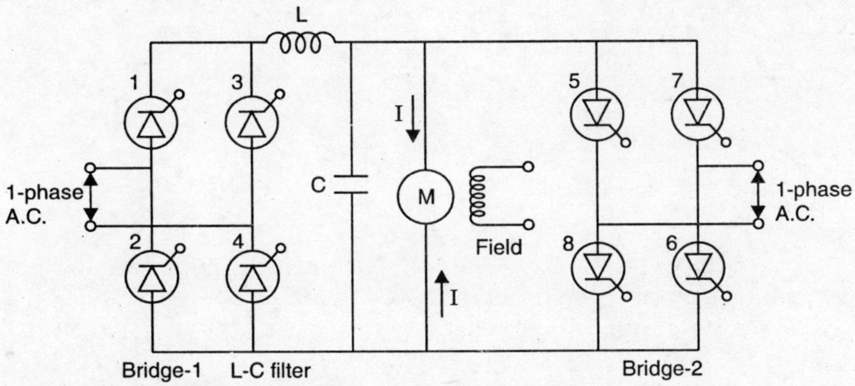

Fig. 2.6. Circuit diagram for speed control of a D.C. shunt motor using a dual converter.

1. Reversible speed control and plugging :

- Four *SCRs*, 1, 2, 3 and 4 form the first bridge (Bridge-1) which serves as a 1-phase full-wave fully-controlled bridge and rectifies the 1-phase A.C. into D.C. This D.C. is filtered by an *L.C. filter* to remove the ripples. In the positive half cycle *SCRs* 1 and 2 conduct simultaneously and in the negative half cycle *SCRs*, 3 and 4 conduct simultaneously. The direction of flow of armature current *I* is clockwise as shown in Fig. 2.6.
- For ***reversing*** the direction of rotation of the motor, the second bridge (Bridge-2) is gated after commutating the first bridge. The Bridge-2 is constituted by the *SCRs* 5, 6, 7 and 8. *SCRs* 5 and 6 conduct simultaneously in the positive half cycle and *SCRs* 7 and 8 conduct simultaneously in the negative half cycle. Thus, the direction of flow of armature current is reversed in this case and the motor tries to rotate in the opposite direction *i.e.* in the anticlockwise direction.

Because the motor was originally running in the clockwise direction, the inertia would oppose the torque developed in the anticlockwise direction. When the *two torques become equal, the motor becomes stationary provided bridge-2 is commutated.* This process of stopping the motor is called ***plugging.*** If the bridge-2 further continues to conduct, the motor would start-running in the opposite direction resulting in speed reversal. In the opposite direction of rotation of the motor, the speed can be controlled by varying the *firing angle* of the second bridge.

2. Regenerative braking :

In this case, after bridge-1 is commutated and bridge-2 is triggered the counter e.m.f. generated in the armature of the motor acts as input for bridge-2 which is connected in the *inverter mode.* The output of bridge-2 which is 1-phase A.C. may be fed back to the mains supply. Thus, we see that bridge-1 acts as a rectifier and bridge acts as inverter. Therefore, in regenerative braking the K.E. of the motor is converted into electrical energy and fed back to the supply system thereby saving energy.

3. EDDY CURRENT DRIVES

- An eddy current drive consists of an *eddy current clutch placed between an induction motor running at a fixed speed and the variable speed load.* Speed is controlled by controlling D.C. excitation to magnetic circuit of the clutch. Since motor itself runs at a fixed speed it can be fed directly from A.C. mains.
- The principle of an eddy-current clutch is identical to an induction motor in which both stator and rotor are allowed to rotate. Stator, which is coupled to driving induction motor, has D.C. winding which produces magnetic field rotating at the speed of rotor. Rotor has a metal drum coupled to the load. Eddy currents are induced in rotor drum by stator magnetic field. Interaction between the stator field and eddy currents produces a torque which causes rotor to move with stator with a slip. Slip, and therefore, the load speed, can be controlled by controlling D.C. current through rotor winding. Speed torque characteristics are identical to an induction motor.
  - — Speed reduction is obtained by wasting a power equal to sP_{in} in the rotor drum. Minimum speed is usually restricted to 30 percent below the synchronous speed, because efficiency becomes too low and cooling of the rotor drum becomes difficult below this speed.
  - — Load can be decoupled from induction motor by setting D.C. winding current to zero. Motor can now be started on no load. Load can be smoothly started by slowly increasing D.C. winding excitation.
- Eddy current clutch is available in different constructions and sizes ranging from fraction of a kW to MW.

Advantages :

The *advantages of eddy current drives* are :

(*i*) Rugged in construction.
(*ii*) Easy to maintain.
(*iii*) Reliable in operation.
(*iv*) Stepless speed control with good speed regulation.
(*v*) Controlled acceleration and soft start.
(*vi*) High starting torque.
(*vii*) High overload capacity.
(*viii*) Ability to handle impact loads.

Applications :

They are widely used in :

- Blowers ;
- Conveyors ;
- Dredges ;
- Winders ;
- Compressors ;
- Cranes ;
- Elevators ;
- Line shafts and paper machines.

However, due to *poor efficiency and cooling,* they are *rarely* used in new installations.

4. ELECTRONIC CONTROL OF A.C. (INDUCTION) MOTORS

4.1. INTRODUCTION

The speed of a D.C. motor can be controlled by varying the field current or the armature voltage through a phase controlled rectifier or by a D.C.—D.C. converter if the input supply is D.C. Also, in a D.C. machine the torque is developed due to the interaction of field flux and the D.C. armature flux which remains stationary in space. Whereas in A.C. machine, a 3-phase supply to the stator winding *produces rotating magnetic field of constant magnitude and which reacts with the rotor m.m.f. to develop the torque.* The rotor m.m.f. in case of an induction motor is created by the stator induction effect, whereas in case of synchronous motor the rotor m.m.f. is created by a separate field winding which carries D.C. current.

The speed of an A.C. machine depends upon the stator supply frequency which produces the synchronously rotating magnetic field. If the frequency of the stator supply is increased to increase the speed of the motor, the magnitude of air gap flux is reduced due to increased magnetising reactance and correspondingly the developed torque is reduced. This shows that the *speed and torque of an A.C. motor can not be controlled independently by the conventional methods of speed control.* For this reason, an *A.C. motor requires 'variable voltage variable frequency' power supply for its speed control.* A *'D.C. link converter system'* consisting of a rectifier and an inverter or a *'cycloconverter'* can be used as a variable voltage variable frequency source.

However, it may be noted that the *voltage and current waves obtained by solid state devices are rich in harmonics and cause problems of harmonic heating torque pulsation.*

4.2. SPEED CONTROL OF A SINGLE-PHASE INDUCTION MOTOR

The most common method for speed control of a single-phase induction motor is the *stator voltage control method.*

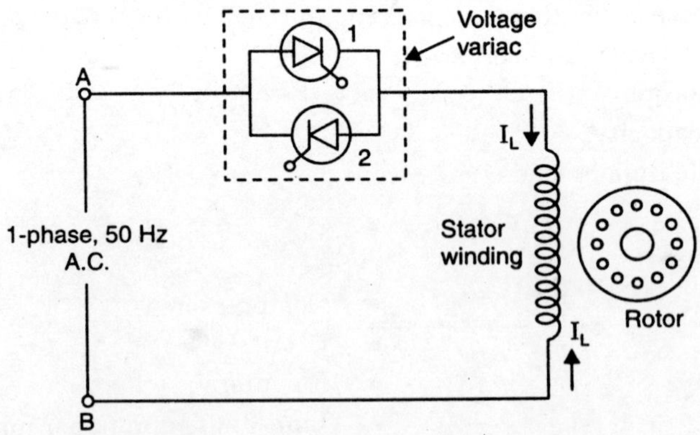

Fig. 4.1. Circuit diagram for speed control of a single-phase induction motor by stator voltage control method.

Fig. 4.1 shows the circuit diagram for speed control of a single-phase induction motor by stator voltage control method. The circuit uses two *SCRs* connected in anti parallel. In the +ve half cycle when point A is positive and point B is negative SCR_1 is triggered. The direction of flow of current in the stator winding is from the top to bottom. In the –ve half-cycle point A becomes negative and B becomes positive. SCR_1 is turned OFF and SCR_2 is triggered. The direction of flow of current in the stator winding is reversed. In other words, the alternating current supply becomes available across the stator winding of the motor. By varying the firing angles of *SCRs* 1 and 2 the magnitude of the A C. voltage across the stator winding of the motor can be *varied* ; this in turn will *vary the speed of the motor.*

Different schemes under this method are discussed below.

1. Speed control by using triac :

By using a triac, *very smooth speed control* of a single-phase induction motor can be obtained. A diac is used as a triggering agent for the triac in the circuit. Fig. 4.2 shows the circuit diagram for this arrangement. A diac-triac pair can provide the widest range of control.

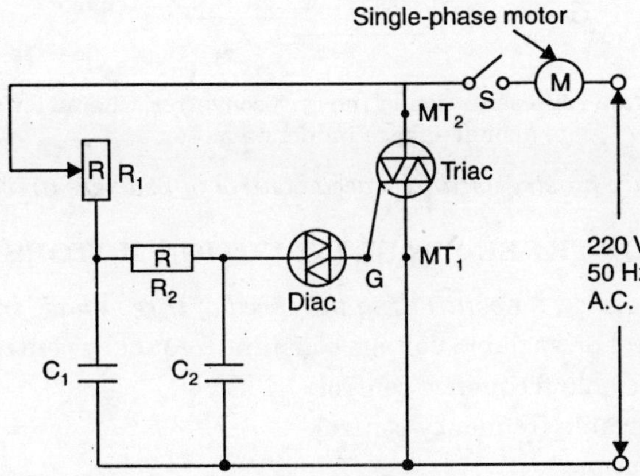

Fig. 4.2. Circuit diagram for the speed control of a single-phase induction motor using a triac.

There are two *R-C* networks. R_1-C_1 form the triggering circuit whereas R_2-C_2 along with C_1 form the π-network (filter) which would bypass any spike in the A.C. mains supply. The values of R_2 and C_2 are lower than the values of R_1 and C_1. R_2 also works as a current limiting resistance for the diac. The *R-C* triggering gate control process adopted in this circuit provides a very wide and smooth speed control for 1-phase induction motor.

- This circuit may be effectively employed for fabricating *fan regulators and illumination controllers.*

2. Speed control using single-phase inverter circuit :

With the help of an 'inverter circuit' we can obtain variable voltage fixed frequency A.C. supply which can be fed to the motor for speed control. Fig. 4.3 shows the block diagram representation for this scheme.

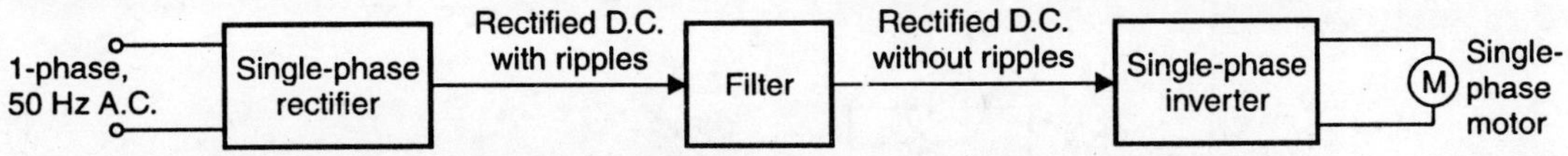

Fig. 4.3. Block diagram representation for the scheme of speed control of a single-phase induction motor using an inverter circuit.

- Single-phase A.C. is rectified with the help of a single-phase full-wave rectifier and then filtered to minimise the ripple content.

- The inverter output, which is a *fixed frequency variable A.C. voltage*, is fed to the motor whose speed is to be controlled. Inverter output (A.C. in nature) is *made variable by changing the firing time (angle) of SCRs*.

This process is known as *fixed frequency variable voltage control.*

3. Speed control by using cycloconverter circuits :

Basically, this is a *variable frequency method for speed control*. By controlling the firing sequence of the *SCRs* connected in a cycloconverter the frequency of the A.C. input voltage can be changed. The variable frequency A.C. supply available at the point of the cycloconverter circuit may be fed to the motor for speed control. A block diagram representation of such a scheme is shown in Fig. 4.4.

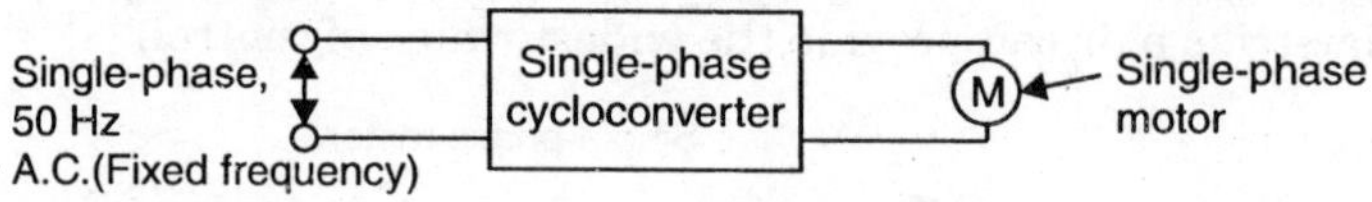

Fig. 4.4. Block diagram representation of the cycloconverter scheme for speed control of a single-phase induction motor.

- *"Cycloconverters" are mostly used for speed control of gearless drives.*

4.3. SPEED CONTROL OF THREE-PHASE INDUCTION MOTORS

Following methods are used for *controlling the speed of three-phase induction motors :*

1. Stator voltage control or variable voltage constant frequency control.
2. Variable voltage variable frequency control.
3. Variable current variable frequency control.
4. Regulation of slip power.

The basic principles of operation of these methods are given as follows :

1. Stator voltage control :

By using a thyristor A.C. controller circuit, constant frequency variable voltage supply can be generated. Connection diagram for such scheme is shown in Fig. 4.5.

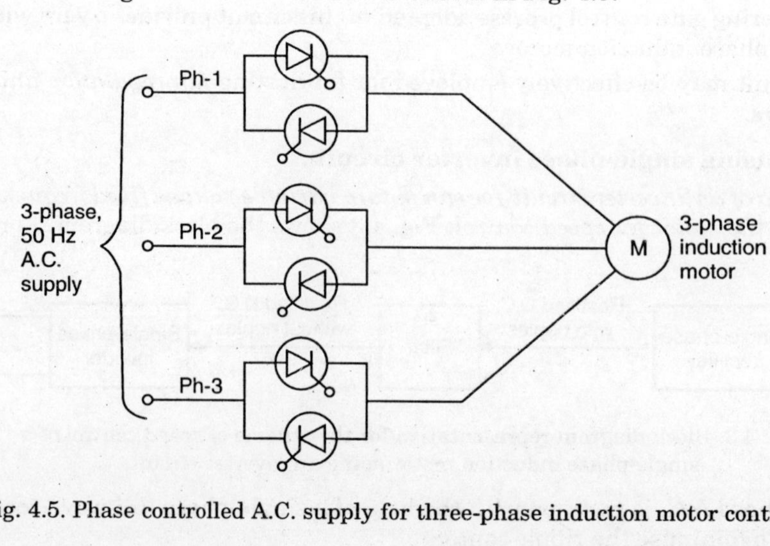

Fig. 4.5. Phase controlled A.C. supply for three-phase induction motor control.

By changing the applied voltage, air gap flux can be changed so also the slip, and motor speed can be altered. To obtain a reasonable control a full thyristor controller is used. Two *SCRs* connected in antiparallel per phase are used to form three such bridges. *SCRs* 1, 2 form the bridge for phase-1, similarly, 3, 4 form the bridge for phase-2 and 5, 6 for phase-3. The controlled (variable) three-phase voltage, when fed to the 3-phase induction motor, will result in the desired speed control of the motor.

- This arrangement is *quite costly* and *its firing circuit will also be quite complicated.*

2. Variable voltage variable frequency control :

Fig. 4.6 shows the basic block diagram for a speed control scheme of a 3-phase induction motor. This is basically a *variable voltage variable frequency supply scheme* for the speed control.

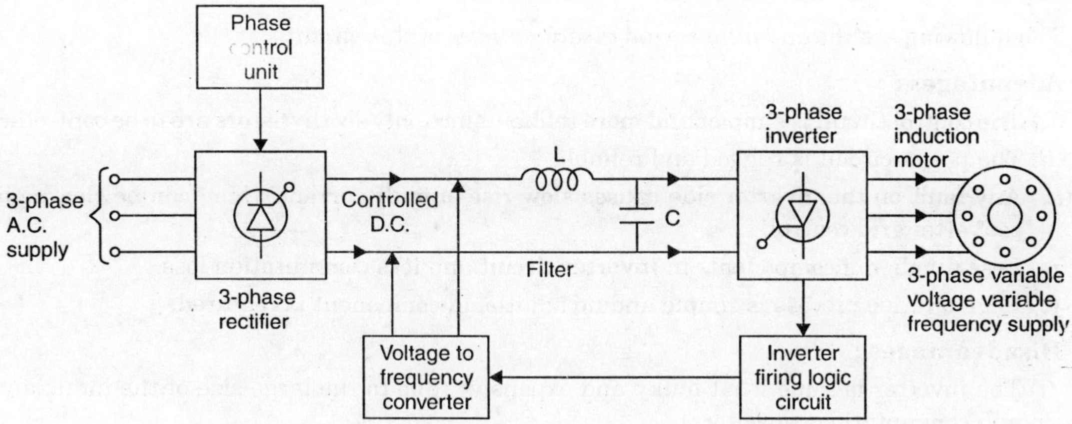

Fig. 4.6. Block diagram for basic scheme (variable voltage variable frequency control) for speed control of a 3-phase induction motor.

- 3-phase A.C. is rectified into D.C. and then filtered to minimise the ripple content. *L-C.* filter is generally used for this purpose. This controlled D.C. is converted into controlled pulses by means of a voltage to frequency converter. These controlled pulses are fed to the inverter bridge for producing the variable voltage variable frequency output. This output is fed to the 3-phase induction motor for controlling its speed.
- The 'phase control circuit' is employed for triggering and logic sensing of 3-phase rectifier circuit. This circuit *controls the firing angle of the rectifier bridge.* The *'inverter firing logic circuit' controls the firing angle of the inverter bridge.*

3. Variable current variable frequency control :

Fig. 4.7, shows a variable current variable frequency control circuit for an induction motor.

A phase controlled rectifier produces variable D.C. voltage which is converted to a current source by connecting a large inductor in series.

A diode rectifier followed by a D.C. chopper can also be used as a variable voltage D.C. source. It can be shown that the voltage at the terminals of 3-phase induction motor is almost sinusoidal with superimposed voltage spikes due to commutation.

The *'converter'* used is a *line commutated* whereas the *'inverter'* is *forced commutated* as the induction motor is a lagging p.f. load.

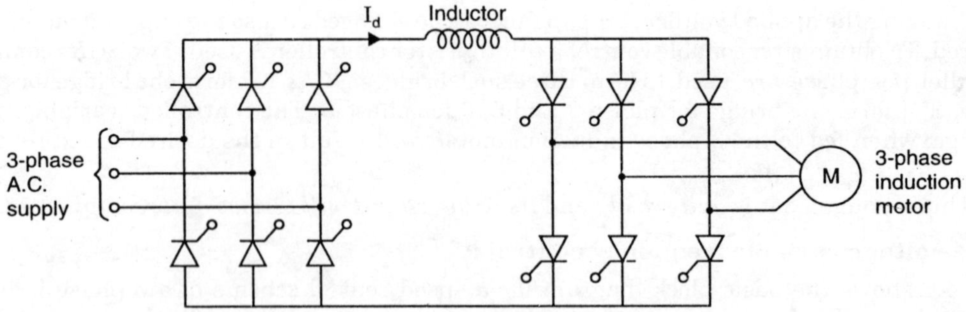

Fig. 4.7. Variable current variable frequency control circuit.

The following are the *advantages* and *disadvantages* of this circuit :

Advantages :

(*i*) The control circuit is simpler and more reliable since only six thyristors are to be controlled.

(*ii*) The power circuit is rugged and reliable.

(*iii*) Any fault on the inverter side causes slow rise of fault current which can be cleared by converter grid control.

(*iv*) Less number of components in inverter circuit and less commutation loss.

(*v*) Regenerative process is simple and no additional component is required.

Disadvantages :

(*i*) The inverter is some what bulky and expensive (due to the large size of the inductance and commutation capacitors).

(*ii*) The response of the drive is somewhat sluggish.

(*iii*) The frequency range of the inverter is low and it cannot operate under no load condition as some minimum load current is required to commutate the inverter satisfactorily.

4. Slip power recovery method :

When the supply frequency and the voltage are fixed, the speed of an induction motor can be *varied by injecting a counter e.m.f. into the rotor circuit of the motor*. This method is, therefore, *used for a wound rotor induction motor.* The inefficiency of the drive system because of large slip power dissipation can be overcome by this method.

In this method the slip power of the motor is rectified by a diode rectifier and is then pumped back to the A.C. line through a line commutated inverter. Fig. 4.8 shows the schematic diagram for this method.

The required power handling capability of the converter corresponds to the slip power and hence if the range of speed control is small the rating of converter is also small. However, if speed control upto standstill is the required, the converters should be rated for the full machine rating. The initial in-rush current in the converters can be avoided by connecting separate starting resistances in the slip ring circuit.

The torque pulsation and additional heating must be considered while designing the drive system.

The ***disadvantage*** of this system is that *regeneration and speed reversal are not possible in the drive system.*

- The draw back of this method that it results in very poor p.f.
- This drive system is used in *large H.P. pump and blower type applications where limited range of speed control is required.*

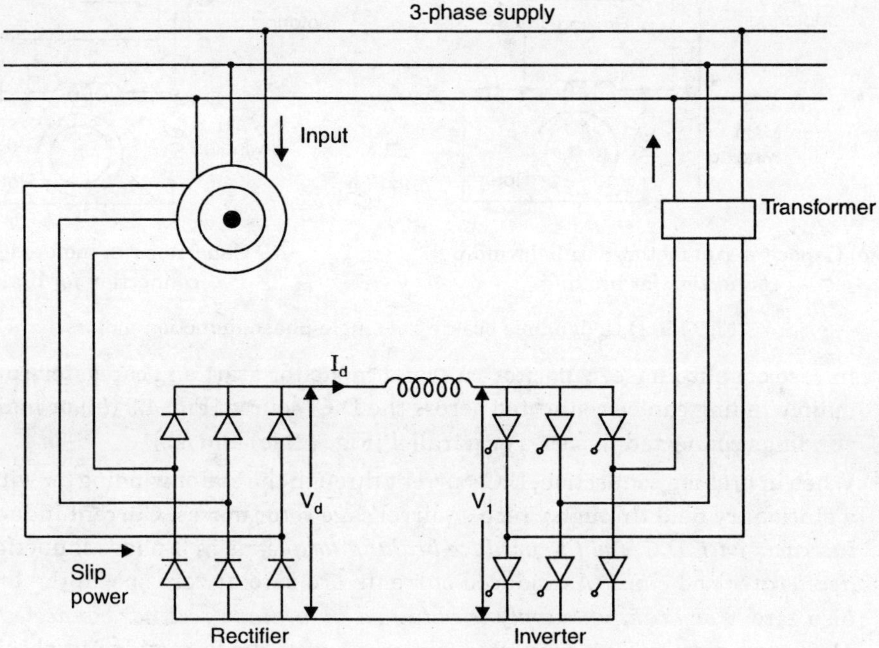

Fig. 4.8. Slip power recovery method.

4.4. BRAKING OF SINGLE-PHASE MOTORS

These motor can be braked by :

(*i*) D.C. dynamic braking (*ii*) Plugging.

(*i*) **D.C. dynamic braking :**

This method is commonly used for braking of single-phase induction motors. With the help of a double pole double throw switch or triple pole double throw switch, motor connection is shifted from A.C. (motoring) to D.C. source for braking. These connections for various single-phase induction motors are shown in Fig. 4.9.

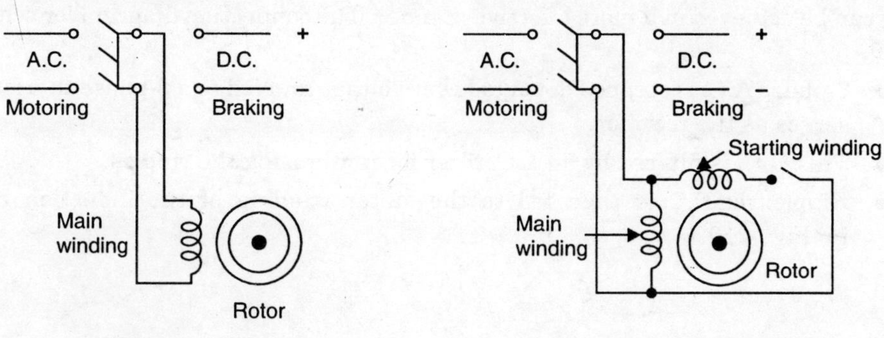

(*a*) Shaded pole motor (*b*) Split-phase motor.

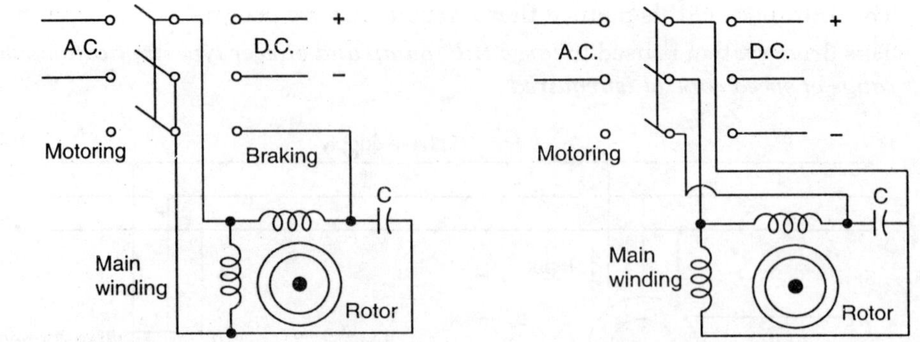

(*c*) Capacitor-run motor, parallel winding connection for braking.

(*d*) Capacitor run motor, series winding connection for braking.

Fig. 4.9. D.C. dynamic braking of single-phase induction motors.

- In case of split-phase, capacitor-run, and capacitor-start and capacitor-run motors, either main winding can be connected across the D.C. source [Fig. 4.9 (*b*)] or main and auxiliary windings connected in series or parallel [Fig. 4.9 (*c*) and (*d*)].
- When in braking connection, D.C. current through the stator winding (or windings) produces a stationary field through which squirrel cage rotor moves. Current induced in rotor bars *interacts with D.C. field to produce braking torque,* as in 3-phase induction motor. Motor *decelerates* and *stops*. As induced currents are zero at zero speed, the braking torque is also zero. *For braking, supply is obtained by a diode rectifier connected to A.C. mains.* Motor winding can be connected directly across diode rectifier to obtain fast braking. Winding is disconnected from D.C. supply after the motor stops.

(*ii*) **Plugging and reversal :**

Except in case of shaded pole motor, *plugging and reversal is obtained by changing phase sequence by reversing polarity of one of the windings.*

4.5. DYNAMIC BRAKING OF A 3-PHASE INDUCTION MOTOR

The speed of an induction motor can be controlled by injecting D.C. voltage in its stator winding. A variable resistance may be used in the rotor (in case of a slip ring induction motor) for dissipating the required amount of power. Now-a-days *thyristor bridges* are used for supplying D.C. which is controllable in nature. With the help of controlled D.C. from a thyristor bridge the dynamic braking can be achieved in a more effective manner. The connection diagram for scheme is shown in Fig. 4.10.

- 3-phase A.C. is stepped down to lower voltage and fed to a 3-phase thyristor bridge which serves as the rectifier.
- This D.C. is filtered by an *L.C.* filter for minimising the ripples.
- Ripple free D.C. is then fed to the stator winding of the induction motor as shown in Fig. 4.10.

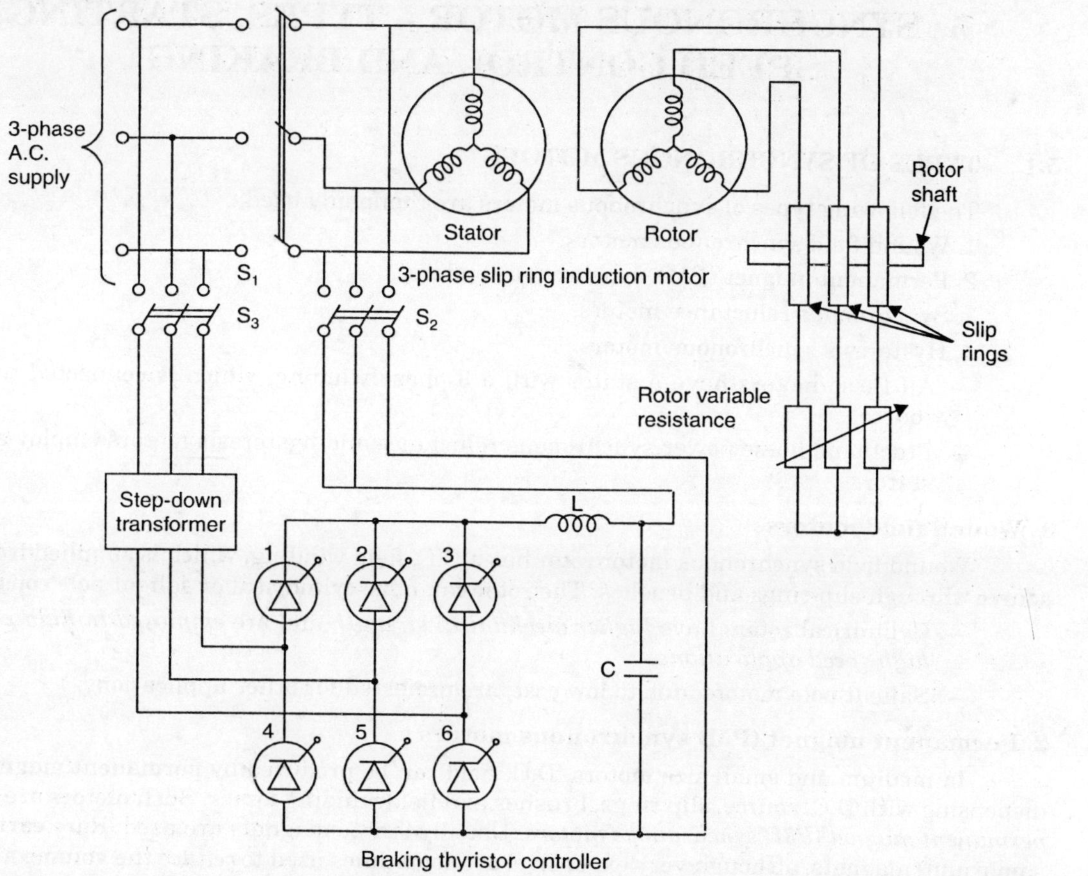

Fig. 4.10. Dynamic braking of a 3-phase slip ring induction motor.

It is to be noted that while feeding D.C. to the stator the *3-phase A.C. input must be disconnected.* A.C. is disconnected with the help of S_1 and D.C. is disconnected with the help of S_2. Since, the input A.C. voltage is stepped down to a lower valve, the thyristor converter may be of lower voltage rating.

5. SYNCHRONOUS MOTOR – TYPES, STARTING, SPEED CONTROL AND BRAKING

5.1. TYPES OF SYNCHRONOUS MOTORS

The following types of synchronous motors are commonly used :

1. Wound field synchronous motors.
2. Permanent magnet (PM) synchronous motors.
3. Synchronous reluctance motors.
4. Hysteresis synchronous motors.

— All these motors have a stator with a 3-phase winding, which is connected to an A.C. source.

— Fractional horse power synchronous reluctance and hysteresis motors employ a 1-phase stator.

1. Wound field motors :

Wound field synchronous motor rotor has a D.C. field winding, which is supplied from a D.C. source through slip-rings and brushes. The rotor can have cylindrical or salient pole construction.

— Cylindrical rotors have *higher mechanical strength* and are *employed in high power and high speed applications*.

— Salient pole motors, due to low cost, are preferred for other applications.

2. Permanent magnet (PM) synchronous motors :

In medium and small size motors, D.C. field can be produced by permanent magnets ; thus dispensing with D.C. source, slip rings, brushes and field winding losses. Such motors are known as *permanent magnet (PM) synchronous motors*. Usually ferrite magnets are used. Rare earth (cobalt-samarium) magnets, although very expensive, are sometimes used to *reduce* the volume and weight of the motor.

PM synchronous motors are classified as follows :

(*i*) ***Surface mounted :***

(*a*) *Projecting type.* In such motors magnets project from the surface of the rotor (Fig. 5.1 (*a*)]

(*b*) *Inset type.* In this case, magnets are inserted into the rotor, providing a smooth rotor surface [Fig. 5.1 (*i*) (*b*)].

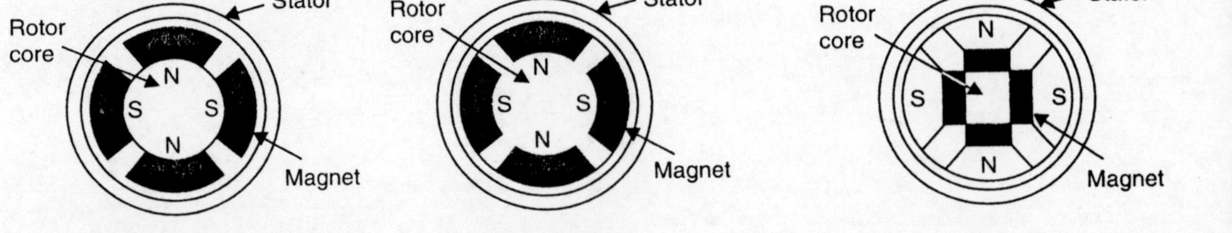

(a) Projecting surface mounting (b) Inset magnet

(i) Surface mounted motor (ii) Interior or buried magnet motors

Fig. 5.1. Types of PM synchronous motors.

- Epoxy glue is used to fix the magnets to the rotor surface in both.
- While these motors are *easy to construct and are less expensive, they are less robust compared to interior type rotors and are not suitable for high speed applications.*

(*ii*) ***Interior (or buried).*** In these motors, magnets are imbedded in the interior of the rotor [Fig. 5.1 (*ii*)].

Features of wound field and permanent magnet synchronous motors :

The wound field and PM synchronous motors have a *higher full load efficiency and power factor than an induction motor.*

Wound field motors can be designed for a higher power rating than induction motors. Since the air-gap flux is not produced solely by the magnetising current drawn from the armature, a larger air-gap suiting the mechanical design can be chosen. The ability to control power factor is an important advantage at higher power levels. *Operating at unity power factor minimizes the inverter rating.*

PM synchronous motor, apart from the *robust construction, has low losses and high efficiency*. Because of low losses, it is possible to make motors with very high power density and torque to inertia ratios. These make them *suitable for servo drives requiring fastest possible dynamic response.*

- One *significant difference* between the wound field and PM motors, which are designed to operate with a source of fixed frequency is discussed below :

When a wound field motor is started as induction motor, D.C. field is kept off. In case of a PM motor, the field *cannot* be 'turned off'. When at a speed below synchronous speed, the rotor field induces a voltage in the stator, which has a frequency different than the frequency of stator supply. The current produced by induced voltage interacts with the rotor field to produce a braking torque, which opposes the induction motor torque due to damper winding. The permanent magnet synchronous motor (PMSM) is designed so that the braking torque is very small compared to induction motor torque. Owing to the capability of starting direct on line these motors are called *line start PMSM*.

— PMSM are available in 3-phase and 1-phase construction.

— Although expensive to induction motors, they have advantages of *high efficiency, high power factor and low sensitivity to supply voltage variations.*

— These motors are preferred for industrial applications with *large duty cycles such as pumps, fans and compressors*.

3. Synchronous reluctance motor :

Single-phase salient-pole synchronous-induction motors, are generally called *reluctance motors*. If the rotor of any uniformly distributed single-phase induction motor is altered so that the laminations tend to produce *salient rotor poles,* as shown in Fig. 5.2, the reluctance of the air-gap flux path will be greater where there are no conductors embedded in slots. Such a motor, coming up to speed as an induction motor, will be pulled into synchronism with the pulsating A.C. single-phase field by the reluctance torque developed at the salient iron poles which have lower-reluctance air gaps.

Working of a reluctance motor. In order to understand the working of such a motor the basic fact which must be kept in mind is that *when a piece of magnetic material is located in a magnetic field, a force acts on the material, tending to bring it into the densest portion of the field. The force tends to align the specimen of material in such a way that the reluctance of the magnetic path that passes through the material will be minimum.*

When supply is given to the stator winding, the revolving magnetic field will exert reluctance torque on the unsymmetrical rotor tending to align the salient pole axis of the rotor with the axis of the revolving magnetic field (because in this position, the reluctance of the magnetic path would be minimum). If the reluctance torque is sufficient to start the motor and its load, the rotor will pull into step with the revolving field and continue to run at the speed of the revolving field. (Actually the motor starts as an induction motor and after it has reached its maximum speed as an induction motor, the reluctance torque pulls its rotor into step with the revolving field so that the motor now runs as synchronous motor by virtue of its saliency).

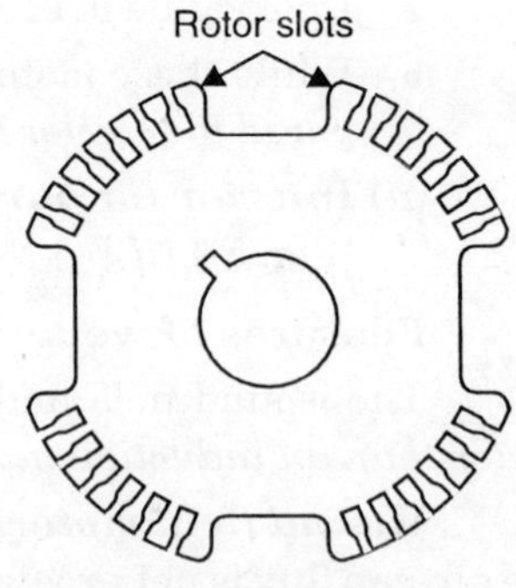

Fig. 5.2. Reluctance-motor lamination.

Reluctance motors have approximately *one-third* the horsepower rating they would have as induction motors with cylindrical rotors, although the ratio may be increased to one-half by proper design of the field windings. *Power factor and efficiency are poorer than for the equivalent induction motor.* Reluctance motors are subject to *'cogging'*, since, the locked-rotor torque varies with the rotor position, but the effect may be *minimized* by *skewing the rotor bars* and by *not* having the number of rotor slots exactly equal to an exact multiple of the number of poles.

Uses. Despite its short-comings, the reluctance motor is widely used for many *constant speed* applications such as *recording instruments, time devices, control apparatus, regulators*, and *phonograph turntables*.

- *Reversing* is obtained as in any single-phase induction motor.

Speed-torque characteristics. Fig. 5.3 shows speed-torque characteristics of a typical single-phase reluctance motor.

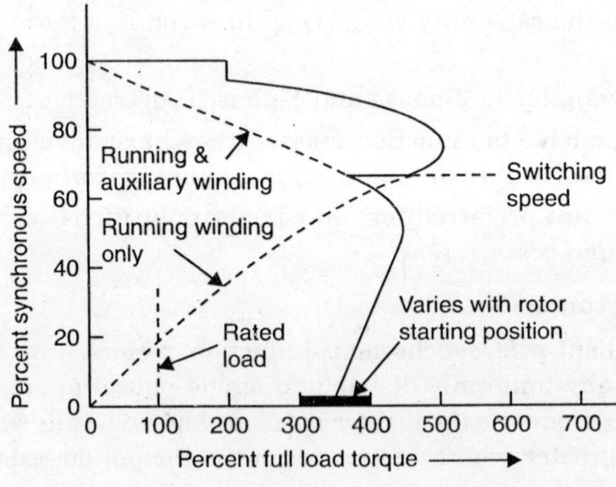

Fig. 5.3. Speed-torque characteristics of a single-phase reluctance motor.

- The motor starts at anywhere from 300 to 400 per cent of its full-load torque (depending on the rotor position of the unsymmetrical rotor with respect to the field windings) as a two-phase motor as a result of the magnetic rotating field created by a starting and running winding (displaced) 90° in both space and time.
- At about 3/4th of the synchronous speed, a centrifugal switch opens the starting winding, and the motor continues to develop a single-phase torque produced by its running winding only.

As it approaches synchronous speed, the reluctance torque (developed as a synchronous motor) is sufficient to pull the rotor into synchronism with the pulsating single-phase field.

- *The motor operates at a constant speed up to a little over 200% of its full-load torque.* If it is loaded beyond the value of pull-out torque, it will continue to operate as a single-phase induction motor up to 500% of its rated output.

5.2. STARTING OF SYNCHRONOUS MOTOR

The purpose of starting method is to bring rotor speed close to synchronous speed. Following methods are used to start synchronous motor :

1. Using damper windings as a squirrel-cage induction motor.
2. Using a low power auxiliary motor.

1. Using damper windings as a squirrel cage induction motor :

One widely used method is to start the synchronous motor as an induction motor with field unexcited and damper winding serving as a squirrel-cage rotor. Regarding this method, following points are worth noting :

(*i*) The current and starting torque can be reduced and increased respectively, by increasing the damper winding resistance. The motor speed while running as an induction motor, for successful pull-in, must be close to synchronous speed. For this the damper winding resistance must be low. Further, for damping hunting oscillation damper winding resistance must be low. The damper winding resistance is so selected as to strike a compromise between these two contradictory requirements.

(*ii*) D.C. field should be applied only after the motor has reached close to full speed.

(*iii*) When the rotor has salient pole construction, the damper winding can have conductors only over the pole arc. This leads to a dip in the speed-torque curve at half of synchronous speed.

(*iv*) On the application of full supply voltage, the starting current in the motor can be 7 to 10 times of full load value. Except in small size motors, such a high starting current causes fluctuations in supply voltage. In case of large size motors, such a high current may cause a large drop in the terminal voltage, thus reducing the already low starting torque further. Starting current can be reduced by employing any one of the reduced voltage starting methods employed for starting induction motors. Reduction in starting current is obtained at the expense of reduction in starting torque. When started at a reduced voltage, the transition to full voltage can be made before or after the pull-in. *Former is preferred as it improves pull-in performance due to following two reasons* :

— With full voltage the speed attained as induction motor is closer to synchronous speed, and

— The pull-in torque increases in proportion to voltage squared, consequently pull-in can be achieved faster and with large motor loads.

2. Using a low power auxiliary motor :

- In this method a low power auxiliary motor is coupled to the synchronous motor shaft. With the help of auxiliary motor, the rotor speed is brought near synchronous speed and then D.C. field is switched-in.
- This method has a very *low starting torque*.

Note. It is practically impossible to start a synchronous motor with its D.C. field energized. Even when left de-energized, the rapidly rotating magnetic field of the stator will induce extremely high voltages in the many turns of the field winding. It is customary, therefore, *to short-circuit the D.C. field winding during the starting period* ; whatever voltage and current are induced in it may then aid in producing induction motor action. In very large synchronous motors, field sectionalising or field-splitting switches are used which short-circuit individual field windings to prevent cumulative addition of induced voltages from pole to pole.

5.3. BRAKING OF SYNCHRONOUS MOTORS

- The motor can work in regenerative braking only at synchronous speed. Therefore, *"regenerative braking" cannot be used for stopping or decelerating a load.*
- *"Dynamic braking"* is obtained by *disconnecting stator from the source and connecting it to 3-phase resistor.* Machine works as a synchronous generator and dissipates generated energy in the braking resistor.

5.4. SPEED CONTROL OF SYNCHRONOUS MOTORS

The speed of synchronous motors can be controlled as follows :

(*i*) By using current-fed delink.

(*ii*) By using cycloconverter.

1. Speed control by current-fed D.C. link :

Fig. 5.4, shows the circuit diagram for speed control of synchronous motor by current-fed D.C. link.

The typical circuit consists of three converters two of which are connected between the 3-phase source and synchronous motor and the third converter (bridge rectifier) supplies D.C. field excitation for the rotor.

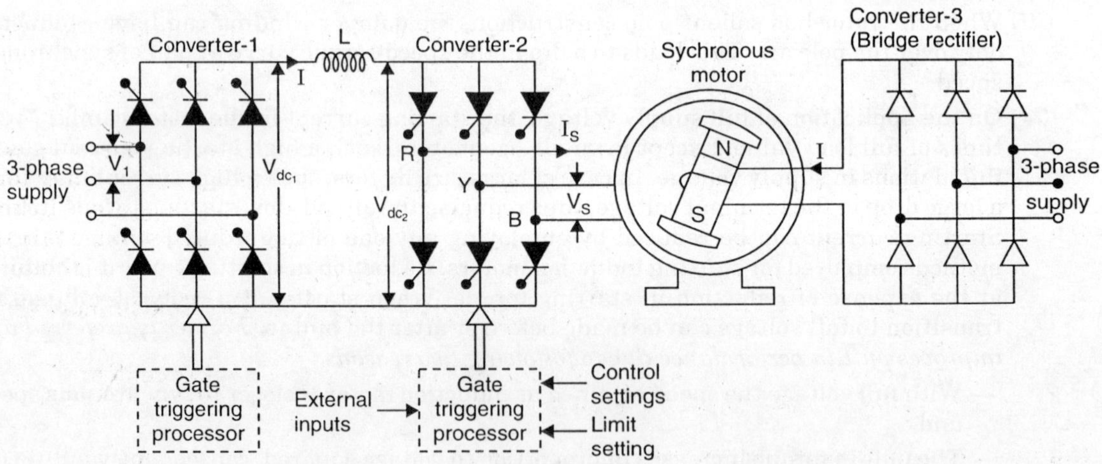

Fig. 5.4. Circuit diagram for speed control of synchronous motor by current-fed D.C. link.

— Converter-1 (C-1) acts as a *controlled rectifier* and feeds D.C. power to converter-2 (C-2). It acts as a current source and controls I.

— Converter-2 (C-2) behaves as a *naturally commutated inverter* whose A.C. voltage and frequency are established by the motor. The converter-2 is naturally commutated by voltage V_S induced across motor terminals by its revolving magnetic flux. The revolving flux which depends on the stator currents and the D.C. field exciting current is usually kept constant. Consequently V_S is proportional to motor speed.

— The function of the smoothing inductor L is to maintain a ripple-free current in the D.C. link between the two converters.

As regards various controls, information picked up from various points is processed in the gate-triggering processors which then send out appropriate gate firing pulses to converters 1 and 2. The processors receive information about the desired rotor speed, its actual speed, instantaneous rotor position, field current, stator voltage and current etc. The processors check whether these inputs represent normal or abnormal conditions and send appropriate gate firing pulses either to correct the situation or meet a specific demand.

Gate triggering of C-1 and C-2 is done at line frequency (50 Hz) and at motor frequency respectively. In fact, gate pulses of C-2 are controlled by rotor position which is pulsed by position transducers mounted at the end of the shaft. The speed of the motor can be increased by increasing either D.C. link I or exciting current I_f.

- This method of speed control is applied to *motors ranging from 1 kW to several MW.*
- *Permanent-magnet synchronous motors used in textile industry and brushless synchronous motors for nuclear reactor pumps are controlled by this method.*

2. Speed control and cycloconverter :

This arrangement consists of *three cycloconverters* connected to the three phases of the synchronous motor and *one controlled rectifier* for supplying field exciting current, I_f to the rotor. Each cycloconverter is composed of two 3-phase bridges and supplies a single-phase output. With a line frequency of 50 Hz, the cycloconverters output frequency can be varied from 0 to 10 Hz (It is well known that a cycloconverter can convert A.C. power at higher frequency to one at a lower frequency).

The cycloconverters and controlled rectifies function as *current sources*. The air-gap flux is kept constant by controlling the magnitude of the stator currents and exciting current I_f. The motor can be made to operate at unity power factor by proper timing of gate pulses.

- The *speed of cycloconverter-driven large slow-speed synchronous motors can be continuously varied from 0 to 15 r.p.m.* Such low speeds permit direct-drive of the ball mill without using a gear reducer. Such high-power low-speed systems are also being used as *propeller drives* on board the ships.

6. DIGITAL CONTROL OF ELECTRIC MOTORS

The speed information can be fed into microcomputer using a D.C. Tacho (speed encoder) and *A/D* converter (speed *I/P* module). The motor current data is usually fed into the computer through a fast *A/D* converter. A synchronizing circuit interface (line synchronizing circuit) is required so that microcomputer can synchronize the generation of the firing pulse data with the supply line frequency. The gate pulse generator is shown as receiving a firing signal from microcomputer. A set of instruction (Program) is stored in the memory and those are executed by computer for proper functioning of a drive.

Advantages of digital control :

1. High reliability
2. Easy software control
3. High precision and accuracy
4. Better speed regulation
5. Faster response
6. Improved performance
7. Economical
8. Flexibility
9. Better time response.
10. *The major advantage of the digital control is that by changing the program, desired control technique can be implemented without any change in the hardware.*

7. TRANSIENT CONDITIONS IN ELECTRICAL DRIVES

7.1. INTRODUCTION

When an electrical drive is in steady state operation and while attaining another steady state operation due to the nature of the drive *transient conditions occur due to the following "reasons/conditions"* :

(*i*) Change of speed, torque and current ;

(*ii*) Change in the load suddenly or during the control operation on the motor during starting, braking or reversal of direction of rotation etc ;

(*iii*) Changes in supply voltage, frequency or unbalance in supply voltage.

The exhaustive study of the transient working conditions of electrical drives helps in the following ways :

(*i*) To select the proper motor power ;

(*ii*) To reduce losses ;

(*iii*) To design control circuit ;

(*iv*) To select starting and braking equipment ;

(*v*) To attain required economy in the driving system etc.

7.2. ACCELERATION TIME

For any electric motor undergoing rotary motion, the equation of motion is given by,

$$T_{motor} - T_{load} = \frac{J\,d\omega}{dt} \text{ Newton metres} \qquad ...(7.1)$$

where, T = Torque (Nm),

J = Moment of inertia (kg m^2),

$d\omega$ = Change in speed (rad/sec), and

dt = Acceleration time (sec).

$\therefore$ The acceleration time required for changing the speed of an electric motor from ω_1 to ω_2 to given by,

$$dt = \frac{J\,d\omega}{(T_{motor} - T_{load})} \qquad ...[7.1\,(a)]$$

Thus, $T_{motor} > T_{load}$ during acceleration time and $T_{load} > T_{motor}$ during deceleration time.

The acceleration time, when the motor accelerates from ω_1 to ω_2, is given by,

$$t_{acc.} = J\int_{\omega_1}^{\omega_2} \frac{d\omega}{T_{motor} - T_{load}} \qquad ...(7.2)$$

- When the motor starts from rest, $\omega_1 = 0$ and ω_2 is the steady state speed ω_n, the acceleration time is given by,

$$t_{acc.} = J\int_{0}^{\omega_n} \frac{d\omega}{T_{motor} - T_{load}} \qquad ...(7.3)$$

As $T_{motor} - T_{load} \to$ zero we have $\dfrac{1}{T_{motor} - T_{load}} \to \infty$ (infinity) and therefore for practical calculations the *acceleration time* ($t_{acc.}$) is considered when $\omega_1 = 0.95\ \omega_2$ and thereby avoiding the complication in calculating the finite time when we take the value as ω_2.

- Now when the dynamic torque (T_{motor}) becomes *negative* the drive will *decelerate*. The torque equation then becomes

$$-T_{motor} - T_{load} = \frac{Jd\omega}{dt}$$

Whenever the motor develops a torque which is less than the load torque of the driven machine, the *deceleration time* ($t_{dec.}$) can be written as,

$$t_{dec.} = \int_{\omega_1}^{\omega_2} \frac{Jd\omega}{-(T_{motor} + T_{load})} = \int_{\omega_2}^{\omega_1} \frac{Jd\omega}{T_{motor} + T_{load}} \quad ...(7.4)$$

Let us now compute the acceleration time (stated above) under different conditions of motor and load torque.

Case 1. *When the motor and load torques are equal :*

When the motor and load torques are constant, then ($T_{motor} - T_{load}$) will also be constant. Since $T_{motor} - T_{load} = \frac{Jd\omega}{dt}$ and as J is a constant we have $\frac{d\omega}{dt}$ as constant. This means the *speed variation will be linear w.r.t. time* and hence

$$\omega = \frac{(T_{motor} - T_{load})\,t}{J} \quad ...(7.5)$$

Hence the time t_1 taken by the motor to reach a speed of ω, rad/sec. is given by

$$t_1 = \frac{J\omega_1}{(T_{motor} - T_{load})} \quad ...(7.6)$$

When the motor starts without load $T_{load} = 0$, then T_{motor} = rated torque (T_{rated}). The moment of inertia J will be due to the rotor of the motor. The motor then attains a speed ω_0 which is the no-load speed of the motor. Hence $t_1 = \frac{J\omega_0}{T_{rated}}$. This is known as the *mechanical constant of the motor* ; for a rated speed of 1000 r.p.m. the time constant varies from 0.4 to 0.6 second.

Case 2. *Load torque remains constant with linearly varying motor torque*

Let us consider the condition when $\omega = 0$, the motor torque being the starting torque T_{start} and when $\omega = \omega_s$ the synchronous speed of induction motor with $T_{motor} = 0$. Then the relation between motor torque and slip (s) can be written as,

$$\frac{T_{start}}{1} = \frac{T_{motor}}{s} \quad \text{or} \quad s = \frac{T_{motor}}{T_{start}} \quad ...(7.7)$$

Hence, *acceleration torque*,

$$\frac{Jd\omega}{dt} = T_{motor} - T_{load} \quad ...(7.8)$$

Also,

$$\omega = \omega_s(1 - s)$$

Differentiating with respect to t, we have

$$\frac{d\omega}{dt} = -\,\omega_s \frac{ds}{dt} \quad ...(7.9)$$

From eqns. (7.7), (7.8) and (7.9), we get

$$-\,\omega_s \frac{ds}{dt} = \frac{s\,T_{start} - T_{load}}{J}$$

or,

$$\frac{ds}{dt} + \frac{sT_{start}}{J\omega_s} - \frac{T_{load}}{J\omega_s} = 0 \quad ...(7.10)$$

The value $\frac{J\omega_s}{T_{start}}$ is having the dimensions of seconds and is known as *starting time constant* (t_{st}) of the drive. Hence J represents the total moment of inertia of the drive system including inertia of any external rotating masses excluding the rotor of the driving motor.

$$\therefore \quad \frac{ds}{dt} + \frac{s}{t_{st}} - \frac{T_{load}}{J\omega_s} = 0 \qquad ...(7.11)$$

Optimum value of $s_{max.\,T}$ when accelerating time is minimum :

When the load torque and friction torque are negligibly small the equation of motion can be written as,

$$T_{motor} = \frac{Jd\omega}{dt} = \frac{J}{dt}\, d\,[\omega_s(1-s)]$$

or
$$T_{motor} = -J\omega_s \frac{ds}{dt} \qquad ...(i)$$

The electromagnetic torque developed by the induction motor neglecting stator resistance is given by

$$T_{motor} = T_{max} \frac{2}{\frac{s}{s_{max\,T}} + \frac{s_{max\,T}}{s}}$$

In order to obtain the time taken by the motor to have slip changing from s_1 to s_2, equating (*i*) and (*ii*), we get

$$-J\omega_s \frac{ds}{dt} = \frac{2}{\frac{s}{s_{max\,T}} + \frac{s_{max\,T}}{s}}$$

$$\therefore \quad \int_0^t dt = \frac{-J\omega_s}{2\,T_{max}}\left[\int_{s_1}^{s_2} \frac{s}{s_{max\,T}}\, ds + \int_{s_1}^{s_2} \frac{s_{max\,T}}{s}\, ds\right]$$

or
$$t = \frac{-J\omega_s}{2\,T_{max}}\left[\left|\frac{s^2}{2s_{max\,T}}\right|_{s_1}^{s_2} + \left|s_{max\,T} \,.\, \log_e s\right|_{s_1}^{s_2}\right]$$

or
$$t = \frac{J\omega_s}{2\,T_{max}}\left[\frac{s_1^2 - s_2^2}{2s_{max\,T}} + s_{max\,T} \log_e\left(\frac{s_1}{s_2}\right)\right] \qquad ...(7.12)$$

This equation indicates that the *time required (t) is a function of slip corresponding to the maximum torque.* To get the value of $s_{max\,T}$ that gives *minimum time* for the slip to change from s_1 to s_2, by differentiating t w.r.t. $s_{max\,T}$ and equating the result to zero, we have

$$(s_{max\,T})_{opt.} = \left[\frac{s_1^2 - s_2^2}{2\log_e(s_1/s_2)}\right]^{1/2} \qquad ...(7.13)$$

Substituting this value of $s_{max\,T}$ in eqn. (7.12), we get the *minimum time* required.

The optimum value of $s_{max\,T}$ to accelerate the motor from $s_1 = 1$ *i.e.*, motor at standstill to any slip s in minimum time is given by,

$$(s_{max\,T})_{opt} = \left[\frac{1-s^2}{2\log_e(1/s)}\right]^{1/2} \qquad ...(7.14)$$

The optimum rotor resistance when the slip changes from s_1 to s_2 in minimum time is obtained by using the relation,

$$(R_2)_{opt} = (s_{max\,T})_{opt}\,(X_1 + X_2)$$

or,
$$(R_2)_{opt} = (X_1 + X_2)\left[\frac{(s_1^2 - s_2^2)}{2\log_e(s_1/s_2)}\right]^{1/2} \quad ...(7.15)$$

7.3. ENERGY LOSSES DURING STARTING

1. D.C. shunt motor :

In a D.C. motor,
$$V = E_b + I_a R_a$$

or,
$$I_a R_a = V - E_b = V - k\omega \quad ...(7.16)$$

This expression is also suitable for separately excited motor.

The equation of motion at no load is given by,

$$T_{motor} = kI_a = \frac{Jd\omega}{dt} \quad ...(7.17)$$

$$\left(\text{or } I_a = \frac{J}{k}\frac{d\omega}{dt}\right)$$

Hence,
$$I_a^2 R_a = I_a(I_a R_a) = \frac{J}{k}\frac{d\omega}{dt}(V - k\omega)$$

or,
$$I_a^2 R_a = \frac{JV}{k}\frac{d\omega}{dt} - \frac{J\omega d\omega}{dt} \quad ...(7.18)$$

As $I_a R_a$ is negligible at no load, $V = k\omega_0$ at no load (ω_0 is the no-load speed of the motor)

$$\therefore \quad I_a^2 R_a dt = J\omega_0 d\omega - J\omega d\omega$$

The energy absorbed (W) by the armature for a change in speed from ω_1 to ω_2 from time t_1 to t_2 is given by,

$$W = \int_{t_1}^{t_2} I_a^2 R_a\, dt = J\omega_0 \int_{\omega_1}^{\omega_2} d\omega - J\int_{\omega_1}^{\omega_2} \omega\, d\omega$$

or,
$$W = J\omega_0(\omega_2 - \omega_1) - \frac{J}{2}(\omega_2^2 - \omega_1^2) \quad ...(7.19)$$

Hence the energy change when the motor changes speed from rest to no load speed ω_0 will be

$$W_{start} = J\omega_0^2 - \frac{J}{2}\omega_0^2$$

or,
$$W_{start} = \frac{J\omega_0^2}{2} \text{ joules} \quad ...(7.20)$$

= K.E. absorbed by the armature in accelerating from standstill to no-load speed.

The *energy loss is independent of armature resistance.*

- The *energy loss at starting, when* the *motor is started with a constant load torque* T_{load}, is calculated as follows :

$$T_{motor} = kJ_a = T_{load} + \frac{J\,d\omega}{dt}$$

$$\therefore \quad I_a^2 R_a = \frac{JV}{k}\frac{d\omega}{dt} - \frac{J\omega\,d\omega}{dt} + \frac{V}{k}(T_{load}) - T_{load}\cdot\omega$$

The energy lost, when the motor speed changes from ω_1 to ω_2, is given by,

$$W = \int_{t_1}^{t_2} I_a^2 R_a dt$$

$$= \frac{JV}{k} \int_{\omega_1}^{\omega_2} d\omega - J \int_{\omega_1}^{\omega_2} \omega \, d\omega + \frac{V}{k} T_{load} \int_{t_1}^{t_2} dt - T_{load} \int_{t_1}^{t_2} \omega (t) \, dt$$

$$= \frac{JV}{k} (\omega_2 - \omega_1) - \frac{J}{2} (\omega_2^2 - \omega_1^2) + \frac{V}{k} T_{load} (t_2 - t_1) - T_{load} \int_{t_1}^{t_2} \omega (t) \, dt$$

Now substituting ω_0 (the no load speed) for $\frac{V}{k}$, the *energy lost at starting on no-load* when the speed changes from 0 to ω_r is given by,

$$W_{start} = J\omega_0\omega_r - \frac{J}{2} \omega_r^2 + T_{load} \, \omega_0 t_{st} - T_{loat} \int_0^t \omega (t) \, dt$$

or

$$W_{start} = J \left(\omega_0\omega_r - \frac{\omega_r^2}{2} \right) + T_{load} \left[\omega_0 t_{st} - \int_0^t \omega (t) \, dt \right] \quad ...(7.21)$$

... where t_{st} is the accelerating time.

This equation shows that the *energy lost during starting depends on the accelerating time and on the speed variation with time during acceleration.*

2. D.C. series motor :

Since a series motor cannot be started at no load, it will be assumed that the motor starts under a constant load torque (T_{load}) to be present while finding the energy lost during starting.

The equation of motor is given by,

$$T_{motor} = kI_a^2 = \frac{J \, d\omega}{dt} + T_{load}$$

The total energy dissipated in the armature circuit resistance is given by,

$$W = \int_{t_1}^{t_2} I_a^2 R_a dt$$

$$= \frac{JR_a}{k} \int_{\omega_1}^{\omega_2} d\omega + \frac{T_{load}}{k} R_a \int_{t_1}^{t_2} dt$$

or

$$W = \frac{JR_a}{k} (\omega_2 - \omega_1) + \frac{T_{load} \, R_a}{k} (t_2 - t_1) \quad ...(7.22)$$

Thus the *energy dissipated* when the motor speed varies from ω_1 to ω_2 will be *proportional to* R_a (this is in contrast to the case of a D.C. shunt motor where the energy dissipated is independent of R_a).

3. Three-phase induction motor :

In a 3-phase induction motor, torque developed is given by,

$$T = \frac{3}{\omega_s} \left(I_2^2 \cdot \frac{R_2}{s} \right) \quad ...(i)$$

Neglecting load and friction torques, we have

$$T = J \frac{d\omega}{dt} = - J\omega_s \frac{ds}{dt} \quad ...(ii)$$

... as $\omega = \omega_s(1 - s)$

The energy loss in the *rotor* of the induction motor when the slip changes from s_1 to s_2 is given by,

$$W = 3\int_{t_1}^{t_2} I_2^2\, R_2\, dt \qquad ...(iii)$$

the slips are s_1 and s_2.

From (*i*), (*ii*) and (*iii*), we have

$$W = 3I_2^2\, R_2 = -J\omega_s^2\, s\,\frac{ds}{dt}$$

i.e.,

$$W = -J\omega_s^2 \int_{s_1}^{s_2} s.ds$$

or,

$$W = \frac{1}{2} J\omega_s^2\, (s_1^2 - s_2^2) \text{ joules} \qquad ...(7.23)$$

At the time of starting the slip changes from 1 to 0 and so the energy lost in the rotor circuit is given by,

$$W_{start} = \frac{1}{2} J\omega_s^2 \text{ joules} \qquad ...(7.24)$$

The eqn. (7.23) indicates that the total energy lost in the rotor circuit depends on the moment of inertia of the rotating masses and initial and final speeds and does not depend on the rotor circuit resistance. This does not mean that the total energy lost is independent of rotor circuit resistance. The total energy lost depends on the stator and rotor resistances.

We know that, $$\frac{\text{Stator copper loss}}{\text{Rotor copper loss}} = \frac{R_1}{R_2}$$

The total energy lost in the motor when the speed changes is given by,

$$W_{motor} = 3\int_{t_1}^{t_2} I_2^2\, (R_1 + R_2)\, dt \qquad ...\text{since } I_1 \simeq I_2$$

Following the same procedure of deriving W, we can write the energy lost in the motor,

$$W_{motor} = \frac{1}{2} J\,\omega_s^2\, (s_1^2 - s_2^2)\left(1 + \frac{R_1}{R_2}\right) \qquad ...(7.25)$$

At starting ($s_1 = 1$; $s_2 = 0$),

$$W_{motor} = \frac{1}{2} J\,\omega_s^2 \left(1 + \frac{R_1}{R_2}\right) \qquad ...(7.26)$$

- When the motor starts under a load torque of T_{load}, we can write

$$T_{motor} - T_{load} = J\,\frac{d\omega}{dt}$$

or,

$$dt = \frac{J\, d\omega}{T_{motor} - T_{load}}$$

Again,

$$d\omega = -\,\omega_s\, ds$$

∴

$$W = 3\int_{t_1}^{t_2} I_2^2\, R_2\, dt$$

$$= \omega_s \int s\, T_{motor}\, dt$$

$$= -\,J\omega^2 \int_{t_1}^{t_2} \frac{T_{motor}}{T_{motor} - T_{load}}\, s.ds$$

or, $$W = -J\omega_r^2 \int_{t_1}^{t_2} \left[1 + \frac{T_{load}}{T_{motor} - T_{load}}\right] s.ds \qquad ...(7.27)$$

— Eqn. (7.27) shows that the energy lost in starting an induction motor on load is much more than that on no load.

— Again when the motor is started using direct on line starter T_{max} will be high at rated voltage and so the energy lost will be minimum. In the reduced voltage starting $V < V_{rated}$ and so the energy losses will be more. The energy loss during starting is wasted as heat in stator and rotor windings of squirrel-cage motor and in external circuit resistance added to the rotor of a wound rotor induction motor. *Whereas the external resistance added to the motor is more compared to stator and rotor resistances to have a low operating temperature, a higher resistance added to the rotor circuit will cause the starting time to increase.*

- *Reduction of energy loss during starting* is achieved by any of the following methods :

(*i*) By reducing the moment of inertia of the rotor.

(*ii*) By smooth variation of supply frequency to induction motors (*V*/*f* control).

(*iii*) By gradual adjustment of applied voltage to D.C. shunt motors.

(*iv*) Starting of multi-speed induction motors in discrete steps of speed.

7.4. ENERGY LOSSES DURING BRAKING

In an electric motor energy losses occur during braking.

1. D.C. motors with shunt or separate excitation :

Let the braking be applied when there is no load on the motor. The energy wasted in the armature circuit during rheostatic braking is calculated using the relation,

$$I_a^2 R_a dt = J\omega_0.d\omega - J\omega.d\omega$$

When rheostatic braking is applied the speed changes from ω_0 to 0 at standstill.

Thus energy wasted in braking

$$= \int_{\omega_0}^{0} -J\omega.d\omega$$

$$= \int_{0}^{\omega_0} J\omega.d\omega$$

$$= \frac{J\omega_0^2}{2} = \text{K.E. of the armature at the start of braking}$$

If the motor *reverses its direction of rotation* and the speed changes from $+\omega_0$ to $-\omega_0$ the energy lost by the armature will be $4\left(\frac{J\omega_0^2}{2}\right)$.

2. Induction motor :

The braking is performed at no load and during reverse current braking (*plugging*) the slip changes from 2 to 1. Substituting in the equation for energy lost, we have

$$W_{braking}\text{(reverse current)} = \frac{1}{2} J\omega_0^2 (s_1^2 - s_2^2)$$

$$= \frac{1}{2} J\omega_0^2 (2^2 - 1^2)$$

$$= \frac{3}{2} J\omega_0^2 \qquad ...(7.28)$$

Eqn. (7.28) indicates that the *power wasted in the rotor circuit is 3 times the K.E. at no load.* Now, if the motor is *reversed* to run at ω_s the *slip varies from 2 to 0*, then the *power loss* will be

$$= \frac{1}{2} J\omega_0^2 (2^2 - 0^2) = 4\left(\frac{J\omega_0^2}{2}\right) \qquad ...(7.29)$$

These losses are in the *rotor alone.*

- The *total energy wasted* by the motor at braking $= \frac{3}{2} J\omega_0^2 \left(1 + \frac{R_1}{R_2}\right)$ joules ...(7.30)
- The *total energy wasted* during the *reversal* of speed $= 4\left(\frac{J\omega_0^2}{2}\right)\left(1 + \frac{R_1}{R_2}\right)$ joules ...(7.31)

The *braking energy is given by the stored energy available at the commencement of the braking process.*

7.5. BRAKING OF MOTORS DURING LOWERING OF LOADS

When load are lowered, as in the case of D.C. shunt (or separately excited) motor, the *torque due to the motor and the load act in the same direction strengthening the total torque available and thereby aiding each other.*

— With the increase in speed the induced e.m.f. in the armature also increases and reaches a value *equal to applied voltage*. At this moment the armature current and therefore the electromagnetic torque becomes *zero*. The downward motion of the hoist will be only due to the downward moving load.

— When the speed *exceeds* the no-load speed resulting in $E_b > V$ the armature current becomes *negative* and the *machine works as an electrical generator* and *provides the braking torque*.

— When the braking torque equals the load torque the *steady speed is obtained.* This type of braking is suitable for "*regenerative braking*" as power is returned to the supply.

— When the *speed is less than the no load speed* braking can be provided by adopting "*rheostatic braking*". Then the motor acts as a generator with its armature terminals short-circuited by an external resistance R_x

Hence, $$E_b = -I(R_a + R_x)$$

and, $$\omega = -T\frac{(R_a + R_x)}{k_1 k_2 \phi^2}, \text{ with } V = 0.$$

D.C. series motor :

'*Regenerative braking*' is *not* practicable with series motors since in a series motor it is not possible to make the field current exceed the armature current and hence to have E_b more than V.

'*Rheostatic braking*', however, is possible if care is taken to interconnect the armature and field windings in such a way as to *make the direction of current in the field remain unaltered although there is a change in the direction of the armature current.* Further, it should be ascertained that the external resistance added *should not* make the total resistance of the motor circuit to *exceed the critical value*. Thus the e.m.f. equation can be written as,

$$E_b = I\,(R_a + R_f + R_x) \qquad \text{...(7.32)}$$

where, R_a = Armature resistance,

R_f = Field resistance, and

R_x = External resistance.

7.6. DYNAMICS OF BRAKING

When a motor is subjected to braking the following torques need to be considered :

(*i*) Torque developed in the motor [– T(ω)] ;

(The –ve sign shows it is the braking torque).

(*ii*) Friction torque (T_f) due to brake friction, which is constant ;

(*iii*) Load torque (T_{load}) ;

(*iv*) Inertia torque $\left(\dfrac{J\,d\omega}{dt}\right)$.

If the motor is arrested in its motion using the electromagnetic torque proportional to the speed and a form of friction torque, we can write the equation (assuming the load torque on the motor to be independent of speed) as :

$$-T(\omega) = T_{load} + T_f + \frac{J\,d\omega}{dt}$$

$$-K\omega = T_{load} + T_f + \frac{J\,d\omega}{dt}, \text{ where } k \text{ is constant}$$

or

$$dt = -\frac{J\,d\omega}{K\omega + T_{load} + T_f}$$

Now, the time required to brake the motor from speed ω_1 to ω_2 is given by,

$$t = -J\int_{\omega_1}^{\omega_2} \frac{d\omega}{K\omega + T_{load} + T_f}$$

or

$$t = \frac{J}{K}\log_e\left(\frac{K\omega_1 + T_{load} + T_f}{K\omega_2 + T_{load} + T_f}\right) \qquad \text{...(7.33)}$$

The time (t_1) required to bring the motor to rest (*i.e.*, $\omega_2 = 0$) is given by,

$$t_1 = \frac{J}{K}\log_e\left(\frac{K\omega_1 + T_{load} + T_f}{T_{load} + T_f}\right) \qquad \text{...(7.34)}$$

The number of revolutions made by the motor before coming to rest is given by,

$$N = \frac{1}{2\pi}\int_0^t \omega_2\,dt$$

$$\omega_2 = \frac{1}{K}\left[(K\omega_1 + T_{load} + T_f)\,e^{-Kt/J} - (T_{load} + T_f)\right]$$

Hence,

$$N = \frac{1}{2\pi K}\left[\frac{J}{K}(K\omega_1 + T_{load} + T_f)(1 - e^{-Kt/J}) - (T_{load} + T_f)\,t\right] \qquad \text{...(7.35)}$$

The number of revolutions (N_r) made by the motor can be obtained by substituting t by t_r in eqn. (7.35), as given below,

$$N_r = \frac{1}{2\pi K}\left[\frac{J}{K}(K\omega_1 + T_{load} + T_f)(1 - e^{-Kt/J}) - (T_{load} + T_f)\,t_r\right] \qquad \text{...(7.36)}$$

Example 7.1. *A 6-pole, 50 Hz synchronous motor coupled to a load has a moment of inertia of 540 kg-m^2. If the load torque is independent of speed and the frictional torque is 300 Nm, calculate the time taken by the motor to come to stop and the number of revolutions made during : (i) Rheostatic braking, the field current and braking resistance being kept constant at values which give initial electrical braking torque of 6000 Nm ; (ii) Plugging which produces a constant braking torque of 3300 Nm.* **(Roorkee University)**

Solution. *Given* : $p = 6$; $f = 50$ Hz ; $J = 540$ kg-m^2 ; $T_f = 300$ Nm.

(*i*) **Rheostatic braking :**

Initial braking torque, $K\omega_1 = 6000$ Nm

$$\omega_1 = \frac{2\pi N}{60} = \frac{2\pi\left(\frac{120\,f}{p}\right)}{60} = \frac{2\pi \times \left(\frac{120 \times 50}{6}\right)}{60} = 104.72 \text{ rad/s.}$$

Hence, $K = \frac{T}{\omega} = \frac{6000}{104.72} = 57.3$

Time taken by the motor to stop

$$t_r = \frac{J}{K} \log_e \left(\frac{K\omega_1 + T_f}{T_f}\right)$$

$$\mathbf{t_r} = \frac{540}{57.3} \log_e \left(\frac{6000 + 300}{300}\right) = \mathbf{28.69 \text{ s.}} \quad \textbf{(Ans.)}$$

Number of revolutions made before coming to rest,

$$N_r = \frac{1}{2\pi K}\left[\frac{J}{K}(K\omega_1 + T_f)(1 - e^{-Kt_r/J}) - T_f \cdot t_r\right]$$

$$= \frac{1}{2\pi \times 57.3}\left[\frac{540}{57.3}(6000 + 300)(1 - e^{-(57.3 \times 28.69/540)}) - 300 \times 28.69\right]$$

$$= 0.002777\,[59371.7\,(1 - e^{-3.044}) - 8607] = 133.12$$

i.e., $\mathbf{N_r = 133.12.}$ **(Ans.)**

(*ii*) **Plugging :**

Total braking torque, $T_R = 3300 + 300 = 3600$ Nm

Since the braking torque is constant, it will produce a uniform retardation of β rad/sec^2.

$$\beta = \frac{T_B}{J} = \frac{3600}{540} = 6.67 \text{ rad/sec}^2$$

$$\omega_1 = 104.72 \text{ rad/s} \quad \text{(as earlier calculated)}$$

Hence time taken to come to rest

$$\mathbf{t_r} = \frac{\omega_1}{\beta} = \frac{104.72}{6.67} = 15.7 \text{ s}$$

∴ Number of revolutions made before coming to rest,

$$N_r = \text{(Average speed in rev./sec.)} \times \text{(Time taken to stop in sec.)}$$

$$= \left[\frac{1}{2} \times \left(\frac{120 \times f}{60 \times p}\right)\right] \times 15.7 = 130.83$$

$$= \frac{1}{2} \times \frac{120 \times 50}{60 \times 6} \times 15.7 = 130.83$$

$\mathbf{N_r = 130.83.}$ **(Ans.)**

8. SIZE AND RATING OF MOTORS

For any particular service the following *factors govern the size and rating of motor :*

(*i*) The maximum temperature rise of the motor under given load conditions ;

(*ii*) The maximum torque required.

It has been observed that a motor which is satisfactory from the point of view of maximum temperature rise usually satisfies the requirement of maximum torque as well.

The maximum permissible temperature rise with insulations *A* and *B* are 40°C and 50°C respectively.

8.1. STANDARD RATING OF MOTORS

The different ratings for electrical motors are as under :

1. **Continuous rating.** It *is an output which a motor can deliver continuously without exceeding the permissible temperature.* It can deliver *25% overload for two hours.*

2. **Continuous maximum rating.** It is similar to continuous rating but *not allowing any overload.*

- It is used for motors of capacity larger than 2 kW per r.p.m.
- These motors are bit inferior to the continuous-rated motors.

3. **Short time rating.** This is an output, which a motor can deliver for a specified period (say 1 hour, $\frac{1}{2}$ hour, $\frac{1}{4}$ hour etc.) without exceeding the specified temperature rise.

4. **Intermittent rating.** This rating indicates the maximum load of the motor for the specified time *followed by a no-load period during which the machine cools down to its original temperature.*

- For intermittent loads short time rated motors can be used if the loading condition corresponds to the short time rating of the motor or if it can be assumed that the heating of the motor is proportional to the square of current, the equivalent continuous rating can be found by taking the r.m.s. value of the load curve and choosing a motor having a continuous rating equal to this. This method is *very frequently used.*

8.2. CLASSES OF DUTY

The rating of a motor selected from the view point of heating depends on the load conditions or duty to which it is subjected. According ISS : 4722, there operating condition are classified into the following *eight classes of duty :*

(*i*) **Continuous duty.** It denotes operation at constant load of sufficient duration for thermal equilibrium to be reached.

Examples. *Centrifugal pumps, fans, compressors* and *conveyors* are some examples of equipment which run continuously with a constant load

(*ii*) **Short time duty.** It denotes the operation at constant load during a given time, *less* than that required to reach thermal equilibrium followed by a rest of sufficient duration to re-establish equality of temperature with the cooling medium.

The recommended values for short time duty are 10, 30, 60 and 90 minutes.

Examples. *Motors used for opening and closing weirs, lock gate and bridges, motor employed in battery-charging units etc.* are rated for such a duty.

(*iii*) **Intermittent periodic duty.** It denotes a sequence of identical duty cycles, each consisting of a period of operation at constant load and a rest period, these periods being too short to attain thermal equilibrium during one duty cycle. The starting current does not affect the temperature rise for this type of duty. Unless otherwise specified, the duration of duty cycle is 10 minutes.

The recommended values for the cycle duration factors are 15, 25, 40 and 60 percent.

Examples. *Motors that are used in different kinds of hoisting mechanisms and those used in trams, trolley buses etc.* are subjected to intermittent periodic duty.

(*iv*) **Intermittent periodic duty with starting.** It indicates a sequence of identical duty cycles each consisting of a period of starting, a period of operation at constant load and a rest period, the operating and rest periods being too short to attain thermal equilibrium during one duty cycle.

— The stopping of motor, in this type of duty, is obtained by natural deceleration after disconnecting the electric supply by means of mechanical brake which does not cause additional heating of the windings.

— This type of duty is characterised by the cyclic duration factor, the number of duty cycles, *factor of inertia* which is the *ratio of the combined inertia of the motor and load to the motor inertia.*

Examples. *Motors that drive metal cutting lathes and certain auxiliary equipment of rolling mills* are subjected to such operating conditions.

(*v*) **Intermittent periodic duty with starting and braking.** It denotes a sequence of identical duty cycles each consisting of a period of starting, a period of operation at a constant load, a period of braking and a rest period. The operating and rest periods are too short to obtain thermal equilibrium during one duty cycle.

In this duty *braking is rapid and is carried out electrically.*

Examples. *Certain auxiliary equipment used in rolling mills and metal cutting lathes* offer such operating conditions to their driving motors.

(*vi*) **Continuous duty with intermittent periodic loading.** It denotes a sequence of identical duty cycles each consisting of a period of operation of constant load and a period of operation at no load, machines with excited windings having normal no-load rated voltage excitation. The *operation and no load periods are too short to attain thermal equilibrium during one duty cycle.*

Unless otherwise specified the duration of the duty cycle is 10 minutes.

The recommended values of cyclic duration factor are 15, 25, 40 and 60 percent.

- This type of duty is distinguished from intermittent periodic duty by the fact that *after a period of operation at constant load follows a period of no-load operation instead of rest.*

(*vii*) **Continuous duty with starting and braking.** It denotes a sequence of identical duty cycles each consisting of a period of starting, a period of operation at constant load and a period of electrical braking. There is *no period of rest.*

This type of duty is also indicated by the *number of cycles per hour and the factor of inertia.*

(*viii*) **Continuous duty with periodic speed changes.** It indicates a sequence of identical duty cycles each consisting of a period of operation at constant load corresponding to a specified speed of rotation, followed immediately by a period of operation at another load corresponding to a different speed of rotation, the operating periods being too short to attain thermal equilibrium during one duty cycle. There is *no rest period.*

For this duty the number of duty cycles per hour and a factor of inertia together with the load have to be indicated. In addition, the cyclic duration factor should be indicated for each speed.

8.3. AMBIENT TEMPERATURE AND RATINGS

- The operating temperature of a motor should *never* exceed the maximum permissible temperature because it will result in *deterioration and breakdown of insulation and will shorten the service life of motors.*
- For simplifying the heating calculations, it is general practice to base the motor power ratings on a standard value of temperature say 35°C. Accordingly, the power given on the name plate of a motor corresponds to the power which the motor is capable of delivering without overheating at an ambient temperature of 35°C. At an ambient temperature considerably *below 35°C*, the motor can safely deliver a *somewhat higher output* than its ratings because there is a greater difference between the ambient temperature and the maximum allowable temperature of insulation used in the motor. Similarly at ambient temperatures greater than 35°C, the load on the motor should be smaller than that indicated on its name plate unless of course special measures are taken to improve the cooling to bring down the temperature rise.
- *The duty cycle is closely related to temperature and is generally taken to include the environmental factors also.*

8.4. OVERLOAD CAPACITY OF MOTORS

The rating of a machine can be determined from heating considerations. However the motor so selected should be *checked for its overload capacity and starting torque*. This is because the motor selected purely on the basis of heating may not be able to meet the mechanical requirements of the load to be driven by it. Table 8.1 lists the coefficients used to determine the permissible instantaneous overload torque of different electric motors.

Table 8.1. Motor instantaneous torque overload co-efficients

S. No.	*Type of motor*	*Torque overload co-efficient*
1.	D.C. motors	2 (for special types upto 3 or 4)
2.	Squirrel cage induction motors	2—2.5
3.	Double cage and deep bar squirrel cage motors	1.8–3
4.	Synchronous motors	2—2.5 (For special types upto 3—4)
5.	A.C. commutator motors	2—3

- In case of *D.C. machines* the maximum overload capacity is determined on the basis of *commutation*, while in case of synchronous and induction motors it is determined by *maximum electromagnetic torque.*

8.5. DETERMINATION OF POWER RATING OF ELECTRIC MOTORS FOR DIFFERENT APPLICATIONS

1. Continuous duty and constant load :

- For *most of the applications*, the rating can be determined from the equation given as under :

$$P = \frac{TN}{975\,\eta}\ \text{kW} \qquad ...(8.1)$$

where, T = Load torque, kg-m,

N = Speed, r.p.m., and

η = Product of the efficiency of the driven equipment and that of transmitting device.

- In case of *linear motion*, the rating of the motor required is given by,

$$P = \frac{F \times v}{102\ \eta}\ \text{kW} \qquad \text{...(8.2)}$$

where, F = Force caused by the load, kg, and

v = Velocity of motion of the load, m/s.

Eqn. (8.2) is directly applicable in case of *hoisting mechanisms*. It is also suitable for *lifts or elevators* ; but since a counterweight which balances the weight of the cage or car as well as one-half of the useful load will be always present, it should be modified as follows :

$$P = \frac{F \times v}{2 \times 102\ \eta}\ \text{kW} \qquad \text{...(8.3)}$$

The velocity of normal passenger lift cabins vary from 0.5 to 1.5 m/s.

- In case of *pumps*, the rating can be determined from the following relation :

$$P = \frac{\rho Q H}{102\ \eta}\ \text{kW} \qquad \text{...(8.4)}$$

where, ρ = Density of liquid pumped, kg/m^3,

Q = Delivery of pump, m^3/s, and

H = Gross head (static head + friction head), *m*.

η varies from 0.8 to 0.9 for *reciprocating pumps* and from 0.4 to 0.8 for *centrifugal pumps*.

- The rating of a *fan motor* is given by,

$$P = \frac{Qh}{102\ \eta}\ \text{kW} \qquad \text{...(8.5)}$$

where, Q = Volume of air or any other gas, m^3/s, and

h = Pressure in mm of water or kg/m^2.

For small power fans, the efficiency η may be taken as 0.6 and for large power ones it may reach a value upto 0.8.

- The rating of a motor used in *metal shearing lathes* can be found from the relation,

$$P = \frac{F \times v}{102 \times 60 \times \eta}\ \text{kW} \qquad \text{...(8.6)}$$

where, F = Shearing force, kg ;

v = Velocity of shearing, m/min, and

η = Mechanical efficiency of the lathe.

Motor Rating for Variable Load :

The following are the commonly used methods for determination of motor rating for variable load drives :

(*i*) Method of average losses ; (*ii*) Equivalent current method ;

(*iii*) Equivalent torque method ; (*iv*) Equivalent power method.

(*i*) **Method of average losses (Q_{av}).** This method consists of finding average loses Q_{av} in the motor when it operates according to the given load diagram. These losses are then compared with Q_{nom}, the losses corresponding to the continuous duty of the machine when operated at its nominal rating. This method presupposes that when $Q_{av} = Q_{nom}$, the motor will operate without temperature rise going above the maximum permissible for the particular class of insulation.

Fig. 8.1 shows a simplified load diagram for a certain drive.

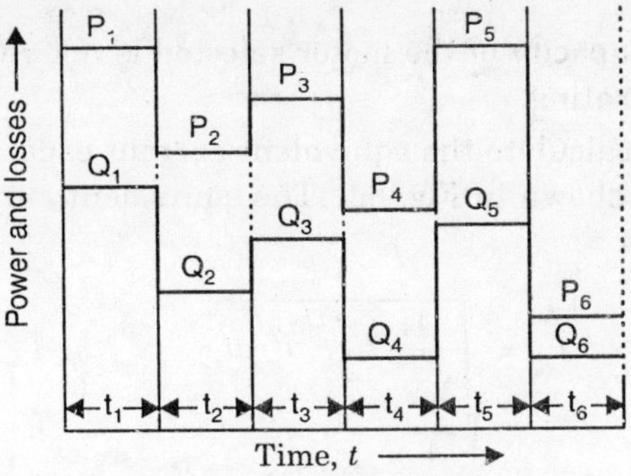

Fig. 8.1. A simplified load diagram for a certain drive.

The loss diagram (loss Q plotted versus time) of the electric motor is shown. The rating of the electric motor can be found from method of successive approximations. To start with a motor is selected in accordance with the required capacity calculated by multiplying average power load a factor of safety of 1.1 to 1.3. The *losses of the motor are calculated for each portion of the load diagram by referring to the efficiency curve of the motor.* The average losses are given by

$$Q_{av} = \frac{Q_1t_1 + Q_2t_2 + Q_3t_3 + \ldots\ldots + Q_nt_n}{t_1 + t_2 + t_3 + \ldots\ldots + t_n} \quad \ldots(8.7)$$

The average losses as found from eqn. (8.7) are compared with losses of selected motor at rated frequency. In case the two losses are equal or differ by a small amount the motor is selected. However, in case the losses differ considerably, another motor is selected and the calculations repeated till the motor having almost the same losses or the average losses is found. It should be checked that the motor selected has a sufficient overload capacity and starting torque.

- This method does *not* take into account the maximum temperature rise under variable load conditions. However, this method is *accurate and reliable for determining the average temperature rise of the motor during one work cycle.*

The *disadvantage* of the method of equal losses is that *it is tedious to work with and also many a times the efficiency curve is not readily available and the efficiency has to be calculated by means of empirical formulae which may not be accurate to work with.*

(ii) **Equivalent current method.** This method is based on the *assumption that the actual variable current may be replaced by an equivalent current I_{eq} which produces the same losses in the motor as the actual current.*

$$\therefore \quad I_{eq} = \sqrt{\frac{I_1^2t_1 + I_2^2\,t_2 + I_2^2t_3 + \ldots + I_n^2t_n}{t_1 + t_2 + t_3 + \ldots\ldots + t_n}} \quad \ldots(8.8)$$

The heating and cooling conditions in self ventilated machines depend upon its speed. At slow speeds the cooling conditions are poorer than at normal speeds. Therefore, if the work cycle involves slow speed operation, it must be taken into consideration when using eqn. (8.8).

The equivalent current as found from eqn. (8.8) should be compared with the rated current of the motor selected and the conditions $I_{eq} \le I_{nom}$ should be met (I_{nom} is the rated current of the machine).

The machine selected should also be checked for its overload capacity.

— For D.C. motor ...... $\dfrac{I_{max}}{I_{nom}} \le 2$ to 2.5

— For induction motors ...... $\frac{T_{max}}{T_{nom}} \le 1.65$ to 2.75

In case the overload capacity of the motor selected is not sufficient it becomes necessary to select a motor of high power rating.

It may not be easy to calculate the equivalent current especially in cases where the current load diagram is irregular as shown in Fig. 8.2. The equivalent current in such cases is calculated from the following expression

$$I_{eq} = \sqrt{\frac{1}{\sum_{1}^{n} t} \int^{\Sigma t} i^2 \, dt} \quad \text{...(8.9)}$$

An easier way is to break up the current load diagram into a series of straight line geometrical figures.

For a *triangular shaped diagram :*

$$I_{eq} = \sqrt{\frac{I_1}{3}} \quad \text{...(8.10)}$$

For a *trapezoidal diagram :*

$$I_{eq} = \sqrt{\frac{I_1^2 + I_1 I_2 + I_2^2}{3}} \quad \text{...(8.11)}$$

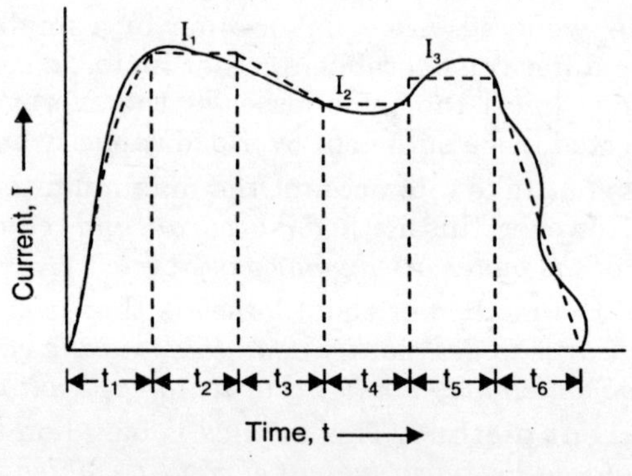

Fig. 8.2. Simplification of irregular current load diagram by short segments of straight lines.

The current values obtained by this method are sufficiently accurate for practical purposes.

(*iii*) and (*iv*) **Equivalent torque and equivalent power method.** For the selection of suitable capacity of the motor it often becomes necessary to use torque or power load diagrams. The equivalent torque or power is found in the same manner as the equivalent current.

Assuming constant flux and constant power factor, the *torque is directly proportional to current* and, therefore, the equivalent torque is :

$$T_{eq} = \sqrt{\frac{T_1^2 t_1 + T_2^2 t_2 + T_3^2 t_3 + \ldots\ldots + T_n^2 t_n}{t_1 + t_2 + t_3 + \ldots\ldots + t_n}} \quad \text{...(8.12)}$$

The equation for equivalent power follows directly from eqn. (8.12) as power is directly proportional to the torque. At constant speed or where the changes in speed are small, the equivalent power is given by :

$$P_{eq} = \sqrt{\frac{P_1^2 t_1 + P_2^2 t_2 + P_3^2 t_3 + \ldots\ldots + P_n^2 t_n}{t_1 + t_2 + t_3 + \ldots\ldots + t_n}} \quad \ldots(8.13)$$

- *The "equivalent current method"* is the *most accurate* out of all the above methods discussed above.

  This method may be used to determine the motor capacity for all uses except where it is necessary to take into account the *changes in so called 'constant losses' i.e.*, the iron and mechanical losses.

- *The "equivalent torque method" cannot be used for cases where equivalent current method cannot be applied.*

  The "equivalent torque method" cannot be used for selection of motor rating for cases in which the field flux does not remain constant like D.C. series motors and for squirrel-cage induction motors under starting and braking conditions.

- The *disadvantage* of the *"equivalent power method"* is that it *cannot be used for motors whose speed varies considerably under load, especially when dealing with starting and running conditions.*

Example 8.1. *Determine the rated current of an electrical machine for the following duty cycle :*

600 A for 3 minutes, a sharp increase to 1200 A and constant at this value for 1 minute, gradually decreasing for 2 minutes to 240 A and constant at this value for 2 minutes, gradually increasing to 600 A during 2 minutes and the repetition of the cycle.

Solution. The load diagram is plotted in Fig. 8.3.

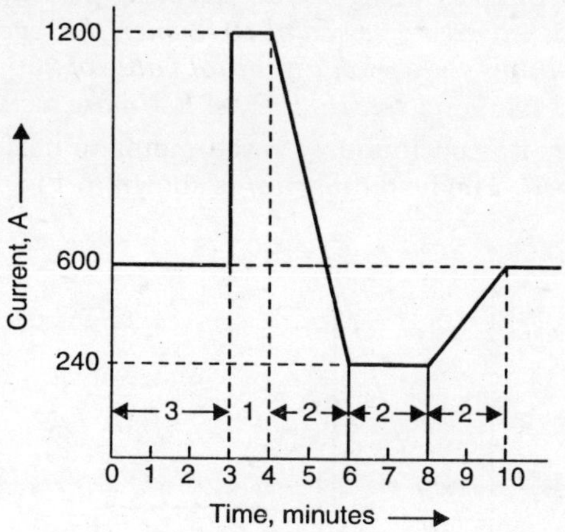

Fig. 8.3

The load diagram can be divided into rectangles and trapeziums (in this case) with the help of line segments. The equivalent current with sides I_1 and I_2 is :

$$I_{eq} = \left(\frac{I_1^2 + I_1 I_2 + I_2^2}{3} \right)^{1/2}$$

∴ The equivalent current for the entire cycle is

$$I_{eq} = \left[\frac{(600)^2 \times 3 + 1200^2 \times 1 + (1/3)\,(1200^2 + 1200 \times 240 + 240^2) \times 2 + 240^2 \times 2 + (1/3)\,(240^2 + 240 \times 600 + 600^2) \times 2)}{(3 + 1 + 2 + 2 + 2)}\right]^{1/2}$$

$$= \left(\frac{1080000 + 1440000 + 1190400 + 115200 + 374400}{10}\right)^{1/2} = \mathbf{648\ A.}\quad \textbf{(Ans.)}$$

Example 8.2. *An induction motor has to perform the following duty cycle, which is repeated indefinitely :*

75 kW for 10 minutes ; No load for 5 minutes ;

45 kW for 8 minutes ; No load for 4 minutes.

Determine a suitable capacity of a continuously rated motor. Motors of standard (continuous) ratings of 45, 55, 75 kW are available. The ratio of maximum torque to nominal torque should be less than 1.8.

Solution. *Given :* $P_1 = 75$ kW, $t_1 = 10$ min ; $P_2 = 0$, $t_2 = 5$ min ; $P_3 = 45$ kW, $t_3 = 8$ min, $P_4 = 0$,

$$t_4 = 4 \text{ min} ; \frac{T_{max}}{T_{nom}} < 1.8$$

The capacity of motor is given by :

$$P = \left[\frac{P_1^2 t_1 + P_2^2 t_2 + P_3^2 t_3 + P_4^2 t_4}{t_1 + t_2 + t_3 + t_4}\right]^{1/2}$$

$$= \left(\frac{75^2 \times 10 + 0^2 \times 5 + 45^2 \times 8 + 0^2 \times 4}{10 + 5 + 8 + 4}\right)^{1/2} = \mathbf{51.8\ kW.}\quad \textbf{(Ans.)}$$

Example 8.3. *A motor driving a colliery winder has to deliver a load rising uniformly from zero to maximum of 1500 kW in 20 s during the accelerating period, 750 kW for 40 s during the full speed period and during the decelerating period of 10 s, when the regenerative braking takes place the kW which is returned to the supply falls from an initial value of 250 to zero. The interval for decking (period of rest) before the next load cycle starts is 20 s. Estimate a suitable rating for the motor.*

Solution. The rating of a continuously rated motor to perform the given duty is found by using *equivalent power method*. The load diagram is shown in Fig. 8.4. It is divided into two triangles and a rectangle.

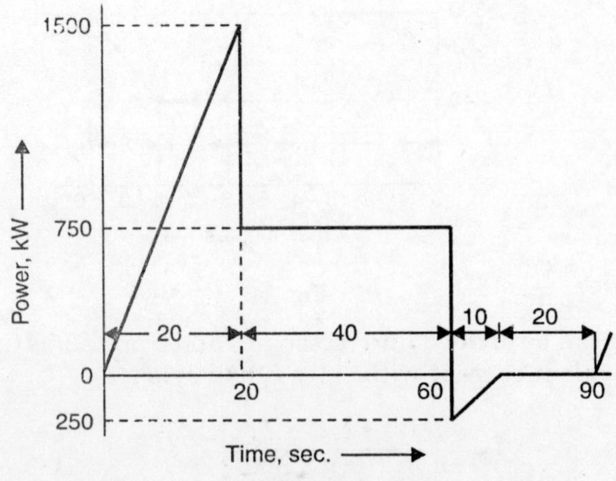

Fig. 8.4

The equivalent rating of motor is given by :

$$= \left[\frac{1/3 \times (1500)^2 \times 20 + (750)^2 \times 40 + 1/3 \times (250)^2 \times 10}{20 + 40 + 10 + 20} \right]^{1/2}$$

= **648 kW. (Ans.)**

Example 8.4. *A motor works on 2 minutes load cycle constituted as follows :*

0 to 15 sec. ...... Load rising from 0 to 1280 H.P. ;

15 to 85 sec. ...... Constant load of 720 H.P. ;

85 to 95 sec. ...... Regenerative braking with the H.P. returned falling uniformly from 240 to 0 H.P. ;

95 to 120 sec. ...... Motor is at rest.

Determine the continuous rating of the motor that would be suitable for the load cycle. Assume the rating to depend upon :

(i) The r.m.s. value of loading ; *(ii) The average value of loading.*

Solution. The loading is shown in Fig. 8.5.

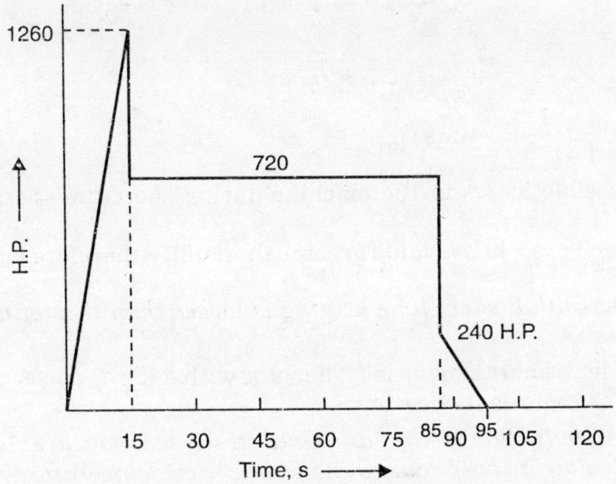

Fig. 8.5

(i) **The r.m.s. value of loading :**

$$(\text{H.P. rating})^2 \times 120 = \int_0^{15} \left(\frac{1260}{15} t \right)^2 dt + (720^2 \times 70) + \int_0^{10} \left(\frac{240}{10} t \right)^2 dt$$

$$= \left[7056 \frac{t^3}{3} \right]_0^{15} + 36288000 + \left[576 \frac{t^3}{3} \right]_0^{10}$$

$$= 7938000 \div 36288000 + 192000 = 44418000$$

$\therefore$ H.P. = **608.4. (Ans.)**

(ii) **The average value of loading :**

$$\text{H.P. rating} \times 120 = \frac{1}{2} \times 1260 \times 15 + 720 \times 70 + \left(\frac{1}{2} \times 240 \times 10 \right)$$

$$= 61050$$

$\therefore$ H.P. $\simeq$ **509. (Ans.)**

8.6. TEMPERATURE RISE WITH SHORT TIME RATINGS

Let, θ_m = Final steady temperature rise with continuous rating. This may be assumed as maximum allowable temperature rise, θ_{per},

T_h = Heating time constant,

θ_m' = Final steady temperature rise that would be attained if the machine were allowed to run for an indefinite time at its short time rating,

t_h = Time for which the machine is run at its short time rating, and

θ_{sh} = Temperature rise with short time rating $t = t_{sh}$.

We know that,
$$\theta_{sh} = \theta_m (1 - e^{-t_h/T_h}) \quad ...(8.14)$$

In order that the insulation of the machine is not operated and the specified temperature limit, the temperature rise $t = t_h$ should not exceed the final steady temperature or θ_{sh} should not exceed the θ_m.

Thus the maximum allowable temperature rise during short time rating is, $\theta_{sh} = \theta_m$

Inserting this value in eqn. (8.14), we get

$$\theta_m = \theta'_m (1 - e^{-t_h/T_h}) \quad ...(8.15)$$

or
$$\frac{\theta'_m}{\theta_m} = \frac{1}{1 - e^{-t_h/T_h}}$$

Obviously θ'_m is $\frac{1}{1 - e^{-t_h/T_h}}$ times θ_m

Thus the permissible losses in the machine during short time rating can also be as many times greater. Since $\frac{1}{1 - e^{-t_h/T_h}}$ has a value greater than unity therefore *with the given dimensions, the output of a machine with a short time heating is larger than that of machine with continuous rating.*

In other words, for a short time load t_h, a motor with a lower power rating should be selected so as to utilize it fully as regards its heating.

Example 8.5. *An induction motor has a final steady temperature rise of 40° C when running at its rated output. Calculate its half-hour rating for the same temperature rise if the copper losses at the rated output are 1.25 times its constant losses. The heating time constant is 90 minutes.*

Solution. *Given :* $\theta_m = 40°C$, $P_c = 1.25 P_i$; $T_h = 90$ min. = 1.5 hour ; $t_h = \frac{1}{2}$ hour.

Half-hour rating :

Let P be the rated output and xP be the half-hour rating. Let P_c be the copper losses at full load and P_i be other constant losses.

$\therefore$ Total losses at full load = $P_c + P_i$

and, total losses with *short time rating* = $x^2P_c + P_i$

Specific heat dissipation and surface remaining the same the temperature rise is proportional to the losses. Suppose θ'_m is the final steady temperature rise with short time rating, then,

$$\frac{\theta'_m}{\theta_m} = \frac{x^2P_c + P_i}{P_c + P_i} = \frac{1.25x^2 + 1}{2.25} \quad \text{as} \quad P_c = 1.25\,P_i$$

The temperature rise after $\frac{1}{2}$ hour of operation at short time rating should not exceed $\theta_m = 40°C$.

Now, using the relation :

$$\theta_{sh} = \theta_m' \, (1 - e^{-t_h/T_h}) \quad \text{...[(Eqn. (8.14)]}$$

$$40 = \theta_m' \, [1 - e^{-1/(2 \times 3/2)}]$$
$$= \theta_m' \, (1 - 0.716) = 0.284 \, \theta'_m$$

or

$$\theta_m' = 141.1°\text{C}$$

$$\therefore \quad \frac{1.25\, x^2 + 1}{2.25} = \frac{141.4}{40} \quad \text{or} \quad x = 2.36$$

Thus half-hour rating of machine is **2.36 times** *its continuous rating.* **(Ans.)**

Example 8.6. *If one hour rating of a machine is $\sqrt{3}$ times the continuous rating and if steady temperature rise for one-hour rating is twice that on normal load, find the ratio of iron to copper loss at full load.*

Solution. Let one-hour rating of transformer be x times its continuous rating.

Losses with continuous rating = $P_c + P_i$

Loss with short time rating = $x^2 P_c + P_i$

Let θ_m and θ'_m be final steady temperature rise with continuous and one-hour rating respectively

$$\therefore \quad \frac{\theta_m'}{\theta_m} = \frac{x^2 P_c + P_i}{P_c + P_i}$$

Now,

$$\frac{\theta'_m}{\theta_m} = 2 \quad \text{and} \quad x = \sqrt{3} \quad \text{...(Given)}$$

$$\therefore \quad 2 = \frac{(\sqrt{3})^2 P_c + P_i}{P_c + P_i} = \frac{3P_c + P_i}{P_c + P_i}$$

or

$$2P_c + 2P_i = 3P_c + P_i$$

$$\therefore \quad P_c = P_i$$

Thus iron *losses are* **equal** *to full-load copper losses for continuous rating.* **(Ans.)**

8.7. LOAD EQUALISATION

In several industrial drives (*e.g.*, electric hammers, presses, rolling mills, reciprocating pumps, planing machines etc.) the load fluctuates between wide limits over a span of few seconds. It is very essential to smooth out the fluctuating load, otherwise during intervals of peak load it will draw a heavy current from the supply *either producing large voltage drop in the system or requiring cables and wires of heavy section. The process of smoothing out the fluctuating load is known as* **load equalisation.**

The process of load equalisation involves the storage of energy during light load periods which can be given out during the peak load period, so that the demand for the system is more or less constant. Load equalisation is often achieved by means of a **flywheel,** which is mounted on the motor-shaft, if the speed of the motor is not to be reversed. In case of reversing rolling mills, the flywheel is mounted on the motor-generator set feeding the driving motor. In order that the flywheel operates effectively, the *driving motor should have a drooping speed characteristic.* During intervals of heavy load, speed of the motor decreases. This enables the flywheel to release a portion of the stored kinetic energy, which together with the energy drawn from supply, will meet the demand of load. During off peak load period, the motor draws more energy than require by the load. The surplus energy taken is again stored as kinetic energy in the flywheel, whose speed, then increases. The load torque required and motor torque developed as well as speed variations with time are shown in Fig. 8.6.

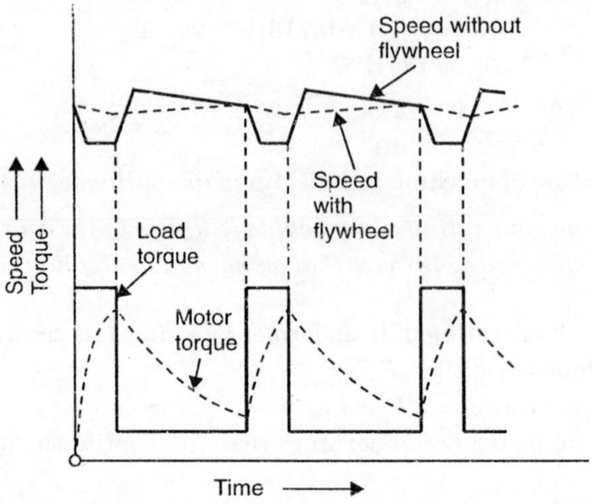

Fig. 8.6. Variations of speed, load torque and motor torque against time.

Flywheel calculations :

Let, T_{load} = Load torque (assumed constant during the period), Nm,

T_{fw} = Flywheel torque, Nm,

T_0 = No-load torque, Nm,

T_{motor} = Motor torque at any instant, Nm,

ω_0 = Motor speed on no-load, rad/s,

ω = Motor speed at any instant, rad/s

$s = (\omega_0 - \omega)$ = Motor slip,

J = Moment of inertia of flywheel, kg-m^2, and

t = Time, s.

Case 1. *Load increasing (flywheel decelerating)* :

During the period the *flywheel decelerates and gives up a part of stored energy in it,* the torque required to be supplied by the motor,

$$T_{motor} = T_{load} - T_{fw} \quad \text{...(8.16)}$$

Energy given out by the flywheel when its speed is reduced from

$$\omega_0 \text{ to } \omega = \frac{1}{2} J (\omega_0^2 - \omega^2) = J \left(\frac{\omega_0 + \omega}{2}\right) (\omega_0 - \omega)$$

But, $\frac{\omega_0 + \omega}{2} = \omega$ (mean speed) and $(\omega_0 - \omega) = s$ (slip)

$\therefore$ Energy given out $= J\omega s$

The power given out by the flywheel

= Rate of energy given up

$$= \frac{d}{dt}(J\,\omega s) = J\omega \frac{ds}{dt}$$

Flywheel torque, $$T_{fw} = \frac{\text{Power}}{\omega} = \frac{J\omega \frac{ds}{dt}}{\omega} = J\frac{ds}{dt} \quad ...(8.17)$$

Eqn. (8.16) thus becomes

$$T_{motor} = T_{load} - J\frac{ds}{dt}$$

As for value of slip upto 10%, the slip is proportional to the torque, *i.e.*, $s = KT_{motor}$.

$$\therefore \quad T_{motor} = T_{load} - JK\frac{dT_{motor}}{dt}$$

or, $$T_{load} - T_{motor} = JK\frac{dT_{motor}}{dt}$$

or, $$\frac{dT_{motor}}{T_{load} - T_{motor}} = \frac{1}{KJ}dt$$

or, $$-\log_e(T_{load} - T_{motor}) = \frac{1}{KJ} + C_1$$

From initial conditions, when $t = 0$; $T_{motor} = T_0$

$$C_1 = -\log_e(T_{load} - T_0)$$

$$\therefore \quad -\log_e(T_{load} - T_{motor}) = \frac{1}{KJ}t - \log_e(T_{load} - T_0)$$

or, $$-\frac{1}{KJ}t = \log_e\left(\frac{T_{load} - T_{motor}}{T_{load} - T_0}\right)$$

or, $$e^{-t/KJ} = \frac{T_{load} - T_{motor}}{T_{load} - T_0}$$

or, $$T_{motor} = T_{load} - (T_{load} - T_0)\,e^{-t/KJ} \quad ...(8.18)$$

Case 2. *Load decreasing (Flywheel accelerating)* :

When *load decreases, the motor accelerates the flywheel to its normal speed, slip is therefore decreased* and $\frac{ds}{dt}$ is –ve.

$$T_{motor} = T_0 + T_{fw}$$

or, $$T_{motor} = T_0 - J\frac{ds}{dt} \quad \text{...since } \frac{ds}{dt} \text{ is –ve}$$

By solving the above expression, we get

$$T_{motor} = T_0 + (T'_{motor} - T_0)\,e^{-t/KJ} \quad ...(8.19)$$

Example 8.7. *A 440 V, 50 H.P., 600 r.p.m. D.C. shunt motor takes 80 A at full load. The current during the starting period varies between the limits 1.1 to 1.5 times the full load current. The moment of inertia of the rotating system is 220 kg-m².*

Calculate the time taken to accelerate the motor to the rated speed against full load torque.

(Gorakhpur University)

Solution. *Given :* Supply voltage = 440 V ; Power = 50 H.P. ; N = 600 r.p.m. ; I_{fl} = 80 A ; I = 220 kg-m².

Time taken to accelerate the motor, t :

Full load torque, $$T_{fl} = \frac{\text{H.P.} \times 735.5}{(2\pi N/60)} = \frac{50 \times 735.5}{(2\pi \times 600/60)} = 585.3 \text{ Nm}$$

The motor is required to accelerate against full-load. The starting current varies from 1.1 to 1.5 times full load current, therefore, starting torque fluctuates between 1.1 and 1.5 times the full-load torque.

$$\therefore \quad \text{Average starting torque} = \frac{1.1 + 1.5}{2} \times \text{full-load torque} = 1.3 \times 585.3 = 760.9 \text{ Nm}$$

Torque available for acceleration,

$$T_{motor} = \text{Average starting torque} - \text{full load torque}$$
$$= 760.9 - 585.3 = 175.6 \text{ Nm}$$

Angular acceleration,
$$\alpha = \frac{T_{motor}}{J} = \frac{175.6}{220} = 0.798 \text{ rad/s}^2$$

Time taken to accelerate the motor to the rated speed,

$$t = \frac{\omega}{\alpha} = \frac{(2\pi \times 600/60)}{0.798} = \mathbf{78.7 \text{ s.}} \quad \textbf{(Ans.)}$$

Example 8.8. *A 60 H.P., 500 r.p.m. shunt motor is used for shop. The stored energy of the machinery averages 4090 Nm / H.P. Assuming the load torque during starting to be equal to the full-load torque, and starting current to be twice the full-load current, estimate the time taken to start the motor.*

(Panjab University)

Solution. *Given :* Power of the motor = 60 H.P. ; N = 500 r.p.m. Average stored energy of the machinery = 4090 Nm/H.P. ; $T_{st} = 2 \times T_{fl}$

Time taken to start the motor, t :

Full-load torque,
$$T_{fl} = \frac{60 \times 735.5}{(2\pi \times 500/60)} = 842.8 \text{ Nm}$$

$$\text{Torque developed during starting} = \frac{\text{Starting current}}{\text{Full-load current}} \times \text{Full-load torque}$$
$$= 2 \times 842.8 = 1685.6 \text{ Nm}$$

Torque available for acceleration,

$$T_{motor} = 1685.6 - 842.8 = 842.8 \text{ Nm}$$

Stored energy of the machinery = 4090 × 60 = 245400 Nm

Since stored energy $= \frac{1}{2} J\omega^2$

$$245400 = \frac{1}{2} \times J \times \left(\frac{2\pi \times 500}{60}\right)^2$$

$$\therefore \quad J = 179 \text{ kg-m}^2$$

Angular acceleration,
$$\alpha = \frac{T_{motor}}{J} = \frac{842.8}{179} = 4.708 \text{ rad/s}^2$$

Starting period,
$$t = \frac{\omega}{\alpha} = \frac{(2\pi \times 500/60)}{4.708} = \mathbf{11.12 \text{ s.}} \quad \textbf{(Ans.)}$$

Example 8.9. *A motor fitted with a flywheel supplies a load of torque 900 Nm for 2 seconds. During no-load period the flywheel regains its original speed. The motor torque is required to be limited to 450 Nm. The no-load speed of the motor is 500 r.p.m. and its full-load slip is 10 percent.*

Determine the moment of inertia of the flywheel.

(Panjab University)

Solution. *Given :*

T_{load} = 900 Nm (for 2 seconds) ; $T_0 = 0$; T_{motor} = 450 Nm ; N = 500 r.p.m. ; s_f = 10%.

Moment of inertia of the flywheel, J :

Slip speed, $s = 0.1 \times 500 = 50 \text{ r.p.m.} = \dfrac{50 \times 2\pi}{60} = 5.236 \text{ rad/s}$

$$K = \frac{s}{T_{motor}} = \frac{5.236}{450} = 0.0116$$

Now, $T_{motor} = T_{load} - (T_{load} - T_o)\, e^{-t/KJ} = T_{load}\,(1 - e^{-t/KJ})$ ...($\because$ $T_0 = 0$)

or, $450 = 900\,[1 - e^{-2/(0.0116 \times J)}]$ or $0.5 = 1 - e^{(-172.4/5)}$

or, $e^{-172.4/J} = 0.5$ or $e^{172.4/J} = 2$

$\therefore$ $J = \mathbf{248.7\ kg\text{-}m^2}$. **(Ans.)**

Example 8.10. *A 25 H.P., 3-phase, 10-pole, 50 Hz induction motor provided with a flywheel has to supply a load torque of 800 Nm for 10 seconds followed by a no-load period during which the flywheel regains the full speed. The full-load slip of the motor is 4 percent and the torque-speed curve may be assumed linear over the working range. Find the moment of inertia of the flywheel if the motor torque is not to exceed twice the full-load torque.* **(AMIE)**

Solution. *Given :* Power of the motor = 25 H.P. ; $p = 10$; $f = 50$ Hz ; $T_{load} = 800$ Nm for 10 seconds ; slip $s_f = 4\%$; $T_{motor} = 2T_{fl}$.

Moment of inertia of flywheel, J :

Synchronous speed, $N_s = \dfrac{120\,f}{p} = \dfrac{120 \times 50}{10} = 600 \text{ r.p.m.}$

Full load speed, $N_f = N_s(1 - s_f) = 600(1 - 0.04) = 576 \text{ r.p.m.}$

Full load torque, $T_{fl} = \dfrac{25 \times 735.5}{(2\pi \times 576/60)} = 304.8 \text{ Nm}$

$$T_{motor} = 2T_{fl} = 2 \times 304.8 = 609.6 \text{ Nm}$$

Slip speed, $s = 0.04 \times 600 = 24 \text{ r.p.m.}$ or $\dfrac{24 \times 2\pi}{60} = 2.5133 \text{ rad/s}$

As the slip at full load is proportional to full-load torque,

$\therefore$ $K = \dfrac{s}{T_{fl}} = \dfrac{2.5133}{304.8} = 0.008246.$

Also, $T_{motor} = T_{load} - (T_{load} - T_0)\, e^{-t/KJ}$

$= T_{load}\,[1 - e^{-t/KJ}]$ ...($\because$ $T_0 = 0$)

or, $609.6 = 800\,[1 - e^{-10/(0.008246 \times J)}]$

or, $0.762 = 1 - e^{(-1212.7/J)}$

or, $e^{-(1212.7/J)} = 0.238$

or, $e^{(1212.7/J)} = 4.2$

$\therefore$ $J = \mathbf{845\ kg\text{-}m^2}$. **(Ans.)**

Example 8.11. *The following data refer to a rolling mill, induction motor equipped with a flywheel :*

Power of the motor	*= 500 H.P.*
No-load speed	*= 40 r.p.m.*
Slip at full-load (torque)	*= 12 percent*

Load torque during actual rolling $= 407115$ *Nm*
Duration of each rolling period $= 10$ *sec.*

Determine inertia of the flywheel required in the above case to limit motor torque to twice its full-load torque. Neglect no-load losses and assume that the rolling mill torque falls to zero between each rolling period. Assume motor slip proportional to full-load torque. **(Nagpur University)**

Solution. *Given :* Power of the motor, $P = 500 \times 735.5 = 367350$ W ; $N_0 = 40$ r.p.m.

i.e., $\omega_0 = \frac{2\pi \times 40}{60} = 4.189$ rad./s ; $s = 12\%$; $T_{load} = 407115$; Duration of each rolling period, $t = 10$ s, $T_{motor} = 2 \times T_{fl}$

Inertia of flywheel J :

Now, full load torque, $T_{fl} = \frac{P}{\omega} = \frac{P}{(1-s)\,\omega_0} = \frac{367750}{(1-0.12) \times 4.189} = 99761$ Nm

$\therefore \quad T_{motor} = 2 \times T_{fl} = 2 \times 99761 = 199522$ Nm

$$s = (\omega_0 - \omega) = \frac{2\pi}{60}(N_0 - N) = \frac{2\pi}{60}(40 - 0.88 \times 40) = 0.503 \text{ rad/s}$$

Also, $\quad s = KT_{fl}$

$$0.503 = K \times 99761 \quad \therefore \quad K = 5.04 \times 10^{-6}$$

We know that, $\quad T_{motor} = T_{load} - (T_{load} - T_0)\, e^{-t/KJ}$ ...(Eqn. 8.18)

Since $\quad T_0 = 0,$

$\therefore \quad T_{motor} = T_{load}\,(1 - e^{-t/KJ})$

$$199522 = 407115\,[1 - e^{-10/(5.04 \times 10^{-6} \times J)}]$$

or $$0.49 = 1 - e^{-10^7/(5.04J)} \quad \text{or} \quad e^{-10^7/(5.05\,J)} = 0.51$$

or $$e^{10^7/(5.04J)} = \frac{1}{0.51} = 1.96$$

$\therefore$ $$\frac{10^7}{5.04\,J} = \log_e 1.96 = 0.673$$

or $$J = \frac{10^7}{5.04 \times 0.673} = \mathbf{2.948 \times 10^6 \ kg\text{-}m^2.} \quad \textbf{(Ans.)}$$

Example 8.12. *A 6-pole, 50 Hz, 3-phase wound rotor induction motor has a flywheel coupled to its shaft. The total moment of inertia is 900 kg-m². Load torque is 900 Nm for 10 seconds followed by a no-load period which is long enough for the motor to reach its no-load speed. Motor has a slip of 5 percent at a torque of 450 Nm. Calculate :*

(i) Maximum torque developed by the motor

(ii) Speed at the end of deceleration period. **(Nagpur University)**

Solution. *Given :* $p = 6$; $f = 50$ Hz ; $J = 900$ kg-m² ; $T_{load} = 900$ Nm

Slip = 5% at torque of 450 Nm (T_{motor}).

(*i*) **Maximum torque developed by the motor :**

$$N_s = \frac{120\,f}{p} = \frac{120 \times 50}{6} = 1000 \text{ r.p.m} \ ; \ s = \frac{2\pi}{60}(1000 \times 0.05) = 5.236 \text{ rad/s}$$

Now, $\quad s = KT_{motor}$

or $\quad 5.236 = K \times 450 \quad \therefore \quad K = 0.0116$

$$(T_{motor})_{max} = T_{load}\,(1 - e^{-t/KJ}) \qquad ...(T_0 = 0)$$

$$= 900\,[1 - e^{-10/(0.0116 \times 900)}] = \mathbf{554.6\ Nm.} \quad \textbf{(Ans.)}$$

(*ii*) **Speed at the end of deceleration period ; N**

$$s = KT_{fl}$$

$$\frac{2\pi}{60}\,(1000 - N) = 0.0116 \times 554.6$$

$$\therefore \qquad N = \mathbf{938.6\ r.p.m.} \quad \textbf{(Ans.)}$$

Example 8.13. *A 6-pole, 50 Hz induction motor has a flywheel of 1200 kg-m² as moment of inertia. Load torque is 980 Nm for 10 seconds. No-load period is long enough for the flywheel to regain its full speed. Motor has a slip of 6 percent at a torque of 490 Nm. Calculate :*

(*i*) *Maximum torque exerted by the motor.*

(*ii*) *Speed at the end of deceleration period.* **(Roorkee University)**

Solution. *Given :* $p = 6$, $f = 50$ Hz, $J = 1200$ kg-m², $T_{load} = 980$ Nm, $t = 10$ s ; $s = 6\%$; $T = 490$ Nm

(*i*) **Maximum torque exerted by the motor, T_m :**

$$T_{motor} = T_{load} - (T_{load} - T_0)\,e^{-t/KJ}$$

Assuming, $T_0 = 0$, we have

$$T_{motor} = T_{load}\,(1 - e^{-t/KJ})$$

$$N_s = \frac{120\,f}{p} = \frac{120 \times 50}{6} = 1000 \text{ r.p.m.} = N_0$$

$$s = (\omega_0 - \omega) = \frac{2\pi}{60}(N_0 - N) = \frac{2\pi}{60}\,(1000 - 940) = 6.283 \text{ rad/s}$$

$$[\because \quad N = N_s(1 - s) = 1000\,(1 - 0.06) = 940 \text{ r.p.m.}]$$

Now, $$K = \frac{s}{T} = \frac{6.283}{490} = 0.0128$$

$$\therefore \qquad T_{motor} = 980\,[1 - e^{-10/(0.0128\,J)}] = 980\,[1 - e^{-10/(0.0128 \times 1200)}]$$

$$= 980\,(1 - e^{-0.651}) = 468.9 \text{ Nm}$$

i.e., $$T_{motor} = \mathbf{468.9\ Nm.} \quad \textbf{(Ans.)}$$

(*ii*) **Speed at the end of deceleration period :**

Slip speed, $$s' = KT_{motor}$$

$$= 0.0128 \times 468.9 = 6.0 \text{ rad/s}$$

$$= \frac{6 \times 60}{2\pi} = 57.3 \text{ r.p.m.}$$

$\therefore$ *Actual speed* = 1000 – 57.3 = **942.7 r.p.m.** **(Ans.)**

Example 8.14. *A 3-phase, 45 kW, 6-pole, 960 r.p.m. induction motor has a constant load torque of 270 Nm and at wide intervals additional torque of 1350 Nm for 10 seconds. Calculate :*

(*i*) *The moment of inertia of the flywheel used for load equalization, if motor torque is not to exceed twice the rated torque.*

(*ii*) *Time taken after removal of additional load, before the motor torque becomes 630 Nm.*

Solution. *Given :* $P = 45$ kW ; $p = 6$; $N = 960$ r.p.m ; $T_{load} = 270 + 1350 = 1620$ Nm ;

$$T_{motor} = 2T_{fl}$$

(*i*) **Moment of inertia of the flywheel, J :**

$$P = T \times \omega \qquad \therefore \quad T = \frac{P}{\omega}$$

$$\therefore \qquad T_{fl} = \frac{45 \times 10^3}{\left(\frac{2\pi \times 960}{60}\right)} = 447.6 \text{ Nm}$$

and, $$T_{motor} = 2T_{fl} = 2 \times 447.6 = 895.2 \text{ Nm}$$

$$N_s = \frac{120 f}{p} = \frac{120 \times 50}{6} = 1000 \text{ r.p.m.}$$

$\therefore$ Full load slip = 1000 – 960 = 40 r.p.m. = 4%

Also, $$s = KT_{fl}$$

$$\frac{2\pi \times 40}{60} = K \times 447.6 \qquad \therefore \quad K = 0.00936$$

$$T_{motor} = T_{load} - (T_{load} - T_0)\, e^{-t/KJ}$$

$$895.2 = 1620 - (1620 - 270)\, e^{-10/(0.00936\,J)} = 1620 - 1350 \times e^{(-1068.4/J)}$$

or, $$e^{-(1068.4)} = 0.537 \quad \text{or} \quad \frac{1068.4}{J} = \log_e \left(\frac{1}{0.537}\right) = 0.622$$

$\therefore$ J = **1717.7 kg-m². (Ans.)**

(*ii*) **Time taken after removal of additional load :**

$$T_{motor} = T_0 + (T'_{motor} - T_0)\; e^{-t/KJ}$$

$$630 = 270 + (895.2 - 270)\, e^{-t/(0.00936 \times 1717.7)}$$

$$(630 - 270) = 625.2 \times e^{-t/16.08}$$

or, $$e^{t/16.08} = 1.737$$

$\therefore$ $t = 16.08 \log_e (1.737) =$ **8.88 s. (Ans.)**

Example 8.15. *An induction motor equipped with a flywheel is driving a rolling mill which requires a load torque of 1900 Nm for 10 sec. followed by 280 Nm for 30 sec. This cycle being repeated indefinitely. The synchronous speed of motor is 750 r.p.m. and it has a slip of 10 percent when delivering 1400 Nm torque. The total moment of inertia of the flywheel and other rotating parts is 2100 kg-m². Draw the curve showing the torque exerted by the motor and the speed for five complete cycles, assuming that initial torque is zero.* **(Nagpur University Summer, 2001)**

Solution. *Given :* T_{load} = 1900 Nm for 10 s ; T_{load} = 280 Nm for 30 s ; N_s = 750 r.p.m. ; slip = 10% at 1400 Nm torque (T_{motor}) ; J = 2100 kg-m².

$$s = 0.1 \times 750 = 75 \text{ r.p.m.} = \frac{75 \times 2\pi}{60} = 7.85 \text{ rad/s}$$

Also, $$s = K\, T_{motor}$$

$$\therefore \qquad K = \frac{s}{T_{motor}} = \frac{7.85}{1400} = 0.0056$$

(*i*) **During 1st cycle**

(*a*) *Flywheel decelerating (on-load period)*

$$T_{motor} = T_{load} - (T_{load} - T_0)\, e^{-t/KJ}$$

$$= 1900 - (1900 - 0)\, e^{-10/(0.0056 \times 2100)} \qquad \text{...assuming } T_0 = 0$$

$$= 1088 \text{ Nm}$$

$$\text{Slip} = 0.0056 \times 1088 = 6.09 \text{ rad/s} = \frac{6.09 \times 60}{2\pi} = 58 \text{ r.p.m.}$$

$\therefore$ Speed = 750 – 58 = 692 r.p.m.

(*b*) *Flywheel accelerating (off-load period)*

$$T_{motor} = T_0 + (T'_{motor} - T_0)\, e^{-t/KJ}$$

$$T_0 = \text{no load torque} = 280 \text{ Nm}$$

T'_{motor} = 1088 Nm (T_{motor} at the beginning of the period *i.e.*, the motor torque at the instant when the load is removed)

After 30 sec., $T_{motor} = 280 + (1088 - 280)\, e^{-30/(0.0056 \times 2100)}$

$$= 280 + (1088 - 280) \times 0.078 = 343 \text{ Nm}$$

$\therefore$ Slip at this T_{motor} = 0.0056 × 343 = 1.92 rad/s = 18.33 r.p.m.

$\therefore$ Speed = 750 – 18.33 $\simeq$ 731 r.p.m.

(*ii*) **During 2nd cycle :**

(*a*) *Flywheel decelerating :*

$$T_0 = 343 \text{ Nm}$$

$$T_{motor} = 1900 - (1900 - 343)\, e^{-10/(.0056 \times 2100)} = 1235 \text{ Nm}$$

$$\text{Slip at this } T_{motor} = 0.0056 \times 1235 = 6.92 \text{ rad/s} = \frac{6.92 \times 60}{2\pi} = 66 \text{ r.p.m.}$$

$\therefore$ Speed = 750 – 66 = 684 r.p.m.

(*b*) *Flywheel accelerating (off-load period) :*

$$T_{motor} = 280 + (1235 - 280) \times e^{-30/(0.0056 \times 2100)} = 354.5 \text{ Nm}$$

$$\therefore \quad \text{Slip at this } T_{motor} = 0.0056 \times 354.5 = 1.99 \text{ rad/s} = \frac{1.99 \times 60}{2\pi} = 19 \text{ r.p.m.}$$

Speed = 750 – 19 = 731 r.p.m.

(*iii*) **During 3rd cycle**

(*a*) *Flywheel decelerating*

$$T_0 = 354.6 \text{ Nm}$$

$$T_{motor} = 1900 - (1900 - 354.6)\, e^{-10/(0.0056 \times 2100)} = 1239 \text{ Nm}$$

$$\therefore \quad \text{Slip at this } T_{motor} = 0.0056 \times 1239 = 6.94 \text{ rad/s} = \frac{6.94 \times 60}{24} = 66 \text{ r.p.m.}$$

Speed = 750 – 66 = 684 r.p.m.

(*b*) *Flywheel accelerating (off-load period) :*

$$T_{motor} = 280 + (1239 - 280)\, e^{-30/(0.0056 \times 2100)} = 354.8 \text{ Nm}$$

$$\text{Slip at this } T_{motor} = 0.0056 \times 354.8 = 1.99 \text{ rad/s} = \frac{1.99 \times 60}{2\pi} = 19 \text{ r.p.m.}$$

Speed = 750 – 19 = 731 r.p.m.

Initial condition at the beginning of the 3rd peak load are thus practically the same as that at the beginning of the 2nd. Therefore motor torque in this and all succeeding load cycles will follow a similar curve to that in second period (Fig. 8.7).

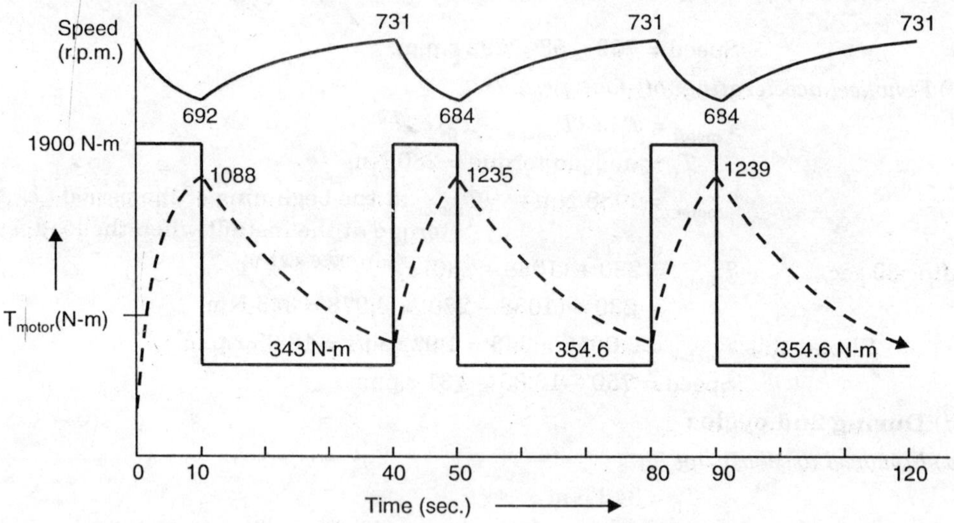

Fig. 8.7

9. ENCLOSURES FOR ROTATING ELECTRICAL MACHINES

The problem of ventilation in rotating electrical machines is closely linked with the types of enclosures used. The various types of enclosures used are :

1. Open machine. One in which there is no restriction to ventilation other than that necessitated by good mechanical construction.

2. Open pedestal machine (OP). An open machine which has pedestal bearing supported independently of the machine frame.

3. Open end-bracket machine (OEB). An open machine having end brackets of which the bearings form an integral part.

4. Protected machine (P). A machine in which the internal rotating parts and line parts are protected mechanically from accidental or inadvertent contact, while ventilation is not materially impeded.

5. Screen-protected machine (SP). A protected machine in which the ventilating openings are not less than 64.5 mm^2 in area. Such protection may be provided by screens of wire mesh, expanded metal, perforated metal or other suitable covers. The use of openings smaller than 64.5 mm^2 is not recognized, as such openings are liable to become closed in service.

6. Drip proof machine (DP). A protected machine in which the opening for ventilation are so protected as to exclude vertically falling water or dist.

7. Splash-proof machine (SPLP). A protected machine in which the ventilating openings are so constructed that drops of liquid or solid particles falling on or reaching any part of the machine at any angle between the vertically downward direction and 100° from that direction cannot enter the machine, whether the machine is running or at rest, by splashing, or otherwise, either directly or by striking and running along a surface.

8. Hose-proof machine (HSP). A protected machine so enclosed as to exclude water whether the machine is running or at rest, when washed by a hose having 9.5 mm diameter nozzle with a maximum pressure of 3.5 bar for a period not exceeding 30 seconds, from a minimum distance of 1.8 metres.

9. Pipe-ventilated or Duct-ventilated machine. A machine in which there is a continuous supply or fresh ventilating air, the frame being so arranged that the ventilating air may be conveyed to and/or from the machine through pipes or ducts attached to the enclosing case.

10. Totally enclosed machine (TEM). A machine so constructed that the enclosed air has no connection with the external air but is not necessarily 'air tight'.

11. Totally enclosed Fan-cooled machine (TEFC). A totally enclosed machine with cooling augmented by a fan, driven by the motor itself, blowing external air over the cooling surface and/or through the cooling passages, if any, incorporated in the machine.

12. Totally enclosed separately air-cooled machine (TESAC). A totally enclosed machine with cooling augmented by a separately-driven fan blowing external air over the cooling surfaces and/or through the cooling passages, if any, incorporated in the machine.

13. Totally enclosed water or other liquid-cooled machine (TEWC). A totally enclosed machine with cooling augmented by water-cooled or other liquid-cooled surfaces embodied in the machine itself.

14. A totally enclosed closed air circuit machine. A totally enclosed machine having special provision for cooling the enclosed air by passing it through its own cooler, usually external to machine.

15. Totally enclosed closed gas circuit machine. A totally enclosed machine cooled by gas other than air, the cooling gas being circulated through associated water-cooled gas coolers.

16. Weath-proof machine (WP). A machine so constructed that it can work without further protection from weather conditions specified by the purchaser.

17. Water tight machine. That it will withstand without damage or sign of leakage, complete immersion in water to a depth of not less than 1 m, or subjection to an external water pressure of 0.1 bar for a period of one hour. The test for water-tightness shall be made with the machine stationary and the temperature of the machine shall not exceed the temperature of the water in which it is immersed.

18. Submersible machine. A machine capable of working for an indefinitely long period when submerged under a specified head of water.

19. Flame proof machine (FPM). A machine which complies with the requirements of IS : 2148-1962 "Specifications for flame-proof enclosures of electrical apparatus".

10. SELECTION OF ELECTRIC MOTOR FOR ANY APPLICATION

The selection of an electric motor for any application depends on the following factors :

1. Electrical Characteristics

(*i*) Running characteristics
(*ii*) Starting characteristics
(*iii*) Speed control
(*iv*) Braking characteristics

II. Mechanical Characteristics

(*i*) Power transmission
(*ii*) Cooling
(*iii*) Noise
(*iv*) Type of enclosure

III. Size and Rating of motors

(*i*) Load cycle

(*a*) Continuous

(*b*) Intermittent

(*c*) Variable

(*ii*) Overload capacity

IV. Cost

(*i*) Initial cost
(*ii*) Cost of control gear
(*iii*) Running cost

Characteristics of electric motors is given in Table 10.1.

Table 10.1. Characteristics of Electric Motors

Motors	*Methods of starting*	*Methods of speed control*	*Voltage limit*	*H.P. limit*	I_{st}/I_{fl}	T_{st}/T_{fl}
D.C. Motors						
1. *Shunt*	Series resistance in armature	(*i*) Variable resistance in field circuit (*ii*) Variable resistance in armature circuit	3 kV	25000	2	2
2. *Series*	(*i*) Series resistance (*ii*) Series-parallel method	(*i*) By tapping the field (*ii*) By field diverter (*iii*) Variable resistance in series (*iv*) Series-parallel control	1.5 kV	3000	2	3
3. *Compound*	Series resistance in armature	(*i*) Variable resistance in shunt field (*ii*) Series field diverter	1.5 kV	3000	2	2-3

Motors	*Methods of starting*	*Methods of speed control*	*Voltage limit*	*H.P. limit*	I_{st}/I_{fl}	T_{st}/T_{fl}
A.C. Motors						
1. *Single phase induction motor*	(*i*) Repulsion start (*ii*) Pole shading (*iii*) Phase splitting by *C*, *L* or *C* and *R*	Voltage drop in series impedance	250 V	1	2	1.5-3
2. *Single phase series motor*	Variable voltage supply	Voltage variation	500 V	3000	2	3
3. *Squirrel cage induction motor*	(*i*) Star delta starter (*ii*) Auto-transformer (*iii*) Resistance in stator circuit	(*i*) Pole changing (*ii*) Variation of frequency	11 kV	300	1	1
4. *Slip ring induction motor*	Resistance in rotor circuit	(*i*) Variation in applied voltage (*ii*) Cascading (*iii*) Pole changing (*iv*) Variation of frequency (*v*) Motor reheostat control (*vi*) Injecting e.m.f. in rotor	11 kV	11000	2	2
5. *Pole changing induction motor*	(*i*) Direct switching (*ii*) Start-delta starter (*iii*) Auto-transformer	(*i*) Pole changing (*ii*) Variation of frequency	11 kV	300	6	1
6. *Schrage motor*	(*i*) Low voltage application (*ii*) Direct switching	(*i*) Shifting the brushes (*ii*) Inserting impedance in secondary winding	1 kV	1000	2	2.6
7. *Synchronous motor*	(*i*) Damper winding (*ii*) Pony brake	Fixed speed	15 kV	10000	2	3

11. MOTORS FOR PARTICULAR APPLICATIONS

For a particular application, the type of electric drive and control gear are determined by the following considerations :

1. Starting torque
2. Duty (whether light, medium, heavy)
3. Limitations on starting current
4. Speed control range and its nature
5. Need for automatic control
6. Conditions of environment.

S. No.	Applications	Type of motor(s) used
1.	***Domestic uses.*** *(e.g. Vacuum cleaners, sewing machines, mixy, electric shavers, fans, cloth-washing machines, refrigerators etc.)*	• *Small universal motors, usually series type.*
2.	***Cranes***	• *D.C. series or compound motors* (on account of their high starting torque and smooth speed control)...... for cranes — The motors used for lifting, travelling and reversing, and for conveying and hoisting are D.C. compound wound, D.C. series wound and A.C. slip ring induction motors respectively.
3.	***Lifts***	• *D.C. compound wound and A.C. slip-ring induction motors.* • Induction repulsion and variable speed A.C. commutator motors are *also* used.
4.	***Lathes, milling and grinding machines***	• *D.C. shunt or squirrel cage induction motors.*
5.	***Planers***	• *Shunt or compound wound motors* (for D.C. supply) • *Slip-ring induction motor* (for A.C. supply).
6.	***Punches, presses and shears***	• *High slip squirrel-cage motors, slip ring induction motors or D.C. cumulative compound motors.*
7.	***Shapers and slotters***	• *Constant speed squirrel-cage induction motors.*
8.	***Drilling machines***	• *Constant speed squirrel-cage or D.C. shunt or multi-speed A.C. motors.*
9.	***Blowers and fans***	• *Squirrel cage induction motors* • *Synchronous motors.*
10.	***Air compressors***	• *Squirrel cage inductions motors*—for small compressors only • *Slip-ring induction motors and synchronous motors*—for large units.

11.	***Electric traction***	• *D.C. series motors*—for all types of services but more particularly for suburban services where high rate of acceleration is essential. • *Single-phase A.C. compensated series motors*—extensively used for main-line work. Due to poor starting these motors are *not* suitable for suburban series where services are frequent.
12.	***Hoist work***	• *3-phase slip-ring induction motors.*
13.	***Pumps***	• *Polyphase induction motors* (Drip proof or totally enclosed surface cooled). — *Squirrel-cage motors*—for centrifugal pumps where T_{st} = 40 to 50% T_{fl} *is required.* — *Slip-ring motors*—for reciprocating pumps where T_{st} = 100 to 200% T_{fl} is required.
14.	***Refrigeration and air-conditioning***	• *1-phase, 230 V, capacitor type induction motor* with D.O.L. starters—for small units. • *Synchronous motor*, driving turbo-compressor may suit especially if power factor correction is also required.
15.	***Machine tools***	• *Squirrel-cage motors.*
16.	***Belt conveyors***	• *Double-cage induction motor* with D.O.L. or *wound-rotor induction motor* — These motors are totally enclosed, surface cooled type as grit and dust remain present in the atmosphere.
17.	***Flour mills***	• *Pipe ventilated or totally enclosed type motors.*
18.	***Mines***	• *Flame proof squirrel-cage induction motor* • *Slip ring induction motor* • *D.C. motor* • *D.C. cumulative motor.*
19.	***Rolling mill drives***	• *D.C. motors* • *A.C. motors with speed control.*
20.	***Kiln drives***	• *Slip-ring induction motor* • *Three-phased shunt wound commutator motor* • *Cascade-controlled A.C. motor* • *Ward Leonard control D.C. motor.*
21.	***Paper industry***	• *Squirrel-cage motor.*
22.	***Iron and steel industry***	• *D.C. shunt wound motor* with stabilizing, compensating and interpole windings.

23.	*Textile industry*	• *Three-phase A.C. motor.*
24.	*Drives for numerical control of machine tools*	• Numerical control is used to keep a part or tool under continuous specific varying control, producing a family of cuttings. Numerical control enables a cutter through a programmed path in the work to give a specific shape of accurate dimensions. • Programmed controlled drives are used to get a command signal from punched card, punched tape or magnetic tape and feed the information to a digital computer which controls the feed drives. Accurate positioning and position measurement is carried out by using *transducers* which provide digital position data by a current of pulses which are compared with the instructions punched on a tape. If there is any change in the two a feed back signal corrects the position and adjusts the drive. D.C. motors are used for numerical control (NC) positioning and they are controlled by SCR in electronic amplifier circuits. These days stepper motors are used to move a motor by pulses through instrumental angular changes to drive a tool or slide through gearing. The use of numerical controlled machines has greatly helped in getting great precision in machine tool works.

Motors of Different Industrial Drives :

Type of motor	Characteristics	Fields of application
1. *D.C. shunt motor*	— Medium starting torque. — Nearly constant speed.	• Constant-speed line shafts • Lathes • Vacuum cleaners • Wood-working machines • Laundry washing machines • Elevators • Conveyors • Small printing presses • Grinders etc.
2. *D.C. series motor*	— High starting torque. — Variable speed.	• Electric locomotives • Steel rolling mills • Lifts and cranes • Hoists.

3. ***D.C. comulative compound motor***	— High starting torque. — Varying speed motor.	• Rotary presses • Compressors • Circular saws • Shearing machines • Elevators • Continuous conveyors • Variable-head centrifugal pumps etc.
4. ***Three-phase synchronous motor***	— Speed remains constant. under varying loads.	• Ammonia and air compressors • Motor-generator sets • Continuous rolling mills • Paper and cement industries.
5. ***Squirrel-cage induction motor***	— Simple and rugged in construction. — High over-load capacity. — Used for low and medium power drives where speed control is not required.	• Water pumps • Tube wells • Lathes • Drills • Grinders • Polishers • Fans and blowers • Laundry washing machines • Compressors • Wood planers.
6. ***Double squirrel-cage motor***	— High starting torque. — Large overload capacity. — Nearly constant speed. — Used for driving loads which require high starting torque.	• Compressor pumps • Reciprocating pumps • Large refrigerators • Crushers • Textile machines • Lathes • Boring mills • Punches etc.
7. ***Slip-ring induction motor***	— High starting torque. — Large overload capacity. — Its speed can be changed upto 50% of its normal speed. — Used for those drives which require high starting torque and speed control.	• Lifts • Pumps • Winding machines • Line shafts • Printing presses • Compressors • Elevators etc.
8. ***Single-phase synchronous motor***	— Constant speed.	• Teleprinters • Recording instruments • All kinds of timing devices • Clocks etc.
9. ***Single-phase series motor***	— High starting torque. — Speed can be controlled over a wide range.	• Refrigerators • Vacuum cleaners etc.
10. ***Repulsion motor***	— High starting torque. — Wide speed control. — High speed at high loads.	• Coil winding machines.

11. *Capacitor-start induction-run motor*	— Fairly constant speed. — Moderately high starting torque. — Speed control not possible.	• Compressors • Refrigerators • Small portable hoists.
12. *Capacitor-start-and-run motor*	— Operating characteristics similar to above motor except that it has better power factor and higher efficiency.	• Drives requiring quiet operations.

12. ENERGY CONSERVATION IN ELECTRICAL DRIVES

- For keeping the manufacturing cost of a product minimum, and make its price competitive in the market, it is necessary to *minimise energy conservation* at all stages of manufacturing process *including electrical drives.*
- Conservation of energy is also necessary because with the ever increasing demand, need for electrical power can only be met by conserving electrical power in addition to installation of new generating units. A major portion of electrical power in a plant is consumed by electrical drives.

A large amount of electrical energy can be saved by the use of efficient and right-type of electrical drives.

Energy conservation in electrical drive is achieved by *reduction of following losses* in its various parts :

(*i*) *Electrical transmission losses*...These losses depend on the drive power factor and harmonics in the line current.

(*ii*) *Electrical motor losses to convert electrical power into mechanical power*...These are determined by choice of motor and quality of supply.

(*iii*) *Conversion losses in the power modulator or converter.*

(*iv*) *Mechanical losses in the parts of transmission system such as bearings, gears, clutches and belts.*

(*v*) *Losses in the load.*

(*vi*) *Losses caused by throttling* or *by other means that control material flow by absorbing or bypassing excess output.*

(*vii*) *Mechanical transmission losses.*

Energy conservation in electrical drives can be materialised by the following *measures* :

- Use of efficient motors.
- Use of efficient semiconductor converters.
- Use of variable speed drives.
- Using a motor of right rating.
- Improvement of power factor.
- Use of single-to-three-phase semiconductor converters in rural applications.
- Regular and preventive maintenance of motors, transformers and coupled equipment.
- Improvement of quality of supply.

Section:

SHORT ANSWER QUESTIONS

SHORT ANSWER QUESTIONS

PART I: D.C. MACHINES

D.C. Generators

1. *What is the main advantage of electric energy over other forms of energy?*

Ans. The main advantage is that electric energy can be transmitted over long distances with *ease and high efficiency*. Its main use is in the form of a transmitting link for transporting other forms of energy *e.g.*, mechanical, sound, light etc. from one physical location to another.

2. *What is an 'Electromechanical energy conversion device'?*

Ans. It is one which converts electrical energy into mechanical energy and mechanical energy into electrical energy. Electromechanical energy conversion takes place *via* the medium of a magnetic or electric field, *the magnetic field being most suited for practical conversion devices.*

3. *What is energy balance equation?*

Ans. For converting energy from one form to another the *'principle of conservation of energy'* (which states that energy can neither be created nor destroyed, it can merely be converted from one form to another) can be involved. In an energy conversion device, out of total input energy, a large part of energy is converted into useful output energy, some energy is stored and rest is converted into heat (called energy loss).

The *"energy balance equation"*, in general, may be written as:

$$\int dW_{\text{input}} = \int dW_{\text{output}} + \int dW_{\text{magnetic field}} + \int dW_{\text{heat}}$$

4. *The electromechanical energy conversion devices with magnetic field as the coupling medium between electrical and mechanical systems are more common than the devices with electric field. Why?*

Ans. It is so, because *energy storing capacity of magnetic field is much larger* (about 25000 times) than that of electric field.

5. *Why energy storing capacity of magnetic field is much larger than that of electric field?*

Ans. The energy storing capacity of magnetic field is much larger because the *value of permittivity of free space is very small compared to permeability of the space.*

6. *What is an 'Electrical generator'?*

Ans. An ***Electrical generator*** *is a machine which converts mechanical energy (or power) into electrical energy (or power).*

This energy conversion is based on the principle of the production of dynamically induced e.m.f.

7. *What is 'Commutator'?*

Ans. It is a mechanical switching device (actuated by the mechanical rotation of the dynamo/generator shaft) which converts the alternating voltage (A.C.) to *unidirection current* (D.C.).

8. *What is 'Yoke'?*

Ans. The ring-shaped portion which serves as the path for the main and commutating pole fluxes is called the ***Yoke.***

9. *What is a commutating pole?*

Ans. • A commutating pole (also called *interpole*) is similar to a main pole and consists of core terminating in a pole shoe, which may have various shapes, and coil mounted on the core.

• The commutating poles *are arranged strictly midway betweeen the main poles* and are bolted to the yoke.

• *Commutating poles are usually made of solid steel, but for machines operating on sharply varying loads they are made of sheet steel.*

10. *What are the major requirements of an armature winding?*

Ans. The *armature winding* is very important element of a machine, as it directly takes part in the conversion of energy from one form into another. The requirements which a winding must meet are diverse and often of a conflicting nature. Among these requirements the following are of major importance.

• The winding must be designed with the *most advantageous utilisation of the material in respect to weight and efficiency.*

• The winding should *provide the necessary mechanical, thermal and electrical strength of the machine* to ensure the usual service life of 16–20 years.

• For D.C. machines proper current collection at the commutator (*i.e.,* absence of detrimental sparking) must be ensured.

11. *What is the difference between lap and wave windings?*

Ans. The winding may be a lap winding or a wave winding. *The difference between the two consists in the arrangement of the end connections at the front of the armature.*

• *In the* '**lap winding**' *the finish of each coil is connected to the start of the next coil so that winding or commutator pitch is unity.* The lap winding may be progressive, *i.e.,* it progresses in the direction in which the coils are wound or retrogressive (*i.e.,* it progresses in the opposite direction).

• *In* '**wave winding**' *the finish of a coil is connected to the start of another coil well away from the first coil.*

• The *advantage of wave winding* is that, *for a given number of poles and armature conductors, it gives more e.m.f. than the lap winding.* Conversely, for the same e.m.f., lap winding would require large number of conductors which will result in higher winding cost and less efficient utilisation of space in the armature in slots. Hence, wave winding is suitable for small generators especially these meant for 500–600 V circuits. Another *advantage is that in wave winding, equalizing connections are not necessary whereas in lap winding they definitely are.*

• *Lap winding is a necessity when large currents are required, because it gives more parallel paths.*

• Hence, *'lap winding' is suitable for comparatively low-voltage but high current generators whereas 'wave-winding' is used for high voltage, low current machines.*

12. *What are 'Dummy coils'?*

Ans. ***Dummy Coils:***

• These coils are also called *idle coils* and are resorted to when the requirements of the windings are not met by the standard armature punchings available in armature-winding shops.

- *These coils do not influence the electrical characteristics of the winding because they are not connected to the commutator.*
- *These coils are exactly similar to the other coils except that their ends are cut short and tapped.*
- *These are used simply to provide mechanical balance for the armature,* because the armature having some slots without windings would be out of balance mechanically.

13. *In what terms are D.C. machines rated?*

Ans. The D.C. machines are rated in terms of kW output at a given speed and voltage.

14. Why is it necessary to use a large number of coils and commutator segments with the coils evenly distributed around the surface of the armature of a D.C. machine.

Ans. *To have practically constant unidirectional voltage at the brushes.*

15. In large machines fabricated steel yokes are used. Why?

Ans. Because of high permeability, thereby resulting in reduced weight.

16. *Why are field coils provided in a D.C. machine?*

Ans. In a D.C. machine field coils are provided to given under various conditions of operation, the number of ampere-turns (AT) of excitation required to give proper flux through the armature to induce e.m.f.

17. *The pole shoe section of a D.C. machine is made larger than its body. Why?*

Ans. The pole shoe acts as a support to the field coils and spreads out the flux over the armature periphery more uniformly and due to large cross-sectional area reduces the reluctance of the magnetic path.

18. *Name the best suited material for commutator segments.*

Ans. The best suited material is silvered copper (extruded copper) containing about 0.05% silver.

19. *Why are brushes staggered?*

Ans. To prevent ridge formation on the surface of the commutator.

20. *In lap wound D.C. machines equilizer rings are provided. Why?*

Ans. To avoid unequal distribution of current at the brushes thereby assist in achieving sparkless commutation.

21. *What are the causes of bad commutation?*

Ans. Poor or bad commutation means that there is sparking at the brushes and commutator surface. It may be caused by mechanical or electrical condition.

- The *mechanical conditions* include uneven commutator surface, non-uniform brush pressure, vibration of brushes in the brush holders etc.
- The *electrical conditions* include an increase in voltage between the commutator segments, an increase in current density at the trailing edge of the brush etc.

22. *What do you understand by self-excitation mode of D.C. machine? Name two D.C. machines working in this mode.*

Ans. The self-excitation mode of D.C. machine is the one in which field winding(s) is/are self-excited by the *current supplied by the machine itself.* In such machines the field coils are interconnected with the armature winding.

The self-excited machine may be:

(*i*) D.C. series machine;

(*ii*) D.C. shunt machine;

(*iii*) D.C. compound wound machine.

23. *Why is it necessary to wind field coils of a D.C. shunt generator with a large number of turns of fine wire?*

Ans. Field coils are wind to have high resistance in order to keep shunt field current as low as possible, as well as to provide required m.m.f.

24. *Why are commutating poles provided in the construction of a large D.C. machine?*

Ans. In a large D.C. machine armature reaction effect becomes very large and the commutation problem increases due to large armature currents. Commutating poles are, therefore, provided to overcome these problems.

25. *It is not advisable to fully load a D.C. shunt generator until its voltage builds up to the rated value. Why?*

Ans. In a D.C. shunt generator the building up process is a step-by-step process which means that initially when the machine is started, generated voltage is small, circulating small current. In case the generator is loaded fully at this stage, the magnitude of excitation current will become very small and hence resulting into the rate of build up becoming very small. Thus the time taken for the voltage to build up to its rated value will become large. Hence, it is advisable not to fully load the D.C. shunt generator during the process of build up of voltage.

26. *Explain the reason that why a D.C. shunt generator can operate under short-circuit condition but series and cumulative compound generators cannot.*

Ans. In a *'D.C. shunt generator'* the field is connected across the armature and hence under short-circuit condition, *some voltage exists* across the armature because of residual flux and this keeps the shunt generator to function at low voltage. In case of *'series and cumulative component generators'*, the series field under the short-circuit condition carries excessive current and hence terminal voltage starts decreasing due to excessive demagnetising effects of armature reactions; thus the terminal voltage *finally reduces to zero* and, thus, such machines cannot operate under short-circuit condition.

27. *The D.C. shunt generators are most suited for parallel operation. Why?*

Ans. Because of their *slightly drooping voltage characteristics.* This satisfactory operation is due to the fact that any tendency on the part of a D.C. shunt generator to take more or less than the proper share of load results in certain changes of the system voltage which immediately opposes this tendency.

28. *What is the simplest way to shift load from one D.C. shunt generator running in parallel to the other D.C. shunt generator?*

Ans. To reduce the excitation of one and to increase the excitation of the other D.C. shunt generator on which load is to be shifted.

29. *The terminal voltage of a D.C. generator at full load is more than its open circuit voltage. What is the mode of excitation of the generator?*

Ans. The mode of excitation of the generator is 'cumulative compound'.

30. *What is the critical resistance of the field circuit of a D.C. shunt generator?*

Ans. The field circuit's critical resistance of a D.C. shunt generator is the maximum value of the resistance with which the generator will be able to *excite*.

31. *In what respect the separately excited D.C. generator is superior to self excited D.C. generator?*

Ans. The separately excited generator has the *advantage of stable operation with any field excitation* over self-excited generators.

32. *In which type of D.C. generator no load and full load voltages are same?*

Ans. Flat-compounded D.C. generator.

33. *What is linear commutation?*

Ans. It is an ideal commutation which causes change of current in a coil undergoing commutation from + I_C to – I_C uniformly.

34. *What is reactance voltage?*

Ans. Reactance voltage is the self induced e.m.f. in the coil undergoing commutation.

35. *What is 'armature reaction'?*

Ans. The action of the armature m.m.f. on the main m.m.f. is termed the armature reaction.

36. *What are the advantages of carbon brushes?*

Ans. (*i*) They have high contact resistance which is useful for having good commutation.

(*ii*) They lubricate and polish the commutator as it rotates.

(*iii*) In the event of sparking, carbon brushes will damage the commutator less in comparison to copper brushes.

37. *What are the reasons for failure of self-excited shunt generator to build up voltage?*

Ans. (*i*) No residual magnetism;

(*ii*) Field connection reversed;

(*iii*) Field circuit resistance too high.

38. *What is an over-compound generator?*

Ans. An over-compound generator is one whose *terminal voltage rises with the application of load* so that its full-load voltage *exceeds* its no-load voltage (negative regulation).

39. *What is an useful application of differentially compounded generator?*

Ans. As an *arc welding generator* where the generator is practically short-circuited every time the electrode touches the metal plates to be welded.

40. *What are commutating poles?*

Ans.
- A commutating pole (also called *interpole*) is similar to a main pole and consists of core terminating in a pole shoe which way have various shapes and coil mounted on the core.
- The commutating poles are arranged strictly *midway* between the main poles and are bolted to the yoke.
- These are usually made of solid steel, but for machines operating on sharply varying loads they are made of sheet-steel.

D.C. Motors

41. *What is an electric motor?*

Ans. The ***electric motor*** is a machine which converts electrical energy into mechanical energy.

42. *State the principle of motor action.*

Ans. The principle of motor action can be stated as follows:

"Whenever a current carrying conductor is placed in a magnetic field, it experiences a force whose direction is given by Fleming's left hand rule."

43. *Why is a commutor needed in a D.C. motor?*

Ans. In a D.C. motor it is necessary that the current through the coils of armature winding be reversed as a particular coil leaves one pole (say N-pole), crosses the neutral line and comes under the influence of next pole which is of opposite polarity (*i.e.,* S-pole). The commutator serves this purpose.

44. *The e.m.f. induced in the armature of a D.C. motor is called the back e.m.f. Why?*

Ans. Since the induced e.m.f. in a D.C. motor opposes the voltage applied to the armature of the motor, it is called back or counter e.m.f.

45. *When does a D.C. motor produces maximum output power?*

Ans. When voltage drop in armature is equal to back e.m.f. E_b *i.e.*, when back e.m.f. *equals half the applied voltage.*

46. *Why the brushes in a D.C. motor are given a backward shift?*

Ans. To *reduce sparking* at the commutator. The backward shift of brushes will induce an e.m.f. in the required direction in the short-circuited coils of the armature undergoing commutation and hence improves commutation.

47. *D.C. series motor is a highly variable speed motor, why?*

Ans. In a D.C. series motor, the field winding is connected in series with the armature winding, therefore, field flux varies approximately linearly with load ($I_{sc} = I_a = I_L$) and thus the motor speed N being inversely proportional to field flux, varies. That is why a D.C. series motor is said to be highly variable speed motor.

48. *Why D.C. motor should not be allowed to run without load?*

Ans. The D.C. series motor on no-load would attain tremendously high speed and the motor may get damaged due to heavy centrifugal forces set-up in rotating parts.

49. *The starting torque of a D.C. series motor is more than that of a D.C. shunt motor, why?*

Ans. In a D.C. series motor flux ϕ is directly proportional to armature current I_a (before saturation) and, therefore, torque developed $T \propto I_a^2$ which in case of a D.C. shunt motor torque developed $T \propto I_a$. Obviously on the basis of the same kW output and speed, a D.C. series motor develops the torque more than that of a shunt motor.

50. *Why does a shunt motor run away at heavy loads?*

Ans. Because the field may be *weakened* due to armature reaction at heavy loads, to the extent that the motor runs away.

51. *What will happen if the windings of a D.C. shunt motor gets disconnected while in normal operation?*

Ans. The *flux will drop to practically zero value* and, therefore, the speed will increase to an extremely high value and the *motor may get damaged* due to heavy centrifugal forces.

52. *A D.C. shunt generator is used for battery charging, why?*

Ans. Because a D.C. shunt generator can be made to provide *constant output voltage.*

53. *A differentially compounded D.C. generator is used for welding purposes, why?*

Ans. Because in this type of generator, the terminal *voltage drops very rapidly* with *increase in load current* and, therefore, short-circuit *cannot* cause any damage to it.

54. *What is 'reactance voltage'?*

Ans. The main cause which makes the current reversal impossible in the specified period is the production of self-induced e.m.f. in the coil undergoing commutation. This self-induced e.m.f. is called *reactance voltage.*

55. *Carbon brushes have high contact resistance, is it helpful in having good commutation?*

Ans. Yes, high contact resistance is useful for good commutation.

56. *What is the critical resistance of a D.C. generator?*

Ans. The value of the resistance represented by the tangent to the initial portion of the O.C.C is known as *critical resistance* of a D.C. generator.

57. *Which type of a D.C. motor has a tendency towards speed instability with a possibility of motor running away and strong possibility of motor starting in wrong direction.*

Ans. Differentially compound wound D.C. motor.

58. *Name the various methods by which the speed of a D.C. motor can be regulated.*

Ans. (*i*) Field control; (*ii*) Rheostatic control;

(*iii*) Voltage control.

59. *How can variations of flux be brought about in a series motor?*

Ans. By the use of

(*i*) Field divertors; (*ii*) Armature divertor;

(*iii*) Tapped field control; (*iv*) Paralleling field coils.

60. *The 'field control' is considered superior to 'armature resistance control', why?*

Ans. Because *very small* power is wasted in controlling resistance whereas field control method of speed control is independent of load on motor and *permits remote control of speed.*

61. *Why are the D.C. motors mostly gear motors?*

Ans. Because the low speed motors are much more expensive than high speed motors of the same kW output rating.

62. *Which are the factors that limit the increase of speed of D.C. shunt motors in field rheostat control?*

Ans. The factors are:

(*i*) Demagnetising effect of armature reaction on the field,

(*ii*) Commutation problems at high speed, and

(*iii*) Large centrifugal forces developed at high speed.

63. *When an external resistance is introduced in the armature circuit of a D.C. motor, how will its speed be affected?*

Ans. The speed of the motor will increase.

64. *The speed of a D.C. shunt motor is to be increased above the base speed, which method must be adopted?*

Ans. Field control method.

65. *Why cumulative compound wound D.C. motors are not suitable for applications requiring adjustable speed for field control?*

Ans. Because with a weak shunt field the series field will overpower the shunt field, hence the changes in load may produce unstable speeds.

66. *While controlling the speed of D.C. shunt motor what should be done to achieve a 'constant torque drive'?*

Ans. *Applied voltage* should be maintained *constant* so as to *maintain field strength constant.*

67. *In what respect shunted armature control is superior to armature resistance control?*

Ans. More stable operation is obtained.

68. *In the armature circuit of D.C. motor, an external resistance is introduced. How shall it affect the speed of the motor?*

Ans. The speed will *decrease.*

69. *What is the speed range (approx.) for the series parallel system of speed control of series motor widely used in electric traction work?*

Ans. For the above system of speed control, the speed of motor when running in parallel is *4 times* that of motors when running in *series.*

70. *Why the field flux speed control method for a D.C. motor is called a constant power drive method?*

Ans. In case of a D.C. motor, the speed is inversely proportional to the flux per pole while the torque is directly proportional to it, so for a given armature current.

71. *Why large variable speed D.C. motors are fitted with compensating winding?*

Ans. To 'reduce sparking in commutation' and prevent field distortion owing to armature reaction.

72. *Is equalizer bar needed for parallel operation of under-compounded generators?*

Ans. The equalizer bar is *not* required.

73. *Why is it desirable that the generators operating in parallel should have the same voltage regulation?*

Ans. To share the total load in proportion to their respective rated capacity.

74. *The D.C. motors are not operated to develop maximum power in practice, why?*

Ans.
- Because it necessitates a current well *beyond* the normal operating range.
- Also an amount of power equal to the mechanical power developed would be *wasted in heating the armature,* so taking other losses into account the efficiency would be well below 50 per cent.

75. *The parallel operation of two D.C. series generators is inherently unstable. How can it be made stable?*

Ans. The operation can be made stable by connecting the armature of the two machines operating in parallel through a heavy copper bar or by cross-connecting their field windings.

76. *Why the torque developed in a D.C. generator is called the backward torque?*

Ans. Because the electromagnet torque developed in a D.C. generator *opposes* the outside driving torque.

77. *What will happen if a D.C. motor is directly switched on to the supply, without any starter?*

Ans. The motor will draw heavy current about 10 to 12.5 times rated current and the motor armature winding may get burnt.

78. *During starting of D.C. shunt and series motors what precautions should be taken and why?*

Ans. Before starting a D.C. motor it is necessary to ensure that the field circuit is closed, the rheostat in series with the shunt field winding is at zero resistance position and starting resistance in the series with armature circuit is at maximum value. It will restrict armature current to a safer value and develop higher torque at the starting instant.

79. *Why is it desirable that the starting lever of a motor starter should fall back to off position when the power fails?*

Ans. If the starter arm remains in the 'ON' position on power failure, on restoration of power supply no back e.m.f. will be acting in the armature circuit, the *armature* being directly across the supply mains will draw excessive current and will get *damaged.*

80. *In a D.C. motor starter is necessary, why?*

Ans. To protect the motor against the flow of excessive current during starting period, it is imperative to insert high resistance in series with the armature circuit.

81. *Why while stopping the motor, the line switch should always be opened rather than throwing back the starting arm?*

Ans. If the starter arm is thrown back the field circuit is broken at the last contact button and owing to inductive nature of the field, this will cause a *hot arc and burn the contact.*

82. *What are the various types of 'electric braking'?*

Ans. Electric braking is of three types:

(*i*) Plugging or counter current braking;

(*ii*) Rheostatic or dynamic braking;

(*iii*) Regenerative braking.

83. *What is plugging or counter current braking?*

Ans. In this method, connection to the armature terminals are reversed so that motor tends to turn in the opposite direction. Due to reversal of armature connections, applied voltage and E_b start acting in the *same direction* around the circuit. In order to limit the armature current to a reasonable value, it is necessary to *insert a resistor* in the circuit while reversing armature connection.

84. *What is rheostatic or dynamic braking in a shunt motor?*

Ans. In this method of electric braking of shunt motors, the armature of the shunt motor is disconnected from the supply and is *connected across a variable resistance R.* The field winding is, however, left connected across the supply undisturbed. The braking effect is controlled by *varying* the series resistance *R*.

85. *What is dynamic braking?*

Ans. In this type of braking, the motor is disconnected from the supply and *operated as a generator* driven by the kinetic energy of the motor and its driven machines. The kinetic energy of rotation is converted into electrical energy which is dissipated in the external resistance connected across the motor at the braking instant.

86. *Rheostat braking is economical to plugging, why?*

Ans. Because in rheostat braking energy required from the supply to brake the motor is eliminated as compared to the plugging.

87. *Which of the following is the most economical method of electric braking?*

(i) Plugging. *(ii) Rheostatic braking.*

(iii) Regeneration braking.

Ans. Regenerative braking.

88. *What are 'iron losses' in a D.C. machine?*

Ans. ***Iron losses:*** *Iron losses are a function of both flux and speed.*

Hysteresis loss: The hysteresis loss P_h is a *measure of the electric energy required to overcome the retentivity of the iron in the magnetic flux path,* using watts as unit,

$$P_h = K_h B^x f V \qquad ...(i)$$

where, V = Volume of iron in dynamo subject to change of flux,

K_h = Constant for the grade of iron employed,

B^x = Flux density raised to the Steinmetz exponent. With modern values of dynamo x is no longer 1.6 but closer to 2.0. This is not to imply that for a given volume, V, of iron the loss has increased, because K_h has been reduced considerably, and

f = Frequency (hz) of reversal of flux.

Eddy current losses: These losses occur not only in the dynamo iron but in all conductive materials with the flux path of rotating or varying magnetic field of the dynamo. The eddy current loss P_e, in watts is,

$$P_e = K_e t^2 B^2 f^2 V \qquad ...(ii)$$

where, K_e = An eddy current constant for the grade of iron employed,

t = Thickness of the laminations of the pole core and armature,

B = Flux density,

f = Frequency (hz) of reversal of flux, and

V = Volume of iron subject to change of flux.

For a D.C. dynamo the frequency f, reversal of flux varies with speed. Thus the hysteresis loss varies directly with speed, whereas the eddy current loss varies as the square of speed. Both hysteresis loss and eddy current loss vary approximately as the square of the flux density. For this reason core losses are considered a function of both flux and speed.

89. *What are the no-load losses in a D.C. machine?*

Ans. These losses consist of core loss or iron loss, windage loss and loss due to friction and bearing.

90. *Which of the power losses occurring in electrical machines are influenced by the magnitude of flux, the load and the square of the load?*

Ans.
- Core or iron losses are influenced by the magnitude of flux;
- Brush contact loss is proportional to the load;
- Copper losses (armature, series field, interpoles, compensating windings, are proportional to the square of the load.

91. *When does the maximum efficiency of a D.C. machine take place?*

Ans. When variable losses *equal* constant losses.

92. *Why do copper losses occur in D.C. machines?*

Ans. Because power is consumed in forcing a current against the resistance of windings and connections.

93. *A brake test is performed in small machines only, why?*

Ans. Because testing of a large machine entails a considerable loss of energy, in fact in case of large machines there may be no facility for such test.

94. *Swinburne's test cannot be performed on D.C. series machines, why?*

Ans. Because at no-load D.C. series motor will attain such a high speed that it will get *damaged* and secondly, this test is only applicable to those machines in which flux and speed remain *constant*.

95. *What is Hopkinson's test?*

Ans. This test (also called back-to-back test) is a *regenerative test.* By this method, full-load test can be carried out on two shunt machines, preferably identical ones, without wasting their output.

96. *Why Swinburne's test is considered convenient and economical method for testing of D.C. shunt machines?*

Ans. Because *power required* to test a large D.C. machine by Swinburne's test is *quite small.*

97. *Is field test on two identical D.C. series machines is regenerative method?*

Ans. No, because output of generator is not fed back to motor but is dissipitated in a resistor.

98. *What is a 'Booster'?*

Ans. A *booster* may be defined as a dynamo electric machine, the armature of which is connected in series with a circuit, its generated e.m.f. being added to or subtracted from that of the circuit, depending upon the polarity of its excitation.

99. *Name three armature excited or cross-field rotating machines.*

Ans. (*i*) Rosenberg generator;

(*ii*) Amplidyne and metadyne;

(*iii*) Rototrol and regular multiple-field exciters.

100. *What is 'Regulex'?*

Ans. A *'Regulex'* is a rotating power amplifier that also operates on the straight-line portion of the characteristic curve. However, instead of a series field, regulex makes use of a shunt field to provide the self-exciting ampere-turns.

101. *What are the applications of a third-brush generator?*

Ans. A '*third-brush generator*' is used to charge batteries and lighting systems in automobiles and trains.

The main requirement for such a service is that the generator must maintain its load current practically constant regardless of speed variations.

102. *What is the difference between 'Amplidyne' and 'Metadyne'?*

Ans. Fully compensated machine is called *'Amplidyne'* and under compensated machine is called *'Metadyne'*.

PART II: A.C. MACHINES

Transformers

1. *What is a 'transformer'?*

Ans. A *'transformer'* is a static electromagnetic device designed for the transformation of the *(primary)* alternating current system into another *(secondary)* one of the same frequency with other characteristics, in particulars, other voltage and current.

2. *What is the function of a transformer?*

Ans. The 'function of a transformer', as the name implies, is to transform alternating current energy from one voltage into another voltage. It has no rotating parts, hence it is called a *'static transformer'.*

3. *What is a 'tapped transformer'?*

Ans. A *'tapped transformer'* is one whose windings are fitted with special taps for changing its voltage or current ratio.

4. *Why an iron or steel core is provided in a transformer?*

Ans. To ensure a high permeability of the magnetic circuit. Because of high permeability, the magnitude of exciting current necessary to create the required flux in the core is *small*. The presence of steel core also causes hundred per cent of magnetic flux.

5. *What is the application of a one-to-one transformer?*

Ans. A 'one-to-one transformer' has *equal* terms in primary and secondary windings. These transformers are employed for isolating the load from supply.

6. *On what principle does a transformer work?*

Ans. A transformer works on the principle of *mutual inductance* between two (and sometimes more) inductively coupled coils. It consists of two windings in close proximity. The two windings are *coupled by magnetic induction.*

7. *What happens when a power transformer is connected to D.C. supply of the same voltage rating?*

Ans. When D.C. voltage is applied to the primary of the transformer, the flux produced in the transformer core will not vary but remain constant in magnitude and, therefore, there will be *no self induced e.m.f.* in the primary winding to oppose the applied voltage. Consequently, since the resistance of the primary winding is quite low, a heavy current will flow through the primary winding that may burn it.

8. *What do you mean by step-up and step-down transformers?*

Ans. When the transformer receives energy at a given voltage and delivers the same energy at *higher* voltage, then the transformer is termed as ***step-up*** transformer. On the other hand, when it delivers the same energy at a *lower voltage*, then the transformer is termed as ***step-down*** transformer.

9. *What is the necessity of step-up and step-down transformers in transmission system?*

Ans. The 'step-up transformer' is necessary for *transmitting* a given amount of energy at a *high voltage* since more will be the voltage less will be the current and correspondingly the *line* loss. Moreover, it is *not possible to generate high voltage from the alternator itself.*

On the other hand, 'step-down transformer' is necessary for *reducing* the high transmission voltage to *safe and usable values.*

10. *What is the main difference between core and shell type construction?*

Ans. The main difference between core and shell type construction is that in the core type the impression is created that the coils have been wound around the core, on the other hand, in the shell type the core has been built around the coils.

11. *Why is the silicon steel used?*

Ans. The silicon is used due to the fact that it reduces losses and prevents ageing.

12. *What are the types of bushings used in transformer terminals?*

Ans. (*i*) Porcelain insulator bushing;

(*ii*) Oil filled bushing;

(*iii*) Capacitor bushing.

13. *What is the conservator tank? Why is it needed?*

Ans. The *'conservator tank'* is fitted over the transformer tank. It is needed for taking up the cyclic expansion and contraction of oil with the change of temperature without allowing the transformer tank oil to come in contact with the ambient air. Thus the transformer oil will have no chance of absorbing moisture from air.

14. *What is a breather?*

Ans. *'Breather'* is a device which absorbs moisture. It generally contains silica-gel.

The function of a breather is to *extract* moisture from the incoming air. In other words, the displacement of the conservator air by oil takes place through a breather.

15. *Why is it advantageous to make short-circuit test on the high voltage test rather than on low voltage side?*

Ans. It is due to the fact that the *ranges of the meter* at the high voltage (h.v.) side at the short-circuit condition of the low voltage (l.v.) side are *much more suitable.*

16. *Which side of the transformer is kept open in the open-circuit test and why?*

Ans. The h.v. side of the transformer is kept open in the open-circuit of the transformer; the reason being that it is *more suitable* to use metering instruments in l.v. side rather than h.v. side during open-circuit.

17. *What is the Buchholz protection of transformer?*

Ans. The Buchholz protection is nothing but actuating a relay in any incipient fault in the transformer. This relay is placed in the pipeline between the transformer tank and separate oil conservator. Any generation of oil vapour or gas may operate the relay.

18. *What is an auto-transformer? Where is it generally used?*

Ans. A transformer in which part of the winding is common to both the primary and secondary circuits is known as an ***auto-transformer.*** The primary is electrically connected to the secondary, as well as mechanically coupled to it.

- It is generally used in cases where the voltage ratio is not less than 2 : 1 approximately and electrical isolation of the primary and secondary circuit is not essential. Examples: Starters, high voltage transmission, etc.

19. *What is the effect of leakage flux in a transformer? What is meant by magnetising current?*

Ans. The effect of leakage flux is to *develop* into their respective windings *e.m.fs. of self inductance* which are proportional to the current and are, therefore, equivalent in effect to the addition *of an inductive coil in series with each winding.* The component of no-load current used to create the flux in the transformer core is known as *magnetising current.*

20. *Define 'voltage regulation' of a transformer.*

Ans. *'Voltage regulation'* of a transformer is defined as the change in magnitude of terminal or secondary voltage when full load (rated load) of specified power factor supplied at rated voltage is thrown of with primary voltage and frequency held constant, as percentage of the rated terminal voltage.

21. *Does the eddy current loss of a transformer depend on the applied voltage?*

Ans. *Yes,* If the transformer is operated with increased voltage but at the same frequency, the eddy current loss increases due to increased value of peak value of flux density and if applied voltage is increased in the increase in frequency, the proportion of increase being same in both, the eddy current loss increases due to increase in supply frequency.

22. *In a transformer short-circuit test, all losses are attributed only to copper loss, why?*

Ans. Because the applied voltage to the primary of transformer is *very low* (5 to 8% of rated voltage) in the short-circuit test, the flux produced in the core is very small and, therefore, iron losses are very small and can be neglected and hence all the losses are attributed to copper losses.

23. *Why silicon content in electrical sheet-steel is limited to 4.5 to 5 per cent?*

Ans. Silicon content exceeding 5 per cent makes the sheet-steel brittle and so causes problem in punchings.

24. *What is the cause of noise in transformers?*

Ans. '*Noise*' in a transformer, occurs mainly due to loosening of stampings and mechanical forces developed during operation.

25. *Why circular coils are always preferred over rectangular coils for winding a transformer?*

Ans. Because 'circular coils' can be *easily wound* on machines, conductors can be easily *bent* and winding does *not* buldge out due to radial forces developed during operation.

26. *Why are transformers rated in kVA instead of kW?*

Ans. The rating of electrical machine (transformer) is governed by the temperature rise caused by its internal losses.

The copper losses in transformer windings depend upon the strength of the output current and core loss on voltage and thus these losses are *independent* of *p.f.* and the transformers are rated in kVA instead of kW.

27. *The excessive insulation is harmful to a coil, why?*

Ans. Because excessive insulation makes heat transfer poor.

28. *How is a transformer rated?*

Ans. By its kVA (or MVA) output, voltage (primary/secondary) and frequency rating.

29. *In what type of transformers sandwich coils are used? What advantage is gained by the use of sandwich coils?*

Ans. "*Sandwich coils*" are used in shell type transformers for l.v. and h.v. windings. The advantage gained by the use of sandwich coils is very low value of leakage reactance resulting in *low voltage drops.*

30. *What is the condition for maximum efficiency in a transformer?*

Ans. Efficiency of a transformer will be maximum on such a load at which variable loss (copper loss) becomes *equal* to constant loss (iron loss).

31. *When will be the ordinary efficiency of a given transformer minimum?*

Ans. At no-load or when supplying load at zero p.f.

32. *What is a 3-phase transformer?*

Ans. It is that transformer which is equivalent to three single phase transformers but wound on one core and enclosed within one common case.

33. *Secondaries of a 3-ϕ transformer bank are connected in delta. If the phase voltages are each equal to V, what e.m.f. would act around the delta if one phase were reversed?*

Ans. When one phase of secondaries of a 3-ϕ transformer bank connected in delta is reversed, the e.m.f. around the delta would be equal to twice the phase voltage *i.e.,* 2 V, instead of zero voltage.

34. *What is open delta connection?*

Ans. In the delta-delta connections of three single phase transformers, if one transformer is removed, the resulting connection is known as *open delta connection.* This connection has VA rating of $\frac{1}{\sqrt{3}}$ *i.e.,* 0.577 times the rating of normal delta connection.

35. *In a star-star connected transformer, if a delta tertiary is provided, what is major reason?*

Ans. To allow sufficient earth fault current to flow for operation of protective equipment, to suppress harmonic voltages, and to limit voltage unbalance when the main load is unsymmetrical.

Polyphase Induction Motors

36. *What is an induction motor?*

Ans. An ***induction motor*** is motor where no current is conducted but the current results from the induced voltage. That is why, it is termed as an induction motor.

37. *What is the difference between 'squirrel cage' and 'wound rotor' induction motors?*

Ans. The difference between the squirrel cage and wound rotor construction lies in the construction of motor. In a squirrel cage motor a series of rotor conductors are evenly spaced around the periphery of rotor of squirrel cage with their *ends short-circuited* by conductors in the form of *end rings.* But in case of wound rotor machines, the winding in the motor will be just like that of stator and the electrical connection are made between the rotor winding and *slip rings* mounted on the shaft.

38. *Is it possible to have a three-phase stator and two phase motor in an induction motor?*

Ans. *Yes,* it is possible but it should be wound rotor machine.

39. *Why are induction motors called "asychronous"?*

Ans. Because the motor does not run in synchronism with the rotating field developed by stator currents.

40. *Why the number of poles of the stator and rotor of an A.C. motor should be equal?*

Ans. The flux set-up by the stator windings is a rotating flux with a definite number of poles; operation is only possible when the number of poles is *same* on both stator and rotor.

41. *What is a rotating magnetic field?*

Ans. When a 3-phase winding is energised by 3-phase supply, a rotating magnetic field is developed that tends the rotor of the machine to move in the direction of motion. The operation of A.C. rotating machines (such as synchronous machines and induction machines) is based on the production of rotating magnetic field.

42. *How does the slip vary with load?*

Ans. When load on a motor is increased it slows down until the relative motion between the rotor and the rotating magnetic field is just sufficient to result in the development of torque required for a particular load. Thus slip *increases* with the increase in load.

43. *How is developed torque of a 3-phase induction motor related with the supply voltage?*

Ans. $T \propto sV^2$, where s is the slip and V is supply voltage respectively.

44. *What are the main factors on which torque developed by a 3-phase induction motor depend?*

Ans. The main factors are:

Rotor resistance, standstill reactance, slip and applied voltage.

45. *At what slip the torque developed in an induction motor will be maximum?*

Ans. Torque developed in an induction motor will be maximum at a slip given by the ratio of rotor resistance to rotor standstill reactance, *i.e.*, $s = \frac{R_2}{X_2}$.

46. *Why is slip always positive in an induction motor?*

Ans. The induction motor cannot run at synchronous speed because at synchronous speed there will not be relative speed between the revolving field and the rotor conductors and so the basic phenomenon will not occur to develop torque. It means that a 3-phase induction motor always runs at a speed *less* than that of the field. The speed being the difference of synchronous speed, N_s and speed N ($N_s > N$) therefore, slip will be always positive.

47. *When applied rated voltage per phase is reduced to one-half, what will be starting torque of a squirrel cage induction motor in terms of its starting torque with full voltage?*

Ans. $\frac{1}{4}$ th of starting torque with full rated voltage.

48. *How do the starting torque and maximum torque vary with rotor circuit resistance?*

Ans. When the rotor resistance R_2 is increased from zero, power factor is improved but rotor impedance is increased and so the rotor current is reduced. At first, the effect of improved power factor predominates over the effect of decrease is rotor current due to increase in impedance and the *'starting torque'*, therefore, *increases,* but after a certain value of rotor resistance, the effect of increased impedance predominates over the effect of improved power factor and so the *'starting torque' decreases. 'Maximum torque'* is *independent of rotor circuit resistance.*

49. *What is the necessity of short-circuited rotor conductors in a squirrel cage induction motor?*

Ans. In a squirrel cage motor, if the rotor conductors are not short-circuited and when the stator is connected to 3-ϕ AC supply, the rotating magnetic field will be set up in the air-gap and an e.m.f. will be induced in the rotor but no current will flow through the rotor conductors and, therefore *no torque will be developed* and the rotor will remain stationary.

50. *If the slip is either positive greater than one or negative, what will be operating conditions of the motor?*

Ans. If the slip is *'positive' greater than one,* the induction motor acts *as a 'brake'* and its absorbs kinetic energy from the still revolving field causing its speed to fall. When the slip is *'negative'* (*i.e.* when the speed is more than the synchronous speed) the induction motor operates as an *'induction generator'* and feeds back to the supply line.

51. *In an induction motor, the rotor always run in the direction of stator field, why?*

Ans. According to Lenz's law, the direction of induced e.m.f. would be in such a direction that it would try to oppose the very cause for which it is due. Since the cause producing the induced currents in the rotor conductors is the relative speed between the rotating magnetic field and stationary rotor conductors, therefore, they circulate in such a way that a torque is produced in the rotor tending to cause it to *follow* the rotating magnetic field and thus *reducing* the relative speed.

52. *If the air-gap flux is kept constant, will the reduction of supply frequency affect the pull-out torque of an induction motor?*

Ans. When the air gap flux is kept *constant,* the supply frequency and supply voltage will have to be *reduced in the same proportion.* In such conditions the torque speed curves are of the same shape but *displaced horizontally i.e.,* maximum developed torque (pull out torque) *remains approximately constant*, but the speed at which it occurs *changes* in the same ratio as the supply frequency.

53. *Enumerate the different losses which take place in a 3-ϕ induction motor.*

Ans. The various losses are:

Stator copper loss, stator iron loss, rotor copper loss, friction and windage losses, stray load loss.

54. *Why an induction motor can be called a generalised transformer?*

Ans. Because the two fields set up by stator (primary) and rotor (secondary) are stationary relative to each other, so the polyphase induction motor *can be considered equivalent to a transformer* having an air gap, separating the iron portions of the magnetic circuit carrying the primary and secondary windings.

55. *Why low speed 3-ϕ induction motors have lower factors as compared to that of high speed motors?*

Ans. In low speed induction motors there are large number of poles as compared to high speed motors, due to which the magnetising current is more due to increase in leakages, which increases with the increase in number of poles. So the power factor in case of low speed induction motors is low.

56. *Why the magnetising current in a transformer is 2 to 5 per cent whereas in 3-ϕ induction motor it is about 25 to 40 per cent of rated current?*

Ans. In an induction motor, due to air gap, the magnetising current is *large* as compared to that of a transformer. The friction and windage losses are also present in the induction motor. Therefore, no-load current of an induction motor is much higher than that of an equivalent transformer.

57. *What do you mean by the term 'Crawling' in an induction motor?*

Ans. The tendency of a squirrel cage induction motor to run stably at speed as low as $\frac{1}{7}$th of its synchronous speed is called ***crawling*** of an induction motor.

58. *Why are the iron losses in the rotor of an induction motor normally neglected?*

Ans. Because of very low frequency of e.m.f. induced in the rotor.

59. *List the factors/parameters which govern the performance of induction motors.*

Ans. (*i*) Rotor resistance;

(*ii*) Air-gap length;

(*iii*) Shapes of both stator and rotor slots and teeth.

60. *How can the starting torque of a 3-ϕ induction motor be improved?*

Ans. By inserting external resistance in each phase of the rotor circuit, which is possible only in case of wound rotor motors.

61. *Why the inner cage of a double cage 3-ϕ induction motor has lower resistance but higher inductance value as compared to the outer cage?*

Ans. The rotor of a double cage 3-ϕ induction motor carries two squirrel cage windings in two rows of slots. The outer winding bars are usually made of manganese brass while the inner winding bars are of red copper and have great cross-section than that of outer winding, so the lower cage has lower resistance, since the outer cage is *nearer* to the air gap so its leakage reactance is *smaller* than that of inner cage.

62. *What do you mean by the term 'single phasing'?*

Ans. This term refers to operation with one stator terminal accidently disconnected, as by a blown fuse or a faulty contact on the run side of a star delta switch.

63. *What is the necessity of a starter in 3-ϕ induction motor?*

Ans. A 3-ϕ induction motor at the starting instant acts as a polyphase transformer having an air gap with a short circuited secondary. If the motor of higher capacity, larger than 2 kW is connected directly to the supply mains, it will draw heavy current (about 5 to 8 times the full load current) at the starting instant which is objectionable because it will cause large voltage drop in the distribution network and thus affect adversely the operation of other electrical equipment connected to the same distribution network. Hence, *to control the starting current,* use of starter is *essential* for starting of 3-ϕ induction motors, of *large sizes, rating exceeting 2 kW.*

64. *The 'induction motor with voltage controller' is suitable for which type of load and why?*

Ans. It is suitable for loads where the *torque required drops off considerably with the drop in speed* such as for *driving fans,* because a large change in speed and developed torque reduces greatly with the reduction in supply voltage.

65. *What is the name of the method by which electric braking is accomplished in an induction motor?*

Ans. The name of the method is 'Plugging or Counter current braking.'

66. *It is required to control the induction motor speed during deceleration, which type of braking is preferable to employ?*

Ans. D.C. dynamic braking; the speed is controlled by controlling the D.C. excitation.

67. *While controlling induction motor speed, how super synchronous speed is achieved?*

Ans. By injecting a slip frequency voltage in phase with e.m.f. induced in the rotor.

Single Phase Induction Motor

68. *Can a single pulsating field resolved?*

Ans. *Yes.* A single pulsing field can be resolved into two components of half its amplitude and rotating in opposite directions with equal speed.

69. *What is the magnitude of the starting torque of single phase induction motor?*

Ans. Zero.

70. *Why are single phase motors usually set on rubber spring moments?*

Ans. Because a pulsating torque is developed as a direct consequence of the pulsating power in single phase circuit.

71. *Why the starting winding of a single phase induction motor is disconnected once the slip reaches 0.25 per cent or less?*

Ans. In a resistance start motor, the starting winding is not designed for continuous operation and so it is necessary to disconnect it from the motor circuit when the motor attains a definite speed. The starting winding is usually *taken out* of the circuit at a speed of about *75 per cent of synchronous speed* because at this speed the torque developed by the main winding above is *greater* than that developed by both windings.

72. *How is the direction of rotation of a single phase induction motor reversed?*

Ans. By reversing the leads to the main or starting winding, but *not both.*

73. *Single-phase induction motors are not self-starting, why?*

Ans. Constructionally a single-phase motor is similar to a 3-phase induction motor *except that the stator is provided with a single-phase winding.* When this winding is fed from a single phase supply, its stator produces a flux which is only alternating along the space axis. It is not a synchronously revolving or rotating flux as in the case of two-or-three phase stator winding fed from a 2-phase or 3-phase supply. *An alternating flux acting on a stationary squirrel cage rotor cannot produce rotation.* This is *why a single phase motor is not self-starting.*

74. *Single-phase series motors are preferred for fraction service rather than induction or synchronous motors, why?*

Ans. Now-a-days single phase motors are manufactured as fractional horse power motors and mostly they are used with domestic appliances, such as tape recorder, record player, hair drier, electric clock, refrigerator, mixi, fans, pumps etc. As these are *light duty machines, are less costly and hence preferred.*

75. *Why a compensating winding is provided in universal motor for A.C. operation?*

Ans. The *compensating winding* is provided in a universal motor *to reduce the reactance voltage* present in the armature when motor runs on A.C. supply. This reactance voltage is caused by the alternating flux by transformer action because the flux produced by the field winding is alternating.

76. *Why reluctance motor has low frequency?*

Ans. The construction of stator of single phase reluctance motor is similar to single-phase induction motor. To make the motor self-starting the starting winding is provided with the main winding. The rotor of the motor is a modified cage rotor with *some of teeth removed.* This *actually provides variation of air-gap reluctance.* When motor is connected to single phase supply it starts like an induction motor. When motor speed reaches near to synchronous speed it will be pulled into sinchronous speed with the stator field due to the *reluctance tarque, because* of the *variation of the air-gap reluctance,* now the motor continues to run at synchronous speed, as a non-excited synchronous motor.

The *'efficiency' and 'power factor' of this type of motor is poor because of the varying reluctance.*

77. *A single phase induction motor is provided with a main winding and an auxiliary winding. Which of the two windings should be more resistive?*

Ans. Auxiliary winding.

78. *What happens to capacitor start induction motor if the auxiliary winding is left in the circuit after it has been used to start the motor?*

Ans. The auxiliary winding will *burn out* (since it is not designed for continuous operation). In addition to it the short duty rated electrolytic *capacitor* connected in series with starting winding will also get *damaged.*

79. *Which type of motor is used in a ceiling fan?*

Ans. Split-phase capacitor type single-phase induction motor.

80. *What is the advantage of using a capacitor-start motor our a resistance start split-phase motor?*

Ans. The capacitor-start motor develops more starting torque than does an equally rated resistance start split-phase motor with a lower in-rush current.

81. *What is hysteresis motor?*

Ans. It is a fractional horse power synchronous motor producing mechanical torque utilising the phenomenon of hysteresis.

Synchronous Generators or Alternators

82. *What do you mean by a synchronous generator or alternator?*

Ans. A *synchronous generator* is an electrical machine where alternating voltage can be developed. The principle of generating voltage is same with that of a D.C. generator; only the difference lies due to the fact that in case of D.C. generator the alternating voltage set-up in the armature conductors is rectified by means of a commutator. In other words, the D.C. generator having closed coil armature winding may be used to serve as a synchronous generator if it is equipped with slip rings suitably connected to the winding.

83. *How can synchronous be classified?*

Ans. (*i*) Generators with revolving field (general used).

(*ii*) Generators with revolving armature.

(*iii*) Inductor-type generators.

84. *How are revolving field type generators classified?*

Ans. (*i*) Salient poles machines;

(*ii*) Non-salient pole machines.

The salient pole machine consists of laminated *salient or projecting poles* in the field.

The non-salient pole machine consists of *cylindrical rotor* having field *windings placed in slots* around the surface of the rotor.

85. *What is an inductor type generator?*

Ans. It is a type of generator where armature and field windings are stationary. Variation of flux linked with armature winding is caused due to the rotation of the two sets of laminated lugs.

86. *What are the essential elements for generating e.m.f. in alternators?*

Ans. The essential elements are:

(*i*) Prime mover; (*ii*) Stator;

(*iii*) Rotor; (*iv*) Exciter.

87. *Why is an alternator, smooth rotor is used for high speeds and salient pole rotor for low speeds?*

Ans. The '*salient pole*' field structure is used almost entirely for low speed alternators, because it is *least expensive* and provides *ample space* for the field ampere-turns but it cannot be used for high speed alternators on account of high peripheral speed and the difficulty of obtaining sufficient mechanical strength. So non-salient field structure is used for high speed operation.

- Salient-pole alternator is *more stable.*

88. *Which type of alternator will be used for?*

(*i*) *Hydraulic turbines;*

(*ii*) *Steam turbines as prime movers.*

Ans. (*i*) Salient pole type;

(*ii*) Smooth cylindrical rotor type.

89. *Why squirrel cage bars are placed in the pole faces of an alternator?*

Ans. To *reduce* the swinging amplitude (hunting).

90. *What is the role of damper winding in the following machines?*

(*i*) *Synchronous generator;* (*ii*) *Synchronous motor.*

Ans. (*i*) To *suppress the negative sequence of field* and to damp the oscillations *when the 'generator' starts 'hunting'*.

(*ii*) To *develop starting torque* so as to make the *'motor' self-starting* and to *develop damping torque* when the machine *starts hunting.*

91. *Why are distributed windings preferred over concentrated windings?*

Ans. The distributed windings are preferred due to the following *reasons*:

(*i*) Improve waveform of induced e.m.f.;

(*ii*) Provide better utilisation of core;

(*iii*) Diminish armature reaction;

(*iv*) Facilitate cooling.

92. *Why is the chording necessary in the armature winding of an alternator?*

Ans. The chording is done:

(*i*) To improve the waveform of induced e.m.f.;

(*ii*) To save copper in the coil ends due to *less span;*

(*iii*) To reduce reluctance of the winding.

93. *When load on an alternator varies, its terminal voltage also varies, why?*

Ans. The variation in terminal voltage is due to the following *reasons*:

(*i*) Voltage drop due to *armature resistance;*

(*ii*) Voltage drop due to *armature leakage reactance;*

(*iii*) Voltage drop due to *armature reaction.*

94. *The short-circuit characteristic of an alternator is linear, why?*

Ans. Because the net excitation is so small that there is no saturation in the magnetic circuit.

95. *Can an alternator have full load terminal voltage greater than the no-load terminal voltage.*

Ans. *Yes.* It occurs when the load supplied by the alternator is highly capacitive, *i.e.,* the *current* supplied by the alternator *leads* the voltage.

96. *Define voltage regulation of a transformer.*

Ans. *Voltage regulation* of an alternator is defined as the increase in terminal voltage, expressed as percentage (or per unit) of the rated terminal voltage when the load at a given factor is thrown off, with speed and field current remaining the *same*.

Mathematically, voltage regulation $= \frac{E_o - V}{V}$ per unit

$$= \frac{E_o - V}{V} \times 100 \text{ in percentage}$$

(where, E_o = No-load terminal voltage, and V = Full-load terminal voltage).

97. *Enumerate the various indirect methods used to determine voltage regulation.*

Ans. (*i*) Synchronous impedance or e.m.f. method;

(*ii*) The ampere-turn or m.m.f. method;

(*iii*) Zero power factor or Potier method.

98. *Which method of voltage regulation is called "pessimistic" and why?*

Ans. '*Synchronous impedance method*' is called pessimistic method.

The results obtained by this method are *too high,* owing to the fact the synchronous impedance determined at short-circuit conditions is *too large due to a very low degree of saturation.* Hence, this method is called *pessimistic method.*

99. *Which method of voltage regulation is called 'optimistic' and why?*

Ans. '*Ampere-turn an m.m.f. method*' is called the optimistic method.

Regulation given by this method is *much lower* than that given by the synchronous impedance method, but it is *nearer the correct value.* Hence, this method is called *optimistic method.*

100. *Why does the determination of synchronous reactance of an alternator from open-circuit and short-circuit characteristics give practically direct axis synchronous reactance?*

Ans. The determination of synchronous reactance of an alternator from open circuit and short circuit characteristics gives *unsaturated value so practically direct-axis synchronous reactance.*

101. *What is use of Potier triangle?*

Ans. To determine voltage regulation of an alternator.

102. *What do you mean by 'armature reaction' of a synchronous machine?*

Ans. The effect of armature flux on the main field flux is known as '*armation reaction.*' It has distorting effect on unity power factor, wholly demagnetising at zero power factor lagging and wholly magnetising at zero power factor leading.

103. *What is an air-gap line?*

Ans. The tangent of open circuit characteristics (O.C.C.) is called the *air-gap line.*

104. *Why does the two reaction theory apply only to salient pole machines?*

Ans. A '*multipolar machine*' with cylindrical rotor has a *uniform air-gap,* therefore, the *reactance remains the same* irrespective of spatial position of rotor. But, in case of *"salient pole machines,"* the *air-gap is not uniform.* Because of non-uniformity of the reactance of the magnetic paths, the m.m.f. of the armature is divided into two components:

(*i*) A direct-acting component along the field pole axis, called the '*direct axis*', and

(*ii*) Quadrature component along the axis passing through the centre of the interpole space, is called the '*quadrature axis.*' These facts form the basis of the two-reactance theory applied to the salient pole machines.

105. *How can the direction of the stator revolving field be reversed?*

Ans. By reversing *any two* of the three supply lines to the stator winding.

106. *What is a 'rotating magnetic field'?*

Ans. It is that field which is constant in magnitude but whose axis of direction *rotates in space* like field system of a D.C. machines.

107. *Upon what factors does the power angle 'δ' depend?*

Ans. The power angle 'δ' depends upon the following factors:

(*i*) Supply voltage;

(*ii*) Armature current;

(*iii*) Power factor of load;

(*iv*) Quadrature component of synchronous reactance.

108. *What is the effect of reducing excitation on a salient pole alternator? What happens when $E = 0$?*

Ans. When the excitation is reduced the *voltage* induced in the stator winding will be *reduced.* Initially the reduction in voltage is non-linear but with large reduction in excitation current the voltage is also reduced *linearly.* In case there is no excitation, *i.e.,* $E = 0$, the power developed due to excitation will be *zero.*

109. *Why is the 'slip test' performed? Why are there fluctuations in ammeter and voltmeter readings in this test?*

Ans. *'Slip test'* is performed to determine direct and quadrature axis reactances of a salient pole synchronous machine.

The effective reactance offered by the machine goes on changing depending upon the position of field poles. It is found that current and voltages fluctuate at twice the slip frequency and, therefore, the ammeter and voltmeter readings will fluctuate.

110. *Why are alternators put in parallel?*

Ans. The alternators are put in parallel for high efficiency of operation, reliability, convenience and economy in maintenance and repair and possibility of additions to plant capacity with the growth in load on the power station.

111. *What do you mean by synchronising of alternators?*

Ans. The process of connecting an alternator in parallel with another alternator or with the common bus-bars is called the ***"synchronising".***

112. *How can the distribution of load between two alternators operating in parallel be varied?*

Ans. By *adjusting governor control.* The governor of one prime move is opened while the others is closed simultaneously so as to *maintain the system frequency constant.*

113. *What is meant by 'infinite bus-bars'?*

Ans. It is a system of large capacity whose frequency and the phase and magnitude of the voltage are *not affected* even if there is a *variation of excitation or power of a synchronous machine connected to it.*

Synchronous Motor

114. *Is there any difference between synchronous generator and motor from the construction point of view?*

Ans. From construction point of view there is *no difference* between the two machines. Only very few minor differences which exist can be expressed as follows:

(*i*) Synchronous motor except for high speed is made of salient poles.

(*ii*) Synchronous motor has a more effective damping device for preventing hunting.

115. *In the synchronous motor self-starting?*

Ans. *No,* the synchronous motor is not self-starting *because it does not possess any starting torque.*

116. *How does the synchronous motor attain synchronous speed?*

Ans. (*i*) Starting by auxiliary motor.

(*ii*) Starting by the principle of the induction motor utilising dampers as squirrel cage windings of induction motors.

117. *What is the advantage of using D.C. motor as auxiliary motor for starting the synchronous motor?*

Ans. D.C. motor can be used as an auxiliary motor for starting and also as D.C. generator for providing field excitation to the synchronous motor after synchronous speed has been attained.

118. *What is the basic principle by virtue of which an induction-start synchronous motor speeds up?*

Ans. When the rotor of a synchronous motor is at rest or revolving at a speed less than the synchronous speed, a relative motion between the poles and the field produced by the armature reaction may occur and that causes currents to be induced in the damper. Thus, these currents produce the same result as that is done by induced current in the squirrel cage winding of the induction motor. Thus the synchronous motor speeds up.

119. *What is the effect of harmonics in the starting of synchronous motor?*

Ans. The motor may be *locked* at some sub-synchronous speed.

120. *Why is the transformer used with the synchronous motor at high voltage supply?*

Ans. When the synchronous motor is induction start, the voltage applied on it should be reduced at the time of starting in case the voltage of line is above 13.5 kV. This voltage is reduced in general by the transformer. That is why transformer is used with synchronous motor at high voltage supply.

121. *What do you mean by the 'V-curves' of synchronous motor?*

Ans. The *'V-curves'* of synchronous motor are the curves which show the *relationship between field current and armature current for constant mechanical output.*

122. *How do you vary the power factor of a synchronous motors at any given load?*

Ans. By *varying the field excitation.*

123. *What is meant by normal excitation, under excitation and over excitation of synchronous motor?*

Ans.
- *'Normal excitation'* of the synchronous motor is that excitation which produces unity power factor.
- *'Under excitation'* is that excitation which causes the motor to take a *lagging current.*
- *'Over excitation'* is that excitation by virtue of which the motor takes *leading current.*

124. *What will be the condition of synchronous machine of the prime mover is uncoupled after it is synchronised with a system?*

Ans. The condition of synchronous machine if the prime mover is uncoupled after it is synchronised can be expressed as follows:

The synchronous machine *will continue to run as a motor at synchronous speed* after the prime mover has been uncoupled. The mechanically losses will be provided electrically. Even with mechanical load on motor, the tendency of showing down will be *compensated with the increase of load angle.* Thus more electrical power will be converted into mechanical power.

125. *What do you mean by 'hunting' of a synchronous motor?*

Ans. A synchronous motor cannot be stable instantaneously if any change in load occurs due to the inertia of the motor.

This oscillation of speed over synchronous speed is termed in 'hunting'.

126. *Can 'hunting' be avoided completely?*

Ans. *No.* However, if the synchronous motor is designed properly and operated under good condition, hunting can be avoided to a great extent.

127. *How can 'hunting' in a synchronous motor be diminished?*

Ans. (*i*) By increasing the moment of inertia by adding a flywheel.

(*ii*) By the use of an amortisseur or damping winding in the pole faces.

128. *What is a super-synchronous motor?*

Ans. The super-synchronous motor is a motor *which can develop very large starting torque with moderate torque.*

129. *What is the 'circle diagram' of a synchronous motor?*

Ans. It is diagram which shows the locus of the extremity of excitation vectors and current vectors for different values of load. Further more, it also indicates the loci of constant developed power, called *power circles*.

130. *Is it practicable to obtain the maximum developed power in synchronous motor?*

Ans. *No,* because the saturation of machine will not allow the degree of excitation needed for obtaining this particular adjustment.

131. *What factors determine the number of poles of a synchronous motor?*

Ans. The supply frequency and speed.

QUESTIONS BANK

(Including Questions from UPSC, GATE, various Universities and AMIE Examinations)

With Answers

PART I : OBJECTIVE TYPE QUESTIONS

- Choose the Correct Answer
- Fill in the Blanks
- Say Yes/No
- Match the Items

PART II : QUESTIONS WITH SHORT ANSWERS

PART I

OBJECTIVE TYPE QUESTIONS

Choose the Correct Answer ; Fill in Blanks ; Say Yes/No ; Match the Items :

1. A 3-phase synchronous motor connected to an infinite bus is operating at half-full load with normal excitation. When the load on the synchronous motor is suddenly increased

(*a*) its speed will first decrease and then become synchronous

(*b*) its speed will first increase and then become synchronous

(*c*) its speed will fluctuate around synchronous speed and then become synchronous

(*d*) its speed will remain unchanged.

2. At a certain current, the energy stored in an iron-cored coil is 1000 J and its copper loss is 2000 W. The time constant (in seconds) of the coil is

(*a*) 0.25 (*b*) 0.5

(*c*) 1.0 (*d*) 2.0.

3. If the applied voltage to a d.c. machine is 230 V, then the back emf, for maximum power developed is

(*a*) 115 V (*b*) 200 V

(*c*) 230 V (*d*) 460 V.

4.

List I *(Methods of control)*	**List II** *(Applications)*
A. Injection of voltage into the rotor circuit of a 3 phase induction motor	1. Steel rolling mills
B. Variation of applied voltage to a 3 phase induction motor at constant frequency	2. Textiles mills
C. Series-parallel control of d.c. series motors	3. Fans and blowers
D. Ward-Leonard control of d.c. motors	4. Traction
	5. Position control in a servo system

Code :	A	B	C	D
(*a*)	1	2	3	4
(*b*)	4	5	2	1
(*c*)	1	5	2	4
(*d*)	2	3	4	1

5.

List I *(Unit)*	**List II** *(Type of rotor)*
A. Synchro-transmitter	1. Dumb-bell rotor
B. Control transformer	2. Drag-cup rotor
C. AC servo-motor	3. Cylindrical rotor
D. Stepper motor	4. Toothed rotor
	5. Phase wound rotor

Codes :	A	B	C	D
(*a*)	1	3	2	4
(*b*)	1	5	3	2
(*c*)	2	4	3	1
(*d*)	3	2	1	5.

6. Consider the following statements regarding the starting of a d.c. shunt motor.
1. External field resistance should be minimum
2. External field resistance should be maximum
3. Reduced voltage should be applied to the armature
4. Increased voltage should be applied to the armature.
Of these statements
(*a*) 1 and 3 are correct (*b*) 1 and 4 are correct
(*c*) 2 and 3 are correct (*d*) 2 and 4 are correct.

7. An induction motor has rotor resistance R_r, stand-still rotor induced voltage E_{ss}, and stator to rotor effective turn-ratio of *m*. In an equivalent circuit of this machine, the rotor circuit resistance is shown in m^2R_r/s, where *s* is the slip. This implies that the value of the equivalent rotor circuit voltage will be
(*a*) E_{ss} (*b*) sE_{ss}
(*c*) $m\ E_{ss}$ (*d*) $m\ SE_{ss}$.

8. An induction motor has a rotor resistance of 0.002 ohm/phase. If the resistance is increased to 0.004 ohm/phase then the maximum torque will
(*a*) reduce to half (*b*) increase by 100%
(*c*) increase by 200% (*d*) remain unaltered.

9. A 5 kVA transformer has a turn ratio of $(N_1/N_2) = 10$. The impedance of the primary winding is $3 + j5$ ohms while that of secondary winding is $0.5 + j0.8$ ohms. The impedance of the transformer when referred to the primary will be
(*a*) $3.5 + j5.8$ ohms (*b*) $8.0 + j13.0$ ohms
(*c*) $53.0 + j85$ ohms (*d*) $3.05 + j5.08$ ohms.

10. Under short-circuit conditions the windings of a transformer having 5% impedance will experience a radial force
(*a*) 50 times of full load value (*b*) 120 times of full load value
(*c*) 250 times of full load value (*d*) 400 times of full load value.

11. Compared to the breaking capacity of circuit breaker its making capacity should normally be
(*a*) more (*b*) equal
(*c*) less (*d*) the two are unrelated to each other.

12. Power despatch through a line can be increased by
(*a*) installing series capacitors (*b*) installing shunt capacitors
(*c*) installing series reactor (*d*) installing shunt reactor.

13. Addition of tubes to the transformer tank improves heat dissipation capacity because of
(*a*) additional cooling surface
(*b*) additional dissipation by radiation only
(*c*) additional dissipation by convection only
(*d*) additional dissipation by radiation and convection both.

14. Air gap at the pole tips of a d.c. machine is kept more than that at the centre of the pole mainly to reduce
(*a*) reactance voltage (*b*) effect of armature reaction
(*c*) losses in armature core (*d*) noise of machine.

15. The quality factor of *R-L-C* circuit will increase if
(*a*) *R* decreases (*b*) *R* increases
(*c*) voltage increases (*d*) none of the above.

16. A d.c. shunt motor drives a load at rated speed and supply voltage. If both the load and voltage are halved, the speed of the motor will be

(*a*) almost doubled (*b*) almost halved
(*c*) the rated speed (*d*) slightly less than the rated speed.

17. A 3-phase induction motor runs at super synchronous speed. For self excitation the machine

(*a*) draws real power from the mains (*b*) draws reactive power from the mains
(*c*) feeds reactive power to the mains
(*d*) generates e.m.f. at the expense of residual magnetism.

18. T is the load torque of a d.c. series motor having linear magnetization and negligible armature resistance. Speed of the motor is

(*a*) inversely proportional to $\sqrt{T}$ (*b*) directly proportional to $\sqrt{T}$
(*c*) inversely proportional to T^2 (*d*) directly proportional to T^2.

19. In a series *R-L-C* circuit at resonance

(*a*) impedance is maximum (*b*) admittance is maximum
(*c*) impedance is purely reactive (*d*) current is minimum.

20. A d.c. series motor develops a torque of 20 Nm at 3A of load current. If the current is increased to 6A, the torque developed will be

(*a*) 10 Nm (*b*) 20 Nm
(*c*) 80 Nm (*d*) 40 Nm.

21. In a synchronous generator delivering lagging power factor load

(*a*) the excitation e.m.f. leads terminal voltage by the power angle
(*b*) the excitation e.m.f. lags the terminal voltage by the power angle
(*c*) excitation e.m.f. leads the terminal voltage by the power factor angle
(*d*) none of the above.

22. In an a.c. circuit if voltage $V = (a + jb)$ and current $I = (c + jd)$, then the power is given by

(*a*) $ac + ad$ (*b*) $ac + bd$
(*c*) $bc - ad$ (*d*) $bc + ad$.

23. A 3-phase, 400 V, 4-pole induction motor is fed from a 3-phase, 400 V, 50 Hz supply and runs at 1440 rpm. The frequency of rotor e.m.f. is

(*a*) 2 Hz (*b*) 50 Hz
(*c*) 48 Hz (*d*) zero Hz.

24. If two synchronous generators are connected, loss of synchronism will result in

(*a*) stalling generators (*b*) wild fluctuations in current
(*c*) wild fluctuations in current and voltage (*d*) none of the above.

25. For stable operation of inter-connected system, the passive element that can be used as the inter-connecting element is

(*a*) reactor (*b*) resistor
(*c*) capacitor (*d*) resistor and capacitor.

26. A *p*-pole lap wound d.c. machine has an armature current I_a. The conductor current in the armature winding is

(*a*) I_a (*b*) I_a/p
(*c*) pI_a (*d*) none of above.

27. While performing short-circuit test on a transformer the impressed voltage magnitude is kept constant but the frequency is increased. The short circuit current will

(*a*) increase (*b*) decrease
(*c*) remain the same (*d*) none of the above.

28. The stator referred resistance in the equivalent circuit of an induction motor, representing mechanical output is

(*a*) $\frac{r_2'}{s}$ (*b*) $r_2\left(\frac{1}{s}-1\right)$

(*c*) $\frac{r_2}{s}$ (*d*) $r_2^2\left(\frac{1}{s}-1\right)$.

29. In a transformer, the exciting current will be in phase quadrature with the impressed voltage provided

(*a*) only the leakage impedance drop is ignored

(*b*) only the core loss is ignored

(*c*) both the leakage impedance drop and the core loss are ignored

(*d*) only the no-load copper loss is ignored.

30. Armature reaction in a synchronous motor at rated voltage and zero power factor (lead) is

(*a*) magnetising (*b*) cross-magnetising

(*c*) both magnetising and cross-magnetising (*d*) demagnetising.

31. In a single phase hysteresis motor

(*a*) starting torque is caused by both eddy current and hysteresis while the running torque is caused by hysteresis

(*b*) starting as well as running torques are caused by both eddy current and hysteresis

(*c*) starting torque is caused by only eddy current while running torque is caused by only hysteresis

(*d*) starting as well as running torques both are caused by only hysteresis.

32. A transformer having constant flux and constant current density, designed for minimum cost must satisfy the following relation :

(*a*) iron loss = copper loss (*b*) weight of iron = weight of copper

(*c*) $\frac{\text{weight of iron}}{\text{weight of copper}} = \frac{\text{specific cost of copper}}{\text{specific cost of iron}}$

(*d*) $\frac{\text{weight of iron}}{\text{weight of copper}} = \frac{\text{specific cost of iron}}{\text{specific cost of copper}}$.

33. Compensating winding in a d.c. machine

(*a*) counteracts armature mmf in the interpolar zone

(*b*) prevents large speed drop

(*c*) shunts most of the armature current

(*d*) performs none of the above.

34. A transformer has hysteresis loss of 30 W, at 240 V, 60 Hz. The hysteresis loss at 200 V, 50 Hz will be

(*a*) 28 W (*b*) 25 W

(*c*) 30 W (*d*) 36 W.

35. A d.c. shunt motor does not operate on a.c. due to

(*a*) low resistance of armature circuit (*b*) high resistance of field circuit

(*c*) high inductance of field circuit (*d*) reversal of polarity.

36. A lap wound d.c. machine has 400 conductors and 8 poles. The voltage induced per conductor is 2 volts. The machine generates a voltage of

(*a*) 100 V (*b*) 200 V

(*c*) 400 V (*d*) 800 V.

37. An induction motor works with
(*a*) d.c. only (*b*) a.c. only
(*c*) a.c. and d.c. both (*d*) none of these.

38. If the prime-mover of an alternator supplying load to an infinite bus is suddenly shut down, then it will
(*a*) stop
(*b*) continue to run as an alternator
(*c*) continue to run as a synchronous motor in the reverse direction
(*d*) continue to run as a synchronous motor in the same direction.

39. A transformer steps up the voltage by a factor 100. The ratio of current in the primary to that in the secondary is
(*a*) 1 (*b*) 100
(*c*) 0.01 (*d*) 0.1.

40. Which of the plants is suitable for peak load ?
(*a*) Diesel engine plant (*b*) Steam power plant
(*c*) Nuclear power plant (*d*) Hydro-electric plant
(*e*) All of them.

41.

List I	List II
A. Conversion of a milliammeter into a voltmeter	1. Use of a shunt
B. Extension of an ammeter range	2. Use of a non-inductive series resistance
C. Extension of an a.c. voltmeter range	3. Use of a potential transformer
D. Conversion of an a.c. voltmeter into an a.c. ammeter	4. Use of a current transformer

Codes :	A	B	C	D
(*a*)	1	2	3	4
(*b*)	2	3	4	1
(*c*)	3	4	1	2
(*d*)	2	1	3	4

42. The power factor of a squirrel cage induction motor is
(*a*) low at light loads only (*b*) low at heavy loads only
(*c*) low at light and heavy loads both (*d*) low at rated load only.

43. A d.c. motor is running with a certain load. The effect of adding an external resistance in the shunt field circuit is
(*a*) to increase the motor speed (*b*) to reduce the motor speed
(*c*) to reduce the armature current of motor (*d*) to stop the motor.

44. The full load slip of a 60 Hz, 12-pole squirrel cage induction motor is 5%. Its full load speed is
(*a*) 600 r.p.m. (*b*) 570 r.p.m.
(*c*) 500 r.p.m. (*d*) 475 r.p.m.

45. The frequency of rotor currents at stand still is equal to
(*a*) zero (*b*) $2f$
(*c*) f (*d*) sf.

46. Silicon content in iron laminations is kept within 5% as it
(*a*) increases hysteresis loss (*b*) increases cost
(*c*) makes the material brittle (*d*) reduces the curie point.

47. An electric train employing a d.c. series motor is running at a fixed speed, when a sudden slight drop in the mains voltage occurs. This would result in

(*a*) drop in speed and rise in current (*b*) rise in speed and drop in current

(*c*) rise in speed and rise in current (*d*) drop in speed with current unaltered.

48. For an ideal transformer shown in the Fig. 1.

(*a*) $V_1 = nV_2, I_2 = -nI_1$ (*b*) $V_2 = nV_1, I_2 = -nI_1$

(*c*) $V_1 = nV_2, I_1 = 1/n\ I_2$ (*d*) $V_1 = nV_2, I_2 = -1/n\ I_1$.

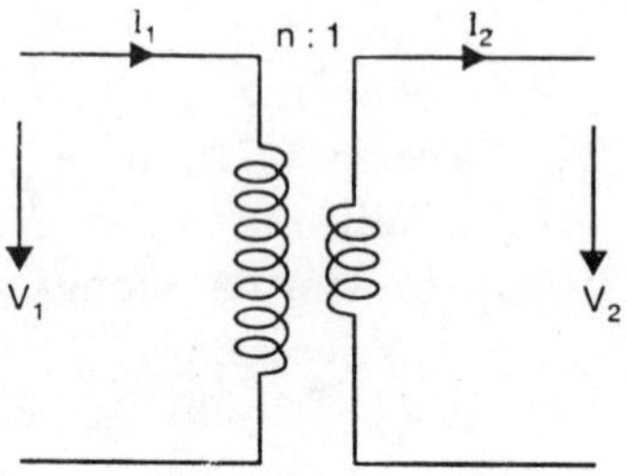

Fig. 1

49. Consider the following statements :

In a 3-phase induction motor connected to a 3-phase supply ; if one of the lines suddenly gets disconnected, then the

1. motor will come to a standstill
2. motor will continue to run at the same speed, with line current unchanged
3. motor will continue to run at a slightly reduced speed with increase in line current
4. rotor current will have both of sf and $(2 - s)f$ component frequencies where s is the slip and f is the supply frequency.

Of these statements

(*a*) 1 and 4 are correct (*b*) 1 and 2 are correct

(*c*) 3 and 4 are correct (*d*) 2 and 3 are correct.

50. The efficiency of a d.c. shunt generator is maximum when the

(*a*) magnetic losses are equal to mechanical losses

(*b*) field copper losses are equal to copper losses

(*c*) stray losses equal constant losses

(*d*) armature copper losses equal constant losses.

51. Synchronous reactance of alternator represents

(*a*) armature reactance and leakage reactance

(*b*) a reactance operating at synchronous speed

(*c*) field winding reactance

(*d*) a reactance connected in series with a synchronous generator.

52. A 210 MW generator should be cooled by

(*a*) water (*b*) air

(*c*) ice (*d*) hydrogen

(*e*) none of them.

53. $A - \overline{A}$ has the logic value

(*a*) 0 (*b*) 1

(*c*) A (*d*) $\overline{A}$.

54. A 220 V shunt motor develops a torque of 54 Nm at an armature current of 10 A. The torque developed when the armature current is 20 A is

(*a*) 54 Nm (*b*) 81 Nm

(*c*) 108 Nm (*d*) none of the above.

55. For a transformer to be tested at full load conditions but consuming only losses from the mains, we perform

(*a*) Load test (*b*) Open circuit and short-circuit tests

(*c*) Back-to-back test (*d*) None of the above.

56. In a transformer operating at constant voltage if the input frequency increases, the core loss

(*a*) increases (*b*) decreases

(*c*) remains constant (*d*) increases as square of the frequency.

57. A synchronous motor is connected to a constant voltage, constant supply frequency. The motor is initially operated at lagging power factor, when the field current is increased

(*a*) the load angle increases (*b*) the power input increases

(*c*) the power factor decreases

(*d*) the armature current decreases upto a certain limit and then increases.

58. 1 kVA, 230 V, 50 Hz, single phase transformer has an eddy current loss of 30 watts. The eddy current loss when the transformer is excited by a d.c. source of same voltage will be

(*a*) 30 W (*b*) more than 30 W

(*c*) less than 30 W (*d*) zero W.

59. In the transformer circuit mode, the core loss is represented as a

(*a*) series resistance (*b*) series inductance

(*c*) shunt resistance (*d*) shunt inductance.

60. Three 1 : 5 single phase transformers have their primaries connected in delta and the secondaries in star to supply a 3-phase load from a 400 V, 3-phase source. The line voltage on the load side is

(*a*) 2000 V (*b*) 80 V

(*c*) 3464 V (*d*) $80\sqrt{3}$ V.

61. A synchronous generator has its field winding on the rotor and armature winding on the stator. When running under steady state conditions its air gap field is

(*a*) stationary with respect to stator

(*b*) rotating at synchronous speed with respect to rotor

(*c*) rotating at synchronous speed against the direction of rotor rotation

(*d*) rotating at synchronous speed in the direction of rotor rotation.

62. For d.c. voltage an inductor

(*a*) is virtually a short-circuit (*b*) is virtually an open-circuit

(*c*) depends on polarity (*d*) depends on magnitude of voltage.

63. For speeds higher than say 3000 r.p.m., the machine used is

(*a*) induction motor (*b*) synchronous motor

(*c*) universal motor (*d*) none of the above.

64. In a shaded pole squirrel cage induction motor the flux in the shaded part always

(*a*) leads the flux in the unshaded pole

(*b*) is in phase with the flux in the unshaded pole segment

(*c*) lags the flux in the unshaded pole segment

(*d*) none of the above.

65. A 210 MW generator should be cooled by

(*a*) water (*b*) air

(*c*) ice (*d*) hydrogen

(*e*) none of them.

66. $A - \overline{A}$ has the logic value

(*a*) 0 (*b*) 1

(*c*) A (*d*) $\overline{A}$.

67. A 220 V shunt develops a torque of 60 Nm at an armature current of 10 A. The torque developed when the armature current is 20 A is
(*a*) 60 Nm (*b*) 90 Nm
(*c*) 120 Nm (*d*) none of the above.

68. Damper winding is used to
(*a*) reduce air gap harmonic flux (*b*) reduce oscillations
(*c*) increase stability limits (*d*) resist the moisture.

69. In an hysteresis motor, the rotor core must have
(*a*) retentivity (*b*) susceptibility
(*c*) resistivity (*d*) none of these.

70. The iron loss in a 100 kVA transformer is 1 kW and the full load copper losses are 2 kW, then maximum efficiency occurs at a load of
(*a*) 141.4 kVA (*b*) 70.7 kVA
(*c*) 50 kVA (*d*) 70.7 kW.

71. In d.c. machines, the armature windings are placed on the rotor because of the necessity for
(*a*) electro-mechanical energy conversion (*b*) generation of voltage
(*c*) commutation (*d*) development of torque.

72. The rotor of a 4-pole, 50 Hz, 3-phase slipring induction motor runs in clockwise direction when its stator terminals 1, 2, 3 are connected to supply terminals *A, B, C* respectively. If 1, 2, 3 are connected to *A, C, B*, respectively of supply terminals and rotor runs in a clockwise direction at synchronous speed, then the frequency of induced e.m.f. across the open circuited rotor terminals is
(*a*) 50 Hz (*b*) zero
(*c*) 25 Hz (*d*) 100 Hz.

73. The dynamic resistance of a parallel resonant circuit is given by
(*a*) $\dfrac{LC}{R_L}$ (*b*) LCR_L
(*c*) $\dfrac{C}{LR_L}$ (*d*) $\dfrac{L}{CR_L}$.

74. An instantaneous change in voltage is not possible in
(*a*) a resistor (*b*) an inductor
(*c*) a capacitor (*d*) a current source.

75. The d.c. generator works on the principle of
(*a*) Fleming's left hand rule (*b*) Fleming's right hand rule
(*c*) Lenz's law (*d*) None of these.

76. The efficiency of a power transformer is around
(*a*) 50% (*b*) 60%
(*c*) 80% (*d*) 95%.

77. An ideal transformer does not change
(*a*) voltage (*b*) power
(*c*) current (*d*) none of these.

78. A leading p.f. load on an alternator implies that its voltage regulation shall be
(*a*) positive (*b*) negative
(*c*) zero (*d*) any one of these.

79. A synchronous motor is supplying a load at unity p.f. If the load on the motor is increased keeping its excitation and terminal voltage constant, the power factor
(*a*) will remain the same (*b*) will become leading
(*c*) will become lagging (*d*) none of the above.

80. Reluctance torque is the
(*a*) torque developed in a given synchronous motor as a function of field excitation and the sine of the torque angle
(*b*) torque developed by salient pole synchronous or synchronous induction motors due to variations in air gap produced by armature reaction
(*c*) torque developed by non-salient pole asynchronous motor
(*d*) none of the above.

81. For the protection of a 3-phase star/delta transformer the CTs for the differential relay should be connected
(*a*) delta/star
(*b*) delta/delta
(*c*) star/delta
(*d*) star/star.

82. Tappings of a transformer are provided
(*a*) at the phase end of *lv* side
(*b*) at the phase end of *hv* side
(*c*) at the neutral end of *hv* side
(*d*) at the middle of the *hv* side.

83. Consider the following statements :
1. In a d.c. series generator, full load voltage is more than the no load voltage
2. In a separately excited d.c. generator full load voltage is more than the no load voltage
3. In a d.c. shunt generator, full load voltage is less than the no load voltage
Of these statements
(*a*) 1, 2 and 3 are correct
(*b*) 1 and 2 are correct
(*c*) 2 and 3 are correct
(*d*) 1 and 3 are correct.

84. A lap wound armature winding, fitted with a commutator and a pair of brushes on it is rotated at a speed N_r in a rotating magnetic field having *p* poles and rotating at a speed N_f in space, N_f and N_r both being in the same direction. The frequency of induced voltage across the brushes on the commutator is
(*a*) $\frac{N_f p}{120}$
(*b*) $\frac{(N_f - N_r)}{120} p$
(*c*) $\frac{(N_f + N_r)p}{120}$
(*d*) $\frac{N_r p}{120}$.

85. When a balanced 3-phase distributed type armature winding is carrying 3-phase, balanced currents, the strength of the resultant rotating magnetic field is
(*a*) three times the amplitude of each constituent of pulsating magnetic field
(*b*) equal to the amplitude of each constituent of pulsating magnetic field
(*c*) half the amplitude of each constituent of pulsating magnetic field
(*d*) one and half times the amplitude of each constituent of pulsating magnetic field.

86. Compensating winding in a d.c. machine
(*a*) counteracts armature mmf in the interpolar zone
(*b*) prevents large speed drop
(*c*) shunts most of the armature current
(*d*) performs none of the above.

87. A transformer has hysteresis loss of 60 W, at 240 V, 60 Hz. The hysteresis loss at 200 V, 50 Hz will be
(*a*) 48 W
(*b*) 50 W
(*c*) 60 W
(*d*) 75 W.

88. Drop in terminal voltage of an alternator due to armature reaction is countered by
(*a*) damper winding
(*b*) effect of saliency
(*c*) increased prime-mover output
(*d*) automatic voltage regulator.

89. A 3-phase, 400 V, 4-pole induction motor is fed from a 3-phase, 400 V, 50 Hz supply and runs at 1440 r.p.m. The frequency of rotor e.m.f. is

(*a*) 2 Hz (*b*) 50 Hz

(*c*) 48 Hz (*d*) zero Hz.

90. A p-pole lap wound d.c. machine has an armature current I_a. The conductor current in the armature winding is

(*a*) I_a (*b*) I_a/p

(*c*) pI_a (*d*) none of above.

91. While performing short-circuit test on a transformer the impressed voltage magnitude is kept constant but the frequency is increased. The short circuit current will

(*a*) increase (*b*) decrease

(*c*) remain the same (*d*) none of the above.

92. The stator referred resistance in the equivalent circuit of an induction motor, representing mechanical output is

(*a*) $\frac{r_2'}{s^2}$ (*b*) $r_2\left(\frac{1}{s^2}-1\right)$

(*c*) $\frac{r_2}{s}$ (*d*) $r_2^2\left(\frac{1}{s^2}-1\right)$.

93. In a transformer, the exciting current will be in phase quadrature with the impressed voltage provided

(*a*) only the leakage impedance drop is ignored

(*b*) only the core loss is ignored

(*c*) both the leakage impedance drop and the core loss are ignored

(*d*) only the no-load copper loss is ignored.

94. Armature reaction in a synchronous motor at rated voltage and zero power factor (lead) is

(*a*) magnetising (*b*) cross-magnetising

(*c*) both magnetising and cross-magnetising (*d*) demagnetising.

95. A separately excited d.c. generator is feeding a d.c. shunt motor. If the load torque on the motor is halved approximately

(*a*) armature currents of both motor and generator are halved

(*b*) armature current of motor is halved and that of the generator is unaltered

(*c*) armature current of generator is halved and that of motor is unaltered

(*d*) armature currents of both motor and generator are unaltered.

96. A 3-phase 4-pole alternator has 48 stator slots carrying the 3-phase distributed winding. Each coil of the winding is short chorded by one slot pitch. The winding factor is given by

(*a*) $\frac{1}{16}\cot(7.5°)$ (*b*) $\frac{1}{8}\cot(7.5°)$

(*c*) $\frac{1}{16}\sin(7.5°)$ (*d*) $\cos(7.5°)$.

97. In a 3-phase induction machine, motoring, generating and braking operations take place in the range of slip "s" given by

(*a*) motoring : $1 > s > 0$; generating : $0 > s > -1$; braking : $s > 1$

(*b*) motoring : $s > 1$; generating : $1 > s > 0$; braking : $0 > s > -1$

(*c*) motoring : $s > 1$; generating : $0 > s > -1$; braking : $1 > s > 0$

(*d*) motoring : $0 > s > -1$; generating : $s > 1$; braking : $1 > s > 0$.

98. The main reason for using a hysteresis motor for high quality tape-recorders and record players is that
(*a*) its speed is constant (synchronous) (*b*) it develops extremely steady torque
(*c*) it requires no centifugal switch
(*d*) its operation is not affected by mechanical vibrations.

99. The efficiency of a d.c. shunt generator is maximum when the
(*a*) magnetic losses are equal to mechanical losses
(*b*) field copper losses are equal to copper losses
(*c*) stray losses equal constant losses
(*d*) armature copper losses equal constant losses.

100. Synchronous reactance of alternator represents
(*a*) armature reactance and leakage reactance
(*b*) a reactance operating at synchronous speed
(*c*) field winding reactance
(*d*) a reactance connected in series with a synchronous generator.

101. The e.m.f. induced in a conductor rotating in bipolar field is
(*a*) d.c. (*b*) a.c.
(*c*) d.c. and a.c. both (*d*) none of these.

102. The power factor of a squirrel cage induction motor is
(*a*) low at light loads only (*b*) low at heavy loads only
(*c*) low at light and heavy loads both (*d*) low at rated load only.

103. A d.c. motor is running with a certain load. The effect of adding an external resistance in the shunt field circuit is
(*a*) to increase the motor speed (*b*) to reduce the motor speed
(*c*) to reduce the armature current of motor (*d*) to stop the motor.

104. The full load slip of a 60 Hz, 12-pole squirrel cage induction motor is 5%. Its full load speed is
(*a*) 600 r.p.m. (*b*) 570 r.p.m.
(*c*) 500 r.p.m. (*d*) 475 r.p.m.

105. The frequency of rotor currents at stand still is equal to
(*a*) zero (*b*) $2f$
(*c*) f (*d*) sf.

106. The direction of rotation of a d.c. shunt motor can be reversed by inter-changing
(*a*) the supply terminals (*b*) the field terminals only
(*c*) the armature terminals only (*d*) either field or armature terminals.

107. Resonant frequency f_r of a series *R-L-C* circuit is related to half power frequencies f_1 and f_2 as
(*a*) $f_r = \frac{f_1 + f_2}{2}$ (*b*) $f_r = \sqrt{f_1 f_2}$
(*c*) $f_r = f_2 - f_1$ (*d*) $f_r = (\sqrt{f_1} + \sqrt{f_2})^2$.

108. An induction motor works with
(*a*) d.c. only (*b*) a.c. only
(*c*) a.c. and d.c. both (*d*) none of these.

109. If the prime-mover of an alternator supplying load to an infinite bus is suddenly shut down, then it will
(*a*) stop (*b*) continue to run as an alternator
(*c*) continue to run as a synchronous motor in the reverse direction
(*d*) continue to run as a synchronous motor in the same direction.

110. A transformer steps up the voltage by a factor 100. The ratio of current in the secondary to that in the primary is
(*a*) 1 (*b*) 100
(*c*) 0.01 (*d*) 0.1.

111. In a double revolving field theory of single-phase induction motor, the slip of the forward motor is s then the slip of the backward motor is
(*a*) $2s$ (*b*) s
(*c*) $2 - s$ (*d*) $s - 2$.

112. A 2 kVA transformer has iron loss of 150 W and full load copper loss of 250 W. The maximum efficiency of the transformer would occur when the total loss is
(*a*) 500 W (*b*) 400 W
(*c*) 300 W (*d*) 275 W.

113. Consider the following statements :
1. In a d.c. series generator, full load voltage is more than the no load voltage
2. In a separately excited d.c. generator full load voltage is more than the no load voltage
3. In a d.c. shunt generator, full load voltage is less than the no load voltage
Of these statements
(*a*) 1, 2 and 3 are correct (*b*) 1 and 2 are correct
(*c*) 2 and 3 are correct (*d*) 1 and 3 are correct.

114. Semi-closed slots or totally closed slots are used in induction motors essentially to
(*a*) improve starting torque (*b*) increase pull-out torque
(*c*) increase efficiency
(*d*) reduce magnetizing current and improve power factor.

115. A lap wound armature winding, fitted with a commutator and a pair of brushes on it is rotated at a speed N_r in a rotating magnetic field having p poles and rotating at a speed N_f in space, N_f and N_r both being in the same direction. The frequency of induced voltage across the brushes on the commutator is
(*a*) $\frac{N_f p^2}{120}$ (*b*) $\frac{(N_f - N_r)}{120} p$
(*c*) $\frac{(N_f + N_r)p^2}{120}$ (*d*) none of the above.

116. When a balanced 3-phase distributed type armature winding is carrying 3-phase, balanced currents, the strength of the resultant rotating magnetic field is
(*a*) three times the amplitude of each constituent of pulsating magnetic field
(*b*) equal to the amplitude of each constituent of pulsating magnetic field
(*c*) half the amplitude of each constituent of pulsating magnetic field
(*d*) one and half times the amplitude of each constituent of pulsating magnetic field.

117. The response of a series *R-L-C* circuit fed from a fixed r.m.s. voltage and variable frequency source is represented graphically in the Fig. 2. Match List I with List II and select the correct answer

List I	**List II**
(*Curve*)	(*Quantity*)
A. AA	1. Current
A. BB	2. Impedance
C. CC	3. Capacitive reactance

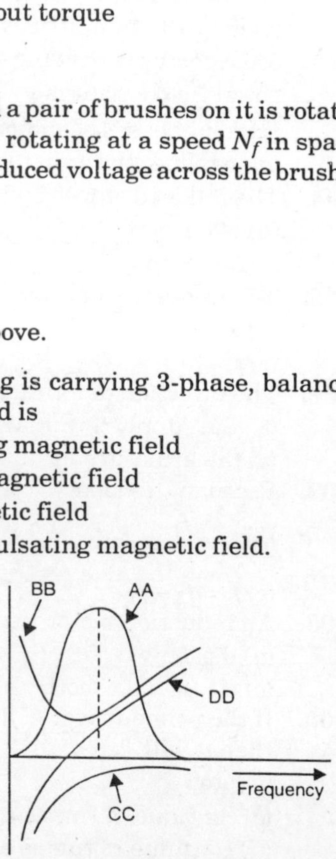

Fig. 2

D. DD 4. Net reactance
5. Inductive reactance

Codes :	A	B	C	D
(*a*)	2	1	3	5
(*b*)	1	2	3	5
(*c*)	1	2	3	4
(*d*)	1	2	4	3

118. A 3-phase synchronous motor connected to an infinite bus is operating at half-full load with normal excitation. When the load on the synchronous motor is suddenly increased
(*a*) its speed will first decrease and then become synchronous
(*b*) its speed will first increase and then become synchronous
(*c*) its speed will remain unchanged
(*d*) none of the above.

119. In a d.c. machine if the number of slots in the armature are more
(*a*) cooling is likely to be poor
(*b*) commutation will be poor
(*c*) cost will be less
(*d*) none of the above.

120. The presence of a dominant 7th harmonic in the winding distribution of a 3-phase, 6-pole, 50 Hz induction motor may cause the motor to crawl at a speed of about
(*a*) 750 rpm
(*b*) 500 rpm
(*c*) 242 rpm
(*d*) 143 rpm.

121. An 11 kV/400 V, 1000 kVA, *Y*/*Y* transformer is reconnected as Δ/*Y* with the high voltage side connected in delta. The rating for new connection will be
(*a*) $11/\sqrt{3}$ kV/400 V, 1000 kVA
(*b*) 11 kV/400 V, 1732 kVA
(*c*) $11\sqrt{3}$ kV/400 V, 1000 kVA
(*d*) 11 kV/$400\sqrt{3}$ V, $1000/\sqrt{3}$ kVA.

122. A round rotor synchronous generator has a leakage reactance of 10%, armature reaction reactance of 90% and negligible armature resistance. With the machine initially running at rated speed and terminal voltage of 1.0 pu, a 3-phase short-circuit is applied. The sustained armature current will be
(*a*) 1.25 pu
(*b*) 1.11 pu
(*c*) 1.0 pu
(*d*) 0.9 pu.

123. A d.c. shunt motor drives a load at rated speed and supply voltage. If both the load and voltage are halved, the speed of the motor will be
(*a*) almost doubled
(*b*) almost halved
(*c*) the rated speed
(*d*) none of the above.

124. A 3-phase induction motor runs at super synchronous speed. For self excitation the machine
(*a*) draws real power from the mains
(*b*) draws reactive power from the mains
(*c*) feeds reactive power to the mains
(*d*) generates e.m.f. at the expense of residual magnetism.

125. T is the load torque of a d.c. series motor having linear magnetization and negligible armature resistance. Speed of the motor is
(*a*) inversely proportional to $\sqrt{T}$
(*b*) directly proportional to $\sqrt{T}$
(*c*) inversely proportional to T^2
(*d*) directly proportional to T^2.

126. In the transformer circuit mode, the core loss is represented as a
(*a*) series resistance
(*b*) series inductance
(*c*) shunt resistance
(*d*) shunt inductance.

127. Three 1 : 5 single phase transformers have their primaries connected in delta and the secondaries in star to supply a 3-phase load from a 400 V, 3-phase source. The line voltage on the load side is

(*a*) 2000 V (*b*) 80 V

(*c*) 3464 V (*d*) $80\sqrt{3}$ V.

128. A synchronous generator has its field winding on the rotor and armature winding on the stator. When running under steady state conditions its air gap field is

(*a*) stationary with respect to stator

(*b*) rotating at synchronous speed with respect to rotor

(*c*) rotating at synchronous speed against the direction of rotor rotation

(*d*) rotating at synchronous speed in the direction of rotor rotation.

129. When a synchronous generator is designed with lower SCR it

(*a*) will give higher stability limit (*b*) will have lower SC current

(*c*) will give better voltage regulation (*d*) will have high synchronizing power.

130. For a soft magnetic material there will be

(*a*) steeply rising magnetization curve

(*b*) relatively narrow and small hysteresis loop

(*c*) less energy loss per cycle of magnetization

(*d*) all the above.

131. In a double revolving field theory of single-phase induction motor, the slip of the forward motor is *s* then the slip of the backward motor is

(*a*) 2*s* (*b*) *s*

(*c*) 3 – *s* (*d*) none of the above.

132. A 2 kVA transformer has iron loss of 300 W and full load copper loss of 500 W. The maximum efficiency of the transformer would occur when the total loss is

(*a*) 1000 W (*b*) 800 W

(*c*) 600 W (*d*) 550 W.

133. Speed of the universal motor is

(*a*) dependent on frequency of supply (*b*) proportional to frequency of supply

(*c*) independent of frequency of supply (*d*) none of the above.

134. If all the dimensions of a transformer are doubled its iron loss will be

(*a*) half (*b*) double

(*c*) four times (*d*) eight times.

135. What is the arrangement of windings in a core type single phase transformer ?

(*a*) Half *LV* inside and half *HV* outside on each core limb

(*b*) *LV* on one core limb and *HV* on the other

(*c*) Sandwiched *LV* and *HV* discs on each core limb

(*d*) Half *HV* inside and half *LV* outside on each core limb.

136. In a synchronous generator delivering lagging power factor load

(*a*) the excitation e.m.f. lags terminal voltage by the power angle

(*b*) the excitation e.m.f. lags the terminal voltage by the power angle

(*c*) excitation e.m.f. lags the terminal voltage by the power factor angle

(*d*) none of the above.

137. Consider the following single-phase motors :

I. Capacitor start motor

II. Capacitor start and run motor

III. Permanent split capacitor motor
IV. Shaded pole motor
The correct sequence of the increasing order of their cost is
(*a*) IV, III, II, I (*b*) IV, III, I, II
(*c*) III, IV, II, I (*d*) III, IV, I, II.

138. A synchronous motor is connected to a constant voltage, constant supply frequency. The motor is initially operated at lagging power factor, when the field current is increased
(*a*) the load angle increases (*b*) the power input increases
(*c*) the power factor decreases
(*d*) the armature current decreases upto a certain limit and then increases.

139. LVDT windings are wound on
[illegible] sheets (laminated) (*b*) aluminium
(*c*) ferrite (*d*) copper.

140. A synchronous machine connected to a power system grid bus-bar is operating as a generator. To make the machine operate as a motor, the
(*a*) direction of rotation is to be reversed (*b*) phase-sequence is to be changed
(*c*) field excitation is to be decreased
(*d*) mechanical input is to be less than the losses at the shaft.

141. Cascade method of speed control involves the use of two coupled induction motors. The necessary conditions for speed control is that
(*a*) both the motors are of the wound rotor type having the same number of poles
(*b*) both the motors are of the squirrel cage rotor type having different number of poles
(*c*) one motor is of the slip-ring type but both the motors have the same number of poles
(*d*) one motor is of the slip-ring type and the two motors have different number of poles.

142. An induction motor has rotor resistance R_r, stand-still rotor induced voltage E_{ss}, and stator to rotor effective turn-ratio of m. In an equivalent circuit of this machine, the rotor circuit resistance is shown as m^2R_r/s, where s is the slip. This implies that the value of the equivalent rotor circuit voltage will be
(*a*) E_{ss} (*b*) sE_{ss}
(*c*) m^2E_{ss} (*d*) none of the above.

143. An induction motor has a rotor resistance of 0.002 ohm/phase. If the resistance is increased to 0.004 ohm/phase then the maximum torque will
(*a*) reduce to half (*b*) increase by 100%
(*c*) increase by 200% (*d*) remain unaltered.

144. A 5 kVA transformer has a turn ratio of $(N_1/N_2) = 10$. The impedance of the primary winding is $3 + j5$ ohms while that of secondary winding is $0.5 + j0.8$ ohms. The impedance of the transformer when referred to the primary will be
(*a*) $3.5 + j5.8$ ohms (*b*) $8.0 + j13.0$ ohms
(*c*) $53.0 + j85$ ohms (*d*) $3.05 + j5.08$ ohms.

145. The power factor for a circuit is of the order of
(*a*) 0.3 (*b*) 0.45
(*c*) 0.6 (*d*) 0.9.

146. The four methods of calculation of voltage regulation of a 3-phase alternator are
1. EMF method
2. Saturated synchronous reactance method

3. New ASA method
4. MMF method

The correct sequence of the ascending order of the values of regulation obtained by these methods is

(*a*) 3, 4, 2, 1 (*b*) 4, 3, 1, 2
(*c*) 3, 4, 1, 2 (*d*) 4, 3, 2, 1.

147. Consider the following statements regarding the starting of a d.c. shunt motor :

1. External field resistance should be minimum
2. External field resistance should be maximum
3. Reduced voltage should be applied to the armature
4. Increased voltage should be applied to the armature.

Of these statements

(*a*) 1 and 2 are correct (*b*) 1 and 4 are correct
(*c*) 2 and 3 are correct (*d*) none of the above.

148. For a specific open circuit voltage of a d.c. generator, the short-circuit current will be maximum when it is

(*a*) separately excited (*b*) shunt connected
(*c*) cumulatively compounded (*d*) differentially compounded.

149. To eliminate the 5th harmonic from the e.m.f. generated in an alternator, the pitch fraction will be

(*a*) 4/5 (*b*) 5/4
(*c*) 5/6 (*d*) 6/5.

150. The maximum possible speed at which an alternator can be driven to generate 50 Hz and 4000 V is

(*a*) 4000 r.p.m. (*b*) 3600 r.p.m.
(*c*) 3000 r.p.m. (*d*) 1500 r.p.m.

151. When the supply terminals of a d.c. shunt motor are interchanged

(*a*) the motor will stop
(*b*) the motor will run at its normal speed in the same direction as before
(*c*) the direction of rotation will reverse
(*d*) the motor will run much faster in the same direction.

152. Under short-circuit conditions the windings of a transformer having 5% impedance will experience a radial force

(*a*) 5 times of full load value (*b*) 20 times of full load value
(*c*) 25 times of full load value (*d*) none of the above.

153. Power despatch through a line can be increased by

(*a*) installing shunt reactor (*b*) installing shunt capacitors
(*c*) installing series reactor (*d*) none of the above.

154. Addition of tubes to the transformer tank improves heat dissipation capacity because of

(*a*) additional cooling surface (*b*) additional dissipation by radiation only
(*c*) additional dissipation by convection only
(*d*) additional dissipation by radiation.

155. Buchholz relay is a

(*a*) oil actuated relay (*b*) current actuated relay
(*c*) gas actuated relay (*d*) oil temperature-actuated relay.

156. The brushes of a d.c. motor suitable for rotation in any direction should be placed as
(*a*) radial (*b*) reaction
(*c*) trailing (*d*) any of these.

157. Minimum number of coils for 400 V, 6-pole lap connected d.c. machine for a maximum voltage of 15 volts between adjacent commutator segments should be
(*a*) 80 (*b*) 160
(*c*) 200 (*d*) 400.

158. Consider the following statements :
In a 3-phase induction motor connected to a 3-phase supply ; if one of the lines suddenly gets disconnected, then the
1. motor will come to a standstill
2. motor will continue to run at the same speed, with line current unchanged
3. motor will continue to run at a slightly reduced speed with increase in line current
4. rotor current will have both of sf and $(2 - s)f$ component frequencies where s is the slip and f is the supply frequency

Of these statements
(*a*) 1 and 4 are correct (*b*) 1 and 2 are correct
(*c*) 3 and 4 are correct (*d*) 2 and 3 are correct.

159. When a 3-phase induction motor is designed with higher value of B_m it will give
(*a*) better full-load power factor (*b*) a higher starting torque
(*c*) higher full load efficiency (*d*) higher over-load capacity.

160. The power input in blocked-rotor test performed on a three-phase induction motor is approximately equal to
(*a*) hysteresis loss in the core (*b*) I^2R-loss in the windings
(*c*) eddy-current loss in the core (*d*) iron-loss in the core.

161. The power input in no-load test performed on a three-phase induction motor is approximately equal to
(*a*) hysteresis loss in the core (*b*) I^2R-loss in the winding
(*c*) eddy-current loss in the core (*d*) iron-loss in the core.

162. In the equivalent circuit of a three-phase induction motor the mechanical load on the motor can be represented by a resistance of value
(*a*) R_2 (*b*) R_2/s
(*c*) $R_2 \frac{(1-s)}{s}$ (*d*) $\frac{R^2}{s} + 1$.

163. A d.c. series motor when connected across an a.c. supply will
(*a*) develop torque in the same direction (*b*) not develop any torque
(*c*) draw dangerously high current (*d*) develop a pulsating torque.

164. To enable a d.c. series motor work satisfactorily with an a.c. supply, the following modifications should be done
(*a*) the yoke and the poles should be completely laminated
(*b*) only the poles should be made of laminated sheets
(*c*) the air-gap between the stator and the rotor be reduced
(*d*) compensating poles should be introduced.

165. Cooling of transformers is necessary to
(*a*) increase the efficiency (*b*) dissipate the heat generated in the windings
(*c*) reduce the losses (*d*) reduce humming.

166. The e.m.f. induced in the windings of a transformer will
(*a*) lag the core flux by 90° (*b*) be in-phase with the core flux
(*c*) be out-of-pulse with the core flux (*d*) be independent of the core flux.

167. A 400 V, 50 Hz, 4-pole, three-phase induction motor cannot run at 1500 r.p.m. because
(*a*) at 1500 r.p.m. there will be no e.m.f. induced in the rotor circuit and hence no torque will be developed
(*b*) an induction motor can run only at a speed higher than its synchronous speed
(*c*) at 1500 r.p.m., torque developed by the rotor may not be sufficient to rotate the rotor
(*d*) at 1500 r.p.m., the rotor will draw excessive current and may be harmful to the motor.

168. The phenomenon of squirrel-cage motors sometimes showing a tendency to run at a very low speed is known as
(*a*) cogging (*b*) crawling
(*c*) damping (*d*) skewing.

169. The speed of revolving field for a 50 Hz, 8-pole machine will be
(*a*) 1500 r.p.m. (*b*) 1440 r.p.m.
(*c*) 1000 r.p.m. (*d*) 750 r.p.m.

170. The torque developed by an induction motor is
(*a*) directly proportional to the square of the rotor resistance
(*b*) directly proportional to the square of the supply voltage
(*c*) inversely proportional to the supply voltage
(*d*) inversely proportional to the slip.

171. The following data pertain to two alternators working in parallel and supplying a total load of 80 MW :
Machine 1 : 40 MVA with 5% speed regulation
Machine 2 : 60 MVA with 5% speed regulation
The load sharing between machine 1 and 2 will be

	P1	P2
(*a*)	48 MW	32 MW
(*b*)	40 MW	40 MW
(*c*)	30 MW	50 MW
(*d*)	32 MW	48 MW.

[U.P.S.C. I.E.S. Elec. Engineering II, 1997]

172. Two similar synchronous generators are working in parallel to supply a common load demand with identical excitations and steam supplies to their prime-movers. Now, if the steam supply to the prime-mover of one of the generators is increased compared to the other, with field excitation kept unchanged, then
(*a*) its active power component will remain the same but the reactive power contribution will increase
(*b*) its active power will decrease while the reactive power will increase
(*c*) both active and reactive components of power will increase
(*d*) its active power contribution will increase but reactive power contributions of both will remain unchanged. **[U.P.S.C. I.E.S. Elec. Engg. II, 1993]**

173. The main reason for using a hysteresis motor for high quality tape recorders and record players is that
(*a*) its speed is constant (synchronous) (*b*) it develops extremely steady torque
(*c*) it requires no centrifugal switch
(*d*) its operation is not affected by mechanical vibrations.

[U.P.S.C. I.E.S. Elec. Engineering II, 1995]

174. If the supply frequency of a transformer is increased, keeping the supply voltage constant, then

(*a*) both magnetising component and core loss component of current will decrease
(*b*) both magnetising component and core loss component of current will increase
(*c*) magnetising component of current will increase while core loss component of current will decrease
(*d*) magnetising component of current will decrease while core loss component of current will increase. **[U.P.S.C. I.E.S. Elec. Engineering II, 1996]**

175. In a transformer operating at constant voltage if the input frequency increases, the core loss

(*a*) increases
(*b*) decreases
(*c*) remains constant
(*d*) increases as square of the frequency.

176. A universal motor runs at

(*a*) higher speed with d.c. supply and with less sparking
(*b*) higher speed with a.c. supply and with less sparking
(*c*) same speed with both a.c. and d.c. supplies
(*d*) higher speed with a.c. supply but with increased sparking at the brushes.

[U.P.S.C. I.E.S. Elec. Engineering II, 1994]

177. When the induction motor is standstill the slip is

(*a*) Zero
(*b*) 1
(*c*) infinity
(*d*) none of the above.

178. A separately excited d.c. generator is feeding a d.c. shunt motor. If the load torque on the motor is halved approximately

(*a*) armature currents of both motor and generator are halved
(*b*) armature current of motor is halved and that of the generator is unaltered
(*c*) armature current of generator is halved and that of motor is unaltered
(*d*) armature currents of both motor and generator are unaltered.

179. A 3-phase 4-pole alternator has 48 stator slots carrying the 3-phase distributed winding. Each coil of the winding is short chorded by one slot pitch. The winding factor is given by

(*a*) $\frac{1}{16}$ cot (7.5°)
(*b*) $\frac{1}{12}$ cot (7.5°)
(*c*) $\frac{1}{16}$ sin (7.5°)
(*d*) none of the above.

180. A 220 V d.c. machine has an armature resistance of 1 Ω. If the full load current is 20 A, the difference in the induced voltage when the machine is running as a motor, and as a generator is

(*a*) 20 V
(*b*) zero
(*c*) 40 V
(*d*) 50 V.

181. A centre zero ammeter connected in the rotor circuit of a 6-pole, 50 Hz induction motor makes 30 oscillations in one minute. The rotor speed is

(*a*) 970 r.p.m.
(*b*) 990 r.p.m.
(*c*) 1010 r.p.m.
(*d*) 1030 r.p.m.

182. The type of d.c. generator used for arc welding purposes is a

(*a*) series generator
(*b*) shunt generator
(*c*) cumulatively compounded generator
(*d*) differentially compounded generator.

183. For the protection of stator winding of an alternator against internal fault involving ground, the relay used is a

(*a*) biased differential relay
(*b*) directional over-current relay
(*c*) plain impedance relay
(*d*) Buchholz relay.

184. Two 3-phase transformers are to be connected for parallel operation. Which one of the following arrangements is impossible ?

(*a*) Transformer A ; primary Y ; secondary Y
Transformer B ; primary Δ ; secondary Δ
(*b*) Transformer A ; primary Δ ; secondary Y
Transformer B ; primary Δ ; secondary Δ
(*c*) Transformer A ; primary Y ; secondary Δ
Transformer B ; primary Δ ; secondary Y
(*d*) Transformer A ; primary Δ ; secondary Δ
Transformer B ; primary Δ ; secondary Y.

185. In a 3-phase induction machine, motoring, generating and braking operations take place in the range of slip "*s*" given by

(*a*) motoring : $1 > s > 0$; generating : $0 > s > -1$ braking : $s > 1$
(*b*) motoring : $s > 1$; generating : $1 > s > 0$ braking : $0 > s > -1$
(*c*) motoring : $s > 1$; generating : $0 > s > -1$ braking : $1 > s > 0$
(*d*) motoring : $0 > s > -1$; generating : $s > 1$ braking : $1 > s > 0$.

186. The main reason for using a hysteresis motor for high quality tape-recorders and record players is that

(*a*) its speed is constant (synchronous)
(*b*) it develops extremely variable torque
(*c*) it requires no centrifugal switch
(*d*) none of the above.

187. A belt-driven cumulative compounded d.c. generator is delivering power to the d.c. main. If the belt snaps then the machine will run as a

(*a*) cumulative compounded motor in the same direction
(*b*) differentially compounded motor in the same direction
(*c*) cumulative compounded motor in the opposite direction
(*d*) differentially compounded motor in the opposite direction.

188. An *R-L-C* series circuit has f_1 and f_2 as the half power frequencies and f_0 as the resonant frequency. The *Q*-factor of the circuit is given by

(*a*) $\frac{f_1 + f_2}{2f_0}$ (*b*) $\frac{f_1 - f_0}{f_2 - f_0}$
(*c*) $\frac{f_0}{f_1 - f_2}$ (*d*) $\frac{f_1 - f_2}{f_0}$.

189. A coil having a resistance of 5 Ω and inductance of 0.1 H is connected in series with a condenser of capacitance 50 μF. A constant alternating voltage of 200 V is applied to the circuit. The voltage across the coil at resonance is

(*a*) 200 volts (*b*) 1788 volts
(*c*) 1800 volts (*d*) 2000 volts.

190. Two mechanically coupled alternators deliver power at 50 Hz and 60 Hz respectively. The highest speed of the alternators is

(*a*) 3600 r.p.m. (*b*) 3000 r.p.m.
(*c*) 600 r.p.m. (*d*) 500 r.p.m.

191. A cylindrical rod of magnetic material fits tightly when it is inserted into a long solenoid, half way through. The nature of the force on the rod and the type of material of the rod are listed below. Match List 1 and List 2 and select the correct answer using codes given below the lists :

List 1 (*Material*)	List 2 (*Force*)
A. Diamagnetic	1. Nil
B. Paramagnetic	2. Repulsive
C. Ferromagnetic	3. Attractive

Codes :	A	B	C
(*a*)	2	2	3
(*b*)	1	3	3
(*c*)	2	3	3
(*d*)	3	2	2

192.

List 1 (*D.C. motors*)	List 2 (*Characteristics*)
A. Cumulatively compounded motor	1. Fairly constant speed irrespective of the load
B. Differentially compounded motor	2. It may start in reverse direction
C. Series motor	3. Define no-load speed
D. Shunt motor	4. Never started without load

Codes :	A	B	C	D
(*a*)	2	3	1	4
(*b*)	3	2	4	1
(*c*)	1	3	2	4
(*d*)	4	1	3	2

193. Consider the following single-phase motors :

I. Capacitor start motor
II. Capacitor start and run motor
III. Permanent split capacitor motor
IV. Shaded pole motor

The correct sequence of the increasing order of their cos

(*a*) IV, III, II, I
(*b*) II, III, I, IV
(*c*) III, IV, II, I
(*d*) none of the above.

194. An effect of phase-lag compensation on servo-system performance is that

(*a*) for a given relative stability, the velocity constant is increased
(*b*) for a given relative stability the velocity constant is decreased
(*c*) the bandwidth of the system is increased
(*d*) the time response is made faster.

195. The e.m.f. induced in the secondary winding of a 50 Hz single-phase transformer having 1000 turns on its secondary is 222 V. The maximum flux density in the core is 0.1 Wb/m^2. The cross-sectional area of the core is

(*a*) 0.1 m^2
(*b*) 0.01 m^2
(*c*) 1 m^2
(*d*) 0.001 m^2.

196. The copper-loss and core-loss of a transformer at various loads are as shown below :

Load	*Core-loss*	*Copper-loss*
(*a*) 50 kVA	320 W	500 W
(*b*) 40 kVA	320 W	320 W
(*c*) 30 kVA	320 W	180 W
(*d*) 20 kVA	320 W	80 W

At what load will the efficiency of the transformer be maximum ?

197. A synchronous motor and an alternator are delivering about 5% of their rated power at unity p.f. If these machines are loaded to their rated capacity, then their armature currents
(*a*) decrease, motor operates at lagging p.f. and alternator at leading p.f.
(*b*) increase, motor operates at leading p.f. and alternator at lagging p.f.
(*c*) increase, both operate at lagging p.fs.
(*d*) increase, motor operates at lagging p.f. and alternator at leading p.f.

198. The starting torque of an induction motor can be increased by
(*a*) increasing the rotor reactance
(*b*) increasing the rotor resistance
(*c*) increasing the supply frequency
(*d*) giving supply through a star-delta starter.

199. Which of the following can be reduced by skewing in induction motor ?
(*a*) Noise
(*b*) Vibration
(*c*) Synchronous cusps
(*d*) All of the above.

200. The speed of a three-phase induction motor will increase, if the
(*a*) number of poles of the stator winding is increased
(*b*) number of poles of the stator winding is decreased
(*c*) frequency of the stator supply is decreased
(*d*) resistance of the rotor circuit is increased.

201. In a resistance split-phase type single-phase induction motor, a time-phase difference between the currents in the main and auxiliary winding is achieved by
(*a*) placing the two windings at an angle of 90° electrical in the stator slots
(*b*) applying two-phase supply across the two windings
(*c*) having different ratio of resistance to inductive reactance for the two windings supplied from a single-phase supply system
(*d*) connecting the two windings in series across a single-phase supply.

202. In a split-phase capacitor-start induction motor, a time-phase difference between the currents in the main and auxiliary winding is achieved by
(*a*) placing the two windings at an angle of 90° electrical in the stator slots
(*b*) applying two-phase supply across the two windings
(*c*) introducing capacitive reactance in the auxiliary winding circuit
(*d*) connecting the two windings in series across a single-phase supply.

203. If any two leads from slip-rings are interchanged in a 3-phase induction motor, the motor will
(*a*) continue running in the same direction as before
(*b*) run in a direction opposite to previous one
(*c*) not run
(*d*) get damaged.

204. If all the stator coils of an induction motor are connected for the same magnetic polarity, there will be formed an equal number of
(*a*) rotor poles with same polarity
(*b*) rotor poles with opposite polarity
(*c*) consequent poles with opposite polarity
(*d*) consequent poles with same polarity.

205. The polyphase induction motors are, by a very considerable margin, the most widely used a.c. motors as
(*a*) these are cheaper in cost, simple and rugged in construction and require little maintenance
(*b*) they are capable of operation under a wide range of power factors both leading and lagging
(*c*) they have maximum efficiency and maximum starting torque
(*d*) they have very good speed regulation.

206. If two induction motors A and B are identical except that the air-gap of motor 'A' is 50% greater than that of motor 'B' then
(*a*) the no-load power factor of A will be better than that of B
(*b*) the no-load power factor of A will be poorer than that of B
(*c*) the core losses of A will be more than those of B
(*d*) the operating flux of A will be smaller than that of B. **[U.P.S.C. I.E.S., 1997]**

207. When the firing angle of a single phase, fully controlled rectifier feeding constant d.c. current into a load is 30°, the displacement power factor of the rectifier is
(*a*) 1 (*b*) 0.5
(*c*) $\frac{1}{\sqrt{3}}$ (*d*) $\frac{\sqrt{3}}{2}$. **[GATE, 1998]**

208. Two transformers of different kVA ratings are working in parallel, share the load in proportion to their ratings when their
(*a*) per unit leakage impedance on the same kVA base are the same
(*b*) per unit leakage impedance on their respective ratings are equal
(*c*) ohmic values of leakage impedances are inversely proportional to their ratings
(*d*) ohmic values of the magnetising reactances are the same. **[GATE, 1992]**

209. A double cage induction motor is a
(*a*) slip-ring type (*b*) squirrel cage type
(*c*) combination of squirrel cage and slip ring induction motors
(*d*) commutator type.

210. For successful parallel operation of two single phase transformers, the most essential condition is that their
(*a*) percentage impedances are equal (*b*) polarities are properly connected
(*c*) turn-ratios are exactly equal (*d*) kVA ratings are equal.
[U.P.S.C. I.E.S., 1994]

211. Two single phase transformers with equal turns have impedances of $(0.5 + j3)$ ohms and $(0.6 + j10)$ ohms with respect to the secondary. If they operate in parallel, how will they share a load of 100 kW at 0.8 pf lagging ?
(*a*) 50 kW, 50 kW (*b*) 62 kW, 38 kW
(*c*) 78.2 kW, 21.8 kW (*d*) 85.5 kW, 14.5 kW. **[U.P.S.C. I.E.S., 1992]**

212. The armature winding of a 2-pole 3-phase alternator for each phase is distributed in a number of slots per phase. The r.m.s. value of the voltage per phase is less than the r.m.s. value of the voltage per coil multiplied by the number of coils in series because the
(*a*) r.m.s. value of the voltage in different coils of the phase is different
(*b*) equal r.m.s. voltages in different coils of the phase has mutual phase difference
(*c*) maximum values of the induced voltage in different coils of the phase are different
(*d*) different coils of the phase pass through different saturated regions of the magnetic circuit.

213. A d.c. shunt generator having a shunt field of 50 Ω was generating normally at 1000 r.p.m. The critical resistance of the machine was 80 Ω. Due to some reasons, the speed of the prime-mover became such that the generator just failed to generate. The speed at that time must have been
(*a*) 1600 r.p.m. (*b*) 800 r.p.m.
(*c*) 625 r.p.m. (*d*) 500 r.p.m.

214. If the armature current in a d.c. machine is increased to double its previous value and the time of commutation is halved, then the reactance voltage will

(*a*) be halved (*b*) remain the same
(*c*) be doubled (*d*) become four times.

215. The presence of a dominant 7th harmonic in the winding distribution of a 3-phase 6-pole, 50 Hz induction motor may cause the motor to crawl at a speed of about

(*a*) 750 rpm (*b*) 500 rpm
(*c*) 242 rpm (*d*) 143 rpm.

216. An 11 kV/400 V, 1000 kVA, *Y*/*Y* transformer is reconnected as Δ/*Y* with the high voltage side connected in delta. The rating for new connection will be

(*a*) 11 kV/400 V, 1000 kVA (*b*) 11 kV/400 V, 1732 kVA
(*c*) $11\sqrt{3}$ kV/400 V, 1000 kVA (*d*) none of these.

217. A round rotor synchronous generator has a leakage reactance of 10%, armature reaction reactance of 90% and negligible armature resistance. With the machine initially running at rated speed and terminal voltage of 1.0 pu, a 3-phase short-circuit its applied. The sustained armature current will be

(*a*) 1.25 pu (*b*) 1.11 pu
(*c*) 1.0 pu (*d*) 0.9 pu.

218. If two meters *X* and *Y* require 40 mA and 50 mA respectively, to give full scale deflection, then

(*a*) *X* is more sensitive (*b*) *Y* is more sensitive
(*c*) both *X* and *Y* are equally sensitive
(*d*) it would not be possible to assess the sensitivity on the basis of the given data.

219. One of the control springs of a permanent magnet moving coil ammeter is broken. If connected in a circuit, the meter would read

(*a*) zero (*b*) half the correct value of the current
(*c*) twice the correct value of the current (*d*) a definite figure.

220. A line voltage regulator is to be used in a single phase 200 V, 5 kVA system to keep the voltage constant for voltage variations within ± 10%.

The rating (in kVA) of the voltage regulator is

(*a*) 0.05 (*b*) 0.5
(*c*) 5 (*d*) 50.

221. Consider the following statements regarding measurement of power in single phase a.c. circuits :

It is possible to measure power without using a wattmeter by

1. using one voltmeter and one ammeter
2. using two voltmeters and two ammeters
3. using three voltmeters
4. using three ammeters.

Of these statements

(*a*) 1 and 2 are correct (*b*) 2 and 3 are correct
(*c*) 3 and 4 are correct (*d*) 1 and 4 are correct.

222. A 35 V d.c. supply is connected across a resistance of 600 Ω in series with an unknown resistance *R*. A voltmeter having a resistance of 1.2 kΩ is connected across the 600 Ω resistance and reads 5 V. The value of resistance *R* will be

(*a*) 120 Ω (*b*) 500 Ω
(*c*) 1.7 kΩ (*d*) 2.4 kΩ.

223. Corona loss can be reduced by the use of hollow conductors because

(*a*) the current density is reduced

(*b*) the eddy current in the conductor is eliminated

(*c*) for a given cross-section, the radius of the conductor is increased

(*d*) of better ventilation in the conductor.

224. A universal motor runs at

(*a*) higher speed with d.c. supply and with less sparking

(*b*) higher speed with a.c. supply and with less sparking

(*c*) same speed with both a.c. and d.c. supplies

(*d*) higher speed with a.c. supply but with increased sparking at the brushes.

225.

List I (*Methods of control*)	**List II** (*Applications*)
A. Injection of voltage into the rotor circuit of a 3 phase induction motor	1. Steel rolling mills
B. Variation of applied voltage to a 3 phase induction motor at constant frequency	2. Textiles mills
C. Series-parallel control of d.c. series motors	3. Fans and blowers
D. Ward-Leonard control of d.c. motors	4. Traction
	5. Position control in a servo system

Codes :	A	B	C	D
(*a*)	1	2	3	4
(*b*)	4	5	2	1
(*c*)	1	5	2	4
(*d*)	2	3	4	1

226. Poles of d.c. machines are often laminated to

(*a*) reduce pulsation loss
(*b*) reduce armature reaction
(*c*) reduce iron weight
(*d*) dissipate more heat.

227. A thermocouple arrangement is to be used to measure temperature in the range of 700-800°C. Point out the pair that would be most suitable for this application :

(*a*) Copper-constantan
(*b*) Iron-constantan
(*c*) Chromel-alumel
(*d*) Platinum-platinum rhodium.

228. Width of carbon brush should be equal to

(*a*) less than the width of one commutator segment

(*b*) the width of 1 and 2 commutator segments

(*c*) the width of 2 and 3 commutator segments

(*d*) the width of more than 3 commutator segments.

229. The inertia constant of a 100 MVA, 11 kV water wheel generator is 4. The energy stored in the rotor at the synchronous speed is

(*a*) 400 MJ
(*b*) 400 kJ
(*c*) 25 MJ
(*d*) 25 kJ.

230. When a 400 V, 50 Hz, 6-pole induction motor is rotating at 960 r.p.m. on no-load, its slip is

(*a*) 1 per cent
(*b*) 2 per cent
(*c*) 3 per cent
(*d*) 4 per cent.

231. The torque-slip characteristic for a three-phase induction motor is such that
(*a*) for lower values of slip, torque is directly proportional to slip and for higher values of slip, torque in inversely proportional to slip
(*b*) for lower values of slip, torque is inversely proportional to slip and for higher values of slip, torque in directly proportional to slip
(*c*) for lower values of slip, torque is directly proportional to the square of the slip and for higher values of slip, torque is inversely proportional to slip
(*d*) for lower values of slip, torque is directly proportional to slip and for higher values of slip, torque is inversely proportional to square of slip.

232. Motor *A* has deeper and narrow slots, whereas motor *B* has shallow and wide slots. Induction motor *A*, as compared to motor *B*, has
(*a*) more starting torque
(*b*) more pull-out torque
(*c*) less starting torque
(*d*) more operating slip.

233. The ratio of the primary to secondary voltage of a transformer is 2 : 1. The saving in terms of weight of copper required if an autotransformers is used instead of a two-winding transformer will be
(*a*) 50%
(*b*) 33.33%
(*c*) 66.67%
(*d*) 97%.

234. For satisfactory parallel operation of two single-phase transformers a number of conditions are to be filled. A number of conditions are written below. Indicate which of these is not required to be filled.
(*a*) kVA ratings of the two transformers should be equal
(*a*) Transformers should be properly connected with regard to their polarity
(*c*) Voltage ratings of the primary windings should be suitable for supply system voltage and frequency. The turn ratio of the transformers must be equal
(*d*) The percentage impedance of the two transformers should be equal.

235. Frequency of a single phase series motor is lowered
(*a*) to reduce reactance drop and improve commutation
(*b*) to reduce reactance only
(*c*) to reduce reactance and iron loss and improve commutation
(*d*) to reduce iron loss only.

236. The core of a transformer is made of
(*a*) silicon steel
(*b*) annealed copper
(*c*) seasoned wood
(*d*) aluminium.

237. The core of a transformer is assembled with laminated sheets to reduce
(*a*) hysteresis loss
(*b*) eddy-current loss
(*c*) magnetic noise
(*d*) magnetising current.

238. The use of grain-oriented laminated sheets in building a transformer core
(*a*) reduces magnetising current
(*b*) reduces eddy-current loss in the core
(*c*) reduces hysteresis loss
(*d*) increases the no-load power factor angle.

239. In a single-phase repulsion motor, torque is developed on the rotor when the brush axis is fixed
(*a*) at 90° electrical with the stator field axis
(*b*) in alignment with the stator field axis

(*c*) at an acute angle with the stator field axis
(*d*) at 90° mechanical with the stator field axis.

240. If the centrifugal switch of a resistance split-phase induction motor fails to close when the motor is de-energised, then
(*a*) no starting torque will be developed when supply is connected again across the motor terminals
(*b*) a dangerously high current will flow through the main winding when supply is connected again across the motor terminals
(*c*) starting torque developed may not be sufficient to enable the motor to restart
(*d*) the motor will develop high starting torque when an attempt is made to restart.

241. The term cogging is associated with ...... motors.
(*a*) induction (*b*) repulsion
(*c*) d.c. series (*d*) d.c. shunt.

242. Distributed winding has the advantage(s) of
(*a*) better utilization of core as a number of small slots evenly spaced are used
(*b*) improved wave-form as harmonic e.m.fs. are reduced
(*c*) diminished armature reactance and armature reaction
(*d*) efficient cooling
(*e*) all of the above.

243. The negative phase sequence in a 3-phase synchronous motor exists when the motor is
(*a*) under-loaded (*b*) over-loaded
(*c*) supplied with unbalanced voltage (*d*) hot. **[U.P.S.C. I.E.S., 1992]**

244. Presence of 5th harmonics in induction motor causes
(*a*) cogging (*b*) crawling
(*c*) small reverse braking torque (*d*) hunting.

245. The number of turns on the primary of current transformer is usually
(*a*) 1 to 5 (*b*) 10 to 20
(*c*) 20 to 100 (*d*) 100 to 500.

246. AC servo-motor is basically a
(*a*) universal motor (*b*) capacitor motor
(*c*) 2-phase induction motor (*d*) 3-phase induction motor.

247. In a 3-phase slip-ring induction motor high starting torque is achieved by
(*a*) increasing supply voltage V
(*b*) increasing supply frequency f
(*c*) connecting a capacitor across the motor terminals
(*d*) connecting a star-connected resistance across the slip-ring terminals of the motor.

248. For an n-pulse rectifier, the r.m.s. value of the a.c. current is related to the d.c. load current as
(*a*) $I_{rms} = \dfrac{I_d}{n}$ (*b*) $I_{rms} = \dfrac{I_d}{\sqrt{n}}$
(*c*) $I_{rms} = I_d$ (*d*) $I_{rms} = \dfrac{2}{\pi} I_d$. **[U.P.S.C. I.E.S., 1995]**

249. The rotor copper losses of a synchronous motor are met by
(*a*) armature input (*b*) d.c. source
(*c*) a.c. supply mains (*d*) none of these.

250. Two mechanically coupled alternators deliver power at 50 Hz and 60 Hz respectively. The highest speed of the alternators is

(*a*) 3600 r.p.m. (*b*) 3000 r.p.m.
(*c*) 600 r.p.m. (*d*) 500 r.p.m. **[U.P.S.C. I.E.S., 1994]**

251. In an induction motor

(*a*) stator current, stator flux and rotor e.m.f. are in phase
(*b*) rotor current and rotor flux are in phase and both lag behind the rotor e.m.f. by 90°
(*c*) rotor current, rotor flux and rotor e.m.f. are in phase
(*d*) both (*a*) and (*b*) are correct.

252. For the same power rating, an alternator operating at lower voltage will be

(*a*) larger in size (*b*) smaller in size
(*c*) less noisy (*d*) more efficient.

253. "In all cases of electromagnetic induction, an induced voltage will cause a current to flow in a closed circuit in such a direction that the magnetic field which is caused by that current will oppose the change that produces the current", is the original statement of

(*a*) Lenz's law (*b*) Faraday's law of magnetic induction
(*c*) Fleming's law of induction (*d*) Ampere's law.

254. In order to reduce hysteresis loss

(*a*) core may be laminated
(*b*) silicon steel may be used as the core material
(*c*) core may be constructed with any permanent magnet material such as alnico
(*d*) core may be impregnated with varnish.

255. Phase splitting can be accomplished in a single phase induction motor

(*a*) only by adding a capacitor in series with the auxiliary winding
(*b*) only by causing the auxiliary winding to have high reactance
(*c*) only by causing the auxiliary winding to have low resistance
(*d*) by any one of the above three method.

256. A fluctuating voltage supply is detrimental to a refrigerator motor but not to a ceiling fan, although both are single phase induction motors because, the refrigerator motor

(*a*) is made more robust than the fan motor
(*b*) is subjected to short duty cycle but the fan motor is subjected to continuous duty
(*c*) is enclosed in sealed unit while the fan motor is open to the environment
(*d*) load is constant, but the fan motor load is voltage dependent.

257. The current and potential coils of a wattmeter were accidentally interchanged while connecting. After energising the circuit, it was observed that the wattmeter did not show the reading. This could be due to

(*a*) damage due to the potential coil
(*b*) damage due to the current coil
(*c*) damage to both potential and current coils
(*d*) loose contact.

258. The electrical conductivity of metals is typically of the order of (in $ohm^{-1}\ m^{-1}$)

(*a*) 10^7 (*b*) 10^{-4}
(*c*) 10^5 (*d*) 10^{-6}.

259. Given ϕ_{1m}, ϕ_{2m} = the fluxes produced by the two portions of the shaded pole,
θ = the angle between ϕ_{1m} and ϕ_{2m},
R = resistance of the disc, the torque developed in an induction relay would be proportional to which of the following ?

(1) ϕ_{1m} and ϕ_{2m}
(2) $1/R$
(3) R
(4) $\sin \theta$.

Select the correct answer using the codes given below

Codes :

(*a*) 1, 2 and 4
(*b*) 1, 3 and 4
(*c*) 1 and 2
(*d*) 2 and 4.

260. Three single phase transformers, each of 100 kVA rating, are connected in a closed delta arrangement. If one of them is taken out, it would be possible to load the bank in such a manner that each one is loaded to the extent of

(*a*) 86.6 kVA
(*b*) 66.7 kVA
(*c*) 57.7 kVA
(*d*) 33.33 kVA.

261. In a d.c. machine "contraction co-efficient" is used to take into account the reduction of

(*a*) air-gap area due to armature slots
(*b*) iron losses in the teeth due to lower tooth density
(*c*) armature mmf due to armature slots
(*d*) torque due to ventilating ducts.

262. The ratio of readings of two wattmeters connected to measure power in a balanced 3-phase load is 5 : 3 and the load is

(*a*) 0.917 lead
(*b*) 0.917 lag
(*c*) 0.6 lead
(*d*) 0.6 lag.

263. Given that D = rotor diameter and L = axial length, a high performance a.c. servomotor is characterised by

(*a*) large D, large L
(*b*) large D, small L
(*c*) small D, small L
(*d*) small D, large L.

264. Magnetizing reactance of a transformer is determined by

(*a*) short-circuit test
(*b*) open-circuit test
(*c*) both open-circuit and short-circuit tests
(*d*) other than the above tests.

265. A 200 V d.c. machine has an armature resistance of 0.5 Ω. If the full load armature current is 30 A, the induced e.m.f. when the machine runs (*i*) as generator and (*ii*) as motor will be respectively

(*a*) 230 V, 170 V
(*b*) 225 V, 175 V
(*c*) 185 V, 215 V
(*d*) 215 V, 185 V.

266. When a synchronous generator is designed with lower SCR it

(*a*) will give higher stability limits
(*b*) will have lower SC current
(*c*) will give better voltage regulation
(*d*) will have high synchronizing power.

267. When an alternator is delivering a balanced load at unity power factor the phase angle between line voltage and line current is

(*a*) 90°
(*b*) 60°
(*c*) 30°
(*d*) 0°
(*e*) none of the above.

268. Speed of the universal motor is

(*a*) dependent on frequency of supply
(*b*) proportional to frequency of supply
(*c*) independent of frequency of supply
(*d*) none of the above.

269. The rotor field of synchronous generators is supplied with
(*a*) d.c. supply (*b*) a.c. supply
(*c*) a.c or d.c. supply (*d*) any of the above.

270. The deflecting torque of a moving iron instrument is proportional to
(*a*) I (*b*) I^2
(*c*) $I^{1/2}$ (*d*) $I^{3/2}$.

271. Silicon content in iron laminations is kept within 5% as it
(*a*) increases hysteresis loss (*b*) increases cost
(*c*) makes the material ductile (*d*) none of these.

272. Which motor can conveniently operate at lagging as well as leading power factor ?
(*a*) Squirrel cage induction motor (*b*) Wound rotor induction motor
(*c*) Synchronous motor (*d*) D.C. shunt motor.

273. Most of the generators in thermal power plants run at
(*a*) 15000 r.p.m. (*b*) 3000 r.p.m.
(*c*) 1500 r.p.m. (*d*) 1000 r.p.m.
(*e*) 750 r.p.m.

274. Three-phase induction motors are widely used for industrial applications because
(*a*) they are rugged in construction, require less maintenance and are less expensive than other motors
(*b*) their speed can be controlled very smoothly over a wide range
(*c*) their operating characteristics are superior over other electrical motors
(*d*) they can be manufactured easily for any hp rating.

275. In a three-phase induction motor
(*a*) three-phase supply is to be given to stator winding and d.c. supply to the rotor winding
(*b*) only three-phase supply is to be given to stator winding
(*c*) three-phase supply is to be given to both stator and rotor windings
(*d*) three-phase supply is to be given to rotor winding.

276. To achieve higher starting torque in a three-phase slip-ring type induction motor
(*a*) extra resistance should be connected across the slip-ring terminals
(*b*) the phase sequence of the supply to the motor should be reversed
(*c*) the supply voltage should be increased
(*d*) the windings should first be connected in star and then in delta.

277. For a three-phase induction motor having rotor circuit resistance of 6 Ω, maximum torque occurs at a slip of 0.6. The value of standstill rotor circuit reactance is
(*a*) 4.44 Ω (*b*) 0.36 Ω
(*c*) 1 Ω (*d*) 10 Ω.

278. In wound-rotor type induction motors the rotor terminals are brought out through slip rings
(*a*) to enable extra resistance to be connected across them during starting
(*b*) to enable closing the rotor circuit externally
(*c*) to enable three-phase supply to be applied across the rotor winding through brush and slip-ring arrangement
(*d*) to enable connecting the rotor windings either in star or in delta depending upon the need.

279. A pulsating (alternating) magnetic field will be produced when
(*a*) two-phase supply is applied across a two-phase stator winding
(*b*) three-phase supply is applied across a three-phase stator winding

(*c*) a polyphase supply is applied across a polyphase stator winding
(*d*) a single-phase supply is applied across a single-phase stator winding.

280. Maximum efficiency of a transformer occurs when
(*a*) hysteresis loss and eddy-current loss are minimum
(*b*) the sum of hysteresis loss and eddy-current loss is equal to copper loss in the windings
(*c*) power factor of the load is leading
(*d*) hysteresis loss is equal to eddy-current loss.

281. The all-day efficiency of a transformer is the ratio of
(*a*) kWh output and kWh input per day
(*b*) kWh input and kWh output in a day
(*c*) output power and input power
(*d*) input power and output power.

282. To make the simplest 6-pole single-phase winding, at least
(*a*) two coils are needed
(*b*) one coil is needed
(*c*) three coils are needed
(*d*) six coils are needed.

283. The direction of the rotating magnetic field produced by the stator ampere turns of a three-phase induction motor changes, if
(*a*) the sequence to supply to the stator terminals is changed
(*b*) a variable frequency voltage is applied across the stator terminals
(*c*) the supply voltage is changed
(*d*) supply to any one phase is disconnected.

284. To obtain sinusoidal flux distribution, chamfering of poles is done in
(*a*) salient-pole alternators only
(*b*) smooth cylindrical pole alternators only
(*c*) both of the above
(*d*) none of these.

285. Which of the following statements regarding skewing of rotor bars in a squirrel cage induction motor are *correct* ?
(1) It prevents cogging
(2) It produces more uniform torque
(3) It increases starting torque
(4) It reduces motor 'hum' during its operation.
Select the answer using the codes given below
(*a*) 2, 3, 4
(*b*) 1, 2, 3
(*c*) 1, 2, 4
(*d*) 1, 3 and 4.

[U.P.S.C. I.E.S., 1997]

286. A 100/5 CT is used in conjunction with a 5 A ammeter. If the latter reads 3.5 A, the line current will be
(*a*) 350 A
(*b*) 175 A
(*c*) 120 A
(*d*) 70 A. **[B.T.E. M.P. Elec. Machines, 1993]**

287. The armature of a single phase alternator is completely wound with T single turn coils distributed uniformly. The induced voltage in each turn is 2 V (r.m.s.). The e.m.f. of the whole winding is
(*a*) $2T$ volt
(*b*) $1.11T$ volt
(*c*) $1.414T$ volt
(*d*) $1.273T$ volt. **[GATE, 1998]**

288. A synchronous motor is operating with normal excitation. With the increase in load the armature current drawn from the supply main increases due to
(*a*) increase in back e.m.f.
(*b*) fall in motor speed
(*c*) increase in resultant voltage across the armature
(*d*) increase in power factor.

289. Which of the following statements associated with repulsion motor is *wrong* ?
(*a*) Its direction of rotation depends upon the position of brushes

(*b*) Its characteristics are similar to those of series motor
(*c*) Its power factor is high
(*d*) It is used where sturdy motor with large starting torque and adjustable but constant speed is required.

290. A power station consists of two synchronous generators *A* and *B* of ratings 250 MVA and 500 MVA with inertia 1.6 pu and 1 pu respectively on their own base MVA ratings. The equivalent per unit inertia constant for the system on 100 MVA common base is
(*a*) 2.6 (*b*) 0.615
(*c*) 1.625 (*d*) 9.0. **[GATE, 1998]**

291. The speed of a universal motor can be controlled by
(*a*) introducing a variable resistance in series with the motor
(*b*) tapping the field at various points
(*c*) centrifugal mechanism
(*d*) all of the above steps. **[U.P.S.C. I.E.S., 1993]**

292. Repulsion induction motor has
(*a*) single phase winding on its stator
(*b*) two separate windings on its rotor in common slots
(*c*) squirrel cage rotor and two separate windings on stator in common slots
(*d*) both (*a*) and (*b*).

293. The volt-ampere equation for a long shunt compound motor is given by
(*a*) $V_t = E_a + I_a r_a$ (*b*) $V_t = E_a + I_a(r_a + r_s)$
(*c*) $V_t = E_a - I_a r_a$ (*d*) $V_t = E_a + I_a r_a + I_L r_s$.

294. The stator and rotor cores of an induction motor are made up of laminated sheets
(*a*) to reduce the hysteresis loss in the core
(*b*) to reduce the eddy-current loss in the core
(*c*) to make the rotor and stator mechanically strong
(*d*) to enable the stator and rotor cores dissipate heat more effectively.

295. The stator and rotor cores of an induction motor are made up of magnetic material
(*a*) to keep the cost of construction low
(*b*) to reduce the magnetising current
(*c*) to make the parts strong
(*d*) because the reluctance of the magnetic material is high.

296. Induction motors now-a-days use die-cast aluminium rotor because
(*a*) aluminium is lighter than copper
(*b*) aluminium is cheaper than copper
(*c*) aluminium is easy to cast because of low melting point and is easily available
(*d*) aluminium has less resistivity than copper.

297. When the rotor circuit resistance of a polyphase induction motor is increased
(*a*) the starting torque increases
(*b*) the maximum value of torque decreases
(*c*) the slip at which maximum torque occurs remains unchanged
(*d*) maximum torque is developed at starting.

298. A d.c. shunt generator has full-load voltage regulation of 10% at rated speed of 1000 r.p.m. If it is now driven at 1250 r.p.m. then its voltage regulation at full load would
(*a*) be more than 10% (*b*) be less than 10%
(*c*) remain unchanged (*d*) be 12.5%.

299. Cogging of induction motors occurs at

(*a*) high voltage and when the number of stator teeth and rotor teeth are equal
(*b*) high voltage and when the number of stator teeth and rotor teeth are not equal
(*c*) low voltage and when the number of stator teeth and rotor teeth are equal
(*d*) low voltage and when the number of stator teeth and rotor teeth are not equal.

300. A hysteresis motor

(*a*) is a self starting motor (*b*) is a constant speed motor
(*c*) does not need d.c. excitation (*d*) all of the above.

301. A transformer when supplying a load maintained 11 kV across load terminals. When the load was switched off, the terminal voltage became 11550 V. What is the voltage regulation at this load ?

(*a*) 11.55% (*b*) 5.5%
(*c*) 5% (*d*) 55%.

302. Power lost in the open-circuit and short-circuit tests on a transformer gives approximately an account of the following losses :

(*a*) core-losses and copper-losses respectively
(*b*) copper-losses and core-losses respectively
(*c*) eddy-current loss and hysteresis loss respectively
(*d*) hysteresis loss and eddy-current loss respectively.

303. With DOL start of an induction motor the heating during acceleration of inertia load is proportional to

(*a*) slip at maximum torque (*b*) maximum torque
(*c*) stored kinetic energy (*d*) electro-magnetic power developed on the rotor.

304. The rating of an induction voltage regulator for regulating the voltage on a 3-phase feeder dealing with 1,000 kVA between the limits of 4,750 and 5,250 V is

(*a*) 50 kVA (*b*) 100 kVA
(*c*) 25 kVA (*d*) 50 kVA.

305. In a repulsion induction motor, the rotor has a

(*a*) squirrel cage winding with rotor bars permanently short-circuited and placed at the bottom of the slots
(*b*) repulsion winding similar to d.c. armature winding placed over the squirrel cage winding
(*c*) repulsion winding placed at the bottom of slots and squirrel cage winding placed above the repulsion winding
(*d*) both (*a*) and (*b*).

306. In a reluctance motor, when the load increases to an extent that it cannot maintain synchronous speed, it will

(*a*) fall out of synchronism and come to rest
(*b*) run as an induction motor
(*c*) draw excessive armature current and burn out
(*d*) become unstable.

307. In comparison to a resistance-start motor, the capacitor-start motor has

(*a*) high starting torque (about double) (*b*) low starting current (about half)
(*c*) high running torque
(*d*) larger phase difference between two winding currents
(*e*) more number of turns on auxiliary winding
(*f*) all of the above statements except (*c*) are correct.

308. In a 3-phase bridge rectifier fed from a star-connected secondary winding of a transformer, let the voltage to the neutral of the *A*-phase (phase sequence *A*, *B*, *C*) be $V_m \sin \omega t$. At the instant when the voltage of *A*-phase is maximum, the output voltage at the rectifier terminals will be

(*a*) $\frac{V_m}{\sqrt{2}}$ (*b*) V_m

(*c*) $1.5V_m$ (*d*) $\sqrt{3}\ V_m$. **[U.P.S.C. I.E.S., 1997]**

309. The power factor of a lightly loaded induction motor is quite low because

(*a*) the current drawn is largely a magnetising current due to air-gap

(*b*) of the current due to air-gap

(*c*) the current drawn is largely a magnetising component due to laminated core

(*d*) the current drawn is largely an energy component due to laminated core.

310. If P_i be the iron loss and P_c be the copper loss on full load, then which of the following conditions had to be satisfied to obtain maximum efficiency at 3/4 full load ?

(*a*) $P_c = \frac{3P_i}{4}$ (*b*) $P_c = \frac{4P_i}{3}$

(*c*) $P_c = \frac{16P_i}{9}$ (*d*) $P_c = \frac{9P_i}{16}$. **[U.P.S.C. I.E.S., 1993]**

311. Armature reaction mmf and leakage reactance of a synchronous machine are determined by

(*a*) open-circuit and short-circuit tests (*b*) open-circuit and zero p.f. tests

(*c*) open-circuit test only (*d*) zero p.f. test only.

312. In a repulsion motor by shifting the brush position

(*a*) speed and direction of rotation can be controlled

(*b*) only speed can be controlled

(*c*) speed, torque and direction of rotation can be controlled

(*d*) only torque can be controlled by varying the speed at constant output power.

313. An alternator running at 1500 r.p.m. delivers rated current of 30 A to a purely inductive load. With the field current remaining constant, if the speed is reduced to 1000 r.p.m., the load current becomes

(*a*) 20 A (*b*) 30 A

(*c*) a little less than 20 A (*d*) a little more than 20 A.

314. A 100 kVA, 1100/400 V, 50 Hz single-phase transformer has 100 turns on the secondary winding. The number of turns on its primary will be

(*a*) 550 (*b*) 275

(*c*) 2750 (*d*) 5500.

315. In a synchronous motor, the synchronizing power comes into action when

(*a*) rotor speed either exceeds or falls below the synchronous speed

(*b*) rotor speed is equal to synchronous speed

(*c*) rotor speed falls below the synchronous speed

(*d*) rotor speed exceeds the synchronous speed.

316. If a 3-phase, 4-pole, 50 Hz induction motor runs at a speed of 1440 r.p.m., then the slip is

(*a*) 0.03 (*b*) 0.10

(*c*) 0.04 (*d*) 0.05.

317. A transformer has N_1 and N_2 turns in primary and secondary windings respectively. Its secondary winding reactance of $x_2\ \Omega$, when referred to primary, is

(*a*) $x_2\left(\frac{N_2}{N_1}\right)^2$ (*b*) $x_2\frac{N_2}{N_1}$

(*c*) $x_2\left(\frac{N_1}{N_2}\right)^2$ (*d*) $x_2\frac{N_1}{N_2}$.

318. The kVA rating of an isolation transformer increases when connected as an auto-transformer due to

(*a*) establishment of conductive link between the primary and secondary

(*b*) rise in secondary terminal voltage

(*c*) increase in secondary current

(*d*) reduced copper losses.

319. Electrical machines are designed to have maximum efficiency at

(*a*) full load (*b*) 50% of full load

(*c*) near about full load (*d*) no load.

320. A 3-phase squirrel-cage induction motor, designed to operate with stator in star, requires *W* kg of copper for its stator winding. If this motor is to be designed to operate with stator in delta, then weight of stator copper would be

(*a*) more than *W* kg (*b*) less than *W* kg

(*c*) $\sqrt{3}\ W$ kg (*d*) $\frac{1}{\sqrt{3}}\ W$ kg

(*e*) *W* kg.

321. Transformer action requires a

(*a*) constant magnetic flux (*b*) increasing magnetic flux

(*c*) alternating magnetic flux (*d*) alternating electric flux.

322. In all rotating electrical machines, electrical torque is developed when relative speed between stator field and rotor field is

(*a*) zero (*b*) equal to rotor speed

(*c*) equal and opposite to rotor speed (*d*) either (*b*) or (*c*).

323. For a 1.15 kVA, 230/460 V transformer connected to a standard 1-phase, 230 V, 50 Hz supply, the no load current is likely to be :

(*a*) 2.5 A (*b*) 0.2 A

(*c*) 1.00 A (*d*) 4.00 A.

324. Speed control by varying the armature-circuit resistance offers

(*a*) constant power drive (*b*) variable power drive

(*c*) constant torque drive (*d*) variable torque drive.

325. When a synchronous motor is started, the field winding is

(*a*) short-circuited (*b*) open-circuited

(*c*) excited from a d.c. source (*d*) excited from a 3-phase a.c. source.

326. In a repulsion motor, commutator provides a means of connecting the

(*a*) stator windings (*b*) armature windings

(*c*) starting winding (*d*) none of these.

327. Two transformers operating in parallel have different quality of their leakage impedances. For a load power factor of 0.8

(*a*) both would operate at p.fs. less than 0.8
(*b*) both would operate at p.fs. more than 0.8
(*c*) both would operate at the same p.fs.
(*d*) one would operate at p.f. > 0.8 and other at p.f. < 0.8.

328. Four identical alternators each rated for 20 MVA, 11 kV having a subtransient reactance of 16% are working in parallel. The short-circuit level at the bus-bars is
(*a*) 500 MVA (*b*) 400 MVA
(*c*) 125 MVA (*d*) 80 MVA.
[U.P.S.C. I.E.S. Elec. Engg. II, 1996]

329. The torque-speed characteristic of a repulsion motor resembles which of the following d.c. motor characteristics ?
(*a*) Separately excited (*b*) Shunt
(*c*) Series (*d*) Compound. **[GATE, 1996]**

330. A 3-phase synchronous motor will not start if
(*a*) voltage is too low (*b*) one phase is open
(*c*) connections are faulty (*d*) any of the above.

331. The function of oil in a transformer is
(*a*) to provide insulation and cooling (*b*) to provide protection against lightning
(*c*) to provide protection against short-circuit
(*d*) to provide lubrication. **[GATE, 1996]**

332. The effect of leading power factor on the voltage regulation of an alternator is
(*a*) increasing in nature (*b*) decreasing in nature
(*c*) maintained at constant value (*d*) oscillating in nature.

333. Electromagnetic torque in rotating machines is present when
(*a*) air gap is uniform (*b*) stator winding alone carries current
(*c*) rotor winding alone carries current (*d*) both stator and rotor windings carry current.

334. The burden of current transformers is expressed in
(*a*) volt-ampere rating (*b*) current rating of secondary winding
(*c*) current and voltage of secondary (*d*) watt rating.

335. A synchronous motor may not run at its rated speed due to
(*a*) low supply voltage (*b*) excessive load
(*c*) low field excitation (*d*) any of these.

336. Electromagnetic force or torque, developed in any physical system, acts in such a direction as to tend to
(*a*) increase both the field energy and co-energy at constant current
(*b*) increase the field energy and decrease the co-energy at constant current
(*c*) decrease both the field energy and co-energy at constant current
(*d*) decrease the field energy and increase the co-energy at constant current.

337. In an electrical machine, if L = slot length and D = slot depth, then slot leakage flux is proportional to
(*a*) $1/LD$ (*b*) LD
(*c*) D/L (*d*) L/D.

338. In an electrical machine, if Z = conductors per slot and W = slot width, then the slot leakage flux is proportional to
(*a*) Z/W (*b*) ZW
(*c*) W/Z (*d*) $1/ZW$.

339. The armature m.m.f. wave of a d.c. machine is
(*a*) sinusoidal and depends on the speed (*b*) square and independent of speed
(*c*) sinusoidal and independent of speed (*d*) triangular and depends on speed
(*e*) triangular and independent of speed.

340. If both the supply voltage and frequency are reduced to 90% of their previous values in 3-phase induction motor, then the new torque T_e in terms of its previous value is
(*a*) 0.90 (*b*) 1.11
(*c*) 2.00 (*d*) 1.00.

341. If the rotor of a 3-phase slip-ring induction motor is fed from balanced 3-phase supply with the stator short circuited, then frequency of stator currents would be
(*a*) sf (*b*) $(2 - s)f$
(*c*) $(1 - s)f$ (*d*) $(s - 1)f$.

342. Induction motors with open slots have
(*a*) better p.f. (*b*) improved efficiency
(*c*) more breakdown torque (*d*) less starting torque.

343. Transformer cores are laminated to reduce
(*a*) eddy current loss (*b*) hysteresis loss
(*c*) both eddy-current and hysteresis losses (*d*) ohmic loss.

344. In d.c. and synchronous machines, pole-face losses consist of
(*a*) hysteresis loss (*b*) I^2R loss in pole-face windings
(*c*) eddy-current loss (*d*) both (*a*) and (*c*).

345. Efficiency of electrical machines should be calculated by measuring
(*a*) output and input (*b*) losses and output
(*c*) losses and input (*d*) losses.

346. The short-circuit test on a transformer is conducted to obtain
(*a*) core losses at any load (*b*) ohmic losses at any load
(*c*) hysteresis losses (*d*) both core losses and ohmic losses.

347. A synchronous motor, connected to an infinite bus, is working at a leading p.f. Its excitation e.m.f. E_f and alternator terminal voltage V_t are related as follows :
(*a*) $E_f > V_t$ and E_f lags V_t (*b*) $E_f < V_t$ and E_f lags V_t
(*c*) $E_f > V_t$ and E_f leads V_t (*d*) $E_f < V_t$ and E_f leads V_t.

348. In a 3-phase induction motor, the maximum torque
(*a*) is proportional to rotor resistance r_2 (*b*) does not depend on r_2
(*c*) is proportional to $\sqrt{r_2}$ (*d*) is proportional to $(r_2)^2$.

349. In a simple magnetic relay, armature is in open position. As the armature moves towards the yoke under the influence of magnetic force f_e, air-gap flux density B_g and f_e vary as under
(*a*) B_g increases and f_e decreases (*b*) Both decrease
(*c*) Both increase (*d*) B_g decreases and f_e increases.

350. If a 3-phase induction motor is to be operated on unbalanced power supply, then it should be operated at
(*a*) higher loads (*b*) lower loads
(*c*) higher speeds (*d*) lower speeds.

351. In polyphase cylindrical-rotor synchronous machines, the synchronous reactance is made up of
(*a*) mutual and leakage reactances (*b*) self and leakage reactances
(*c*) magnetizing and leakage reactances (*d*) magnetizing and mutual reactances.

352. In a 3-phase induction motor, the mechanical power developed, in terms of air gap power P_g is

(*a*) $(s-1)P_g$ (*b*) $\frac{P_g}{1-s}$

(*c*) $\frac{1}{s}P_g$ (*d*) $(1-s)P_g$.

353. The possible armature-circuit resistance of a d.c. machine is

(*a*) 0.01 Ω (*b*) 1 Ω

(*c*) 50 Ω (*d*) 200 Ω.

354. Distribution transformers have core losses

(*a*) > copper loss (*b*) < copper losses

(*c*) = copper loss (*d*) = $\frac{1}{2}$ (copper losses). **[B.T.E. U.P., 1996]**

355. In a 3-phase synchronous motor, the magnitude of back e.m.f. set up in the stator depends on

(*a*) rotor excitation (*b*) supply voltage

(*c*) rotor speed (*d*) load on motor.

356. While operating on variable frequency supplies, the a.c. motor requires variable voltage as well in order to

(*a*) protect the insulation (*b*) avoid the effect of saturation

(*c*) improve the capabilities of the inverter (*d*) protect the thyristor from $\frac{dv}{dt}$.

[U.P.S.C. I.E.S., 1993]

357. A dual voltage single phase a.c. motor can be operated on either

(*a*) 230 or 400 V (*b*) 30 or 400 V

(*c*) 115 or 230 V (*d*) 160 or 320 V.

358. In a synchronous machine, if the field flux axis is ahead of the armature field axis in the direction of rotation, the machine operating is

(*a*) synchronous motor (*b*) synchronous generator

(*c*) asynchronous motor (*d*) asynchronous generator.

359. In 3-phase induction motors, sometimes copper bars are placed deep in the rotor. It is done in order to improve

(*a*) efficiency (*b*) starting torque

(*c*) power factor (*d*) none of these.

360. The pull-out torque for a normal squirrel cage induction motor usually occurs at a slip ranging from

(*a*) 0.1 to 0.15 (*b*) 0.2 to 0.25

(*c*) 0.4 to 0.5 (*d*) 0.5 to 0.6.

361. In the transformer circuit mode, the core loss is represented as a

(*a*) series resistance (*b*) series inductance

(*c*) shunt resistance (*d*) shunt inductance.

362. The primary (220 V side) of a 220/6 V, 50 Hz transformer is connected to a 110 V, 60 Hz supply. The secondary output voltage is :

(*a*) 3.6 V (*b*) 3.0 V

(*c*) 2.5 V (*d*) 1.667 V.

363. For a 3-phase, 50 Hz, squirrel cage induction motor, rotor leakage reactance at standstill is twice of its resistance. The frequency of the supply at which maximum torque is obtained at starting is :
(*a*) 50 Hz (*b*) 25 Hz
(*c*) 100 Hz (*d*) 75 Hz.

364. The full-load voltage regulation of an alternator is 6% at 0.8 p.f. lagging and at rated speed of 1200 r.p.m. Its full-load voltage regulation at 0.8 p.f. lagging and at 1100 r.p.m. would be
(*a*) less than 6% (*b*) more than 6%
(*c*) 6% (*d*) 5.5%.

365. If the rotor of a 3-phase induction motor is assumed purely resistive, then electromagnetic torque is
(*a*) minimum with load angle of 0° (*b*) minimum with load angle of 90°
(*c*) optimum with load angle of 90° (*d*) optimum with load angle of 0°
(*e*) maximum with load angle of 90°.

366. The no-load speed of a 3-phase, induction motor connected to a 3-phase 50 Hz supply is likely to be
(*a*) 1400 r.p.m. (*b*) 1480 r.p.m.
(*c*) 1450 r.p.m. (*d*) 1498 r.p.m.

367. When a single phase supply is connected across a single-phase winding, the nature of the magnetic field produced is
(*a*) pulsating in nature (*b*) rotating in nature
(*c*) constant in magnitude but rotating at synchronous speed
(*d*) constant in magnitude and direction.

368. In modern alternators, the rotating part is
(*a*) field system (*b*) armature
(*c*) armature as well as field system (*d*) none of the above.

369. A single 3-phase alternator in the laboratory is not run at synchronous speed. It would produce no effect on its
(*a*) occ (*b*) scc
(*c*) occ and scc (*d*) zpfc.

370. For a 3-phase uniformly distributed and narrow-spread winding, the distribution factor is
(*a*) 0.9 (*b*) 0.955
(*c*) 0.965 (*d*) 0.97.

371. The capacitor employed in a capacitor-start induction motor has no
(*a*) voltage rating (*b*) polarity marking
(*c*) dielectric rating (*d*) definite capacitance value.

372. For providing a load power of 100 MW at 0.8 p.f., ratings of alternator, its prime-mover and transformer should respectively be
(*a*) 100 MW, 125 MVA, 125 MVA (*b*) 125 MVA, 100 MW, 125 MVA
(*c*) all 125 MVA (*d*) 100 MW, 100 MW, 125 MVA.

373. The field system of an alternator is usually excited at
(*a*) 1250/250 V d.c. (*b*) 110/220 V a.c.
(*c*) 3-phase, 50 Hz, 400 V (*d*) 230 V a.c.

374. The purpose of using interpoles in d.c. machines is to nullify
(*a*) the demagnetizing effect of armature m.m.f.
(*b*) the cross-magnetizing effect of armature m.m.f.

(*c*) the reactance voltage
(*d*) both the demagnetizing m.m.f. and the reactance voltage
(*e*) both the cross-magnetizing m.m.f. and the reactance voltage.

375. In a 3-phase induction motor reactance under running condition is less than its standstill value. This is due to reduction in
(*a*) rotor inductance
(*b*) stator magnetic flux
(*c*) frequency of rotor e.m.f.
(*d*) mutual flux linking the stator and rotor.

376. In a capacitor-start single phase induction motor, capacitor of suitable value is connected, through a centrifugal switch
(*a*) in series with the auxiliary winding
(*b*) across the auxiliary winding
(*c*) across the main winding
(*d*) in series with the main winding.

377. The output of an alternator depends on
(*a*) generated emf.
(*b*) power factor
(*c*) terminal voltage
(*d*) all of these.

378. In comparison to mercury-arc rectifiers metal rectifiers
(*a*) operate on low temperatures
(*b*) can operate on high voltages
(*c*) can operate on high loads
(*d*) give poor voltage regulation.

379. In an alternator, the use of short pitch coils having pitch of 160° will eliminate ...... harmonic component for the e.m.f. generated.
(*a*) 3rd
(*b*) fifth
(*c*) 7th
(*d*) ninth.

380. The ratio error in the current transformer is attributed to
(*a*) magnetising component of no-load current
(*b*) energy component of excitation current
(*c*) leakage flux
(*d*) power factor of the primary.

381. The method that can be used conveniently for speed control of a squirrel cage induction motor is
(*a*) pole changing control
(*b*) rotor resistance control
(*c*) cascade operation
(*d*) secondary foreign voltage control.

382. A salient pole synchronous motor continues running even if its excitation current is reduced to zero. This is due to
(*a*) rotating magnetic field of stator
(*b*) rotating magnetic field of rotor
(*c*) magnetization of rotor salient poles by stator magnetic field
(*d*) interlocking action between stator and rotor rotating magnetic fields.

383. Cascade method of speed control involves the use of two coupled induction motors. The necessary condition for speed control is that
(*a*) both the motors are of the wound rotor type having the same number of poles
(*b*) both the motors are of the squirrel cage rotor type having different number of poles
(*c*) one motor is of the slip-ring type but both the motors have the same number of poles
(*d*) one motor is of the slip-ring type and the two motors have different number of poles.

[U.P.S.C. I.E.S., 1993]

384. Hysteresis motor is a
(*a*) synchronous induction motor
(*b*) single phase induction motor
(*c*) single phase synchronous motor without any salient pole and without d.c. excitation
(*d*) single phase synchronous motor with salient poles and without d.c. excitation.

385. A 400/200 V transformer has its l.v. resistance of 0.02 per unit. The resistance when referred to h.v. side is
(*a*) 0.02 p.u. (*b*) 0.04 p.u.
(*c*) 0.01 p.u. (*d*) 0.08 p.u.

386. E_1, E_3, E_5 are the fundamental, third harmonic and fifth harmonic per phase voltages generated in a star-connected alternator. Its line voltage is
(*a*) $3\sqrt{E_1^2 + E_3^2 + E_5^2}$ (*b*) $\sqrt{3}\sqrt{E_1^2 + E_3^2 + E_5^2}$
(*c*) $\sqrt{3}\sqrt{E_1^2 + E_3^2}$ (*d*) $\sqrt{3}\sqrt{E_1^2 + E_5^2}$.

387. The most advantageous number of anodes for a mercury-arc rectifier is six, if more than six anodes are used, then
(*a*) voltage regulation will increase
(*b*) the cost increases rapidly without a comparative increase in output
(*c*) power factor falls
(*d*) all of the above.

388. In d.c. generators, the polarity of interpoles is
(*a*) same as that of main pole behind (*b*) same as that of main pole ahead
(*c*) opposite to that of the main pole ahead (*d*) none of these.

389. DC series motor should never be switched on at no load, because
(*a*) the field current is initially zero (*b*) the motor does not pick up
(*c*) the speed becomes dangerously high (*d*) it will take too long to accelerate.

390. A pole-pitch in electrical machines is equal to
(*a*) 180 electrical degrees (*b*) 90 electrical degrees
(*c*) 120 electrical degrees (*d*) 360 electrical degrees.

391. For transformers employing CRGOS, the area of yoke is taken ...... that of core.
(*a*) 80 to 85% of (*b*) 115 to 125% of
(*c*) equal to (*d*) half of.

392. A 3-phase, 50 Hz, 4-pole squirrel cage induction motor has its stator rewound for 6 poles without any alterations in the rotor. The motor would now run at a speed
(*a*) < 1000 r.p.m. (*b*) < 1500 r.p.m.
(*c*) < 3000 r.p.m. (*d*) zero r.p.m.

393. In a single-phase repulsion motor power factor is
(*a*) always leading (*b*) high at low speed
(*c*) high at high speed (*d*) always unity.

394. Decrease in supply voltage to a 3 phase induction motor will not cause
(*a*) any difficulty in accelerating the load
(*b*) increase in torque
(*c*) decrease in speed as well as in core loss
(*d*) increase in stator as well as rotor currents
(*e*) operating temperature abnormally high owing to increase in copper loss.

395. In a 3-phase induction motor, electrical torque T_e in terms of supply voltage V_1 is proportional to
(*a*) V_1 (*b*) $\sqrt{V_1}$
(*c*) V_1^2 (*d*) V_1^4.

396. The rotor of a 3-phase induction motor is running at a speed of n_r r.p.s. The slip frequency currents in the rotor would produce a field which is

(*a*) stationary with respect to rotor
(*b*) rotating at synchronous speed w.r.t. stator
(*c*) rotating at n_r r.p.s. w.r.t. stator
(*d*) rotating at small slip w.r.t. stator.

397. In concentrated winding, the number of slots is equal to the number of

(*a*) coil sides
(*b*) poles
(*c*) conductors
(*d*) coils.

398. Which of the following is not the advantage of an induction generator ?

(*a*) It does not hunt or drop out of synchronism
(*b*) It is simple and rugged in construction, cheaper in cost and needs little maintenance
(*c*) It delivers only leading current
(*d*) When short-circuited it delivers little or no sustained power.

399. Skew of rotor bar eliminates

(*a*) the effect of space harmonics
(*b*) the entire effect of crawling
(*c*) magnetic noise
(*d*) vibration due to unequal force developed on rotor.

400. In a 3-phase induction motor, the electrical representation of the variable mechanical load is the resistance

(*a*) $r_2(s-1)$
(*b*) $r_2\left(\frac{1}{s}-1\right)$
(*c*) $r_2\left(1-\frac{1}{s}\right)$
(*d*) $r_2(1-s)$.

401. The speed of a synchronous motor can be varied by varying its

(*a*) excitation
(*b*) supply voltage
(*c*) supply frequency
(*d*) load.

402. The rated secondary voltage of 220 V for a transformer means the terminal voltage of 220 V on

(*a*) no-load
(*b*) rated full load
(*c*) half-full load
(*d*) none of these.

403. The torque-slip characteristic of an induction motor is approximately a/an

(*a*) straight line
(*b*) rectangular parabola
(*c*) parabola
(*d*) hyperbola.

404. A universal motor is one which has

(*a*) constant speed
(*b*) constant output
(*c*) capability of operating both on a.c. and d.c. with comparable performance
(*d*) maximum efficiency.

405. An infinite bus-bar has

(*a*) constant voltage
(*b*) constant frequency
(*c*) infinite voltage
(*d*) both (*a*) and (*b*)
(*e*) both (*b*) and (*c*).

406. The load shared between two alternators operating in parallel may be adjusted by varying

(*a*) their speed
(*b*) their excitation
(*c*) power input to their prime-movers
(*d*) any of the above.

407. In case of a split-phase motor, the phase shift between currents in the two windings is around

(*a*) 30°
(*b*) 70°
(*c*) 90°
(*d*) 120°. **[GATE, 1996]**

408. Efficiency of a transformer is maximum if
(*a*) transformer is unloaded (*b*) copper losses are equal to iron losses
(*c*) eddy current losses are equal to hysteresis losses
(*d*) it is maximally loaded.

409. In an induction motor under running condition, the rotor reactance per phase is its standstill phase reactance
(*a*) *s* times (*b*) equal to
(*c*) 1/*s* times (*d*) (1 – *s*) times.

410. Short circuit test on a single-phase transformer gave the following data :
30 V at 50 Hz, 20 A, p.f. = 0.2 lag.
If a.c. test is performed on 30 V, 25 Hz, then short-circuit current is
(*a*) decreased at p.f. < 0.2 (*b*) increased at p.f. < 0.2
(*c*) increased at p.f. > 0.2 (*d*) decreased at p.f. > 0.2
(*e*) same at p.f. = 0.2.

411. The m.m.f. produced by a single phase winding is
(*a*) pulsating and stationary (*b*) pulsating and rotating
(*c*) constant in amplitude and stationary (*d*) constant in amplitude and rotating.

412. A d.c. shunt motor is running with a certain load. The effect of adding an external resistance in the field circuit is to
(*a*) reduce the motor speed (*b*) increase the motor speed
(*c*) reduce the armature current of the motor
(*d*) stop the motor.

413. For the purpose of analysis exact equivalent circuit of a transformer is usually replaced by an approximate equivalent circuit. In doing so errors introduced due to winding copper loss and core loss are of different nature. Due to this, the analysis by approximate equivalent circuit gives fairly satisfactory results. Under the circumstances, which one of the following statements is correct in respect of losses referred to approx. equivalent circuit as compared to exact equivalent circuit ?
(*a*) This accounts for somewhat lesser primary winding copper loss and less core loss
(*b*) This accounts for somewhat lesser primary winding copper loss and more core loss
(*c*) This accounts for somewhat greater secondary winding copper loss and less core loss
(*d*) This accounts for somewhat lesser secondary winding copper loss and core loss.
[U.P.S.C. I.E.S., 1995]

414. The outer cage of a double squirrel rotor consists of
(*a*) manganese brass (*b*) red copper
(*c*) bronze (*d*) steel. **[B.T.E. M.P. 1995]**

415. CRGO laminations in a transformer are used to minimise
(*a*) eddy current loss (*b*) hysteresis loss
(*c*) both eddy current and hysteresis losses (*d*) ohmic loss.

416. The chording angle for eliminating 5th harmonics should be
(*a*) 30° (*b*) 34°
(*c*) 36° (*d*) 35°.

417. The efficiency of electrical machines is maximum when
(*a*) constant losses = *x* (variable losses) (*b*) constant losses = variable losses
(*c*) constant losses = $\frac{1}{x}$ (variable losses) (*d*) constant losses = $\frac{1}{2}$ (variable losses).

418. A d.c. shunt motor is running at normal speed at rated voltage. If the supply voltage is reduced to half then its speed
(*a*) remains unchanged (*b*) is reduced to half
(*c*) increase slightly (*d*) decreases slightly.

419. A transformer can have zero regulation at
(*a*) zero p.f. (*b*) unity p.f.
(*c*) lagging p.f. (*d*) leading p.f. **[S.C.E. & T.E.W.B., 1996]**

420. In an induction motor, if the air-gap is increased
(*a*) speed will reduce (*b*) efficiency will improve
(*c*) power factor will be lowered (*d*) breakdown torque will reduce. **[GATE, 1996]**

421. A repulsion motor is equipped with
(*a*) slip-rings (*b*) centrifugal switch
(*c*) commutators and brushes (*d*) all of the above.

422. Two alternators *A* and *B* are sharing an inductive load equally. If the excitation of alternator *A* is increased
(*a*) alternator *B* will deliver more current and alternator *A* will deliver less current
(*b*) alternator *B* will deliver less current and alternator *A* will deliver more current
(*c*) both will continue to share load equally
(*d*) both will deliver more current. **[U.P.S.C. I.E.S., 1992]**

423. A 3-phase slip-ring induction motor is fed from the rotor side with stator winding short-circuited. The frequency of the currents in the short-circuited stator is
(*a*) slip frequency (*b*) supply frequency
(*c*) frequency corresponding to rotor speed (*d*) zero. **[GATE, 1993]**

424. In a shaded-pole motor, shading coils are used to
(*a*) reduce winding losses (*b*) reduce friction losses
(*c*) produce rotating magnetic field (*d*) protect against sparking.
[U.P.S.C. I.E.S., 1992]

425. Beyond the point of maximum torque any further increase in load will cause
(*a*) increase in torque developed and so the motor will run at a high speed
(*b*) no change in the torque developed and so the motor will run at the same speed as before
(*c*) decrease in torque developed and the motor will slow down
(*d*) increase in current drawn from the supply mains resulting in damage to it.

426. Under which of the following starting methods, an induction motor draws largest starting current ?
(*a*) Star-delta starting (*b*) Auto-transformer starting
(*c*) Direct-on-line starting (*d*) Reduced voltage starting.

427. Which one of the following capacitor-start split-phase motor will have the largest value of capacitance ?
(*a*) $\frac{1}{2}$ hp ; 3,450 r.p.m. (*b*) $\frac{1}{4}$ hp ; 1,725 r.p.m.
(*c*) $\frac{1}{2}$ hp ; 1,140 r.p.m. (*d*) $\frac{3}{4}$ hp ; 1,140 r.p.m. **[U.P.S.C. I.E.S., 1993]**

428. The flux in transformer core
(*a*) increases with load (*b*) decreases with load
(*c*) remains constant irrespective of load (*d*) none of the above.

429. When the speed becomes more than the synchronous speed during hunting, the damper bars develop

(*a*) synchronous motor torque (*b*) single-phase induction motor torque

(*c*) induction motor torque (*d*) induction generator torque.

430. In a transformer, the exciting current will be in phase quadrature with the impressed voltage provided

(*a*) only the leakage impedance drop is ignored

(*b*) only the core loss is ignored

(*c*) both the leakage impedance drop and the core loss are ignored

(*d*) only the no-load copper loss is ignored.

431. A 3-phase induction motor is operating at its rated torque. If the supply voltage falls to 75% of its normal value,

(*a*) the motor will get heated up to inadmissible extent after some time

(*b*) the motor will stop

(*c*) the slip will decrease

(*d*) the stator and rotor current will decrease.

432. A 3-phase squirrel cage induction motor has a full-load efficiency of 0.8 and a maximum efficiency of 0.9. It is operated at a slip of 0.6 by applying a reduced voltage. The efficiency of the motor at this operating point is

(*a*) less than 0.4 (*b*) greater than 0.6

(*c*) in the range of 0.8 ± 1 (*d*) none of the above. **[GATE, 1998]**

433. Major insulation in a transformer is the insulation between the

(*a*) lv winding and core (*b*) lv winding and hv winding

(*c*) turns of the winding turns (*d*) both (*a*) and (*b*).

434. A 40 kVA transformer has a core loss of 400 W and a full-load copper loss of 800 W. The proportion of full load at maximum efficiency is

(*a*) 50% (*b*) 62.3%

(*c*) 70.7% (*d*) 100%. **[U.P.S.C., 1997]**

435. Which of the following synchronous motors will be smallest in size ?

(*a*) 5 kW, 375 r.p.m. (*b*) 5 kW, 500 r.p.m.

(*c*) 5 kW, 750 r.p.m. (*d*) any of these.

436. In a synchronous motor

(*a*) the rotor mmf and stator mmf are stationary with respect to each other

(*b*) rotor mmf rotates slightly faster in comparison to stator mmf

(*c*) stator mmf rotates slightly faster than rotor mmf

(*d*) none of the above.

437. The low-voltage winding of a core type transformer is subdivided into two equal halves, each of half the original width of the single winding with the high-voltage winding in between (instead of having the usual construction of low voltage winding adjacent to the core and surrounded by the high-voltage winding). Such an interlacing of coils would make the combined primary and secondary leakage reactance (in terms of the primary) nearly

(*a*) twice (*b*) equal

(*c*) half (*d*) one-fourth. **[U.P.S.C. I.E.S., 1997]**

438. Two 3-limb, 3-phase delta-star connected transformers are supplied from the same source. One of the transformers is Dy1 and the other is Dy11 connection. The phase difference between the corresponding phase voltage of the secondaries would be

(*a*) 0° (*b*) 30°

(*c*) 60° (*d*) 120°. **[U.P.S.C. I.E.S., 1997]**

439. Single phase a.c. motors are used in largest number. This is due to

(*a*) their great variety of useful services in the home, office, the factory, in business establishment and on the farm

(*b*) their wide applications in automatic control devices of various types

(*c*) lower cost

(*d*) both (*a*) and (*b*).

440. Synchronous motors are not used

(*a*) in power houses and substations in parallel to bus bars for p.f. correction

(*b*) in factories having a large number of induction motors and other power apparatus operating at lagging power factor, for p.f. improvement

(*c*) for driving electric trains, cranes, machine tools and line shaft drives

(*d*) in rubber mills, textile mills, cement factories, mining industries and other big industries for power applications

(*e*) for driving continuously operating and constant speed equipment such as centrifugal pumps, fans, blowers, ammonia and air compressors, motor generator sets etc.

441. In salient pole field structure, the pole shoes cover about ...... of pole pitch.

(*a*) one-third (*b*) one half

(*c*) two-third (*d*) whole.

442. Induction motors, when directly switched on across their normal rated supply voltage, the torque developed at starting is about ...... of their full load torque.

(*a*) half (*b*) 1.5 to 2.5 times

(*c*) 5 times (*d*) 10 times.

443. When the supply voltage to an induction motor is reduced by 10%, the maximum torque will decrease by approximately

(*a*) 5% (*b*) 10%

(*c*) 20% (*d*) 40%. **[GATE, 1996]**

444. Single phase preventer

(*a*) suppresses negative sequence currents

(*b*) compensates for voltage drops

(*c*) provides protection in the event of non-availability of one of the phases

(*d*) none of the above.

445. Increase in supply voltage to a 3-phase induction motor will not cause

(*a*) decrease in input current and so reduction in copper losses

(*b*) increase in torque

(*c*) decrease in speed as well as in core loss

(*d*) operating temperature abnormally higher owing to increase in core loss.

446. The interaction torque in a synchronous motor depends upon

(*a*) stator field strength alone

(*b*) rotor field strength alone

(*c*) torque angle alone

(*d*) all of the above *i.e.* on stator field and rotor field strengths and torque angle.

447. Electrical noise cannot be heard with
(*a*) power on (*b*) power off
(*c*) rotor locked (*d*) defective bearings.

448. In a 3-phase cylindrical rotor synchronous machine, the effect of armature reaction is taken care of by
(*a*) synchronous reactance (*b*) magnetizing reactance
(*c*) self reactance (*d*) mutual reactance
(*e*) leakage reactance.

449. There are two 2-pole d.c. machines. Main field axis of machine *A* is horizontal and that of *B* is vertical. It is preferable to purchase
(*a*) machine *A* (*b*) machine *B*
(*c*) any of the two machines (*d*) none of these.

450. A 10 kW, 400 V, 3-phase induction motor is started with direct-on line starter. The motor current and p.f. at starting will respectively be
(*a*) 120 A, 0.2 (*b*) 20 A, 0.2
(*c*) 120 A, 0.6 (*d*) 20 A, 0.6.

451. The hunting in synchronous machines can be guarded against by
(*a*) using a flywheel
(*b*) designing the synchronous machines with suitable synchronizing power
(*c*) by damper bars
(*d*) all of (*a*), (*b*) and (*c*)
(*e*) only (*a*) and (*c*).

452. An air-core transformer, as compared to iron-core transformer, has
(*a*) less magnetic core loss (*b*) more magnetic core loss
(*c*) no magnetic core loss (*d*) less ohmic loss.

453. If a transformer primary is energised from a square-wave voltage source, then its output voltage will be :
(*a*) zero (*b*) a sine wave
(*c*) a triangular wave (*d*) a pulsed wave.

454. In a self excited induction generator, to keep the frequency of generated voltage constant with the increase in load, the speed of the induction machine should be
(*a*) increased (*b*) decreased
(*c*) maintained less than the rated synchronous speed
(*d*) maintained more than the rated synchronous speed. **[U.P.S.C. I.E.S., 1995]**

455. The secondary of a CT is never left open-circuited because otherwise
(*a*) heat dissipation in the core will be very large
(*b*) the core will be saturated and permanently magnetised rendering it useless
(*c*) dangerously high e.m.fs. will be induced in the secondary
(*d*) all of the above.

456. In d.c. machines, the air-gap length under the poles is not kept uniform so as to
(*a*) obtain a sinusoidal main-field flux
(*b*) obtain better cooling
(*c*) minimise the effect of armature mmf on main field
(*d*) obtain sinusoidal armature mmf wave.

457. With core type transformers, the limbs are stepped so as to
(*a*) reduce the iron material and, therefore, iron loss
(*b*) reduce the conductor material and, therefore, I^2R loss
(*c*) provide better cooling
(*d*) provide more mechanical strength to the core.

458. Single phase synchronous motors
(*a*) are self starting motors
(*b*) do not need d.c. excitation
(*c*) operate at constant speed
(*d*) are built for a wider range of output
(*e*) all of the above.

459. In a 3-phase induction motor, the rotor current is produced by
(*a*) induction effect
(*b*) Lenz's law
(*c*) rotor supply
(*d*) none of these.

460. An induction motor having full-load torque of 60 Nm when delta connected develops a starting torque of 120 Nm. For the same supply voltage, if the motor is changed to star-connection, the starting torque developed will be
(*a*) 40 Nm
(*b*) 60 Nm
(*c*) 90 Nm
(*d*) 120 Nm. **[GATE, 1996]**

461. When an alternator designed for operation at 60 Hz is operated at 50 Hz
(*a*) kVA rating will increase in the ratio of 1.2
(*b*) operating voltage will reduce in the ratio of 5/6
(*c*) operating voltage will increase in the ratio of 1.2
(*d*) operating voltage will reduce in the ratio of $\left(\frac{5}{6}\right)^2$

462. Synchronous impedance of a synchronous machine is given as phasor sum of
(*a*) armature resistance and armature reactance
(*b*) armature resistance and leakage reactance
(*c*) armature resistance and synchronous reactance
(*d*) armature reactance and leakage reactance.

463. Linear induction motor is used in
(*a*) traction
(*b*) magnetic attraction
(*c*) mechanical workshops
(*d*) textile mills. **[B.T.E. M.P., 1995]**

464. In dynamic braking of 3-phase induction motors
(*a*) the supply terminals of any two stator phases are interchanged
(*b*) any two stator terminals are earthed
(*c*) the stator terminals are switched over to a d.c. source from the a.c. supply
(*d*) a d.c. voltage is injected in the rotor circuit.

465. While braking a 3-phase induction motor, its terminals were switched-over to a d.c. source from a.c. supply. The method of braking is
(*a*) regenerative braking
(*b*) counter-current braking
(*c*) eddy current braking
(*d*) none of the above.

466. Induction generators deliver power at
(*a*) leading power factor only
(*b*) lagging power factor only
(*c*) leading as well as lagging power factor
(*d*) unity power factor only.

467. The thickness of transformer core laminations is
(*a*) 3 mm to 5 mm
(*b*) 0.35 mm to 0.5 mm
(*c*) 30 mm to 50 mm
(*d*) 0.035 mm to 0.05 mm.

468. The drag cup rotor is employed in a two phase induction motor to give
(*a*) low inertia (*b*) high inertia
(*c*) low torque (*d*) high torque.

469. In a double cage induction motor outer cage used for starting has
(*a*) high resistance and low leakage reactance
(*b*) high leakage reactance and low resistance
(*c*) high resistance and high leakage reactance
(*d*) low resistance and low leakage reactance.

470. For production of a rotating magnetic field
(*a*) a single-phase supply is to be connected across a single-phase winding
(*b*) a two-phase supply should be connected across a two-phase winding
(*c*) a d.c. supply is to be connected across a single-phase winding
(*d*) the polarities of the d.c. supply across a single phase winding should be continuously reversed through a suitable switching device.

471. A 3-phase, 50 Hz induction motor takes a power input of 30 kW at its full-load speed of 1440 r.p.m. Total stator losses are 1 kW. The slip and rotor ohmic losses at full load are
(*a*) 0.02, 600 W (*b*) 0.04, 580 W
(*c*) 0.04, 1160 W (*d*) 0.04, 1200 W.

472. The regulation obtained by synchronous impedance method is always higher than actual value because
(*a*) synchronous reactance is assumed variable while it is not
(*b*) effective armature resistance is assumed constant
(*c*) synchronous reactance is assumed constant while it is not
(*d*) field current is increased to give short-circuit current about twice of full-load current.

[B.T.E. M.P., 1991]

473. In an oil-filled transformer, oil is provided for
(*a*) cooling (*b*) insulation
(*c*) lubrication (*d*) both cooling and insulation
(*e*) preventing accumulation of dust.

474. Synchronizing power comes into play when rotor speed is
(*a*) = synchronous speed n_s (*b*) > n_s
(*c*) < n_s (*d*) either more or less than n_s.

475. Which of the following synchronous motors has cost comparable to that of an induction motor ?
(*a*) High kW output high speed (*b*) High kW output low speed
(*c*) Low kW output low speed (*d*) Low kW output high speed.

476. The torque-speed characteristic of two phase induction motor is largely affected by
(*a*) voltage (*b*) speed
(*c*) $\frac{X}{R}$ ratio (*d*) supply frequency. **[U.P.S.C. I.E.S., 1992]**

477. An open-circuit test on a transformer is performed basically to determine
(*a*) copper loss (*b*) iron loss
(*c*) regulation (*d*) leakage impedance.

478. The iron loss in a 100 kVA transformer is 1 kW and the full-load copper losses are 2 kW, then maximum efficiency occurs at a load of

(*a*) 141.4 kVA (*b*) 70.7 kVA
(*c*) 50 kVA (*d*) 70.7 kW.

479. For a concentrated winding the polar group of each phase is ...... electrical degrees apart.
(*a*) 360 (*b*) 180
(*c*) 120 (*d*) 60.

480. In a 230-V d.c. machine, the brushes are made up of
(*a*) copper (*b*) copper-carbon
(*c*) copper-graphite (*d*) electrographite.

481. If the supply frequency in a transformer is doubled
(*a*) hysteresis loss also doubles (*b*) eddy current loss also doubles
(*c*) iron losses double (*d*) copper losses double.

482. A transformer possesses a percentage resistance and percentage reactance of 1% and 4% respectively. Its voltage regulation at power factor 0.8 lagging and 0.8 leading would be
(*a*) 2.4% and – 0.8% (*b*) 3.2% and – 1.6%
(*c*) 3.2% and – 3.2% (*d*) 4.8% and – 1.6%. **[U.P.S.C. I.E.S., 1996]**

483. The voltage regulation of a transformer at full load and 0.8 power factor lagging is 2.5%. The voltage regulation at full load, 0.8 power factor leading will be
(*a*) – 2.5% (*b*) zero
(*c*) 1.0% (*d*) 2.5%. **[U.P.S.C. I.E.S., 1994]**

484. The principle of operation of a 3-phase induction motor closely resembles to that of a
(*a*) two winding transformer with its secondary short-circuited
(*b*) two winding transformer with its secondary open-circuited
(*c*) synchronous motor
(*d*) d.c. shunt motor.

485. A synchronous motor installed at the end of a transmission line is operating at lagging power factor. With the fall in supply voltage, the power factor of the synchronous motor will
(*a*) go down (*b*) improve
(*c*) remain unchanged (*d*) none of these.

486. Which of the following statements associated with a Scharge motor is/are correct ?
(*a*) The magnitude of the slip frequency voltage injected into the secondary circuit depends upon the spacing between the two sets of brushes, and the phase depends on their angular position with respect to centre of the rotor winding
(*b*) When the two sets of brushes of each phase are making contact with the same commutator bar, the secondary is short-circuited and the motor operates as an ordinary induction motor
(*c*) Its power factor can be varied by rocking the brushes as a whole round the commutator
(*d*) All of the above.

487. A 3-phase wound rotor induction motor, when started with load connected to its shaft, was found to start but settle down at about half synchronous speed. If the rotor winding as well as stator winding were star-connected, the cause of the malfunctioning could be attributed to
(*a*) one of the stator phase winding being short-circuited
(*b*) one of the supply fuses being blown
(*c*) one of the rotor phases being open-circuited
(*d*) two of the rotor phases being open-circuited. **[U.P.S.C. I.E.S., 1997]**

488. The stator of a $\frac{2}{4}$ pole changing cage motor is initially wound for 2-poles. The reconnection of the stator winding to 4-poles through a change over switch, while the motor is running would result in

(*a*) constant torque drive (*b*) constant hp drive
(*c*) plugging to standstill
(*d*) regenerative braking to half the original speed. **[U.P.S.C. I.E.S., 1992]**

489. If F_1 is the constant amplitude of fundamental rotating m.m.f. wave, then for the space harmonics of the order of $6k + 1$, the harmonic m.m.f. wave is of

(*a*) constant amplitude and stationary in space
(*b*) constant amplitude and rotates against F_1
(*c*) varying amplitude and rotates along F_1
(*d*) constant amplitude and rotates along F_1.

490. A large synchronous generator gives output at the standard frequency. This frequency can be increased by

(*a*) increasing the field current (*b*) increasing the speed of the armature
(*c*) increasing the speed of the field (*d*) decreasing the field current.

491. A d.c. machine is fitted with both interpole winding *IW* and a compensating winding *CW*. With respect to the armature circuit

(*a*) both *IW* and *CW* are in parallel (*b*) *IW* is in series and *CW* is in parallel
(*c*) *IW* is in parallel and *CW* is in series (*d*) both *IW* and *CW* are in series.

492. Two single phase 100 kVA transformers, each having different leakage impedances are connected in parallel. When a load of 150 kVA at 0.8 p.f. lagging is applied

(*a*) both transformers will operate at power factor more than 0.8 lagging
(*b*) both transformers will operate at power factor less than 0.8 lagging
(*c*) one of transformers will operate at p.f. more than 0.8 lagging and other will operate at p.f. less than 0.8 lagging
(*d*) both transformers will operate at identical power factors. **[U.P.S.C. I.E.S., 1992]**

493. The operation of an induction motor is based on

(*a*) Lenz's law (*b*) Ampere's law
(*c*) principle of self induction (*d*) none of these.

494. Shaded pole induction motor does not have the advantage of

(*a*) high starting torque (*b*) rugged construction
(*c*) low initial as well as maintenance costs (*d*) comparatively small starting current
(*e*) use of concentrated winding and absence of centrifugal switch.

495. Which type of motor is most suitable for a computer printer drive ?

(*a*) Reluctance motor (*b*) Hysteresis motor
(*c*) Shaded pole motor (*d*) Stepper motor. **[GATE, 1996]**

496. A d.c. shunt motor has external resistance R_f in the field circuit and R_a in the armature circuit. The starting armature current can be reduced by keeping

(*a*) maximum R_f and minimum R_a (*b*) maximum R_f and maximum R_a
(*c*) minimum R_f and minimum R_a (*d*) minimum R_f and maximum R_a.

497. The stator and rotor pole number may be different in a

(*a*) pole changing induction motor (*b*) reluctance motor
(*c*) repulsion motor (*d*) synchronous motor. **[U.P.S.C. I.E.S., 1993]**

498. Repulsion motor
(*a*) receives its armature current by induction
(*b*) is similar to a series motor except that rotor and stator windings are inductively coupled
(*c*) can be directly switched on to supply line
(*d*) has high starting torque and high speed on no load
(*e*) all of the above.

499. Supply to one terminal of a delta-wye connected three phase core type transformer which is on no-load fails. Assuming magnetic circuit symmetry, voltages on the secondary side will be
(*a*) 230, 230, 115 (*b*) 230, 115, 115
(*c*) 345, 115, 115 (*d*) 345, 0, 345. **[GATE, 1995]**

500. The efficiency of a 100 kVA transformer is 0.98 at full as well as half load. For this transformer at full load the copper loss
(*a*) is less than core less (*b*) is equal to core loss
(*c*) is more than core loss (*d*) none of the above. **[GATE, 1998]**

501. A cylindrical rotor synchronous motor is switched on to the d.c. supply with its field windings shorted on themselves. It will
(*a*) not start (*b*) start but not run at synchronous speed
(*c*) start as an induction motor and then run as a synchronous motor
(*d*) start and run as a synchronous motor. **[GATE, 1996]**

502. The efficiency of two identical transformers under load conditions can be determined by
(*a*) back to back test (*b*) oper-circuit test
(*c*) short-circuit test (*d*) any of the above. **[U.P.S.C. I.E.S., 1992]**

503. In phase shift control the control is carried out by varying the ...... of grid voltage.
(*a*) magnitude (*b*) polarity
(*c*) phase (*d*) any of the above.

504. In 36 slot, 4-pole, 3-ϕ alternator the winding pitch is 7. The electrical angle by which the winding is chorded is equal to
(*a*) 40° (*b*) 30°
(*c*) 60° (*d*) 15°. **[B.T.E. M.P. Elec. Machines, 1992]**

505. The direction of rotation of a shaded pole induction motor
(*a*) cannot be reversed unless there is a provision for shifting of shading coil from one half to the other half of the pole
(*b*) can be reversed by interchanging main winding terminals
(*c*) can be reversed by interchanging the supply terminals
(*d*) can be reversed by open-circuiting the shading rings.

506. Open-circuit test in a transformer is performed with
(*a*) rated transformer voltage (*b*) rated transformer current
(*c*) direct current (*d*) high frequency supply.

507. Which of the following statements about a single phase induction regulator is *incorrect* ?
(*a*) The primary winding is usually wound on the stator and the secondary is wound on rotor
(*b*) Both primary and secondary windings are usually two poles and spread over approximately 120°
(*c*) The primary winding is connected across the circuit to be regulated and the secondary is connected in series with it.
(*d*) The rotor slots carry an additional short-circuited winding placed on the rotor at right angle to the primary winding.

508. Transformer-core laminations are made of
(*a*) cast iron (*b*) wrought iron
(*c*) silicon steel (*d*) cast steel
(*e*) stainless steel.

509. For a d.c. series motor with saturated condition of the field circuit, the torque developed is proportional to
(*a*) I_a^2 (*b*) I_a
(*c*) speed (*d*) voltage.

510. The electromagnetic force, developed in any physical system, acts in such a direction as to tend to
(*a*) decrease the coenergy at constant mmf (*b*) increase the coenergy at constant flux
(*c*) decrease the coenergy at constant flux (*d*) increase the coenergy at constant mmf.

511. For terminal voltage V_t = 11 kV, the field current required is 20 A for unity p.f. load on an alternator. For maintaining constant V_t, the field current required for loads of 0.8 p.f. lag and 0.8 p.f. lead would respectively be :
(*a*) > 20 A, < 20 A (*b*) > 20 A, > 20 A
(*c*) < 20 A, < 20 A (*d*) < 20 A, > 20 A.

512. Two transformers *A* and *B*, with identical ratings, are to be designed with flux densities of 1.2 and 1.4 Wb/m^2 respectively. The weight of transformer *A* per kVA is :
(*a*) less than that of transformer *B* (*b*) more than that of transformer *B*
(*c*) equal to that of transformer *B* (*d*) none of these.

513. The magnetising current of a three-phase induction motor is much higher than an equivalent transformer because
(*a*) the size of an induction motor is larger than an equivalent transformer
(*b*) of the presence of air-gap between stator and rotor in an induction motor
(*c*) gain oriented magnetic material is used for the core of an induction motor
(*d*) inferior magnetic material is used for the core of an induction motor.

514. Smooth speed control of a three-phase induction motor over a wide range is possible, by
(*a*) pole changing method (*b*) frequency control method
(*c*) using consequent pole technique (*d*) by voltage control method.

515. A d.c. shunt motor draws an armature current of I_a amperes under normal rated conditions. For constant power output, if both the field flux and armature terminal voltage are halved, then the armature current becomes
(*a*) I_a (*b*) $2I_a$
(*c*) $\frac{1}{2} I_a$ (*d*) $4I_a$
(*e*) $\frac{1}{4} I_a$.

516. The power factor of an induction motor at no load is around
(*a*) 0.2 lead (*b*) 0.2 lag
(*c*) 0.5 lag (*d*) 0.8 lag.

517. In comparison to three phase induction motors, single phase induction motors have
(*a*) lower efficiency (*b*) lower pf
(*c*) poor speed regulation (*d*) more slip at full load
(*e*) all of the above.

518. If the rotor circuit resistance is increased in an induction motor, the maximum torque will occur at
(*a*) lower speed (*b*) high speed
(*c*) the same speed (*d*) none of these.

519. The absolute speed of magnetic field in space of a 3-phase rotor fed induction motor is
(*a*) synchronous speed N_s (*b*) rotor speed N_r
(*c*) $(N_s - N_r)$ (*d*) $(N_s + N_r)$. **[U.P.S.C. I.E.S., 1996]**

520. A voltmeter gives 120 oscillations per minute when connected to the rotor of an induction motor. The stator frequency is 50 Hz. The slip of the motor is
(*a*) 2% (*b*) 2.5%
(*c*) 4% (*d*) 5%. **[U.P.S.C. I.E.S., 1995]**

521. Two transformers of identical voltages but of different capacities are operating in parallel. For satisfactory load sharing
(*a*) impedances must be equal
(*b*) per unit impedance must be equal
(*c*) per unit impedances and X/R ratio must be equal
(*d*) impedance and X/R ratio must be equal. **[GATE, 1994]**

522. The value of capacitor for a capacitor-start induction motor is determined by its
(*a*) speed (*b*) efficiency
(*c*) starting torque (*d*) power factor.

523. Which of the following statements regarding repulsion induction motor is *wrong* ?
(*a*) At start squirrel cage winding does not produce any torque
(*b*) At start and at low speeds, major part of the torque is developed by repulsion winding
(*c*) At rated load it almost runs at synchronous speed
(*d*) It employes centrifugally operated mechanism for short-circuiting the commutator.

524. A synchronous machine connected to a power system grid bus-bar is operating as a generator. To make the machine operate as a motor, the
(*a*) direction of rotation is to be reversed
(*b*) phase-sequence is to be changed
(*c*) field excitation is to be decreased
(*d*) mechanical input is to be less than the losses at the shaft. **[U.P.S.C. I.E.S., 1993]**

525. The two windings provided on the stator of a single phase induction motor, one main winding and the other auxiliary winding are connected
(*a*) in parallel (*b*) in series
(*c*) either in series or in parallel depending on the design of the motor
(*d*) through inductive coupling.

526. The armature current of a synchronous motor on no load
(*a*) leads the applied voltage by 90° (*b*) lags behind the applied voltage by 90°
(*c*) is in phase with applied voltage (*d*) zero.

527. A 3-phase synchronous generator is operating at 0.8 p.f. lagging with respect to the excitation voltage. The nature of armature reaction m.m.f. produced by the armature currents is
(*a*) magnetizing (*b*) demagnetizing
(*c*) cross-magnetizing and partly demagnetizing
(*d*) partly demagnetizing and partly cross-magnetizing.

528. A 3-phase induction motor is running at half-full load. If the fuse in one of phases gets burnt, the motor would
(*a*) come to standstill
(*b*) continue running at the same speed

(c) continue running at slightly reduced speed
(d) continue running at an increased speed.

529. The name plate of an alternator indicates an operating p.f. of 0.85. The nature of this p.f. is
(a) lagging because this requires less field current I_f
(b) lagging because I_f required is more
(c) leading because I_f required is less
(d) leading because I_f required is more.

530. Scott-connections are used for
(a) single phase to three phase transformation
(b) three phase to single phase transformation
(c) three phase to two phase transformation
(d) any of the above. **[U.P.S.C. I.E.S., 1992]**

531. For slip-ring induction motors, rotor resistance starting is preferred over reduced voltage starting because it
(a) limits the speed (b) limits the starting current
(c) increases the starting torque (d) improves the starting power factor
(e) limits starting current, increases starting torque and also improves starting power factor.

532. In a 3-phase synchronous motor, the field poles
(a) lead the resultant air-gap flux were ϕ_r (b) lag ϕ_r
(c) are along ϕ_r (d) are opposite to ϕ_r.

533. When a synchronous motor is running at synchronous speed, the damper winding produces
(a) damping torque (b) eddy current torque
(c) torque aiding the developed torque (d) no torque. **[U.P.S.C. I.E.S., 1997]**

534. The synchronous impedance of a synchronous induction motor is much larger than that of a synchronous motor due to
(a) its larger air-gap (b) presence of damper bars in it
(c) its less magnetic reluctance (d) supply of d.c. excitation to its 3-phase rotor.

535. The efficiency of a d.c. shunt generator is maximum when
(a) magnetic loss is equal to the mechanical loss
(b) field ohmic loss is equal to the constant loss
(c) stray load loss is equal to the armature circuit loss
(d) armature circuit loss is equal to the sum of rotational-loss and field circuit loss.

536. Speed control by the variation of field flux results in
(a) constant power drive (b) variable power drive
(c) constant torque drive (d) variable torque drive.

537. The main advantage(s) of an auto-transformer over a conventional two winding transformer is/are
(a) continuously varying voltage is available
(b) considerable saving in winding material
(c) iron losses are reduced
(d) copper losses are considerably reduced
(e) both (a) and (b).

538. Single phase induction motors are made self starting by
(a) increasing rotor resistance (b) using an external starting device
(c) providing an additional winding on the stator called the auxiliary winding
(d) any of the above methods.

539. Two transformers are operating in parallel, their load sharing will depend upon
(*a*) cooling methods
(*b*) all day efficiency
(*c*) maximum efficiency
(*d*) per unit impedance.
[B.T.E.M.P. Elec. Machines, 1995]

540. Which of the following connections of transformer will give the highest secondary voltage ?
(*a*) Delta primary, delta secondary
(*b*) Delta primary, star secondary
(*c*) Star primary, star secondary
(*d*) Star primary, delta secondary.
[U.P.S.C. I.E.S., 1992]

541. The stator core of a 3-phase induction motor is laminated in order to reduce the
(*a*) eddy current loss
(*b*) hysteresis loss
(*c*) both eddy current and hysteresis loss
(*d*) weight of the stator
(*e*) windage and frictional losses.

542. When a transformer winding suffers a short-circuit, the adjoining turns of the same winding experience
(*a*) an attractive force
(*b*) a repulsive force
(*c*) no force
(*d*) any of the above. **[GATE, 1994]**

543. ...... losses have the high proportion of losses occuring in synchronous machines.
(*a*) Iron
(*b*) Stator copper
(*c*) Friction and windage
(*d*) Eddy current.

544. Increase in length of air-gap in an induction motor causes
(*a*) decrease in power factor
(*b*) decrease in pulsation losses
(*c*) increase in ampere-turns required to produce same flux density in the air gap
(*d*) all of the above.

545. Phase advancers are employed with large induction motors to
(*a*) regulate their speeds
(*b*) reduce copper losses
(*c*) improve power factor
(*d*) reduce noise and vibrations.

546. A 3-phase synchronous motor is exclusively used where
(*a*) load fluctuates frequency
(*b*) the speed is to be maintained absolutely constant
(*c*) very high speed is required
(*d*) high torque loads are to be driven
(*e*) speed regulation is required.

547. Transformer zero voltage regulation occurs at
(*a*) unity p.f.
(*b*) leading p.f.
(*c*) lagging p.f.
(*d*) zero p.f. leading.

548. The direction of rotation of an ordinary shaded pole single-phase induction motor
(*a*) can be reversed by reversing the supply terminal connections to the stator winding
(*b*) cannot be reversed
(*c*) can be reversed by open-circuiting the shading rings
(*d*) can be reversed by short-circuiting the shading rings.

549. Direction of rotation of split-phase type single-phase induction motor can be reversed by
(*a*) reversing the supply terminal connections
(*b*) reversing the connections of only the auxiliary winding across the supply terminals
(*c*) reversing the connections of either the main winding or the auxiliary winding terminals
(*d*) reversing the connections of only the main winding across the supply terminals.

550. Open-circuit test on a single-phase transformer gave the following data : 230 V at 50 Hz, 2 A, p.f. = 0.2 lag.
If open-circuit test is performed at 230 V, 45 Hz, then no-load current is
(*a*) decreased at p.f. > 0.2
(*b*) decreased at p.f. < 0.2
(*c*) increased at p.f. > 0.2
(*d*) increased at p.f. < 0.2
(*e*) same at p.f. = 0.2.

551. In Swinburne's method of testing d.c. machines, the shunt machine is run as a
(*a*) motor at full load
(*b*) generator at full load
(*c*) motor at no load
(*d*) generator at no load.

552. The terminal voltage of a d.c. series generator running at rated speed and at no load is equal to
(*a*) rated voltage
(*b*) half of its rated voltage
(*c*) zero
(*d*) a very little voltage.

553. If the rotor of an induction motor is assumed purely inductive, then electromagnetic torque is
(*a*) optimum with load angle $\delta = 0°$
(*b*) optimum with $\delta = 90°$
(*c*) maximum with $\delta = 90°$
(*d*) maximum with $\delta = 0°$
(*e*) zero.

554. For core type power transformers, both primary and secondary windings have circular coil sections, because this section
(*a*) is easier to wind
(*b*) requires minimum copper and, therefore, less I^2R loss
(*c*) has the strongest mechanical shape
(*d*) results in less core material and, therefore, less core loss.

555. A d.c. shunt motor is running at 900 r.p.m. It is required to run this motor at ± 150 r.p.m. The method suitable for this purpose is
(*a*) Ward-Leonard method
(*b*) field control method
(*c*) armature resistance method
(*d*) armature voltage control method.

556. In a 3-phase synchronous generator, the field poles
(*a*) lead the resultant air gap flux wave
(*b*) lag the resultant air-gap flux wave
(*c*) are along the resultant air-gap flux wave
(*d*) are opposite to the resultant air-gap flux wave.

557. The commutator segments of d.c. machines are made up of
(*a*) brass
(*b*) copper
(*c*) hard-drawn copper
(*d*) aluminium.

558. Ripple frequency in a single phase full-wave rectifier operating at 50 Hz frequency supply is
(*a*) 25
(*b*) 50
(*c*) 100
(*d*) 200.

559. The armature core of a d.c. machine is laminated to minimise
(*a*) hysteresis loss
(*b*) eddy-current loss
(*c*) copper loss
(*d*) mechanical loss
(*e*) temperature rise.

560. The windings for an alternator are
I. 36 slots, 4-poles, span 1 to 3
II. 72 slots, 6-poles, span 1 to 10
III. 96 slots, 6-poles, span 1 to 12

The windings having pitch factor of more than 0.9 are

(*a*) I and II only (*b*) II and III only

(*c*) I and III only (*d*) I, II and III. **[U.P.S.C. I.E.S., 1992]**

561. A starting torque of 80 Nm is developed in an induction motor by an auto-transformer starter with a tapping of 30%. If the tapping of auto-transformer is 60%, then the starting torque will be

(*a*) 40 Nm (*b*) 160 Nm

(*c*) 240 Nm (*d*) 320 Nm. **[U.P.S.C. I.E.S., 1994]**

562. When a balanced 3-phase distributed type armature winding is carrying 3-phase, balanced currents, the strength of the resultant rotating magnetic field is

(*a*) three times the amplitude of each constituent pulsating magnetic field

(*b*) equal to the amplitude of each constituent pulsating magnetic field

(*c*) half the amplitude of each constituent pulsating magnetic field

(*d*) one and half times the amplitude of each constituent of pulsating magnetic field. **[U.P.S.C. I.E.S., 1994]**

563. An induction motor has rotor resistance R_r, stand-still rotor induced voltage E_{ss}, and stator to rotor effective turn-ratio of m. In an equivalent circuit of this machine, the rotor circuit resistance is shown as m^2R_r/s, where s is the slip. This implies that the value of the equivalent rotor circuit voltage will be

(*a*) E_{ss} (*b*) sE_{ss}

(*c*) mE_{ss} (*d*) msE_{ss}. **[U.P.S.C. I.E.S., 1993]**

564. The equivalent circuit per phase of a three phase transformer is similar to that of a three phase induction motor but the transformer does not develop any torque. This is due to

(*a*) insufficient voltage (*b*) low supply frequency

(*c*) non-fulfilment of condition of space condition of winding

(*d*) none of the above.

565. Which of the following statements about reluctance-start motor is/are *correct* ?

(*a*) It is basically an induction motor and not a synchronous motor

(*b*) The air-gap between rotor and stator is uniform

(*c*) Its working principle is similar as that of a shaded pole motor

(*d*) Both (*a*) and (*c*).

566. A hysteresis motor

(*a*) has high starting torque due to its high rotor hysteresis loss

(*b*) is extremely quite in operation due to absence of teeth and winding on rotor

(*c*) accelerates from rest to full speed almost instantaneously

(*d*) all of the above.

567. If the capacitor of a capacitor-start induction motor fails to open when it picks up the speed

(*a*) the motor will stop (*b*) the auxiliary winding will get damaged

(*c*) the capacitor will get damaged (*d*) the main winding will get damaged.

568. The polarity of the output voltage in a cross-field generator can be altered by

(*a*) reversing the direction of rotation

(*b*) reversing both the direction of rotation and field current

(c) reversing the direction of field current
(d) reversing either the direction of rotation or the field current.

569. The electromagnetic torque, developed in any physical system and with magnetic saturation neglected, acts in such a direction as to tend to
(a) decrease both the reluctance and inductance
(b) increase both the reluctance and inductance
(c) decrease the reluctance and increase the inductance
(d) increase the reluctance and decrease the inductance.

570. In a transformer, eddy current loss is 100 watts which is half of the total core loss. If both the thickness of laminations and frequency are increased by 10%, the new core loss would be :
(a) 256.41 W (b) 231 W
(c) 267.41 W (d) 242 W.

571. The balanced short-circuit current of a polyphase alternator is 20 A at a speed of 1000 r.p.m. For the same field current, the balanced short-circuit current at 900 r.p.m. would be
(a) less than 20 A (b) more than 20 A
(c) 20 A (d) 18 A.

572. In a 3-phase induction motor, if stator impedance is neglected, then the slip at maximum torque is equal to
(a) x_2/r_2 (b) r_2/x_2
(c) $\sqrt{\dfrac{r_2}{x_2}}$ (d) $\sqrt{\dfrac{x_2}{r_2}}$.

573. In d.c. motors, the polarity of interpoles is
(a) same as that of main pole behind (b) same as that of main pole ahead
(c) opposite to that of the main pole behind (d) none of these.

574. An amplidyne develops an output voltage of 100 volts at an armature speed of n r.p.m. If its armature is now driven at $2n$ r.p.m., then its output voltage is
(a) 100 volts (b) 200 volts
(c) 400 volts (d) 50 volts
(e) 25 volts.

575. One transformer has leakage impedance of $1 + j4\ \Omega$ and $3 + j11\ \Omega$ for its primary and secondary windings respectively. This transformer has
(a) h.v. primary (b) medium voltage primary
(c) l.v. primary (d) none of these.

576. A 3-phase induction motor should have small air-gap length so that it has
(a) more starting torque (b) more pull-out torque
(c) better p.f. (d) improved efficiency.

577. Transformer maximum efficiency, for a constant load current, occurs at
(a) 0.8 p.f. (b) zero p.f. leading
(c) zero p.f. lagging (d) unity power factor.

578. A d.c. shunt motor is a
(a) variable-speed motor (b) adjustable-speed motor
(c) constant-speed motor (d) constant-speed-adjustable-speed motor.

579. The object of using compensating winding in d.c. machines is to neutralize the
(a) armature reaction in the interpolar zone
(b) armature reaction in the commutating zone

(*c*) armature reaction under the pole faces
(*d*) cross-magnetizing armature reaction.

580. Rotational losses in electrical machines consist of
(*a*) friction and windage losses
(*b*) stator core, friction and windage losses
(*c*) rotor core, friction and windage losses
(*d*) stray load losses and friction and windage losses.

581. In a synchronous motor, squirrel cage winding is provided for making motor
(*a*) noise free
(*b*) self starting
(*c*) cheap
(*d*) quick start. **[B.T.E.M.P., 1996]**

582. The salient pole rotors have
(*a*) two or four projecting poles carrying the field coils
(*b*) a large number of projecting poles carrying the field coils
(*c*) thick steel laminations riveted together and fixed to the rotor by a dove-tail joint
(*d*) pole faces usually provided with slots for damper windings
(*e*) all of the above except (*a*).

583. A 3-phase, 400 V, 50 Hz synchronous motor is operating normally. If the load on the motor is doubled keeping the excitation constant, the new load angle will become nearly
(*a*) double
(*b*) half
(*c*) four times
(*d*) $\sqrt{2}$ times.

584. The torque angle of a synchronous machine operating from a constant voltage bus, is usually defined as the space angle between
(*a*) rotor mmf wave and stator mmf wave
(*b*) rotor mmf wave and resultant flux density wave
(*c*) stator mmf wave and resultant flux density wave
(*d*) stator mmf wave and resultant mmf wave. **[GATE, 1992]**

585. The direction of shaded pole motor can be reversed by
(*a*) reversing supply terminals
(*b*) shifting any shading coil to the other half of the pole
(*c*) shifting two shading coils to the other half of the pole
(*d*) shifting all the shading coils to the other half of the pole. **[B.T.E.M.P., 1992]**

586. In a repulsion motor
(*a*) stator usually carries a distributed winding similar to the main winding of an ordinary single phase induction motor
(*b*) the rotor or armature is similar to a d.c. motor armature with a drum type winding connected to the commutator
(*c*) the brushes fixed directly opposite to each other are connected to each other or short-circuited
(*d*) the speed is function of load
(*e*) all of the above.

587. The desirable properties of transformer core material are
(*a*) low permeability and low hysteresis loss
(*b*) high permeability and high hysteresis loss
(*c*) high permeability and low hysteresis loss
(*d*) low permeability and high hysteresis loss. **[U.P.S.C. I.E.S., 1992]**

588. Consider the circuit shown in the given Fig. 3. For maximum power transfer to the load, the primary to secondary turn-ratio must be

(*a*) 9 : 1 (*b*) 3 : 1

(*c*) 1 : 3 (*d*) 1 : 9.

[U.P.S.C. I.E.S., 1995]

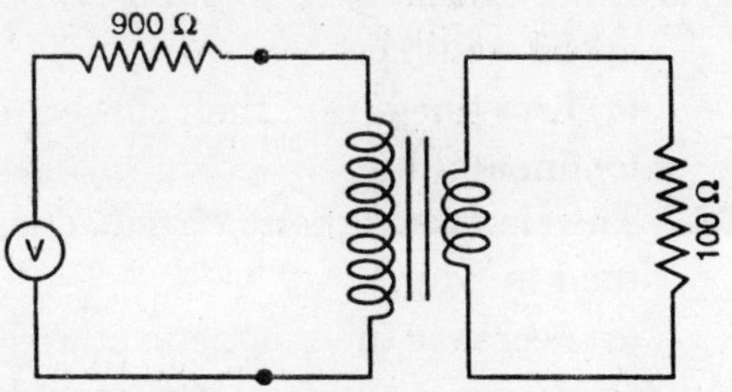

Fig. 3

589. A salient-pole synchronous motor runs under steady state conditions with armature current I_a. If the field-circuit gets suddenly open-circuited, then

(*a*) the motor stops, I_a becomes zero

(*b*) it continues to run at the same speed, I_a increases

(*c*) it continues to run at the same speed, I_a decreases

(*d*) it runs at a slower speed, I_a increases.

590. A centre-zero ammeter connected in the rotor circuit of a 6-pole, 50 Hz slip-ring induction motor makes 45 complete oscillations in one minute. In r.p.m., the rotor speed and speed, of stator field with respect to rotor are respectively

(*a*) 985, 15 (*b*) 970, 1000

(*c*) 985, 985 (*d*) 985, 970.

591. An alternator and a synchronous motor are connected to an infinite bus. Both are working at unity p.f. and reactive power handled by them is as under :

(*a*) both deliver it to the bus

(*b*) both absorb it from the bus

(*c*) alternator is delivering it and motor is absorbing it

(*d*) neither of these machines is delivering nor absorbing it.

592. One 200 V, 100 W bulb is connected in series with the primary of a 200 V, 10 kVA transformer. If its secondary is left open circuited, then the bulb would have

(*a*) full brightness (*b*) poor brightness

(*c*) a little less than full brightness (*d*) more than full brightness.

593. In case of a power transformer, the no-load current in terms of rated current is

(*a*) 10 to 20% (*b*) 2 to 6%

(*c*) 15 to 30% (*d*) 30 to 50%.

594. A d.c. shunt motor runs at rated speed. If its field circuit gets open-circuited, then soon after this the motor speed would tend to

(*a*) decrease (*b*) remain unchanged

(*c*) increase (*d*) fluctuate around its previous speed.

595. In a transformer

(*a*) both a.c. and s.c. tests are conducted on l.v. side

(*b*) o.c. test is conducted on h.v. side and s.c. test on l.v. side

(*c*) o.c. test is conducted on l.v. side and s.c. test on h.v. side

(*d*) both o.c. and s.c. tests are conducted on h.v. side.

596. The field windings of a d.c. compound motor are connected in short shunt configuration. If the series and shunt field windings are interchanged by mistake and the motor is connected to rated voltage,

(*a*) series winding gets overloaded (*b*) shunt winding gets overloaded

(*c*) both windings get overloaded (*d*) neither of the two windings gets overloaded.

597. If the excitation is kept constant in an alternator, then the excitation voltage variation with speed would be
(*a*) first linear and then curved upwards (*b*) first linear and then curved downwards
(*c*) linear (*d*) first curved and then linear.

598. The electromagnetic torque, developed in any physical system, acts in such a direction as to tend to
(*a*) decrease the magnetic stored energy at constant current
(*b*) decrease the magnetic stored energy at constant flux
(*c*) increase the magnetic stored energy at constant flux
(*d*) increase the magnetic stored energy at constant current.

599. The possible resistance of the shunt field winding of a d.c. machine is
(*a*) 1 Ω (*b*) 2 Ω
(*c*) 200 Ω (*d*) 20 Ω.

600. In a 3-phase induction motor if the leakage reactance is reduced by using open slots
(*a*) starting torque and starting current will decrease but power factor will increase
(*b*) starting torque and starting current both will increase but power factor will decrease
(*c*) pull-out torque will decrease
(*d*) starting current will increase but starting torque will decrease.

601. In a phase-shift control, the phase shift between anode and grid voltages can be achieved by means of
(*a*) an induction regulator (*b*) a synchronous motor
(*c*) a synchronous generator (*d*) any of the above.

602. A single phase transformer has specifications as 250 kVA, 11000 V/415 V, 50 Hz. What are the approximate values of primary and secondary currents ?
(*a*) Primary = 22.7 A, Secondary = 602 A (*b*) Primary = 602 A, Secondary = 22.7 A
(*c*) Primary = 301 A, Secondary = 22.7 A (*d*) Primary = 11.4 A, Secondary = 301 A.

603. The no-load current in a transformer lags the applied voltage by
(*a*) 90° (*b*) 0°
(*c*) 75° (*d*) about 115°. **[B.T.E.U.P., 1996]**

604. The method that gives continuous speed control of a slip-ring induction motor is
(*a*) rotor resistance control (*b*) secondary foreign voltage control
(*c*) concatenation or cascade operation (*d*) line voltage control.

605. For the applications requiring speed ratio other than 2 : 1, the speed control can be affected by varying the number of stator poles employing
(*a*) multiple stator winding (*b*) pole amplitude modulation technique
(*c*) consequent pole technique (*d*) any of the above.

606. A transformer has at full-load iron loss of 900 watts and copper loss of 1600 watts. At what % of the load transformer will have maximum efficiency ?
(*a*) 133% (*b*) 125%
(*c*) 75% (*d*) 66.6%. **[S.C.E. & T.E. W.B., 1996]**

607. The torque developed by a 3-phase induction motor depends upon
(*a*) rotor phase resistance (*b*) rotor standstill phase reactance
(*c*) applied voltage (*d*) slip
(*e*) all of the above.

608. The impedance value of a generator is 0.2 pu on a base value of 11 kV, 50 MVA. The impedance value for a base value of 22 kV, 150 MVA is

(*a*) 0.15 pu. (*b*) 0.2 pu.
(*c*) 0.3 pu. (*d*) 0.24 pu. **[U.P.S.C. I.E.S., 1996]**

609. Booster transformers should never .

(*a*) have fuse in the hv side (*b*) be left open-circuited
(*c*) be left closed circuited (*d*) none of the above.

610. The primary current in a *CT* is dictated by

(*a*) the secondary burden (*b*) the core of the transformer
(*c*) the load current (*d*) none of the above. **[GATE, 1997]**

611. A 400 V, 50 Hz three-phase induction motor rotates at 1440 r.p.m. on full-load. The motor is wound for

(*a*) 2 poles (*b*) 4 poles
(*c*) 6 poles (*d*) 8 poles.

612. The slip of 400 V, three phase, 4-pole induction motor when rotating at 1440 r.p.m. is

(*a*) 2 per cent (*b*) 3 per cent
(*c*) 4 per cent (*d*) 5 per cent.

613. The relation between synchronous speed, stator supply frequency and stator number of poles of a three-phase induction motor is given by

(*a*) $N_s = \dfrac{P}{120f}$ (*b*) $f = \dfrac{120N_s}{P}$
(*c*) $f = \dfrac{PN_s}{120}$ (*d*) $N_s = \dfrac{120P}{f}$.

614. Two 3-phase induction motors have efficiencies of 82% and 92% and their no-load p.f.s. are 0.1 and 0.3. These two motors *A* and *B* have the following values for their p.f.s. and efficiencies :

(*a*) Bigger motor (*BM*) – 0.1, 82% ; smaller motor (*SM*) – 0.3, 92%
(*b*) *BM A* – 0.1, 9.2% ; *SM B* – 0.3, 82%
(*c*) *BM A* – 0.3, 92% ; *SM B* – 0.1, 82%
(*d*) *BM A* – 0.3, 82% ; *SM B* – 0.1, 92%.

615. The rotor of an induction motor can't run at synchronous speed, because if it did so then

(*a*) rotor e.m.f. would be zero (*b*) rotor current would be zero
(*c*) rotor torque would be zero (*d*) none of (*a*), (*b*) and (*c*)
(*e*) all of (*a*), (*b*) and (*c*).

616. The torque power factor of an induction motor will be high when

(*a*) running at no-load (*b*) running at full-load
(*c*) rotor is blocked (*d*) the rotor is crawling.

617. A 5 hp, three-phase, 400 V star-connected squirrel-cage induction motor meant to drive a milling machine, at starting takes about

(*a*) 40 A (*b*) 100 A
(*c*) 150 A (*d*) 200 A.

618. While starting a three-phase induction motor, a star-delta starter is used to

(*a*) reduce the starting current to a safe value
(*b*) achieve higher starting torque
(*c*) enable the motor to start in the right direction
(*d*) be able to reverse the direction of rotation of the rotor as and when necessary.

619. A delta-connected 400 V, 50 Hz, three-phase induction motor when started direct-on-line takes a starting current of 30 A. When the motor is started through a star-delta starter, the starting current will be

(*a*) 3 A (*b*) 10 A

(*c*) 15 A (*d*) 30 A.

620. Torque developed by a three-phase, 400 V, induction motor is 100 N-m. If the applied voltage is reduced to 200 V, the developed torque will be

(*a*) 50 N-m (*b*) 25 N-m

(*c*) 200 N-m (*d*) 62.5 N-m.

621. If the rotor circuit resistance of a three-phase induction motor is increased

(*a*) its starting torque will increase and the maximum torque developed will also increase

(*b*) both the starting torque and maximum torque developed will decrease

(*c*) its starting torque will increase but the maximum torque developed will decrease

(*d*) its starting torque will increase but the maximum torque will remain unchanged.

622. Which of the following statements about a phase advancer is *incorrect* ?

(*a*) It is a particular type of a.c. exciter which may be connected in the rotor circuit of an induction motor to improve power factor

(*b*) It uses the principle of injection through slip-rings of the motor a current leading with regard to the rotor voltage

(*c*) It resembles with an induction motor in construction

(*d*) Its kVA capacity needs be only 5% or less of kVA correction affected in the main supply circuit

(*e*) It has compensating windings on the stator to ensure good commutation.

623. An inverted *V*-curve of synchronous motor shows the variation of

(*a*) power factor and d.c. excitation at constant load

(*b*) supply voltage and field current at constant excitation

(*c*) power factor and supply voltage during hunting

(*d*) supply voltage and excitation current at constant load. **[U.P.S.C. I.E.S., 1992]**

624. The torque angle δ is defined as the space angle between

(*a*) stator field axis and resultant field axis

(*b*) rotor field axis and resultant field axis

(*c*) stator field axis and rotor field axis

(*d*) stator field axis and mutual field axis.

625. A double squirrel cage induction motor has

(*a*) two series windings in stator (*b*) two series windings in rotor

(*c*) two parallel windings in rotor (*d*) two parallel windings in stator.

626. For a given developed power, synchronous motor operating from a constant voltage and constant frequency supply, will draw the minimum and maximum armature currents, I_{min} and I_{max} respectively, corresponding to

(*a*) I_{min} at unity p.f., but I_{max} at zero p.f. (*b*) I_{max} at unity p.f., but I_{min} at zero p.f.

(*c*) both I_{min} and I_{max} at unit p.f. (*d*) both I_{min} and I_{max} at zero p.f.

[U.P.S.C. I.E.S., 1997]

627. Cogging and crawling are phenomenon associated with

(*a*) cage induction machines and they are essentially the same

(*b*) squirrel cage induction machines, the former during starting and the latter at a fraction of its rated speed (*c*) squirrel cage induction machines, the former at a fraction of its rated speed and the latter during starting

(*d*) wound rotor induction machines and they are reduced by skewing, chording and distribution of windings. **[U.P.S.C. I.E.S., 1995]**

628. Breakdown torque of a 3-phase induction motor of negligible stator impedance is

(*a*) directly proportional to the rotor resistance

(*b*) inversely proportional to the rotor resistance

(*c*) directly proportional to the reactance

(*d*) inversely proportional to the rotor leakage reactance. **[U.P.S.C. I.E.S., 1993]**

629. In a 3-phase a.c. machine, phase e.m.f. is E_1 and output is P_1 for a phase spread of 60° ; whereas phase e.m.f. is E_2 and output is P_2 for a phase spread of 120°. The terms E_1, E_2, P_1 and P_2 are related as under :

(*a*) $E_1 = E_2, P_1 = 1.15\ P_2$
(*b*) $E_1 = 1.15E_2, P_1 = 1.15P_2$
(*c*) $E_2 = 1.15E_1, P_1 = 1.15P_2$
(*d*) $E_2 = 1.15E_1, P_2 = 1.15P_1$.

630. The no-load speed of a 3-phase 50 Hz induction motor is 1485 r.p.m. The number of alterations per minute which the rotor e.m.f. will make and the speed of rotor mmf with respect to rotor would respectively be

(*a*) 0.5, 1485 r.p.m.
(*b*) 30, zero r.p.m.
(*c*) 15, 15 r.p.m.
(*d*) 30, 15 r.p.m.

631. The stator of a 3-phase, 4-pole, slip-ring induction motor is fed from 50 Hz source and its rotor from 30 Hz source. The motor will run at

(*a*) 1500 r.p.m.
(*b*) 900 r.p.m.
(*c*) zero r.p.m.
(*d*) 2400 r.p.m.

632. In d.c. machines, the armature reaction m.m.f. is

(*a*) stationary with respect to armature
(*b*) rotating with respect to stator
(*c*) stationary with respect to stator
(*d*) rotating with respect to brushes.

633. Reluctance torque in rotating machines is present when

(*a*) air gap is not uniform
(*b*) reluctance seen by the stator mmf varies
(*c*) reluctance seen by the rotor mmf varies
(*d*) reluctance seen by the working mmf varies.

634. In a single-phase transformer, with subscripts 1 and 2 for primary and secondary windings :

(*a*) $E_1N_2 = E_2N_1$ and $I_1N_1 = I_2N_2$
(*b*) $E_1N_1 = E_2N_2$ and $I_1N_1 = I_2N_2$
(*c*) $E_1N_2 = E_2N_1$ and $I_1N_2 = I_2N_1$
(*d*) $E_1N_1 = E_2N_2$ and $I_1N_2 = I_2N_1$.

635. For successful parallel operation of two single-phase transformers, the essential condition is that their

(*a*) percentage impedances should be equal
(*b*) turns ratios should be exactly equal
(*c*) polarities must be properly connected
(*d*) kVA ratings should be equal.

636. The flux involved in the e.m.f. equation of a transformer has

(*a*) r.m.s. value
(*b*) average value
(*c*) total value
(*d*) maximum value.

637. The no-load current in a transformer lags the applied voltage by

(*a*) 90°
(*b*) about 75°
(*c*) 0°
(*d*) about 115°.

638. The two windings of a transformer are designated as :

(*a*) primary and secondary windings
(*b*) primary and h.v. windings
(*c*) secondary and l.v. windings
(*d*) secondary and h.v. windings
(*e*) h.v. and l.v. windings.

639. Chording and distribution of armature windings for machines result in :
(*a*) reduction in air-gap mmf harmonics
(*b*) reduction in fundamental component of induced voltage
(*c*) increase in effective turns
(*d*) both (*a*) and (*b*)
(*e*) both (*a*) and (*c*).

640. In a transformer, the tappings are provided on
(*a*) h.v. side at one end of the winding (*b*) l.v. side at one end of the winding
(*c*) h.v. side at the middle (*d*) l.v. side at the middle.

641. In d.c. machines, constant loss is composed of
(*a*) iron loss and mechanical loss
(*b*) friction, windage (*FW*) and iron loss and field circuit loss
(*c*) iron loss and field circuit loss
(*d*) *FW* and stray load losses.

642. A d.c. shunt motor drives a load requiring constant power. If both the field flux and armature terminal voltage are halved, then
(*a*) both speed w_m and armature current I_a remain constant
(*b*) w_m remains constant and I_a is doubled
(*c*) w_m increases slightly and I_a is doubled
(*d*) w_m increases slightly and I_a remains constant.

643. Which of the following motor is preferred for tape recorder ?
(*a*) Shaded pole motor (*b*) Hysteresis motor
(*c*) Two-value capacitor motor (*d*) Universal motor.

644. The main objection to the use of constant current transformer is that
(*a*) its power factor falls as the distance between the coils is increased
(*b*) its power factor falls as the distance between the coils is reduced
(*c*) its efficiency reduces as the distance between the coils is reduced
(*d*) none of the above.

645. In case of a *PT* with the increase in load on secondary side
(*a*) both of the ratio error and phase angle increase
(*b*) the ratio error increases but phase angle decreases
(*c*) the ratio error decreases but phase angle increases
(*d*) both of the ratio error and phase angle error decrease.

646. The armature m.m.f. wave for a d.c. machine is
(*a*) sinusoidal and rotates w.r.t. armature (*b*) sinusoidal and rotates w.r.t. field
(*c*) triangular and rotates w.r.t. armature (*d*) triangular and stationary w.r.t. armature.

647. The noise, in a transformer, caused by vibrations of laminations set by magnetic forces, is termed as
(*a*) zoom (*b*) hum
(*c*) buzz (*d*) none of these.

648. For successful parallel operation of two single phase transformers, the most essential condition is that their
(*a*) percentage impedances are equal (*b*) polarities are properly connected
(*c*) turn-ratios are exactly equal (*d*) kVA ratings are equal.

649. Most of the generators in thermal power plants run at
(*a*) 15000 r.p.m. (*b*) 3000 r.p.m.
(*c*) 1500 r.p.m. (*d*) 1000 r.p.m.
(*e*) 750 r.p.m.

650. Given that D = rotor diameter and L = axial length, a high performance a.c. servometer is characterized by
(*a*) large D, large L (*b*) large D, small L
(*c*) small D, small L (*d*) small D, large L.

651. In a d.c. machine "contraction co-efficient" is used to take into account the reduction of
(*a*) air-gap area due to armature slots
(*b*) iron losses in the teeth due to lower tooth density
(*c*) armature mmf due to armature slots
(*d*) torque due to ventilating ducts.

652. For d.c. voltage an inductor
(*a*) is virtually a short-circuit (*b*) is virtually an open-circuit
(*c*) depends on polarity (*d*) depends on magnitude of voltage.

653. For speeds higher than say 3000 r.p.m., the machine used is
(*a*) induction motor (*b*) synchronous motor
(*c*) universal motor (*d*) none of the above.

654. Air gap at the pole tips of a d.c. machine is kept more than that at the centre of the pole mainly to reduce
(*a*) reactance voltage (*b*) effect of armature reaction
(*c*) losses in armature core (*d*) noise of machine.

655. In a d.c. machine if the number of slots in the armature are more
(*a*) cooling is likely to be poor (*b*) commutation will be poor
(*c*) cost will be more (*d*) the flux pulsation will be more.

656. Three single phase transformers, each of 100 kVA rating, are connected in a closed delta arrangement. If one of them is taken out, it would be possible to load the bank in such a manner that each one is loaded to the extent of
(*a*) 86.6 kVA (*b*) 66.7 kVA
(*c*) 57.7 kVA (*d*) 33.33 kVA.

657. The power-slip characteristics of a 3-phase induction machine, shown in the (Fig. 4), has three regions of operation

A—Braking region
B—Motor region
C—Generator region

Which of these are correctly represented in the given Fig. 4 ?
(*a*) *A* and *B* (*b*) *A* and *C*
(*c*) *B* and *C* (*d*) *A*, *B* and *C*.

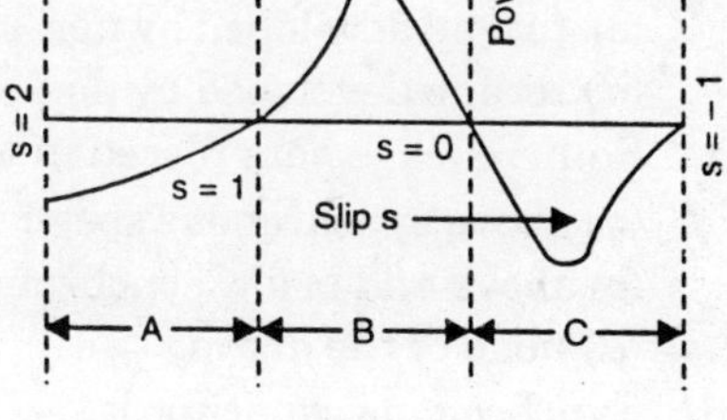

Fig. 4

AMIE EXAMINATIONS QUESTIONS

WINTER, 1993

658. The e.m.f. induced in a conductor rotating in a bipolar field is

(*a*) d.c. (*b*) a.c.

(*c*) d.c. and a.c. both (*d*) none of the above.

659. The field coils of a dynamo have an inductance of 8 henries and have a resistance of 60 ohms and are connected to 120 V d.c. source. The steady state current in the field coil is

(*a*) 15 amp. (*b*) 2 amp.

(*c*) Zero amp. (*d*) none of the above.

660. Approximate physical size of the 50 kW, 1800 r.p.m. dynamo as a function of a 25 kW, 900 r.p.m. dynamo will be

(*a*) 1 (*b*) 1/2

(*c*) 1/4 (*d*) none of the above.

661. A 220 V shunt motor develops a torque of 54 Nm at an armature current of 10 A. The torque produced when the armature current is 20 A, is

(*a*) 54 Nm (*b*) 81 Nm

(*c*) 108 Nm (*d*) None of the above.

662. We understand by dynamic braking a control function that

(*a*) provides braking by reversing motor line voltage polarity so that the motor developes a counter torque and exerts a retarding force

(*b*) brakes a drive (or motor) by dissipating its stored energy in a resistor

(*c*) brakes a drive by converting the kinetic energy of the motor and its driven machinery and returning to the power system

(*d*) none of the above.

663. Armature reaction in polyphase and d.c. dynamos is

(*a*) pulsating (*b*) constant

(*c*) pulsating and constant both (*d*) none of the above.

664. Reluctance torque is the

(*a*) torque developed in a given synchronous motor as a function of field excitation and the sine of the torque angle

(*b*) torque developed by salient pole synchronous or synchronous induction motors due to variations in air gap produced by armature reaction

(*c*) torque developed by non-salient pole asynchronous motor

(*d*) torque developed by none of the above.

665. Squirrel-cage bars placed in the rotor pole faces of an alternator help reduce hunting

(*a*) above synchronous speed only (*b*) below synchronous speed only

(*c*) above and below synchronous speeds both

(*d*) none of the above.

666. Synchronous capacitor is

(*a*) an ordinary static capacitor bank

(*b*) an over-excited synchronous motor driving mechanical load

(*c*) an over-excited synchronous motor running without mechanical load

(*d*) none of the above.

667. An eight-pole wound rotor induction motor operating on 60 Hz supply is driven at 1800 r.p.m. by a prime mover in the opposite direction of revolving magnetic field. The rotor current is at

(*a*) 60 Hz
(*b*) 120 Hz
(*c*) 180 Hz
(*d*) none of the above.

668. The power factor of a squirrel-cage induction motor is

(*a*) low at light loads only
(*b*) low at heavy loads only
(*c*) low at high and heavy loads both
(*d*) low at rated load only.

669. The full load slip of 60 Hz, 12 pole, squirrel cage induction motor is 5 percent. Its full load speed is

(*a*) 600 r.p.m.
(*b*) 570 r.p.m.
(*c*) 500 r.p.m.
(*d*) 475 r.p.m.

670. In the shaded pole squirrel cage induction motor the flux in the shaded part always

(*a*) leads the flux in the unshaded pole segment
(*b*) is in phase with the flux in the unshaded pole segment
(*c*) lags the flux in the unshaded pole segment
(*d*) none of the above.

671. A shunt d.c. motor works on a.c. mains

(*a*) unsatisfactorily
(*b*) satisfactorily
(*c*) not at all
(*d*) none of the above.

672. Speed of the universal motor is

(*a*) dependent on frequency of supply
(*b*) proportional to frequency of supply
(*c*) independent of frequency of supply
(*d*) none of the above.

673. In operating a 400 Hz transformer at 50 Hz

(*a*) only torque is reduced in the same proportion as the frequency
(*b*) only kVA rating is reduced in the same proportion as the frequency
(*c*) both voltage and kVA rating are reduced in the same proportion as the frequency
(*d*) none of the above.

674. In performing the short circuit test of a transformer

(*a*) high voltage side is usually short circuited
(*b*) low voltage side is usually short circuited
(*c*) any side is short circuited with preference
(*d*) none of the above.

675. In properly connected Y-Δ transformer, a voltmeter connected in the secondary (opening a node of delta) the voltmeter will read

(*a*) zero voltage
(*b*) line voltage of supply
(*c*) phase voltage in the secondary
(*d*) none of the above.

676. In double-revolving field theory of single phase induction motor, the slip of the forward motor is s then the slip of the backward motor is

(*a*) $2s$
(*b*) s
(*c*) $2 - s$
(*d*) $s - 2$.

677. For a machine on infinite bus active power can be varied

(*a*) changing field excitation
(*b*) changing of prime mover speed
(*c*) both (*a*) and (*b*) above
(*d*) none of the above.

SUMMER, 1994

678. **If a current-carrying coil is placed in a uniform magnetic field with its plane perpendicular to the direction of magnetic induction, then**
(*a*) the net force and torque on the coil are both zero
(*b*) the net force is zero but torque is finite
(*c*) the net force is finite but torque is zero
(*d*) the net force and torque are both finite.

679. **A d.c. shunt motor is operating at no load. If the armature is shunted by a resistance, its speed will**
(*a*) increase **(*b*) decrease**
(*c*) remain the same **(*d*) be unpredictable.**

680. **If the prime mover of an alternator supplying load to an infinite bus is suddenly shut down, then it will**
(*a*) stop
(*b*) continue to run as an alternator
(*c*) continue to run as a synchronous motor running in the reverse direction
(*d*) continue to run as a synchronous motor running in the same direction.

681. **Synchronous motor is supplying a load at unity p.f. If the load on the motor is increased keeping its excitation and the terminal voltage constant, the power factor**
(*a*) will remain the same **(*b*) will become leading**
(*c*) will become lagging **(*d*) none of the above.**

682. **W_1 and W_2 are the readings of two wattmeters used to measure power of a 3-phase balanced load. The reactive power drawn by the load is**
(*a*) $W_1 + W_2$ **(*b*) $W_1 - W_2$**
(*c*) $\sqrt{3}\,(W_1 + W_2)$ **(*d*) $\sqrt{3}\,(W_1 - W_2)$.**

683. **Compared to a resistor split-phase motor, a capacitor start motor has**
(*a*) higher starting torque **(*b*) lower starting torque**
(*c*) higher running torque **(*d*) lower running torque.**

684. **The synchronous speed of an induction motor can be increased by**
(*a*) reducing mechanical friction **(*b*) increasing the no. of poles**
(*c*) increasing the supply voltage **(*d*) increasing the supply frequency.**

685. **Transformer core generally uses grain oriented laminated sheets. The grain orientation reduces**
(*a*) copper loss **(*b*) eddy current loss**
(*c*) hysteresis loss **(*d*) none of the above.**

686. **In a particular motor there is a 3-phase stator winding and a short-circuited rotor winding. The motor runs at synchronous speed but does not necessitate any d.c. excitation. The motor is a**
(*a*) 3-phase induction motor **(*b*) 3-phase synchronous motor**
(*c*) 3-phase reluctance motor **(*d*) hysteresis motor.**

687. **While braking a 3-phase induction motor, its stator terminals were switched over to a d.c. source from a.c. supply. The method of braking is**
(*a*) regenerating braking **(*b*) counter current braking**
(*c*) eddy current braking **(*d*) none of the above.**

688. A circular metallic disc is placed in a vertical magnetic field of constant induction in the downward direction. If the disc is rotated uniformly in a horizontal plane, the e.m.f. induced will be

(*a*) zero (*b*) constant independent of disc size
(*c*) increasing radially in the outward direction
(*d*) decreasing radially in the outward direction.

689. A *p*-pole lap wound d.c. machine has an armature current I_a. The conductor current in the armature winding is

(*a*) I_a (*b*) I_a/p
(*c*) pI_a (*d*) None of the above.

690. Compensating winding in a d.c. machine

(*a*) counteracts armature m.m.f. in the interpolar zone
(*b*) prevents large speed drop
(*c*) shunts most of the armature current
(*d*) performs none of the above.

691. For a specified open circuit voltage of a d.c. generator, the short-circuit current will be maximum when it is

(*a*) separately excited (*b*) shunt connected
(*c*) cumulatively overcompounded (*d*) differentially compounded.

692. T is the load torque of a d.c. series motor having linear magnetisation and negligible armature resistance. Speed of the motor is

(*a*) inversely proportional to $\sqrt{T}$ (*b*) directly proportional to $\sqrt{T}$
(*c*) inversely proportional to T^2 (*d*) directly proportional to T^2.

693. A d.c. shunt motor drives a load at rated speed and supply voltage. If both the load and voltage are halved, the speed will be

(*a*) almost double (*b*) almost halved
(*c*) the rated speed (*d*) slightly less than the rated speed.

694. In the transformer circuit model, the core loss is represented as a

(*a*) series resistance (*b*) series inductance
(*c*) shunt resistance (*d*) shunt inductance.

695. While performing short-circuit test on a transformer the impressed voltage magnitude is kept constant but the frequency is increased. The short-circuit current will

(*a*) increase (*b*) decrease
(*c*) remain the same (*d*) none of the above.

696. To eliminate the 5th harmonic from the e.m.f. generated in an alternator, the pitch fraction will be

(*a*) 4/5 (*b*) 5/4
(*c*) 5/6 (*d*) 6/5.

697. Armature reaction in a synchronous motor at rated voltage and zero power factor (lead) is

(*a*) magnetising (*b*) cross-magnetising
(*c*) both magnetising and cross-magnetizing
(*d*) demagnetising.

698. A transformer has a hysteresis loss of 30 W at 240 V, 60 Hz. The hysteresis loss at 200 V, 50 Hz will be

(*a*) 20.8 W (*b*) 25 W
(*c*) 30 W (*d*) 36 W.

699. A 3-phase induction motor runs at supersynchronous speed. For self-excitation the machine
(*a*) draws real power from the mains
(*b*) draws reactive power from the mains
(*c*) feeds reactive power to the mains
(*d*) generates e.m.f. at the expenses of residual magnetism.

700. A d.c. shunt motor does not operate on a.c. due to
(*a*) low resistance of armature circuit (*b*) high resistance of field circuit
(*c*) high inductance of field circuit (*d*) reversal of polarity.

701. In a single-phase hysteresis motor
(*a*) starting torque is caused by both the eddy current and hysteresis while the running torque is caused by hysteresis
(*b*) starting as well as running torque are caused by both eddy current and hysteresis
(*c*) starting torque is caused by only eddy current while the running torque is caused by hysteresis
(*d*) starting as well as running torque are both caused by only hysteresis.

702. In a transformer the exciting current will be in phase quadrature with the impressed voltage provided
(*a*) only the leakage impedance drop is ignored
(*b*) only the core loss is ignored
(*c*) both the leakage and impedance drop and the cores loss are ignored
(*d*) only the no-load copper loss is ignored.

703. Drop in terminal voltage of an alternator due to armature reaction is countered by
(*a*) damper winding (*b*) effect of saliency
(*c*) increased prime mover input (*d*) automatic voltage regulator.

704. The stator referred resistance in the equivalent circuit of an induction motor representing mechanical output is
(*a*) r_2'/s (*b*) $r_2'\left(\frac{1}{s}-1\right)$
(*c*) r_2/s (*d*) $r_2^2\left(\frac{1}{s}-1\right)$.

705. For maximum current during 'slip test' on a synchronous machine, the armature m.m.f. aligns
(*a*) along q-axis (*b*) along d-axis
(*c*) along 45° to q-axis (*d*) 45° to q-axis.

706. Phase splitting can be accomplished in a single-phase induction motor
(*a*) only by adding the capacitor in series with the auxiliary winding
(*b*) only by causing the auxiliary winding to have high reactance
(*c*) only by causing the auxiliary winding to have high resistance
(*d*) by any one the above three methods.

707. A single-phase reluctance motor
(*a*) has salient-pole rotor structure and runs at subsynchronous speed
(*b*) has salient-pole rotor structure and runs at supersynchronous speed
(*c*) has salient-pole rotor structure and runs at synchronous speed
(*d*) has non-salient-pole rotor structure and runs at synchronous speed.

SUMMER, 1995

708. Magnetising impedance of a transformer is determined by
(*a*) short circuit test (*b*) open circuit test
(*c*) both open circuit and short circuit tests (*d*) other than the above.

709. In a transformer operating at constant voltage if the input frequency increases the core loss will
(*a*) increase (*b*) decrease
(*c*) remain constant (*d*) increase as the square of the frequency.

710. For a transformer to be tested at full load conditions but consuming only losses from the mains, we do
(*a*) load test (*b*) open circuit and short circuit tests
(*c*) back to back test (*d*) none of the above.

711. In a two-wattmeters method of measuring power in a 3-phase system one of the wattmeters reads negatively implying
(*a*) wattmeter connection is faulty (*b*) load is unbalanced
(*c*) power flow is in the reverse direction (*d*) p.f. is less than 0.5.

712. Pole shoe of a d.c. machine is laminated for the purpose of
(*a*) decreasing hysteresis loss (*b*) decreasing eddy current loss
(*c*) decreasing both hysteresis and eddy current loss
(*d*) manufacturing ease.

713. Core loss in a d.c. machine occurs in
(*a*) armature only (*b*) pole shoe only
(*c*) yoke only (*d*) in all the above three.

714. Efficiency of a d.c. machine is maximum when the variable losses are equal to
(*a*) constant losses (*b*) square of the constant losses
(*c*) square root of the constant losses (*d*) none of the above.

715. In a synchronous generator delivering lagging power factor load
(*a*) the excitation e.m.f. leads terminal voltage by the power angle
(*b*) excitation e.m.f. lags terminal voltage by the power angle
(*c*) excitation e.m.f. leads terminal voltage by the power factor angle
(*d*) none of the above.

716. For speeds higher than say 3000 r.p.m., the machine used is
(*a*) induction motor (*b*) synchronous motor
(*c*) universal motor (*d*) none of the above.

717. The relationship between rotor frequency f_2, slip s and the slator frequency f_1 is given by
(*a*) $f_1 = sf_2$ (*b*) $f_2 = sf_1$
(*c*) $f_2 = f_1(1 - s)$ (*d*) $f_2 = \sqrt{sf}$.

WINTER, 1995

718. The wattage rating for a ceiling fan motor will be in the range :
(*a*) 500 to 250 W (*b*) 250 to 500 W
(*c*) 50 to 150 W (*d*) 10 to 20 W.

719. Which of the following motors is preferred for tape recorder ?
(*a*) Shaded pole motor (*b*) Hysteresis motor
(*c*) Two value capacitor motor (*d*) Universal motor.

720. In single-phase repulsion motor power factor is
(*a*) always leading (*b*) high at low speeds
(*c*) high at high speed (*d*) always unity.

721. In case one-phase of a 3-phase synchronous motor is short-circuited the motor will :
(*a*) not start (*b*) run at $\frac{2}{3}$ of synchronous speed
(*c*) run with excessive vibration (*d*) take less than the rated load.

722. A pony motor is basically a
(*a*) small induction motor (*b*) d.c. series motor
(*c*) d.c. shunt motor (*d*) double winding a.c./d.c. motor.

723. Number of slip rings in 3-phase synchronous motor will be
(*a*) 0 (*b*) 1
(*c*) 2 (*d*) 3 or 4.

724. The highest voltage for transmitting electrical power in India is :
(*a*) 33 kV (*b*) 66 kV
(*c*) 132 kV (*d*) 400 kV.

725. A shell type transformer has :
(*a*) high eddy current loss (*b*) reduced magnetic leakage
(*c*) negligible hysteresis loss (*d*) none of the above.

726. When speed control in higher range is required which starter must be preferred ?
(*a*) 3 point starter (*b*) 4 point starter
(*c*) None of the above.

727. A sinusoidal voltage of 5 Hz is applied to the field of a d.c. shunt generator running at *N* rps. The frequency of armature voltage wave
(*a*) will be zero (*b*) will be 5 Hz
(*c*) will be 5 × *N* Hz (*d*) will be $\frac{N}{5}$ Hz.

WINTER, 1996

728. A frog-leg winding is used in a d.c. machine to eliminate
(*a*) the use of equilizer's in lap winding (*b*) the use of dummy coils in armature
(*c*) the use of compensating winding
(*d*) reactance voltage produce due to commutation.

729. Growler is used to detect
(*a*) inter turn short or open condition of winding
(*b*) condition of insulation used
(*c*) working condition of bearing
(*d*) None of the above.

730. Laminated yoke in d.c. motor can reduce
(*a*) speed regulation (*b*) iron loss
(*c*) temperature rise (*d*) sparking on load.

731. Skew of rotor bars eliminates
(*a*) the effect of space harmonic (*b*) the entire effect of crawling
(*c*) magnetic noise
(*d*) vibration due to unequal force developed on rotor.

732. The effect of leading power factor on the voltage regulation of an alternator is
(*a*) increasing in nature (*b*) decreasing in nature
(*c*) maintained at constant value (*d*) oscillating in nature.

733. A synchronous motor will deliver maximum power when
(*a*) load angle is equal to input power factor angle
(*b*) input power factor is unity
(*c*) load angle is 45°
(*d*) load angle is 0°.

734. Damper winding is used in synchronous motor for
(*a*) power factor improvement (*b*) efficiency improvement
(*c*) eliminating hunting of the rotor (*d*) minimising temperature rise

735. Centrifugal switch fitted on the rotor will operate when
(*a*) rotor speed reaches at its rated condition
(*b*) rotor speed exceeds 70 percent of its rated value
(*c*) rotor speed exceeds synchronous speed
(*d*) rotor speed exceeds 40 percent of its rated value.

736. In repulsion motor by shifting the brush position
(*a*) speed and direction of rotation can be controlled
(*b*) only speed can be controlled
(*c*) speed, torque and direction of rotation can be controlled
(*d*) only torque can be controlled by varying the speed at constant out power.

SUMMER, 1997

737. Squirrel-cage bars placed in the rotor pole faces of an alternator help reducing hunting
(*a*) above synchronous speed only (*b*) below synchronous speed only
(*c*) above and below synchronous speeds both
(*d*) none of the above.

738. An eight-pole wound rotor induction motor operating on 60 Hz supply is driven at 1800 r.p.m. by a prime mover in the opposite direction of revolving magnetic field. The rotor current frequency is
(*a*) 60 Hz (*b*) 120 Hz
(*c*) 180 Hz (*d*) none of the above.

739. Reluctance torque is the
(*a*) torque developed in a given synchronous motor as a function of field excitation and the sine of the torque angle
(*b*) torque developed by salient pole synchronous or synchronous induction motors due to variations in air gap produced by armature reaction
(*c*) torque developed by non-salient pole asynchronous motor
(*d*) torque developed by none of the above.

740. Grain oriented laminated sheet steel in transformer reduces
(*a*) copper loss (*b*) eddy current loss
(*c*) hysteresis loss (*d*) none of the above.

741. Compared to a resistor split phase motor a capacitor start motor has
(*a*) higher starting torque (*b*) lower starting torque
(*c*) higher running torque (*d*) lower running torque.

742. W_1 and W_2 are readings of two wattmeters used to measure power of a 3-phase balanced load. The reactive power drawn by the load is

(*a*) $W_1 + W_2$ (*b*) $W_1 - W_2$
(*c*) $\sqrt{3}\ (W_1 + W_2)$ (*d*) $\sqrt{3}\ (W_1 - W_2)$.

743. A 220 V shunt motor develops a torque of 54 Nm at an armature current of 10 A. The torque developed when the armature current is 20 A, is

(*a*) 54 Nm (*b*) 81 Nm
(*c*) 108 Nm (*d*) none of the above.

744. A synchronous capacitor is

(*a*) an ordinary static capacitor bank
(*b*) an overexcited synchronous motor driving mechanical load
(*c*) an overexcited synchronous motor without much load
(*d*) none of the above.

745. If the prime mover of an alternator supplying load to an infinite bus is suddenly shut down, then it will

(*a*) stop (*b*) continue to run as an alternator
(*c*) continue to run as a synchronous motor running in the reverse direction
(*d*) continue to run as synchronous motor running in the same direction.

746. If a current carrying coil is placed in a uniform magnetic field with its plane perpendicular to direction of magnetic induction, then

(*a*) the net force and torque on the coil are both zero
(*b*) the net force is zero but torque is finite
(*c*) the net force is finite but torque is zero
(*d*) the net force and torque are both finite.

WINTER, 1997

747. During measurement of power in a 3-phase load by two-wattmeters method, one wattmeter reads negative, the power factor of the circuit is

(*a*) 0.5 (*b*) less than 0.5
(*c*) greater than 0.5 (*d*) unity.

748. The armature of a d.c. machine is laminated to reduce

(*a*) eddy current loss (*b*) hysteresis loss
(*c*) copper loss (*d*) friction and windage loss.

749. For a given d.c. motor, the speed depends upon

(*a*) flux only (*b*) applied voltage alone
(*c*) back e.m.f. alone (*d*) back e.m.f. and flux.

750. An eight-pole, lap wound d.c. shunt generator is supplying 295 A to a load. The field current of the generator is 5 A. The current per parallel path is

(*a*) 300 A (*b*) 37.5 A
(*c*) 150 A (*d*) none of the above.

751. R_1 is the resistance of the primary winding of the transformer. The turn ratio in terms of primary to secondary is K. Then the equivalent resistance of the primary referred to secondary is

(*a*) R_1/K (*b*) R_1/K^2
(*c*) R_1K^2 (*d*) R_1K.

752. Copper losses at half full load in a transformer is 300 W. Then the copper losses at 3/4 full load will be

(*a*) 450 W (*b*) 675 W
(*c*) 225 W (*d*) 533.3 W.

753. If the full load slip of a 3-phase induction motor is 5 percent, the frequency of the rotor induced e.m.f. is

(*a*) 2.5 Hz (*b*) 50 Hz
(*c*) 5 Hz (*d*) 25 Hz.

754. For high starting torque, the most suited 3-phase induction motor is

(*a*) squired cage type (*b*) slip ring type
(*c*) deep bar squirrel cage type (*d*) double cage induction motor.

755. A 3-phase double layer stator winding has a coil span of 150° electrical degrees. Its coil span factor will be

(*a*) 0.96 (*b*) 1.0
(*c*) 0.966 (*d*) 0.98.

756. A 3-phase synchronous motor can operate over a wide range of power factors, *i.e.*, from lagging to leading power factors. This is achieved by

(*a*) changing the field excitation (*b*) varying the speed
(*c*) varying the applied voltage (*d*) changing the load.

SUMMER, 1998

757. Two wattmeters are used for the measurement of power factor of a three-phase balanced load. If one wattmeter reads positive and the other as negative, then the power factor is

(*a*) 0.8 (*b*) 0.6
(*c*) 0.5 (*d*) 0.3.

758. Choose the *correct* statement

(*a*) e.m.f. per turn in h.v. winding is more than the e.m.f. per turn in l.v. winding
(*b*) e.m.f. per turn in both the windings is equal
(*c*) e.m.f. per turn in h.v. winding is less than the e.m.f. per turn in l.v. winding

759. In a transformer

(*a*) both *OC* and *SC* tests are conducted on l.v. side
(*b*) *OC* test is conducted on h.v. side and *SC* test on l.v. side
(*c*) *OC* test is conducted on l.v. side and *SC* test on h.v. side
(*d*) both *OC* and *SC* tests are conducted on h.v. side.

760. The armature m.m.f. wave in a d.c. machine is

(*a*) sinusoidal (*b*) triangular
(*c*) rectangular (*d*) square.

761. In d.c. generators, the polarity of interpoles is

(*a*) same as that of the main pole behind (*b*) same as that of the main pole ahead
(*c*) opposite to that of the main pole ahead (*d*) None of the above.

762. A star-delta starter is equivalent to an autotransformer starter with tapping of

(*a*) 86.6% (*b*) 57.73%
(*c*) 57% (*d*) 58%.

763. A three-phase induction motor is running at slip *s*. If its two supply leads are interchanged, then its slip at that instant is equal to

(*a*) 2*s* (*b*) 1 – *s*
(*c*) 2 – *s* (*d*) zero.

764. For simplex wave and lap windings, the back pitch Y_b and the front pitch Y_f are as follows
(*a*) Y_b is odd, Y_f is even (*b*) Y_b is even, Y_f is odd
(*c*) both Y_b and Y_f are odd (*d*) both Y_b and Y_f are even.

WINTER, 1998

765. The armature m.m.f. of a d.c. machine is
(*a*) sinusoidal and depends on speed (*b*) triangular and depends on speed
(*c*) triangular and independent of speed (*d*) trapezoidal and independent of speed.

766. The chording angle for eliminating 5th harmonic should be
(*a*) 30° (*b*) 34°
(*c*) 36° (*d*) 35°.

767. The direction of rotation of a d.c. shunt motor can be reversed by interchanging
(*a*) the supply terminals (*b*) the field terminals only
(*c*) the armature terminals only (*d*) either field or armature terminals.

768. A cumulatively compounded d.c. motor drives a constant power load. If series field winding gets short-circuited then
(*a*) armature current I_a increases and speed ω_m rises
(*b*) I_a remains constant and ω_m rises
(*c*) I_a remains constant and ω_m falls
(*d*) I_a decreases and ω_m rises.

769. The power in a 3-phase system is given by $\sqrt{3}\, V_L I_L \cos \phi$ where ϕ is the p.f. angle between
(*a*) line voltage and line current (*b*) line voltage and phase current
(*c*) phase voltage and phase current (*d*) phase voltage and line current.

770. The stator of a 3-phase, 4-pole, slip-ring induction motor is fed from 50 Hz source and its rotor from 30 Hz source. The motor will run at
(*a*) 1500 r.p.m. (*b*) 900 r.p.m.
(*c*) 0 r.p.m. (*d*) 2400 r.p.m.

771. A salient-pole synchronous motor is fed from an infinite bus and is running at no load. It is field current is reduced to zero, then the motor would
(*a*) stop (*b*) run at a reduced speed
(*c*) run at synchronous speed (*d*) run above synchronous speed.

772. Voltage regulation of a transformer is given by
(*a*) $\dfrac{E_2 - V_2}{V_2}$ (*b*) $\dfrac{E_2 - V_2}{E_2}$
(*c*) $\dfrac{V_2 - E_2}{V_2}$ (*d*) $\dfrac{V_2 - E_2}{E_2}$.

773. The direction of rotation of a 1-phase induction motor can be reversed by
(*a*) reversing the leads of main winding only
(*b*) reversing the leads of auxiliary winding only
(*c*) reversing the supply leads
(*d*) Either (*a*) or (*b*).

774. One transformer has leakage impedance of $1 + j4\ \Omega$ and $3 + j14\ \Omega$ for its primary and secondary windings respectively. This transformer has

(*a*) l.v. primary (*b*) h.v. primary
(*c*) medium voltage primary
(*d*) insufficient data to find out about the primary winding.

SUMMER, 1999

775. An 8-pole, 50 Hz, single-phase induction motor is running at 690 r.p.m. Its slip w.r.t. forward field is (in pu)
(*a*) + 0.08 (*b*) + 1.92
(*c*) – 0.08 (*d*) 1.00.

776. For a constant torque load, the supply voltage of a squirrel cage induction motor is reduced by a factor of $\left(\frac{1}{\sqrt{2}}\right)$, its rotor current is modified by a factor of
(*a*) 2 (*b*) $\frac{1}{\sqrt{2}}$
(*c*) $\sqrt{2}$ (*d*) none.

777. A synchronous motor is operating at constant load while its excitation is adjusted to give UPF current. If the excitation is now increased the p.f. will
(*a*) lead (*b*) lag
(*c*) remain at unity (*d*) become zero.

778. A synchronous generator is operating at constant load while its excitation is adjusted to give UPF current. If the excitation is now increased the p.f. will
(*a*) lead (*b*) lag
(*c*) remain at unity (*d*) become zero.

779. The function of brush and commutator in a d.c. motor is
(*a*) to convert a.c. into d.c.
(*b*) to produce unidirectional torque
(*c*) to produce unidirectional current in the armature
(*d*) to reduce demagnetising effect of armature reaction.

780. The motor best suited for crane is the
(*a*) induction motor (*b*) synchronous motor
(*c*) d.c. series motor (*d*) d.c. shunt motor.

781. For a cyclic load variation motor rating is determined by the
(*a*) average load (*b*) peak load
(*c*) r.m.s. load (*d*) $\frac{3}{4}$ th of peak load.

782. A single-phase transformer has its maximum efficiency at 60% of full-load. At full-load copper loss will be
(*a*) equal to core loss (*b*) less than core loss
(*c*) more than core loss (*d*) zero.

783. The flux in a magnetic core is alternating sinusoidally at a frequency of 600 Hz. The maximum flux density is 2 tesla and eddy current loss is 15 watt. If the frequency is raised to 800 Hz and the maximum flux density reduced to 1.5 tesla, the eddy current loss will be
(*a*) same (*b*) reduced to half
(*c*) doubled (*d*) increased by 50 per cent.

784. The core flux in transformer depends mainly on
(*a*) supply voltage (*b*) supply voltage and frequency

(*c*) supply voltage-frequency and load
(*d*) supply voltage and load but independent of frequency.

WINTER, 1999

785. In a d.c. armature winding the back pitch and front pitch in terms of coil sides must be :
(*a*) even and odd respectively (*b*) odd and even respectively
(*c*) both even (*d*) both odd.

786. For building up of voltage, residual magnetism is essential in field systems of all types of d.c. generators *except* :
(*a*) series (*b*) shunt
(*c*) compound (*d*) separately excited.

787. A cumulative compound motor runs at 1200 r.p.m. on no-load. When the flux increases by 14% on full load, the full load drop in armature comes to be 5%. What is the full load speed ?
(*a*) 1000 r.p.m. (*b*) 1332 r.p.m.
(*c*) 1428 r.p.m. (*d*) 972 r.p.m.

788. For large transformers best utilisation of available core space can be made by using
(*a*) square core section (*b*) rectangular core section
(*c*) stepped core section (*d*) none of the above.

789. The transformer exciting current has two components, magnetising component and core loss component. Neglecting impedance drop
(*a*) both of them are in phase with the supply voltage
(*b*) the former lags the supply, voltage by 90° while the latter is in phase with the supply voltage
(*c*) both of them lag the supply voltage by 90°
(*d*) the former is in phase with the supply voltage while the latter lags the supply voltage by 90°.

790. In a 3-phase induction motor, slip for maximum torque in terms of rotor resistance r_2 is :
(*a*) independent of r_2 (*b*) inversely proportional to r_2
(*c*) proportional to r_2^2 (*d*) directly proportional to r_2.

791. In a slip ring induction motor if the rotor resistance is increased then
(*a*) starting torque and efficiency increase
(*b*) starting torque and efficiency decrease
(*c*) starting torque increases but efficiency decreases
(*d*) starting torque decreases but efficiency increases.

792. In a synchronous generator
(*a*) the open circuit voltage leads the terminal voltage by an angle known as power angle
(*b*) the open circuit (o.c.) voltage leads the terminal voltage by an angle known as overlap angle
(*c*) the o.c. voltage lags the terminal voltage by an angle known as power angle
(*d*) the o.c. voltage lags the terminal voltage by an angle known as overlap angle.

793. A 3-phase synchronous motor having constant excitation drives a load taking power from infinite bus at leading p.f., if the shaft load decreases :
(*a*) the power angle decreases while p.f. increases
(*b*) the power angle increases while p.f. decreases
(*c*) both power angle and p.f. decreases
(*d*) both power angle and p.f. increase.

ANSWERS

PART I
OBJECTIVE TYPE QUESTIONS

1. (*c*)	**2.** (*c*)	**3.** (*a*)	**4.** (*d*)	**5.** (*a*)	**6.** (*a*)
7. (*c*)	**8.** (*d*)	**9.** (*c*)	**10.** (*d*)	**11.** (*a*)	**12.** (*a*)
13. (*d*)	**14.** (*a*)	**15.** (*a*)	**16.** (*d*)	**17.** (*b*)	**18.** (*a*)
19. (*b*)	**20.** (*c*)	**21.** (*a*)	**22.** (*b*)	**23.** (*a*)	**24.** (*a*)
25. (*a*)	**26.** (*b*)	**27.** (*b*)	**28.** (*b*)	**29.** (*c*)	**30.** (*d*)
31. (*a*)	**32.** (*c*)	**33.** (*d*)	**34.** (*b*)	**35.** (*c*)	**36.** (*a*)
37. (*b*)	**38.** (*d*)	**39.** (*b*)	**40.** (*a*)	**41.** (*d*)	**42.** (*a*)
43. (*a*)	**44.** (*b*)	**45.** (*c*)	**46.** (*c*)	**47.** (*a*)	**48.** (*a*)
49. (*c*)	**50.** (*d*)	**51.** (*a*)	**52.** (*d*)	**53.** (*b*)	**54.** (*c*)
55. (*c*)	**56.** (*b*)	**57.** (*d*)	**58.** (*d*)	**59.** (*c*)	**60.** (*c*)
61. (*d*)	**62.** (*a*)	**63.** (*c*)	**64.** (*c*)	**65.** (*d*)	**66.** (*b*)
67. (*c*)	**68.** (*b*)	**69.** (*a*)	**70.** (*b*)	**71.** (*c*)	**72.** (*d*)
73. (*d*)	**74.** (*b*)	**75.** (*b*)	**76.** (*d*)	**77.** (*b*)	**78.** (*d*)
79. (*c*)	**80.** (*b*)	**81.** (*a*)	**82.** (*d*)	**83.** (*d*)	**84.** (*a*)
85. (*d*)	**86.** (*d*)	**87.** (*b*)	**88.** (*d*)	**89.** (*a*)	**90.** (*b*)
91. (*b*)	**92.** (*b*)	**93.** (*c*)	**94.** (*d*)	**95.** (*a*)	**96.** (*b*)
97. (*a*)	**98.** (*b*)	**99.** (*d*)	**100.** (*a*)	**101.** (*b*)	**102.** (*a*)
103. (*a*)	**104.** (*b*)	**105.** (*c*)	**106.** (*d*)	**107.** (*a*)	**108.** (*b*)
109. (*d*)	**110.** (*c*)	**111.** (*c*)	**112.** (*c*)	**113.** (*d*)	**114.** (*d*)
115. (*d*)	**116.** (*d*)	**117.** (*c*)	**118.** (*d*)	**119.** (*d*)	**120.** (*d*)
121. (*a*)	**122.** (*c*)	**123.** (*d*)	**124.** (*b*)	**125.** (*a*)	**126.** (*c*)
127. (*c*)	**128.** (*d*)	**129.** (*b*)	**130.** (*d*)	**131.** (*d*)	**132.** (*c*)
133. (*c*)	**134.** (*d*)	**135.** (*a*)	**136.** (*d*)	**137.** (*b*)	**138.** (*d*)
139. (*c*)	**140.** (*d*)	**141.** (*d*)	**142.** (*d*)	**143.** (*d*)	**144.** (*c*)
145. (*c*)	**146.** (*d*)	**147.** (*d*)	**148.** (*c*)	**149.** (*a*)	**150.** (*c*)
151. (*d*)	**152.** (*d*)	**153.** (*d*)	**154.** (*d*)	**155.** (*c*)	**156.** (*d*)
157. (*b*)	**158.** (*c*)	**159.** (*d*)	**160.** (*b*)	**161.** (*d*)	**162.** (*c*)
163. (*a*)	**164.** (*a*)	**165.** (*b*)	**166.** (*a*)	**167.** (*a*)	**168.** (*b*)
169. (*d*)	**170.** (*b*)	**171.** (*d*)	**172.** (*d*)	**173.** (*b*)	**174.** (*a*)
175. (*b*)	**176.** (*a*)	**177.** (*b*)	**178.** (*a*)	**179.** (*d*)	**180.** (*c*)
181. (*a*)	**182.** (*d*)	**183.** (*a*)	**184.** (*b*)	**185.** (*a*)	**186.** (*d*)
187. (*b*)	**188.** (*c*)	**189.** (*c*)	**190.** (*c*)	**191.** (*c*)	**192.** (*b*)
193. (*d*)	**194.** (*a*)	**195.** (*b*)	**196.** (*b*)	**197.** (*d*)	**198.** (*b*)
199. (*d*)	**200.** (*b*)	**201.** (*c*)	**202.** (*c*)	**203.** (*a*)	**204.** (*c*)
205. (*a*)	**206.** (*b*)	**207.** (*a*)	**208.** (*a*)	**209.** (*b*)	**210.** (*b*)
211. (*c*)	**212.** (*b*)	**213.** (*c*)	**214.** (*d*)	**215.** (*d*)	**216.** (*d*)
217. (*c*)	**218.** (*a*)	**219.** (*a*)	**220.** (*b*)	**221.** (*c*)	**222.** (*d*)
223. (*c*)	**224.** (*c*)	**225.** (*d*)	**226.** (*a*)	**227.** (*b*)	**228.** (*c*)
229. (*a*)	**230.** (*d*)	**231.** (*a*)	**232.** (*c*)	**233.** (*a*)	**234.** (*a*)
235. (*c*)	**236.** (*a*)	**237.** (*b*)	**238.** (*a*)	**239.** (*c*)	**240.** (*a*)
241. (*a*)	**242.** (*e*)	**243.** (*c*)	**244.** (*c*)	**245.** (*a*)	**246.** (*c*)
247. (*d*)	**248.** (*b*)	**249.** (*b*)	**250.** (*c*)	**251.** (*d*)	**252.** (*a*)
253. (*a*)	**254.** (*b*)	**255.** (*d*)	**256.** (*d*)	**257.** (*b*)	**258.** (*a*)

259. (*a*)	**260.** (*a*)	**261.** (*a*)	**262.** (*b*)	**263.** (*d*)	**264.** (*b*)
265. (*d*)	**266.** (*b*)	**267.** (*c*)	**268.** (*c*)	**269.** (*a*)	**270.** (*b*)
271. (*d*)	**272.** (*c*)	**273.** (*b*)	**274.** (*a*)	**275.** (*b*)	**276.** (*a*)
277. (*d*)	**278.** (*a*)	**279.** (*d*)	**280.** (*b*)	**281.** (*a*)	**282.** (*c*)
283. (*a*)	**284.** (*a*)	**285.** (*c*)	**286.** (*d*)	**287.** (*d*)	**288.** (*c*)
289. (*c*)	**290.** (*d*)	**291.** (*b*)	**292.** (*d*)	**293.** (*b*)	**294.** (*b*)
295. (*b*)	**296.** (*c*)	**297.** (*a*)	**298.** (*a*)	**299.** (*c*)	**300.** (*d*)
301. (*c*)	**302.** (*a*)	**303.** (*d*)	**304.** (*a*)	**305.** (*d*)	**306.** (*b*)
307. (*f*)	**308.** (*b*)	**309.** (*a*)	**310.** (*c*)	**311.** (*b*)	**312.** (*c*)
313. (*b*)	**314.** (*b*)	**315.** (*a*)	**316.** (*c*)	**317.** (*c*)	**318.** (*a*)
319. (*c*)	**320.** (*e*)	**321.** (*c*)	**322.** (*a*)	**323.** (*b*)	**324.** (*c*)
325. (*a*)	**326.** (*a*)	**327.** (*d*)	**328.** (*a*)	**329.** (*c*)	**330.** (*d*)
331. (*a*)	**332.** (*b*)	**333.** (*d*)	**334.** (*a*)	**335.** (*d*)	**336.** (*a*)
337. (*b*)	**338.** (*a*)	**339.** (*e*)	**340.** (*d*)	**341.** (*a*)	**342.** (*c*)
343. (*a*)	**344.** (*d*)	**345.** (*d*)	**346.** (*b*)	**347.** (*a*)	**348.** (*b*)
349. (*c*)	**350.** (*b*)	**351.** (*c*)	**352.** (*d*)	**353.** (*b*)	**354.** (*b*)
355. (*a*)	**356.** (*b*)	**357.** (*c*)	**358.** (*b*)	**359.** (*b*)	**360.** (*b*)
361. (*c*)	**362.** (*b*)	**363.** (*b*)	**364.** (*c*)	**365.** (*c*)	**366.** (*b*)
367. (*a*)	**368.** (*a*)	**369.** (*b*)	**370.** (*b*)	**371.** (*b*)	**372.** (*b*)
373. (*a*)	**374.** (*e*)	**375.** (*c*)	**376.** (*a*)	**377.** (*b*)	**378.** (*a*)
379. (*b*)	**380.** (*b*)	**381.** (*a*)	**382.** (*c*)	**383.** (*d*)	**384.** (*c*)
385. (*a*)	**386.** (*d*)	**387.** (*b*)	**388.** (*b*)	**389.** (*c*)	**390.** (*a*)
391. (*b*)	**392.** (*a*)	**393.** (*c*)	**394.** (*b*)	**395.** (*c*)	**396.** (*b*)
397. (*b*)	**398.** (*c*)	**399.** (*a*)	**400.** (*b*)	**401.** (*c*)	**402.** (*b*)
403. (*b*)	**404.** (*c*)	**405.** (*d*)	**406.** (*c*)	**407.** (*a*)	**408.** (*b*)
409. (*a*)	**410.** (*c*)	**411.** (*a*)	**412.** (*b*)	**413.** (*b*)	**414.** (*a*)
415. (*b*)	**416.** (*c*)	**417.** (*b*)	**418.** (*d*)	**419.** (*d*)	**420.** (*c*)
421. (*c*)	**422.** (*b*)	**423.** (*a*)	**424.** (*c*)	**425.** (*c*)	**426.** (*c*)
427. (*d*)	**428.** (*c*)	**429.** (*d*)	**430.** (*c*)	**431.** (*a*)	**432.** (*a*)
433. (*d*)	**434.** (*c*)	**435.** (*a*)	**436.** (*a*)	**437.** (*c*)	**438.** (*c*)
439. (*d*)	**440.** (*c*)	**441.** (*c*)	**442.** (*b*)	**443.** (*c*)	**444.** (*c*)
445. (*c*)	**446.** (*d*)	**447.** (*b*)	**448.** (*b*)	**449.** (*a*)	**450.** (*a*)
451. (*d*)	**452.** (*c*)	**453.** (*d*)	**454.** (*a*)	**455.** (*d*)	**456.** (*c*)
457. (*b*)	**458.** (*e*)	**459.** (*a*)	**460.** (*a*)	**461.** (*b*)	**462.** (*c*)
463. (*a*)	**464.** (*c*)	**465.** (*d*)	**466.** (*a*)	**467.** (*b*)	**468.** (*a*)
469. (*a*)	**470.** (*b*)	**471.** (*c*)	**472.** (*c*)	**473.** (*d*)	**474.** (*d*)
475. (*b*)	**476.** (*c*)	**477.** (*b*)	**478.** (*b*)	**479.** (*a*)	**480.** (*d*)
481. (*a*)	**482.** (*b*)	**483.** (*c*)	**484.** (*a*)	**485.** (*b*)	**486.** (*d*)
487. (*c*)	**488.** (*d*)	**489.** (*d*)	**490.** (*c*)	**491.** (*d*)	**492.** (*c*)
493. (*d*)	**494.** (*a*)	**495.** (*d*)	**496.** (*d*)	**497.** (*b*)	**498.** (*e*)
499. (*b*)	**500.** (*c*)	**501.** (*b*)	**502.** (*a*)	**503.** (*c*)	**504.** (*a*)
505. (*a*)	**506.** (*a*)	**507.** (*a*)	**508.** (*c*)	**509.** (*b*)	**510.** (*d*)
511. (*a*)	**512.** (*b*)	**513.** (*b*)	**514.** (*b*)	**515.** (*b*)	**516.** (*b*)
517. (*e*)	**518.** (*a*)	**519.** (*c*)	**520.** (*c*)	**521.** (*c*)	**522.** (*c*)
523. (*d*)	**524.** (*d*)	**525.** (*a*)	**526.** (*a*)	**527.** (*d*)	**528.** (*c*)
529. (*b*)	**530.** (*c*)	**531.** (*c*)	**532.** (*b*)	**533.** (*d*)	**534.** (*c*)

535. (*d*)	**536.** (*a*)	**537.** (*e*)	**538.** (*c*)	**539.** (*d*)	**540.** (*b*)
541. (*a*)	**542.** (*a*)	**543.** (*a*)	**544.** (*d*)	**545.** (*c*)	**546.** (*b*)
547. (*b*)	**548.** (*b*)	**549.** (*c*)	**550.** (*d*)	**551.** (*c*)	**552.** (*d*)
553. (*e*)	**554.** (*c*)	**555.** (*a*)	**556.** (*a*)	**557.** (*c*)	**558.** (*c*)
559. (*b*)	**560.** (*a*)	**561.** (*d*)	**562.** (*d*)	**563.** (*c*)	**564.** (*c*)
565. (*d*)	**566.** (*d*)	**567.** (*c*)	**568.** (*c*)	**569.** (*c*)	**570.** (*a*)
571. (*c*)	**572.** (*b*)	**573.** (*a*)	**574.** (*c*)	**575.** (*c*)	**576.** (*c*)
577. (*d*)	**578.** (*d*)	**579.** (*c*)	**580.** (*c*)	**581.** (*b*)	**582.** (*e*)
583. (*a*)	**584.** (*a*)	**585.** (*d*)	**586.** (*e*)	**587.** (*c*)	**588.** (*b*)
589. (*b*)	**590.** (*a*)	**591.** (*d*)	**592.** (*b*)	**593.** (*b*)	**594.** (*c*)
595. (*c*)	**596.** (*d*)	**597.** (*c*)	**598.** (*b*)	**599.** (*c*)	**600.** (*b*)
601. (*a*)	**602.** (*a*)	**603.** (*c*)	**604.** (*b*)	**605.** (*b*)	**606.** (*c*)
607. (*e*)	**608.** (*a*)	**609.** (*c*)	**610.** (*c*)	**611.** (*b*)	**612.** (*c*)
613. (*c*)	**614.** (*b*)	**615.** (*e*)	**616.** (*b*)	**617.** (*a*)	**618.** (*a*)
619. (*b*)	**620.** (*b*)	**621.** (*d*)	**622.** (*c*)	**623.** (*a*)	**624.** (*c*)
625. (*c*)	**626.** (*a*)	**627.** (*b*)	**628.** (*d*)	**629.** (*b*)	**630.** (*d*)
631. (*d*)	**632.** (*c*)	**633.** (*d*)	**634.** (*a*)	**635.** (*c*)	**636.** (*d*)
637. (*b*)	**638.** (*e*)	**639.** (*d*)	**640.** (*c*)	**641.** (*b*)	**642.** (*c*)
643. (*b*)	**644.** (*a*)	**645.** (*a*)	**646.** (*c*)	**647.** (*b*)	**648.** (*b*)
649. (*b*)	**650.** (*d*)	**651.** (*a*)	**652.** (*a*)	**653.** (*c*)	**654.** (*a*)
655. (*c*)	**656.** (*a*)	**657.** (*c*)			

AMIE EXAMINATIONS QUESTIONS

658. (*b*)	**659.** (*b*)	**660.** (*a*)	**661.** (*c*)	**662.** (*b*)	**663.** (*a*)
664. (*b*)	**665.** (*c*)	**666.** (*c*)	**667.** (*c*)	**668.** (*a*)	**669.** (*b*)
670. (*c*)	**671.** (*a*)	**672.** (*a*)	**673.** (*c*)	**674.** (*b*)	**675.** (*a*)
676. (*c*)	**677.** (*b*)	**678.** (*a*)	**679.** (*a*)	**680.** (*d*)	**681.** (*c*)
682. (*d*)	**683.** (*a*)	**684.** (*d*)	**685.** (*b*)	**686.** (*c*)	**687.** (*c*)
688. (*a*)	**689.** (*b*)	**690.** (*a*)	**691.** (*a*)	**692.** (*a*)	**693.** (*c*)
694. (*c*)	**695.** (*b*)	**696.** (*a*)	**697.** (*a*)	**698.** (*b*)	**699.** (*a*)
700. (*d*)	**701.** (*d*)	**702.** (*b*)	**703.** (*d*)	**704.** (*b*)	**705.** (*b*)
706. (*b*)	**707.** (*c*)	**708.** (*b*)	**709.** (*a*)	**710.** (*c*)	**711.** (*d*)
712. (*b*)	**713.** (*d*)	**714.** (*a*)	**715.** (*a*)	**716.** (*c*)	**717.** (*b*)
718. (*c*)	**719.** (*a*)	**720.** (*c*)	**721.** (*d*)	**722.** (*b*)	**723.** (*d*)
724. (*d*)	**725.** (*b*)	**726.** (*c*)	**727.** (*b*)	**728.** (*b*)	**729.** (*a*)
730. (*b*)	**731.** (*a*)	**732.** (*b*)	**733.** (*a*)	**734.** (*c*)	**735.** (*b*)
736. (*a*)	**737.** (*d*)	**738.** (*c*)	**739.** (*b*)	**740.** (*c*)	**741.** (*c*)
742. (*d*)	**743.** (*c*)	**744.** (*c*)	**745.** (*d*)	**746.** (*a*)	**747.** (*b*)
748. (*a*)	**749.** (*d*)	**750.** (*b*)	**751.** (*b*)	**752.** (*b*)	**753.** (*a*)
754. (*b*)	**755.** (*c*)	**756.** (*a*)	**757.** (*d*)	**758.** (*b*)	**759.** (*c*)
760. (*a*)	**761.** (*b*)	**762.** (*c*)	**763.** (*c*)	**764.** (*c*)	**765.** (*c*)
766. (*c*)	**767.** (*d*)	**768.** (*a*)	**769.** (*c*)	**770.** (*d*)	**771.** (*a*)
772. (*b*)	**773.** (*d*)	**774.** (*a*)	**775.** (*a*)	**776.** (*a*)	**777.** (*a*)
778. (*b*)	**779.** (*b*)	**780.** (*c*)	**781.** (*c*)	**782.** (*c*)	**783.** (*a*)
784. (*b*)	**785.** (*d*)	**786.** (*d*)	**787.** (*a*)	**788.** (*c*)	**789.** (*b*)
790. (*d*)	**791.** (*b*)	**792.** (*a*)	**793.** (*c*)		

PART II

QUESTIONS WITH SHORT ANSWERS

AMIE SUMMER, 1994

1. A particular application employing a D.C. generator requires high voltage and low current. Should be machine be lap wound or wave wound ?
   Ans. Wave wound.
2. Iron loss in a D.C. machine increases with increasing load. What is the mode of excitation ?
   Ans. Series excitation.
3. Why is that the V/F ratio is kept constant while controlling the speed of a 3-phase induction motor by varying the supply frequency ?
   Ans. To keep the flux constant.
4. A full pitch coil in an alternator has a span of 18 slots. What should be the span if the third harmonic component of the e.m.f. is to be eliminated ?
   Ans. A span of 12 slots.
5. A 3-phase induction motor is to be started first using an autotransformers with 80% tapping and next direct on-line. What will be the ratio of the starting torques ?
   Ans. $\frac{T_{auto}}{T_{d.o.l.}} = 0.64.$
6. Why does synchronous impedance method give a poorer voltage regulation ?
   Ans. It does not account for saturation.
7. A synchronous motor is to be used as a synchronous condenser. Under what condition should it operate ?
   Ans. It should be over-excited.
8. Can a repulsion-induction motor ever run at super-synchronous speed ?
   Ans. Yes, at no load.
9. Does the power factor of a single phase series motor increase or decrease with increasing load torque ?
   Ans. Power factor decreases with increase in load torque.
10. What is the dimension of $\sqrt{LC}$?
    Ans. Reciprocal of velocity *viz.*, LT^{-1}.

AMIE SUMMER, 1995

11. Is maximum torque of a 3-phase induction motor dependent on the rotor resistance ?
    Ans. Independent of rotor resistance.
12. What is the relationship of development torque of a 3-phase induction motor with the supply voltage ?
    Ans. Directly proportional.
13. In a 3-phase synchronous motor field m.m.f. lags the air gap flux, then what is the position of air gap flux vis-a-vis armature m.m.f. ?
    Ans. It also lags.
14. By which test armature reaction m.m.f. and leakage reactance of a synchronous machine are determined ?
    Ans. O.C. and S.C. tests.

15. What happens to the speed and p.f. of a single-phase series motor when the load increases ?
Ans. Speed decreases and p.f. increases.

16. Two single-phase transformers A and B of identical voltage and turns ratio operate in parallel. If the impedances of the two transformers are equal while resistance/reactance ratio of A is higher than that B, what would be the magnitudes of I_A and I_B and their phase positions ?
Ans. Magnitudes of I_A and I_B are equal but phase angle of I_A is less than that of I_B.

17. A 2-winding transformer is used to step down the voltage from 220 V to 110 V. If an auto-transformer is used for the same purpose, what would be the ratio of copper weights in them ?

Ans. $\dfrac{\text{Weight of copper of autotransformer}}{\text{Weight of copper of two winding transformer}} = 0.5.$

18. While controlling the speed of a d.c. shunt motor what should be done to achieve a constant torque drive ?
Ans. Supply voltage is to be controlled.

19. How plugging of a d.c. motor is done ?
Ans. By cutting the field supply.

20. Alternators have no commutators, why ?
Ans. They have to supply electrical energy with an alternating voltage.

AMIE SUMMER, 1996

21. An alternator can be used as a synchronous condenser. Under what conditions it should operate ?
Ans. Over excited.

22. Which test gives the copper-loss of a transformer ?
Ans. S.C. test.

23. What is dimension of L/R ?
Ans. T.

24. Stator of an induction motor has full pitch coil span of 18 slots. What must be short pitch angle in degree if the 5th harmonics components of the e.m.f. is to be eliminated ?
Ans. 36°.

25. A 3-phase, 50 Hz induction motor runs at 960 r.p.m. on no load and 940 r.p.m. at full-load. Find full-load slip.
Ans. 6%.

26. What is critical speed in a d.c. shunt generator ?
Ans. Speed for which shunt field resistance represents critical resistance.

27. Which method we must adopt to control the speed of a d.c. shunt motor above the base speed ?
Ans. Field control method.

28. What are the advantages of two wattmeters method over single wattmeter for 3-phase power measurement ?
Ans. Can be used for balanced as well as unbalanced system.

29. Name the motor being used in the ceiling fan.
Ans. Single phase induction motor with split phase.

30. Which is the cheapest method of starting a 3-phase induction motor ?
Ans. D.O.L. starting.

AMIE WINTER, 1996

31. When a d.c. motor produces maximum output power ?

Ans. When back e.m.f. is equal to half the applied voltage.

32. What type of load should be connected to a transformer for getting maximum voltage regulation ?

Ans. Load required to be connected for maximum voltage regulation is inductive and for maximum regulation $\phi = \tan^{-1} (X_1/R_1)$.

33. What is the condition for producing maximum torque in a 3-phase induction motor ?

Ans. $R_2 = sX_2$.

34. What measures can be taken for minimising the effect of crawling ?

Ans. Effect of crawling can be minimised by chording, skewing and integral slot winding.

35. Which parameter of the equivalent circuit should change the diameter of the circle diagram and how it effects the pull-out torque ?

Ans. Total standstill leakage reactance per phase referred to stator. Pull-out torque is reduced by increased value of X.

36. What parameter of load influences the armature reaction of an alternator ?

Ans. Power factor of the load.

37. What is the use of Potier's triangle ?

Ans. To find voltage regulation.

38. How the power factor of a synchronous motor is changed keeping the shaft load undisturbed ?

Ans. By changing the excitation.

39. What are the advantages of synchronous motor ?

Ans. It runs at constant synchronous speed and is capable of being operated under a wide range of power factor both lagging and leading.

40. How the direction of rotation of a capacitor start 1-phase induction motor is changed ?

Ans. By changing the connection of one of the windings.

AMIE SUMMER, 1997

41. Does the armature m.m.f. of an alternator delivering a capacitive load magnetise or demagnetise the main field ?

Ans. Will demagnetise the field.

42. The percentage leakage impedance of a 1-phase 2000 V/100 V, 5 kVA transformer is 2.5. What voltage should be applied to the HV side for carrying out SC test at rated current ?

Ans. 125 V.

43. A particular application employing a d.c. generator requires high voltage and low current. Should the machine be lap wound or wave wound ?

Ans. Wave wound.

44. A full pitch coil in an alternator has a span of 18 slots. What should be the span if the third harmonic component of the e.m.f. is to be eliminated ?

Ans. 12 slots.

45. Can a repulsion-induction motor ever run at super-synchronous speed ?

Ans. Yes, at no load.

46. What is dimension of $\sqrt{LC}$?

Ans. LT^{-1}.

47. On switching a d.c. motor it is found to rotate in direction opposite to that for which it is designed. Identify the motor.

Ans. D.C. series motors.

48. A synchronous motor runs at N_s r.p.m. at full load. What will be its speed if the load is reduced to half the full load ?

Ans. N_s.

49. A 3-phase induction motor is to be started first using an auto-transformer with 80% tapping and next direct-on-line. What will be the ratio of the starting torques ?

Ans. $\frac{T_{auto}}{T_{d.o.l.}} = 0.64$.

50. Three identical impedances having the same $\frac{X}{R}$ ratio are first connected in star and then in delta to a 400 V, 50 Hz supply. What is the ratio of active powers drawn ?

Ans. $\frac{1}{\sqrt{3}}$.

AMIE WINTER, 1997

51. What is the power drawn by a 3-phase balanced load ?

Ans. $\sqrt{3}\, V_L I_L \cos\phi$.

52. A dummy coil is in the armature winding of a d.c. machine. Is the machine lap wound or wave wound ?

Ans. Lap wound.

53. The air gap reluctance of a resistance motor is uniform. Is the statement true or false ?

Ans. False.

54. Is the interpole AT required for a compensated d.c. machine more or less than that of an uncompensated one ?

Ans. Less.

55. A 3-phase induction motor is to be started first using an auto transformer with 80% tapping and next direct-on-line. What will be the ratio of the starting torques ?

Ans. 0.64.

56. Why does synchronous impedance method give a poorer voltage regulation ?

Ans. Saturation conditions are lower.

57. A synchronous motor is to be used as a synchronous condenser. Under what condition should it operate ?

Ans. Over excited.

58. Can a repulsion-induction motor ever run at super synchronous speed ?

Ans. Yes.

59. A full pitch coil in an alternator has a span of 18 slots. What should be the span if the third harmonic component of the e.m.f. is to be eliminated ?

Ans. 12 slots.

60. If the applied voltage to the primary of a transformer is reduced to 50 percent of its initial value, at the same time the number of turns in the secondary are doubled. What will be the secondary voltage ?

Ans. Same.

AMIE SUMMER, 1998

61. Derive the relation between electrical and mechanical angle in case of rotating machine.

Ans. Mechanical degrees in rotation = 360°

Electrical degrees = No. of pairs of poles times 360°

$$\therefore \text{ Mechanical degrees} = \frac{\text{electrical degrees}}{\text{pair of poles}} \quad i.e.\ Q_{mech} = \frac{Q_{elect.}}{(p/2)}.$$

62. A 6-pole d.c. generator has a rated speed of 600 r.p.m. Calculate the frequency of generated e.m.f. in armature conductors and at its brushes.

Ans.
$$N = \frac{120f}{p}$$

$$\therefore \quad f = \frac{pN}{120} = \frac{6 \times 600}{120} = 30 \text{ Hz.}$$

63. Why does an induction motor always operate at lagging power factor ?

Ans. Induction motor has large magnetising current at very low power factor, hence even at load it will always work at lagging power factor.

64. Can the excitation level give the idea of operating power factor of synchronous motor ?

Ans. Excitation decides operating power factor of synchronous motor, for example over excited motor has leading power factor.

65. Explain the advantages of using closed slots for the rotor of small rating squirrel cage induction motor.

Ans. Close side slot gives low air gap reluctance and hence has reduced value of magnetising current. This provides high value of leakage reactance, due to which starting current is limited.

66. Transformers, belonging to phase group *DY*1 and *YY*0 cannot be operated in parallel, state why.

Ans. There exists a phase angle difference between the two secondary voltages, hence cannot be connected in parallel.

67. Comment on the statement, 'triplex order harmonics current can flow in a delta connected winding but are absent in the line current'.

Ans. Triplex harmonic currents are cophased, hence they can flow in delta local circuit, there resultant in line becomes zero.

68. A d.c. shunt motor may run away at heavy loads. Comment.

Ans. At heavy loads the internal drops become very large and hence terminal voltages are reduced to a very low value, finally resulting into run away.

69. Why, the pole changing method of speed control, can be used in squirrel cage induction motor ?

Ans. By providing different sets of windings in the stator of squirrel cage induction motor and selecting a particular winding at a time the speed of the motor can be controlled.

70. An induction generator is running at a slip of 0.03 and is feeding power to the infinite bus. What will happen if the input is suddenly removed ?

Ans. The induction generator will start functioning as induction motor by taking power from the infinite bus.

AMIE WINTER, 1998

71. Draw the equivalent circuit of a d.c. compound generator for long shunt and mark the various voltages, currents and polarities.

Ans. Fig. 5 shows the equivalent circuit.

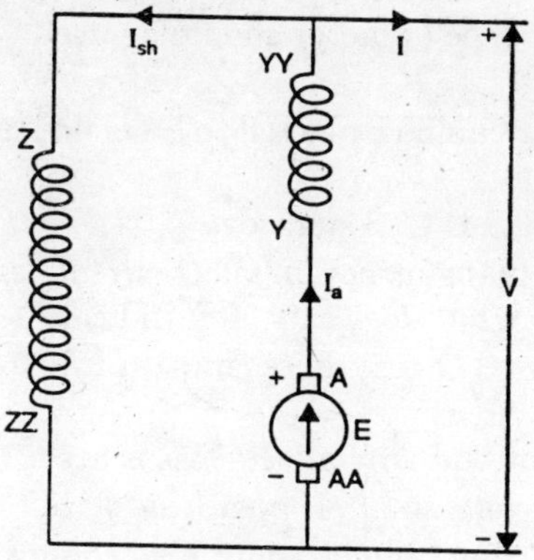

Fig. 5

72. An air gap is introduced in the core of a transformer. With this change, what happens to the magnetizing current ?

Ans. By the introduction of air gap magnetising current will be increased largely.

73. Write the relation between electrical and mechanical degrees.

Ans. Electrical degrees = $\frac{p}{2}$ × mechanical degrees.

74. Write the function of interpole in a d.c. machine.

Ans. The function of interpoles is to neutralise the reactance voltage in the coil undergoing commutation.

75. Give the components of synchronous reactance of a synchronous machine.

Ans. Leakage reactance X_L and armature reactance X_a (*i.e.* $X_S = X_L + X_a$).

76. A centre-zero galvanometer in the rotor circuit of a 6-pole, 50 Hz wound-rotor induction motor makes 90 complete oscillations in one minute. Calculate the rotor speed.

Ans. Rotor speed = $\frac{120 \times 50}{6} - 90 = 910$ r.p.m.

77. A 3-phase, 50 Hz induction motor has a full-load speed of 960 r.p.m. Calculate speed of the rotor field with respect to rotor structure and also with respect to stator field.

Ans. Rotor field speed with respect to rotor structure = 40 r.p.m. and with respect to stator field 960 r.p.m.

78. How can the direction of rotation be reversed in a repulsion motor ?

Ans. By changing the direction of lead of brushes.

79. A transformer is operated at rated frequency but at a voltage higher than its rated value. How will the following quantities change ?

(*a*) No-load current (*b*) Hysteresis loss.

Ans. (*a*) No load current will increase

(*b*) Hysteresis losses will increase.

80. Primary winding of a 1-phase transformer is energized from a triangular wave source. Give the nature of the output voltage wave obtained from its secondary.

Ans. Rectangular wave.

WINTER, 1999

81. In a D.C. traction use, what type of motor should be used ?

Ans. D.C. series motor.

82. What kind of D.C. generator can be expected to have no load voltage and full load voltage to be same ?

Ans. Cumulative compounded D.C. generator.

83. A 1000 V/240 V, 1-ϕ, 50 Hz transformer on short circuit test takes 50 V to circulate full load current in the transformer. What does this 50 V indicate ?

Ans. 50 V required to perform S.C. test represents the internal drops in primary and secondary windings of transformer.

84. In a transformer open circuit test why all the losses are contributed to iron loss ?

Ans. No load current being very small, copper losses are considered negligible.

85. In alternator, the coils are short pitched and also conductors of the individual phases are distributed within their phase spread. What is the reason ?

Ans. To avoid harmonic voltage, short pitching is done. Distribution of phases is done to get required phase spread.

86. In terms of terminal voltage/phase V, open circuit voltage per phase E and V, the internal angle of a 3-ϕ synchronous motor, when the motor power factor will be leading ?

Ans. For an over-excited synchronous motor.

87. In a star-delta starter, what is the amount of reduction in the line current ?

Ans. Line current is reduced to $\frac{1}{3}$ in star-delta starter at starting.

88. What is the relationship between rotor frequency f_2, stator frequency f_1 and slip s.

Ans. $f_2 = sf_1$.

89. What are the characteristics of repulsion induction motor ?

Ans. Characteristics of repulsion induction motor are shown in Fig. 6.

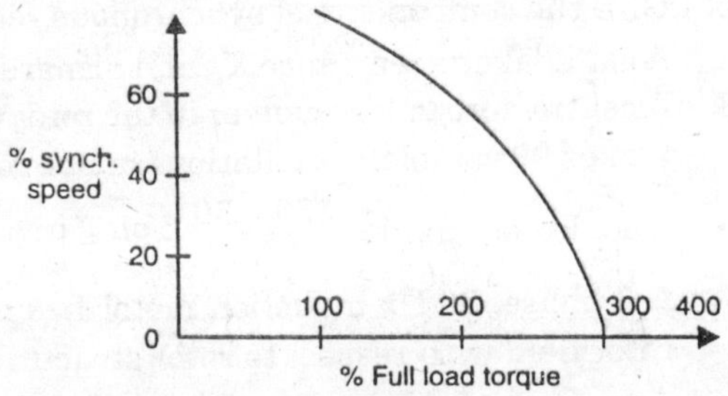

Fig. 6

90. In a measurement of 3-phase power of a load by two wattmeters, is it possible to measure the reactive power ?

Ans. Yes.

SUMMER, 2000

91. Fill in the blanks, justifying your answer in not more than three or four sentences :

(*i*) In d.c. machines, the field-flux axis and armature mmf axis are respectively along ...... axis and ...... axis.

(*ii*) A differentially, compounded d.c. generator is delivering full load current at a terminal voltage of 220 V. If its series field winding is short circuited, then its terminal voltage will ....

(*iii*) The full-load voltage regulation of an alternator is 6% at 0.8 p.f. lagging and at rated speed of 1200 r.p.m. Its full-load voltage regulation at 0.8 p.f. lagging and at 1100 r.p.m. would be .... Neglect armature resistance.

(*iv*) A synchronous motor, connected to an infinite bus, is working at a leading power factor. Its excitation e.m.f. E_f and alternator terminal voltage V_t are related as follows :
E_f is ...... (less/more) than V_t, and E_f ...... (lags/leads) V_t.

(v) For satisfactory performance of a 3-phase 420 V, 60 Hz induction motor, the supply voltage at 50 Hz should be equal to ...... V.

(vi) The stator of a 3-phase 6-pole wound-rotor induction motor is connected to 50 Hz source but its rotor is energised from 20 Hz source. The rotor of this motor would run at speeds of ...... r.p.m. and ...... r.p.m.

(vii) The direction of rotation of a single-phase repulsion motor can be reversed by ....

(viii) If a single-phase series motor is operated at a frequency lower than its rated frequency, then its speed would ...... (decrease/increase) and torque would ...... (decrease/increase).

(ix) Two windings of a single transformer are indicated by terminals *AB* and *CD* as shown in Fig. 7. When a voltage of 100 V is applied across *AB* with *BD* short-circuited, a voltage of 200 V appears across *AC*. The turns ratio from *CD* to *AB* is ...... or ....

(x) One transformer has leakage impedance of $1 + j4\ \Omega$ and $3 + j11\ \Omega$ for its primary and secondary windings respectively. This transformer has ...... (l.v./h.v.) primary and ...... (l.v./h.v.) secondary.

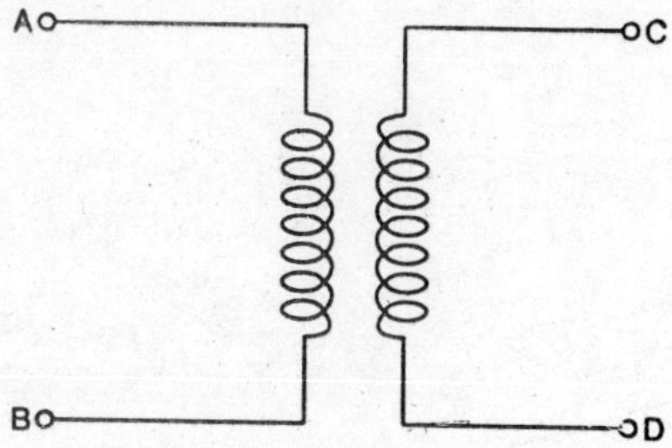

Fig. 7

Ans. (i) Pole axis and magnetic neutral axis.

Field flux is produced by pole excitation hence will act along pole axis. Armature mmf is produced by armature current and hence acts along the direction of the axis produced by coil formation of armature conductors.

(ii) Terminal voltage will increase because of series field winding being shorted, series field becomes zero and hence the net field will increase.

(iii) Voltage regulation at 1100 r.p.m. will be also 6% because with reduction in speed induced voltage will reduce by the same amount as the reactive drop, because both are proportional to frequency or speed.

(iv) $E_f > V_t$ and E_f leads V_r.

(v) For satisfactory performance at 50 Hz, the supply voltage should be equal to

$$420 \times \frac{50}{60} = 350 \text{ volts.}$$

(vi) The rotor will run at speeds corresponding to frequencies (50 + 20) and (50 – 20) Hz. The respective speeds would be 1400 r.p.m. and 600 r.p.m.

(vii) The direction of rotation of a single phase repulsion motor can be reversed by changing the direction of shift of brushes.

(viii) When operated at lower than the rated frequency then its speed increases and torque would decrease.

(ix) Turns ratio from *CD* to *AB* is 1 or 3.

Depending upon the polarity of connection either both the voltages will be additive or substractive.

(x) The transformer has l.v. primary and h.v. secondary.

Index

PART I : DIRECT CURRENT MACHINES

PART II : ALTERNATING CURRENT MACHINES